Handbook of
NATURAL TOXINS

HANDBOOK OF NATURAL TOXINS

Series Editor
Anthony T. Tu
Department of Biochemistry
Colorado State University
Fort Collins, Colorado

Volume 1 Plant and Fungal Toxins, *edited by Richard F. Keeler and Anthony T. Tu*

Other Volumes in Preparation

Volume 2 Insect Poisons, Allergens, and Other Invertebrate Venoms, *edited by Anthony T. Tu*
Volume 3 Bacterial Toxins, *edited by M. Carolyn Hardegree, William H. Habig, and Anthony T. Tu*
Volume 4 Reptile and Amphibian Venoms, *edited by Anthony T. Tu*
Volume 5 Marine Venoms and Toxins, *edited by Anthony T. Tu*

Handbook of
NATURAL TOXINS

Volume 1
PLANT AND FUNGAL TOXINS

Edited by

RICHARD F. KEELER
Poisonous Plant Research Laboratory
U.S. Department of Agriculture
Logan, Utah

ANTHONY T. TU
Department of Biochemistry
Colorado State University
Fort Collins, Colorado

MARCEL DEKKER, INC.   New York and Basel

Library of Congress Cataloging in Publication Data
Main entry under title:

Plant and fungal toxins.

(Handbook of natural toxins ; v. 1)
Includes indexes.
1. Mycotoxins—Physiological effect—Handbooks,
manuals, etc. 2. Poisonous plants—Toxicology—
Handbooks, manuals, etc. I. Keeler, Richard F.
II. Tu, Anthony T., [date]. III. Series.
QP632.M9P55 1983 615.9'52 83-15442
ISBN 0-8247-1893-3

COPYRIGHT © 1983 by MARCEL DEKKER, INC. ALL RIGHTS RESERVED

Neither this book nor any part may be reproduced or transmitted in any form
or by any means, electronic or mechanical, including photocopying, micro-
filming, and recording, or by any information storage and retrieval system,
without permission in writing from the publisher.

MARCEL DEKKER, INC
270 Madison Avenue, New York, New York 10016

Current printing (last digit):
10 9 8 7 6 5 4 3 2 1
PRINTED IN THE UNITED STATES OF AMERICA

C. D. Marsh

This volume is dedicated to the memory of Dr. C. D. Marsh, early pioneer in the study of the effects on domestic livestock of poisonous plants and their toxins. He directed much of the work on this subject in the U.S. Department of Agriculture in the early part of this century from laboratories and offices in Washington, D.C., and field stations throughout the United States and at the principal Salina, Utah, station. He sits here astride his favorite mare at a Colorado field location in a photo dated September 11, 1911.

PREFACE TO THE HANDBOOK

Natural toxins are unique toxins which possess some common properties, whether they are obtained from plants, microorganisms, or animals. One common characteristic is that they exert a pronounced effect on the metabolism and biological functions of the intoxicated animals with just a minute quantity. Since ancient times human beings have pondered the physiological effects of various toxins and venoms. How do these natural poisons work? Despite possessing some common nature, each toxin, however, has its unique mode of action and its own characteristic structure.

Drugs are compounds that have specific beneficial effects with a minute quantity. Usually, natural toxins also have very specific effects. Therefore, it is not suprising that many natural toxins are potentially good drugs.

Heretofore, the study of each field of toxins has been taking an independent pathway. Scientists in a specific toxin field are often unaware of the activity in other toxin fields. It is thus desirable to have a primary source of information on all natural toxins so that scientists in a specific discipline of toxin research can easily obtain useful information from other toxin researchers.

This five-volume handbook of toxins will include the following volumes:

1. Plant and Fungal Toxins
2. Insect Poisons, Allergens, and Other Invertebrate Venoms
3. Bacterial Toxins
4. Reptile and Amphibian Venoms
5. Marine Venoms and Toxins

The editor expresses sincere thanks to Maurits Dekker, Chairman of the Board of Marcel Dekker, Inc., for initiating this project.

Anthony T. Tu

FOREWORD

Natural toxic agents abound in agricultural environments. Even though widely distributed, they form a vast reservoir in western America and they renew readily. They are synthesized by fungi or by plants and are eaten by animals. They create scientific and economic problems that are studied and solved at endemic regions.

During the early decades of the twentieth century, a research pioneer of the U.S. Department of Agriculture, C. D. Marsh, studied both poisonous plants and poisoned animals. Subsequently, other investigators—botanists, chemists, geneticists, managers, nutritionists, pathologists, physiologists, and veterinarians—followed the Marsh lead, persevered in the research, and solved many problems. In 1915 the USDA established a research station at Salina Canyon, Utah, and in about 1954 transferred it to Logan, Utah. At both locations the station was active and productive.

Research programs, state, federal, and in other nations, have steadily and significantly increased the information specific to plant poisonings. Early investigations identified species of toxic plants, categories of toxic agents, and species of susceptible animals. Such findings gradually concentrated scientific attention on specific aspects of the problems. As a result, chemists, for example, found specific culpable compounds and determined their pharmacological actions. Other investigators found target systems, organs, tissues, and cells, as well as the nature of changes produced in affected animals. Of great importance was the finding that some edible plants produce teratogenic compounds that are generally nontoxic to consuming maternal animals but are severely toxic to embryos in critically specific stages of development. Such poisoned embryos may die early or develop fatal deformities. Another heuristic finding was the indication that under specific environmental conditions, combined actions, or perhaps interactions between some host plants, particularly grasses and some epiphytic fungi, may cause serious toxicologic problems to consuming livestock.

These and all other cogent discoveries now constitute a cognitive base on which rational principles and pragmatic programs can be built for abating and preventing injuries and deaths to food animals. Research administrators, research scientists, and taxpayers, however, must not be deluded by past achievements that all problems are solved. Among the numerous aspects requiring further study are determinations of the full extent of plant-fungus interactions in the production of toxins, and the development of materials and technologies for controlling and possibly eradicating some noxious plants of pastures and ranges.

This volume presents the collated and interpreted results from field observations and laboratory investigations on plant and fungal poisoning. It fixes the perspective of plant and fungal toxins as a major category of pathogens. As a member of a series of related volumes, it should become a useful and convenient library resource to research scientists, practicing veterinarians, range and pasture managers, students, and livestock producers.

Now, I return to C. D. Marsh. At each ingress to his work domain, I see the species of plants he saw, suspicioned, and verified as poisonous. These dangerous species include death camus, larkspur, locoweed, milkweed, and wild tobacco. Then I see bleached bones possibly from poisoned animals, and I remember the nation's gratitude for the truths he found and gave.

Rue Jensen
University of Wyoming, Laramie

PREFACE TO VOLUME 1

Plant and fungal toxins produce in humans and other animals a great variety of toxic effects, from transient illness with little or no apparent histopathologic or biochemical evidence to debilitation, death, or teratogenic sequela with profound evidence of toxicosis. These effects arise from a vast divergence of toxin types, from simple, small molecules through very high molecular weight compounds such as proteins. Application of the term "toxin" was once restricted to the latter, but modern usage includes the entire spectrum of compound types responsible for this vast divergence of toxicoses. We have included in this volume samplings of the broad range selected to provide a view of the diversity of effects in animal systems. Also included are examples of toxins whose toxic forms represent compounds resulting from digestive tract, microsomal enzyme, or some other type of conversion from nontoxic naturally occurring forms.

Because few toxins affect single organs, any classification can be arbitrary. But in this volume toxins are grouped to illustrate the following: effects on the cardiovascular or pulmonary system, carcinogenic effects, reproductive effects, psychic or neurotoxic effects, gastrointestinal or hepatic effects, effects of toxins on species interactions, and the usefulness of toxins in medicine.

Authors have provided information on chemistry of the toxins, source, gross and histopathologic effects, and mechanism of action. Coverage has varied according to information available and the inclination of each author, but one can grasp the magnitude of information available in each instance.

Plant and fungal toxins play a significant role in the health of humans and animals. Domestic livestock are particularly at risk to both toxin types. U.S. Department of Agriculture estimates suggest that up to 5% of all livestock grazing arid western North American ranges have a serious encounter with poisonous plants and their toxins each year and that perhaps half of these are debilitated, die, abort, or give birth to deformed offspring. Fungal toxins are most troublesome in the moist eastern areas, where moldy pasture grasses, hays, or improperly dried grains are the usual culprits. Humans are less frequently affected by both types of toxins. Even so, Poison Control Centers note that poisonous plant ingestion, for example, usually ranks in the top five or six among toxicoses reaching their attention each year.

Plant and fungal toxins consequently are of great importance to us, and
this volume provides contemporary information on selected significant examples.

The responsibility for the scientific content of each chapter in this volume
rests with the individual authors. Any opinions, findings, and conclusions
or recommendations expressed in this publication are those of the authors and
do not necessarily reflect the views of the U.S. Department of Agriculture,
Colorado State University, or any other organization.

Richard F. Keeler
Anthony T. Tu

CONTENTS

PART I. TOXINS INDUCING EFFECTS IN THE CARDIOVASCULAR OR PULMONARY SYSTEMS

PART II. TOXINS THAT ARE TERATOGENIC

PART VII.   TOXINS IN EVOLUTION AND SPECIES INTERACTION

PART VIII.   TOXINS USEFUL IN MEDICINE

CONTRIBUTORS

HAROLD BAER   *Allergenic Products Branch, Office of Biologics, National Center for Drugs and Biologics, Food and Drug Administration, Bethesda, Maryland*

JANET M. BENSON   *Lovelace Inhalation Toxicology Research Institute, Albuquerque, New Mexico*

ROGER G. BREEZE   *Department of Veterinary Microbiology and Pathology, College of Veterinary Medicine, Washington State University, Pullman, Washington*

LEO T. BURKA   *Department of Chemistry, Vanderbilt University, Nashville, Tennessee*

JAMES R. CARLSON   *Department of Animal Sciences, Washington State University, Pullman, Washington*

ALEX CIEGLER*   *Southern Regional Research Center, U.S. Department of Agriculture, New Orleans, Louisiana*

C. C. J. CULVENOR   *Commonwealth Scientific and Industrial Research Organization, Division of Animal Health, Victoria, Melbourne, Australia*

MEMORY P. F. ELVIN—LEWIS   *Department of Biomedicine, Washington University, School of Dental Medicine, St. Louis, Missouri*

SELWYN L. EVERIST[†]   *Department of Veterinary Pathology and Public Health, University of Queensland, Brisbane, Australia*

---

*Present affiliation: Computer Science Corporation, NASA National Space Technology Laboratory, Building 2204, NSTL Station, Mississippi

[†]Deceased.

HEINZ FAULSTICH   *Max-Planck-Institute for Medical Research, D-6900 Heidelberg, Federal Republic of Germany*

STEPHEN M. HAMMOND   *Department of Microbiology, Leeds University, Leeds, England*

JEFFREY B. HARBORNE   *Phytochemical Unit, Plant Science Laboratories, University of Reading, England*

RONALD D. HOOD*   *Developmental Biology Section, Department of Biology, The University of Alabama, Tuscaloosa, Alabama*

PETER T. HOOPER   *Department of Tropical Veterinary Science, James Cook University of North Queensland, Queensland, Australia*

JANA HRDLICKA   *Department of Veterinary Pathology and Public Health, University of Queensland, Brisbane, Australia*

G. WAYNE IVIE   *Veterinary Toxicology and Entomology Research Laboratory, Agricultural Research Service, U.S. Department of Agriculture, P.O. Drawer GE, College Station, Texas*

LYNN F. JAMES   *Poisonous Plant Research Laboratory, U.S. Department of Agriculture, Science and Education Administration, Logan, Utah*

A. EARL JOHNSON   *Poisonous Plant Research Laboratory, U.S. Department of Agriculture, Logan, Utah*

RICHARD F. KEELER   *Poisonous Plant Research Laboratory, U.S. Department of Agriculture, Logan, Utah*

BONG—SEOP KIL[†]   *Department of Biology, University of Massachusetts, Amherst, Massachusetts*

A. DOUGLAS KINGHORN   *Department of Medicinal Chemistry and Pharmacognosy, College of Pharmacy, University of Illinois at Chicago, Chicago, Illinois*

JOHN M. KINGSBURY   *Department of Clinical Sciences, New York State College of Veterinary Medicine, Cornell University, Ithaca, New York*

PETER A. LAMBERT   *Department of Pharmacy, University of Aston in Birmingham, Birmingham, England*

S. MARK LEE   *Department of Environmental Toxicology, University of California, Davis, California*

---

*Also affiliated with: Ronald D. Hood and Associates, Consulting Toxicologists, Tuscaloosa, Alabama.

[†]Present affiliation: Won Kwang University, Iri, Korea

WALTER H. LEWIS *Department of Biology, Washington University, Campus Box 1137, St. Louis, Missouri*

PETER H. MORTIMER *Ministry of Agriculture and Fisheries, Ruakura Agricultural Research Center, Hamilton, New Zealand*

MURIEL S. PALMGREN *School of Public Health and Tropical Medicine, Tulane University, New Orleans, Louisiana*

J. E. PETERSON *Commonwealth Scientific and Industrial Research Organization, Division of Animal Health, Victoria, Melbourne, Australia*

JONATHAN E. POULTON *Department of Botany, University of Iowa, Iowa City, Iowa*

JOHN W. RONALDSON *Ministry of Agriculture and Fisheries, Ruakura Agricultural Research Center, Hamilton, New Zealand*

ALAN A. SEAWRIGHT *Department of Veterinary Pathology and Public Health, University of Queensland, Brisbane, Australia*

JAMES N. SEIBER *Department of Environmental Toxicology, University of California, Davis, California*

LAWRENCE G. STOWE* *Department of Botany, University of Massachusetts, Amherst, Massachusetts*

GEORGE M. SZCZECH *Burroughs Wellcome Company, Toxicology and Experimental Pathology Department, Research Triangle Park, North Carolina*

COY W. WALLER *Research Institute of Pharmaceutical Sciences, University of Mississippi, University, Mississippi*

THEODOR WIELAND *Max-Planck-Institute for Medical Research, D-6900 Heidelberg, Federal Republic of Germany*

BENJAMIN J. WILSON *Biochemistry Department, School of Medicine, Vanderbilt University, Nashville, Tennessee*

DONALD A. WITZEL *Veterinary Toxicology and Entomology Research Laboratory, U.S. Department of Agriculture, College Station, Texas*

---

*Present affiliation: Choate Rosemary Hall, Wallingford, Connecticut

Handbook of
NATURAL TOXINS

Part I

TOXINS INDUCING EFFECTS IN THE
CARDIOVASCULAR OR PULMONARY SYSTEMS

1

SWEET POTATO TOXINS AND RELATED TOXIC FURANS

BENJAMIN J. WILSON and LEO T. BURKA

Vanderbilt University, Nashville, Tennessee

I. INTRODUCTION

Ipomoea batatas (L.) Lam., the common sweet potato, is a member of the Convolvulaceae family of plants. It has served as a food source for many peoples over several centuries. Although apparently unknown to certain of the more advanced ancient civilizations, primitive groups in tropical countries used the sweet potato root as food long before written historical records came into existence, and consumption has since become widespread throughout much of the world (Wilson, 1973). It is now grown in the United States, tropical America, the warmer islands of the Pacific, Japan, and since 1930 in the USSR. The ethnobotanical history of the sweet potato in relation to Oceania has been extensively reviewed by Yen (1974).

Some reports suggest that the sweet potato was introduced into the United States by Sir John Hawkins in 1565 (Jones et al., 1958). At present many thousands of acres, largely in southern regions, are set aside for its cultivation. It is used primarily as a human food (60-70% of the annual crop) but has also served as a source of starch production and certain industrial products as well as an item of livestock feed, especially for cattle (Steinbauer and Cushman, 1971).

Many different varieties of sweet potato are recognized, and new strains having disease resistance properties are continuously being developed by plant geneticists. The interior root color may vary from a light buff to light yellow, orange, and various shades of purple. In certain southern states sweet potatoes are sometimes referred to as "yams." However, the true yam is a different plant, belonging to the genus *Dioscorea*.

Sweet potatoes are subject to many different disease agents, including insect and fungus pathogens. One estimate suggests that losses to each year's crop in the United States can amount to as much as 35% of total production (Steinbauer and Cushman, 1971). Losses may be reduced by use of plant fungicides and careful mechanical harvesting methods that eliminate trauma to the roots. Other useful measures that diminish postharvest spoilage consist of washing the roots in hypochlorite solution, curing at 30-31°C at a relative humidity of 85-90% (which promotes surface blemish healing), and packing in boxes using paper to separate the layers of roots during storage at approximately 15°C. Lower-temperature storage results in biochemical processes that are detrimental (Lieberman et al., 1958). Storage may be extended for as long as 6-7 months under optimal conditions.

The sweet potato is but one among several plants capable of forming stress metabolites (phytoalexins) when stimulated by a variety of factors, including mechanical injury, insect invasion, exogenous chemicals, and microbial pathogens. The abnormal metabolites apparently have only limited inhibitory effect against certain microbial pathogens but can render the root unpalatable or even toxic to consuming animals.

The 3-substituted furans have received considerable research emphasis beginning with that of Japanese investigators predating World War II. Most of the early reports dealt mainly with ipomeamarone, a hepatotoxic furanosesquiterpene whose production is stimulated by several of the previously mentioned factors (Takeuchi, 1946). Hydroxycinnamic acid derivatives (including coumarins) are accumulated in sound tissue adjacent to sections undergoing fungal invasion (Uritani and Miyano, 1955). At least one of these compounds, chlorogenic acid, is present in sound tissue in the absence of fungal infection (Rudkin and Nelson, 1947). Chlorogenic acid (Uritani and Muramatsu, 1953), caffeic acid (Uritani and Miyano, 1955), isochlorogenic acid (Uritani and Miyano, 1955), umbelliferone (Uritani and Hoshiya, 1953), and scopoletin (Uritani and Hoshiya, 1953), have been found in sweet potatoes infected with the fungal black-rot pathogen, *Ceratocystis filmbriata*.

In both the United States (Hansen, 1928) and Japan (Hiura, 1943), mold-damaged sweet potato roots have been recognized for several decades as a dietary agent causing fatal repiratory disease in cattle. Sporadic enzootics of the disease in the United States have been reported, since at least 1928, under such names as pulmonary adenomatosis, sweet potato poisoning, atypical interstitial pneumonia, and acute bovine pulmonary edema and emphysema (ABPE) (Peckham et al., 1972). The latter designated syndrome is now recognized by some investigators as having several causative agents, including moldy sweet potatoes, perilla mint plant, lush pasture, and occasionally other food sources (Wilson et al., 1978).

A devastating outbreak of bovine pulmonary disease, described as atypical interstitial pneumonia, occurred in the summer of 1969 in Tift County, Georgia. A total of 69 head of Hereford cattle were lost out of a herd of 275 that had been given access to cull (moldy) sweet potatoes (Peckham et al., 1972). Other outbreaks of sweet potato poisoning have since been recorded. The Tift County outbreak, however, was the stimulus for intensive and successful efforts to isolate and describe the lung-toxic factors. It now appears that the lung-toxic principle is of primary importance in spite of the fact that hepatotoxic furans are usually present in moldy sweet potatoes in somewhat larger quantities than those having lung specificity (Boyd and Wilson, 1971). Moreover, no cases of liver disease due to blemished sweet potatoes have been reported.

Many other lung-toxic, hepatotoxic, and nephrotoxic 2- and 3-substituted furans and related compounds occurring naturally in plants, or used as therapeutic agents, have been studied as well as several chemically synthesized congeners of these toxins. For example, among the naturally occurring analogs of sweet potato lung toxins are certain lung-edemagenic, normal metabolites of the common mint plant, *Perilla frutescens* (Wilson et al., 1977). Another natural source of lung toxicity is L-tryptophan, a normal amino acid constituent of pasture grass which, under certain conditions, may be converted by ruminal microorganisms to 3-methylindole, considered by some investigators to be the specific causative agent of ABPE (Hammond et al., 1980). 3-Methylfuran occurs as a natural contaminant of the atmosphere in certain urban areas (Saunders et al., 1974), and could conceivably play a role in respiratory diseases in view of its toxicity to the repiratory tract of experimental animals (Boyd et al., 1978b). Nitrofurantoin, a commonly used chemotherapeutic agent for urinary tract infections, may also produce pulmonary edema in patients treated with excessively large doses (Boyd et al., 1977).

This chapter consists of a compilation of chemical, biochemical, and toxicological data on a selected number of toxic furans, most of which are naturally occurring. The reader is referred to other reviews referenced in the text for additional information on these and other compounds that may be of interest.

II. HEPATOTOXIC 3-SUBSTITUTED FURANOTERPENES

A. Isolation, Characterization, and Synthesis

A variety of agents will cause stress metabolite production in sweet potatoes. In studies designed to isolate and characterize the compounds, three stressing agents have been used most frequently. Uritani, Kubota, and earlier Japanese groups have used the fungus *C. fimbriata* as the stressing agent. Most of the early work at Vanderbilt University was done with *Fusarium solani*; however, mercuric chloride has been used as the stressing agent for recent studies of hepatotoxic metabolites. In general, sweet potato slices are treated with the agent and placed in a moist environment at room temperature for 3-7 days. After this time, the slices are either extracted directly or after desiccation. The usual solvents are acetone, chloroform-methanol, or methylene chloride-methanol mixtures. Wilson and co-workers found it expedient to steam distill directly the treated sweet potato slices to obtain quantities of fairly pure ipomeamarone. Early investigators relied on repeated fractional distillation to separate the mixture of compounds obtained. Most recent work has employed column, thin-layer, or high-pressure liquid chromatography (HPLC) for separation.

1. *Ipomeamarone*

[1]

Ipomeamarone [1] was reported as a stress metabolite of unknown structure in mold-damaged sweet potatoes in 1943 (Hiura, 1943). Early structural studies were performed by at least three groups in Japan (Takeuchi, 1946; Ohno and Toyao, 1952; Watanabe and Nishiyana, 1952). Determination of the accepted structure by Kubota and co-workers was based on a series of oxidative degradations; this work was reviewed in 1958 (Kubota, 1958) and will not be detailed here.

Of related historical interest is the investigation of toxic principles from the Ngaio tree (*Myoporum laetum*) of New Zealand begun by McDowall in the 1920s (McDowall, 1925, 1927, 1928). Structural studies on ngaione, the major toxic principle, were carried out in Australia and New Zealand contemporaneously with the ipomeamarone studies in Japan. Ohno noted the similarity between ngaione and ipomeamarone (Ohno, 1952). At one time it was thought that ngaione and ipomeamarone were diastereomers with ipomeamarone having a cis configuration about the tetrahydrofuran ring and ngaione having a trans configuration (Birch et al., 1954). Kubota's synthesis (to be described later) and subsequent work by Kubota and Matsuura (1958) established ngaione and ipomeamarone to be enantiomers, both having a cis configuration. This was the first of several examples of sweet potato stress metabolites which have the same structure as normal secondary metabolites in other plants. In fact, ipomeamarone, ngaione, and their epimers, epiipomeamarone and epigaione, are fairly common secondary plant metabolites. One or more of them have been found in plants of the genera *Myoporum* (McDowall, 1928), *Eremophilia* (Birch et al., 1954), *Anthanasia* (Bohlmann and Rao, 1972), *Eumorphia* (Bohlmann and Zdero, 1978), and *Gymnopentzia* (Bohlmann and Zdero, 1978), which are widespread geographically.

The first synthesis of racemic ipomeamarone, outlined in Scheme 1, was carried out by Kubota and his co-workers (Kubota, 1958). The critical step in this synthesis, addition of diisobutylcadmium to the mixture of acid chlorides [8], gave a tricyclic ketone [9] as the major product and a minor product that was similar to ipomeamarone but had a different infrared spectrum. It was rationalized that the cis acid chloride cyclized to form the tricyclic ketone and that the trans acid chloride gave the epimer [10] of

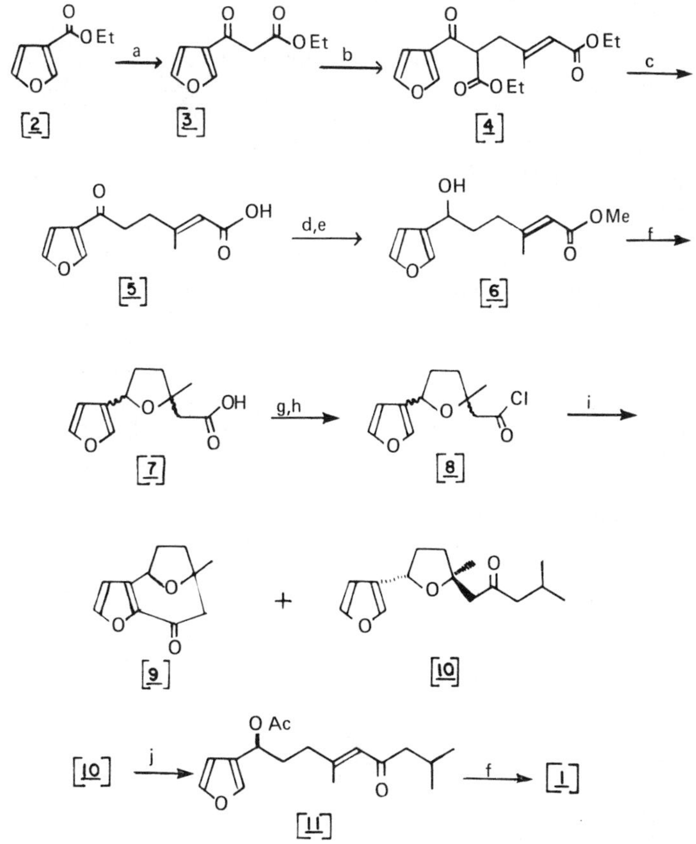

SCHEME 1 (a) Na/EtOAc, (b) Na/benzene/BrCH$_2$(CH$_3$)C=CHCO$_2$Et, (c) H$_2$SO$_4$/HOAc, (d) CH$_2$N$_2$, (e) Al$_2$[OCH(CH$_3$)$_2$]$_3$, (f) NaOH, (g) NaOMe, (h) (COCl)$_2$, (i) Cd[CH$_2$CH(CH$_3$)$_2$]$_2$, (j) KOAc/Ac$_2$O.

ipomeamarone. Ipomeamarone, therefore, must have a cis configuration
about the tetrahydrofuran ring as shown in structure [1].

The synthesis of ipomeamarone was completed by treating the trans
epimer with potassium acetate and acetic anhydride to give *seco*-enone [11].
Treatment of [11] with base effected saponification and ring closure. The
material resulting from this sequence had properties differing from the
trans product but similar to the material obtained by the same ring-opening
and saponification sequence performed either on natural ngaoine or ipomeamarone. Thus ipomeamarone was assumed to have a cis configuration about
the tetrahydrofuran ring.

Two syntheses of racemic ipomeamarone have been reported since Kubota's
work. One involved a shorter synthesis of *seco*-enone [11] and is outlined
in Scheme 2 (Burka et al., 1974c). The starting material, 1-ipomeanyl acetate [12], was prepared from 1-ipomeanol, the synthesis of which will be
described later. Hydrolysis of the *seco*-enone was shown to give a 1:1 mixture of ipomeamarone and epiipomeamarone. It was found that the epimers
were formed at about the same rate during hydrolysis and that the relative
concentrations did not change with equilibration. Apparently, the ring closure is not stereospecific.

The other reported synthesis of ipomeamarone (Kondo and Matsumoto,
1976) involves synthesis of dithiane [13], an umpolung version of Kubota's
acid chloride [8]. This synthesis is shown in Scheme 3. A notable feature
of this sequence is the formation of the furan ring from acyclic precursors.
It is one of the few syntheses described in this chapter in which the starting
material is not derived from 3-furoic acid.

For each of the stress metabolites sufficient spectral information is given
to identify the compounds and to provide information that can be used to
help identify similar compounds that may be isolated in the future. Ultraviolet and infrared absorption spectra are not included, but proton and
carbon nuclear magnetic resonance (NMR) spectra, major electron impact
mass spectral fragments, and other physical constants that seem appropriate are given. The numbering system used in NMR assignments (see below)
is retained throughout, even for degraded furanosesquiterpenes. Many of
the carbon NMR spectra have not been published previously.

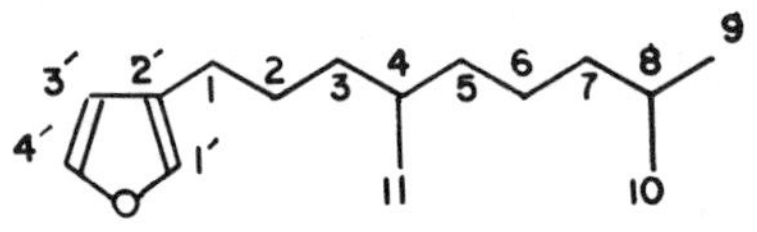

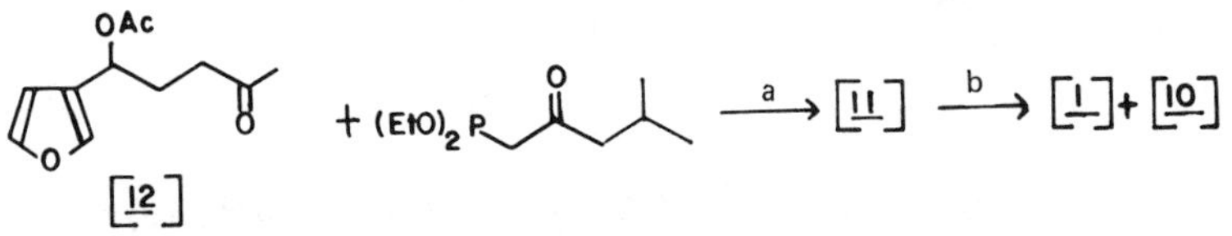

SCHEME 2   (a) NaH/MeOCH$_2$CH$_2$OMe, (b) NaOH/MeOH/H$_2$O.

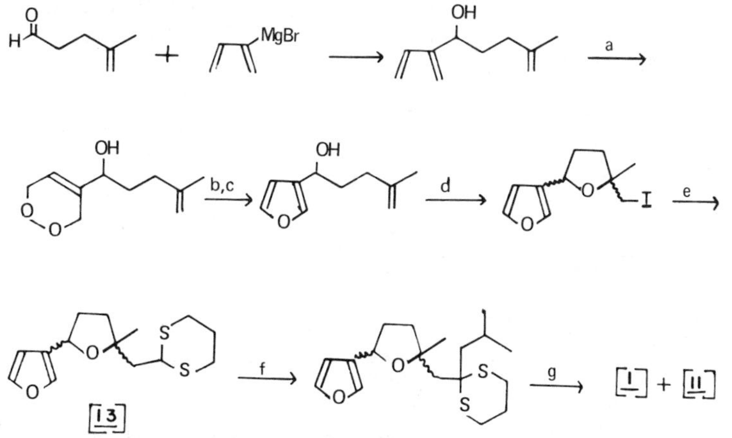

SCHEME 3 (a) $^1O_2$, (b) LiOC(CH$_3$)$_3$, (c) H$_2$SO$_4$/THF, (d) NaHCO$_3$/dioxane/ N-iodosuccinimide, (e) 1,3-dithiane/n-BuLi/HMPA/diglyme/THF, (f) n-BuLi/ ClCH$_2$CH(CH$_3$)$_2$/HMPA/THF, (g) AgNO$_3$/N-chlorosuccinimide/CH$_3$CN/H$_2$O.

Both the proton and carbon NMR spectra are available for ipomeamarone: ^{1}H-NMR (CDCl$_3$) δ 0.88 (d, J = 6 Hz, C-9, C-10), 1.32 (c, C-11), 1.8-2.4 (m, C-2, C-3, C-7, C-8), 2.68 (s, C-5), 4.90 (m, C-1), 6.36 (br s, C-3'), and 7.36 (br s, C-1', C-4') ^{13}C-NMR (CDCl$_3$) δ 22.5 (C-9, C-10), 24.4 (C-8), 26.7 (C-11), 33.1 (C-2 or C-3), 37.2 (C-3 or C-2), 53.6 (C-7 or C-5), 54.2 (C-5 or C-7), 72.6 (C-1), 81.7 (C-4), 108.9 (C-3'), 127.4 (C-2), 139.1 (C-1' or C-4'), 143.2 (C-4' or C-1'), and 209.0 (C-6). Major mass spectral fragments are at m/e 151 and 85. The optical rotation at the D line in ethanol was reported to be +28° (Kubota, 1958).

2. Dehydroipomeamarone

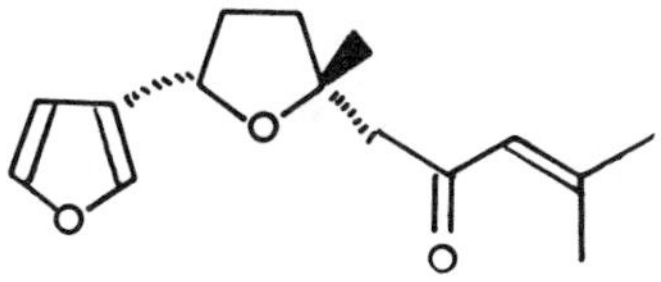

Dehydroipomeamarone [14] was isolated from *C. fimbriata*-infected sweet potatoes (Oguni and Uritani, 1973, 1974a,b). This compound and its epimer were previously isolated from *Myoporum deserti* (Hamilton et al., 1973) and from *Anthanasia* spp. (Bohlmann and Rao, 1972). Identification was made by comparison of spectra data. The proton NMR has been reported (Bohlmann and Rao, 1972): ^{1}H-NMR δ (CDCl$_3$), 1.26 (s, C-11), 1.87 (d, C-9), 2.0 (m, C-2, C-3), 2.05 (d, C-10), 2.75 (s, C-5), 4.88 (m, C-1), 6.11 (m, C-7), 6.26 (dd, J = 1 and 1 Hz, C-3'), 7.17 (d, J = 1 Hz, C-4'), and 7.20

(d, J = 1 Hz, C-1'). Major mass spectral fragments are at m/e 151, 108 and
83 (Bohlman and Rao, 1972). Optical rotation at the D line in ethanol was
reported to be +14.7° (Oguni and Uritani, 1973, 1974).

3. Ipomeamaronol

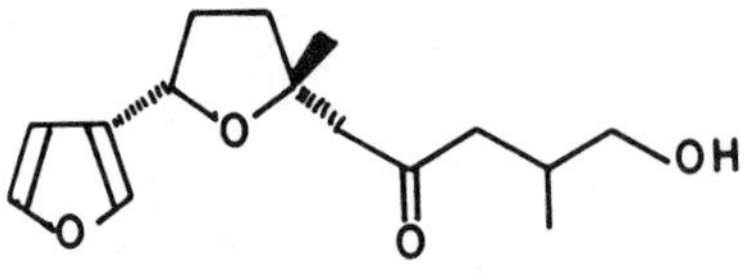

$$\left[\underline{15}\right]$$

Ipomeamaronol [15] was isolated from *C. fimbriata*-infected sweet potatoes
(Kato et al., 1971, 1973) and from roots infected by *F. solani* (Yang et al.,
1971). The compound is apparently very difficult to obtain in a pure state
by silica gel chromatography. For final purification Kato and co-workers
(1971, 1973) resorted to formation of the 3,5-dinitrobenzoate ester (mp 87°C)
and recrystallization. Yang and co-workers (1971) derivatized with bis(trimethylsilyl)trifluoroacetamide to form the trimethylsilyl ether, which was isolated by preparative layer chromatography. The proton NMR spectrum of
the trimethylsilyl ether was reported (Yang et al., 1971): ^{1}H-NMR (CDCl$_3$)
δ 0.08 (OTMS), 0.86 (d, C-10), 1.33 (s, C-11), 2.71 (s, C-5), 3.39 (d,
C-9), 4.91 (m, C-1), 6.37 (t, C-3'), and 7.37 (d, C-1', C-4'). Major mass
spectral fragments of the free alcohol are at m/e 151 and 101.

4. Myoporone, 1-Myoporol, and 6-Myoporol

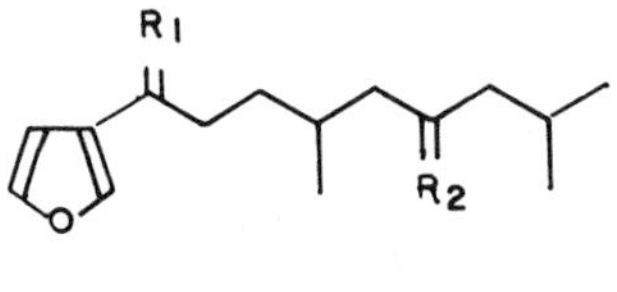

[16] $R_1 = R_2 = O$

[17] $R_1 = H, OH; R_2 = O$

[18] $R_1 = O; R_2 = H, OH$

Myoporone [16] has been isolated from *Myoporum* (Kubota and Matsuura,
1957; Blackburne et al., 1972), *Eumorphia* (Bohlmann and Zdero, 1978) and
Eremophila (Blackburne et al., 1972) species. It was recently isolated from
sweet potatoes treated with mercuric chloride (Burka and Iles, 1979). A
combination of column chromatography and HPLC was used to isolated 39 mg
of [16] per kilogram of treated sweet potato slices. In the same paper the
isolation of related keto alcohols 1-myoporol [17] (53 mg/kg) and 6-myoporol
[18] (109 mg/kg) were reported. The compounds were identified by spectral characteristics. Oxidation of [17] and [18] with pyridinium chlorochromate gave myoporone.

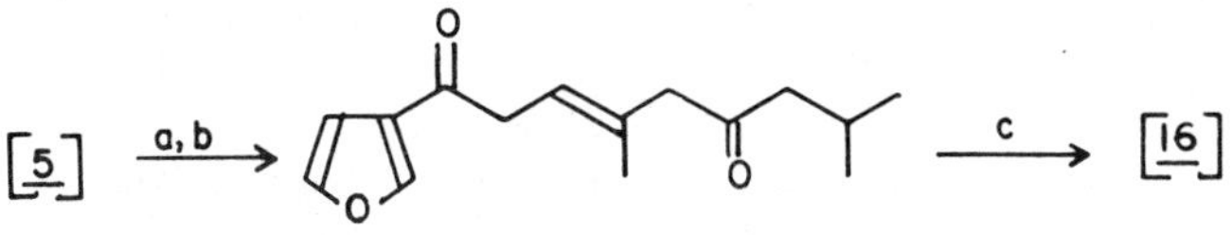

SCHEME 4 (a) $(COCl)_2$, (b) $Cd[CH_2CH(CH_3)_2]_2$, (c) H_2/Pd-C.

A synthesis of myoporone was reported by Kubota and Matsuura (1957) (Scheme 4). Ketoacid [5] from his ipomeamarone synthesis was utilized.

The synthesis of 6-myoporol was described by Still and Darst (1980) as part of their work on remote asymmetric induction. Thus, as shown in Scheme 5, *trans*-2,6-dimethyl-3,6-heptadiene [19] was hydroborated and oxidized to give diol [20] as the major product. This diol was then homologated to 6-myoporol.

Both the proton and carbon NMR spectra are available for myoporone: ^{1}H-NMR (CDCl$_3$) δ 0.90 (m, C-9, C-10, C-11), 1.4-2.2 (m, C-3, C-4, C-8), 2.28 (m, C-5, C-7), 2.76 (t, J = 7 Hz, C-2), 6.76 (m, C-3'), 7.44 (m, C-4'), and 8.04 (m, C-1'); ^{13}C-NMR (CDCl$_3$) δ 19.8 (C-11), 22.6 (c-9, C-10), 24.6 (C-8), 28.3 (C-4), 31.2 (C-3), 38.3 (C-2), 50.6 (C-5 or C-7), 52.5 (C-7 or C-5), 108.8 (C-3'), 127.8 (C-2'), 144.3 (C-1' or C-4'), 147.2 (C-4' or C-1'), 195.1 (C-1), and 210.7 (C-6). Major mass spectral fragments are at m/e 150, 110, and 95. The optical rotation was found to be -8.5°.

The proton NMR is available for 1-myoporol: ^{1}H-NMR (CDCl$_3$) δ 0.90 (d, J = 7 Hz, C-9, C-10, C-11), 1.34 (m, C-3), 1.5-2.5 (m, C-2, C-4, C-5,

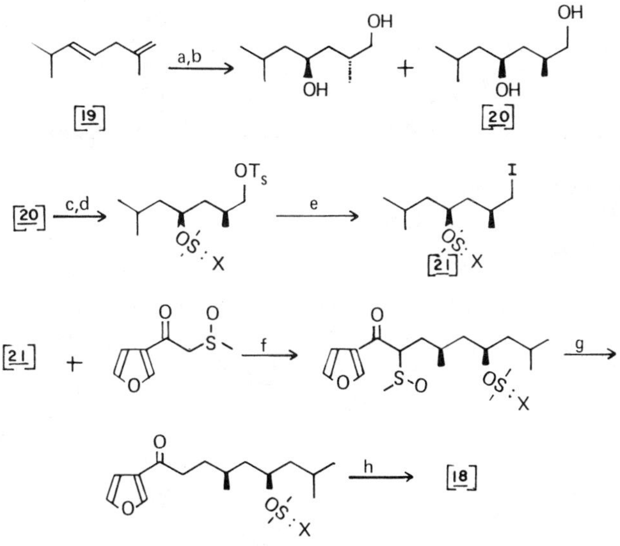

SCHEME 5 (a) Thexylborane, (b) NaOOH, (c) p-toluenesulfonyl chloride/ pyridine, (d) t-butylchlorodimethylsilane/imidazone/DMF, (e) NaI/acetone, (f) NaH/DMV, (g) Al[Hg]/THF/H$_2$O, (h) AcOH/H$_2$O/50°.

C-7, C-8), 4.66 (t, J = 7 Hz, C-1), 6.42 (m, C-3'), and 7.42 (m, C-1' and C-4'). The major mass spectral peaks are at m/e 252, 127, and 97. The optical rotation is -13°, not -6.4° as reported (Burka and Iles, 1979).

Both the proton and carbon NMR spectra are available for 6-myoporol: ^{1}H-NMR (CDCl$_3$) δ 0.91 and 0.93 (2d, C-9, C-10, C-11), 1.2-2.2 (m, C-3, C-4, C-5, C-7, C-8), 2.80 (t, J = 7 Hz, C-2), 3.88 (m, C-6), 6.80 (m, C-3'), 7.46 (m, C-4'), and 8.07 (m, C-1'); ^{13}C-NMR (Still and Darst, 1980) (CDCl$_3$) δ 20.33, 22.0, 23.4, 24.5, 29.1, 30.0, 37.8, 45.5, 47.0, 67.4, 108.6, 144.1, 147.0, and 195.6. Major mass spectral fragments are at m/e 123, 110, and 95. The optical rotation is -6.4°.

5. *4-Hydroxymyoporone*

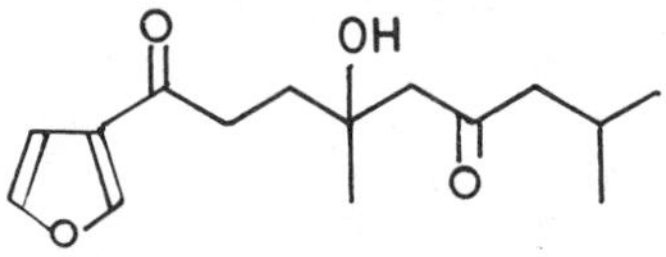

[22]

4-Hydroxymyoporone [22] was found in sweet potatoes infected either with *C. fimbriata* or *F. solani* or treated with mercuric chloride (Burka et al., 1974a). In one report 42 mg of [22] per kilogram of sweet potatoes treated with mercuric chloride was isolated using a combination of column chromatography and HPLC (Burka et al., 1974b). The compound can be easily recognized by the brilliant magenta color formed on reaction with Ehrlich's reagent. The compound was identified primarily by spectra, and its conversion to ipomeanine by base-catalyzed retro-aldol reaction, as discussed later.

No synthesis of 4-hydroxymyoporone has been reported. However, oxidation of ipomeamarone with t-butyl perbenzoate and cuprous chloride gives a good yield of 4-hydroxymyoporone (Burka and Kuhnert, 1977). Therefore, the ipomeamarone syntheses are also formally syntheses of 4-hydroxymyoporone.

The proton and carbon NMR spectra are available for 4-hydroxymyoporone: ^{1}H-NMR (CDCl$_3$) δ 0.92 (d, C-9, C-10), 1.22 (s, C-11), 1.5-2.4 (m, C-3, C-7, C-8), 2.55 (s, C-5), 2.90 (m, C-2), 4.09 (br s, OH), 6.77 (m, C-3'), 7.43 (m, C-4'), and 8.08 (m, C-1'); ^{13}C-NMR (CDCl$_3$) δ 22.5 (C-9, C-10), 24.4 (C-8), 26.7 (C-11), 34.8 (C-2 or C-3), 35.6 (C-3 or C-2), 52.4 (C-5 or C-7), 53.6 (C-7 or C-5), 71.1 (C-4), 108.7 (C-3'), 127.6 (C-2'), 144.2 (C-1' or C-4'), 147.4 (C-4' or C-1'), 195.2 (C-1), and 213.1 (C-6). Major mass spectral fragments are at m/e 166, 148, and 95. The optical rotation is +12.8°.

6. *Dehydro-4-hydroxymyoporone*

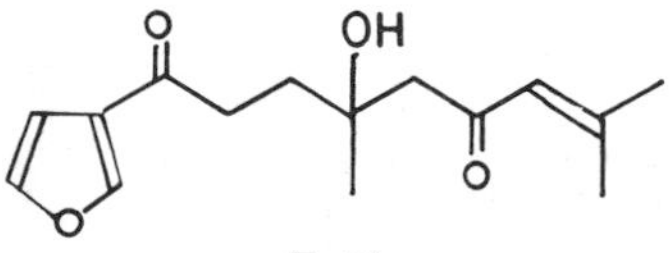

[23]

The isolation of dehydro-4-hydroxymyoporone [23] by a combination of column and thin-layer chromatography from *C. fimbriata*-infected sweet potatoes, was reported by Inoue and co-workers (1977). The identification was primarily by comparison of spectral data from 4-hydroxymyoporone to that of the new compound: ^{1}H-NMR (CDCl$_3$) δ 1.24 (s, C-11), 1.85 (m, C-3), 1.91 (d, J = 1.2 Hz, C-9), 2.17 (d, J = 1 Hz, C-10), 2.61 (s, C-5), 2.93 (m, C-2), 4.45 (br s, OH), 6.05 (m, C-7), 6.78 (q, C-3'), 7.43 (q, C-4'), and 8.07 (q, C-1'); mass spectral peaks occurred at m/e 246, 148, and 141.

7.   7-Hydroxymyoporone

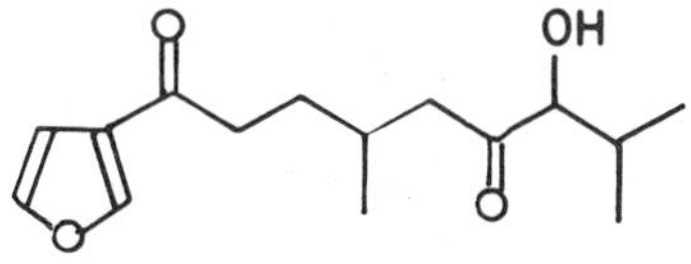

7-Hydroxymyoporone [24] was obtained from *C. fimbriata*-treated sweet potatoes by column chromatography followed by preparative gas chromatography of the trimethylsilyl ether. The ether was then cleaved using an alkylammonium fluoride (Burka et al., 1974b). Approximately 20 mg of [24] per kilogram of sweet potato slices was isolated in this way. The assignment of structure was based on spectra and degradation. Cleavage of the acyloin functionality by sodium periodate gave homobatatic acid [25] and, presumably, isobutyraldehyde. Homobatatic acid was subsequently synthesized (Scheme 6).

The synthesis of 7-hydroxymyoporone, reported by Reich and co-workers (1979, 1981), is shown in Scheme 7. This sequence uses organoselenium dianion [26] in an unusual approach to the construction of the acyloin moiety.

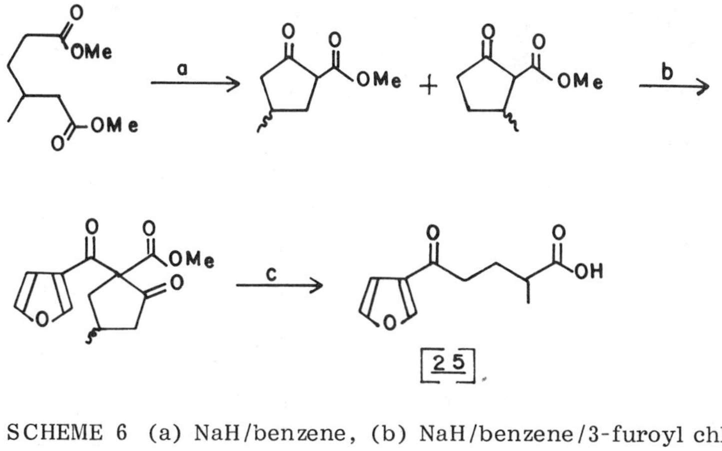

SCHEME 6   (a) NaH/benzene, (b) NaH/benzene/3-furoyl chloride, (c) NaOH.

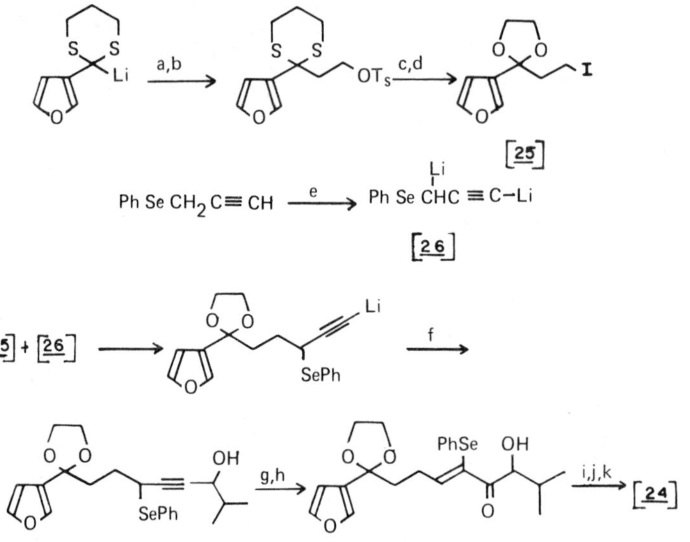

SCHEME 7 (a) Ethylene oxide, (b) p-toluenesulfonyl chloride/pyridine, (c) ethylene glycol/chloramine T, (d) NaI/acetone, (e) 2LDA, (f) iso-butyraldehyde, (g) peracid, (h) pyridine, (i) Me₂CuLi, (j) thiophenol, (k) AcOH/H₂O.

The synthesis gives a 30:70 mixture of compounds epimeric at C-4 and C-7. The spectra of the minor isomer correspond to those of the natural product.

Both the proton and carbon NMR spectra are available: ^{1}H-NMR (CDCl₃) δ 0.72 (d, J = 7 Hz, CH₃), 0.96 (d, J = 7 Hz, CH₃), 1.12 (d, J = 7 Hz, CH₃), 1.5-1.85 (m, C-4, C-8), 1.95-2.30 (m, C-3), 2.41 (m, C-5), 2.78 (t, J = 7 Hz, C-2), 3.37 (br s, OH), 4.01 (m, C-7), 6.72 (m, C-3'), 7.41 (m, C-4'), and 8.00 (m, C-1'); ^{13}C-NMR (CDCl₃) δ 14.7 (C-11), 19.4 (C-10 or C-9), 19.7 (C-9 or C-10), 28.7 (C-4), 30.8 (C-3 and C-8), 37.7 (C-2), 45.1 (C-5), 81.1 (C-7), 108.4 (C-3'), 127.4 (C-2'), 144.0 (C-4'), 146.9 (C-1'), 194.4 (C-1), and 211.6 (C-6). There are several intense peaks in the mass spectrum, at m/e 195, 194, 166, 147, 123, 110, and 95. The optical rotation is +12.3°.

8. 1-(3'-Furyl)-6,7-Dihydroxy-4,8-Dimethylnonan-1-One

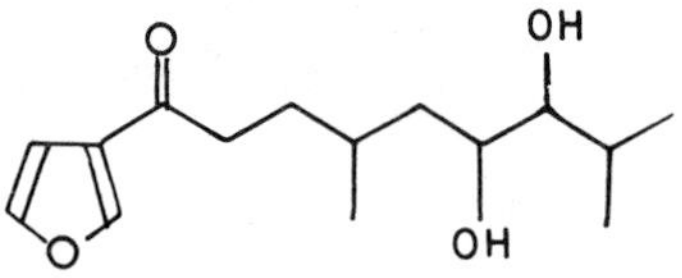

1-(3'-Furyl)-6,7-dihydroxy-4,8-dimethylnonan-1-one [27] was isolated from mercuric chloride-treated sweet potatoes (26 mg/kg) by column chromatography and HPLC (Burka, 1978). The structure was derived from spectra, oxidative degradation to homobatatic acid [25], and comparison to the products of partial reduction of 7-hydroxymyoporone. Compound [27] is a solid, mp 70-71°C: ^{1}H-NMR (CDCl$_3$) δ 0.94 and 0.97 (superimposed d, J = 7 Hz, C-9, C-10, C-11), 1.1-2.1 (m, C-3, C-4, C-5, C-8), 2.20 (br s, OH), 2.82 (t, J = 7 Hz, C-2), 3.14 (br t, J = 5 Hz, C-7), 3.75 (m, C-6), 6.84 (m, C-3'), 7.56 (m, C-4'), and 8.16 (m, C-1'); ^{13}C-NMR (CDCl$_3$) δ 16.7 (C-11), 19.1 (C-9 or C-10), 19.7 (C-10 or C-9), 29.1 (C-4 or C-8), 30.0 (C-8 or C-4), 32.0 (C-3), 38.0 (C-2), 41.0 (C-5), 69.7 (C-6), 79.6 (C-7), 108.6 (C-4'), 127.6 (C-2'), 144.0 (C-4'), 147.0 (C-1'), and 195.4 (C-1). Major mass spectral fragments are at m/e 196, 195, 123, 110, and 95. The optical rotation is +17°.

9. 6-Oxodendrolasin and 6-Hydroxydendrolasin

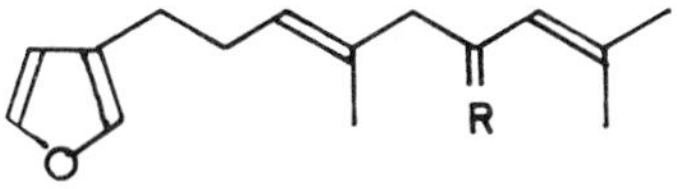

[28] R = O
[29] R = H, OH

6-Oxodendrolasin [28] was recently isolated from mercuric chloride-treated sweet potatoes (8.5 mg of [28] per kilogram) by a combination of column chromatography and HPLC (Burka et al., 1981). The structure was elucidated from spectral data and synthesis. Dimitriadis and Massy-Westropp (1980) isolated [28] from *Eremophila rotundifolia*. The compound has also been found in *Lasiospermum radiatum* and *Anthanasia crithmifolia* (Bohlmann and Rao, 1972).

6-Hydroxydendrolasin [29] was isolated from mercuric chloride-treated sweet potatoes (1.2 mg of [29] per kg) (Burka et al., 1981). This compound was also found in *E. rotundifolia* (Dimitriadis and Massy-Westropp, 1980).

The synthesis of [28] is shown in Scheme 8 (Burka et al., 1981). It depends on a tin-lithium exchange reaction to generate 3-furylmethyl lithium, which is then alkylated. The perillene [31] thus formed is regioselectively oxidized with selenium dioxide. Lithium aluminum hydride reduction of [28] gives 6-hydroxydendrolasin.

Proton and carbon NMR spectra were reported for both compounds: [28] ^{1}H-NMR (CDCl$_3$) δ 1.62 (s, C-11), 1.88 (s, C-10), 2.17 (s, C-9), 2.2-2.6 (m, C-1, C-2), 3.05 (s, C-5), 5.33 (br t, C-3), 6.13 (br s, C-7), 6.33 (m, C-3'), 7.27 (m, C-1'), and 7.38 (m, C-4'); ^{13}C-NMR (CDCl$_3$) δ 16.5 (C-11), 20.7 (C-10), 24.8 (C-1 or C-2), 25.7 (C-9), 28.7 (C-2 or C-1), 55.4 (C-5), 111.1 (C-3'), 123.1 (C-7), 124.8 (C-2'), 128.6 (C-3), 130.6 (C-4), 139.0 (C-1'), 142.6 (C-4'), 155.4 (C-8), and 199.0 (C-6); [29] ^{1}H-NMR (CDCl$_3$) δ 1.65 (s, CH$_3$), 1.68 (s, CH$_3$), 1.72 (s, CH$_3$), 2.14 (d, J = 6 Hz, C-5), 2.1-2.6 (m, C-1, C-2), 4.42 (m, C-6), 5.0-5.5 (m, C-3, C-7), 6.28 (m, C-3'), 7.23 (m, C-4'), and 7.35 (m, C-1'); ^{13}C-NMR (CDCl3) δ 16.3 (C-11), 18.8 (C-10), 24.8 (C-1 or C-2), 25.7 (C-9), 28.7 (C-2 or C-1), 48.2 (C-5), 66.1 (C-6), 110.9 (C-3'), 124.7 (C-2'), 127.7 (C-3 and C-7), 132.5 (C-4), 134.7

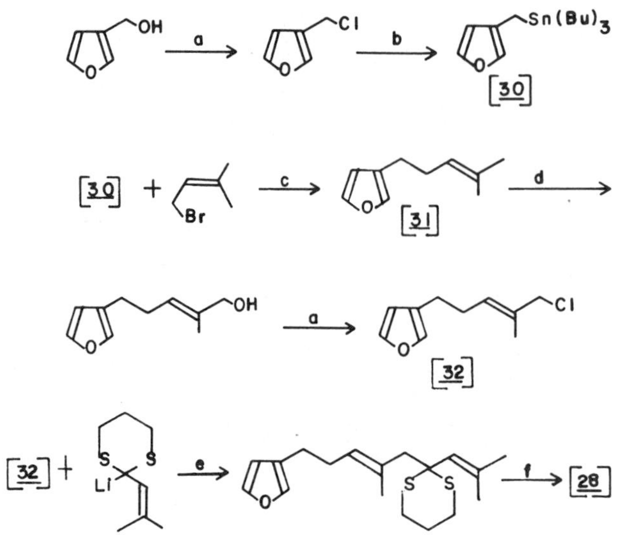

SCHEME 8 (a) LiCl/DMF/CH$_3$SO$_2$Cl/collidine, (b) n-Bu$_3$SnLi, (c) n-BuLi, (d) SeO$_2$/(CH$_3$)$_3$COOH, (e) LDA, (f) AgNO$_3$/N$_2$O/EtOH.

(C-8), 138.9 (C-1'), and 142.7 (C-4'). Major mass spectral fragments are at m/e 151, 134, and 83 for [28] and at m/e 150, 136, and 135 for [29]. The optical rotation of [29] is -2.4°.

Inspection of the proton NMR spectra for the stress metabolites allows a generalization to be made as to the appearance of the furan region as a function of the oxidation state of C-1. If C-1 is a ketone, the proton signals appear as three more or less equally spaced multiplets at δ 6.8, 7.4, and 8.1. If C-1 is in the alcohol state of oxidation as in [1] or [17], the signals for C-1' and C-4' are not well resolved (at 100 MHz) and one obtains a one-proton multiplet at about 6.4 and a two-proton multiplet at 7.4. If C-1 is a methylene group, one observes signals at 6.3, 7.25, and 7.35.

B. Stereochemistry

As a result of synthetic and degradative studies on the preceding furanosesquiterpenes, considerable information has been gathered concerning the configurations of asymmetric centers in the molecules. The fact that ngaione and ipomeamarone are cis compounds was discussed earlier. Additional information garnered by Hegarty and co-workers (1970) allows assignment of absolute configuration at C-1 and C-4. These authors found (-)ngaione to have a 1(S),4(R)-configuration. It follows that (+)ngaione (ipomeamarone), the enantiomer, is 1(R),4(S). It was observed that the copper-catalyzed t-butyl perbenzoate oxidation of ipomeamarone leads to 4-hydroxy-myoporone [22] with the same sign and magnitude of rotation as naturally occurring [22] (Burka and Kuhnert, 1977). Thus C-4 in [22] also has the S-configuration.

$$\boxed{33}$$

Several degradations have led to isolation of an unoxygenated chiral center at C-4. Batatic acid, [33], probably derived by catabolism of furanosesquiterpenes, was isolated from *C. fimbriata*-infected sweet potatoes by
Kubota (1958). Ozonolysis of [33] gave S-(+)-2-methylglutaric acid (Kubota,
1958). Blackburne and co-workers (1972) ozonized (-)myoporone (same sign
as the sweet potato isolate) to dimethyl S-(+)-2-methylglutarate. Homobatatic acid from 7-hydroxymyoporone and the related diol [27] was ozonized to
S-(-)-3-methyladipic acid (L. T. Burka and L. J. Felice, unpublished observations, 1979). An S-configuration at C-4 seems a general occurrence
in these stress metabolites.

The hydride reduction of 7-hydroxymyoporone [24], part of the structure
proof of [27], gave an intriguing result. The diol obtained was identical in
all respects to the natural product except that the optical rotation was opposite in sign (Burka, 1978). Since both [24] and [27] have the 4(S)-configura-
tion, they must be epimeric at C-7. This being true, they must result from
divergent biosynthetic pathways, and neither is the precursor of the other.

The stereospecific synthesis of 6-myoporol by Still and Darst proved C-4
and C-6 in the natural product to have a threo stereochemistry.

C. Biosynthesis

The biosynthesis of furanosesquiterpenes in the sweet potato has received
considerable attention. All the furans isolated thus far have a normal,
unrearranged terpene skeleton. Early studies demonstrated the incorporation of usual terpene precursors; acetate (Akazawa, 1964; Hyodo et al.,
1968; Akazawa and Uritani, 1962; Akazawa et al., 1962; Oba et al., 1970),
leucine (Oshima-Oba et al., 1969), mevalonate (Akazawa, 1964; Akazawa et
al., 1962; Oshima and Uritani, 1968), and farnesol (Oguni and Uritani, 1970,
1971). In most of these studies, and in later ones as well, ipomeamarone [1]
was the compound isolated for study. It is a major constituent (about 1 mg
of [1] per gram of sweet potato), and it is easily isolated and derivatized.
Most of the recent work has centered on the oxidation and reduction processes that are required to convert farnesol pyrophosphate to ipomeamarone.
Very early in these studies it was found that dehydroipomeamarone was
efficiently incorporated into [1] (Oguni and Uritani, 1973, 1974a,b).
Recently, an incorporation of about 1% for [^{14}C]-6-oxodendrolasin into [1]
has been demonstrated (Burka et al., 1981). Evidence for oxidation of the
side chain before furan ring formation has also been found since a 1% incorporation of 9-hydroxyfarnesol into [1] was also observed (Burka et al., 1981).
Only 0.1% or less of the radioactivity from [^{14}C]dendrolasin was incorporated
(L. T. Burka, J. Illes, and L. J. Felice, unpublished observations, 1978).

9-Hydroxyfarnesol has been isolated as a stress metabolite; dendrolasin has not been detected in sweet potatoes (L. T. Burka, J. Iles, and L. J. Felice, unpublished observations, 1978). The percent incorporation for dendrolasin is quite high for complex precursors in higher plants, but since it is sufficiently low in comparison with the other sweet potato feeding experiments, it has been discounted as a precursor.

Furanosesquiterpenes beyond [1] in the biosynthetic pathway have also been studied. Ipomeamarone is oxidized initially to two compounds, 4-hydroxy-myoporone [22] and ipomeamaronol. About 1% incorporation of ^{14}C-[1] into [22] (Burka and Kuhnert, 1977) and nearly 7% incorporation of ^{14}C-[1] into ipomeamaronol was found (L. T. Burka and L. J. Felice, unpublished observations, 1979). Inoue and Uritani reported the reduction of dehydro-4-hydroxymyoporone [23] to [22] (Inoue and Uritani, 1980). Metabolite [23] probably results from oxidation of dehydroipomeamarone in much the same way as [22] arises from [1].

The radiolabel incorporation results are summarized in Scheme 9. The heavy arrows are direct experimental results, whereas the dotted arrows delineate one possible pathway leading from farnesol to ipomeamarone. A trioxygenated intermediate between 6-oxodendrolasin and dehydroipomeamarone has not been found.

D. Toxic Properties of 3-Substituted Furanosesquiterpenes

The toxic effects of stress metabolites having an undegraded sesquiterpene structure have received little attention compared to the lung toxic compounds discussed later. Some detailed studies have been carried out on ngaione, the enantiomer of ipomeamarone, by Seawright and co-workers. This work has been reviewed recently (Seawright et al., 1978) and in Chap. 16. Ngaione, epingaione, and ipomeamarone all have an LD$_{50}$ value of about 200 mg/kg by intraperitoneal administration in mice and all produce the same

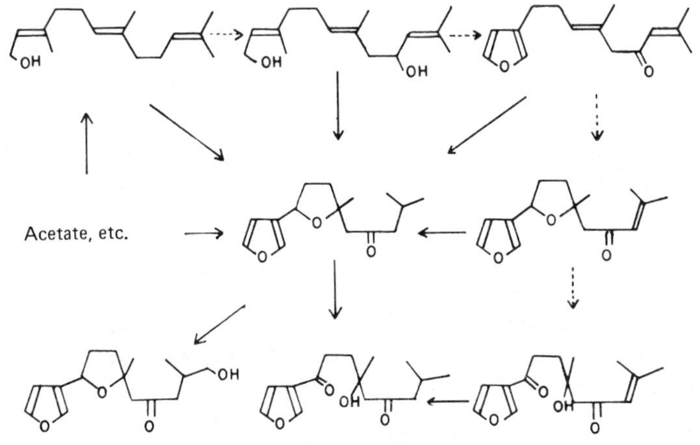

SCHEME 9

liver lesion (Seawright and Mattocks, 1973). Rats and mice dosed with
ngaione develop midzonal necrosis of the hepatic parenchymal cells (Seawright
et al., 1978). Pretreatment of mice with phenobarbital results in periportal
necrosis, whereas pretreatment with SKF525A increases the LD_{50} value to
about 500 mg/kg and results in centrilobular necrosis (Seawright and
Hrdlicka, 1972). These results suggest that ngaione metabolism by hepatic
mixed-function oxidase is required for toxicity (Seawright et al., 1978).

Wilson and Burka have reported toxicity data for mice on five of the furanosesquiterpenes from sweet potatoes (Wilson and Burka, 1979). The compounds tested and their LD_{50} values for intraperitoneal administration were
as follows: ipomeamaronol [15] (266 mg/kg), 6-myoporol [18] (84 mg/kg),
4-hydroxymyoporone [22] (235 mg/kg), 7-hydroxymyoporone [24] (200 mg/
kg), and dihydro-7-hydroxymyoporone [27] (184 mg/kg). All five compounds caused marked neurological effects shortly after injection, but these
reactions usually disappeared within an hour. Death usually occurred in
less than 24 hr but not more than 48 hr after administration of the compound.
Postmortem examination of animals dying in less than 24 hr revealed an enlarged and congested liver. Microscopically, liver damage consisted of extensive necrosis of hepatocytes extending from the centrilobular zone nearly
to the periportal zone. No evidence of pulmonary edema or damage to the
kidneys was observed by light microscopy.

Myoporone [16] from *Myoporum* spp. was reported to have an LD_{50} value
of 180 mg/kg by intraperitoneal administration in mice; the pathology was
typical of ngaione poisoning (Blackburne et al., 1972). A dose of 230 mg/kg
resulted in no effect on a sheep (Blackburne et al., 1972).

6-Oxodendrolasin has an LD_{50} value of 230 mg/kg in mice by intraperitoneal administration (L. T. Burka and A. Thorsen, unpublished observations, 1980). The gross pathology is the same as that of the other furanosesquiterpenes.

III. LUNG-TOXIC 3-SUBSTITUTED FURANS FROM SWEET POTATOES

A. Initial Detection in Toxic Sweet Potatoes

The 1969 outbreak of sweet potato poisoning in a herd of Hereford cattle in
Tift County, Georgia (Peckham et al., 1972), mentioned in the introduction,
provided Vanderbilt University investigators with a ready source of toxic
sweet potato roots for investigations on the causative agent of the bovine
pulmonary disease. Samples exhibited variable states of decomposition and
yielded more than 150 fungal isolates (Peckham et al., 1972). Roots showing
most advanced deterioration usually were least toxic, however. Ether extracts of specimens with moderately invasive blemishes were found to contain
the toxic principle as demonstrated by injection of the extracts into laboratory mice. After a few hours they became dyspneic and died within 24 hr,
apparently of anoxia. Marked pleural effusion and intraalevolar edema were
prominent (Wilson et al., 1970).

Several of the more prominent fungal isolates were cultured on heat-
sterilized sweet potato preparations and on various laboratory media.

Extracts of the cultures failed to reproduce the respiratory syndrome in mice. Only one of the many isolates studied, *Fusarium solani (javanicum)*, grown on viable sweet potato slices, stimulated the production of ether soluble, lung-toxic material as determined by oral and parenteral injections of mice (Wilson et al., 1971) and by feeding artificially infected sweet potatoes to cattle (Peckham et al., 1972). The same *F. solani* isolate grown on corn and fed to day-old chicks by other investigators caused reduced weight gains and deaths (Peckham et al., 1972). The avian pathology was not described, and any toxic factor present must have been unrelated to the lung toxins obtained from the living (infected) sweet potato slices.

Since most Japanese reports of sweet potato stress metabolism had employed the black-rot fungus, *Ceratocystis fimbriata*, this organism was also tested as a stimulator of lung toxin formation. In this case considerable quantities of ipomeamarone (Boyd and Wilson, 1971) and other hepatotoxic metabolites were obtained but without any trace of lung toxins.

Polar solvent extracts of *F. solani*-infected sweet potato slices were partially separated on thin-layer silica gel plates developed in 10% methanol in benzene. These plates were sprayed with modified Ehrlich reagent (dimethylaminobenzaldehyde containing HCl) to reveal numerous brightly colored spots varying from light pink to purple to dark gray. Ipomeamarone and other hepatotoxins previously described were detected. A particular purple spot at $R_f = 0.4$, containing several compounds detectable by gas-liquid chromatography, included the lung-toxic substances 4-ipomeanol and 1-ipomeanol, described later in this chapter (Boyd et al., 1974).

The pulmonary toxic response obtained with sweet potato material devoid of ipomeamarone and other hepatotoxic metabolites led Wilson and co-workers to postulate the presence of a distinct lung-toxic factor in *F. solani*-infected sweet potato root tissue (Wilson et al., 1970). This assumption was sustained in subsequent work which led to isolation of the four closely related 1,4-dioxygenated-1-(3-furyl)pentanes [34-37].

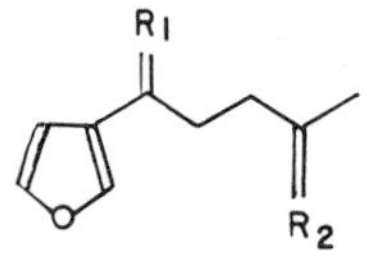

[34] $R_1 = 0;\ R_2 = H, OH$

[35] $R_1 = H, OH;\ R_2 = 0$

[36] $R_1 = R_2 = H, OH$

[37] $R_1 = R_2 = 0$

B. Bioproduction and Isolation

The lung-toxic compounds were bioproduced by inoculating slices of unblemished sweet potatoes with liquid shake cultures of *F. solani* and allowing the fungus to grow for approximately 6 days at room temperature (22-25°C). Acetone extracts of the infected slices were then partitioned on a column of silica gel using a hexane-ethyl ether gradient. Ipomeamarone and ipomeanine were contained in the hexane fractions; the ipomeanols eluted with the 10-15% ether fractions, and the diol was found in the 75-100% ether eluates (Boyd et al., 1974).

Preparative gas-liquid chromatographic separation of the hydroxylated compounds was accomplished following silylation to prevent dehydration on the column. The lung toxins were further purified using HPLC. Spectral and physical data on the naturally occurring toxins were compared with synthetic compounds subsequently prepared. Details of the foregoing methodology are described in a report by Boyd and co-workers (1974).

Ipomeanine [37] was usually obtained in very small quantities. It had been isolated from sweet potatoes earlier by Kubota and Ichikawa (1954), but its toxicity had not been reported. 4-Ipomeanol [34] was most abundant and probably most responsible for the sweet potato lung toxicity. Its isomer, 1-ipomeanol [35], was somewhat less stable and less toxic. The presence of the diketone and the ketols suggested the possibility that the corresponding diol, 1,4-ipomeadiol [36], might also be present. The diol was obtained by reduction of ipomeanine, which provided an authentic standard and led to its identification as a naturally occurring member of the group (Boyd et al., 1974).

C. Synthesis

Syntheses of ipomeanine (Kubota and Ichikawa, 1954), 4-ipomeanol (Boyd et al., 1972), 1-ipomeanol (Boyd et al., 1974), and 1,4-ipomeadiol (Boyd et al., 1974) using ethyl 3-furoylacetate [3] as a common intermediate have been reported. They are outlined in Scheme 10. The syntheses are straightforward and provide an easy access to these interesting compounds. A somewhat shorter and more efficient synthesis of 1-ipomeanol has been described by Woesser (1979) (Scheme 11).

Proton NMR spectra are available for all four compounds (Boyd et al., 1974). Carbon spectra are available for [34] and [35] only: [34] ^{1}H-NMR (CCl$_4$) δ 1.15 (d, J = 6 Hz, C-11), 1.67 (m, C-3), 2.50 (br s, OH), 2.83 (t, J = 7 Hz, C-2), 3.73 (m, C-4), 6.75 (m, C-3'), 7.42 (m, C-4'), and 8.07 (m, C-1'); ^{13}C-NMR (CDCl$_3$) δ 23.3 (C-11), 33.2 (C-3), 36.4 (C-2), 66.5 (C-4), 108.5 (C-3'), 127.6 (C-2'), 144.1 (C-4'), 147.7 (C-1'), and 195.3 (C-1); [35] ^{1}H-NMR (CCl$_4$) δ 2.22 (s, C-11), 2.0-2.2 (m, C-2), 2.70 (m, C-3), 4.86 (t, J = 7 Hz, C-1), 6.64 (m, C-3'), and 7.67 (m, C-1', C-4'); ^{13}C-NMR (CDCl$_3$) δ 29.4 (C-11), 31.5 (C-2), 39.2 (C-3), 65.4 (C-1), 108.7 (C-3'), 129.3 (C-2'), 138.7 (C-1'), 142.8 (C-4'), and 208.7 (C-4). 4-Ipomeanol has major peaks in the mass spectrum at m/e 110 and 95; the natural product is optically active, [α] +8°. The trimethysilyl ether of 1-ipomeanol has major peaks in the mass spectrum at m/e 182, 169, 75, and 73; the natural product is also optically active, [α] +22°. The proton NMR (CDCl$_3$) of [36] has absorptions at δ 1.17 (d, J = 6 Hz, C-11), 1.59 (m, C-3), 1.82 (m, C-2), 3.94 (m, OH, C-4), 4.75 (m, C-1), 6.55 (m, C-3'), and 7.55 (m, C-1', C-4'). Naturally occurring 1,4-ipomeadiol has a zero or slightly positive rotation at the D line of sodium. The proton NMR (CCl$_4$) of [37] has absorptions at δ 2.12 (s, C-11), 2.87 (m, C-2, C-3), 6.75 (m, C-3'), 7.44 (m, C-4'), and 8.08 (m, C-1'). Ipomeanine is a solid, mp 41-42°C.

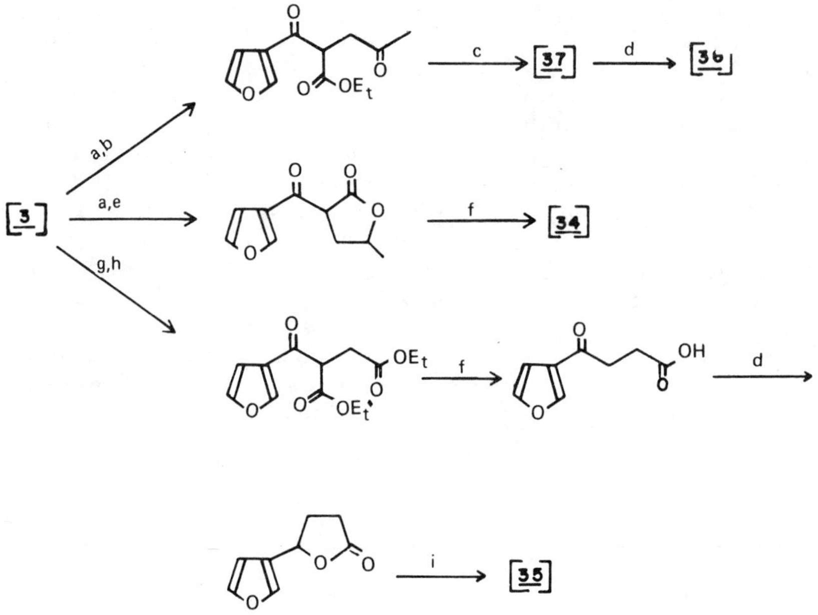

SCHEME 10 (a) NaOEt/EtOH, (b) $BrCH_2COCH_3$, (c) dil. NaOH, (d) $NaBH_4$, (e) propylene oxide, (f) H^+, (g) NaH/DME, (h) $BrCH_2CO_2Et$, (i) CH_3Li/ ether/-78°.

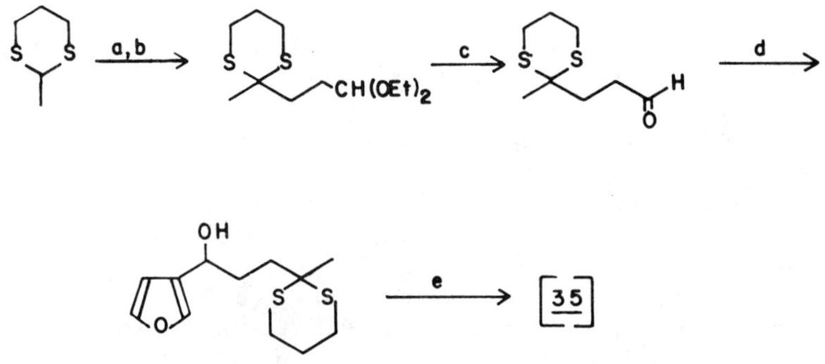

SCHEME 11 (a) n-BuLi/THF, (b) $ClCH_2CH_2CH(OEt)_2$, (c) H^+, (d) 3-furyl-lithium, (e) $Hg\bar{C}l_2/CaCO_3$/acetone/H_2O.

D. Biosynthesis

The 1,4-dioxygenated-1-(3-furyl)pentanes [34]-[37] have an interesting biogenesis. They are not found in sweet potatoes infected with many common
fungal pathogens nor are they present in sweet potatoes treated with mercuric chloride. The compounds are produced if *F. solani* or *F. oxysporum*
and certain other species of *Fusarium* are grown on raw, but not on autoclaved, sweet potatoes. It was postulated that catabolism of one or more of
the sesquiterpene stress metabolites by the *Fusaria* was the biogenesis of
[34]-[37] (Burka and Wilson, 1976). It was shown that incubation of [^{14}C]-
4-hydroxymyoporone with fungal mats of *F. solani* led to [^{14}C]-4-ipomeanol
and [^{14}C]ipomeanine with no dilution of radiolabel (Burka et al., 1974a, 1977).
F. oxysporum was also able to convert 4-hydroxymyoporone to [34]-[37],
although *C. fimbriata* did not degrade the sesquiterpene.

E. Toxicity

The outbreak of bovine pulmonary disease mentioned in the introduction,
attributed to the feeding of moldy sweet potatoes (Peckham et al., 1972),
was given the descriptive name "atypical interstitial pneumonia" by veterinary pathologists who conducted the postmortem examinations.

Clinical findings, which began a few days after the cattle were given access to the sweet potatoes, consisted of severe respiratory distress with a
rapid respiratory rate and frothy exudate around the mouth shortly before
death.

Gross lesions were confined to the respiratory tract in most cases. The
lungs were wet, firm, and large and did not collapse when the thorax was
opened. All lobes of the lungs had marked interstitial and interlobular
emphysema.

Light microscopy revealed that the lungs were quite edematous with
alveolar and interstitial emphysema, proliferation of the alveolar epithelium,
and hyaline membrane formation. Foci of conjestion and hemorrhage were
also prominent.

Doster and co-workers (1978), using a small number of cross-bred, 18-
month-old heifers, found the bovine to be among the most susceptible species
to 4-ipomeanol. The minimum nonlethal dose given intraruminally was estimated to be between 7.5 and 9.0 mg/kg body weight. Animals receiving 9-14
mg/kg toxin died within 3-4 days after dosing and at necropsy exhibited
many of the features of the natural outbreaks of sweet potato poisoning.
Two animals receiving doses of 6.0-7.5 mg/kg had proportionally fewer pulmonary changes, although alveolar septa were thickened and hypertrophy of
alveolar epithelium was multifocal.

Insight into the ultrastructural series of events occurring in the bovine
lung following intraperitoneal injection of 4-ipomeanol was provided by recent
work of Doster and co-workers using 3-week-old Holstein calves (Doster,
personal communication, 1982). Pulmonary tissues were examined both by
transmission and scanning electron microscopy. Results of this study will
be published at a later time.

Each of the lung-toxic 1,4-dioxygenated-1-(3-furyl)pentanes produce
identical reactions in the lungs of mice whether the doses (in 50% propylene

glycol-water) are given orally, intraperitoneally, or by intravenous injection. After receiving lethal doses mice became increasingly dyspneic and die, usually within 24 hr, following brief convulsive seizures apparently induced by anoxia. Postmortem examination reveals changes confined to the lungs and pleural cavity. The lungs, which show mild congestion, are surrounded by clear pleural fluid which clots upon standing. Intrapulmonary fluid is demonstrable by holding an animal with its nose downward, which allows fluid to drip from the nostrils (Wilson et al., 1971).

Microscopic sections of mouse lung show widespread perivascular, peribronchiolar, and intraalveolar edema. Most blood vessels are congested and the bronchiolar lining cells are eroded. No significant gross or microscopic pathology is seen in the other organs. LD_{50} figures derived from the format prescribed by Balazs (1970) and calculated by the method of Miller and Tainter (1944) are shown in Table 1.

More detailed studies on microscopic aspects of the lung lesions in mice and rats revealed that minimal doses of 4-ipomeanol produced necrosis of nonciliated (Clara) cells in the smallest bronchioles as the only significant reaction. Larger or lethal doses, however, affected nonciliated cells in the larger airways as well. It has been suggested that the bronchiolar effects represent primary lesions, with other pulmonary effects being secondary or tertiary events (Boyd, 1977).

In addition to the pulmonary effects in mice, adult male animals surviving near-lethal lung-toxic levels of the toxins often became cachectic over a few days and died of nonpulmonary disease. The kidneys were found to be enlarged and pale. Microscopically, the tubular epithelium was necrotic and calcium stain revealed deposits in the cellular debris. Immature male mice were highly resistant to this effect, as were adult females. Nephrotoxicity was not evident in any other species of laboratory animal tested (Boyd and Dutcher, 1978).

The rat, rabbit, and guinea pig were also quite susceptible to 4-ipomeanol (Dutcher and Boyd, 1978). Lethal doses were in the range 10-60 mg/kg. In the rat pleural effusion was rarely noted, although perivascular and intra-alveolar edema were prominent. A positive relationship was noted between increasing intraperitoneal doses in this species and ratios of lung wet weight to lung dry weight. In addition to intraalveolar edema, the guinea pig also exhibited extensive vascular hemorrhage, resulting in dark red, fluid-filled lungs at death. The hamster was somewhat more resistant to the pulmonary

TABLE 1 LD_{50} Values (mg/kg) in Mice for Synthetic Lung Toxins

	Intravenous	Intraperitoneal	Oral
Ipomeanine	14 ± 1	25 ± 1	26 ± 1
4-Ipomeanol	21 ± 1	36 ± 4	38 ± 3
1-Ipomeanol	34 ± 6	49 ± 2	79 ± 9
1,4-Ipomeadiol	68 ± 8	67 ± 8	104 ± 12

effects of 4-ipomeanol with the intraperitoneal LD$_{50}$ value approximating 150 mg/kg. Centrilobular liver lesions were also noted occasionally at the high dosage levels (Dutcher and Boyd, 1979).

In birds, including the Japanese quail and the chicken, 4-ipomeanol failed to produce pulmonary or renal lesions; instead, liver appeared to be the main target organ. This phenomenon was related to the fact that these avian species lack both ciliated and nonciliated lining cells found in other verte-brates (Buckpitt and Boyd, 1978; Boyd, 1980).

F. Distribution and Excretion of 4-Ipomeanol

1. Synthesis of [^{14}C] 4-Ipomeanol

Syntheses of [^{14}C]-labeled 4-ipomeanol are based on the synthesis of un-labeled compounds as outlined in Scheme 11. In one preparation, 2-[^{14}C]- or 1,3-[^{14}C]acetone, was brominated and the product used to synthesize [^{14}C]ipomeanine (Boyd et al., 1975). Partial reduction of the ipomeanine gave [^{14}C]4-ipomeanol as a mixture with the other dioxygenated furylpen-tanes. The mixture was then separated by chromatography. [^{14}C]-4-Ipomeanol has also been synthesized from [^{14}C]propylene oxide (Boyd, 1977). [^{3}H]-4-Ipomeanol has been prepared by an exchange reaction with tritium oxide using rhodium on alumina as a catalyst (Boyd, 1977).

2. Disposition of 4-Ipomeanol in Rats

Distribution and excretion of [^{14}C]-4-ipomeanol has been determined using male Sprague-Dawley rats (Boyd et al., 1975). A characteristic pattern emerged which was consistent regardless of the dose administered, which radioactive compound was used, or whether the dose was given intraperi-toneally or orally. Ten milligrams per kilogram, a nonlethal but toxic intra-peritoneal dose, was chosen for detailed studies.

Within 2-4 hr approximately half the administered radioactivity appeared in the urine as unmetabolized toxin. The total urinary excretion increased only slightly by 96 hr. Much smaller amounts of radioactivity were detected in expired air and feces within 24 hr; the fecal excretion, containing un-identified metabolites, increased to 15-20% of the administered dose by 96 hr, whereas the amount in expired air was less than 5% at that time.

Determinations of tissue radioactivity were expressed as relative specific activities for several organs and tissues. Lung, liver, and kidney reached peak levels within 1-2 hr after intraperitoneal injection, and at 24 hr they had established relatively high plateaus of radioactivity, 70-90% of which was apparently covalently bound to tissue macromolecules. The lung contained highest bound activity followed by the liver and kidney. Intravenous injec-tion of toxin established that peak covalent binding to lung tissue occurred in 40 min.

G. Toxic Mechanisms

1. Covalent Binding and Toxicity of 4-Ipomeanol

All subcellular fractions of lung and liver in [^{14}C]-4-ipomeanol-injected rats contained significant levels of bound radioactivity after 24 hr, which were roughly equivalent to those found in the respective crude homogenates

(Boyd and Burka, 1978). Sephadex chromatography of solubilized residues revealed that the radioactivity eluted primarily with protein. A similar predilection for target organ alkylation exists for lungs of rabbits and guinea pigs (Dutcher and Boyd, 1979). In keeping with earlier studies showing marked toxicity of 4-ipomeanol for the kidneys of adult male mice (Boyd et al., 1974), radioactive toxin was noted to bind covalently to the tubular epithelium of this species. The same was true for hepatic cells of hamsters. This preferential binding to liver or kidney was in addition to that seen in the lungs of those animals in which pulmonary edema is also a prominent feature of the toxic response. However, chickens and the Japanese quail bind 4-ipomeanol preferentially to the liver, which is the major target organ in these avian species (Buckpitt and Boyd, 1978).

Rat experiments by Boyd and Burka (1978) demonstrated that covalent binding, pulmonary edemagenesis, and lethal effects were related events that also correlated with the quantities of toxin administered. Doses in the range 2-10 mg/kg elicited a mild increase in lung wet weight, whereas doses beyond 10-15 mg/kg evoked a marked increment in lung fluid accumulation that could lead to death of the animals. These observations suggested that saturation of normal pulmonary mechanisms for removal of excess fluid occurs, resulting in fatal anoxia.

2. Role of Microsomal Enzymes in Covalent Binding of 4-Ipomeanol

In vitro studies: The conclusion that cytochrome P_{450}-dependent monooxygenase enzymes were required for activation and binding of 4-ipomeanol was based on several in vitro studies in which [^{14}C]-4-ipomeanol was incubated with microsomes of target organ tissues. Oxygen and nicotinamide adenine dinucleotide phosphate dehydrogenase (NADPH) were required for optimal covalent binding of toxin to microsomal protein. The presence of carbon monoxide or cytochrome c resulted in greatly decreased binding. When heat-denatured microsomes were used or when the incubation was carried out at 1-2°C, little or no covalent binding was observed. Cyanide ion did not affect binding. Microsomes for these experiments were obtained from both lung and liver of mice, hamsters, guinea pigs, and rabbits, from kidneys of adult male mice, and from chicken liver, all of which are capable of binding the toxin. Microsomes of chicken lung were inactive in binding toxin.

Experiments employing recognized inhibitors of monooxygenase enzymes demonstrated decreased covalent binding of 4-ipomeanol to both rat lung and liver microsomes in vitro (Boyd et al., 1978a). The inhibitors included pyrazole, piperonyl butoxide, and SKF525A. Cobaltous chloride, presumably as a result of its ability to decrease cytochrome P_{450}, also decreased binding, as did antibody to purified rat liver NADPH-cytochrome c reductase.

Guengerich (1977a,b), employing highly purified cytochromes P_{450}, plus purified NADPH-cytochrome P_{450} reductase and phospholipid, succeeded in demonstrating covalent binding of 4-ipomeanol to protein using enzymes from rat and rabbit liver and rabbit lung. He noted that reconstituted enzyme mixtures containing rat liver cytochrome P_{450} induced by either 3-methylcholanthrene or phenobarbital could catalyze in vitro covalent binding of 4-ipomeanol. This finding was consistent with previous experiments by other investigators (Boyd et al., 1978a) which demonstrated the activation of 4-ipomeanol by liver microsomes of rats given the same pre- treatments and further correlated with an increase in liver microsomal cyto- chrome P_{450} and increased covalent binding of 4-ipomeanol.

Rat lung cytochrome P_{450} content was not increased by phenobarbital pretreatment, and the in vitro lung microsomal alkylation by 4-ipomeanol was unaffected. On the other hand, although 3-methylcholanthrene increased lung cytochrome P_{450} levels, it caused no significant change, or in some cases only a slight lowering of the covalently bound toxin to lung microsomes (Boyd et al., 1980).

Attempts to elucidate the foregoing differences between lung and liver alkylation by 4-ipomeanol led to experiments (Boyd et al., 1978a) showing that the K_m value for lung microsomal alkylation was more than 10-fold lower than for microsomes of the liver. The V_{max} values for lung and liver micro- somal alkylation were approximately equal when expressed in quantity of toxin bound per unit of protein per minute. However, maximal rates in terms of microsomal P_{450} content were more than 10-fold greater for lung micro- somes than for those of the liver. Other studies (Sasame and Boyd, 1978) suggested that the high rate of lung alkylation was not caused by a defi- ciency in lung microsomal detoxication of 4-ipomeanol.

Studies by Sasame and co-workers (Sasame and Boyd, 1978; Sasame et al., 1978) investigated the role of cytochrome b_5 on the alkylation of 4-ipomeanol for lung tissue. Using antibody to cytochrome b_5, an inhibitory effect on nucleotide adenine dinucleotide hydrogenase (NADH)-mediated reduction of cytochrome c and the NADPH-mediated metabolic activation of 4-ipomeanol by lung microsomes was noted. This was in contrast to the situation with liver microsomes, in which inhibition of NADH-mediated cytochrome c reduction effected only small decreases in the rate of 4-ipomeanol activation. The cytochrome b_5 antibody significantly inhibited the disappearance and covalent binding of 4-ipomeanol using either NADPH or NADH or a mixture of the two cofactors. NADPH was a more effective cofactor than NADH; the two sub- stances were neither synergistic nor additive in their action. Boyd (1980) has proposed a mechanism for the in vitro pulmonary microsomal activation of 4-ipomeanol which stipulates a two-electron transfer in the lung cyto- chrome P_{450} scheme; one or both electrons could be obtained from NADPH- cytochrome c reductase. However, transfer of the second electron by way of cytochrome b_5 is the rate-limiting step in lung microsomes. The possible role of this overall mechanism for in vivo activation of the toxin or other xenobiotics in lung tissue must await further study.

In vivo experiments: Experiments have demonstrated that metabolic acti- vation and covalent binding of 4-ipomeanol also occur in vivo and that lung is the main target organ (Boyd and Burka, 1978). Pyrazole, piperonyl butoxide, and cobaltous chloride, as in the in vitro studies, decreased the covalent binding and hence the toxicity of the toxin. SKF525A, however, had no significant effect on the binding or toxicity. Piperonyl butoxide in decreasing pulmonary binding also increased both blood and pulmonary levels of unmetabolized toxin, which were found to be nearly equal. Comparable levels of toxin in blood and lungs, which also attains in control animals, sug- gests that marked pulmonary binding of 4-ipomeanol is not due to selective pulmonary uptake or accumulation of nonactivated toxin.

As pretreatment agents for in vivo rat toxicity studies, the metabolic inducers phenobarbital and 3-methylcholanthrene, showed striking differences in their action (Boyd and Burka, 1978). Pretreatment with the latter agent resulted in a marked increase in covalent binding of 4-ipomeanol in the liver, resulting in centrilobular hepatic necrosis but diminished lung binding and the lung edemagenic response. This latter phenomenon was interpreted to be the result of increased hepatic clearance and shortened biological half-life of the administered toxin. Studies in adult male mice also supported the idea of enhanced in situ activation and binding of toxin in the livers of 3-methylcholanthrene-pretreated mice, with a corresponding decrease in both lung and kidney binding and the respective toxic responses. This phenomenon was not seen in a strain of noninducible mice.

Phenobarbital pretreatment, which was shown to increase the rate of covalent binding of toxin in liver microsomes in vitro, did not alter the target organ in vivo even though the lethal dose level was increased. Since this inducer decreased covalent binding in both the lung and liver in vivo compared to control rats, it seems likely that detoxifying mechanisms were predominant over toxifying activity.

Glutathione as a protective nucleophile: Enhancement of glutathione levels have been shown both in vivo (Boyd et al., 1979; Buckpitt and Boyd, 1979) and in vitro to decrease covalent binding and animal toxicity of 4-ipomeanol. It did not, however, prevent formation of activated toxin but served as an alternate nucleophile which reacted with activated toxin to form nontoxic, water-soluble conjugates in vitro, as was also true for cysteine. Depletion of glutathione levels by diethylmaleate pretreatment of animals resulted in increased tissue covalent binding and enhanced the toxicity of 4-ipomeanol. Toxic doses of 4-ipomeanol tended to deplete lung glutathione levels, an effect that could be prevented by piperonyl butoxide pretreatment. In these studies there was no evidence to suggest that glutathione was preferentially depleted prior to 4-ipomeanol alkylation of lung tissue.

Binding to bronchiolar cells: The availability of both [^{14}C]- and tritiated 4-ipomeanol enabled the use of autoradiographic techniques for determining the localization of toxin in the respiratory tract of rats, mice, and hamsters (Boyd, 1977). In all three species the toxin was found to be covalently bound to nonciliated bronchiolar (Clara) cells of the terminal airways, which subsequently became necrotic. Lethal doses of tritiated toxin resulted in binding of toxin with subsequent necrosis in noncilicated cells of the larger bronchioles as well. Pretreatment of animals with piperonyl butoxide diminished the covalent binding of toxin. These studies with living animals were extended to in vitro systems composed of intact lung cells, including mouse lung slices and isolated whole lungs perfused with oxygenated Krebs solution (Longo and Boyd, 1979). The covalent binding to Clara cells was time dependent and also concentration dependent up to the range 0.25-0.50 mM 4-ipomeanol, which gave maximal binding rates. Heat-denatured lung did not bind the toxin. Lungs from animals pretreated with piperonyl butoxide showed diminished binding, whereas depletion of glutathione levels by pretreatment with diethylmaleate (in vivo) enhanced the in vitro cellular binding. These observations, together with those previously described for the bovine lung, indicate that the nonciliated bronchiolar cell, which is known to have prominent endoplasmic reticulum, is a prime cellular target of 4-ipomeanol attack. This phenomenon can be accounted for on the basis of cellular cyto-chrome P_{450} activation of toxin with in situ covalent binding that results in cell death. This fact points to the potential vulnerability of the Clara cell to other xenobiotic agents, including inhaled carcinogens, which might require activation by mixed-function oxygenases.

Tolerance induction: Repeated daily injections of sublethal doses of 4-ipomeanol result in development of tolerance on the part of laboratory rodents to otherwise lethal challenge doses of the toxin (Boyd et al., 1981). Intra-peritoneal pretreatment of rats for a week with 7 mg/kg daily increased the LD_{50} value nearly threefold. Even though the main target organ (lungs) remained the same, covalent binding of radioactive toxin to lung and liver tissue in toxin-pretreated rats was markedly decreased in challenged animals.

A somewhat *greater* degree of tolerance was induced in mice if the daily pretreatment doses were gradually increased over the 7-day period. Lethal challenge doses of toxin resulted in the characteristic lung edema reaction except in those pretreatment regimens in which the doses were increased gradually or prolonged beyond 7 days. In the latter animals challenge with lethal doses of 4-ipomeanol gave rise to hepatic and/or renal necrosis.

A single intraperitoneal 10-mg/kg dose of 4-ipomeanol induced a detect-able amount of resistance to toxin challenge provided that the challenge was given more than 8 hr later. Maximal tolerance to a 40 mg/kg, i.p. challenge was attained within 24 hr, and some tolerance persisted for at least 20 days after pretreatment.

The methyl and phenyl (nonfuran) analogs of 4-ipomeanol (which are not lung toxic) were ineffective for inducing tolerance, indicating the furan moiety requirement for tolerance induction, as is also the case for toxicity. The closely related lung toxin, ipomeanine, induced tolerance to 4-ipomeanol; conversely, 4-ipomeanol induced tolerance to ipomeanine. However, the well-known lung toxin α-naphthylthiourea (ANTU) failed to induce tolerance to 4-ipomeanol even though the latter toxin pretreatment induced tolerance to ANTU. Secondary pretreatments with CCl_4 or SKF525A modified the induced tolerance to 4-ipomeanol in 4-ipomeanol-pretreated animals, but CCl_4 pre-treatment alone had no effect on the susceptibility to 4-ipomeanol. Other experiments were carried out to show that tolerance to 4-ipomeanol induced by phenobarbital and 3-methylcholanthrene pretreatments are by different mechanisms than that induced by 4-ipomeanol itself (Boyd et al., 1981).

These studies suggest that tolerance to 4-ipomeanol in mice induced by repeated nonlethal doses is based on an alteration in its metabolism in aspects of toxifying or detoxifying pathways. The final determination as to which of these situations attains must await further work.

In experiments with rats Doster (1977) demonstrated by electron micro-scopy that injection of near-lethal doses of 4-ipomeanol daily resulted in Clara cell necrosis (at 12 hr) followed by destruction of ciliated cells within 24 hr after injection of the first dose. In spite of continued daily injections of toxin, which resulted in destruction of much of the bronchiolar epithelium within 48 hr of the first dose, groups of proliferating nonciliated cells were noted at 72 hr which by 96 hr had progressed to form single layers of cu-boidal epithelial cells covered with stubby microvilli. Restoration of normal bronchiolar lining was complete by the seventh day.

IV. LUNG TOXINS FROM *PERILLA FRUTESCENS*

A. *Perilla frutescens* as a Causative Agent of Animal Lung Disease

Perilla frutescens (L.) Britt. is now a common annual weed in much of the
Eastern United States, although some varieties (mostly deep purple) are
used as ornamentals. The wild plant has predominantly serrated green
leaves with slightly purple colored reverse sides. Crushed leaves have a
pronounced minty scent characteristic of the essential oil. Square stems
support plants that may grow up to 1 m or more in height. The plant is
known by a variety of common names, including wild coleus, perilla mint,
beefsteak plant, and other descriptive names. Its history indicates it was
an Asian import to this country that garden-escaped (Fernald, 1950). Oil
from the seeds has been considered for use as a drying oil in paints, lac-
quers, and varnishes. The extracted seed cake has been fed to livestock
as a high-protein food source (Gardner, 1936).

Several varieties of *Perilla* are grown in Japan, and the essential oil of
leaves and stems has been used for food flavoring (Arctander, 1960). In
china and other countries the pungent oil was apparently the main constituent of "soyo," a folk medicine used in a variety of illnesses (Nagao et al.,
1974). *Perilla frutescens* and related species are also found in certain
European countries and have been used extensively in biochemical and
physiological studies in the United States (Mounts and Dutton, 1964), the
USSR (Rushkovskii, 1967), and elsewhere (Krzywacka and Rylska, 1960).

The essential oil of *P. frutescens* contains at least three potent lung
edemagenic agents, known as perilla ketone [38], egomaketone [39], and
isoegomaketone [40], whose structural similarities to the sweet potato lung
toxins are quite apparent.

Field investigations by Peterson in the years 1961-1963 (Peterson, 1965)
provided strong evidence that wild perilla mint was another cause of extensive seasonal outbreaks of the syndrome known as acute bovine pulmonary
emphysema (ABPE) in Oklahoma. A review of similar outbreaks in several
other states, including Arkansas and Texas, suggests that such outbreaks
are not uncommon, occurring especially in the late summer or fall when
normal forage may be scarce.

In more recent years outbreaks have been reported in Tennessee following late summer drought (Linnabary et al., 1978a). Incrimination of perilla
as a causative agent of ABPE was the result of finding remnants of perilla
seeds in the forestomachs or droppings of animals, and noting the typical
perilla mint odor upon opening the gastrointestinal tract at autopsy.
Generally, cattle will avoid grazing perilla unless driven by hunger when

pastures are low. The same is probably true of other livestock, including sheep, goats, and horses, all of which are susceptible to the experimental lung disease caused by administration of perilla ketone (Linnabary et al., 1978b).

Perilla lung toxins may be extracted from the leaves, stems, and roots of the plant using a variety of organic solvents (Wilson et al., 1977). The respective compounds can be separated using HPLC and high-vacuum distillation procedures. Both HPLC and gas-liquid chromatography may be utilized for quantitative analyses.

Suitable quantities of perilla ketone for experimental studies are best obtained by chemical synthesis.

B. Synthesis of Perilla Ketone

Many syntheses of perilla ketone [38] have been devised. Some, which begin with a preformed 3-substituted furan, are shown in Scheme 12. Arata and Achiwa (1958) acylated a malonate ester with 3-furoyl chloride and after hydrolysis and decarboxylation obtained [38]. Matsuura (1957) obtained [38]

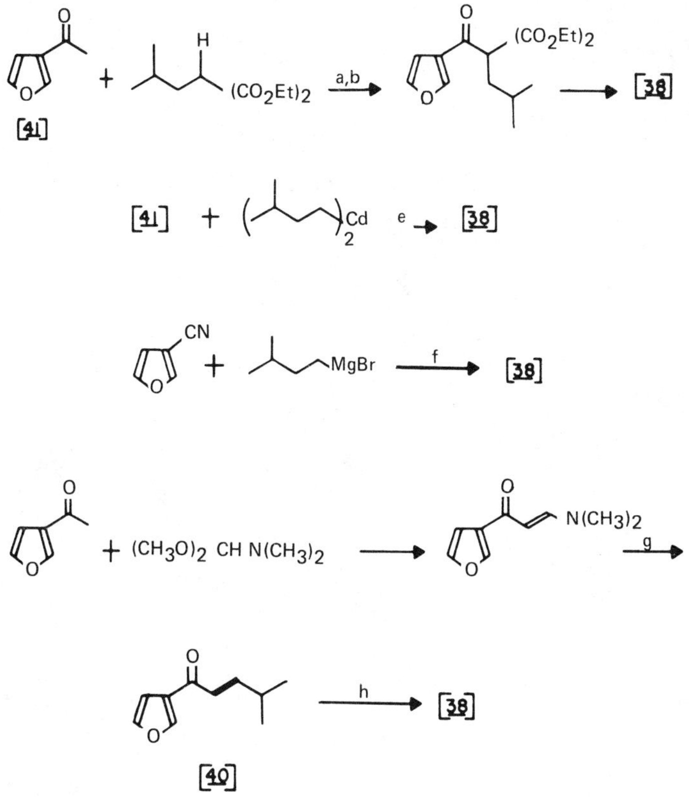

SCHEME 12 (a) $Mg/EtOH/CCl_4/Et_2O$, (b) H^+, (c) $AcOH/H_2SO_4/H_2O$, (d) OH^-, (e) Et_2O, (f) benzene, (g) $(CH_3(_2CHLI$, (h) H_2/Pd-$C/EtOH$.

directly by reacting diisoamylcadmium with 3-furoyl chloride. Another direct approach was developed by Garst, who reacted isoamylmagnesium bromide with 3-cyanofuran, conveniently prepared by copyrolysis of 3-furoic acid and 1,3-dicyanobenzene (J. E. Garst, unpublished observations, 1978). In a multistep synthesis, Abdulla and Fuhr (1978) reacted 3-acetylfuran with the dimethyl acetal of dimethylformamide and converted the resulting ketoenamine via isoegomaketone [40] to [38].

Several syntheses of [38] from acryclic precursors or furan itself are shown in Scheme 13. Inomata et al. (1979) reacted the anion of the protected acrolein-thiophenol adduct [42] with isocapraldehyde. Oxidation, addition of an additional carbon as formaldehyde, and deprotection gave [38]. Kondo and Matsumoto (1976) reacted the dienol prepared from 2-(1,3-butadienyl)magnesium chloride and isocapraldehyde with singlet oxygen and then converted the adduct to [38]. Two groups have reported the conversion of the photoadduct of furan and isocapraldehyde to [38] (Zamojski and Koźluk, 1977; Kitamura et al., 1977).

C. Animal Toxicity of Perilla Ketone

Toxicity studies indicate that perilla ketone, egomaketone, and isoegomaketone are approximately equal in intraperitoneal LD_{50} values and produce identical pulmonary effects in laboratory mice. Gross pulmonary lesions are

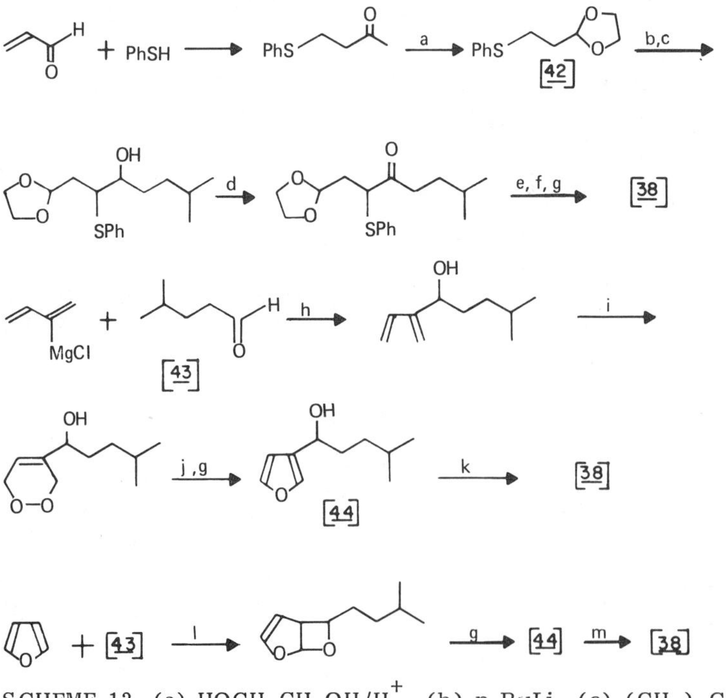

SCHEME 13 (a) $HOCH_2CH_2OH/H^+$, (b) n-BuLi, (c) $(CH_3)_2CHCH_2CH_2CHO$, (d) pyridinium chlorochromate, (e) LDA, (f) H_2CO, (g) H^+, (h) THF, (i) $^1O_2$, (j) $(CH_3)_3COLi$, (k) MnO_2, (l) hν, (m) Jones' reagent.

indistinguishable from the disease caused by 4-ipomeanol; however, unlike 4-ipomeanol, perilla ketone does not affect the kidneys of adult male mice. The intraperitoneal LD_{50} figures for perilla ketone in male Notre Dame strain mice is approximately 6 mg/kg and 2.5 mg/kg for females. The per os LD_{50} value for females is approximately 16 mg/kg. In male Sprague-Dawley rats the intraperitoneal LD_{50} value is 10 mg/kg, while the per os LD_{50} value is 25 mg/kg (Wilson et al., 1978; 1977).

Although laboratory rodents appear to be more susceptible to perilla ketone than to 4-ipomeanol, limited bovine experiments suggest that this species is less susceptible. Injections of 3 and 9 mg/kg, respectively, of perilla ketone directly into the rumen elicited no responses in Angus heifers. Another heifer receiving 30 mg/kg intravenously, however, developed severe respiratory distress beginning within 10 hr and died on the third day after injection. Extensive pleural effusion was present together with other pulmonary disease signs. Another animal receiving 40 mg/kg per os was unaffected.

An adult male sheep receiving 19 mg/kg of perilla ketone intravenously developed respiratory distress on day 3 but had improved somewhat by the fifth day postinjection, when it was sacrificed. The lungs, as primary targets of the toxin, showed focal capillary congestion and intraalveolar edema. Two female goats injected intravenously with 18 and 40 mg/kg, respectively, died within 12 hr with severe pulmonary edema. A dose of 10 mg/kg, i.v. killed another female goat in 36 hr. A dose of 150 mg/kg injected into the rumen of a goat resulted in death of the recipient in 48 hr. Ponies injected intravenously with 18 mg of perilla ketone per kilogram developed a severe respiratory disease beginning a few hours after dosing which became progressively more severe and ended in death about 6 days later. The lung lesions histologically were similar to those of ABPE in cattle and goats. Comparable doses of 4-ipomeanol produced a much milder disease, suggesting that this species, unlike the bovine, is more susceptible to perilla ketone than to the sweet potato toxin.

Breeze and Carlson (1982) have reviewed the lung injury induced by perilla ketone and other chemical agents in livestock and have chosen to limit the term "acute bovine pulmonary edema" (ABPE) to the syndrome caused by 3-methylindole in cattle.

Studies on the mechanism of perilla ketone toxicity are under way which, although incomplete, suggest that with few exceptions, the mechanism of its toxic action is very similar to that of 4-ipomeanol.

V. OTHER LUNG-TOXIC FURANS AND RELATED COMPOUNDS

A. 3-Methylfuran

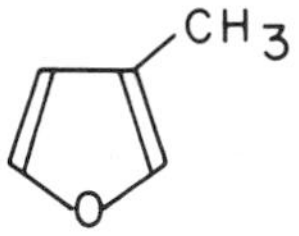

45

It has been suggested that 3-methylfuran [45] may originate naturally from atmospheric photooxidation and degradation of volatile compounds such as terpenes and isoprene, released from deciduous trees. It has been reported as a major component of smog occurring in Washington, D.C., in 1973 (Saunders et al., 1974). Boyd and co-workers (Boyd et al., 1978b; Statham et al., 1978) have shown that 3-methylfuran, although less toxic than 4-ipomeanol, behaves toxicologically in much the same way in bronchiolar cells. Both inhalation and parenteral injection of the compound result in cytochrome P_{450} enzyme activation and covalent binding to nonciliated bronchiolar epithelium. The covalent binding is diminished by piperonyl butoxide pretreatment. High concentrations of 3-methylfuran have also been observed to become covalently bound and to cause necrosis of both nonciliated and ciliated bronchiolar cells. It is not evident whether the ciliated cell effects are mediated by these same cells or are the result of activated material emanating from the Clara cells.

B. Furan Carbamates

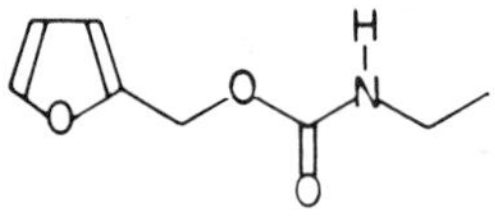

Both 2- and 3-(N-ethylcarbamoylhydroxymethyl)furan [46,47] have been synthesized and tested for toxicity in laboratory rodents. The 3-substituted compound, reported by Seawright and Mattocks (1973), is the more toxic (LD_{50} value comparable to that of 4-ipomeanol), causing pulmonary edema and renal tubular necrosis (the latter presumably occurred in adult male mice). The hepatotoxicity of the 3-substituted furan [47] was enhanced by phenobarbital pretreatment in mice while decreasing the pulmonary and renal effects in these animals. The 2-substituted furan [46], prepared and tested by Guengerich, had an LD_{50} value of 80 mg/kg in rats (Guengerich, 1977c). It is metabolically activated in vitro by microsomal enzymes. Covalently bound radioactive toxin was found in significant quantities in all subcellular fractions of rat liver; toxin was also bound in lesser quantities in the lung, kidney, and other tissues. Toxin was observed to be covalently bound in vivo, but it is not known whether this phenomenon required metabolic activation. However, phenobarbital pretreatment enhanced the hepatotoxicity in rats.

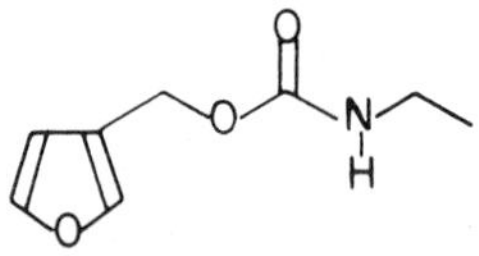

REFERENCES

Abdulla, R. F., and Fuhr, K. H. (1978). An efficient conversion of ketones to α,β-unsaturated ketones. *J. Org. Chem. 43*:4248-4250.

Akazawa, T. (1964). Biosynthesis of ipomeamarone. II. Synthetic mechanism. *Arch. Biochem. Biophys. 105*:512-516.

Akazawa, T., and Uritani, I. (1962). Biosynthesis of ipomeamarone. The incorporation of acetate-2-C^{14} into ipomeamarone. *Agric. Biol. Chem.*, *26*:131-133.

Akazawa, T., Uritani, I., and Akazawa, Y. (1962). Biosynthesis of ipomeamarone. I. The incorporation of acetate-2-C^{14} and mevalonate-2-C^{14} into ipomeamarone. *Arch. Biochem. Biophys.* *99*:52-59.

Arata, Y., and Achiwa, K. (1958). Perillaketone. *Kanazawa Daigaku Yakugakubu Kenkyu Nempô* 8:29-31. *Chem. Abstr.* 53:5228a (1959).

Arctander, S. (1960). *Perfume and Flavor Materials of Natural Origin.* Rutgers University Press, Elizabeth, N.J., p. 520.

Balazs, T. (1970). Measurement of acute toxicity. In *Methods in Toxicology*, G. E. Paget (Ed.). Davis, Philadelphia, pp. 49-131.

Birch, A. J., Massy-Westropp, R., Wright, S. E., Kubota, T., Matsuura, T., and Sutherland, M. D. (1954). Ipomeamarone and ngaione. *Chem. Ind.*, p. 902.

Blackburne, I. D., Park, R. J., and Sutherland, M. D. (1972). Terpenoid chemistry. XX. Myoporone and dehydromyoporone, toxic furanoid ketones from *Myoporum* and *Eremophila* species. *Aust. J. Chem. 25*: 1787-1796.

Bohlmann, F., and Rao, N. (1972). Neue Furansesquiterpene aus *Athanasia*-arten. *Tetrahedron Lett.*, pp. 1039-1044.

Bohlmann, F., and Zdero, C. (1978). New furanosesquiterpenes from *Eumorphia* species. *Phytochemistry* 17:1155-1159.

Boyd, M. R. (1977). Evidence for the Clara cell as a site of cytochrome P450-dependent mixed-function oxidase activity in lung. *Nature (Lond.) 269*:713-715.

Boyd, M. R. (1980). Biochemical mechanisms in chemical-induced lung injury: role of metabolic activation. In *CRC Critical Review in Toxicology*, CRC Press, Boca Raton, Fla., pp. 103-176.

Boyd, M. R., and Burka, L. T. (1978). In vivo studies on the relationship between target organ alkylation and the pulmonary toxicity of a chemically reactive metabolite of 4-ipomeanol. *J. Pharmacol. Exp. Ther. 207*: 687-697.

Boyd, M. R., and Dutcher, J. S. (1978). Role of renal metabolism in the pathogenesis of renal corticol necrosis produced by 4-ipomeanol in the mouse. *Toxicol. Appl. Pharmacol. 45*:229.

Boyd, M. R., and Wilson, B. J. (1971). Preparative and analytical gas chromatography of ipomeamarone, a toxic metabolite of sweet potatoes. *J. Agric. Food Chem. 19*:547-550.

Boyd, M. R., Wilson, B. J., and Harris, T. M. (1972). Confirmation by chemical synthesis of the structure of 4-ipomeanol, a lung-toxic metabolite of the sweet potato, *Ipomoea batatas. Nature New Biol. 236*:158-159.

Boyd, M. R., Burka, L. T., Harris, T. M., and Wilson, B. J. (1974). Lung-toxic furanoterpenoids produced by sweet potatoes (*Ipomoea batatas*) following microbial infection. *Biochim. Biophys. Acta 337*:184-195.

Boyd, M. R., Burka, L. T., and Wilson, B. J. (1975). Distribution, excretion, and binding of radioactivity in the rat after intraperitoneal administration of the lung-toxic furan, [^{14}C]4-ipomeanol. *Toxicol. Appl. Pharmacol. 32*:147-157.

Boyd, M. R., Sasame, H., Mitchell, J. R., and Catignani, G. (1977). Dose-dependent pulmonary toxicity by nitrofurantoin (NF), and modification by vitamin E, dietary fat, and oxygen. *Fed. Proc. 36*:405.

Boyd, M. R., Burka, L. T., Wilson, B. J., and Sasame, H. A. (1978a). In vitro studies on the metabolic activation of the pulmonary toxin, 4-ipomeanol, by rat lung and liver microsomes. *J. Pharmacol. Exp. Ther. 207*:677-685.

Boyd, M., Statham, C., Franklin, R., and Mitchell, J. (1978b). Pulmonary bronchiolar alkylation and necrosis by 3-methylfuran, a naturally occurring potential atmospheric contaminant. *Nature (Lond.) 272*:270-271.

Boyd, M. R., Statham, C., Stiko, A., Mitchell, J., and Jones, R. (1979). Possible protective role of glutathione in pulmonary toxicity by 4-ipomeanol. *Toxicol. Appl. Pharmacol. 48*:A66.

Boyd, M. R., Sasame, H. A., and Franklin, R. B. (1980). Comparison of ratios of covalent binding to total metabolism of the pulmonary toxin, 4-ipomeanol, in vitro in pulmonary and hepatic microsomes, and the effects of pretreatments with phenobarbital or 3-methylcholanthrene. *Biochem. Biophys. Res. Commun. 93*:1167-1172.

Boyd, M. R., Burka, L. T., Wilson, B. J., and Sastry, B. V. R. (1981). Development of tolerance to the pulmonary toxin, 4-ipomeanol. *Toxicology 19*:85-100.

Breeze, R. G., and Carlson, J. R. (1982). Chemical-induced lung injury in domestic animals. *Adv. Vet. Sci. and Comp. Med. 26*:201-231.

Buckpitt, A. R., and Boyd, M. R. (1978). Xenobiotic metabolism in birds, species lacking pulmonary Clara cells. *Pharmacologist 20*:181.

Buckpitt, A. R., and Boyd, M. R. (1979). Determination of electrophilic metabolites produced during microsomal metabolism of 4-ipomeanol by high pressure anion exchange chromatography of the glutathione adducts. *Fed. Proc. 38*:692.

Burka, L. T. (1978). 1-(3'-Furyl)-6,7-dihydroxy-4,8-dimethylnonan-1-one, a stress metabolite from sweet potatoes (*Ipomoea batatas*). *Phytochemistry 17*:317-318.

Burka, L. T., and Iles, J. (1979). Myoporone and related keto alcohols from stressed sweet potatoes. *Phytochemistry 18*:873-874.

Burka, L. T., and Kuhnert, L. (1977). Biosynthesis of furanosesquiterpenoid stress metabolites in sweet potatoes (*Ipomoea batatas*). Oxidation of ipomeamarone to 4-hydroxymyoporone. *Phytochemistry 16*:2022-2023.

Burka, L. T., and Wilson, B. J. (1976). Toxic furanosesquiterpenoids from mold-damaged sweet potatoes (*Ipomoea batatas*). In *Mycotoxins and Other Fungal Related Food Problems*, J. V. Rodricks (Ed.). American Chemical Society, Washington, D.C., pp. 387-399.

Burka, L. T., Kuhnert, L., Wilson, B. J., and Harris, T. M. (1977). Biogenesis of lung-toxic furans produced during microbial infection of sweet potatoes (*Ipomoea batatas*). *J. Am. Chem. Soc. 99*:2302-2305.

Burka, L. T., Kuhnert, L., Wilson, B. J., and Harris, T. M. (1974a). 4-Hydroxymyoporone, a key intermediate in the biosynthesis of pulmonary toxins produced by *Fusarium solani* infected sweet potatoes. *Tetrahedron Lett.*. pp. 4017-4020.

Burka, L. T., Bowen, R. M., Wilson, B. J., and Harris, T. M. (1974b). 7-Hydroxymyoporone, a new toxic furanosesquiterpene from mold-damaged sweet potatoes. *J. Org. Chem. 39*:3241-3244.

Burka, L. T., Wilson, B. J., and Harris, T. M. (1974c). A synthesis of racemic ipomeamarone and epiipomeamarone. *J. Org. Chem.* *39*:2212-2214.

Burka, L. T., Felice, L. J., and Jackson, S. W. (1981). 6-Oxodendrolasin, 6-hydroxydendrolasin, 9-oxofarnesol and 9-hydroxyfarnesol., stress metabolites of the sweet potato (*Ipomoea batatas*). *Phytochemistry 20*: 647-652.

Dimitriadis, E., and Massy-Westropp, R. A. (1980). Furanosesquiterpenes from *Eremophila rotundifolia*. *Aust. J. Chem.* *33*:2729-2736.

Doster, A. R. (1977). An ultrastructural study of bronchiolar lesions in rats induced by 4-ipomeanol, a product from mold-damaged sweet potatoes. Doctoral dissertation submitted to faculty of the College of Veterinary Medicine, University of Georgia, Athens.

Doster, A. R., Mitchell, F. E., Ferrell, R. L., and Wilson, B. J. (1978). Effects of 4-ipomeanol, a product from mold-damaged sweet potatoes, in the bovine lung. *Vet. Pathol.* *15*:367-375.

Dutcher, J. S., and Boyd, M. R. (1978). Species and strain differences in tissue alkylation and toxicity by 4-ipomeanol: predictive value of covalent binding in the study of target organ toxicity. *Toxicol. Appl. Pharmacol.* *45*:267.

Dutcher, J. S., and Boyd, M. R. (1979). Species and strain differences in target organ alkylation and toxicity by 4-ipomeanol. Predictive value of covalent binding in studies of target organ toxicities by reactive metabolites. *Biochem. Pharmacol.* *28*:3367-3372.

Fernald, M. L. (1950). Family 15.1 labiatae (mint family), *Perilla frutescens*. In *Gray's Manual of Botany*, 8th cent. ed., A. Gray (Ed.). American Book, New York, p. 1251.

Gardner, H. A. (1936). Scientific section. In *National Paint, Varnish and Lacquer Association, Inc., Circ. 506*, Washington, D.C., pp. 178-198.

Guengerich, F. P. (1977a). Separation and purification of multiple forms of microsomal cytochrome P-450. Activities of different forms of cytochrome P-450 towards several compounds of environmental interest. *J. Biol. Chem.* *252*:3970-3979.

Guengerich, F. P. (1977b). Preparation and properties of highly purified cytochrome P-450 and NADPH-cytochrome P-450 reductase from pulmonary microsomes of untreated rabbits. *Mol. Pharmacol.* *13*:911-923.

Guengerich, F. P. (1977c). Studies on the activation of a model furan compound-toxicity and covalent binding of 2-(N-ethyl-carbamoylhydroxy-methyl)-furan. *Biochem. Pharmacol.* *26*:1909-1915.

Hamilton, W. D., Park, R. J., Perry, G. J., and Sutherland, M. D. (1973). Terpenoid chemistry. XXI. (-)-Epingaione, (-)-dehydrongaione, (-)-dehydroepingaione, and (-)-deisopropylngaione, toxic furanoid sesqui-terpenoid ketones from *Myoporum deserti*. *Aust. J. Chem.* *26*:375-387.

Hammond, A. C., Carlson, J. R., and Breeze, R. G. (1980). Prevention of tryptophan-induced acute bovine pulmonary oedema and emphysema (fog fever). *Vet. Rec.* *107*:322-325.

Hansen, A. A. (1928). Potato poisoning. *N. Am. Vet.* *9*:31-34.

Hegarty, B. F., Kelly, J. R., Park, R. J., and Sutherland, M. D. (1970). Terpenoid chemistry. XVII. (-)-Ngaione, a toxic constituent of *Myoporum deserti*. The absolute configuration of (-)-ngaione. *Aust. J. Chem. 23*: 107-117.

Hiura, M. (1943). Studies on storage and rot of sweet potato. *Rep. Gifu Agric. Coll. 50*:1-5.

Hyodo, H., Uritani, I., and Akai, S. (1968). Formation of a callus tissue in sweetpotato stems in response to infection by an incompatible strain of *Ceratocystis fimbriata*. *Phytopathology 58*:1032-1033.

Inomata, K., Sumita, M., and Kotake, H. (1979). A convenient method for the preparation of 3-acylfurans. A new synthesis of perilla ketone. *Chem. Lett.*, pp. 709-712.

Inoue, H., and Uritani, I. (1980). Conversion of 4-hydroxydehydromyoporone to other furanoterpenes in *Ceratocystis fimbriata*-infected sweet potato. *Agric. Biol. Chem. 44*:1935-1936.

Inoue, H., Kato, N., and Uritani, I. (1977). 4-Hydroxydehydromyoporone from infected *Ipomoea batatas* root tissue. *Phytochemistry 16*:1063-1065.

Jones, H. A., Stevenson, F. J., and Smith, O. (1958). Potato. In *Encyclopaedia Britannica*, Vol. 18. Encyclopaedia Britannica, Chicago, p. 325.

Kato, N., Imaseki, H., Nakashima, H., and Uritani, I. (1971). Structure of a new sesquiterpenoid, ipomeamaronol, in diseased sweet potato root tissue. *Tetrahedron Lett.*, pp. 843-846.

Kato, N., Imaseki, H., Nakashima, N., and Uritani, I. (1973). Isolation of a new phytoalexin-like compound, ipomeamaronol, from black-rot fungus infected sweet potato root tissue, and its structural elucidation. *Plant Cell Physiol. 14*:597-606.

Kitamura, T., Kawakami, Y., Imagawa, T., and Kawanisi, M. (1977). One-pot synthesis of 3-substituted furan: a synthesis of perillaketone. *Synth. Commun. 7*:521-528.

Kondo, K., and Matsumoto, M. (1976). Synthesis of furanoterpenes: perillaketone, α-clausenane, (+)-ipomeamarone, and (+)-epiipomeamarone. *Tetrahedron Lett.*, pp. 4363-4366.

Krzywacka, T., and Rylska, T. (1960). Studies on photoperiodism in *Perilla ocimoides* L. II. Morphological response of plants to prolonged night. *Acta Soc. Bot. Pol. 29*:11-56.

Kubota, T. (1958). Volatile constituents of black-rotted sweet potato and related substances. *Tetrahedron 4*:68-86.

Kubota, T., and Ichikawa, N. (1954). On the chemical constitution of ipomeanine, a new ketone from the black-rotted sweet potato. *Chem. Ind.*, pp. 902-903.

Kubota, T., and Matsuura, T. (1957). The constitution of myoporone, a new furano-terpene from myoporum. *Chem. Ind.*, pp. 491-492.

Kubota, T., and Matsuura, T. (1958). The synthesis of (+)-ipomeamarone [(+)-ngaione] and its steric isomers. *J. Chem. Soc.*, pp. 3667-3673.

Lieberman, M., Craft, C. C., Audia, W. V., and Wilcox, M. S. (1958). Biochemical studies of chilling injury in sweetpotatoes. *Plant Physiol. 33*:307-311.

Linnabary, R. D., Warren, J., Wilson, B. J., and Byerly, C. S. (1978a). Acute bovine pulmonary emphysema produced by *Perilla frutescens*. *Mod. Vet. Pract.*, September, pp. 687-688.

Linnabary, R. D., Wilson, B. J., Garst, J. E., and Holscher, M. A. (1978b). Acute pulmonary emphysema (ABPE): perilla ketone as another cause. *Vet. Hum. Toxicol. 2*:325-326.

Longo, N., and Boyd, M. (1979). In vitro metabolic activation of the pulmonary toxin, 4-ipomeanol, by lung slices and isolated whole lungs. *Toxicol. Appl. Pharmacol.* 48:A130.

McDowall, F. H. (1925). Constituents of *Myoporum laetum*, Forst. (the "Ngaio"). Part I. *J. Chem. Soc.* 127:2200-2207.

McDowall, F. H. (1927). Constituents of *Myoporum laetum*, Forst. (the "Ngaio"). Part II. Hydrogenation of ngaione and ngaiol and dehydration of ngaiol. *J. Chem. Soc.*, pp. 731-740.

McDowall, F. H. (1928). Constituents of *Myoporum laetum*, Forst. (the "Ngaio"). Part III. The oxide rings of ngaione. *J. Chem. Soc.*, pp. 1324-1331.

Matsuura, T. (1957). Natural furan derivatives. Part I. The synthesis of perillaketone. *Bull. Chem. Soc. Jpn.* 30:430-431.

Miller, L. C., and Tainter, M. L. (1944). Estimation of the ED_{50} and its error by means of logarithmic-probit graph paper. *Proc. Soc. Exp. Biol. Med.* 57:261-264.

Mounts, T. L., and Dutton, H. J. (1964). Efficient production of biosynthetically labeled fatty acids. *J. Am. Oil Chem. Soc.* 41:537-539.

Nagao, Y., Komiya, T., Fujioka, S., and Matsuoka, T. (1974). Quality of the Chinese drug soyo and the cultivation of the original plants. I. *J. Takeda Res. Lab* 33:111-118. *Chem. Abstr.* 81:117072 (1974).

Oba, K., Shibuta, H., and Uritani, I. (1970). The mechanism supplying acetyl-CoA for terpene biosynthesis in sweet potato with black rot: incorporation of acetate-2-^{14}C, pyruvate-3-^{14}C and citrate-2,4-^{14}C into ipomeamarone. *Plant Cell Physiol.* 11:507-510.

Oguni, I., and Uritani, I. (1970). The incorporation of farnesol-2-^{14}C into ipomeamarone. *Agric. Biol. Chem.* 34:156-158.

Oguni, I., and Uritani, I. (1971). Participation of farnesol in the biosynthesis of ipomeamarone. *Plant Cell Physiol.* 12:507-515.

Oguni, I., and Uritani, I. (1973). Isolation of dehydro-ipomeamarone, a new sesqui-terpenoid from the black rot fungus infected sweet potato root tissue and its relation to the biosynthesis of ipomeamarone. *Agric. Biol. Chem.* 37:2443-2444.

Oguni, I., and Uritani, I. (1974a). Dehydro-ipomeamarone from infected *Ipomoea batatas* root tissue. *Phytochemistry* 13:521-522.

Oguni, I., and Uritani, I. (1974b). Dehydroipomeamarone as an intermediate in the biosynthesis of ipomeamarone, a phytoalexin from sweet potato root infected with *Ceratocystis fimbriata*. *Plant Physiol.* 53:649-652.

Ohno, T. (1952). The bitter substance, produced in black-rotten sweet potato. II. On the constitution of ipomoeamarone. Part 1. *Bull. Chem. Soc. Jpn.* 25:222-225.

Ohno, T., and Toyao, M. (1952). The bitter substance, produced in black-rotten sweet potato. III. On the constitution of ipomoeamarone. Part 2. *Bull. Chem. Soc. Jpn.* 25:414-418.

Oshima, K., and Uritani, I. (1968). Participation of mevalonate in the biosynthetic pathway of ipomeamarone. *Agric. Biol. Chem.* 32:1146-1152.

Oshima-Oba, K., Sugiura, I., and Uritani, I. (1969). The incorporation of leucine-U-^{14}C into ipomeamarone. *Agric. Biol. Chem.* 33:586-591.

Peckham, J. C., Mitchell, F. E., Jones, O. H., Jr., and Doupnik, B., Jr. (1972). Atypical interstitial pneumonia in cattle fed moldy sweet potatoes. *J. Am. Vet. Med. Assoc.* 160:169-172.

Peterson, D. T. (1965). Bovine pulmonary emphysema caused by the plant
    Perilla frutescens. *Proc. Symp. Acute Bovine Pulmonary Emphysema
    (ABPE)*, University of Wyoming, Laramie, pp. R1-R13.
Reich, H. J., Gold, P. M., and Chow, F. (1979). Synthetic applications of
    organoselenium reagents: synthesis of epi-7-hydroxymyoporone. *Tetra-
    hedron Lett.*, pp. 4433-4434.
Reich, J. J., Shah, S. K., Gold, P. M., and Olson, R. E. (1981). Selenium-
    stabilized anions. Preparation of α,β-unsaturated carbonyl compounds
    using propargyl selenides. Synthesis of (+)-7-hydroxymyoporone. *J.
    Am. Chem. Soc. 103*:3112-3120.
Rudkin, G. O., and Nelson, J. M. (1947). Chlorogenic acid and respiration
    of sweet potatoes. *J. Am. Chem. Soc. 69*:1470.
Rushkovskii, S. V. (1967). Biology and biochemistry of perilla. *Biokhim.
    Fiziolog. Maslich. Rast. 2*:15-28.
Sasame, H. A., and Boyd, M. R. (1978). Possible role of cytochrome b_5
    as a rate-limiting factor in metabolic activation of 4-ipomeanol (IPO) by
    lung microsomes. *Fed. Proc. 37*:464.
Sasame, H. A., Gillette, J. R., and Boyd, M. R. (1978). Effects of anti-
    NADPH-cytochrome $\underline{c}$ reductase and anti-cytochrome b_5 antibodies on the
    hepatic and pulmonary microsomal metabolism and covalent binding of the
    pulmonary toxin 4-ipomeanol. *Biochem. Biophys. Res. Commun. 84*:389-
    395.
Saunders, R., Griffith, J., and Saalfed, F. (1974). Identification of some
    organic smog components based on rain water analysis. *Biomed. Mass
    Spectrom. 1*:192-194.
Seawright, A. A., and Hrdlicka, J. (1972). The effect of prior dosing with
    phenobarbitone and β-diethylaminoethyl diphenylpropyl acetate (SKF525A)
    on the toxicity and liver lesion caused by ngaione in the mouse. *Br. J.
    Exp. Pathol. 53*:242-252.
Seawright, A. A., and Mattocks, A. R. (1973). The toxicity of two syn-
    thetic 3-substituted furan carbamates. *Experientia 29*:1197-1200.
Seawright, A. A., Lee, J. S., Allen, J. G., and Hrdlicka, J. (1978).
    Toxicity of *Myoporum* spp. and their furanosesquiterpenoid essential
    oils. In *Effects of Poisonous Plants on Livestock*, R. F. Keeler, K. R.
    Van Kampen, and L. F. James (Eds.). Academic Press, New York, pp.
    241-250.
Statham, C. N., Franklin, R. B., and Boyd, M. R. (1978). Pulmonary
    bronchiolar alkylation and necrosis by 3-methylfuran, a potential atmo-
    spheric contaminant derived from natural sources. *Toxicol. Appl.
    Pharmacol. 45*:267.
Steinbauer, C. E., and Cushman, L. J. (1971). *Sweet Potato Culture and
    Diseases. Agric. Handbook 388*. Agric. Res. Serv., U.S. Dept. Agric.,
    Washington, D.C., p. 1.
Still, W. C., and Darst, K. P. (1980). Remote asymmetric induction. A
    stereoselective approach to acyclic diols via cyclic hydroboration. *J. Am.
    Chem. Soc. 102*:7385-7387.
Takeuchi, T. (1946). Bitter substance, produced in black-rotted sweet
    potatoes. *Sci. Insect Control 12*:26-29. *Chem. Abstr. 43*:8453 (1949).
Uritani, I., and Hoshiya, I. (1953). Phytopathological chemistry of black-
    rotted sweet potato. VI. Coumarin substances from sweet potato and

their physiology. *J. Agric. Chem. Soc. Jpn.* 27:161-164. *Chem. Abstr.* 47:10634 (1953).

Uritani, I., and Miyano, M. (1955). Derivatives of caffeic acid in sweet potato attacked by black rot. *Nature (Lond.)* 175:812.

Uritani, I., and Muramatsu, K. (1953). Phytopathological chemistry of black-rotted sweet potato. IV. Isolation and identification of polyphenols from the injured sweet potato. *J. Agric. Chem. Soc. Jpn.* 27:29-33. *Chem. Abstr.* 47:10634 (1953).

Watanabe, H., and Nishiyana, S. (1952). The black-rotten sweet potato. III. Chemical properties of ipomeamarone. *J. Agric. Chem. Soc. Jpn.* 26:200-202. *Chem. Abstr.* 47:2361 (1953).

Wilson, B. J. (1973). Toxicity of mold-damaged sweetpotatoes. *Nutr. Rev.* 31:73-78.

Wilson, B. J., and Burka, L. T. (1979). Toxicity of novel sesquiterpenoids from the stressed sweet potato (*Ipomoea batatas*). *Food Costmet. Toxicol.* 17:353-355.

Wilson, B. J., Yang, D. T. C., and Boyd, M. R. (1970). Toxicity of mould-damaged sweet potatoes (*Ipomoea batatas*). *Nature (Lond.)* 227:521-522.

Wilson, B. J., Boyd, M. R., Harris, T. M., and Yang, D. T. C. (1971). A lung oedema factor from mouldy sweet potatoes (*Ipomoea batatas*). *Nature (Lond.)* 231:52-53.

Wilson, B. J., Garst, J. E., Linnabary, R. D., and Channell, R. B. (1977). Perilla ketone: a potent lung toxin from the mint plant, *Perilla frutescens* Britton. *Science* 197:573-574.

Wilson, B. J., Garst, J. E., and Linnabary, R. D. (1978). Pulmonary Toxicity of naturally occurring 3-substituted furans. In *Effects of Poisonous Plants on Livestock,* R. F. Keeler, K. R. Van Kampen, and L. F. James (Eds.). Academic Press, New York, pp. 311-323.

Woesser, W. D. (1979). Protected γ-ketoaldehyde synthesis. 1-Ipomeanol. *Synth. Commun.* 9:147-149.

Yang, D. T. C., Wilson, B. J., and Harris, T. M. (1971). The structure of ipomeamaronol: a new toxic furanosesquiterpene from moldy sweet potatoes. *Phytochemistry* 10:1653-1654.

Yen, D. E. (1974). *The Sweet Potato and Oceania: An Essay in Ethnobotany,* Bernice P. Bishop Museum Bull. 236. Bishop Museum Press, Honolulu.

Zamojski, A., and Koźluk, T. (1977). Synthesis of 3-substituted furans. *J. Org. Chem.* 42:1089-1090.

2

CARDIAC GLYCOSIDES (CARDENOLIDES) IN SPECIES OF *ASCLEPIAS* (ASCLEPIADACEAE)

JAMES N. SEIBER and S. MARK LEE

University of California, Davis, California

JANET M. BENSON

Lovelace Inhalation Toxicology Research Institute, Albuquerque, New Mexico

I. INTRODUCTION

The milkweed plant family, Asclepiadaceae, comprises some 200 genera and 2500 species of perennial shrubs, herbs, and vines distributed throughout the tropics and extending to temperate areas of the world. Milkweeds are generally characterized by the milky latex they exude when a leaf, pod, or stem is ruptured. Some species have found medicinal uses in the treatment of cancers, tumors, and warts (Kupchan et al., 1964, and references therein; Koike et al., 1980), as emetics and to treat bronchitis (Mittal et al., 1962 and references therein), and for their strong action on heart muscle (Hesse and Reicheneder, 1936, and references therein). These activities are associated with cardenolides, most notably in species of *Asclepias*, *Pergularia*, *Gomphocarpus*, and *Calotropis*. It should be noted that reported chemical examinations of genera of Asclepiadaceae—by no means exhaustive—have led to the isolation of cardenolides from 12 genera (Hoch, 1961; Singh and Rastogi, 1970), while their presence has been noted in several additional ones (Abisch and Reichstein, 1962).

The poisonous nature of some milkweeds is particularly well documented for *Calotropis procera*, which was used to treat arrow tips in Africa (Hesse and Reicheneder, 1936; Hesse et al., 1939), and for several species of *Asclepias* which are sporadic causes of death among sheep and cattle, particularly in the western U.S. range (Marsh and Clawson, 1924; Kingsbury, 1964). The evidence, again, points to cardenolides as the active agents (Benson et al., 1978). Aside from livestock poisoning, *Asclepias* cardenolides are involved in more subtle plant-herbivore-predator interactions based on their toxicity to vertebrates. The best studied example involves sequestration of the poisons from *Asclepias* by larvae of the monarch butterfly, *Danaus plexippus* L. (Parsons, 1965; Reichstein et al., 1968). If the monarch caterpillar has accumulated sufficient cardenolide, then the caterpillar, pupa, or adult will cause a bird which eats it to vomit—a response which may be learned by the predator such that it avoids the prey in subsequent encounters (Brower et al., 1968; Brower, 1969). Sequestration for defense has been demonstrated in a number of other insects that feed on Asclepiadaceae, nearly all of which are brightly colored and unpalatable to some vertebrates (Duffey and Scudder, 1972, 1974; von Euw et al., 1967).

The dominant milkweed genus in North America is *Asclepias* L., which comprises 108 described species distributed from Mexico northward to Canada through virtually all states of the United States, in some islands of the Caribbean, and, infrequently, in South America (Woodson, 1941, 1954). *Asclepias syriaca* and *A. speciosa* are among the more widespread and abundant broadleaf species, the former in the northeastern, north-central, and midwestern United States and the latter from the Pacific states and Canadian provinces to, roughly, the Missouri-Mississippi river floodplains. These perennials are easily recognizable as principal weeds along highways and railroads and in open fields and dry creek beds throughout the summer. Prolific lacticifers, they have been proposed for cultivation for rubber production and energy-producing biomass (Nielsen et al., 1977). Floss from their seed pods has been explored for use in pillows (Whiting, 1943), while their young shoots are, to some, an asparagus-like delicacy (Gibbons, 1972).

Localized and/or rare species include *A. masonii* (Baja California) and *A. labriformis* (eastern Utah). As little as 0.8 oz of dry *A. labriformis* leaves may be fatal to a 100-lb sheep, making this plant one of the most poisonous of the milkweeds (Kingsbury, 1964).

Asclepias species of specialized habitat include *A. incarnata*, found only in marshy areas of the eastern United States; *A. exaltata*, a deciduous forest border species; and *A. amplexicaulis*, found on sandy shores or associated with old alluvial deposits. *A. curassavica*, apparently a native of the

Caribbean Islands, is the only species to have been cultivated widely as an ornamental and is now found in semitropical and temperate areas worldwide. It is quite likely that other species, such as *A. syriaca* and *A. speciosa*, have been spread far beyond their native confines relatively recently, as a result of the colonization of the North American continent (Fink and Brower, 1981).

It is noteworthy that virtually all *Asclepias* species examined to date (Duffey, 1970; Duffey and Scudder, 1972; Roeske et al., 1976; Isman et al., 1977a,b) contain at least small amounts of cardenolides. It is our objective to describe the chemistry, plant distribution, biological activity, and analysis of this group of compounds as they are found in *Asclepias*, but with some examples for cardenolides in other genera of the Asclepiadaceae included for analogy.

II. CARDENOLIDES IN *ASCLEPIAS* SPECIES

A. Chemistry

The cardenolides are a group of C_{23} steroids which conform to the general structure

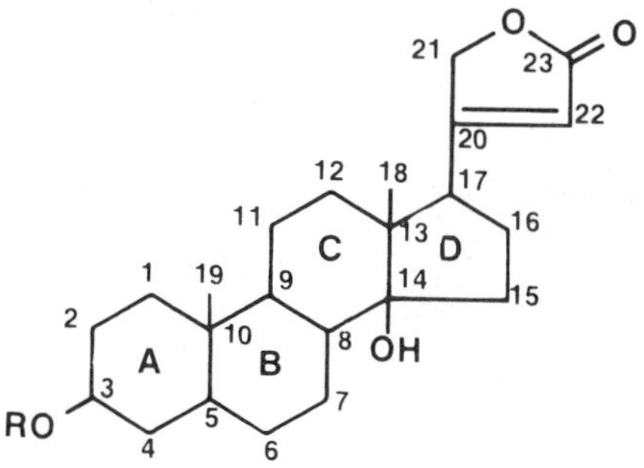

They are characterized by the presence in the "genin" (aglycone, R = H) of (1) an α,β-unsaturated γ-lactone (butenolide) ring attached at C-17, (2) a cis juncture of rings C and D, and (3) a 14β-hydroxy group. The configurations at C-3, C-5, and C-17 are differentiating features, as are oxygenation patterns (usually hydroxy or carbonyl O) at positions from among C-1, C-2, C-5, C-11, C-12, C-15, C-16, and C-19. Some cardenolides have an additional olefinic double bond, and others have an epoxy group in the steroid ring.

The cardenolides usually occur in nature as glycosides ("cardiac glycosides"), attached generally through an OH at C-3 (C-1, C-2, and C-11 are other positions of glycosidation) to one or more sugar moieties. Over 20 sugars have been isolated from hydrolysis of cardenolides, and only three occur with other classes of natural products. A common, but by no means inclusive pattern is for the genin to be attached to one or more rare sugars, and then to one or more glucose molecules. *Digitalis* cardenolides are examples. Unless enzymatic hydrolysis is inhibited during extraction and isolation procedures, the glycosides may retain only the rare sugars (Fieser and Fieser, 1959; Reichstein, 1967a; Singh and Rastogi, 1970).

The configuration at the A:B ring juncture primarily distinguishes cardenolides of the Asclepiadaceae from the clinically useful cardenolides of the
Apocyanaceae and Scrophulariaceae, being 5α(trans-A/B) in the former and
5β(cis-A/B) in the latter. For example, uzarigenin is widely distributed in
the Asclepiadaceae while its 5β isomer, digitoxigenin, is not found in this
plant family. In addition, a number of milkweed cardenolide genins have
hydroxy groups at both C-2 and C-3, both of which may be involved in a
cyclic bridge to a single sugar moiety. This feature, which is not found
among the clinically useful cardiac glycosides, produces cardenolides which
are markedly resistant to acid hydrolysis (Hesse and Reicheneder, 1936;
Crout et al., 1963, 1964; Kupchan et al., 1964; Coombe and Watson, 1964;
Carman et al., 1964; Brüschweiler et al., 1969a; Brown et al., 1979). The
structures of some representative cardenolide genins found in Asclepiadaceae
are shown in Fig. 1.

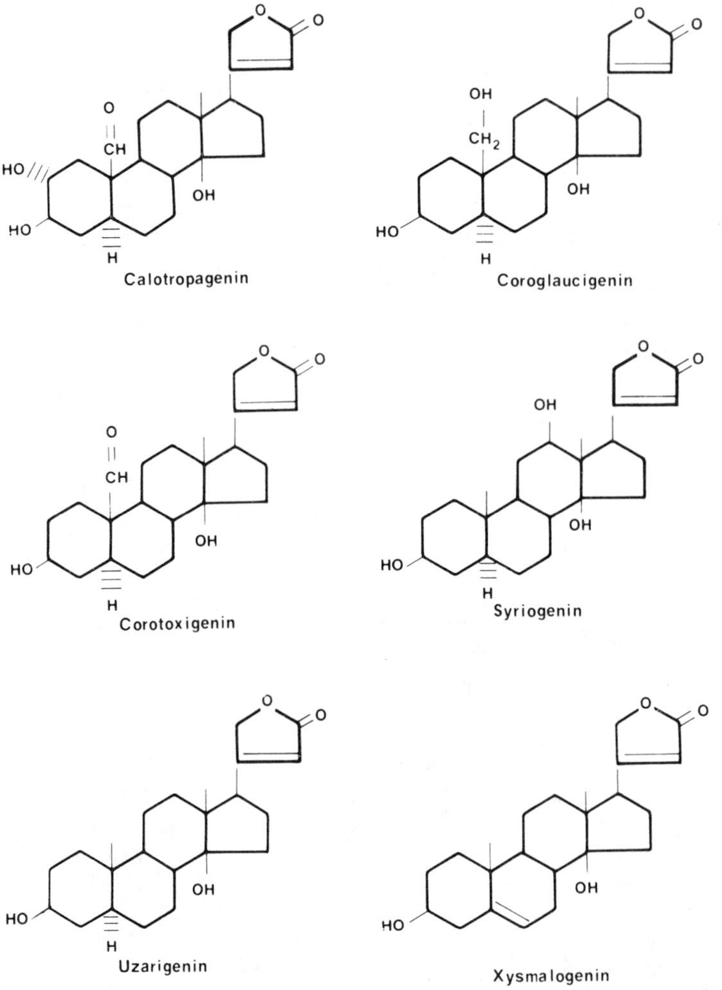

FIGURE 1 Representative cardenolide genins found in Asclepiadaceae.

Calotropis procera, a shrub native to Africa, was the first source for isolation of the cyclic bridged glycosides. From *Calotropis* latex, Hesse and co-workers separated calotropin, calactin, calotoxin, uscharidin, uscharin, and voruscharin—a new series of cardenolides with a common genin, calotropagenin. After several false starts (Hesse et al., 1950; Hassall and Reyle, 1959; Crout et al., 1963, 1964), the structure of calotropagenin and carbon structure of its glycosides were determined (Brüschweiler et al., 1969a,b; Lardon et al., 1970) by comparison with gomphoside isolated and identified earlier from *Asclepias fruticosa* (Coombe and Watson, 1964; Watson and Wright, 1954, 1956, 1957). The evidence, summarized by Brown et al. (1979) with additional stereochemical detail added by Cheung and Watson (1980) and Cheung et al. (1981), led to the following structures for gomphoside, afroside, the six latex cardenolides of *C. procera*, and the more recently isolated proceroside (Brüschweiler et al., 1969a) and asclepin (Singh and Rastogi, 1969, 1972):

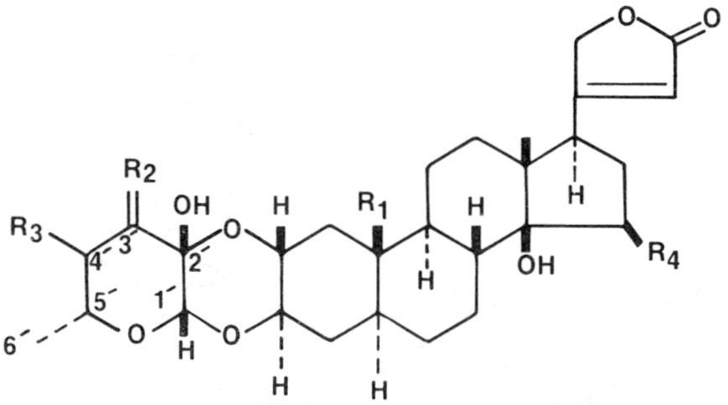

	R_1	R_2	R_3	R_4
Gomphoside	Me	α-H, β-OH	H	H
Afroside	Me	α-H, β-OH	H	OH
Calactin	CHO	α-H, β-OH	H	H
Calotropin	CHO	β-H, α-OH	H	H
Asclepin	CHO	β-H, α-OAc	H	H
Uscharidin	CHO	O	H	H
Uscharin	CHO	N=CH / S-CH$_2$	H	H
Voruscharin	CHO	NH-CH$_2$ / S-CH$_2$	H	H
Calotoxin	CHO	ξ-H, ξ-OH	ξ-OH	H

Gomphoside and the *Calotropis* cardenolides possess adjacent *trans*-hydroxy groups at C-2 and C-3, which together bond to the hexosone through hemiketal (C-2) and acetal (C-3) bonds. Uscharidin has a carbonyl oxygen at C-3', and may be converted to a mixture of calactin and calotropin by partial reduction (Hesse et al., 1950, 1959). Uscharin and voruscharin generate uscharidin by acid or mercuric salt-induced hydrolysis of the spiro thiazoline and thiazolidine rings at C-3' (Hesse and Gampp, 1952; Hesse and Lettenbauer, 1957; Hesse and Mix, 1959; Hesse and Ludwig, 1960). Proceroside and calotoxin have an additional OH, in steroid rings C or D in the former and at C-4' in the latter. Gomphoside and afroside are differentiated from the *Calotropis* cardenolides by the oxidation state of C-19.

Asclepias curassavica from Trinidad, West Indies, analyzed by Santavy et al. (cited by von Euw et al., 1967) and a cultivated sample analyzed by us (Seiber et al., 1980) contained a nearly identical array of cardenolides as *C. procera*. Several additional *Asclepias* species also contain *Calotropis* cardenolides, that is, glycosides of calotropagenin (Roeske et al., 1976; Seiber et al., 1982). It appears that the *Calotropis* cardenolides form one of at least two major cardenolide groups in *Asclepias* species. Leaves and stems of *A. curassavica* from Brazil afforded uzarigenin, its bis-D-glucose conjugate uzarin, coroglaucigenin (OH at C-19), corotoxigenin (aldehyde carbonyl at C-19), and calotropagenin (Tschesche et al., 1958, 1959). These results vary substantially with those of others working with *A. curassavica* from the locations cited previously (von Euw et al., 1967; Seiber et al., 1980) and India (Singh and Rastogi, 1969).

A second major group of 2,3-dihydroxylated cardenolides which form cyclic bridged glycosides has been found in *Asclepias* species. From *A. syriaca* Bauer et al. (1961) and Masler et al. (1962a,b) isolated uzarigenin, syriogenin, desglucouzarin, syrioside, and syriobioside. The latter two were incorrectly formulated as rhamnose-glucose conjugates of syriogenin. Brown et al. (1979) showed that syrioside, syriobioside, and an additional component of *A. syriaca*, desglucosyrioside, have a methyl at C-10, epoxide at C-7 and C-8, and oxygen functions at C-11 and C-12. The hexosulose is attached to hydroxy groups at C-2 and C-3, as in the *Calotropis* cardenolides. The 7,8-epoxy function was deduced by analogy with other cardenolides possessing this function, such as sarverogenin.

One of the three cardenolides with unusually high O/C ratios isolated from *A. labriformis* and *A. eriocarpa*, "eriocarpin" (Seiber et al., 1978) was shown subsequently to be identical to desglucosyrioside (Cheung et al., 1980). A second component of these two species, labriformidin, was formulated as the 3'-keto based, in part, on the finding that desglucosyrioside was among the reduction products. Similar reasoning established the structure of the third cardenolide, labriformin, which yielded labriformidin on acid hydrolysis. The structures of the five epoxy cardenolides so far identified from *A. syriaca, A. eriocarpa, A. labriformis*, and *A. erosa* are as follows (Brown et al., 1979; Cheung et al., 1980):

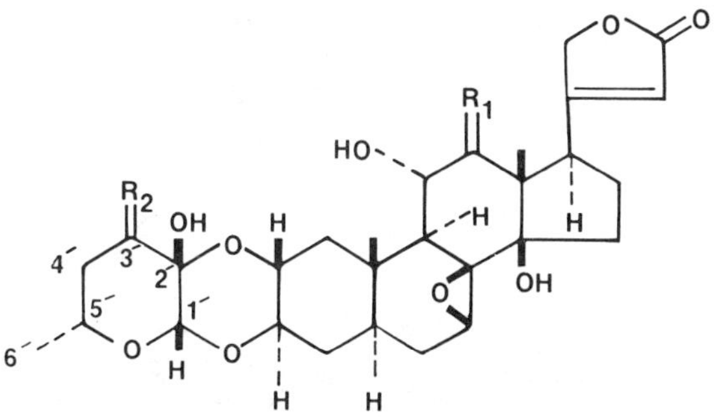

	R_1	R_2
Desglucosyrioside	O	β-OH, α-H
Labriformidin	O	O
Labriformin	O	S-CH$_2$ / N=CH (cyclic)
Syrioside	O	β-O-glu, α-H
Syriobioside	α-OH, β-H	β-OH, α-H

A. erosa has virtually identical cardenolide components as *A. eriocarpa* and *A. labriformis* (Seiber et al., 1978). *A. speciosa* contains desglucosyrioside and several additional cardenolides in common with *A. syriaca*, and apparently lacks labriformidin and labriformin, as does *A. syriaca* (Seiber et al., 1982). Derivatives of desglucosyrioside thus form a second major group of cyclic bridged cardiac glycosides in *Asclepias* species.

There is presently no evidence on the possible occurrence of afroside and gomphoside, isolated from *A. fruticosa* (≡ *Gomphocarpus fructicosa*), in other *Asclepias* species. Time may prove that these C-10 methyl cardenolides constitute still a third major group of cyclic bridged cardiac glycosides in this genus. It should be noted that 2,3-cyclic bridged cardenolides were recently identified from *Elaeodendron glaucum* (Celastraceae) (Kupchan et al., 1977), and from *Anodendron affine* (Yamauchi et al., 1979), indicating a more widespread distribution than Asclepiadaceae alone.

In summary, *Asclepias* species contain several cardenolides of the 5α series, including simple cardenolide genins (uzarigenin, syriogenin) and their sugar conjugates, and at least two groups of 2,3-dihydroxy cardenolide derivatives with cyclic bridged sugars. One of these groups comprises glycosides of calotropagenin, and the second contains epoxy cardenolides— desglucosyrioside and its derivatives. However, there are typically many unidentified cardenolides in extracts of those species which have been examined, and a large number of species yet to be looked at. It is possible,

then, that still other major cardenolide groups are present in *Asclepias*
species, and a number of derivatives of the two groups thus far characterized which await identification.

B. Distribution

Virtually all of the *Asclepias* species examined to date contain cardenolides,
but with extensive differences in quantities and types of cardenolides.
Quantitative data for some leaf/aerial and seed samples of several species
are given in Table 1. These data are primarily from Roeske et al. (1976),
with some additional values from Isman et al. (1977b), Benson et al. (1978),
Nishio (1980), Nelson et al. (1981), and Brower et al. (1982). Analyses
were performed by comparable spectrophotometric procedures using digitoxin as a reference. Species of subgenus I (Asclepias), series 8 (Roseae),
uniformly contain substantial quantities of cardenolides, and include two
species (*A. masonii* and *A. albicans*) with the highest reported amounts of
cardenolides. The four species of subgenus I, series 5 (Syriaca) and 6
(Purpurascentes) examined to data, which include the widely distributed
A. syriaca (midwest-eastern United States), *A. speciosa* (western United
States), *A. humistrata* (southeastern United States), and *A. linaria* (Mexico),
are also relatively enriched in cardenolides. It is, however, premature to
match taxonomic classification or geographical distribution with cardenolide
quantity until additional species are examined. Furthermore, it should be
noted that many of the data in Table 1 are for single samples, from single
locations and dates.

Preliminary qualitative analyses by thin-layer chromatography (TLC)
were carried out for several of the species in Table 1 (Roeske et al., 1976).
Similarity in TLC patterns were found for at least three groups of species,
but the groupings did not correspond simply with taxonomic classification.

Recent studies have provided information on the effect of spatial and
temporal factors, and of plant part on cardenolide quantity and quality
within a single *Asclepias* species. Samples of root, stem, leaf, and latex
were collected from three plots in one population of *A. eriocarpa* at 12
monthly intervals (Nelson et al., 1981). The samples were assayed for total
cardenolide by spectroassay and for individual cardenolides by TLC. From
May to September mean milligram equivalents of digitoxin per gram of dried
plant were: latices (latex), 56.8 >> stems, 6.12 > leaves, 4.0 > roots, 2.5.
With the exception of the roots, which were relatively constant in content
over a 12-month period, significant changes in gross cardenolide content
occurred for each sample type with time of collection during the growing season (Fig. 2). The proportions of different cardenolides also varied with
sample type (Fig. 3), but not significantly with time of collection. Latex
samples were enriched in labriformin, a thiazoline ring-containing cardenolide of low polarity; leaf samples contained principally labriformin, labriformidin, desglucosyrioside, and other predominately more polar cardenolides; stems contained most of the same cardenolides as the leaves, but
among all the sample types were unique in their content of uzarigenin; and
the roots predominated in desglucosyrioside and polar cardenolides, including syriogenin and cardenolide glucosides. Significantly, there was little
qualitative variation within a single sample type with time of collection.

TABLE 1 Gross Cardenolide Content of *Asclepias* Species Listed as Enumerated by Woodson (1954)

	Cardenolide content (mg digitoxin equivalent/g dried material)[a]		
	Leaf or aerial		Seeds
Subgenus I. *Asclepias*			
Series 1. *Incarnata*			
1. *incarnata* L.	nil - 0.28	(L/a, 2)	0.70-5.42 (82)
7. *curassavica* L.	2.08-4.14	(L/a, 10)	
8. *nivea*	5.44	(L/a, p)	
10. *fascicularis* Dcne	0.04-0.12	(1 + 2p)	0.51-0.73 (2)
fascicularis Dcne[b]	0.22	(L, p)	0.25-0.44 (3p)
13. *verticillata* L.	nil	(L/a, p)	nil (p)
Series 2. *Tuberosa*			
17. *tuberosa* L.	nil-0.06	(L/a, p)	nil-0.03 (38)
Series 3. *Exaltata*			
23. *exaltata* L.	nil-0.70	(L/a, 3)	1.59-7.24 (49)
26. *amplexicaulis* Sm.	nil-0.22	(L/a, 5)	2.64 (p)
Series 5. *Syriaca*			
36. *syriaca* L.	0.06-2.64	(L/a, 16)	1.34-9.03 (160)
syriaca L.[c]	1.34	(L, p)	
syriaca L.[c]	3.40	(s, p)	
37. *humistrata* Walt[d]	5.13-8.42	(L, 6)	
humistrata Walt[d]	9.95	(s, 3)	
42. *linaria* cav.	7.78	(L/a, p)	
Series 6. *Purpurascentes*			
49. *speciosa* Torr.	0.149	(L/a, p)	2.58-3.89 (2)
speciosa Torr.[b]			1.8-2.7 (2p)
Series 8. *Roseae*			
63. *labriformis* M. E. Jones	2.14	(L/a, p)	
labriformis M. E. Jones[e]	7.2	(L/a, p)	
64. *erosa* Torr.	1.30	(L/a, p)	
erosa Torr.[b]	3.0	(L, p)	1.9-2.4 (2p)
65. *eriocarpa* Benth.	1.13-5.20	(L/a, 3 + 2p)	1.66-5.13 (2 + p)
eriocarpa Benth.[b]	3.7	(L, p)	1.3-3.4 (3p)
eriocarpa Benth.[f]	1.02-9.19	(L, 172)	
66. *masonii* Woods.	11.2-147.0	(L/a, p + 1)	
67. *subaphylla* Woods.	7.65-18.8	(L/a, 2p)	
68. *albicans* S. Wats.	1.79-55.1	(L/a, 3p)	
69a. *vestita vestita*	3.05	(L/a, p)	
vestita vestita[b]	6.8	(L, p)	7.1 (p)
69b. *vestita parishii* (Jeps) Woods.	1.25	(L/a, p)	

TABLE 1 (Continued)

	Cardenolide content (mg digitoxin equivalent/g dried material)[a]		
	Leaf or aerial		Seeds
Subgenus II. *Podostema*			
74. *subulata*	3.21-29.6	(L/a, p)	
Subgenus IV. *Asclepiodella*			
86. *cordifolia* (Benth) Jeps.	<0.57	(L/a, p)	3.06 (p)
cordifolia (Benth) Jeps.[b]			1.3-3.1 (2p)
Subgenus V. *Acerates*			
90. *viridiflora* Raf.	nil	(L/a, p)	0.51 (p)
Subgenus VI. *Solanoa*			
92a. *californica californica*	0.145-0.271	(L/a, 2p)	
92b. *californica greenei* Woods.	0.166	(L/a, p)	
californica greenei Woods.[b]	0.27	(L, p)	3.1-4.3 (2p)
93a. *cryptoceras cryptoceras*	0.252	(L/a, p)	
94. *solanoana* Woods.	0.244	(L/a, p)	
solanoana Woods.[b]			1.2 (p)

[a]Minimum and maximum values (range) are given where possible, each representing a different plant or seed sample; plant part (L/a, leaf/aerial; L, leaf; s, stem) and number of samples is given in parentheses; if an individual sample is from more than one plant it is indicated as pooled (p).
[b]Isman et al., 1977b.
[c]Nelson et al., 1981.
[d]Nishio, 1980.
[e]Benson et al., 1978.
[f]Brower et al., 1982.
Source: Values are from Roeske et al. (1976) except when noted otherwise.

The high concentration of cardenolides in the latices of *A. eriocarpa*—an order of magnitude greater than in the tissues—may represent a defensive adaptation. Based on the structure of laticifers (Esau, 1965), and considering that in other plant species the latex serves as a storage depot for toxic secondary products (Fairbairn et al., 1974; Matile, 1976; Evans, 1977; Evans and Schmidt, 1976; Haupt, 1976), the laticifers may afford plant species such as *A. eriocarpa* with storage sites for cardenolides isolated from the plant's vascular system. As the latex permeates through most tissues, a pressurized secretion of latex rich in toxic metabolites is thus made available at any point of attack or injury. As this secretion oozes, collects, and dries at the site of injury, the high concentration of cardenolides in the immediate area might discourage further attack and/or secondary infection by microorganisms.
That is, the laticifers may provide a means of avoiding autotoxicity via physical isolation of cardenolides in a system that distributes cardenolides throughout the plant, making them available for defense. This is in addition to the

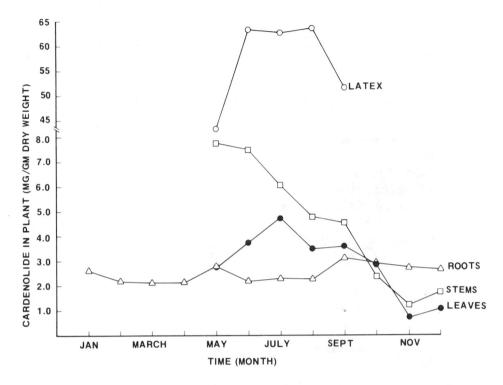

FIGURE 2 Variation in cardenolide content in plant tissues and latex of *Asclepias eriocarpa* during one year. (From Nelson et al., 1981.)

physical barrier presented by the sticky latex oozing at a point of injury, which may trap or deter some potential herbivores.

The propensity of the relatively nonpolar, thiazoline ring-containing cardenolide labriformin in *A. eriocarpa*, and of its chemical counterparts uscharin and voruscharin in *A. curassavica* and *Calotropis procera* (Brüschweiler et al., 1969a) to accumulate in the latex is striking. The presence of nitrogen and sulfur in an additional ring may aid their stabilization within the latex. Also, their ready transformation via hydrolysis and reduction to an array of cardenolide products could further enhance their defensive role for plants under herbivore attack. Finally, uzarigenin in *A. eriocarpa* may act as a deterrent to the monarch's feeding on the stem, as this is the only cardenolide found to date which manifests toxicity (melanism) in developing monarch larvae (Seiber et al., 1980).

In sum, the experiments with *A. eriocarpa* plants led to a more complete understanding of cardenolide localization and distribution within plants of this toxicologically important species, and a possible explanation of how plants of this and presumably other milkweed species can store and deploy cardenolides for defense while avoiding toxicity in the process.

In a related study, the latex and leaf cardenolide content of a number of *Asclepias* species and *Calotropis procera* were compared (Seiber et al., 1982). Most of the samples were from natural or cultivated stands typical of the

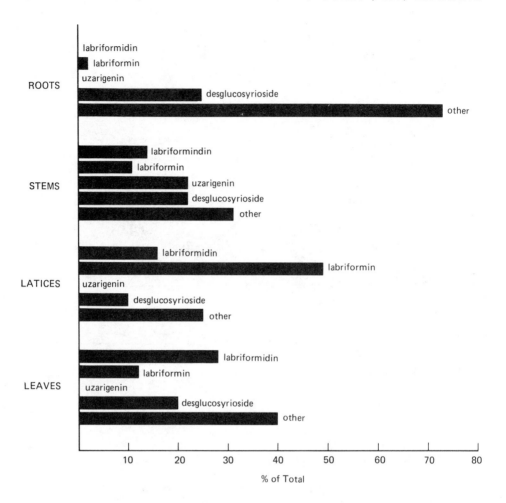

FIGURE 3 Proportions of cardenolides in June *Asclepias eriocarpa* roots, stems, latex, and leaves. (From Nelson et al., 1981.)

plant's habitat. Latex-leaf ratios for samples subjected to lead acetate clean-up were as follows: *A. curassavica*, 51.4 and 54.6 for two sample sets; *C. procera*, 79.4; *A. vestita* subsp. *vestita*, 3.93 and 10.0 for two sample sets from different locations; *A. vestita* subsp. *parishii*, 28.1; *A. eriocarpa*, 8.75; *A. cordifolia*, 1.54; *A. speciosa*, <0.07 and <0.29 for two sample sets; *A. californica*, <0.30; and *A. fascicularis*, less than detectable cardenolide in both leaf and latex. It appeared from these results that latex storage may be relatively more important in plants with a generally high cardenolide content (e.g., *A. curassavica, C. procera, A. vestita,* and *A. eriocarpa*) and that in plants with low concentrations of cardenolides the physical isolation afforded by the laticifers may not be necessary. This argument assumes that cardenolides may be harmful to milkweed plants if present in vital cells and tissues, for which there is experimental evidence in the studies of MacRobbie (1962) and Brown et al. (1965).

Among the species with high cardenolide concentrations in the latex, it was cardenolides of low polarity (high TLC R_f) which predominated. Among *Asclepias* species and *C. procera*, it was primarily cardenolides containing the thiazoline-thiazolidine (NS) ring system at C-3' which were particularly adapted to latex storage. Those species with a very low latex-leaf ratio, such as *A. speciosa* and *A. fascicularis*, did not contain NS-substituted cardenolides or contained only very small quantities of them. Whether some species lack the appropriate cardenolide precursor to NS-ring formation (e.g., a keto group on the sugar moiety), the NS-ring amino acid precursor (i.e., cysteine), or the appropriate enzyme system to effect NS-ring condensation with decarboxylation is not known.

The latex-leaf study provided further evidence for the hypothesis advanced by Roeske et al. (1976) that many western U.S. milkweeds contain cardenolides belonging to one of two structural groups. *A. vestita* and *A. cordifolia* contain cardenolides in common with *A. curassavica* and *C. procera*, that is, derivatives of calotropagenin; *A. eriocarpa*, *A. speciosa*, and from other studies, *A. erosa* and *A. labriformis* contain the epoxy cardenolides, desglucosyrioside and its derivatives.

Brower et al. (1982) reported the gross cardenolide content and cardenolide TLC profile of the leaves of approximately 150 individual *A. eriocarpa* plants from six geographically separate and ecologically different areas in California. The gross cardenolide content of the plants exhibited normal variation with a large variance, ranging from 1.02 to 9.19 mg of digitoxin equivalents per gram of dry material with a mean of 4.21 mg/g. Geographic variation was not a major contributor to this variation. Twenty discernible cardenolides of varying intensity were resolved in the plants by TLC analysis; the overall plant cardenolide profile was remarkably consistent over the geographic locations sampled. From this study it appears that plant location and habitat did not substantially influence either cardenolide quantity or quality for *A. eriocarpa*. Additional western *Asclepias* species now under study by similar methodology conform to these generalizations for *A. eriocarpa*, but it should be noted that for some species, such as *A. syriaca*, location and habitat may have a pronounced effect on plant cardenolide content (Roeske et al., 1976).

III. PLANT-HERBIVORE INTERACTIONS

The potent biological activity of milkweed-derived cardenolides is manifested in animals that feed directly on milkweeds, and in those that prey on milkweed feeders. The first case is exemplified by poisoning among livestock that graze milkweed-infested forage or consume milkweed-contaminated feed. Evidence linking cardenolides to these poisoning episodes comes from toxicity testing of whole plants, plant extracts, and isolated cardenolides in vivo to a range of animal species and in vitro to isolated enzymes. Milkweeds may be included among those plants which gain a competitive advantage by virtue of their poison content, to the extent that these poisons prevent or limit consumption of the plants by at least some herbivores (see, e.g., Fraenkel, 1959; Freeland and Janzen, 1974; Feeny, 1976; Swain, 1977; Laycock, 1978). The second case involves herbivores that have evolved a tolerance to some plant poisons; of particular interest are insects that are not only tolerant of the poisons, but additionally have the capability to sequester them from their plant host and store them for defense against their predators. The herbivore in this situation may enjoy an adaptive advantage by virtue of its complement of sequestered poison (see, e.g., Eisner, 1970; Schildknecht, 1971; Rothschild, 1972; Duffey, 1980). This evolutionary tour de force, originally termed ecological chemistry (Brower and Brower, 1964), has been most thoroughly studied with *Asclepias* milkweeds; monarch butterflies, which feed upon *Asclepias* during larval development; and bluejays, experimental predators of monarchs (Brower, 1969; Roeske et al., 1976). Defended prey, monarchs in this example, frequently adopt distinctive warning coloration (i.e., become aposematic) such that predators learn to avoid them on sight following one or more unpleasant encounters. Protection may further be extended to nonpoisonous prey species through their adoption of similar coloration—the phenomenon of Batesian mimicry (Brower, 1969).

A. Livestock

Because of the general unpalatability of milkweeds, range animals do not normally eat them. However, in cases of overgrazing of rangeland, or in instances of drought, milkweeds may provide the only forage available.

Several species of *Asclepias* have been responsible for sporadic livestock poisonings, particularly in the western United States. Among species commonly associated with livestock poisonings are *A. mexicana* (frequently confused with *A. fascicularis*) and *A. speciosa* (Marsh et al., 1920; Fleming and Peterson, 1920), *A. subverticillata* (=*A. verticillata*, *A. galioides*) (Glover et al., 1918; Marsh et al., 1920; Clark, 1979), *A. labriformis* (Kingsbury, 1964; Huffman et al., 1956; Holmgren, 1945; Stoddart et al., 1949), *A. eriocarpa* (Marsh and Clawson, 1924; Baxter, 1944), and *A. pumila* (Huffmann et al., 1956). Milkweed poisoning in sheep and cattle remains a problem to ranchers, particularly when adequate quantities of potable forage are unavailable or when milkweeds are incorporated in cut feed. For example, in a 1975 episode, 250 sheep in Sonoma County, California, died after eating hay contaminated with *A. eriocarpa*. Toxicities of several milkweed species to sheep are compared in Table 2.

Toxic effects on all range animals are similar, although differences in intensity of effects with different plants may be seen. Nervous and gastrointestinal symptoms include depression and apathy, weakness, loss of muscle control, falling, dilated pupils, respiratory paralysis, intestinal stasis and fermentation, and occasionally, diarrhea. Sometimes tetanic seizures are seen. Depending on the toxicity and amount of plant eaten, symptoms may begin as early as 2 hr after ingestion of the plant and may continue for hours or days. A poisoned animal may or may not experience a period of coma before dying. Lesions at death include acute catarrhal gastroenteritis and congestion of the lungs and sometimes of the kidneys (Smith et al., 1972).

Limited work has been done to determine the toxic principles of various milkweeds. Couch (1929) reported the isolation of a toxic, nonalkaloidal, nonglycosidal resinoid from *A. eriocarpa* which was believed to be the plant's source of toxicity. Toxic effects, toxic doses, and pathology were

TABLE 2   Comparative Toxicity of Several Milkweed Species to Sheep[a]

Species	Minimum toxic dose (lb/100-lb sheep)	Minimum lethal dose (lb/100-lb sheep)
A. labriformis[b]	—	0.05-0.2
A. eriocarpa[c]	0.1-0.2	0.22
A. galioides[c] (= *A. subverticillata*)	0.22	0.22
A. pumila[d]	0.79	2.16
A. mexicana[c]	0.88	1.32
A. verticillata[c]	2.20	—
A. speciosa[e]	2.60	—

[a]Data are expressed as pounds of whole green plant required to produce a given effect in a 100-lb sheep.
[b]Kingsbury, 1964; Benson et al., 1979.
[c]Marsh and Clawson, 1921. Note that in the early literature, *A. mexicana* is frequently confused with *A. fascicularis* (Woodson, 1954).
[d]Marsh and Clawson, 1924.
[e]Fleming and Peterson, 1920. Only green leaves were fed.

not described. Roeske (1971) investigated four milkweed species, *A. labriformis*, *A. eriocarpa*, *A. speciosa,* and *A. fascicularis*, for their poisonous principles. A positive correlation was found between the cardenolide content of the plants and their toxicity to mice. *A. labriformis* and *A. eriocarpa* were highly toxic and had a high cardenolide content. *A. speciosa* was slightly toxic and contained less cardenolides. *A. fascicularis* was the least toxic and contained no detectable cardenolides. Cardenolides were further implicated as the toxic principles of *A. eriocarpa* and *A. labriformis* in studies conducted by Benson et al. (1979). Four preparations were tested on sheep: dried, ground *A. eriocarpa* and *A. labriformis* plants, a crude ethanolic *A. eriocarpa* extract, and a partially purified extract from which pigments and fats had been removed. In addition, labriformin (a purified cardenolide present in *A. labriformis* and *A. eriocarpa*) and digitoxin were administered. Toxic symptoms and gross histopathological lesions were qualitatively similar regardless of whether plants, extracts, labriformin, or digitoxin were administered. Symptoms of poisoning and some pathological changes were similar to those observed in sheep and goats chronically fed *Calotropis procera*, a member of the Asclepiadaceae plant family which also has a high cardenolide content (Mahmoud et al., 1979a,b).

Cardenolides may not be the chief toxic principle in all toxic *Asclepias* species or the sole toxic principle in cardenolide-containing species. For example, *A. subverticillata* has little, if any, cardenolide (Roeske et al., 1976), yet it is listed as one of the more toxic species (Table 1). *A. mexicana* has a low cardenolide content and moderate toxicity. Marsh et al. (1920) isolated and partially characterized a resinoid, galitoxin, from *A.*

subverticillata and *A. mexicana* which was believed to be the toxic principle
of these species. The effects of galitoxin on guinea pigs resembled effects
produced in both guinea pigs and sheep by administration of *A. subverticillata*. Alkaloids and glycosides were also isolated from *A. subverticillata*
and *A. mexicana*, but did not produce effects similar to those produced by
the intact plants. Further investigation is clearly needed to substantiate
these early studies, and to ascertain whether, in fact, noncardenolide toxicants are of importance in poisonings attributed to *Asclepias* species.

To summarize, while *Asclepias* species vary considerably in their toxicity to livestock, they include some members which are among the most toxic
plants encountered in the open range. For several species there exists
strong evidence that cardenolides are the primary poisons. For others, however, it has been suggested that noncardenolide poisons are the principal
toxic agents. Circumstantial evidence, including the avoidance of *Asclepias*
by livestock under usual forage conditions, suggests that cardenolides, and
perhaps other *Asclepias* constituents, provide a competitive advantage to
these plants in nature by limiting their consumption by large herbivores.

B. Insects

Early studies on cardenolide sequestration from milkweeds were carried out
with the grasshopper, *Poekilocerus bufonius* (Reichstein, 1967b; von Euw
et al., 1967). This aposematic North African species, which probably feeds
exclusively upon milkweeds in nature, has a poison gland from which the
contents may be ejected as a spray during the immature hopper stage, or
as a foamy froth in the adult stage. The secretion contains histamine and
one cat-lethal dose of cardenolide. When reared on *Calotropis procera* or
Asclepias curassavica, the grasshoppers secretion contained only calactin
and calotropin in appreciable quantities of the seven or so related cardenolides present in the plant. Some experimental predators, such as
white mice and the European jay, learned to avoid *P. bufonius* only after
experiencing disagreeable side effects following contact with the insect.
These included, in the case of the jay, the characteristic vomiting response
elicited by cardenolides. Predators that are less sensitive to cardenolides,
such as the European hedgehog, consumed *P. bufonius* with impunity.

A similar situation exists with monarch butterflies (*Danaus plexippus* L.).
Monarch larvae feed exclusively on *Asclepias* species from the time of hatching to pupation (Urquhart, 1960). Adults fed as larvae on *A. curassavica*
contained principally calactin and calotropin and lacked uscharin, voruscharin, and uscharidin—principal foliage cardenolides of *A. curassavica*
(Parsons, 1965; Reichstein, 1967b; Reichstein et al., 1968). Several cardenolides more polar than calactin and calotropin were also present in the adult
butterflies.

Selective storage of some cardenolides from among several in the food
plant was confirmed in a later study with *A. curassavica*, and was also found
to occur with different groups of cardenolides in *Gomphocarpus physocarpus*
(Roeske et al., 1976) and *A. eriocarpa* (Brower et al., 1982). That metabolic conversion might be involved in this selectivity (von Euw et al., 1967;
Roeske et al., 1976) was proven by separate administration of graded doses
of individual cardenolides to monarch larvae, with subsequent analysis of the

cardenolide content of the larvae and adults (Seiber et al., 1980). Calactin and calotropin were stored intact, but uscharidin was metabolized to a mixture of calactin and calotropin, which were the cardenolides stored. This conversion, a reduction, occurred very rapidly during feeding, apparently in the gut. Reasons offered for the storeability of calactin and calotropin, but not uscharidin, included chemical instability of the latter in the gut milieu, polarity (calactin and calotropin are more polar than uscharidin), and the possibility that the C-3' OH in calactin and calotropin was needed for binding to a carrier and/or tissue storage site. Uscharin is also not stored, apparently giving way to uscharidin by enzymatic hydrolysis and then to calactin and calotropin by the reduction mentioned above. This series of conversions, and that for the cardenolides labriformin and labriformidin when *A. eriocarpa* was fed to monarchs (Brower et al., 1982), are summarized in Fig. 4.

Uzarigenin, a milkweed cardenolide, and digitoxigenin, the 5β isomer of uzarigenin not found in milkweeds, were also metabolized by monarch larvae, giving way to products much more polar than the administered chemicals (Seiber et al., 1980). The products have tentatively been identified as 3-β-D-glucosides (C. J. Nelson and J. N. Seiber, unpublished observations, 1980). It was hypothesized that glycoside formation was a detoxication reaction for the larvae, circumventing the melanism caused when uzarigenin was fed at higher dose levels. Glycoside formation was apparently reversed later in the monarch's development, for only unconjugated uzarigenin and digitoxigenin were present in adults from the rearing experiments.

Although both *P. bufonius* and *D. plexippus* apparently sequester cardenolides for defense, the grasshopper uses active deployment through a defensive secretion, whereas the butterfly's defensive load is distributed in its body tissue. Certain lygaeids that feed on milkweeds more resemble *P. bufonius* in that particularly high cardenolide concentrations accumulated in secretions of a special series of dorsolateral glands (Scudder and Duffey, 1972). *Oncopeltus fasciatus* reared on *A. syriaca* seeds selectively stored polar cardenolides in the dorsolateral space fluid, while smaller amounts of less polar components were present in tissues and hemolymph (Duffey and Scudder, 1974). Metabolism was apparently involved in this segregation, from the evidence that digitoxin was converted by *O. fasciatus* to two unidentified metabolites of greater polarity. Ouabain, a very polar cardenolide, was not metabolized under the same conditions. Digitoxin and ouabain were used as models for cardenolide disposition, although neither occurs naturally in milkweeds.

Selective cardenolide storage was also observed for the aposematic aphid, *Aphis nerii*, reared on *Nerium oleander* and *A. curassavica* (Rothschild et al., 1970). In this case, it is apparently the selective feeding habits of the aphid, rather than a sorting or metabolism following ingestion, which results in the sequestering of a cardenolide mixture differing from that in the plants. The aphid feeds selectively on the phloem, which may have a different cardenolide complement than the tissue or latex.

In addition to the qualitative regulation of its stored cardenolides referred to above, monarchs apparently regulate the quantities they store both from among plants of a single species and those derived from different species. From collections of adults at several field locations of *A. eriocarpa* in

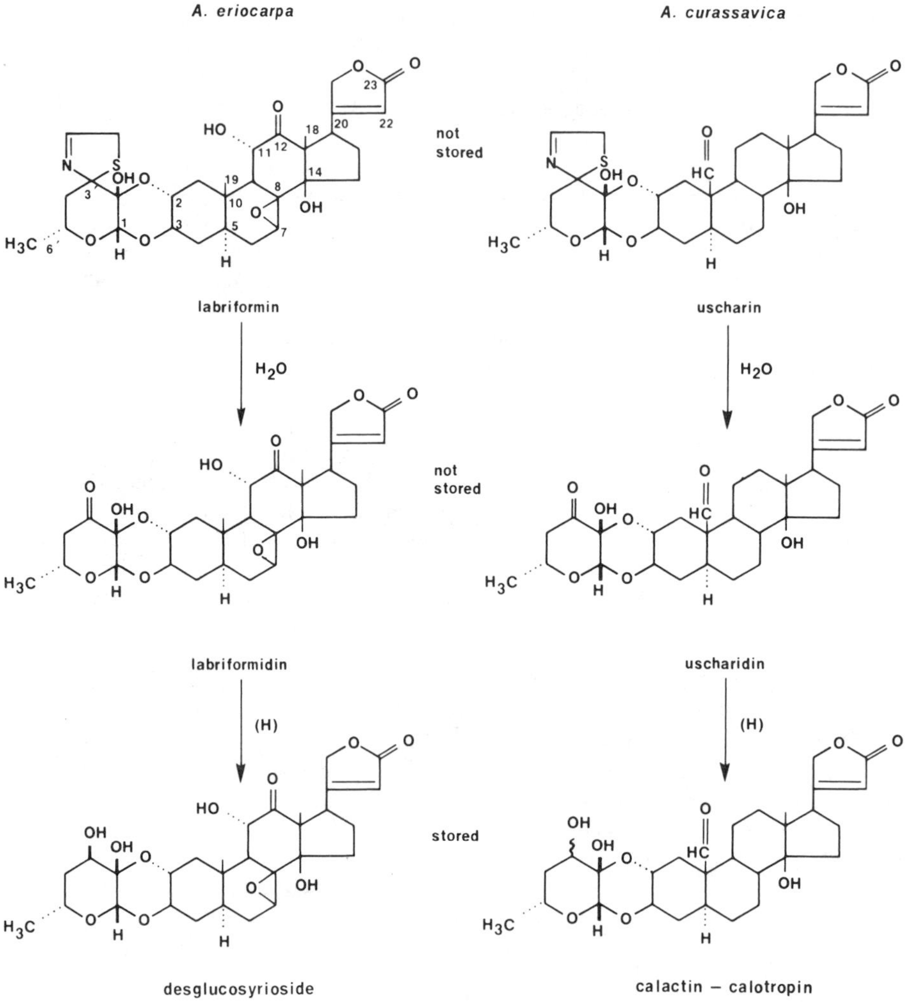

FIGURE 4 Chemical and metabolic conversions for some major cardenolides of *Asclepias curassavica* and *A. eriocarpa*. (From Seiber et al., 1981; Brower et al., 1982.)

California, Brower et al. (1982) found that, while there was a large variation in the total cardenolide content of the plants (1.02-9.19 mg/g), less variation occurred in total cardenolide for the corresponding butterflies (1.36-6.06 mg/g). Adult butterflies reared on plants low in cardenolide concentrated the chemicals, whereas those reared on cardenolide-enriched plants excluded more cardenolide relative to the amount stored, resulting in a butterfly-plant cardenolide relationship virtually independent of plant cardenolide concentration. This quantitative leveling over broad swings in the cardenolide content of the food plant provided evidence for cardenolide regulation to achieve a storage capacity, at least for cardenolides of a specific structural type (e.g., desglucosyrioside and its derivatives in *A. eriocarpa*). That this storage capacity may differ for cardenolides of different structural groups was indicated by two findings. (1) Adult monarchs had 13% of the total cardenolide content of consumed leaves when reared as larvae on *A. curassavica*, but only 3% for *Gomphocarpus physocarpus*, despite both milkweed species having nearly identical concentrations of total cardenolide. The cardenolides of these two plants, while similar in polarity, are structurally dissimilar (Roeske et al., 1976). (2) Cardenolide storage levels were higher for larvae fed the glycosides uscharidin and calotropin than for those administered the genins uzarigenin and digitoxigenin (Seiber et al., 1980).

A consequence of quantitative regulation is that monarchs reared as larvae on milkweed species which have the same structural groups of cardenolies but differing total amounts of them, sequester and store very similar quantities of cardenolides. This has been observed for *A. cordifolia, A. californica, A. vestita,* and *A. curassavica* (L. P. Brower, C. J. Nelson, and J. N. Seiber, unpublished observations, 1982)—all of which contain cardenolides of the *Calotropis* series (Seiber et al., 1982).

Even though the cardenolide profile in adult monarchs does not simply mirror that in the larvae's food plant, insofar as sorting and metabolism have altered the latter's profile, it is nevertheless likely that the adults will show a profile (or "fingerprint") which is related to its *Asclepias* host plant. The fingerprint relationship can be determined experimentally by TLC analysis of cardenolide extracts from adults and the foliage of their corresponding food plants. For *A. eriocarpa* the fingerprint relationship—a composite of 85 individual plant-butterfly pairs collected from six widely separate locations in California—was presented by Brower et al. (1982). Similar analysis of individual butterflies taken from migratory populations may allow for matching them to their larval food plant (Roeske et al., 1976). For example, the differing profiles for overwintering butterflies from California and migratory individuals collected in Massachusetts indicated substantial variations in the cardenolide content of the *Asclepias* flora on which they were reared. This may explain the differences in emetic potencies of butterflies from the two populations (California specimens were much more emetic than those from Massachusetts, after normalizing for cardenolide concentration) and the bimodal distribution in cardenolide content in the California specimens (Brower and Moffitt, 1974). These fingerprinting techniques may show interesting contrasts between populations of monarchs from the spectacular overwintering sites in Mexico, which originate largely from east of the Rockies (Urquhart and Urquhart, 1976, 1979; Brower et al., 1977), and those from the California overwintering sites, which are composed of individuals originating west of the Sierras (Tuskes and Brower, 1978). At any rate, a palatability spectrum reflecting both the quantities and types of cardenolides in natural butterfly populations, with its origins in the larvae's food sources, seems reasonable and defensible based on the accumulating evidence (Brower, et al., 1968; Brower et al., 1972; Brower and Moffitt, 1974; Roeske et al., 1976; Fink and Brower, 1981; Brower et al., 1982).

Underlying this argument is the hypothesis that cardenolides confer unpalatability to butterflies toward their natural enemies—an activity which can be quantified by the "emetic dose fifty (ED_{50})" assay with bluejays as the experimental predator (Brower et al., 1972; Brower and Moffitt, 1974). Although objections have been raised as to the ecological relevance of this assay (Duffey, 1977), recent evidence has shown that cardenolides induce emesis in 12 species of birds belonging to nine families, some of which have been observed selectively feeding on palatable butterflies in the wild (Fink and Brower, 1981). Another criticism (Dixon et al., 1978) of sole reliance on the bluejay ED_{50} assay is that one may overlook other potential defensive substances in butterflies, such as pyrrolizidine alkaloids derived by Danaidae from adult nectar gathering (Edgar et al., 1979; Boppré et al., 1978).

Some work has been done on the distribution of cardenolides within the monarch's body. Brower and Glazier (1975) showed that cardenolides of differing emetic potencies were unevenly distributed in body parts of the monarch. The concentration, number of ED_{50} units, and relative emetic potency are given in Fig. 5 for the abdomen, wings, and thorax of male and female monarchs reared on *A. curassavica*. Butterflies had the highest concentration of cardenolides in the wing, but these were the least emetic. It was argued that a high concentration of bitter cardenolides in the wings could induce a predator to release its prey unharmed after seizing the prey by its wings (Brower and Glazier, 1975; Fink and Brower, 1981). If the predator persists in consuming the prey, it will encounter the more emetic cardenolides in the abdomen, become sick if the abdomen contains a sufficient

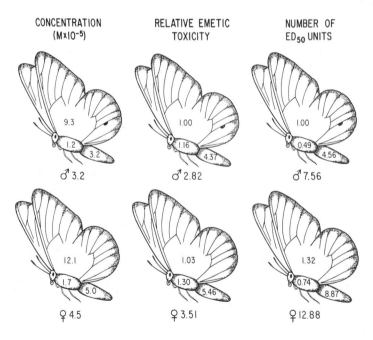

FIGURE 5 Cardenolide concentration (left), relative emetic potency (center), and ED_{50} units (right) for monarch butterflies reared on *Asclepias curassavica*. (From Brower and Glazier, 1975. Copyright 1975 by the American Association for the Advancement of Science.)

quantity of cardenolides, and thereby learn to avoid poisonous monarchs on sight or by first tasting the wings thereafter. Support for this argument lies in a number of beak-marked monarchs (i.e., live individuals with the imprint of bird beaks in their wings) captured in the wild.

The physicochemical basis for cardenolide storage in *Oncopeltus* has been studied by several workers. Yoder et al. (1976) showed that digitoxin and ouabain diffuse across gut cell membranes in proportion to their concentration, but that ouabain's rate was only about 10% that of digitoxin, presumably because of ouabain's lower lipophilicity. Duffey et al. (1978) found that, once in the hemolymph, ouabain is rapidly sequestered into the dorsolateral glands, where its concentration was about 75 times that in the hemolymph. By contrast, the less polar digitoxin gave a concentration ratio of less than 5 for dorsolateral space fluid relative to hemolymph. For both cardenolides, uptake into the dorsolateral space fluid increased linearly with dose administered in the feed—seeds of *A. syriaca*. A linear increase with seed concentration of native cardenolides was also noted for seeds of several *Asclepias* species (Isman et al., 1977b). The evidence suggests that polar cardenolides, such as ouabain, are concentrated by a physical emulsion-phase transfer process which transports these chemicals to the dorsolateral gland so rapidly that hemolymph concentration is low at all times (Duffey, 1980). Less polar cardenolides, such as digitoxin, must diffuse into the dorsolateral space fluid and do not interact efficiently with the emulsion-phase sequestration system.

The relevance of the processes of sequestration in *Oncopeltus* to monarchs is questionable, in that monarchs sequester and store cardenolides only during their larval stages and, in the adult, it is tissue or exoskeletal binding which is involved rather than storage in a defensive fluid. Although the underlying physicochemical basis controlling these processes in monarchs remains to be elucidated, the commonality of a free OH group at C-3' of the sugar, or C-3 of the genin in those cardenolides which are efficiently stored, suggests that specific binding sites exist in tissues which become cardenolide storage depots (Seiber et al., 1980; Brower et al., 1982).

In sum, although many mechanistic details of sequestration and storage are unknown, the accumulation of evidence strongly suggests that cardenolides are primary protective agents for several insects which are adapted to feed on cardenolide-containing milkweeds. Adapted insects are apparently relatively insensitive to cardenolide intoxication (Jungreis and Vaughan, 1977; Vaughan and Jungreis, 1977), unlike their vertebrate predators.

IV. COMPARATIVE PHARMACOLOGICAL-TOXICOLOGICAL ACTIVITY IN VERTEBRATES

Median lethal doses (LD_{50}) of eriocarpin (= desglucosyrioside), labriformidin, and labriformin, isolated from *A. eriocarpa*, of uscharidin and calotropin from *A. curassavica* (a milkweed also known for its toxicity to cattle, Tokarnia et al., 1972), and of ouabain were determined using male Swiss Webster mice (Table 3). A threefold range in toxicity exists in the LD_{50} values of labriformidin and labriformin from *A. eriocarpa*, and this range encompasses the LD_{50} value of ouabain. The *A. curassavica* cardenolides

TABLE 3 Median Lethal Doses (LD_{50}) of Cardenolides to Male Swiss Webster Mice[a]

Cardenolide	LD_{50} (mg/kg)	LD_{50} Relative to ouabain
Labriformidin	3.1 (4)[b]	0.46
Ouabain	6.8 (4)	1.0
Eriocarpin (= desglycosyrioside)	6.5 (2)	1.03
Labriformin	9.2 (1)	1.35
Calotropin[c]	9.8 (2)	1.44
Uscharidin	11.8 (2)	1.74

[a]Administration was intraperitoneally in Emulphor 719-water-ethanol (1:8:2).
[b]Numbers in parentheses refer to the number of times an experiment was carried out. LD_{50} values are an average of the results obtained in all experiments with a given compound.
[c]Calotropin contained approximately 20% of its isomer, calactin, as measured by high-performance liquid chromatography.
Source: Modified from Benson et al., 1978.

calotropin and uscharidin appear to be slightly less toxic than the *A. eriocarpa* cardenolides and ouabain.

Affected animals became quiet and ataxic. Respiration became rapid but labored, and animals laid flat on their abdomens with their limbs spread. There was some trembling, and severely affected animals convulsed. Symptoms appeared within a few minutes after the mice were dosed; and death, accompanied by clonic convulsions, usually occurred within 1-2 hr. Mice that exhibited symptoms of poisoning, but survived, did not appear ill on the days following treatment.

No lesions were clearly visible in mice that died within 24 hr of being dosed. Hemorrhages were seen in the lungs of animals euthanized on the days following treatment, but the incidence of this lesion was not significantly greater than the incidence in the control group.

The median lethal doses of several milkweed cardenolides, and digitoxin and ouabain, to cats are given in Table 4. Details on toxic effects are noted in the references. Cats are several times more sensitive than mice to cardenolide intoxication and, generally speaking, are more sensitive to glycosides than to genins.

(Na^+, K^+)-ATPase present in cardiac muscle has been proposed as the pharmacological receptor for cardiac glycosides. Inhibition of this enzyme system is thought to be responsible for the therapeutic or toxic effects of cardiac glycosides on the heart of vertebrates. The ability of natural and semisynthetic cardiac glycosides to inhibit (Na^+, K^+)-ATPase preparations has been used to predict the extent of their effect on the myocardium. Benson et al. (1978) studied the ability of several milkweed cardenolides to bind lamb myocardium (Na^+, K^+)-ATPase. Results, expressed in terms of

TABLE 4 Acute Toxicity (LD_{50}) of Several Cardenolides to Cats (Intravenous)

Cardenolide	LD_{50} (mg/kg)	Reference
Ouabain	0.11	Detweiler, 1967
	0.107	Chen et al., 1942
Calotropin	0.118	Brüschweiler et al., 1969a
	0.105	Chen et al., 1942
Calactin	0.11	Detweiler, 1967
	0.118	Chen, 1970
	0.118	Brüschweiler at al., 1969a
Uscharin	0.144	Brüschweiler et al., 1969a
G-strophanthin	0.132	Patnaik and Köhler, 1978
Proceroside	0.199	Brüschweiler et al., 1969a
Asclepin	0.236	Patnaik and Köhler, 1978
Digitoxin	0.38	Parsons and Summers, 1971
	0.40	Detweiler, 1967
	0.43	Patnaik and Köhler, 1978
Digitoxigenin	0.58	Patnaik and Köhler, 1978
Uzarigenin	1.38	Chen, 1970
Uscharidin	1.4	Brüschweiler et al., 1969a
Calotropagenin	1.52	Chen, 1970
	2.57	Brüschweiler et al., 1969a

the concentration of cardenolide in the enzyme incubation medium necessary to inhibit enzyme activity by 50% (I_{50}), are shown in Table 5. All of the compounds studied had I_{50} values very close to that of ouabain, despite considerable differences in their structures.

The evidence cited here and by Chen (1970) and other authors indicates that all cardenolides may be regarded as highly toxic, that sensitivity to cardenolides may vary considerably among species and with route of administration, and that the principal toxic effects are associated with the myocardium. Aside from the therapeutic cardenolides of the 5β series, very little pharmacokinetic and metabolic fate information exists for cardenolides in vertebrates.

V. METHODS OF ANALYSIS

Cardenolide analysis in biological samples may be carried out to find (1) the total amount or total concentration of cardenolides, (2) the types of cardenolides present, and (3) the amounts, concentrations, and/or proportions of

TABLE 5 Cardenolide I_{50} Values Inhibition of Lamb Cardiac (Na^+, K^+)-ATPase[a]

Cardenolide	Experimental (mean I_{50} μM ± S.D.)	Cardenolide I_{50} relative to ouabain I_{50} (I_{50} cardenolide/I_{50} ouabain) [mean (n = 3) ± S.D.][b]
Digitoxin	0.36 ± 0.12	0.20 ± 0.13
Labriformidin	0.78 ± 0.07	0.75 ± 0.09
Eriocarpin	0.81 ± 0.24	0.77 ± 0.18
Labriformin	0.96 ± 0.18	0.92 ± 0.10
Uscharidin	0.77 ± 0.20	0.92 ± 0.17
Ouabain	0.89 ± 0.2	1.00
Calotropin	0.68 ± 0.16	1.04 ± 0.18

[a] Values for ouabain are an average of six determinations, while those for the remaining cardenolides are an average of three determinations.
[b] I_{50} relative to ouabain was derived during separate assays for each compound.
Source: Data from Benson et al., 1978.

individual cardenolides when mixtures are present. The total amount of cardenolide is generally determined by spectrophotometric assay of a colored derivative based on conversion of the steroid or butenolide rings. Determination of the types of cardenolides present and their individual concentrations or proportions involves a differential analysis based on chromatography, often following one or more cleanup steps to reduce noncardenolide coextractives to a tolerable level. Identifications are based on comparison of chromatographic mobilities of unknowns with those of standards—under more than one chromatographic condition to increase certainty in the assignments--or by isolation of resolved and purified components for spectrometric (infrared, IR; nuclear magnetic resonance, NMR; and mass spectrometry, MS) measurement. Quantitation is achieved by comparing a response [TLC spot density or area, high-performance liquid chromatography (HPLC) peak area] with that of a standard. The following sections are meant to illustrate these general principles with a few examples.

A. Spectrophotometric Assay

A characteristic feature of cardenolides is the presence of an α, β-unsaturated γ-lactone ring attached to C-17 of the steroid nucleus. The maximum ultraviolet (UV) absorption band of the lactone ring, at ca. 217 nm, is not suitable for determination of cardenolides in most sample matrices because of large numbers of potential interferences with similar absorption characteristics. To circumvent this, derivatization is generally required, most frequently through formation of charge-transfer complexes formed with polynitrated aromatic compounds in the presence of a strong base. An example is the base-catalyzed reaction of cardenolides with 2,2',4,4'-tetranitrodiphenyl (TNDP) to form a blue-colored derivative, λ_{max} = ca. 620 nm. This procedure formed the basis for measuring the total cardenolide content in ethanol extracts of plant and insect samples (Brower et al., 1972, with modifications by Brower et al., 1975). Applications have been to wild-caught *Danaus plexippus* (Brower and Moffitt, 1974; Fink and Brower, 1981) and *D. chrysippus* samples (Brower et al., 1975), to *Oncopeltus fasciatus*, *Lygaeus kalmii kalmii* and their milkweed hosts (Isman et al., 1977a,b; Duffey et al., 1978), *D. plexippus* reared on *A. curassavica* and *Gomphocarpus physocarpus* (Roeske et al., 1976), uscharidin, calotropin, digitoxigenin, and uzarigenin-dosed monarch larvae (Seiber et al., 1980), plant parts and latex of *A. eriocarpa* (Nelson et al., 1981) and other *Asclepias* species (Seiber et al., 1982), and leaves of *A. eriocarpa* and monarchs reared as larvae on them (Brower et al., 1982).

Variations in the methodology include the conditions of extraction and cleanup steps employed, if any. Isman et al. (1977a) described extraction procedures for insects by soaking in chloroform-methanol solvent. For milkweed seeds, these authors defatted the samples by extraction with petroleum ether, chloroform-methanol, and 95% ethanol in sequence. Tuskes and Brower (1978) found that defatting extracts of wild-caught monarchs facilitated cardenolide analysis and allowed for fat determination at the same time. Rothschild et al. (1970) defatted aphids with petroleum ether prior to cardenolide determinations. For plant leaves, stems, roots, and so on, most authors have used 95% ethanol at an elevated temperature for cardenolide extraction (Isman et al., 1977a,b; Nelson et al., 1981). Reichstein et al. (1968) performed extraction under a CO_2 atmosphere, to minimize oxidation of cardenolides. We have found that normal precautions (high-quality solvents, evaporation under nitrogen, and storing all extracts in a freezer) guard sufficiently against breakdown.

No cleanup is required for determination of cardenolides at ca. 10 μM in most samples, and as little as 0.1 μM in some (Isman et al., 1977a,b). The results may be expressed in absolute concentrations if standards of the cardenolide(s) in the sample are available as a reference, or in mass of cardenolide equivalent to digitoxin per unit mass of sample (Brower et al., 1982). The latter data cannot be converted to the exact amount of cardenolides present since the molar absorptivities of the TNDP complex of all cardenolides present in a sample are generally not known. However, it is acceptable for most applications because the variation in extinction coefficients is relatively minor (Moffitt and Brower, as cited by Brower and Glazier, 1975, footnote 17). Also, using digitoxin as a reference standard allows for comparison of results from many sample types, including those whose cardenolide composition is not completely characterized (Roeske et al., 1976).

B. Chromatography

1. Cleanup

While spectroassay may be performed on crude extracts of most samples, or on samples provided with very minimal cleanup, subsequent analysis by TLC or HPLC may require more extensive cleanup. Roeske et al. (1976)

described a cleanup procedure based on solvent partitioning which eliminated most TLC interferences, but gave cardenolide recoveries of only 13-30% for plant, butterfly, and butterfly frass samples. It also yielded higher recoveries for less polar cardenolides than for more polar ones. A more satisfactory method for plant and butterfly samples was subsequently described by these same workers (Nelson et al., 1981; Brower et al., 1982; Seiber et al., 1982). The method, described in detail by Brower et al. (1982) (Fig. 6), is a modification of Rowson's (1952) procedure for *Digitalis* leaf tinctures; it reduces the amounts of some potentially interfering pigments, lipids, and so on, through their precipitation with lead(II) acetate, with subsequent removal of excess lead from the cardenolide-containing supernatant by precipitation with sulfate. This method gave cardenolide recoveries of 94% for *A. eriocarpa* leaves and 82% for *A. eriocarpa*-reared butterflies. Some hydrolysis of a thiazoline ring-containing cardenolide (labriformin) occurred, and

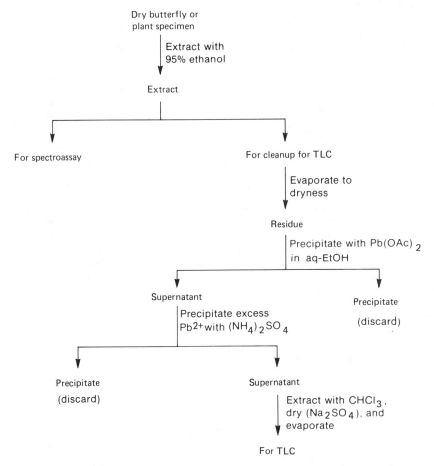

FIGURE 6 Sample extraction and cleanup for spectroassay and thin-layer chromatography of *Asclepias* cardenolides in plants and butterflies. (From Brower et al., 1982.)

was minimized by reducing the time and temperature of contact with $Pb(OAc)_2$. Other applications gave recoveries in excess of 70% for most plant samples, with removal of TLC interferences to acceptable levels (Nelson et al., 1982; Seiber et al., 1982).

2. *Thin-Layer Chromatography*

Differential cardenolide analyses have employed paper, thin-layer, or gas-liquid chromatography (Heftmann, 1976). TLC methods are the most common, and a variety of conditions have been developed for samples of plants, insects, drug preparations, and drugs in animal excreta, fluids, and tissues (see, e.g., Carvalhas and Figueira, 1973; Faber, 1977). Chromogenic reagents include dinitrobenzoic acid (Kedde reagent), 2,2',4,4'-tetranitrodiphenyl (TNDP), and picric acid, while treatment with mineral acids produces fluorescent products. Quantitation is principally by densitometry or through the use of radiolabeled tracers.

Roeske et al. (1976) described two solvent systems that provide good resolution of most cardenolides found in milkweeds when employed with multiple development on nonactivated TLC plates. Digitoxin and/or digitoxigenin were used as reference standards, and TNDP was the chromogenic reagent. A compilation of R_f values for several cardenolides in the solvent systems of Roeske et al., but with heat-activated TLC plates, is presented in Table 6. TLC analysis has been applied to samples cleaned up by solvent partition (Roeske et al., 1976) or lead acetate (Nelson et al., 1981; Brower et al., 1982). Examples include a variety of plant and butterfly samples (Seiber et al., 1978, 1980, 1982). Quantitation was achieved in one example by spectroassay of bands recovered from preparative TLC (Roeske et al., 1976), while in other studies relative proportions of cardenolides were measured by densitometry (Seiber et al., 1980) or subjectively (Brower et al., 1982; Seiber et al., 1982). The problem of rapid color fading after visualization with TNDP can be overcome by photographing the TLC plates just after spraying, and making all subsequent measurements from colored photographic prints (Seiber et al., 1980).

3. *High-Performance Liquid Chromatography*

The development of HPLC as an analytical tool during the 1970s has spurred the application of this technique to cardenolides and related compounds which are generally not stable to gas-liquid chromatography (GLC) conditions. A recent review has appeared summarizing the applications of HPLC to this class of steroids (Seiber et al., 1981).

HPLC was used to resolve and quantify cardenolides in extracts cleaned up by the same lead acetate procedure employed for TLC (Benson and Seiber, 1978). Conditions included a μBondapak C_{18} reverse-phase column eluted with aqueous acetonitrile, detection at 220 nm, and quantitation against external standards. With this procedure the major cardenolides in *Asclepias eriocarpa* and *A. curassavica* plant and butterfly samples were clearly evident when compared with these from *Gonolobus rostratus*, a milkweed that lacks cardenolides (Brower, 1969). Qualitatively, HPLC provided evidence that *A. eriocarpa* butterflies concentrated primarily desglycosyrioside ("eriocarpin") and lacked the principal leaf cardenolides labriformin and labriformidin.

TABLE 6 TLC R_f and R_{dig} (R_f Relative to Digitoxin) Values for *Asclepias* Cardenolides and Reference Standards[a]

| | TLC solvent system | | | |
| | EtOAc:MeOH (97:3), 2× | | CHCl$_3$:MeOH:HCONH$_2$ (90:6:1), 4× | |
Cardenolide	R_f	R_{dig}	R_f	R_{dig}
Afroside	0.493	1.48	0.273	0.95
Ascleposide	0.233	0.70	0.240	0.84
Calactin	0.493	1.48	0.280	0.98
Calotoxin	0.162	0.86	0.173	0.60
Calotrapagenin	0.167	0.50	0.227	0.79
Calotropin	0.387	1.16	0.287	1.00
Coroglaucigenin	0.22	0.66	0.22	0.77
Coroglaucigenin α-L-rhamnoside	0.093	0.28	0.093	0.33
Corotoxigenin	0.413	1.24	0.320	1.12
Corotoxigenin α-L-rhamnoside	0.133	0.40	0.153	0.53
Desglucosyrioside	0.513	1.54	0.313	1.09
Digitoxigenin	0.613	1.84	0.633	2.21
Digitoxin	0.333	1.00	0.287	1.00
Frugoside	0.087	0.26	0.107	0.37
Gofruside	0.173	0.52	0.167	0.58
Gomphoside	0.553	1.66	0.360	1.26
Labriformidin	0.667	2.00	0.713	2.49
Labriformin	0.520	1.56	0.700	2.44
Mallogenin α-L-rhamnoside	0.173	0.52	0.127	0.44
Oleandergenin	0.593	1.78	0.620	2.16
Panogenin α-L-rhamnoside	0.333	0.10	—	—
Proceroside[b]	0.347	1.04	0.287	1.00
	0.280	0.84	0.267	0.93
Syriobioside	0.287	0.86	0.180	0.63

TABLE 6 (Continued)

| Cardenolide | TLC solvent system | | | |
| | EtOAc:MeOH (97:3), 2× | | CHCl$_3$:MeOH:HCONH$_2$ (90:6:1), 4× | |
	R_f	R_{dig}	R_f	R_{dig}
Syriogenin	0.313	0.94	0.260	0.91
Syrioside	0.027	0.08	0.033	0.12
Uscharidin	0.633	1.90	0.513	1.79
Uscharin	0.507	1.52	0.447	1.56
Uzarigenin	0.567	1.70	0.533	1.93
Uzarin	0.00	0.00	0.040	0.14
Voruscharin	0.567	1.70	0.580	2.02

[a]All R_f values were measured on heat-activated (100°C, 3 hr) 20 × 20 cm Silica Gel G plates in lined, preequilibrated solvent chambers. Plates were developed twice (2×) or four times (4×) in the indicated solvent systems.
[b]Standard had two spots of equal intensity.

Similarly, *A. curassavica* butterflies sequestered calactin and calotropin but not uscharin or uscharidin.

Detectability with HPLC was about 0.1 µg, permitting analysis of the major cardenolides in a portion of a single butterfly and in as little as 0.1 g of dried plant. This makes possible the quantitation of sequestering on an individual butterfly-leaf basis. Resolution was generally superior to TLC. The major drawback lay in the lack of specific detection; at 220 nm many noncardenolide chemicals may be detected and thus interfere. By contrast, TLC visualization with TNDP or Kedde reagent is very selective, so that one may be reasonably certain that a blue (positive) spot is indeed a cardenolide even if no standard is available for it. Retention data for some cardenolides on reverse phase HPLC columns are given in Table 7.

In unpublished work, we have used reverse-phase HPLC to monitor stages of purification during cardenolide isolation. Diluted latex and all cardenolide positive fractions in typical isolations (Seiber et al., 1978, 1980, 1982) can be injected directly on HPLC. We have also used HPLC for preparative purification, particularly for radiolabeled cardenolides from precursor-dosed plants. When coupled with scintillation counting or a flow-through radiomonitor, HPLC may well be the separation technique of choice for biosynthetic and metabolic studies with radiolabeled cardenolides.

TABLE 7 HPLC Retentions for Some *Asclepias* Cardenolides

Cardenolide	Retention time (min)	HPLC condition
Solvent (MeOH)	2.10	a
Calotropagenin	2.70	a
Calotoxin	3.30	a
Calotropin	4.05	a
Uscharidin	4.80	a
Uzarigenin	7.65	a
Uscharin	10.95	a
Voruscharin	12.00	a
Desglucosyrioside	7.79	b
Labriformidin	14.55	b

[a]Altex ultrasphere ODS column, 4.6 mm × 25 cm, flow rate 1.2 ml/min; solvent: $MeOH:H_2O:THF$ (65:31.5:3.5).
[b]Waters Associates, μBondapak C_{18}, column 6.2 mm × 30 cm, flow rate 1.8 ml/min; solvent: acetonitrile:H_2O (30:70). A UV detector, at ca. 220 nm, was used for all determinations.

C. Spectrometric identification

Structural determination and identification of cardenolides have generally been carried out by one or more spectrometric methods, particularly by IR, ^{1}H- and ^{13}C-NMR, and mass spectrometry. References to the spectra of some cardenolides from *Asclepias* species are given in Table 8.

Although it is beyond the scope of this review to discuss details of individual spectra, some salient features may be summarized as follows. Cardenolide IR spectra show characteristic absorption bands at ca. 1780, 1745, and 1630 cm^{-1} corresponding to the butenolide ring, and OH stretch of variable intensities at ca. 3500 (free) and 3300 cm^{-1} (H-bonded). Complex bands from ca. 1300-900 cm^{-1} and those from functional groups (C=N, C=O, etc.) present in some compounds differentiate individual cardenolides.

Some characteristic ^{1}H-NMR bands for cardenolides are at ca. 4.8 ppm (C-21H), 5.8-5.9 ppm (C-22H), and 0.8-1.2 ppm (C-18H and, if present as CH_3, C-19H). The presence of the hexosulose ring in *Asclepias* cardenolides is signaled by bands at ca. 4.5 (C-1'H) and a doublet at ca. 1.3-1.4 ppm (C-6'H). The thiazoline ring (e.g., uscharin) shows bands at ca. 7.5 (CH) and 3.8 (CH_2) ppm, while an aldehyde at C-19 is easily recognized by a band at ca. 10 ppm.

Although electron impact mass spectra of cardenolides are complex, loss of the elements H_2O (18 amu) and CO (28 amu), frequently in alternating sequence, may form distinctive patterns for some cardenolides. Molecular ion peaks may be visible for cardenolide genins but not for glycosides unless the latter are acylated or run under soft ionization conditions.

TABLE 8   References to Spectra of Some Cardenolides Found in *Asclepias* Species, and Their Acetylated Derivatives

Cardenolide	IR	^{1}H-NMR	^{13}C-NMR	Mass spectrometry
Afrogenin		Cheung et al., 1981	Cheung et al., 1981	
Afroside		Cheung and Watson, 1980	Cheung et al., 1981	Brüschweiler et al., 1969a
		Cheung et al., 1981		Cheung et al., 1981
Asclepin		Singh and Rastogi, 1972		
Calactin	Mittal et al., 1962	Brown et al., 1979	Brown et al., 1979	Brüschweiler et al., 1969a
		Singh and Rostogi, 1972		
Calotoxin		Singh and Rastogi, 1972		Reichstein, 1967b
Calotropin	Mittal et al., 1962	Brown et al., 1979		Reichstein, 1967b
		Singh and Rastogi, 1972		
Calotropagenin	Mittal et al., 1962	Mittal et al., 1962		Brüschweiler et al., 1969a
Desglucosyrioside	Brown et al., 1979	Cheung et al., 1980	Cheung et al., 1980	Brown et al., 1979
	Seiber et al., 1978	Brown et al., 1979	Brown et al., 1979	Seiber et al., 1978
		Seiber et al., 1978		
Desglucouzarin	Koike et al., 1980	Koike et al., 1980		Koike et al., 1980
Gomphogenin	Lardon et al., 1969	Cheung et al., 1981	Cheung et al., 1981	Lardon et al., 1969
		Carman et al., 1964		
Gomphoside		Cheung et al., 1981	Cheung et al., 1981	Brüschweiler et al., 1969a
		Brown et al., 1979	Brown et al., 1979	

TABLE 8 (Continued)

Cardenolide	IR	1H-NMR	13C-NMR	Mass Spectrometry
Gomphoside (Continued)		Cheung and Watson, 1980		
		Singh and Rostogi, 1972		
Labriformidin	Seiber et al., 1978	Seiber et al., 1978	Cheung et al., 1980	Seiber et al., 1978
Labriformin	Seiber et al., 1978	Seiber et al., 1978		Seiber et al., 1978
Panogenin	Roberts et al., 1967	Roberts et al., 1967		
Proceroside	Brüschweiler et al., 1969b	Brüschweiler et al., 1969b		Brüschweiler et al., 1969b
		Singh and Rostogi, 1972		
Syriobioside	Brown et al., 1979	Brown et al., 1979		Brown et al., 1979
Syriogenin	Masler et al., 1962b	Casagrande et al., 1974		Brüschweiler et al., 1969b
Syrioside	Brown et al., 1979	Brown et al., 1979		Brown et al., 1979
Uscharidin	Brüschweiler et al., 1969a	Brüschweiler et al., 1969a	Cheung et al., 1980	Reichstein, 1967b
		Singh and Rostogi, 1972		
Uscharin	Brüschweiler et al., 1969b	Brüschweiler et al., 1969a		
Uzarigenin	Kuritzkes et al., 1959	Roberts et al., 1967	Cheung and Watson, 1980	
Voruscharin		Seiber et al., 1982		

Summarizing, a positive color response with Kedde's or TNDP reagents, together with characteristic butenolide absorption bands in the IR and ^{1}H-NMR spectra, will serve to classify unknowns as cardenolides. Identification may then proceed through examination of spectral details and comparison of spectral and chromatographic properties with those of known standards.

For de novo structure determination, the use of ^{13}C-NMR (Brown et al., 1979), field ionization mass spectrometry (Brown et al., 1971) or x-ray cyrstallography (Kupchan et al., 1977; Yamauchi et al., 1979) may be required. Considerable assistance in data interpretation may be gained by comparison of underivatized cardenolides with acetate (Brown et al., 1979, Cheung et al., 1980) or isopropylidene derivatives (Cheung et al., 1981).

VI. SUMMARY AND CONCLUSIONS

The *Asclepias* genus, named for the Greek god of medicine, is an important source of cardenolides of the 5α series. Nearly all species examined to date contain these chemicals, although the amounts, types, and proportions of cardenolides vary considerably with species. Compounds identified to date include relatively simple genins, such as uzarigenin and syriogenin, and at least three series of cyclic bridged glycosides, two of which—derivatives of calotropogenin and desglycosyrioside—form major cardenolide groups in several species. The latex of some species is a rich source of the less polar glycosides, particularly those which contain spiro thiazoline and thiazolidine rings at C-3' of the sugar moiety.

Cardenolides are important mediators of plant-herbivore interactions, forming the toxic principles in several *Asclepias* species—particularly in the western U.S. range—which have caused livestock intoxication, and as chemical agents which are sequestered and stored by insects of several orders for protection against their vertebrate predators. In the latter case, sequestration and storage is generally selective in terms of both quantities and types of cardenolides taken from available cardenolides in plants. This phenomenon has been most thoroughly studied with monarch butterflies, which utilize *Asclepias* species as their sole larval host plants. All the cardenolides examined to date are extremely toxic to vertebrates but, apparently, much less so to insects.

Analytical methods are available for identifying and quantifying cardenolides in biological samples. It may be expected that these methods will form the basis for a more comprehensive study of cardenolides in *Asclepias* species, from which chemotaxonomic correlations and biosynthetic pathways may be derived, and of cardenolide pharmacodynamics in a variety of vertebrate and invertebrate species--a study that is presently in its infancy.

ACKNOWLEDGMENTS

The experimental work of the authors cited in this chapter was supported in part by NIH Grant ES00125 and NSF Grant DEB 75-14266. We would especially like to thank Lincoln Brower and Carolyn Nelson for reviewing the manuscript, and Sandy Wendland and Luanne Clayton for their excellent typing of it.

REFERENCES

Abisch, E., and Reichstein, T. (1962). Chemical orientation investigation of some Asclepiadaceae and Periplocaceae. *Helv. Chim. Acta 45*:2090-2116.

Bauer, S., Masler, L., Bauerova, O., and Sikl, D. (1961). Uzarigenin and desglucouzarin from *Asclepias syriaca* L. *Experientia 17*:15.

Baxter, C. M. (1944). *Cornell Vet. 34*:256.

Benson, J. M., and Seiber, J. N. (1978). High speed liquid chromatography of cardiac glycosides in milkweed plants and monarch butterflies. *J. Chromatogr. 148*:521-527.

Benson, J. M., Seiber, J. N., Keeler, R. F., and Johnson, A. E. (1978). Studies on the toxic principle of *Asclepias eriocarpa* and *Asclepias labriformis*. In *Effects of Poisonous Plants on Livestock*, R. F. Keeler, K. R. Van Kampen, and L. F. James (Eds.). Academic Press, New York, pp. 273-284.

Benson, J. M., Seiber, J. N., Bagley, C. V., Keeler, R. F., Johnson, A. E., and Young, S. (1979). Effects on sheep of the milkweeds *Asclepias eriocarpa* and *A. labriformis* and of cardiac glycoside-containing derivative material. *Toxicon 17*:155-165.

Boppré, M., Petty, R. L., Schneider, D., and Meinwald, J. (1978). Behaviorally mediated contacts between scent organs: another prerequisite for pheromone production in *Danaus chrysippus* males (Lepidoptera). *J. Comp. Physiol. 126*:97-103.

Brower, L. P. (1969). Ecological chemistry. *Sci. Am. 220*:22-29.

Brower, L. P., and Brower, J. V. Z. (1964). Birds, butterflies and plant poisons: a study in ecological chemistry. *Zoologica 49*:137-159.

Brower, L. P., and Glazier, S. C. (1975). Localization of heart poisons in the monarch butterfly. *Science 188*:19-25.

Brower, L. P., and Moffitt, C. M. (1974). Palatability dynamics of cardenolides in the monarch butterfly. *Nature (Lond.) 249*:280-283.

Brower, L. P., Ryerson, W. N., Coppinger, L. L., and Glazier, S. C. (1968). Ecological chemistry and the palatability spectrum. *Science 161*:1349-1351.

Brower, L. P., McEvoy, P. B., Williamson, K. L., and Flannery, M. A. (1972). Variation in cardiac glycoside content of monarch butterflies from natural populations in eastern North America. *Science 177*:426-429.

Brower, L. P., Edmunds, M., and Moffitt, C. M. (1975). Cardenolide content and palatability of a population of *Danaus chrysippus* butterflies from West Africa. *J. Entomol. (A) 49*:183-196.

Brower, L. P., Calvert, W. H., Hedrick, L. E., and Christian, J. (1977). Biological observations on an overwintering colony of monarch butterflies (*Danaus plexippus*, Danaidae) in Mexico. *J. Lepid. Soc. 31*:232-242.

Brower, L. P., Seiber, J. N., Nelson, C. J., Lynch, S. P., and Tuskes, P. M. (1982). Plant determined variation in the cardenolide content and emetic potency of monarch butterflies, *Danaus plexippus*, reared on the milkweed *Asclepias eriocarpa* in California. *J. Chem. Ecol. 8*:579-633.

Brown, H. D., Neucere, N. J., Altschul, A. M., and Evans, W. J. (1965). Activity patterns of purified ATPase from *Arachis hypogaea*. *Life Sci. 4*:1439-1447.

Brown, P., Brüschweiler, F., and Pettit, G. R. (1971). Field ionization mass spectrometry. III. Cardenolides. *Org. Mass Spectrom.* 5:573-597.

Brown, P., von Euw, J., Reichstein, T., Stöckel, K., and Watson, T. R. (1979). Cardenolides of *Asclepias syriaca* L., probable structure of syrioside and syriobioside. *Helv. Chim. Acta 62*:412-441.

Brüschweiler, F., Stöckel, K., and Reichstein, T. (1969a). *Calotropis*-glycosides, presumed partial structure. *Helv. Chim. Acta 52*:2276-2303.

Brüschweiler, F., Stöcklin, W., Stöckel, K., and Reichstein, T. (1969b). Glycosides of *Calotropis procera* R. Br. *Helv. Chim. Acta 52*:2086-2106.

Carman, R. M., Coombe, R. G., and Watson, T. R. (1964). The cardiac glycosides of *Gomphocarpus fruticosus* (R. Br.). IV. The nuclear magnetic resonance spectrum of gomphoside. *Aust. J. Chem. 17*:573-577.

Carvalhas, M. L., and Figueira, M. A. (1973). Comparative study of thin-layer chromatographic techniques for separation of digoxin, digitoxin, and their main metabolites. *J. Chromatogr. 86*:254-260.

Casagrande, C., Ronchetti, F., and Russo, G. (1974). The structure of syriogenin. *Tetrahedron 30*:3587-3589.

Chen, K. K. (1970). Newer cardiac glycosides and aglycones. *J. Med. Chem. 13*:1029-1034.

Chen, K. R., Bliss, C. I., and Robbins, E. G. (1942). Digitalis-like principles of *Calotropis* compared with other cardiac substances. *J. Pharmacol. Exp. Ther. 74*:223.

Cheung, H. T. A., and Watson, T. R. (1980). Stereochemistry of the hexosulose in cardenolide glycosides of the Asclepiadaceae. *J. Chem. Soc. Perkin Trans. 1*, pp. 2162-2168.

Cheung, H. T., Watson, T. R., Seiber, J. N., and Nelson, C. J. (1980). 7β,8β-Epoxy-cardenolide glycosides of *Asclepias eriocarpa*. *J. Chem. Soc. Perkin Trans. 1*, pp. 2169-2173.

Cheung, H. T. A., Coombe, R. G., Sidwell, W. T. L., and Watson, T. R. (1981). Afroside, a 15β-hydroxycardenolide. *J. Chem. Soc. Perkin Trans. 1*, pp. 64-72.

Clark, J. G. (1979). Whorled milkweed poisoning. *Vet. Hum. Toxicol. 21*(6):431.

Coombe, R. G., and Watson, T. R. (1964). The cardiac glycosides of *Gomphocarpus fruticosus* R. Br. III. Gomphoside. *Aust. J. Chem. 17*: 92-100.

Couch, J. F. (1929). Experiments with extracts from the woolly-pod milkweed (*Asclepias eriocarpa*). *Am. J. Pharm. 101*:815-821.

Crout, D. H. G., Curtis, R. F., Hassall, C. H., and Jones, T. L. (1963). The cardiac glycosides of *Calotropis procera*. *Tetrahedron Lett. 1963*: 63-67.

Crout, D. H. G., Hassall, C. H., and Jones, T. L. (1964). Cardenolides. Part VI. Uscharidin, calotropin, and calotoxin. *J. Chem. Soc. 1964*: 2187-2194.

Detweiler, D. K. (1967). Comparative pharmacology of cardiac glycosides. *Fed. Proc. 26*:1119-1124.

Dixon, C. A., Erickson, J. M., Kellett, D. N., and Rothschild, M. (1978). Some adaptions between *Danaus plexippus* and its food plant, with notes on *Danaus chrysippus* and *Euploea core* (Insecta: Lepidoptera). *J. Zool. Lond. 185*:437-467.

Duffey, S. S. (1970). Cardiac glycosides and distastefulness: some observations on the palatability spectrum of butterflies. *Science 169*: 78-79.

Duffey, S. S. (1977). Arthropod allomones: Chemical effronteries and antagonists. *15th Int. Congr. Entomol. 1976*, Washington, D.C., pp. 323-394.

Duffey, S. S. (1980). Sequestration of plant secondary products by insects. *Annu. Rev. Entomol. 25*: 447-477.

Duffey, S. S., and Scudder, G. G. E. (1972). Cardiac glycosides in North America Asclepiadaceae, a basis for unpalatability in brightly coloured Hemiptera and Coleoptera. *J. Insect Physiol. 18*: 63-78.

Duffey, S. S., and Scudder, G. G. E. (1974). Cardiac glycosides in *Oncopeltus fasciatus* (Dallas) (Hemiptera:Lygaeidae). I. The uptake and distribution of natural cardenolides in the body. *Can. J. Zool. 52*: 283-290.

Duffey, S. S., Blum, M. S., Isman, M. B., and Scudder, G. G. E. (1978). Cardiac glycosides: a physical system for their sequestration by the milkweed bug. *J. Insect Physiol. 24*: 639-645.

Edgar, J. A., Boppré, M., and Schneider, D. (1979). Pyrrolizidine alkaloid storage in African and Australian danaid butterflies. *Experientia 35*: 1447-1448.

Eisner, T. (1970). Chemical defense against predation in arthropods. In *Chemical Ecology*, E. Sondheimer and J. G. Simeone (Eds.). Academic Press, New York, Chap. 8.

Esau, K. (1965). *Plant Anatomy*, 2nd ed. Wiley, New York.

Evans, F. J. (1977). A new phorbol triester from the latices of *Euphorbia frankiana* and *E. coerulescens*. *Phytochemistry 16*: 395-396.

Evans, F. J., and Schmidt, R. J. (1976). Two new toxins from the latex of *Euphorbia poisonii*. *Phytochemistry 15*: 333-335.

Faber, D. B. (1977). Quantitation with high-performance thin-layer chromatography and programmed multiple development with high performance micro-thin-layer material for drug analyses in biological fluids. *J. Chromatogr. 142*: 421-430.

Fairbairn, J. W., Hakim, F., and El Kheir, Y. (1974). Alkaloidal storage, metabolism and translocation in the vesicles of *Papaver somniferum* latex. *Phytochemistry 13*: 1133-1139.

Feeny, P. (1976). Plant apparency and chemical defense. *Recent Adv. Phytochem. 10*: 1-40.

Fieser, L. F., and Fieser, M. (1959). *Steroids*. Reinhold, New York, Chap. 20.

Fink, L. S., and Brower, L. P. (1981). Birds can overcome the cardenolide defense of monarch butterflies in Mexico. *Nature (Lond.) 291*: 67-70.

Fleming, C. E., and Peterson, N. F. (1920). The narrow-leaved milkweed (*Asclepias mexicana*) and the broad-leaved or showy milkweed (*Asclepias speciosa*); plants poisonous to livestock in Nevada. *Univ. Nev. Agric. Exp. Stn. Bull. 99*.

Fraenkel, G. S. (1959). The raison d'etre of secondary plant substances. *Science 129*: 1466-1470.

Freeland, W. J., and Janzen, D. H. (1974). Strategies for herbivory by mammals: the role of plant secondary compounds. *Am. Nat. 108*: 245-269.

Gibbons, E. (1972). How to milk a milkweed. *Org. Gard. Farm. 1972* (January): 148-153.

Glover, G. H., Newsom, I. E., and Robbins, W. W. (1918). A new poisonous plant. The whorled milkweed (*Asclepias verticillata*). *Agric. Exp.
Stn., Colo. Agric. Coll., Bull. 246.*

Hassall, C. H., and Reyle, K. (1959). Cardenolides, Part III. Constitution
of calotropagenin. *J. Chem. Soc. 1959*: 85.

Haupt, I. (1976). Separation of the sites of synthesis and accumulation of
3,4-dihydroxyphenylalanine in *Euphorbia lathyris* L. *Nova Acta Leopold.
Suppl.* 7: 129-132.

Heftmann, E. (1976). *Chromatography of Steroids.* Elsevier, New York, pp.
121-123.

Hesse, G., and Gampp, H. W. (1952). African arrow poisons. VI. The
heterocyclic part of uscharin. *Chem. Ber. 85*: 933-936.

Hesse, G., and Lettenbauer, G. (1957). A second sulfur-containing compound from the milky-juice of *Calotropis procera. Angew. Chem. 69*: 392.

Hesse, G., and Ludwig, G. (1960). African arrow poisons. XIV. Voruscharin, a second sulfur-containing heart poison from *Calotropis procera* ·
L. *Liebigs Ann. Chem. 632*: 158-171.

Hesse, G., and Mix, K. (1959). African arrow poisons. IX. Structure and
partial synthesis of uscharin. *Liebigs Ann. Chem. 625*: 146-156.

Hesse, G., and Reicheneder, F. (1936). African arrow poison calotropin. I.
Liebigs Ann. Chem. 526: 252-276.

Hesse, G., Reicheneder, F., and Eysenbach, H. (1939). African arrow
poisons. II. Heart poisons in *Calotropis* latex. *Liebigs Ann. Chem. 537*:
67-86.

Hesse, G., Heuser, L. J., Hutz, E., and Reicheneder, F. (1950). African
arrow poisons. V. Relationships between the most important poisons of
Calotropis procera. Liebigs Ann. Chem. 566: 130-139.

Hesse, G., Fasold, H., and Geiger, W. (1959). African arrow poisons.
X. Calotropin from uscharidin. *Liebigs Ann. Chem. 625*: 157-160.

Hoch, J. H. (1961). *A Survey of Cardiac Glycosides and Genins.* University
of South Carolina Press, Charleston, S.C.

Holmgren, A. (1945). Important poisonous plants of Utah. *Utah Exp. Stn.
Farm. Home Sci.* 6: 11.

Huffman, W. T., Moran, E. A., and Binns, W. (1956). Poisonous plants.
In *Yearbook of Agriculture 1956. Animal Diseases*, A. Stefferud (Ed.).
U.S. Dept. Agric., U.S. Government Printing Office, Washington, D.C.,
p. 118.

Isman, M. B., Duffey, S. S., and Scudder, G. G. E. (1977a). Cardenolide
content of some leaf- and stem-feeding insects on temperate North American milkweeds (*Asclepias* spp.). *Can. J. Zool. 55*: 1024-1028.

Isman, M. B., Duffey, S. S., and Scudder, G. G. E. (1977b). Variation in
cardenolide content of the lygaeid bugs, *Oncopeltus fasciatus* and
Lygaeus kalmii kalmii and of their milkweed hosts (*Asclepias* spp.) in
central California. *J. Chem. Ecol. 3*: 613-624.

Jungreis, A. M., and Vaughan, G. L. (1977). Insensitivity of lepidopteran
tissues to ouabain: absence of ouabain binding and Na^+-K^+ ATPases in
larval and adult midgut. *J. Insect Physiol. 23*: 503-509.

Kingsbury, J. M. (1964). *Poisonous Plants of the United States and
Canada*, Prentice-Hall, Englewood Cliffs, N.J., pp. 267-270.

Koike, K., Bevelle, C., Talapatra, S. K., Cordell, G. A., and Farnsworth, N. R. (1980). Potential anticancer agents. V. Cardiac glycosides of *Asclepias albicans* (Asclepiadaceae). *Chem. Pharm. Bull. 28*:401-405.

Kupchan, S. M., Knox, J. R., Kelsey, J. E., and Renauld, J. A. S. (1964). Calotropin, a cytotoxic principle isolated from *Asclepias curassavica* L. *Science 146*:1685-1686.

Kupchan, S. M., Uchida, I., Shimada, K., Fei, B. Y., Stevens, D. M., Sneden, A. T., Miller, R. W., and Bryan, R. F. (1977). Elaeodendroside A: a novel cytotoxic cardiac glycoside from *Elaeodendron glaucum*. *J. Chem. Soc. Chem. Commun.*, pp. 255-256.

Kuritzkes, A., von Euw, J., and Reichstein, T. (1959). 3-Epi-uzarigenin and 3-epi-17α-uzarigenin. *Helv. Chim. Acta 42*:1502-1515.

Lardon, A., Stöckel, K., and Reichstein, T. (1969). Gomphogenin partial synthesis and structure of calotropagenin. *Helv. Chim. Acta 52*:1940-1954.

Lardon, A., Stöckel, K., and Reichstein, T. (1970). Partial synthesis of 2α,3β,19-triacetoxy-14β-hydroxy-5α-card-20(22)-enolide. Additional proof of calatropagenin structure. *Helv. Chim. Acta 53*: 167-170.

Laycock, W. A. (1978). Coevolution of poisonous plants and large herbivores on rangelands. *J. Range Manage. 31*:335-343.

MacRobbie, E. A. C. (1962). Ionic relations of *Nitella translucens*. *J. Gen. Physiol. 45*:861-878.

Mahmoud, O. M., Adam, S. E. I., and Tartour, G. (1979a). The effects of *Calotropis procera* on small ruminants. I. Effects of feeding sheep with the plant. *J. Comp. Pathol. 89*:241-250.

Mahmoud, O. M., Adam, S. E. I., and Tartour, G. (1979b). The effects of *Calotropis procera* on small ruminants. II. Effects of administration of the latex to sheep and goats. *J. Comp. Pathol. 89*:251-263.

Marsh, C. D., and Clawson, A. B. (1921). The mexican whorled milkweed (*Asclepias mexicana*) as a poisonous plant. *U.S. Dept. Agric. Bull. 969*.

Marsh, C. D., and Clawson, A. B. (1924). The woolly-pod milkweed (*Asclepias eriocarpa*) as a poisonous plant. *U.S. Dept. Agric. Bull. 1212*.

Marsh, C. D., Clawson, A. B., Couch, J. F., and Eggleston, W. W. (1920). The whorled milkweed (*Asclepias galioides*) as a poisonous plant. *U.S. Dept. Agric. Bull. 800*.

Masler, L., Bauer, S., Bauerova, O., and Sikl, D. (1962a). Cardiac glycosides from *Asclepias syriaca*. I. Isolation of cardiac active steroids. *Collect. Czech. Chem. Commun. 27*:872-881.

Masler, L., Bauer, S., Bauerova, O., and Sikl, D. (1962b). Cardiac glycosides from *Asclepias syriaca*. II. Structure of syriogenin and its glycosides. *Collect. Czech. Chem. Commun. 27*:895-901.

Matile, P. (1976). Localization of alkaloids and mechanism of their accumulation in vacuoles of *Chelidonium majus* laticifers. *Nova Acta Leopold. Suppl. 7*:139-156.

Mittal, O. P., Tamm, Ch., and Reichstein, T. (1962). Glycosides of *Pergularia extensa* (Jacq) N. E. Br. *Helv. Chim. Acta 45*:907-924.

Nelson, C. J., Seiber, J. N., and Brower, L. P. (1981). Seasonal and intraplant variation of cardenolide content in the California milkweed, *Asclepias eriocarpa*, and implications for plant defense. *J. Chem. Ecol. 7*:981-1010.

Nielsen, D. E., Nishimura, H., Otos, J. W., and Calvin, M. (1977). Plant crops as a source of fuel and hydrocarbon-like materials. *Science 198*: 942-944.

Nishio, S. (1980). The fates and adaptive significance of cardenolides sequestered by larvae of *Danaus plexippus* (L.) and *Cycnia inopinatus* (Hy. Edwards). Ph. D. thesis, University of Georgia, Athens.

Parsons, J. A. (1965). A digitalis-like toxin in the monarch butterfly, *Danaus plexippus* L. *J. Physiol. 178*: 290-304.

Parsons, J. A., and Summers, R. J. (1971). Cat assay for the emetic action of digitalis and related glycosides (digitoxin, digoxin, lanatoside C, ouabain, and calactin). *Br. J. Pharmacol. 42*: 143-152.

Patnaik, G. K., and Köhler, E. (1978). Pharmacological investigation on *Asclepias*—a new cardenolide from *Asclepias curassavica*. *Arzneim.-forsch. /Drug Res. 28*: 1365-1372.

Reichstein, T. (1967a). Cardenolide- and pregnanglycosides. *Naturwissenschaften 54*: 53-67.

Reichstein, T. (1967b). Cardiac glycosides as defensive substances in insects. *Naturwiss. Rundsch. 20*: 499-511.

Reichstein, T., von Euw, J., Parsons, J. A., and Rothschild, M. (1968). Heart poisons in the monarch butterfly. *Science 161*: 861-866.

Roberts, K. D., Weiss, E., and Reichstein, T. (1962). The cardenolides of the seeds of *Mallotus paniculatus* Müll-Arg (Euphorbiaceae). *Helv. Chim. Acta 50*: 1645-1664.

Roeske, C. N. (1971). Correlation of cardenolide content and mammalian toxicity in four species of milkweed (*Asclepias* spp.). M.S. thesis, University of California, Davis.

Roeske, C. N., Seiber, J. N., Brower, L. P., and Moffitt, C. M. (1976). Milkweed cardenolides and their comparative processing by monarch butterflies (*Danaus plexippus* L.). *Recent Adv. Phytochem. 10*: 93-167.

Rothschild, M. (1972). Secondary plant substances and warning colouration in insects. In *Insect Plant Relationships*, H. F. van Emden (Ed.). Blackwell, Oxford, pp. 59-83.

Rothschild, M., von Euw, J., and Reichstein, T. (1970). Cardiac glycosides in the oleander aphid, *Aphis nerii*. *J. Insect Physiol. 16*: 1141-1145.

Rowson, J. M. (1952). Studies in the genus *Digitalis*, Part I. The colorimetric estimation of digitoxin and preparations of *Digitalis purpurea*. *J. Pharm. Pharmacol. 4*: 814-830.

Schildknecht, H. (1971). Evolutionary peaks in the defensive chemistry of insects. *Endeavor 30*: 136.

Scudder, G. G. E., and Duffey, S. S. (1972). Cardiac glycosides in the Lygaeinae (Hemiptera:Lygaeidae). *Can. J. Zool. 50*: 35-42.

Seiber, J. N., Roeske, C. N., and Benson, J. M. (1978). Three new cardenolides from the milkweeds *Asclepias eriocarpa* and *A. labriformis*. *Phytochemistry 17*: 967-970.

Seiber, J. N., Tuskes, P. M., Brower, L. P., and Nelson, C. J. (1980). Pharmacodynamics of some individual milkweed cardenolides fed to larvae of the monarch butterfly. *J. Chem. Ecol. 6*: 321-339.

Seiber, J. N., Nelson, C. J., and Benson, J. M. (1981). HPLC analysis of cardiac glycosides and related steroids. In *Steroid Analysis by HPLC*, M. P. Kautsky (Ed.). Marcel Dekker, New York, pp. 41-80.

Seiber, J. N., Nelson, C. J., and Lee, S. M. (1982). Cardenolides in the latex and leaves of some *Asclepias* species and *Calotropis procera*. *Phytochemistry* 21:2343-2348.

Singh, B., and Rastogi, R. P. (1969). Chemical investigation of *Asclepias curassavica* Linn. *Indian J. Chem.* 7:1105-1110.

Singh, B., and Rastogi, R. P. (1970). Cardenolides–glycosides and genins. *Phytochemistry* 9:315-331.

Singh, B., and Rastogi, R. P. (1972). Structure of asclepin and some observations on the nmr spectra of *Calotropis* glycosides. *Phytochemistry* 11:757-762.

Smith, H. A., Jones, T. C., and Hunt, R. D. (1972). *Veterinary Pathology*, 4th ed. Lea & Febiger, Philadelphia, pp. 863-864.

Stoddart, L., Holmgren, A., and Cook, C. (1949). Important poisonous plants of Utah. *Utah Agric. Exp. Sta. Spec. Rep.* 2:13.

Swain, T. (1977). Secondary compounds as protective agents. *Annu. Rev. Plant Physiol.* 28:479-501.

Tokarnia, C. H., Dobereiner, J., and Canella, D. F. F. (1972). *Pesqui. Agropecu. Bras. Ser. Vet.* 7:31.

Tschesche, R., Forstmann, D., and Rao, V. K. M. (1958). Cardenolide components of *Asclepias curassavica* L. *Chem. Ber.* 91:1204-1211.

Tschesche, R., Snatzke, G., and Grimmer, G. (1959). Calotropagenin from *Asclepias curassavica* L. *Naturwissenschaften* 46:263-264.

Tuskes, P. M., and Brower, L. P. (1978). Overwintering ecology of the monarch butterfly, *Danaus plexippus* L., in California. *Ecol. Entomol.* 3:141-153.

Urquhart, F. A. (1960). *The Monarch Butterfly*. University of Toronto Press, Toronto.

Urquhart, F. A., and Urquhart, N. R. (1976). The overwintering site of the eastern population of the monarch butterfly (*Danaus plexippus*; Danaidae) in southern Mexico. *J. Lepid. Soc.* 30:153-158.

Urquhart, F. A., and Urquhart, N. R. (1979). Vernal migration of the monarch butterfly (*Danaus plexippus*, Lepidoptera:Danaidae) in North America from the overwintering site in the neo-volcanic plateau of Mexico. *Can. Entomol.* 111:15-18.

Vaughn, G. L., and Jungreis, A. M. (1977). Insensitivity of lepidopteran tissues to ouabain: physiological mechanisms for protection from cardiac glycosides. *J. Insect Physiol.* 23:585-589.

von Euw, J., Fishelson, L., Parsons, J. A., Reichstein, T., and Rothschild, M. (1967). Cardenolides (heart poisons) in a grasshopper feeding on milkweeds. *Nature (Lond.)* 214:35-39.

Watson, T. R., and Wright, S. E. (1954). The cardiac glycosides of *Gomphocarpus fruticosus* (R. Br.). *Chem. Ind. 1954*:1178.

Watson, T. R., and Wright, S. E. (1956). The cardiac glycosides of *Gomphocarpus fruticosus* R. Br. I. Afroside. *Aust. J. Chem.* 9:497.

Watson, T. R., and Wright, S. E. (1957). The cardiac glycosides of *Gomphocarpus fruticosus* R. Br. II. Gomphoside. *Aust. J. Chem.* 10:79.

Whiting, A. G. (1943). A summary of the literature on milkweeds (*Asclepias* spp.) and their utilization. *USDA Biogr. Bull.* 2, Washington, D.C., October 15.

Woodson, R. E., Jr. (1941). The North American Asclepiadaceae. I. Perspective of the genera. *Ann. Mo. Bot. Gard.* 28:193.

Woodson, R. E., Jr. (1954). The North American species of *Asclepias* L. *Ann. Mo. Bot. Gard. 41*:1-211.

Yamauchi, T., Miyahara, K., Abe, F., and Kawasaki, T. (1979). Affino-side B, a cardiac glycoside with a diosphenol system in the aglycone, from *Anodendron affine* (Anodendron. I). *Chem. Pharm. Bull. 27*:2463-2467.

Yoder, C. A., Leonard, D. E., and Lerner, J. (1976). Intestinal uptake of ouabain and digitoxin in the milkweed bug, *Oncopeltus fasciatus*. *Experienta 32*:1549-1550.

3

CAUSE AND PREVENTION OF ACUTE PULMONARY EDEMA AND EMPHYSEMA IN CATTLE

JAMES R. CARLSON and ROGER G. BREEZE

Washington State University, Pullman, Washington

I. NATURALLY OCCURRING ACUTE BOVINE PULMONARY EDEMA AND EMPHYSEMA

A. Abrupt Change to Lush Pasture

Acute bovine pulmonary edema and emphysema (ABPE) is a naturally occurring disease of cattle that can cause respiratory distress and death after an abrupt change in pasture (Schofield, 1948; Hyslop, 1969). In the United Kingdom the disease is referred to as "fog fever," reflecting its common occurrence when cattle are given access to lush regrowth (foggage) pasture. In this pasture-induced disease, clinical signs and pulmonary lesions usually occur within 2-10 days after change from sparse, poor-quality forage to lush, green, rapidly growing pasture (Selman et al., 1974). Although the specific history and nature of pasture change can vary, aburpt exposure to improved grazing or feeding conditions is the most consistent observation associated with the onset of ABPE.

Several factors may be important in the development of ABPE after pasture change. In many cases, sparse grazing conditions prior to pasture change result in the consumption of inadequate quantities of energy, protein, and other nutrients in low-quality forage. Although there is no direct experimental evidence available, it is possible that this low plane of nutrition has a preconditioning effect on the cattle which influences the onset and severity of ABPE. Another possibility is that lower consumption of dry matter and nutrients may contribute to hunger in the cattle and excessive consumption of lush forage during the first few days after pasture change. Forage consumption has not been measured experimentally under these conditions, but this concept is consistent with casual observations by ranchers that the largest, most vigorous cows may consume larger quantities of lush forage, and these may be the individuals that develop ABPE. It is tempting to speculate that individual cows in the herd that consume the largest quantities of lush forage may be subjected to the most drastic dietary change and increased risk of ABPE.

The change in feeding conditions from sparse, dry forage to lush green pasture also results in a drastic change in the composition of feed consumed. Many overripe dry grasses contain much higher crude fiber (33% versus 18%), lower crude protein (5% versus 26%) and lower digestible energy than fresh, immature, rapidly growing grass (National Academy of Sciences, 1971). In addition, the digestibility coefficient of crude protein may be approximately twice as high in lush grass than in dry, overripe grass of the same species. These differences in feed composition will result in intake of lush forage with lower fiber and higher energy, but the greatest change in composition is the higher content of digestible protein. Earlier workers (Leslie, 1949) have surmised that the protein content of lush grass may have some relationship to ABPE. An abrupt change to feed with large differences in composition, compounded by a possible excessive intake, represents a drastic nutritional change that would be expected to cause significant alterations in ruminal fermentation.

The term "lush pasture" is an arbitrary term used in an attempt to describe a transition in grazing conditions associated with ABPE. Obviously, extreme variation exists in the quality of forage described as "lush forage" by scientific investigators, veterinarians, and ranchers describing the occurrence of ABPE; but the transition from poor-quality forage to better grazing conditions is a consistent feature associated with ABPE. ABPE has been reported after grazing kale, rape, alfalfa, turnip tops, small grains, rye grass, Bermuda grass, mixed meadow grass, reed canary grass, and many other rapidly growing forages (Hyslop, 1969). No particular species of lush forage has been specifically implicated as a causative agent in ABPE.

B. Season of Year

Most reports of ABPE indicate that the highest incidence occurs in late summer and early fall, especially during late August, September, and October (Selman et al., 1974; Heron and Suther, 1979; Blake and Thomas, 1971). Although spring outbreaks have been reported, the predominant occurrence in late summer and fall may be related to customary cattle management systems involving the movement of cattle from dry summer pastures to more accessible meadows or fields containing regrowth or seeded forage for fall and winter grazing. Spring outbreaks of ABPE usually involve a similar management pattern, resulting in a transition at that time of year. These animal management systems and possible environmental effects on plant composition could explain the higher incidence of ABPE in late summer and fall. Van Soest (1968) and Deinum (1966) present evidence that plant composition is influenced by light, temperature, and maturity, with individual variation among plant species. Light intensity has a positive effect on soluble carbohydrate content and a negative effect on crude protein content. Higher temperature decreases soluble carbohydrates and increases lignin and cell wall constituents, resulting in a lower nutritional quality. It is conceivable that lower light intensity and cooler temperatures in the fall could lower soluble carbohydrates, increase protein content, and lower lignin content, possibly improving digestibility of young regrowth pasture.

C. Clinical Signs and Pathology

The majority of cattle affected by ABPE are adult females over 2 years of age, and they may often be nursing calves. The disease occurs in animals of both sexes and at younger ages but with lower frequency (Selman et al., 1974; Blake and Thomas, 1971). Although reports of clinical cases of ABPE may predominate in Hereford or crossbred cattle, other breeds of beef and dairy cattle are susceptible and there is no evidence of a breed difference in susceptibility to ABPE (Selman et al., 1974; Breeze et al., 1976; Monlux et al., 1970). The reason for more cases of ABPE reported in Hereford or Hereford-crossbred cattle may reflect greater numbers of these cattle subjected to risk. It is clear that a variety of respiratory diseases may be present in cattle diagnosed as having ABPE (fog fever) (Breeze et al., 1975b; Selman et al., 1977b), and accurate diagnosis is a prerequisite for meaningful evaluation of the incidence and susceptibility of cattle to ABPE.

Clinical signs of respiratory distress usually appear from 1 to 14 days after pasture change and death may follow within 2-4 days (Selman et al., 1974). The morbidity varies from 0 to about 50%, and mortality is approximately 30% of severely affected animals. Affected animals have an increased respiration rate and labored breathing, with an apparent increase in expiratory effort. The signs are progressive and in severe cases there is an expiratory "grunt," frothing at the mouth, the head is extended and lowered, and the animal may breath through the mouth. There is generally no significant elevation in body temperature. Although there are respiratory sounds on auscultation in advanced stages, coughing is not common. If affected cattle do not die within 2-7 days after onset of clinical signs, they usually recover uneventfully without permanent lung injury.

At death the pathological lesions are limited to the lungs, which are inflated, heavier than normal, dark red in color, and firm and rubbery. The airways are filled with frothy edema fluid and the lungs have a glistening appearance and contain large interlobular gas bullae. Interlobular septa and perivascular connective tissue contain edema fluid and fluid runs from cut surfaces of the lungs. Histologically, the alveoli and small airways contain edema, hyaline membranes, and cellular infiltration. Animals that die or are killed for necropsy 3 or more days after onset of clinical signs have thickened alveolar septa, resulting from diffuse alveolar epithelial cell hyperplasia. The alveoli are lined by cuboidal type 2 pneumocytes with frequent mitotic figures (Pirie et al., 1974; Breeze et al., 1975a).

II. THEORIES ON ETIOLOGY OF ABPE

The occurrence of ABPE has been recognized for over 150 years (Knowlson, 1819; Maki, 1963), and many theories on the cause and prevention have been proposed. Most can be classified as being related to an anaphylactic or hypersensitivity reactivity reaction or to a toxicity from feed or microorganisms.

A. Anaphylaxis and Hypersensitivity

The rapid onset of clinical signs and the presence of lung lesions, including edema and interstitial emphysema, have contributed to the view that ABPE results from anaphylaxis (Aitken and Sanford, 1969). Although some similarities exist between ABPE and anaphylaxis described by Aitken and Sanford (1969, 1972), there are difference in time of onset, mortality, and in lung lesions. Anaphylaxis develops much more rapidly than ABPE and results in pulmonary congestion, edema, intraalveolar hemorrhage, and interstitial emphysema. In contrast, cattle with ABPE have progressive clinical signs resulting in death in 2-4 days, and the lungs contain diffuse alveolar epithelial cell hyperplasia and hyaline membranes.

Hypersensitivity to moldy hay has also be suggested in the etiology of ABPE (Jenkins and Pepys, 1965) since sera from cattle with ABPE (fog fever) formed precipitins to *Micropolyspora faeni* associated with "farmer's lung" in humans. The clinical signs and pathology of ABPE (fog fever) in cattle are clearly different from those in cattle that develop farmer's lung in response to *M. faeni*, and there is no known association between these diseases (Breeze et al., 1978b).

Another form of hypersensitivity related to an allergic reaction to lungworms (*Dictyocaulus viviparus*) has been suggested (Michel, 1953, 1954;

Aitken and Sanford, 1973). The experimental production of ABPE (fog fever) on lungworm-free pastures and the absence of clinical response of recovered fog fever cases to a large challenge with viable lungworm larvae indicate that hypersensitivity to *D. viviparus* is not the cause of ABPE (fog fever) (Breeze et al., 1974; Selman et al., 1977a).

It has not been possible to duplicate the time course, clinical signs, or pathology changes of ABPE with any experimental regimen involving hypersensitivity or anaphylactic response. The similarities are superficial and there is no direct evidence that ABPE results from the above-mentioned or other hypersensitivity reactions.

B. Toxins

Consumption of pneumotoxic chemicals in the feed can cause acute pulmonary disease similar to those in cattle after abrupt pasture change. Sweet potatoes infected with the fungus *Fusarium solani* produce several pneumotoxic chemicals; and the furan 4-ipomeanol is the most toxic. Cattle that eat moldy sweet potatoes develop acute respiratory distress, pulmonary lesions, and high mortality (Wilson, 1973; Peckham et al., 1972; Doster et al., 1978). 4-Ipomeanol causes pulmonary edema and other lesions in several species, and the mechanism of lung damage has been extensively studied and recently reviewed (Boyd, 1980a,b; Wilson et al., 1978). The pulmonary toxicity of 4-ipomeanol results from metabolic activation by a cytochrome P_{450}-dependent mixed-function oxidase (MFO) system. A reactive metabolite binds covalently to cellular macromolecules, primarily in the nonciliated bronchiolar epithelial cells, which contain the appropriate MFO activity and are the target cells for 4-ipomeanol toxicity.

Other pneumotoxic 3-substituted furans are present in *Perilla frutescens* (purple mint), a common weed in the United States (Wilson et al., 1978). Wilson et al. (1977) isolated perilla ketone as a pneumotoxic constituent that is closely related chemically to 4-ipomeanol. Perilla ketone causes pulmonary edema in mice, rats, cattle, and sheep. Peterson (1965) suggested that *P. frutescens* was the cause of several outbreaks of acute pulmonary disease in grazing cattle.

It is clear that acute pulmonary disease in cattle can result from the consumption of moldy sweet potatoes and *P. frutescens*, which contain pneumotoxic furans. But these compounds cannot account for outbreaks of ABPE where these plant materials are not present and where ABPE occurs as a result of pasture change. Even though outbreaks of lung disease from moldy sweet potatoes and *P. frutescens* often occur after a change in pasture, it is unlikely that these or other toxic plants are responsible for most cases of ABPE. In the western United States and in the United Kingdom, most outbreaks of ABPE occur in pastures that do not contain poisonous plants or materials; and ABPE has been produced experimentally on reseeded rye grass pasture that did not contain lung worms or extraneous plant species (Selman et al., 1977a).

The involvement of *Clostridium* exotoxins has been suggested in the etiology of ABPE but has not been verified (Schofield, 1948; Maki and Tucker, 1962; Hyslop, 1969, Tucker and Maki, 1962). Inhalation of nitrogen dioxide, ammonia, and other substances of ruminal origin have been suggested but not implicated.

A new theory on the cause of ABPE arose from a study designed to determine the effect of tryptophan (TRP) on liver enzyme activity in cattle and sheep (Johnson and Dyer, 1966). Five of eight steers died from acute pulmonary edema and intestitial emphysema within 1-7 days after an oral dose of 0.57 g of D,L-TRP/kg body weight (BW). The clinical signs and pulmonary lesions were similar to those that had been described for naturally occurring ABPE (Dickinson et al., 1967). The fact that TRP is a component of protein in lush grass was the basis for a working hypothesis to determine whether TRP-induced acute pulmonary edema and emphysema was related to the naturally occurring disease. This theory is consistent with previous observations that the disease results as a consequence of consuming lush grass, and the change in digestible protein content of forage is one of the greatest differences in diet composition after abrupt pasture change. The remainder of this review will describe the evidence for the involvement of TRP in the development of ABPE after abrupt pasture change and summarize current information on the formation, mechanism of action, and effects on the lungs of the causative agent. In addition, experimental approaches and recommendations for prevention of ABPE are described.

III. TRYPTOPHAN-INDUCED ABPE

A. Pathogenesis

After an oral dose of TRP to cattle, clinical signs of respiratory distress appear within 24-48 hr after dosing and become progressively more severe until 2-6 days. Approximately one-half of the treated cows are usually severely affected and may die from acute pulmonary edema and interstitial emphysema (Carlson et al., 1968; Dickinson et al., 1967; Yokoyama et al., 1975). Cattle that survive for 1 week after dosing usually recover without complications. At necropsy the lungs are dark red, firm, and heavier than normal. They contain diffuse edema and interstitial emphysema. Predominant microscopic changes include alveolar edema, hyaline membranes, hyperplasia of alveolar epithelial cells, and thickened alveolar septa. The clinical signs, time course, and pulmonary lesions of TRP-induced ABPE are indistinguishable from those associated with naturally occurring ABPE.

The initial experiments utilized oral doses of 0.57-0.70 g of D,L-TRP/kg BW or 0.26 to 0.35 g of L-TRP/kg BW (Johnson and Dyer, 1966; Dickinson, 1970; Carlson et al., 1968; Yokoyama et al., 1975). Oral doses of D-TRP do not cause lung injury in cattle (Carlson et al., 1972). The minimum effective dose of L-TRP has not been established, but the development of ABPE is not strictly dose dependent. A similar incidence and severity of lung damage is observed with doses of 0.26-0.35 g of L-TRP/kg BW, suggesting that lower doses would be effective and that the dose of TRP is not the only factor involved in the development of lung injury (Carlson and Dickinson, 1978).

Since acute lung disease had not been reported after TRP doses in simple-stomached animals, the involvement of ruminal fermentation of TRP seemed possible. In contrast to the lung injury caused by oral TRP doses, there were no clinical signs in cows after intraperitoneal or intravenous administration of TRP indicating that ruminal fermentation of TRP was a prerequisite for the development of TRP-induced ABPE.

B. Ruminal Fermentation of TRP

Deamination and decarboxylation of amino acids leading to the formation of ammonia, CO_2, and volatile fatty acids is the normal process of amino acid degradation in the rumen. Amino acids are resynthesized by ruminal microorganisms from these precursors, and they are available to the microorganisms and host animal (Blackburn, 1965; Allison, 1969). TRP can be degraded to indole, indoleacetic acid (IAA), indolepropionic acid, tryptamine, 3-methylindole (3MI), and other end products by ruminal bacteria (Lacoste, 1961; Lewis and Emery, 1962a,b; Schatzmann and Gerber, 1972; Scott et al., 1964). In our studies, in vitro incubation of TRP with ruminal fluid resulted in the formation of 3MI and indole as the primary end products. The widely studied tryptophanase enzyme in many bacteria is probably responsible for the formation of indole (DeMoss and Moser, 1969; Roth et al., 1971; Newton and Snell, 1964). The presence of IAA was shown to be an intermediate in the formation of 3MI (Yokoyama and Carlson, 1974). Several indole compounds were converted to 3MI through the intermediates indolepyruvic and indoleacetic acids. The formation of 3MI was a two-step process involving the initial formation of IAA from TRP and subsequent decarboxylation of IAA to 3MI. Diverse ruminal microorganisms have been shown to convert TRP to IAA and indolepropionic acid (Stowe, 1955; Chung et al., 1975; Elsden et al., 1976). The decarboxylation of IAA to 3MI is catalyzed by a *Lactobacillus* sp., recently isolated from ruminal fluid (Yokoyama et al., 1977). This organism is a gram-positive, nonmotile, non-spore-forming rod. It is an obligate anaerobe that ferments a limited number of sugars to D-(-)-lactic acid. The reaction mechanism or the form in which the carbon is eliminated during the decarboxylation of IAA is not known. It is conceivable that either CO_2 or formic acid are eliminated as products, but preliminary attempts to collect $^{14}CO_2$ from $1-[^{14}C]IAA$ were unsuccessful (Wong and Carlson, 1980). The elimination of formic acid in the decarboxylation process is postulated (M. T. Yokoyama, personal communication, 1981), which could provide an ecological advantage to the organism as an electron sink for reducing equivalents. In addition to IAA, the *Lactobacillus* sp. decarboxylates p-hydroxyphenylacetic acid to p-cresol, 5-hydroxyindolacetic acid to 5-hydroxy-3-methylindole, and 3,4-dihydroxyphenylacetic acid to methylcatechol (Yokoyama and Carlson, 1981). Competition studies with IAA and p-hydroxyphenylacetic suggested the involvement of different enzymes in the production of 3MI and p-cresol by this organism.

Little is known about the ecological distribution, requirements, and biochemistry of the ruminal *Lactobacillus* sp. responsible for 3MI formation. Wide distribution is suggested by the fact that decarboxylation of IAA to 3MI has been demonstrated in mixed fecal contents of various exotic and domestic animals and humans (Yokoyama and Carlson, 1979b), and 3MI has been demonstrated in ruminal fluid of cattle in the United States and the United Kingdom (Hammond et al., 1979a; Selman et al., 1977a; Yokoyama et al., 1975). Wong (1979) demonstrated the highest conversion of TRP to 3MI by ruminal fluid, in vitro, at pH 6.5-7.0, which probably reflects the combined effect of pH on at least two organisms involved in this conversion. In vitro incubation of ruminal fluid from cattle fed barley converted only 9% of the TRP to 3MI compared to 45% conversion in ruminal fluid from cattle fed hay. In addition, after an oral dose of TRP, ruminal fluid 3MI concentrations reached a maximum of approximately 100-120 μg of 3MI/ml in cattle fed hay compared to only 5 μg/ml in cattle fed a concentrate ration containing barley. As expected, ruminal pH was lower in the cattle fed the concentrate ration compared to the hay ration (J. Kowalczyk and J. R. Carlson, unpublished data, 1980). It is clear that the type of diet, ruminal pH, and perhaps other ruminal fermentation conditions can dramatically influence the proportion of TRP converted to 3MI in the rumen and that ruminal 3MI production is not necessarily dependent on the TRP dose. Further studies are needed to evaluate the fermentation conditions and biochemical factors associated with ruminal 3MI production.

C. Absorption of 3MI and Route of Exposure

Adverse effects of intestinal and ruminal fermentation products of TRP in animals depends on absorption of these end products into the blood. Fordtran et al. (1964) observed almost quantitative absorption of both indole and skatole (3MI) from the jejunum, ileum, and colon of human subjects. Horvath et al. (1975) reported peak plasma 3MI concentrations within 1.5-3 hr after giving an oral dose to goats, suggesting rapid ruminal absorption of 3MI. More recently, Carlson et al. (1981) used goats fitted with ruminal fistulas and reentrant duodenal cannulas to demonstrate rapid and quantitative absorption of 3MI from the forestomachs of goats after 3MI administration into the rumen. Ruminal 3MI concentrations decreased rapidly and peak plasma concentrations were present at 2 hr after dosing with 0.05 g of 3MI/kg BW. After 24 hr less than 2% of the dose remained in the rumen and less than 0.03% of the dose had exited the abomasum through the duodenal cannula, indicating that more than 98% of the dose had been absorbed from the rumen and/or forestomachs. The lipophilic properties of 3MI probably contribute to its capacity to penetrate biological membranes (Rogers et al., 1969) and enhance absorption. It is clear that 3MI and indole are efficiently absorbed from both the rumen and intestinal tract, resulting in exposure of animals to these bacterial fermentation products of TRP.

It is likely that the blood is the route of exposure. Even though a transient odor of 3MI can sometimes be detected on the breath, the potent odor of 3MI would suggest that this represents only minute quantities of 3MI. The fact that the inhalation is not the route of exposure was demonstrated in a study by Williams et al. (1977) in which goats given intraruminal doses of 3MI developed acute lung disease, even though they were tracheotomized to eliminate the inhalation of eructed gases.

D. Effect of TRP Fermentation Products on Lungs of Cattle and Goats

1. General Effects of Amino Acids and Fermentation Products of TRP

Administration of tyrosine, histidine, and arginine to cattle in doses equivalent to those of TRP did not cause lung injury, which indicates that the effect of TRP is not a general effect of large quantities of amino acids in the rumen (Dickinson, 1970). Administration of indole to cattle resulted

in severe hemolysis and hemoglobinuric nephrosis, but there were no significant pulmonary lesions (Hammond et al., 1980a; Jarvie et al., 1977). In
contrast, 3MI caused pulmonary injury in cattle and goats after both oral and
intravenous administration (Carlson et al.; 1975). The clinical signs and
pulmonary lesions were identical to those of TRP-induced ABPE and 3MI was
present in the blood plasma and ruminal fluid of cattle given TRP (Yokoyama
et al., 1975). As stated previously, intraruminal doses of TRP that have
been used to induce acute pulmonary edema and emphysema in cattle ranged
from 0.26 to 0.35 g of L-TRP/kg BW. Intraruminal doses of 0.075-0.2 g of
3MI/kg BW and intravenous doses of 0.03-0.06 g of 3MI-kg BW are effective
in cattle and goats, and clinical disease has been detected at 0.02 g of 3MI/kg
BW. Lung injury is present in goats within 30 min after beginning infusion,
during which time only 0.01 g of 3MI/kg BW had been administered. It is
apparent that 3MI is a potent pneumotoxic agent in ruminants, and that its
effects are related to dose and duration of exposure (Carlson et al., 1972;
Dickinson et al., 1976; Bradley and Carlson, 1980; Bray and Carlson, 1979a;
J. R. Turk, 1981).

Other ruminal fermentation products of TRP have been tested for possible
pneumotoxic effects. Oral doses of IAA cause lung injury as a consequence
of ruminal decarboxylation to 3MI (Dickinson, 1970; Carlson et al., 1972).
Equimolar oral doses of indole, tryptamine, 5-hydroxytryptamine, and 3MI
were given to calves, but only those given 3MI developed lung disease. 3MI
is the only fermentation product of TRP that causes acute respiratory disease
typical of ABPE when given orally to cattle (Jarvie et al., 1977), which
contradicts the hypothesis that tryptamine or 5-hydroxytryptamine cause
ABPE (Schatzmann and Gerber, 1972; Eyre, 1972, 1975). There is substantial
evidence for the conclusion that 3MI is the ruminal metabolite of TRP that is
absorbed into the blood, resulting in the development of acute pulmonary
edema and emphysema.

2. Effects of 3MI on Lungs

The effects of 3MI are limited to the lungs. The clinical signs of respiratory disease in cattle after administration of 3MI are identical to those observed
after TRP doses or in naturally occurring ABPE, except that the onset is
earlier. Peak ruminal 3MI concentrations are present 12-18 hr after a TRP
dose, which represents the time necessary for ruminal microorganisms to convert significant quantities of TRP to 3MI (Carlson et al., 1975; Yokoyama et
al., 1975). It is interesting to note that younger cattle are apparently less
susceptible to 3MI-induced lung damage (Cornelius et al., 1979; R. G. Breeze,
unpublished data, 1981), which is also consistent with observations on
naturally occurring ABPE.

Pulmonary edema is the first morphological change seen in ruminants after
3MI adminstration. Within a few hours after 3MI administration the lungs are
inflated, dark red in color, heavier than normal, and contain edema fluid in
interstitial spaces and airways. Large gas bullae may be present, especially
in cattle. Except for anatomical differences, such as the more easily separated
intralobular septa in cattle, the gross lesions are similar in cattle, goats, and
sheep. Two to four days after 3MI, pulmonary edema is reduced but the lungs
remain inflated, heavy, and firm (Dickinson et al., 1976; Carlson et al., 1975).
Focal interstitial pulmonary fibrosis may be present in cattle and goats after

recovery from repeated 3MI treatment, but these changes are not extensive (J. R. Turk, 1981).

Examination of the lungs with the light microscope reveals alveolar edema within a few hours after 3MI dosing. This is followed by formation of hyaline membranes, interstitial emphysema, and proliferation of alveolar and bronchiolar epithelial cells at 2-4 days. At this time there is also infiltration of neutrophils, eosinophils, and macrophages (Carlson et al., 1975; Dickinson et al., 1976; Huang et al., 1977; Bradley et al., 1977; Pirie et al., 1976). Proliferation of alveolar and bronchiolar epithelial cells is probably a general response to lung injury, reflecting repopulation of alveolar type 1 and bronchiolar epithelial cells that were desquamated during earlier stages of the disease.

Scanning and transmission electron microscopic studies during the early stages of 3MI-induced lung injury demonstrated the most severe effects of 3MI on alveolar type 1 and nonciliated bronchiolar epithelial cells in goat lungs (Huang et al., 1977; Bradley et al., 1977; Bradley and Carlson, 1980). Swollen mitochondria, dilated intracellular vesicles and other changes were present in these cells within 1/2 hr after beginning 3MI infusion. There were some reversible changes in capillary endothelial cells, but they appeared normal at 4 hr, which was after the 3MI infusion was completed. At 2 hr interstitial and alveolar edema was severe, and there were areas of denuded alveolar and bronchiolar basement membranes from sloughing of alveolar type 1 and nonciliated bronchiolar epithelial cells. Physiological studies suggest that the alveolar epithelium is the barrier that restricts passage of water-soluble molecules from the capillary lumen into the alveolar space (Taylor and Gaar, 1970; Hurley, 1978). Although the mechanism of edema formation is not known, 3MI administration did not cause pulmonary artery hypertension (Olcott, 1981), and it is possible that alveolar edema resulted from 3MI-induced type 1 cell damage. But other mechanisms of edema formation involving changes in lymph flow or membrane permeability cannot be ruled out.

By 4 hr the surviving alveolar type 1 and nonciliated bronchiolar cells contained larger and more prominent clusters of smooth endoplasmic reticulum (SER) compared to controls. The proliferation of SER continued in these cells for at least 48 hr. These studied indicate that 3MI has rapid and selective toxic effects on alveolar type 1 and nonciliated bronchiolar epithelial cells. These cells become necrotic, or if they survive there is extensive proliferation of SER. This SER proliferation suggests the involvement of mixed-function oxidase (MFO) enzymes (Staubli et al., 1969) and a role of metabolism in 3MI-induced lung injury.

There are dose sensitivity and species differences in the effects of 3MI on lungs and lung cells. In cattle and goats given a lower oral dose of 3MI (0.075 g of 3MI/kg BW) in an attempt to induce mild lung injury without mortality (J. R. Turk, 1981), clinical signs were induced, but these animals had only minimal damage to the bronchiolar epithelium. Since alveolar type 1 cell damage was extensive, the results suggest a slightly lower sensitivity of nonciliated bronchiolar epithelial cells to 3MI in cattle and goats compared to alveolar type 1 cells. Administration of 3MI to horses causes severe obstructive pulmonary disease resulting from damage to the bronchiolar epithelium, but alveolar type 1 cell lesions and alveolar edema are absent (Breeze et al., 1978a; M. A. M. Turk, 1981). There is an increase in rate and depth of respiration from 24 to 72 hr followed by labored breathing and severe respiratory distress in horses given 3MI. Hypoxemia may be severe and death may occur within a few days after dosing. In contrast to cattle and goats, which can recover from 3MI treatment, clinical signs persist for at least 30 days in horses. Physiological changes reflect airway obstruction, which is consistent with pathological findings. Epithelial necrosis is present in all small airways and desquamated epithelial cells fill the small airways at 24 hr. Later, there is proliferation of bronchiolar epithelial cells, bronchiolar exudates, plugs of desquamated cells, and bronchiolitis obliterans that result in severe acinar overinflation. After 30 days, repair is incomplete and narrow airways and bronchiolar hyperplasia are still present. There may be panlobar and centrilobar emphysema along the ventral edges of caudal lung lobes.

Unpublished preliminary experiments indicate that a variety of other animals are not susceptible to 3MI-induced lung injry. Doses of 3MI have been given to rats (Watanabe and Aviado, 1974), mice, rabbits, hamsters, ferrets, swine, cats, and cynomologus monkeys without success in producing acute lung lesions. These species appear to be less susceptible to 3MI-induced lung injury.

E. Mechanism of Action of 3MI

1. *Physical and Chemical Properties of 3MI*

The common name for 3MI is skatole, a name denoting the intensely unpleasant fecal odor. It is planar and composed of an aromatic and pyrrole ring with a methyl group substituted at the 3-position (C_9H_9N). The molecular weight of 3MI is 131.17. It forms white to white-brown scaly crystals with a melting point of 95±C and boiling point of 265-266°C. A relatively low vapor pressure contributes to volatility and the permeating fecal odor. It is soluble in organic solvents but only slightly soluble in distilled water at room temperature (solubility in water is approximately 450-500 μg/ml). The natural logarithm of the partition coefficient of 3MI in n-octanol:aqueous sodium phosphate buffer, pH 7.4, is approximately 6.0, indicating that it is more lipophilic than benzene and indole and less lipophilic than diphenylamine, naphthalene, pyrene, and similar compounds (Rogers, 1969). Rogers et al. (1969) suggest that compounds such as 3MI with intermediate lipophilicity penetrate nonpolar:polar interfaces (membranes) more readily than do compounds that are either more or less lipophilic. The ultraviolet absorption spectrum of 3MI is similar to other indoles (Bakri and Carlson, 1970). The amine proton is only slightly ionizable. 3MI is neutral at physiological pH, but it is subject to destruction by strong acid.

1. *Direct Effects of 3MI*

The direct effects of 3MI on biological membranes have been known for many years. High concentrations of 3MI cause hemolysis of erythrocyte membranes (Rogers, 1969; Bray and Carlson, 1975), inflammation and arthritic damage in rabbit knee joints (Nakoneczna et al., 1969; Rogers et al., 1969), and disintegration of ruminal protozoa (Eadie and Oxford, 1954). Indole and 3MI interact with lecithin micelles by intercalating between the fatty acids near the polar head. The amine proton hydrogen bonds with the phosphate group, and micelle structure and water mobility are altered

(Bray et al., 1974). 3MI also induces structural perturbations and alters ATPase activity in bovine erythrocyte membranes (Bray and Carlson, 1975; Bray et al., 1975). Both indole and 3MI cause the release of lysosomal enzymes from rabbit lung lavage cells (Yang and Carlson, 1977). These results indicate that both indole and 3MI cause disruptive effects on biological membranes at the concentrations used in these studies (>100 µg/ml). It is not known whether indole and 3MI weaken membranes at lower concentrations. The effects of indole and 3MI are qualitatively similar, but indole concentrations of approximately two times those of 3MI are required for equivalent membrane damage in vitro.

The dramatic effects of indole and 3MI on biological membranes could be interpreted as the cause of acute pulmonary edema and emphysema through direct alteration of lung capillary and cell membrane permeability or release of lysosomal enzymes. If this were the case, it would seem that both indole and 3MI would cause lung damage as well as injury to other biological membranes in the body. The qualitatively similar lipophilic properties of indole and 3MI, and the fact that only 3MI causes lung injury without similar effects in other organs, suggests some other basis for the toxic effect. Another factor that argues against direct effects of 3MI as the primary mechanism of lung injury is the fact that 3MI does not concentrate in the lung during the onset of lung 3MI-induced lung injury, and that the concentrations are much lower than those associated with membrane disruption (Bradley and Carlson, 1982). It is our current belief that direct effects of 3MI on biological membranes are not primarily responsible for lung injury. The data do not rule out the possibility that direct effects could contribute to a mild onset of clinical signs or to reversible ultrastructural changes in capillary endothelial cells during 3MI infusion (Bradley and Carlson, 1980).

Eyre (1980) recently suggested that the effects of 3MI in 3MI-induced lung disease were examples of an anaphylactoid reaction. A direct vasoconstrictive effect of 3MI and other indolic compounds was demonstrated in pulmonary veins of cattle and was suggested to contribute to pulmonary edema (Eyre, 1975). In these studies, 3MI was 1000 times less potent than serotonin in contracting calf pulmonary vein. Sheep pulmonary vein was relaxed by these agents, with 3MI being 8000 times less potent than serotonin. Constriction of pulmonary veins by 3MI would possibly enhance edema formation through a postulated elevation in blood pressure in the lung. Olcott (1981) recently demonstrated little elevation in pulmonary artery and pulmonary wedge pressure after rapid injection of 3MI into anesthetized goats, even though lung injury was evident. The lack of response in sheep pulmonary veins was considered as a possible explanation for species differences in the response of cattle and sheep to a TRP dose. Since 3MI does induce pulmonary injury in sheep (Bradley et al., 1978), it is apparent that the extent of ruminal conversion of TRP to 3MI is the primary reason for differences in response to TRP between cattle and sheep rather than differences in the susceptibility to 3MI. The results of Jarvie et al. (1977) indicate that 5-hydroxytryptamine and tryptamine do not cause the acute pulmonary lesions, even though they are much more potent than 3MI in causing the constriction of bovine pulmonary vein preparations (Eyre, 1975).

Atkinson et al. (1977) demonstrated that 3MI produced an anaphylactoid reaction in calves given rapid intravenous injections of 1-16 mg of 3MI/kg BW in cremophor EL. Immediate cardiopulmonary effects, including pulmonary hypertension, systemic hypotension, dyspnea, and apnea, were observed. The response decreased with subsequent injections and was abolished by sodium meclofenamate, an antagonist to mediators of anaphylaxis in cattle. Since sodium meclofenamate does not reduce the severity of 3MI-induced lung injury in cattle (Breeze, 1978), the anaphylactoid effects of 3MI are not primarily responsible for the development of this lung disease. This conclusion was confirmed by the observation that rapid injection of 3MI caused similar cardiopulmonary effects in goats pretreated with piperonyl butoxide (BT), which reduces the severity of 3MI-induced pulmonary lesions (Olcott, 1981; Bray and Carlson, 1979a). The pharmacological effects of 3MI were reduced in goats pretreated with phenobarbital (PB) that developed the most severe lung injury.

Current evidence indicates that 3MI can have a variety of biophysical and pharmacological effects on biological membranes and systems. Nevertheless, none of the effects reviewed above can be demonstrated to be directly responsible for the acute lung injury associated with 3MI administration to cattle and goats.

3. Role of 3MI Metabolism in Pulmonary Injury

Tissue distribution and metabolism. Prior to the discovery that 3MI caused lung injury, there was little information available on its body distribution and metabolism. Skatole (3MI) has long been known to be present in feces, with increased concentrations associated with malabsorption syndromes (Herter, 1907). Hydroxyskatoles were reported in the urine of normal patients and patients with forms of mental illness and malabsorption syndrome (Mori et al., 1978; Sprince, 1962; Heacock and Mahon, 1963, 1964; Mahon and Mattok, 1967; Horning and Dalgliesh, 1958). King et al. (1966) studied the metabolism of radioactive indole in rats and reported two metabolic pathways: one via indoxyl, $\underline{N}$-formylanthranilic acid, and anthranilic acid, and the other via oxindole to 5-hydroxyoxindole, $\underline{o}$-aminophenylacetic acid, and anthranilic acid. The majority of the indole was oxidized through the indoxyl pathway, emphasizing the greater reactivity of the indole nucleus at the 3-position; this position is occupied by a methyl group in the case of 3MI.

More recently, the action of rat liver pyrrolooxygenases on 3MI has been reported to catalyze the formation of 3-methyloxindole, 2-formimidoacetophenone, 2-aminoacetophenone, and 3-hydroxy-2-aminoacetophenone (Frydman et al., 1971, 1972, 1973). These end products of 3MI metabolism suggest preferential oxidation at the 2-position and subsequent pyrrole ring cleavage in rats. The enzyme(s) appeared to be mixed-function oxidase (MFO) dependent, requiring oxygen and NADPH. It is widely distributed in plant and animal tissues, but the activity is inhibited in crude tissue preparations because of a powerful macromolecular inhibitor(s).

We investigated the tissue distribution, plasma concentration, and metabolism of 3MI in goats. A remarkable feature is the relatively low 3MI concentration in the plasma after intraruminal or intravenous administration. Infusion of 0.06 g of 3MI/kg BW over 2 hr in cattle and 0.03-0.04 g of 3MI/kg BW over 2 hr in goats resulted in peak 3MI concentrations in the plasma of about 8-10 μg of 3MI/ml. The total blood volume did not contain more than 1-1.5% of the dose at any time, and 3MI was cleared from the plasma with a half-life of 20-25 min (Carlson et al., 1975; Bradley and Carlson, 1982; Bray and Carlson, 1979a; Carlson and Dickinson, 1978). Plasma 3MI concentrations apparently reflect the balance between absorption (or infusion rate) and metabolism and/or body accumulation. Since infusion of 0.04 g/kg BW over 2 hr would result in a dose of approximately 1 g of 3MI/hour in a 50-kg goat, rapid disposition of 3MI is indicated.

Our results also indicate that 3MI does not accumulate in body tissues, suggesting that metabolism and/or excretion is the mechanism of disposition (Bradley, 1977; Bradley and Carlson, 1982). The lungs of goats infused for 2 hr with 0.04 g of 3MI/kg BW containing [methyl-^{14}C]-3MI contained only 1.3 and 7.5 μg of 3MI per gram of lung tissue at 0.5 and 2 hr, respectively, after beginning the infusion. Concentrations of less than 1.0 μg/g were present at all other times during a 24-hr experiment. Similarly, concentrations of 3MI in liver, kidney, heart, skeletal muxcle, and brain did not exceed 9 μg of 3MI/g. The data clearly indicate that 3MI does not accumulate in lung or other body tissues during infusion or during the development of lung injury. In contrast to data on 3MI concentration, the radioactivity exceeded equivalent 3MI concentrations in the lung by 13- to 254-fold during the experiment. Radioactivity also exceeded equivalent 3MI concentrations in liver, kidney, heart, and skeletal muscle, but the highest proportions of radioactivity were in the lung. These data indicate that disposition of 3MI results from rapid metabolism and the metabolites are present in tissues (particularly in the lung) prior to excretion.

[Methyl-^{14}C]-3-methylindole was used to characterize 3MI metabolism in goats. Goats excreted 90% of the radioactivity in the urine, less than 1% in the expired air, and negligible quantities in the feces (Hammond et al., 1979c). Accumulation of urinary radioactivity was rapid with 50% of the radioactivity excreted in 2-7 hr (Bray and Carlson, 1979a). Ion-exchange chromatography of goat urine results in separation of at least 10 radioactive metabolites, and unmetabolized 3MI was not present. The majority of the metabolites represent 3-methyloxindole (3MOI) and derivatives and a minor route of metabolism leads to the formation of indole-3-carboxylic acid (Hammond et al., 1979c). Rats appear to excrete a larger quantity of indole-3-carboxylic acid after 3MI administration (Carlson and Markey, 1975). A tentative scheme representing these two metabolic pathways is shown in Fig. 1. Preliminary unpublished data also suggest the presence of small quantities of acetoaminophenone derivatives and hydroxyskatoles. These compounds have been reported in other species. All data currently available confirm that 3MI undergoes rapid biotransformation and the products are excreted in the urine. Since tissue concentrations of 3MI appear to be too low to elicit direct membrane effects, the role of metabolism in pulmonary injury was investigated further.

Role of metabolism in lung injury. We investigated whether 3MI was the toxic agent or whether a pneumotoxic metabolite(s) was generated. Tentative identification of 3MI metabolites suggested that 3-methyloxindole and indole-3-carbinol were early metabolites in the 3MOI and indole-3-carboxylic acid pathways, respectively (Fig. 1). Organic synthesis of 3MOI and tritium exchange labeling provided [^{3}H]-3MOI and [^{3}H]indole-3-carbinol for double-isotope metabolism studies with [^{14}C]-3MI. Doses of 3MOI or indole-3-carboxylic acid equivalent to those of 3MI that cause severe lung disease

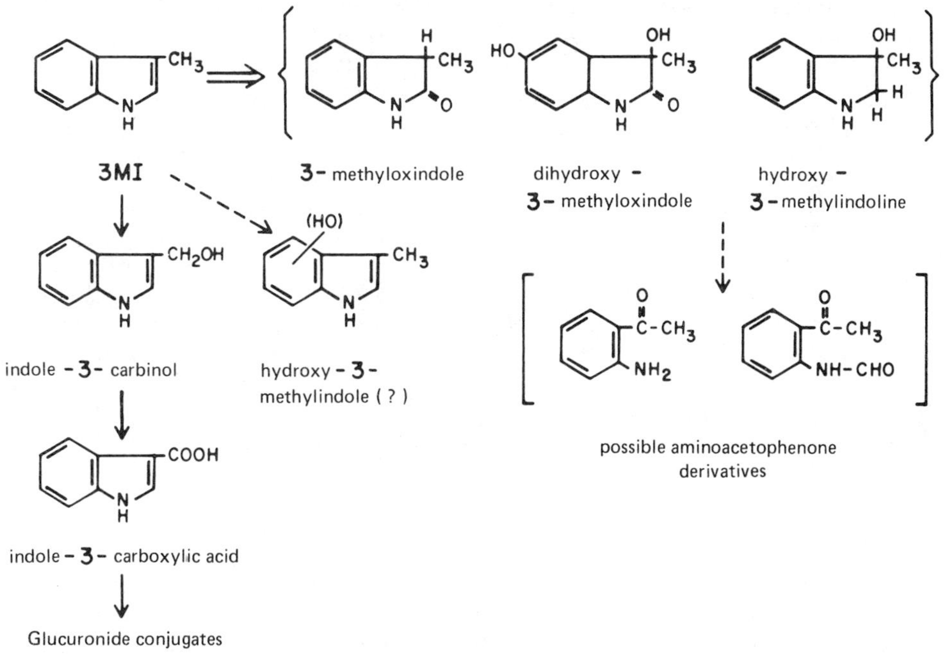

FIGURE 1 Tentative pathways of 3-methylindole (3MI) metabolism in goats.

failed to cause any lung injury, indicating that 3MOI, indole-3-carbinol, or their subsequent metabolites were not pneumotoxic (Potchoiba et al., 1982). In addition, ion-exchange chromatograms of [^{3}H]-3MOI metabolites were similar to those of [^{14}C]-3MI, indicating again that the majority of 3MI excretion products are nontoxic 3MOI or its degradation products. There was little indication of tritium exchange or the existence of unidentified pathways since 3MI metabolite peaks could be accounted for by metabolism of 3MOI or indole-3-carbinol.

Accumulating evidence pointed toward the possibility that the MFO system may be involved in the metabolism of 3MI. This evidence included the observed proliferation of SER in lung cells after 3MI administration, the metabolism of 3MI by MFO-linked pyrrolooxygenases in rat liver, and the chemical nature of oxidized and hydroxylated 3MI metabolites. To test the hypothesis that MFO enzymes were involved in the metabolism of 3MI, groups of goats were pretreated with a known inducer (phenobarbital, PB) or inhibitor (piperonyl butoxide, BT) of the MFO system (Bray and Carlson, 1979a). Goats pretreated with PB developed more severe lung injury compared to controls given only 3MI, and goats pretreated with BT were protected from the effects of 3MI infusion. A shorter half-life for clearnace of 3MI from plasma in PB goats compared to BT goats indicated more rapid metabolism of 3MI in PB-induced goats.

Since lung injury was enhanced when 3MI metabolism was enhanced, this experiment suggested that the MFO system was involved in 3MI metabolism and implicated metabolism in the mechanism of pneumotoxicity. It also rules out the parent compound (3MI) as the causative agent since pneumotoxicity was prevented when metabolism of 3MI was inhibited. This information and the double-isotope studies support the concept that pneumotoxicity of 3MI results from an early reaction in 3MI metabolism, prior to the formation of 3MOI or indole-3-carbinol.

It is well known that the MFO system results in biotransformation of foreign compounds, yielding products that are often more polar and less toxic than the parent compound. There are also numerous examples of metabolic activation of nontoxic parent compounds by the MFO system to reactive inter-mediates with potent cytotoxic, carcinogenic, and mutagenic effects (Jollow and Smith, 1977). These effects may be mediated by unstable electrophiles or nucleophiles which interact and in some cases covalently bind with cellular macromolecular and disrupt critical cellular process (Brooks, 1977). A widely used indicator of the formation of reactive intermediates has been the measurement of in vitro and in vivo covalent binding of radioactivity from labeled substrates to DNA, protein, lipids, and other macromolecules. Although covalent binding studies can establish the formation of reactive intermediates, they implicate the reactive intermediate in the mechanism of toxicity only if there is a close correlation between in vitro and in vivo co-valent binding and severity of response under a variety of experimental con-ditions. These conditions may include the use of metabolic inducers and in-hibitors, time and dose studies, as well as cellular, organ, and species-specific damage (Boyd, 1980a,b; Dutcher and Boyd, 1979; Gillette, 1974, 1979; Gillette and Pohl, 1977). The involvement of reactive intermediates has been implicated in the toxicity of several compounds, including the hepatotoxins and carcino-gens, acetaminophen, aflatoxins and bromobenzene, and the pneumotoxin 4-ipomeanol (Hinson, 1980; Reid, 1973; Reid et al., 1973; Gillette, 1979; Boyd, 1980a,b; Swenson, 1981).

The toxicity of a reactive intermediate depends on both its rate of forma-tion by the MFO or other system, and its rate of detoxification by conjugation, enzymatic transformation, or other means (Jollow and Smith, 1977). Conse-quently, conjugating agents such as glutathione, and the activity of conju-gating enzymes such as glutathione-S-transferase, often function to protect against the toxic effects of reactive intermediates (Boyd, 1980a,b; Hinson, 1980; Jerina and Bend, 1977).

The formation of a reactive intermediate of 3MI was investigated in in vitro and in vivo covalent binding experiments in cattle using [^{3}H]-3MI (Hanafy and Bogan, 1980), and in in vitro studies with goats using [^{14}C]-3MI (Bray and Carlson, 1979b, 1980; Hammond and Carlson, 1980a). The lungs of calves given [^{3}H]-3MI contained more bound radioactivity than any other tissue studied, including liver. In vitro binding in bovine microsomal preparations was inhibited by cytochrome c and carbon monoxide and re-quired oxygen and NADPH, suggesting that a cytochrome P_{450}-dependent MFO was responsible for the formation of the reactive intermediate. Goat lung microsomes resulted in significantly more covalent binding from [^{14}C]-3MI than any other tissue and a nontoxic metabolite, [^{3}H]-3MOI, did not result in significant covalent binding. In addition, in vitro covalent binding in goat lung microsomes was inhibited by the addition of piperonyl butoxide, glutathione, and cysteine. In vitro covalent binding also showed some species specificity, with significant binding in lung microsomes of goat, cattle, and horses but not in hamsters or monkeys. In in vivo studies, the severity of lung injury and covalent binding are altered by changes in tissue glutathione concentration (Nocerini and Carlson, 1983). All these results clearly indicate that a reactive intermediate is formed by lung microsomes, and that a cytochrome P_{450}-dependent mixed-function oxidase is probably involved in the mechanism of lung injury. The chemical nature of the reactive intermediate remains to be determined.

　　Metabolism of 3MI in various tissues. The organ and cellular selective 3MI-induced lung injury in cattle, goats, and horses could result from differences in substrate specificity of metabolic characteristics of particular forms of cytochrome P_{450}-dependent mixed-function oxidases in lungs and lung cells of these species. It could also result from differences in the concentration of conjugating agents or activities of detoxifying enzymes among organs or cell types. In the case of the lung-toxic furan 4-ipomeanol, autoradiography demonstrated preferential binding of reactive intermediate in nonciliated bronchiolar epithelial (Clara) cells in lungs, which led to the conclusion that these cells contained the appropriate mixed-function oxidase activity (Boyd, 1977, 1980b).

　　Incubation of [^{14}C]-3MI with crude microsomal preparations of lung, liver, and kidney results in the formation of metabolites in all tissues, indicating that 3MI is not exclusively metabolized in the lung (Winston et al., 1980). Nevertheless, the higher proportion of radioactivity in the lung during infusion (Bradley and Carlson, 1982) and higher in vitro covalent binding in lung microsomes (Bray and Carlson, 1979b, 1980; Hammond and Carlson, 1980a) suggest active metabolism of 3MI by lung tissue. The fact that 3MI can cause acute lung injury without involvement of other tissues and organs was demonstrated in isolated perfused horse lungs. A 3-hr perfusion of horse lungs with [^{14}C]-3MI resulted in bronchiolar necrosis with no such changes in vehicle- or indole-perfused lungs. Covalently bound radioactivity was present in the perfused lung tissue and metabolites were present in the perfusate (Carlson et al., 1980a). Metabolism of 3MI and acute lung injury in isolated horse lungs indicates that the lung is capable of carrying out the reactions that lead to lung damage. The experiment does not rule out the possibility that other tissues or organs may contribute to lung injury in the event that reactive intermediates would escape detoxification, enter the circulation, and adversely affect the lung.

F. Other Sources of 3MI

This review has focused on the ruminal formation of 3MI and lung injury in ruminants and horses. 3MI has long been known to be a constituent of feces in animals, and it is produced by anaerobic fermentation of TRP. Human feces may contain 5-100　µg of 3MI/g and appreciable quantities are also present in feces of rats, pigs, and other animals (Herter, 1907; Horning and Dalgliesh, 1958; Yoshihara and Maruta, 1977; Spoelstra, 1977). Traces of 3MI have been detected in milk of cows fed *Lepidium* sp. (Conochie, 1953). The quantity of 3MI in the feces is affected by diet, resulting in a 100-fold increase in rats fed a meat diet compared to a chow diet (Anderson, 1975). In view of the rapid absorption of 3MI from the gastrointestinal tract, it is apparent that all animals are probably exposed to varying amounts of 3MI as a result of intestinal fermentation.

Another source of 3MI is cigarette smoke. Significant quantities of 3MI are formed by pyrolysis of TRP in tobacco leaves (Hoffman and Rathkamp, 1970) and smoke contains approximately 4-50 μg of 3MI per cigarette (Wynder and Hoffman, 1967; Hoshika, 1977). Considering the potent effects of 3MI in ruminants, the quantities of 3MI in cigarette smoke may represent a significant exposure, compounded by the fact that inhalation would probably result in direct contact of 3MI with lung cell populations that may be most susceptible to its effects. There is no direct evidence regarding possible risk of exposure to 3MI in humans, but fortunately, preliminary studies with some monogastric species tend to suggest less susceptibility to 3MI-induced lung injury in the species tested.

Since 3MI causes lung injury as a result of metabolism by the MFO system, identification of the specific pneumotixc agent and reactions responsible for its formation and detoxification may aid in future assessment of the risk of exposure to 3MI in humans. It is important to keep in mind that the presence of other xenobiotics or drugs in the environment and dietary alterations (Bray and Atwal, 1980; Campbell, 1978; Campbell and Hayes, 1974) could conceivably alter specific activities of the MFO system and the severity and species susceptibility to lung injury from 3MI and other chemicals.

G. Role of 3MI in Naturally Occurring ABPE

The involvement of 3MI as the etiologic agent in naturally occurring ABPE after pasture change is postulated, based on the many similarities between the naturally occurring and experimentally induced syndromes. Administration of 3MI is the only experimental model that reproduces the clinical signs and pulmonary lesions of naturally occurring ABPE. As previously reviewed (Carlson and Dickinson, 1978), calculations suggest that cattle can consume at least 0.13 g of TRP/kg BW per day in lush pasture, which exceeds by two- to threefold the intravenous dose of 3MI that can cause acute lung injury in cattle and goats. This indicates that sufficient TRP is available in lush forage; and because of the total TRP content and relatively high protein digestibility of lush forage, it is likely that TRP is the primary precursor for ruminal formation of 3MI. Other indolic compounds such as indoleacetic acid are present in amounts of nanograms per gram compared to milligram to gram quantities of TRP (Knegt and Bruinsma, 1973; Kamisaka and Larsen, 1977), and it is unlikely that they could account for more than a minor fraction of the substrate.

Even though TRP is probably the major precursor of 3MI in the rumen, it is not essential that pastures where outbreaks of ABPE occur have higher TRP content than other pastures. Mackenzie (1975) found no significant difference in the TRP content of pastures where ABPE occurred compared to other pastures (TRP content = 0.33-0.45% of dry matter). As discussed above, a primary consideration in the development of ABPE is an abrupt change to lush pasture, which undoubtedly causes drastic alterations in ruminal fermentation. Cattle are usually not affected by ABPE after they become adapted to the new feeding regimen, even though the TRP content of the forage would not be expected to change significantly. We believe that abnormal fermentation associated with the abrupt change in pasture influences the proportion of TRP converted to 3MI and that adequate substrate (TRP) is available in lush forage. The concept that ruminal fermentation characteristics affect the proportion of TRP converted to 3MI is an important concept in understanding the etiology of ABPE, and it offers clues to possible prevention strategies.

More direct evidence that 3MI is the causative agent in ABPE comes from field trials in which cattle have been subjected to abrupt pasture change. A Scottish study (Selman et al., 1977a) demonstrated the presence of 3MI in ruminal contents in 11 cattle that developed clinical signs of ABPE and were slaughtered 15-18 days after pasture change. In another study (Hammond et al., 1979a), mean ruminal 3MI concentrations increased for 3-4 days after pasture change and then gradually decreased until the end of the experiment at 7 days. These experiments verify the presence of 3MI in the ruminal contents after pasture change and indicate that these levels increase during the first few days and then gradually decrease.

The foregoing facts present evidence for the hypothesis that 3MI is the cause of ABPE, but proof of the hypothesis has been elusive. In our field trials (Hammond et al., 1978; Carlson et al., 1980b), the clinical incidence of ABPE and ruminal 3MI concentrations have been lower than expected, which is attributed to the necessity to disturb the cattle on a daily basis to collect samples. More data are needed in well-controlled field trials where larger numbers of animals are used, and a greater incidence of clinical illness is achieved. These studies will allow rigorous assessment of the effectiveness of prophylactic agents to reduce ruminal 3MI concentrations and the incidence of ABPE, which is needed for further proof of the 3MI hypothesis.

IV. PREVENTION STRATEGIES FOR ABPE

Approaches to the prevention of ABPE could involve alteration of ruminal fermentation or animal management schemes to reduce the formation of 3MI, or systemic intervention to inhibit the effects of 3MI on the lung. Experimentally, systemic administration of BT (Bray and Carlson, 1979a), or cysteine (Nocerini and Carlson, 1983), which maintains tissue glutathione concentrations and dietary manipulation (Bray and Atwal, 1980), can protect goats from effects of 3MI administration. Even though systemic intervention may be possible experimentally, it does not seem feasible as a practical prophylactic strategy because of regulatory policies and difficulty in administration of the prophylactic agent. We have focused on the control of ruminal fermentation and animal management practices as the most practical approaches to prevention of ABPE.

A. Inhibition of Ruminal 3MI Formation

More than 27 compounds representing a variety of classifications of antibiotics and metabolic inhibitors were screened in an in vitro procedure to assess their effects on ruminal conversion of TRP to 3MI (Hammond and Carlson, 1980b). Two iodonium compounds with known inhibitory effects on ruminal amino acid deamination (Chalupa, 1977, 1980), inhibited 3MI formation in vitro at 25 µg/ml but not at 5 µg/ml. All of the polyether antibiotics tested inhibited 3MI production by more than 80%, and they were the most potent compounds tested. Monensin and lasalocid inhibited by 87% and 90%, respectively, at 5 μg/ml. Monensin also inhibited 3MI formation effectively at 0.2 μg/ml, but the inhibition from lasalocid was lost at concentrations tested below 5 μg/ml, indicating approximately a 30-fold difference in the potency of monensin over lasalocid. Monensin inhibited the conversion of IAA to 3MI (Hammond and Carlson, 1980b) and inhibited the growth of the ruminal *Lactobacillus* sp. responsible for 3MI formation (Yokoyama and Carlson, 1979a). Schelling et al. (1978) demonstrated that monensin also inhibited the degradation of TRP in in vitro systems, indicating that it can effect both steps in the formation of 3MI from TRP. There is currently no information on which of the two steps in the formation of 3MI from TRP is most sensitive to monensin or which is the rate-limiting reaction.

Since monensin was already a widely used feed additive for cattle, its effect on TRP-induced ABPE was tested. A dose of 200 mg of monensin per day, given 1 day before until 4 days after a TRP dose to induce ABPE, significantly reduced ruminal and plasma 3MI concentrations and completely prevented lung injury in the treated cows. A similar dose of lasalocid had no effect on ruminal and plasma 3MI concentrations and the incidence or severity of pulmonary lesions (Hammond et al., 1978, 1979b, 1980b).

The promising results with in vitro and in vivo experiments encouraged us to examine the effects of monensin treatment under field conditions. Although there were no cases of ABPE that developed in control or treatment groups after pasture change, monensin significantly reduced ruminal 3MI concentrations and ruminal fluid from treated cows converted less TRP to 3MI, in vitro. The effect of monensin was greatest during the first 4 days of the experiment (Carlson et al., 1980b). This experiment confirmed that monensin was a potential prophylactic agent for ABPE.

At the present time, monensin is not approved by the U.S. Food and Drug Administration for use in breeding cows, but it is licensed for use in feedlot cattle and to slaughter stocker and feeder cattle on pasture. If approved by the FDA and if field research continues to indicate that it is effective as a prophylactic agent for ABPE, a number of commercial preparation should make it readily available for supplementation in various forms. Since monensin is toxic to horses, it must be used carefully.

Other approaches to the control of ruminal 3MI production have been proposed, including the use of sodium bicarbonate, a ruminal buffering agent. In our hands, sodium bicarbonate is ineffective in reducing TRP-induced ruminal 3MI production or clinical signs of ABPE (Breeze et al., 1981).

B. Management Approaches to Prevention of ABPE

In the past few years a great deal of information has been generated on the etiology and prevention of ABPE, but definitive recommendations based on proven effectiveness are not yet possible. There are a number of animal management recommendations that can be expected to reduce the incidence and severity of ABPE after abrupt pasture change. The primary focus of these procedures should be to minimize the degree of change in grazing conditions and limit the intake of lush forage for the first 10-14 days to facilitate ruminal adaptation to the new feeding conditions. Examples of some possible procedures are as follows. Whenever possible, move cows to new pasture before the previous feed is entirely utilized. If necessary, temporarily pregraze excessively lush pasture with less susceptible species, such as horses, or with younger cattle. Avoid giving hungry cattle access to lush pasture by prefeeding with hay or pregrazing lower-quality forage. Supply supplementary hay or other feed to cattle during the first 10-14 days after pasture change. Interrupt the grazing pattern by gathering animals from lush pasture at intervals of 1 or 2 days and feeding hay or other dry rough-age before returning them to the pasture. In some cases, gradually exposing cattle to lush forage by movable fence and supplementary hay feeding may be effective if care is taken to ensure that cattle do not have access to too much lush forage at one time. There are obviously a variety of other possible pro-cedures that may also be effective in reducing the incidence of ABPE by facilitating a smoother transition in ruminal fermentation after pasture change.

When an outbreak of ABPE occurs, the first step should be to verify the diagnosis and rule out other possible respiratory diseases (Breeze et al., 1975b). All cattle and their calves should be removed, without unnecessary excitement, to dry lot, dry pasture, or other facilities and be fed hay or other forage. After recovery of affected animals, it may be possible to re-introduce the herd gradually to the pasture.

C. Treatment of ABPE

Many drugs have been suggested in the treatment of ABPE but none have been tested under controlled conditions. Antagonists of anaphylaxis and hypersensitivity do not alter the course of 3MI-induced pulmonary disease and would not be expected to be effective in naturally occurring ABPE. These compounds include acetylsalicylic acid, mepyramine maleate, diethyl-carbamazine citrate, sodium meclofenamate, and betamethasone (Breeze, 1978; Breeze and Carlson, 1981). Even though there is an absence of proven effective therapy, it should be noted that no benefit is to be expected from monensin if given after clinical signs appear. Because of the known rapid effects of 3MI, monensin's only potential value in ABPE would be as a pro-phylactic agent to prevent formation of 3MI.

V. SUMMARY

Over the past 15 years substantial progress has been made in understanding the etiology, mechanism of action, and prevention of ABPE. Prior to that time a multitude of theories were being applied to a poorly characterized disease, the result being that little progress was possible. The hypothesis that the disease was of dietary origin resulting from ingestion of TRP arose from an unanticipated observation and was a controversial idea because of the prevailing view that the disease originated from some type of hypersensi-tivity reaction. Nevertheless, the TRP provided an opportunity for systematic, controlled experiments in which the clinical signs and pulmonary lesions of ABPE could be reproduced. These studies led to the knowledge that rumen fermentation was involved, and that 3MI was the only ruminal fermentation product of TRP that resulted in pulmonary injury. Rapid absorption of 3MI into the blood and further metabolism by the MFO system was responsible for lung damage. The lungs of cattle, sheep, goats, and horses are susceptible to the effects of 3MI; and these effects are rapid and cellular specific. Since 3MI itself and most metabolites of 3MI are not pneumotoxic, the causative agent appears to be a reactive intermediate formed in an early step of metabolism by the MFO system in the lung. All of these and other observations support the hypothesis that 3MI is the cause of naturally occurring ABPE, and it is currently the only hypothesis in which experimental observations are identical and can account for observations in naturally occurring ABPE. Approaches to prevention of ABPE based on the TRP hpothesis show promise and hopefully will lead to reduced incidence of ABPE and more efficient beef production in the United States and other countries.

ACKNOWLEDGMENTS

Much of the work reviewed was supported by NIH Grant HL-13645, USDA-SEA Grant 901-15-169, and Washington State University Agricultural Experiment Station Project 1893.

REFERENCES

Aitken, M. M., and Sanford, J. (1969). Experimental anaphylaxis in cattle. *J. Comp. Pathol. 79*:131-139.

Aitken, M. M., and Sanford, J. (1972). Modification of acute systemic anaphylaxis in cattle by drugs and by vagotomy. *J. Comp. Pathol. 82*: 247-256.

Aitken, M. M., and Sanford, J. (1973). "Fog fever" and lungworms. *Vet. Rec. 93*:209-210.

Allison, M. J. (1969). Biosynthesis of amino acids by ruminal microorganisms. *J. Anim. Sci. 29*:797-807.

Anderson, G. M. (1975). Quantitation of tryptophan metabolites in rat feces by thin-layer chromatography. *J. Chromatogr. 105*:323-328.

Atkinson, G., Bogan, J. A., Breeze, R. G., and Selman, I. E. (1977). Effects of 3-methylindole in cattle. *Br. J. Pharmacol. 61*:285-290.

Bakri, M., and Carlson, J. R. (1970). Chromatographic separation of tryptophan metabolites. *Anal. Biochem. 34*:46-65.

Blackburn, T. H. (1965). Nitrogen metabolism in the rumen. In *Physiology of Digestion in the Ruminant*, R. W. Dougherty (Ed.). Butterworth, Washington, D.C., pp. 322-334.

Blake, J. T., and Thomas, D. W. (1971). Acute bovine pulmonary emphysema in Utah. *J. Am. Vet. Med. Assoc. 158*:2047-2052.

Boyd, M. R. (1977). Evidence for the Clara cell as a site of cytochrome P-450 dependent mixed function oxidase activity in the lung. *Nature (Lond.) 269*:713-715.

Boyd, M. R. (1980a). Biochemical mechanisms in chemical-induced lung injury: roles of metabolic activation. *CRC Crit. Rev. Toxicol. 7*:103-176.

Boyd, M. R. (1980b). Biochemical mechanisms in pulmonary toxicity of furan derivatives. In *Biochemical Reviews in Toxicology*, Vol. 2, E. Hodgson, J. R. Bend, and R. M. Philpot (Eds.). Elsevier/North-Holland, New York, pp. 71-101.

Bradley, B. J. (1977). Electron and light microscopy of lungs and body distribution of ^{14}C-3-methylindole in experimentally induced pulmonary edema and emphysema in goats. Ph.D. thesis, Washington State University, Pullman.

Bradley, B. J., and Carlson, J. R. (1980). Ultrastructural pulmonary changes induced by intravenously administered 3-methylindole in goats. *Am. J. Pathol. 99*:551-560.

Bradley, B. J., and Carlson, J. R. (1982). Concentration of 3-methylindole (3MI) and distribution of radioactivity from ^{14}C-3MI in goat tissue associated with acute pulmonary edema. *Life Sci. 30*:455-463.

Bradley, B. J., Carlson, J. R., and Dickinson, E. O. (1977). Light and scanning electron microscope observations of goat lungs following 3-methylindole infusion. In *Pulmonary Macrophage and Epithelial Cells*, ERDA Symp. 43, Tech. Inf. Ctr., Energy Res. Dev. Adm., Pub. pp. 590-602.

Bradley, B. J., Carlson, J. R., and Dickinson, E. O. (1978). 3-Methylindole-induced pulmonary edema and emphysema in sheep. *Am. J. Vet. Res. 39*: 1355-1358.

Bray, T. M., and Atwal, O. S. (1980). Effect of energy restriction on mixed function oxidase (MFO) activities and the development of acute pulmonary edema (APE) in goats. *Fed. Proc. 39*:891.

Bray, T. M., and Carlson, J. R. (1975). The effects of 3-methylindole on hemolysis, transport of ^{22}Na$^+$, and ATPase activities of bovine erythrocytes. *Proc. Soc. Exp. Biol. Med. 148*:875-879.

Bray, T. M., and Carlson, J. R. (1979a). Role of mixed function oxidase in 3-methylindole-induced acute pulmonary edema in goats. *Am. J. Vet. Res. 40*:1268-1272.

Bray, T. M., and Carlson, J. R. (1979b). Covalent binding of 3-methyl-indole metabolites in goat lung microsomal preparations. *Fed. Proc. 38*: 1329.

Bray, T. M., and Carlson, J. R. (1980). Tissue and subcellular distribution and excretion of [^{14}C]-3-methylindole in rabbits after intratracheal infusion. *Can. J. Physiol. Pharmacol. 58*:1399-1405.

Bray, T. M., Magnuson, J. A., and Carlson, J. R. (1974). Nuclear magnetic resonance studies of lecithin-skatole interaction. *J. Biol. Chem. 249*:914-918.

Bray, T. M., Sandberg, H. E., and Carlson, J. R. (1975). An EPR study of structural perturbations induced by 3-methylindole in the protein and lipid regions of erythrocyte membranes. *Biochim. Biophys, Acta 382*:534-541.

Breeze, R. G. (1978). Fog fever and heaves: studies on the respiratory diseases of adult cattle and horses. *Proc. Am. Coll. Vet. Int. Med.*, Dallas, TX, pp. 87-119.

Breeze, R. G., and Carlson, J. R. (1981). Acute bovine pulmonary edema and emphysema. In *Current Veterinary Therapy, Food Animal Practice*, J. L. Howard (Ed.). W. B. Saunders, Philadelphia, pp. 832-835.

Breeze, R. G., Pirie, H. M., Selman, I. E., and Wiseman, A. (1974). Fog fever: provocation tests with *Dictyocaulus viviparus*. *J. Comp. Pathol. 84*:577-587.

Breeze, R. G., Pirie, H. M., Dawson, C. O., Selman, I. E., and Wiseman, A. (1975a). The pathology of respiratory diseases of adult cattle in Britain. *Folia Vet. Lat.* 5:95-128.

Breeze, R. G., Pirie, H. M., Selman, I. E., and Wiseman, A. (1975b). Acute respiratory distress in cattle. *Vet. Rec.* 97:226-229.

Breeze, R. G., Pirie, H. M., Selman, I. E., and Wiseman, A. (1976). Fog fever (acute pulmonary emphysema) in cattle in Britain. *Vet. Bull.* 46: 243-251.

Breeze, R. G., Lee, H. A., and Grant, B. D. (1978a). Toxic lung disease. *Mod. Vet. Pract.* 59:301.

Breeze, R. G., Selman, I. E., Wiseman, A., and Pirie, H. M. (1978b). A reappraisal of atypical interstitial pneumonia in cattle. *Bovine Pract.* 13: 75-81.

Breeze, R. G., Carlson, J. R., Potchoiba, M. J., Nocerini, M. R., and Hammond, A. C. (1981). Sodium bicarbonate does not prevent acute bovine pulmonary edema and emphysema. *Bovine Pract.* 16:40-43.

Brooks, P. (1977). Role of covalent binding in carcinogenicity. In *Biological Reactive Intermediates*, D. J. Jollow, J. J. Kocsis, R. Snyder, and H. Vainio (Eds.). Plenum Press, New York, pp. 470-480.

Campbell, T. C. (1978). Effect of dietary protein on drug metabolism. In *Nutrition and Drug Interrelations*, J. N. Hathcock and J. Coon (Eds.). Academic Press, New York, pp. 409-422.

Campbell, T. C., and Hayes, J. R. (1974). Role of nutrition in the drug-metabolizing system. *Pharmacol. Rev.* 26:171-197.

Carlson, J. R., and Dickinson, E. O. (1978). Tryptophan-induced pulmonary edema and emphysema in ruminants. In *Effects of Poisonous Plants on Livestock*, R. F. Keeler, K. R. Van Kampen, and L. F. James (eds.). Academic Press, New York, pp. 261-272.

Carlson, J. R., and Markey, S. L. (1975). Metabolism and excretion of 3-methylindole (skatole) by rats. *Fed. Proc.* 34:900.

Carlson, J. R., Dyer, I. A., and Johnson, R. J. (1968). Tryptophan-induced interstitial pulmonary emphysema in cattle. *Am. J. Vet. Res.* 29:1983-1989.

Carlson, J. R., Yokoyama, M. T., and Dickinson, E. O. (1972). Induction of pulmonary edema and emphysema in cattle and goats with 3-methylindole. *Science* 176:298-299.

Carlson, J. R., Dickinson, E. O., Yokoyama, M. T., and Bradley, B. J. (1975). Pulmonary edema and emphysema in cattle after intraruminal and intravenous administration of 3-methylindole. *Am. J. Vet. Res.* 36:1341-1347.

Carlson, J. R., Breeze, R. G., Campbell, K. B., Wakao, Y., Klavano, P. A., and Nocerini, M. R. (1980a). Effect of 3-methylindole (3MI, skatole) on isolated perfused horse lungs. *Fed. Proc.* 39:306.

Carlson, J. R., Hammond, A. C., Breeze, R. G., Potchoiba, M. J., and Nocerini, M. R. (1980b). Effect of monensin on ruminal 3-methylindole production after abrupt pasture change. *Abstr., 72nd Meet. Am. Soc. Anim. Sci.*, Cornell University, Ithaca, N.Y., p. 233.

Carlson, J. R., Kowalczyk, J., and Honeyfield, D. C. (1981). Absorption of 3-methylindole (3MI) from the rumen of goats and partitioning of 3MI in ruminal contents. *Abstr., 73rd Meet. Am. Soc. Anim. Sci.*, Raleigh, N.C., p. 283.

Chalupa, W. (1977). Manipulating rumen fermentation. *J. Anim. Sci.* 45: 585-599.

Chalupa, W. (1980). Chemical control of rumen microbial metabolism. In *Digestive Physiology and Metabolism in Ruminants,* Y. Ruckebusch and P. Thivend (Eds.). AVI Publishing Co., Westport, Conn., pp. 325-347.

Chung, K. T., Anderson, G. M., and Fulk, G. E. (1975). Formation of indoleacetic acid by intestinal anaerobes. *J. Bacteriol.* 124:573-575.

Conochie, J. (1953). Indole and skatole in the milk of the ruminant feeding on *Lepidium* sp. *Aust. J. Exp. Biol. Med. Sci.* 31:373-384.

Cornelius, L. M., Coulter, D., Doster, A., and Rawlings, C. (1979). Pathophysiologic studies of calves given 3-methylindole intraruminally. *Am. J. Vet. Res.* 40:571-575.

Deinum, B. (1966). Influence of some climatological factors on the chemical composition and feeding value of herbage. *Proc. 10th Int. Grassl. Congr.,* Helsinki, pp. 415-418.

DeMoss, R. D., and Moser, K. (1969). Tryptophanase in diverse bacterial species. *J. Bacteriol.* 98:167-171.

Dickinson, E. O. (1970). Effect of tryptophan on bovine lungs. Ph.D. thesis, Washington State University, Pullman.

Dickinson, E. O., Spencer, G. R., and Gorham, J. R. (1967). Experimental induction of an acute respiratory syndrome resembling bovine pulmonary emphysema. *Vet. Rec.* 80:487-489.

Dickinson, E. O., Yokoyama, M. T., Carlson, J. R., and Bradley, B. J. (1976). Induction of pulmonary edema and emphysema in goats by intra-ruminal administration of 3-methylindole. *Am. J. Vet. Res.* 37:667-672.

Doster, A. R., Mitchell, F. E., Farrell, R. L., and Wilson, B. J. (1978). Effects of 4-ipomeanol, a product from mold damaged sweet potatoes on the bovine lung. *Vet. Pathol.* 15:367-375.

Dutcher, J. S., and Boyd, M. R. (1979). Species and strain differences in target organ alkylation and toxicity by 4-ipomeanol: predictive value of covalent binding in studies of target organ toxicities by reactive metabolites. *Biochem. Pharmacol.* 28:3367-3372.

Eadie, J. M., and Oxford, A. E. (1954). A remarkable disintegrative effect of skatole upon certain rumen ciliate protozoa. *Nature (Lond.)* 174:973.

Elsden, S. R., Hilton, M. G., and Waller, J. M. (1976). The end products of the metabolism of aromatic amino acids by *Clostridia. Arch. Microbiol.* 107:293-288.

Eyre, P. (1972). Acute bovine pulmonary emphysema. *Vet. Rec.* 91:38-40.

Eyre, P. (1975). Comparison of reactivity of pulmonary vascular and airway smooth muscle of sheep and calf to tryptamine analogues, histamine, and antigen. *Am. J. Vet. Res.* 36:1081-1084.

Eyre, P. (1980). Pharmacological aspects of hypersensitivity in domestic animals: a review. *Vet. Res. Commun.* 4:83-98.

Fordtran, J. S., Scroggie, W. B., and Polter, D. E. (1964). Colonic absorption of tryptophan metabolites in man. *J. Lab. Clin. Med.* 64:125-132.

Frydman, R. B., Tomaro, M. L., and Frydman, B. (1971). Pyrrolooxygenases: the biosynthesis of 2-aminoacetophenone. *FEBS Lett.* 17:273-276.

Frydman, R. B., Tomaro, M. L., Frydman, B. (1972). Pyrrolooxygenases: isolation, properties and products formed. *Biochim. Biophys. Acta 284*: 63-79.

Frydman, B., Frydman, R. B., and Tomaro, M. L. (1973). Pyrrolooxygenases: a new type of oxidases. *Mol. Cell. Biochem. 2*:121-136.

Gillette, J. R. (1974). A perspective on the role of chemically reactive metabolites of foreign compounds in toxicity. I. Correlation of changes in covalent binding of reactive intermediate with changes in the incidence and severity of toxicity. *Biochem. Pharmacol. 23*:2785-2794.

Gillette. J. R. (1979). Effects of induction of cytochrome P-450 enzymes on the concentration of foreign compounds and their metabolites and on the toxicological effects of these compounds. *Drug. Metab. Rev. 10*:59-87.

Gillette, J. R., and Phol, L. R. (1977). A prospective on covalent binding and toxicity. *J. Toxicol. Environ. Health 2*:849-871.

Hammond, A. C., and Carlson, J. R. (1980a). Factors affecting binding of 3-methylindole metabolites in crude microsomal preparations. *Fed. Proc. 39*:306.

Hammond, A. C., and Carlson, J. R. (1980b). Inhibition of ruminal degradation of L-tryptophan to 3-methylindole, in-vitro. *J. Anim. Sci. 51*:207-214.

Hammond, A. C., Carlson, J. R., and Breeze, R. G. (1978). Monensin and the prevention of tryptophan-induced acute bovine pulmonary edema and emphysema. *Science 201*:153-155.

Hammond, A. C., Bradley, B. J., Yokoyama, M. T., Carlson, J. R., and Dickinson, E. O. (1979a). 3-Methylindole and naturally-occurring acute bovine pulmonary edema and emphysema. *Am. J. Vet. Res. 40*:1398-1401.

Hammond, A. C., Carlson, J. R., Breeze, R. G., and Selman, I. E. (1979b). Progress in the prevention of acute bovine pulmonary emphysema. *Bovine Pract. 14*:9-14.

Hammond, A. C., Carlson, J. R., and Willett, J. D. (1979c). The metabolism and disposition of 3-methylindole in goats. *Life Sci. 25*:1301-1306.

Hammond, A. C., Carlson, J. R., and Breeze, R. G. (1980a). Indole toxicity in cattle. *Vet. Rec. 107*:344-346.

Hammond, A. C., Carlson, J. R., and Breeze, R. G. (1980b). Prevention of tryptophan-induced acute bovine pulmonary edema and emphysema (fog fever). *Vet. Rec. 107*:322-325.

Hammond, A. C., Carlson, J. R., and Breeze, R. G. (1980c). Effect of monensin pretreatment on tryptophan-induced acute bovine pulmonary emphysema. *Abstr., 72nd Meet. Am. Soc. Anim. Sci.*, Cornell University, Ithaca, N.Y., p. 365.

Hanafy, M. S. M., and Bogan, J. A. (1980). The covalent binding of 3-methylindole metabolites to bovine tissue. *Life Sci. 27*:1225-1231.

Heacock, R. A., and Mahon, M. E. (1963). The hydroxylation of skatole. *Can. J. Biochem. Physiol. 41*:2381-2390.

Heacock, R. A., and Mahon, M. E. (1964). Conjugated hydroxyskatoles in human urine. *Can. J. Biochem. 42*:813-819.

Heron, B. R., and Suther, D. E. (1979). A retrospective investigation and random sample survey of acute bovine pulmonary emphysema in northern California. *Bovine Pract. 14*:2-8.

Herter, C. A. (1907). *The Common Bacterial Infections of the Digestive Tract and the Intoxications Arising from Them.* Macmillan, New York, p. 241.

Hinson, J. A. (1980). Biochemical toxicology of acetaminophen. In *Reviews in Biochemical Toxicology,* Vol. 2, E. Hodgson, J. R. Bend, and R. M. Philpot (Eds.). Elsevier/North-Holland, New York, pp. 103-130.

Hoffman, D., and Rathkamp. G. (1970). Quantitative determination of 1-alkylindoles in cigarette smoke. *Anal. Chem. 42:*366-370.

Horning, E. C., and Dalgliesh, D. E. (1958). The association of skatole-forming bacteria in the small intestine with the malabsorption syndrome and certain anaemias. *Biochem. J. 70:*13P-14P.

Horvath, H. A., Carlson, J. R., and Dickinson, E. O. (1975). Induction of pulmonary edema and emphysema in goats. *J. Anim. Sci. 41:*404.

Hoshika, Y. (1977). Simultaneous gas chromatographic analysis of lower fatty acids, phenols, and indoles using a glass capillary column. *J. Chromatogr. 144:*181-189.

Huang, T. W., Carlson, J. R., Bray, T. M., and Bradley, B. J. (1977). 3-Methylindole-induced pulmonary injury in goats. *Am. J. Pathol. 87:* 647-666.

Hurley, J. V. (1978). Current views on the mechanisms of pulmonary edema. *J. Pathol. 125:*59-79.

Hyslop, N. St. G. (1969). Bovine pulmonary emphysema. *Can. Vet. J. 10:* 251-257.

Jarvie, A., Breeze, R. G., Selman, I. E., and Wiseman, A. (1977). Effects of oral dosage with tryptamine analogues in cattle. *Vet. Rec. 101:*267-268.

Jenkins, P. A., and Pepys, J. (1965). Fog fever precipitin (FLH) reactions to mouldy hay. *Vet. Rec. 77:*464-466.

Jerina, D. M., and Bend, J. R. (1977). Glutathione-S-transferases. In *Biological Reactive Intermediates,* D. J. Jollow, J. J. Kocsis, R. Snyder, and H. Vainio (Eds.). Plenum Press, New York, pp. 207-236.

Johnson, R. J., and Dyer, I. A. (1966). Effect of orally administered tryptophan on tryptophan pyrrolase activity in ovine and bovine. *Life Sci. 5:*1121-1124.

Jollow, D. J., and Smith, C. (1977). Biochemical aspects of toxic metabolites: formation, detoxification, and covalent binding. In *Biological Reactive Intermediates,* D. J. Jollow, J. J. Kocsis, R. Snyder, and H. Vainio (Eds.). Plenum Press, New York, pp. 42-59.

Kamisaka, S., and Larsen, P. (1977). Improvement of the indole-α-pyrone fluorescence method for quantitative determination of endogenous indole-3-acetic acid in lettuce seedlings. *Plant Cell Physiol. 18:*595-602.

King, L. J., Parke, D. V., and Williams, R. T. (1966). The metabolism of [2-^{14}C]-indole in the rat. *Biochem. J. 98:*266-277.

Knegt, E., and Bruinsma, J. (1973). A rapid, sensitive, and accurate determination of indolyl-3-acetic acid. *Phytochemistry 12:*753-756.

Knowlson, J. (1819). *The Yorkshire Cattle-Doctor and Farrier,* William Walker, London.

Lacoste, A. M. (1961). Dégradation du tryptophane par les bactéries de la panse de ruminants. *C. R. Acad. Sci. 252:*1233-1235.

Leslie, V. J. S. (1949). Fog fever. *Vet. Rec. 61:*228-229.

Lewis, T. R., and Emery, R. S. (1962a). Intermediate products in the catabolism of amino acids by rumen microorganisms. *J. Dairy Sci.* 45: 1363-1368.

Lewis, T. R., and Emergy, R. S. (1962b). Metabolism of amino acids in the bovine rumen. *J. Dairy Sci.* 45:1487-1492.

Mackenzie, A. (1975). Pasture levels of tryptophan in relation to outbreaks of fog fever. *Res. Vet. Sci.* 19:227-228.

Mahon, M. E., and Mattock, G. L. (1967). The differential determination of conjugated hydroxyskatoles in human urine. *Can. J. Biochem.* 45: 1317-1322.

Maki, L. R. (1963). A review of the literature on the history and occurrence of bovine pulmonary emphysema. *Proc. Symp. Acute Bovine Pulmonary Emphysema,* University of Wyoming, Laramie, pp. 1-17.

Maki, L. R., and Tucker, J. O. (1962). Acute pulmonary emphysema of cattle. II. Etiology. *Am. J. Vet. Res.* 23:824-826.

Michel, J. F. (1953). Fog fever syndrom in parasitic bronchitis. *Nature (Lond.)* 171:940.

Michel, J. F. (1954). A contribution to the aetiology of fog fever. *Vet. Rec.* 66:381-384.

Monlux, W. S., Cutlip, R. C., and Estes, P. C. (1970). Breed susceptibility to tryptophan-induced pulmonary adenomatosis in cattle. *Cornell Vet.* 60:547-551.

Mori, A., Yasaka, Y., Masamoto, K., Hiramatsu, H. (1978). Gas chromatography of 5-hydroxy-3-methylindole in human urine. *Clin. Chim. Acta* 84:63-68.

Nakoneczna, I., Forbes, J. C., and Rogers, K. S. (1969). The arthritogenic effect of indole, skatole, and other tryptophan metabolites in rabbits. *Am. J. Pathol.* 57:523-532.

National Academy of Sciences (1971). *Atlas of Nutritional Data on United States and Canadian Feeds.* Washington, D.C.

Newton, W. A., and Snell, E. E. (1964). Catalytic properties of tryptophanase, a multifunctional pyridoxal phosphate enzyme. *Proc. Natl. Acad. Sci. USA* 51:382-389.

Nocerini, M. R., Carlson, J. R., and Breeze, R. G. (1983). Effect of glutathione status on covalent binding and pneumotoxicity of 3-methylindole in goats. *Life Sci.* 32:449-458.

Olcott, B. (1981). Cardiopulmonary effects of 3-methylindole following inhibition and induction of mixed function oxidation in goats. M.S. thesis, Washington State University, Pullman.

Peckham, J. C., Mitchell, F. E., Jones, O. H., Jr., and Doupnik, B., Jr. (1972). Atypical interstitial pneumonia in cattle fed moldy sweet potatoes. *J. Am. Vet. Med. Assoc.* 160:169-172.

Peterson, D. R. (1965). Bovine pulmonary emphysema caused by the plant *Perilla frutescens. Proc. 3rd Symp. Acute Bovine Pulmonary Emphysema,* University of Wyoming, Laramie, pp. R1-R13.

Pirie, H. M., Breeze, R. G., Selman, I.E., and Wiseman, A. (1974). Fog fever in cattle: pathology. *Vet. Rec.* 95:479-483.

Pirie, H. M., Breeze, R. G., Selman, I. E., and Wiseman, A. (1976). Indoleacetic acid, 3-methylindole and type 2 pneumonocyte hyperplasia in a proliferative alveolitis of cattle. *Vet. Rec.* 98:259-260.

Potchoiba, M. J., Carlson, J. R., and Breeze, R. G. (1982). Metabolism and pneumotoxicity of 3-methylindole, indole-3-carbinol and 3-methyloxindole, in goats. *Am. J. Vet. Res. 43*:1418-1423.

Reid, W. D. (1973). Mechanism of renal necrosis induced by bromobenzene or chlorobenzene. *Exp. Mol. Pathol. 19*:197-214.

Reid. W. D., Ilett, K. F., Glick, J. M., and Krishna, G. (1973). Metabolism and binding of aromatic hydrocarbons in the lung. *Am. Rev. Respir. Dis. 107*:539-551.

Rogers, K. S. (1969). Rabbit erythrocyte hemolysis by lipophilic, aryl molecules. *Proc. Soc. Exp. Biol. Med. 130*:1140-1142.

Rogers, K. S., Forbes, J. C., and Nakoneczna, I. (1969). Arthritogenic properties of lipophilic aryl molecules. *Proc. Soc. Exp. Biol. Med. 131*: 670-672.

Roth, C. W., Hock, J. A., and DeMoss, R. D. (1971). Physiological studies of biosynthetic indole excretion in *Bacillus alvei*. *J. Bacteriol. 106*:97-106.

Schatzmann, H. J., and Gerber, H. (1972). Production of tryptamine from tryptophan by ruminal fluid, in-vitro. *Zentralbl. Veterinaermed. Reihe A 19*:482-489.

Schelling, G. T., Spires, H. R., Mitchell, G. E., and Tucker, R. E. (1978). The effect of various antimicrobials on amino acid degradation rates by rumen microbes. *Fed. Proc. 37*:411.

Schofield, F. W. (1948). Acute pulmonary emphysema of cattle. *J. Am. Vet. Med. Assoc. 112*:254-259.

Scott, T. W., Ward, P. F. V., and Dawson, R. M. C. (1964). The formation and metabolism of phenyl-substituted fatty acids in the ruminant. *Biochem. J. 90*:12-24.

Selman, I. E., Wiseman, A., Pirier, H. M., and Breeze, R. G. (1974). Fog fever in cattle: clinical and epidemiological features. *Vet. Rec. 95*:139-146.

Selman, I. E., Breeze, R. G., Bogan, J. A., Wiseman, A., and Pirie, H. M. (1977a). Experimental production of fog fever by change to pasture free from *Dictyocaulus viriparus* infection. *Vet. Rec. 101*:278-283.

Selman, I. E., Wiseman, A., Breeze, R. G., and Pirie, H. M. (1977b). Differential diagnosis of pulmonary disease in adult cattle in Britain. *Bovine Pract. 12*:63-74.

Spoelstra, S. F. (1977). Simple phenols and indoles in anaerobically stored piggery wastes. *J. Sci. Food Agric. 28*:415-423.

Sprince, H. (1962). Biochemical aspects of indole metabolism in normal and schizophrenic subjects. *Ann. N.Y. Acad. Sci. 96*:399-418.

Staubli, W., Hess, R., and Weiberl, E. R. (1969). Correlated morphometric and biochemical studies on the liver cell. II. Effect of phenobarbital on rat hepatocytes. *J. Cell Biol. 42*:92-112.

Stowe, R. W. (1955). The production of indoleacetic acid by bacteria. *Biochem. J. 61*:ix-x.

Swenson, D. H. (1981). Metabolic activation and detoxification of aflatoxins. In *Reviews in Biochemical Toxicology*, Vol. 3, E. Hodgson, J. R. Bend, and R. M. Philpot (Eds.). Elsevier/North-Holland, New York, pp. 155-192.

Taylor, A. E., and Gaar, K. A. (1970). Estimation of equivalent pore radii of pulmonary capillary and alveolar membranes. *Am. J. Physiol. 218*: 1133-1140.

Tucker, J. O., and Maki, L. R. (1962). Acute pulmonary emphysema of cattle. I. Experimental production. *Am. J. Vet. Res. 23*:821-823.

Turk, J. R. (1981). Ultrastructural characterization of bovine fibrosing alveolitis and evaluation of 3-methylindole as an experimental model. Ph.D. thesis, Washington State University, Pullman.

Turk, M. A. M. (1981). Effect of 3-methylindole on equine lungs. Ph.D. thesis, Washington State University, Pullman.

Van Soest, P. J. (1968). Composition, maturity, and the nutritive value of forages. In *Cellulases and Their Application*. American Chemical Society, Washington, D.C., pp. 262-277.

Watanabe, T., and Aviado, D. M. (1974). Functional and biochemical effects on the lung following inhalation of cigarette smoke and constituents. II. Skatole, acrolein, and acetaldehyde. *Toxicol. Appl. Pharmacol. 30*:201-209.

Williams, G. D., Hatkin, J., and Bridges, C. H. (1977). Tissue distribution of 3-methlyindole (3MI) in goats following oral administration of 3MI. Abstr. 37, *58th Meet. Conf. Res. Workers Anim. Dis.*, Pick-Congress Hotel, Chicago.

Wilson, B. J. (1973). Toxicity of mold damaged sweet potatoes. *Nutr. Rev. 31*:73-78.

Wilson, B. J., Garst, J. E., Linnabary, R. D., and Channel, T. B. (1977). Perilla ketone: a potent lung toxin from the mint plant, *Perilla frutescens* Britton. *Science 197*:573-574.

Wilson, B. J., Garst, J. E., Linnabary, R. D., and Doster, A. R. (1978). Pulmonary Toxicity of naturally-occurring 3-substituted furans. In *Effects of Poisonous Plants on Livestock*, R. F. Keeler, K. R. Van Kampen, and L. F. James (Eds.). Academic Press, New York, pp. 311-323.

Winston, G. W., Carlson, J. R., and Hammond, A. C. (1980). Metabolism of 3-methylindole (3MI) by a crude goat lung microsomal preparation. *Fed. Proc. 39*:1751.

Wong, L. P. (1979). Factors affecting 3- methylindole production by *Lactobacillus* sp. from the rumen. M. S. thesis, Washington State University, Pullman.

Wynder, E. L., and Hoffman, D. (1967). *Tobacco and Tobacco Smoke*. Academic Press, New York, pp. 377-379.

Yang, N. Y. J., and Carlson, J. R. (1977). Release of marker lysosomal enzymes by 3-methylindole and indole from rabbit lung lavage cells. *Proc. Soc. Exp. Biol. Med. 154*:269-273.

Yokoyama, M. T., and Carlson, J. R. (1974). The dissimilation of tryptophan related indolic compounds by ruminal compounds by ruminal microorganisms, in vitro. *Appl. Microbiol. 27*:540-548.

Yokoyama, M. T., and Carlson, J. R. (1979a). Further studies on the production of skatole and para-cresol by a ruminal *Lactobacillus* sp. *XV Rumen Function Conf.*, Chicago.

Yokoyama, M. T., and Carlson, J. R. (1979b). Microbial metabolites of tryptophan in the intestinal tract with special reference to skatole. *Am. J. Clin. Nutr. 32*:173-178.

Yokoyama, M. T., and Carlson, J. R. (1981). Production of skatole and para-cresol by a rumen *Lactobacillus* sp. *Appl. Environ. Microbiol. 41*: 71-76.

Yokoyama, M. T., Carlson, J. R., and Dickinson, E. O. (1975). Ruminal and plasma concentrations of 3-methylindole associated with tryptophan-induced pulmonary edema and emphysema in cattle. *Am. J. Vet. Res. 36*:1349-1352.

Yokoyama, M. T., Carlson, J. R., and Holdeman, L. V. (1977). Isolation and characterization of a skatole producing *Lactobacillus* sp. from the bovine rumen. *Appl. Environ. Microbiol. 34*:837-842.

Yoshihara, I., and Maruta, K. (1977). Gas chromatographic microdetermination of indole and skatole in gastrointestinal contents of domestic animals. *Agric. Biol. Chem. 41*:2083-2085.

4

CYANOGENIC COMPOUNDS IN PLANTS AND THEIR TOXIC EFFECTS

JONATHAN E. POULTON

University of Iowa, Iowa City, Iowa

I. INTRODUCTION

Cyanogenesis, the ability of plants and other living organisms to release hydrogen cyanide (HCN) under certain circumstances, has been known for several centuries. In 1803, the German pharmacist Schrader (1803) crushed leaves of cherry laurel and other plants and detected the odor of hydrogen cyanide. A few years later, Bergemann (1812) reported that an oil, obtained from the bark of *Prunus padus* by distillation, killed a medium-size dog in 10 min, but, on the other hand, proved beneficial to gout sufferers when given in small doses! The phenomenon of cyanogenesis is now recognized in over 2050 species of higher plants, distributed throughout 110 different families. The list of species includes ferns, gymnosperms, and both monocotyledonous and dicotyledonous angiosperms. Among the families most noted for cyanogenesis are the Rosaceae (150 species), Leguminosae (125), Gramineae (100), Araceae (50), Compositae (50), Euphorbiaceae (50), and Passifloraceae (30) (Gibbs, 1974). Of special concern in this chapter is the recognition that many plants used as food sources by humans and domestic animals may also be markedly cyanogenic (Table 1). The aim of this chapter is to review the general nature of cyanogenesis in higher plants and to describe the toxic effects, both acute and chronic, that HCN may have upon animals.

Although it is clear that cyanogenesis is widespread within the plant kingdom, the source of the HCN remains unknown in many cases. Indeed, only in approximately 200 cases has the precursor been characterized. The release of HCN results from the hydrolysis of two classes of "secondary" plant products: the cyanogenic glycosides and the cyanolipids. Hydrolysis generally follows the disruption of cyanophoric tissue either by crushing, mastication, or fungal injury; endogenous hydrolases are thereby allowed access to these glycosides or cyanolipids, whose rapid catabolism to HCN is promoted.

II. CHEMICAL NATURE OF CYANOGENIC GLYCOSIDES AND THEIR OCCURRENCE

The 26 cyanogenic glycosides currently known to occur in plants are all β-glycosidic derivatives of α-hydroxynitriles. The great majority are derived by multistep biosynthetic sequences from the five common protein amino acids L-phenylalanine, L-tyrosine, L-leucine, L-isoleucine, and L-valine. The glycosides gynocardin, tetraphyllin A and B, and epitetraphyllin B are based on the cyclopentene ring structure and presumably originate from L-2-cyclopentene-1-glycine. Many excellent reviews concerning the isolation, chromatography, and characterization of these compounds are available (Seigler, 1977; Conn, 1980a; Nahrstedt, 1981). Detailed information describing the distribution of cyanogenic glycosides within the plant kingdom may be found in the invaluable chemotaxonomic treatises of Hegnauer (1963-1973) and Gibbs (1974) and regional lists compiled by Seigler (1976a,b). Readers interested in poisonous plants, including cyanogenic species, are directed to excellent books by Kingsbury (1964), Watt and Breyer-Brandwijk (1962), and Everist (1974).

Depending on the species, L-phenylalanine may serve as the starting point for the biosynthesis of several cyanogenic glycosides; these include the monosaccharides prunasin and sambunigrin and the disaccharides amygdalin,

TABLE 1 Yield of HCN Released from Food Plants

Plant	HCN yield (mg/100 g)
Bitter almond	
Seed	290
Young leaves	20
Wild cherry, leaves	90-360
Apricot, seed	60
Peach	
Seed	160
Leaves	125
Sorghum	
Mature seed	0
Etiolated shoot tips	240
Young green leaves	60
Bamboo	
Stem, unripe	300
Tops of unripe sprouts	800
Linen flax	
Seedling tops	910
Linseed cake	50
Bitter cassava	
Leaves	104
Bark of tuber	84
Inner part of tuber	33
Vicia sativa, seed	52
Lima bean, mature seed	
Puerto Rico, small black	400
Puerto rico, black	300
Java, colored	312
Burma, white	210
Jamaica, speckled white	17
Arizona, colored	17
American, white	10

Source: Modified from Montgomery (1965),
Baumeister et al. (1975), and Conn (1979).

vicianin, and lucumin (Fig. 1). Prunasin [the O-β-D-glucopyranoside of (R)-mandelonitrile] is widespread in many families, including the Myrtaceae, Polypodiaceae, Saxifragaceae, and Scrophulariaceae. It is also present at variable levels in leaves of numerous members of the domesticated stone fruits (Rosaceae) (Table 1). The poisoning of livestock after eating leaves of several species of wild cherry is ascribed to their prunasin content (Kingsbury, 1964). The presence of a chiral center in mandelonitrile provides the opportunity for a second epimeric cyanogenic glycoside having the (S)-configuration, namely sambunigrin, which occurs in the Caprifoliaceae, Oleaceae, and Leguminosae.

Undoubtedly the best known of all cyanogenic glycosides is amygdalin, which was the first cyanogen to be isolated and fully characterized (see Robinson, 1930, for a review). It differs from the majority of cyanogenic glycosides in being a disaccharide, in which β-gentiobiose is attached to the aglycone (R)-mandelonitrile. A general feature of cyanophoric plants is that the cyanogen(s) may not necessarily be distributed equally throughout all tissues and organs of the plant. A clear illustration is afforded by the fruits of the domesticated rosaceous species (e.g., cherry, apple, peach, apricot, and pear). The fleshy portion of these fruits, which have long been enjoyed by humans, is noncyanogenic; in contrast, the seeds enclosed within these fruits may be highly cyanogenic (Table 1), with amygdalin being the sole or major cyanogen. Ingestion of such seeds has often resulted in cases of accidental poisoning (Lewis, 1977), which will be discussed later with regard to the Laetrile controversy.

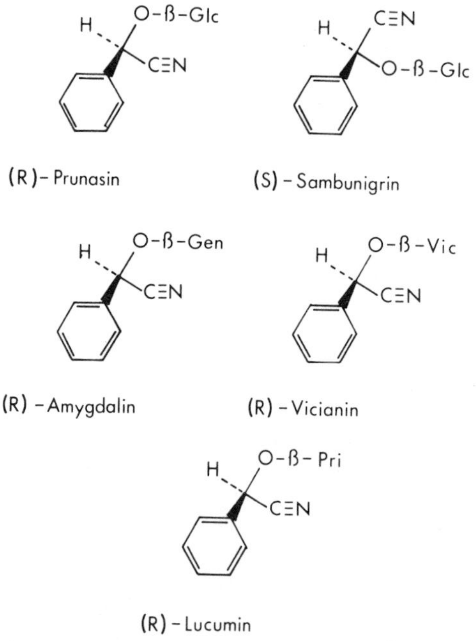

FIGURE 1 Structures of several cyanogenic glycosides derived from L-phenylalanine. The abbreviations Glc, Gen, Vic, and Pri stand for the sugars glucose, gentiobiose, vicianose, and primaverose, respectively.

It is fortunate that the dry starch-rich *Sorghum* seed, used as a food source by millions of humans in India and Africa and by cattle in North America, is noncyanogenic. By contrast, young sorghum seedlings may be highly cyanogenic (Table 1) and have frequently caused accidental poisoning of livestock (Gibb et al., 1974). Indeed, the poisoning of the transport animals of Kitchener's army in the Sudan by *Sorghum* species native to the Upper Nile Valley led to the identification of the cyanogenic glycoside dhurrin (4-hydroxy-S-mandelonitrile-β-D-glucopyranoside) as the toxic agent responsible (Dunstan and Henry, 1902) (Fig. 2). A dramatic synthesis of dhurrin from L-tyrosine occurs within the first week of germination in either light or darkness. Dhurrin may constitute up to 30% of the dry weight of the leaves and coleoptiles of 2-day-old dark-grown seedlings (Saunders and Conn, 1977). The biosynthetic pathway from L-tyrosine to this glycoside has been elucidated by Conn and co-workers (Conn, 1980a), and involves the unusual intermediates N-hydroxytyrosine, p-hydroxyphenylacetaldoxime, p-hydroxyphenylacetonitrile, and p-hydroxymandelonitrile (Fig. 3). McFarlane et al. (1975) demonstrated the conversion of L-tyrosine to p-hydroxymandelonitrile using a microsomal fraction isolated from dark-grown sorghum seedlings. This multienzyme system exhibited the phenomenon of metabolic channeling, in which negligible quantities of intermediates accumulated during product formation. Subsequent glucosylation by a soluble stereospecific UDPG:aldehyde cyanohydrin β-glucosyltransferase yielded dhurrin (Reay and Conn, 1974). Dhurrin and its 2-epimer taxiphyllin (4-hydroxy-(R)-mandelonitrile β-D-glucopyranoside) were once considered to have a limited distribution, but several recent reports have indicated their occurrence in many plant families (Seigler, 1980). Taxiphyllin, present in young bamboo shoots, has been responsible for several cases of human cyanide poisoning (Baggchi and Ganguli, 1943).

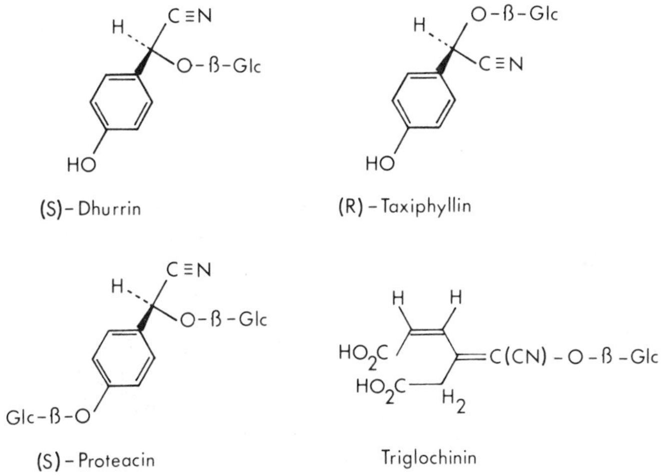

(S)-Dhurrin   (R)-Taxiphyllin

(S)-Proteacin   Triglochinin

FIGURE 2 Structures of cyanogenic glycosides derived from L-tyrosine. Proteacin is a diglucoside of p-hydroxy-(S)-mandelonitrile isolated from bitter nuts of *Macadamia ternifolia*.

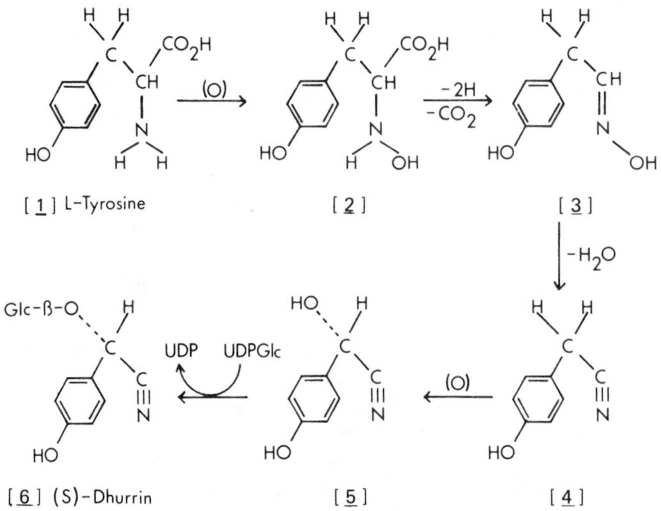

[**1**] L-Tyrosine   [**2**]   [**3**]

[**6**] (S)-Dhurrin   [**5**]   [**4**]

FIGURE 3 Biosynthesis of dhurrin [**6**] from L-tyrosine [**1**] in *Sorghum*,
involving the intermediates N-hydroxytyrosine [**2**], p-hydroxyphenylacetaldoxime [**3**], p-hydroxyphenylacetonitrile [**4**], and p-hydroxymandelonitrile [**5**].

A further cyanogenic glycoside receiving special attention recently is triglochinin, whose structure is O-[β-D-glucopyranosyl]-1-cyano-1-hydroxy-2-
methylcarboxy-4-carboxy-Δ-1,2(Z)-3,4(Z)-butadiene (Majak et al., 1980a). The
related derivatives isotriglochinin and triglochinin monomethylester may be
artifacts of the extraction procedure (Nahrstedt, 1981). Triglochinin is the
major cyanogen in arrowgrass (*Triglochin maritima* L.) and small arrowgrass
(*T. palustris*) and, in association with lesser quantities of taxiphyllin, is
responsible for the toxicity of arrowgrass to cattle and sheep (Muenscher,
1945; Kingsbury, 1964). Majak et al. (1980a), investigating the seasonal
variation in the cyanide potential of arrowgrass, noted that cyanogen levels
in leaves were substantially elevated during a period of severe moisture
deficit; this correlates with the higher toxicity of this plant to sheep during
extreme drought (Clawson and Moran, 1937). In view of its chemical structure and of the fact that its taxonomic distribution is linked to that of taxiphyllin and/or dhurrin, it seems likely that triglochinin is derived from L-
tyrosine, possibly via a 3,4-dihydroxyphenolic intermediate (Jaroszewski and
Ettlinger, 1981). The in vitro biosynthesis of taxiphyllin by extracts of *T.
maritima* seedlings was recently described by Hösel and Nahrstedt (1980).
Cutler et al. (1981) have extended these studies, demonstrating that microsomes from dark-grown arrowgrass seedlings catalyze the multistep conversion of tyrosine to p-hydroxymandelonitrile, the immediate precursor of
taxiphyllin. Several interesting kinetic differences are observed between the
Sorghum and *T. maritima* microsomal systems as regards cyanogenic glycoside
biosynthesis. In arrowgrass, N-hydroxytyrosine and p-hydroxyphenylacetaldoxime are apparently channeled, whereas in *Sorghum* the aldoxime is
freely exchangeable, and N-hydroxytyrosine and p-hydroxyphenylacetonitrile

are channeled. The possible relationship between the biosyntheses of taxiphyllin and triglochinin is a challenging problem that awaits elucidation. A thorough description of the biosyntheses of known cyanogenic glycosides is beyond the purpose of this review. Recent reviews by Conn (1980a,b) provide detailed information for the interested reader.

In addition to the foregoing glycosides of aromatic α-hydroxynitriles, significant concentrations of glycosides of aliphatic α-hydroxynitriles may also be present in foodstuffs. Linamarin (α-hydroxyisobutyronitrile-β-D-glucopyranoside) and its homolog lotaustralin [(R)-2(2-hydroxy-2-methylbutyronitrile)-β-D-glucopyranoside] (Fig. 4) are widely distributed among cyanophoric plants, with both compounds invariably being present within the same plant (Butler, 1965). This suggests that linamarin and lotaustralin share a common set of biosynthetic enzymes controlling their formation from the precursor amino acids L-valine and L-isoleucine, respectively. The observed tissue ratio of linamarin to lotaustralin varies with the species involved; in *Lotus arenarius* Brot., linamarin predominates (ratio 99:1), whereas the same ratio in *Lotus tenuis* L. is 1:24. (S)-Lotaustralin is not known to occur naturally. Studies on the ability of linseed to protect animals against the toxic effects of ingested selenium have revealed two new cyanogenic glycosides in linseed meal (Smith et al., 1980); these are the disaccharides linustatin [2-[(6-O-β-D-glucopyranosyl-β-D-glucopyranosyl)oxy]-2-methylpropane nitrile] and neolinustatin [(2R)-[(6-O-β-D-glucopyranosyl-β-D-glucopyranosyl)oxy]-2-methylbutanenitrile], corresponding to the monosaccharides linamarin and lotaustralin, respectively. Although flax seed possesses relatively little cyanogenic glycoside (∿0.1% dry weight), the young green *Linum usitatissimum* seedlings may contain 5% (dry weight) of linamarin and lotaustralin (ratio 55:45). Accordingly, germinating flax seedlings have provided a valuable system for investigating the biosynthesis of these two major aliphatic cyanogenic glycosides. The biosynthetic pathway

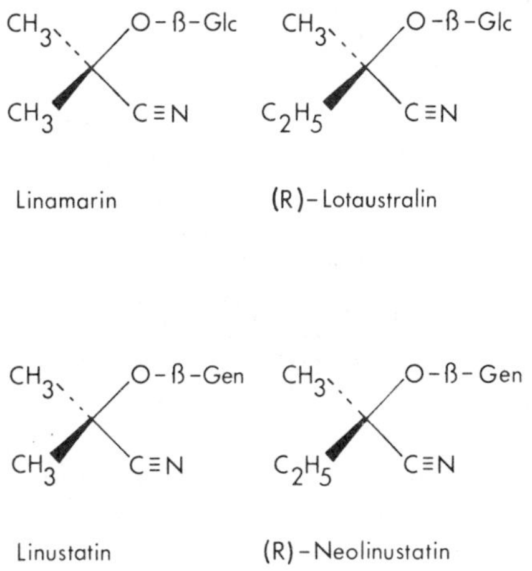

Linamarin   (R)-Lotaustralin

Linustatin   (R)-Neolinustatin

FIGURE 4 Structures of cyanogenic glycosides derived from L-valine and L-isoleucine.

from L-valine to linarmarin, which had been suggested by earlier isotopic tracer studies (Conn and Butler, 1969), has received considerable support using microsomal preparations from etiolated flax seedlings (Cutler and Conn, 1981). A specific glucosyltransferase, catalyzing the final step in this pathway, was purified 120-fold from flax seedlings (Hahlbrock and Conn, 1970). Linamarin and lotaustralin were formed from their corresponding α-hydroxynitriles with UDPG serving as glucose donor. This enzyme was inactive with aromatic α-hydroxynitriles such as mandelonitrile.

Cassava (*Manihot esculenta* Crantz) serves as the basic source of carbohydrates for approximately 300 million people in tropical areas of the world and alone meets 8-10% of the daily global human caloric needs. Cassava also finds increasing usage as a livestock feed and source of starch in the textile, paper, and food industries (Nestel, 1973). The cyanogen concentration in edible tubers varies in the range 30-150 mg of HCN per kilogram of fresh weight depending on the plant variety and climatic and cultural conditions (Coursey, 1973). During the processing of cassava for eating, these cyanogens may be removed or broken down to release HCN. That this process may not always be quantitative is indicated by reports of pathological conditions resulting from prolonged ingestion of cassava products presumably containing sublethal concentrations of HCN and cyanogenic glycosides (see Sec. VIII.B).

Linamarin and lotaustralin are also found in varieties of white clover (*Trifolium repens*). This plant is used extensively as an important cover crop in many parts of the world due to its ability to fix nitrogen and thereby increase the nitrogen content of the soil. In young leaves, cyanogen levels may reach 350 mg of HCN per 100 g of tissue, with lotaustralin predominating over linamarin. The genetics of cyanogenesis in *Trifolium* are discussed in Sec. V.

Viehoever (1940), in a comprehensive review of the edible and poisonous beans of the lima type (*Phaseolus lunatus* L.), discussed the origin, distribution, and toxicity of this important edible legume. All parts of the plant, including the developing seed pods, are cyanogenic, possessing predominantly linamarin. The cyanogenic potential of the mature bean varies with its color, shape, and size. The white lima beans of American origin ("butter beans") yield only low amounts (less than 0.01%) of HCN on analysis and may be considered desirable and safe for consumption. On the other hand, the seed of the small black wild lima bean native to most of Central America is highly toxic, containing up to 400 mg of HCN per 100 g of seeds, and has caused frequent human and livestock poisoning (Montgomery, 1965). Contrary to data of Viehoever (1940), small amounts of cyanogen are also detectable in the kidney bean *Phaseolus vulgaris* (2.0 mg of HCN per 100 g). Additionally, the following legume seeds are also cyanogenic: common pea (*Pisum sativum*, 2.3 mg per 100 g); black-eyed pea (*Vigna sinensis*, 2.1 mg per 100 g); chick pea (*Cicer arietinum*, 0.8 mg per 100 g); and common vetch (*Vicia sativa*, 52 mg per 100 g) (Montgomery, 1965). The toxicity of vetch is due not only to its content of the cyanogenic glycoside vicianin but also because it possesses the lathyrism factor β-cyanoalanine. Hydrogen cyanide released during the turnover of vicianin is converted to β-cyanoalanine by reaction with cysteine in the presence of β-cyanoalanine synthase (Rosenthal and Bell, 1979).

III. THE CATABOLISM OF CYANOGENIC GLYCOSIDES

Under normal physiological conditions, the tissues of a cyanogenic species contain no detectable free HCN, but HCN may be rapidly liberated from the cyanogenic glycoside when that plant tissue is crushed or otherwise disrupted. The mechanism by which cyanogenic glycosides are catabolized appears to be a general one (Fig. 5). Initially, the glycoside is hydrolyzed by a β-glycosidase, producing D-glucose and the respective α-hydroxynitrile. The latter may dissociate spontaneously, releasing HCN and the corresponding aldehyde or ketone. However, cyanogenic tissues generally possess a second enzyme, an hydroxynitrile lyase, which catalyzes dissociation of the cyanohydrin.

A substantial amount of literature exists describing the occurrence of β-glycosidases in plants (Nisazawa and Hashimoto, 1970; Hösel, 1981). These enzymes were often regarded by early workers as being rather unspecific, each capable of cleaving a large variety of glycosidic substrates. This viewpoint was somewhat encouraged by the common practice of testing the substrate specificity of relatively crude glycosidase preparations using model substrates such as p-nitrophenyl-β-D-glucopyranoside (PNPG), arbutin, and salicin, for ease of assay, rather than the natural substrates present within the tissue in question. Consequently, the literature remains regrettably inadequate as regards the action of β-glycosidases on the cyanogenic glycosides.

The substrate specificities of a few substantially purified β-glycosidase preparations have been investigated. Two β-glucosidases have been partially purified and characterized from *Sorghum* seed and vegetative tissues (Mao and Anderson, 1967). Glucosidase I, purified 10-fold from dried stems and leaves, hydrolyzed salicin and the chromogenic substrate PNPG but not dhurrin, the cyanogenic glycoside found in *Sorghum* shoots. Glucosidase II was purified

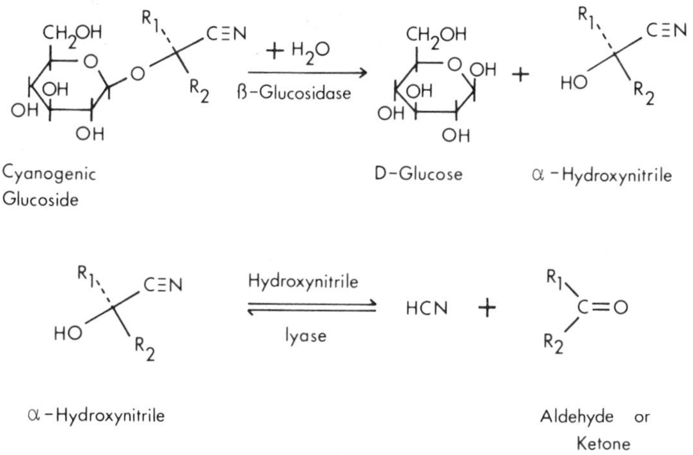

FIGURE 5 Enzymatic hydrolysis of cyanogenic glycosides. Initially, the cyanogenic glucoside is hydrolyzed by a β-glucosidase, releasing D-glucose and an α-hydroxynitrile. The latter compound may dissociate either enzymatically or nonenzymatically to yield HCN and the corresponding aldehyde or ketone.

85-fold from vegetative tissues and 4-fold from seeds but could not be obtained free of glucosidase I. By contrast, it hydrolyzed dhurrin (seed enzyme K_m = 0.31 mM; vegetative enzyme K_m = 0.55 mM), taxiphyllin, prulaurasin, and benzyl-β-glucoside.

β-Glycosidases active toward linamarin have been detected in several tissues that possess both linamarin and lotaustralin, although no general picture has emerged as to their substrate specificity. Butler et al. (1965) used column electrophoresis on Sephadex G-25 to obtain a linamarase preparation from linseed meal which was 105-fold purified with respect to the initial crude extract. Highest activity was observed with linamarin, but the enzyme could also hydrolyze arbutin and salicin as well as the aromatic cyanogenic glycosides prunasin, taxiphyllin, and dhurrin. The disaccharide amygdalin was not accepted as substrate. A 350-fold purification of linamarase was achieved by Cooke et al. (1978) from cassava parenchymal tissue. Enzyme activity was shown toward both linamarin (K_m = 1.45 mM) and PNPG (K_m = 0.46 mM), but its activity toward other substrates was unfortunately not investigated. Preliminary studies indicated that cassava peel possesses a greater linamarase concentration than parenchymal tissue but that the peel enzyme is more unstable.

More recent findings support the proposal that within every cyanophoric plant there exists a β-glycosidase specific for the particular cyanogen present in that plant. Hösel and Nahrstedt (1975) have purified a β-glucosidase(s) from *Alocasia macrorrhiza* which exhibits maximum catalytic activity toward triglochinin, the cyanogen of this species. Dhurrin and several chromogenic substrates could be hydrolyzed at far lower rates, but little or no detectable activity was shown toward seven other cyanogenic glycosides. It therefore showed remarkable specificity for triglochinin. Both taxiphyllin and triglochinin have been detected in seedlings of *Triglochin maritima*. Interestingly, two β-glucosidases have been characterized in these seedlings (Nahrstedt et al., 1979). One (glucosidase B) shows a pronounced specificity for triglochinin (V_{max}/K_m = 72); the corresponding V_{max}/K_m values for PNPG and dhurrin were 0.55 and 0.5. Activity toward taxiphyllin was too low for accurate determination. The second enzyme (glucosidase C) was active with taxiphyllin (V_{max}/K_m = 9) but also hydrolyzed PNPG, triglochinin, and dhurrin with V_{max}/K_m values of 31, 10, and 1.75, respectively.

The enzymic hydrolysis of cyanogenic disaccharides may be more complex than that of the monosaccharides since their α-hydroxynitriles could be released either by the successive cleavage of the two sugars (i.e., with a cyanogenic monosaccharide as intermediate) or by the elimination of both sugars as a disaccharide. This question has been addressed in depth only for amygdalin (Haisman and Knight, 1967) and vicianin (Bertrand and Weisweiler, 1910).

The kernels of *Prunus* species are a rich source not only of amygdalin but also of the complex enzyme system commonly known as "emulsin," which attacks a wide variety of β-glycosidic bonds. Since Wohler and Liebig (1837) first described the action of emulsin on amygdalin, many workers have attempted to determine the enzymic nature of emulsin and exactly how many enzymes are involved in amygdalin catabolism. Haisman and Knight (1967) reported kinetic data favoring the probability that two distinct β-glucosidases were involved in amygdalin hydrolysis: (1) an "amygdalin lyase," which produces prunasin and D-glucose, and (2) a "prunasin lyase," which cleaves prunasin to mandelonitrile and D-glucose. With the advent of modern protein purification techniques, several attempts were made to answer these fundamental questions, but for the most part, these efforts have failed to clarify the situation. Almond emulsin was found to contain three (Grover et al., 1977) or at least four (Helferich and Kleinschmidt, 1968) isoenzymes. Unfortunately, since synthetic substrates were used to assay β-glucosidase activities, the relationship of these proteins to the enzymes in almonds that hydrolyze amygdalin and prunasin remains unknown. A preliminary report by Haisman et al. (1967) represents the sole case in which almond emulsin β-glucosidases have been purified with reference to their capacity to hydrolyze cyanogenic glycosides. Evidence was given for the identification of amygdalin lyase, prunasin lyase, and hydroxynitrile lyase by continuous electrophoresis in a free buffer film. However, complete resolution of these enzymes was not achieved, thus preventing detailed study of their kinetic and molecular properties. In summary, it is generally accepted, but not yet rigorously proven, that the hydrolysis of amygdalin to mandelonitrile proceeds via the monosaccharide prunasin, requiring the cooperation of two β-glucosidases, each specific for one hydrolytic stage. That a similar mechanism might occur in mammalian tissues is supported by the partial purification of a β-glucosidase from feline kidney, which cleaves the terminal glucose molecule of amygdalin to yield prunasin (Freese et al., 1980); a distinct β-glycosidase appears responsible for prunasin hydrolysis. Amygdalin hydrolysis to prunasin is also effected by a β-glucosidase from yeast (Fischer, 1895). In contrast to the almond, yeast, and feline β-glycosidases, an enzyme from snail intestines cleaved amygdalin at the linkage between the cyanohydrin and sugar moieties, releasing β-gentiobiose (Giaja, 1919). Similarly, Bertrand and Weisweiller (1910) reported that a crude enzyme preparation from seeds of *Vicia angustifolia* cleaved vicianin at the aglycone-vicianose bond, yielding vicianose [6-(α-L-arabinosido-D-glucose)] and (R)-mandelonitrile.

It is an interesting feature of cyanogenic members of the Prunoideae that, in general, prunasin is the cyanogen within the leaves, whereas the seeds possess amygdalin (Karrer, 1958; Hegnauer, 1963-1973). This situation allows one the unique opportunity to test whether the β-glycosidase(s) in different organs within one plant is (are) specific for the cyanogen(s) within those tissues. Preliminary studies in this laboratory have indicated that this may be the case. Leaf extracts of almond (*Prunus amygdalus*), apricot (*Prunus armeniaca*) and of a *Prunus lyonii* X *Prunus ilicifolia* hybrid displayed high β-glucosidase activity toward prunasin but were virtually inactive toward amygdalin. By contrast, extracts from pits of almond and apricot showed high β-glucosidase activity toward both amygdalin and prunasin. Similar data were reported by Armstrong et al (1912) with cherry laurel leaves and by Heuser (1972), who showed that an enzyme preparation from stems of the hybrid plum "Marianna" (*Prunus cerasifera* X *P. munsoniana*) hydrolyzed prunasin but not amygdalin.

Hydroxynitrile lyase (mandelonitrile benzaldehyde lyase, E.C.4.1.2.10) was first observed in almond emulsin by Rosenthaler (1908) and has now been extensively purified from several members of the Rosaceae (Gerstner and Pfeil, 1972, 1975). This enzyme catalyzes the reaction between HCN and a great number of aliphatic, aromatic, and heterocyclic aldehydes, yielding optically active α-hydroxynitriles. In almonds, three isozymes (in the ratio 7:53:40) are recognizable by slight differences in behavior during ion-exchange and gel filtration chromatography and isoelectric focusing (Aschhoff and Pfeil, 1970). Recent investigations (Jorns, 1979) have centered around the curious finding that these isozymes possess bound flavine-adenine dinucleotide (FAD), despite not catalyzing an oxidation-reduction reaction. In contrast, the purified hydroxynitrile lyases from etiolated *Sorghum* seedlings and cassava (E.E. Conn, personal communication, 1981) lack the flavin prosthetic group; other notable differences in physical and kinetic properties of the almond and sorghum lyases have been reported (Seely et al., 1966). It will be important to investigate oxynitrilases from other species having cyanogenic glycosides derived from nonaromatic amino acids. Stevens and Strobel (1968) described a fungal hydroxynitrile lyase which catalyzes the reversible dissociation of the cyanohydrins of acetone and methyl ethyl ketone, but which is inactive with aromatic cyanohydrins.

The hydrolysis of a cyanogenic glycoside occurs at a significant rate only after the tissue of the cyanogenic plant has been macerated and the glycoside is thereby brought into contact with its catabolic enzymes. This strongly suggests that these two components of the hydrolytic system exist in different compartments within the intact plant. Several possibilities regarding compartmentation under normal physiological conditions may be envisaged. First, the glycoside and its degradative enzymes may be sequestered in different tissues. Alternatively, the glycoside and the β-glycosidase may exist within the same cell, but in separate intracellular compartments. Finally, all components could exist even within the same intracellular compartment (e.g., the vacuole) if the activity of the β-glycosidase could be controlled by endogenous inhibitors.

Kojima et al. (1979) have investigated these alternatives using epidermal and mesophyll protoplasts and bundle-sheath strands isolated by cellulase digestion from light-grown *Sorghum* leaves. The glycoside dhurrin was located entirely in the epidermal layers of the leaf blade; its vacuolar location was demonstrated by Saunders and Conn (1977). By contrast, the two enzymes responsible for dhurrin catabolism, dhurrin β-glucosidase and hydroxynitrile lyase, resided almost exclusively in the mesophyll tissue. It was concluded that the separation of dhurrin and its catabolic enzymes in different tissues prevents its large-scale hydrolysis under normal physiological conditions, but cyanogenesis would be expected to proceed rapidly when the contents of the ruptured epidermal and mesophyll cells are allowed to mix. Whether this mode of compartmentation applies to other cyanogenic species remains to be clarified.

IV. CHEMICAL NATURE AND CATABOLISM OF CYANOLIPIDS

An alternative source of HCN in plants are the cyanolipids, which appear restricted to the seed oils of the family Sapindaceae. The distribution, biosynthesis, and chemistry of these unusual compounds have been excellently reviewed by Mikolajczak (1977). All four known cyanolipids possess the same branched, five-carbon nitrile skeleton, derived presumably from leucine (Fig. 6), but variation exists in the position of the double bond and the number and location of hydroxyl groups. Only two of these ([7] and [8]) may release

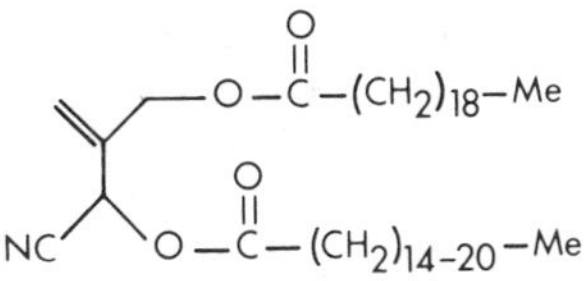

FIGURE 6 Structures of cyanolipids found in seed oils of the Sapindaceae.

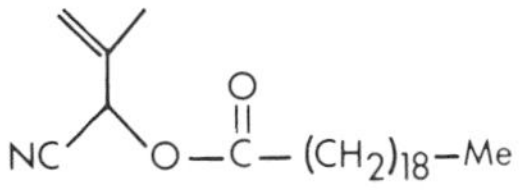

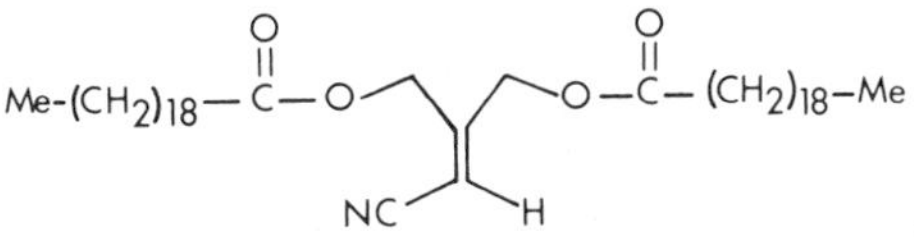

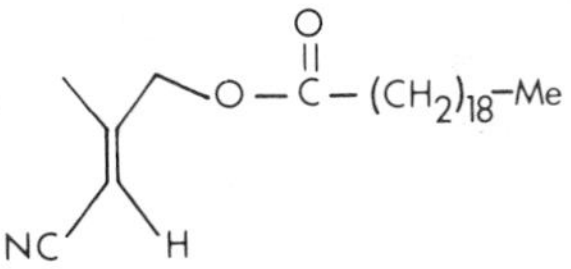

HCN by enzymic hydrolysis; endogenous lipases remove the acyl groups, forming an unstable cyanohydrin, which subsequently decomposes to HCN and a carbonyl compound. The lipase substrate specificity and the subcellular localizations of both lipases and cyanolipids are currently unknown. Cyanolipids may occur in copious amounts in species of economic interest. For example, kusum oil, the seed oil of *Schleichera trijuga*, may be quite toxic if untreated, but after processing it finds usage as an edible and medicinal oil, a hair dressing, and as a raw material for soap production. Extensive studies of the toxic properties of cyanolipids have not been conducted (Seigler, 1979).

V. FACTORS AFFECTING CYANOGEN LEVELS IN PLANTS

Knowledge of the tissue distribution of cyanogens and of factors affecting the levels of cyanogenic glycosides in plants is critical in reducing or eliminating the potential toxicity of cyanogenic plants used as foodstuffs. It is difficult to generalize regarding the tissue distribution of cyanogenic glycosides within cyanophoric plants. Leaves and seeds frequently contain the highest concentration but, as exemplified by Table 1, these compounds have been found in probably all plant organs. In general, young shoots are more dangerous than older tissue (e.g., sorghum, wild cherries, white clover). In a given plant, significant amounts of cyanogen may be present in one or more organs. However, it should be borne in mind that the tissues where cyanogenic glycosides accumulate may not necessarily be the site of synthesis since translocation of glycosides may be possible, as suggested by preliminary evidence of De Bruijn (1973) and Clegg et al. (1979).

The amount of glycoside present at a particular time may be influenced not only by the age of the plant but also by a variety of factors, including diurnal, seasonal, environmental, nutritional, and genetic factors. The existence of diurnal variation in HCN levels has been demonstrated in white clover (De Waal, 1942). Prunasin levels in the California toyon (*Heteromeles arbutifolia*) vary with the season, being higher in the spring and late autumn

(Dement and Mooney, 1974). Frosts may increase cyanogenesis in sorghum. Cyanogen levels were increased by temperatures above -2°C, which did not kill the plant (Wattenbarger et al., 1968), but lethal temperatures of -5°C and below caused a decrease in cyanogenesis, partly by loss of HCN into the atmosphere. The effect of moisture stress is less clear. While some authors reported a small increase in cyanogenesis by sorghum in water-deficient plants (Heinrichs and Anderson, 1947; Nelson, 1953), Boyd et al. (1938) maintained that drought did not increase the cyanide content but that the plants remained small and in the high-cyanide stage. Serious drought increased glycoside content in cassava (De Bruijn, 1973). The effect of fertilizers on cyanogen levels has been investigated by several groups (Oke, 1969). Physiological and genetic factors affecting cyanogen content in cassava have been reviewed (Butler et al., 1973; De Bruijn, 1973).

The polymorphic nature of cyanogenesis, indicated by the existence of both cyanogenic and noncyanogenic plants in a population of a single species, has been most extensively studied in *Trifolium repens* and *Lotus corniculatus* (Corkill, 1942; Nass, 1972; Hughes, 1973). In *T. repens*, cyanogenesis is controlled by two independently inherited genes, whose relationship is shown in Fig. 7. Alleles of the gene Ac determine the biosynthesis of the cyanogenic glycosides linamarin and lotaustralin from their respective amino acid precursors valine and isoleucine. The presence or absence of the enzyme linamarase is governed by alleles of the Li gene. Only plants possessing at least one dominant functional allele of both genes liberate HCN when damaged. Of the four homozygous genotypes AcAcLiLi, AcAclili, acacLiLi, and acaclili obtained by selective breeding, solely the first genotype is cyanogenic, since it possesses both the glycoside and β-glycosidase. The second genotype contains the cyanogenic glycoside but fails to release HCN upon tissue disruption, because it lacks the β-glycosidase. It should be realized, however, that even these individuals may be potentially toxic if eaten with other food materials possessing β-glycosidases capable of hydrolyzing linamarin or lotaustralin. Individuals heterozygous for Ac or Li have intermediate levels of cyanogenic glucosides or linamarase, respectively, compared with homozygous individuals. Comparison of the linamarin biosynthetic pathway between *Trifolium* plants producing the cyanogenic glycosides and plants which do not has suggested that the Ac locus may control more than one step in this biosynthetic pathway (Hughes and Conn, 1976). The genetic significance of this unexpected finding remains unclear.

Fikenscher and Hegnauer (1977) reported cyanogenic polymorphism in *Achillea macrophylla*, *A. millefolium*, *Centaurea scabiosa*, *Taxus baccata*, *Sedum* spp., *Sorbus aucuparia*, *Campanula cochlearifolia*, and *C. rotundifolia*. *Prunus amygdalus*, *Manihot esculenta*, *Sorghum bicolor*, and *Macadamia ternifolia* are also undoubtedly polymorphic, since one can find individual

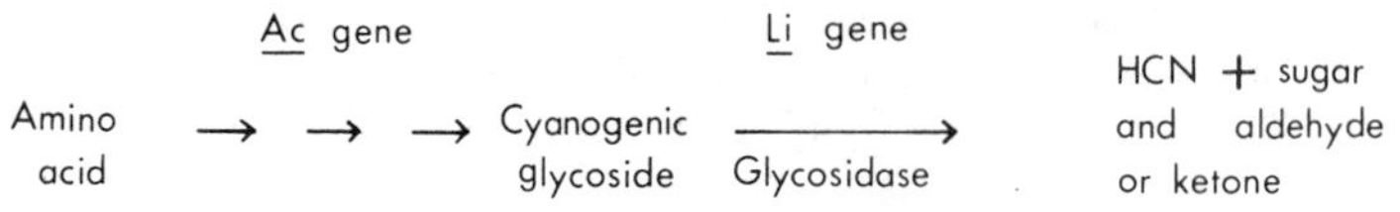

FIGURE 7 Genetics of cyanogenesis in *Trifolium repens*.

plants within these species that are either strongly cyanogenic or only weakly so (Jones, 1972).

Selective breeding over decades has enabled plant breeders to obtain low-cyanogen varieties of several important food plants, including almonds (*Prunus amygdalus*), cassava, lima beans, and sorghum. Amygdalin is apparently responsible both for the bitterness and toxicity of seeds of several rosaceous species, such as bitter almonds (100 μmol/g) and apricots (20-80 μmol/g). Selective breeding has produced the sweet almond, whose kernel is edible, sweet, and of commercial value (Heppner, 1926). Its amygdalin concentration is far lower than in bitter almonds but still detectable by qualitative tests for HCN. Bitter almonds are grown only as a source of the bitter almond oil used as a flavoring agent. Tissue of all cassava cultivars so far examined contains cyanogenic glycosides, usually in the range 30-150 mg of HCN per kilogram of tissue. Both "sweet" and "bitter" cassava cultivars are recognized; in general, bitter cassava has a high cyanide content while sweet cassava tends to have lower values, but a great deal of overlapping exists between these classes (Sinha and Nair, 1968). Factors contributing to the sweetness or bitterness of cassava, other than the cyanogen content, require further examination (Coursey, 1973). It should be noted that the bitter variety may sometimes be preferred because of the generally bland, starchy taste of the sweet form (Conn, 1979). Since the grass species *Sorghum sudanense* (sudangrass) and *S. bicolor* and their hybrid sorghum-sudangrass find usage in many countries as animal feed, continuing efforts are being made by plant breeders to produce low-HCN strains. The genetics of cyanogenesis in *Sorghum* appear more complex than exist in *T. repens*, and authors differ as to the number of genes involved and the dominance relationship of these genes (Snyder, 1950; Barnett and Caviness, 1968; Nass, 1972).

VI. REDUCTION OF POTENTIAL TOXICITY OF CYANOGENIC PLANT MATERIAL BY FOOD PROCESSING

The potential toxicity of cyanogenic plants used as foodstuffs by humans and domestic animals may be significantly reduced during food processing. Action is thereby taken by appropriate means to reduce the HCN potential of the plant material either by removal of the cyanogenic glycosides, by inactivation of the β-glycosidases, or by a combination of both goals. Several effective techniques are discussed here with special reference to cassava, which provides a high-carbohydrate, low-protein staple food in many tropical countries. Similar arguments apply to the detoxication of other cyanogenic plants, such as lima beans.

As described earlier, cyanogen levels within a particular species may be affected by genetic and environmental factors. Where climatic and soil conditions permit their cultivation, low-HCN varieties are obviously desirable. Knowledge of the tissue distribution of the cyanogenic glycosides in a particular species is an essential aid to predicting the potential toxicity of different organs. In the case of cassava, both the tuber and leaves are cyanogenic due to their content of linamarin and lotaustralin (Table 1) but they may be eaten following appropriate processing. As in several other cyanogenic plants, the glucoside concentration in cassava leaves decreases with age

(De Bruijn, 1973). The tuber is considered more palatable. In most varieties, the concentration of cyanogenic glycoside is appreciably higher in the tuber bark than in the flesh, the ratio being about 5 or 10:1. Usually, the highest cyanogen concentration is found at the proximal end of the root. In a horizontal section, there is an increase in glucoside concentration from the center outwards. No evidence for a direct relationship between the concentration of linamarase in the plant and the level of glucoside was revealed by studies probing the distribution of this enzyme in different parts of four cassava clones (De Bruijn, 1973). On average, the bark contained 30 times more linamarase than the inner part of the tuberous roots.

At least in theory, the release of HCN from cyanophoric plants could be prevented by inactivation or removal of the endogenous β-glycosidases, which catabolize the cyanogenic glycosides upon tissue disruption. This supposition may be false, however, since, if treatment of the food does not simultaneously remove or destroy the glycosides, toxicity may still result from the catabolism of these cyanogens in the gastrointestinal tract either by microfloral β-glycosidases or by β-glycosidases contributed by other components of the diet. Similar arguments would apply following ingestion of plants which, due to their genetic character, possess cyanogen but lack the β-glycosidase. Seddon and King (1930), noting these hazards, performed model experiments using sweet almonds to release HCN from foliage of *Acacia glaucescens* and *Eremophila maculata*, which contained a cyanogenic glycoside but no β-glucosidase. One should also be aware of the ability of some noncyanogenic plants to hydrolyze cyanogenic glycosides. Extracts derived from common components of a typical salad, namely lettuce, celery, and mushrooms, catalyzed amygdalin hydrolysis (Conn, 1979), so clearly this calls for some caution in consuming these plants and a source of cyanogenic glycosides simultaneously.

Few effective means are available for removal of cyanogenic glycosides from food material. In general, these glycosides are heat stable. Thus cooking cassava or other cyanophoric plants by boiling, roasting, or frying before consumption would tend to inactivate the endogenous β-glycosidases but would have little or no effect on the cyanogen itself. In cassava roots, linamarase but not the glucoside linamarin was destroyed by cooking at 72°C (Joachim and Pandittesekere, 1944). DeBruijn (1973) reported that during cooking the glycoside could not get out of the starchy product and up to 90% of the original content remained. This thermostability of linamarin may explain the cases of people poisoned by eating colored lima beans after boiling and draining them (Viehoever, 1940). Taxiphyllin, being unusually thermolabile (Schwarzmaier, 1976), constitutes an exception. This glycoside is present in young bamboo shoots but should be destroyed by boiling the shoots in water for 35-40 min; boiling should also drive off the liberated HCN, if the cooking pot remains uncovered.

Alternative food processing techniques which have been successful in reducing the acute toxicity of cyanogenic plants are exemplified by the wide variety of traditional ways devised to detoxicate the more poisonous varieties of cassava. These procedures are designed to allow cyanogens to come into contact with endogenous β-glycosidases, causing their hydrolysis. The liberated HCN may be eliminated by solution in water or by volatilization. Coursey (1973) describes how cassava tubers are grated or ground to initiate release of HCN, which may be subsequently removed by soaking in running or static water or by drying in hot air. Alternatively, linamarin breakdown may be assisted by fermentation processes (Akinrele, 1964). The interested reader is directed to several detailed articles covering the production of cassava-based food products in various parts of the world (Oke, 1968; Normanha, 1965; Maner, 1972).

The amount of analytical data available on the efficacy of these processes is limited and generally unreliable (see Coursey, 1973, for a review). There are conflicting reports as to efficiency of drying in detoxication. Several authors state that HCN levels may be greatly reduced by simple boiling of the roots. For example, a variety originally containing 332 mg of HCN/kg possessed only 10 mg of HCN/kg after boiling (Raymond et al., 1941). The success of this treatment may be more attributable to the leaching of cyanogens rather than to their thermal degradation, since Joachim and Pandittesekere (1944), during investigation of the effect of boiling on the cyanogen content of several cassava varieties, noted that those varieties losing least toxicity did not become soft and floury upon boiling.

One must conclude that cassava food products prepared by these traditional detoxication methods may still contain appreciable traces of HCN and unhydrolyzed cyanogenic glycosides. Oke (1968) reported the following HCN levels (in mg of HCN/kg tissue) in samples of several African cassava-based food products: fresh cassava, 380; gari, 19; fufu, 25; lafun, 10; and kpokpo-gari, 11. Kokonte, a flour prepared from sun-dried cassava chips, contains 20 mg of HCN/kg (Wood, 1965). In some areas of southern Nigeria, where ataxic neuropathy exists in almost epidemic proportions, purupuru, which contains 0.1 μmol of cyanide per gram of tissue, is commonly eaten. Assuming the consumption of 3 kg of purupuru per day, the total cyanide intake per day would be about 50 mg, a figure close to the lethal dose of cyanide for humans (60 mg). Similarly, the linamarin content of gari, together with its large intake (0.8 kg/day) by some populations, means that such individuals are potentially exposed to approximately 20 mg of HCN daily. The incidence of several neurological conditions exhibited by cultures where cassava is intensely cultivated and consumed is discussed later (see Sec. IX).

VII. TOXICOLOGY OF CYANOGENIC GLYCOSIDES IN ANIMALS, INCLUDING HUMANS

The literature contains numerous examples of acute poisoning of animals believed caused by ingestion of cyanogenic plants. There are reports of human poisoning following ingestion of amygdalin-containing seeds of several rosaceous species (see Lewis, 1977, for a review), notably bitter almonds (Pack et al., 1972), apricot kernels (Gunders et al., 1969; Sayre and Kaymakcalan, 1964), and choke cherry seeds (Pijoan, 1942). The leaves of these species may also be toxic due to their prunasin content. Cattle fed saskatoon service-berry (*Amelanchier alnifolia*), an important if not preferred browse for range-land livestock and wildlife, showed symptoms of cyanide poisoning attributed to prunasin breakdown (Majak et al., 1980b). Accidental poisoning of children after drinking "tea" made from peach leaves is documented (Siegler, 1977). Among the grasses (Gramineae), young sorghum (Gibb et al., 1974) and arrowgrass (Muenscher, 1945; Moran, 1954; Beath, 1939) have been cited for much loss of livestock in the United States. Young bamboo shoots, a delicacy in many countries, may also cause cyanide poisoning (Baggchi and Ganguli, 1943). Cyanophoric species of acacia have been blamed for the death of sheep and cattle in Australia (Hurst, 1942), South Africa (Steyn and Rimington, 1935), and the United States (Kingsbury, 1964).

Whether the unhydrolyzed cyanogenic glycosides themselves are toxic to animals remains a difficult question to answer. It appears more likely that the aforementioned poisonings are instead caused by the HCN released upon catabolism of these glycosides. This belief is undoubtedly supported by the clinical symptoms of acute cyanide poisoning which may follow ingestion of cyanogenic glycosides. Consequently, the potential toxicities of the cyanogenic glycosides per se and of the noncarbohydrates produced by their catabolism, notably the α-hydroxynitriles, aldehydes, and ketones corresponding to these glycosides, have been largely ignored. While aliphatic ketones should be metabolized without difficulty, the formation of aromatic aldehydes may well be harmful and their detoxication may require increased energy expenditure by the animal.

Several factors will determine whether an animal will actually suffer cyanide poisoning following the ingestion of plant material which contains potentially toxic levels of cyanogenic glycosides (Kingsbury, 1964). These include the nature and size of the animal, the level of β-glycosidases in the plant, the length of time between tissue disruption and ingestion, the presence and nature of other components of the meal, and the rate of detoxication of cyanide by the animal. For acute toxicity, enough plant material must be ingested in a sufficiently short time period such as to accumulate lethal cyanide levels. The minimum lethal dose of HCN in sheep is 2.4 mg/kg (Coop and Blakley, 1949). However, sheep could well tolerate 15-20 mg of HCN/kg body weight per day in the normal grazing situation when the ingestion of forage is relatively slow.

In cases where acute cyanide poisoning occurs rapidly, one can probably safely assume that the plant material has supplied both the glycoside and active endogenous β-glycosidases capable of hydrolyzing it. Alternative mechanisms for cyanide release from cyanogenic glycosides have been considered. For example, it has been suggested that acid hydrolysis of cyanogenic glycosides might occur in the stomach of humans and of other monogastric organisms, where the pH of the contents may be approximately 2. This appears improbable since the glycosides are stable to such acid conditions. In confirmation of earlier studies (Caldwell and Cortauld, 1907), Dunn and Conn demonstrated that neither linamarin nor amygdalin was hydrolyzed by incubation at 37°C for 18 hr in 0.1 N H_2SO_4 (Conn, 1979). Furthermore, direct incubation of amygdalin with human gastric juices possessing a pH range of 2.2-4.1 failed to release HCN; some amygdalin hydrolysis occurred when the juice acidity exceeded pH 5.0, a level at which colonization of the stomach by bacteria has been documented (Newton et al., 1981). Plant β-glycosidases exhibit highest activity in the pH range around pH 5-6. The acidity of the monogastric stomach ($\sim$pH 2) would therefore be expected to cause inactivation or at least inhibition of ingested β-glycosidases. A significant point, which has received little attention, is whether this inactivation/inhibition is reversible. If reversibility were possible, the breakdown of cyanogens to HCN by plant β-glycosidases could resume when the stomach contents pass to the duodenum and are neutralized by the bile juices. It should be noted that the inactivation of plant β-glycosidases might be incomplete in hypochlorhydria. In contrast to the monogastric stomach, the rumen of ruminants maintains a pH near neutrality, which would favor the continuing hydrolysis of cyanogenic glycosides by ingested β-glycosidases. This may explain, at least partially, the observation that ruminants are more susceptible to poisoning by cyanogenic plants than are nonruminants (Moran, 1954).

The ability of mammalian digestive enzymes to hydrolyze cyanogenic glycosides has also be questioned, but this too seems unlikely since, being α-glycosidases, they would in theory be incapable of cleaving the β-glycosidic linkage in the known cyanogenic glycosides. Whether purified intestinal carbohydrases can indeed hydrolyze cyanogenic glycosides has not yet been reported.

Cycasin (methylazoxymethanol-β-D-glucoside), an unrelated β-glucoside of plant origin, depends on the presence of the gastrointestinal flora for its metabolism and toxicity (Spatz et al., 1967). Thus it seems feasible that the release of cyanide from cyanogenic glycosides in vivo might also be effected by gut microfloral enzymes. Barrett et al. (1977) demonstrated that when a nonlethal dose of linamarin was administered orally to rats, a portion (19%) was absorbed by the animal and excreted unchanged within the urine in the first 24 hr. An additional 24% was excreted within 72 hr as urinary thiocyanate, but linamarin was absent from the feces. Although thiocyanate excretion constitutes evidence for linamarin hydrolysis within the rat, these experiments cannot distinguish whether this catabolism occurred within the body of the rat following linamarin absorption or instead was effected by intestinal microflora with subsequent cyanide detoxication by rat cells.

Early literature on the hydrolysis of cyanogenic glycosides by bacteria seems confusing (Winkler, 1951), but the rumen flora are apparently capable of this hydrolysis (Coop and Blakley, 1949; Smith, 1971). The participation of intestinal microflora was further suggested by the studies of Newton et al. (1981), who fed amygdalin (250 mg/kg) orally to control rats by stomach tube. Urinary thiocyanate levels rose immediately following this administration, indicating rather extensive amygdalin hydrolysis. When amygdalin was administered to experimental rats that had been treated with neomycin to reduce the bacterial population of the digestive tract, much less thiocyanate was excreted. Similar data were obtained by Carter et al. (1980), who compared the toxicity and metabolism of amygdalin when administered orally to germ-free and conventional rats. Conventional rats given single oral doses of amygdalin (600 mg/kg) showed symptoms characteristic of acute cyanide poisoning and they usually died within 2-5 hr. High cyanide concentrations (2.6-4.5 µg/ml) were detected in their blood. By contrast, identical doses administered to germ-free rats produced no visible signs of toxicity; moreover, their blood cyanide levels were less than 0.4 µg/ml, a level displayed by control rats that had not received amygdalin. The authors concluded that the gastrointestinal flora are obligatory for reactions leading to the release of toxic amounts of cyanide from amygdalin. As additional support for this conclusion, when amygdalin was administered orally at a nontoxic level, it could be recovered in the feces of germ-free rats but not of conventional rats. Moreover, the stomach and cecal contents of conventional rats, but not of germ-free rats, catabolized amygdalin to benzaldehyde and presumably cyanide under aerobic conditions. It is puzzling that amygdalin hydrolysis did not proceed under anaerobic conditions, since this is the physiological environment under which most reactions attributed to the flora occur (Goldman, 1978).

The participation of intestinal microflora in amygdalin catabolism in humans may be inferred from distinct differences observed following oral and intravenous amygdalin administration (Moertel et al., 1981). In contrast to data from intravenous administration, human patients treated with oral amygdalin (0.5 g thrice daily) displayed extremely low peak plasma amygdalin levels (less than 1 μg/ml), and far less unmetabolized amygdalin was excreted in the urine. Blood thiocyanate levels also showed delayed but significant increases in these patients.

The nature of the gut microflora capable of hydrolyzing cyanogenic glycosides is currently under investigation. Ten strains of two bacterial genera isolated from mouse intestine, *Enterobacteria* and *Enterococci*, released HCN from amygdalin when incubated together in vitro (Smith, 1971). Newton et al. (1981) demonstrated that *Bacteroides fragilis*, a major constituent of human stool, efficiently hydrolyzed amygdalin in vitro. Several other human fecal bacteria, including *Enterobacter aerogenes, Streptococcus fecalis,* and *Clostridium perfringens*, but not *Escherichia coli*, were also active. When amygdalin was inclubated with approximately 1 g of human feces, 53% of the available HCN was released by hydrolysis. Taken together, these data probably explain why a Laetrile (amygdalin) enema was nearly fatal to one patient (Ortega and Creek, 1978). It should be noted that hydrolysis of amygdalin forms the basis of a biochemical diagnostic test for the identification of several strains of gastrointestinal bacteria (Holdeman et al., 1977).

Only recently has the toxicity of cyanogenic glycosides following parenteral administration been considered in depth; this route of administration presumably avoids the participation of intestinal microflora. Smith (1971) showed that far larger cyanogen doses may be administered parenterally than orally; mice were resistant to high doses of amygdalin (up to 5 g/kg) when administered intraperitoneally, but were killed by relatively low oral doses (LD_{50} value of 350 mg/kg). Confirming earlier reports (Ames et al., 1978), Moertel et al. (1981) demonstrated that amygdalin given intravenously (4.5 g/m^2 per day) to human patients with advanced cancer was largely excreted unchanged in the urine. High peak plasma levels of amygdalin (up to 1.16 mg/ml) were recorded, but whole-blood cyanide levels remained essentially undetectable. That some amygdalin breakdown with cyanide release had indeed occurred was suggested by a slight rise in plasma thiocyanate level over a period of several days in two out of three patients investigated. However, the rate of amygdalin hydrolysis must have been small enough for the resulting HCN to be detoxified, since no detectable side effects were observed over the 21-day treatment period.

It should be emphasized, however, that others have observed more significant hydrolysis of amygdalin following parenteral administration, leading in some cases to acute cyanide poisoning and death. Newton et al. (1981) reported a pronounced rise in 24-hr urine output of thiocyanate following intravenous doses (250 mg/kg) of amygdalin to rats; approximately 6% of the available cyanide from amygdalin was released. Khandekar and Edelman (1979) reported that amygdalin given intraperitoneally to rats for 5 days in doses of 250, 500, and 750 mg/kg caused mortalities of 30.8%, 44.1% and

56.8%, respectively. Mortality was not a result of cumulative toxicity. The acute mode of death in these animals, together with the demonstration of increased blood cyanide levels, indicated that cyanide poisoning was the most probable cause of death. It remains unknown whether a β-glycosidase capable of amygdalin hydrolysis exists in the rat peritoneal cavity.

Whereas the digestive juices of mammals contain only α-glycosidases, β-glycosidases are widely distributed within mammalian tissues (Cohen et al., 1952; Neufeld et al., 1975). Unfortunately, little is known about their capacity to hydrolyze cyanogenic glycosides. A partially purified steroid β-D-glucosidase preparation from rabbit liver could also hydrolyze prunasin, amygdalin, dhurrin, taxiphyllin, and linamarin (Ng, 1975). However, the K_m values for these cyanogens were large and their rates of hydrolysis were meager compared with values for the steroidal substrate. Recently, Freese et al. (1980) have detected a β-glucosidase in cat, rat, and rabbit (but not in mouse or human) kidney tissue, which catalyzes the hydrolytic cleavage of the terminal glucose residue of amygdalin yielding prunasin. Although primarily confined to renal tissue, lesser amounts are detectable in intestinal mucosa preparations. Following 61-fold purification from feline kidney, the enzyme displayed optimum activity at pH 7.5 and had a K_m value for amygdalin of 3.8 mM. The natural substrate for this enzyme remains unknown. Preliminary evidence indicated that two further enzymes with different tissue distributions catalyze the hydrolysis of prunasin and glucocerebroside, respectively.

In conclusion, it appears that routes of administration which provide the most direct contact of the cyanogenic glycoside with the gastrointestinal flora maximize the possibility of cyanide release and subsequent toxicity. This danger is increased dramatically if sources of β-glucosidase are simultaneously ingested with the cyanogenic glycoside (Schmidt et al., 1978; Moertel et al., 1981).

VIII. TOXICITY OF CYANIDE

A. Acute Cyanide Toxicity

The occurrence of intoxication symptoms depends on the velocity of increase of cyanide in tissues. This decisive factor is determined by the kind of the cyanide compound administered, by the intoxication mode, by the ingested dose, and by the ability of the organism to detoxicate cyanide. HCN may enter the organism via the lungs, gastrointestinal tract, and the skin. In humans, the minimum lethal dose (MLD) of HCN taken orally is approximately 0.5-3.5 mg/kg body weight (Chen et al., 1934; Gettler and Baine, 1938; Halstrom and Moller, 1945). The MLD of alkaline cyanides is approximately twice that of HCN (Montgomery, 1969). Christensen (1976) cites the following values for oral lethal doses of HCN: cat, 2.0 mg/kg; mouse, 3.7 mg/kg; dog, 4.0 mg/kg; and rat, 10 mg/kg. The lethal HCN dose for cattle and sheep is 2.0 mg/kg body weight (Moran, 1954). Bearing in mind that HCN comprises only 5-10% of the molecular weight of the known cyanogenic glycosides, these figures give some indication of the amount of cyanogenic plant that must be consumed to be hazardous.

Cyanide exerts its acute toxic effect by combining with metalloporphyrin-containing enzyme systems. Most important, it has an extremely strong affinity for cytochrome oxidase. Cyanide concentrations of only 33 µM can completely block electron transfer through the mitochondrial electron transport chain, thus swiftly preventing the utilization of O_2 by the cell. Death ensues from generalized cytotoxic anoxia, the brain, heart, and central nervous system being most rapidly affected. The sequence of events in acute cyanide poisoning in humans are: hyperventilation, headache, nausea and vomiting, generalized weakness, collapse and coma, perhaps with convulsions, and then respiratory depression. With a large dose, collapse, coma, and death may occur within a few minutes, but with smaller doses, a much slower response is seen and death may not supervene for 15-20 min.

If acute toxicity does not lead to immediate death, it is critical that the condition be quickly recognized and that treatment be initiated. Several successful antidotes to acute cyanide poisoning, whose therapeutic aim is to rapidly lower the cyanide concentration in blood and tissues, have been employed in recent years (Baumeister et al., 1975; Oke, 1969). Unspecific measures such as artificial respiration (Friedberg and Schwarzkopf, 1969) and administration of 100% O_2 (Sheehy and Way, 1968) have been recommended. More specific measures against cyanide toxicity have included the provision of a sulfur donor, such as sodium thiosulfate, which promotes the detoxication of cyanide to thiocyanate by rhodanese (see Sec. VIII.C). Sodium thiosulfate has a very high detoxication capacity, but suffers from a relatively slow rate of action, and only after several minutes is a therapeutic effect seen (Friedberg, 1968). An additional procedure is to convert some of the blood hemoglobin into methemoglobin (ferrihemoglobin); the methemoglobin competes with cytochrome oxidase for the binding of cyanide ion, producing the nontoxic form, cyanomethemoglobin. As the plasma cyanide concentration later decreases, cyanide may dissociate from cyano-methemoglobin and eventually be metabolized or excreted. The hemoglobin-methemoglobin conversion can be effected by primary nitrites, such as sodium nitrite (for injection), amyl nitrite (for inhalation), and by methylene blue. Assuming a total hemoglobin amount in an adult person of approximately 1000 mg and a 10% hemoglobin-methemoglobin conversion, theoretically 156 mg of cyanide can be bound; this approximately corresponds to a lethal dose (Friedberg, 1968). According to Chen and Rose (1952), the combination of nitrites with thiosulfate has proven therapeutically successful in animal experiments and in cases of human cyanide poisoning.

The usefulness of nitrite in treating acute cyanide toxicity in humans has been recently questioned (Kiese and Weger, 1969). Nitrite reacts only slowly with human hemoglobin; consequently, safe doses of nitrite produce only small amounts of ferrihemoglobin, which accumulate too late to detoxify lethal cyanide doses. Similar criticisms have been expressed for amyl nitrite administration (Bastian and Mercker, 1959). Higher nitrite doses, which produce effective amounts of ferrihemoglobin, cause cardiovascular collapse (Kiese and Weger, 1969). Aminophenols have been suggested as being superior to nitrite in the treatment of cyanide poisoning. Both 4-dimethyl-aminophenol (4-DMAP) and 4-methylaminophenol (4-MAP) rapidly produced controlled amounts of ferrihemoglobin in the blood of various species in vitro and in vivo. The rapid reaction of these compounds was also observed after intravenous injection in humans. Doses of 4-DMAP and 4-MAP, which oxidized 30-40% of the hemoglobin, produced the half-maximal ferrihemoglobin concentration in 1 and 2 min, respectively, without any immediate effect on the cardiovascular system. The optimum concentration of ferrihemoglobin for the treatment of cyanide poisoning is about 50% (Lörcher and Weger, 1971).

Several authors advocated use of cobalt compounds, such as cobalt-histidine, cobalt-ethylenediaminetetraacetic acid (EDTA), and hydroxocobalamine, which are believed to combine directly with cyanide within the plasma rather than indirectly by ferrihemoglobin formation (for references, see Rose et al., 1965; Isom and Way, 1973). However, model cyanide intoxication studies with cats showed that dimethylaminophenol-HCl administration was far superior to treatment with hydroxocobalamine, cobalt-EDTA, cobalt-histidine, or $NaNO_2$ (Offterdinger and Weger, 1969). Although administration of cobalt-containing compounds seems unlikely to replace the classic cyanide antidote ($Na_2S_2O_3$-$NaNO_2$), they may be used beneficially to supplement it. Burrows and Way (1977) demonstrated that either cobaltous chloride or oxygen with the thiosulfate-sodium nitrite antidote resulted in dramatic improvement of the protection against cyanide intoxication in mice, but not in sheep. The administration of cobalt compounds is not, however, without potential danger (Offterdinger and Weger, 1969; Isom and Way, 1973).

B. Chronic Cyanide Toxicity

With education, with plant selection, and with government regulation of plant importation, the problem of acute poisoning by HCN of plant origin has been largely overcome in many countries. However, the possible dangers of long-term exposure to low cyanide levels that in single doses do not produce clinical signs of poisoning are not well understood. Only recently have possible correlations been implicated between chronic cyanide uptake and specific diseases such as Nigerian nutritional neuropathy (Osuntokun, 1973), tobacco amblyopia (Heaton et al., 1958), retrobulbar neuritis in pernicious anemia (Freeman and Heaton, 1961), and Leber's optic atrophy (Wilson, 1965; Baumeister et al., 1975). In addition to cyanide originating from the breakdown of dietary cyanogens, humans are continually exposed to low levels from other sources. Exposure to cigarette smoke, in which HCN levels may reach 1600 ppm (Towill et al., 1978), is probably the most common mode of cyanide uptake. Bacterial action in the urinary tract and intestines may also release cyanide (Adams et al., 1966). Cyanide and related compounds (e.g., acrylonitrile, phthalonitrile) are common industrial chemicals used in electroplating, metal refining, and in the production of synthetic fibers, adhesives, plastics, and insecticides. Workers in such industries may be exposed to higher cyanide levels than those encountered by members of the general population. Although some disagreement exists on this question (NIOSH, 1976), a variety of signs and symptoms have been grouped to form what can be presently described as a "cyanide syndrome," with toxic effects involving the central nervous system, gastrointestinal tract, and thyroid.

Given the availability of sufficient sulfur amino acids (methionine, cysteine, and cystine), cyanide originating from these sources does not normally accumulate but is detoxified principally by rhodanese to thiocyanate (see Sec. VIII.C). Thus smokers exhibit higher thiocyanate levels in plasma and other biological fluids than do nonsmokers, whereas cyanide levels are not significantly different (Baumeister et al., 1975). It is questionable whether chronic cyanide poisoning might be accurately assessed by monitoring the levels of thiocyanate in body fluids; these levels are not a simple index of the amount of cyanide ingested, since they are directly affected by several other factors (Wilson, 1965). The great majority of thiocyanate is excreted in the urine and saliva, but this metabolite may also undergo slow oxidation to sulfate (Clemedson et al., 1960). In normal humans, a dynamic equilibrium exists between cyanide and thiocyanate, the concentration of cyanide present in body fluids being approximately 0.25-0.5 mg per 100 ml of cells. These trace amounts may act as a brake on cellular oxidative processes. A minor conversion of thiocyanate back to cyanide is catalyzed by an endogenous oxidase from red blood cells (Goldstein and Rieders, 1953). Thiocyanate levels may also increase following consumption in the diet of thiocyanate itself, which is present in significant amounts in several green vegetables (notably *Brassica*), derived from glucosinolates, and in beer, milk, and eggs (Van Etten and Tookey, 1979; Oke, 1969). Finally, it should be noted that the availability of sulfur amino acids (which may be deficient in chronic malnutrition), the tissue levels of vitamin B_{12}, and minor degrees of dehydration may all affect thiocyanate levels in body fluids (Montgomery, 1979). Thus it may be difficult to detect, or assess the importance of, fluctuations in the thiocyanate pool following consumption of cyanogenic glycosides.

C. Detoxication of Cyanide

Cyanide circulating in the bloodstream may be metabolized by various mechanisms (for a review, see Oke, 1973). The principal metabolic pathway is the production of the less toxic thiocyanate from cyanide and thiosulfate, catalyzed by the enzyme rhodanese [thiosulfate sulfur transferase; E.C.2.8.1.1]. The distribution, structure, and biochemistry of this enzyme are reviewed elsewhere (Westley, 1973). Although rhodanese is present in almost all animal tissues, highest concentrations are observed in the liver and kidney (Himwich and Saunders, 1948). In general, while the rhodanese content in liver, kidney, and suprarenal glands varied markedly from species to species, the level of this enzyme in the brain did not. This may account for the fact that the rate at which small doses of cyanide are detoxified appears related to the liver rhodanese contents in dogs and rabbits, whereas the LD_{50} value for intravenously injected NaCN was approximately the same for all species tested (Mukerji and Smith, 1943).

Several questions arise when one attempts to relate in vitro work on rhodanese to the situation where an intact animal has been poisoned with cyanide. Based on observed in vitro activities, Himwich and Saunders (1948) estimated that in 15 min the whole liver and total skeletal muscle of one dog could have detoxified 4015 and 1763 g of cyanide, respectively. Thus the known toxicity of much lower quantities of cyanide clearly indicates that parameters other than the level of rhodanese present may constitute the limiting factors in the in vivo detoxication process. Rhodanese operates efficiently in vitro with a thiosulfate/CN^- concentration ratio of at least 3; that such a concentration of thiosulfate exists normally in the cell is doubtful. The likelihood that the availability of sulfur is a limiting factor in vivo is further suggested by Chen et al. (1934) and Gilman et al. (1946).

Following the observation that acetone powder extracts of rat liver converted cyanide to thiocyanate as readily with β-mercaptopyruvate as with thiosulfate, it was suggested that β-mercaptopyruvate might be a sulfur-donor substrate for rhodanese (Wood and Fiedler, 1953). The optimal pH range and the sulfur-cyanide ratios closely paralleled values obtained with rhodanese for thiosulfate. Sörbo (1954) subsequently showed that β-mercaptopyruvate would not serve as a donor for crystalline rhodanese from beef liver. This alternative pathway to thiocyanate is instead catalyzed by the copper protein mercaptopyruvate sulfurtransferase, which greatly resembles rhodanese (Westley, 1973).

Voegtlin et al. (1926) observed that animals could be protected from a minimum lethal dose of cyanide by prior injection of cystine. This may be explained by the spontaneous reaction of cystine with cyanide, forming cysteine and β-thiocyanoalanine. The latter tautomerizes to 2-aminothiazoline-4-carboxylic acid or the equivalent 2-amino-4-thiazolidinecarboxylic acid. Since 2-imino-4-thiazolidinecarboxylic acid is metabolically inert and is excreted in the urine, this pathway provides an independent mechanism in animals for cyanide detoxication (Wood and Cooley, 1956). Its quantitative significance may be relatively minor as the recovery of intraperitoneally injected cyanide (in the absence of cystine injection) as thiocyanate and as the thiazolidine was 80% and 15%, respectively. The prophylactic action of cystine against cyanide poisoning is a direct detoxication reaction, apparently potentiated by raising the blood level of sulfur amino acids.

Vitamin B_{12} (cyanocobalamin) is converted to vitamin B_{12a} (hydroxocobalamin) in the presence of light. The latter may react with cyanide, producing vitamin B_{12} (Wokes and Picard, 1955). Several lines of evidence suggest that this interconversion may play an active role in cyanide detoxication in animals. Injection of sublethal doses of cyanide to rats caused a significant depletion of the liver store of vitamin B_{12} (Braekkan et al., 1957). Moreover, dietary deficiency of vitamin B_{12} leads to increased thiocyanate excretion; from this, it has been suggested that competition exists between hydroxocobalamin and rhodanese for the cyanide to be detoxicated. Stress conditions that increase the vitamin B_{12} requirements, such as pregnancy, lactation, and menstruation, also increase thiocyanate excretion. Finally, administration of large doses of hydroxocobalamin, but not of cyanocobalamin, prevented and reversed the toxic effects of cyanide in mice (Mushett et al., 1952). This cyanide detoxication method may well be limited, however, since even if one assumes that all the liver vitamin B_{12} occurs in the hydroxo form, the total amount would be sufficient to detoxify only approximately 25 μg of cyanide.

IX. PATHOLOGICAL CONDITIONS RELATING TO CHRONIC CYANIDE INTOXICATION

Complex neuropathies believed to be of nutritional origin have been described from many parts of the world during the past century (Montgomery, 1979). Many have arisen in communities with low standards of nutrition, especially in the tropics or in prisoner-of-war camps. It seems unlikely that the special etiological factors are the same or occur in the same proportions in every circumstance where such syndromes have been reported; similar clinical syndromes may well be the consequence of different intoxications, dietary deficiencies, and other pathogenetic mechanisms. Of principal interest here is the syndrome tropical ataxic neuropathy (TAN) common in certain areas of Nigeria (Osuntokun, 1973) and Tanzania (Makene and Wilson, 1972), which is characterized by myelopathy, bilateral optic atrophy, bilateral perceptive deafness, and polyneuropathy. Stomatoglossitis, motor neurone disease, Parkinson's disease, cerebellar disintegration, psychosis, and dementia may also be associated with this disease. The sexes are equally affected, as are all age groups with the exception of children under 10 years old. While familial cases accounted for 41% of the patients, analysis revealed no evidence of a genetically determined predisposition. Instead, the data reflected a similar environment and dietary history of the individuals. The experimental findings of Osuntokun (1973) strongly support the proposal of Moore (1934), Clark (1936), and Oluwole (1935) that cyanide intoxication resulting from a protein-poor cassava diet is a major factor in the etiology of TAN in Nigeria. TAN patients display elevated levels of plasma thiocyanate and cyanide (both free and bound) and of urinary thiocyanate. Foodstuffs containing high thiocyanate concentrations, such as *Brassica* and dairy products, are unlikely sources of this thiocyanate, since they are rarely consumed by Nigerian TAN sufferers. Instead, considerable evidence links TAN with consumption of large amounts of cassava derivatives (e.g., gari, eba, and purupuru) known to be cyanogenic. When TAN patients received hospital diets which included cassava only twice per week, their elevated plasma cyanide and thiocyanate levels fell, but they increased again when the patient returned to a high-cassava diet. The prevalence of TAN in villagers eating cassava as staple diet was approximately 3%; in the age group 51-60, even 8% prevalence was reached in one village. By contrast, the disease was absent in a village where predominantly yam was eaten.

Biochemical studies indicated that liver rhodanese levels were normal in TAN patients and that a deficiency in vitamin B_{12} at the tissue or cellular level was unlikely. A major role of riboflavin deficiency in the etiology of TAN was also discounted. Biochemical evidence for protein-calorie deficiency as seen in kwashiorkor was not obtained. However, a poor intake of sulfur amino acids (methionine, cysteine, and cystine) was confirmed by the important finding that plasma levels of these acids were absent or severely reduced in TAN patients. Alternatively, these results could be interpreted as a conditioned deficiency resulting from repeated and excessive cyanide detoxication.

Neuropathological symptoms of TAN patients include posterior column myelopathy, optic atrophy, perceptive deafness, and peripheral nerve demyelination. That these symptoms may be linked to chronic cyanide intoxication is suggested by numerous experiments with animals administered repeated injections of sublethal cyanide doses (Montgomery, 1979; Wilson, 1965). Only in certain of these trials, however, can the results be reasonably attributed to chronic rather than acute anoxic effects of cyanide. Smith et al. (1963) administered rats frequent small doses of potassium cyanide over 5 months and described myelin pallor, with cellular changes in the cortex and cerebellum, apparently preceded by oligodendroglial proliferation. These changes were associated with a rise in plasma thiocyanate concentration (Smith and Duckett, 1965). The demyelination of peripheral nerves induced in rodents by cyanide injection bore a striking resemblance to the lesions in biopsy specimens of peripheral nerves of Nigerian TAN patients (Williams and Osuntokun,

1969). Interestingly, Wistar rats given cassava diets for 18-24 months exhibited ataxic and segmental demyelination and elevated thiocyanate levels (Osuntokun, 1973). Changes in optic nerve tracts have been described in cyanide poisoning in primates (Ferraro, 1933; Hurst, 1940). Rats receiving cyanide in sublethal doses for up to 3 weeks exhibited demyelination in optic nerves and retina, in addition to callosed lesions (Lessell, 1971). Possible mechanisms by which the observed patterns of visual field defects in TAN patients might be generated in chronic cyanide poisoning are discussed in detail elsewhere (Montgomery, 1979; Osuntokun, 1979). The nature of perceptive deafness in TAN patients, which indicated both receptor organ and neuronal lesions (Hinchcliffe et al., 1972), is compatible with described effects of cyanide on the chick embryo otocyst (Friedmann, 1972).

Endemic goiter, a widespread disease affecting more than 200 million people throughout the world, is principally caused by iodine deficiency, but the additional role of natural goitogens has been suggested for several geographical regions (Delange and Ermans, 1971). Osuntokun (1979) reported a higher prevalence (2-5%) of goiter in TAN patients in Nigeria than in the general population, which appeared related to cassava diet and high plasma thiocyanate levels. Thiocyanate, a common component of foods and feeds from crucifers, is a well-established goitrogen, which competes directly with the iodide-trapping process of the thyroid and at higher concentrations, with the intrathyroidal processing of iodine (Delange et al., 1976). The quantity of iodine available for hormone synthesis is thereby reduced, while increases occur in urinary iodine excretion. This situation could induce or aggravate a state of iodine deficiency and thus promote goiter.

In recent years, the goiter endemic of Idjwi Island on Lake Kivu, eastern Zaire, has been an active area of research. The presence of a dietary goitrogen in this region was suspected when studies indicated that goiter prevalence was 10 times higher in the north of the island than in the southwest, although a severe and uniform iodine deficiency had been described throughout the whole island. Dietary studies revealed that the major foods eaten in these two areas were identical; cruciferous plants were not consumed in appreciable quantities, but larger amounts of cassava were eaten in goitrous regions. The latter fact appears significant, as Ekpechi (1973) has demonstrated in rats the goitrogenic and thionamide-like antithyroid activity of unfermented cassava from endemic areas of eastern Nigeria. Ingestion of cassava grown in goitrous areas of Idjwi resulted in decreased thyroidal radioiodine uptake with concomitant increases in renal excretion of stable and labeled iodide. Curiously, iodine uptake was not affected after ingestion of equally cyanogenic cassava roots grown in the nongoitrous areas. Cassava consumption by humans and experimental animals is followed by increases in both serum and urine thiocyanate levels, presumably resulting from the endogenous catabolism of cassava cyanogenic glycosides to HCN, with its subsequent detoxication by sulfurtransferases (Van der Velden et al., 1973; Osuntokun, 1979). Although additional factors may well be involved, it has been proposed that these elevated levels of thiocyanate with their known antithyroid effects (Ermans et al., 1973) play a critical role in goiter development under the conditions of iodine deficiency existing on Idjwi Island (Delange et al., 1973).

Endemic cretinism of the myxedematous type reaches 1.1% prevalency in goitrous regions of Idjwi Island. This distinct epidemological distribution suggests that, at least for Idjwi, iodine deficiency alone can no more account for the prevalence of cretinism than it can for that of endemic goiter. Although no firm conclusion has yet been reached, the ingestion of cassava or the action of a factor liberated during its catabolism are suspected to account for the difference in prevalency of cretinism on Idjwi (Pharoah et al., 1980). Indeed, studies with neonatal iodine-deficient rats support the proposition that cyanogenic glycosides and thiocyanate derived therefrom may affect thyroid function (Kreutler et al., 1978).

Comparison of the world distribution of type J and pancreatic diabetes with the consumption of cassava and other cyanide-containing foods supports the concept that low protein uptake, combined with ingestion of cyanide or a cyanide precursor, may lead to exocrine and endocrine pancreatic changes (McMillan and Greevarghese, 1979). Due to reduced detoxication rates prevailing under these conditions, it is proposed that cyanide might accumulate in the plasma and be capable of pancreatic damage and diabetes. Preliminary experiments showed that oral and parenteral cyanide caused marked hyperglycemia in rats, although the mechanism involved remains unknown. However, despite evidence of recurrent hyperglycemia, cyanosis, and epidermal changes during chronic ingestion of cyanide, rats did not develop diabetes. Age and species difference in the susceptibility of the pancreatic islets to hyperglycemic damage may account for this failure.

The use of cassava as a food source is projected to increase in future years (Nestel, 1973). One should expect that the chronic toxicity observed in humans and animals on high-cassava diets may become a problem of mounting significance unless effective means are taken to reduce its toxicity. Screening for acyanogenesis and low cyanide levels in the extensive collections of cultivated lines of cassava which are available and also of uncultivated "wild" material is undoubtedly desirable. In addition, preventive work should strive to improve the nutritional balance among the exposed population, while reducing exposure to cyanogen.

X. ANTITUMOR ACTIVITY OF AMYGDALIN AND RELATED COMPOUNDS

Amygdalin was first used in cancer chemotherapy in 1845, but was abandoned due to marked toxicity and lack of specific canceristatic effects (Viehoever and Mack, 1935). Krebs and Krebs later proposed that L-mandelonitrile-β-glucuronide, a derivative of amygdalin which they named Laetrile, might be beneficial in cancer therapy. According to the patents issued (E. T. Krebs and E. T. Krebs, Jr., U.S. Patent No. 2,985,664, issued in 1961; British Patent No. 788,855, issued in 1958), this compound may be produced by (1) the hydrolysis of amygdalin and subsequent oxidation of prunasin with platinum black, or (2) by condensation of mandelonitrile with glucose followed by oxidation, or (3) by condensation of mandelonitrile with glucuronic acid. Subsequent attempts to prepare Laetrile by the patented procedures were unsuccessful, but a two-step procedure employing an immobilized glucuronyltransferase from rat liver has recently become vailable, which should produce L-mandelonitrile-β-glucuronide in quantities sufficient for proper clinical assessment (Fenselau et al., 1977). Careful pharmacological assessment has now shown that the material known in the United States as "Laetrile" is not L-mandelonitrile-β-glucuronide but principally amygdalin. Furthermore, analytical tests demonstrated that certain oral and injectable forms of amygdalin of Mexican origin were chemically subpotent, microbially contaminated, and unfit as pharmaceutical products for human use when evaluated by U.S. crieria (Davignon et al., 1978; Cairns et al., 1978).

The proposed anticancer mechanism of Laetrile has changed several times over past years, the compound being promoted initially as a drug, but more recently, as a vitamin, namely vitamin B_{17} (for a review, see Dorr and Paxinos, 1978). In an early proposal, cancer cells were believed to possess abundant amounts of a β-glucosidase that hydrolyzes amygdalin to release HCN, but they were purportedly deficient in rhodanese, the enzyme detoxifying HCN to thiocyanate. By contrast, normal cells were assumed to have low levels of β-glucosidases but were rich in rhodanese. Thus the theoretical basis for amygdalin use was that cancer cells are preferentially killed by amygdalin hydrolysis while normal cells are spared. This selective target toxicity has not received firm experimental support. Both normal and cancer cells had similar rhodanese contents (Gal et al., 1952). Furthermore, knowledge of the levels of amygdalin hydrolase (in contrast to PNPGase) activity of cancer cells is not yet available. Fishman and Anlyan (1947) observed that certain tumors exhibited higher β-glucuronidase activity than did the corresponding uninvolved tissue, but these data remain difficult to interpret since the model substrate phenolphthalein mono-β-glucuronide was utilized in enzyme assays rather than Laetrile or amygdalin.

Assuming that HCN were released in vivo from amygdalin by suitable mechanisms, the question still remains whether it could selectively kill cancer cells. Burk et al. (1971) reported that Ehrlich ascites cells, treated in vitro, are sensitive to combined treatment with amygdalin and β-glucosidase because of the synergistic effects of the HCN-benzaldehyde mixture, which produces a decreased respiration and an increased aerobic glycolysis. In contrast, Levi et al. (1965) demonstrated that amygdalin had no significant effect on aerobic and anaerobic glycolysis of cancer cells from primary human tumors or from animal ascites. Protein, RNA, and DNA synthesis were also not inhibited by amygdalin. These authors confirmed earlier reports that cyanide is not cancerocidal unless cells are totally deprived of glucose. Their conclusion that amygdalin would be ineffective as a clinical anticancer agent, by virtue of cyanide release, is supported by experimental evidence. Direct intratumoral administration of cyanide caused no measurable improvement of uterine cancer (Brown et al., 1960). Amygdalin and its breakdown products had no preferential cytotoxicity in vitro to the colony-forming cells from human acute myelogenous leukemia cell lines compared to normal human bone marrow granulocyte-monocyte clonogenic cells (Koeffler et al., 1980).

Little information exists in the scientific literature and lay press supporting either Laetrile or amygdalin for cancer therapy (Morrone, 1962; see Lewis, 1977, and Dorr and Paxinos, 1978, for reviews). Preliminary trails conducted at the Sloan-Kettering Institute indicated that amygdalin given in doses of 1-2 g/kg, i.p. caused significant inhibition of spontaneous mammary tumors, inhibition of lung metastases, and prevention, to some extent, of the formation of new tumors in CD_8F_1 mice. Unfortunately, attempts to reproduce these encouraging experiments proved unsuccessful (Stock et al., 1978b). Amygdalin was ineffective also against a wide range of transplantable tumors (Laster and

Schabel, 1975; Wodinsky and Swiniarski, 1975; Hill et al., 1976; Stock et al., 1978a; Ovejera et al., 1978). Furthermore, no antitumor activity was observed when amygdalin and β-glucosidase were administered simultaneously, but the lethal toxicity of amygdalin was potentiated (Laster and Schabel, 1975; Wodinsky and Swiniarski, 1975; Ovejera et al., 1978). Thus amygdalin has apparently failed as an active chemotherapeutic agent in the standard animal tumor screens utilized to identify potentially active anticancer drugs. Similarly, review of data for human patients treated with amygdalin has not provided convincing evidence of efficacy (Lewis, 1977; U.S. DHEW, 1977; Ellison et al., 1978; Mehta, 1980; Moertel et al., 1981). In response to widespread public interest and a long and bitter dispute between Laetrile advocates and the scientific community, amygdalin was clinically tested by the National Cancer Institute, who judged it ineffective against cancer, whether administered alone or with huge doses of vitamins (Sun, 1981).

Concern has been expressed about potential dangers resulting from amygdalin administration. The contentions of Laetrile proponents that amygdalin is totally nontoxic (Krebs, 1970) or, at worst, a nontoxic placebo are without doubt contradicted by reports of acute cyanide poisoning following ingestion of amygdalin itself or of seeds of certain rosaceous species (Lewis, 1977; Sadoff et al., 1978; Smith et al., 1978; Humbert et al., 1977; U.S DHEW, 1977). Amygdalin toxicity is clearly accentuated when sources of β-glycosidase are ingested with the glycoside (Schmidt et al., 1978; Moertel et al., 1981). Parenteral rather than oral administration appears preferable, but this too may be dangerous (Khandekar and Edelman, 1979). The possibility of chronic toxic reaction in Laetrile users has been suggested but requires further examination (Newton et al., 1981). Precautionary measures for Laetrile users against cyanide toxicity have been described (Moertel et al., 1981). They should also be warned that amygdalin may be teratogenic (Lewis, 1977) and, along with Laetrile and mandelonitrile, mutagenic (Fenselau et al., 1977).

REFERENCES

Adams, J. H., Blackwood, W., and Wilson, J. (1966). Further clinical and pathological observations on Leber's optic atrophy. *Brain 89:*15-26.

Akinrele, I. A. (1964). Fermentation of cassava. *J. Sci. Food Agric. 15:* 589-594.

Ames, M. M., Kovach, J. S., and Flora, K. P. (1978). Initial pharmacologic studies of amygdalin (Laetrile) in man. *Res. Commun. Chem. Pathol. Pharmacol. 22:*175-185.

Armstrong, H. E., Armstrong, E. F., and Horton, E. (1912). Studies on enzyme action. XVI. The enzymes of emulsin. (I) Prunase, the correlate of prunasin. *Proc. R. Soc. B 85:*359-362.

Aschhoff, H-J., and Pfeil, E. (1970). Auftrennung und Charakterisierung der Isoenzyme von D-Hydroxynitril-Lyase (D-Oxynitrilase) aus Mandeln. *Hoppe-Seyler's Z. Physiol. Chem. 351:*818-826.

Baggchi, K. N., and Ganguli, H. D. (1943). Toxicology of young shoots of common bamboos (*Bambusa arundinacea* Willd). *Indian Med. Gaz. 78:* 40-42.

Barnett, R. D., and Caviness, C. E. (1968). Inheritance of hydrocyanic acid production in two sorghum × sudangrass crosses. *Crop Sci. 8:*89-91.

Barrett, M. D., Hill, D. C., Alexander, J. C., and Zitnak, A. (1977). Fate of orally dosed linamarin in the rat. *Can. J. Physiol. Pharmacol. 55:*134-136.

Bastian, G., and Mercker, H. (1959). On the question of the usefulness of amylnitrite in the treatment of cyanide poisoning. *Arch. Exp. Pathol. Pharmakol. 237:*285-295.

Baumeister, R. G. H., Schievelbein, H., and Zickgraf-Rudel, G. (1975). Toxicological and clinical aspects of cyanide metabolism. *Drug Res. 25:* 1056-1064.

Beath, O. A. (1939). Poisonous plants and livestock poisoning. *Univ. Wyo. Agric. Exp. Stn., Laramie, Bull. 231.*

Bertrand, G., and Weisweiller, G. (1910). Sur la constitution du vicianose et de la vicianine. *C. R. Acad. Sci. 151:*884-886.

Bergemann, M. (1812). De l'existence de l'acid prussique dans les écorces d'arbre. *Ann. Chim. 83:*215-216.

Boyd, F. T., Aamodt, O. S., Bohstedt, G., and Truog, E. (1938). Sudan grass management for control of cyanide poisoning. *J. Am. Soc. Agron. 30:*569-582.

Braekkan, O., Njaa, L. R., and Utne, F. (1957). The effect of cyanide on liver reserves of vitamin B_{12}. *Acta Pharmacol. (Kbh.) 13:*228-232.

Brown, W. E., Wood, C. D., and Smith, A. N. (1960). Sodium cyanide as a cancer chemotherapeutic agent. Laboratory and clinical studies. *Am. J. Obstet. Gynecol. 80:*907-918.

Burk, D., McNaughton, A. R. L., and Von Ardenne, M. (1971). Hyper-thermy of cancer cells with amygdalin-glucosidase, and synergistic action of derived cyanide and benzaldehyde. *Panminerva Med. 13:*520-522.

Burrows, G. E., and Way, J. L. (1977). Cyanide intoxication in sheep: therapeutic value of oxygen or cobalt. *Am. J. Vet. Res. 38:*223-227.

Butler, G. W. (1965). The distribution of the cyanoglucosides linamarin and lotaustralin in higher plants. *Phytochemistry 4:*127-131.

Butler, G. W., Bailey, R. W., and Kennedy, L. D. (1965). Studies on the glucosidase "linamarase." *Phytochemistry 4:*369-381.

Butler, G. W., Reay, P. F., and Tapper, B. A. (1973). Physiological and genetic aspects of cyanogenesis in cassava and other plants. In *Chronic Cassava Toxicity*, B. Nestel and R. MacIntyre (Eds.) International Development Research Centre, Ottawa, pp. 65-71.

Cairns, T., Froberg, J. E., Gonzales, S., Langham, W. S., Stamp, J. J., Howie, J. K., and Sawyer, D. T. (1978). Analytical chemistry of amygdalin. *Anal. Chem. 50:*317-322.

Caldwell, R. J., and Cortauld, S. L. (1907). The hydrolysis of amygdalin by acids. *J. Chem. Soc. 91:*666-671.

Carter, J. H., McLafferty, M. A., and Goldman, P. (1980). Role of the gastrointestinal microflora in amygdalin (Laetrile)-induced cyanide toxicity. *Biochem. Pharmacol. 29:*301-304.

Chen, K. K., and Rose, C. L. (1952). Nitrite and thiosulfate therapy in cyanide poisoning. *JAMA 149:*113-119.

Chen, K. K., Rose, C. L., and Clowes, G. H. A. (1934). Comparative values of several antidotes in cyanide poisoning. *Am. J. Med. Sci. 188*: 767-781.

Christensen, H. E. (1976). Registry of Toxic Effects of Chemical Substances. *DHEW Publ. (NIOSH) 76-191.* National Institute of Occupational Safety and Health, U.S. Public Health Service, Rockville, Md.

Clark, A. (1936). Report on effects of certain poisons contained in food plants of West Africa upon health of native races. *J. Trop. Med. Hyg. 39*: 269-276, 285-295.

Clawson, A. B., and Moran, E. A. (1937). Toxicity of arrowgrass for sheep and remedial treatment. *U.S. Dept. Agric. Tech. Bull. 580.* Washington, D.C.

Clegg, D. O., Conn, E. E., and Janzen, D. H. (1979). Developmental fate of the cyanogenic glycoside linamarin in Costa Rican wild lima bean seeds. *Nature (Lond.) 278*: 343-344.

Clemedson, C. J., Sorbo, B., and Ullberg, S. (1960). Autoradiographic observations on injected S^{35}-thiocyanate and C^{14}-cyanide in mice. *Acta Physiol. Scand. 48*: 382-389.

Cohen, R. B., Rutenberg, S. H., Tsou, K-C., Woodbury, M. A., and Seligman, A. M. (1952). The colorimetric estimation of β-D-glucosidase. *J. Biol. Chem. 195*: 607-614.

Conn. E. E. (1979). Cyanogenic glycosides. In *International Review of Biochemistry*, Vol. 27: *Biochemistry of Nutrition*. A. Neuberger and T. H. Jukes (Eds.). University Park Press, Baltimore, pp. 21-43.

Conn, E. E. (1980a). Cyanogenic glycosides. In *Secondary Plant Products, Encyclopedia of Plant Physiology, New Series*, Vol. 8, E. A. Bell and B. V. Charlwood (Eds.). Springer-Verlag, New York, pp. 461-492.

Conn. E. E. (1980b). Cyanogenic compounds. *Annu. Rev. Plant Physiol. 31*: 433-451.

Conn. E. E., and Butler, G. W. (1969). The biosynthesis of cyanogenic glycosides and other simple nitrogen compounds. In *Perspectives in Phytochemistry*, J. B. Harborne and T. Swain (Eds.). Academic Press, London, pp. 47-74.

Cooke, R. D., Blake, G. G., and Battershill, J. M. (1978). Purification of cassava linamarase. *Phytochemistry 17*: 381-383.

Coop, I. E., and Blakley, R. L. (1949). The metabolism and toxicity of cyanides and cyanogenic glycosides in sheep. I. Activity in the rumen. *N.Z. J. Sci. Technol. Sect. A 30*: 277-291.

Corkill, L. (1942). Cyanogenesis in white clover (*Trifolium repens*). V. The inheritance of cyanogenesis. *N.Z. J. Sci. Technol. Sect. B 23*: 178-193.

Coursey, D. G. (1973). Cassava as food: toxicity and technology. In *Chronic Cassava Toxicity*, B. Nestel and R. MacIntyre (Eds.). International Development Research Centre, Ottawa, pp. 27-36.

Cutler, A. J., and Conn, E. E. (1981). The biosynthesis of cyanogenic glycosides in *Linum usitatissimum* (Linum flax) in vitro. *Arch. Biochem. Biophys. 212*: 468-474.

Cutler, A. J., Hosel, W., Sternberg, M., and Conn, E. E. (1981). The in vitro biosynthesis of taxiphyllin and the channeling of intermediates in *Triglochin maritima*. *J. Biol. Chem. 256*: 4253-4258.

Davignon, J. P., Trissel, L. A., and Kleinman, M. A. (1978). Pharmaceutical assessment of amygdalin (Laetrile) products. *Cancer Treat. Rep. 62*: 99-104.

DeBruijn, G. H. (1973). The cyanogenic character of cassava (*Manihot esculenta*). In *Chronic Cassava Toxicity*, B. Nestel and R. MacIntyre (Eds.). International Development Research Centre, Ottawa, pp. 43-48.

Delange, F. and Ermans, A. M. (1971). Role of a dietary goitrogen in the etiology of endemic goiter on Idjwi Island. *Am. J. Clin. Nutr. 24*:1354-1360.

Delange, F., Van der Velden, M., and Ermans, A. M. (1973). Evidence of an antithyroid action of cassava in man and in animals. In *Chronic Cassava Toxicity*, B. Nestel and R. MacIntyre (Eds.). International Development Centre, Ottawa, pp. 147-151.

Delange, F., Bourdoux, P., Camus, M., Gerard, M., Mafuta, M., Hanson, A., and Ermans, A. M. (1976). The toxic effect of cassava on human thyroid. *Proc. 4th Symp. Int. Soc. Trop. Root Crops*, Kali, Columbia. *IDRC Publ. 080E*, pp. 237-242.

De Waal, D. (1942). Het Cyanophore Karakter van Witte Klaver. Ph.D. thesis, Agricultural University, Wageningen, The Netherlands.

Dement, W. A., and Mooney, H. A. (1974). Seasonal variation in the production of tannins and cyanogenic glucosides in the chaparral shrub *Heteromeles arbutifolia. Oecologia 15*:65-76.

Dorr, R. T., and Paxinos, J. (1978). The current status of Laetrile. *Ann. Intern. Med. 89*:389-397.

Dunstan, W. R., and Henry, T. A. (1902). Cyanogenesis in Plants. Part II. The great millet, *Sorghum vulgare. Philos. Trans. R. Soc. Lond. Ser. A 199*:399-410.

Ekpechi, O. L. (1973). Endemic goitre and high cassava diets in eastern Nigeria. In *Chronic Cassava Toxicity*, B. Nestel and R. MacIntyre (Eds.). International Development Research Center, Ottawa, pp. 139-145.

Ellison, N. M., Byar, D. P., and Newell, G. R. (1978). Special report on Laetrile: the NCI Laetrile review. *N. Engl. J. Med. 299*:549-552.

Ermans, A. M., Van der Velden, M., Kinthaert, J., and De Lange, F. (1973). Mechanism of the goitrogenic action of cassava. In *Chronic Cassava Toxicity*, B. Nestel and R. MacIntyre (Eds.). International Development Research Centre, Ottawa, pp. 153-157.

Everist, S. L. (1974). *Poisonous plants of Australia*. Angus & Robertson, Sydney.

Fenselau, C., Pallante, S., Batzinger, R. P., Benson, W. R., Barron, R. P., Sheinin, E. B., and Maienthal, M. (1977). Mandelonitrile β-glucuronide: synthesis and characterization. *Science 198*:625-627.

Ferraro, A. (1933). Experimental toxic encephalomyelopathy: diffusion sclerosis following subcutaneous injections of potassium cyanide. *Arch. Neurol. Psychiatry (Chicago) 29*:1364-1367.

Fikenscher, L. H., and Hegnauer, R. (1977). Über die cyanogen Verbindungen bei einigen Compositae, bei den Olinaceae und in der *Rutaceen*-Gattung, *Zieria. Pharm. Weekbl. 112*:11-20.

Fischer, E. (1895). Ueber ein neues dem Amygdalin ähnliches Glucosid. *Chem. Ber. 28*:1508-1511.

Fishman, W. H., and Anlyan, A. J. (1947). Comparison of the β-glucuronidase activity of normal, tumor, and lymph node tissues of surgical patients. *Science 106*: 66-67.

Freeman, A. G., and Heaton, J. M. (1961). The aetiology of retrobulbar neuritis in addisonian pernicious anaemia. *Lancet 1*: 908-911.

Freese, A., Brady, R. O., and Gal, A. E. (1980). A β-glucosidase in feline kidney that hydrolyses amygdalin (Laetrile). *Arch. Biochem. Biophys. 201*: 363-368.

Friedberg, K. D. (1968). Antidote bei Blausaurevergiftungen. *Arch. Toxikol. 24*: 41-48.

Friedberg, K. D., and Schwarzkopf, H. A. (1969). Blausäureexhalation bei der Cyanidvergiftung (The exhalation of hydrocyanic acid in cyanide poisoning). *Arch. Toxicol. 24*: 235-248.

Friedmann, I. (1972). The effect of sodium cyanide on the chick-embryo otocyst in vitro. *Acta Oto-laryngol. 73*: 280-289.

Gal, E. M., Fung, F.-H., and Greenberg, D. M. (1952). Studies on the biological action of malononitriles. II. Distribution of rhodanese (transulfurase) in the tissues of normal and tumor-bearing animals and the effect of malononitriles thereon. *Cancer Res. 12*: 574-579.

Gerstner, E., and Pfeil, U. (1972). Zur Kenntnis des Flavinenzyms Hydroxynitril-Lyase (D-Oxynitrilase). *Hoppe-Seyler's Z. Physiol. Chem. 353*: 271-286.

Gerstner, E., and Pfeil, E. (1975). Eine neue Mandelsäurenitril-Lyase (D-Oxynitrilase) aus *Prunus laurocerasus* (Kirschlorbeer). *Hoppe-Seyler's Z. Physiol. Chem. 356*: 1853-1857.

Gettler, A. O., and Baine, J. O. (1938). The toxicology of cyanide. *Am. J. Med. Sci. 195*: 182-198.

Giaja, R. (1919). Sur l'action successive des deux genres d'émulsines sur l'amygdaline. *C. R. Soc. Biol. (Paris) 82*: 1196-1198.

Gibb, M. C., Carbery, J. T., Carter, R. G. and Catalinac, S. (1974). Hydrocyanic acid poisoning of cattle associated with sudan grass. *N. Z. Vet. J. 22*: 127.

Gibbs, R. D. (1974). *Chemotaxonomy of Flowering Plants*, Vols. I-IV. McGill-Queens University Press, Montreal.

Gilman, A., Philips, F. S. and Koelle, E. S. (1946). The renal clearance of thiosulfate with observations on its volume distribution. *Am. J. Physiol. 146*: 348-357.

Goldman, P. (1978). Biochemical pharmacology of the intestinal flora. *Annu. Rev. Pharmacol. Toxicol. 18*: 523-539.

Goldstein, F., and Rieders, R. (1953). Conversion of thiocyanate to cyanide by erythrocytic enzyme. *Am. J. Physiol. 173*: 287-290.

Grover, A. K., MacMurchie, D. D., and Cushley, R. J. (1977). Studies on almond emulsin β-D-glucosidase. I. Isolation and characterization of a bifunctional enzyme. *Biochim. Biophys. Acta 482*: 98-108.

Gunders, A. E., Abrahamov, A., and Weisenberg, E. (1969). Cyanide poisoning following ingestion of apricot (*Prunus armeniaca*) kernels. *J. Isr. Med. Assoc. 76*: 536-538.

Hahlbrock, K., and Conn, E. E. (1970). The biosynthesis of cyanogenic glycosides in higher plants. I. Purification and properties of a uridine diphosphate glucose:ketone cyanohydrin β-glucosyltransferase from *Linum usitatissimum* L. *J. Biol. Chem. 245*: 917-922.

Haisman, D. R., and Knight, D. J. (1967). The enzymic hydrolysis of amygdalin. *Biochem. J. 103*:528-534.

Haisman, D. R., Knight, D. J., and Ellis, M. J. (1967). The electrophoretic separation of the β-glucosidases of almond "emulsin." *Phytochemistry 6*:1501-1505.

Halstrom, F., and Moller, K. D. (1945). Content of cyanide in human organs from cases of poisoning with cyanide taken by mouth, with contribution to toxicology of cyanides. *Acta Pharmacol. Toxicol. 1*:18-28.

Heaton, J. M., McCormick, A. J. A., and Freeman, A. G. (1958). Tobacco amblyopia: a clinical manifestation of vitamin B_{12} deficiency. *Lancet 2*: 286-290.

Hegnauer, R. (1963-1973). *Chemotaxonomie der Pflanzen*, Vols. I-VI. Birkhäuser-Verlag, Basel.

Heinrichs, D. H., and Anderson, L. J. (1947). Toxicity of sorghum in southwestern Saskatchewan. *Sci. Agric. 27*:186-191.

Helferich, B., and Kleinschmidt, T. (1968). Zur Kenntnis des Süssmandelemulsins. *Hoppe-Seyler's Z. Physiol. Chem. 349*:25-28.

Heppner, M. J. (1926). Further evidence on the factor for bitterness in the sweet almond. *Genetics 11*:605-606.

Heuser, C. W. (1972). β-Glucosidase from 'Marianna' plum. *Phytochemistry 11*:2455-2457.

Hill, G. J., Shine, T. E., Hill, H. Z., and Miller, C. (1976). Failure of amygdalin to arrest B_{16} melanoma and BW5147 AKR leukemia. *Cancer Res. 36*:2102-2107.

Himwich, W. A., and Saunders, J. P. (1948). Enzymatic conversion of cyanide to thiocyanate. *Am. J. Physiol. 153*:348-354.

Hinchcliffe, R., Osuntokun, B. O., and Adeuja, A. O. C. (1972). Hearing levels in Nigerian ataxic neuropathy. *Audiology 49*:111-137.

Holdeman, L. V., Cato, E. P., and Moore, W. E. C. (1977). In *Anaerobe Laboratory Manual*, 4th ed. VPI Anaerobe Laboratory, Blacksburg, Va., p. 144.

Hösel, W. (1981). Glycosylation and glycosidases. In *The Biochemistry of Plants*, Vol. 7, P. K. Stumpf and E. E. Conn (Eds.). Academic Press, New York, pp. 725-753.

Hösel, W., and Nahrstedt, A. (1975). Spezifische Glucosidasen für das Cyanglucosid Triglochinin: Reinigung und Charakterisierung von β-Glucosidasen aus *Alocasia macrorrhiza* Schott. *Hoppe-Seyler's Z. Physiol. Chem. 356*:1265-1275.

Hösel, W., and Nahrstedt, A. (1980). In vitro biosynthesis of the cyanogenic glycoside taxiphyllin in *Triglochin maritima*. *Arch. Biochem. Biophys. 203*:753-757.

Hughes, M. A. (1973). The genetics of cyanogenesis. In *Chronic Cassava Toxicity*, B. Nestel and R. MacIntyre (Eds.). International Development Research Centre, Ottawa, pp. 49-54.

Hughes, M. A., and Conn, E. E. (1976). Cyanoglucoside biosynthesis in white clover (*Trifolium repens*). *Phytochemistry 15*:697-701.

Humbert, J. R., Tress, J. H., and Braico, K. T. (1977). Fatal cyanide poisoning: accidental ingestion of amygdalin [letter]. *JAMA 238*:482.

Hurst, E. W. (1940). Experimental demyelination of the central nervous system. *Aust. J. Exp. Biol. Med. Sci. 18*:201-223.

Hurst, E. (1942). *The Poison Plants of New South Wales.* Snelling Printing Works, Sydney.

Isom, G. E., and Way, J. L. (1973). Cyanide intoxication: protection with cobaltous chloride. *Toxicol. Appl. Pharmacol. 24*:449-456.

Jaroszewski, J. W., and Ettlinger, M. G. (1981). Ring cleavage of phenols in higher plants: biosynthesis of triglochinin. *Phytochemistry 20*:819-821.

Joachim, A. W. R., and Pandittesekere, D. G. (1944). Investigations on the hydrocyanic acid content of manioc. *Trop. Agric. 100*:150-163.

Jones, D. A. (1972). Cyanogenic glycosides and their function. In *Phytochemical Ecology*, J. B. Harborne (Ed.). Academic Press, New York, pp. 103-124.

Jorns, M. S. (1979). Mechanism of catalysis by the flavoenzyme oxynitrilase. *J. Biol. Chem. 254*:12145-12152.

Karrer, W. (1958). *Konstitution and Vorkommen der organischen Pflanzenstoffe.* Birkhäuser Verlag, Basel-Stuttgart.

Khandekar, J. D., and Edelman, H. (1979). Studies of amygdalin (Laetrile) toxicity in rodents. *JAMA 242*:169-171.

Kiese, M., and Weger, N. (1969). Formation of ferrihaemoglobin with aminophenols in the human for treatment of cyanide poisoning. *Eur. J. Pharmacol. 7*:97-105.

Kingsbury, J. M. (1964). *Poisonous Plants of the United States and Canada.* Prentice-Hall, Englewood Cliffs, N.J., pp. 364-370.

Koeffler, H. P., Lowe, L., and Golde, D. W. (1980). Amygdalin (Laetrile): effect on clonogenic cells from human myeloid leukemia cell lines and normal human marrow. *Cancer Treat. Rep. 64*:105-109.

Kojima, M., Poulton, J. E., Thayer, S. S., and Conn, E. E. (1979). Tissue distributions of dhurrin and of enzymes involved in its metabolism in leaves of *Sorghum bicolor. Plant Physiol. 63*:1022-1028.

Krebs, E. T., Jr. (1970). The nitrilosides (vitamin B_{17}), their nature, occurrence and metabolic significance. Antineoplastics vitamin B_{17}. *J. Appl. Nutr. 22*:75-86.

Kreutler, P. A., Varbanov, V., Goodman, W., Olaya, G., and Stanbury, J. B. (1978). Interactions of protein deficiency, cyanide, and thiocyanate on thyroid function in neonatal and adult rats. *Am. J. Clin. Nutr. 31*:282-289.

Laster, W. R., and Schabel, F. M. (1975). Experimental studies of the antitumor activity of amygdalin MF (NSC-15780) alone and in combination with β-glucosidase (NSC-128056). *Cancer Chemother. Rep. 59*:951-965.

Lessell, S. (1971). Experimental cyanide optic neuropathy. *Arch. Ophthalmol. 84*:194-204.

Levi, L., French, W. N., Bickis, I. J., and Henderson, I. W. D. (1965). Laetrile: a study of its physicochemical and biochemical properties. *Can. Med. Assoc. J. 92*:1057-1061.

Lewis, J. (1977). Laetrile. [Informed Opinion.] *West. J. Med. 127*:55-62.

Lörcher, W., and Weger, N. (1971). Optimal concentration of ferrihaemoglobin for the treatment of cyanide poisoning. *Arch. Exp. Pathol. Pharmakol. Suppl. 270*:R 88.

Makene, W. J., and Wilson, J. (1972). Biochemical studies in Tanzanian patients with ataxic tropical neuropathy. *J. Neurol. Neurosurg. Psychiatry 35*:31-33.

Majak, W., McDiarmid, R. E., Hall, J. W., and Van Ryswyk, A. L. (1980a). Seasonal variation in the cyanide potential of arrowgrass (*Triglochin maritima*). *Can. J. Plant Sci. 60*:1235-1241.

Majak, W., Udenberg, T., Clark, L. J., and McLean, A. (1980b). Toxicity of saskatoon serviceberry to cattle. *Can. Vet. J. 21*:74-76.

Maner, J. H. (1972). Feeding swine with rations based on cassava. Document prepared by Centro Internacional de Agricultura Tropical (CIAT).

Mao, C.-H., and Anderson, L. (1967). Cyanogenesis in *Sorghum vulgare*. III. Partial purification and characterization of two β-glucosidases from *Sorghum* tissues. *Phytochemistry 6*:473-483.

McFarlane, I., Lees, E., and Conn, E. E. (1975). The in vitro biosynthesis of dhurrin, the cyanogenic glycoside of *Sorghum bicolor*. *J. Biol. Chem. 250*:4708-4713.

McMillan, D. E., and Geevarghese, P. J. (1979). Dietary cyanide and tropical malnutrition diabetes. *Diabetes Care 2*:202-208.

Mehta, P. (1980). Ineffectiveness of Laetrile in the treatment of acute lymphoblastic leukemia. *Clin. Pediatr. 19*:363-364.

Mikolajczak, K. L. (1977). Cyanolipids. *Prog. Chem. Fats Other Lipids 15*: 97-130.

Moertel, C. G., Ames, M. M., Kovach, J. S., Moyer, T. P., Rubin, J. R., and Tinker, J. H. (1981). A pharmacological and toxicological study of amygdalin. *JAMA 245*:591-594.

Montgomery, R. D. (1965). The medical significance of cyanogen in plant foodstuffs. *Am. J. Clin. Nutr. 17*:103-113.

Montgomery, R. D. (1969). Cyanogens. In *Toxic Constituents of Plant Foodstuffs*, I. E. Liener (Ed.) Academic Press, New York, pp. 143-157.

Montgomery, R. D. (1979). Cyanogenetic glycosides. In *Handbook of Clinical Neurology*, Vol. 36, J. N. Vinken and W. Bruyn (Eds.). North-Holland, Amsterdam, pp. 515-527.

Moore, D. F. (1934). Retrobular neuritis and partial optic atrophy as sequelae of avitaminosis. *Ann. Trop. Med. 28*:295-303.

Moran, E. A. (1954). Cyanogenetic compounds in plants and their significance in animal industry. *Am. J. Vet. Res. 15*:171-176.

Morrone, J. A. (1962). Chemotherapy of inoperable cancer—preliminary report of 10 cases treated with laetrile. *Exp. Med. Surg. 20*:299-308.

Muenscher, W. C. (1945). *Poisonous Plants of the United States*. Macmillan, New York.

Mukerji, B., and Smith, R. G. (1943). Cyanide detoxication in rabbit and dog as measured by urinary thiocyanate excretion. *Ann. Biochem. Exp. Med. 3*:23-34.

Mushett, C. W., Kelley, K. L., Boxer, G. E., and Rickards, J. C. (1952). Antidotal efficacy of viatmin B_{12a} (hydroxocobalamin) in experimental cyanide poisoning. *Proc. Soc. Exp. Biol. Med. 81*:234-237.

Nahrstedt, A. (1981). Isolation and structure elucidation of cyanogenic glycosides. In *Cyanide in Biology*, B. Vennesland, E. E. Conn, C. J. Knowles, J. Westley, and F. Wissing (Eds.). Academic Press, London.

Nahrstedt, A., Hösel, W., and Walther, A. (1979). Characterization of cyanogenic glycosides and β-glucosidases in *Triglochin maritima* seedlings. *Phytochemistry 18*:1137-1141.

Nass, H. G. (1972). Cyanogenesis: its inheritance in *Sorghum bicolor, S. sudanense, Lotus,* and *Trifolium repens*—a review. *Crop Sci. 12*:503-506.

National Institute for Occupational Safety and Health. (1976). Occupational exposure to hydrogen cyanide and cyanide salts. Criteria Document. NIOSH Publ. 77-108.

Nelson, C. E. (1953). HCN content of certain sorghums under irrigation as affected by nitrogen fertilizer and soil moisture stress. *Agron. J. 45:* 615-617.

Nestel, B. (1973). Current utilization and future potential for cassava. In *Chronic Cassava Toxicity,* B. Nestel and R. MacIntyre (Eds.). International Development Research Centre, Ottawa, pp. 11-26.

Neufeld, E. F., Lim, T. L., and Shapiro, L. J. (1975). Inherited disorders of lysosomal metabolism. *Annu. Rev. Biochem. 44:* 357-376.

Newton, G. W., Schmidt, E. S., Lewis, J. P., Conn. E. E., and Lawrence, R. (1981). Amygdalin toxicity studies in rats predict chronic cyanide poisoning in humans. *West. J. Med. 134:* 97-103.

Ng, S. J. (1975). A comparative study of the enzymic hydrolysis of cyanogenic glycosides. M. S. thesis, University of California, Davis.

Nisazawa, K., and Hashimoto, Y. (1970). Glycoside hydrolases and glycosyltransferances. In *The Carbohydrates,* Vol. IIA, W. Pigman and D. Horton (Eds.). Academic Press, New York, pp. 241-300.

Normanha, E. S. (1965). Analise de HCN em mandioca. *Cienc. Cult. (Sao Paulo) 17(2):* 197.

Offterdinger, H., and Weger, N. (1969). Circulation and respiration in cyanide poisoning and its therapy with ferrihemoglobin forming and cobalt containing agents. *Arch. Exp. Pathol. Pharmakol. 264:* 289.

Oke, O. L. (1968). Cassava as food in Nigeria. *World Rev. Nutr. Diet. 9:* 227-250.

Oke, O. L. (1969). The role of hydrocyanic acid in nutrition. *World Rev. Nutr. Diet. 11:* 170-198.

Oke. O. L. (1973). The mode of cyanide detoxication. In *Chronic Cassava Toxicity,* B. Nestel and R. MacIntyre (Eds.). International Development Research Centre, Ottawa, pp. 97-104.

Oluwole, I. L. (1935). Quoted by Clark (1935). On the aetiology of pellagra and allied nutritional diseases. *W. Afr. Med. J. 8:* 7-9.

Ortega, J. A., and Creek, J. E. (1978). Acute cyanide poisoning following administration of Laetrile enemas. *J. Pediatr. 93:* 1059.

Osuntokun, B. O. (1973). Ataxic neuropathy associated with high cassava diets in West Africa. In *Chronic Cassava Toxicity,* B. Nestel and R. MacIntyre (Eds.). International Development Research Centre, Ottawa, pp. 127-138.

Osuntokun, B. O. (1979). A degenerative neuropathy with blindness and chronic cyanide intoxication of dietary origin: the evidence in the Nigerians. *Int. Symp. Toxicol. Trop.,* Ibadan, Nigeria, February 1979.

Ovejera, A. A., Houchens, D. P., Barker, A. D., and Venditti, J. M. (1978). Inactivity of DL-amygdalin against human breast and colon tumor xenografts in athymic (nude) mice. *Cancer Treat. Rep. 62:* 576-578.

Pack, W. K., Raudonat, H. W., and Schmidt, K. (1972). Über eine tödliche Blausäurevergiftung nach dem Genuss bitterer Mandeln (*Prunus amygdalus*). *Z. Rechtsmed. 70:* 53-54.

Pharoah, P., Delange, F., Fierro-Benitez, R., and Stanbury, J. B. (1980). Endemic cretinism. In *Endemic Goiter and Endemic Cretinism,* J. B. Stanbury and B. S. Hetzel (Eds.). Wiley, New York, pp. 395-421.

Pijoan, M. (1942). Cyanide poisoning from choke cherry seed. *Am. J. Med. Sci. 204*:550-553.

Raymond, W. D., Jojo, W., and Nicodemus, Z. (1941). The nutritive value of some Tanganyka foods. II. Cassava. *E. Afr. Agric. J. 6*:154-159.

Reay, P. F., and Conn. E. E. (1974). The purification and properties of a uridine diphosphate glucose:aldehyde cyanohydrin β-glucosyltransferase from sorghum seedlings. *J. Biol. Chem. 249*:5826-5830.

Robinson, M. E. (1930). Cyanogenesis in plants. *Biol. Rev. 5*:126-141.

Rose, C. L., Worth, R. M., Kikuchi, K., and Chen, K. K. (1965). Cobalt salts in acute cyanide poisoning. *Proc. Soc. Exp. Biol. Med. 120*:780-783.

Rosenthal, G. A., and Bell, E. A. (1979). Naturally occurring, toxic non-protein amino acids. In *Herbivores—Their Interactions with Secondary Plant Metabolites*, G. A. Rosenthal and D. H. Janzen (Eds.). Academic Press, New York, pp. 353-385.

Rosenthaler, L. (1908). Durch Enzyme bewirkte asymmetrische Synthesen. *Biochem. Z. 14*:238-253.

Sadoff, L., Fuchs, K., and Hollander, J. (1978). Rapid death associated with laetrile ingestion. *JAMA 239*:1532.

Saunders, J., and Conn, E. E. (1977). Subcellular localization of the cyanogenic glucoside of *Sorghum* by autoradiography. *Plant Physiol. 59*:647-652.

Sayre. J. W., and Kaymakcalan, S. (1964). Hazards to health—cyanide poisoning from apricot seeds among children in central Turkey. *N. Engl. J. Med. 270*:1113-1115.

Schmidt, E. S., Newton, G. W., Saunders, S. M., Lewis, J. P., and Conn, E. E. (1978). Laetrile toxicity studies in dogs. *JAMA 239*:943-947.

Schrader, C. C. (1803). Neue Wahrnegmungen über die Blausäure. *Gilbert Ann. 13*:503-504.

Schwarzmaier, U. (1976). Über die Cyanogenese von *Bambusa vulgaris* und *B. gradua. Chem. Ber. 109*:3379-3389.

Seddon, H. R., and King, R. O. C. (1930). The fatal dose for sheep of cyanogenetic plants containing sambunigrin or prunasin. *J. Counc. Sci. Ind. Res. Aust. 3*:14-24.

Seely, M. K., Criddle, R. S., and Conn, E. E. (1966). The metabolism of aromatic compounds in higher plants. VIII. On the requirement of hydroxynitrile lyase for flavin. *J. Biol. Chem. 241*:4457-4462.

Seigler, D. S. (1976a). Plants of the northeastern United States that produce cyanogenic compounds. *Econ. Bot. 30*:395-407.

Seigler, D. S. (1976b). Plants of Oklahoma and Texas capable of producing cyanogenic compounds. *Proc. Okla. Acad. Sci. 56*:95-110.

Seigler, D. S. (1977). The naturally occurring cyanogenic glycosides. *Prog. Phytochem. 4*:83-120.

Seigler, D. S. (1979). In *Herbivores—Their Interaction with Secondary Plant Metabolites*, G. A. Rosenthal and D. H. Janzen (Eds.). Academic Press, New York, pp. 449-470.

Seigler, D. S. (1981). Recent developments in the chemistry and biology of cyanogenic glycosides and lipids. *Rev. Latinoam. Quim. 12*:39-48.

Sheehy, M., and Way, J. L. (1968). Effect of oxygen on cyanide intoxication. III. Mithridate. *J. Pharmacol. Exp. Ther. 161*:163-168.

Sinha, S. K., and Nair, T. V. R. (1968). Studies on the variability of cyanogenic glucoside content in cassava tubers. *Indian J. Agric. Sci. 38*:958-963.

Smith, R. L. (1971). The role of gut flora in the conversion of inactive compounds to active metabolites. In *A Symposium on Mechanisms of Toxicity*, W. N. Aldridge (Ed.). Macmillan, New York, pp. 229-247.

Smith, A. D. M., and Duckett, S. (1965). Cyanide, vitamin B$_{12}$, experimental demyelination and tobacco amblyopia. *Br. J. Exp. Pathol. 46*:615-622.

Smith, A. D. M., Duckett, S., and Waters, A. H. (1963). Neuropathological changes in chronic cyanide intoxication. *Nature (Lond.) 200*:179-181.

Smith, C. R., Weisleder, D., Miller, R. W., Palmer, I. S., and Olson, O. E. (1980). Linustatin and neolinustatin: cyanogenic glycosides of linseed meal that protect animals against selenium toxicity. *J. Org. Chem. 45*: 507-510.

Smith, F. P., Butler, T. P., Cohan, S., and Schein, P. S. (1978). Laetrile toxicity: a report of two patients. *Cancer Treat. Rep. 62*:169-171.

Snyder, F. B. (1950). Inheritance and association of hydrocyanic acid potential, disease reactions and other characters in sudangrass. Ph.D. thesis, Univ. of Wisconsin, Madison.

Sörbo, B. (1954). β-Mercaptopyruvate as a substrate of rhodanese. *Acta Chem. Scand. 8*:694-695.

Spatz, M., Smith, D. W., McDaniel, E. G., and Laquer, L. (1967). Role of intestinal microorganisms in determining cycasin toxicity. *Proc. Soc. Exp. Biol. Med. 124*:691-697.

Stevens, D. L., and Strobel, G. A. (1968). Origin of cyanide in cultures of a psychrophilic basidiomycete. *J. Bacteriol. 95*:1094-1102.

Steyn, D. G., and Rimington, C. (1935). The occurrence of cyanogenetic glucosides in South African species of *Acacia*. *Onderstepoort J. Vet. Sci. Anim. Ind. 4*:51-63.

Stock, C. C., Tarnowski, G. S., Schmid, F. A., Hutchison, D. J., and Teller, M. N. (1978a). Antitumor tests of amygdalin in transplantable animal tumor systems. *J. Surg. Oncol. 10*:81-88.

Stock, C. C., Martin, D. S., Sugiura, K., Fugmann, R. A., Mountain, I. M., Stockert, E., Schmid, F. A., and Tarnowski, G. S. (1978b). Antitumor tests of amygdalin in spontaneous animal tumor systems. *J. Surg. Oncol. 10*:89-123.

Sun, M. (1981). Commentary article: "Laetrile brush fire is out, scientists hope." *Science 212*:758-759.

Towill, L. E., Drury, J. S., Whitfield, B. L., Lewis, E. B., Galyan, E. L., and Hammons, A. S. (1978). Reviews of the environmental effects of pollutants. V. Cyanide. *U.S. EPA Doc. EPA-600/1-78-027*. U.S. Environmental Protection Agency, Cincinnati, Ohio.

U.S. Department of Health, Education and Welfare. (1977). Laetrile: the commissioner's decision. DHEW Publ. 77-3056, U.S. Government Printing Office, Washington, D.C.

Van der Velden, M., Kinthaert, J., Orts, S., and Ermans, A. M. (1973). A preliminary study on the action of cassava on thyroid iodine metabolism in rats. *Br. J. Nutr. 30*:511-517.

Van Etten, C. H., and Tookey, H. L. (1979). In *Herbivores—Their Interaction with Secondary Plant Metabolites*, G. A. Rosenthal and D. H. Janzen (Eds.). Academic Press, New York, pp. 471-500.

Viehoever, A. (1940). Edible and poisonous beans of the lima type (*Phaseolus lunatus* L.). *Thai Sci. Bull. 2*:1-99.

Viehoever, A., and Mack, H. (1935). Biochemistry of amygdalin. *Am. J. Pharmacol.* *107*:397-450.

Voegtlin, C., Johnson, J. M., and Dyer, H. A. (1926). Biological significance of cystine and glutathione. (I) Mechanism of cyanide action. *J. Pharmacol. Exp. Ther.* *27*:467-483.

Watt, J. M., and Breyer-Brandwijk, M. G. (1962). *The Medicinal and Poisonous Plants of Southern and East Africa*, 2nd ed. Churchill Livingstone, Edinburgh.

Wattenbarger, D. W., Gray, E., Rice, J. S., and Reynolds, J. H. (1968). Effects of frost and freezing on hydrocyanic acid potential of sorghum plants. *Crop. Sci.* *8*:526-528.

Westley, J. (1973). Rhodanese. *Adv. Enzymol.* *39*:327-368.

Williams, A. O., and Osuntokun, B. O. (1969). Light and electron microscopy of peripheral nerves in tropical ataxic neuropathy. *Arch. Neurol. (Chicago)* *21*:475-492.

Wilson, J. (1965). Leber's hereditary optic atrophy: a possible defect of cyanide metabolism. *Clin. Sci.* *29*:505-515.

Winkler, W. O. (1951). Report on hydrocyanic glucosides. *J. Assoc. Off. Agric. Chem.* *34*:541.

Wodinsky, I., and Swiniarski, J. K. (1975). Antitumor activity of amygdalin MF (NSC-15780) as a single agent and with beta-glucosidase (NSC-128056) on a spectrum of transplantable rodent tumors. *Cancer Chemother. Rep.* *59*:939-950.

Wohler, F., and Liebig, J. (1837). Ueber die Bildung des Bittermandelöls. *Ann. Phys. Chem.* *11*:345-366.

Wokes, F., and Picard, C. W. (1955). The role of vitamin B_{12} in human nutrition. *Am. J. Clin. Nutr.* *3*:383-390.

Wood, T. (1965). The cyanogenic glucoside content of cassava and cassava products. *J. Sci. Food Agric.* *16*:300-305.

Wood, J. L., and Cooley, S. L. (1956). Detoxication of cyanide by cystine. *J. Biol. Chem.* *218*:449-457.

Wood, J. L., and Fiedler, H. (1953). β-Mercaptopyruvate, a substrate for rhodanese. *J. Biol. Chem.* *205*:231-234.

Part II
TOXINS THAT ARE TERATOGENIC

5

NATURALLY OCCURRING TERATOGENS FROM PLANTS

RICHARD F. KEELER

U.S. Department of Agriculture, Logan, Utah

I. INTRODUCTION

Congenital deformities of humans and animals were speculated in antiquity to arise from causes ranging from maternal mental impressions at the time of conception or pregnancy to association with devils or witches (Warkany, 1971). More recently, through epidemiologic investigation and experimental trials, investigators have secured more reliable information on causes. Now, most congenital defects can be ascribed to naturally occurring compounds from plants or fungi (teratogens as they are called), human-made compounds (teratogens) of various types, viral diseases, or genetic defects. In this volume on natural toxins, consideration is given to both plant and fungal teratogens—the former in this chapter, the latter in Chap. 6.

Information on teratogenic compounds from plants has become available generally from two sources: (1) investigations of epidemics of congenital deformities in livestock, and (2) investigations of hazards of plant derived compounds used as drugs, in foods, or in other items of commerce.

In some instances, experimental work has progressed to the point that teratogen structures, configurations, and mechanisms of effect are known together with practical methods for elimination of the problems. In some instances, teratogens are yet to be identified from plants positively incriminated as teratogenic sources or plants speculated to be sources.

Although there are a wide variety of compounds found in plants that are known to be teratogenic, for purposes of this review we will exclude those that are common constituents of mammalian systems and that are teratogenic under unusual circumstances. Thus we will not include vitamins, inorganic elements, sugars, common purines and pyrimidines, or common amino acids; in other words, compounds excluded are those generally considered to be nontoxic. The reader is referred to excellent reviews covering teratogenic effects of such compounds and the unusual circumstances giving rise to these effects (Shephard, 1980; Kalter, 1968; and Wilson and Fraser, 1977).

Considered in this chapter are the naturally occurring plant-derived teratogens not commonly found in mammalian systems—the secondary plant constituents that in many cases are quite toxic as well as teratogenic. Considered also are those plants known or thought to give rise to teratogenic

effects whose teratogens are yet to be isolated. The organization of the text
arranges teratogens according to compound class, with a final section on
plants with teratogens of unknown structure.

II. ALKALOIDS

A. Protoalkaloids

1. *Colchicine*

The protoalkaloid colchicine [1] long used to relieve the pain associated
with gout, and that held some promise in cancer chemotherapy, is found in
seeds and corms of *Colchicum autumnale* (autumn crocus). The plant is a
native of the United Kingdom and Europe and is present in the United States
as a garden escape. Livestock are occasionally poisoned by the plant because
of its presence in hay or pastures. Colchicine has been used extensively in
genetic investigations as a polyploiding agent because of its ability to inhibit
mitosis at the metaphase, thereby preventing cell division (Kingsbury, 1964).

Colchicine or derivatives are teratogenic in laboratory animals. In a hamster embryo study on mitosis inhibition, Ferm (1963a) found an unexpected
teratogenic effect on the embryo and investigated teratogenicity in pregnant
hamsters by intravenous injection of colchicine on the eighth day of gestation.
The dose produced high fetal mortality and gross deformities that included
microphthalmia, anophthalmia, umbilical hernia, exencephaly, and skeletal
anomalies. Shoji and Makino (1966) showed that SC injections of colchicine
during the first 2 weeks of pregnancy produced a variety of congenital deformities in mice, including exencephaly, polydactyly, cleft palate, agnathia,
facial defects, syndactyly, and umbilical hernia.

Teratogenic effects of colchicine and derivatives have been studied by
many workers (Ingalls et al., 1968; Szabo and Kang, 1969; and Sieber et al.,
1978), including those who wondered whether some human abnormalities may
have been caused by colchicine therapy during pregnancy. For example,
Ferreira and Frota-Pessoa (1969) reported that children from 2 of 54 parents
undergoing colchicine therapy had Down's syndrome.

2. *Mescaline*

The protoalkaloid mescaline [2] is found in the peyote cactus *Lophophora
williamsii* and is one of the many psychoactive alkaloids of the plant. The
peyote cactus was used in religious rites by native Americans long before

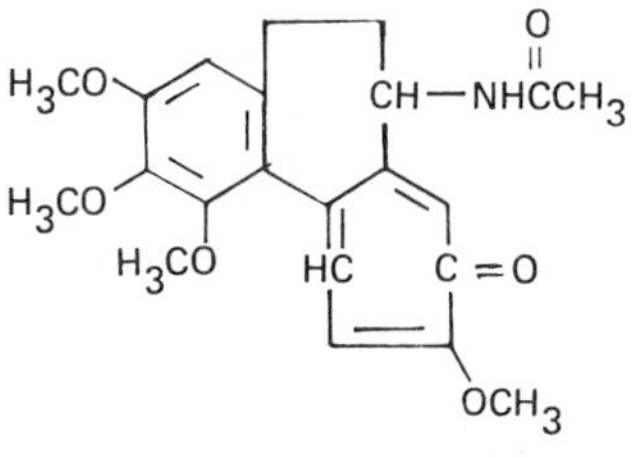

[1]

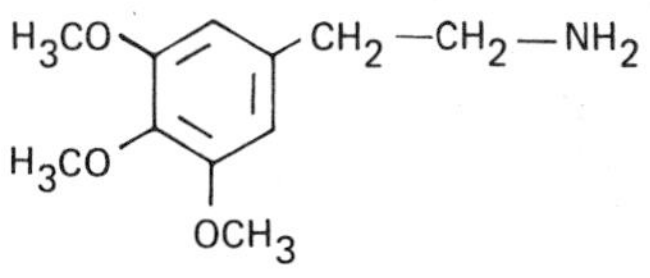

[2]

Spaniards came to this hemisphere and is still in use by some members of the Native American Church in religious ceremonies.

Mescaline was shown by Gerber (1967) to be teratogenic in hamsters when injected subcutaneously on the eighth day of pregnancy. Abnormalities included exencephaly, spina bifida, meningocele, omphalocele, hydrocephalus, myelocele, edema, and hemorrhage.

Mescaline is a rather simple phenethylamine related in structure to certain pressor amines such as L-dopa [3], epinephrine [4], and norepinephrine [5]. These three compounds, common constituents of mammalian systems involved in nerve transmission, have all been reported to be teratogenic (see Shephard, 1980). L-Dopa is a constituent of *Vicia faba* (broad bean) (Liener, 1980), a common food and feed in some parts of the world. Norepinephrine has been reported in bananas and plantain (Strong, 1966). The mechanism of action of the mescaline teratogenesis has not been studied, but it may share teratogenic mechanisms with the pressor amines L-dopa, epinephrine, and norepinephrine and may relate to the potent vasoconstrictive propensities of these compounds. Constriction of uterine blood vessels with impairment of placental and fetal blood supply could lead to teratogenic effects.

B. Imidazole Alkaloids

1. *Pilocarpine*

Members of the *Pilocarpus* genus are sources of the imidazole alkaloid pilocarpine [6]. The compound acts on cholinergic receptor sites, as does acetylcholine. Because of its potent cholinergic properties, it has been used in treatment of glaucoma, where it promptly reduces intraocular pressure and thereby helps to prevent damage to the optic nerve and retina.

Landauer showed in 1953 that pilocarpine was teratogenic in the chick. He injected pilocarpine hydrochloride into the yolk sac of developing eggs at various dosages. Treatment at 16 hr of incubation produced rumplessness,

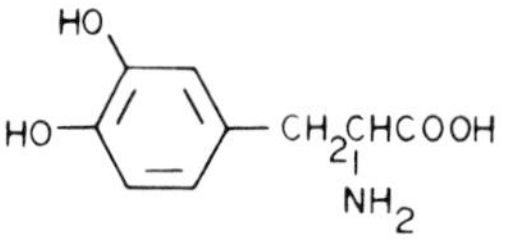

[3]

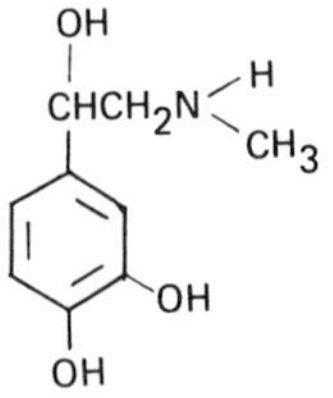

[4]

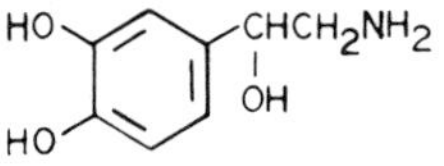

[5]

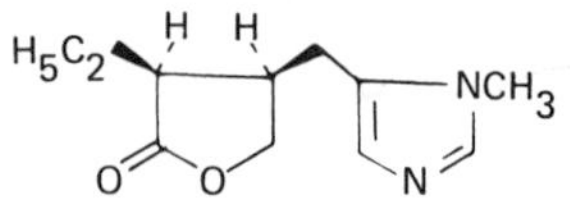

[6]

beak defects, bent tibia, tarsometatarus defects, short toes, and syndacty-lism. Nicotinamide protected a large proportion of the embryos against damage, leading Landauer to speculate that the mechanism of effect must be an interference with carbohydrate utilization.

C. Indole Alkaloids

1. *Reserpine*

Rauwolfia spp. have been used for centuries in Indian folk medicine. Scientific examination of the folk claims resulted in isolation of the alkaloids reserpine [7], deserpidine, and others. These alkaloids find extensive use as hypotensive agents and tranquilizers. They deplete norepinephrine, the neurotransmitter, from sympathetic nerves, relax blood vessels, and thus lower blood pressure.

Two reports suggest possible teratogenicity for these alkaloids. Tuchmann-Duplessis and Mercier-Parot (1961) reported that deserpidine at high dose induced a low incidence of limb, tail, and skeletal defects in rats. Goldman and Yakovac (1965) treating pregnant cats with reserpine found anopthalmia and other defects in a small percentage of the survivors. Other species seem resistant to teratogenesis by these alkaloids (see the summary by Schardein, 1976).

2. *Vincristine*

The periwinkle plant (*Vinca rosea*) is the source of the tumor chemothera-peutic indole alkaloid vincristine [8] and related alkaloids. Some of these alkaloids have proven teratogenic in a number of species of laboratory ani-mals—hamsters, rats, rabbits, and monkeys (Schardein, 1976; Kalter, 1968). Perhaps terata might be induced under grazing conditions in cattle because in India and Australia at least, *Vinca* has been reported to be poisonous to cattle—so clearly they are ingesting it (see Gardner and Bennetts, 1956). Ferm (1963b) showed that microphthalmia, anophthalmia, spina bifida, and various skeletal defects were common in hamsters born to mothers treated on the eighth day of gestation with vincristine. Joneja and Ungthavorn (1970) reported defects of the central nervous system, eyes, limbs, tail, and ribs in mice born to vincristine-treated mothers. Demyer (1965) found that vincristine by intramuscular injection in rats on the ninth day of pregnancy

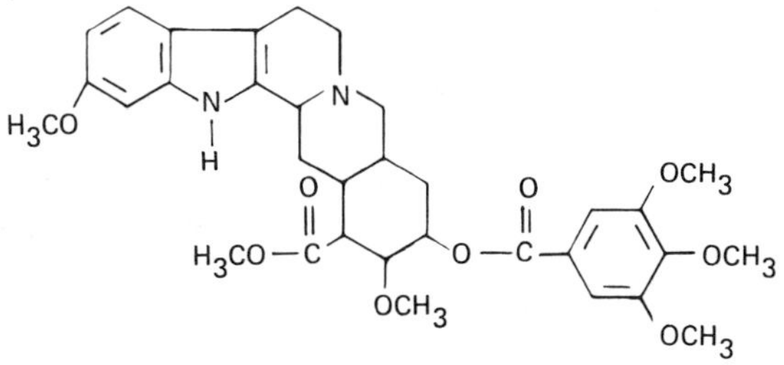

[7]

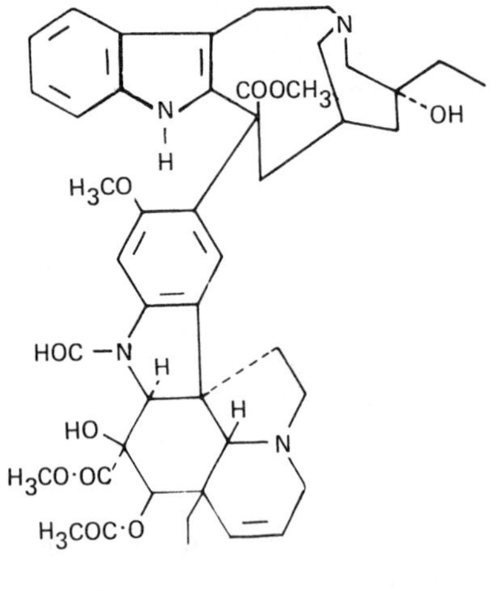

[8]

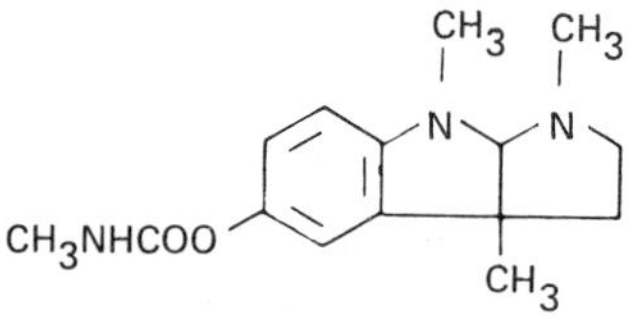

[9]

produced numerous offspring with microcephaly, encephalocele, exencephaly, or anencephaly.

3. Physostigmine

Physostigmine (eserine) [9], an indole alkaloid from *Physostigma veneno-sum*, is considered in drug parlance as an anticholinesterase because it inhibits the enzyme cholinesterase, thereby raising the acetylcholine level at neuroskeletal junctions, giving rise to stimulation of voluntary muscles. For this reason it has been used in treatment of myasthenia gravis and has also been used for glaucoma.

Physostigmine has long been recognized as teratogenic in chicks, having been shown to produce brachymelia and spina bifida (Ancel, 1946); syndacty-lism and micromelia (Landauer, 1954); and severe vertebral defects and micro-melia (Bueker and Platner, 1956). According to Arcuri and Gautieri (1973), physostigmine is also teratogenic in CF1 mice. Landauer (1975) found that administration of gallium together with physostigmine moderated somewhat cer-tain of the teratogenic effects of physostigmine alone, specifically the verte-bral defects. Acetylcholine itself at high doses was also teratogenic but pro-duced entirely different effects. Landauer speculated that the mechanism of the teratogenic effect of physostigmine was a displacement of acetylcholine from its receptors or a formation of complexes with acetylcholine.

4. Indole-3-acetic Acid

An ubiquitous naturally occurring plant growth hormone, indole-3-acetic acid [10], was found to produce teratogenic effects in mice and rats by John et al. (1979). Although it is somewhat beyond the stated scope of this chap-ter, we include it because of its structural relationship to the indole serotonin (another of the common pressor amines), which is also teratogenic. Sero-tonin is found in bananas, plantain, pineapple, and watermelon (Liener, 1980; Strong, 1966). The work of John et al. (1979) showed that indole-3-acetic

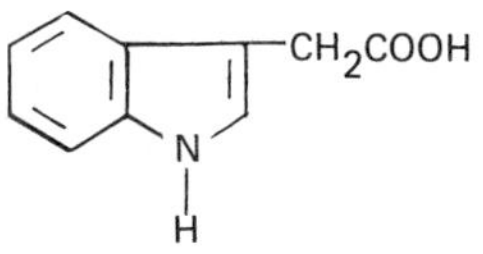

[10]

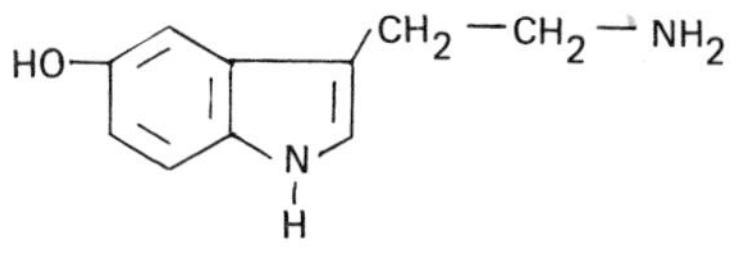

[11]

acid produced cleft palate in CF1 mice and Sprague-Dawley rats. In addition, treated mice sometimes had exencephaly, ablepharia, dilated cerebral ventricles, and crooked tail.

5. Serotonin

Subcutaneous injection of serotonin [11] in mice produced about 25% deformed fetuses (Thompson and Gautieri, 1969). Deformities included gastroschisis, hydrocephalus, and hydronephrosis. A possible mechanism by which serotonin (and possibly indole-3-acetic acid) could exert teratogenic effects relates to vasoconstriction. Robson and Sullivan (1968) suggested that vasoconstriction of uterine blood vessels immediately adjacent to and supplying the placenta could occur, thereby depriving the placenta and the fetus of oxygen and various nutrients.

D. Isoquinoline Alkaloids

1. D-Tubocurarine

This complex alkaloid [12] derived from *Chondrodendron tomentosum* is a useful muscle relaxant. But it has produced clubfoot in both chick embryos (Drachman and Coulombre, 1962) and rat fetuses (Shoro, 1972) when given during late incubation or gestation. Its ability to paralyze skeletal muscles suggests that its action as a teratogen may simply result from temporarily immobilizing prenatal movement during gestation, which Shephard (1980) suggests emphasizes that movement of joints is essential during fetal development.

E. Piperidine Alkaloids

1. Coniine

Although most crooked calf disease in western United States is due to maternal ingestion of lupine plants during gestation (see Sec. II.J), *Conium maculatum* is the cause of some cases not due to lupine ingestion. Deformities include limb rotation, permanent carpal or elbow joint flexure, spinal curvature, or cleft palate and lip (Keeler, 1974; Keeler and Balls, 1978). We found that both coniine [13] and γ-coniceine [14], two of the main piperidine alkaloids of the plant, were teratogenic (Keeler, 1974). Evidently, introduction of a double bond in the ring between the nitrogen and the α-carbon atom (i.e., producing a tertiary nitrogen) as in γ-coniceine did not result in loss of activity.

A number of coniine analogs were tested to determine what changes could be introduced in the side chain and ring of coniine and still retain teratogenicity (Keeler and Balls, 1978). None caused congenital deformities. Both

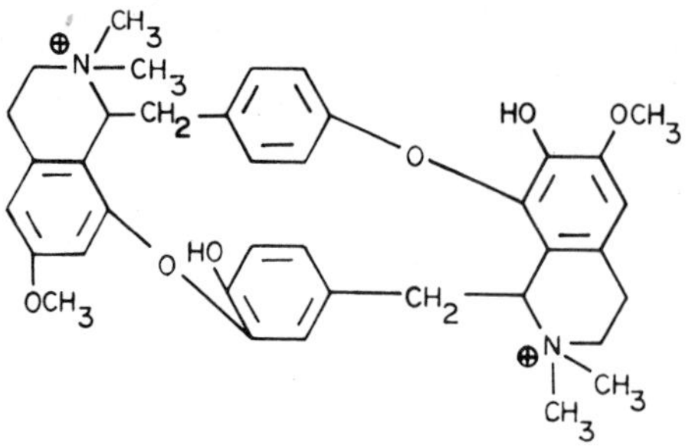

[12]

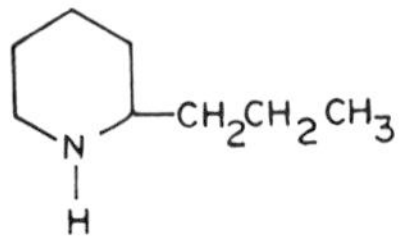

[13]

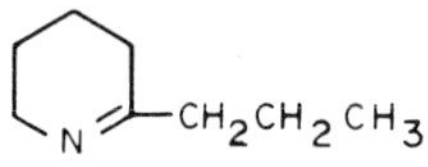

[14]

chain length and degree of ring unsaturation appeared to affect teratogenicity. Neither 2-ethylpiperidine nor 2-methylpiperidine was teratogenic, even
though coniine (2-propylpiperidine) was. Coniine with a fully saturated ring
and γ-coniceine with one double bond were active, but the coniine analog,
conyrine, with the ring fully unsaturated was not active.

2. Anabasine

From the coniine analog results above, one would expect to be teratogenic
naturally occurring piperidines with a side chain α to the nitrogen at least
propyl in length and with the piperidine ring saturated. One such possibility is the naturally occurring piperidine alkaloid in tobacco, anabasine [15].
We have speculated that it may account for teratogenic effects of tobacco on
pigs (Keeler, 1979), a condition originally noticed by Crowe (1969).

Crowe (1969) described malformed pigs from five Kentucky farms. During
gestation sows that delivered 64 litters with 300 skeletally deformed pigs had
access to and had apparently ingested tobacco stalks during gestation. Crowe
speculated that the tobacco produced the deformities. Subsequently, Menges
et al. (1970) reported a tobacco-related epidemic in pigs of similar deformities.
Twisting of the fore or hind limb digits was common. In 1974, Crowe and
Swerczek produced congenital deformities in pigs by feeding alkaloid containing tobacco leaf filtrates to pregnant sows. They did not determine whether
the alkaloids were responsible, but they reported negative results with nicotine suggesting that nicotine [16] was probably not the teratogen. Feeding
trials in sheep and cows with free-base nicotine at our laboratory in cooperation with Dr. Crowe produced no discernible congenital defects (but see Sect.
II.G for teratogenic effects in laboratory animals).

The suspect teratogen in tobacco, anabasine [15], is a known teratogen
in chicks (Upshall, 1972; Sanatina et al., 1969/1971; Landauer, 1960; Tsumo
et al., 1960) and therefore likely teratogenic in livestock.

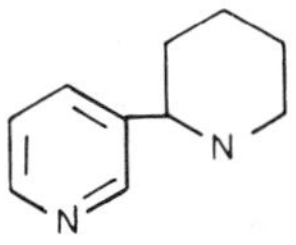

[15]

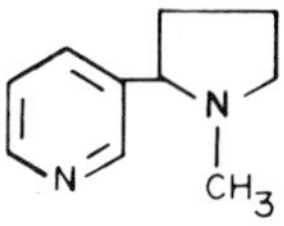

[16]

One of the best naturally occurring sources of anabasine is *Nicotiana glauca*, in which anabasine is reported to occur as the principal or possibly the sole alkaloid (Smith, 1935; Rindle and Sapiro, 1949). We conducted teratogenicity trials with *N. glauca* and found it to be teratogenic (Keeler, 1979; Keeler et al., 1981). Seven calves born to seven cows fed the plant from gestation days 50 to 75 or 40 to 75 were deformed at birth. Daily doses produced clinical signs of toxicity (irregular gait, excessive salivation, recumbency) in the dams. Abnormalities in the calves at birth were principally arthrogryposis of fore limbs and spinal curvature. Teratogenicity trials in pigs showed that *N. glauca* produces deformities in that species too. Deformed offspring were found in four litters from four sows fed *N. glauca* from gestation days 18 to 68 at daily doses that produced moderate to severe toxicosis signs (irregular gait, tremors, recumbency) in dams during the feeding period. Abnormalities in newborn pigs included arthrogryposis of front or rear limbs.

Anabasine was present in the plant material fed at a concentration of 0.113% and represented well over 99% of the total alkaloids as measured by gas chromatographic separations and as identified by comparative infrared spectra with authentic anabasine (Keeler, 1979) and by mass spectral fragmentation (Keeler et al., 1981).

Clearly, if an alkaloid from *N. glauca* is responsible for the teratogenicity of that plant, it must be anabasine. If anabasine is responsible, it becomes a prime suspect to account for the teratogenicity in pigs fed common tobacco because the alkaloid-containing fraction from common tobacco produces deformities (Crowe and Swerczek, 1974) and it contains anabasine (Keeler, 1979).

F. Xanthine Alkaloids

The xanthine alkaloids caffeine [17], theophylline [18], and theobromine [19] derived from members of the *Coffea*, *Camellia*, and *Theobroma* species are common constituents of certain beverages such as coffee, tea, and chocolate. Use of these beverages and others containing xanthine alkaloids is nearly worldwide, so exposure of humans to xanthine alkaloids is nearly universal. These beverages are popular because of the stimulating properties of the contained xanthine alkaloids. But the purine-type structure of the alkaloids suggests that they may pose some hazard as mutagens or teratogens.

Indeed, a number of reports suggest that both caffine and theophylline are teratogenic in laboratory animals. But epidemiologic evidence for teratogenicity in humans at generally used dosages is conflicting.

Nishimura and Nakai (1960) using single intraperitoneal dosages of caffeine in pregnant SMA swine found offspring with digital defects, cleft palate, or hematoma. Knoche and Konig (1964), using daily oral doses in NMRI mice

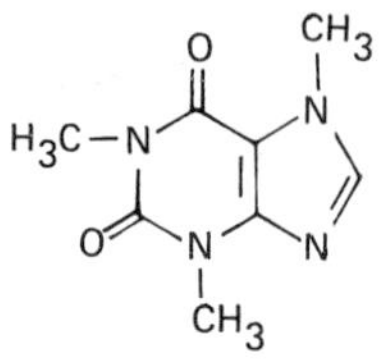

[17]

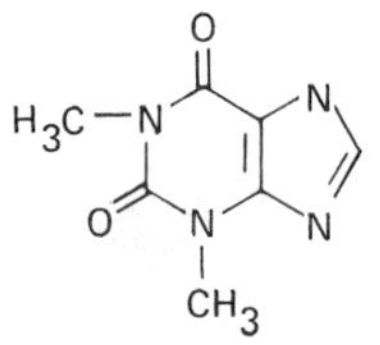

[18]

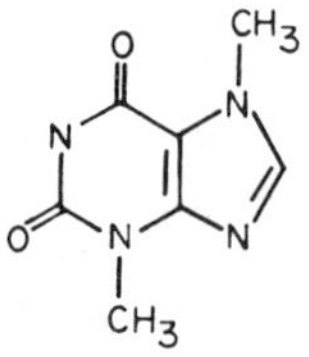

[19]

before conception and throughout gestation, found a very low incidence of cleft palate or other facial defects in offspring. Bertrand et al. (1965) reported 20% limb reductions in Wistar rat fetuses whose dams received daily oral doses before conception and throughout gestation. Fujii et al. (1969) found cleft palate, hematoma, clubfoot, and digital reduction at an incidence of about 20% in ICR-JCL mice when dams received caffeine subcutaneously or intraperitoneally on the gestation day 12. Bertrand et al. (1970) reported deformities in some breeds of rats and rabbits from maternal oral intake of caffeine. Fujii and Nishimura (1972) found cranial and sacral defects in Sprague-Dawley rat fetuses following daily oral administration throughout gestation. Gilani and Giovinazzo (1981) injected chick embryos with caffeine and found particularly high teratogenicity and toxicity on day 2. Terata included micropthalmia, exencephaly, short neck, abnormal beak, reduced body size, and hemorrhage.

Ishikawa et al. (1978) reported that 1 ml of theophylline topically applied to the surface of the chorioallantoic membrane of chick eggs in the vicinity of the embryonic heart caused aortic anurisms in 63% of the chicks and, of these 48% had ventricular septal defects.

These two xanthine alkaloids (caffeine and theophylline) certainly seem to be teratogenic in some strains of laboratory animals. Mulvihill (1973) and Shephard (1980) reviewed the literature, most of which reports caffeine to be teratogenic; but some investigators, using other laboratory animal strains, have found no teratogenicity.

G. Pyridine Alkaloids

Alkaloids of the pyridine class are found in members of the *Lobelia* and *Nicotiana* genera. They contain a number of pyridine alkaloids, the most common of which is nicotine [16]. Vara and Kinnunen (1951) showed that nicotine

produced fetal resorptions in rabbits. Injected subcutaneously or intraperi-
tonally into pregnant mice (Nishimura and Nakai, 1958), nicotine produced
resorptions, decreased litter size, and congenital defects of the skeletal system. When given at most any time during the second week of gestation,
nicotine produced malformations in fetal mice, including defects of the digits,
of the elbow and wrist joints, of the spine, and of the palate. The injection
of nicotine sulfate into fertilized chicken eggs at various stages of incubation
resulted in teratogenic effects in the hatched chicks, according to Landauer
(1960), ranging from a shortening and twisting of the neck that was due to
incomplete and irregular formation and fusion of cervical vertebrae to dwarf-
ing.

Upshall (1972) found nicotine injected into the yolk sac of chicken eggs on
day 4 of incubation to be severely embryolethal and teratogenic. Mosier and
Jansons (1972), using tritiated nicotine, found that fetal plasma radioactivity
exceeded that of maternal plasma, within 30 min after intravenous injection of
nicotine in the dam, showing a very quick and prolonged exposure of fetal
tissue to the compound.

H. Pyrrolizidine Alkaloids

Pyrrolizidine alkaloids are found in *Senecio, Amsinckia, Crotalaria, Echium,
Heliotropium, Trichodesma*, as well as other genera. Some of these alkaloids
are hepatotoxic (Kingsbury, 1964; Johnson, 1978; Culvenor, 1973). Among
the most toxic are the allylic esters of heliotridine and retronecine such as
the compound heliotrine [20]. Hepatotoxic alkaloids of the pyrrolizidines are
all esterified and have 1,2-double bonds. Toxicosis, however, is not due to
the alkaloids, but rather to metabolites produced by liver microsomes when
they oxidize the compounds to the dehydroform as the dehydro alkaloids or
as further hydrolyzed to dehydramino alcohols. Pyrrolic dehydro alkaloids
(e.g., dehydroheliotrine) are alkylating agents, which may account for
their hepatotoxicity (Kingsbury, 1964; Culvenor, 1973).

Chronic liver lesions from low pyrrolizidine doses are esssentially irreversible and progressive; they may not be evident until months or years
after ingestion of the alkaloids. Livers often have enlarged parenchymal
cells; they may be deformed and nodular and have varying degrees of fibrosis. Acute poisoning from high doses is characterized by liver necrosis
(Kingsbury, 1964; Johnson, 1978). Certain structural features conferring
toxicity have been defined by Mattocks (1978).

[20]

Schoental (1959) found that lasiocarpine and retrorsine, two pyrrolizidine alkaloids from *Senecio*, fed to lactating rats, in doses having no apparent effects on the mothers, caused fetal liver lesions in the young. The hepatotoxic alkaloids can pass into the milk, so Green and Christie (1961) assessed the teratogenic effects of pyrrolizidine alkaloids. Pregnant rats were administered heliotrine intraperitoneally during gestation in a single dose, usually during the second week of gestation. Offspring were deformed and development was generally retarded, with malformations of ribs, vertebra, lower jaw, and occasionally, palate, limbs, and tail. Persaud and Hoyte (1974) showed that fulvine was fetotoxic and highly teratogenic in rats. Intraperitoneal injection on days 9-12 of gestation induced exencephaly, cleft palate, microphthalmia, limb and tail abnormalities, hydrocephalus, ectopic kidneys, and other defects. The dehydroalkaloid metabolites of pyrrolizidine alkaloids are so reactive that they generally exert their toxic effects on the dam in the organ of origin, the liver, or the next organ, the lung. They seldom circulate to other organs. Effects elsewhere are usually caused by the less reactive metabolites, the dehydroamino alcohols. For this reason Peterson and Jago (1980) assessed the teratogenicity of the dehydroamino alcohol of heliotrine, dehydroheliotridine, in rats. Upon day 14 dosing, it produced a number of abnormalities, including retarded ossification, distented rib and long bones, cleft palate, and feet and visceral defects. Johnson (1983) fed two groups of pregnant cows chronic lethal or near-lethal doses of *Senecio jacobaea* on gestation days 15 through 30 or 30 through 45, but failed to produce teratogenic effects or definitive liver changes in the offspring.

The mechanism is unknown by which deformities are produced by pyrrolizidine alkaloids but may be due to the alkylating properties of the metabolites, probably the dehydroaminoalcohols.

I. Quinoline Alkaloids

The quinoline alkaloid quinine [21] from *Cinchona ledgeriana*, which found widespread use as an antimalarial drug early in the ninetheenth century, may possess teratogenic properties.

Covell (1936) found congenital defects in the ear of fetal guinea pigs after maternal dosages of quinine, as did Mosher (1938). Whether quinine has caused congenital defects in humans is equivocal, however (Robinson et al., 1963; Tanimura and Lee, 1972).

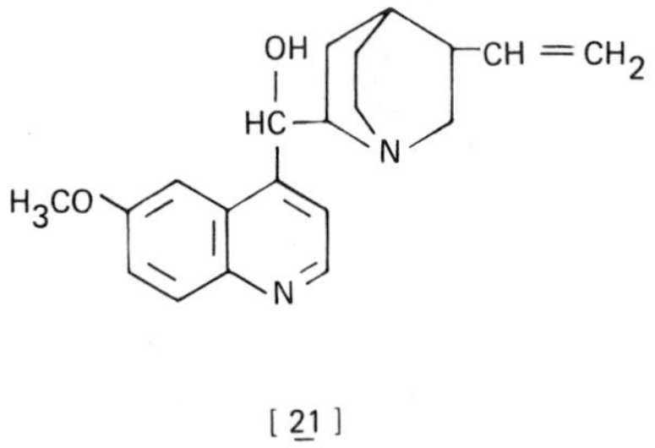

[21]

J. Quinolizidine Alkaloids

A congenital deformity in calves called crooked calf disease (Fig. 1) is prevalent in Alaska, Canada, and the United States. The disease is expressed by deformities that include twisted or bowed limbs, spinal or neck curvature, and cleft palate. The disease is caused by maternal ingestion of certain members of the *Lupinus* genus by the pregnant cow from gestation day 40 to 70 (Shupe et al., 1968). Sometimes as many as 30% of calves born in a given herd will be affected. The skeletal deformities are permanent and may increase in severity because of the stress of added body weight.

Epidemiologic evidence (Keeler, 1973a,b) suggested that the teratogen was probably anagyrine [22] because of correlation between presence of anagyrine and the ability of a given lupine to produce the disease. Results from feeding trial experiments (Keeler, 1976) established that alkaloid extracts from teratogenic lupines could produce the disease. Preparations rich in anagyrine were teratogenic, but other alkaloids present in the semipurified anagyrine preparations did not produce the disease when fed to cows. There-

[22]

FIGURE 1 Arthrogrypotic manifestation of crooked calf disease from maternal *Lupinus* consumption.

fore, both feeding trials and epidemiologic evidence suggest that the terato-genicity was due to anagyrine. Severity of malformations was directly re-lated to the level of anagyrine present in the preparations fed (Keeler, 1976).

In *L. sericeus* and *L. caudatus*, two teratogenic lupines, concentration of all alkaloids, including anagyrine, in aboveground parts was high early in growth and decreased markedly as the plants matured. Concentration in mature, intact seeds was also very high (Keeler et al., 1976a). Consequently, a pregnant cow in the period from gestation day 40 to 70 is at greatest hazard when grazing teratogenic lupine early in growth or during the seeding stage. The hazard is low when flowering and postseeding stage plants are grazed.

K. Steroidal Alkaloids

1. *C-Nor-D-homosteroidal Alkaloids (Jerveratrum Series)*

Before the teratogenic cause was understood, epidemics of cyclopia and related congenital deformities in sheep were common in Idaho. Hundreds of lambs were afflicted with these deformities each year (Binns et al., 1963). Deviations from normal included a single or double globe cyclopia (Fig. 2) and a shortened upper jaw and protruding and curved lower jaw, occasionally with a peculiar skin-covered proboscis above the single eye. Some lambs had normal eyes and a shortening of the upper jaw or cebocephaly. *Veratrum californicum* caused the condition when pregnant ewes ingested it on gestation day 14 (Binns et al., 1965). Results from feeding trials suggested that the teratogen was an alkaloid. A number of veratrum alkaloids were tested in pregnant sheep for teratogenicity. The compounds jervine [23], cyclopamine (11-deoxojervine) [24], and cycloposine (3-glucosyl-11-deoxo-jervine) [25] produced deformities similar to natural cases (Keeler and Binns, 1968).

The three teratogenic compounds are closely related steroidal furano-piperidines, but cyclopamine is the teratogen of natural importance because of plant concentration. Closely related compounds devoid of the furan ring did not produce cyclopia in sheep, suggesting that an intact furan ring was required for activity (Keeler, 1970a), perhaps conferring some essential configuration on the molecule.

Cattle and goats as well as sheep (all ruminant animals) proved suscepti-ble to the teratogenic action of orally administered *V. californicum* (Binns et al., 1972).

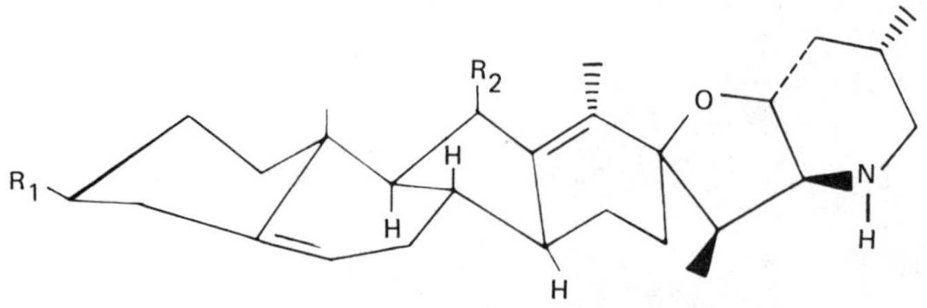

[23] , R_1 = OH, R_2 = O
[24] , R_1 = OH, R_2 = 2H
[25] , R_1 = GLU, R_2 = 2H

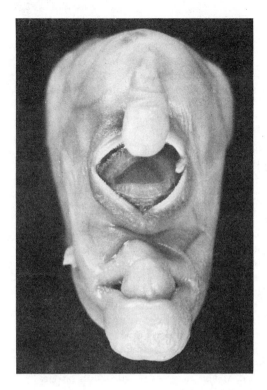

FIGURE 2 Cyclopia in a lamb from maternal consumption of *Veratrum*.

Rabbits were also susceptible (Keeler, 1970b) when the compound was administered orally on gestation day 7, even though there was some aromatization of the administered cyclopamine to veratramine as a result of the lability of cyclopamine to stomach acid. Veratramine does not produce cephalic deformities but is very toxic. Anatomical deviations in rabbits included double globe cyclopia and cebocephalia with a single or closely spaced double nostril (Keeler, 1970b; 1971). Cyclopamine also produced deformities in rats, mice, and hamsters gavaged during the same primitive streak/neural plate stage of development (Keeler, 1975), but cyclopics were not evident. In rats, microphthalmia and cebocephalia predominated; in mice, a few exencephalics resulted; whereas in hamsters, cebocephalia, encephalocele (cranial bleb), exencephaly, and hare lip were common. Malformations in hatched chicks were produced by direct application of 1-2 mg of cyclopamine to the embryonic shield of windowed chicken eggs (Bryden et al., 1973; Bryden and Keeler, 1974). This showed that maternal metabolic alteration of cyclopamine was not necessary. Further, in other experiments in sheep, intrauterine injection of as little as 1-2 mg of cyclopamine produced deformities. Direct treatment of the embryo with cyclopamine produced teratogenic effects (Bryden and Keeler, 1974).

2. Spirosolane Furanoalkaloids

Solasodine and its glycosides are found in certain plants, including eggplant. These spirosolane alkaloids are based on the C_{27}-carbon steroid skeleton of cholestane with a furan ring fused to ring D of the steroid and a piperidine ring attached spiro to the furan at carbon 22. Solasodine [26] was first reported to be teratogenic in rats by Seifulla and Ryzhova in 1972. Keeler et al. (1976b) showed that the compound produced teratogenic effects in hamsters dosed orally on days 7 and 8 of gestation. Deformities produced in hamsters were exencephaly, cranial bleb, and spina bifida, each occurring with about equal incidence. Keeler et al. (1976b) compared the teratogenicity in hamsters of solasodine [26]; tomatidine [27]; diosgenin [28], which is the non-nitrogen-containing analog of solasodine; and cyclopamine [24]. Solasodine was teratogenic and produced the deformities mentioned above when dams were gavaged a dose nearly 10 times that required to produce terata with cyclopamine. Neither diosgenin nor tomatidine was teratogenic. Neither the C-nor-D-homo steroid skeleton nor the fused furanopiperidine was apparently essential to teratogenicity. Presence of a nitrogen in ring F appeared to be essential, as well as a certain configurational position of that nitrogen conferred by virtue of the spiro connection between rings D and E or E and F. In the teratogens cyclopamine and solasodine the essential nitrogen projects α with respect to the plane of the steroid. In nonteratogenic tomatidine the nitrogen of the piperidine ring projects in the β direction with respect to the plane of the steroid. Cyclopamine is almost 10 times as teratogenic as solasodine. The data suggested as one possibility that teratogenicity might be a function of the extent of the α projection of the nitrogen, inasmuch as the extent of α projection is much greater in cyclopamine than in solasodine.

3. Solanidan Alkaloids

These alkaloids, common to potatoes and tomatoes (*Solanum* and *Lycopersicon* genera), are based on the C_{27}-carbon steroid skeleton of cholestane with a tertiary nitrogen shared by rings E and F, the nonsteroidal rings of the molecule. Important examples include solanidine [29], demissidine, and their glycosides. Because of interest in an hypothesized teratogenicity of potatoes (Renwick, 1972, and see below), many workers investigated whether solanidan alkaloids might be teratogenic. A few workers reported teratogenicity of solanidine in chick embryos and hamsters, but most workers found no terata— rats, rabbits, and mice seemed resistant.

Swinyard and Chaube (1973) injected pure solanine, the glycoside of solanidine, in pregnant rat and rabbit dams but produced no neural tube defects. Nishie et al. (1975) reported that the glycosides tomatine, α-chaconine, and α-solanine did not produce significant teratogenic effects in chicks. Intraperitoneal injection of solanine to pregnant mice failed to produce terata, according to Bell et al. (1976). Chaube and Swinyard (1976) found no neural tube defects in offspring from rats subject to acute or chronic intraperitoneal injection of α-chaconine or α-solanine. Mun et al. (1975), however, showed that pure solanine produced "rumplessness or trunklessness" in chick embryos when eggs were treated during early development. Jelinek et al. (1976) verified and extended those observations.

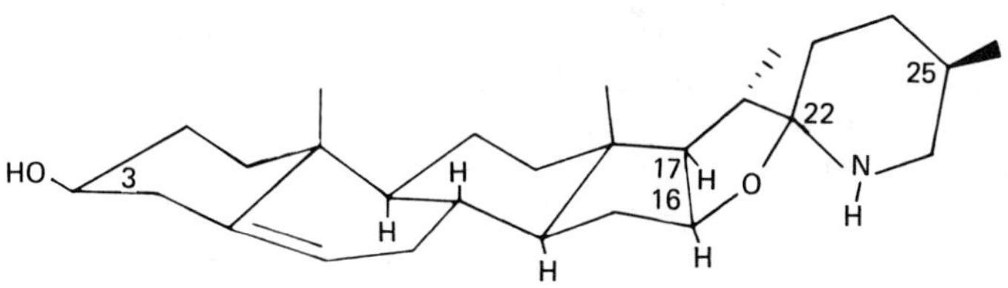
HO
3
H
H
H
17
16
H
O
H
22
25
N
H

[26]

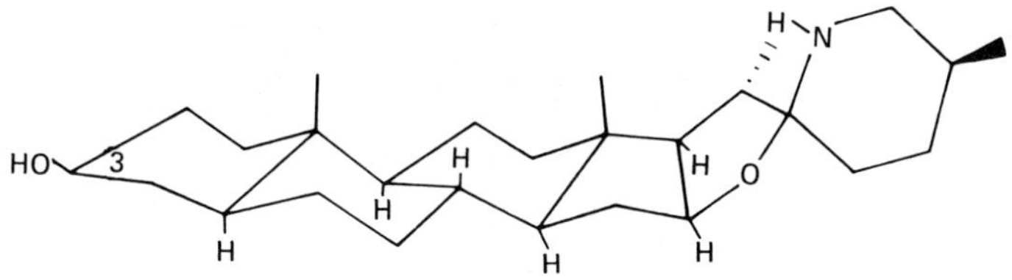
HO
3
H
H
H
H
H
O
H
H
N

[27]

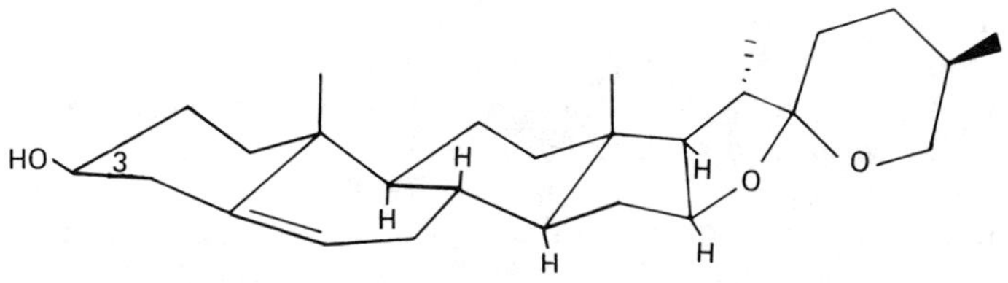
HO
3
H
H
H
H
O
H
O

[28]

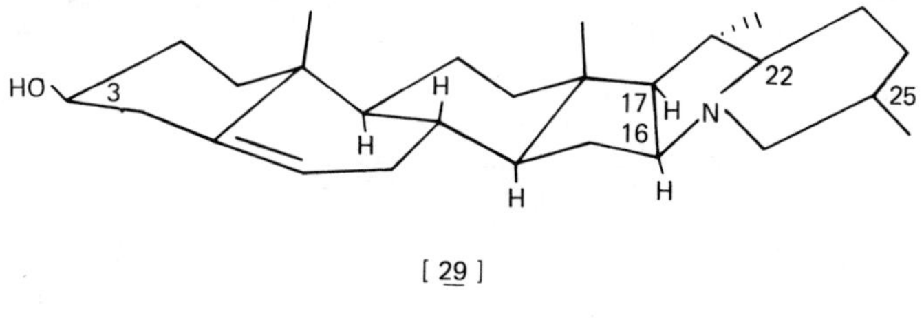
HO
3
H
H
H
17
16
H
H
N
22
25

[29]

Solanidan analogs of solanidine and demissidine that were R-S isomers at
C-22 and C-25, providing a series of conformational isomers with the lone
electron pair on the nitrogen projecting in either the α or β direction with
respect to the steroid plane, were tested for teratogenicity in hamsters
(Brown and Keeler, 1978). In the 22R,25S isomer, the lone electron pair
projects β, whereas in the 22S,25R isomer, the pair projects α to the steroid
plane. The 22R,25S isomer with its β-projecting pair was not teratogenic in
hamsters, but the 22S,25R isomer was. Naturally occurring solanidine, the
aglycone of the common potato alkaloidal glycoside solanine, is the 22R,25S
or nonteratogenic form.

L. Tropane Alkaloids

Among the plants containing tropane alkaloids, *Atropa belladonna* (deadly
nightshade), *Datura stramonium* (jimsonweed), and *Hyoscyamus niger* (hen-
bane) are perhaps the most common. *D. stramonium* causes frequent and
serious toxicosis in domestic livestock. It is a large annual with tubular
flowers. All *Datura stramonium* plant parts are highly toxic because of high
alkaloid levels. Thirst, impaired vision, flushed skin, hyperirritability of
the central nervous system, delirium, rapid and weak heartbeat, and convulsions are among the symptoms of intoxication.

Hysocyamine [30] is the most common alkaloid of the tropane derivatives
found in these plants. It is hydrolyzed to tropine and tropic acid, two products devoid of the physiologic properties of the parent compound (Swan,
1967). The racemic form of hyoscyamine, atropine, is used as a drug in
ophthalmology for pupillary dilation. Isolated atropine probably results from
racemization of hyoscyamine during isolation procedures. The alkaloids are
hallucinogenic and were used by sorcerers in the Middle Ages. The Babyloni-
ans used *Hyoscyamus niger* seeds for relief of asthma (Swan, 1967).

Atropine has produced skeletal abnormalities in CF1 mice given single subcutaneous injections of the compound on day 9 of gestation, according to
Arcuri and Gautieri (1973). McBride (1980) reported that hyoscine (sco-
polamine) produces interior abdominal wall and reduction deformities in chick
embryos by injection of the compound at 96 hr into the yolk sac. But Bueker
and Platner (1956) did not produce deformities in chick eggs using atropine.

[30]

M. Benzylisoquinoline Alkaloids

1. Papaverine

Papaver somniferum, the opium poppy, is the source of a great many medically useful alkaloids which in excess are highly toxic. Among them is papaverine [31], useful as an antispasmodic, which is known to relax smooth muscle and lower the intracellular level of free Ca^{2+} ions (Imai and Takeda, 1967).

Lee and Nagele (1979) demonstrated that papaverine produced neural tube closure defects in explanted chick embryos. The embryos (stage 8 of development) were cultured in a nutrient medium containing papaverine (50 mg/ml of medium). The compound inhibited the uplifting of neural folds and the neuroepithelial cells lost their characteristic wedge shape and elongated appearance. The investigators believed that the mechanism whereby papaverine induced the neural tube closure defects was by lowering free Ca^{2+} levels, which relaxed contracted apical microfilaments in neuroepithelial cells. This mechanism seemed plausible because the effect could be reversed by ionophore A23187, a carboxylic antibiotic known to raise the intracellular level of free Ca^{2+} by promoting release of bound Ca^{2+} (Reed and Lardy, 1972).

N. Phenanthrene Alkaloids

In addition to benzylisoquinoline alkaloids, the opium poppy contains a number of phenanthrene alkaloids of high toxicity that have, nonetheless, found use as analgesic agents in medicine. They are among the most potent and useful of all analgesics for intolerable pain. They include such alkaloids as morphine [32], heroin, thebain, codeine, and others.

Harpel and Gautieri (1968) reported a study made to assess the teratogenicity of morphine in CF1 mice on gestation day 8 or 9. High-level subcutaneous doses induced exencephaly and axial skeletal fusions. Following up on this work, Iuliucci and Gautieri (1971) sought to determine whether the release of histamine and consequent constriction of placental vessels caused by morphine administration was the responsible mechanism. No enhanced incidence of deformities was caused, however, by concomitant histamine administration; nor was there a reduction in deformity incidence produced by concomitant antihistamine administration. Arcuri and Gautieri (1973) in an additional investigation from the same laboratory found evidence to suggest that morphine-induced malformations were related to reduced oxygen concentration and hyperglycemia in pregnant mice administered the compound. In

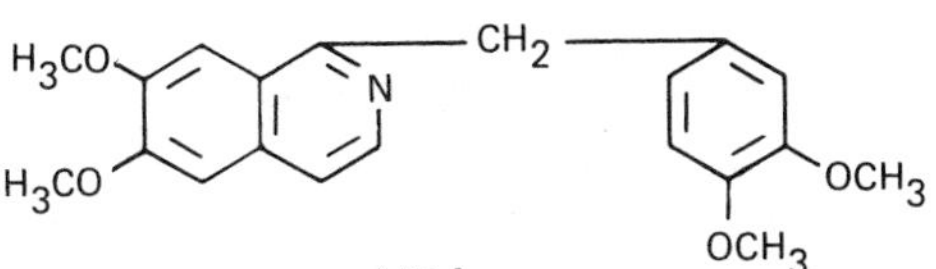

[31]

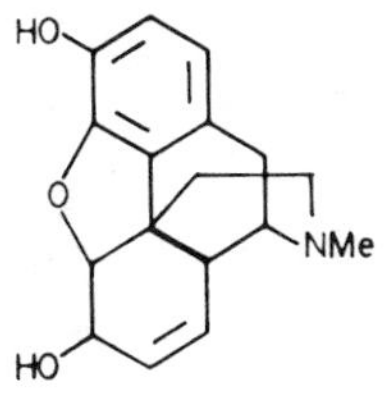

[32]

an extension of this work, Gerber and Schramm (1975) tested teratogenicity of a number of narcotic analgesics, including morphine, heroin, thebain, and codeine in LAK:LVG golden hamsters injected intravenously on day 8 of pregnancy. Malformations were particularly prevalent in the morphine and heroin groups and included exencephaly and crainioschisis. Zellers and Gautieri (1977) reported the teratogenic effects of codeine in the offspring of treated CF1 mice. Various skeletal abnormalities resulted when dams were dosed subcutaneously on any of days 7-12 of gestation.

It seems doubtful to some (Santomauro, 1976) that these laboratory animal studies indicate potential terata problems in humans since laboratory animal doses have been so much higher than doses administered to humans for analgestic purposes.

III. AZOXY COMPOUNDS

A. Methylazoxymethanol

Cycas circinalis and related tropical palms are valuable food plants because of the high concentration of starch they contain; however, without processing, cycads are quite toxic. Cycasin from cycads is hydrolyzed by β-glucosidases in the gut to yield methylazoxymethanol [33], which is both a neurotoxin and a carcinogen (Yang and Mickelsen, 1969).

Human toxicoses are infrequent from methylazoxymethanol; but extensive grazing of toxin-containing palms of the *Bowenia, Cycas,* and *Macrozamia* spp. by livestock, particularly in Australia, produces massive outbreaks of toxicoses (Hooper, 1978).

Spatz et al. (1967) showed that hydrocephalus, microcephalus, cranioschisis, rachischisis, anophthalmia, and microphthalmia occurred in offspring from golden hamsters dosed intravenously with methylazoxymethanol on day 8 of gestation. Other research reported by Spatz (1969) described microcephaly production in offspring from rats given methylazoxymethanol on day 14 or 15 of gestation. At 4 months of age offspring from treated rats had an intellectual deficit as shown by performance in a maze. Fisher et al. (1972) confirmed and extended the experiments on microcephaly production in rats by using methylazoxymethanol acetate.

The mechanism by which methylazoxymethanol produces liver, central nervous system, and teratogenic lesions is not known. Carcinogenic effects may result from methylation of deoxyribonucleic acid (Matsumoto and Higa, 1966), inasmuch as some protection against the effect is afforded by the radioprotective compound cysteamine (Miwa, 1975).

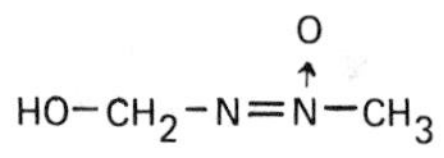

$$HO-CH_2-N\overset{O}{\underset{\uparrow}{=}}N-CH_3$$

[33]

IV. COUMARINS

Two general types of toxic coumarins from plants have received some attention by researchers because of teratogenic possibilities. They are coumarins
related to dicoumarol used as blood anticoagulants and the coumarin rotenone,
the naturally occurring pesticide.

Huebner and Link (1941) and Overman et al. (1944) reported the anticoagulant properties of dicoumarol in a study on the etiology of hemorrhagic sweet
clover disease in cattle. Dicoumarol is found as a metabolite of coumarins in
spoiled sweet clover. A number of coumarin compounds are now used medically as anticoagulant drugs, including warfarin [34] (coumarin). Quite a
controversy has arisen over whether anticoagulant coumarins are responsible
for certain terata in humans (e.g., Warkany, 1976). Pauli and Hall (1979),
reviewing all the literature up to the time of their report, concluded that for
humans "use of coumarin derivatives in the 1st trimester is associated with
the (so called) warfarin embryopathy in some infants, while use in the 2nd
and 3rd trimester is associates with an increased incidence of central nervous
system defects." They believe that only two-thirds of the children are born
normal where mothers received anticoagulent therapy during gestation. The
resolution of the controversy has not been helped greatly by animal experiments—placental hemorrhages and abortion occur in rabbits and mice, but
there is little if any evidence for terata (Kraus et al, 1949; Kronick et al.,
1974; McCallion et al., 1971).

Rotenone [35] is an interesting naturally occurring toxin used as an insecticide. The compound was originally classed as a oddity used by natives of
South America to poison small streams or pools, thereupon allowing harvesting
of the floating killed fish for food. The natives recognized that rotenone was
highly selective in toxicity. It killed fish, but they could safely eat the fish
(Schery, 1952). The high toxicity extends to certain insects, which makes
rotenone a useful insecticide dust on vegetables without subsequent toxicosis
in those who eat the vegetables (Schery, 1952). The low mammalian toxicity
makes rotenone useful as an insecticide to kill ticks, lice, and fleas on livestock and useful in lotions and ointments for the treatment of humans for
chiggers and scabies (Schery, 1952).

Rotenone has a teratogenic propensity at least in explanted chick embryos.
Rao and Chauhan (1971) found that the compound caused a variety of defects,

[34]

[35]

ranging from complete arrest of development, through failure of neural tube
closure or failure to develop distinct brain divisions, to simple retardation in
development. Defects were prevented by administration of adenosine triphosphate (ATP), an expected effect since rotenone is a known inhibitor of the
mitochondrial respiratory chain between flavoprotein and cytochromes. Khera
et al. (1981) reported that high maternal oral doses of rotenone to pregnant
Wistar rats increased resorptions, lowered birth weight, increased bone
mineralization, and produced extra ribs in offspring.

V. LIGNANS

A cytotoxic compound from *Podophyllum peltatum* (may apple), the lignan
podophyllotoxin [36], considered for use as an antitumor agent has been examined for teratogenicity in laboratory animals. Thiersch (1963) found the
compound nonteratogenic in rats, but litter resorption occurred. Litter resorption had been noted earlier when the compound was given to mice by
Wiesner and Yudkin (1955). Joneja and LeLiever (1974) examined the effect
of intraperitoneal injection of podophyllotoxin in DBA/2J mice dosed on days
7-11 of gestation at various doses. There was a high incidence of total resorption and of weight retardation in offspring but no significant increase in
terata over controls. The investigators concluded that podophyllotoxin can
be considered "a strong embryocidal and growth retarding agent rather than
a teratogenic agent."

The one report that showed teratogenic propensity dealt with teratogenicity of two derivatives of podophyllotoxin of lower general toxicity, but higher
antitumor activity than the parent podophyllotoxin. These two derivatives,
VP-16-213 and VM-26, were tested by Sieber et al. (1978) for teratogenicity
in mice by intraperitoneal injection on day 6, 7, or 8 of gestation. Terata
produced included various cranial abnormalities, including exencephaly,
hydrocephalus, anopthalmia, and microtia and other major skeletal malformations.

VI. MACROLIDES

Maytenus serrata and *M. bucuananii* are the usual sources of the macrolide,
maytansine [37]. Maytansine is one among various compounds with high potential usefulness as antitumor agents. According to Chabner et al. (1977),
it has shown promising results in patients with lymphoblastic leukemia and
ovarian carcinoma. Sieber et al. (1978) tested maytansine and other plant-
derived antitumor agents for teratogenicity. They found maytansine to be
quite teratogenic in Swiss albino mice injected intraperitoneally particularly
on day 7 of gestation and to be quite embryolethal on day 6, 7, or 8 of gestation. Among the more common defects were exencephaly, spina bifida, hydrocephalus, gastroschisis, cleft palate, microtia, and axial skeletal defects.
Fortunately, the dose of maytansine effective against P-388 leukemia (0.025
mg/kg) is 10-fold lower than the teratogenic dose of 0.25 mg/kg (Sieber et
al., 1978). The mechanism by which maytansine exerts its teratogenic effect
is not known, but its antitumor potency is thought to result from an effect on
the mitotic spindle apparatus of dividing cells (Sieber et al., 1976).

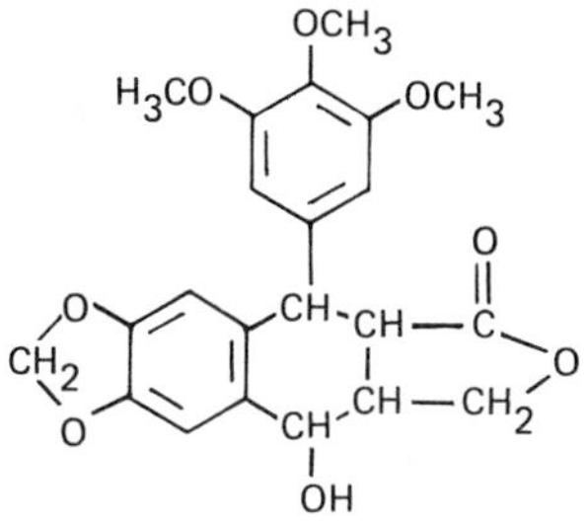

[36]

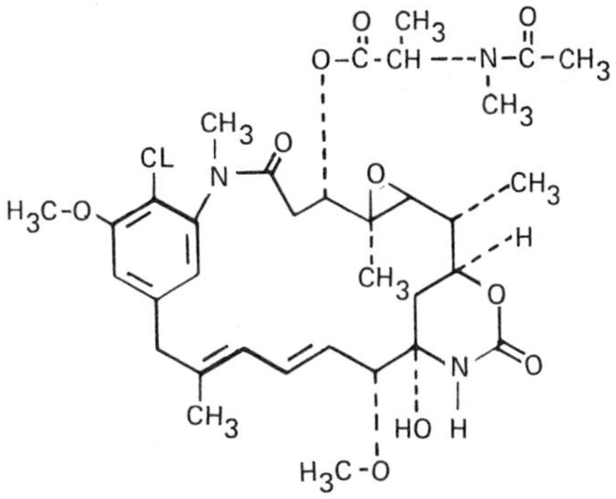

[37]

VII. NITRILES

Nitriles from certain members of the *Lathyrus* and *Vicia* genera are teratogenic in laboratory animals. The teratogens β-aminopropionitrile (βAPN) or the γ-glutamyl dipeptide of βAPN [38] are classed as osteolathyrogens and give rise to spinal or limb curvature, among other effects, when ingested free or from plant sources by young animals. The same compounds produce similar and other types of teratogenic effects in offspring from dams that ingest the compounds during gestation.

Dislocation of rear limb joints and loosening of intersegmental septa were noted by Chang et al. (1954) in *Xenopus* larvae treated with *Lathryus odoratus* extracts. Long bone curvature occurred in chicks treated in ovo with βAPN (Chang et al., 1955). Rosenberg (1957) found teratogenic effects in chicks treated with βAPN in ovo, including scoliosis, hemothorax, abnormal beaks, buckled mandible, and bowed tibia.

Many investigators studied teratogenic effects in treated mammals. Frequently observed have been curvature of the spine or limbs in rats, mice, lambs, and calves (e.g., Azumis, 1955; Ferm, 1960; McCallum, 1958; Keeler et al., 1967; Keeler et al., 1969) or aortic anurisms in rats and mice (e.g., Stamler, 1955; Herd and Orbison, 1966; McCallum, 1958). These effects have usually been seen when the compounds were administered after the embryonic period and well into the fetal period. Other types of terata induced by *Lathyrus* or which contained nitrile lathyrogens when administered earlier in gestation to various mammals have included cleft palate, microagnathia, ectopic heart, and gastroschisis (e.g., Abramovich and DeVoto, 1968; Barrow and Steffek, 1974; Pratt et al., 1971; Steffek et al., 1972; Steffek and Hendrickx, 1972).

The mechanism by which osteolathyrogens exert their effect on young developing postnatal animals as well as prenatally as teratogens is believed to center on production of defects in the cross-linking of elastin and collagen. That mechanism of effect and other aspects of osteolathyrogens have been reviewed extensively. Two of many good reviews include those by Barrow

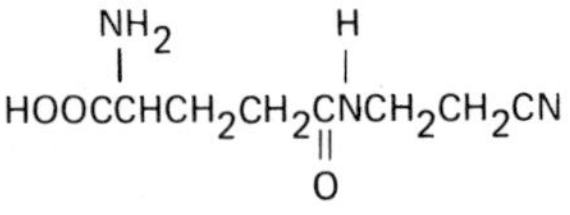

$$\text{HOOCCHCH}_2\text{CH}_2\overset{\overset{\displaystyle H}{|}}{\underset{\underset{\displaystyle O}{||}}{C}}\text{NCH}_2\text{CH}_2\text{CN}$$

$$\overset{\displaystyle NH_2}{|}$$

[38]

et al. (1974) and Padmanaban (1980). Essentially, osteolathyrogens exert their effects by inhibiting the initial reaction in cross-link formation. The enzyme lysyl oxidase, which catalyzes the oxidative deamination of peptide-bound lysine giving rise to allysine, is irreversibly inhibited by βAPN. βAPN thereby stops the initial cross-link reaction since the enzyme acts on specific lysyl residues involved in cross-linking in both elastin and collagen substrates (Pinnel and Martin, 1968).

VIII. TERPENOIDS

Elsewhere in this volume are discussions of plants with toxins that have neurotoxic effects. Marihuana (*Cannabis sativa*) is one of them. Because of its widespread use as the "opium of the people" throughout the world, it has been investigated in great detail. Among those studies are many that have reported research on whether marihuana or associated compounds such as Δ^9-tetrahydrocannabinol [39] are teratogenic. Much controversy has been generated partly because of the use of such a wide variety of experimental protocols. Fleischman et al. (1975) reviewed that controversy and tabulated the species and strain of experimental animals, preparations, dosages, routes of administration, and days of treatment during gestation used by the various investigators. Considering the known general "principles of teratology" (Wilson and Fraser, 1977) relating to variation in species and strain susceptibility, dosage, gestational susceptibility, and so on, it is little wonder that there is controversy. The greater number of investigators have reported no teratogenic effects, but at least four reports describing terata produced by *Cannabis* preparations or compounds seem incontrovertible if one is cautious in claiming direct correlation to human marihuana usage. Reference to those reports follows.

In 1968, Persaud and Ellington reported that 4.2 mg/kg of *Cannabis* resin injected intraperitoneally in pregnant inbred albino rats once per day on days 1-6 of gestation produced malformations in 57% of offspring from implanted sites and resorptions from 29% of implanted sites. Those figures compare to 0% malformed and 17% resorbed in controls. Deformities of high incidence

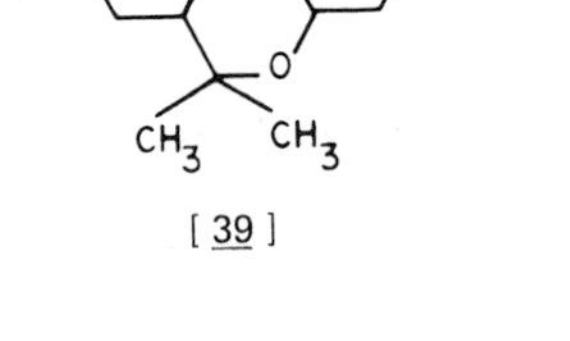

[39]

included general stunting, syndactyly, encephalocele, phycomelia, and eventration of abdominal viscera.

Gerber and Schramm (1969) injected a marihuana extract subcutaneously in pregnant New Zealand white rabbits (up to 500 mg/kg) and golden hamsters (up to 300 mg/kg) and found a significant increase in dead fetuses, runts, and abnormalities in treated dams at high doses compared to controls. Abnormalities included phocomelia, omphalocele, spina bifida, exencephaly, and myelocele.

Mantilla-Plata et al. (1975) assessed the teratogenic effects of Δ^9-tetrahydrocannabinol in Swiss-Webster mice on various gestational days. No skeletal anomalies were found that could be attributed to the Δ^9-tetrahydrocannabinol, but cleft palate occurred in offspring from dams receiving 75, 80, and 100 mg/kg during the days 6 to 15 of gestation or after a single 300 mg/kg dose on days 12 and 14. Resorptions were also common in treated animals.

Joneja (1976) also reported studies on teratogenic effects in offspring from Δ^9-tetrahydrocannabinol-treated pregnant Swiss-Webster and DBA mice after single exposure to the compound during gestation. An oral dose in Swiss-Webster mice of 400 mg/kg given on day 9 of gestation induced a 12% incidence of malformation among live fetuses. Among DBA mice given 200 mg/kg, p.o. on day 10 of gestation, 23% were malformed.

IX. TOXIC AMINO ACIDS

A. Mimosine

The amino acid mimosine [40], β-N-(3-hydroxyl-4-pyridone)-α-aminopropionic acid, is found in *Leucaena leucocephala* and other legumes, including certain *Mimosa* spp. *Leucaena* is toxic, but even so, is considered adequate forage for certain classes of livestock as a source of protein. Ingestion of large quantities results in toxicosis—mimosine being the toxic compound. It produces alopecia, inanition, decreased weight gain, cataracts, and infertility. Monogastric animals are more susceptible to mimosine toxicosis than are ruminants. Rats and swine are extremely susceptible.

When pregnant rats were fed diets containing mimosine, a variety of fetal deformities were produced (Dewreede and Wayman, 1970). Uterine perforations produced fetal constriction of cranium, thorax, and pelvis because these parts, still in fetal membranes, protruded during gestation through the uterine perforations. Abortions were frequent. Wayman et al. (1970) found a significant increase in fetal resorptions and in polypodia of forelimbs in offspring from gilts fed *Leucaena* during gestation. The uterine perforations may have been caused by formation of a stable complex between mimosine

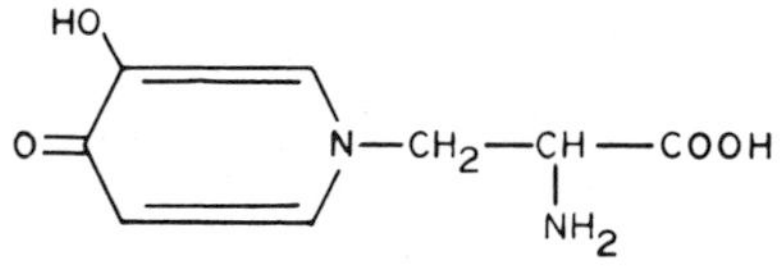

[40]

and pyridoxal phosphate, as shown by Linn et al. (1962), in which the aldehyde function of the coenzyme is lost. Embryonic events susceptible to decreased pyridoxal phosphate levels would be affected.

B. Indospicine

Pearn (1967) found that an extract of *Indigofera spicata* produced cleft
palate and a high incidence of intrauterine deaths and resorptions in rats
when given on day 13 of gestation. Hegarty and Pound (1968) isolated an
amidine-containing amino acid from this extract that they called indospicine
[41]. Pearn and Hegarty (1970) demonstrated that indospicine was responsible for the cleft palate condition and embryolethality. They believed that
the mechanism of the teratogenicity was either a competitive inhibition with
arginine (i.e., incorporation of indospicine into a structurally incompetent
protein) or a direct toxic effect on the rotation and subsequent fusion of the
palatine shelves. Recent reports by this group (Madsen et al., 1970; Madsen
and Hegarty, 1970) reported that indospicine in the L form is a competitive
inhibitor of arginase. They speculated that other enzymes for which arginine is substrate may be similarly affected. They showed that indospicine
competes with arginine for binding to arginyl-tRNA and inhibits DNA synthesis (Hegarty and Peterson, 1973).

   Widespread occurrence of plants of this genus, particularly in Australian
ranges (Gardner and Bennetts, 1956), suggests that the plant could be responsible for production of malformations in domestic livestock.

C. Cyclopropyl Amino Acids

Blighia sapida is a tree native to Africa but cultivated in the tropics of this
hemisphere. The fruit is commonly eaten after parboiling and frying, but
improperly prepared fruit retains a hypoglycemic compound, hypoglycin-A
[42], a cyclopropane amino acid [β-(methylene-cyclopropyl)alanine], which
produces severe hypoglycemia in humans consuming it (Kingsbury, 1964;
Hylin, 1969). Persaud has shown that hypoglycin-A is teratogenic in rats
(1968a), and produces severe stunting and fetal resorptions in rabbits
(Persaud, 1968b).

   The mechanism of teratogenesis may be identical to that proposed for the
hypoglycemia effect—a reduction in rate of fatty acid β-oxidation (Von Holt
et al., 1966). Persaud (1971) has shown that hypoglycin-A-induced deformities can be prevented by riboflavin phosphate. This suggests that the
effect is due to hypoglycin-A inhibition of the same acyl dehydrogenase
flavin-dependent oxidation reaction involved in the hypoglycemic effect.

$$CH_2-C-CH-CH_2-CH-COOH$$
$$\diagdown\diagup\qquad\qquad|$$
$$CH_2\qquad\quad NH_2$$

$$H_2N-C-CH_2-CH_2-CH_2-CH_2-CH-COOH$$
$$\quad\|\qquad\qquad\qquad\qquad\qquad\quad|$$
$$\quad NH\qquad\qquad\qquad\qquad\qquad NH_2$$

[41]   [42]

X. PLANTS WITH UNIDENTIFIED TERATOGENS

A. Loco Plants

Certain loco plants, for example, *Astragalus lentiginosus* and *Astragalus pubentissimus*, induce deformities and abortions in lambs from ewes ingesting these plants during gestation. Affected lambs may have excessive flexure of carpal joints, tendon contracture, or anterior flexure and hypermobility of the hock joints. Natural incidence is sporadic because of the variable abundance of the plant. In years of abundant plant growth, incidence of abortions may approach 60%, and up to 30% or more of the live lambs born may be deformed. Feeding trials have demonstrated that deformities are due to ingestion of loco by the pregnant ewe at almost any period during gestation (James et al., 1967). *Astragalus lentiginosus* is particularly potent. The teratogen and mechanism are unknown.

B. Jimsonweed

Leipold et al. (1973) speculated that an outbreak of arthrogryposis in newborn pigs in Kansas in 1971 was due to ingestion by sows of the plant *Datura stramonium* (jimsonweed) during the second and third months of pregnancy. They based their speculation on two observations. First, the cases observed did not appear to fit an hereditary pattern. Second, signs of jimsonweed intoxication (see Sec. II.L) were observed in sows during the second and third months of gestation, and there was evidence that the jimsonweed had been foraged. After the jimsonweed was eradicated, the owners had no cases the following year using the same breeding groups. During the intervening years since the speculation that jimsonweed is teratogenic, no reports have appeared describing experimental testing of the plant for teratogenicity in livestock. But reference is made that gives the impression that the possibility is more than speculation (Wilson and Fraser, 1977).

Consequently, we sought to establish by experimental feeding whether jimsonweed was, in fact, teratogenic in pigs through trials in purebred Hampshire sows. They were fed levels of the plant that induced clinical signs of intoxication (due to the tropane alkaloids present) during the second and third months of pregnancy. No arthrogypotically deformed newborn pigs were farrowed by the sows, suggesting that it is quite unlikely that jimsonweed is teratogenic in pigs (Keeler, 1981), contrary to the epidemiological speculation by Leipold et al. (1973) and despite the presence of teratogenic tropane alkaloids (see above).

C. Cyanogenic Glycoside Containing Plants

Three otherwise unrelated plants that are believed to cause livestock deformities have in common the ability to synthesize and store cyanogenic glycosides. The plants are sudangrass (*Sorghum sudanense*), wild black cherries (*Prunus serotina*), and bamboo (*Bambusa arundinacea*). If the plants prove teratogenic by feeding trials, perhaps the cyanide could be responsible in view of the known teratogenic propensity of hypoxia (see Shephard, 1980) and the ability of cyanide to induce the hypoxic state.

In 1971, Selby et al. reported an outbreak of swine malformations in Missouri that they believed to be associated with consumption by pregnant sows of wild black cherries. Deformed piglets were born alive and had no tail, no arms, very small external sex organs, and limb deformities. Thirteen of 184 piglets were born deformed to 13 farrowing sows bred and maintained in a pasture where there was evidence that wild black cherries had been eaten by the sows during early gestation. Six sows, bred by the same boar and in an adjacent pasture free of cherries, delivered normal pigs. The workers believed the cherries responsible and speculated that a cyanogenic glycoside was the teratogen.

In 1978, Alikutty and Aleyas reported a mammary gland anomaly in a calf born to a dam that had ingested liberal amounts of bamboo leaves during gestation. The calf had only two teats at birth. The workers speculated that the cause might be estrogenic compounds in the bamboo, but another possibility is caynide.

In 1967, Prichard and Voss reported that foals from four mares that had been bred to the same stallion had anklosis of the joints. The stallion had bred 200 mares previously and sired no deformed foals. The mares in question, however, had grazed on a hybrid sudangrass pasture during gestation, leading the researchers to speculate that the sudangrass was responsible.

D. *Pteridium aquilinum*

Because of reported toxicity and carcinogenicity of *Pteridium aquilinum* (bracken fern), Yasuda et al. (1974) examined the embryotoxicity of the plant in pregnant mice. A diet containing 33% dried bracken fern fed the mice throughout gestation caused growth suppression, increased rib variations, and retarded ossification of sternebrae in fetuses compared to controls. The authors speculate that the effect may be related to the known antihistamine activity of bracken.

E. *Trachymene* spp.

A condition in lambs called bent leg occurs in Australia and is thought to be due to the wild parsnips *Trachymene cyanantha* and *Trachymene ochracea* (Clark et al., 1975). The condition in lambs involves an uneven growth of epiphyseal plates during fetal or neonatal development from a teratogen in the maternal circulation, ultimately reaching the fetus or from maternal milk. The deformities are usually characterized by limb bone curvature or alteration in articulation of joints. Identity of the teratogen is unknown.

F. Potatoes

Renwick (1972) hypothesized that maternal consumption of blighted potatoes by pregnant humans produced congenital anencephaly and spina bifida (ASB). He speculated that the teratogen might be elaborated by potatoes as a consequence of blight.

Many researchers (Elwood and MacKenzie, 1973; Emanuel and Sever, 1973; Field and Kerr, 1973; Kinlen and Hewitt, 1973; McMahon et al., 1973; Spiers et al., 1974) found that the epidemiology upon which Renwick based his hypothesis was best interpreted other than as Renwick suggested. In addition, the birth of ASB children to mothers on potato-avoidance trials during pregnancy (Nevin and Merrett, 1975) was convincing evidence against his hypothesis.

Other investigators examined whether laboratory animals would give birth to deformed offspring if mothers were fed during pregnancy with various potato preparations. Poswillo et al. (1972) found cranial osseus defects in neonatal marmosets born to dams dosed with certain potato preparations, but they were unable to confirm the observation (Poswillo et al., 1973). Ruddick et al. (1974) found that freeze-dried, blighted "Kennebec" potatoes produced no terata when fed to rats. We found (Keeler et al., 1974, 1975) that neither early nor late blighted 'Russet Burbank' potatoes, control healthy tubers, nor extracts therefrom were teratogenic in rats, rabbits, hamsters, or mice.

Keeler et al. (1978) tested the teratogenicity in hamsters of dried, ground potato sprout preparations from seven potato cultivars. Deformities were produced and included exencephaly, encephalocele, microphthalmia, and spina bifida. Neither peel nor tuber material from sprouted or control tubers was teratogenic. Oral doses of the sprout material in hamsters ranged from 2500 to 3500 mg/kg. The average 50- to 70-kg pregnant woman would have to eat from 125 to 250 g of dry sprout material (900-1750 g of fresh weight) in a single dose to consume comparable amounts. Peel and tuber material were not teratogenic at three to five times that dose level, and sprouts are not generally consumed by humans in any amount. Consequently, potato tubers seem perfectly safe for human consumption despite the teratogenicity of very large doses of sprouts in one hamster strain. The identity of the sprout teratogen is unknown.

XI. SUMMARY

It is evident from the wide variety of natural teratogens cited in this review that animals, including humans, are at some risk from plant-derived teratogens. The list grows rapidly, and we can speculate that many more plant-derived natural teratogens will be found.

REFERENCES

Abramovich, A., and DeVoto, F. C. H. (1968). Anomalous maxillofacial patterns produced by maternal lathyrism in rat foetuses. *Arch. Oral Biol. 13*:823-826.

Alikutty, K. M., and Aleyas, N. M. (1978). Amastia of rear quarters in a cow. *Mod. Vet. Pract. 59*:623.

Ancel, P. (1946). Recherche expérimentale sur le spina bifida. *Arch. Anat. Microsc. 36*:45-68.

Arcuri, P. A., and Gautieri, R. F. (1973). Morphine induced fetal malformations. 3. Possible mechanisms of action. *J. Pharm. Sci. 62*:1626-1634.

Azumis, S. (1955). Experimentally produced scoliosis of vertebrae. *J. Osaki City Med. Cent. 4*:19-27.

Barrow, M. V., and Steffek, A. J. (1974). Teratologic and other embryotoxic effects of β-aminopropionitrile in rats. *Teratology 10*:165-172.

Barrow, M. V., Simpson, C. F., and Miller, E. J. (1974). Lathyrism: a
review. *Q. Rev. Biol. 49*:101-128.

Bell, D. P., Gibson, J. G., McCarroll, A. M., McClean, G. A., and
Geraldine, A. (1976). Embryotoxicity of solanine and aspirin in mice.
J. Reprod. Fertil. 46:257-259.

Bertrand, M., Schwam, E., Frandon, A., Vagne, A., and Alary, J. (1965).
Sur un effect tératogène systématique et spécifique de la caféine chez les
rongeurs. *C. R. Soc. Biol. 159*:2199-2201.

Bertrand, M., Girod, J., and Rigaud, M. F. (1970). Ectrodactylie provoquée
par la caféine chez les rongeurs. Rôle des facteurs spécifiques et généti-
ques. *C. R. Soc. Biol. 164*:1488-1489.

Binns, W., James, L. F., Shupe, J. L., and Everett, G. (1963). A congenital cyclopian-type malformation in lambs induced by maternal ingestion of
a range plant, *Veratrum californicum*. *Am. J. Vet. Res. 24*:1164-1175.

Binns, W., Shupe, J. L., Keeler, R. F., and James, L. F. (1965). Chrono-
logic evaluation of teratogenicity in sheep fed *Veratrum californicum*. *J.
Am. Vet. Med. Assoc. 147*:839-842.

Binns, W., Keeler, R. F., and Balls, L. D. (1972). Congenital deformities
in lambs, calves, and goats resulting from maternal ingestion of *Veratrum
californicum*: hare lip, cleft palate, ataxia, and hypoplasia of metacarpal
bones. *Clin. Toxicol. 5*:245-261.

Brown, D., and Keeler, R. F. (1978). Structure-activity relation of steroid
teratogens. 3. Solanidan epimers. *J. Agric. Food Chem. 26*:566-569.

Bryden, M. M., and Keeler, R. F. (1974). Effects of alkaloids of *Veratrum
californicum* on developing embryos. *J. Anat. 116*:464.

Bryden, M. M., Perry, C., and Keeler, R. F. (1973). Effects of alkaloids
of *Veratrum californicum* on the developing chick embryo. *Teratology 8*:
19-25.

Bueker, E. D., and Platner, W. S. (1956). Effect of cholinergic drugs on
development of chick embryos. *Proc. Soc. Exp. Biol. Med. 91*:538-543.

Chabner, B., Levine, A., Adamson, R., Johnson, B., Whang-Peng, J., and
Young, R. (1977). Initial clinical and pharmacological studies of maytansine. *Proc. Am. Assoc. Cancer Res. 18*:129.

Chang, C. Y., Witschi, E., and Ponseti, I. V. (1954). Teratogenic development in *Xenopus* larvae caused by the sweet pea seed (*Lathyrus odoratus*)
and their extracts. *Anat. Rec. 120*:816.

Chang, C. Y., Witschi, E., and Ponseti, I. V. (1955). Teratogenic effects
of *Lathyrus odoratus* seeds on the development and regeneration of
vertebrate limbs. *Proc. Soc. Exp. Biol. Med. 90*:45-50.

Chaube, S., and Swinyard, C. A. (1976). Teratological and toxicological
studies of alkaloids and phenolic compounds from *Solanum tuberosum*. *J.
Toxicol. Appl. Pharmacol. 36*:227-237.

Clark, L., Carlisle, C. H., and Beasley, P. S. (1975). Observations on the
pathology of bent leg of lambs in south-western Queensland. *Aust. Vet.
J. 51*:4-10.

Covell, W. P. (1936). A cytotoxic study of the effects of drugs on the
cochlea. *Arch. Otolaryngol. 23*:633-641.

Crowe, M. W. (1969). Skeletal anomalies in pigs associated with tobacco.
Mod. Vet. Pract. 69:54-55.

Crowe, M. W., and Swerczek, T. W. (1974). Congenital arthrogryposis in offspring of sows fed tobacco (*Nicotiana tabacum*). *Am. J. Vet. Res. 35*: 1071-1073.

Culvenor, C. C. J. (1973). Alkaloids. In *Chemistry and Biochemistry of Herbage*, Vol. 1, G. W. Butler and R. W. Bailey (Eds.). Academic Press, New York, pp. 375-446.

Demyer, W. (1965). Cleft lip and jaw induced in fetal rats by vincristine. *Arch. Anat. 48*: 181-186.

Dewreede, S., and Wayman, O. (1970). Effect of mimosine on the rat fetus. *Teratology 3*: 21-28.

Drachman, D. B., and Coulombre, A. J. (1962). Experimental club foot and arthrogryposis multiplex congenita. *Lancet 2*: 523-526.

Elwood, J. H., and MacKenzie, G. (1973). Associations between the incidence of neurological malformations and potato blight outbreaks over 50 years in Ireland. *Nature (Lond.) 243*: 476-477.

Emanuel, I. S., and Sever, L. E. (1973). Questions concerning the possible association of potatoes and neural tube defects, and an alternative hypothesis relating to maternal growth and development. *Teratology 8*: 325-331.

Ferm, V. H. (1960). Osteolathyrogenic effects on the developing rat foetus. *J. Embryol. Exp. Morphol. 8*: 94-97.

Ferm, V. H. (1963a). Colchicine teratogenesis in hamster embryos. *Proc. Soc. Exp. Biol. Med. 112*: 775-778.

Ferm, V. H. (1963b). Congenital malformations in hamster embryos after treatment with vinblastine and vincristine. *Science 141*: 426.

Ferreira, N. R., and Frota-Pessoa, O. (1969). Trisomy after colchicine therapy. *Lancet 1*: 1160-1161.

Field, B., and Kerr, C. (1973). Potato blight and neural tube defects. *Lancet 2*: 507-508.

Fisher, M. H., Welker, C., and Waisman, H. A. (1972). Generalized growth retardation in rats induced by prenatal exposure to methylazoxymethyl acetate. *Teratology 5*: 223-232.

Fleischman, R. W., Hayden, D. W., Rosenkrantz, H., and Braude, M. C. (1975). Teratologic evaluation of delta-9-tetrahydrocannibinol in mice, including a review of the literature. *Teratology 12*: 47-50.

Fujii, T., and Nishimura, H. (1972). Adverse effects of prolonged administration of caffeine on rat fetus. *Toxicol. Appl. Pharmacol. 22*: 449-457.

Fujii, T., Sasaki, H., and Nishimura, H. (1969). Teratogenicity of caffeine in mice related to its mode of administration. *Jpn. J. Pharmacol. 19*: 134-138.

Gardner, C. A., and Bennetts, H. W. (1956). *The Toxic Plants of Western Australia*. Western Australia Newspaper Ltd., Perth.

Gerber, W. F. (1967). Congenital malformation induced by mescaline, lysergic acid diethylamide, and bromolysergic acid in the hamster. *Science 158*: 265-267.

Gerber, W. F., and Schramm, L. C. (1969). Effect of marihuana extract on fetal hamsters and rabbits. *Toxicol. Appl. Pharmacol. 14*: 276-282.

Gerber, W. F., and Schramm, L. C. (1975). Congenital malformations of the central nervous system produced by narcotic analgesics in the hamster. *Am. J. Obstet. Gynecol. 123*: 705-713.

Gilani, S. H., and Giovinazzo, J. (1981). Experimental studies on the effects of caffeine on chick embryonic development. *Teratology 23*:36A.

Goldman, A. S., and Yakovac, W. C. (1965). Teratogenic action in rats of reserpine alone and in combination with salicylate and immobilization. *Proc. Soc. Exp. Biol. Med. 118*:857-862.

Green, C. R., and Christie, G. S. (1961). Malformations in foetal rats induced by the pyrrolizidine alkaloid heliotrine. *Br. J. Exp. Pathol. 42*: 369-378.

Harpel, H. S., Jr., and Gautieri, R. F. (1968). Morphine-induced fetal malformations. I. Exencephaly and axial skeletal fusions. *J. Pharm. Sci. 57*:1590-1597.

Hegarty, M. P., and Peterson, P. J. (1973). Free amino acids, bound amino acids, amines, and ureides. In *Chemistry and Biochemistry of Herbage*, Vol. 1, G. W. Butler and R. W. Bailey (Eds.). Academic Press, New York, pp. 1-62.

Hegarty, M. P., and Pound, A. W. (1968). Indospicine, a new hepatotoxic amino-acid from *Indigofera spicata*. *Nature (Lond.) 217*:354-355.

Herd, J. K., and Orbison, J. L. (1966). Lathyrism in the fetal rat. *Arch. Pathol. 81*:60-66.

Hooper, P. T. (1978). Cycad poisoning in Australia—etiology and pathology. In *Effects of Poisonous Plants on Livestock*, R. F. Keeler, K. R. Van Kampen, and L. F. James (Eds.). Academic Press, New York, pp. 337-347.

Huebner, C. F., and Link, K. P. (1941). Studies on the hemorrhagic sweet clover disease. VI. The synthesis of the δ-diketone derived from the hemorrhagic agent through alkaline degradation. *J. Biol. Chem. 138*: 529-534.

Hylin, J. W. (1969). Toxic peptides and amino acids in foods and feeds. *J. Agric. Food Chem. 17*:492-496.

Imai, S., and Takeda, K. (1967). Action of calcium and certain multivalent cations on potassium contracture of guinea-pigs, *Tainia coli*. *J. Physiol. 190*:155-169.

Ingalls, T. H., Curley, F. J., and Zappasodi, P. (1968). Colchicine-induced craniofacial defects in the mouse embryo. *Arch. Environ. Health 16*:326-332.

Ishikawa, S., Gilbert, E. F., Bruyere, H. J., and Cheung, M. O. (1978). Aortic aneurysm associated with cardiac defects in theophylline stimulated chick embryos. *Teratology 18*:23-30.

Iuliucci, J. D., and Gautieri, R. F. (1971). Morphine-induced fetal malformations. II. Influence of histamine and diphenylhydramine. *J. Pharm. Sci. 60*:420-425.

James, L. F., Shupe, J. L., Binns, W., and Keeler, R. F. (1967). Abortive and teratogenic effects of loco plants. *Am. J. Vet. Res. 28*:1379-1388.

Jelinek, R., Kyzlink, V., and Blattný, C., Jr. (1976). An evaluation of the embryo-toxic effects of blighted potatoes on chicken embryos. *Teratology 14*:335-342.

John, J. A., Blogg, C. D., Murray, F. J., Schwetz, B. A., and Gehring, P. J. (1979). Teratogenic effects of the plant hormone indole-3-acetic acid in mice and rats. *Teratology 19*:321-326.

Johnson, A. E. (1978). Tolerance of cattle to tansy ragwort. *Am. J. Vet. Res. 39*:1542-1544.

Johnson, A. E. (1983). The effects on cattle and their calves of tansy rag-wort (*Senecio jacoboea*) fed in early gestation. *Am. J. Vet. Res.* (accepted for publication).

Joneja, M. G. (1976). A study of teratological effects of intravenous sub-cutaneous, and intragastric administration of Δ^9-tetrahydrocannabinol in mice. *Toxicol. Appl. Pharmacol. 36*:151-162.

Joneja, M. G., and LeLiever, W. C. (1974). Effects of vinblastine and podo-phyllin in DBA mouse fetuses. *Toxicol. Appl. Pharmacol. 27*:408-414.

Joneja, M. G., and Ungthavorn, S. (1970). Teratogenic effects of vincris-tine in three lines of mice. *Teratology 2*:235-240.

Kalter, H. (1968). *Teratology of the Central Nervous System.* University of Chicago Press, Chicago.

Keeler, R. F. (1970a). Teratogenic compounds of *Veratrum californicum* (Durand). IX. Structure activity relationships. *Teratology 3*:169-174.

Keeler, R. F. (1970b). Teratogenic compounds of *Veratrum californicum* (Durand). X. Cyclopia in rabbits produced by cyclopamine. *Teratology 3*:175-180.

Keeler, R. F. (1971). Teratogenic compounds of *Veratrum californicum* (Durand). XI. Gestational chronology and compound specificity in rabbits. *Proc. Soc. Exp. Biol. Med. 136*:1174-1189.

Keeler, R. F. (1973a). Lupin alkaloids from teratogenic and non-teratogenic lupins. I. Correlation of crooked calf disease incidence with alkaloid distribution determined by gas chromatography. *Teratology 7*:23-30.

Keeler, R. F. (1973b). Lupin alkaloids from teratogenic and non-teratogenic lupins. II. Identification of the major alkaloids by tandem gas chromatography-mass spectrometry in plants producing crooked calf disease. *Teratology 7*:31-35.

Keeler, R. F. (1974). Coniine, a teratogenic principle from *Conium macula-tum* producing congenital malformations in calves. *Clin. Toxicol. 7*:195-206.

Keeler, R. F. (1975). Teratogenic effects of cyclopamine and jervine in rats, mice, and hamsters. *Proc. Soc. Exp. Biol. Med. 149*:302-306.

Keeler, R. F. (1976). Lupin alkaloids from teratogenic and non-teratogenic lupins. III. Identification of anagyrine as the possible teratogen by feed-ing trials. *J. Toxicol. Environ. Health 1*:887-898.

Keeler, R. F. (1979). Congenital defects in calves from maternal ingestion of *Nicotiana glauca* of high anabasine content. *Clin. Toxicol. 15*:417-426.

Keeler, R. F. (1981). Absence of arthrogryposis in newborn Hampshire pigs from sows ingesting toxic levels of jimsonweed during gestation. *Vet. Hum. Toxicol. 23*:413-415.

Keeler, R. F., and Balls, L. D. (1978). Teratogenic effects in cattle of *Conium maculatum* and conium alkaloids and analogs. *Clin. Toxicol. 12*:49-64.

Keeler, R. F., and Binns, W. (1968). Teratogenic compounds of *Veratrum californicum* (Durand). V. Comparison of cyclopian effects of steroidal alkaloids from the plant and structurally related compounds from other sources. *Teratology 1*:5-10.

Keeler, R. F., James, L. F., Binns, W., and Shupe, J. L. (1967). An in-vestigation of the apparent relationship between locoism and lathyrism. *Can. J. Comp. Med. Vet. Sci. 31*:334-341.

Keeler, R. F., James, L. F., Binns, W., and Shupe, J. L. (1969). Preliminary investigation of the relationship between bovine congenital
lathyrism induced by aminoacetonitrile and the lupine induced crooked
calf disease. *Can. J. Comp. Med.* *33*:89-92.

Keeler, R. F., Douglas, D. R., and Stallknecht, G. F. (1974). Failure of
blighted Russet Burbank potatoes to produce congenital deformities in
rats. *Proc. Soc. Exp. Biol. Med.* *146*:284-286.

Keeler, R. F., Douglas, D. R., and Stallknecht, G. F. (1975). The testing
of blighted, aged, and control Russet Burbank potato tuber preparations
for ability to produce spina bifida and anencephaly in rats, rabbits,
hamsters, and mice. *Am. Potato J.* *52*:125-132.

Keeler, R. F., Cronin, E. H., and Shupe, J. L. (1976a). Lupin alkaloids
from teratogenic and non-teratogenic lupins. IV. Concentration of total
alkaloids, individual major alkaloids, and the teratogen anagyrine as a
function of plant part and stage of growth and their relationship to
crooked calf disease. *J. Toxicol. Environ. Health* *1*:899-908.

Keeler, R. F., Young, S., and Brown, D. (1976b). Spina bifida, exencephaly, and cranial bleb produced by the solanum alkaloid solasodine.
Res. Commun. Chem. Pathol. Pharmacol. *13*:723-730.

Keeler, R. F., Young, S., Brown, D., Stallknecht, G. F., and Douglas, D.
(1978). Congenital deformities produced in hamsters by potato sprouts.
Teratology *17*:3.

Keeler, R. F., Balls, L. D., and Panter, K. (1981). Teratogenic effects of
Nicotiana glauca and concentration of anabasine, the suspect teratogen in
plant parts. *Cornell Vet.* *71*:47-53.

Khera, K. S., Whalen, C., and Angers, G. (1981). Teratogenicity study on
pyrethrins and rotenone (of natural origin) and ronnel in pregnant rats.
Teratology *23*:45A.

Kingsbury, J. M. (1964). *Poisonous Plants of United States and Canada.*
Prentice-Hall, Englewood Cliffs, N.J.

Kinlen, L., and Hewitt, A. (1973). Potato blight and anencephaly in Scot-
land. *Br. J. Prev. Soc. Med.* *27*:208-213.

Knoche, C., and Konig, J. (1964). Zur prävatalen Toxizität von Diphenyl
Pyralin-8-chloro-thophyllinat unter Bërucksichtigung von Erfahrungen
mit Thalidomid und Coffein. *Arzneim-forch.* *14*:415-424.

Kraus, A. P., Perlow, A., and Singer, K. (1949). Danger of dicoumarol
treatment in pregnancy. *JAMA* *139*:758-762.

Kronick, J., Phelps, N. E., McCallion, D. J., and Hirst, J. (1974).
Effects of sodium warfarin administered during pregnancy in mice. *Am.
J. Obstet. Gynecol.* *118*:819-823.

Landauer, W. (1953). On teratogenic effects of pilocarpine in chick development. *J. Exp. Zool.* *122*:469-483.

Landauer, W. (1954). On the chemical production of developmental abnormalities and of phenocopies in chicken embryos. *J. Cell. Comp. Physiol.*
43:261-305.

Landauer, W. (1960). Nicotine-induced malformations of chicken embryos
and their bearing on the phenocopy problem. *J. Exp. Zool.* *143*:107-122.

Landauer, W. (1975). Cholinomimetic teratogens: studies with chicken
embryos. *Teratology* *12*:125-146.

Lee, H., and Nagele, R. G. (1979). Neural tube closure defects caused by papaverine in explanted early chick embryos. *Teratology 20*:321-332.

Leipold, H. W., Oehme, F. W., and Cook, J. E. (1973). Congenital arthrogryposis associated with ingestion of jimsonweed by pregnant sows. *J. Am. Vet. Med. Assoc. 162*:1059-1060.

Liener, I. E. (1980). Miscellaneous toxic factors. In *Toxic Constituents of Plant Foodstuffs*, 2nd ed., I. E. Liener (Eds.). Academic Press, New York, pp. 430-467.

Linn, J. Y., Shih, Y. M., and Ling, K. H. (1962). Studies on the mechanism of toxicity of mimosine, β-N-3-hydroxypyridone-α-amino propionic acid. (1) Studies of the reactions of mimosine and pyridoxal-5-phosphate using the spectrophotometric method. *J. Formosan Med. Assoc. 61*:997-1003.

Madsen, N. P., and Hegarty, M. P. (1970). Inhibition of rat liver homogenate arginase activity in vitro by the hepatotoxic amino acid indospicine. *Biochem. Pharmacol. 19*:2391-2393.

Madsen, N. P., Christie, G. S., and Hegarty, M. P. (1970). Effect of indospicine on incorporation of L-arginine-^{14}C into protein and transfer ribonucleic acid by cell-free systems from rat liver. *Biochem. Pharmacol. 19*:853-857.

Mantilla-Plata, B., Clewe, G. L., and Harbison, R. D. (1975). Δ^9-Tetrahydrocannabinol-induced changes in prenatal growth and development of mice. *Toxicol. Appl. Pharmacol. 33*:333-340.

Matsumoto, H., and Higa, H. H. (1966). Studies on methylazoxymethanol, the aglycone of cycasin: methylation of nucleic acids in vitro. *Biochem. J. 98*:20c.

Mattocks, A. R. (1978). Recent studies on mechanisms of cytotoxic action of pyrrolizidine alkaloids. In *Effects of Poisonous Plants on Livestock*, R. F. Keeler, K. R. Van Kampen, and L. F. James (Eds.). Academic Press, New York, pp. 177-187.

McBride, W. G. (1980). Hyoscine in the chick embryo. *Lancet 2(8191)*: 430.

McCallion, D. J., Phelps, N. E., Hirst, J., and Cade, J. F. (1971). Effects of Coumadin administered during pregnancy in rabbits and mice. *Teratology 4*:235-236.

McCallum, H. M. (1958). Lathyrism in mice. *Nature (Lond.) 182*:1169-1170.

McMahon, B. S., Yen, S., and Rotham, K. J. (1973). Potato blight and neural tube defects. *Lancet 1*:598-599.

Menges, R. W., Selby, L. A., Marienfeld, C. J., Aue, W. A., and Greer, D. L. (1970). A tobacco related epidemic of congenital limb deformities in swine. *Environ. Res. 3*:285-302.

Miwa, T. (1975). Protective effect of cysteamine and glutathione against the toxicity and carcinogenicity of methylazoxymethanol. *Gifu Daigaku Igakubu Kiyo 23*:495-500.

Mosher, H. P. (1938). Does animal experimentation show similar changes in the ear of the mother and fetus after the ingestion of quinine by the mother? *Laryngoscope 48*:361-395.

Mosier, H. D., and Jansons, R. A. (1972). Distribution and fate of nicotine in the rat fetus. *Teratology 6*:303-312.

Mulvihill, J. J. (1973). Caffeine as a teratogen and mutagen. *Teratology 8*: 69-72.

Mun, A. M., Barden, E. S., Wilson, J. M., and Hogan, J. M. (1975).
Teratogenic effects in early chick embryos of solanine and glycoalkaloids
from potatoes infected with late-blight *Phytophthora infestans*. *Teratology*
11:73-78.

Nevin, N. C., and Merrett, J. D. (1975). Potato avoidance during pregnancy in women with a previous infant with either anencephaly and/or
spina bifida. *Br. J. Prev. Soc. Med. 29*:111-115.

Nishie, K., Norred, W. P., and Swain, A. P. (1975). Pharmacology and
toxicology of chaconine and tomatine. *Res. Commun. Chem. Pathol.
Pharmacol. 12*:657-668.

Nishimura, H., and Nakai, K. (1958). Developmental anomalies in offspring
of pregnant mice treated with nicotine. *Science 127*:877-878.

Nishimura, H., and Nakai, K. (1960). Congenital malformations in offspring
treated with caffeine. *Proc. Soc. Exp. Bio. Med. 104*:140-142.

Overman, R. S., Stahmann, M. A., Huebner, C. F., Sullivan, W. R., Spero,
L., Doherty, D. G., Ikawa, M., Graf, L., Roseman, S., and Link, K. P.
(1944). Studies on the hemorrhagic sweet clover disease. XIII. Anticoagulant activity and structure in the 4-hydroxycoumarin group. *J. Biol.
Chem. 153*:5-24.

Padmanaban, G. (1980). Lathyrogens. In *Toxic Constituents of Plant Foodstuffs*, 2nd ed., I. E. Liener (Eds.). Academic Press, New York, pp.
239-263.

Pauli, R. M., and Hall, J. G. (1979). Warfarin embryopathy. *Lancet 2(8134)*:
144.

Pearn, J. H. (1967). Studies on a site specific cleft palate teratogen. The
toxic extract from *Indigofera spicata* Forssk. *Br. J. Exp. Pathol. 48*:620-
626.

Pearn, J. H., and Hegarty, M. P. (1970). Indospicine—the teratogenic
factor from *Indigofera spicata* extract causing cleft palate. *Br. J. Exp.
Pathol. 51*:34-36.

Persaud, T. V. N. (1968a). Teratogenic effects of hypoglycin-A. *Nature
(Lond.) 217*:471.

Persaud, T. V. N. (1968b). Hypoglycin-A and foetal development in rabbits.
West Indian Med. J. 17:52-56.

Persaud, T. V. N. (1971). Mechanism of teratogenic action of hypoglycin-A.
Experientia 27:414.

Persaud, T. V. N., and Ellington, A. C. (1968). Teratogenic activity of
cannabis resin. *Lancet 2*:406-407.

Persaud, T. V. N., and Hoyte, D. A. N. (1974). Pregnancy and progeny in
rats treated with the pyrrolizidine alkaloid fulvine. *Exp. Pathol. 9*:59-63.

Peterson, J. E., and Jago, M. V. (1980). Comparison of the toxic effects of
dehydroheliotridine and heliotrine in pregnant rats and their embryos.
J. Pathol. 131:339-355.

Pinnel, S. R., and Martin, G. R. (1968). Cross-linking of collagen and elastin. The enzymatic conversion of lysine in peptide linkage to alpha-
aminoadipic-5-semialdehyde by an extract from bone. *Proc. Natl. Acad.
Sci. USA 61*:708-716.

Poswillo, D. E., Sopher, D., and Mitchell, S. J. (1972). Experimental induction of foetal malformation with "blighted" potato: a preliminary report.
Nature (Lond.) 239:462-464.

Poswillo, D. E., Sopher, D., Mitchell, S. J., Coxon, D. T., Curtis, R. F., and Price, K. R. (1973). Investigations into the teratogenic potential of imperfect potatoes. *Teratology* 8:339-348.

Pratt, R. M., Wilk, A. L., Horigan, E. A., and King, C. T. G. (1971). Inhibition of cross linking in embryonic collagen and β-aminopropionitrile (βAPN) induced cleft palate in the rat. *Teratology* 4:239.

Prichard, J. T., and Voss, J. L. (1967). Fetal ankylosis in horses associated with hybrid sudangrass pasture. *J. Am. Vet. Med. Assoc.* 150:871-873.

Rao, K. V., and Chauhan, S. P. S. (1971). Teratogenic effects of rotenone on the early development of chick embryos in vitro. *Teratology* 4:191-198.

Reed, P., and Lardy, H. (1972). A23187: a divalent cation ionophore. *J. Biol. Chem.* 247:6970-6977.

Renwick, J. H. (1972). Hypothesis: anencephaly and spina bifida are usually preventable by avoidance of a specific but unidentified substance present in certain potato tubers. *Br. J. Prev. Soc. Med.* 26:67-88.

Rindle, M., and Sapiro, M. L. (1949). A chemical investigation of the consituents of *Nicotiana glauca* R. Grah. (Solanaceae) (wild tobacco). *Onderstepoort J. Vet. Sci. Anim. Ind.* 22:301-311.

Robinson, G. C., Brommitt, J. R., and Miller, J. R. (1963). Hearing loss in infants and preschool children. II. Etiologic considerations. *Pediatrics* 32:115-124.

Robson, J. M., and Sullivan, F. M. (1968). Effect of 5-hydroxytryptamine on maintenance of pregnancy, congenital abnormalities, and the development of toxemia. In *Advances in Pharmacology*, Vol. 6, Part B, pp. 187-189, Academic Press, New York.

Rosenberg, E. E. (1957). Teratogenic effects of beta-aminopropionitrile in the chick embryo. *Nature (Lond.)* 180:706-707.

Ruddick, J. A., Harwig, J., and Scott, P. M. (1974). Non-teratogenicity in rats of blighted potatoes and compounds contained in them. *Teratology* 9:165-168.

Sanatina, K. G., Lazutka, F. A., Gefen, S. H. G., and Vaitekuniene, D. (1969). Effect of anabasine sulfate on embryogeny and sexual function of rats. *Vopr. Epidemiol. Geg. Liton. SSR Mater. Nauch. Sess.*, pp. 175-181. *Chem. Abstr.* 74:110840j (1971).

Santomauro, A. G. (1976). Central nervous system congenital malformations produced by narcotic analgesics. *Am. J. Obstet. Gynecol.* 125:1158-1159.

Schardein, J. L. (1976). *Drugs as Teratogens*. CRC Press, Boca Raton, Fla.

Schery, R. W. (1952). *Plants for Man*. Prentice-Hall, Englewood Cliffs, N.J.

Schoental, R. (1959). Liver lesions in young rats suckled by mothers treated with pyrrolizidine (*Senecio*) alkaloids, lasciocarpine and retrorsine. *J. Pathol. Bacteriol.* 77:485-495.

Seifulla, Kh. I., and Ryzhova, K. E. (1972). Effect of hydrocortisone, solasodine, and solanine on the growth and development of fetuses of pregnant rats. *Mater. Vses. Konf. Issled. Lek. Rast. Perspekt. Ikh. Ispol'z. Proizvoid. Lek. Prep.* 1970:160. *Chem. Abstr.* 83:22383t (1975).

Selby, L. A., Menges, R. W., Hauser, E. C., Flatt, R. E., and Case, A. A. (1971). Outbreak of swine malformations associated with the wild black cherry, *Prunus serotina*. *Arch. Environ. Health* 22:496-501.

Shephard, T. H. (1980). *Catalog of Teratogenic Agents.* Johns Hopkins University Press, Baltimore.

Shoji, R., and Makino, S. (1966). Preliminary notes on the teratogenic and embryocidal effects of colchicine on mouse embryos. *Proc. Jpn. Acad. 42:* 822-827.

Shoro, A. A. (1972). Club foot and intrauterine growth retardation produced by tubocurarine in the rat fetus. *J. Anat. 111:*506-508.

Shupe, J. L., Binns, W., James, L. F., and Keeler, R. F. (1968). A congenital deformity in calves induced by the maternal consumption of lupine. *Aust. J. Agric. Res. 19:*335-340.

Sieber, S. M., Mead, J. A. R., and Adamson, R. H. (1976). Pharmacology of antitumor agents from higher plants. *Cancer Treat. Rep. 60:*1127-1140.

Sieber, S. M., Whang-Peng, J., Botkin, C., and Knutsen, T. (1978). Teratogenic and cytogenic effects of some plant-derived antitumor agents (vincristine, colchicine, maytansine, VP-16-213, and VM-26) in mice. *Teratology 18:*31-48.

Smith, C. R. (1935). Occurrence of anabasine in *Nicotiana glauca* R. Grah. (Solanaceae). *J. Am. Chem. Soc. 57:*959-960.

Spatz, M. (1969). Toxic and carcinogenic alkylating agents from cycads. *Ann. N. Y. Acad. Sci. 163:*848-855.

Spatz, M., Dougherty, W. J., and Smith, S. W. E. (1967). Teratogenic effects of methylazoxymethanol. *Proc. Soc. Exp. Biol. Med. 124:*467-478.

Spiers, P. S., Pietrzyk, J. J., Piper, J. M., and Glebatis, D. M. (1974). Human potato consumption and neural malformation. *Teratology 10:*125-128.

Stamler, F. W. (1955). Reproduction in rats fed *Lathyrus* peas or aminonitriles. *Proc. Soc. Exp. Biol. Med. 90:*294-298.

Steffek, A. J., and Hendrickx, A. G. (1972). Lathyrogen-induced malformations in baboons: a preliminary report. *Teratology 5:*171-179.

Steffek, A. J., Verrusio, A. C., and Watkins, C. A. (1972). Cleft palate in rodents after maternal treatment with various lathyrogenic agents. *Teratology 5:*33-40.

Strong, F. M. (1966). Pressor amines in foods. In *Toxicants Occurring Naturally in Foods.* National Academy of Sciences Publ. 1354. National Research Council, Washington, D.C.

Swan, G. A. (1967). *An Introduction to the Alkaloids.* Wiley, New York.

Swinyard, C. A., and Chaube, S. (1973). Are potatoes teratogenic for experimental animals? *Tetratology 8:*349-357.

Szabo, K. T., and Kang, J. Y. (1969). Comparative teratogenic studies with various therapeutic agents in mice and rabbits. *Teratology 2:*270.

Tanimura, T., and Lee, S. (1972). Discussions on the suspected teratogenicity of quinine to humans. *Teratology 6:*122.

Thiersch, J. B. (1963). Effect of podophyllin and podophyllotoxin on the rat litter in utero. *Proc. Soc. Exp. Biol. Med. 113:*124-127.

Thompson, R. S., and Gautieri, R. F. (1969). Comparison and analysis of the teratogenic effects of serotonin, angiotensin-II, and bradykinin in mice. *J. Pharm. Sci. 58:*406-412.

Tsumo, S., Takahashi, K., Kato, S., and Kubota, A. (1960). The malformation induced by 2-(3-pyridyl)piperidine (neonicotine) in chick embryos. *Nippon Yakurigaku Zasshi 56:*659-667.

Tuchmann-Duplessis, H., and Mercier-Parot, L. (1961). Malformations foetales chez le rat traité par de fortes doses de déserpidine. *C. R. Seances Soc. Biol. 155*:2291-2293.

Upshall, D. G. (1972). Correlation of chick embryo teratogenicity with the nicotinic activity of a series of tetrahydropyrimidines. *Teratology 5*: 287-294.

Vara, P., and Kinnunen, O. (1951). The effect of nicotine on the female rabbit and developing foetus. *Ann. Med. Exp. Biol. Fenn. 29*:202-213.

Von Holt, C., Von Holt, C. M., and Bohn, H. (1966). Metabolic effects of hypoglycin and methylenecyclopropaneacetic acid. *Biochim. Biophys. Acta 125*:11-21.

Warkany, J. (1971). *Congenital Malformations*. Yearbook Medical Publishers, Chicago.

Warkany, J. (1976). Warfarin embryopathy. *Teratology 14*:205-210.

Wayman, O., Iwanaga, I. I., and Hugh, W. I. (1970). Fetal resorption in swine caused by *Leucaena leucocephala* (Lam.) DeWitt in the diet. *J. Anim. Sci. 30*:583-588.

Wiesner, B. P., and Yudkin, J. (1955). Control of fertility by antimitotic agents. *Nature (Lond.) 176*:249-250.

Wilson, J. G., and Fraser, F. C. (1977). *Handbook of Teratology*, Vol. 1. Plenum Press, New York.

Yang, M. G., and Mickelsen, O. (1969). Cycads. In *Toxic Constituents of Plant Foodstuffs*, I. E. Liener (Eds.). Academic Press, New York, pp. 159-167.

Yasuda, Y., Kihara, T., and Nishimura, H. (1974). Embryotoxic effects of feeding bracken fern (*Pteridium aquilinum*) to pregnant mice. *Toxicol. Appl. Pharmacol. 28*:264-268.

Zellers, J. E., and Gautieri, R. F. (1977). Evaluation of teratogenic potential of codeine sulfate in CF-1 mice. *J. Pharm. Sci. 66*:1727-1731.

6

TERATOGENICITY OF FUNGAL TOXINS AND FUNGAL - PRODUCED ANTIMICROBIAL AGENTS

RONALD D. HOOD*

The University of Alabama, Tuscaloosa, Alabama

GEORGE M. SZCZECH

Burroughs Wellcome Company, Research Triangle Park, North Carolina

---

*
Also affiliated with Ronald D. Hood and Associates, Consulting Toxicologists, Tuscaloosa, Alabama.

I. INTRODUCTION

Knowledge of the potential of mycotoxins for inducing birth defects is of recent origin and has largely been obtained from laboratory experimentation.
The first such reports appeared in 1964, when Le Breton et al. described
increased prenatal mortality following exposure of pregnant rats to aflatoxin
B_1, and Verrett et al. reported embryotoxicity of aflatoxin in the chick. Not
until 1973 were additional mycotoxins, such as rubratoxin B (Hood et al.,
1973) and zearalenone (Miller et al., 1973) named as teratogenic agents. A
number of additional mycotoxins have subsequently been screened for developmental toxicity, but the majority of even the currently known mycotoxins have
not yet been so tested.

Tests for teratogenicity of antibiotics appeared just after those with mycotoxins. The initial study on streptonigran by Warkany and Takacs appeared
in 1965, but few such papers involving significant antibiotics other than
griseofulvin have appeared subsequently.

The remainder of this chapter is devoted primarily to the results of exposure of developing mammals and birds to fungal metabolites. Prenatal effects
of mycotoxins and of antibiotics are briefly summarized in Tables 1 and 2,
respectively.

Timing of the days of gestation has been adjusted to avoid confusion. The
day on which a copulation plug or sperm-positive vaginal smear was seen (mice
or rats), the day following mating the prior evening (hamsters), or the day of
mating (rabbits) has been designated gestation day 1.

II. MYCOTOXINS

A. Aflatoxins and Related Compounds

Aflatoxin B_1 is produced by various molds, particularly *Aspergillus flavus*
(Lillehoj et al., 1976), and is found in numerous human foods and livestock
feeds (Stoloff, 1980). Because of their widespread natural occurrence and
potent toxicity, aflatoxins have been the best studied mycotoxins. Of the
aflatoxins, aflatoxin B_1 (AB_1) has been the subject of the majority of studies.

In the initial study involving a mammal, Le Breton et al. (1964) reported
that AB_1 (300-400 µg per rat) given to pregnant rats caused increased prenatal death and hemorrhage at the uteroplacental junction. Fetal growth
retardation followed chronic administration of smaller doses. Butler and
Wigglesworth (1966) soon reported on the effects of oral aflatoxin doses of
approximately 5.6 mg/kg on one of gestation days 6-22 in Wistar rats. Animals treated before day 17 showed little effect, but fetuses from rats treated

TABLE 1 Summary of Teratogenicity Test Results for Mycotoxins and Their Metabolites[a]

Mycotoxin	Test species	Pre- or postnatal death	Retarded growth	Teratogenic effects	Reference
Aflatoxin B_1	Rat	+	+	−	Le Breton et al., 1964
		+	+	?	Butler and Wigglesworth, 1966
		−	+	−	Butler, 1971
		?	?	?	Elegbe et al., 1974
		+	*	*	Goerttler et al., 1980
	Hamster	+	*	+	Elis and DiPaolo, 1967
		−	+	−	Schmidt and Panciera, 1980
	Mouse	+	−	−	DiPaolo et al., 1967
	Pig	−	−	−	Hintz et al., 1967
	Chick embryo	+	*	+	Verrett et al., 1964
		+	?	?	Clegg, 1964
		+	+	−	Shibko et al., 1968
		+	+	+	Bassir and Adekunle, 1970a
		+	+	−	Stoloff et al., 1972
Aflatoxin P_1	Chick embryo	−	−	−	Stoloff et al., 1972
Aflatoxin Q_1	Chick embryo	+	*	*	Hsieh et al., 1974
Palmotoxin B_o	Chick embryo	+	+	+	Bassir and Adekunle, 1970a
Palmotoxin G_o	Chick embryo	+	+	?	Bassir and Adekunle, 1970a
Alternariol	Mouse	+	+	+	Pero et al., 1973
Alternariol monomethyl ether	Mouse	+	+	?	Pero et al., 1973

TABLE 1 (Continued)

Mycotoxin	Test species	Pre- or postnatal death	Retarded growth	Teratogenic effects	Reference
Citrinin	Mouse	+	+	-	Hood et al., 1976a
	Chick embryo	+	*	+	Ciegler et al., 1977
Cytochalasin B	Chick embryo	+	+	+	Linville and Shepard, 1972
		*	*	+	Karfunkle, 1972
		*	*	+	Messier and Auclair, 1974, 1977
		*	*	+	Lee and Kalmus, 1976
		*	*	+	Greenaway et al., 1977
	Hamster	+	*	+	Wiley, 1980
	Rat	?	-	-	Ruddick et al., 1974
	Mouse	+	*	+	Hayakawa et al., 1968
		-	-	-	Austin and Brown, 1979
Cytochalasin A	Chick embryo	*	*	+	Greenaway et al., 1977
Cytochalasin D	Chick embryo	*	+	+	Gilani and Kreshel, 1980
		*	*	+	Greenaway et al., 1977
	Mouse	*	*	+	Foerder, 1978
		+	+	+	Hayakawa et al., 1968
		+	+	+	Shepard and Greenaway, 1977
		+	*	+	Austin and Brown, 1979
	Hamster	+	*	+	Wiley, 1980
Cytochalasin E	Chick embryo	*	*	+	Greenaway et al., 1977
	Mouse	+	*	+	Austin and Brown, 1979

Compound	Species				Reference
Ergot alkaloids					
Ergotamine and others	Rat	+	+	−	Sommer and Buchanan, 1955
	Rat	+	*	*	Mantle, 1969
Ergotamine	Mouse, rabbit	+	+	−	Grauwiler and Schön, 1973
		−	+	−	Grauwiler and Schön, 1973
	Sheep	+	+	−	Greatorex and Mantle, 1974
Agroclavine	Rat	+	*	*	Edwardson, 1968
	Mouse	+	*	*	Mantle, 1969
Elymoclavine	Mouse	+	*	+	Witters et al., 1975
Crude ergot	Mouse	+	*	*	Campbell and Burfening, 1972
	Gilt	−	*	*	Campbell and Burfening, 1972
Fusarenon-X	Mouse	+	+	−	Ito et al., 1980
Moniliformin	Mouse	−	−	−	Hayes and Hood, 1974
	Chick embryo	+	*	−	Burmeister et al., 1979
Ochratoxin A	Mouse	+	+	+	Hayes et al., 1974
	Mouse	+	+	+	Hood et al., 1978
	Mouse	+	*	+	Szczech and Hood, 1981
	Rat	+	+	+	Moré and Galtier, 1974
	Rat	+	+	+	Brown et al., 1976
	Hamster	+	+	+	Hood et al., 1976b
	Japanese quail	+	*	*	Prior et al., 1978
	Japanese quail	+	*	*	Prior et al., 1979
	Chick embryo	+	*	+	Gilani et al., 1978

TABLE 1 (Continued)

Mycotoxin	Test species	Pre- or postnatal death	Retarded growth	Teratogenic effects	Reference
Patulin	Mouse	-	-	-	Hayes and Hood, 1974
		+	+	-	Reddy et al., 1978
		+	*	-	Osswald et al., 1978
	Rat	-	-	-	Dailey et al., 1977
		-	-	-	Gallo et al., 1977
	Chick	+	*	+	Ciegler et al., 1976
		+	*	*	Ciegler et al., 1977
Penicillic acid	Mouse	?	?	-	Hayes and Hood, 1978
Penitrem A	Mouse	-	-	+	Hayes and Hood, 1978
Rubratoxin B	Mouse	+	+	+	Hood et al., 1973
		+	+	+	Koshakji et al., 1973
		+	+	?	Hayes and Hood, 1976
		+	+	+	Evans and Harbison, 1977
	Hamster	+	+	?	Hayes and Hood, 1976
	Chick embryo	+	+	+	Gilani et al., 1979
Sclerotinia toxins	Rat	-	?	-	Ruddick and Harwig, 1975
Secalonic acid D	Chick	-	*	-	Yamazaki et al., 1971
		-	*	-	Ciegler et al., 1980
	Mouse	+	+	+	Reddy et al., 1981
Stachybotrys toxins	Mouse	+	+	-	Korpinen, 1974
Sterigmatocystin	Mouse	-	-	-	Hayes and Hood, 1976

Toxin	Species			Reference
T-2 toxin	Mouse	+	+	Stanford et al., 1975
		+	+	Lary and Hood, 1977
		+	-	Hood et al., 1978
	Chick	-	+	Wyatt et al., 1975
Viriditoxin	Mouse	-	+	Hood et al., 1976a
Zearalenone (F-2)	Pig	+	-	Miller et al., 1973
		-	+	Étienne and Jemmali, 1979
	Rat	-	+	Ruddick et al., 1976
		+	-	Bailey et al., 1976
		+	?	Hidy et al., 1977
Combined effects				
Ochratoxin A + T-2 toxin	Mouse	+	+	Hood et al., 1978
T-2 toxin + rubratoxin B		+	+	R. D. Hood, unpublished observations, 1982
Rubratoxin B + ochratoxin A		+	+	R. D. Hood, unpublished observations, 1982
Citrinin + patulin	Chick embryo	+	*	Ciegler et al., 1977

[a]Test responses are given as follows: +, presence of the indicated response; -, absence of the indicated response; *, not tested or not reported; ?, equivocal or unclear results.

TABLE 2 Summary of Teratogenicity Test Results for Antibiotics[a]

Antibiotic	Test species	Pre- or postnatal death	Retarded growth	Teratogenic effects	Reference
Actinomycin D	Rat	+	+	+	Tuchmann-Duplessis, 1975
		+	+	+	Wilson, 1966
		+	-	+	Harvey and Srebnik, 1978
	Mouse	+	+	+	Lary et al., 1977
	Rabbit	+	*	+	Tuchmann-Duplessis, 1975
	Hamster	+	+	+	Tuchmann-Duplessis et al., 1973
Adriamycin	Rat	+	+	-	Ogura et al., 1978
		+	+	+	Thompson et al., 1978
	Rabbit	+	-	-	Thompson et al., 1978
Carminomycin	Mouse	+	+	?	Damjanov and Celluzzi, 1980
Daunorubicin	Rat	+	*	+	Roux and Taillemite, 1969
		+	*	-	Chaube and Murphy, 1968
		+	*	?	Thompson et al., 1978
	Rabbit	-	-	-	Julou et al., 1967
		-	*	-	Thompson et al., 1978
	Mouse	-	-	-	Julou et al., 1967
	Chick	-	-	-	Julou et al., 1967

					Reference
Griseofulvin	Rat	+	+	+	Slonitskaya, 1969
		+	+	+	Klein and Beal, 1972
		+	+	+	Steelman and Kocsis, 1978
	Mouse	+	+	+	Aujezaska et al., 1978
	Cat	*	+	*	Gillick and Bullmar, 1972
		+	+	*	Cott et al., 1974
Mitomycin C	Mouse	+	+	+	Tanimura, 1968
	Rat	+	-	*	Chaube and Murphy, 1968
Streptonigrin	Rat	+	+	+	Warkany and Takacs, 1965
		+	+	*	Chaube et al., 1969

[a]Test responses are given as follows: +, presence of the indicated response; -, absence of the indicated response; *, not tested or not reported; ?, equivocal or unclear results.

on day 17 were stunted. No malformations were reported, but the authors described some fetuses as having "loose, wrinkled skin and unduly large heads." Histological examination of fetuses and placentas were negative. When additional dams were treated with a more crude aflatoxin preparation, maternal liver damage was correlated with decreased fetal weight. Aflatoxin was detected in the fetal livers, however, so a direct toxic effect on the fetus could not be ruled out. Such an effect seems likely, because when day 17-treated rats were allowed to litter, their pups either died or failed to thrive.

According to Butler (1971), administration of AB_1 (5-7 mg/kg) to rats on gestation day 17 retarded fetal growth. Aflatoxin-treated dams had reduced food intakes. When controls were pair-fed with these animals, their fetuses were similarly reduced in weight. The author suggested that the fetal stunting seen in his previous study (Butler and Wigglesworth, 1966) may also have been due to reduced diet consumption.

In an attempt to determine the toxicological mechanism responsible for neonatal mortality in rat pups exposed to aflatoxin in utero, Das et al. (1978) looked for effects on the lung. Pregnant Sprague-Dawley rats were injected intraperitoneally with 5 mg/kg aflatoxin in dimethyl sulfoxide (DMSO) and examined on day 21. Treatment resulted in increased lipid synthesis but decreased phosphatidyl choline production in the fetal lungs. Treatment was associated with ultrastructural changes suggestive of a toxic effect.

AB_1 treatment late in gestation was also employed by Goerttler et al. (1980). Sprague-Dawley rats were injected intraperitoneally on gestation days 15-18 or 18-21; some pups were given an additional injection on days 2-5 or 14-17 postpartum. Aflatoxin treatment at a dose of 0.5 mg/kg resulted in increased fetal deaths. When additional treatments were given postnatally, mortality increased further.

The hamster has also been used as an animal model for the study of aflatoxicosis. According to Elis and DiPaolo (1967), intraperitoneal injection of the dams with 2 mg/kg AB_1 on gestation day 8 had no effect on their offspring, while 6 mg/kg killed half the dams. Treatment with 4 mg/kg resulted in increased prenatal mortality and malformation. AB_1 treatment also resulted in maternal and fetal liver damage. In a more recent study, Schmidt and Panciera (1980) also treated pregnant hamsters with AB_1. Treatment was given on day 8 or 9 with an intraperitoneal dose of 4 or 6 mg/kg. Treated dams exhibited renal and hepatic necrosis and fetal growth was often stunted. Contrary to the report of Elis and DiPaolo (1967), however, neither an increased incidence of fetal malformations nor fetal liver necrosis was observed, although a largely transient occurrence of subcutaneous hemorrhage, edema, and ascites was seen in several fetuses.

Mammals other than the rat and hamster have been largely ignored in tests of aflatoxin teratogenicity. An exception is the use of the C3H/He mouse by DiPaolo et al. (1967). Treatment by intraperitoneal injection on multiple gestation days during the period of major organogenesis with AB_1 doses of 1-4 mg/kg per day did not cause malformation or decreased fetal growth. Prenatal deaths appeared to be increased in several cases, and a single dose of 8 or 12 mg/kg on gestation day 9 resulted in 90% resorptions and deaths, but no malformed survivors.

One additional report indicated no adverse effects on the offspring of Duroc pigs fed a ration containing 450 ppb AB_1 for 5 months prior to pregnancy and during pregnancy and lactation (Hintz et al., 1967). Since it is likely that these animals were metabolizing aflatoxin more rapidly during gestation than would animals naive for the toxin, however, the results may have differed if the dosing had begun during pregnancy.

The foregoing reports suggest that while AB_1 may be fetotoxic, it has only modest teratogenic potential. Whether this finding is generally applicable to mammals, however, remains to be seen.

In addition to rodents, the chick embryo has been the target of extensive testing with AB_1 and related compounds. For example, Verrett et al. (1964) noted the production of limb malformations following administration of AB_1. Also, Clegg (1964) reported that the percentage of chicks hatching from eggs injected with "crude aflatoxin" increased with increasing age of the embryo through the eleventh day of incubation.

When hen's eggs were injected with various amounts of AB_1, 0.12 µg per egg was found to be lethal to 98% of treated eggs (Shibko et al., 1968). Eggs treated with 0.05 µg per egg (the approximate LD_{70}) showed decreased growth after day 12 of incubation, but no gross malformations or significant changes in enzyme levels.

Bassir and Adekunle (1970a) injected AB_1 or palmotoxins B_0 (PB_0) or G_0 (PG_0), also from *A. flavus*, into the yolk sac of chicks. PB_0 treatment at doses of 0.2-0.6 µg per egg produced beak, skull, eye, and limb malformations, while similar doses of AB_1 produced limb defects only. PG_0 was less effective. Chicks exposed to doses of up to 6.3 µg per egg exhibited a small number of eye defects, but there was no apparent dose-response relationship. Birds treated with AB_1 or PB_0 had rough plumage, and all toxins were associated with stunting of growth and decreased viability. The same workers also found fatty infiltration and cellular necrosis in the livers of chick embryos treated with toxin doses of 0.3 µg per egg (Bassir and Adekunle, 1970b). AB_1 and PB_0 were more potent in their effects than was PG_0.

PB_0 and PG_0 were tested by injecting them into chicken eggs of five different strains (Smith et al., 1974). Their results were said to suggest a strain difference in response to AB_1. While exposure to AB_1 or PB_0 resulted in histopathological lesions, PG_0 had little effect. All three toxins were detected in various tissues of chicks sampled after hatching (Adekunle and Bassir, 1975).

Further studies with aflatoxins in developing chick embryos have examined the role of metabolism in altering AB_1 toxicity. For example, Stoloff et al. (1972) compared the effects of aflatoxin P_1 with AB_1 in the chick. The parent compound (B_1) was embryotoxic, but the metabolite was not, at doses of up to 0.19 µg per egg. Similarly, when the metabolite aflatoxin Q_1 was tested, it proved much less toxic, as measured by percent mortality, than was AB_1 (Hsieh et al., 1974).

In another approach, Howarth and Wyatt (1976) added AB_1 to the diets of hens at levels of 5 or 10 µg/g. Although fertility was unaffected by the mycotoxin treatment, hatchability of the fertile eggs declined for eggs collected only 1 week after the start of the trial, at both dose levels. No effects of the aflatoxin were seen in chicks that hatched, however.

Such findings with nonmammalian species confirm the toxicity of aflatoxins to developing systems. They also support the view that AB_1 is teratogenic only in specific organisms. Such data imply that AB_1 often displays neither specificity of action nor preferential concentration in specific tissues. At least one of these characteristics is probably necessary for the production of malformations (rather than more general toxic effects, such as growth retardation).

B. *Alternaria* Toxins

Pero et al. (1973) tested two out of the variety of mycotoxins elaborated by members of the genus *Alternaria*. These toxins were the dibenzopyrones alternariol (AOH) and alternariol monomethyl ether (AME) isolated from *Alterneria tenuis*. The toxins were given subcutaneously to DBA/2 mice. AOH on gestation days 9-12 was associated with prenatal deaths and stunting at a dose of 100 mg/kg. AME was associated with similar effects, although they were statistically nonsignificant at doses up to 100 mg/kg. When given together at a level of 25 mg/kg each, the combination of AOH and AME was fetotoxic. Neither toxin was fetotoxic when given on days 13-16 of pregnancy, although the result of 100 mg/kg AOH was suggestive of an effect. Early and late treatment regimens were associated with suggestive but nonsignificant incidences of gross malformations; in the case of the high dose of AOH on days 13-16 the effect was significant. These data are difficult to evaluate, as only small numbers of litters were treated in most groups, and the compounds were tested in two difference solvents (DMSO and 1:1 honey and water). Nevertheless, it does appear that AOH and possibly AME are fetotoxic and teratogenic in mice.

C. Citrinin

Citrinin, a toxic product of several *Penicillia* and *Aspergilli* (Jordan et al., 1977), has been tested for prenatal effects by Hood et al. (1976a). Citrinin was given intraperitoneally to CD-1 mice at doses of 10-40 mg/kg on one of gestation days 7-10. The high doses tended to result in fetal stunting and decreased survival. No increases were observed in malformations, however.

Although in the study above, single doses of citrinin did not prove teratogenic to mice, Ciegler et al. (1977) reported teratogenic effects in the chick following citrinin injection of the egg. "Malformations of the extremities," as well as exencephaly, exophthalmia, and crossed beaks followed treatment of 4-day-old embryos. The percentage of malformed chicks increased as the citrinin dose rose from 10 to 150 µg per egg. The LD_{50} value for citrinin was calculated to be 80.5 µg per egg.

According to the limited data available, citrinin thus appears to be a teratogen for the chick embryo and was found to be fetotoxic in one strain of mice. Little appears to be known of citrinin's mechanism of action, however, and much more extensive testing would be required to establish teratogenicity or its lack in mammals.

D. Cytochalasins

Cytochalasins are metabolites of various molds, such as species of *Helmintho-sporia* and *Hormiscia* (Hood, 1979). The effects of these mycotoxins, particularly cytochalasins B (CB) and D (CD), have been studied extensively by teratologists and developmental biologists. This interest is due to the biological activities of cytochalasins, such as disruption of microfilaments and interference with cleavage, cell movements, and activities of the cell membrane (Granholm, 1979).

Linville and Shepard (1972) were among the first to use CB as a tool for the study of development. They treated explanted early somite stage chick embryos with CB in the culture medium. Treatment resulted in neural tube closure defects and retarded development, and at the higher doses (1 or 2 µg per explant) frequently resulted in death. The authors speculated that the effects of CB on neural tube morphogenesis may have been due to interference with intracellular microfilaments.

Karfunkel (1972) cultured stage 8 chick embryos in the presence of 10 µg/ml CB. In treated embryos, the neural tube failed to close and intracellular microfilaments in the neural cells were said to be largely absent; microtubules appeared unaffected. The anterior wedge-shaped cells lost that shape, while the more posterior cells failed to achieve the normal wedge shape.

In 1974, Messier and Auclair described the effects of exposure of explanted chick embryos to 0.5 µg/ml CB. They confirmed the observations of Linville and Shepard (1972) and Karfunkel (1972) regarding failure of neural tube closure. These authors, however, observed bands of microfilaments, although their arrangement was atypical. Inhibition of interkinetic nuclear migration was also observed in neural cells of treated embryos.

Lee and Kalmus (1976) performed similar experiments with explanted chicken embryos and by use of electron microscopy confirmed the disruption of microfilaments. They also noted reduction in the frequency of cytoplasmic extensions but no visible effect on microtubules or other subcellular structures.

Later work by Messier and Auclair (1977) with cultured chick embryos determined that CB altered intercellular permeability, presumably by altering cellular attachment. Treatment with trypsin, however, had the same effect. Both trypsin and CB also inhibit interkinetic nuclear migration. The authors suggest that both agents may disrupt apical junctions necessary for nuclear migration and for closure of the neural tube.

The assumption that CB exerts its inhibitory effect on certain cell functions by affecting the plasmalemma was supported by Stagno and Low (1978). They exposed 3- to 4-day-old chick embryos to CB for brief periods. Observations by both scanning and transmission electron microscopy revealed that endodermal cells tended to become dissociated, with large surface blebs associated with an atypical cell membrane. Microfilaments were also affected, being unevenly dispersed, rather than arranged parallel to the cell surface. Nagela and Lee (1978) observed similar results in embryonic chick neuroepithelial cells, which were less adherent when exposed to CB and had an altered distribution of extracellular coat material.

A related study by De-Voy et al. (1979) also employed chick embryos. They examined the migrating presumptive mesodermal cells of early embryos after exposure to CB. Treatment resulted in rounding up of the cells, wider separation from other mesodermal cells and from the adjacent ectoderm, and an altered surface appearance.

In the studies described previously, the explanted chick embryo has been the test system used. In a study by Gilani and Kreshel (1980), however, CD was injected into chicken eggs on one of days 0-3 of incubation. Treatment resulted in gross malformations, such as gastroschisis, microphthalmia, and

anophthalmia. Retarded growth, edema, and hemorrhage were also noted.
According to these results, although failure of neural tube closure was noted
in cultured chick embryos, if treated in the egg, the embryos displayed
other defects after further development.

Foerder (1978), using 8-day-old mouse embryos, reported that cytochalasin D affected the presumptive epidermal cells overlying the neural folds.
Treated epidermis had less of the ruffles indicative of migrating epithelium
and was inhibited from advancing toward the dorsal midline and participating
in closure together with the underlying neural cells. Thus the stage may be
set for malformations such as exencephaly.

According to the results of Wiley (1980) with developing hamsters, both
CD and CB were teratogenic, with the former being the more potent. Exencephalies and encephaloceles were indeed seen, and survival was decreased
at the highest dose levels (5-7 mg/kg for CB; 2 mg/kg for CD). Treatment
with either mycotoxin resulted in failure of approximation and closure of the
neural folds. Ultrastructural changes in the apical membranes of the neu-
rectodermal cells were observed, but no changes in the apical microfilament
bundles were seen at doses that resulted in teratogenesis. The author thus
concluded that the cytochalasins were affecting primarily target cell membranes.

Although cytochalasins B and D have been the most studied, Greenaway
et al. (1977) compared their effects with those of cytochalasins A and E (CA
and CE). They used cultured chick embryos and found that all four toxins
inhibited neural tube closure, with CD the most potent, CA and CB intermediate in effect, and CE the least active. When these compounds were
tested for interference with cytokinesis (resulting in binucleation of mouse
embryo cells in culture), CE was the most effective, A and D intermediate,
and B last in activity. Such results indicate that while microfilament disruption may be important in causing binucleation, another mechanism may be
influential in causing the observed open neural tubes.

The rat has also been treated with CB, but the results were negative with
regard to teratogenicity (Ruddick et al., 1974). Doses of 0.25 or 1.0 mg/kg
per day were given by gavage on gestation days 6-15. Only three litters
per dose level were examined, however, so these results are not conclusive.

Shepard and Greenaway (1977) tested CD for teratogenicity. They injected CD intraperitoneally at doses of 0.4-0.9 mg/kg on days 8-12 of gestation. CD induced exencephaly, hypognathia, and skeletal malformations in
C57BL/6J and BALB/c mice but not in a Swiss-Webster strain. Resorptions
were increased in all three strains, however. Oral dosing with 7 mg/kg CD
was also teratogenic in BALB/c mice, even when the CD was autoclaved.

Strain-specific effects of cytochalasins were also seen by Austin and
Brown (1979), who compared the effects of CB, CD, and CE in mice of the
A/J, C57BL/6J (B6), OEL-N, and aa strains. The adult LD_{50} value for
single intraperitoneal doses of CD ranged from 2.3 to 2.7 mg/kg, for CE from
2.7-2.8 mg/kg, and for CB from 30-32 mg/kg, in nonpregnant mice. When
mice were injected intraperitoneally on gestation days 8-10, CB had no effect
at a total dose up to the adult LD_{50} value. In contrast, CD and CE caused
prenatal deaths at the adult LD_{50} value. CD treatment at 1/6 of the LD_{50}
value resulted in gross defects in OEL-N and aa mice, while similar effects
were seen at 5/6 of the LD_{50} value in A/J and B6 mice. Defects seen were

also strain specific. B6 X aa offspring were intermediate in sensitivity compared with the parental strains, while CE was significantly teratogenic only
in the A/J strain.

Additional studies with CB in mice were published by Granholm (1979),
who treated pre- and early postimplantation mouse embryos, and who reviewed similar work with early mouse and rabbit embryos and the use of
cytochalasins to induce polyploidy.

E. Ergot Toxins

Ergot fungi, members of the genus *Claviceps*, synthesize a number of toxic
alkaloids. Since *C. purpurea* and others live on food grains, they are of
significances to humans and have been implicated in human disease (Hood,
1979).

1. Ergotamine and Related Alkaloids

In 1955, Sommer and Buchanan described the effects of administration of
each of the following ergot-derived alkaloids: ergotamine methansulfonate,
ergotoxine ethansulfonate, methylergonovine, and Hydergine (a mixture of
equal amounts of dihydroergocornine, dihydroergokryptine, and dihydro-
ergocristine). Doses of 0.5 mg/kg were given intraperitoneally twice daily
from gestation days 12-21 to Sprague-Dawley rats. The rats were allowed to
litter and the offspring observed for 21 days postpartum. Both ergotamine
and ergotoxine treatments often resulted in fetal death, while Hydergine was
less lethal, and methylergonovine least toxic of all. Postnatal survival and
growth were decreased in ergotamine-treated litters; only growth was depressed in the ergotoxine-exposed group. The effects on growth, however,
were probably due at least in part to decreased maternal milk production.

Grauwiler and Schön (1973) subsequently administered ergotamine tartrate orally on gestation days 7-16 to mice and rats and days 7-19 to rabbits.
Dose levels employed (in mg/kg) were: mice, 30-300; rats, 1-100; rabbits,
1-30. No teratogenic effects were noted, but development was retarded in
all three species and prenatal mortality was enhanced in the rats.

Ergotamine administration during gestation has also been investigated in
sheep (Greatorex and Mantle, 1974). Ground ergot sclerotia (0.4 g/kg)
containing 0.25% ergotamine (0.29% total alkaloids), alkaloid-free ergot, and
alkaloid-free ergot plus ergotamine tartrate at a dose of 1 mg/kg were administered orally or through a rumen cannula each day during pregnancy to
Cheviot ewes. Treatment was started at various times during pregnancy.
Whole sclerotia and the alkaloid-free sclerotia plus ergotamine were toxic to
the adults and induced fetal deaths and abortions.

2. Clavine Alkaloids

Among the clavine alkaloids elaborated by various *Claviceps* species, agroclavine was the first to be examined for prenatal effects. Edwardson (1968)
fed agroclavine (10 mg per 100 g of diet) to rats on gestation days 1-5 or
6-10. Early treatment totally prevented implantation, while the later treatment had a lesser effect. Other rats given 2 mg of agroclavine per day on

days 1-5 all had litters. Edwardson's (1968) study was followed by that of Mantle (1969). Feeding a diet containing agroclavine (250 µg/day) to mice during early pregnancy prevented implantation. A total oral dose of 350 µg/day (days 3-5) of other alkaloids, such as ergotoxine ethanesulfonate, lysergic acid α-hydroxyethylamide, or ergosine/ergosinine (60:40), also interrupted pregnancy. Subsequently, Campbell and Burfening (1972) fed 0.53% crude ergot in the diet to ICR mice through gestation day 10. Both pre- and postimplantation losses were noted, in contrast to the prior reports. Dissimilar results were obtained when pregnant gilts were fed the same ergot-containing ration from the day after breeding for a total of 38 days, as no significant decrease in embryo survival was noted.

Another clavine alkaloid, elymoclavine, was administered to pregnant mice by Witters et al. (1975). Single intraperitoneal doses of 3, 30, or 60 mg/kg were given on gestation day 10, and the mice were kept in an environmental chamber at a temperature of 20, 30, 35, or 40°C for 24 hr. On other gestation days, environmental temperature was 20°C. Treatment with the high dose was associated with skeletal defects and maternal toxic effects. Maternal effects as well as skeletal defects and prenatal deaths were exacerbated by elevated environmental temperature, in what appeared to be a potentiative effect of temperature on elymoclavine toxicity.

F. Fusarenon-X

Fusarenon-X, a trichothecene mycotoxin produced by *Fusarium nivale*, was administered to pregnant DDD mice by addition to the diet (5, 10, or 20 ppm) or by subcutaneous injection (0.63-2.6 mg/kg) by Ito et al. (1980). Chronic toxin feeding throughout gestation or for shorter periods resulted in interruption of pregnancy or fetal stunting. Single or repeated fusarenon-X injections had similar effects. Such effects may be related to the ability of the mycotoxin to inhibit synthesis of protein and DNA (Ohtsubo and Saito, 1970) and preferentially damage dividing cells (Ueno et al., 1971).

G. Moniliformin

Moniliformin, a toxin derived from *Fusarium moniliforme*, was given intraperitoneally to mice on one of gestation days 8 or 10 at a dose of 15, 20, or 25 mg/kg (Hayes and Hood, 1974). The highest dose was lethal to a portion of the dams, but no effect was seen on the conceptuses. This report was followed by that of Burmeister et al. (1979). Chicken eggs during day 4 of incubation were injected with various doses of moniliformin. A LD_{50} value of 2.8 µg per egg was obtained, but no teratogenic effects were seen.

The basis of the toxic effects of moniliformin appears to be unknown, as no studies of its mechanism of action were found in the literature.

H. Ochratoxin A

Ochratoxin A (OA) and the less toxic ochratoxin B are widely observed toxins of *Aspergillus ochraceus* and *Penicillium viridicatum* (Scott et al., 1972).

OA was assayed in CD-1 mice by Hayes et al. in 1974. Pregnant mice were treated intraperitoneally on one of gestation days 7-12 with a 5-mg/kg dose. Except on day 9, treatment resulted in fetal stunting. Treatment was associated with increased prenatal mortality and gross and skeletal malformations. Exencephalies median facial clefts, open or missing eyes, limb defects, and malformed ribs or vertebrae were common.

More recent experiments involving transplacental exposure of fetal mice to OA were performed by Szczech and Hood (1981). They treated CD-1 or ICR mice by gavage or intraperitoneal injection with 2-5 mg/kg OA on single or 2-11 consecutive days, at various times during gestation. The dams were killed on day 15, 17, 18, or 20, or were allowed to give birth and the pups sacrificed on postpartum day 7, 21, or 35. The brains of the fetuses or pups were then fixed and prepared for histopathological examination. When treatment was given on days 15-17, many fetuses exhibited prominent, widespread cerebral necrosis, although they were grossly normal in appearance. Treatment at midgestation, although often teratogenic, did not result in such necrosis. Single doses, regardless of pre- or postnatal treatment day, failed to cause brain necrosis. Similar results were obtained following either intraperitoneal or oral treatment in both mouse strains. In addition, decreased postnatal viability of pups transplacentally exposed to OA was reported.

Moré and Galtier (1974) confirmed the teratogenicity of OA when they treated pregnant rats with doses of 4 (i.p.) or 5 (p.o.) mg/kg. Treatments were given on from 2 to 8 days and resulted in lowered fetal weights and decreased viability. The only malformation observed was coelosomy.

A second study involving the rat was performed by Brown et al. (1976). OA was given orally daily on days 7-16 of pregnancy at doses of 0.25-8 mg/kg. The highest doses resulted in totally resorbed litters, as well as some maternal deaths at 4 and 8 mg/kg. Gross and skeletal malformations were noted, mainly at doses above 0.25 mg/kg. Treatment also generally resulted in fetal stunting.

Teratogenic effects of OA in the hamster were described by Hood et al. (1976b). They injected pregnant hamsters intraperitoneally with OA doses of from 2.5 to 20 mg/kg on one of gestation days 7-10, with higher doses employed on the later days. The higher doses given on day 7, 8, or 9 decreased fetal survival and on day 9 retarded growth. Gross malformations were associated with OA treatment, and included hydrocephalus, short jaws, cleft lip, and limb and tail defects. Visceral defects included heart malformations and ectopic gonads, but no skeletal defects were seen.

Japanese quail (*Coturnix*) have also been used as a test organism for OA. Feeding 16 ppm OA to the parents was reported to result in no eggs hatching (Prior et al., 1978). In a follow-up study, Prior et al. (1979) concluded that chronic OA feeding of the parent quail did not affect their fertility but resulted in early death of the embryos from treated females, whether or not the male parents were treated. When fertile eggs from untreated birds were injected with 1600 or 5000 ng of OA per egg (50 or 100 ppb OA in the egg), early embryonic death was also observed. These results caused the authors to conclude that the mortality seen in the embryos of treated parents was due to a direct effect of OA on the embryo.

Malformations in chick embryos followed treatment with 0.5-7 µg per egg by Gilani et al. (1978). Injections were made into the air sacs at 48, 72, or

96 hr of incubation. A number of gross and heart malformations were seen, and survival was decreased, particularly at the higher doses.

According to the reviewed studies, ochratoxin A is embryotoxic and teratogenic in a variety of species. This mycotoxin must therefore be viewed with suspicion as a potential human teratogen. Human exposure to significant levels of OA is presumably rare in most areas. Contamination of bread and cereals in certain geographical areas, however, such as parts of Croatia in Yugoslavia, has recently been documented (Pavlović et al., 1979).

The mechanism of OA teratogenesis is unknown, but could be related to interference with enzyme activity and energy metabolism (Warren and Hamilton, 1980), or to inhibition of protein synthesis (Creppy et al., 1980).

I. Patulin

Patulin is a carcinogenic toxin produced by several species of molds, such as *Penicilium expansum* (Ciegler et al., 1977), *Byssochlamys fulva*, and *B. nivea* (Rice et al., 1977).

In 1974, Hayes and Hood reported that intraperitoneal injections of patulin (4 mg/kg) in CD-1 mice had no discernible embryotoxic or teratogenic effects when given on days 8 or 9. A 6-mg/kg dose on day 10 was lethal to five of six pregnant females. These results were inconclusive, however, as only small numbers of litters were examined, and treatments were given only during a brief period. Reddy et al. (1978) tested patulin further in CD-1 mice by giving intraperitoneal doses of 1.5 or 2 mg/kg on days 6-17 of gestation. They also failed to detect teratogenicity, but reported total resorption of the high-dose litters. The low dose resulted in decreased fetal weights.

Osswald et al. (1978) administered 2-mg/kg doses of patulin twice daily to Swiss mice on gestation days 14-19. Increased mortality of the pups 2-6 days postpartum resulted.

Dailey et al. (1977) gave male and female rats chronic doses of patulin, mated them, and gave the females daily patulin doses of 1.5-15 mg/kg, p.o. Mortality of the dams was high in the highest dose groups, and only the low dose was evaluated for teratogenicity. A second generation of rats, exposed in utero and later to chronic patulin at the low dose, was mated, treated during pregnancy, and evaluated. No teratogenic effects were seen in offspring from either generation. A similar two-generation rat study by Gallo et al. (1977) also yielded negative results with doses of up to 1.5 mg/kg three times per week.

Ciegler et al. (1976) found the LD_{50} value for patulin to be 68.7 μg per egg following injection of unincubated chicken eggs. The LD_{50} value for 4-day embryos was much lower—2.35 μg. Patulin treatment with 10 μg prior to incubation, or 1-2 μg on incubation day 4, resulted in abnormalities, particularly leg defects. Use of patulin reacted with cysteine, to determine if reaction with SH groups detoxifies the mycotoxin, resulted in teratogenesis at dose levels of 15-100 μg per egg, but no embryolethality. A similar LD_{50} value for 4-day-old chick embryos, 2.4 μg per egg, was obtained by Ciegler et al. (1977).

Patulin has not yet been found to be teratogenic in mammals, although it can be fetotoxic, and it is teratogenic to the chick. Since it is both a highly toxic inhibitor of RNA polymerase and possibly carcinogenic as well (Tashiro et al., 1979), patulin exposure should obviously be avoided regardless of whether or not it is a likely teratogen in humans.

J. Penicillic Acid

The only report involving prenatal exposure to penicillic acid, a product of various *Penicillia* and *Aspergilli* (Harwig and Munro, 1975), appears to be that of Hayes and Hood (1978). In this study, CD-1 mice were given 30 or 50 mg/kg penicillic acid, i.p. on one of gestation days 7-10. No adverse effects on the conceptus were noted. Treatment of eight pregnant mice with 90 mg/kg patulin on day 10 resulted in six maternal deaths. Fetuses in the remaining two litters were stunted and had decreased viability. These results indicate that penicillic acid can be fetotoxic in mice, but at a dose level highly toxic to the mother.

K. Penitrem A

Penitrem A is a tremorigen produced by a number of *Penicillia*, such as *P. crustosum, P. simplicissimum*, and *P. commune* (Pitt, 1979; Wagener et al., 1980). Hayes and Hood (1978) injected CD-1 mice with 1 to 3-mg/kg doses of penitrem A on one of gestation days 7-10. No dose-related fetotoxic effects were observed, but skeletal abnormalities were noted after day 8, 9, or 10 treatment. A few gross malformations were also seen, mainly open eyes. Thus penitrem A appears to have at least a limited potential for teratogenicity.

The mechanism of action of penitrem A is not yet certain, although Hayes et al. (1977) reported effects on hepatic lipid and glycogen metabolism and DNA content in mice.

L. Rubratoxin B

Rubratoxin B, a toxic metabolite of *Penicillium rubrum* (Moss and Hill, 1970), was one of the first mycotoxins investigated for prenatal effects. In 1973, Hood et al. administered rubratoxin B (RB) intraperitoneally to CD-1 mice on one of days 6-12 of pregnancy. Treatment with 0.6 or 0.9 mg/kg on day 6, 7, or 9 resulted in gross malformations and caused completely resorbed litters if given on day 8. A dose of 0.4 mg/kg on day 8 allowed some grossly malformed survivors. All except day 9 treatments decreased both survival and growth. Malformations included exencephaly, malformed pinnae, blunt jaws, umbilical hernia and open eye.

Results of the preceding study were confirmed by Koshakji et al. (1973), who treated Swiss-Webster mice with RB. Doses of 0.5 mg/kg on one of gestation days 8-10 resulted in 90% prenatal death; 0.3 mg/kg was much less effective, as was the high dose on any of days 12-14. RB doses of 1 mg/kg on one of days 8-14 resulted in more prenatal death following treatment on the earlier than on the later days. Both multiple 0.5-mg/kg treatments and single 1-mg/kg doses caused stunting. Gross malformations but no skeletal defects were observed, as in the previous study.

Teratogenicity of RB in the chick was reported by Gilani et al. (1979). They injected 0.5-7 µg per egg on one of incubation days 1-3. Treatment on day 1 or 2 decreased survival; on day 3, it did so only at the higher doses. LD_{50} values were 1 and 5 µg/egg on days 1 and 2, respectively. Treatment caused gross malformations, particularly limb and neck defects, exencephaly, microphthalmia, and gastroschisis, and resulted in stunting.

The literature on rubratoxin B suggests that it is a potent teratogen and embryotoxin, although it has been tested primarily on mice and on chick embryos. Preliminary results with RB in the hamster indicate that it is at least fetotoxic in that species as well (Hayes and Hood, 1976). In addition, RB has been administered by gavage to pregnant mice during early organogenesis at doses of 100-125 mg/kg (Hayes and Hood, 1976). This protocol resulted in embryotoxicity and a few malformations. The high dose required is due to the greatly decreased toxicity of RB when given orally, as reported previously by Wogan et al. (1971). Rose and Moss (1970) stated that a number of modifications of the RB molecule reduce its toxicity. Such evidence indicates that RB, while an interesting environmental toxin, is less hazardous than more stable toxins when present in the food supply.

As is commonly the case with mycotoxins, the mechanism of toxicity of RB is controversial, making speculation about the mechanism of teratogenesis difficult. Effects of RB on mitochondrial electron transport (Hayes, 1976), polyribosome disaggregation (Watson and Hayes, 1977), and ATP, cAMP, and adenosine triphosphatase (Hayes et al., 1978; Desaiah et al., 1977) have been investigated, but no definitive picture has yet emerged.

M. *Sclerotinia* Toxins

When growing on the proper host, *Sclerotinia sclerotiorum* can produce phototoxic agents, such as 5- or 8-methoxypsoralen, and 4,5,8-trimethylpsoralen, although on other hosts, toxin production may not occur (Ciegler et al., 1978). To assess the teratogenicity potential of *Sclerotinia* toxins, air-dried sclerotia from sunflower seeds were added at levels of 0.5-8% to the diet of pregnant Wistar rats on gestation days 1-21 by Ruddick and Harwig (1975). Maternal food intake and weight gain were reduced at high dose levels, but the only fetal effect was delayed ossification. Thus *S. sclerotiorum* sclerotia infesting sunflowers did not appear hazardous to the mouse fetus, although types or amounts of toxins present were not determined.

N. Secalonic Acid D

Secalonic acid D (SAD), a metabolite of *Penicillium oxalicum*, was investigated by Ciegler et al. (1980). Doses of up to 150 µg in chicken eggs caused no deaths or gross malformations, a result similar to that obtained earlier by Yamazaki et al. (1971). Ciegler et al. (1980) speculated that the lack of toxicity may have resulted because the major effect of SAD is on the lung, an organ unimportant to the embryo. This speculation was not supported by the findings of Reddy et al. (1981), however. They gave the toxin intraperitoneally to CD-1 mice on days 7-15 of gestation at doses of 5-10 mg/kg per day in 5% $NaHCO_3$ or 5-15 mg/kg per day in 5% $NaHCO_3$ plus 10% DMSO. Treatment resulted in reductions in maternal weight, and at doses over 5 mg/kg, increased prenatal deaths. Fetal growth was decreased due to all treatments. Teratogenesis was observed at doses of 6 mg/kg or above for SAD in the 5% NaHCO$_3$ solution and at 10 mg/kg for SAD in the solvent with added DMSO. Gross malformations included cleft palate, cleft lip, open eye, and micrognathia. Skeletal alterations were also reported, consisting largely of conditions considered to be toxicity-induced variations or delayed ossification.

The foregoing results suggest that secalonic acid D, produced by a fungus commonly found on corn (Ciegler et al., 1980), should be regarded as a potential teratogen, a possibility that warrants further investigation.

O. *Stachybotrys* Toxins

Korpinen (1974) tested toxins from *S. alternans* because of reports by Forgacs and Carll (1962) and Vachev et al. (1970) that such toxins may have been involved in field cases of abortion in livestock. Grain infested with *S. alternans* was added to the diet of pregnant mice. Similar mice were given toxin-containing fungal growth medium or a "partly purified preparation" by gavage. Treatment occurred on gestation day 3 or 5 or for various 5-day periods, and tended to result in implantation failure, prenatal death, or reduced fetal growth. Uteroplacental hemorrhage was detected, but no fetal malformations were found. The author's explanation of the embryotoxic effects as being secondary to placental damage is plausible; however, the supporting evidence invoked is less so. The author contends that the presence of both affected and uneffected fetuses in the same litter argues against a direct effect of *Stachybotrys* toxins on the fetus. Such observations of heterogeneous effects are, however, the rule rather than the exception in teratological studies involving polytocous mammals.

P. Sterigmatocystin

Sterigmatocystin is a carcinogenic metabolite of *Aspergillus versicolor*, *A. nidulans* (Athnasios and Kuhn, 1977), *Chaetomium udagawae*, and *C. thielavioideum* (Udagawa et al., 1979). Hayes and Hood (1976) gave 30 mg/kg sterigmatocystin to pregnant CD-1 mice on gestation day 8 or 10 and saw no toxic effects. It appears that more teratogenicity data on sterigmatocystin are needed, as no additional reports were found in our literature search.

Q. T-2 Toxin

T-2 toxin (T-2) is one of the more potent and common toxins synthesized by *Fusarium tricinctum* and other *Fusaria*, as well as *Gibberella zea* (Burmeister et al., 1972). Nevertheless, few studies of its prenatal effects have been reported.

Stanford et al. (1975) administered 0.5-1.5 mg/kg of T-2, i.p., to CD-1 mice on one of gestation days 7-11. They reported gross malformations, particularly those of the tail and limbs, exencephaly, open eye, and blunt jaws, following treatment on day 9, 10, or 11. Rib and vertebral defects were seen after day 9 or 10 treatment. Fetal weights were also reduced in some cases, and survival was decreased, particularly at the highest doses.

More recently, Lary and Hood (1977) reported the results of in utero T-2 treatment of mice that were homozygous wild-type (+/+) or heterozygous for the brachyury (*T*) gene (*T*/+). Treatment increased the frequency of short-tailed or tailless fetuses in litters containing heterozygotes. T-2 also increased the incidence of fetuses with abnormal skeletons and enhanced the severity of the defects, as well as decreasing fetal viability and growth rate. No treatment-related differences were observed, however, among genotypes in fetal survival or weight gain, or in other malformations. These results constitute evidence that the mycotoxin increased the expressivity of the brachyury gene with regard to tail length and exacerbated the adverse effects of T-2 on skeletal formation.

Such results suggest that teratogenicity may be added to the list of hazards expected from exposure to T-2. Teratogenicity and other toxic effects may be due to T-2 toxin's known ability to inhibit protein synthesis (Cundliffe et al. (1974), although DNA synthesis inhibition has also been reported (Munsch and Müller, 1980).

R. Viriditoxin

Viriditoxin, a product of certain *Aspergilli* (Lillehoj and Ciegler, 1972), was administered by Hood et al. (1976a) in 2- to 3.5-mg/kg doses to CD-1 mice on one of days 7-10 of pregnancy. The highest doses were often fatal to the dams. No adverse effect on the litters of survivors was detected, with the exception of decreased weight of fetuses treated on day 10 at the high dose. Since this dose resulted in 62% maternal deaths, however, the maternal toxicity observed was of more significance than effects on the conceptus. These results indicate that viriditoxin may not be teratogenic, but more extensive testing must be done to confirm or deny this contention.

S. Zearalenone (F-2)

Zearalenone (or F-2 toxin) is an estrogenic metabolite of *Gibberella zeae*, a plant pathogen widely distributed on cereal grains, and is also produced by other *Fusaria* (Hidy et al., 1977).

In 1973, Miller et al. implicated F-2 as the causative agent in a field outbreak of stillbirths, small litters, and neonatal deaths. Liveborn pigs had a "splayleg" condition, with weakness and uncoordinated hind limbs. Grain fed to the pregnant sows contained F-2 at levels of 50-75 μg per 100 g. One sow and one gilt were then bred and treated during pregnancy with intramuscular injections of 5 mg of F-2 daily in the last gestational month. The litters produced by the treated pigs contained offspring with the splayleg condition, as well as some dead piglets. Thus it is likely that F-2 caused the original problem.

Reproductive problems in swine fed F-2-contaminated grain (3.6 mg/kg) were also described by Étienne and Jemmali (1979), who noted blocking of estrous cycles. Gilts that conceived had normal numbers of fetuses by gestation day 80, but fetal weights were reduced.

Ruddick et al. (1976) administered F-2 to Wistar rats orally on days 6-15 of pregnancy. Dose levels were 1-10 mg/kg per day. Fetal weights were decreased somewhat in the high-dose group, and skeletal ossification was delayed. No gross or visceral anomalies were reported.

Bailey et al. (1976) also tested F-2 in the rat in a two-generation study. F-2 was given at dose levels of 0.1, 1, or 10 mg/kg per day, with dietary levels adjusted weekly. The high dose inhibited food consumption and growth in the F_1 generation, and mated F_1 and F_2 females had increased resorptions and deaths per litter. No fetal abnormalities were observed, but there was an alteration of the medullary trebeculae of the marrow cavity of the femur.

Hidy et al. (1977) evaluated breeding performance, embryotoxicity, teratogenicity, and peri- and postnatal effects of F-2 in the rat. They reported an oral no-effect dose of 1-3 mg/kg per day. Doses of 5-10 mg/kg per day were associated with some embryotoxic effects, together with undescribed "skeletal anomalies."

Such data indicate that zearalenone has a potential for prenatal toxicity and possibly for teratogenicity. This compound can also affect reproductive cycles, presumably due to its estrogenic activity, as it competes with estradiol for estrogen receptors (Greenman et al., 1979). Although relatively large amounts may be present in feed samples (Mirocha et al., 1971), the acute toxicity of F-2 is low (Hidy et al., 1977). Nevertheless, chronic low doses had adverse biological effects, according to the studies previously discussed, and should be avoided.

T. Combinations of Toxins

Hood et al. (1978) exposed pregnant CD-1 mice to both ochratoxin A and T-2 toxin, concurrently or separately, as a test for additive or synergistic effects. The mycotoxins were given intraperitoneally on gestation day 8 or 10. T-2 was given at a dose of 0.5 mg/kg, a level only marginally teratogenic, particularly with day 10 treatment. OA was given at 2 or 4 mg/kg, levels that were teratogenic on day 10, but not on day 8.

An apparent potentiative effect was observed following treatment with the high combined dose (4 mg/kg OA + 0.5 mg/kg T-2) on day 10. Malformations known to be associated with T-2 treatment were present, but none typical only of OA effects were seen. Thus simultaneous exposure to the two mycotoxins appeared to have resulted in exacerbation of only one toxin's effects. Effects of combination treatments were never more than additive on fetal weights or mortality, and few skeletal or visceral defects were seen. Of course, results would be dependent on such factors as the combinations of toxins used, dose levels, test species, and time of administration. These suggestions are supported by results obtained in mice with additional combinations of mycotoxins (T-2 or OA plus rubratoxin B), where synergistic effects were not seen (R. D. Hood and M. H. Kuczuk, unpublished observations, 1982).

One additional study where embryos were exposed to two different mycotoxins was that of Ciegler et al. (1977). Here, 4-day-old chick embryos were injected simultaneously with citrinin (1-3 µg per egg) and patulin (20-76 µg per egg). Examination of the results indicated only additive effects.

III. ANTIBIOTICS

Although it is common knowledge that many fungal metabolites have antimicrobial effects, there are still many gaps in our understanding of the teratogenic potential of fungal-produced antimicrobials. Except where antibiotics have unique properties of research interest or where there is significant human exposure, teratogenic potential of these agents is seldom reported in the open literature. The remainder of this review deals primarily with reports of teratogenic effects produced in animals by less toxic fungal metabolites that have prominent antimicrobial activity, although the division between mycotoxin and antibiotic is necessarily somewhat arbitrary.

Owing to limitations of space, this chapter does not include every antibiotic that can be found in the teratological literature. For reviews of additional compounds, see Tuchmann-Duplessis (1975, pp. 127-132), Wilson (1977, p. 331), and Shepard (1980).

A. Actinomycin D

Actinomycin D, a product of various *Streptomyces* spp. (Woodruff and Waksman, 1960), has become a tool of cell biologists due to its well-known effects on nucleic acid synthesis. Tuchmann-Duplessis (1975) found that actinomycin D treatment of rats (50-75 µg/kg) between gestation days 2 and 6 or 7 and 9 decreased prenatal viability, but only the later treatment was dysmorphogenic. Treatment after day 10 was only lethal to the fetus. Higher doses (300 µg/kg) resulted in more defects, and in defects after day 10 treatment as well. Wilson (1966) also found actinomycin D to be teratogenic in the rat, as did Harvey and Srebnik (1978), who gave doses of 200 µg/kg, s.c., on gestation day 13. Treatment caused talipes, exomphalos, crooked tail, edema, and stunting.

Lary et al. (1977) found that actinomycin D (0.1-0.15 mg/kg) increased the incidence of skeletal defects in litters of mice containing fetuses heterozygous for brachyury (*T*) in comparison with wild-type litters. Thus the *T* gene appeared to have enhanced the effects of actinomycin D on skeletal development. Prenatal mortality, stunting, and gross malformations were also increased in all genotypes.

In the rabbit, actinomycin D doses of 50-75 µg/kg resulted in resorptions and in spina bifida, encephalocele, and eye defects (Tuchmann-Duplessis, 1975), while in the hamster (Tuchmann-Duplessis et al., 1973), actinomycin D (100 µg/kg) was embryotoxic, and was teratogenic to the surviving offspring.

B. Adriamycin

Adriamycin is a glycosidic, anthracycline antibiotic produced by *Streptomyces peucetius* (Thompson et al., 1978). Ogura et al. (1973) treated pregnant rats with adriamycin doses of up to 1.0 mg/kg on days 8-14 of pregnancy and reported no teratogenic effect. Their results disagreed with those of Thompson et al. (1978), however, who gave adriamycin at doses between 1 and 2 mg/kg to Sprague-Dawley rats on various days of gestation. Treatment on days 6-9 or 6-15 resulted in gross and visceral abnormalities. Of particular interest was the high rate of esophageal atresia with tracheoesophageal fistula. Retarded skeletal ossification and extra ribs were also seen. Resorption rates increased, and fetal weights decreased in treated groups. When adriamycin was given intravenously to Dutch Belted rabbits at dose levels of up to 6 mg/kg per day, no teratogenicity or effect on fetal growth was observed, but there was a high incidence of abortion. These results indicate that adriamycin is embryotoxic and teratogenic to mammals, with species-specific effects. The effects seen are not surprising in view of the antibiotic's ability to inhibit nucleic acid synthesis (Lee and Byfield, 1976).

C. Carminomycin

Carminomycin is also an anthracycline antibiotic, structurally similar to adriamycin and to duanorubicin (Crooke, 1977). Damjanov and Celluzzi (1980) gave the antibiotic at dose levels of 0.5-3.5 mg/kg, s.c., or 0.3-0.5 mg/kg, i.p. Treatments were given on one of days 8-11 (s.c.) or 8-10 (i.p.), or chronically on days 8-11 (s.c.). Carminomycin tended to decrease prenatal survival and fetal weight, and only a few possibly treatment-related malformations were seen.

The prenatal effects of carminomycin in the mouse were mainly typical of toxicity rather than teratogenicity. Nevertheless, carminomycin is believed to inhibit DNA and RNA synthesis (Crooke, 1977) and could prove teratogenic in other species or strains.

D. Daunorubicin (Daunomycin, Rubidomycin)

Daunorubicin is an anthracycline antibiotic isolated from *Streptomyces peuceteus*, and has been called daunomycin and rubidomycin (Windholz, 1976).

Roux and Taillemite (1969) gave 1-3 mg/kg daunorubicin from *Streptomyces coeruleo rubidus* intraperitoneally to rats from gestation days 7 to 9 and reported congenital defects affecting the eyes, heart, kidneys, and brain, as well as decreased viability. Their results differed, however, from those of Julou et al. (1967), who found no teratogenicity in the rabbit, mouse, or chick, but who used lower doses, and from the findings of Chaube and Murphy (1969), who injected rats on gestation days 5-12 and saw no congenital defects.

Daunorubicin was also studied by Thompson et al. (1978). They gave doses of 1-4 mg/kg at intervals during pregnancy to Sprague-Dawley rats. Several treated dams delivered their litters early. When others were examind, their uteri were found to be in a state of tonic contraction, resulting in mechanical deformation of the fetuses. The resorption rate was increased at the high dose, and fetal growth was inhibited at all doses. Some abnormalities, mostly involving the viscera, were noted. Litters from Dutch Belted rabbits treated with 0.15 to 0.6-mg/kg doses on days 7-19 were unaffected.

Such findings indicate that daunorubicin is a teratogen, although not a universal one. More studies are needed to clarify its potential for prenatal effects. Since it is similar to other anthracycline antibiotics in its effects on nucleic acids (Wang et al., 1972), it may have a similar mechanism of action on the fetus.

E. Griseofulvin

Penicillum griseofulvum produces griseofulvin (GFN), one of the most widely used antibiotics for the treatment of human and animal dermatophytic infections.

In 1969, Slonitskaya gave pregnant rats 50-500 mg/kg GFN, p.o., at several stages of major organogenesis. Embryolethal effects were most apparent when GFN was given on day 9 of pregnancy, whereas teratogenic effects predominated in fetuses taken from dams treated daily on days 11-14. Anophthalmia, hydrocephalus, positional abnormalities of brain and eye, cerebral hypoplasia, deformed or missing tails, umbilical hernia, genital hypoplasia, subcutaneous hemorrhage, and fetal edema were reported.

Klein and Beall (1972) reproduced adverse effects in Sprague-Dawley rats given griseofulvin orally on days 6-15 after mating. Dose levels of 125-1500 mg/kg per day represented approximately 6- to 75-fold multiples of the recommended human dose (0.02 g/kg). The increase in number of resorbing fetuses in dams given 1250 or 1500 mg/kg GFN per day was significant, and fetal body weights were decreased at most dose levels. Also, dams treated with the largest doses had body weights smaller than those of controls. Skeletal ossification was retarded in fetuses from dams given high GFN doses, and offspring viability was reduced by all but the lowest dose. There was a low but increased incidence of gross (mainly tail defects) and visceral anomalies. However, major osseous anomalies were seen following the highest doses.

Steelman and Kocsis (1978) reported similar results in rats treated with smaller doses (50-500 mg/kg per day) of GFN on gestation days 6-15. Embryolethal effects predominated at the high dose, but osseous malformations were prevalent in the lower-dose groups.

Aujezaska et al. (1978) mixed GFN in diets fed to ICR mice on days 6 and 7, 8 and 9, or 10 and 11 of gestation. They estimated a daily antibiotic intake of 5000 mg/kg. Teratogenic effects were most prominent in offspring exposed on days 8 and 9, consisting mainly of skeletal malformations involving the spine and ribs.

Since ringworm is a problem in domestic cats and is treated with oral griseofulvin, it is not surprising that reports of malformed kittens began to appear (Gillick and Bullmer, 1972, Scott et al., 1974). There appeared to be no consistent syndrome of malformations; however, the teratogenic effect could not be doubted. Many of the gross, visceral, and skeletal malformations reported in field cases were reproduced by giving comparable dosages (500-1000 mg per cat per week) of griseofulvin to pregnant laboratory cats by Gillick and Bulmer (1972) and by Scott et al. (1974).

Although large doses of griseofulvin were required to produce overt teratogenic effects in laboratory animals, there were also signs of teratogenic effects at doses comparable to those used therapeutically. Since griseofulvin has proven to be teratogenic and prenatally toxic in at least three mammalian species, pregnant women should be considered at risk until reliable evidence should prove otherwise. Low levels of teratogenicity or toxicity to the conceptus are difficult to detect in the human population. Therefore, unequivical evidence of the safety of griseofulvin therapy during pregnancy will likely not be soon forthcoming.

F. Mitomycin C

Mitomycin C, a product of *Streptomyces caespitosus* (Hata et al., 1956) was given to mice by Tanimura (1968) and produced abnormalities. Doses of 5-10 mg/kg were used on one of gestation days 7-13. Chaube and Murphy (1968) reported that doses of 0.5-2 mg/kg in the rat were not teratogenic, and the maternal and fetal LD_{50} values were similar (2-2.5 mg/kg). They related the effects of mitomycin C to its reputed ability to block DNA synthesis.

G. Streptonigrin

Streptonigrin is a quinoid antibiotic produced by *Streptomyces flocculus* (Rao and Cullen, 1959-1960). Warkany and Takacs (1965) injected Long-Evans rats intraperitoneally on gestation day 9, 10, or 11 with a streptonigrin dose of 0.25 mg/kg. Many day 10-treated fetuses were abnormal, with a variety of defects observed. These included omphalocele, exencephaly, hydrocephalus, and eye and skeletal defects.

Chaube et al. (1969) also treated streptonigrin for teratogenic potential. Pregnant Wistar strain rats were given streptonigrin intraperitoneally at levels ranging from 0.02 to 0.7 mg/kg as a single dose on one of gestation days 6-18. Streptonigrin killed 50 to 100% of the dams when multiple doses of 0.4-0.7 mg/kg were given on days 10 and 13 or days 13, 15, and 18 of gestation. Single doses of 0.02-0.3 mg/kg streptonigrin were not lethal to any dams regardless of the day on which exposure occurred. Only the lowest dose (0.02 mg/kg) of streptonigrin had no adverse effects on the fetus. The level of 0.05 mg/kg was the lowest dose at which malformations were found and was the largest dose that was not embryolethal. At the larger dose levels, most litters were destroyed. Anomalies produced included cleft palate, cleft lip, exencephaly, encephalocoele, shortened maxilla, mandible, and trunk, deformed appendages, severe skeletal defects, and a variety of visceral anomalies.

From such results it can be seen that the teratogenic potential of streptonigrin must be appreciated, although it has been tested only in the rat.

REFERENCES

Adekunle, A. A., and Bassir, O. (1975). Studies with aflatoxin B_1 and palmotoxins B_0 and G_0 in chicken embryos. *Toxicol. Appl. Pharmacol.* 31:384-389.

Athnasios, A. K., and Kuhn, G. O. (1977). Improved thin layer chromatographic method for the isolation and estimation of sterigmatocystin in grains. *J. Assoc. Off. Anal Chem.* 60:104-106.

Aujezdska, A., Jindra, J., and Janousek, V. (1978). Influence of griseofulvin on the mouse fetus skeleton. *Cesk. Hyg.* 23:55-57.

Austin, W. L., and Brown, K. S. (1979). Differential lethality and CNS teratogenicity of cytochalasins among mouse strains. *Proc. Fed. Am. Soc. Exp. Biol.* 38:438.

Bailey, D. E., Cox, G. E., Morgareidge, K., and Taylor, J. (1976). Acute and subacute toxicity of zearalenone in the rat. *Toxicol. Appl. Pharmacol.* 37:144.

Bassir, O., and Adekunle, A. (1970a). Teratogenic action of aflatoxin B_1, polmotoxin B_0 and palmotoxin G_0 on the chick embryo. *J. Pathol.* 102:49-51.

Bassir, O., and Adekunle, A. A. (1970b). The histopathological effects of aflatoxin B_1 and the palmotoxins B_0 and G_0 on the liver of the developing chick embryo. *FEBS Lett.* 10:198-201.

Brown, M. H., Szczech, G. M., and Purmalis, B. P. (1976). Teratogenic and toxic effects of ochratoxin A in rats. *Toxicol. Appl. Pharmacol. 37*:331-338.

Burmeister, H. R., Ellis, J. J., and Hesseltine, C. W. (1972). Survey for *Fusaria* that elaborate T-2 toxin. *Appl. Microbiol. 23*:1165-1166.

Burmeister, H. R., Ciegler, A., and Vesonder, R. F. (1979). Moniliformin, a metabolite of *Fusarium moniliforme* NRRL 6322: purification and toxicity. *Appl. Environ. Microbiol. 37*:11-13.

Butler, W. H. (1971). The effect of maternal liver injury and dietary reduction on fetal growth in the rat. *Food Cosmet. Toxicol. 9*:57-63.

Butler, W. H., and Wigglesworth, J. S. (1966). The effects of aflatoxin B$_1$ on the pregnant rat. *Br. J. Exp. Pathol. 47*:242-247.

Campbell, C. W., and Burfening, P. J. (1972). Effects of ergot on reproductive performance in mice and gilts. *Can. J. Anim. Sci. 52*:567-569.

Chaube, S., and Murphy, M. L. (1968). The teratogenic effects of the recent drugs active in cancer chemotherapy. In *Advances in Teratology*, Vol. 3, D. H. M. Woollam (Ed.). Logos/Academic Press, New York, pp. 181-237.

Chaube, S., Kuffer, F. R., and Murphy, M. L. (1969). Comparative teratogenic effects of sterptonigrin (NSC-45383) and its derivatives in the rat. *Cancer Chemother. Rep. 53*:23-31.

Ciegler, A., Beckwith, A. C., and Jackson, L. K. (1976). Teratogenicity of patulin and patulin adducts formed with cysteine. *Appl. Environ. Microbiol. 31*:664-667.

Ciegler, A., Vesonder, R. F., and Jackson, L. K. (1977). Production and biological activity of patulin and citrinin from *Penicillium expansum*. *Appl. Environ. Microbiol. 33*:1004-1006.

Ciegler, A., Burbridge, K. A., Ciegler, J., and Hesseltine, C. W. (1978). Evaluation of *Sclerotinia sclerotiorum* as a potential mycotoxin producer on soybeans. *Appl. Environ. Microbiol. 36*:533-535.

Ciegler,, A., Hayes, A. W., and Vesonder, R. F. (1980). Production and biological activity of secalonic acid D. *Appl. Environ. Microbiol. 39*:285-287.

Clegg, D. J. (1964). The hen egg in toxicity and teratogenicity studies. *Food Cosmet. Toxicol. 2*:717-727.

Creppy, E. E., Schlegel, M., Röschenthaler, R., and Dirheimer, G. (1980). Phenylalanine prevents acute poisoning by ocratoxin A in mice. *Toxicol. Lett. 6*:77-80.

Crooke, S. T. (1977). A review of carminomycin: a new anthracycline developed in the U.S.S.R. *J. Med. 8*:295-316.

Cundliffe, E., Cannon, M., and Davies, J. (1974). Mechanism of inhibition of eukaryotic protein synthesis by trichothecene fungal toxins. *Proc. Natl. Acad. Sci. USA 71*:30-34.

Dailey, R. E., Brouwer, E., Blaschka, A. M., Reynaldo, E. F., Green, S. G., Monlux, W. S., and Ruggles, D. I. (1977). Intermediate duration toxicity study of patulin in rats. *J. Toxicol. Environ. Health 2*:713-725.

Damjanov, I., and Celluzzi, A. (1980). Embryotoxicity and teratogenicity of the anthracycline antibiotic carminomycin in mice. *Res. Commun. Chem. Pathol. Pharmacol. 28*:497-504.

Das, S. K., Nair, R. C., Patthey, H. L., and Mgbodile, V. K. (1978). The effects of aflatoxin B$_1$ on rat fetal lung lipids. *Biol. Neonate 33*:283-288.

Desaiah, D., Hayes, A. W., and Ho, I. K. (1977). Effect of rubratoxin B on adenosine triphosphatase activities in the mouse. *Toxicol. Appl. Pharmacol. 39*:71-79.

De-Voy, K., England, M. A., and Wakely, J. (1979). The effect of cyto-chalasin B on chick mesoderm cells as studied by scanning electron micro-scopy. *Anat. Embryol. 158*:63-73.

DiPaolo, J. A., Elis, J., and Erwin, H. (1967). Teratogenic response by hamsters, rats and mice to aflatoxin B_1. *Nature (Lond.) 215*:638-639.

Edwardson, J. A. (1968). The effects of agroclavine, an ergot alkaloid, on preg-nancy and lactation in the rat. *Br. J. Pharmacol. Chemother. 33*:215-216.

Elegbe, R. A., Amure, B. O., and Adekunle, A. A. (1974). Induction of im-plantation by aflatoxin B_1 in the rat. *Acta Embryol. Exp. 2*:199-201.

Elis, J., and DiPaolo, J. A. (1967). Aflatoxin B_1, induction of malformations. *Arch. Pathol. 83*:53-57.

Etienne, M., and Jemmali, M. (1979). Conséquences de l'ingestion de mais fusarié la truie reproductrice. *C. R. Acad. Sci. Ser. D 288*:779-782.

Evans, M. A., and Harbison, R. D. (1977). Prenatal toxicity of rubratoxin B and its hydrogenated analog. *Toxicol. Appl. Pharmacol. 39*:13-22.

Foerder, B. A. (1978). Mechanisms in mouse neural tube closure: an ultra-structural study of normal and cytochalasin D treated embryos. *Anat. Rec. 190*:396-397.

Forgacs, J., and Carll, W. T. (1962). Mycotoxicoses. *Adv. Vet. Sci. 7*: 273-293.

Gallo, M. A., Bailey, D. E., Babish, J. G., Taylor, J. M., and Daily, R. E. (1977). Toxicity and reproduction studies with patulin in the rat. *Toxicol. Appl. Pharmacol. 41*:139.

Gilani, S. H., and Kreshel, C. (1980). The effects of cytochalasin D on chick embryogenesis. *Anat. Rec. 196*:62A.

Gilani, S. H., Bancroft, J., and Reily, M. (1978). Teratogenicity of ochra-toxin A in chick embryos. *Toxicol. Appl. Pharmacol. 46*:543-546.

Gilani, S. H., Bancroft, J., and Reilly, M. (1979). Rubratoxin B and chick embryogenesis: an experimental study. *Environ. Res. 20*:199-204.

Gillick, A., and Bulmer, W. S. (1972). Griseofulvin, a possible teratogen. *Can. Vet. J. 13*:244.

Goerttler, K., Löhrke, H., Schweizer, H.-J., and Hesse, B. (1980). Effects of aflatoxin B_1 on pregnant inbred Sprague-Dawley rats and their F_1 generation. A contribution to transplacental carcinogenesis. *J. Natl. Cancer Inst. 64*:1349-1354.

Granholm, N. H. (1979). Effects of cytochalasin B on preimplantation and early postimplantation mouse embryos in vitro. In *Advances in the Study of Birth Defects*, Vol. 3; *Abnormal Embryogenesis: Cellular and Molecu-lar Aspects*, T. V. N. Persaud (Ed.). University Park Press, Baltimore, pp. 71-94.

Grauwiler, J., and Schön, H. (1973). Teratological experiments with ergota-mine in mice, rats, and rabbits. *Teratology 7*:227-236.

Greatorex, J. C., and Mantle, P. G. (1974). Effect of rye ergot on the pregnant sheep. *J. Reprod. Fertil. 37*:33-41.

Greenaway, J. C., Shepard, T. H., and Kuc, J. (1977). Comparison of cyto-chalasins (A, B, D, and E) in chick explant teratogenicity and tissue culture systems. *Proc. Soc. Exp. Biol. Med. 155*:239-242.

Greenman, D. L., Mehta, R. G., and Wittliff, J. L. (1979). Nuclear inter-action of *Fusarium* mycotoxins with estradiol binding sites in the mouse uterus. *J. Toxicol. Environ. Health 5*:593-598.

Harvey, J. E., and Srebnik, H. H. (1978). Inductions of malformations by actinomycin D at mid-pregnancy in the rat and protection by L-thyroxine. *Proc. Soc. Exp. Biol. Med. 157*:553-555.

Harwig, J., and Munro, I. C. (1975). Mycotoxins of possible importance in diseases of Canadian farm animals. *Can. Vet. J. 16*:125-141.

Hata, T., Sano, Y., Sugawara, R., Matsumae, A., Kanamori, K., Shima, T., and Hoshi, T. (1956). Mitomycin C, a new antibiotic from *Streptomyces. Int. J. Antibiot., Ser. A9*:141.

Hayakawa, S., Matsushima, T., Kimura, T., Minato, H., and Katagiri, K. (1968). Zygosporin A, a new antibiotic from *Z. masonii. J. Antibiot. (Tokyo) 21*: 523-524.

Hayes, A. W. (1976). Action of rubratoxin B on mouse liver mitochondria. *Toxicology 6*:253-261.

Hayes, A. W., and Hood, R. D. (1974). Mycotoxin induced developmental abnormalities. In *Proceedings of the Western Hemisphere Nutrition Congress IV*, P. L. White and N. Selvey (Eds.). Publishing Sciences Group, Acton, Mass., pp. 397-402.

Hayes, A. W., and Hood, R. D. (1976). Effect of prenatal exposure to mycotoxins. In *Proceedings of the European Society of Toxicology*, Vol. 17. Excerpta Medica Int. Congr. Ser. 376. Excerpta Medica Press, Amsterdam, pp. 209-219.

Hayes, A. W., and Hood, R. D. (1978). Effects of prenatal administration of penicillic acid and penitrem A to mice. *Toxicon 16*:92-96.

Hayes, A. W., Hood, R. D., and Lee, H. L. (1974). Teratogenic effects of ochratoxin A in mice. *Teratology 9*:93-98.

Hayes, A. W., Phillips, R. D., and Wallace, L. C. (1977). Effect of penitrem A on mouse liver composition. *Toxicon 15*:293-300.

Hayes, A. W., Hoskins, B., Pfaffman, M. A., and Watson, S. A. (1978). Effects of rubratoxin B on hepatic ATP and cyclic AMP levels in mice. *Toxicol. Lett. 1*:241-246.

Hidy, P. H., Baldwin, R. S., Greasham, R. L., and McMullen, J. R. (1977). Zearalenone and some derivatives: production and biological activities. In *Advances in Applied Microbiology*, Vol. 22, D. Perlman (Ed.). Academic Press, New York, pp. 59-82.

Hintz, H. F., Heitman, H., Jr., Booth, A. N., and Gagné, W. E. (1967). Effects of aflatoxin on reproduction in swine. *Proc. Soc. Exp. Biol. Med. 126*:146-148.

Hood, R. D. (1979). Effects of mycotoxins on development. In *Advances in the Study of Birth Defects*, Vol. 2: *Teratological Testing*, T. V. N. Persaud (Ed.). MTP Press, Lancaster, England, pp. 191-210.

Hood, R. D., Innes, J. E., and Hayes, A. W. (1973). Effects of rubratoxin B on prenatal development in mice. *Bull. Environ. Contam. Toxicol. 10*: 200-207.

Hood, R. D., Hayes, A. W., and Scamell, J. G. (1976a). Effects of prenatal administration of citrinin and viriditoxin to mice. *Food Cosmet. Toxicol. 14*:175-178.

Hood, R. D., Naughton, M. J., and Hayes, A. W. (1976b). Prenatal effects of ochratoxin A in hamsters. *Teratology 13*:11-14.

Hood, R. D., Kuczuk, M. H., and Szczech, G. M. (1978). Effects in mice of simultaneous prenatal exposure to ochratoxin A and T-2 toxin. *Teratology 17*:25-30.

Howarth, B., Jr., and Wyatt, R. D. (1976). Effect of dietary aflatoxin on fertility, hatchability, and progeny performance of broiler breeder hens.

Appl. Environ. Microbiol. 31:680-684.

Hsieh, D. P. H., Salhab, A. S., Wong, J. J., and Yang, S. L. (1974). Toxicity of aflatoxin Q_1 as evaluated with the chicken embryo and bacterial auxotrophs. *Toxicol. Appl. Pharmacol. 30*:237-242.

Ito, Y., Ohtsubo, K., and Saito, M. (1980). Effects of fusarenon-X, a trichothecene produced by *Fusarium nivale*, on pregnant mice and their fetuses. *Jpn. J. Exp. Med. 50*:167-172.

Jordan, W. H., Carlton, W. W., and Sansing, G. A. (1977). Citrinin mycotoxicosis in the mouse. *Food Cosmet. Toxicol. 15*:29-34.

Julou, L., Ducrot, R., Fournel, J., Ganter, P., Maral, R., Popularie, P., Koenig, F., Myon, J., Pascal, S., and Pasqnet, J. (1967). Un nouvel antibiotique doué d'activité antitumorale: la rubidomycine (13.057 R.P.). *Arzneim.-Forsch. 17*:948-954.

Karfunkel, P. (1972). The activity of microtubules and microfilaments in neurulation in the chick. *J. Exp. Zool. 181*:289-302.

Klein, M. F., and Beall, J. K. (1972). Griseofulvin: a teratogenic study. *Science 175*:1483-1484.

Korpinen, E. (1974). Studies on *Stachybotrys alternans*. IV. Effect of low doses of *Stachybotrys* toxins on pregnancy of mice. *Acta pathol. Microbiol. Scand. Sect. B 82*:457-464.

Koshakji, R. P., Wilson, B. J., and Harbison, R. D. (1973). Effect of rubratoxin B on prenatal growth and development in mice. *Res. Commun. Chem. Pathol. Pharmacol. 5*:584-592.

Lary, J. M., and Hood, R. D. (1977). Developmental interactions between T-2 toxin and the brachyury (*T*) gene in mice. *Teratology 15*:19A.

Lary, J. M., Hood, R. D., Scott, W. T., Gearhart, A. M., and Molay, P. M. (1977). Effects of actinomycin D on skeletal abnormalities in mice heterozygous for the brachyury (*T*) gene. *ASB Bull. 24*:65.

LeBreton, E., Frayssinet, C., Lafarge, C., and de Recondo, A. M. (1964). Aflatoxine-mécanisme de l'action. *Food Cosmet. Toxicol. 2*:675-678.

Lee, H. Y., and Kalmus, G. W. (1976). Effects of cytochalasin B on the morphogenesis of explanted early chick embryos. *Growth 40*:153-162.

Lee, J. C., and Byfield, J. E. (1976). Induction of DNA degradation in vivo by adriamycin. *J. Natl. Cancer Inst. 57*:221-224.

Lillehoj, E. B., and Ciegler, A. (1972). A toxic substance from *Aspergillus viride-nutans*. *Can. J. Microbiol. 18*:193-197.

Lillehoj, E. B., Fennell, D. I., and Kwolek, W. F. (1976). *Aspergillus flavus* and aflatoxin in Iowa corn before harvest. *Science 193*:495-496.

Linville, G. P., and Shepard, T. H. (1972). Neural tube closure defects caused by cytochalasin B. *Nature New Biol. 236*:246-247.

Mantle, P. G. (1969). Interruption of early pregnancy in mice by oral administration of agroclavine and sclerotia of *Claviceps fusiformis* (Loveless). *J. Reprod. Fertil. 18*:81-88.

Messier, P.-E., and Auclair, C. (1974). Effect of cytochalasin B on interkinetic nuclear migration in the chick embryo. *Dev. Biol. 36*:218-223.

Messier, P.-E., and Auclair, C. (1977). Alteration of apical junctions and inhibition of interkinetic nuclear migration by cytochalasin B and trypsin. *Acta Embryol. Ex. 3*:341-356.

Miller, J. K., Hacking, A., Harrison, M. A., and Gross, V. J. (1973). Stillbirths, neonatal mortality and small litters in pigs associated with the ingestion of *Fusarium* toxin by pregnant sows. *Vet. Rec. 93*:555-559.

Mirocha, C. J., Christensen, C. M., and Nelson, G. H. (1971). F-2 (zearale-none) estrogenic mycotoxin from *Fusarium*. In *Microbial Toxins*, Vol. VI: *Algal and Fungal Toxins*, S. Kadis, A. Ciegler, and S. J. Ajl (Eds.). Academic Press, New York, pp. 107-138.

Moré, J., and Galtier, P. (1974). Toxicité de l'ochratoxine A. I. Effet embryotoxique et tératogène chez le rat. *Ann. Rech. Vet.* 5:167-178.

Moss, M. O., and Hill, I. W. (1970). Strain variation in the production of rubratoxins by *Penicillum rubrum* Stoll. *Mycopathol. Mycol. Appl.* 40: 81-88.

Munsch, N., and Müller, W. E. G. (1980). Effects of T$_2$ toxin on DNA polymerases and terminal deoxynucleotidyl transferase of Molt$_4$ and Nu$_8$ cell lines. *Immunopharmacology* 2:313-318.

Nagele, R. G., Jr., and Lee, H.-Y. (1978). Re-examination of neural tube defects caused by 5-bromodeoxyuridine and cytochalasin B in chick embryos. *Bull. N.J. Acad. Sci.* 23:2-5.

Ogura, T., Hatanao, M., Imamura, T., and Shimizu, M. (1973). A study on the safety of adriamycin HCL. Report No. 4: Deformity-inducing (terato-logical) experiment. *Yakubutsu Ryoho* 6:1152-1164.

Ohtsubo, K., and Saito, M. (1970). Cytotoxic effects of scirpene compounds, fusarenon-X produced by *Fusarium nivale*, dihydronivalenol and dihydro-fusarenon-X, on Hela cells. *Jpn. J. Med. Sci. Biol.* 23:217-225.

Osswald, H., Frank, H. K., Komitowski, D., and Winter, H. (1978). Long-term testing of patulin administered orally to Sprague-Dawley rats and Swiss mice. *Food Cosmet. Toxicol.* 16:243-247.

Pavlović, M., Plestina, R., and Krogh, P. (1979). Ochratoxin A contamina-tion of foodstuffs in an area with Balkan (endemic) nephropathy. *Acta Pathol. Microbiol. Scand.* 87:243-246.

Pero, R. W., Posner, H., Blois, M., Harvan, D., and Spalding, J. W. (1973). Toxicity of metabolites produced by the "*Alternaria*." *Environ. Health Perspect.* 4:87-94.

Pitt, J. I. (1979). *Penicillium crustosum* and *P. simplicissimum*, the correct names for two common species producing tremorgenic mycotoxins. *Mycologia* 71:1166-1177.

Prior, M. G., O'Neil, J. B., and Sisodia, C. S. (1978). Effects of ochratoxin A on production characteristics and hatchability of Japanese quail. *Can. J. Anim. Sci.* 58:29-33.

Prior, M. G., Sisodia, C. S., O'Neil, J. B., and Hrudka, F. (1979). Effect of ochratoxin A on fertility and embryo viability of Japanese quail (*Coturnix coturnix japonica*). *Can. J. Anim. Sci.* 59:605-609.

Rao, K. V., and Cullen, W. P. (1959-1960). Streptonigrin, an antitumor substance. I. Isolation and characterization. *Antibiot. Ann.* 7:950-953.

Reddy, C. S., Chan, P. K., and Hayes, A. W. (1978). Teratogenic and dominant lethal studies of patulin in mice. *Toxicology* 11:219-223.

Reddy, C. S., Reddy, R. V., Hayes, A. W., and Ciegler, A. (1981). Terato-genicity of secalonic acid D in mice. *J. Toxicol. Environ. Health* 7:445-455.

Rice, S. L., Beuchat, L. R., and Worthington, R. E. (1977). Patulin pro-duction by *Byssochlamys* spp. in fruit juices. *Appl. Environ. Microbiol.* 34:791-796.

Rose, H. M., and Moss, M. O. (1970). The effect of modifying the structure of rubratoxin B on the acute toxicity to mice. *Biochem. Pharmacol.* 19: 612-615.

Roux, C., and Taillemite, J. L. (1969). Action teratogène de la rubidomycine chez le rat. *C. R. Soc. Biol. 163*:1299-1302.

Ruddick, J. A., and Harwig, J. (1975). Prenatal effects caused by feeding sclerotia of *Sclerotinia sclerotiorum* to pregnant rats. *Bull. Environ. Contam. Toxicol. 13*:524-526.

Ruddick, J. A., Harwig, J., and Scott, P. M. (1974). Nonteratogenicity in rats of blighted potatoes and compounds contained in them. *Teratology 9*: 165-168.

Ruddick, J. A., Scott, P. M., and Harwig, J. (1976). Teratological evaluation of zearalenone administered orally to the rat. *Bull. Environ. Contam. Toxicol. 15*:678-681.

Schmidt, R. E., and Panciera, R. J. (1980). Effects of aflatoxin on pregnant hamsters and hamster foetuses. *J. Comp. Pathol. 90*:339-347.

Scott, F. W., DeLahunta, A., Schultz, R. D., Bistrer, S. I., and Riis, R. C. (1974). Teratogenesis in cats associated with griseofulvin therapy. *Teratology 11*:79-86.

Scott, P. M., von Walbeek, W., Kennedy, B., and Anyeti, D. (1972). Mycotoxins (ochratoxin A, citrinin, and sterigmatocystin) and toxigenic fungi in grains and other agricultural products. *Agric. Food Chem. 20*:1103-1109.

Shepard, T. H. (1980). *Catalog of Teratogenic Agents*, 3rd ed. Johns Hopkins, Baltimore.

Shepard, T. H., and Greenaway, J. C. (1977). Teratogenicity of cytochalasin D in the mouse. *Teratology 16*:131-136.

Shibko, S. I., Arnold, D. L., Morningstar, J., and Friedman, L. (1968). Studies on the effect of aflatoxin B_1 on the development of the chick embryo. *Proc. Soc. Exp. Biol. Med. 127*:835-839.

Slonitskaya, N. N. (1969). Teratogenic effects of griseofulvin-forte on rat fetuses. *Antibiotiki 14*:44-48.

Smith, J. A., Adekunle, A. A., and Bassir, O. (1974). Comparative histopathological studies with aflatoxin B_1 and palmotoxins B_0 and G_0 on some organs of the newly hatched chick. *Egypt. J. Microbiol. 9*:97-104.

Sommer, A. F., and Buchanan, A. R. (1955). Effects of ergot alkaloids on pregnancy and lactation in the albino rat. *Am. J. Physiol. 180*:296-300.

Stagno, P. A., and Low, F. N. (1978). Effects of cytochalasin B on the fine structure of organized endodermal cells of the early chick embryo. *Am. J. Anat. 151*:159-172.

Stanford, G. K., Hood, R. D., and Hayes, A. W. (1975). Effect of prenatal administration of T-2 toxin to mice. *Res. Commun. Chem. Pathol. Pharmacol. 10*:743-746.

Steelman, R. L., and Kocsis, J. J. (1978). Determination of the teratogenic and mutagenic potential of griseofulvin. *Toxicol. Appl. Pharmacol. 45*: 343.

Stoloff, L. (1980). Aflatoxin control; past and present. *J. Assoc. Off. Anal. Chem. 63*:1067-1073.

Stoloff, L., Verrett, M. J., Dantzman, J., and Reynaldo, E. F. (1972). Toxicological study of aflatoxin P_1 using the fertile chicken egg. *Toxicol. Appl. Pharmacol. 23*:528-531.

Szczech, G. M., and Hood, R. D. (1981). Brain necrosis in mouse fetuses transplacentally exposed to the mycotoxin ochratoxin A. *Toxicol. Appl. Pharmacol. 57*:127-137.

Tanimura, T. (1968). Effects of mitomycin C administered at various stages of pregnancy upon mouse fetuses. *Okajimas Folia Anat. Jap.* 44:337-355.

Tashiro, F., Hirai, K., and Ueno, Y. (1979). Inhibitory effects of carcinogenic mycotoxins on deoxyribonucleic acid-dependent ribonucleic acid polymerase and ribonuclease H. *Appl. Environ. Microbiol.* 38:191-196.

Thompson, D. J., Molello, J. A., Strebing, R. J., and Dyke, I. L. (1978). Teratogenicity of adriamycin and daunomycin in the rat and rabbit. *Teratology* 17:151-158.

Tuchmann-Duplessis, H. (1975). *Drug Effects on the Fetus.* ADIS Press, Seaforth, Australia.

Tuchmann-Duplessis, H., Hiss, D., Mottot, C., and Rosner, I. (1973). Embryotoxic and teratogenic effect of actinomycin D in the Syrian hamster. *Toxicology* 1:131-133.

Udagawa, S., Muroi, T., Kurata, H., Sekita, S., Yoshihira, K., and Natori, S. (1979). *Chaetomium udagawae*: a new producer of sterigmatocystin. *Trans. Mycol. Soc. Jpn.* 20:475-480.

Ueno, Y., Ueno, I., Iitoi, Y., Tsunoda, H., Enomoto, M., and Ohtsubo, K. (1971). Toxicological approaches to the metabolites of *Fusaria.* III. Acute toxicity of fusarenon-X. *Jpn. J. Exp. Med.* 41:521-539.

Vachev, V., Dyakov, L., Peichev, P., and Tabakov, B. (1970). Abortion caused by *Stachybotrys* toxin in swine. *Vet. Sbir.* 67:7-10.

Verrett, M. J., Marliac, J., and McLaughlin, J., Jr. (1964). The use of the chicken embryo in the assay of aflatoxin toxicity. *J. Assoc. Off. Anal. Chem.* 47:1003-1006.

Wagener, R. E., Davis, N. D., and Diener, U. L. (1980). Penitrem A and roquefortine production by *Penicillium commune. Appl. Environ. Microbiol.* 39:882-887.

Wang, J., Chervinsky, D. S., and Rosen, J. M. (1972). Comparative biochemical studies with Adriamycin and Daunomycin in leukemic cells. *Cancer Res.* 32:511-515.

Warkany, J., and Takacs, E. (1965). Congenital malformations in rats from streptonigrin. *Arch. Pathol. (Chicago)* 79:65-79.

Warren, M. F., and Hamilton, P. B. (1980). Inhibition of the glycogen phosphorylase system during ochratoxicosis in chickens. *Appl. Environ. Toxicol.* 40:522-525.

Watson, S. A., and Hayes, A. W. (1977). Evaluation of possible sites of action of rubratoxin B-induced polyribosomal disaggregation in mouse liver. *J. Toxicol. Environ. Health* 2:639-650.

Wiley, M. J. (1980). The effects of cytochalasins on the ultrastructure of neurulating hamster embryos in vivo. *Teratology* 22:59-69.

Wilson, J. G. (1966). Effects of acute and chronic treatment with actinomycin D on pregnancy and the fetus in the rat. *Harper Hosp. Bull.* 24:109-118.

Wilson, J. G. (1977). Embryotoxicity of drugs in man. In *Handbook of Teratology*, Vol. 1: *General Principles and Etiology*, J. G. Wilson and F. C. Fraser (Eds.). Plenum, New York, pp. 309-355.

Windholz, M. (Ed.) (1976). *The Merck Index*, 9th ed. Merck, Rahway, N.J., p. 371.

Witters, W. L., Wilms, R. A., and Hood, R. D. (1975). Prenatal effects of elymoclavine administration and temperature stress. *J. Anim. Sci. 41*: 1700-1705.

Wogan, G. N., Edwards, G. S., and Newberne, P. M. (1971). Acute and chronic toxicity of rubratoxin B. *Toxicol. Appl. Pharmacol. 19*:712-720.

Woodruff, H. B., and Waksman, S. A. (1960). Historical background: the actinomycins and their importance in the treatment of tumors in animals and man. *Ann. N.Y. Acad. Sci. 89*:287-298.

Wyatt, R. D., Hamilton, P. B., and Burmeister, H. R. (1975). Altered feathering of chicks caused by T$_2$ toxin. *Poult. Sci. 54*:1042-1045.

Yamazaki, M., Maebayashi, Y., and Miyaki, K. (1971). The isolation of secalonic acid A from *Aspergillus ochraceus* cultured on rice. *Chem. Pharm. Bull. 19*:199-201.

Part III

TOXINS INDUCING
CARCINOGENIC OR RELATED EFFECTS

7

CARCINOGENIC AND COCARCINOGENIC TOXINS FROM PLANTS

A. DOUGLAS KINGHORN

University of Illinois at Chicago, Chicago, Illinois

I. INTRODUCTION

While it has been known for about 200 years that humans are susceptible to
the carcinogenic effects of chemicals, it is only within the last 30 years or so
that it has been recognized that carcinogens occur in the natural environment
as constituents of plants (Shimkin and Triolo, 1969). In humans these toxins
may be ingested or otherwise contacted, and occur, for example, in foods,
beverages, herbal remedies, and smoking and chewing mixtures. Carcinogenic principles may either be inherently present in a particular plant-derived
item in the diet, or foods may become contaminated with carcinogenic plant,
animal, or microorganism biosynthetic products, or with synthetic additives.
Compounds in some foods may become carcinogenic as a result of cooking or
other processing, or by metabolic transformation in the digestive tract. It is
now estimated that between 70 and 90% of all human cancers are environmental
in origin (Weisburger, 1976). Plant-derived carcinogenic principles are suspected to account for a discernible proportion of human cancers, although
relatively few plant extracts have been shown to induce the formation of
tumors in laboratory animals. In many cases the chemical nature of the active
principle(s) of plant with demonstrated carcinogenic activity is still uncertain.
    Carcinogens have been defined as "agents causative for the generation of
neoplasia whose origin, nature and identity are unequivocally clarified"
(Hecker, 1976a). Hecker (1976a) terms such compounds "solitary carcinogens." About 30 compounds have been demonstrated unambiguously to be
carcinogenic for humans, and most of these, including all the plant carcinogens known to date, require biological activation from a procarcinogen form
to a corresponding reactive ultimate carcinogen form (Weisburger, 1976).
J. A. Miller and E. C. Miller (1971) have proposed that the ultimate reactive
forms of carcinogens are highly electrophilic and may attack tissue nucleophiles such as DNA, RNS, and proteins.
    The term "cocarcinogenesis" was first used to describe the enhancement
of the carcinogenic effects on mouse skin of 3,4-benz[*a*]pyrene by a basic
fraction of creosote (Shear, 1938). Cocarcinogenesis has now come to express
a wide variety of physical, chemical, and biological phenomena in which tumors
are produced by the action of more than one agent, applied either concomitantly or sequentially (Salaman, 1958; Scribner and Süss, 1978). The "two-
stage" theory of chemical carcinogenesis, which is based on the work in the
1940s of Berenblum and others, was developed after repetitive doses of
croton oil, applied subsequent to a single subthreshold dose of a carcinogen,
were found to produce papillomas on mouse skin (Berenblum, 1941a,b;
Mottram, 1944; Berenblum and Shubik, 1947). In this type of system, the
solitary carcinogen is referred to as a "tumor initiator," and the cocarcinogen
as a "tumor promoter," with the whole process being divisible into initiation
and promotion stages (Hecker, 1976b). Not all authorities regard the terms
"tumor promoter" and "cocarcinogen" as being synonymous. Van Duuren
(1976) regards promoting agents as compounds applied repeatedly after a
single dose of an initiating agent, while cocarcinogens are repeatedly administered simultaneously with the initiator. Tumor promoters all have weak
tumorigenic activity, and tumor-promotion experiments result in the production of both benign and malignant tumors in laboratory animals (Van Duuren,
1976). However, since the most potent tumor promoters, inclusive of the

phorbol esters, anthralin, and *n*-dodecane, are also active when used in cocarcinogenic experimental protocols (Sivak, 1979), the two terms will be used
interchangeably in this chapter.

There is a pressing need to identify the chemical nature of plant carcinogenic and cocarcinogenic toxins and to relate their distribution in the plant
kingdom. This is no small task, since many thousands of plant primary and
secondary metabolites are known, and the number is increasing rapidly as a
result of increased analytical capacity in phytochemical laboratories all over
the world. It has been estimated that 100,000 low-molecular-weight organic
compounds are known to be biosynthesized by plants, animals, and unicellular
organisms (E. C. Miller and J. A. Miller, 1979a), and at least 286,000 species
of higher plants alone are recognized (Grant, 1963). However, several approaches to this problem have been made. Valuable information about toxic
agents, which have later been found to be carcinogens or tumor promoters, has
been gained from investigations of livestock toxicoses. Examples include cattle
poisoning from *Macrozamia* species (cycasin-related compounds) (Whiting, 1963);
Senecio and *Crotalaria* species (pyrrolizidine alkaloids) (Bull et al., 1968) and
Pimelea simplex (daphnane esters) (Roberts et al., 1975). Epidemiological observations of high cancer incidences in human populations in restricted geographical locations may also lead to the identification of carcinogenic plants
and possibly their active principles. For example, high rates of buccal cancer
have been associated with betel quid chewing in the lowland populations of
Papua and New Guinea (Atkinson et al., 1964). Also, the observation of high
levels of esophageal cancer associated with the use of *Croton flavens* L. as a
herbal tea in Curaçao (Morton, 1968), resulted in the eventual isolation of
potent cocarcinogenic diterpene esters from this plant (Weber and Hecker,
1978). Finally, plants may be screened for carcinogenic or tumor-promoting
activity using long-term tests in laboratory animals (Berenblum and Shubik,
1947; Hecker, 1963, 1968a; O'Gara, 1968; Wynder and Hoffmann, 1969;
Weisburger, 1975; E. C. Miller and J. A. Miller, 1979b) or short-term mutagenicity assays (E. C. Miller et al., 1979). Experimental protocols in long-term
carcinogenicity testing on rodents, such as selection of test animal, dosing,
route of administration, and evaluation of results, have been discussed by
Weisburger (1975), Weisburger (1976), and Miller and Miller (1979b). The response of the host or target tissue in these tests may be quantified in terms
of the latent period (the time that elapses after the application of test compound or extract until the first tumor appears), the cancer rate (the number
of individuals bearing at least one tumor), and/or the cancer yield (the number of tumors observed per individual test animal) (Hecker, 1976a). Short-
term tests for the detection of potential carcinogens, including mammalian
tissue-mediated mutagenicity tests and assays for the malignant transformation of cells in culture, have become more widely used as prescreens in recent
years (E. C. Miller and J. A. Miller, 1979b). Tazima (1974) has reviewed the
area of naturally occurring mutagens, including plant metabolites.

The task of compiling information for this chapter has been considerably
aided by the publication of several major works on natural product carcinogens
in the last few years (Schoental, 1976; Farnsworth et al., 1976; IARC, 1976;
Wogan, 1977; Hirono, 1979, 1981; E. C. Miller et al., 1979). Also, a number
of recent review articles on carcinogens in foods have devoted large portions
of their content to plant carcinogens (J. A. Miller and E. C. Miller, 1976;

E. C. Miller and J. A. Miller, 1979b; Bababunmi et al., 1978; Austwick and
Mattocks, 1979). In addition, useful reviews specific to individual classes of
plant carcinogens have been published in the last 10 years, including articles
on bracken fern carcinogen (Evans, 1976; Pamakcu and Bryan, 1979; Cooper,
1980); cycasin (Yang et al., 1972; Laquer and Spatz, 1975; IARC, 1976;
Laquer, 1977; Matsumoto, 1979); pyrrolizidine alkaloids (Mattocks, 1972;
McLean, 1974; Crout, 1976; Hirono et al., 1979b); betel nut carcinogen
(Arjungi, 1976), and safrole and related compounds (J. A. Miller et al.,
1979). The area of diterpenoid tumor promoters from plants of the Euphorbiaceae and Thymelaeaceae has been regularly reviewed over the last decade,
and some of the more recent articles include works by Hecker (1977, 1978,
1979, 1981), Adolf and Hecker (1977), Evans and Soper (1978), and Kinghorn
(1979a).

The plant carcinogens and cocarcinogens discussed in this chapter will be
restricted to those compounds biosynthesized by plants. Plant-derived carcinogens that are produced by combustion, such as the active principles of
tobacco smoke condensate (Van Duuren, 1968; Wynder and Hoffmann, 1979)
and heated cooking oils (O'Gara et al., 1969), or by acid hydrolysis, as in
degraded carrageenan (Wakabayashi et al., 1979), will not be included because of space constraints. Also, only those *N*-nitroso compounds thought
to be actual plant constituents will be dealt with here. Carcinogenic fungal
metabolites, and a more extensive treatment of the toxicological actions of
pyrrolizidine alkaloids, will be covered in Chaps. 9 and 19, respectively.
This chapter covers the literature up to the end of March 1981.

II. PLANT CARCINOGENS

A. Bracken Fern Carcinogen

Bracken fern [*Pteridium aquilinum* (L.) Kuhn; *Pteris aquilina*] (Polypodiaceae)
is indigenous to temperate latitudes in the northern hemisphere (Farnsworth
et al., 1976). It has long attracted the attention of veterinarians, since it
produces lethal radiomimetic symptoms in cattle, including bone marrow hyperplasia, hemorrhaging, and severe gastrointestinal damage (Evans, 1968). In
horses, bracken fern induces vitamin B_1 deficiency, because it contains thi-
aminase (Hirono, 1979).

While a possible relationship of bracken fern ingestion with an increased
incidence of bovine urinary bladder cancer was suggested in the early 1960s
(Rosenberger and Heeschen, 1960; Pamakcu, 1963), its carcinogenic activity
in laboratory animals was first demonstrated in 1965 by Evans and Mason,
when rats fed bracken fern developed intestinal adenocarcinomas. An
American-Turkish team confirmed the carcinogenic nature of bracken fern by
producing neoplasms of the small intestine in the rat (Price and Pamakcu,
1968; Pamakcu and Price, 1969; Pamakcu et al., 1970b). The carcinogen in
bracken fern also produced intestinal carcinomas in the Japanese quail (Evans,
1968) and cancer of the bladder in rats (Pamakcu and Price, 1969; Pamakcu et
al., 1970b), cows (Pamakcu et al., 1967; Price and Pamakcu, 1968); guinea
pigs (Evans, 1968), and mice (Pamakcu et al., 1966; Miyakawa and Yoshida,
1975). Pulmonary tumors and lymphatic leukemias were experienced by female
Swiss mice fed bracken fern orally (Pamakcu et al., 1972).

Pteridium aquilinum and related ferns are used in the human diet in Japan, Canada, and the northeastern United States in salads (Pamakcu and Bryan, 1979). Bracken fern has thus been suggested as a possible human carcinogenic hazard (Evans et al., 1971), and the risk of esophageal carcinoma is especially high when hot tea gruel and bracken fern are consumed together (Pamakcu and Bryan, 1979). In Japan, bracken fern is collected between April and June, when the young frond is in the fiddlehead or crosier stage of growth (Hirono, 1979). Bracken fern is processed in Japan either by boiling in water and seasoning, or by storage in salt prior to immersion in boiling water before use (Hirono, 1979). It has been observed that the carcinogenic potency in rats of bracken fern used in human food is reduced, but not obviated, by treatment before testing in boiling water (Hirono et al., 1970b) and by wood ash, sodium bicarbonate, and sodium chloride (Hirono et al., 1972a). The carcinogenic activity is not affected by mincing to a paste before administration to rats (Mori et al., 1977). However, it has recently been established that a decreased carcinogenic potency of bracken fern for rats results after storage at 4°C or room temperature for up to 2 years (Kawai et al., 1981).

A further health hazard for humans by bracken fern has been suggested by the observation that the carcinogen is passed into cows' milk intended for human consumption (Evans et al., 1972; Pamakcu et al., 1978). Feeding experiments have shown that all parts of the bracken fern may produce carcinogenic effects in rats, with the stalks, young fronds, and rhizomes exhibiting, respectively, increasing carcinogenic potency (Hirono et al., 1973a).

There have been many attempts to elucidate the chemical nature of the bracken carcinogen. Leach et al. (1971) isolated a water-soluble active principle with a molecular formula $C_7H_8O_4$ that was positive for acute toxicity, mutagenicity, and carcinogenicity in mice. The molecular formula was revised in a later study when the carcinogenic agent was identified as the common plant primary metabolite, shikimic acid (Fig. 1) (Evans and Osman, 1974). In a later study by a Japanese group no tumorigenic activity by shikimic acid was detected for rats (Hirono et al., 1977). Pterolactam (Fig. 1), a methanol-soluble lactone from bracken fern (Takatori et al., 1972), was also later found to be noncarcinogenic for mice (Hirono et al., 1975). The carcinogen in bracken fern was more recently confirmed as being extractable in boiling water, and probably water soluble (Hirono et al., 1978b).

In a major study on the constituents of bracken fern (*Pteridium aquilinum* Kuhn var. *latiusculum* Underwood), over 30 compounds with the 1-indanone nucleus have been isolated, and two of these, pterosin B, and its glycoside, pteroside B (Fig. 1), shown to be inactive as carcinogens in the rat (Saito et al., 1975; Yoshihira et al., 1978; Fukuoka et al., 1978; Kuroyanagi et al., 1979). Pteroside B was originally isolated by a different Japanese group (Hikino et al., 1970).

Meanwhile, at the University of Wisconsin, Pamacku and co-workers showed that the carcinogenic fractions of bracken fern are methanol soluble (Pamacku et al., 1970a), and that tannin from the plant produced urinary bladder cancers in mice (Wang et al., 1976), but not in rats (Pamakcu et al., 1980a). The mutagen quercetin (Fig. 1) was recently shown to be a rat intestinal and bladder carcinogen, although not so active a bladder carcinogen in rats as bracken fern extracts (Pamakcu et al., 1980b). Quercetin, however, is inactive as a carcinogen in mice (Wang et al., 1976; Saito et al., 1980).

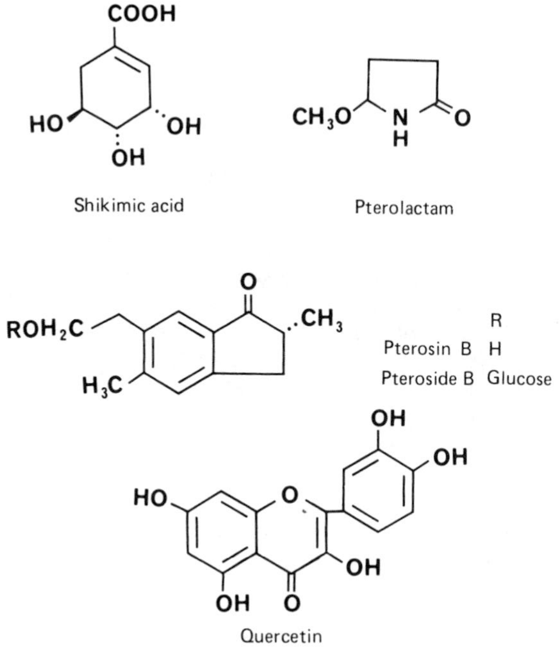

COOH

HO⟋ ⟍OH
        ÖH

Shikimic acid

CH₃O—⟨N⟩=O
         H

Pterolactam

O

ROH₂C ⟋⟍ ⟍CH₃

H₃C⟋

| R
Pterosin B   H
Pteroside B Glucose

OH

OH

HO

OH

OH   O

Quercetin

FIGURE 1   Bracken fern constituents tested for carcinogenic activity in laboratory animals.

According to Jarrett et al. (1978), bracken fern carcinogenicity may be due to an inhibition of the normal immunological reactions to the papilloma virus. Hirono (1981) states that the chemical nature of bracken fern carcinogen is still uncertain due, in part, to difficulties in isolation because no acute toxicity symptoms, analogous to those experienced by cattle, are observed in small laboratory animals, and hence there is a consequent need to wait for 6 months or more to observe its carcinogenic effects on rats or mice.

B. Cycasin and Related Compounds

Cycads belong to the gymnosperm order Cycadales, which consists of 10 genera representing about 100 species found mainly in the tropics and subtropics (Birdsey, 1972). While various toxic effects of cycads to cattle and sheep have long been recognized (Whiting, 1963), the isolation of a glycoside, toxic orally but not parenterally to guinea pigs, from an Australian cycad, *Macrozamia spiralis* (Salisb.) Miq., was not achieved until 1941 (Cooper, 1941). The structure of this compound, macrozamin (Fig. 2), was established by Lythgoe and co-workers (Lythgoe and Riggs, 1949; Langley et al., 1951). Cycasin (Fig. 2), a structurally related compound, was isolated as a toxic principle of *Cycas revoluta* Thunb. (Nishida et al., 1955) and *C. circinalis* (Riggs, 1956). Other azoxyglycosides from *C. revoluta*, neocycasins A, B, C, and E (Fig. 2), were structurally characterized by Nishida

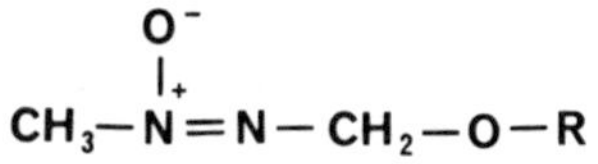

$$O^-$$
$$CH_3-\overset{\underset{|}{+}}{N}=N-CH_2-O-R$$

R

Macrozamin	β-D-Primeverosyloxy-
Cycasin	β-D-Glucosyloxy-
Neocycasin A	β-Laminaribiosyloxy-
Neocycasin B	β-Gentiobiosyloxy-
Neocycasin C	β-Laminaritetraoxyloxy-
Neocycasin E	β-Cellobiosyloxy-
Methylazoxymeth- anol (MAM)	H

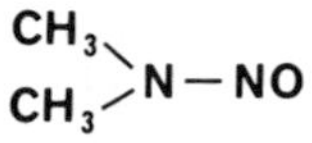

$$\underset{CH_3}{\overset{CH_3}{>}}N-NO \qquad CH_2N_2$$

Dimethylnitrosamine   Diazomethane

$$\left[\begin{array}{c} CH_3\overset{+}{N}\equiv N \\ ^-OH \end{array}\right]$$

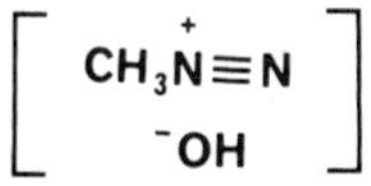

Methyl diazonium hydroxide

FIGURE 2 Cycasin-related plant constituents and some postulated carcinogenic metabolites.

and associates (Nishida et al., 1959; Nagahama et al., 1959, 1960, 1961). All these compounds are glycosides of the same aglycone, methylazoxymethanol (MAM) (Fig. 2), which was itself isolated from *C. circinalis* (Matsumoto and Strong, 1963). MAM is liberated from its glycosides in cycads by the action of β-glycosidase, and is an unstable compound that degrades spontaneously into equimolar amounts of nitrogen gas, methanol, and formaldehyde (Yagi et al., 1980). Cycasin has been found in representatives of all 10 genera in the Cycadales, and is considered to be characteristic of, and exclusive to, the cycads (de Luca et al., 1980).

Cycas circinalis seeds, when incorporated into the diet of rats, were first shown to possess carcinogenic properties by Laquer et al. (1963), when benign and malignant tumors of the liver, kidneys, intestine, and lungs were observed. Oral administration of cycad meal produced neoplasms in guinea pigs (Spatz, 1964), and an aqueous extract of the seeds applied to the skin of young adult mice led to the development of hepatic and renal tumors (O'Gara et al., 1964). Cycasin has been repeatedly shown to be carcinogenic for rats (Laquer, 1964, 1965; Kawaji et al., 1968; Hirono et al., 1968a,b; Gusek and Mestwerdt, 1969; Williams and Murphy, 1971; Fukunishi et al., 1971, 1972; Watanabe et al., 1975). Other laboratory animals for which cycasin is carcinogenic include mice (Hirono et al., 1969; Hirono and Shibuya, 1970; Williams and Murphy, 1971; Shibuya and Hirono, 1971), fish (Spatz, 1968); hamsters (Laquer and Spatz, 1968; Hirono et al., 1971), and monkeys (Seiber et al., 1980). Hirono (1972) has summarized the observed differences in carcinogenic effect of cycasin on various laboratory animals depending on such factors as dosing regimen and route of administration. Other aspects in regard to cycasin carcinogenicity and toxicity have been considered by Laquer and Spatz (1968, 1975), Spatz (1969), and Laquer (1977).

Laquer (1964, 1965) showed that while cycasin is carcinogenic only after passage through the gastrointestinal tract, its aglycone, MAM, induced tumors in rats independent of the route of administration. The hydrolysis of cycasin occurs only when given orally to rats older than 28 days (Laquer, 1977), and does not take place in germ-free rats (Laquer, 1964, 1965). Therefore, it has been postulated that its proximate carcinogen is MAM, which is liberated by the action of bacterial β-glucosidase in the intestinal tract (Laquer et al., 1967). The levels of β-D-glucosidase vary within the first month after birth in rats, and there is good correlation between β-glucosidase activity in the small intestine and the incidence of tumors in young rats of different ages when injected with cycasin (Matsumoto, 1979). β-D-Glucosidase activity has also been found in the skin of newborn rats (Spatz, 1967, 1968), which may help to explain the carcinogenic activity of cycasin when administered subcutaneously to newborn laboratory animals (Laquer and Spatz, 1968, 1975; Hirono and Shibuya, 1970; Hirono et al., 1971). Several other reports have described the carcinogenic effects of MAM or its semisynthetic acetate (Laquer and Matsumoto, 1966; Narisawa and Nakano, 1973; Zedeck et al., 1972, 1977; Zedeck and Sternberg, 1974; Matsubara et al., 1978; Seiber, 1980). MAM, when injected, produces colon cancers in rats, and apparently reaches the colon via the vascular system (Zedeck et al., 1977; Matsubara et al., 1978; Matsumoto, 1979).

The observed similarities between the chemical structures and biological activities of cycasin and dimethylnitrosamine (Fig. 2) are suggestive of a common metabolic pathway. Shank and Magee (1967) postulated that, while cycasin and dimethylnitrosamine are metabolized to the same biologically active compound, the observed differences in their carcinogenic effects can be explained by variations in the ability of tissues to metabolize the two carcinogens. J. A. Miller (1964) has suggested that both are metabolized to diazomethane (Fig. 2), a compound known to possess alkylating properties and as a strong carcinogen. Matsumoto (1979) has rejected this possibility on the basis of proton magnetic resonance studies, and has suggested that methyl diazonium hydroxide (Fig. 2) is a more likely intermediate. Schoental (1973) suggested that MAM might be oxygenated by alcohol dehydrogenase to methylazoxyformaldehyde, which would react at similar cellular sites to the aldehyde derivatives of nitrosamines. MAM is known to methylate nucleic acids in vitro (Matsumoto and Higa, 1966). The effects of MAM acetate on DNA and RNA synthesis have been studied by Zedeck and associates (Zedeck et al., 1970, 1972).

Cycads have a long history of use by humans as sources of food and medicine (Whiting, 1963). *Cycas revoluta* is used in Japan as a source of bean paste, a sample of which was shown by Kobayashi (1972) to be free of cycasin. However, small toy dolls made from raw, untreated cycad nuts in Japan did contain considerable amounts of cycasin, and hence represent a carcinogenic risk to humans coming into contact with them (Kobayashi, 1972). In Guam, *C. circinalis* seeds are used for the production of flour, in part because the plant often survives adverse climatic events such as typhoons (Campbell et al., 1966). Cycad flour used by Guamanians produced no carcinogenic effects when fed to rats, and it has been suggested that the local treatment of soaking the cycad flour effectively removes any carcinogenic substances present (Yang et al., 1966). Although the natives of the Miyako Islands, Okinawa, Japan, subsisted mainly on cycads in 1959, their incidence of cancer mortality due to hepatomas did not differ significantly from that of the population of the rest of Japan during the period 1961-1970 (Hirono et al., 1970a).

C. Pyrrolizidine Alkaloids

Pyrrolizidine alkaloids are widely distributed in the plant kingdom, especially in the families Boraginaceae (all genera), Compositae (tribes Senecioneae and Eupatorieae), and Leguminosae (genus *Crotalaria*) (Smith and Culvenor, 1981). The first observation of the carcinogenic nature of these compounds was reported by Cook et al. (1950), who produced liver tumors in rats fed with intermittent doses of the crude alkaloids of *Senecio jacobaea*, a cattle-poisoning plant. Of well over a 100 such compounds known, relatively few have been demonstrated to be carcinogenic for laboratory animals (Figs. 3 and 4). These carcinogens structurally exhibit a variety of amino alcohols (necine bases), and may be unesterified, or be mono- or diesters, or contain a macrocylic ester function. Compounds with known carcinogenic activity are listed in Table 1, and have been grouped in increasing order of complexity of their ester moieties. While the mechanism of the carcinogenic activity of pyrrolizidine alkaloids is not known, it should be noted that one requirement for activity is a carbonyl group allylic to the 1,2-double bond of the pyrrolizidine nucleus. It has been suggested that the proximate carcinogen from monocrotaline is dehydroretronecine (Fig. 3), a pyrrole, which is known to bind macromolecules in vivo and in vitro (Shumaker et al., 1976; Robertson et al., 1977; Johnson et al., 1978). Knowledge of the structural requirements for the hepatotoxicity of pyrrolizidine alkaloids, however, is much more detailed (Culvenor et al., 1969, 1976; McLean, 1970). It is interesting to note that pyrrolizidine alkaloids also exhibit activity as anticancer agents (Culvenor, 1968).

A number of plants that contain pyrrolizidine alkaloids have been found to be carcinogenic for rats and other laboratory animals. Such plants comprise those, in addition to *Senecio jacobaea* L., that are poisonous to livestock, such as *S. longibus* Benth. (Harris and Chen, 1970) and *Amsinckia intermedia* Fisch. and Mey (Schoental et al., 1970), as well as plants important in human ecology as herbal remedies, including *Heliotropium supinum* L. (Schoental et al., 1970); *S. longibus* (Harris and Chen, 1970; Smith and Culvenor, 1981); *Petasites japonicus* Maxim. (Hirono et al., 1973b); *Tussilago farfara* L. (Hirono et al., 1976), and *Symphytum officinale* L. (Hirono et al., 1978a). Several of the alkaloids listed in Table 1 are either derived from the plants listed above or from other plants with a history of human use. Alkaloids of

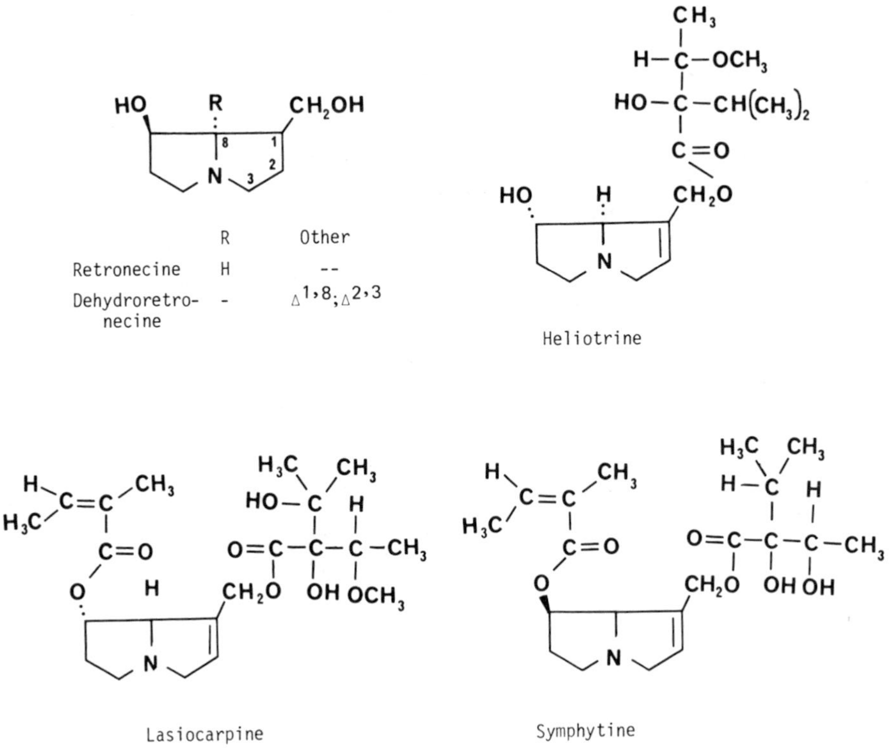

FIGURE 3 Carcinogenic pyrrolizidine alkaloids: necine bases, monoesters, and acyclic diesters.

S. jacobaea have been found in honey produced from its nectar (Deinzer et al., 1977) and in milk from cows grazing on this species (Dickinson et al., 1976). Pyrrolizidine alkaloids therefore present a significant carcinogenic risk to humans and livestock alike.

D. Betel Quid Carcinogen

It has been estimated that 10% of the world's population indulges in the habit of chewing betel or areca nuts (*Areca catechu* L., Palmae), especially in oriental countries (Fendell and Smith, 1970). Since the betel nut is highly astringent and acidic, calcium hydroxide (lime) may be added for neutralization purposes, and is one of several possible agents found in betel quid, as chewed in India. Other typical ingredients are the leaf of *Piper betle* L.; aqueous extracts of *Acacia catechu* (L. fil.) Willd. or *A. suma* (Roxb.) Buch.-Hum. ex Voigt and *Uncaria gambir* (Hunter) Roxb.; tobacco (*Nicotiana tabacum* L.), and various spices (Muir and Kirk, 1960; Jussawalla and Deshpande, 1971; Arjungi, 1976). Very early clinical investigations on the incidence of oral cancer in Southeast Asia, according to Muir and Kirk (1960), did not all show a definite relationship with the chewing of betel nut-tobacco

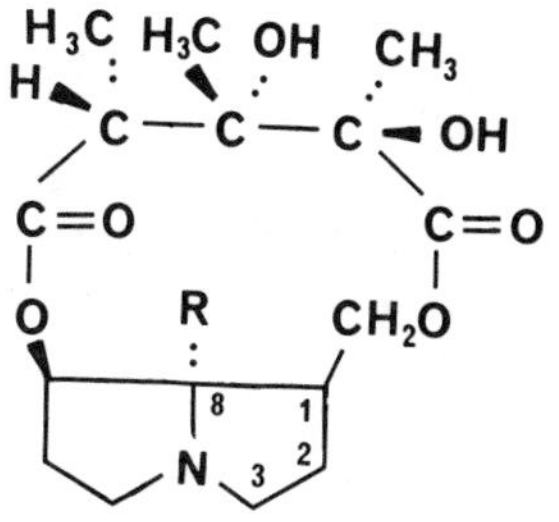

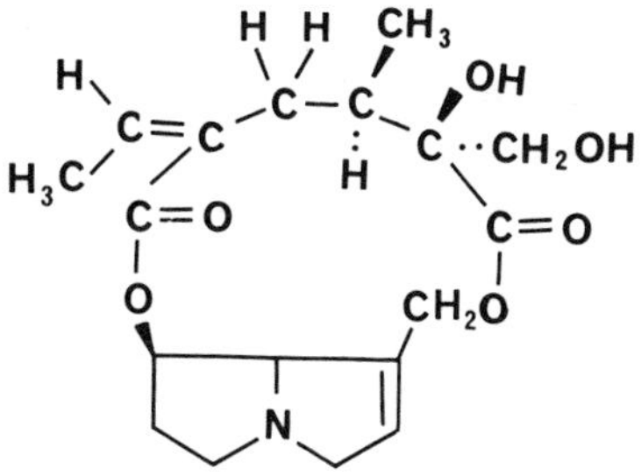

	R	Other
Monocrotaline	H	$\Delta^{1,2}$
Monocrotaline pyrrole	-	$\Delta^{1,8}$;$\Delta^{2,3}$

	Other
Retrorsine	--
Isatidine	N→O

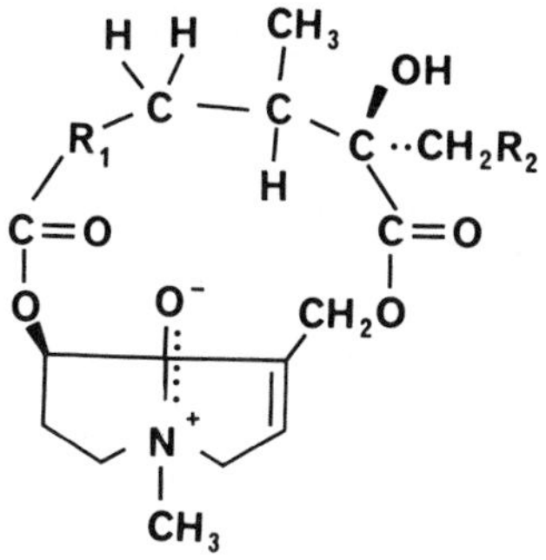

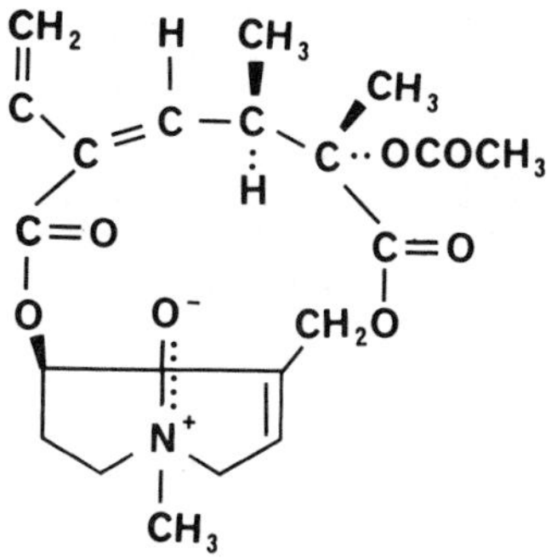

Clivorine

	R_1	R_2
Petasitenine	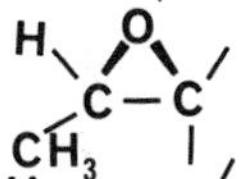	H
Senkirkine	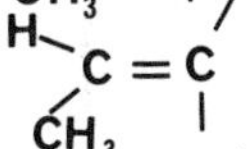	H
Hydroxysenk-irkine	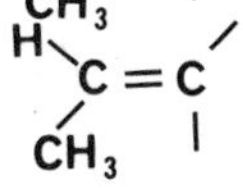	OH

FIGURE 4 Carcinogenic pyrrolizidine alkaloids: macrocyclic diesters.

mixtures. Later articles have concluded that there is such a correlation (Mendelson and Ellis, 1924; Orr, 1933; Muir and Kirk, 1960; Atkinson et al., 1964; Boyland, 1968; Dunham, 1968; Fendell and Smith, 1970; Jussawalla and Deshpande, 1971; Ramanthan and Lakshimi, 1976; Arjungi, 1976). Populations who chew betel quid formulations not containing tobacco may run an increased risk of cancer of the oropharynx and esophagus since the juice of mastication is often swallowed in such circumstances. Therefore, betel quid excluding tobacco may be suspected as containing weak carcinogens or cocarcinogens (Jassawalla and Deshpande, 1971).

Several groups have produced carcinogenic and other toxic effects in laboratory animals from extracts of betel nut and betel quid. Woelfel and

TABLE 1 Pyrrolizidine Alkaloids with Carcinogenic Activity for Rats

Compound	Route of administration	Dosing method	Type of tumor produced	References
Retronecine	s.c.	Single	Spinal cord	Schoental and Cavanagh, 1972
Dehydroretronecine[a]	s.c.	Multiple	Rhabdomyosarcoma	Allen et al., 1975; Schumaker et al., 1976
	s.c., topical	Multiple	Skin	Johnson et al., 1978
Heliotrine	i.p.[b]	Single, double	Pancreatic islet cell; hepatoma; urinary bladder; testicular	Schoental, 1975
Lasiocarpine	i.p.	Multiple	Hepatoma; squamous cell carcinoma of skin	Svoboda and Reddy, 1972, 1974
	Oral	Multiple	Angiosarcoma of liver; hepatocellular carcinoma; skin; lymphoma	Rao and Reddy, 1978
Symphytine	i.p.	Multiple	Liver cell adenoma	Hirono et al., 1979a
Monocrotaline	Oral	Single	Hepatoma	Schoental and Head, 1955
	i.p.	Multiple	Hepatoma	Schoental and Head, 1955
	s.c.	Multiple	Rhabdomyosarcoma; leukemia; hepatocellular carcinoma; pulmonary adenoma	Allen et al., 1975; Schumaker et al., 1976
Monocrotaline pyrrole[a]	s.c.	Multiple	Skin sarcoma	Hooson and Grasso, 1976
Retrorsine	Oral	Multiple	Hepatoma	Schoental et al., 1954
Isatidine	Oral	Multiple	Hepatoma	Schoental et al., 1954
Petasitenine	Oral	Multiple	Liver	Hirono et al., 1977
Senkirkine	i.p.	Multiple	Liver	Hirono et al., 1979a
Hydroxysenkirkine	i.p.	Single	Brain	Schoental and Cavanagh, 1972
Clivorine	Oral	Multiple	Liver	Kuhara et al., 1980

[a]Not naturally occurring.
[b]Administered with nicotinamide.

co-workers generated severe ulcerations on mouse skin with ethanolic fractions of *A. catechu* seeds. Ether-soluble fractions did not produce tumors
after 15 months of application, and the ethanol extract from which tannins
had been precipitated out also gave negative results (Woelfel et al., 1941).
Extracts of betel containing tobacco, when painted on the ears or instilled
in the vagina of mice, produced a low incidence of papillomas and squamous
cell carcinomas at the sites of treatment (Muir and Kirk, 1960; Reddy and
Anguli, 1967). Suri and associates found that repeated topical applications
of dimethyl sulfoxide extracts of betel nut to the mucosa of the buccal pouch
of hamsters resulted in the development of leukoplakia and tumors. The
incidence of tumors increased when extracts were made of a mixture of betel
nut and cured tobacco, but extracts of tobacco alone caused leukoplakia but
no tumors (Suri et al., 1971). Subcutaneous applications of betel nut aqueous extracts produced fibrosarcomas in Swiss mice (Ranadive et al., 1976)
and malignant tumors in NIH black rats (Kapadia et al., 1978). Mori et al.
(1979) observed epidermal thickening in the upper digestive tract of rats
fed a diet containing betel nut, lime, and *Piper betel* leaves, as well as a
forestomach papilloma in one rat given *P. betel* leaves in the diet. Dunham
et al. (1966) demonstrated epithelial atypia but no tumors in hamster cheek
pouches treated repeatedly with calcium hydroxide. Arecoline (Fig. 5), an
alkaloidal constituent of *A. catechu*, when applied after calcium hydroxide to
the cheek pouch of hamsters, produced proliferative lesions in two out of
nine animals, and one papilloma. However, when hamsters were exposed to
arecoline in dimethyl sulfoxide, no lesions developed in the pouch or esophagus of the test animals (Dunham et al., 1974). Betel quid containing lime
and tobacco produced epithelial atypia in the buccal mucosa of baboons and a
malignant growth in one animal (Hamner, 1972; Hamner and Reed, 1972).

So far the carcinogenic principle of betel nut has not been identified, and
the difficulty in so doing is compounded because betel nut is rarely chewed
alone. The presence of lime and/or tobacco has been suspected as being the
carcinogenic agent of betel quid (Mendelson and Ellis, 1924; Orr, 1933;
Atkinson et al., 1964; Hamner, 1972). However, these proposals were made
before the more recent experiments which indicate that betel nut extract per
se is carcinogenic. Boyland and Nery (1969) showed that arecoline and its
metabolite arecaidine (Fig. 5) from betel nut react at the 3,4-olefinic bond
with thiol groups, and are thus biological alkylating agents, and hence possibly carcinogenic. A somewhat different hypothesis was proposed by
Shivapurkar and associates, who showed that the chewing of betel quid

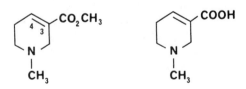

Arecoline   Arecaidine

FIGURE 5 Two alkylating agents from betel nut.

(excluding tobacco) enhanced levels of nitrite in the oral cavity. They projected that these increases in nitrite concentrations may then promote increased generation of carcinogenic nitrosamines (Shivapurkar et al., 1980). Studies on the tumor-promoting agents of unburned tobacco are mentioned later in this chapter.

E. Safrole and Other Alkenylbenzene Derivatives

Safrole (4-allyl-1,2-methylenedioxybenzene) (Fig. 6) is the principal constituent of sassafras oil [*Sassafras albidum* (Nutt.) Nees] (Lauraceae), Brazilian sassafras oil (*Ocotea cymbarum* HBK. and *O. pretiosa* Benth. and Hook. f.), and star anise oil (*Illicium anisatum* L.). It is a major component of the essential oils of several spices, including nutmeg (*Myristica fragrans* Houtt.), cinnamon (*Cinnamomum zeylanica* Nees), and camphor (*Cinnamomum camphora* Seib.) (Hickey, 1948; Guenther and Althausen, 1949). The compound is widely distributed in the plant kingdom, having been recorded in 53 species and varieties, representing 10 families (Farnsworth et al., 1976). Safrole is used in perfumery and soaps, and oil of sassafras, containing this ingredient, was formerly used as a flavoring agent in the soft-drink industry before being removed from the market in the United States in 1960 (IARC, 1976). However, it was recently detected in a number of flavored alcoholic and nonalcoholic beverages, including vermouth (Salvatore et al., 1980).

Safrole has been found to be a liver carcinogen for rats by several groups when administered orally (Homberger et al., 1961a,b; Long et al., 1961, 1963; Homberger et al., 1965; Borchert et al., 1973a). Since levels of 0.5-5% in the diet are necessary to elicit this activity, safrole is considered to be a weak carcinogen. The compound is carcinogenic for young mice after oral and subcutaneous administration, producing pulmonary tumors and hepatomas (Innes et al., 1969; Epstein et al., 1970). Adult male mice fed safrole (0.4-0.5%) for 13 months also developed liver tumors (Borchert et al., 1973a).

The discovery of the carcinogenicity of safrole, coupled with its wide distribution in the plant kingdom, led to the investigation of the tumorigenicity of other essential oils and their constituents. Oil of calamus, from *Acorus calamus* L., has as major constituent, β-asarone (Fig. 6), and, when fed to rats at high dose levels, produced mesenchymal tumors of the small intestine (Gross et al., 1967). The safrole isomer, isosafrole (Fig. 6), a constituent of ylang-ylang oil (*Cananga odorata* Hook. f. et Thomson), and the semisynthetic compound dihydrosafrole (Fig. 6) are weak hepatocarcinogens when fed to mice (Long and Jenner, 1963; Innes et al., 1969). The latter compound has also been shown to evoke esophageal tumors in rats (Hagan et al., 1965). Estragole (Fig. 6) (1-allyl-4-methoxybenzene), a major component of the oils of the spices tarragon (*Artemisia draculuncus* L.), basil (*Ocimium basilicum* L.), and fennel (*Foeniculum vulgare* Miller), is also a weak hepatocarcinogen when administered by subcutaneous injection to young mice (Drinkwater et al., 1976).

J. A. Miller and E. C. Miller, at the University of Wisconsin, have investigated the formation of electrophilic metabolites of safrole and estragole. They found 1'-hydroxysafrole and 1'-hydroxyestragole (Fig. 7) to be more potent hepatocarcinogens for rodents than their respective parent allylbenzenes (Borchert et al., 1973; Drinkwater, et al., 1976). A low percentage conver-

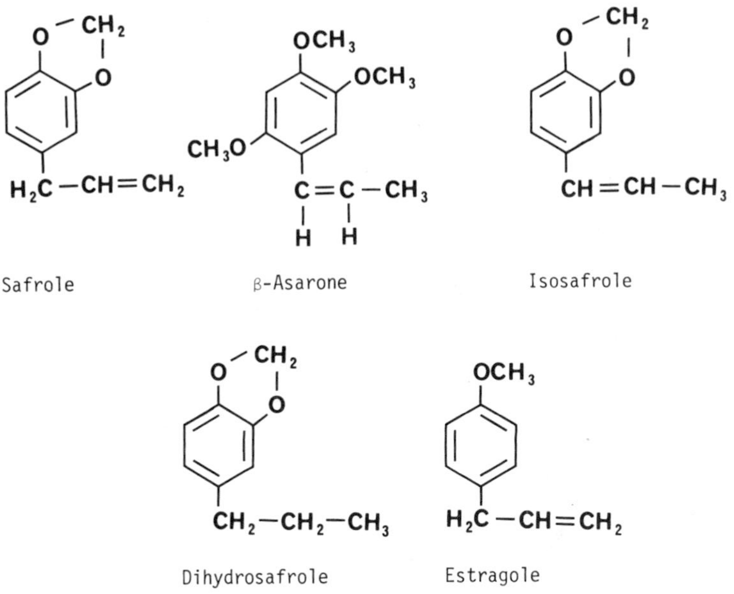

FIGURE 6 Safrole and some related compounds.

sion to 1'-hydroxysafrole, excreted in the glucuronide form, occurred in rats, hamsters, and guinea pigs after intraperitoneal administration of pure safrole (Borchert et al., 1973b). Similarly, about 20% of estragole administered to rats was found to be excreted in the urine as a conjugate of 1'-hydroxyestragole (Drinkwater et al., 1976). The metabolism of 1'-hydroxysafrole was further studied by Peele and Oswald (1978). Other possible in vivo metabolites of safrole include safrole-2',3'-epoxide, 1'-hydroxysafrole-2',3'-epoxide, and 1'-oxosafrole (Fig. 7) (Stillwell et al., 1974; J. A. Miller et al., 1979). Repeated topical applications of 1'-hydroxysafrole-2',3'-epoxide and cocarcinogenesis with croton oil resulted in the formation of papillomas on mouse skin (Wislocki et al., 1977). 1'-Oxosafrole, however, did not produce any carcinogenic activity after subcutaneous injection to adult rats or weanling mice (Wislocki et al., 1977). The electrophilic reactivity, expressed by reaction with guanosine, and the mutagenicity against several *Salmonella typhimurium* strains, have been determined for several known and speculative metabolites of safrole and estragole (Wislocki et al., 1976, 1977; J. A. Miller et al., 1979).

F. Tannins

The vegetable tannins are a complex group of polyphenolic compounds, with a molecular weight range of 500-3000, that are widely distributed in plants (Haslam, 1965). Tannins are divided into two groups: condensed tannins, which are derivatives of flavonols, and hydrolyzable tannins, which are esters of a sugar and one or more trihydroxybenzenecarboxylic acids (Windholz, 1976).

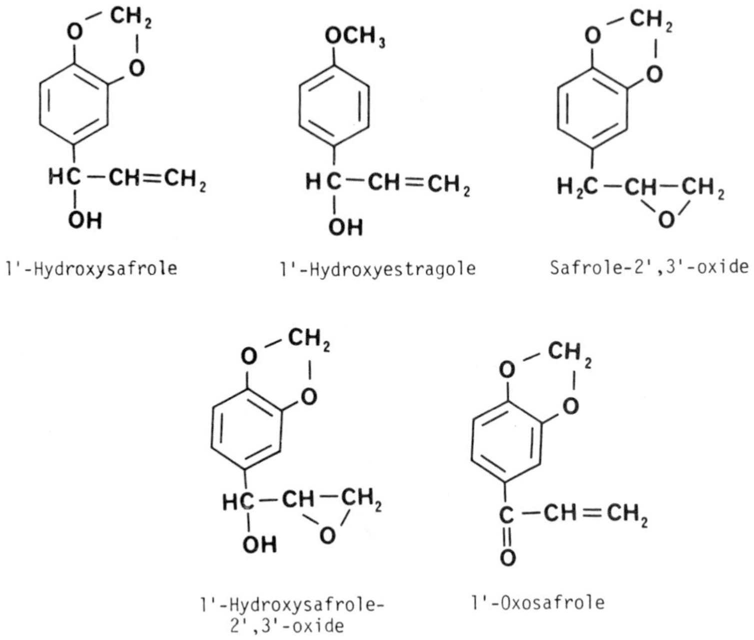

1'-Hydroxysafrole   1'-Hydroxyestragole   Safrole-2',3'-oxide

1'-Hydroxysafrole-   1'-Oxosafrole
2',3'-oxide

FIGURE 7 Structures of some metabolites of safrole and estragole.

Tannic acid is a hydrolyzable tannin from the galls of many oak (*Quercus*) species and from *Rhus semilata* leaves (Haslam, 1965). The empirical formula of this pentadigalloyl glucoside is usually given as $C_{76}H_{52}O_{46}$, and it has a molecular weight of 1701.2 (IARC, 1976). This compound was shown to produce cirrhosis and hepatomas in rats on prolonged subcutaneous administration (Korpássy and Mosonyi, 1950). Also, it exhibited a synergistic effect in the production of malignant hepatomas in young rats when administered subcutaneously with acetylaminofluorene (Mosonyi and Korpássy, 1953).

Other compounds in this group have also been tested for their carcinogenic effects. Kirby (1960) showed that parenteral administration of extracts of both condensed and hydrolyzable tannins were carcinogenic for rats and mice. In general, condensed tannins evoked liver tumors as well as sarcomas at the site of injection, while liver tumors only were induced by extracts of hydrolyzable tannins (Kirby, 1960). In the last decade, Kapadia and co-workers have demonstrated the carcinogenic effects in laboratory animals of a variety of tannin-containing extracts from plants, including a fraction from tea (*Camellia sinensis*) (Table 2). The earlier data obtained by Kirby (1960) are omitted from this tabulation since they have recently been summarized in a review article (Farnsworth et al., 1976).

Tannic acid produces polyribosome disaggregation in rat liver, and inhibition of [14C]leucine incorporation into liver protein (Reddy et al., 1970). Tannins have been determined to have anticancer activity, especially against certain solid tumors such as the Walker 256 carcinosarcoma (Fong et al., 1972).

TABLE 2 Some Tannin-Containing Plant Extracts Carcinogenic for Rats

Family	Species	Plant part	Tumor-bearing[a] animals	Reference
Altingiaceae	*Liquidambur styriciflua*	Leaf[b-d]	26/30	Kapadia et al., 1976
Anacardiaceae	*Rhus copallina*	Root[b-d]	10/30	Kapadia et al., 1978
Ebenaceae	*Diospyros virginiana*	Leaf[b-d]	17/30	Kapadia et al., 1978
		Unripe fruit[b-d]	21/30	Kapadia et al., 1976
Fagaceae	*Quercus falcata pagodaefolia*	Bark[b-d]	28/30	Kapadia et al., 1976
Krameraceae	*Kramera ixina*	Aerial[b-d]	28/30	Pradhan et al., 1974
		Herb[e-g]	15/15	O'Gara et al., 1971
		Aerial[b,c,g]	12/15	O'Gara et al., 1974
Leguminosae	*Acacia villosa*	Root[b-d]	29/30	Pradhan et al., 1974
		Root[c,e,g]	17/18	O'Gara et al., 1974
Myricaceae	*Myrica cerifera*	Bark[b-g]	8/30	Kapadia et al., 1976
Plumbaginaceae	*Limonium nashii*	Root[b-d]	13/30	Kapadia et al., 1976
Sterculiaceae	*Melochia tomentosa*	Root[c,e,g]	11/11	O'Gara et al., 1974
Theaceae	*Camellia sinensis*	Leaf[b-d]	21-30	Kapadia et al., 1976

[a]Numerator: number of treated animals bearing tumors; denominator: number of test animals.
[b]Tannin fraction.
[c]Administered subcutaneously.
[d]Fibrous histiocytomas produced.
[e]Hot aqueous extract.
[f]Administered intramuscularly.
[g]Sarcomas produced.

Tannins have been pinpointed as possible carcinogenic constituents of plants suspected of causing esophageal and other types of human cancer (Morton, 1970, 1972, 1980; Segi, 1975). Of all the compound classes detailed in this chapter, perhaps tannins are ingested to the greatest extent by human populations, primarily as a result of the imbibition of beverages such as tea and wine (Morton, 1980). It has been pointed out that the tannin levels of black (fermented) teas produced in Sri Lanka and India are markedly higher than those in green (unfermented) teas from China and Japan (Morton, 1980). Also, the British habit of adding milk to tea has some scientific basis, since the contained casein fixes the tannin constituents of tea, thereby restricting possible harmful effects in the gastrointestinal tract (Morton, 1972). This practice almost certainly was responsible for the lower levels of esophageal cancer recorded in the nineteenth-century British population than in their Dutch counterparts, since in Holland tea was widely consumed without milk at that time (Morton, 1980). A number of tannin-containing herbal remedies have also been cited for their possible carcinogenic effects (Morton, 1980).

In view of the widespread occurrence of tannins in the human diet, and the demonstrated carcinogenic activity of this class of compounds in laboratory animals, their further investigation is warranted, preferably using pure compounds of known structure, rather than unpurified plant extracts.

G. *N*-Nitrosamines

In a search for a possible cause of the high levels of esophageal cancer experienced by certain Bantu populations in the Transkei, Du Plessis et al. (1969) isolated the known carcinogen dimethylnitrosamine (Fig. 8) from *Solanum incanum* L. fruits. Juice from this plant is used to curdle milk (Du

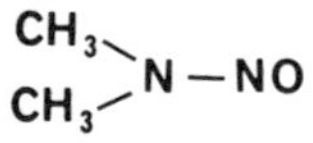

Dimethylnitrosamine

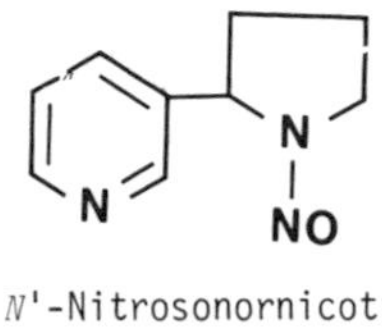

N'-Nitrosonornicotine

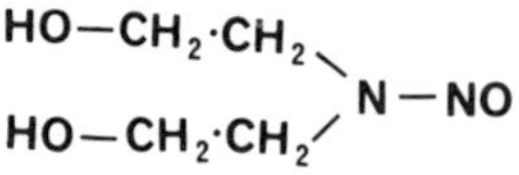

N-Nitrosodiethanolamine

FIGURE 8 Plant-derived *N*-nitrosamines.

Plessis et al., 1969). Dimethylnitrosamine has been shown to produce hepatomas, lung adenomas, and renal tumors in mice (Terracini et al., 1966).

It was later found that N-nitrosamines, including N'-nitrosonornicotine, and N-nitrosodiethanolamine (Fig. 8) occur not only in tobacco smoke, but also at lower levels in unburnt tobacco (*Nicotiana tabacum* L.) (Hoffmann et al., 1974; Schmeltz and Hoffmann, 1977). Such compounds, which are derived from the tobacco alkaloids, may be formed during the curing of tobacco (Hecht et al., 1978), or may occur in the fresh, unfermented leaves of tobacco (Bharadwaj et al., 1975). N'-Nitrosonornicotine has caused nasal cavity and esophageal cancers in rats (Hoffmann et al., 1975; Singer and Taylor, 1976), respiratory tract tumors in hamsters (Hilfrich et al., 1977), and pulmonary tumors in mice (Boyland et al., 1964). N-Nitrosodiethanolamine has exhibited hepatic carcinogenicity for rats (Druckrey et al., 1967).

N-Nitrosamines, which are formed by the reaction of naturally occurring amines with nitrites, are considered potential health hazards at concentrations in the parts per billion region (Hoffmann et al., 1974; Issenberg, 1976; Kawabata et al., 1979). Since the majority of these compounds are not naturally occurring plant constituents, they are not considered further in this chapter.

H. Polycyclic Aromatic Hydrocarbons

Several polycyclic aromatic hydrocarbons (PAHs), which include some of the most potent carcinogens known to humans, have been demonstrated to occur in a variety of plants and plant products (Grasso and O'Hare, 1976). The potent carcinogen benz[a]pyrene (3,4-benzpyrene) (Fig. 9) has been detected in green vegetables and tree leaves (Gräf and Diehl, 1966), unrefined vegetable oils including sunflower seed, palm kernel, and coconut oils

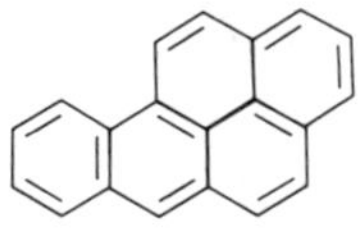

Benz[*a*]pyrene

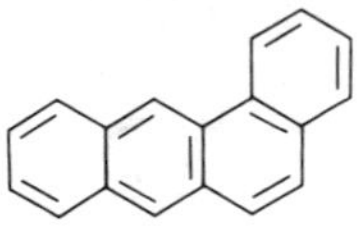

Benz[*a*]anthracene

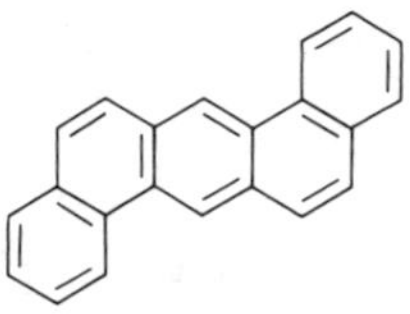

Didenz[*a*,*h*]anthracene

FIGURE 9 Some plant-derived carcinogenic polycyclic aromatic hydrocarbons.

(Grimmer and Hildebrandt, 1967, 1968; Biernoth and Rost, 1967), as well as coffee substitutes based on chicory (Maier and Stender, 1969). Benz[a]-anthracene (1,2-benzanthracene) and dibenz[a,h]anthracene (1,2:5,6-dibenzanthracene) (Fig. 9) are examples of other PAHs known to occur in vegetables (Gräf and Diehl, 1966; Grasso and O'Hare, 1976). Plant sources of carcinogenic hydrocarbons, including PAHs, have recently been reviewed by Il'initskii et al. (1978) and Zamfir and Paladi (1979).

The origin of PAHs in plants has been suggested as being due to one or more of several factors: endogenous synthesis, atmospheric pollution, and contamination from the soil (Grasso and O'Hare, 1976). While it has been asserted that benz[a]pyrene is a product of plant biosynthesis (Gräf and Diehl, 1966; Petrun, 1977), there is evidence that the levels of PAHs appear to increase in plant products as a result of processing. For example, while only small amounts of PAHs occurred in sun-dried copra, smoke-dried copra was found to contain distinctly higher amounts (Grimmer and Hildebrandt, 1967). The subject of the occurrence of PAHs in cooked food has been discussed at length by Grasso and O'Hare (1976).

I. Miscellaneous Carcinogenic Toxins Biosynthesized by Plants

A variety of plant secondary metabolites, many widely distributed in the plant kingdom, have been shown to evoke carcinogenic effects in laboratory animals. The sources and types of tumor produced by a number of plant constituents are listed in Table 3. Structural formulas for these compounds are presented in Fig. 10.

Although coumarin (Fig. 10) when fed at a level of 0.5% produced liver cancers in 12 out of 14 surviving rats (Griepentrog, 1973) (Table 3), Hagan and co-workers were unable to show that coumarin produced benign or malignant tumors in any rat organ when fed in the diet at up to 5000 ppm for 2 years (Hagan et al., 1967). In addition, it was recently reported that male and female Syrian golden hamsters fed 0.1 or 0.5% coumarin for up to 2 years did not develop tumors or other changes in the bile duct (Ueno and Hirono, 1981). Therefore, evidence for the carcinogenicity of coumarin is contradictory. The coumarin derivative 8-methoxypsoralen (methoxalen) (Fig. 10) is currently used in photochemotherapy in the treatment of psoriasis (Towers, 1980), and the consequent risk of cutaneous carcinoma in patients treated in this manner has been pointed out (Stern et al., 1979) (Table 3).

The carcinogenic effect of rotenone (Fig. 10) reported in Table 3 is of particular concern because of the massive consumption of this compound as an insecticide, piscicide, and parasitic control agent in the United States and other Western countries (Gosálvez and Díaz-Gil, 1978). This compound, which is a respiratory inhibitor, has been shown to increase cerebral glutamate and somatomedin concentrations in rats (Gosálvez et al., 1979). Similarly, the widespread use of *Argemone mexicana* L. seed oil (argemone oil) for cooking in parts of India is cause for environmental concern because there are high levels of the carcinogenic benzophenanthridine alkaloid sanguinarine (Fig. 10) present (Table 3). Ranadive et al. (1972) have commented on the

TABLE 3 Tumors Produced in Experimental Animals by Some Miscellaneous Plant Constituents

Compound	Plant origin	Test animal	Route of administration	Tumor type	Reverences
Coumarin	Plant essential oils, including *Dipteryx odorata* and *Cinnamoeim cassia*	Rat	Oral	Bile-duct carcinoma	IARC, 1976; Griepentrog, 1973
8-Methoxypsoralen	Many plants in the families Umbelliferae, Rutaceae, and Leguminosae	Mouse	Oral, s.c.[a]	Skin (ear)	Towers, 1980; Hakim et al., 1960
		Mouse (hairless)	Topical[b]	Skin	Forbes et al., 1976
Parasorbic acid	*Sorbus acuparia* L.	Rat	s.c.	Sarcoma (local)	IARC, 1976; Dickens and Jones, 1963
Rotenone	*Derris*, *Tephrosia*, and *Lonchocarpus* species (Leguminosae)	Rat	Oral, i.p.	Mammary adenoma	Farnsworth et al., 1976; Gosálves and Merchán, 1973; Gosálves et al., 1977
Sanguinarine	*Argemone mexicana* L. and other species in the Papaveraceae	Rat	Bladder implantation	Bladder tumor	Farnsworth et al., 1976; Sarkar, 1948; Hakim, 1968

[a]Treated animals exposed to fluorescent light.
[b]Treated animals exposed to solar simulator.

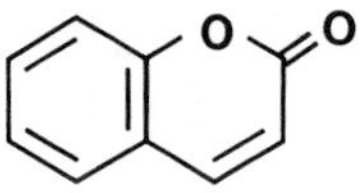

Coumarin

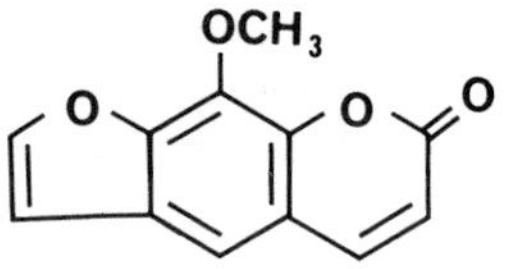

8-Methoxypsoralen

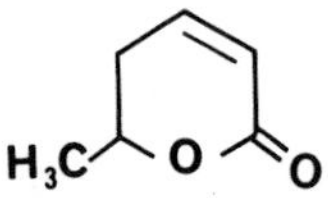

Parasorbic Acid

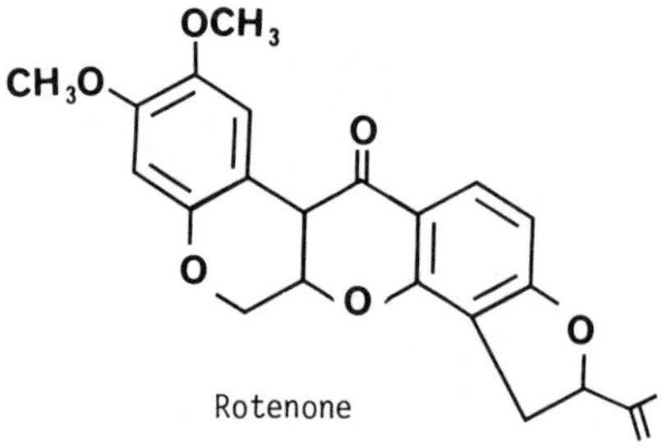

Rotenone

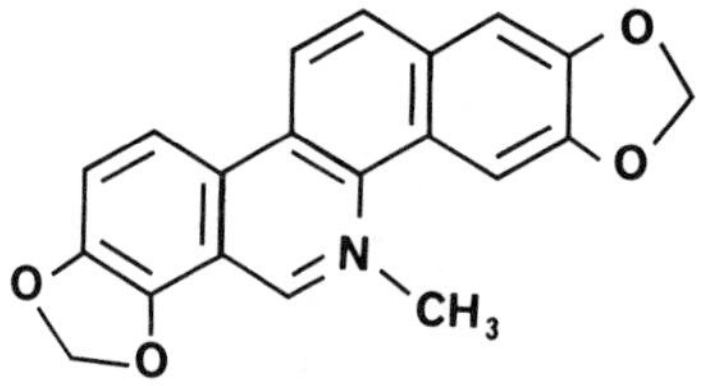

Sanguinarine

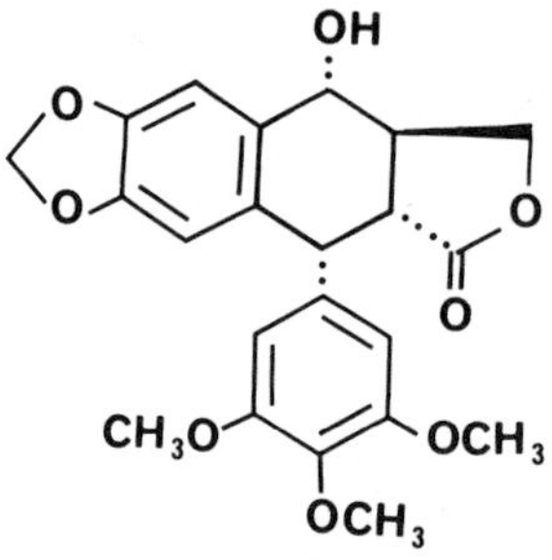

Podophyllotoxin

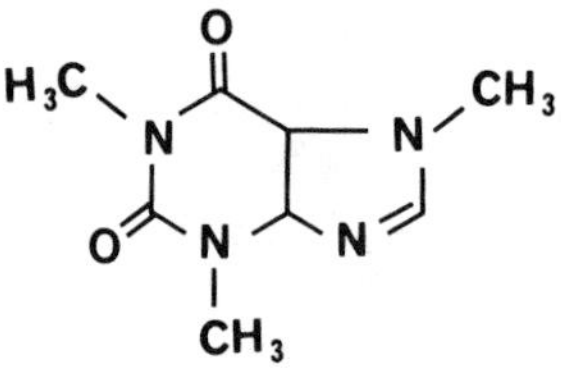

Caffeine

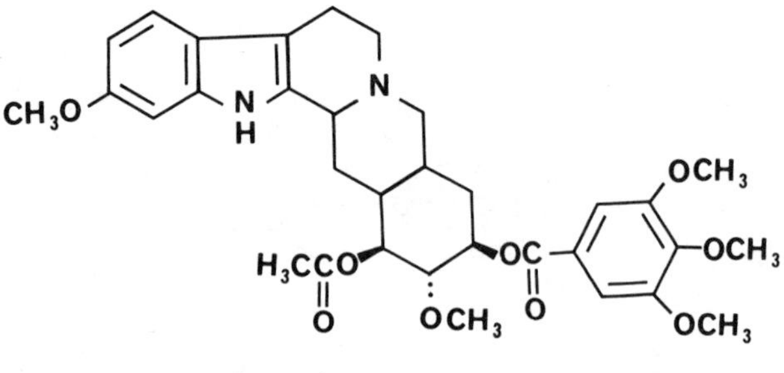

Reserpine

FIGURE 10 Miscellaneous plant constituents with known or suspected carcinogenic activity.

possible health hazards presented by argemone oil, which, although not carcinogenic for mice at the doses used, tended to potentiate the carcinogenic
effects of a contaminated sample of mustard oil. Sanguinarine has been found
to be transmitted from the diet into the milk of goats and rabbits (Hakim et
al., 1961). A postulated metabolite of this compound, 3,4-benzacridine, has
also been demonstrated to be a carcinogen for rodents (Hakim et al., 1961).

Several plant constituents, excluded from Table 3, may be regarded as
potential carcinogens. Three such examples, podophyllotoxin, caffeine, and
reserpine (Fig. 10), will be discussed in turn.

Podophyllotoxin is a lignan that is a major constituent of podophyllin, a
resin obtained from *Podophyllum peltatum* L. (Berberidaceae) (Schoental,
1974). Podophyllin has been implicated as the cause of epidermoid carcinomas
of the cervix and vaginal wall of a strain A mouse (Kaminetzky and McGrew,
1962), and when administered on a prolonged basis in the diet of BALB/c
mice, three hepatomas and three lymphomas were observed (O'Gara, 1968).
Schoental (1974, 1976) has called for the examination of the carcinogenicity
of podophyllotoxin, since it is also a constituent of certain *Juniperus* species
used as bedding materials for laboratory animals, and may be a contributory
factor in the generation of "spontaneous" tumors in carcinogenesis experiments.

Caffeine (1,3,7-trimethylxanthine) is mutagenic in test systems involving
Drosphila melanogaster and mice, and causes chromatid breakage in human
cells in vitro (Kuhlmann et al., 1968). Despite the existence of negative
evidence for carcinogenicity in the literature, caffeine remains a suspicious
substance in this regard (*British Medical Journal*, 1976).

Reserpine, a constitutent of *Rauvolfia serpentina* (L.) Benth. ex Kurz, and
many other species in the family Apocynaceae (Farnsworth et al., 1976), is a
major therapeutic agent in the United States for the control of essential hypertension (IARC, 1976). Early work indicated the development of mammary
tumors in female C3H mice, and lymphosarcomas and hepatomas in male and
female Wistar rats, on administration of reserpine (IARC, 1976). Recently,
results released by the National Cancer Institute (NCI) indicated that reserpine caused cancers of the adrenal gland in male rats, the mammary gland
in female mice, and the seminal vesicles in male mice (*Chemical Engineering
News*, 1980). In a recent journal editorial, Feldmann (1980) strongly criticized the NCI test protocols with respect to the evaluation of the carcinogenicity of reserpine, since these experiments involved the administration of
maximally tolerated doses of test compound to two species of rodents for only
one generation. The role of reserpine as a carcinogen must be regarded as
uncertain, since a substantial body of negative laboratory and clinical data
exists. For example reserpine failed to significantly affect the development
of tumors induced by 7,12-dimethylbenz[a]anthracene in ovariectomized female
rats (Welsch and Meites, 1970), and did not induce malignant transformation
of mouse fibroblasts (line M2) in cell culture (Marquardt, 1975). Several
retrospective clinical studies have shown a lack of association of increased
breast cancer incidence in women with previous ingestion of reserpine or other
Rauwolfia alkaloid prescription products (Armstrong et al., 1974; O'Fallon and
Larbarthe, 1975; Mack et al., 1975).

A number of other natural products of plant origin have been classified as
suspected carcinogens by the National Institute of Occupational Safety and

Health (Christensen et al., 1976), including several amino acids, lipids, sugars, polysaccharides, essential oil components, and alkaloids such as colchicine and vinblastine.

J. Miscellaneous Plants with Uncharacterized Carcinogenic Principles

There have been several preliminary reports in the literature of plant extracts found to exhibit carcinogenic activity in experimental animals, for which effects no known constituents have been attributed as being responsible. In Table 4, extracts of a number of plants used in human ecology in both primitive and advanced cultures that have been found to produce unambiguous carcinogenic effects in laboratory animals are listed. While pyrrolizidine alkaloids have been found in many species in the genus *Heliotropium*, no representatives of this class of carcinogens have been reported to occur in *H. angiospermum* (Table 4) (Smith and Culvenor, 1981). It may be inferred from data presented in Table 4 that carcinogens other than safrole occur in *Sassafras albidum*.

A study on the carcinogenic nature of several plants indigenous to Curaçao produced somewhat less clear-cut conclusions (O'Gara et al., 1971), and hence has not been incorporated into Table 4. Boiling aqueous extracts of sorsaka (*Annona muricata*), sali (*Heliotropium ternatum*), ration (*Gliricidia sepium*), and laraha (*Citrus aurantium*) were shown by O'Gara et al. (1971) to cause subcutaneous sarcomas in 1 or 2 out of 15 rats used in each experiment. No known carcinogens occur in any of these taxa.

In view of the concern in Japan about the carcinogenicity of bracken fern, because of its use in the human diet, other edible Japanese plants have also been tested for carcinogenic effects. Young leaves of artemisia (*Artemisia princeps* Pamp.), horsetail (*Equisetum arvense* L.), osmund (*Osmunda japonica* Thunb.), and seeds of ginkgo (*Ginkgo biloba* L.) did not produce statistically significant evidence of carcinogenic activity when fed to rats for 7-337 days, although a few tumors were observed during testing (Hirono et al., 1972b). Similarly, fiddlehead greens, from the ostrich fern, *Matteucia struthiopteris*, when fed in 1-10% concentrations for 70-73 weeks, did not result in the development of tumors in rats (Newberne, 1976).

III. PLANT COCARCINOGENS AND TUMOR PROMOTERS

A. Tigliane and Related Diterpenoids

Much has been learnt about the isolation techniques, structural characteristics, and biological activities of the tumor-promoting principles of certain species of the Euphorbiaceae and Thymelaeaceae, since the first demonstration of the cocarcinogenic activity of the seed oil of *Croton tiglium* L. (croton oil) on mouse skin in 1941 (Berenblum, 1941a). The isolation of the pure, biologically active constituents of croton oil was achieved independently by three groups in the mid-1960s (Hecker et al., 1964a; Van Duuren et al., 1963b; Van Duuren and Orris, 1965; Arroyo and Holcomb, 1965; Hecker, 1968b), after the efforts of many previous workers had proven inconclusive (Hecker,

TABLE 4 Some Miscellaneous Plant Extracts Used to Produce Carcinogenic Effects in Rats

Family	Species and common name	Plant part	Extract	Route of administration	Tumor type	Reference
Boraginaceae	*Heliotropium angiospermum*[a] (cocolade)	Aerial	Boiling aqueous	s.c.	Sarcoma	O'Gara et al., 1974
Chenopodiaceae	*Chenopodium ambrosiodes*[a] (sagrado)	Aerial	Hot aqueous	s.c.	Mesenchymal	Kapadia et al., 1978
Lauraceae	*Sassafras albidum* (sassafras)	Root	Petroleum[b] ether	s.c.	Mesenchymal	Kapadia et al., 1978
Leguminosae	*Glycine max* L. (soybean)	Defatted seed	Methanol	Oral[c]	Thyroid	Kimura et al., 1979
Solanaceae	*Solanum tuberosum* L. (potato)	Tuber, green	Undiluted sap	i.p.	Stomach	Ivankovic, 1978

[a]Author citation not given in reference quoted.
[b]This extract was further partitioned into methylene chloride and ethanol to eliminate safrole.
[c]Fed under thyroid-deficiency conditions.

1971a). The correct structure of phorbol, the parent diterpene alcohol of
the tumor-promoting principles of croton oil, as determined by x-ray crystallography, appeared in the literature in 1967 (Hoppe et al., 1967; Pettersen
et al., 1967). The tumor-promoting principles of croton oil, which are also
potent skin irritants, have been established as phorbol diesters, of varying
acyl chain lengths (Hecker, 1971a).

In 1961, Roe and Peirce showed that the latices of 10 *Euphorbia* species,
a genus related to *Croton* in the Euphorbiaceae, also exhibited tumor-
promoting activity on mouse skin (Roe and Peirce, 1961). Following this preliminary observation, bioassay-guided phytochemical studies on related
euphorbiaceous species, and on plants in the family Thymelaeaceae, have led
to the characterization of over 30 polyfunctional diterpene alcohols structurally related to phorbol (Hecker, 1978, 1981; Kinghorn, 1979a). All such
compounds are based on three diterpene hydrocarbons, tigliane, daphnane,
and ingenane, with the parent alcohols representing each skeleton being
phorbol, resiniferonol, and ingenol, respectively (Fig. 11) (Stout et al.,
1970; Zechmeister et al., 1970; Kupchan et al., 1976a; Hecker, 1978). These
diterpene alcohols must be esterified at certain positions to be biologically
active. In this chapter the numbering system for tigliane and related diterpenoids adopted by Hecker (1968a) is employed, rather than the less widely
used *Chemical Abstracts* nomenclature preferred by Van Duuren (1976).

About 130 biologically active tigliane, daphnane, and ingenane esters are
now known. Of these, over 65 are tigliane-based, while about 35 daphnane
and 30 ingenane derivatives have also been structurally characterized.
Literature reports of such compounds indicate that to date, they are restricted to only 13 genera in the Euphorbiaceae: *Aleurites, Baliospermum,
Croton, Elaeophorbia, Excoecaria, Euphorbia, Hippomane, Hura, Jatropha,
Micrandra, Sapium, Stillingia,* and *Synadenium* (Hecker, 1978; Evans and
Soper, 1978; Kinghorn, 1979a, 1980; Gunasekera et al., 1979; Adolf and
Hecker, 1980). [A daphnane ester was obtained by Gunasekera et al. (1979)
from plant material originally identified as *Cunuria spruceana* Baill. and
later reidentified as *Micrandra elata* (Didr.) Muell.-Arg.] All these genera
are classified in two (Crotonoideae and Euphorbioideae) of the five sub-
families recently proposed by Webster (Webster, 1975; Kinghorn, 1979a).
Tigliane and/or daphnane derivatives have been identified in seven genera
of the Thymelaeaceae so far (*Aquilaria, Daphne, Daphnopsis, Gnidia, Lasio-
syphon, Pimelea,* and *Synaptoleptis*) (Hecker, 1978; Gunasekera et al., 1981).
It is interesting to note that in a contemporary classification of the Angiosperms, Thorne (1968) placed both the Euphorbiaceae and the Thymelaceaceae
in the same order, and hence there is a chemotaxonomic basis for the occurrence of these diterpene esters in the two families.

Not all of the known biologically active tigliane, daphnane, and ingenane
derivatives have been tested for tumor-promoting activity on mouse skin.
Perhaps the majority of such isolates have been tested only for their skin-
irritant activity. While all tumor promoters of this type are skin irritants,
the converse does not seem to be the case (Fürstenberger and Hecker, 1972).
Compounds of this type have also been isolated as a result of bioassays directed toward piscicidal activity against killifish (*Oryzias latipes*) (Hirota et
al., 1980); abortifacient activity (Ying et al., 1977), and in vivo and in
vitro activity against the P-388 mouse lymphocytic leukemia test system

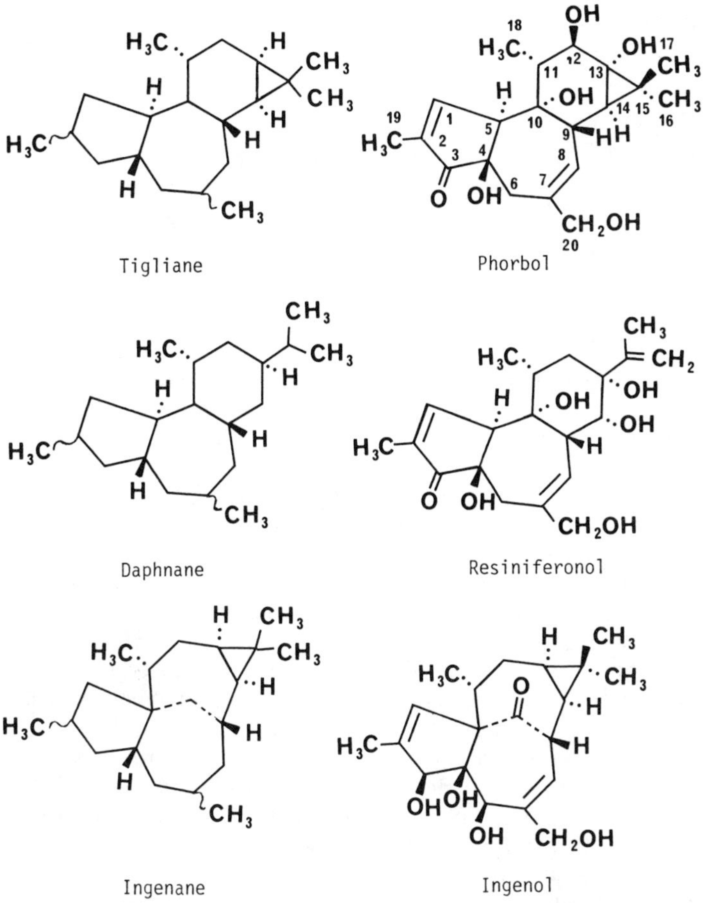

FIGURE 11 Structures of the hydrocarbon skeleta and corresponding parent diterpene alcohols of the cocarcinogens of plants of the Euphorbiaceae and Thymelaeaceae.

(Ogura et al., 1978). It is thus of some interest to note that the same compound, 12-O-deca-2,4,6-trienoylphorbol-13-acetate (Fig. 12), has been isolated as a piscicidal constituent of the twigs and bark of *Sapium japonicum* Pax and Hoffman (Ohigashi et al., 1972), a highly irritant principle of *Euphorbia tirucalli* L. latex (Fürstenberger and Hecker, 1977a), and a cytotoxic principle of *Aquilaria malaccensis* Lamk. stem bark (Gunasekera et al., 1981).

Diterpene esters based on the tigliane, daphnane, and ingenane skeleta, although potent biologically, tend to occur in the plant in low yields as complex natural mixtures. For example, the yields (w/w) of the major skinirritant principles of the latices of *Euphorbia tirucalli* L. (grown in Colombia) and *Synadenium grantii* Hook. f., which are both 4-deoxyphorbol diesters, were determined as 0.1% and 0.05%, respectively (Kinghorn, 1979b, 1980).

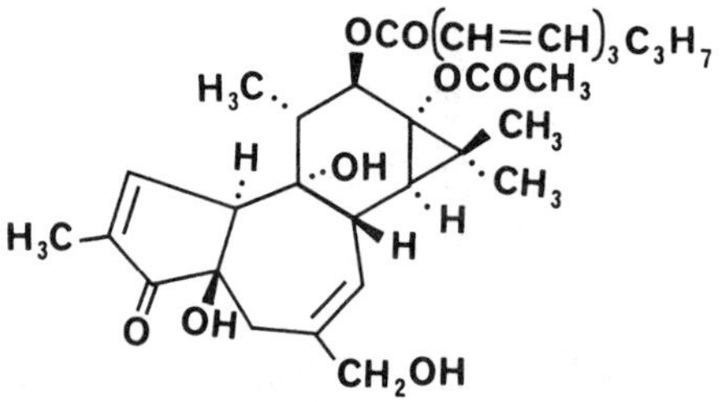

12-*O*-Deca-2,4,6-trienoyl-
phorbol-13-acetate

FIGURE 12 Piscicidal, skin-irritant, and cytotoxic phorbol diester of diverse botanical origin.

Compounds of the phorbol ester type are sensitive to acids and alkalis (Hecker and Schmidt, 1974), and undergo autooxidation even in the solid form (Schmidt and Hecker, 1975). Ingenol esters seem to be the most difficult of this group of biogenetically related diterpenoids to isolate in their naturally occurring form, since they may undergo transesterification to inactive derivatives when chromatographed on silica gel columns or thin-layer plates (Hecker, 1971b; Upadhyay et al., 1976). There seems to be some evidence for the existence of chemical races in regard to the diterpenoid profiles of euphorbiaceous plants, as exemplified by the variations of structure of irritant compounds isolated from *E. tirucalli* grown in three different geographical locations (Fürstenberger and Hecker, 1977a,b; Kinghorn, 1979b).

In the following paragraphs the results of tumor-promoting experiments on isolates of the diterpenoid type from plants of the Euphorbiaceae and Thymelaeaceae will be discussed. Active principles will be dealt with under the categories tigliane, daphnane, and ingenane derivatives, respectively. It should be pointed out that relatively few of the irritant, piscicidal, or antileukemic diterpenoids isolated from these plants have been tested for tumor-promoting activity, and that the majority of the tumor-promotion (Berenblum) experiments performed on these classes of diterpenoids have been carried out by the Hecker group at Heidelberg. Hecker and his colleagues have used a routine bioassay for this purpose on the skin of NMRI mice (Hecker and Bresch, 1965; Hecker, 1968b).

Fourteen tumor-promoting phorbol diesters (Table 5) from croton oil have been isolated, characterized, and biologically tested (Hecker, 1962, 1971a; Hecker et al., 1964b; Hecker and Schmidt, 1974). These compounds occur in two groups, A and B factors, which are short-chain and long-chain 13-*O*-acylates, respectively (Clarke and Hecker, 1965; Hecker and Schairer, 1967). As can be seen from Table 5, the most potent tumor-promoting agent from croton oil on mouse skin (and indeed the most potent cocarcinogen known in this system) is 12-*O*-tetradecanoylphorbol-13-acetate (TPA), which is compound A_1 of Hecker (Hecker et al., 1964a; Hecker and Bresch, 1965) and compound C of Van Duuren (Van Duuren et al., 1963c; Hecker, 1968b). Compounds A_5', B_8', and B_9' (Table 5) are not natural products, but were obtained

TABLE 5 Tumor-Promoting Activity on Mouse Skin of Phorbol Diesters
Obtained from Croton Oil (*Croton tiglium* L.)[a]

Compound	Ester group at 12-position	Ester group at 13-position	Tumor-promoting activity[b]	
			Tumor rate %	Average tumor yield[c] (tumors/survivor)
A_1	Tetradecanoate	Acetate	82	3.6
A_2	Decanoate	Acetate	8	0.3
A_3	Dodecanoate	Acetate	29	0.6
A_4	Hexadecanoate	Acetate	64	2.6
A_5'	Tiglate	Butyrate	14	0.3
B_1	(+)-S-2-Methyl- butyrate	Dodecanoate	71	3.1
B_2	(+)-S-2-Methyl- butyrate	Decanoate	86	6.2
B_3	Tiglate	Decanoate	61	3.0
B_4	Acetate	Dodecanoate	64	2.3
B_5	(+)-S-2-Methyl- butyrate	Octanoate	32	2.0
$^aB_6$	Tiglate	Octanoate	29	1.0
B_7	Acetate	Decanoate	57	2.4
B_8'	Tiglate	Dodecanoate	42	1.0
B_9'	Butyrate	Dodecanoate	32	0.8

[a]The structure of phorbol is presented in Fig. 11.
[b]Doses used: initiator, 0.1 μM 7,12-dimethylbenz[a]anthracene, single dose; promoters, A_1-A_4, B_8', B_9' 0.02 μM, A_5' 0.06 μM, B_1-B_7 10 μg per application.
[c]After 12 weeks (24 applications of tumor promoters).
Source: Hecker, 1971a.

from croton oil by the selective hydrolysis of the C_{20}-acyl groups of their respective naturally occurring phorbol triester "cryptic cocarcinogens" (Hecker, 1971a, 1976b). Experiments that have studied the tumor-promoting activity of phorbol (Fig. 11) and TPA (Table 5) in tissues other than mouse skin have been summarized by Diamond et al. (1980). Interestingly, only 12-*O*-tigloylphorbol-13-decanoate (B_3, Table 5) of the croton oil tumor-promoting factors showed antileukemic activity in mice in the dose ranges used (Kupchan et al., 1976b).

Additional tigliane esters from four species of the Euphorbiaceae have been tested by the Hecker group for cocarcinogenic activity. Several 12-deoxyphorbol esters from *Euphorbia triangularis* Desf. (Gschwendt and Hecker, 1974), and 12-deoxy-16-hydroxyphorbol esters from *E. cooperi* N.E. Br. (Gschwendt and Hecker, 1973) were shown to exhibit marginal or no tumor-promoting activity on mouse skin. In contrast, two irritant factors, F_1 (12-O-hexadecanoyl-16-hydroxyphorbol-13-acetate) and F_2 (12-O-hexadecanoyl-4-deoxy-16-hydroxyphorbol-13-acetate) (Fig. 13), from *Croton flavens* L. were found to demonstrate equivalent cocarcinogenic potency on mouse skin to TPA (Weber and Hecker, 1978.

As over 25 representatives of the tigliane class of diterpenoids have been tested for tumor-promoting activity on mouse skin in Berenblum experiments (Hecker, 1978; Diamond et al., 1980), it is possible to make preliminary conclusions on structural features necessary for activity. Essential functionalities for this biological property appear to include the presence of a 4β-substituent, a C_1, C_2-double bond, a primary hydroxy group at C_{20}, and saturated ester groups at C_{12} or C_{13} of moderate chain length (8-16 carbons) (Hecker, 1978).

Only a limited number of Berenblum experiments have been carried out by the Hecker group on daphnane derivatives, and the data obtained have been reported entirely in various review articles. While the 5β-hydroxyresiniferonol-6α,7α-oxide derivatives huratoxin and simplexin (Fig. 14), and the 5β,12β-dihydroxyresiniferonol-6α,7α-oxide derivative mezerein (Fig. 14) exhibit moderate to considerable mouse-skin tumor-promoting activity relative to TPA, the resiniferonol ester resiniferatoxin (Fig. 14) is inactive in this respect (Hecker, 1971b, 1977, 1978; Adolf and Hecker, 1977). The 1α-alkyldaphnane derivative Pimelea factor P_2 (Fig. 14) also has been found to exhibit a comparable tumor rate to TPA in Berenblum experiments on mouse skin (Adolf and Hecker, 1977; Hecker, 1978). Paradoxically, mezerein exhibits in vivo antileukemic activity in mice (Kupchan and Baxter, 1975). The plants of origin of the compounds mentioned above with tumor-promoting activity on mouse skin are presented in Table 6.

Irritant compounds based on the ingenane skeleton have so far been isolated only from plants of the large euphorbiaceous genus *Euphorbia* (Hecker,

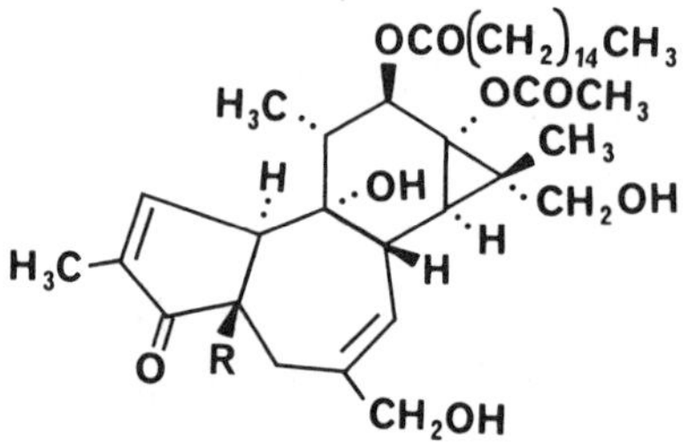

FIGURE 13 Potent mouse skin tumor-promoting agents from *Croton flavens*.

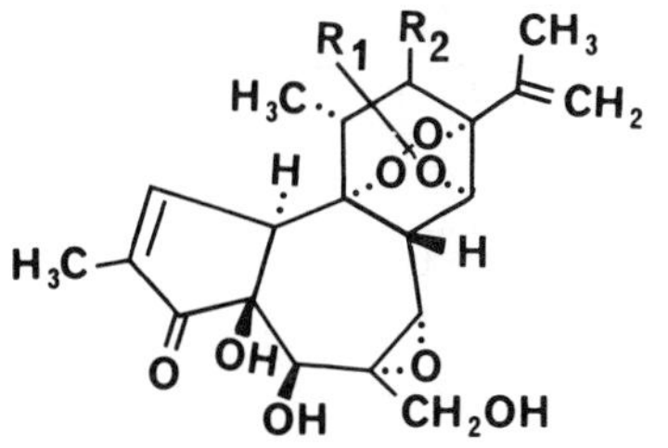

	R_1	R_2
Huratoxin	$-(CH{=}CH)_2(CH_2)_8CH_3$	H
Simplexin	$-(CH_2)_8CH_3$	H
Mezerein	$-(CH\overset{E}{=}CH)_2(CH_2)_8CH_3$	$-OCO(CH\overset{E}{=}CH)_2C_6H_5$

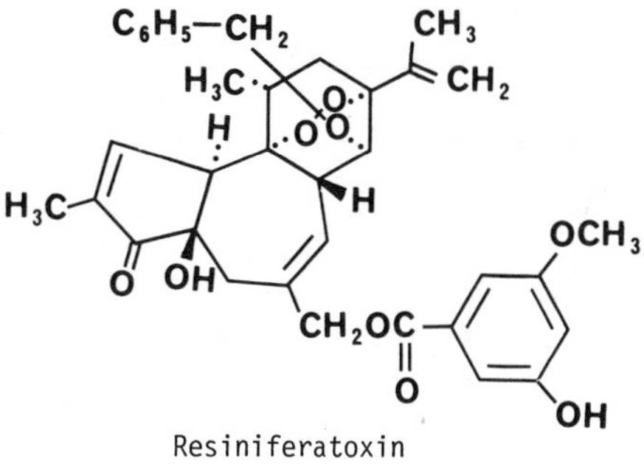

Resiniferatoxin

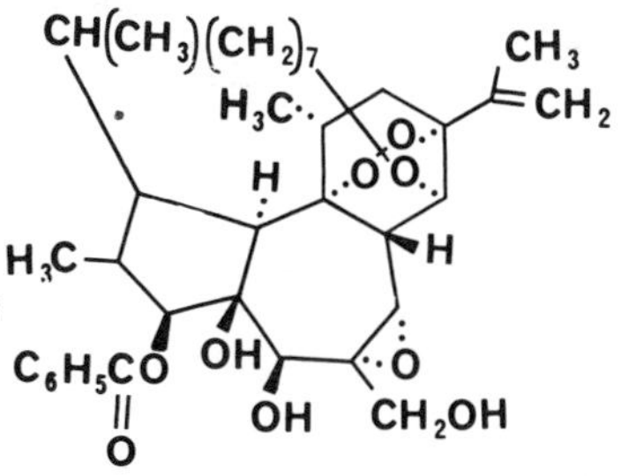

Pimelea factor P_2

FIGURE 14 Daphnane derivatives that have been tested for tumor-promoting activity in Berenblum experiments on mouse skin.

1978). Within this genus, ingenol is the most commonly recorded diterpene, having been reported to occur in nearly 50 species to date (Hecker, 1977, 1978; Evans and Kinghorn, 1977; Upadhyay et al., 1980). The plant origin of the ingenol esters that have been shown to be tumor promoters is indicated in Table 7. Ingenol-3-hexadecanoate (Hecker's compound $L_5 = I_1$) (Fig. 15) has been determined to possess a tumor-promoting potency on mouse skin of about one-tenth that of TPA (Adolf and Hecker, 1975b; Hecker, 1977, 1978). Interestingly, its positional isomer, ingenol-20-hexadecanoate (Hecker's compound $L_4 = I_2$), is devoid of this activity at comparable applied doses (Adolf and Hecker, 1975b; Hecker, 1977, 1978).

Phorbol esters have now become important pharmacological tools, and produce a vast array of cellular and biochemical responses, often at very low dose levels (Sivak, 1978; Diamond et al., 1980; Blumberg, 1980, 1981). The subject of the mechanism of action of cocarcinogens was considered in depth in a recent volume (Slaga et al., 1978). While it is considered that TPA is biologically active in an unmetabolized form (Hecker, 1978), the questions of whether there is a specific cellular receptor for phorbol esters, and how the responses of the cell to the phorbol esters are mediated, remain uncertain

TABLE 6 Plant Origin of Daphnane Esters with Tumor-Promoting Activity on Mouse Skin

Compound	Plant(s) of origin	References
Huratoxin	*Hura crepitans* L.[a]	Sakata et al., 1972
	Hippomane mancinella L.[a]	Adolf and Hecker, 1975a
	Pimelea simplex F. Muell.[b]	Freeman et al., 1979
Simplexin (= pimelea factor P_1)	*P. simplex* F. Muell.[b]	Freeman et al., 1979; Roberts et al., 1975; Zayed et al., 1977b
	P. prostrata Willd.[b]	Zayed et al., 1977b
	P. trichostachya Lindl.[b] Form B	Freeman et al., 1979
Mezerein	*Daphne mezereum* L.[b]	Ronlán and Wickberg, 1970; Kupchan and Baxter, 1975
Pimelea factor P_2 (= daphnopsis factor R_1 = linifolin b)	*P. prostrata* Willd.[b]	Zayed et al., 1977a; Hecker, 1978
	Daphnopsis racemosa[b] Griseb.	Zayed et al., 1977a; Hecker, 1978
	P. linifolia[b,c]	Tyler and Howden, 1981
	P. ligustrina[b,c]	Tyler and Howden, 1981

[a]Euphorbiaceae.
[b]Thymelaeaceae.
[c]Author citation not given in original research report.

TABLE 7 Ingenol Esters Isolated from *Euphorbia* Species with Tumor-Promoting Activity on Mouse Skin

Compound	Plant of origin	References
Ingenol-3-hexadecanoate	*E. lathyris* L.	Adolf and Hecker, 1975b
	E. ingens E. Mey	Opferkuch and Hecker, 1974
	E. serrata L.	Upadhyay et al., 1976
	E. esula L.	Upadhyay et al., 1978
Mixture of ingenol-3-: 2',6'-dimethyloctanoate, 2'-methylnonanoate, 2',6'-dimethylnonoate, 2'-methyldecanoate, 2',6'-dimethyldecanoate, 2'-methylundecanoate	*E. resinifera* Berg.	Hergenhahn et al., 1975

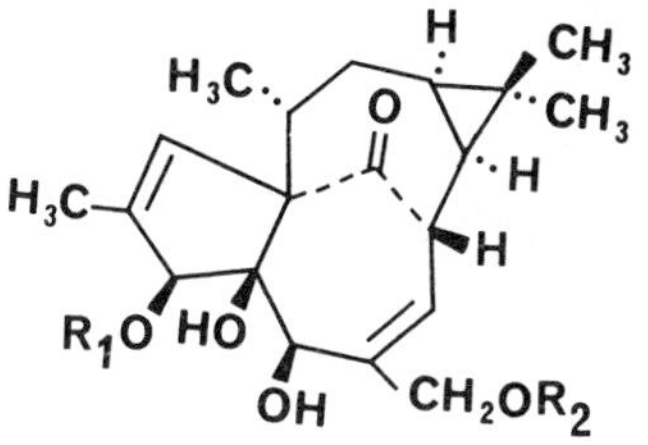

	R_1	R_2
Factor $L_5 = I_1$	Hexadecanoate	H
Factor $L_4 = I_2$	H	Hexadecanoate

FIGURE 15 Two ingenol esters from *Euphorbia lathyris* and *E. ingens*.

(Weinstein and Troll, 1977). Boutwell and co-workers (1978, 1979), Hecker (1979), and Diamond et al. (1980) have recently commented on possible mechanisms of tumor promotion by phorbol esters. Slaga et al. (1980) have provided evidence for the existence of several stages in the phenomenon of tumor promotion.

Hecker (1981) has pointed out that cocarcinogenic diterpenoids of the tigliane, daphnane, and ingenane classes may occur in herbal teas, official plant drugs, in household and garden ornamental plants, and in honey collected from bees that have fed on nectar from euphorbiaceous plants. The possible carcinogenic risk to humans from these compounds may be minimized by refraining from ingesting plants or plant products containing them, and by avoiding extensive contact of these plant parts with the skin (Hecker, 1981).

B. Cyclopropenoid Fatty Acids

A group at Oregon State University first reported the cocarcinogenic effect of cyclopropenoid fatty acids (CPFA) in aflatoxin B_1-induced hepatomas in rainbow trout (*Salmo gairdneri*) in 1966 (Sinnhuber et al., 1966). The incidence and rate of growth of liver tumors after 6 months were greatly increased in fish fed a diet incorporating 200 ppm of the triglycerides of CPFAs from *Sterculia foetida* oil and 4 ppb aflatoxin B_1, compared to the effects of a diet incorporating only the aflatoxin (Sinnhuber et al., 1968). *S. foetida* oil (49% sterculic acid, 7% malvalic acid) (Fig. 16) is a more potent cocarcinogen than *Hibiscus syriacus* oil (2% sterculic acid, 19% malvalic acid) in this system (Lee et al., 1968). Since diets containing 200 ppm epoxyoleic acid (Fig. 16) (in *Vernonia anthelimintica* oil) did not alter the carcinogenic potency of aflatoxin B_1, the need for a cyclopropene ring is indicated, rather than an epoxide functionality, for the expression of cocarcinogenic activity of this type (Lee et al., 1968). CPFAs are also known to enhance hepatic carcinogenesis in rainbow trout initiated by aflatoxin M (Sinnhuber et al., 1974), and methyl sterculate has been shown to act as a tumor promoter when repeatedly fed to rainbow trout that initially received a diet incorporating 20 ppb aflatoxin B_1 for 30 days (Lee et al., 1970).

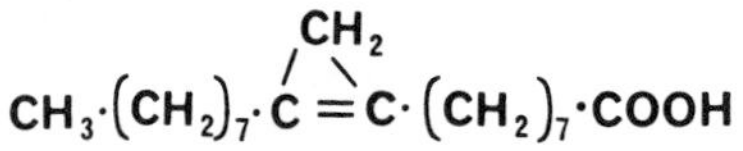

$$CH_3 \cdot (CH_2)_7 \cdot \overset{\displaystyle \overset{CH_2}{\diagup \diagdown}}{C} = C \cdot (CH_2)_7 \cdot COOH$$

Sterculic Acid

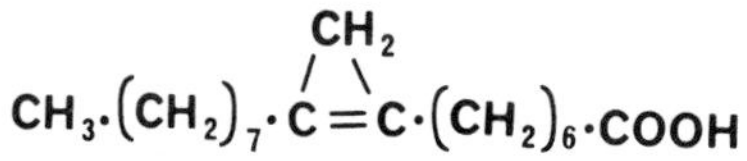

$$CH_3 \cdot (CH_2)_7 \cdot \overset{\displaystyle \overset{CH_2}{\diagup \diagdown}}{C} = C \cdot (CH_2)_6 \cdot COOH$$

Malvalic Acid

$$CH_3 \cdot (CH_2)_4 \cdot \overset{\displaystyle \overset{O}{\diagup \diagdown}}{CH} - CH \cdot CH_2 \cdot CH = CH \cdot (CH_2)_7 \cdot COOH$$

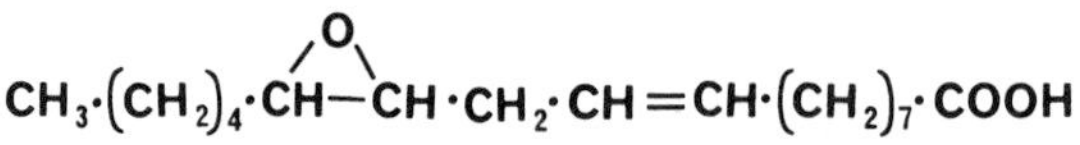

12,13-Epoxyoleic Acid
(Vernolic Acid)

FIGURE 16 Some fatty acids used in cocarcinogenesis experiments on the rainbow trout.

Sterculic acid was more recently found to be a carcinogen per se for rainbow trout at levels in the diet of 45, 135, and 405 ppm, after 8 months of feeding (Sinnhuber et al., 1976). A similar observation was made in rainbow trout fed diets containing glandless cottonseed (*Gossypium hirsuitum*) kernels and lightly processed cottonseed oil. The CPFA glyceride content is about 0.5-1.0% of the total lipids in the seeds of this species (Hendricks et al., 1980).

Neither the synergism in aflatoxin-induced carcinogenesis, nor the carcinogenic effect of sterculic acid, has been convincingly demonstrated in rats (Lee et al., 1969; Wells et al., 1974).

CPFAs have been subjected to a considerable number of biochemical experiments in an attempt to explain the mechanism of their cocarcinogenic activity for trout. For example, these compounds cause significant reductions in liver protein and in the activity of several hepatic dehydrogenase enzymes of rainbow trout (Taylor et al., 1973). They produce cytoplasmic alternations in hepatocytes of *S. gairdneri* (Scarpelli et al., 1974), and increased plasma and liver cholesterol levels, as well as a high incidence of aortic atherosclerosis, in rabbits (Ferguson et al., 1976). Sterculic acid possesses mitogenic activity in rainbow trout hepatocytes, a fact that may explain its cocarcinogenic activity, since mitogenic stimuli are known to potentiate the carcinogenic effect of certain chemical carcinogens (Scarpelli, 1974).

While some commercial salad oils and margarine consumed by humans may contain CPFAs, the human implications of the long-term ingestion of these compounds have not been established, since no data exist on their effects on subhuman primates (Scarpelli, 1974). It is to be noted that the cyclopropene ring is labile during normal cooking procedures, and that biological activity of CPFAs is thus eliminated (Scarpelli, 1974; Hendricks et al., 1980).

C. Essential Oils and Their Constituents

In the first of a series of experiments carried out at the London Hospital Medical College, Roe (1959) demonstrated that the essential oil of sweet orange (*Citrus sinensis*) produced epidermal hyperplasia and papillomas on mouse skin, when applied after a subthreshold dose of 7,12-dimethylbenz[*a*]-anthracene (DMBA). Similar results were obtained using the essential oils of lime, lemon, and grapefruit in a follow-up study (Roe and Peirce, 1960). Lime oil was found to be also an active tumor-promoting agent for the epithelium of the mouse forestomach, after an initiating dose of DMBA or benz-[*a*]pyrene (Peirce, 1961), and the active tumor-promoting principle(s) found to be thermostable (Field and Roe, 1965). Orange oil and d-limonene (Fig. 17), its major constituent, were without apparent tumor-promoting activity on mouse forestomach epithelium (Field and Roe, 1965). The active tumor-promoting agent in sweet orange oil has been hypothesized as the d-limonene impurity, 1-hydroperoxy-1-vinyl-cyclohex-3-ene (Fig. 17) by Homberger and Boger (1968). The latter compound has been shown to be weakly carcinogenic when painted on mouse skin (Van Duuren et al., 1963a).

Further results on the tumor-promoting properties of various essential oils were reported in a review article by Roe and Field (1965). Weak tumor promotion on mouse skin was observed with undiluted turpentine and eucalyptus oils after initiation with DMBA. The major constituents of each oil, α-pinene and β-phellandrene (Fig. 17), respectively, when applied at a 40% concentration in acetone, proved to be weak tumor promoters in this test system (Roe and Field, 1965). Bergamot oil, in contrast, which does not contain high levels of terpene hydrocarbons, was found to be inactive on mouse skin as a promoter after DMBA initiation (Field and Roe, 1965). Cashew nut shell oil (3-5% in acetone), applied once weekly after an initiating dose of DMBA, produced papillomas in mice after 20 weeks of treatment (Roe and Field, 1965).

D. Miscellaneous Plant Tumor-Promoting Agents

A number of structurally unrelated plant constituents have been shown to exhibit tumor-promoting activity on laboratory animals. Several simple widely occurring lipid constituents of plants, including *n*-dodecane, *n*-tetradecane, decanol, dodecanol, dodecanoic acid, and oleic acid (Fig. 18), have been

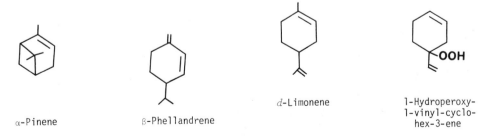

α-Pinene   β-Phellandrene   *d*-Limonene   1-Hydroperoxy-1-vinyl-cyclo-hex-3-ene

FIGURE 17 Constituents of essential oils tested for tumor-promoting or carcinogenic activity.

found to promote the carcinogenic effects of 7,12-dimethylbenz[*a*]anthracene (DMBA) on mouse skin (Horton et al., 1957; Holsti, 1959; Siće, 1966; Opdyke, 1975). *n*-Dodecane and *n*-tetradecane, in addition to *n*-decane (Fig. 18) also exhibited a cocarcinogenic effect on the carcinogenic potential of ultraviolet (UV) light of 254 nm on mouse skin (Bingham and Nord, 1977). In a series of experiments on analogs of the synthetic tumor promoter anthralin, the naphthoquinone, juglone (Fig. 18), was found to exhibit tumor-promoting activity on female ICR/Ha Swiss mouse skin, after a single subcarcinogenic dose of DMBA was applied (Van Duuren et al., 1978). Juglone is found in the green parts of species of the genus *Juglans*, and is readily isolated from green walnut shells (Thompson, 1971). Asiaticoside (Fig. 18), a triterpene glycoside constituent of *Centella asiatica* (L.) Urban, when dissolved in the carcinogen benzene, showed weak tumor-promoting effects on the hairless mouse epidermis, by increasing the yield of papillomas over that obtained using solvent alone (Laernum and Iversen, 1972). The addition of undegraded carrageenan to the diet of F344 rat treated with the carcinogens azoxymethane or methylnitrosourea caused an enhancing effect on the colorectal carcinogenesis produced (Watanabe et al., 1978). Carrageenan is a complex sulfated polysaccharide obtained by extraction with water from members of the red seaweed families Gigartinaceae and Solieraceae (IARC, 1976), that is used in the U.S. dairy industry as a stabilizer, gelling agent, and viscosity control agent (Watanabe et al., 1978).

Several substances used in beverages have been tested for or associated with tumor-promoting activity. The tumor-promoting activity of tea [*Camellia sinensis* (L.) Kuntze.] was demonstrated in young Swiss mice that developed epithelial cell carcinoma of the neck region, after an aqueous extract was repeatedly painted over an area of skin, following a single dose of 1% benz-[*a*]pyrene (Kaiser and Bartone, 1966). Kaiser (1967) later ascribed this activity to the presence of phenols in tea. Chlorogenic acid (Fig. 18), a major constituent of coffee (*Coffea arabica* L.), which under mild conditions is a very powerful catalyst for *N*-nitrosamine formation, has thus been postulated as a possible cocarcinogen (Challis and Bartlett, 1975). However, the distillation residues of sake and Scotch whisky, when used in two-stage carcinogenesis experiments with DMBA on Wistar rats and CF_1 and ddN mice, did not statistically increase tumor development over controls (Kuratsune et al., 1971).

Finally, the tumor-promoting activities of tobacco and hashish have been studied. Mention has been made earlier in this chapter of the chewing of tobacco (*Nicotiana tabacum* L.) as a constituent of betel quid. There is now ample experimental evidence that extracts of unburned tobacco possess tumor-promoting activity (Van Duuren et al., 1966; Bock, 1968). For example, Bock et al. (1964) showed that over one-third of a group of ICR Swiss mice, treated sequentially with a tumor-initiating dose of DMBA and repeated doses of an aqueous extract of unburned American tobacco, developed skin tumors within 26 weeks. Bock and his co-workers have found that a combination of two constituents of tobacco extract, nicotine (Fig. 18), and an uncharacterized compound with a high molecular weight (>13,000), are necessary to elicit this tumor-promoting effect (Bock, 1980; Bock and Clausen, 1980). Since many of the cocarcinogenic and tumor-promoting agents of tobacco smoke (Van Duuren and Goldschmidt, 1976; Hecht et al., 1975) may not be natural products, their discussion is outside the scope of this chapter. In the case of hashish

$CH_3 \cdot (CH_2)_8 \cdot CH_3$ $CH_3 \cdot (CH_2)_{10} \cdot CH_3$ $CH_3 \cdot (CH_2)_{12} \cdot CH_3$

n-Decane   *n*-Dodecane   *n*-Tetradecane

$CH_3 \cdot (CH_2)_8 \cdot CH_2OH$ $CH_3 \cdot (CH_2)_{10} CH_2OH$

1-Decanol   1-Dodecanol

$CH_3 \cdot (CH_2)_{10} \cdot COOH$ $CH_3 \cdot (CH_2)_7 \cdot CH:CH \cdot (CH_2)_7 \cdot COOH$

Dodecanoic Acid   Oleic Acid

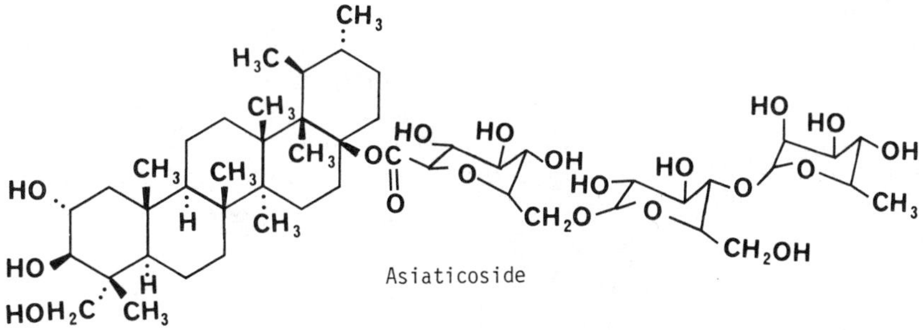

Asiaticoside

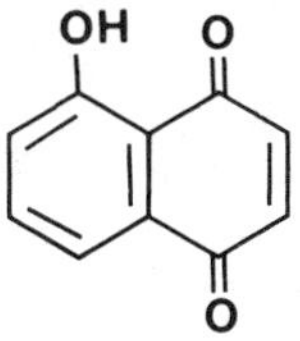

Juglone

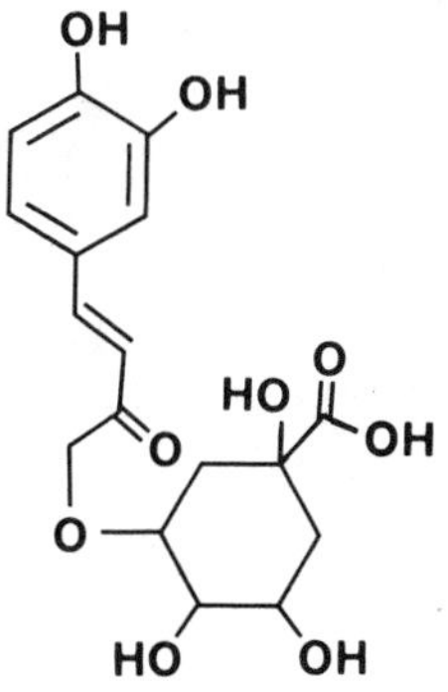

Chlorogenic Acid

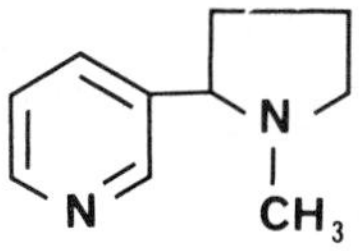

Nicotine

FIGURE 18 Miscellaneous plant-derived known or suspected cocarcinogens.

(*Cannabis sativa* L.), an acetone extract, repeatedly applied to mouse skin after a subcarcinogenic dose of DMBA, caused a high incidence of fibro-epithelial and papillomatous neoplasms (Procter et al., 1974).

IV. CONCLUSIONS

In this chapter over 20 carcinogens and about 30 cocarcinogenic toxins that are plant constituents have been reported. Many of these compounds are potent, as expressed by their activity in long-term tests on rodents, as

exemplified by the carcinogens cycasin, certain polycyclic aromatic hydrocarbons, and the pyrrolizidine alkaloids, and the cocarcinogenic diterpenoids
of the tigliane and daphnane types. There is every reason to believe that
many of the presently known pyrrolizidine alkaloids and diterpenes of the
Euphorbiaceae and Thymelaeaceae will also prove to be active if subjected to
the necessary testing procedures.

Considering that plant carcinogens were unknown just over 30 years ago,
there has been remarkable progress in their chemical recognition, which has
been aided by several valuable separatory and spectroscopic techniques
routinely available in many laboratories. However, much remains to be done
in this respect, especially the identification of the carcinogens in betel quid
and bracken fern, and the cocarcinogens in unburned tobacco and hashish.
Also, the precise structural identification of carcinogenic tannins, of tea and
other plant sources, needs to be performed. The mechanisms of action of
certain plant carcinogens are now better understood, and carcinogenic metabolites of several carcinogens, notably of safrole, have been identified.

Since some plant constituents have recently been shown to be carcinogenic
for nonhuman primates, the carcinogenic risk that the plants mentioned in
this chapter present to humans cannot be understated. A number of plant
products which contain carcinogenic and cocarcinogenic toxins have been
described herein. Contact with, and ingestion of, these plants should therefore be avoided, to minimize the potential health hazards afforded by these
materials.

NOTE ADDED IN PROOF

Since this chapter was written, significant papers have appeared in the literature detailing the structures and biological activities of two classes of tumor-
promoting agent from specimens of the blue-green alga, *Lyngbya majuscula*, a
causative organism of "swimmer's itch" in Hawaii, where it is indigenous
(Fujiki et al., 1981; Sugimura, 1982). Thus, lyngbyatoxin A, an indole alkaloid, and the polyacetate derivative aplysiatoxin, have been found to be as
potent as TPA in mouse-skin two-stage carcinogenesis experiments after initiation with DMBA (Sugimura, 1982). Lyngbyatoxin A is closely related in structure to the isomeric compounds teleocidins A and B, of bacterial origin, which
have also recently been established as potent tumor-promoting agents in
mouse skin (Sugimura, 1982).

REFERENCES

Adolf, W., and Hecker, E. (1975a). On the irritant and cocarcinogenic
    principles of *Hippomane mancinella*. *Tetrahedron Lett.* *1975*:1587-1590.
Adolf, W., and Hecker, E. (1975b). On the active principles of the spurge
    family. III. Skin irritant and cocarcinogenic factors from the caper
    spurge. *Z. Krebsforsch.* *84*:325-344.
Adolf, W., and Hecker, E. (1977). Diterpenoid irritants and cocarcinogens
    in Euphorbiaceae and Thymelaeaceae: structural relationships in view of
    their biogenesis. *Isr. J. Chem.* *16*:75-83.

Adolf, W., and Hecker, E. (1980). New irritant diterpene-esters from roots of *Stillingia sylvatica* L. (Euphorbiaceae). *Tetrahedron Lett. 1980*:2887-2890.

Allen, J. R., Hsu, I.-C., and Carstens, L. A. (1975). Dehydroretronecine-induced rhabdomyosarcomas in rats. *Cancer Res. 35*:997-1002.

Arjungi, K. N. (1976). *Areca* nut. A review. *Arzneim.-Forsch. 26*:951-956.

Armstrong, B., Stevens, N., and Doll, R. (1974). Retrospective study of the association between use of *Rauwolfia* derivatives and breast cancer in English women. *Lancet 2*(7882):672-675.

Arroyo, E. R., and Holcomb, J. (1965). Structural studies of an active principle from *Croton tiglium* L. *J. Med. Chem. 8*:672-675.

Atkinson, L., Chester, I. C., Smyth, F. G., and ten Seldam, R. E. J. (1964). Oral cancer in New Guinea. A study in demography and etiology. *Cancer 17*:1289-1298.

Austwick, P., and Mattocks, R. (1979). Naturally occurring carcinogens in food. *Chem. Ind. (Lond.) 1979*:76-83.

Bababunmi, E. A., Uwaifo, A. O., and Bassir, O. (1978). Hepatocarcinogens in Nigerian foodstuffs. *World Rev. Nutr. Diet. 28*:188-209.

Berenblum, I. (1941a). The cocarcinogenic action of croton resin. *Cancer Res. 1*44-48.

Berenblum, I. (1941b). The mechanism of carcinogenesis. A study of the significance of cocarcinogenic action and related phenomena. *Cancer Res. 1*:807-814.

Berenblum, I., and Shubik, P. (1947). A new quantitative approach to the study of the stages of chemical carcinogenesis in the mouse's skin. *Br. J. Cancer 1*:383-391.

Bharadwaj, V. P., Takayama, S., Yamada, T., and Tanimura, A. (1975). N'-Nitrosonornicotine in Japanese tobacco products. *Gann 66*:585-586.

Biernoth, G., and Rost, H. E. (1967). The occurrence of polycyclic aromatic hydrocarbons in coconut oil and their removal. *Chem. Ind. (Lond.) 1967*:2002-2003.

Bingham, E., and Nord, P. J. (1977). Cocarcinogenic effects of *n*-alkanes and ultraviolet light on mice. *J. Natl. Cancer Inst. 58*:1099-1101.

Birdsey, M. R. (1972). A brief description of the cycads. *Fed. Proc. 31*:1467-1469.

Blumberg, P. M. (1980). In vitro studies on the mode of action of the phorbol esters, potent tumor promoters: Part 1. *CRC Crit. Rev. Toxicol. 8*:153-197.

Blumberg, P. M. (1981). In vitro studies on the mode of action of the phorbol esters, potent tumor promoters: Part 2. *CRC Crit. Rev. Toxicol. 8*:199-234.

Bock, F. G. (1968). The nature of tumor-promoting agents in tobacco products. *Cancer Res. 28*:2363-2368.

Bock, F. G. (1980). Cocarcinogenic properties of nicotine. *Banbury Rep.: A Safe Cigarette? 3*:129-139.

Bock, F. G., and Clausen, D. F. (1980). Further fractionation and co-promoting activity of the large molecular weight compounds of aqueous tobacco extracts. *Carcinogenesis (N.Y.) 1*:317-321.

Bock, F. G., Moore, G. E., and Crouch, S. K. (1964). Tumor-promoting activity of extracts of unburned tobacco. *Science 145*:831-833.

Boutwell, R. K. (1978). Biochemical mechanism of tumor promotion. In *Carcinogenesis*, Vol. 2: *Mechanisms of Tumor Promotion and Cocarcinogenesis*, T. J. Slaga, A. Sivak, and R. K. Boutwell (Eds.). Raven Press, New York, pp. 49-58.

Boutwell, R. K., O'brien, T. G., Verma, A. K., Weekes, R. G., De Young, L. M., Ashendel, C. L., and Astrup, E. G. (1979). On the biochemical mechanisms of tumor promotion in mouse skin. In *Naturally Occurring Carcinogens—Mutagens and Modulators of Carcinogenesis*, E. C. Miller, J. A. Miller, I. Hirono, T. Sugimura, and S. Takayama (Eds.). University Park Press, Baltimore, pp. 287-300.

Borchert, P., Miller, J. A., Miller, E. C., and Shires, T. K. (1973a). 1'-Hydroxysafrole, a proximate carcinogenic metabolite of safrole in the rat and mouse. *Cancer Res. 33*:590-600.

Borchert, P., Wislocki, P. G., Miller, J. A., and Miller, E. C. (1973b). The metabolism of the naturally occurring hepatocarcinogen safrole to 1'-hydroxysafrole and the electrophilic reactivity of 1'-acetoxysafrole. *Cancer Res. 33*:575-589.

Boyland, E. (1968). The possible carcinogenic action of alkaloids of tobacco and betel nut. *Planta Med. 16*(Suppl.):13-23.

Boyland, E., and Nery, R. (1968). Mercapturic acid formation during the metabolism of arecoline and arecaidine in the rat. *Biochem. J. 113*:123-130.

Boyland, E., Roe, F. J. C., and Gorrod, J. W. (1964). Induction of pulmonary tumors in mice by nitrosonornicotine, a possible constituent of tobacco smoke. *Nature (Lond.) 202*:1126.

British Medical Journal (1976). Caffeine, coffee, and cancer. *1*(1607): 1031-1032.

Bull, L. B., Culvenor, C. C. J., and Dick, A. T. (1968). *The Pyrrolizidine Alkaloids*. North-Holland, Amsterdam.

Campbell, M. E., Mickelsen, O., Yang, M. G., Laquer, G. L., and Kereszetesy, J. C. (1966). Effects of strain, age and diet on the response of rats to the ingestion of *Cycas circinalis*. *J. Nutr. 88*:115-124.

Challis, B. C., and Bartlett, C. D. (1975). Possible cocarcinogenic effects of coffee constituents. *Nature (Lond.) 254*:532-533.

Chemical Engineering News (1980). Science/technology concentrates. *September 22*, p. 17.

Christensen, H. E., Fairchild, E. J., and Lewis, R. J. (Eds.) (1976). *Suspected Carcinogens*, 2nd ed. A Subfile of the NIOSH Registry of Toxic Effects of Chemical Substances. U.S. Government Printing Office, Washington, D.C.

Clarke, E., and Hecker, E. (1965). On the active principles of croton oil. V. Purification and characterisation of further irritant and cocarcinogenic compounds of the B group. *Z. Krebsforsch. 67*:197-204.

Cook, J. W., Duffy, E., and Schoental, R. (1950). Primary liver tumors in rats following feeding with alkaloids of *Senecio jacobaea*. *Br. J. Cancer 4*: 405-410.

Cooper, J. M. (1941). Isolation of a toxic principle from the seeds of *Macrozamia spiralis*. *J. Proc. R. Soc. N.S. Wales 74*:450-454 (through *Chem. Abstr. 35*:4917-1, 1941).

Cooper, P. (1980). More chapters on the bracken saga. *Food Cosmet. Toxicol. 18*:311-312.

Crout, D. H. C. (1976). The pyrrolizidine alkaloids, their physiological activity and biosynthesis in higher plants. *Chimia 30*:270-271.

Culvenor, C. C. J. (1968). Tumor-inhibitory activity of pyrrolizidine alkaloids. *J. Pharm. Sci. 57*:1112-1117.

Culvenor, C. C. J., Downing, D. T., Edgar, J. A., and Jago, M. V. (1969). Pyrrolizidine alkaloids as alkylating and antimitotic agents. *Ann. N.Y. Acad. Sci. 163*:837-847.

Culvenor, C. C. J., Edgar, J. A., Jago, M. V., Outteridge, A., Peterson, J. E., and Smith, L. W. (1976). Hepato- and pneumotoxicity of pyrrolizidine alkaloids and derivatives in relation to molecular structure. *Chem. Biol. Interact. 12*:299-324.

Deinzer, M. L., Thomson, P. A., Burgett, D. M., and Isaacson, D. L. (1977). Pyrrolizidine alkaloids: their occurrence in honey from tansy ragwort (*Senecio jacobaea* L.). *Science 195*:497-499.

De Luca, P., Moretti, A., Sabato, S., and Gigliano, G. S. (1980). The ubiquity of cycasin in cycads. *Phytochemistry 19*:2230-2231.

Diamond, L., O'Brien, T. G., and Baird, W. M. (1980). Tumor promoters and the mechanism of tumor promotion. *Adv. Cancer Res. 32*:1-74.

Dickens, F., and Jones, H. E. H. (1963). Further studies on the carcinogenic and growth-inhibitory activity of lactones and related substances. *Br. J. Cancer 17*:100-108.

Dickinson, J. O., Cooke, M. P., King, R. R., and Mohamed, P. A. (1976). Milk transfer of pyrrolizidine alkaloids in cattle. *J. Am. Vet. Med. Assoc. 169*:1192-1196.

Drinkwater, N. R., Miller, E. C., Miller, J. A., and Pitot, H. C. (1976). Hepatocarcinogenicity of estragole (1-allyl-4-methoxybenzene) and 1'-hydroxyestragole in the mouse and mutagenicity of 1'-acetoxyestragole in bacteria. *J. Natl. Cancer Inst. 57*:1323-1331.

Druckrey, H., Preussmann, R., Ivankovic, S., and Schmäll, D. (1967). Organotrope carcinogene Wirkungen bie 65 verschiedenen N-Nitroso-Verbindungen an BD-Ratten. *Z. Krebsforsch. 69*:103-201.

Dunham, L. J. (1968). A geographic study of the relationship between oral cancer and plants. *Cancer Res. 28*:2369-2371.

Dunham, L. J., Muir, C. S., and Hamner, J. E., III (1966). Epithelial atypia in hamster cheek pouches repeatedly treated with calcium hydroxide. *Br. J. Cancer 20*:588-593.

Dunham, L. J., Sheets, R. H., and Morton, J. F. (1974). Proliferative lesions in cheek pouch and esophagus of hamsters treated with plants from Curacao, Netherland Antilles. *J. Natl. Cancer Inst. 53*:1259-1269.

Du Plessis, L. S., Nunn, J. R., and Roach, W. A. (1969). Carcinogen in a Transkeian food additive. *Nature (Lond.) 222*:1198-1199.

Epstein, S. S., Fujii, K., Andrea, J., and Mantel, N. (1970). Carcinogenicity testing of selected food additives by parenteral administration to infant Swiss mice. *Toxicol. Appl. Pharmacol. 16*:321-334.

Evans, F. J., and Kinghorn, A. D. (1977). A comparative phytochemical study of the diterpenes of some species of the genera *Euphorbia* and *Elaeophorbia* (Euphorbiaceae). *Bot. J. Linn. Soc. (Lond.) 74*:23-25.

Evans, F. J., and Soper, C. J. (1978). The tigliane, daphnane and ingenane diterpenes, their chemistry, distribution and biological activities. *Lloydia 41*:193-233.

Evans, I. A. (1968). The radiomimetic nature of bracken toxin. *Cancer Res. 28*:2252-2261.

Evans, I. A. (1976). The bracken carcinogen. In *Chemical Carcinogens*, C. E. Searle (Ed.). ACS Monogr. 173. American Chemical Society, Washington, D.C., pp. 690-700.

Evans, I. A., and Mason, J. (1965). Carcinogenic activity of bracken. *Nature (Lond.) 208*:913-914.

Evans, I. A., and Osman, M. A. (1974). Carcinogenicity of bracken fern and shikimic acid. *Nature (Lond.) 250*:348-349.

Evans, I. A., Widdop, B., Jones, R. S., Barber, G. D., Leach, H., Jones, D. L., and Mainwaring-Burton, R. (1971). The possible human hazard of the naturally occurring bracken carcinogen. *Biochem. J. 124*:28P-29P.

Evans, I. A., Jones, R. S., and Mainwaring-Burton, R. (1972). Passage of bracken fern toxicity into milk. *Nature (Lond.) 273*:107-108.

Farnsworth, N. R., Bingel, A. S., Fong, H. H. S., Saleh, A. A., Christenson, G. M., and Saufferer, S. M. (1976). Oncogenic and tumor-promoting spermatophytes and pteridophytes and their active principles. *Cancer Treat. Rep. 60*:1171-1214.

Feldmann, E. G. (1980). Carcinogenicity testing—a national tragedy [editorial]. *J. Pharm. Sci. 69(2)*:I.

Fendell, L. D., and Smith, J. R. (1970). Betel-nut-associated cancer: report of case. *J. Oral Surg. 28*:455-456.

Ferguson, T. L., Wales, J. H., Sinnhuber, R. O., and Lee, D. J. (1976). Cholesterol levels, atherosclerosis and liver morphology in rabbits fed cyclopropenoid fatty acids. *Food Cosmet. Toxicol. 14*:15-18.

Field, W. E. H., and Roe, F. J. C. (1965). Tumor promotion in the forestomach epithelium of mice by oral administration of citrus oils. *J. Natl. Cancer Inst. 35*:771-787.

Fong, H. H. S., Bhatti, W., and Farnsworth, N. R. (1972). Antitumor activity of certain plants due to tannins. *J. Pharm. Sci. 61*:1818.

Forbes, P. D., Davies, R. E., and Urbach, F. (1976). Phototoxicity and photocarcinogenesis: comparative effects of anthracene and 8-methoxypsoralen in the skin of mice. *Food Cosmet. Toxicol. 14*:303-306.

Freeman, P. W., Ritchie, E., and Taylor, W. C. (1979). The constituents of Australian *Pimelea* spp. I. The isolation and structure of the toxin of *Pimelea simplex* and *P. trichostachya* form B responsible for St. George disease of cattle. *Aust. J. Chem. 32*:2495-2506.

Fujiki, H., Mori, M., Nakayasu, M., Terada, M., Sugimura, T., and Moore, R. E. (1981). Indole alkaloids: Dihydroteleocidin B, teleocidin, and lyngbyatoxin A as members of a new class of tumor promoters. *Proc. Natl. Acad. Sci. USA 78*:3872-3876.

Fukunishi, R., Watanabe, K., Terashi, S.-I., and Kawaji, K. (1971). Shift of target organs in chemical carcinogenesis. *Gann 62*:353-358.

Fukunishi, R., Terashi, S.-I., Watanabe, K., and Kawaji, K. (1972). High yield of hepatic tumors in rats by cycasin. *Gann 63*:575-578.

Fukuoka, M., Kuroyanagi, M., Yoshihira, K., and Natori, S. (1978). Chemical and toxicological studies on bracken fern, *Pteridium aquilinum* var. *latiusculum*. II. Structures of pterosins, sesquiterpenes having 1-indanone skeleton. *Chem. Pharm. Bull. 26*:2365-2385.

Fürstenberger, G., and Hecker, E. (1972). Zum Wirkungsmechanismus cocarcinogener Pflanzeninhaltsstoffe. *Planta Med. 22*:241-266.

Fürstenberger, G., and Hecker, E. (1977a). New highly irritant euphorfactors from latex of *Euphorbia triucalli* L. *Experientia 33*:986-988.

Fürstenberger, G., and Hecker, E. (1977b). The new diterpene 4-deoxy-phorbol and its highly unsaturated irritant diesters. *Tetrahedron Lett.* *1977*:925-928.

Gosálvez, M., and Merchán, J. (1973). Induction of rat mammary adenomas with the respiratory inhibitory rotenone. *Cancer Res.* *33*:3047-3050.

Gosálvez, M., and Díaz-Gil, J. J. (1978). Rotenone: a possible environmental carcinogen? *Eur. J. Cancer* *14*:1403-1404.

Gosálvez, M., and Díaz-Gil, J., Coloma, J., and Salganicoff, L. (1977). Spectral and metabolic characteristics of mitochondrial fractions from rotenone-induced tumours. *Br. J. Cancer* *36*:243-253.

Gosálvez, M., Díaz-Gil, J., Alcañiz, J., and Borrell, J. (1979). A possible pathogenic mechanism for the induction of rotenone tumours. *Biochem. Soc. Trans.* *7*:113-115.

Gräf, W,and Diehl,H.(1966). Über den naturbedingten Normalpegel Kanzerogener polcyclischer Aromate und sein Ursache. *Arch. Hyg.* *150*:49-59.

Grant, V. (1963). *The Origin of Adaptions.* Columbia University Press, N.Y.

Grasso, P., and O'Hare, C. (1976). Carcinogens in food. In *Chemical Carcinogens*, C. E. Searle (Ed.). ACS Monogr. 173. American Chemical Society, Washington, D.C., pp. 701-728.

Griepentrog, F. (1973). Pathological-anatomical results on the effect of coumarin in animal experiments. *Toxicology 1*:93-102.

Grimmer, G., and Hildebrandt, A. (1967). Content of polycyclic hydrocarbons in crude vegetable oils. *Chem. Ind. (Lond.).* *1967*:2000-2002.

Grimmer, G., and Hildebrandt, A. (1968). Kohlenwasserstoffe in der Umgebung des Menschen. VI. Mitteilung: Der Gehalt polycyclischer Kohlenwasserstoffe in rohen Pflanzenölen. *Arch. Hyg.* *152/3*:255-259.

Gross, H. A., Jones, W. I., Cook, E. L., and Boone, C. C. (1967). Carcinogenicity of oil of calamus. *Proc. Am. Assoc. Cancer Res.* *8*:24.

Gschwendt, M., and Hecker, E. (1973). Über die Wirkstoffe der Euphorbiaceen. I. Hautreizende und cocarcinogene Faktoren aus *Euphorbia cooperi* N.E. Br. *Z. Krebsforsch.* *80*:335-350.

Gschwendt, M., and Hecker, E. (1974). Über die Wirkstoffe der Euphorbiaceen. II. Hautreizende und cocarcinogene Faktoren aus *Euphorbia triangularis* Desf. *Z. Krebsforsch.* *81*:193-210.

Guenther, E., and Althausen, D. (1949). *The Essential Oils*, Vol. 2. *The Constituents of Volatile Oils.* D. Van Nostrand, Princeton, N.J.

Gunasekera, S. P., Cordell, G. A., and Farnsworth, N. R. (1979). Potential anticancer agents. XIV. Isolation of spruceanol and montanin from *Cunuria spruceana* (Euphorbiaceae). *J. Nat. Prod.* *42*:658-662.

Gunasekera, S. P., Kinghorn, A. D., Cordell, G. A., and Farnsworth, N. R. (1981). Potential anticancer agents. XIX. Constituents of *Aquilaria malaccensis*. *J. Nat. Prod.* *44*:569-572.

Gusek, W., and Mestwerdt, W. (1969). Cycasin-induced renal tumors in the Wistar rat, with special consideration of adenomas. *Beitr. Pathol. Anat.* *139*:199-218.

Hagan, E. C., Jenner, P. M., Jones, W. I., Fitzhugh, O. G., Long, E. L., Brouwer, J. G., and Webb, W. K. (1965). Toxic properties of compounds related to safrole. *Toxicol. Appl. Pharmacol.* *7*:18-24.

Hagan, E. C., Hansen, W. H., Fitzhugh, O. G., Jenner, P. M., Jones, W. I., Taylor, J. M., Long, E. L., Nelson, A. A., and Brouwer, J. G. (1967). Food flavorings and compounds of related structure. II. Subacute and chronic toxicity. *Food Cosmet. Toxicol.* *5*:141-157.

Hakim, R. E., Griffin, A. C., and Knox, J. M. (1960). Erythema and tumor formation in methoxsalen-treated mice exposed to fluorescent light. *Arch. Dermatol.* 82:572-577.

Hakim, S. A. E. (1968). Sanguinarine—a carcinogenic contaminant in Indian edible oils. *Indian J. Cancer* 5:183-197.

Hakim, S. A. E., Miković, V., and Walker, J. (1961). Experimental transmission of sanguinarine in milk: detection of a metabolic product. *Nature (Lond.)* 189:201-204.

Hamner, J. E., III (1972). Betel quid inducement of epithelial atypia in the buccal mucosa of baboons. *Cancer* 30:1001-1005.

Hamner, J. E., III, and Reed, O. M. (1972). Betel quid carcinogenesis in the baboon. *J. Med. Primatol.* 1:75-85.

Harris, P. N., and Chen, K. K. (1970). Development of hepatic tumors in rats following ingestion of *Senecio longilobus*. *Cancer Res.* 30:2882-2886.

Haslam, E. (1976). *Chemistry of Vegetable Tannins*. Academic Press, London.

Hecht, S. S., Thorne, R. L., Marinpot, R. R., and Hoffmann, D. (1975). A study of tobacco carcinogenesis. XIII. Tumor-promoting substances of the weakly acidic fraction. *J. Natl. Cancer Inst.* 55:1329-1336.

Hecht, S. S., Chen, C.-H. B., Hirota, N., Ornaf, R. M., Tso, T. C., and Hoffmann, D. (1978). Tobacco-specific nitrosamines: formation from nicotine in vitro and during tobacco curing and carcinogenicity in strain A mice. *J. Natl. Cancer Inst.* 60:819-824.

Hecker, E. (1962). The toxic, inflammatory, and co-carcinogenic components of croton oil. *Angew Chem. Int. Ed.* 1:602.

Hecker, E. (1963). Über die Wirkstoffe des Crotonöls. I. Biologische Teste zur quantitativen Messung der entzündlichen, cocarcinogenen und toxischen Wirkung. *Z. Krebsforsch.* 65:325-333.

Hecker, E. (1968a). Cocarcinogenic principles from the seed oil of *Croton tiglium* and from other Euphorbiaceae. *Cancer Res.* 28:2338-2349.

Hecker, E. (1968b). Cocarcinogene Wirkstoffe aus Euphorbiaceen. *Planta Med.* 16(Suppl.):24-45.

Hecker, E. (1971a). Isolation and characterization of the cocarcinogenic principles from croton oil. In *Methods of Cancer Research*, Vol. 6, H. Busch (Ed.). Academic Press, New York, pp. 439-484.

Hecker, E. (1971b). Cocarcinogens from Euphorbiaceae and Thymelaeaceae. In *Pharmacognosy and Phytochemistry*, H. Wagner and L. Hörhammer (Eds.). Springer-Verlag, Berlin, pp. 147-165.

Hecker, E. (1976a). Definitions and terminology in cancer (tumor) etiology. An analysis aiming at proposals for a current internationally standardized terminology. *Z. Krebsforsch.* 86:219-230; *Int. J. Cancer* 18:122-129; *Gann* 67:471-481; *Bull. WHO* 54:1-10.

Hecker, E. (1976b). Aspects of cocarcinogenesis. In *Scientific Foundations of Oncology*, T. Symington and R. L. Carter (Eds.). Heinmann, London, pp. 310-318.

Hecker, E. (1977). New toxic, irritant and cocarcinogenic diterpene esters from Euphorbiaceae and from Thymelaeaceae. *Pure Appl. Chem.* 49:1423-1431.

Hecker, E. (1978). Structure-activity relationships in diterpene esters irritant and cocarcinogenic to mouse skin. In *Carcinogenesis*, Vol. 2: *Mechanisms of Tumor Promotion and Cocarcinogenesis*, T. J. Slaga, A. Sivak, and R. K. Boutwell (Eds.). Raven Press, New York, pp. 11-48.

Hecker, E. (1979). Diterpene ester type modulators of carcinogenesis—new findings in the mechanism of chemical carcinogenesis and in the etiology of human tumors. In *Naturally Occurring Carcinogens—Mutagens and Modulators of Carcinogenesis*, E. C. Miller, J. A. Miller, I. Hirono, T. Sugimura, and S. Takayama (Eds.). University Park Press, Baltimore, pp. 263-286.

Hecker, E. (1981). Cocarcinogenesis and tumor promoters of the diterpene ester type as possible carcinogenic risk factors. *J. Cancer Res. Clin. Oncol. 99*:103-124.

Hecker, E., and Bresch, H. (1965). Über die Wirkstoffe des Crotonöls. III. Reindarstellung und Charakterisierung eines toxisch, entzündlich und cocarcinogen hochaktiven Wirkstoffes. *Z. Naturforsch. 20B*:216-226.

Hecker, E., and Schairer, H. U. (1967). Über die Wirkstoffe des Corotonöls. VIII. Verbessertes Isolierungsverfahren für die Wirkstoffgruppen A und B sowie-Isolierung und Charakterisierung weiterer Wirkstoffe der Gruppe A. *Z. Krebsforsch. 70*:1-12.

Hecker, E., and Schmidt, R. (1974). Phorbolesters—the irritants and cocarcinogens of *Croton tiglium* L. *Fortsch. Chem. Org. Naturst. 31*:377-467.

Hecker, E., Bresch, H., and v. Szczepanski, C. (1964a). Cocarcinogen A_1—the first pure, highly active constituent of croton oil. *Angew Chem. Int. Ed. 3*:227.

Hecker, E., Kubinyi, H., and Bresch, H. (1964b). A new group of co-carcinogens from croton oil. *Angew. Chem. Int. Ed. 3*:747-748.

Hendricks, J. D., Sinnhuber, R. O., Loveland, P. M., Pawlowski, N. E., and Nixon, J. E. (1980). Hepatocarcinogenicity of glandless cottonseeds and cottonseed oil to rainbow trout (*Salmo gairdnerii*). *Science 208*:309-311.

Hergenhahn, M., Kusumoto, S., and Hecker, E. (1974). Diterpene esters from *Euphorbium* and their irritant and cocarcinogenic activity. *Experientia 30*:1438-1440.

Hickey, M. J. (1948). Investigation of the chemical constituents of Brazilian sassafras oil. *J. Org. Chem. 13*:443-446.

Hikino, H., Takahashi, T., Arihara, S., and Takemoto, T. (1970). Structure of pteroside B, glycoside of *Pteridium aquilinum* var. *latiusculum*. *Chem. Pharm. Bull. 18*:1488-1489.

Hilfrich, J., Hecht, S. S., and Hoffmann, D. (1977). Effects of N'-nitrosonornicotine and N'-nitrosoanabasine in Syrian golden hamsters, *Cancer Lett. 2*:169-176.

Hirono, I. (1972). Carcinogenicity and neurotoxicity of cycasin with special reference to species differences. *Fed. Proc. 31*:1493-1499.

Hirono, I. (1979). Naturally-occurring carcinogenic substances. *Gann Monogr. Cancer Res. 24*:85-102.

Hirono, I. (1981). Natural carcinogenic products of plant origin. *CRC Crt. Rev. Toxicol. 8*:235-277.

Hirono, I., and Shibuya, C. (1970). High incidence of pulmonary tumors in dd mice by a single injection of cycasin. *Gann 61*:403-407.

Hirono, I., Laquer, G. L., and Spatz, M. (1968a). Tumor induction in Fischer and Osborne-Mendel rats by a single administration of cycasin. *J. Natl. Cancer Inst. 40*:1003-1010.

Hirono, I., Laquer, G. L., and Spatz, M. (1968b). Transplantability of cycasin-induced tumors in rats, with emphasis on nephroblastomas. *J. Natl. Cancer Inst. 40*:1011-1025.

Hirono, I., Shibuya, C., and Fushimi, K. (1969). Tumor induction in
C57BL/6 mice by a single administration of cycasin. *Cancer Res.* 29:
1658-1662.

Hirono, I., Kachi, H., and Kato, T. (1970a). A survey of acute toxicity of
cycads and mortality rate from cancer in the Miyaho Islands, Okinawa.
Acta Pathol. Jpn. 20:327-337.

Hirono, I., Shibuya, C., Fushimi, K., and Haga, M. (1970b). Studies on
carcinogenic properties of bracken, *Pteridium aquilinum*. *J. Natl. Cancer
Inst.* 45:179-188.

Hirono, I., Hayashi, K., Mori, H., and Miwa, T. (1971). Carcinogenic effects
of cycasin in Syrian golden hamsters and the transplantability of induced
tumors. *Cancer Res.* 31:283-287.

Hirono, I., Shibuya, C., Shimizu, M., and Fushimi, K. (1972a). Carcinogenic activity of processed bracken used as human food. *J. Natl. Cancer
Inst.* 48:1245-1250.

Hirono, I., Shibuya, C., Shimizu, M., Fushimi, K., Mori, H., and Miwa, T.
(1972b). Carcinogenicity examination of some edible plants. *Gann* 63:383-
386.

Hirono, I., Fushimi, K., Mori, H., Miwa, T., and Haga, M. (1973a). Comparative study and carcinogenic activity in each part of bracken. *J. Natl.
Cancer Inst.* 50:1367-1371.

Hirono, I., Shimizu, M., Fushimi, K., Mori, H., and Kato, K. (1973b).
Carcinogenic activity of *Petasites japonicus*, a kind of coltsfoot. *Gann* 64:
527-528.

Hirono, I., Sasaoka, I., Shibuya, C., Shimizu, M., Fushimi, K., Mori, H.,
Kato, K., and Haga, M. (1975). Natural carcinogenic products of plant
origin. *Gann Monogr. Cancer Res.* 17:205-217.

Hirono, I., Mori, H., and Culvenor, C. C. J. (1976). Carcinogenic activity
of coltsfoot, *Tussilago farfara* L. *Gann* 67:125-129.

Hirono, I., Fushimi, K., and Matsubara, N. (1977). Carcinogenicity test of
shikimic acid in rats. *Toxicol. Lett.* 1:9-10.

Hirono, I., Mori, H., and Haga, M. (1978a). Carcinogenic activity of *Symphytum officinale*. *J. Natl. Cancer Inst.* 61:865-869.

Hirono, I., Ushimaru, Y., Kato, K., Mori, H., and Sasaoka, I. (1978b).
Carcinogenicity of boiling water extract of bracken, *Pteridium aquilinum*.
Gann 69:383-388.

Hirono, I., Haga, M., Fujii, M., Matsuura, S., Matsubara, N., Nakayama,
M., Furuya, T., Hikichi, M., Takanashi, H., Uchida, E., Hosaka, S.,
and Ueno, I. (1979a). Induction of hepatic tumors in rats by senkirkine
and symphytine. *J. Natl. Cancer Inst.* 63:469-472.

Hirono, I., Mori, H., Haga, M., Fujii, M., Yamada, K., Hirata, Y.,
Takanashi, H., Uchida, E., Hosaka, S., Ueno, I., Matsushima, T.,
Umezama, K., and Shirai, A. (1979b). Edible plants containing carcino-
genic pyrrolizidine alkaloids in Japan. In *Naturally Occurring Carcinogens
—Mutagens and Modulators of Carcinogenesis*, E. C. Miller, J. A. Miller,
I. Hirono, T. Sugimura, and S. Takayama (Eds.). University Park Press,
Baltimore, pp. 79-87.

Hirota, M., Ohigashi, H., Oki, Y., and Koshimizu, K. (1980). New ingenol-
esters as piscicidal constituents of *Euphorbia cotinifolia* L. *Agric. Biol.
Chem.* 44:1351-1356.

Hoffmann, D., Hecht, S. S., Ornaf, R. M., and Wynder, E. L. (1974). N'-Nitrosonornicotine in tobacco. *Science 286*:265-276.

Hoffmann, D., Raineri, R., Hecht, S. S., Moronpot, R., and Wynder, E. L. (1975). A study of tobacco carcinogenesis. XIV. Effects of N'-nitrosonornicotine and N'-nitrosoanabasine in rats. *J. Natl. Cancer Inst. 55*: 977-981.

Holsti, P. (1959). Tumor promoting effects of some long chain fatty acids in experimental skin carcinogenesis in the mouse. *Acta Pathol. Microbiol. Scand. 46*:51-58.

Homberger, F., and Boger, E. (1968). The carcinogenicity of essential oils, flavors, and spices: a review. *Cancer Res. 28*:2372-2374.

Homberger, F., Friedler, G., Kelley, T., Jr., and Russfield, A. B. (1961a). 4-Allyl-1,2-methylenedioxybenzene (safrole) as a new tool for the study of pathogenesis of toxic hepatic cirrhosis and adenomatosis. *Fed. Proc. 20*: 288.

Homberger, F., Kelley, T., Jr., Friedler, G., and Russfield, A. B. (1961b). Toxic and possible carcinogenic effects of 4-allyl-1,2-methylenedioxybenzene (safrole) in rats on deficient diets. *Med. Exp. 4*:1-11.

Homberger, F., Kelley, T. F., and Bogdonoff, P. D. (1965). Effects of diet on toxicity of safrole in rats. *Toxicol. Appl. Pharmacol. 7*:485.

Hooson, J., and Grasso, P. (1976). Cytotoxic and carcinogenic response to monocrotaline pyrrole. *J. Pathol. 118*:121-128.

Hoppe, W., Brandl, F., Strell, I., Röhrl, M., Gassmann, I., Hecker, E., Bartsch, H., Kreibich, G., and v. Szczepanski, C. (1967). X-Ray structure analysis of neophorbol. *Angew Chem. Int. Ed. 6*:809-810.

Horton, A. W., Denman, D. T., and Trosset, R. P. (1957). Carcinogenesis of the skin. II. The accelerating properties of aliphatic and related hydrocarbons. *Cancer Res. 17*:758-766.

IARC (1976). *IARC Monographs on the Evaluation of the Carcinogenic Risk of Chemicals to Man*, Vol. 10; *Some Naturally Occurring Substances*. International Agency for Research on Cancer, Lyon.

Il'initskii, A. P., Mischenko, V. S., and Shabad, L. M. (1978). Natural sources of carcinogenic hydrocarbons. *Gig. Sanit. 8*:39-43 (through *Chem. Abstr. 89*:17443z, 1978).

Innes, J. R. M., Ulland, B. M., Valerio, M. G., Petrucelli, L., Fishbein, L., Hart, E. R., Pallota, A. J., Bates, R. R., Falk, H. L., Gart, J. J., Klein, M., Mitchell, I., and Peters, J. (1969). Bioassay of pesticides and industrial chemicals for tumorigenicity in mice: a preliminary note. *J. Natl. Cancer Inst. 42*:1101-1114.

Issenberg, P. (1976). Nitrite, nitrosamines, and cancer. *Fed. Proc. 35*: 1322-1326.

Ivankovic, S. (1978). Gastric cancer in rats after chronic intraperitoneal application of sap of green parts of potatoes (*Solanum tuberosum* L.). *Experientia 34*:645.

Jarrett, W. F. H., McNeil, P. E., Grimshaw, W. T. R., and Selman, I. E. (1978). High incidence of cattle cancer with a possible interaction between an environmental carcinogen and a papilloma virus. *Nature (Lond.) 274*: 215-217.

Johnson, W. D., Robertson, K. A., Pounds, J. G., and Allen, J. R. (1978). Dehydroretronecine-induced skin tumors in mice. *J. Natl. Cancer Inst. 61*:85-89.

Jussawalla, D. J., and Deshpande, V. A. (1971). Evaluation of cancer risk in tobacco chewers and smokers: an epidemiologic assessment. *Cancer* 28:244-252.

Kaiser, H. E. (1967). Cancer-promoting effects of phenols in tea. *Cancer* 20:614-616.

Kaiser, H. E., and Bartone, J. C. (1966). The carcinogenic activity of ordinary tea. A preliminary note. *J. Natl. Med. Assoc.* 58:361.

Kaminetzky, H. A., and McGrew, E. A. (1962). Podophyllin and mouse cervix. *Arch. Pathol.* 73:481-485.

Kapadia, G. J., Paul, B. D., Chung, E. B., Ghosh, B., and Pradhan, S. N. (1976). Carcinogenicity of *Camellia sinensis* (tea) and some tannin-containing folk medicinal herbs administered subcutaneously in rats. *J. Natl. Cancer Inst.* 57:207-209.

Kapadia, G. J., Chung, E. B., Ghosh, B., Shukla, Y. N., Basak, S. P., Morton, J. F., and Pradhan, S. N. (1978). Carcinogenicity of some folk medicinal herbs in rats. *J. Natl. Cancer Inst.* 60:683-686.

Kawabata, T., Ohshima, H., Uibu, J., Nakamura, M., Matsui, M., and Hamano, M. (1979). Occurrence, formation, and precursors of N-nitroso compounds in Japanese diet. In *Naturally Occurring Carcinogens— Mutagens and Modulators of Carcinogenesis*, E. C. Miller, J. A. Miller, I. Hirono, T. Sugimura, and S. Takayama (Eds.). University Park Press, Baltimore, pp. 195-209.

Kawai, T., Takanashi, H., Nakayama, M., Mori, H., and Hirono, I. (1981). Effect of storage on carcinogenic activity of bracken fern. *Cancer Lett.* 12:29-35.

Kawaji, K., Fukunishi, R., Terashi, S.-I., Higashi, J., and Watanabe, K. (1968). Induction of mammary cancer of rats with cycasin: a preliminary report. *Gann* 59:361.

Kimura, S., Suwa, J., Ito, M., and Sato, H. (1979). Experimental studies on the role of defatted soybean in the development of malignant goiter. In *Naturally Occurring Carcinogens—Mutagens and Modulators of Carcinogenesis*, E. C. Miller, J. A. Miller, I. Hirono, T. Sugimura, and S. Takayama (Eds.). University Park Press, Baltimore, pp. 101-110.

Kinghorn, A. D. (1979a). Cocarcinogenic irritant Euphorbiaceae. In *Toxic Plants*, A. D. Kinghorn (Ed.). Columbia University Press, New York, pp. 137-159.

Kinghorn, A. D. (1979b). Characterization of an irritant 4-deoxyphorbol diester from *Euphorbia tirucalli*. *J. Nat. Prod.* 42:112-115.

Kinghorn, A. D. (1980). Major skin-irritant principle from *Synadenium grantii*. *J. Pharm. Sci.* 69:1446-1447.

Kirby, K. S. (1960). Induction of tumors by tannin extracts. *Br. J. Cancer* 14:147-150.

Kobayashi, A. (1972). Cycasin in cycad materials used in Japan. *Fed. Proc.* 31:1476-1477.

Korpássy, B., and Mosonyi, M. (1950). The carcinogenic activity of tannic acid. Liver tumors induced in rats by prolonged subcutaneous administration of tannic acid solutions. *Br. J. Cancer* 4:411-420.

Kuhara, K., Takanashi, H., Hirono, I., Furuya, T., and Asada, Y. (1980). Carcinogenic activity of clivorine, a pyrrolizidine alkaloid isolated from *Ligularia dentata*. *Cancer Lett.* 10:117-122.

Kuhlmann, W., Fromme, H.-G., Heege, E.-M., and Ostertag, W. (1968). The mutagenic action of caffeine in higher organisms. *Cancer Res.* 28:2375-2389.

Kupchan, S. M., and Baxter, R. L. (1975). Mezerein: antileukemic principle isolated from *Daphne mezereum* L. *Science 187*:652-653.

Kupchan, S. M., Shizuri, Y., Murae, T., Sweeny, J. G., Haynes, H. R., Shen, M.-S., Barrick, J. C., Bryan, R. F., van der Helm, D., and Wu, K. K. (1976a). Gnidimacrin and gnidimacrin 20-palmitate, novel macrocyclic antileukemic diterpenoid esters from *Gnidia subcordata*. *J. Am. Chem. Soc. 98*:5719-5720.

Kupchan, S. M., Uchida, I., Branfman, A. R., Dailey, R. G., Jr., and Yu Fei, B. (1976b). Antileukemic principles isolated from Euphorbiaceae plants. *Science 191*:571-572.

Kuratsune, M., Kohchi, S., Horie, A., and Nishizumi, N. (1971). Test of alcoholic beverages and ethanol solutions for carcinogenicity and tumor-promoting activity. *Gann 62*:395-405.

Kuroyanagi, M., Fukuoka, M., Yoshihira, K., and Natori, S. (1979). Chemical and pharmacological studies on bracken fern, *Pteridium aquilinum* var. *latiusculum*. III. Further characterization of pterosins and pterosides, sesquiterpenes and their glycosides having a 1-indanone skeleton, from the rhizomes. *Chem. Pharm. Bull. 27*:592-601.

Laerum, O. D., and Iversen, O. H. (1972). Reticuloses and epidermal tumors in hairless mice after topical skin applications of cantharidin and asiaticoside. *Cancer Res. 32*:1463-1469.

Langley, B. W., Lythgoe, B., and Riggs, N. V. (1951). Macrozamin. Part II. The aliphatic azoxy structure of the aglycone part. *J. Chem. Soc. 1951*:2309-2316.

Laquer, G. L. (1964). Carcinogenic effects of cycad meal and cycasin, methylazoxymethanol glycoside, in rats and effects of cycasin in germ-free rats. *Fed. Proc. 23*:1386-1388.

Laquer, G. L. (1965). The induction of intestinal neoplasms in rats with the glycoside cycasin and its aglycone. *Virchows Arch. Pathol. Anat. 340*:151-163.

Laquer, G. L. (1977). Oncogenicity of cycads and its implications. In *Advances in Modern Toxicology*, Vol. 3,: *Environmental Cancer*, H. F. Kraybill and M. A. Mehlman (Eds.). Hemisphere, Washington, D.C., pp. 231-261.

Laquer, G. L., and Matsumoto, H. (1966). Neoplasms in female Fischer rats following intraperitoneal injection of methylazoxymethanol. *J. Natl. Cancer Inst. 37*:217-232.

Laquer, G. L., and Spatz, M. (1968). Toxicology of cycasin. *Cancer Res. 28*: 2262-2267.

Laquer, G. L., and Spatz, M. (1975). Oncogenicity of cycasin and methylazoxymethanol. *Gann Monogr. Cancer Res. 17*:189-204.

Laquer, G. L., Mickelsen, O., Whiting, M. G., and Kurland, L. T. (1963). Carcinogenic properties of nuts from *Cycas circinalis* L. indigenous to Guam. *J. Natl. Cancer Inst. 31*:919-951.

Laquer, G. L., McDaniel, E. G., and Matsumoto, H. (1967). Tumor induction in germfree rats with methylazoxymethanol (MAM) and synthetic MAM acetate. *J. Natl. Cancer Inst. 39*:355-371.

Leach, H., Barber, G. D., Evans, I. A., and Evans, W. C. (1971). Isolation of an active principle from the bracken fern that is mutagenic, carcinogenic and lethal to mice on intraperitoneal injection. *Biochem. J. 124*:13P-14P.

Lee, D. J., Wales, J. H., Ayres, J. L., and Sinnhuber, R. O. (1968). Synergism between cyclopropenoid fatty acids and chemical carcinogens in rainbow trout (*Salmo gairdneri*). *Cancer Res. 28*:2312-2318.

Leach, H., Barber, G. D., Evans, I. A., and Evans, W. C. (1971). Isolation of an active principle from the bracken fern that is mutagenic, carcinogenic and lethal to mice on intraperitoneal injection. *Biochem. J. 124*:13P-14P.

Lee, D. J., Wales, J. H., Ayres, J. L., and Sinnhuber, R. O. (1968). Synergism between cyclopropenoid fatty acids and chemical carcinogens in rainbow trout (*Salmo gairdneri*). *Cancer Res. 28*:2312-2318.

Lee, D. J., Wales, J. H., and Sinnhuber, R. O. (1969). Hepatoma and renal tubule adenoma in rats fed aflatoxin and cyclopropenoid fatty acids. *J. Natl. Cancer Inst. 43*:1037-1044.

Lee, D. J., Wales, J. H., Sinnhuber, R. O., and Roehm, J. N. (1970). Promotion of aflatoxin-induced hepatoma growth by cyclopropenoid fatty acids. *Fed. Proc. 29*:568.

Long, E. L., and Jenner, P. M. (1963). Esophageal tumors produced in rats by the feeding of dihydrosafrole. *Fed. Proc. 22*:275.

Long, E. L., Hansen, W. H., and Nelson, A. A. (1961). Liver tumors produced in rats by feeding safrole. *Fed. Proc. 20*:287.

Long, E. L., Nelson, A. A., Fitzhugh, O. G., and Hansen, W. H. (1963). Liver tumors produced in rats by feeding safrole. *Arch. Pathol. 75*:595-604.

Lythgoe, B., and Riggs, N. V. (1949). Macrozamin. Part I. The identity of the carbohydrate component. *J. Chem. Soc. 1949*:2716-2718.

Mack, T. M., Henderson, B. E., Gerkins, V. R., Arthur, M., Baptista, J., and Pike, M. C. (1975). Reserpine and breast cancer in a retirement community. *N. Engl. J. Med. 292*:1366-1371.

Maier, H. G., and Stender, W. (1969). Carcinogenic hydrocarbons in coffee substitutes. *Dtsch. Lebensm.-Rundsch. 65*:341-343.

Marquardt, H. (1975). Reserpine and chemical carcinogenesis. *Lancet 1* (7912):925-926.

Matsubara, N., Mori, H., and Hirono, I. (1978). Effect of colostomy on intestinal carcinogenesis by methylazoxymethanol acetate in rats. *J. Natl. Cancer Inst. 61*:1161-1164.

Matsumoto, H. (1979). Carcinogenicity of cycasin, its aglycone methylazoxymethanol, and methylazoxymethyl-glucosiduronic acid. In *Naturally Occurring Carcinogens—Mutagens and Modulators of Carcinogenesis*, E. C. Miller, J. A. Miller, I. Hirono, T. Sugimura, and S. Takayama (Eds.). University Park Press, Baltimore, pp. 67-77.

Matsumoto, H., and Higa, H. H. (1966). Studies on methylazoxymethanol, the aglycone of cycasin: methylation of nucleic acids in vitro. *Biochem. J. 98*:20C-22C.

Matsumoto, H., and Strong, F. M. (1963). The occurrence of methylazoxymethanol in *Cycas circinalis* L. *Arch. Biochem. Biophys. 101*:299-310.

Mattocks, A. R. (1972). Toxicity and metabolism of *Senecio* alkaloids. In *Phytochemical Ecology*, J. B. Harborne (Ed.). Academic, London, pp. 179-200.

McLean, E. K. (1970). The toxic actions of pyrrolizidine (*Senecio*) alkaloids. *Pharmacol. Rev. 22*:430-483.

McLean, E. K. (1974). *Senecio* and other plants as liver poisons. *Isr. J. Med. Sci. 10*:436-440.

Mendelson, R. W., and Ellis, A. G. (1924). Cancer as a public health problem in Siam. *J. Trop. Med. Hyg. 27*:274-278.

Miller, E. C., and Miller, J. A. (1979a). Overview of the relevance of naturally occurring carcinogens, promotes, and modulators of carcinogenesis in human cancer. In *Naturally Occurring Carcinogens—Mutagens and Modulators of Carcinogenesis*, E. C. Miller, J. A. Miller, I. Hirono, T. Sugimura, and S. Takayama (Eds.). University Park Press, Baltimore, pp. 1-17.

Miller, E. C., and Miller, J. A. (1979b). Naturally occurring chemical carcinogens that may be present in foods. In *International Review of Biochemistry*, Vol. 27: *Biochemistry of Nutrition 1A*, A. Neuberger and T. H. Jukes (Eds.). University Park Press, Baltimore, pp. 123-165.

Miller, E. C., Miller, J. A., Hirono, I., Sugimura, T., and Takayama, S. (Eds.) (1979). *Naturally Occurring Carcinogens—Mutagens and Modulators of Carcinogenesis*. University Park Press, Baltimore.

Miller, J. A. (1964). Comments on chemistry of cycads. *Fed. Proc. 23:* 1361-1362.

Miller, J. A., and Miller, E. C. (1971). Chemical carcinogenesis: mechanisms and approaches to its control [editorial]. *J. Natl. Cancer Inst. 47(3):*V-XIV.

Miller, J. A, and Miller, E. C. (1976). Carcinogens occurring naturally in foods. *Fed. Proc. 35:*1316-1321.

Miller, J. A., Swanson, A. B., and Miller, E. C. (1979). The metabolic activation of safrole and related naturally occurring alkenylbenzenes in relation to carcinogenesis by these agents. In *Naturally Occurring Carcinogens—Mutagens and Modulators of Carcinogenesis*, E. C. Miller, J. A. Miller, I. Hirono, T. Sugimura, and S. Takayama (Eds.). University Park Press, Baltimore, pp. 111-125.

Miyakawa, M., and Yoshida, O. (1975). Induction of tumors of the urinary bladder in female mice following surgical implantation of glass beads and feeding of bracken fern. *Gann 66:*437-439.

Mori, H., Kato, K., Ushimaru, Y., Kato, T., and Hirono, I. (1977). Effect of drying with hot forced draft and of mincing bracken fern on its carcinogenic activity. *Gann 68:*517-520.

Mori, H., Matsubara, N., Ushimaru, Y., and Hirono, I. (1979). Carcinogenicity examination of betel nuts and piper betel leaves. *Experientia 35:* 384-385.

Morton, J. F. (1968). Plants associated with esophageal cancer cases in Curaço. *Cancer Res. 28:*2268-2271.

Morton, J. F. (1970). Tentative correlations of plant usage and esophageal cancer zones. *Econ. Bot. 24:*217-226.

Morton, J. F. (1972). Further associations of plant tannins and human cancer. *Q. J. Crude Drug Res. 12:*1829-1841.

Morton, J. F. (1980). Search for carcinogenic principles. In *Recent Advances in Phytochemistry*, Vol. 14: *The Resource Potential in Phytochemistry*, T. Swain (Ed.). Plenum, New York, pp. 53-73.

Mosonyi, M., and Korpássy, B. (1953). Rapid production of malignant hepatomas by simultaneous administration of tannic acid and 2-acetylaminofluorene. *Nature (Lond.) 171:*791.

Mottram, J. C. (1944). A developing factor in experimental blastogenesis. *J. Pathol. 56:*181-186.

Muir, C. S., and Kirk, R. (1960). Betel, tobacco, and cancer of the mouth. *Br. J. Cancer 14:*597-608.

Nagahama, T., Numata, T., and Nishida, K. (1959). Azoxy glycosides of *Cycas revoluta*. II. Neocycasin B and macrozamin. *Bull. Agric. Chem. Soc. Jpn. 23:*556-557 (through *Chem. Abstr. 54:*9006d, 1960).

Nagahama, T., Nishida, K., and Numata, T. (1960). Azoxy glycosides of *Cycas revoluta*. IV. The structure of neocycasin C formed by trans-glycosylation with cycad emulsion. *Bull. Agric. Chem. Soc. Jpn. 24*: 536-537 (through *Chem. Abstr.* 55:3448c, 1961).

Nagahama, T., Nishida, K., and Numata, T. (1961). Azoxy glycosides of *Cycas revoluta*. VII. Neocycasin E, β-cellobiosyloxyazoxymethane. *Agric. Biol. Chem. 25*:937-938.

Narisawa, T., and Nakano, H. (1973). Carcinoma of the large intestine of rats induced by rectal infusion of methylazoxymethanol. *Gann 64*:93-95.

Newberne, P. M. (1976). Biologic effects of plant toxins and aflatoxins in rats. *J. Natl. Cancer Inst. 56*:551-555.

Nishida, K., Kobayashi, A., and Nagahama, T. (1955). Cycasin, a toxic glycoside of *Cycas revoluta*. I. Isolation and structure of cycasin. *Bull. Agric. Chem. Soc. Jpn. 19*:77-84 (through *Chem. Abstr.* 50:1376g, 1956).

Nishida, K., Kobayashi, A., Nagahama, T., and Numata, T. (1959). Some new azoxy glycosides of *Cycas revoluta*. I. Neocycasin A. *Bull. Agric. Chem. Soc. Jpn. 23*:460-464 (through *Chem. Abstr.* 54:3609d, 1960).

O'Fallon, W. M., and Larbarthe, D. R. (1975). *Rauwolfia* derivatives and breast cancer. *Lancet 2*(7938):773.

O'Gara, R. W. (1968). Biologic screening of selected plant material for carcinogens. *Cancer Res. 28*:2272-2275.

O'Gara, R. W., Brown, J. M., and Whiting, M. G. (1964). Induction of hepatic and renal tumors by topical application of aqueous extract of cycad nut to artificial skin ulcers in mice. *Fed. Proc. 23*:1383.

O'Gara, R. W., Stewart, L., Brown, J., and Hueper, W. C. (1969). Carcinogenicity of heated fats and fat fractions. *J. Natl. Cancer Inst. 42*: 275-287.

O'Gara, R. W., Lee, C., and Morton, J. F. (1971). Carcinogenicity of extracts of selected plants from Curaçao after oral and subcutaneous administration to rodents. *J. Natl. Cancer Inst. 46*:1131-1137.

O'Gara, R. W., Lee, C. W., Morton, J. F., Kapadia, G. J., and Dunham, L. J. (1974). Sarcoma induced in rats by extracts of plants and by fractionated extracts of *Krameria ixina*. *J. Natl. Cancer Inst. 52*:445-448.

Ogura, M., Koike, K., Cordell, G. A., and Farnsworth, N. R. (1978). Potential anticancer agents. VIII. Constituents of *Baliospermum montanum* (Euphorbiaceae). *Planta Med. 33*:128-143.

Ohigashi, H., Kawazu, K., Koshimizu, K., and Mitsui, T. (1972). A piscicidal constituent of *Sapium japonicum*. *Agric. Biol. Chem. 36*:2529-2537.

Opdyke, D. L. J. (1975). Fragrance raw materials. Alcohol C-14 myristic. *Food Cosmet. Toxicol. 13*(Suppl.):699-700.

Opferkuch, H. J., and Hecker, E. (1974). New diterpenoid irritants from *Euphorbia ingens*. *Tetrahedron Lett. 1974*:261-264.

Orr, I. M. (1933). Oral cancer and betel nut chewers in Travancore. *Lancet 2*:575-580.

Pamakcu, A. M. (1963). Epidemiologic studies on urinary bladder tumors in Turkish cattle. *Ann. N.Y. Acad. Sci. 168*:938-947.

Pamakcu, A. M., and Price, J. M. (1969). Induction of intestinal and urinary bladder cancer in rats by feeding bracken fern (*Pteris aquilina*). *J. Natl. Cancer Inst. 43*:275-281.

Pamakcu, A. M., and Bryan, G. T. (1979). Bracken fern, a natural urinary bladder and intestinal carcinogen. In *Naturally Occurring Carcinogens— Mutagens and Modulators of Carcinogenesis*, E. C. Miller, J. A. Miller, I. Hirono, T. Sugimura, and S. Takayama (Eds.). University Park Press, Baltimore, pp. 89-99.

Pamakcu, A. M., Olson, C., and Price, J. M. (1966). Assay of fractions of bovine urine for carcinogenic activity after feeding bracken fern (*Pteris aquilina*). *Cancer Res.* 26(Part 1):1745-1753.

Pamakcu, A. M., Göksoy, S. K., and Price, J. M. (1967). Urinary bladder neoplasms induced by feeding bracken fern (*Pteris aquilina*) to cows. *Cancer Res.* 27(Part 1):917-924.

Pamakcu, A. M., Price, J. M., and Bryan, G. T. (1970a). Assay of fractions of bracken fern (*Pteris aquilina*) for carcinogenic activity. *Cancer Res.* 30:902-905.

Pamakcu, A. M., Yalçiner, Ş., Price, J. M., and Bryan, G. T. (1970b). Fffects of coadministration of thiamine on the incidence of urinary bladder carcinomas in rats fed bracken fern. *Cancer Res.* 30:2671-2674.

Pamakcu, A. M., Ertürk, E., Price, J. M., and Bryan, G. T. (1972). Lymphatic leukemia and pulmonary tumors in female Swiss mice fed bracken fern (*Pteris aquilina*). *Cancer Res.* 32:1442-1445.

Pamakcu, A. M., Ertürk, E., Yalçiner, Ş., Milli, U., and Bryan, G. T. (1973). Carcinogenic and mutagenic activities of milk from cows fed bracken fern (*Pteridium aquilinum*). *Cancer Res.* 38:1556-1560.

Pamakcu, A. M., Wang, C. Y., Hatcher, J., and Bryan, G. T. (1980a). Carcinogenicity of tannin and tannin-free extracts of bracken fern (*Pteridium aquilinum*) in rats. *J. Natl. Cancer Inst.* 65:131-136.

Pamakcu, A. M., Yalçiner, Ş., Hatcher, J. F., and Bryan, G. T. (1980b). Quercetin, a rat intestinal and bladder carcinogen present in bracken fern (*Pteridium aquilinum*). *Cancer Res.* 40:3468-3472.

Peele, J. D., Jr., and Oswald, E. O. (1978). Metabolism of the proximate carcinogen 1'-hydroxysafrole and the isomer 3'-hydroxyisosafrole. *Bull. Environ. Contam. Toxicol.* 19:396-402.

Peirce, W. E. H. (1961). Tumour-promotion by lime oil in the mouse fore-stomach. *Nature (Lond.)* 189:497-498.

Petrun, A. S. (1977). Level of carcinogenic polycyclic hydrocarbons in cultivated plants. *Ratsion. Pitan.* 12:71-75; 98-102 (through *Chem. Abstr.* 93:108600e, 1980).

Pettersen, R. C., Ferguson, G., Crombie, L., Games, M. L., and Pointer, D. J. (1967). The structure and stereochemistry of phorbol, diterpene parent of co-carcinogens of croton oil. *J. Chem. Soc. Chem. Commun.* 1967:716-717.

Pradhan, S. N., Chung, E. B., Ghosh, B., Paul, B. D., and Kapadia, G. J. (1974). Potential carcinogens. I. Carcinogenicity of some plant extracts and their tannin-containing fractions in rats. *J. Natl. Cancer Inst.* 52:1579-1582.

Price, J. M., and Pamakcu, A. M. (1968). The induction of neoplasms of the urinary bladder of the cow and the small intestine of the rat by feeding bracken fern (*Pteris aquilina*). *Cancer Res.* 28:2247-2251.

Procter, B. G., Dussault, P., Rona, G., and Chappel, C. I. (1974). Studies on the carcinogenicity of an acetone extract of hashish. *Toxicol. Appl. Pharmacol.* 29:76.

Ramanathan, K., and Lakshimi, S. (1976). Oral carcinoma in peninsular Malaysia: racial variations in the Indians, Malays, Chinese, and Caucasians. *Gann Monogr. Cancer Res.* *18*:27-36.

Ranadive, K. J., Gothoskar, S. V., and Tezabwala, B. U. (1972). Carcinogenicity of contaminants in indigenous edible oils. *Int. J. Cancer 10*:652-666.

Ranadive, K. J., Gothoskar, S. V., Rao, A. R., Tezabwalla, B. U., and Ambaye, R. Y. (1976). Experimental studies on betel nut and tobacco carcinogenicity. *Int. J. Cancer 17*:469-476.

Rao, M. S., and Reddy, J. K. (1978). Malignant neoplasms in rats fed lasiocarpine. *Br. J. Cancer 37*:289-293.

Reddy, D. G., and Anguli, V. C. (1967). Experimental production of cancer with betel nut, tobacco, and slaked lime mixture. *J. Indian Med. Assoc.* *49*:315-318.

Reddy, J. K., Chiga, M., Harris, C. C., and Svoboda, D. J. (1970). Polyribosome disaggregation in rat liver following administration of tannic acid. *Cancer Res. 30*:58-65.

Riggs, N. V. (1956). Glucosyloxyazoxymethane, a constituent of the seeds of *Cycas circinalis* L. *Chem. Ind. (Lond.) 1956*:926.

Roberts, H. B., McClure, T. J., Ritchie, E., Taylor, W. C., and Freeman, P. W. (1975). The isolation and structure of the toxin of *Pimelea simplex* responsible for St. George disease of cattle. *Aust. Vet. J. 51*:325-326.

Robertson, K. A., Seymour, J. L., Hsia, M.-T., and Allen, J. R. (1977). Covalent interaction of dehydroretronecine, a carcinogenic metabolite of the pyrrolizidine alkaloid monocrotaline, with cysteine and glutathione. *Cancer Res. 37*:3141-3144.

Roe, F. J. C. (1959). Oil of sweet orange: a possible role in carcinogenesis. *Br. J. Cancer 13*:92-93.

Roe, F. J. C., and Peirce, W. E. H. (1960). Tumor promotion by citrus oils: tumors of the skin and urethral orifice in mice. *J. Natl. Cancer Inst. 24*:1389-1403.

Roe, F. J. C., and Peirce, W. E. H. (1961). Tumor promotion by *Euphorbia* latices. *Cancer Res. 21*:338-344.

Roe, F. J. C., and Field, W. E. H. (1965). Chronic toxicity of essential oils and certain other products of natural origin. *Food Cosmet. Toxicol.* *3*:311-324.

Ronlán, A., and Wickberg, B. (1970). The structure of mezerein, a major toxic principle of *Daphne mezereum* L. *Tetrahedron Lett. 1970*:4621-4624.

Rosenberger, C., and Heeschen, W. (1960). Alderfarn (*Pteris aquilina*) die Ursache des sog. Stallrotes der Rinder (Haematurie vesicalis bovis chronica). *Dtsch. Tieraerztl. Wochenschr. 67*:201-208.

Saito, D., Shirai, A., Matsushima, T., Sugimura, T., and Hirono, I. (1980). Test of carcinogenicity of quercetin, a widely distributed mutagen in food. *Teratog. Carcinog. Mutagen. 1*:213-221 (through *Chem. Abstr. 94*: 59538a, 1981).

Saito, M., Umeda, M., Enomoto, M., Hatanaka, Y., Natori, S., Yoshihira, K., Fukuoka, M., and Kuroyanagi, M. (1975). Cytotoxicity and carcinogenicity of pterosins and pterosides, 1-indanone derivatives from bracken (*Pteridium aquilinum*). *Experientia 31*:829-831.

Sakata, K., Kawazu, K., Mitsui, T., and Masaki, N. (1971). Structure and stereochemistry of huratoxin, a piscicidal constituent of *Hura crepitans*. *Tetrahedron Lett. 1971*:1141-1144.

Salaman, M. H. (1958). Cocarcinogenesis. *Br. Med. Bull 14*:116-120.

Salvatore, G., Stacchini, A., and Di Marzio, S. (1980). Presence of aromatic substances in food. II. Safrole: identification and determination by head-space gas chromatography. *Riv. Soc. Ital. Sci. Aliment. 9*:253-264 (through *Chem. Abstr. 94*:45680d, 1981).

Sarkar, S. N. (1948). Isolation from argemone oil of dihydrosanguinarine and sanguinarine: Toxicity of sanguinarine. *Nature (Lond.) 162*:265-266.

Scarpelli, D. G. (1974). Mitogenic activity of sterculic acid, a cyclopropenoid fatty acid. *Science 185*:958-960.

Scarpelli, D. G., Lee, D. J., Sinnhuber, R. O., and Chiga, M. (1974). Cytoplasmic alterations of hepatocytes in rainbow trout (*Salmo gairdneri*) induced by cyclopropenoid fatty acids. *Cancer Res. 34*:2984-2990.

Schmeltz, I., and Hoffmann, D. (1977). Nitrogen-containing compounds in tobacco and tobacco smoke. *Chem. Rev. 77*:295-311.

Schmidt, R., and Hecker, E. (1975). Autoxidation of phorbol esters under normal storage conditions. *Cancer Res. 35*:1375-1377.

Schoental, R. (1973). The mechanisms of action of the carcinogenic nitroso and related compounds. *Br. J. Cancer 28*:436-439.

Schoental, R. (1974). The role of podophyllotoxin in the bedding and dietary zearalenone on incidence of spontaneous tumors in laboratory animals. *Cancer Res. 34*:2419-2420.

Schoental, R. (1975). Pancreatic islet-cell and other tumors in rats given heliotrine, a monoester pyrrolizidine alkaloid, and nicotinamide. *Cancer Res. 35*:2020-2024.

Schoental, R. (1976). Carcinogens in plants and microorganisms. In *Chemical Carcinogens*, C. E. Searle (Ed.). ACS Monogr. 173. American Chemical Society, Washington, D.C., pp. 626-689.

Schoental, R., and Cavanagh, J. B. (1972). Brain and spinal cord tumors in rats treated with pyrrolizidine alkaloids. *J. Natl. Cancer Inst. 49*: 665-671.

Schoental, R., and Head, M. A. (1955). Pathological changes in rats as a result of treatment with monocrotaline. *Br. J. Cancer 9*:229-237.

Schoental, R., Head, M. A., and Peacock, P. R. (1954). *Senecio* alkaloids: primary liver tumors in rats as a result of treatment with (1) a mixture of alkaloids from *S. jacobaea* Lin., (2) retrosine; (3) isatidine. *Br. J. Cancer 8*:458-465.

Schoental, R., Fowler, M. E., and Coady, A. (1970). Islet cell tumors of the pancreas found in rats given pyrrolizidine alkaloids from *Amsinckia intermedia* Fisch. and Mey and from *Heliotropium supinum* L. *Cancer Res. 30*:2127-2131.

Scribner, J. D., and Süss, R. (1978). Tumor initiation and promotion. *Int. Rev. Exp. Pathol. 18*:137-198.

Segi, M. (1975). Tea-gruel as a possible factor for cancer of the esophagus. *Gann 66*:199-202.

Seiber, S. M., Correa, P., Dalgard, D. W., McIntire, K. R., and Adamson, R. H. (1980). Carcinogenicity and hepatotoxicity of cycasin and its aglycone methylazoxymethanol acetate in nonhuman primates. *J. Natl. Cancer Inst.* 65:177-189.

Shank, R. C., and Magee, P. N. (1967). Similarities between the biochemical actions of cycasin and dimethylnitrosamine. *Biochem. J.* 105:521-527.

Shear, M. J. (1938). Studies in carcinogenesis. V. Methyl derivatives of 1:2-benzanthracene. *Am. J. Cancer* 33:499-537.

Shibuya, C., and Hirono, I. (1973). Relations between postnatal days of mice and carcinogenic effect of cycasin. *Gann* 64:109-110.

Shimkin, M. B., and Triolo, V. A. (1969). History of chemical carcinogenesis: Some prospective remarks. *Prog. Exp. Tumor Res.* 11:1-20.

Shivapurkar, N. M., D'Souza, A. V., and Bhide, S. V. (1980). Effect of betel-quid chewing on nitrite levels in saliva. *Food Cosmet. Toxicol.* 18: 277-281.

Shumaker, R. C., Robertson, K. A., Hsu, I. C., and Allen, J. R. (1976). Neoplastic transformation in tissues of rats exposed to monocrotaline or dehydroretronecine. *J. Natl. Cancer Inst.* 56:787-790.

Sicé, J. (1966). Tumor-promoting activity of *n*-alkanes and 1-alkanols. *Toxicol. Appl. Pharmacol.* 9:70-74.

Singer, G. M., and Taylor, H. W. (1976). Carcinogenicity of *N'*-nitrosonornicotine in Sprague-Dawley rats. *J. Natl. Cancer Inst.* 57:1275-1276.

Sinnhuber, R. O., Wales, J. H., and Lee, D. J. (1966). Cyclopropanoids, co-carcinogens for aflatoxin-induced hepatoma in trout. *Fed. Proc.* 25: 255.

Sinnhuber, R. O., Lee, D. J., Wales, J. H., and Ayres, J. L. (1968). Dietary factors and hepatoma in rainbow trout (*Salmo gairdneri*). II. Cocarcinogenesis by cyclopropenoid fatty acids and effect of gossypol and altered lipids on aflatoxin-induced liver cancer. *J. Natl. Cancer Inst.* 41: 1293-1301.

Sinnhuber, R. O., Lee, D. J., Wales, J. H., Landers, M. K., and Keyl, A. C. (1974). Hepatic carcinogenesis of aflatoxin M_1 in rainbow trout (*Salmo gairdneri*) and its enhancement by cyclopropene fatty acids. *J. Natl. Cancer Inst.* 53:1285-1288.

Sinnhuber, R. O., Hendricks, J. D., Putnam, G. B., Wales, J. H., Pawlowski, N. E., Nixon, J. E., and Lee, D. J. (1976). Sterculic acid, a naturally occurring cyclopropene fatty acid, a liver carcinogen to rainbow trout, *Salmo gairdneri. Fed. Proc.* 35:505.

Sivak, A. (1978). Mechanisms of tumor promotion and cocarcinogenesis: a summary from one point of view. In *Carcinogenesis*, Vol. 2: *Mechanisms of Tumor Promotion and Cocarcinogenesis*, T. J. Slaga, A. Sivak, and R. K. Boutwell (Eds.). Raven Press, New York, pp. 553-564.

Sivak, A. (1979). Cocarcinogenesis. *Biochim. Biophys. Acta* 560:67-69.

Slaga, T. J., Sivak, A., and Boutwell, R. K. (Eds.) (1978). *Carcinogenesis*, Vol. 2: *Mechanisms of Tumor Promotion and Cocarcinogenesis*. Raven Press, New York.

Slaga, T. J., Fischer, S. M., Nelson, K., and Gleason, G. L. (1980). Studies on the mechanism of skin tumor promotion: evidence for several stages in promotion. *Proc. Natl. Acad. Sci. USA* 77:3659-3663.

Smith, L. W., and Culvenor, C. C. J. (1981). Plant sources of pyrrolizidine alkaloids. *J. Nat. Prod.* 44:129-152.

Spatz, M. (1964). Carcinogenic effect of cycad meal in guinea pigs. *Fed. Proc.* 23:1384-1385.

Spatz, M. (1968). Hydrolysis of cycasin by β-D-glucosidase in skin of new-born rats. *Proc. Soc. Exp. Biol. Med. 128:*1005-1008.

Spatz, M. (1969). Toxic and carcinogenic alkylating agents from cycads. *Ann. N.Y. Acad. Sci. 163:*848-859.

Stern, R. S., Thibodeau, L. A., Kleinerman, R. A., Parrish, J. A., Fitz-patrick, T. B., and 22 participating investigators (1979). Risk of cutaneous carcinoma in patients treated with oral methoxsalen photo-chemotherapy for psoriasis. *N. Engl. J. Med. 300:*809-813.

Stillwell, W. G., Carman, M. J., Bell, L., and Horning, M. G. (1974). The metabolism of safrole and 2',3'-epoxysafrole in the rat and guinea pig. *Drug Metab. Dispos. 2:*489-498.

Stout, G. H., Balkenol, W. G., Poling, M., and Hickernell, G. L. (1970). The isolation and structure of daphnetoxin, the poisonous principle of *Daphne* species. *J. Am. Chem. Soc. 92:*1070-1071.

Sugimura, T. (1982). Potent tumor promoters other than phorbol ester and their significance. *Gann 73:*499-507.

Suri, K., Goldman, H. M., and Wells, H. (1971). Carcinogenic effect of a dimethyl sulphoxide extract of betel nut on the mucosa of the hamster buccal pouch. *Nature (Lond.) 230:*383-384.

Svoboda, D. J., and Reddy, J. K. (1972). Malignant tumors in rats given lasiocarpine. *Cancer 32:*908-913.

Svoboda, D. J., and Reddy, J. K. (1974). Lasiocarpine induced, transplant-able squamous cell carcinoma of rat skin. *J. Natl. Cancer Inst. 53:*1415-1418.

Takatori, K., Nakano, S., Nagata, S., Okumura, K., Hirono, I., and Shimizu, M. (1972). Pterolactam, a new compound isolated from bracken. *Chem. Pharm. Bull. 20:*1087.

Taylor, S. L., Montgomery, M. W., and Lee, D. J. (1973). Liver dehydro-genase levels in rainbow trout, *Salmo gairdneri*, fed cyclopropenoid fatty acids and aflatoxin B_1. *J. Lipid Res. 14:*643-646.

Tazima, Y. (1974). Naturally occurring mutagens of biological origin. A review. *Mutat. Res. 26:*225-234.

Terracini, B., Palestro, G., Gigliardi, M. R., and Montesano, R. (1966). Carcinogenicity of dimethylnitrosamine in Swiss mice. *Br. J. Cancer 20:* 871-876.

Thomson, R. H. (1971). *Naturally Occurring Quinones*, 2nd ed. Academic Press, London.

Thorne, R. F. (1968). Synopsis of a putatively phylogenetic classification of the flowering plants. *Aliso 6:*57-66.

Towers, G. H. N. (1980). Photosensitizers from plants and their photo-dynamic action. In *Progress in Phytochemistry*, Vol. 6, L. Reinhold, J. B. Harborne, and T. Swain (Eds.). Pergamon, Oxford, pp. 183-202.

Tyler, M. L., and Howden, M. E. H. (1981). Piscicidal constituents of *Pimelea* species. *Tetrahedron Lett. 1981:*689-690.

Ueno, I., and Hirono, I. (1981). Non-carcinogenic response to coumarin in Syrian golden hamsters. *Food Cosmet. Toxicol. 19:*353-355.

Upadhyay, R. R., Ansarin, M., Zarintan, M. H., and Shakui, P. (1976). Tumor promoting constituent of *Euphorbia serrata* L. latex. *Experientia 32:*1196-1197.

Upadhyay, R. R., Bakhtavar, F., Ghaisarzadeh, M., and Tilabi, J. (1978). Cocarcinogenic and irritant factors of *Euphorbia esula* L. latex. *Tumori 64:*99-102.

Upadhyay, R. R., Bakhtavar, F., Mohseni, H., Sater, A. M., Saleh, N.,
    Tafazuli, A., Dizaji, F. N., and Mohaddes, G. (1980). Screening of
    Euphorbia from Azarbaijan for skin irritant activity and for diterpenes.
    Planta Med. 38:151-154.
Van Duuren, B. L. (1968). Tobacco carcinogenesis. *Cancer Res. 28*:2357-
    2362.
Van Duuren, B. L. (1976). Tumor-promoting and co-carcinogenic agents in
    chemical carcinogenesis. In *Chemical Carcinogens*, C. E. Searle (Ed.).
    CAS Monogr. 173. American Chemical Society, Washington, D.C., pp.
    24-51.
Van Duuren, B. L., and Goldschmidt, B. M. (1976). Cocarcinogenic and
    tumor-promoting agents in tobacco carcinogenesis. *J. Natl. Cancer Inst.
    56*:1237-1242.
Van Duuren, B. L., and Orris, L. (1965). The tumor-enhancing principles
    of *Croton tiglium* L. *Cancer Res. 25*:1871-1875.
Van Duuren, B. L., Nelson, N., Orris, L., Palmes, E. D., and Schmitt,
    F. L. (1963a). Carcinogenicity of epoxides, lactones, and peroxy com-
    pounds. *J. Natl. Cancer Inst. 31*:41-55.
Van Duuren, B. L., Orris, L., and Arroyo, E. (1963b). Tumor-enhancing
    activity of the active principles of *Croton tiglium* L. *Nature (Lond.)
    200*:1115-1116.
Van Duuren, B. L., Arroyo, E., and Orris, L. (1963c). The tumor-enhancing
    and irritant principles from *Croton tiglium* L. *J. Med. Chem. 6*:616-617.
Van Duuren, B. L., Sivak, A., Segal, A., Orris, L., and Langseth, L.
    (1966). The tumor-promoting agents of tabacco leaf and tobacco smoke
    condensate. *J. Natl. Cancer Inst. 37*:519-526.
Van Duuren, B. L., Segal, A., Tseng, S.-S., Rusch, G. M., Loewengart,
    G., Maté, U., Roth, D., Smith, A., Melchionne, S., and Seidman, I.
    (1978). Structure and tumor-promoting activity of analogues of anthralin
    (1,8-dihydroxy-9-anthrone). *J. Med. Chem. 21*:26-31.
Wakabayashi, K., Fujimoto, Y., Oohashi, Y., Kuwabara, N., and Fukuda, Y.
    (1979). Induction of colorectal tumors and its early lesions by degraded
    carrageenan in mice. In *Naturally Occurring carcinogens—Mutagens and
    Modulators of Carcinogenesis*, J. C. Miller, J. A. Miller, I. Hirono, T.
    Sugimura, and S. Takayama (Eds.). University Park Press, Baltimore,
    pp. 127-138.
Wang, C. Y., Chiu, C. W., Pamakcu, A. M., and Bryan, G. T. (1976).
    Identification of carcinogenic tannin isolated from bracken fern (*Pteridium
    aquilinum*). *J. Natl. Cancer Inst. 56*:33-36.
Watanabe, K., Yoshii, H., Iwashita, H., Muta, K., Hamada, Y., Hamada, K.,
    Isaka, H., and Nishi, M. (1975). Intestinal tumors of rats by gastric or
    intestinal administration of cycad extract and cycasin. *Gann 66*:449-453.
Watanabe, K., Reddy, B. S., Wong, C. Q., and Weisburger, J. H. (1978).
    Effect of dietary undegraded carrageenan on colon carcinogenesis in F344
    rats treated with azoxymethane or methylnitrosurea. *Cancer Res. 38*:
    4427-4430.
Weber, J., and Hecker, E. (1978). Cocarcinogens of the diterpene ester
    type from *Croton flavens* L. and esophageal cancer in Curaçao. *Experi-
    entia 34*:679-682.
Webster, G. L. (1975). Conspectus of a new classification of the Euphorbi-
    aceae. *Taxon 24*:593-601.

Weinstein, I. B., and Troll, W. (1977). National Cancer Institute workshop on tumor promotion and cofactors in carcinogenesis (meeting report). *Cancer Res.* 37:3461-3463.

Weisburger, J. H. (1976). Bioassays and tests for chemical carcinogens. In *Chemical Carcinogens*, C. E. Searle (Ed.). ACS Monogr. 173. American Chemical Society, Washington, D.C., pp. 1-23.

Weisburger, E. K. (1975). A critical evaluation of the methods used for determining carcinogenicity. *J. Clin. Pharmacol.* 15:5-15.

Wells, P., Aftergood, L., Alfin-Slater, R. B., and Straus, R. (1974). Effect of sterculic acid upon aflatoxicosis in rats fed diets containing saturated and unsaturated fat. *J. Am. Oil Chem. Soc.* 51:456-460.

Welsch, C. W., and Meites, J. (1970). Effects of reserpine on development of 7,12-dimethylbenzanthracene induced mammary tumors in female rats. *Experientia* 26:1133-1134.

Whiting, M. G. (1963). Toxicity of cycads. *Econ. Bot.* 17:271-302.

Williams, P. D., and Murphy, G. P. (1971). Dose related effects of cycasin induced renal and hepatic tumors. *Res. Commun. Chem. Pathol.* 2:627-632.

Windholz, M. (Ed.). (1976). *The Merck Index*, Merck, Rahway, N.J.

Wislocki, P. G., Borchert, P., Miller, J. A., and Miller, E. C. (1976). The metabolic activation of the carcinogen 1'-hydroxysafrole in vivo and in vitro and the electrophilic reactivities of possible ultimate carcinogens. *Cancer Res.* 36:1686-1695.

Wislocki, P. G., Miller, E. C., Miller, J. A., McCoy, E. C., and Rosenkrantz, H. S. (1977). Carcinogenic and mutagenic activities of safrole, 1'-hydroxy-safrole, and some known or possible metabolites. *Cancer Res.* 37:1883-1891.

Woelfel, W. C., Spies, J. W., and Cline, J. K. (1941). Cancer of the mouth. I. Some chemical aspects of buyo-cheek cancer. *Cancer Res.* 1:748-749.

Wogan, G. N. (1977). Mycotoxins and other naturally occurring carcinogens. In *Advances in Modern Toxicology*, Vol. 3: *Environmental Cancer*, H. F. Kraybill and M. A. Mehlman (Eds.). Hemisphere, Washington, D.C., pp. 263-290.

Wynder, E. L., and Hoffmann, D. (1969). Bioassays in tobacco carcino-genesis. *Prog. Exp. Tumor Res.* 11:163-193.

Wynder, E. L., and Hoffmann, D. (1979). Tobacco and health. A societal challenge. *N. Engl. J. Med.* 300:894-903.

Yagi, F., Tadera, K., and Kobayashi, A. (1980). Simultaneous determination of cycasin, methylazoxymethanol and formaldehyde by high performance liquid chromatography. *Agric. Biol. Chem.* 44:1423-1425.

Yang, M. G., Mickelsen, O., Campbell, M. E., Laquer, G. L., and Keresztesy, J. C. (1966). Cycad flour used by Guamanians: effects produced in rats by long-term feeding. *J. Nutr.* 90:153-156.

Yang, M. G., Kobayashi, A., and Mickelsen, O. (1972). Bibliography of cycad research. *Fed. Proc.* 31:1543-1546.

Ying, B.-P., Wang, C.-S., Chou, P.-N., Pan, P.-C., and Liu, J.-S. (1977). Studies on the active principles of the root of Yuan-hua (*Daphne genkwa*). *Hua Hsueh Hsueh Pao* 35:103-108.

Yoshihira, K., Fukuoka, M., Kuroyanagi, M., Natori, S., Umeda, M., Morohashi, T., Enomoto, M., and Saito, M. (1978). Chemical and toxico-logical studies on bracken fern, *Pteridium aquilinum* var. *latiusculum*.

I. Introduction, extraction and fractionation of constituents, and toxicological studies including carcinogenicity tests. *Chem. Pharm. Bull. 26*:
2346-2364.

Zamfir, G., and Paladi, L. (1979). Natural synthesis of some carcinogenic
hydrocarbons. *Rev. Med.-Chir. 83*:197-200 (through *Chem. Abstr. 92*:
160466j, 1980).

Zayed, S., Adolf, W., Hafez, A., and Hecker, E. (1977a). New highly
irritant 1-alkyldaphnane derivatives from deveral species of Thymelaeaceae.
Tetrahedron Lett. 1977:3481-3482.

Zayed, S., Hafez, A., Adolf, W., and Hecker, E. (1977b). New tigliane and
daphnane derivatives from *Pimelea prostrata* and *Pimelea simplex*. *Experientia 33*:1554-1555.

Zechmeister, K., Brandl, F., Hoppe, W., Hecker, E., Opferkuch, H. J.,
and Adolf, W. (1970). Structure determination of the new tetracyclic
diterpene ingenol-triacetate with triple product methods. *Tetrahedron
Lett. 1970*:4075-4078.

Zedeck, M. S., and Sternberg, S. S. (1974). A model system for studies
of colon carcinogenesis: tumor induction by a single injection of methylazoxymethanol acetate. *J. Natl. Cancer Inst. 53*:1419-1421.

Zedeck, M. S., Sternberg, S. S., Poynter, R. W., and McGowan, J. (1970).
Biochemical and pathological effects of methylazoxymethanol acetate, a
potent carcinogen. *Cancer Res. 30*:801-812.

Zedeck, M. S., Sternberg, S. S., McGowan, J., and Poynter, R. W. (1972).
Methylazoxymethanol acetate: induction of tumors and early effects on
RNA synthesis. *Fed. Proc. 31*:1485-1492.

Zedeck, M. S., Grab, D. J., and Sternberg, S. S. (1977). Differences in
the acute response of the various segments of rat intestine to treatment
with the intestinal carcinogen, methylazoxymethanol acetate. *Cancer
Res. 37*:32-36.

8

AFLATOXINS

MURIEL S. PALMGREN

Tulane University, New Orleans, Louisiana

ALEX CIEGLER*

U.S Department of Agriculture, New Orleans, Louisiana

I. INTRODUCTION (HISTORICAL REVIEW)

Aflatoxins are a group of structurally related secondary fungal metabolites that are carcinogenic, hepatoxic, teratogenic, and immunosuppressive (Ciegler, 1975; Thaxton et al., 1974). *Aspergillus flavus* Link ex Fries and *Aspergillus parasiticus* Speare have been demonstrated to produce aflatoxins on a variety of grains and peanuts, but have also been reported on numerous other agricultural commodities.

The current era of mycotoxin research developed as a direct result of concurrent outbreaks of disease in poultry and fish in diverse geographical locations during 1960. Most notoriety was given to the report of severe losses of turkey poults in the United Kingdom (Blount, 1961) as a result of

*Present affiliation: Computer Science Corporation, Nasa National Space Technology Laboratory, Building 2204, NSTL Station, Mississippi

an unknown etiological agent, the disorder thereby being dubbed turkey X disease. The disease was characterized by acute hepatic necrosis with bile duct hyperplasia. Acute manifestiations included loss of appetite, lethargy, wing weakness, and a distinctive attitude of the head and neck at the time of death (Spensley, 1963). A common factor in the outbreaks was the incorporation of Brazilian groundnut (peanut) meal into the feed (Blount, 1961; Carnaghan and Sargeant, 1961). Symptoms analogous to those of turkey X disease were also reported in outbreaks of farm animals that had consumed feed containing non-Brazilian peanut meal (Asplin and Carnaghan, 1961). The toxic factors in the peanut meal were extracted, isolated, identified, and named aflatoxins (Sargeant et al., 1961; Smith and McKernan, 1962; Nesbitt et al., 1962; van der Zijden et al., 1962; Asao et al., 1963). *A. flavus* was isolated from the contaminated meal and produced the toxic factors upon re-inoculation into peanut meal (Sargeant et al., 1961). These factors were named aflatoxins B_1, B_2, G_1, and G_2; the letters referred to their fluorescent colors (B, blue; G, green) and the subscripts to their relative position on chromatography plates.

A. flavus was associated with previous outbreaks of toxic syndromes prior to 1961. Seibold and Bailey (1952) described a disease in dogs that they termed "hepatitis X." Later, the etiology of the disease was determined to be the peanut meal in their diet and was reproduced by feeding the toxic meal (Newberne et al., 1955). Moldy corn poisoning of swine and cattle, characterized by liver lesions, was associated with the presence of *A. flavus* and another fungus, *Penicillium rubrum* (Burnside et al., 1957). Later, moldy corn poisoning in swine was demonstrated to be the same disease as hepatitis X in dogs (Bailey and Groth, 1959).

Poultry also were affected by dietary components that were contaminated with *A. flavus*. Toxigenic strains of fungi including *A. flavus* were associated with "hemorrhagic syndrome" in poultry (Forgacs et al., 1958). These early episodes of toxic syndromes are difficult to attribute solely to the presence of aflatoxin because the toxin was not isolated from the feed, there were other toxigenic fungi present, and aflatoxin alone did not reproduce all facets of the syndromes. Therefore, aflatoxin was probably responsible in part for the symptoms, but other toxins may also have contributed to produce the toxic effects. The importance of *A. flavus* and aflatoxin in toxic syndromes remained undefined until the 1960s and the occurrence of turkey X disease.

Symptoms analogous to those of turkey X disease were reported in outbreaks of other farm animals that had consumed feed whose peanut meal had originated from non-Brazilian sources (Asplin and Carnaghan, 1961). Sources of toxic peanuts and peanut meal were reported from Kenya (Carnaghan and Sargeant, 1961), Uganda (Sargeant et al., 1961), and West Africa (Sargeant et al., 1961). Cottonseed meal in the United States was also reported as being contaminated with aflatoxin (Jackson et al., 1968). The scope of the problem escalated from an isolated incident concerning primarily British farmers to a problem of worldwide concern. Additional alarm was generated when the toxic meal was demonstrated to be capable of producing hepatomas in laboratory animals (Lancaster et al., 1961; Dickens and Jones, 1963). Aflatoxins were subsequently found to be widespread, representing a toxic and carcinogenic threat to a variety of farm animals, and potentially a threat to humans consuming contaminated commodities.

Prior to detection of toxins, in feed or food, certain features are useful in identifying potential outbreaks of aflatoxicosis or mycotoxicoses in general. Although these features were described some time ago (Feuell, 1966), they are still relevant:

1. Mycotoxicoses are not transmissible.
2. Drug and antibiotic treatment have little or no effect on the disease.
3. In field outbreaks, the trouble is often seasonal.
4. The outbreak is usually associated with a specific food or feedstuff.
5. The degree of toxicity is often influenced by the age, sex, and nutritional status of the host.
6. The examination of the suspected food or feed reveals signs of fungal activity.

Patterns of both animal and human outbreaks of aflatoxicosis exhibit these features.

II. PRODUCTION

A. flavus was initially implicated with turkey X disease and the production of aflatoxins. Since then, aflatoxin production by *A. flavus* and *A. parasiticus* has been well documented. *A. parasiticus* appears to be prevalent primarily in tropical and semitropical areas and all isolates to date have been toxin producers (Hesseltine et al., 1968, 1970). However, this is not so with *A. flavus* isolates. Whether aflatoxin plays any role for the fungus or as a selective factor is unknown and has puzzled investigators.

There have been other reports that attribute aflatoxin production to various other fungi. However, these reports have not been confirmed. On the contrary, extensive studies that tested isolates of these fungi have failed to demonstrate that any fungi other than *A. flavus* or *A. parasiticus* could produce aflatoxins (Bullerman and Ayres, 1968; Hesseltine et al., 1966; Mislivec et al., 1968; Parrish et al., 1966; Wilson et al., 1968). Wilson and his colleagues (1968) suggested several ways in which false positive results could have been obtained by inadvertent contamination of substrates with prior fungal growth or glassware contaminated with aflatoxins. Another possibility was that other fluorescent metabolites had R_f values on thin-layer chromatography (TLC) plates that were similar to aflatoxin (Detroy et al., 1971). After the development of confirmation tests for aflatoxins (Andrellos and Reid, 1964; Przybylski, 1975; Ashoor and Chu, 1975; Stack and Pohland, 1975), it was possible to distinguish between aflatoxins and other metabolites that appear similar on TLC plates. The general consensus now is that only *A. flavus* and *A. parasiticus* are capable of producing aflatoxins.

Widespread contamination of a variety of food and feed components reflects the ubiquitous distribution of *A. flavus* in nature. Corn and peanuts represent commodities that provide the greatest possibility for contamination of food and feed, while soybeans appear resistant to toxin production but not to fungal contamination. Formation of aflatoxin under laboratory conditions on various substrates should be distinguished from natural occurrence of the toxin. Many substrates will support toxin production under laboratory conditions, but are not contaminated with the toxins naturally in the field.

Initially, production of aflatoxin in the field was considered to be a problem of improper storage. However, investigations proved that *A. flavus* invaded corn and produced aflatoxin before harvest (Anderson et al., 1975;

Lillehoj et al., 1975a, 1976a). Similar results were obtained for cottonseed and peanuts (Marsh et al., 1969; Pettit and Taber, 1968). Thus the scope of the problem now includes cultivation as well as storage practices.

Natural occurrence and production in the field are affected by numerous parameters, including dissemination of the fungus, predisposing stresses on the plant, and cultivation techniques. Insect damage has been linked with field occurrence, which implicates insects as vectors (Ashworth et al., 1971; Fennell et al., 1975; Lillehoj et al., 1976b; McMeans and Brown, 1975; Hamsa and Ayres, 1977). Other parameters investigated include: variety and region in which planted (McMeans et al., 1977; Zuber et al., 1976; Lillehoj et al., 1975b); moisture, drought, and irrigation stresses (Dickens and Pattee, 1966; Dickens and Satterwhite, 1971; Diener et al., 1965; Pettit et al., 1971; Hamsa and Ayres, 1977; Russell et al., 1976); stage of fruit development and weather (Lillehoj and Hesseltine, 1977); harvesting practice, crop sequence, and weather, including temperature cycles (Pettit and Taber, 1968; Stutz and Krumperman, 1976); strain of fungus and composition of a mixed microbial mileau (Lillehoj et al., 1976b; Naguib and El-Khadem, 1976; Ashworth et al., 1965). Development of varieties more resistant to fungal invasion and toxin production may effectively alter many of these parameters.

Laboratory production studies have defined the optimum and range of conditions under which aflatoxins are produced. One requirement for toxin production that was recognized in early studies was zinc (Mateles and Adye, 1965; Lee et al., 1966; Davis et al., 1967). Production on liquid medium and solid substrates are described in detail by Detroy and colleagues (1971) and by Heathcote and Hibbert (1978). For the biosynthetic pathway of aflatoxin production, the reader is referred to the excellent review by Steyn et al. (1980).

III. DETECTION

Numerous methods exist for qualitative and quantitative analysis of aflatoxins. Generally, they employ extraction with an organic solvent, removal of or separation from interfering compounds, and observation of the characteristic fluorescence at 254 nm.

A number of "minicolumn" procedures have been developed (Holaday, 1968; Velasco, 1972; Pons et al., 1973; Shannon et al., 1973; Holaday and Lansden, 1975). Basically, they pass an aliquot of a solvent extract through a small glass column containing one or more layers of absorbent. The fluorescing bands are compared to those on a column on which an aflatoxin standard has been added. The procedures vary depending on the commodities being tested.

Official quantitative methods for aflatoxins are established for various commodities by collaborative testing supervised by the Association of Official Analytical Chemists (AOAC) and their official referees. The AOAC procedures (AOAC, 1975) employ a representative sampling technique (the amount varies from commodity to commodity), solvent extraction (one phase and two phase systems), use of precipitating agents (removal of protein and other interfering substances such as pigments), cleanup column chromatography (Silica gel, cellulose) and TLC separation of the aflatoxins with measurement by fluorodensitometry or visual methods.

Modifications of AOAC procedures have been adapted for use with various commodities. The commodities for which procedures exist include: peanut, peanut meal, cottonseeds and cottonseed meal, copra (AOAC, 1975; Ayres et al., 1970; Engelbrecht et al., 1965; Marsh et al., 1973; McKinney, 1975; Pons and Goldblatt, 1975; Robertson et al., 1966; Velasco and Whitaker, 1975), corn (Shotwell, 1977), roasted corn (Shannon and Shotwell, 1975), mixed feeds, nuts (Ayres et al., 1970; Romer, 1975), cocoa beans (IUPAC, 1973), groundnuts and groundnut products (Coomes and Sanders, 1963), coffee beans (Levi and Borker, 1968), and cereal grains (Hesseltine, 1974).

Adequate sampling is the key to accurate measurements of aflatoxins in agricultural commodities. Often, aflatoxin contamination is localized within a lot with individual kernels containing high levels of toxin and others, none at all. Evaluation of the true level of contamination of the lot would necessitate extraction of the entire lot. Since this is impossible, selection of large representative samples is achieved by probes or stream splitters. Samples then are ground and blended for uniform toxin distribution prior to 50-g aliquots being removed for analysis. In this manner, each aliquot represents a close approximation of the entire lot.

The use of high-pressure liquid chromatography (HPLC) is an attractive alternative to the TLC fluorodensitometry and visual analytical procedures, because of the high resolution potential, rapidity of separations, and potentially improved quantitative accuracy and precision (Garner, 1975; Seitz, 1975; Hsieh et al., 1976). Its limitations are that large numbers of samples cannot be quantitated as rapidly as with thin-layer chromatography and that interfering substances contained in extracts obtained from naturally contaminated foods and feeds impede the measurement of aflatoxins. However, for the detection of aflatoxins in smaller numbers of urine, blood, or other tissue samples, HPLC offers distinct advantages.

IV. STRUCTURAL DIVERSITY OF AFLATOXINS

Originally, the toxic factors isolated from feed were separated chromatographically into four distinct compounds: aflatoxins B_1, B_2, G_1, and G_2 (Nesbitt et al., 1962; Sargeant et al., 1961) (Fig. 1). The molecular formulas indicated that aflatoxins B_2 and G_2 (AFB_2 and AFG_2) were dihydro derivatives of the parent AFB_1 and G_1, respectively (Asao et al., 1963; Chang et al., 1963; Cheung and Sim, 1964; van der Merwe et al., 1963; van Dorp et al., 1963; van Soest and Peerdeman, 1964). A subsequent report proposed structural formulas, which were later verified by chemical synthesis of aflatoxin B_1 (Büchi et al., 1966; Knight et al., 1966). Aflatoxins contain a coumarin nucleus fused to a bifuran and either a pentanone (AFB_1 and B_2) or a six-membered lactone (AFG_1 and G_2). AFB_1 and AFG_1 were more toxic to ducklings, rats, and fish than either AFB_2 or AFG_2, with AFB_1 being the most toxic (Wogan et al., 1971; Abedi and Scott, 1969). A similar pattern holds for its carcinogenic potency, $AFB_1 > AFG_1 > AFB_2$ (Wogan et al., 1971; Ayres et al., 1971).

Hydroxylated aflatoxin derivatives, called AFM_1 and AFM_2, were reported in the milk of cows fed toxic rations (Allcroft and Carnaghan, 1963; DeIongh et al., 1964, 1965) (Fig. 2). These derivatives were subsequently isolated from lactating, rats, rat liver, and sheep urine, liver, and kidneys, as well

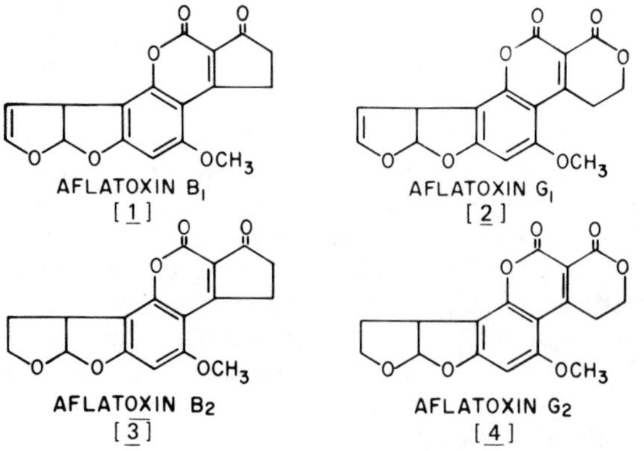

FIGURE 1 Structures of aflatoxins B_1, B_2, G_1, and G_2.

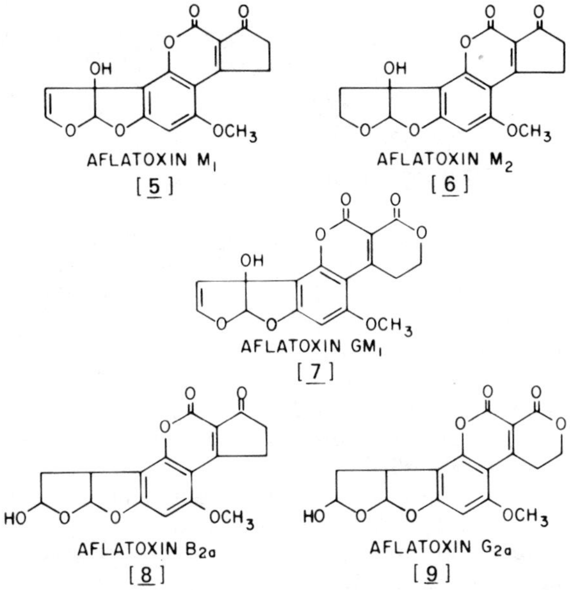

FIGURE 2 Hydroxylated aflatoxin derivatives.

as from moldy peanuts and corn (DeIongh et al., 1964, 1965; Butler and Clifford, 1965; Allcroft et al., 1966; Masri et al., 1967; Holzapfel et al., 1966; Shotwell et al., 1976). AFM_1 was established to be as toxic but less carcinogenic than AFB_1 (Purchase, 1967; Wogan and Paglialunga, 1974; Canton et al., 1975).

$AFGM_1$, a hydroxylated derivative of AFG_1, was isolated from *A. flavus* cultures (Heathcote and Dutton, 1969) (Fig. 2). However, $AFGM_1$ is a very minor natural metabolite. Other derivatives produced by *A. flavus*, the 2-hydroxyaflatoxins, have been reported under several names by various scientists. They are known as AFB_{2a} and AFG_{2a} (Dutton and Heathcote, 1966, 1968), but have also been described as AFB_1 hemiacetal (Büchi et al., 1966; Pohland et al., 1968), aflatoxin-W (Andrellos and Reid, 1964), and hydroxydihydroaflatoxin B_1 (Ciegler and Peterson, 1968) (Fig. 2). AFB_{2a} and G_{2a} are relatively nontoxic (Dutton and Heathcote, 1966).

A possible precursor in the biosynthesis of aflatoxins was isolated from rice and wheat inoculated in the laboratory (Stubblefield et al., 1970). In place of the terminal pentanone ring was an ethanol moiety. It was described as 6-methoxy-7-(2'-hydroxyethyl)difurocoumarin and named parasiticol (Stubblefield et al., 1970) or AFB_3 (Heathcote and Dutton, 1969 (Fig. 3). AFB_3 is a natural metabolite of *A. flavus* and *A. parasiticus*. *Rhizopus* species are also capable of metabolizing AFG_1 to form AFB_3 (Cole and Kirksey, 1971). Ducklings appear to be as sensitive to AFB_3 as AFB_1, while for chick embryos, AFB_3 is much less toxic than AFB_1 (Stubblefield et al., 1970).

Reduction of the pentanone of AFB_1 by microorganisms yielded aflatoxicol (AFL or AFR_0) (Detroy and Hesseltine, 1970; Robertson et al., 1970). The importance of AFL increased dramatically when AFL was demonstrated to be produced by animals and its highly toxic and carcinogenic activities became apparent (Patterson and Roberts, 1971; Schoenhard et al., 1974, 1976).

Animals also metabolize AFB_1 by O-demethylation to produce AFP_1 (Fig. 4). It has been detected in the urine of several animal species (Wogan et al., 1967; Dalezios et al., 1971; Bassir and Emafo, 1970). AFP_1 is less toxic than AFB_1 (Büchi et al., 1973). In vitro metabolism by liver microsomes from monkeys and humans produced a major metabolite of AFB_1 called AFQ_1 that contained a hydroxyl group at the C-3 position (Masri et al., 1974; Steyn et al., 1974; Büchi et al., 1974) (Fig. 4).

Detoxification procedures using ammonia produced two major derivatives of AFB_1. They were called AFD_1 and AFD_2 (Lee et al., 1974; Kiermeier and

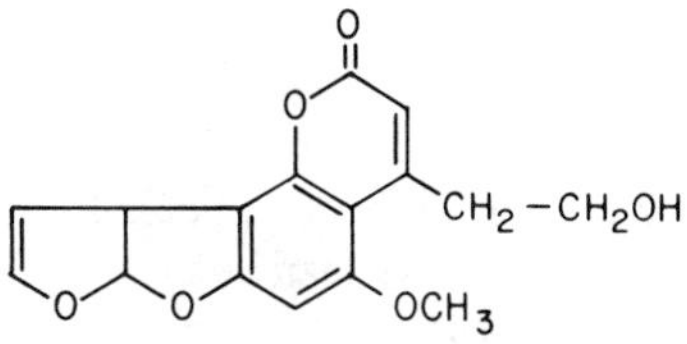

AFLATOXIN B_3
(Parasiticol)
[10]

FIGURE 3 Parasiticol.

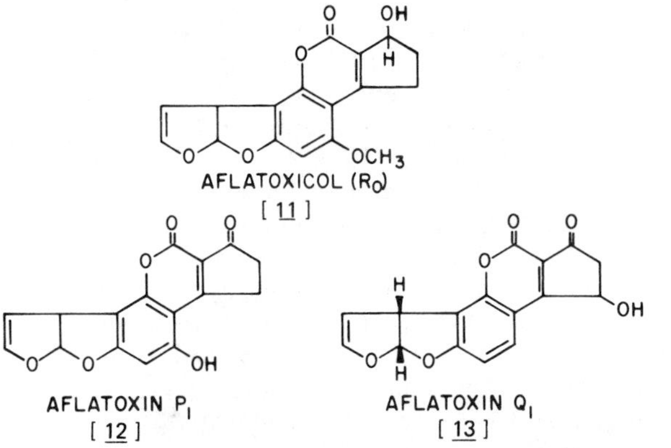

AFLATOXICOL (R₀)
[11]

AFLATOXIN P₁
[12]

AFLATOXIN Q₁
[13]

FIGURE 4 Metabolized forms of aflatoxin B_1.

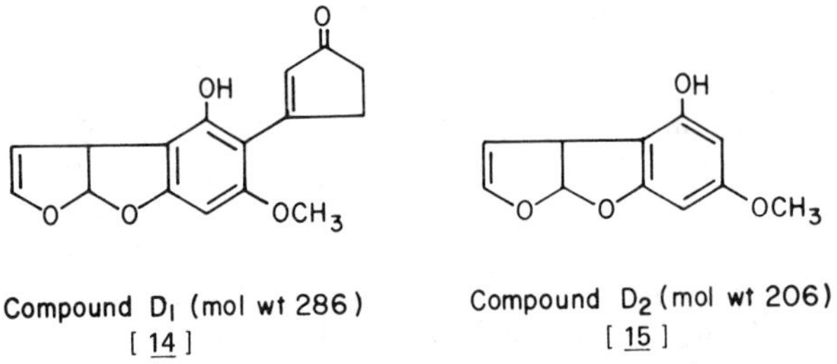

Compound D₁ (mol wt 286)
[14]

Compound D₂ (mol wt 206)
[15]

FIGURE 5 Degradation products of aflatoxin B_1.

Ruffer, 1974; Cucullu et al. , 1976). The six-membered lactone ring in AFB_1 was disrupted by treatment with ammonia (Fig. 5).

The structures described represent most of the major metabolites and de-rivatives of the four original aflatoxins.

V. CARCINOGENESIS OF AFLATOXINS

Carcinogenic effects of aflatoxins have been documented in several species: rat, mouse, trout, duck, and monkey (Ashley et al. , 1964; Adamson et al. , 1976). Aflatoxins are suspected to have a role in human cancers as well (Kraybill and Shimkin, 1964; Alpert et al. , 1971). Most frequently, the tar-get organ is the liver. However, in some circumstances, primary tumors have also been observed in the kidney (Merkow et al. , 1973) of animals given AFB_1.

In early studies, crude extracts containing mixtures of the naturally occurring aflatoxins were fed to animals. The results from these studies cannot be attributed solely to AFB_1. When procedures for increased yields and improved isolation of AFB_1 were developed, some studies were conducted in which pure AFB_1 was incorporated into diets or injected. The first evidence that aflatoxin was carcinogenic was reported in 1961 when 9 of 11 weanling rats fed for 7 months on a purified diet containing 20% toxic peanut meal developed multiple liver tumors, with lung metastases in two animals. In one study, a single oral dose of 12.7 mg/kg of a crude extract was given that contained 40% AFB_1 and 60% G_1. After 26 months, 7 of 18 rats developed liver tumors (Carnaghan, 1967). There are numerous other studies using crude extracts, which have been reviewed extensively (Enomoto and Saito, 1972; Heathcote and Hibbert, 1978; Linsell and Peers, 1972; Mirocha et al., 1968; Newberne and Rogers, 1973; Ong, 1975; Wogan, 1973).

The rat is a species that is highly susceptible to the carcinogenic effects of aflatoxin B_1. The target organ is the liver. In the rat, aflatoxin B_1 induces multicentric hepatocellular carcinoma without cirrhosis or much fibrosis (Carnaghan, 1967). The nodules are soft, friable, and a yellowish gray color. In later stages, tumors may be hemorrhagic and necrotic. Metastasis to the lung is common. The development of liver lesions progresses through various types of damage. Early lesions include bile duct hyperplasia, parenchymal cell damage, necrosis, and megalocytosis. This damage usually regresses. Focal hyperplasia of parenchymal cells occurs and is followed by nodule formation and hepatocellular carcinoma (Newberne and Wogan, 1968).

Extrinsic factors may modify the effects of aflatoxins (i.e., diet, preexisting cirrhosis, and hepatitis B virus). Diets low in protein enhanced toxic effects but not carcinogenicity in rats (Madhavan and Gopalan, 1965). In areas with a high incidence of primary liver carcinoma, there often is a high incidence of hepatitis B virus antigen (HB Ag) in patients' serum. These same areas often were ones with high levels of aflatoxin in food. When serum and liver tissue from patients with primary liver cancer were examined, there was a good correlation with the presence of HB Ag and/or aflatoxins.

Evidence for the role of aflatoxins in human carcinogenicity is very difficult to gather. The data consist of epidemiological studies and clinical examinations of people in developing nations where agricultural commodities are contaminated with aflatoxins. Uganda, Indonesia, and South Africa have a high incidence of primary liver cancer (Alpert et al., 1971; Pang et al., 1974; Kew, 1978). In Uganda, hepatoma incidence was 15 in 100,000. In a northern province of Uganda, an extremely high incidence, 15%, correlated well with high frequency and high concentrations of aflatoxin in the food. It was estimated that the per capita ingestion of aflatoxin B_1 in that province could be in the range of 0.02-2 mg daily (Alpert et al., 1971), a level that was carcinogenic in nonhuman primates.

Aflatoxin B_1 has been detected in the serum and liver tissue of patients with primary liver carcinoma. Four of five cases from Nigeria had detectable levels, while 17 controls and 3 other patients with cirrhosis rather than carcinoma had no detectable level of aflatoxin (Onyemelukwe et al., 1980). In the United States, aflatoxin B_1 was detected in two cases of primary liver carcinoma (Philips et al., 1976; Wray and Hayes, 1980).

VI. TOXICITY OF AFLATOXINS

Many animal species are susceptible to the acute, toxic effects of aflatoxins. Among the more susceptible species of mammals and birds to AFB_1 are rabbits and ducklings (LD_{50} value approximately 0.3 mg/kg), while female adult rats of the Porton-Wistar strain are among the less susceptible species (LD_{50} approximately 18 mg/kg) (Lijinsky and Butler, 1966; Carnaghan, 1967; Butler, 1964). For a comprehensive review of the acute toxicity of both laboratory and farm animals, the reader is referred to the recent work of Wyllie and Morehouse (1978).

The teratogenic action of AFB_1 has been reported in chickens (Bassir and Adekunle, 1970). However, the mutagenic action has been investigated more extensively. In HeLa cells, both single- and double-stranded DNA breaks occurred in cultures exposed to AFB_1 (Umeda et al., 1972). In the Ames mutagenicity assay, which uses *Salmonella typhimurium*, AFB_1 was one of the most potent compounds of the 300 they tested (McCann et al., 1975). Microsomal activation was necessary for AFB_1 to be mutagenic.

Suspected cases of acute aflatoxicosis in humans have been reported from India (Krishnamachari et al., 1974, 1975). Over 200 villages in western India experienced an outbreak of a disease affecting humans and dogs that was characterized by jaundice, rapidly developing ascites, portal hypertension, and a high mortality rate, with death usually resulting from massive gastrointestinal tract bleeding. The disease was confined to the very poor, who were forced by economic circumstances to consume badly molded corn containing aflatoxin between 6.25 and 15.6 ppm, an average daily intake per victim of 2-6 mg of aflatoxin. Analyses of liver, sera, and urine were inconclusive, probably because of a time lag between corn consumption and sample collection.

Cases of a children's disease may also be linked with acute aflatoxin ingestion. Reports of encephalopathy and fatty degeneration of the viscera (EFDV) in northeast Thailand have suggested that aflatoxins may be involved (Bourgeois et al., 1971; Olson et al., 1971; Shank et al., 1971). In northeast Thailand, where there is a high aflatoxin incidence in the food, EFDV is a common cause of death among children. Almost all cases occur in children from rural areas, with the incidence increasing during the latter part of the rainy season. The symptoms of EFDV were described originally by Reye and his colleagues (1963) in Australia, who characterized the features of the illness in children. EFDV is also referred to as Reye's syndrome (RS). Symptoms included disturbed consciousness, fever, convulsions, vomiting, disturbed respiratory rhythm, altered muscle tone, and altered reflexes. Levels of serum glutamic-pyruvic acid transaminase and serum glutamic-oxalacetic acid transaminase were elevated. Often hypoglycaemia and low cerebrospinal fluid glucose were noted. The onset was usually associated with coughing, rhinorrhea, sore throat, or earache. Seventeen of twenty-one cases (81%) diagnosed were fatal. At necropsy, there was cerebral swelling, a slightly enlarged, firm yellow liver, and a pale, slightly widened renal cortex.

The criteria for diagnosing RS that the Centers for Disease Control (CDC) use are as follows:

1. Acute onset of encephalopathy
2. Evidence of hepatic involvement from liver biopsy autopsy, or serum glutamic-oxalic transaminase (SGOT) and serum glutamic-pyruric transaminase (SGPT) levels twice normal

3. No other reasonable explanation of these findings (CDC, 1974)

Using these criteria, CDC had recorded more than 250 cases of RS in the United States and 139 in Thailand in the period 1963-1974. The number of cases has increased in subsequent years.

There are similarities between the symptoms of acute aflatoxin B_1 poisoning and RS. The pathological, biochemical, and clinical responses of young cynomologus monkeys to aflatoxin included many of those found in RS such as cerebral edema with neuronal degeneration, marked fatty degeneration of the liver, heart, and kidneys, hypoglycemia, and increased levels of transaminases. Cough, vomiting, diarrhea, and coma were also characteristic in the monkeys (Bourgeois et al., 1971).

Various reports have documented the presence of aflatoxins in the food associated with RS cases and in tissues, urine, and blood from RS patients (Dvorackova et al., 1977; Shank et al., 1971; Ryan et al., 1979). AFB_1 has been associated with cases of RS in New Zealand, Thailand, Czechoslavakia, and the United States (Becroft and Webster, 1971; Shank et al., 1971; Dvorackova et al., 1977; Ryan et al., 1979; Chaves-Carballo et al., 1976).

A case control study was conducted recently in the United States (Nelson et al., 1980). Dietary histories, medical information, blood, and urine were collected from hospitalized children demonstrating RS symptoms, from their families, and from neighborhood children. Statistical analysis of food preference questionnaires for 17 patients and 43 control children revealed that corn bread and cornmeal were consumed more frequently by children having measurable levels of aflatoxins. All peanut products, milk, and other corn products were consumed at equivalent rates by all the children. However, no food product was consumed more frequently by RS patients than by control subjects (Nelson et al., 1980). Therefore, corn bread and cornmeal were implicated as likely sources of aflatoxin for those children with aflatoxin in their blood or urine. Blood and urine were examined for aflatoxins. Of 17 patients, 2 had detectable levels of aflatoxin B_1 in their urine only. Of 108 patients, family members and control subjects tested, 23 people (21%) had detectable levels of aflatoxin B_1 or M_1 in either their blood or urine; levels ranged from 2.4 to 170 ng/ml. These results indicated that aflatoxin does enter the diet of some Americans. However, detectable levels of aflatoxins were not statistically different between RS patients and neighborhood controls. There still is the possibility of insult by aflatoxin prior to exposure to a viral agent, which would compromise the immune system, damage the liver, and lead to RS (Mullen, 1978). By the time the RS patient was hospitalized, levels of aflatoxin could be sufficiently reduced to be undetectable. Further investigation in the human population and animal models is warranted.

Infection with influenza viruses type A and type B, as well as herpes zoster (chickenpox virus) had preceded most if not all cases of RS (Reye et al., 1963; CDC, 1974). Viruses have been isolated from some patients with RS (Wilson et al., 1980). However, when cynomogologus monkeys were given aflatoxin B_1, they also exhibited symptoms resembling viral infections, anorexia, cough, vomiting, and in some, diarrhea (Bourgeois et al., 1971). Neither virus injection nor AFB_1 alone appears to produce the complete cymptom complex of RS. Therefore, RS may be a result of a combined assault of

these two and/or other agents on an immature liver. There is also some epidemological evidence indicating that pesticides may induce hepatic damage resulting in RS-like symptoms. RS or RS-like disease may therefore result from
a complex array of etiological agents and circumstances.

Although the physiological cause of central nervous system damage has not
been determined, hyperammonemia has been implicated, since it has been observed in RS patients. Elevated blood ammonia may be responsible for the
encephalopathy, coma, and brain damage. This elevation of ammonia was suggested to be caused by the reduced activities of mitochrondrial enzymes in
the urea cycle. Damage to mitochrondria (e.g., swelling and pleomorphism)
was visible in electron micrographs of liver biopsies from RS patients (Partin
et al., 1971). Reduced activity of hepatic carbamyl phosphate synthetase
(CPS) and ornithine transcarbamylase (OTC) was observed in the first days
of clinical symptoms of RS, but returned to normal in the next week (Snodgrass and DeLong, 1976; Sinatra et al., 1975). In view of the clinical and
histological evidence of disturbances in mitochondrial enzymes of the urea
cycle, it appears likely that hyperammonemia is the result of hepatic damage
and subsequent central nervous system effects in RS.

Various agents have been examined for their effect on CPS and OTC activities. Since the onset of RS is frequently preceded by viral infection, the
effect of influenza A and B infections on urea cycle enzymes has been investigated (Pierson et al., 1976). Reductions in the activity of CPS, 12%, and
OTC, 17%, were observed when mice were infected with influenza A. These
reductions were not of the magnitude present with RS nor were they sufficient to produce hyperammonemia. The reductions probably represented a
nonspecific result of pneumonia.

Exogenous toxins, which have been implicated in Jamaican vomiting sickness, have also been considered as potential pathological agents. One such
toxin, a short-chain fatty acid, 4-pentanoic acid, was incubated with rat
liver. It reduced the activity of CPS and also reduced the V_{max} of OTC but
not its activity (Sinatra et al., 1975). These results do not reflect the
pattern typical of RS.

The effect of AFB_1 on urea cycle enzymes was also investigated. The
activities of CPS and OTC in rats were measured at 2 and 24 hr after treatment with AFB_1. The rats had been given AFB_1 at a dosage of 3 mg/kg body
weight. At 24 hr, AFB_1 significantly reduced the activities of both of the
enzymes, CPS 61% of normal and OTC 67% of normal; there was no significant
reduction at 2 hr (Thurlow et al., 1980). Although these reductions are considerably greater than those produced by influenza A, they are still not as
severe as those reported in RS. Species differences or synergism between
these or other agents are possibly explanations for AFB_1 alone not having
greater reductions in enzyme activity. These results offer some evidence for
the involvement of AFB_1 but are not conclusive.

VII. AFLATOXIN METABOLISM

Aflatoxin B_1 can be metabolized by higher mammals and birds via several
pathways, resulting in several transformation products. Various workers
have indicated that aflatoxin B_1 requires activation to become metabolically
active (Garner, 1973a, 1976; Garner et al., 1972; Swenson et al., 1975).

Garner (1976) postulated that of the various derivatives formed, aflatoxin B_{2a} and the epoxide represent the active forms responsible for acute toxicity and carcinogenicity, whereas the other derivatives are probably detoxification products. However, some of these latter products (e.g., aflatoxin M_1 and aflatoxicol) are still toxic and carcinogenic.

Different species exhibit varying degrees of susceptibility to the acutely toxic and carcinogenic actions of aflatoxin B_1, presumably as a result of differences in their metabolism of the toxin. Susceptibility to each action is independent of the other. Mice are resistant to acute toxic effects of aflatoxin B_1 as well as the carcinogenic action. Rats are moderately susceptible to the acute effects but highly susceptible to the carcinogenic insult, whereas guinea pigs are more susceptible to the acute effect than rats, but resistant to the carcinogenic effect (Newberne and Butler, 1969). Monkeys are also susceptible to the acute action of AFB_1, but relatively resistant to the carcinogenic action (Wong and Hsieh, 1980). A thorough comparison of the metabolism and kinetics of AFB_1 in the monkey, rat, and mouse revealed much about susceptibility to the acute and carcinogenic actions of AFB_1 (Wong and Hsieh, 1980). The monkey represents a model species relatively resistant to the carcinogenic action of AFB_1, but sensitive to the acute action. In this animal, AFB_1 was rapidly distributed and taken up by the tissues, resulting in a high tissue concentration of AFB_1. This high tissue concentration, coupled with high metabolic activity of the tissues, would explain the sensitivity of the monkey to the acute effects more than to the chronic carcinogenic effects. In addition, the monkey had the lowest excretion rate of the three species (monkey, rat, and mouse), thereby contributing to higher intrabody levels of toxin during the first 24 hr.

There appears to be some difficulty in finding a consistent pattern of metabolism that correlates with the differences in the toxic and carcinogenic responses of various species. In vitro studies that used liver fractions of several species, including humans, failed to demonstrate a correlation between metabolites of AFB_1 and species susceptibility (Roebuck and Wogan, 1977). The ability of species to produce aflatoxicol (AFR_0) may interfere with such correlations. AFR_0 can be produced by such species as chickens, ducks, turkeys, rabbits, and primates, while none has been detected in mice, rats, or guinea pigs (Patterson and Roberts, 1971, 1972; Salhab and Hsieh, 1975). Loveland et al. (1977) presented important evidence to show that AFR_0 could be converted back to AFB_1 by trout postmitochondrial enzymes; hence AFR_0 could serve as a reservoir for the release of AFB_1 to target cells, thereby enhancing the toxic effect of AFB_1. Other data indicate that the species ability to reduce AFB_1 to AFR_0 is related to the sensitivity of the species to acute aflatoxicosis and that the ratio of reductive and oxidative activities is an index of species susceptibility to the carcinogenic effect of AFB_1 (Hsieh et al., 1977).

The rat represented a species relatively sensitive to the carcinogenic action of AFB_1, but resistant to the acute action. The volume of distribution of AFB_1 was lower and redistribution by release of the toxin from the tissues occurred, which caused a rise in plasma levels. Therefore, the concentration in tissues was not as high. Aflatoxicol (AFR_0) was a major metabolite in the plasma of the rat, but not in the monkey or mouse. The authors (Wong and Hsieh, 1980) suggest that AFR_0 formation may be important in the carcinogenic activity of AFB_1, since AFR_0 is easily converted back to AFB_1, thereby prolonging tissue exposure to AFB_1 but at lower levels.

The mouse was relatively resistant to both carcinogenic and acute actions of AFB_1. It had a low volume of distribution, rapid excretion, and metabolism to primarily AFP_1. The mouse also produced more water-soluble urinary conjugates than did either the monkey or rat. These differences in tissue distribution metabolism and excretion may explain the variation in species susceptibility to AFB_1. Liver protein levels as well as aniline hydroxylase and N-demethylase enzymes levels were statistically lower in the duck, which is more sensitive to both acute toxicity and carcinogenicity than is the rat (Thabrew and Bababunmi, 1980).

Aniline hydroxylase and N-demethylase are enzymes responsible for modifying key structural features of aflatoxins. The rate at which AFB_1 was metabolized to AFB_{2a} correlated to species susceptibility to acute toxic reactions. Rapid production of AFB_{2a} by a species rendered it more susceptible (Paterson, 1976). Thus the levels of protein and enzymes, which would affect the rate of metabolism by these enzymes, correlated with the degree of susceptibility of a species.

Species susceptibility may be also related to the binding capacity of hepatic macromolecules and plasma proteins for AFB_1 and its metabolites. Binding of AFB_1 to nucleic acids and proteins in target organs has been demonstrated and may represent a mechanism for its carcinogenic and acutely toxic actions (Garner, 1973b). Binding capacities in mice, guinea pigs, and rats paralleled their relative susceptibilities (Ueno et al., 1980). In mice, there was less radiolabeled AFB_1 bound to hepatic macromolecules and plasma proteins than in either guinea pigs or rats; this correlates with the greater resistance of mice to the acute toxicity of AFB_1. Guinea pigs had a greater accumulation of radiolabel in hepatic RNA and protein than rats, which correlates with the greater susceptibility of guinea pigs than rats to acute toxicity of AFB_1. The lower level of radiolabel found in guinea pig hepatic DNA inversely correlates with the greater resistance of guinea pigs than rats to the carcinogenic action of AFB_1.

Binding of aflatoxin B_1 metabolites to nucleic acids appears to have an effect on the susceptibility of a species to the carcinogenicity of aflatoxin B_1. In vivo binding levels of AFB_1 metabolites to nucleic acids were greater in the rat, the more susceptible species, than the hamster (Garner and Wright, 1975). Similar results were obtained when liver slices were incubated with AFB_1 and the level of binding to nucleic acids determined (Garner, 1980).

ACKNOWLEDGMENTS

The authors are most grateful to L. S. Lee and A. J. DeLucca II for their assistance in the preparation and editing of this chapter.

REFERENCES

Abedi, A. H., and Scott, P. M. (1969). Detection of toxicity of aflatoxins, sterigmatocystin, and other fungal toxins by lethal action on zebra fish larvae. *Anal. Chem. 52*:963-969.

Adamson, R. H., Correa, P., Sieber, S. M., McIntire, K., and Dalgard, D. W. (1976). Carcinogenicity of aflatoxin B_1 in *Rhesus* monkeys: two additional cases of primary liver cancer. *J. Natl. Cancer Inst. 57*:67-68.

Allcroft, R., and Carnaghan, R. B. A. (1963). Toxic properties in ground-nuts, biological effects. *Chem. Ind. (Lond.)*, pp. 50-53.

Allcroft, R., Rogers, H., Lewis, G., Nabney, J., and Best, P. E. (1966). Metabolism of aflatoxin in sheep: excretion of the "milk toxin." *Nature (Lond.)* 209:154-155.

Alpert, M. E., Hutt, M. S. R., Wogan, G. N., and Davidson, C. S. (1971). Association between aflatoxin content of food and hepatoma frequency in Uganda. *Cancer* 28:253-260.

Anderson, H. W., Nehring, E. W., and Wichser, W. R. (1975). Aflatoxin contamination of corn in the field. *J. Agric. Food Chem.* 23:775-782.

Andrellos, P. J., and Reid, G. R. (1964). Confirmatory tests for aflatoxin B_1. *J. Assoc. Off. Anal. Chem.* 47:801-803.

Asao, T., Büchi, G., Abdel-Kader, M. M., Chang, S. B., Wick, E. L., and Wogan, G. N. (1963). Aflatoxins B and G. *J. Am. Chem. Soc.* 85:1706-1707.

Ashley, L. M., Halver, J. M., and Wogan, G. N. (1964). Hepatoma and aflatoxicosis in trout. *Fed. Proc.* 23:105.

Ashoor, S. H., and Chu, F. S. (1975). New confirmatory test for aflatoxins B_1 and B_2. *J. Assoc. Off. Anal. Chem.* 53:617-618.

Ashworth, L. J., Jr., Schroeder, H. W., and Langley, B. C. (1965). Afla-toxins: environmental factors governing occurrence in peanuts. *Science* 148:1288-1289.

Ashworth, L. J., Jr., Rice, R. E., McMeans, J. L., and Brown, C. M. (1971). The relationship of insects to infection of cotton bolls by *Aspergillus flavus*. *Phytopathology* 61:488-493.

Asplin, F. D., and Carnaghan, R. B. A. (1961). The toxicity of certain groundnut meals for poultry with special reference to their effect on ducklings and chickens. *Vet. Rec.* 73:1215-1219.

Association of Official Analytical Chemists. (1975). Natural poisons. In *Official Methods of Analysis of AOAC*, 12th ed., W. Horowitz (Ed.). AOAC, Arlington, Va., pp. 462-482.

Ayers, J. L., Lillard, H. S., and Lillard, D. A. (1970). Mycotoxins: detection in foods. *Food Technol.* 24:55-60.

Ayres, J. L., Lee, D. J., Wales, J. H., and Sinnhuber, R. O. (1971). Aflatoxin structure and hepatocarcinogenicity in rainbow trout. *J. Natl. Cancer Inst.* 46:561-564.

Bailey, W. S., and Groth, A. H., Jr. (1959). The relationship of hepatitis X of dogs and moldy corn poisoning of swine. *J. Am. Vet. Med. Assoc.* 134:514-516.

Bassir, O., and Adekunle, A. (1970). Tetratogenic action of aflatoxin B_1, palmotoxin B_0 and palmotoxin G_0 on the chick embryo. *J. Pathol.* 102:49-51.

Bassir, O., and Emafo, P. O. (1970). Oxidative metabolism of aflatoxin B_1 by mammalian liver slices and microsomes. *Biochem. Parmacol.* 19:1681-1687.

Becroft, D. M. O., and Webster, D. R. (1971). Aflatoxins and Reye's disease. *Br. Med. J.* 4:117-118.

Blount, W. P. (1961). Turkey "X" disease. *Turkeys (J. Br. Turkey Fed.)* 9:52,55-58, 61, 77.

Bourgeois, C. H., Shank, R. C., Grossman, R. A., Johnson, D. O., Wooding, W. L., and Chandavinol, P. (1971). Acute aflatoxin B_1 toxicity in the macaque and its similarities to Reye's syndrome. *Lab. Invest.* 24:206-216.

Büchi, G., Foulkes, D. M., Kurono, M., and Mitchell, G. F. (1966). The total synthesis of racemic aflatoxin B_1. *J. Am. Chem. Soc.* 88:4534-4536.

Büchi, G., Spitzner, D., Paglialunga, S., and Wogan, G. N. (1973). Synthesis and toxicity evaluation of aflatoxin P_1. *Life Sci.* 13:1143-1149.

Büchi, G., Muller, P. M., Roebuck, B. D., and Wogan, G. N. (1974). Aflatoxin Q_1: a major metabolite of aflatoxin B_1 produced by human liver. *Res. Commun. Chem. Pathol. Pharmacol.* 8:585-592.

Bullerman, L. B., and Ayres, J. C. (1968). Aflatoxin producing potential of fungi isolated from cured and aged meats. *Appl. Microbiol.* 16:1945-1946.

Burnside, J. E., Sippel, W. L., Forgacs, J., Carll, W. T., Atwood, M. B., and Doll, E. R. (1957). A disease of swine and cattle caused by eating moldy corn. *Am. J. Vet. Res.* 18:817-824.

Butler, W. H. (1964). Acute toxicity of aflatoxin B_1 in rats. *Br. J. Cancer* 18:756-762.

Butler, W. H., and Clifford, J. I. (1965). Extraction of aflatoxin from rat liver. *Nature (Lond.)* 206:1045-1047.

Canton, J. H., Kroes, R., van Logten, M. J., van Schothorst, M., Stavenuiter, J. F. C., and Verhülsdonk. (1975). The carcinogenicity of aflatoxin M_1 in rainbow trout. *Food Cosmet. Toxicol.* 13:441-443.

Carnaghan, R. B. A. (1967). Hepatic tumors and other chronic liver changes in rats following a single oral administration of aflatoxin. *Br. J. Cancer* 21:811-814.

Carnaghan, R. B. A., and Sargeant, K. (1961). The toxicity of certain groundnut meals to poultry. *Vet. Rec.* 73:726-727.

Center for Disease Control. (1974). Influenza surveillance—Reye's syndrome and viral infections. *Morbid. Mortal. Wkly. Rep.* 23:58

Chang, S. B., Kader, A., Wick, E. L., and Wogan, G. N. (1963). Aflatoxin B_2: chemical identity and biological activity. *Science* 142:1191-1192.

Chaves-Carballo, E., Ellefson, R. D., and Gomez, M. R. (1976). An aflatoxin in the liver of a patient with Reye-Johnson Syndrome. *Mayo Clin. Proc.* 51:48-50.

Cheung, K. K., and Sim, G. A. (1964). Aflatoxin G_1: direct determination of the structure by the method of isomorphous replacement. *Nature (Lond.)* 201:1185-1188.

Ciegler, A. (1975). Mycotoxins: occurrence, chemistry, biological activity. *Lloydia* 38:21-34.

Ciegler, A., and Peterson, R. E. (1968). Aflatoxin detoxification: hydroxy-dihydro-aflatoxin B_1. *Appl. Microbiol.* 16:665-668.

Cole, R. J., and Kirksey, J. W. (1971). Aflatoxin G_1 metabolism by *Rhizopus* species. *J. Agric. Food Chem.* 19:222-223.

Coomes, T. J., and Sanders, J. C. (1963). The detection and estimation of aflatoxin in groundnuts and groundnut material. *Analyst (Lond.)* 88:209-213.

Cucullu, A. F., Lee, L. S., Pons, W. A., Jr., and Stanley, J. B. (1976). Ammoniation of aflatoxin B_1: isolation and characterization of a product with molecular weight 206. *J. Agric. Food Chem.* 24:408-410.

Dalezios, J., Wogan, G. N., and Weinreb, S. M. (1971). Aflatoxin P_1: a new aflatoxin metabolite in monkeys. *Science* 171:584-585.

Davis, N. D., Diener, U. L., and Agnihotri, V. P. (1967). Production of aflatoxins B_1 and G_1 in chemically defined medium. *Mycopathol. Mycol. Appl. 31*:251-256.

DeIongh, H., Vles, R. O., and van Pelt, J. G. (1964). Milk of mammals fed on aflatoxin-containing diet. *Nature (Lond.) 202*:466-467.

DeIongh, H., Vles, R. O., and deVogel, P. (1965). The occurrence and detection of aflatoxin in food. In *Mycotoxins in Foodstuffs*, G. N. Wogan (Ed.). MIT Press, Cambridge, Mass., pp. 235-245.

Detroy, R. W., and Hesseltine, C. W. (1970). Aflatoxicol: structure of a new transformation product of aflatoxin B_1. *Can. J. Biochem. 48*:830-832.

Detroy, R. W., Lillehoj, E. B., and Ciegler, A. (1971). Aflatoxin and related compounds. In *Microbial Toxins*, Vol. VI: *Fungal Toxins*, A. Ciegler, S. Kadis, and S. J. Ajl (Eds.). Academic Press, New York, pp. 4-178.

Dickens, F., and Jones, H. E. H. (1963). The carcinogenic action of aflatoxin after its subcutaneous injection in the rat. *Br. J. Cancer 17*:691-698.

Dickens, J. W., and Pattee, H. E. (1966). The effects of time, temperature and moisture on aflatoxin production in peanuts inoculated with a toxic strain of *Aspergillus flavus*. *Trop. Sci. 8*:11-12.

Dickens, J. W., and Satterwhite, J. B. (1971). Diversion program for farmers stock peanuts with high concentrations of aflatoxin. *Oleagineux 26*:321-328.

Diener, U. L., Jackson, C. R., Cooper, W. E., Stipes, R. D., and Davis, N. D. (1965). Invasion of peanut pods in the soil by *Aspergillus flavus*. *Plant Dis. Rep. 49*:931-935.

Dutton, M. F., and Heathcote, J. G. (1966). Two new hydroxyaflatoxins. *Biochem. J. 101*:21-22.

Dutton, M. F., and Heathcote, J. G. (1968). The structure, biochemical properties and origin of the aflatoxins B_{2a} and G_{2a}. *Chem. Ind. (Lond.)*, pp. 418-421.

Dvorackova, I., Kusak, V., Vesely, D., Vesela, J., and Nesnidal, P. (1977). Aflatoxin and encephopathy with fatty degeneration of viscera (Reye). *Ann. Nutr. Aliment. 31*:977-990.

Engelbrecht, R. H., Ayres, J. L., and Sinnhuber, R. O. (1965). Isolation and determination of aflatoxin B_1 in cottonseed meals. *J. Assoc. Off. Anal. Chem. 48*:815-818.

Enomoto, M., and Saito, M. (1972). Carcinogens produced by fungi. *Annu. Rev. Microbiol. 26*:279-312.

Fennell, D. I., Lillehoj, E. B., and Kwoles, W. F. (1975). *Aspergillus flavus* and other fungi associated with insect damaged field corn. *Cereal Chem. 52*:314-321.

Feuell, A. J. (1966). Toxic factors of mold origin. *Trop. Sci. 8*:61-70.

Forgacs, J., Koch, H., Carll, W. T., and White-Stevens, R. H. (1958). Additional studies on the relationship of mycotoxicoses to the poultry hemorrhagic syndrome. *Am. J. Vet. Res. 19*:744-753.

Garner, R. C. (1973a). Chemical evidence for the formation of a reactive aflatoxin B_1 metabolite by hamster liver microsomes. *FEBS Lett. 36*:261.

Garner, R. C. (1973b). Microsomal-dependent binding of aflatoxin B_1 to DNA, RNA, polynucleotides and protein in vitro. *Chem. Biol. Interact. 11*:123-131.

Garner, R. C. (1975). Aflatoxin separation by high pressure liquid chroma-
    tography. *J. Chromatogr.* *103*:186-188.
Garner, R. C. (1976). The role of epoxides in bioactivation and carcino-
    genesis. In *Progress in Drug Metabolism*, J. W. Bridge and L. F.
    Chassead (Eds.). Wiley, New York, pp. 77-128.
Garner, R. C. (1980). Carcinogenesis by fungal products. *Br. Med. Bull.*
    36:47-52.
Garner, R. C., and Wright, C. M. (1975). Binding of [14C]aflatoxin B_1 to
    cellular macromolecules in the rat and hamster. *Chem. Biol. Interact.* *11*:
    123-132.
Garner, R. C., Miller, E. C., and Miller, J. A. (1972). Liver microsomal
    metabolism of aflatoxin B_1 to a reactive derivative toxic to *Salmonella*
    typhimurium TA 1530. *Cancer Res.* *32*:2058-2066.
Hamsa, T. A. P., and Ayres, J. C. (1977). Factors affecting aflatoxin con-
    tamination of cottonseed. I. Contamination of cottonseed with *Aspergillus*
    flavus at harvest and during storage. *J. Am. Oil Chem. Soc.* *54*:219-224.
Heathcote, J. G., and Dutton, M. F. (1969). New metabolites of *Aspergillus*
    flavus. *Tetrahedron* *25*:1497-1500.
Heathcote, J. G., and Hibbert, J. R. (1978). *Aflatoxins: Chemical and*
    Biological Aspects, Vol. I: *Developments in Food Science*, Elsevier, New
    York.
Hesseltine, C. W. (1974). Natural occurrence of mycotoxins in cereals.
    Mycopathol. Mycol. Appl. *53*:141-153.
Hesseltine, C. W., Shotwell, O. L., Ellis, J. J., and Stubblefield, R. D.
    (1966). Aflatoxin formation by *Aspergillus flavus*. *Bacteriol. Rev.* *30*:
    795-805.
Hesseltine, C. W., Shotwell, O. L., Smith, M., Ellis, J. J., Vandergraft, E.,
    and Shannon, G. (1968). Production of various aflatoxins by strains of
    the *Aspergillus flavus* series. *Proc. First U.S.-Japan Conf. Toxic*
    Microorganisms, M. Herzberg (Ed.). UJNR Joint Panels on Toxic Micro-
    organisms and U.S. Dept. Interior, Washington, D.C., pp. 202-210.
Hesseltine, C. W., Sorrenson, W. G., and Smith, M. (1970). Taxonomic
    studies of the aflatoxin-producing strains in the *Aspergillus flavus* group.
    Mycologia *62*:123-132.
Holaday, C. E. (1968). Rapid method for detecting aflatoxins in peanuts.
    J. Am. Oil Chem. Soc. *45*:680-682.
Holaday, C. E., and Lansden, J. (1975). Rapid screening method for afla-
    toxin in a number of products. *J. Agric. Food Chem.* *23*:1134-1136.
Holzapfel, C. W., Steyn, P. S., and Purchase, I. F. H. (1966). Isolation
    and structure of aflatoxins M_1 and M_2. *Tetrahedron Lett.*, pp. 2799-
    2803.
Hsieh, D. P. H., Fitzell, D. L., Miller, J. L., and Seiber, J. N. (1976).
    High-pressure liquid chromatography of oxidative aflatoxin metabolites.
    J. Chromatogr. *117*:474-479.
Hsieh, D. P. H., Wong, Z. A., Wong, J. J., Michas, C., and Ruebner, B.
    H. (1977). Comparative metabolism of aflatoxin. In *Mycotoxins in Human*
    and Animal Health, J. V. Rodricks, C. W. Hesseltine, and M. A. Mehlman
    (Eds.). Pathotox, Park Forest South, Ill., pp. 37-50.
International Union of Pure and Applied Chemistry (IUPAC). (1973).
    Recommended method for aflatoxin analysis in cocoa beans. *IUPAC Inf.*
    Bull. Tech. Rep. *8*:1-12.

Jackson, W. E., Wolf, H., and Sinnhuber, R. O. (1968). The relationship of hepatoma in rainbow trout to aflatoxin contamination and cottonseed meal. *Cancer Res. 28*:987-991.

Kew, M. C. (1978). Hepatocellular cancer in southern Africa. In *Primary Liver Tumors*, H. Remmer, H. M. Bolt, P. Bannasch, and H. Pepper (Ed.). University Park Press, Baltimore, pp. 179-183.

Kiermeier, F., and Ruffer, L. Z. (1974). Aflatoxin formation in milk and milk products. XXIV. Changes of aflatoxin B_1 in alkaline solutions. *Z. Lebensm. Unters. Forsch. 155*:129-141.

Knight, J. A., Roberts, J. C., Roffey, P., and Sheppard, A. H. (1966). Synthesis of ($\pm$)-tetrahydrodeoxo-aflatoxin-B_1, a racemic form of the laevoratatory hydrogenation product of aflatoxin B_1. *Chem. Commun.*, pp. 706-707.

Kraybill, H. F., and Shimkim, M. B. (1964). Carcinogenesis realted to foods contaminated by processing and fungal metabolites. *Adv. Cancer Res. 8*: 191-248.

Krishnamachari, K. A., Bhat, R. V., Nagaragan, V., and Tilak, T. B. G. (1974). Investigations into an outbreak of hepatitis in parts of western India. *Indian J. Med. Res. 63*:1036-1040.

Kirishnamachari, K. A., Bhat, R. V., Nagarajan, V., and Tilak, T. B. G. (1975). Hepatitis due to aflatoxicosis. An outbreak in western India. *Lancet*, pp. 1061-1062.

Lancaster, M. C., Jenkins, F. P., and Philip, J. McL. (1961). Toxicity associated with certain samples of groundnuts. *Nature (Lond.) 192*: 1095-1096.

Lee, E. G.-H., Townsley, P. M., and Walden, C. C. (1966). Effect of bivalent metals on the production of aflatoxins in submerged cultures. *J. Food Sci. 31*:432-436.

Lee, L. S., Stanley, J. B., Cucullu, A. F., Pons, W. A., Jr., and Goldblatt, L. A. (1974). Ammoniation of aflatoxin B_1: isolation and identification of the major reaction product. *J. Assoc. Off. Anal. Chem. 57*: 626-631.

Levi, C. P., and Borker, E. (1968). Survey of green coffee for potential aflatoxin contamination. *J. Assoc. Off. Anal. Chem. 51*:600-602.

Lijinsky, W., and Butler, W. H. (1966). Purification and toxicity of aflatoxin G_1. *Proc. Soc. Exp. Biol. Med. 123*:151-154.

Lillehoj, E. B., and Hesseltine, C. W. (1977). Aflatoxin control during plant growth and harvest of corn. In *Mycotoxins in Human and Animal Health*, J. V. Rodricks, C. W. Hesseltine, and M. A. Niehlman (Eds.). Pathotox, Park Forest South, Ill., pp. 107-120.

Lillehoj, E. B., Kwolek, W. F., Shannon, G. M., Shotwell, O. L., and Hesseltine, C. W. (1975a). Aflatoxin occurrence in 1973 corn at harvest. I. A limited survey in the southeastern U.S. *Cereal Chem. 52*:603-611.

Lillehoj, E. B., Kwolek, W. F., Vandergraft, E. E., Zuber, M. S., Calvert, O. H., Widstrom, N., Futrell, M. C., and Bockholt, A. J. (1975b). Aflatoxin production in *Aspergillus flavus* inoculated ears of corn grown at diverse locations. *Crop Sci. 15*:267-270.

Lillehoj, E. B., Fennell, D. I., and Kwolek, W. F. (1976a). *Aspergillus flavus* and aflatoxin in Iowa corn before harvest. *Science 193*:495-496.

Lillehoj, E. B., Kwolek, W. F., Peterson, R. E., Shotwell, O. L., and Hesseltine, C. W. (1976b). Aflatoxin contamination, fluorescence and insect damage in corn infected with *Aspergillus flavus* before harvest. *Cereal Chem.* *53*:505-512.

Linsell, C. A., and Peers, F. G. (1972). The aflatoxins and human liver cancer. In *Current Problems in the Epidemology of Cancer and Lymphomas*, E. Grundmann and H. Tulinius (Eds.). Springer-Verlag, New York, pp. 193-207.

Loveland, P. M., Sinnhuber, R. O., Berggren, K. E., Libbey, L. H., Nixon, J. E., and Pawlowski, N. E. (1977). Formation of aflatoxin B_1 from aflatoxicol by rainbow trout (*Salmo gairdneri*) in vitro. *Res. Commun. Chem. Pathol. Pharmacol.* *16*:167-170.

Madhavan, T. V., and Gopalan, C. (1965). Effect of dietary protein on aflatoxin liver injury in weanling rats. *Arch. Pathol.* *80*:123-126.

Marsh, P. B., Simpson, M. E., Ferretti, R. J., Campbell, T. C., and Donoso, J. (1969). Relation of aflatoxins in cottonseed at harvest to fluorescence in the fiber. *J. Agric. Food Chem.* *17*:462-467.

Marsh, P. B., Simpson, M. E., Craig, H. O., Donoso, D., and Ramey, G. H. (1973). Occurrence of aflatoxins in cotton seeds at harvest in realtion to location of growth and field temperature. *J. Environ. Qual.* *2*:276.

Masri, M. S., Lundin, R. E., Page, J. R., and Garci, V. C. (1967). Crystalline aflatoxin M_1 from urine and milk. *Nature (Lond.)* *215*:753-755.

Masri, M. S., Haddon, W. F., Lundin, R. E., and Hsieh, D. P. H. (1974). Aflatoxin Q_1. A newly identified major metabolite of aflatoxin B_1 in monkey liver. *J. Agric. Food Chem.* *22*:514-555.

Mateles, R. I., and Adye, J. C. (1965). Production of aflatoxins in submerged culture. *Appl. Microbiol.* *13*:208-211.

McCann, J., Choi, E., Yamasaki, E., and Ames, B. N. (1975). Detection of carcinogens as mutagens in the *Salmonella*/microsome test: assay of 300 chemicals. *Proc. Natl. Acad. Sci. USA* *72*:5135-5139.

McKinney, J. D. (1975). Use of zinc acetate in extract purification for aflatoxin assay of cottonseed products. *J. Am. Oil Chem. Soc.* *52*:213.

McMeans, J. L., and Brown, C. M. (1975). Aflatoxin in cottonseed as affected by the pink bollworm. *Crop Sci.* *15*:865-866.

McMeans, J. L., Brown, C. M., McDonald, R. L., and Parker, L. L. (1977). Aflatoxins in cottonseed: a comparison of two cultivars. *Crop Sci.* *17*:707-709.

Merkow, L. P., Epstein, S. M., Slifkin, M., and Pardo, M. (1973). The ultrastructure of renal neoplasms induced by aflatoxin B_1. *Cancer Res.* *33*:1608-1614.

Mirocha, C. J., Christensen, C. M., and Nelson, G. H. (1968). Toxic metabolites produced by fungi implicated in mycotoxicoses. *Biotechnol. Bioeng.* *10*:468-482.

Mislivec, P. B., Hunter, J. H., and Tuite, J. (1968). Assay for aflatoxin production by the genera *Aspergillus* and *Penicillium*. *Appl. Microbiol.* *16*:1053-1055.

Mullen, P. W. (1978). Immunopharmacological considerations in Reye's syndrome: a possible xenobiotic initiated disorder. *Biochem. Pharmacol.* *27*:145-149.

Naguib, M. M., and El-Khadem, M. (1976). Toxicity of *Aspergillus flavus* cultures isolated from Egypt. *Zentralbl. Bakteriol.* *131*:506-509.

Nelson, D. B., Kimbrough, R., Landrigan, P. S., Hayes, A. W., Yang, G. C., and Benanides, J. (1980). Aflatoxin and Reye's syndrome: a case control study. *Pediatrics 66*:865-869.

Nesbitt, B. F., O'Kelly, J., Sargeant, K., and Sheridan, A. (1962). Toxic metabolites of *Aspergillus flavus. Nature (Lond.) 195*:1062-1063.

Newberne, P. M., and Butler, W. H. (1969). Acute and chronic effects of aflatoxin on liver of domestic and laboratory animals. *Cancer Res. 29*: 236-250.

Newberne, P. M., and Rogers, A. E. (1973). Animal model of human disease: primary hepatocellular carcinoma. *Am. J. Pathol. 72*:137-141.

Newberne, P. M., and Wogan, G. N. (1968). Sequential morphological changes in aflatoxin B_1 carcinogenesis in the rat. *Cancer Res. 28*:770-781.

Newberne, P. W., Bailey, W. S., and Siebold, H. R. (1955). Notes on a recent outbreak and experimental reproduction of hepatitis X in dogs. *J. Am. Vet. Med. Assoc. 127*:59-62.

Olson, L. C., Bourgeois, C. H., Jr., Cotton, R. B., Harikul, S., Grossman, R. A., and Smith, T. J. (1971). Encephalopathy and fatty degeneration of the viscera in northeastern Thailand. Clinical syndrome and epidemiology. *Pediatrics 47*:707-716.

Ong, T.-M. (1975). Aflatoxin mutagenesis. *Mutat. Res. 32*:35-53.

Onyemelukwe, C. G., Nirodi, C., and West, C. E. (1980). Aflatoxin B_1 in hepatocellular carcinoma. *Trop. Georgr. Med. 32*:237-240.

Pang, R. T. L., Karyadi, H., and Karyadi, D. (1974). Aflatoxin and primary hepatic cancer in Indonesia. *5th World Congr. Gastroenterol.*, Mexico, October 13-19, 1974, pp. 15-47.

Parrish, F. W., Wiley, B. J., Simmons, E. G., and Long, L. (1966). Production of aflatoxin and kojic acid by *Aspergillus* and *Penicillum. Appl. Microbiol. 14*:139.

Partin, J. C., Schubert, W. K., and Partin, J. S. (1971). Mitochrondrial ultrastructure in Reye's syndrome (encephalopathy and fatty degeneration of the viscera). *N. Engl. J. Med. 285*:1339-1343.

Patterson, D. S. P. (1976). Structure, metabolism and toxicity of the aflatoxins: a review. *Nutr. Diet. Suppl. 2*:71-76.

Patterson, .D S. P., and Roberts, B. A. (1971). The in vitro reduction of aflatoxins B_1 and B_2 by soluble avian liver enzymes. *Food Cosmet. Toxicol. 9*:829-837.

Patterson, D. S. P., and Roberts, B. A. (1972). Aflatoxin metabolism in duck-liver homogenates: the relative importance of reversible cyclopentenone reduction and hemiacetal formation. *Food Cosmet. Toxicol. 10*: 501-512.

Pettit, R. E., and Taber, R. A. (1968). Factors influencing aflatoxin accumulation in peanut kernels and the associated mycoflora. *Appl. Microbiol. 16*:1230-1234.

Pettit, R. E., Taber, R. A., Schroeder, H. W., and Harrison, A. L. (1971). Influence of fungicides and irrigation practice on aflatoxin in peanuts before digging. *Appl. Microbiol. 22*:629-634.

Phillips, D. L., Yourtree, D. M., and Searles, S. (1976). Presence of aflatoxin B_1 in human liver in the United States. *Toxicol. Appl. Pharmacol. 36*:403-406.

Pierson, D., Knight, V., Hansard, P., and Chan, E. (1976). Hepatic carbamyl phosphate synthetase and ornithine transcarbamylase in mouse influenza A and influenza B infection. *Proc. Soc. Exp. Biol. Med. 152*: 67-70.

Pohland, A. E., Cushmac, M. E., and Andrellos, P. J. (1968). Aflatoxin B hemiacetal. *J. Assoc. Off. Anal. Chem. 51*:907-910.

Pons, W. A., Jr., and Goldblatt, L. A. (1975). The determination of aflatoxins in cottonseed products. *J. Am. Oil Chem. Soc. 42*:471-475.

Pons, W. F., Jr., Cucullu, A. F., Franz, A. O., Jr., Lee, L. S., and Goldblatt, L. A. (1973). Rapid detection of aflatoxin contamination of aflatoxins in cottonseed products. *J. Am. Oil Chem. Soc. 56*:803-807.

Przybylski, W. (1975). Formation of aflatoxin derivatives on thin layer chromatographic plates. *J. Assoc. Off. Anal. Chem. 58*:163-164.

Purchase, I. H. F. (1967). Acute toxicity of aflatoxins M_1 and M_2 in one-day-old ducklings. *Food Cosmet Toxicol. 5*:339-342.

Reye, R. D. K., Morgan, G., and Baral, J. (1963). Encephalopathy and fatty degeneration of the viscera—A disease entity in childhood. *Lancet*, pp. 749-752.

Robertson, G. A., Lee, L. S., Cucullu, A. F., and Goldblatt, L. A. (1966). Determination of aflatoxins in individual peanuts and peanut sections. *J. Am. Oil Chem. Soc. 43*:89-92.

Robertson, J. A., Teunisson, D. J., and Boudreaux, G. J. (1970). Isolation and structure of a biologically reduced aflatoxin B_1. *J. Agric. Food Chem. 18*:1090-1091.

Roebuck, B. D., and Wogan, G. N. (1977). Species comparison of aflatoxin B. *Cancer Res. 37*:1649-1656.

Romer, F. R. (1975). Screening method for the detection of aflatoxins in mixed feeds and other agricultural commodities with subsequent confirmation and quantitative measurement of aflatoxins in positive samples. *J. Assoc. Off. Anal. Chem. 58*:500-506.

Russell, T. E., Watson, T. F., and Ryan, G. F. (1976). Field accumulation of aflatoxin in cottonseed as influenced by irrigation termination dates and pink bollworm infestation. *Appl. Environ. Microbiol. 31*:711-713.

Ryan, N. J., Hogan, G. R. Hayes, A. W., Unger, P. D., and Siray, M. Y. (1979). Aflatoxin B_1, its role in the etiology of Reye's syndrome. *Pediatrics 64*:71-75.

Salhab, A. S., and Hsieh, D. P. H. (1975). Aflatoxin H_1: a major metabolite of aflatoxin B_1 produced by human and *Rhesus* monkey livers in vitro. *Res. Commun. Chem. Pathol. Pharmacol. 10*:419-431.

Sargeant, K., Sheridan, A., O'Kelly, J., and Carnaghan, R. B. A. (1961). Toxicity associated with certain samples of groundnuts. *Nature (Lond.) 192*:1096-1097.

Schoenhard, G. L., Lee, D. L., and Sinnhuber, R. O. (1974). Aflatoxin B_1 activation and aflatoxicol toxicity in rainbow trout (*Salmo gairdneri*). *Fed. Proc. 33*:247.

Schoenhard, G. L., Lee, D. J., Nowell, G. E., Pawlowski, N. E., Libbey, L. M., and Sinnhuber, R. O. (1976). Aflatoxin B_1 metabolism to aflatoxicol and derivatives lethal to *Bacillus subtilis*. *Cancer Res. 36*:2040-2045.

Seibold, H. R., and Bailey, W. S. (1952). An epizootic of hepatitis in dogs. *J. Am. Vet. Med. Assoc. 121*:201-206.

Seitz, L. M. (1975). Comparison of methods for aflatoxin analysis by high-pressure liquid chromatography. *J. Chromatogr. 104*:81-89.

Shank, R. C., Bourgeois, C. H., Keschamras, N., and Chandavimol, P. (1971). Aflatoxins in autopsy specimens from Thai children with an acute disease of unknown aetiology. *Food Cosmet. Toxicol. 9*:501-507.

Shannon, G. M., and Shotwell, O. L. (1975). A quantitative method for determination of aflatoxin B_1 in roasted corn. *J. Assoc. Off. Anal. Chem. 58*:743-745.

Shannon, G. M., Stubblefield, R. D., and Shotwell, O. L. (1973). Modified rapid screening method for aflatoxin in corn. *J. Assoc. Off. Anal. Chem. 56*:1024-1025.

Shotwell, O. L. (1977). Aflatoxin in corn. *J. Am. Oil Chem. Soc. 54*:216A.

Shotwell, O. L., Goulden, M. L., and Hesseltine, C. W. (1976). Aflatoxin M_1. Occurrence in stored and freshly harvested corn. *J. Agric. Food Chem. 24*:683-684.

Sinatra, F., Yoshida, T., Applebaum, M., Mason, W., Hoogenraad, J., and Sunshine, P. (1975). Abnormalities of carbarmyl phosphate synthetase and ornithine transcarbamylase in liver of patients with Reye's syndrome. *Pediatr. Res. 9*:829-833.

Smith, R. H., and McKernan, W. (1962). Hepatotoxic action of chromato-graphically separated fractions of *Aspergillus flavus* extracts. *Nature (Lond.) 195*:1301-1303.

Snodgrass, P. J., and DeLong, G. R. (1976). Urea-cycle deficiencies and an increased nitrogen load producing hyperammonemia in Reye's syndrome. *N. Engl. J. Med. 294*:855-860.

Spensley, P. C. (1963). Aflatoxin, the active principle in turkey "X" disease. *Endeavor 22*:75-79.

Stack, M. E., and Pohland, A. E. (1975). Collaborative study of a method for chemical confirmation of the identity of aflatoxins. *J. Assoc. Off. Anal. Chem. 58*:110-113.

Steyn, P. S., Vleggaar, R. M., Pitout, J., Steyn, M., and Thiel, P. G. (1974). 3-Hydroxy-aflatoxin B_1: a new metabolite of in vitro aflatoxin B_1 metabolism by vervet monkey (*Cercopithecus aethiops*) liver. *J. Chem. Soc. Perkin Trans. 1*, pp. 2551-2552.

Steyn, P. S., Vleggaar, R., and Wessels, P. L. (1980). The biosynthesis of aflatoxin and its congeners. In *The Biosynthesis of Mycotoxins: A Study in Secondary Metabolism*, P. S. Steyn (Ed.). Academic Press, New York, pp. 105-155.

Stubblefield, R. D., Shotwell, O. L., Shannon, G. M., Weisleder, D., and Rohwedder, W. K. (1970). Parasiticol: a new metabolite from *Aspergillus parasiticus*. *J. Agric. Food Chem. 18*:391-393.

Stutz, K. K., and Krumperman, P. H. (1976). Effect of temperature cycling on the production of aflatoxin by *Aspergillus parasiticus*. *Appl. Environ. Microbiol. 32*:327-332.

Swenson, D. H., Miller, J. A., and Miller, E. C. (1975). The reactivity and carcinogenicity of aflatoxin B_1-2,3,-dichloride, a model for the putative 2,3-oxide metabolite of aflatoxin B_1. *Cancer Res. 35*:3811-3823.

Thabrew, M. I., and Bababumi, E. A. (1980). Levels of microsomal drug-metabolizing enzymes in animals which are highly susceptible to aflatoxin carcinogenicity: the case of the duck. *Cancer Lett. 9*:333-338.

Thaxton, J. P., Tung, H. T., and Hamilton, P. B. (1974). Immunosuppression in chickens by aflatoxin. *Poult. Sci. 53*:721-725.

Thurlow, P. M., Desai, R. K., Newberne, P. M., and Brown, H. (1980). Aflatoxin B_1 acute effects on three hepatic urea cycle enzymes using semi-automatic methods: a model for Reye's syndrome. *Toxicol. Appl. Pharmacol.* *53*:293-298.

Ueno, I., Friedman, L., and Stone, C. L. (1980). Species difference in the binding of aflatoxin B_1 to hepatic macromolecules. *Toxicol. Appl. Pharmacol.* *52*:177-180.

Umeda, M., Tsutsui, T., and Saito, M. (1972). DNA-strand breakage of HeLa cells induced by several mycotoxins. *Jpn. J. Exp. Med.* *42*:527-530.

van der Merwe, K. H., Fourie, L., and de Scott, B. (1963). On the structures of the aflatoxins. *Chem. Ind. (Lond.)*, pp. 1660-1661.

van de Zijden, A. S. M., Koelensmid, W. A. A. B., Boldingh, J., Barrett, C. B., Ord, W. O., and Philip, J. (1962). Isolation in crystalline form of a toxin responsible for turkey X disease. *Nature (Lond.)* *195*:1060-1062.

van Dorp, D. A., van der Zijden, A. S. M., Berrthuis, R. K., Sparreboom, S., Ord, W. Q., de Jong, K., and Keuning, R. (1963). Dihydro-aflatoxin B, a metabolite of *A. flavus*: remarks on the structure of aflatoxin B. *Recl. Trav. Chim.* *82*:587-592.

van Soest, T. C., and Peerdeman, A. F. (1964). An X-ray study of dihydro-aflatoxin B_1. *Kl. Ned. Akad. Wet. Proc. Ser. B* *67*:469-472.

Velasco, J. (1972). Detection of aflatoxin using small columns of florisil. *J. Am. Oil Chem. Sco.* *49*:141-142.

Velasco, J., and Whitaker, T. B. (1975). Sampling cottonseed lots for aflatoxin contamination. *J. Am. Oil Chem. Soc.* *52*:191-195.

Wilson, B. J., Campbell, T. C., Hayes, A. W., and Hanlin, R. T. (1968). Investigation of reported aflatoxin production by fungi outside the *Aspergillus flavus* group. *Appl. Microbiol.* *16*:819-821.

Wilson, R. Miller, J., Greene, H., Rankin, R., Lumeng, L., Gordon, D., Nelson, D., and Noble, G. (1980). Reye's syndrome in three siblings: association with type A influenza infection. *Am. J. Dis. Child* *134*:1032-1034.

Wogan, G. N. (1973). Aflatoxin carcinogenesis. *Methods Cancer Res.* 7: 309-344.

Wogan, G. N., and Paglialunga, S. (1974). Carcinogenicity of synthetic aflatoxin M_1 in rats. *Food Cosmet. Toxicol.* *12*:381-384.

Wogan, G. N., Edwards, G. S., and Shank, R. C. (1967). Excretion and tissue distribution of radioactivity from aflatoxin B_1-^{14}C in rats. *Cancer Res.* *27*:1729-1736.

Wogan, G. N., Edwards, G. S., and Newberne, P. M. (1971). Structure-activity relationships in toxicity and carcingoenicity of aflatoxins and analogs. *Cancer Res.* *31*:1936-1942.

Wong, Z. A., and Hsieh, D. P. H. (1980). The comparative metabolism and toxicokinetics of aflatoxin B_1 in the monkey, rat, and mouse. *Toxicol. Appl. Pharmacol.* *55*:115-125.

Wray, B. B., and Hayes, A. W. (1980). Aflatoxin B_1 in the serum of a patient with primary hepatic carcinoma. *Environ. Res.* *22*:400-403.

Wyllie, T. D., and Morehouse, L. G. (Eds.) (1978). *Mycotoxic Fungi, Mycotoxins, Mycotoxicoses*, Vol. II: *Mycotoxicoses of Domestic and Laboratory Animals, Poultry, and Aquatic Invertebrates and Vertebrates.* Marcel Dekker, New York, pp. 1-653.

Zuber, M. S., Calvert, O. H., Lillehoj, E. B., and Kwolek, W. F. (1976). Preharvest development of aflatoxin B_1 in corn in the United States. *Phytopathology 66*:1120-1121.

9

TOXICITY AND CARCINOGENICITY OF FUNGAL LACTONES: PATULIN AND PENICILLIC ACID.

MURIEL S. PALMGREN

Tulane University, New Orleans, Louisiana

ALEX CIEGLER*

U.S. Department of Agriculture, New Orleans, Louisiana

I. PATULIN

Patulin is a fungal metabolite that was originally investigated because of its potential as an antibiotic. Its chemical structure was determined to be that of a furopyrone (Birkinshaw et al., 1943) (Fig. 1). Early findings indicated that patulin was toxic in varying degrees to a number of microorganisms. Of more than 75 bacterial species tested, none were completely resistant to patulin. Fungi varied in their susceptibility to patulin, while for protozoa, time of incubation and concentration determined their sensitivity to patulin. Testing of humans produced initial favorable reports of patulin's activity against the common cold (Gye, 1943; Hopkins, 1943). However, these results could not be confirmed (Medical Research Council, 1944; Stansfeld et al., 1944). Tests proved that patulin was rather ineffective against influenza virus in mice (Rubin and Giarman, 1947). Deleterious effects appeared in subsequent testing in humans, for example, dermal irritation after topical application

---

*Present affiliation: Computer Science Corporation, NASA National Space Technology Laboratory, Building 2204, NSTL Station, Mississippi

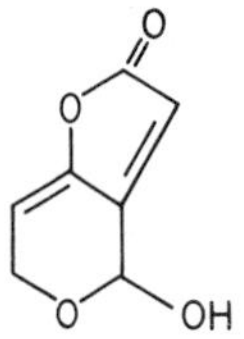

FIGURE 1 Patulin.

(Dalton, 1952) and after oral administration, stomach irritation, nausea, and vomiting (deRosnay et al., 1952; Freerksen and Bönicke, 1951; Walker and Wiesner, 1944). More detailed information on the antibiotic properties of patulin can be obtained from various reviews (Florey et al., 1944; Singh, 1967; Korzybski, 1967).

A. Chemical and Physical Properties

Patulin, 4-hydroxy-4H-furo[3,2-c]pyran-2(6H)-one, is a lactone of fungal origin and has an empirical formula $C_7H_6O_4$ and a molecular weight of 154. It has been given various names, including clavacin, claviformin, expansin, leucopin, mycoin C, penicidin, and tercinin (Singh, 1967). Patulin is a colorless to white crystal that melts at 110.5°C. It is optically inactive and has a single, ultraviolet absorption peak at 276 nm (Katzman et al., 1944). In water and most polar organic solvents, patulin is soluble, but it is insoluble in pentane-hexane. Patulin is unstable in alkali and loses biological activity, but is stable in acid (Chain et al., 1942).

An interesting property of patulin is its reaction with sulfhydryl-containing compounds. It is this adduct formation of patulin with cysteine that has been postulated to be responsible for its antibiotic activity (Geiger and Conn, 1945). The mechanism is complex and can be postulated to involve a Micheal addition, that is, a nucleophilic addition of a carbanion of the sulfhydryl group to the double bond of the unsaturated lactone system of patulin. Analysis of adducts of patulin with cysteine revealed a complex mixture of reaction products, some of which contained more than one atom of sulfur (Ciegler et al., 1976). Data of Ashoor and Chu (1973a,b) indicated that both the sulfhydryl and amino group of cysteine may react with patulin. Adduct formation may account for patulin's disappearance in various commodities and the assumption of some researchers that patulin was inactivated or destroyed (Ashoor and Chu, 1973a,b; Atkinson and Stanley, 1943; Dickens and Cooke, 1965; Hofmann et al., 1971).

B. Production, Detection, and Natural Occurrence

Various species of fungi are capable of producing patulin, including the following species of *Penicillium* and *Aspergillus*: *P. claviforme*, *P. cyclopium*, *P. equinum*, *P. expansum* (= *P. leucopus*), *P. granulatum* (= *P. divergens*), *P. lanosum*, *P. lapidosum*, *P. melinii*, *P. novae-zeelnadiae*, *P. uricae* (= *P.*

patulum and perhaps *P. griseofulvum*), *A. clavatus*, *A. giganteus*, *A. terreus*, and also *Byssochlamys nivea* (= *Gymnoascus* sp.) (Wilson, 1976).

Two synthetic media, Raulin-Thom and Czapek-Dox, and modifications of these, are commonly used for production of patulin by various *Penicillium* and *Aspergillus* spp. The addition of yeast extract or corn-steep liquor to Czapek-Dox medium reduced patulin production by a *Penicillium* sp. (Lochhead et al., 1946) but not with *Aspergillus clavatus* (Katzman et al., 1944). Yields of patulin were higher in stationary cultures than when shaken, higher at 20-25°C than 30°C, and maximum at 8-12 days after inoculation (Lochhead et al., 1946).

Natural substrates that support patulin production include barley; malt; rice (Yamamoto, 1954a); wheat straw (Norstadt and McCalla, 1971); soil-containing roots, leaves, or other residues of apple and other fruit trees (Börner, 1963a,b); and apples (Brian et al., 1956). High yields of patulin were also obtained from *P. urticae* grown on potato-dextrose broth, as high as 2.7 g/liter (Norstadt and McCalla, 1969b).

The biosynthesis of patulin begins with acetate and proceeds through various aromatic intermediates, such as 6-methylsalicyclic acid and gentisaldehyde, to yield patulin. For a discussion of the hypothetical pathways for patulin biosynthesis, the reader is referred to Zamir (1980).

Analyses for patulin in foods have employed primarily thin-layer, gas, and liquid chromatography. Visualization was by colorimetric or fluorometric means. Patulin has been extracted with ethyl acetate, theyl acetate-water mixtures (Scott and Somers, 1968; Pohland et al., 1970), and acetonitrile-hexane mixtures (Pohland and Allen, 1970). Evaporation of the ethyl acetate extract to dryness should be avoided because it causes a loss of patulin (Scott and Kennedy, 1973; Wilson and Nuovo, 1973).

Cleanup procedures have employed silica gel columns (Scott and Somers, 1968), preparative thin-layer chromatography (TLC) (Pohland and Allen, 1970), and solvent partition (Stoloff et al., 1971).

TLC has been used extensively to quantitate patulin. Scott (1974) has presented a thorough review of TLC developing solvents and spray reagents for visualization. The most sensitive reagents were ammonia and phenylhydrazine hydrochloride (0.02-0.05 µg of patulin) (Scott and Kennedy, 1973), O-dianisidine in glacial acetic acid (0.02 mg of patulin) (Reiss, 1971), and N-methylbenzthiazol-one-2-hydrazone (Besthorn's hydrazone) (0.06 µg of patulin) (Reiss, 1973b).

Various methods have been developed using gas chromatography (GC) to detect patulin. Pohland et al. (1970) prepared silyl ether, acetate, and chloroacetate derivatives of patulin. Other methods have used silyl ether or trimethylsilyl derivatives of patulin for gas chromatographic (GC) analysis (Pero and Harvan, 1973; Pero et al., 1972). A method was also developed in which patulin was detected by liquid chromatography using a silica column (Ware et al., 1974).

Patulin has been reported to occur naturally in apple juice made from decaying apples (Brian et al., 1956; Walker, 1969), and in apple juice commercially available in Canada and the United States (Scott et al., 1972; Dubé, 1972; Wilson and Nuovo, 1973; Scott and Kennedy, 1973). As much as 1 mg of patulin per liter of apple juice has been reported in commercial apple juice (Scott et al., 1972). Lesions of pears and stone fruits decayed by *Penicillium expansum* have also contained patulin (Buchanan et al., 1974).

Patulin has also been reported in spontaneously moldy bread and baked goods (Reiss, 1973a), as well as temporarily present during ripening of fermented sausage (Alperden et al., 1973). This toxin has been found in soil and wheat straw residue associated with phytotoxic problems (Norstadt and McCalla, 1963, 1969a) and in animal feed (Ukai et al., 1954).

C. Toxicity and Carcinogenicity

As previously noted, patulin is toxic to a wide range of bacteria, fungi, and protozoa (Ciegler et al., 1971). Several toxicological studies have been conducted with mice. The LD_{50} values for mice varied with the route of administration. In mice, the LD_{50} values by subcutaneous injection ranged from 8 to 15 mg/kg (Katzman et al., 1944; Broom et al., 1944; McKinnen and Carlton, 1980), by intravenous injection, 16-25 mg/kg (Yamamoto, 1954b; Broom et al., 1944); by intraperitoneal injection, 6-30 mg/kg (Ciegler et al., 1976; Hofmann et al., 1971; Broom et al., 1944; Andraud et al., 1964; McKinnen and Carlton, 1980), and orally 35-48 mg/kg (Broom et al., 1944; Andraud et al., 1964; McKinnen and Carlton, 1980). Edema, swelling, and discoloration were evident at the site of subcutaneous injection. The mice became restless, with heavy, labored breathing (Katzman et al., 1944). When injected intravenously, death was preceded by convulsions (Broom et al., 1944). Pulmonary edema and hemorrhage were the main pathological findings. However, ascites and congestion of the liver, kidneys, lungs, and spleen also occurred (Broom et al., 1944; Hopkins, 1943).

In rats, the LD_{50} doses of patulin were slightly higher: 15 to 25 mg/kg, s.c. (Broom et al., 1944; Katzman et al., 1944) and 25 to 50 mg/kg, i.v. (Broom et al., 1944). The signs, symptoms, and pathology were much the same as for mice except for a marked antidiuretic effect observed in rats (Katzman et al., 1944; Broom et al., 1944).

Mutiple doses of patulin produced cumulative toxicity. Cumulative effects were observed by deRosnay et al. (1952) and Lembke and Hahn (1954) in mice after subcutaneous and intraperitoneal injections of 0.1 mg for up to 4 weeks; in chicks, liver lesions occurred when dosed orally daily with 0.2 mg for 6 weeks. In contrast, daily oral doses produced no effects in mice (Freerkesen and Bönicke, 1951; Lembke and Hahn, 1954; McKinnen and Carlton, 1980).

No evidence of tetratogenicity was observed in a two-generation study of rats. Both males and females received sublethal doses of patulin (1.5 mg/kg per day) orally for 10 or 14 weeks. The progeny were normal except for reduced growth rates (Dailey et al., 1966b). However, teratogenicity was observed when low levels of patulin (1-2 µg) were injected into chick embryos. Malformed feet, ankles, and beaks, exencephaly, and exophthalmia were the major abnormalities encountered (Ciegler et al., 1976).

Metabolism, distribution, and excretion of patulin were studied with [^{14}C]-patulin administered orally to rats. Within 7 days, 40% of the total radioactivity in a single dose (3.0 mg/kg) was excreted in the urine and 50% in feces (Daily et al., 1977a). One to two percent of the radiolabel was expired as CO_2 and the remainder was in the animal. Most of the radiolabel was excreted in the urine during the first 24 hr and in the feces during the first 48 hr. Free, radiolabeled patulin was not detected in the urine. Of the radioactivity in the tissues, red blood cells retained the majority of the bound label for 7 days after treatment. Patulin was rapidly excreted and metabolized in the rat.

The evidence for carcinogenicity of patulin is a study in which rats were treated with 0.2 mg of patulin subcutaneously twice a week for 61-64 weeks. Fibrosarcomas, restricted to the site of injection, developed in six of eight rats (Dickens and Jones, 1961). In view of the route of entry from environmental exposure to patulin, carcinogenicity produced by oral administration would be a more valid test than subcutaneous injection. To date, oral administration of patulin has not produced tumors (Enomoto and Saito, 1972; Osswald et al., 1978). Even in the Ames mutagenic assay, patulin did not produce mutations either with or without microsomal activation (Ueno et al., 1978; Kuczuk et al., 1978).

In cell culture systems, various actions have been reported. Patulin inhibited cell division in mouse fibroblasts (Perlman et al., 1959). Based on data from experiments with amphibian eggs, the disruption of the mitotic spindle apparatus was responsible for patulin's inhibition of cell division (Dustin, 1963; Ciegler et al., 1971). Patulin also caused chromosomal breakage in salamander eggs (Sentein, 1955) and polyploid cells in human leukocyte cultures (Withers, 1965). Breaks in both single-and double-stranded DNA of HeLa cells and single-stranded DNA of FM3A mouse carcinoma cells were observed after incubation with patulin (Umeda et al., 1972, 1977). DNA, RNA, and protein syntheses were inhibited in HeLa cells 80% or more with patulin (Kawasaki et al., 1972). In Chang liver cells, reductions in RNA and protein synthesis were only 40-60% (Schaeffer et al., 1975).

Various enzymes are inhibited by patulin. Among those are enzymes containing thiol groups, muscle aldolase, muscle lactic dehydrogenase (LDH), and yeast alcohol dehydrogenase (ADH) (Ashoor and Chu, 1973a,b). Cysteine blocked the inhibition of LDH, but not of aldolase and ADH. The data indicated that in the case of aldolase, amino groups may be reacting with patulin. Mouse brain, kidney, and liver ATPases were also inhibited by patulin, but the inhibition was blocked by pretreating patulin with sulfhydryl compound (Phillips and Hayes, 1976, 1978).

Aerobic respiration was inhibited by patulin 80% in guinea pig brain tissue, and 50% in kidney tissue (Andraud et al., 1963). This inhibition may be a result of inhibition of the active transport mechanism of the membrane, which relies on ATPases (Busby and Wogan, 1981). Most of the levels of patulin used in the in vitro enzyme studies were higher than would be expected in vivo. Thus the effects are not likely to be primary ones.

II. PENICILLIC ACID

Penicillic acid was first isolated by Alsberg and Black (1913). The fungus, *Penicillium puberulum*, which had been recovered from corn, was reinoculated in the laboratory on corn and tested for toxicity to laboratory animals. Even at this early period in mycological chemistry, it was postulated that fungal products might cause animal diseases. Later, the toxicity of penicillic acid was confirmed, but the amount of metabolite produced by the fungus was so low that characterization of the toxic compound was impractical (Birkinshaw and Raistrick, 1932). Later, *Penicillum cyclopium* was found to produce relatively large amounts of pencillic acid, and both structural and chemical properties were determined (Birkinshaw et al., 1936).

A. Chemical and Physical Properties

The structure of penicillic acid was determined to be an open-chain substituted hexanoic acid or its corresponding γ-hydroxylactone (Birkinshaw et al., 1936) (Fig. 2). The pH of the solution determines the predominate form. In alkaline solutions, the open-chain acid [3] predominates; in neutral solutions, the cyclic lactol [2] predominates.

The empirical formula for penicillic acid is $C_8H_{10}O_4$. It has a molecular weight of 170.2 and crystallizes as anhydrous needles from pentane, hexane, or benzene with a melting point of 83-84.8°C. The monohydrate ($C_8H_{12}O_5$) crystallizes from water as large, transparent monoclinic or triclinic rhomboid crystals (m.p. 58-64°C). Penicillic acid is slightly soluble in cold water, soluble in hot water, alcohol, ether, benzene, chloroform, and ethyl acetate, but insoluble in hexane and petroleum ether. Penicillic acid is stable in mild acid and weak alkaline solutions, but unstable in strong alkali (Heatley and Philpot, 1947; Shaw, 1946; Szilagyi et al., 1963; Ford et al., 1950). When dissolved in methanol, penicillic acid has an ultraviolet absorption peak at 221 nm. Using infrared spectrum data, penicillic acid was discovered to strongly self-associate, forming mainly a dimer (Kovac et al., 1969). Exposure to ammonia produces a fluorescent derivative with an excitation maximum at 350 nm and an emission maximum at 440 nm (Ciegler and Kurtzman, 1970a).

The reactions of penicillic acid with sulfhydryl or amino groups have been studied by several investigators. These studies have practical significance in that the stability and toxicity of penicillic acid gradually reacted with primary amines and amino acids (Oxford, 1942). Cysteine reacted with penicillic acid and the reaction products were not toxic (Dickens and Cooke, 1965; Geiger and Conn, 1945). Arginine, histidine, and lysine were found to react with penicillic acid (Ciegler et al., 1972b). The reaction went to completion in 9 days at pH 7.0. In the same study, cysteine and glutathione reactions with penicillic acid were virtually complete after 7 hours at pH 5, 6, or 7. The adducts formed were examined with nuclear magnetic resonance (NMR) and conventional functional group tests. Data indicated that the addition of the sulfhydryl group was to the isolated double bond rather than the conjugated double bond of penicillic acid. Ciegler et al. (1972b), proposed that the adducts formed with cysteine were derivatives of the cyclic form of

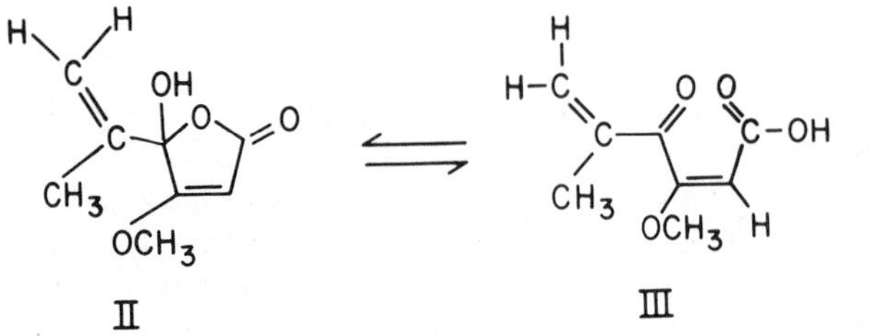

FIGURE 2 Tautomeric forms of penicillic acid.

penicillic acid, while Black (1966) proposed that they were derivatives of the open-chain form. The cysteine and glutathione adducts were not toxic to mice or quail, but were toxic to chick embryos (Ciegler et al., 1972a). Therefore, the toxicity of penicillic acid may be reduced by the reactive amino acids in such foods as sausage (Ciegler et al., 1972b), flour, and orange juice (Scott and Somers, 1968).

B. Production, Detection, and Natural Occurrence

Penicillic acid is produced by a large number of fungi, including the following species of *Penicillium*: *P. aurantio-virens*, *P. baarense*, *P. cannesceans*, *P. chrysogenum*, *P. cyclopium*, *P. fenelliae*, *P. griseum*, *P. janthinellum*, *P. lilacinum*, *P. lividum*, *P. madriti*, *P. martensii*, *P. olivino-viride*, *P. palitans*, *P. puberulum*, *P. roqueforti* (= *P. suavolens*), *P. simplicissimum*, *P. thomii*, and *P. viridicatum*; species of *Aspergillus*, including *A. alliaceus*, *A. melleus*, *A. ochraceus* (= *A. quercinus*), *A. ostianus*, *A. sclerotiorum*, and *A. sulphureus*; and *Paecilomyces ehrlichii* (Ciegler et al., 1971; Wilson, 1976). Several of these fungi are also capable of producing other mycotoxins, such as ochratoxin, citrinin, and patulin, which is important because of possible synergistic toxic effects.

Factors affecting production of penicillic acid have not been well studied. Incubation at low temperatures favored the accumulation of penicillic acid (Ciegler and Kurtzman, 1970b). The temperature range for production of the toxin was between 5 and 32°C, with maximum production between 15 and 20°C. There was a sharp drop in production at 25°C and no mold growth above 32°C (Kurtzman and Ciegler, 1970). CO_2 levels also affect toxin production. By increasing the CO_2 level to 60%, toxin production was reduced at 10°C and blocked at 5°C (Lillehoj et al., 1972). *P. cyclopium* produced no penicillic acid on Czapek-Dox medium with glucose as the sole carbon source and sodium nitrate as the nitrogen source (Birkinshaw et al., 1936), but produced considerable quantities of toxin in Raulin-Thom medium.

The biosynthesis of penicillic acid involves the formation of orsellinic acid by the condensation of one molecule of acetyl coenzyme A with three molecules of malonyl coenzyme A and the loss of three CO_2 molecules (Bentley and Keil, 1961). Several six-membered, cyclic intermediates have been postulated. Patulin biosynthesis has been reviewed by Zamir (1980).

Most analytical methods for the detection of penicillic acid have employed extraction with organic solvents, cleanup with silica gel columns, preparative TLC or GC, and quantitation with TLC or GC. Culture filtrates have been extracted with ether, chloroform, or ethyl acetate, usually after acidification of the filtrate (Birkinshaw et al., 1936; Betina et al., 1969; Suzuki et al., 1971; Sassa et al., 1971; Bentley and Keil, 1962). Penicillic acid has been extracted from grain with various mixtures of chloroform-methanol (Ciegler, 1972). For sausages, acetonitrile-water extracted penicillic acid and a hexane wash removed the lipid material (Ciegler et al., 1972b).

Elution of penicillic acid from silica gel columns with ethyl acetate or benzene-acetone provided partial purification of penicillic acid (Scott and Somers, 1968; Kobayashi et al., 1971). Preparative TLC and transfer into dilute sodium bicarbonate solution are other means to purify penicillic acid partially (Pero et al., 1972; Sassa et al., 1971).

Quantitation of penicillic acid has frequently been performed on TLC plates. Developing solvents that have commonly been used for penicillic acid are toluene:ethyl acetate:90% formic acid (6:3:1), benzene:methanol: acetic acid (24:2:1), chloroform:methanol (97:3), chloroform:ethyl acetate (1:1), and tetrahydrofuran:benzene (2:8) (Scott and Somers, 1968; Ciegler and Kurtzman, 1970b; Scott et al., 1970; Pero et al., 1972). Other systems have been described in a systematic comparison of various TLC systems (Durackova et al., 1976).

Production of a blue fluorescent derivative by exposing penicillic acid to ammonia provided the basis for a sensitive fluorodensitometric assay that could detect and quantitate between 1 and 10 μg of penicillic acid (Ciegler and Kurtzman, 1970a). Phenylhydrazine hydrochloride solution, acidic *p*-anisaldehyde solution, potassium permanganate solution, 50% sulfuric acid, diphenylboric acid ethanolamine salt, *p*-tolualdehyde, and *p*-dimethyl anisaldehyde are other spray reagents that have been used to visualize penicillic acid on TLC plates (Scott and Somers, 1968; Scott et al., 1970; Natori et al., 1970; Ciegler et al., 1972a,b; Neelaktan et al., 1978).

Trimethysilyl and trifluoroacetate derivatives of penicillic acid have been used to quantitate penicillic acid by GC. Columns that have been used include: OV-101, 3% OV-17 or Dexsil 300, and 3% SE-30 (Thorpe and Johnson, 1974; Pero et al., 1972; Bacon et al., 1973; Suzuki et al., 1974, 1975; Fujimoto et al., 1975).

Penicillic acid has been detected in a variety of agriculture commodities. Seven of 20 samples of commercial corn from the United States that had mold problems had levels of penicillic acid ranging from 5 to 230 ng/g and also 5 out of 20 samples of commercial dried beans at levels of 11-179 ng/g (Thorpe and Johnson, 1974). Analysis of moldy tobacco from commercial storage also detected levels of penicillic acid of 110 and 230 ng/kg (Snow et al., 1972). In this same article, up to 4% of penicillic acid that was added to cigarettes was recovered from smoke condensate. Penicillic acid was not detected in meat products overgrown with *Penicillium* spp. (Ciegler et al., 1972b).

C. Toxicity and Carcinogenicity

The toxicity of penicillic acid to laboratory animals was established by Alsberg and Black (1913) and confirmed by Birkinshaw and Raistrick (1932). Its antimicrobial spectrum was also investigated (Oxford, 1942). Penicillic acid was active primarily against gram-negative bacteria and a few gram-positive bacteria (Heatley and Philpot, 1947; Kavanagh, 1947). However, its potential as an antibiotic was diminished when it proved to be too toxic for clinical use (Oxford, 1942). The LD_{50} dose in mice by subcutaneous injection was 100 mg/kg (Spector, 1957) to 110 mg/kg (Murnaghan, 1946); by intraperitoneal injection, 70 mg/kg (Lindenfelser et al., 1973) to 100 mg/kg (Sansing et al., 1976); by intravenous injection, 250 mg/kg (Murnaghan, 1946); and orally, 600 mg/kg (Murnaghan, 1946). When a dose of penicillic acid of 750 mg/kg was administered intravenously, the mice went into convulsions before dying (Murnaghan, 1946). Lesions of the liver, kidney, and thyroid gland were reported in mice by Kobayashi et al. (1971). However, these data should be interpreted with caution, because the subcutaneous LD_{50} value reported in mice was 1/50 dose in the previous studies (Spector, 1957;

Murnaghan, 1946). Damage to the liver has been confirmed in other studies. Ciegler et al. (1972b) reported generalized necrosis of the liver of mice. Dogs that received 20 mg of penicillic acid per kilogram exhibited extensive hepatic changes (Hayes et al., 1977). In postmortum examination of dogs, hemorrhages of the serosal surfaces of the abdomen were reported. Histologically, there was congestion and dilation of hepatic sinusoids.

To trace the distribution and excretion of penicillic acid, an oral dose of [^{14}C]penicillic acid (40 mg/kg) was administered to rats (Park et al., 1978b). The highest level of radioactivity accumulated in red blood cells. Other organs containing radioactivity were, in descending order, the liver, bladder, kidneys, lungs, and heart (Park et al., 1978b). The half-life of the radio-label in the body was relatively short, 18.7 hr for male rats and 17.1 hr for female rats. Within 7 days, 82% of the radiolabel was recovered in the urine, 13% in the feces, and less than 1% as CO_2. There was no free penicillic acid detected in the urine (Park et al., 1978c). A significant route of elimination in the rat was determined to be in bile (Park et al., 1978a). Bile flow was unaffected by penicillic acid in the rat, but was significantly depressed in the mouse on a body weight basis (Chan and Hayes, 1981). In view of these data, additional studies in mice on excretion and elimination of penicillic acid are warranted.

Synergistic effects between penicillic acid and other mycotoxins have been observed. Administration of penicillic acid with either ochratoxin or citrinin increased mortality beyond additive values for joint administration (Lindenfelser et al., 1973; Sansing et al., 1976). A synergistic effect was not apparent in dogs receiving both penicillic acid and rubratoxin B (Hayes et al., 1977). However, patulin and penicillic acid increased mortality and enhanced histological changes synergistically (Reddy et al., 1979). Since more than one fungus may contaminate food and feeds and more than one toxin may be produced by certain fungi, more attention should be directed toward studies of possible synergisms.

Carcinogenicity of penicillic acid was established in rats by Dickens and Jones (1961, 1963). Sarcomas were produced in rats at the site of injection when administered subcutaneously with 1 mg of penicillic acid in oil twice a week for 64 weeks. Two milligrams of penicillic acid in aqueous solution administered in the same regime also produced sarcomas in rats at the site of injection after 52 weeks. A dose as low as 0.1 mg initiated tumor development (Dickens and Jones, 1963). No tumors were induced in rats or mice by oral administration of penicillic acid (Enomoto and Saito, 1972). Penicillic acid was carcinogenic in mice also (Dickens and Jones, 1965). Subcutaneous doses of 0.2 mg of penicillic acid twice a week induced sarcomas at the injection site in 6 of 19 surviving mice. In contrast to its carcinogenic properties, penicillic acid exhibited significant antitumor activity against Ehrlich ascites carcinoma and ascites from leukemia SN-36 (Suzuki et al., 1971).

The mutagenic potential of penicillic acid varied with respect to the test system. Penicillic acid induced single- and double-stranded DNA breaks in HeLa cells (Umeda et al., 1972). Significant DNA damage was observed in the *Bacillus subtilis* recombination-deficient mutagenic assay (Ueno and Kubota, 1976). In FM3A mouse carcinoma cells, penicillic acid was a potent inducer of mutants (Umeda et al., 1977). However, penicillic acid was inactive in both Ames' *Salmonella typhimurium* and *Sacchromyces cerevisiae* mutagenesis assays with and without microsomal activation (Engel and Van Milczenki, 1976; Kuczuk et al., 1978; Ueno et al., 1978).

Morphological distortion and inhibition of cell division has been reported with penicillic acid in HeLa cells (Suzuki et al., 1971). Mitotic poisoning of cells caused by spindle fiber damage was noted by Saito et al. (1971). Damage caused by penicillic acid could be prevented by pretreatment with sulfhydryl compounds (Rondanelli et al., 1968). DNA, RNA, and protein synthesis in HeLa cells was inhibited (Umeda et al., 1972, 1974; Kawaski et al., 1972).

Various enzymes were inhibited by penicillic acid. Among them were enzymes containing thiol groups: muscle lactic dehydrogenase (LDH), yeast alcohol dehydrogenase (ADH), and muscle aldolase (Ashoor and Chu, 1973a,b). Cysteine blocked the inhibition of LDH by penicillic acid but not that of muscle aldolase or ADH. Mouse brain and kidney (Na^+,K^+)-ATPases were inhibited by penicillic acid both in vivo and in vitro (Chan and Hayes, 1981). A linear correlation was demonstrated between free enzyme sulfhydryl groups, ATPase activity, and varying concentrations of penicillic acid (Phillips et al., 1980).

ACKNOWLEDGMENTS

The authors are most grateful to L. S. Lee and A. J. DeLucca II for their assistance in the preparation and editing of this chapter, and to P. Burnaman for the preparation of the manuscript.

REFERENCES

Alperden, I., Mintzlaff, H. J., Tauchmann, F., and Leistner, L. (1973). Untersuchung über die Bilding des Mykotoxins Patulin in Rohwurst. *Fleischwirtschaft 53*:566-568.

Alsberg, C. L., and Black, O. F. (1913). Contribution to the study of maize deterioration; biochemical and toxicological investigations of *Penicillium puberulum* and *Penicillium stoloniferum*. *U.S. Dept. Agric. Bur. Plant Ind. Bull. 270.*

Andraud, G., Aublet-Cuvelier, A. M., Couquelet, J., Curvelier, R., and Tronche, P. (1963). Activaté conparée sur la respiration cellulaire de la patuline naturelle et d'un isomère de synthèse. *C. R. Soc. Biol. 157*: 1444-1447.

Andraud, G., Tronche, P., and Couquelet, J. (1964). Effets cytotoxiques dans la série de la patuline. Étude par la méthode de Warburg. *Ann. Biol. Chem. (Paris) 22*:1067-1074.

Ashoor, S. H., and Chu, F. S. (1973a). Inhibition of alcohol and lactic dehydrogenases by patulin and penicillic acid in vitro. *Food Cosmet. Toxicol. 11*:617-624.

Ashoor, S. H., and Chu, F. S. (1973b). Inhibition of muscle aldolase by penicillic acid and patulin in vitro. *Food Cosmet. Toxicol. 11*:995-1000.

Atkinson, N., and Stanley, N. F. (1943). Antibacterial substances produced by moulds. IV. The detection and occurrence of suppressors of penicidin activity. *Aust. J. Exp. Biol. Med. Sci. 21*:249-254.

Bacon, C. W., Sweeney, J. G., Robbins, J. D., and Burdick, D. (1973). Production of penicillic acid and ochratoxin A on poultry feed by *Aspergillus ochraceus*: temperature and moisture requirements. *Appl. Microbiol. 26*:155-160.

Bentley, R., and Keil, J. G. (1961). Role of acetate and malonate in the biosynthesis of penicillic acid. *Proc. Chem. Soc.*, pp. 111-112.

Bentley, R., and Keil, J. G. (1962). Tetronic acid biosynthesis in moulds. II. Formation of penicillic acid in *Penicillium cyclopium. J. Biol. Chem. 237*:867-873.

Betina, V., Gasparikova, E., and Nemec, P. (1969). Isolation of penicillic acid from *Penicillium simplicissum. Biologia (Bratislava) 24*:482-485.

Birkinshaw, J. H., and Raistrick, H. (1932). Studies in the biochemistry of micro-organisms. Puberulic acid $C_8H_4O_6$, and an acid $C_8H_4O_6$, new products of the metabolism of glucose by *Penicillium puberulum* Bainier and *Penicillium aurantio-virens* Biourge. *Biochem. J. 26*:441-453.

Birkinshaw, J. H., Oxford, A. E., and Raistrick, H. (1936). LXIV. Studies in the biochemistry of micro-organisms. XLVIII. Penicillic acid, a metabolic product of *Penicillium puberulum* Bainier and *P. cyclopium* Westling. *Biochem. J. 30*:394-411.

Birkinshaw, J. H., Bracken, A., Micheal, S. E., and Rainstrick, H. (1943). Patulin in the common cold. II. Biochemistry and chemistry. *Lancet*, pp. 625-630.

Black, D. K. (1966). The addition of L-cysteine to unsaturated lactones and related compounds. *J. Chem. Soc. C*:1123-1127.

Börner, H. (1963a). Untersuchungen über die Bildung antiphytotischer und antimikrobieller Substanzen durch Mikroorganismem im Boden und ihre mögliche Bedeutung für die Bodenmüdigkeit beim Apfel (*Pirus malus* L.). I. Bildung von Patulin und einer phenolischen Verbindung durch *Penicillium expansum* auf Wurzel- und Blattrückständen des Apfels. *Phytopathol. Z. 48*:370-396.

Börner, H. (1963b). Untersuchungen über die antimikrobieller Substanzen durch Mikroorganismen im Boden und ihre mögliche Bedeutung für die Bodenmüdigkeit beim Apfel (*Pirus malus* L.). II. Der Einfluss verschiebner Faktoren auf die Bildung von Patulin und einer phenolischen Verbindung durch *Penicillium expansum* auf Blatt- und Wurzelrückständen des Apfels. *Phytophathol. Z. 49*:1-28.

Brian, P. W., Elson, G. W., and Lowe, D. (1956). Production of patulin in apple fruits by *Penicillium expansum. Nature (Lond.) 178*:263-264.

Broom, W. A., Bülbring, E. Chapman, C. J., Hampton, J. W. F., Thompson, A. M., Ungar, J., Wein, R., and Woolfe, G. (1944). The pharmacology of patulin. *Br. J. Exp. Pathol. 25*:195-207.

Buchanan, J. R., Somer, N. F., Fortlage, R. J., Maxie, E. C., Mitchell, F. G., and Hsieh, D. P. H. (1974). Patulin from *Penicillium expansum* in stone fruits and pears. *J. Am. Soc. Hort. Sci. 99*:262-265.

Busby, W. F., Jr., and Wogan, G. N. (1981). Patulin and penicillic acid. In *Mycotoxins and N-Nitroso Compounds*, Vol. II: *Environmental Risks*, R. C. Shank (Ed.). CRC Press, Boca Raton, Fla., pp. 121-127.

Chain, E. H., Florey, W., and Jennings, M. A. (1942). An antibacterial substance produced by *Penicillium claviforme*. *Br. J. Exp. Pathol. 23*: 202-205.

Chan, P. K., and Hayes, A. W. (1981). Effect of penicillic acid on biliary excretion of indocyanine in the mouse and rat. *J. Toxicol. Environ. Health 7*:169-180.

Ciegler, A. (1972). Bioproduction of ochratoxin A and penicillic acid by members of the *Aspergillus ochaceus* group. *Can. J. Microbiol. 18*:631-636.

Ciegler, A., and Kurtzman, C. P. (1970a). Fluorodensitometer assay of penicillic acid. *J. Chromatogr. 51*:511-516.

Ciegler, A., and Kurtzman, C. P. (1970b). Penicillic acid production by blue-eye fungi on various commodities. *Appl. Microbiol. 20*:761-764.

Ciegler, A., Detroy, R. W., and Lillehoj, E. B. (1971). Patulin, penicillic acid, and other carcinogenic lactones. In *Microbial Toxins*, Vol. VI: *Fungal Toxins*, A. Ciegler, S. Kadis, and S. J. Ajl (Eds.). Academic Press, New York, pp. 409-434.

Ciegler, A., Mintzlaff, H.-J., Machnik, W., and Leistner, L. (1972a). Untersuchungen über das Toxinbildungsvermögen von Rohwürsten iso-lieter Schimmelpilze der Gattung *Penicillium*. *Fleischwirtschaft 52*:1311-1314, 1317-1318.

Ciegler, A., Mintzlaff, H.-J., Weisleder, D., Leistner, L. (1972b). Potential production and detoxification of penicillic acid in mould-fermented sausage (salami). *Appl. Microbiol. 24*:114-119.

Ciegler, A., Beckwith, A. C., and Jackson, L. K. (1976). Teratogenicity of patulin and patulin adducts formed with cysteine. *Appl. Environ. Microbiol. 31*:664-667.

Dailey, R. E., Blaschka, A. M., and Brouwer, E. A. (1977a). Absorption distribution, and excretion of [^{14}C]patulin by rats. *J. Toxicol. Environ. Health 3*:479-490.

Dailey, R. E., Brouwer, E., Blaschka, A. M., Reynaldo, E. F., Green, S., Monlux, W. S., and Ruggles, D. I. (1977b). Intermediate-duration toxicity study of patulin in rats. *J. Toxicol. Environ. Health 2*:713-725.

Dalton, J. E. (1952). Keloid resulting from a positive patch test. *Arch. Dermatol. Syphilol. 65*:53-55.

deRosnay, C. D., Martin-Dupont, C., and Jensen, R. (1952). Étude d'une substance antibiotique, la "Mycoine C." *J. Med. Bord. 129*:189-199.

Dickens, F., and Cooke, J. (1965). Rates of hydrolysis and interaction with cysteine of some carcinogenic lactones and related substrates. *Br. J. Cancer 19*:404-410.

Dickens, F., and Jones, H. E. H. (1961). Carcinogenic activity of a series of reactive lactones and realted substances. *Br. J. Cancer 15*:85-100.

Dickens, F., and Jones, H. E. H. (1963). Further studies on the carcino-genic and growth-inhibitory activity of lactones and related substances. *Br. J. Cancer 17*:100-108.

Dickens, F., and Jones, H. E. H. (1965). Further studies on the carcino-genic action of certain lactones and related substances in the rat and mouse. *Br. J. Cancer 19*:392-403.

Durackova, Z., Betina, V., and Nemec, P. (1976). Systematic analysis of mycotoxins by thin layer chromatography. *J. Chromatogr. 116*:141-154.

Dustin, P., Jr. (1963). New aspects of pharmacology of antimitotic agents. *Pharmacol. Rev. 15*:449-480.

Engel, G., and Van Milczenki, K. E. (1976). Zum Machweis von Myko-
toxinen nach activierung unt. Rattenlerhomogenaten mittles histidinemapel
Mutation von *Salmonella typhimurium*. *Kiel. Milchwiertsch. Forschungsber.*
28:309-366.

Enomoto, M., and Saito, M. (1972). Carcinogens produced by fungi. *Annu.
Rev. Microbiol. 26*:279-312.

Florey, H. W., Jennings, M. A., and Philpot, F. J. (1944). Claviformin from
Aspergillus gigantus Wehm. *Nature (Lond.) 153*:139.

Ford, J. H., Johnson, A. R., and Hinman, J. W. (1950). The structure of
penicillic acid. *J. Am. Chem. Soc. 72*:4529-4531.

Freerksen, E., and Bönicke, R. (1951). Die Inaktivierung des Patulins in
vivo (modell Versuche zur Bestimmung des wertes antibakterieller Substanzer für therapeutische Zwecke). *Z. Hyg. Infektionskr. 132*:274-291.

Fujimoto, Y., Suzuki, T., and Hoshino, Y. (1975). Determination of penicillic acid and patulin by gas-liquid chromatography with an electron-capture
detector. *J. Chromatogr. 105*:99-106.

Geiger, W. B., and Conn, J. E. (1945). The mechanism of antibiotic action
of clavacin and penicillic acid. *J. Am. Chem. Soc. 67*:112-116.

Gye, W. E. (1943). Patulin in the common cold. III. Preliminary trial in the
common cold. *Lancet*, pp. 630-631.

Hayes, A. W., Unger, P. D., and Williams, W. L. (1977). Acute toxicity of
penicillic acid and rubratoxin B in dogs. *Ann. Nutr. Aliment. 31*:711-722.

Heatley, N. G., and Philpot, F. J. (1947). The routine examination for antibiotics produced by moulds. *J. Gen. Microbiol. 1*:232-237.

Hofmann, K., Mintzlaff, H.-J., Alperden, I., and Leistner, L. (1971).
Untersuchung über die Inaktivierung des mykotoxins Patulin durch
Sulfhydrylgruppen. *Fleischwirtschaft 51*:1534-1536, 1539.

Hopkins, W. E. (1943). Patulin in the common cold. IV. Biological properties extended trial in the common cold. *Lancet*, pp. 631-634.

Katzman, P. A., Hays, E. E., Cain, C. K., van Wyk, J. J., Reithel, F. J.,
Thayer, S. A. , Doisy, E. A., Gaby, W. L., Carroll, C. J., Muir, R. D.,
Jones, L. R., and Wade, N. J. (1944). Clavacin, and antibiotic sub-
stance from *Aspergillus clavatus*. *J. Biol. Chem. 154*:475-486.

Kavanagh, F. (1947). Antibacterial substances from fungi and green plants.
Adv. Enzymol. 7:461-511.

Kawasaki, I., Oki, T., Umeda, M., and Saito, M. (1972). Cytotoxic effect
of penicillic acid and patulin on HeLa Cells. *Jpn. J. Exp. Med. 42*:327-
340.

Kobayaski, H., Tsunoda, T., and Tatsuno, T. (1971). Recherches toxicologiques sur les mycotoxines qui polluent le fourrage artificiel du porc.
Chem. Pharm. Bull. 19:839-842.

Korzybski, T. (1967). Patulin, syn. clavicin, clavatin, claviformin, expansin, penicidin. In *Antibiotics*, Vol. I: *Mechanisms of Actions*, D. Gottlieb
and P. D. Shaw (Eds.). Springer-Verlag, New York, pp. 1223-1230.

Kovac, S., Solcaniova, E., and Eglinton, G. (1969). Infrared studies with
terpenoid compounds. VI. Infrared spectra of penicillic acid and its
derivative. *Tetrahedron 25*:3617-3622.

Kuczuk, M. H., Benson, P. M., Health, H., and Hayes, A. W. (1978).
Evaluation of mutagenic potential of mycotoxins using *Saccharomyces
cerevisae*. *Mutat. Res. 53*:11-20.

Kurtzman, C. P., and Ciegler, A. (1970). Mycotoxin from a blue-eyed mold of corn. *Appl. Microbiol. 20*:207-207.

Lembke, A., and Hahn, B. (1954). Action of metabolic products of *Penicillium claviforme* on cells and tissues varying in their stage of development. *Kiel. Milchwirtsch. Forschungsber. 6*:41-58, 219-241.

Lillehoj, E. B., Milburn, M. S., and Ciegler, A. (1972). Control of *Penicillium martensii*, development and penicillic acid production by atmospheric gases and temperature. *Appl. Microbiol. 24*:198-201.

Lindenfelser, L. A., Lillehoj, E. B., and Milburn, M. S. (1973). Ochratoxin and penicillic acid in tumorigenic and acute toxicity tests with white mice. *Dev. Ind. Microbiol. 14*:331-336.

Lochhead, A. G., Chase, F. E., and Landerkin, G. B. (1946). Production of claviformin by soil penicillic acid. *Can. J. Res. 24E*:1-9.

McKinnen, E. R., and Carlton, W. W. (1980). Patulin mycotoxicosis in Swiss ICR mice. *Food Cosmet. Toxicol. 18*:181-187.

Medical Research Council (1944). Patulin clinical trials committee. *Lancet*, pp. 373-375.

Murnaghan, M. F. (1946). The pharmacology of penicillic acid. *J. Pharmacol. Exp. Ther. 88*:119-132.

Natori, S., Sakari, S., Kurata, H., Udagawa, S., Ichinoe, M., and Umeda, M. (1970). Chemical and cytotoxicity survey on the production of ochratoxins and penicillic acid by *Aspergillus ochraceus* Wilk. *Chem. Pharm. Bull. 18*:2259-2268.

Neelaktan, S., Balasurbamanian, J., Balsaraswathi, R., Jamine, G. I., and Swaninathan, R. (1978). Detection of penicillic acid in foods. *J. Food Sci. Technol. 15*:125-126.

Norstadt, F. A., and McCalla, T. M. (1963). Phytotoxic substance from a species of *Penicillium*. *Science 140*:410-411.

Norstadt, F. A., and McCalla, T. M. (1969a). Microbial populations in stubble-mulched soil. *Soil Sci. 107*:188-193.

Norstadt, F. A., and McCalla, T. M. (1969b). Patulin production by *Penicillium urticae* Bainier in batch culture. *Appl. Microbiol. 17*:193-196.

Norstadt, F. A., and McCalla, T. M. (1971). Growth and patulin formation by *Penicillium urticae* Bainier in pure and mixed cultures. *Plant Soil 34*: 97-108.

Osswald, H., Frank, H. K., Komitowski, D., and Winter, H. (1978). Long-term testing of patulin administered orally to Sprague-Dawley rats and Swiss mice. *Food Cosmet. Toxicol. 16*:243-248.

Oxford, A. E. (1942). On the chemical reactions occurring between certain substances which inhibited bacterial growth and constituents of bacteriological media. *Biochem. J. 36*:438-444.

Park, D. L., Brouver, E., and Heath, J. L. (1978a). Biliary excretion of [^{14}C]penicillic acid by rats. *Toxicol. Environ. Health 4*:9-13.

Park, D. L., Dailey, R. E., Friedman, L., and Heath, J. L. (1978b). The absorption, distribution and excretion of [^{14}C]-penicillic acid by rats. *Ann. Nutr. Aliment. 31*:919-934.

Park, D. L., Miller, E., and Heath, J. L. (1978c). Preliminary report of the distribution of [^{14}C]-penicillic acid in rats; autoradiographic technique. *Ann. J. Vet. Res. 39*:1863-1865.

Perlman, D., Giuffre, N. A., Jackson, P. W., and Giardinello, F. E. (1959). Effects of antibiotics on multiplication of L cells in suspension culture. *Proc. Soc. Exp. Biol. Med. 102*:290-292.

Pero, R. W., and Harvan, D. (1973). Simultaneous detection of metabolites from several toxigenic fungi. *J. Chromatogr.* *80*:255-258.

Pero, R. W., Harvan, D., Owens, R. G., and Snow, J. P. (1972). A gas chromatographic method for the mycotoxin penicillic acid. *J. Chromatogr.* *65*:501-506.

Phillips, T. D., and Hayes, A. W. (1976). Effects of patulin on adenosine triphosphatase activities in the mouse. *Toxicol. Appl. Pharmacol.* *42*:175-188.

Phillips, T. D., and Hayes, A. W. (1978). Effects of patulin on the kinetics of substrate and cationic ligand activation of adenosine triphosphatase in mouse brain. *J. Pharmacol. Exp. Ther.* *205*:606-616.

Phillips, T. D., Chan, P. K., and Hayes, A. W. (1980). Inhibitory characteristics of the mycotoxin penicillic acid on (Na^+-K^+)-activated adenosine triphosphatase. *Biochem. Pharmocol.* *29*:19-26.

Pohland, A. E., and Allen, R. (1970). Analysis and chemical confirmation of patulin in grains. *J. Assoc. Off. Anal. Chem.* *53*:686-687.

Pohland, A. E., Sanders, K., and Thorpe, C. W. (1970). Determination of patulin in apple juice. *J. Assoc. Off. Anal. Chem.* *57*:621-625.

Reddy, C. S., Chan, P. K., Hayes, A. W., Williams, W. L., and Ciegler, A. (1979). Acute toxicity of patulin and its interaction with penicillic acid. *Food Cosmet. Toxicol.* *17*:605-609.

Reiss, J. (1971). Detection of patulin by thin-layer chromatography. *Chromatographia* *4*:576.

Reiss, J. (1973a). Mycotoxine in Nahrungsmitteln. III. Bildung von Patulin auf verschieden Schnittbrotarten durch *Penicillium expansum.* *Chem. Microbiol. Technol. Lebensm.* *2*:171-173.

Reiss, J. (1973b). *N*-Methylbenzthiazolone-(2)-hydrazone (Besthorn's hydrazone) as a sensitive spray reagent for patulin. *J. Chromatogr.* *86*:190-191.

Rondanelli, R., Carco, F. P., Fossati, G. C., and Dionisi, D. (1968). Sul meccanismo di, azione antibiotica di alcuni antibiotices minori. *G. Ital. Chemioter.* *15*:78-83.

Rubin, B. A., and Giarman, N. J. (1947). The therapy of experimental influenza in mice with antibiotic lactones and related compounds. *Yale J. Biol. Med.* *19*:1017-1025.

Saito, M., Enomoto, M., Umeda, M., Ohtsub, K., Ishiko, T., Yamamoto, S., and Toyokawa, H. (1971). Field surveys of mycotoxin producing fungi contaminating human foodstuffs in Japan; with epidemiological background. In *Mycotoxins in Human Health,* I. F. H. Purchase (Ed.). Macmillan, New York, pp. 179-183.

Sansing, G. A., Lillehoj, E. B., Detroy, R. W., and Miller, M. A. (1976). Synergistic toxic effects of citrinin, ochratoxin A, and penicillic acid in mice. *Toxicon* *14*:213-220.

Sassa, S., Hayakari, S., Ikeda, M., and Miura, Y. (1971). Plant growth inhibitors produced by fungi. Part I. Isolation and identification of penicillic acid and dihydro-penicillic acid. *Agric. Biol. Chem.* *35*:2130-2131.

Schaeffer, W. I., Smith, N. E., Payne, P. A., and Wilson, D. M. (1975). Pysiological and biochemical effects of the mycotoxin patulin on Chang liver cell cultures. *In Vitro* *11*:69-77.

Scott, P. M. (1974). Patulin. In *Mycotoxins*, I. F. H. Purchase (Ed.). Elsevier, New York, pp. 383-403.

Scott, P. M., and Kennedy, B. P. C. (1973). Improved method for the thin
   layer chromatographic determination of patulin in apple juice. *J. Assoc.
   Off. Anal. Chem. 56*:813-816.

Scott, P. M., and Somers, E. (1968). Stability of patulin in fruit juices and
   flour. *J. Agric. Food Chem. 16*:483-485.

Scott, P. M., Lawrence, J. W., and van Walbeek, W. (1970). Detection of
   mycotoxins by thin-layer chromatography: application to screening of
   fungal extracts. *Appl. Microbiol. 20*:839-842.

Scott, P. M., Miles, W. F., Toft, P., and Dube, J. G. (1972). Occurrence
   of patulin in apple juice. *J. Agric. Food Chem. 20*:450-451.

Sentein, P. (1955). Altération du fuseau mitotique et fragmentation des
   chromosomes par l'action de la patuline sur l'oeuf d'Urodéles en segmen-
   tation. *C. R. Soc. Biol. 149*:1621-1622.

Shaw, E. (1946). Synthesis of protoanemonin. Tautomerism of acetylacrylic
   acid and penicillic acid. *J. Am. Chem. Soc. 68*:2510-2513.

Singh, J. (1967). Patulin. In *Antibiotics*, Vol. I: *Mechanisms of Action*,
   D. Gottlieb and P. D. Shaw (Eds.). Springer-Verlag, New York, pp. 621-
   630.

Snow, J. P., Lucas, G. B., Harvan, D., Pero, R. A., and Owens, R. G.
   (1972). Analysis of tobacco and smoke condensate for penicillic acid.
   Appl. Microbiol. 24:34-36.

Spector, W. S. (1957). *Handbook of Toxicology*, Vol. II. Saunders,
   Philadelphia.

Stansfeld, J. H., Francis, A. E., and Stuart-Harris, C. H. (1944). Laboratory clinical traits of patulin. *Lancet*, pp. 370-372.

Stoloff, L., Nesheim, S., Yin, L., Rodricks, J. V., Stack, M., and Campell,
   A. D. (1971). A multimycotoxin detection method for aflatoxins, ochra-
   toxins, zearalenone, sterigmatocystin, and patulin. *J. Assoc. Off. Anal.
   Chem. 54*:91-97.

Suzuki, S., Kimura, T., Saito, F., and Ando, K. (1971). Antitumor and
   antiviral properties of penicillic acid. *Agric. Biol. Chem. 35*:287-290.

Suzuki, T., Fujimoto, Y., Hoshino, V., and Tanaka, A. (1975). Trimethyl-
   silyation of penicillic acid and patulin, and the stability of the products.
   J. Chromatogr. 105:95-98.

Szilagyi, I., Valyi-Nagy, T., and Galambo, G. (1963). Die penicillium-saure
   und ihre Tautomerie. *Mikrochim. Ichnoanal. Acta 5-6*:864-871.

Thorpe, C. W., and Johnson, R. L. (1974). Analysis of penicillin acid by
   gas-liquid chromatography. *J. Assoc. Off. Anal. Chem. 57*:861-865.

Ueno, Y., and Kubota, K. (1976). DNA-attacking ability of carcinogenic
   mycotoxins in recombination-deficient mutant cells of *Bacillus subtilis*.
   Cancer Res. 36:445-451.

Ueno, Y., Kubota, K., Ito, T., and Nakamura, Y. (1978). Mutagenicity of
   carcinogenic mycotoxins in *Salmonella typhimurium*. *Cancer Res. 38*:536-
   542.

Ukai, T., Yamamoto, Y., and Yamamoto, T. (1954). Studies on the poisonous substance from a strain of *Penicillium* (Hori-Yamamoto strain). II.
   Culture method of Hori-Yamamoto strain and chemical structure of its
   poisonous substance. *J. Pharm. Soc. (Jpn.) 74*:450-454.

Umeda, M., Yamamoto, T., and Saito, M. (1972). DNA-strand breakage of
   HeLa cells induced by several mycotoxins. *Jpn. J. Exp. Med. 42*:527-535.

Umeda, M., Yamashita, T., Saito, M., Sekita, S., Takahashi, C., Yoshihira, K., Natori, S., Kurata, H., and Udagawa, S. (1974). Chemical and cytotoxic survey on the metabolites of toxic fungi. *Jpn. J. Exp. Med.* 44:83-96.

Umeda, M., Tsutsui, T., and Saito, M. (1977). Mutagenicity and inducibility of DNA single-strand breaks and chromosone aberrations by various mycotoxins. *Gann* 68:619-626.

Walker, J. R. L. (1969). Inhibition of the apple phenolase system through infection by *Penicillium expansum*. *Phytochemistry* 8:561-566.

Walker, K., and Wiesner, B. P. (1944). Patulin and clavicin. *Lancet 246*: 294.

Ware, G. M., Thorpe, C. W., and Pohland, A. E. (1974). Liquid chromatographic method for determination of patulin in apple juice. *J. Assoc. Off. Anal. Chem.* 57:1111-1113.

Wilson, D. M. (1976). Patulin and penicillic acid. In *Mycotoxins and Other Related Food Problems, Advances in Chemistry Series 149*, J. V. Rodricks (Ed.). American Chemical Society, Washington, D.C., pp. 90-109.

Wilson, D. M., and Nuovo, G. J. (1973). Patulin production in apples decayed by *Penicillium expansum*. *Appl. Microbiol.* 26:124-125.

Withers, R. F. J. (1965). The action of some lactones and related compounds on human chromosomes. *Symp. Mutat. Process., Mech. Mutat. Inducing Factors*, Prague, pp. 359-364.

Yamamoto, T. (1954a). Studies on the poison-producing mould isolated from dry malt. I. Distribution, isolation, cultivation, and formation of the toxic substance. *J. Pharm. Soc. (Jpn.)* 74:797-801.

Yamamoto, T. (1954b). Studies on the poison-producing mould isolated from dry malt. IV. On the toxicity. *J. Pharm. Soc. Jpn.* 74:810-812.

Zamir, L. O. (1980). The biosynthesis of patulin and penicillic acid. In *The Biosynthesis of Mycotoxins: A Study in Secondary Metabolism*, P. S. Steyn (Ed.). Academic Press, New York, pp. 223-268.

Part IV

TOXINS INDUCING
PHOTOSENSITIVITY OR ALLERGIC EFFECTS

10

PHOTOSENSITIZING TOXINS FROM PLANTS AND THEIR BIOLOGIC EFFECTS

A. EARL JOHNSON

U.S. Department of Agriculture, Logan, Utah

I. INTRODUCTION

In 1898, Oscar Raab, a student in Munich, Germany, very astutely noted that acridine rendered paramecia sensitive to sunlight to the degree that they were rapidly killed. Paramecia exposed to acridine alone or to sunlight alone were not so sorely affected.

Stimulated by Raab's observations on the potentiating effect of this dye in tissue in the presence of light, other investigators soon noted that many dyes and pigments could sensitize living organisms to light and that some of these agents, chlorophyll for example, occurred naturally in plants.

By the 1930s several livestock diseases were recognized to be caused by the ingestion of plants containing substances that rendered animals sensitive to sunlight, and many observations related to these maladies were recorded. Similarly, during this time, there was interest and observations on photosensitizations and light-related diseases in humans, although there was much confusion regarding the etiology of these diseases. The activity through this period prior to 1950 related to diseases caused by light was summarized by Blum (1941), who also gave an excellent discussion of photodynamics and photobiology.

Clare (1952) further updated the information on photosensitization diseases of domestic animals and introduced the classification of these diseases as type I, primary photosensitization; type II, photosensitization due to aberrant pigment synthesis; and type III, secondary or hepatogenous photosensitization. These will be discussed in more detail later.

In the late 1950s and 1960s additional interest in human photosensitization was generated as compounds with photosensitizing properties began to be used in suntan oils to stimulate pigment formation in the skin. Some of these compounds, namely psoralens, are found in plants and are under intensive investigation at the present time.

As might be expected, photosensitizations due to plant toxins are primarily a problem in livestock, particularly of livestock on open rangeland areas, since they are exposed to a wide variety of plants and to longer periods of direct sunlight. Horses, pigs, and fowls have photosensitization problems in localized areas; cattle may be involved to varying degrees, but the greatest economic losses generally occur in sheep herds, where hundreds of animals may be affected at a time.

Present-day environments in which livestock are managed offer a variety of sources of phototoxic compounds, other than plants, ranging from intentionally administered therapeutic compounds to contamination by phototoxins in feed and water. Losses from photosensitizations from nonplant sources, however, are generally small, involving only a few animals, whereas losses from naturally growing range or pasture plants may be devastating.

There are a variety of plants that may cause photosensitization, but many cause only occasional problems and are of little economic importance. These have received little research attention and the conditions under which they may be phototoxic are not all completely understood. Plants responsible for serious economic losses to the livestock industry, even though they are often sporadic in generating photosensitizations, have received the major attention and there are several reviews relating to them in addition to those already mentioned (Spikes, 1968; Giese, 1971; Scheel, 1973). It is not the purpose of this chapter to provide an in-depth study of these well-known plants and toxins and the speculations that have been presented regarding their actions, but rather to provide an overview of the problem so that the reader may have a feeling for the potential of a variety of plants and plant compounds to cause photosensitization and have some understanding of how these compounds act within the animal's body.

II. PHOTODYNAMIC ACTION

Information available on photodynamic action, or the activation of photosensitizers by light and the subsequent chemical reactions generated, is not complete. Mechanisms are complicated and several theories have been advanced to explain the various aspects of photodynamic action (Spikes, 1968); proof of most of these theories still awaits elucidation. For our purposes here, only a brief review of the general mechanisms involved is given.

Compounds that photosensitize are termed photodynamic agents, photo-sensitizers, photosensitizing compounds, or phototoxic compounds and do not belong to a specific class of chemicals. They are generally pigments that fluoresce, but their chemical formulations vary greatly. They do, however, have one common characteristic: they are able to absorb and hold intact for short time periods units of light energy. These units are generally referred to as quanta or photons. Molecules of a photosensitizing agent that have absorbed quanta of light energy are known as activated molecules. An activated molecule may be likened to a loaded, cocked gun; light energy is its bullet, and time pulls the trigger. When it is fired it is ready to be loaded again for a repeat performance.

Each photosensitizing agent can be activated only by light rays in a given wavelength range. The most destructive light is that below 330 nm. There-fore, natural photosensitizing compounds that absorb, or have an absorption spectrum, in the range of about 250 to 330 nm are those most destructive to biological systems. Longer wavelengths can also be destructive but they re-quire more intensity and longer exposure periods. Sunlight produces light in the range 290-1850 nm and is therefore capable of activating a wide variety of photosensitizing compounds.

An activated molecule can hold its absorbed light energy for only a short time. If, when it releases this energy, the activated molecule is in contact with certain receptor molecules, the energy is transferred to the receptor molecule. As stated by Blum: "The molecule to which the energy of activa-tion is transferred may participate in chemical reaction, the nature of which is determined by the chemical properties of that molecule, and those of other molecules in the environment." Thus the substrate or compounds entering into this chemical reaction and the subcellular structures that respond deter-mine the type and degree of tissue damage that may result.

An array of disruptive effects are therefore possible; a few that have been observed are: abnormal cell division, change in permeability of membranes and active transport processes, interference with glycolysis and cellular respiration, disruption of protein and DNA synthesis, and cell death.

Many of the known reactions involving the substrate are oxidations and molecular oxygen must be available for the reaction to proceed; other reac-tions do not require oxygen. These reactions are often classed as those that require oxygen and those that do not require oxygen. For a more de-tailed discussion and review of proposed mechanisms of photodynamic action, the reader is referred to Spikes (1968).

III. CLINICAL MANIFESTATIONS OF PHOTOSENSITIZATION

The clinical manifestations of photosensitization in the affected animal can vary greatly depending on two factors primarily: (1) the amount of light-absorbing material (phototoxin) that reaches the skin by way of peripheral blood circulation, and (2) the degree and time of exposure to the photosen-sitizing rays (sunlight) and the degree of protection from light afforded the animal by hair, wool, or pigmented skin. Other factors, such as the type of phototoxin and the nature of the chemical reaction that it elicits in the tissue, may also be of prime importance.

The first indication that animals may be reacting to phototoxins is that they become restless as they feel discomfort, itching, and burning of the affected parts. In sheep these parts are lips, eyes, and ears, and in some situations, udder and vulva; in cattle, areas where hair is thin or absent, as on the udder, teats, and escutcheon, are affected first. Animals attempt to scratch these areas by rubbing on any available surface. Erythema may then develop, followed by edema. In sheep the entire head may become highly edematous, giving rise to the common names for the disease—big head, swell-head, or geeldikkop (thick yellow head). If the edema is extensive, serum may extrude, resulting in scab formation; often secondary infection develops. The skin may become necrotic, especially on the muzzle, eyes, nostrils, and ears. Exfoliation may also occur in these severely affected areas. In cattle and horses dead skin may peel off in large, dry, leathery pieces and sheep may lose wool from large areas. Icterus may or may not be noted, and in extreme cases, the urine may become a dark, reddish brown. Pregnant animals may abort. If animals are not removed from sunlight, the trauma may result in shock and they may die from this in combination with liver damage (in hepatogenous photosensitization).

Cutaneous lesions may heal in days or weeks depending on the severity of the condition. However, retinas in range sheep are sometimes scarred so severely that they are blind. Also, wool may not grow in denuded areas on some sheep and they are lost as productive animals.

Cutaneous histopathologic changes, as might be expected, are those typical of inflammation, edema, and dermatitis.

IV. CLASSIFICATION OF PHOTOSENSITIZATIONS

Grouping photosensitizing toxins from plants into related classes is difficult because of the dissimilarity of plants in which they occur and the dissimilarity of their chemical formulations. Clare (1952), however, has classified photosensitizations in livestock into three groups and for convenience they will be considered here by his classification. Two of Clare's groups, primary and secondary photosensitization, relate to plant-caused photosensitizations, while the third class is due to a metabolic defect in livestock and is not plant related.

Primary photosensitization is differentiated from secondary photosensitization in that in the primary type, the phototoxic agent is in its toxic form in the plant tissue. As the plant is eaten the toxin is absorbed in an unchanged state into the circulatory system, where it eventually reaches the skin, there to be acted upon by sunlight to cause the toxic reaction.

A. Secondary Photosensitization

In secondary photosensitization the phototoxic agent is phylloerythrin. This porphyrin is an anaerobic fermentation product of chlorophyll produced by microorganisms in the stomachs of ruminants. Both chlorophyll and phylloerythrin are porphyrins; both are photodynamic compounds (Griffiths et al., 1955), and both absorb light in the general porphyrin absorption range 300-450 nm. In a normal animal, intact chlorophyll [1] molecules cannot be

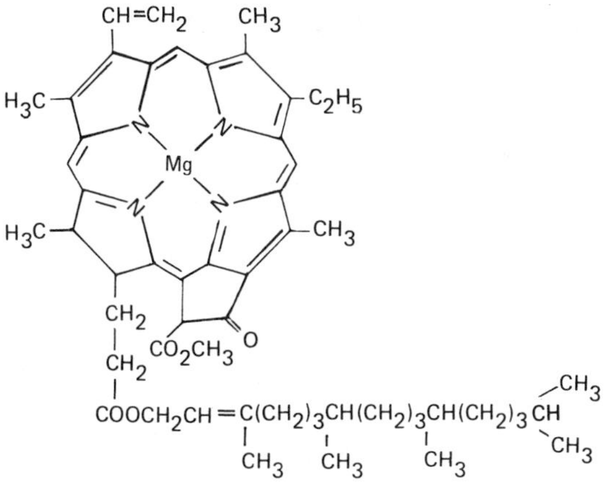

[1]

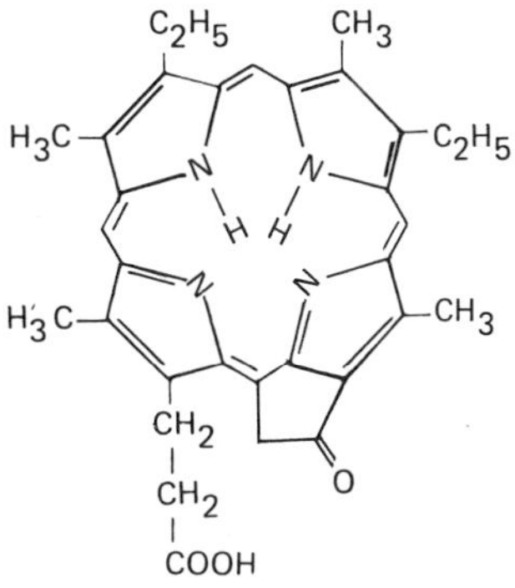

[2]

absorbed from the digestive tract because of their long phytyl hydrocarbon side chains; thus they cause no photodynamic reaction. Phylloerythrin [2], on the other hand, has lost these long side chains to action of digestive micro-organisms and can be absorbed from the gut. Absorbed phylloerythrin is normally removed from the circulation by the liver and is excreted in the bile before it can reach the skin. In an animal that has liver damage of a specific type or biliary stasis wherein the hepatic excretion of phylloerythrin is pre-vented, there is no way this phototoxin can be eliminated from the body even though small amounts are excreted in the urine. Thus phylloerythrin reaches the skin, absorbs its light from sunlight, and produces its phototoxic reac-tion in tissue as it releases its light energy.

Secondary photosensitization is often, and perhaps more appropriately, labeled hepatogenous photosensitization.

Hepatogenous photosensitization is by far the most frequently encountered type in livestock; therefore, chlorophyll is the most important photosensitizing toxin that occurs in plants, even though it must be metabolized to a secondary product to exert its phototoxic action.

Most of the knowledge we have of secondary photosensitization was ob-tained through studies of geeldikkop, a photosensitization syndrome that occurs in Africa and is caused by sheep eating *Tribulus terrestris* (puncture vine) and other *Tribulus* species. It is perhaps the best known example of secondary photosensitization. It was first observed in 1886 (Henning, 1932), and the etiology of this syndrome was so carefully worked out primarily by Quin (1928), by Quin and Rimington (1935), Theiler (1918), and others that little has been added to the knowledge of hepatogenous photosensitization since their work in the 1930s. For a complete history of the early work on geeldikkop, the reader is referred to Blum (1941).

Although the general etiology of hepatogenous photosensitization has been worked out, some of the basic mechanisms have not been elucidated. For instance, a difference of opinion exists as to the importance of the chlorophyll quality of the feed needed to cause hepatogenous photosensitization. Clare (1952) believes that too much emphasis is placed on this factor, and that even

hay with obviously degraded chlorophyll can cause phylloerythrinemia. However, Smith et al. (1972) state that for hepatogenous photosensitization to
occur, the animal must be "on a diet containing liberal amounts of chlorophyll." Thus the quantity of circulating phylloerythrin needed to cause
photosensitization has not been ascertained. It may be small indeed, and its
buildup rate in the peripheral blood may be more related to the type of liver
damage involved than the amount of chlorophyll ingested by the animal.
Rimington and Quin (1934) were able to cause photosensitization in sheep
readily by injecting as little as 41 mg per sheep, but they did not determine
minimum amounts needed.

The actual mechanism of phylloerythrin excretion by the liver is not known;
thus the damage that blocks that excretion by the liver is not understood.
It is known that extrahepatic cholestasis produced by ligation of the common
bile duct can generate hepatogenous photosensitization in animals on a diet
containing chlorophyll (Rimington and Quin, 1934). Intrahepatic cholestasis
produces the same effect (Quin, 1936; Gopinath and Ford, 1969) and it is
entirely possible that phylloerythrin excretion may be blocked by subtle biochemical changes at the hepatocyte or membrane level (Heikel et al., 1960;
Pass et al., 1978).

Smith et al. (1972) state that there is no specific morphologic pattern of
liver changes in hepatogenous photosensitization but say that changes described for acute toxic hepatitis are usually in evidence. They admit, however, that photosensitization is seen in only a small minority of toxic hepatic
illnesses that can occur from a variety of causes.

Certainly, there is much to be learned of both the intrinsic and extrinsic
causes of hepatogenous photosensitization. If this syndrome is as simple and
straightforward as we are tempted to record it, why is it so difficult to produce experimentally with good consistency? This difficulty has been expressed by almost every worker who has attempted to produce experimental
hepatogenous photosensitization and leads one to believe that there are important factors of which we are not yet aware. Furthermore, Johnson (1974a)
has shown that potentiating or preconditioning substances found in black
sagebrush (*Artemesia nova*) are important in the development of tetradymia-
induced hepatogenous photosensitization in sheep but has not pinpointed the
specific compounds or their mode of action. Elucidation of these relationships might provide some important clues.

1. Plants Causing Secondary or Hepatogenous Photosensitization

Although all secondary photosensitizations have a common phototoxic agent,
phylloerythrin, they are designated by the hepatotoxic plant that is actually
the primary causative agent. Thus *Tribulus terrestris* and possibly other
Tribulus species cause tribulosis, or geeldikkop. This toxic syndrome has
been studied extensively in Africa (see above). Attempts to produce this
phototoxic reaction in sheep by experimentally feeding the plant have often
been fruitless (Quin, 1933). It has long been suspected that the instability
of the hepatotoxic agent is at least partially responsible for this difficulty.
The recent work of Bath et al. (1978) seems to confirm this suspicion. They
collected wilted plant, which is reported to be most toxic, quickly froze it,
and stored it frozen for 7 weeks. The plant, which was then rapidly thawed
and immediately fed to sheep, caused photosensitization.

T. terrestris (puncture vine) occurs in the United States and other parts of the world, but reported outbreaks are not as frequent or as serious as those recorded in Africa.

Other South African plants that cause symptoms and photosensitization similar to tribulosis are *Lippia rehmanni* Pears and *L. pretoriensis* Pears. Both cause hepatogenous photosensitization in sheep; *L. rehmanni* is the more important of the two. Rimington and Quin (1935) and Rimington et al. (1937) isolated a compound they named "icterogenin" as the hepatotoxic agent from *Lippia*, and Quin (1936) concluded that the primary effect of this compound was an intrahepatic blockage of bile excretion.

Further work by Rimington and Quin (1937) also incriminated two grass species, *Panicum coloratum* and *P. laerifolium*, growing in South Africa as causing signs of photosensitization identical to those caused by *Tribulus* and *Lippia* species. *P. coloratum*, known as kleingrass in the United States, has in recent years become a major photosensitization problem in lambs in Texas (Bailey et al., 1977). Other species of panic grasses from other parts of the world are also reported to be phototoxic.

Photosensitizations caused by *Tetradymia glabrata* (littleleaf horsebrush) and *T. canescens* (spineless or little gray horsebrush) in sheep are some of the most devastating phototoxicities occurring in the United States. Many thousands of sheep have been lost as a result of the hepatotoxicity and photo-toxicity produced by these plants. *T. glabrata*-caused outbreaks, although sporadic, are more frequent than those caused by *T. canescens*. Losses occur most often in early spring as sheep are trailed in desert areas. In addition to photosensitization losses, abortion and death are common. The photosensitization syndrome is difficult to produce experimentally and instability of the hepatotoxin has been ruled out as the reason for the difficulty (Johnson, 1974b). Johnson (1974a), however, has recently determined that other plants, notably *Artemesia nova* (black sagebrush), can potentiate or precondition sheep for the toxic action of tetradymia. Jennings et al. (1974) and Jennings et al. (1978) have isolated several furanoeremophilanes that are hepatotoxic from *T. glabrata* but have not determined that they cause the type of hepatotoxicity that leads to secondary photosensitization.

Lantana camara (lantana) is a highly toxic plant found in several parts of the world but especially in South Africa, Australia, and the United States. It causes hepatogenous photosensitization in livestock and the hepatotoxic compound is the polycyclic triterpenoid, lantadene. Gopinath and Ford (1969) report that lantadene blocks bile excretion by causing hepatocytes to swell. This, in turn, interferes with canalicular drainage of bile. Pass et al. (1978) believe that the damage to the canaliculi is more of a direct injury caused by the lantana toxin. Other lantana species may also cause phototoxicity.

Nolina texana, known as sacahuiste or bear grass, has been determined by Mathews (1940) to cause secondary photosensitization, particularly in sheep and goats. Mathews states its range to be western Texas, Arizona, and New Mexico. Apparently, buds and flowers are most toxic.

Agave lecheguilla (lechuguilla) is another plant described by Jungherr (1931) and Mathews (1937) that causes hepatogenous photosensitization. This plant occurs in essentially the same areas as sacahuiste. A saponin isolated by Wall et al. (1962) is the suspected hepatotoxic agent.

Many other plants of lesser importance cause hepatogenous photosensitization or at least they cause hepatotoxicity, and photosensitization sometimes follows. Given these circumstances, the inclination is to label the disease

hepatogenous photosensitization when it should not be so labeled unless phylloerythrin can be detected in the blood at the time photosensitization signs
are present. A plant may contain both an hepatotoxic agent and a phototoxic agent.

Other plants cause occasional photosensitizations and hepatoxicity only
under certain unknown conditions, and still other plants reportedly cause
photosensitization, but there is no proof of this. Furthermore, plants known
to cause definitive hepatoses, such as the toxic *Senecios*, occasionally cause
hepatogenous photosensitization.

Some of the plants in the foregoing categories generally listed as responsible for hepatogenous photosensitization are *Kochia scoparia* (summer cyprus
or Mexican fireweed) (Dickie and Berryman, 1979), *Brassica napus* (cultivated rape) (Hansen, 1930; Case, 1957), *Euphorbia maculata* (spotted spurge)
(Case, 1954, 1957), *Vicia* spp. (vetches) (Forsyth, 1954), and *Narthecium
ossifragum* (Ender, 1955; Dishington and Laksesvela, 1976).

Information available on the toxicity of these and other plants that cause
infrequent photosensitizations in livestock in various parts of the world may
be found in Clare (1952) and Kingsbury (1964).

Additionally, some of the common feed plants for livestock occasionally
cause photosensitization generally thought to be of the hepatogenous type.
Clover, notably *Trifolium hybridum* (alsike clover), causes photosensitization
primarily in horses (Fincher and Fuller, 1942) but also in cattle and sheep.
Oats have been reported to be a common cause of photosensitization in pigs
(Sippel and Burnside, 1954), and there are numerous reports of alfalfa
(*Medicago sativa*) causing mild photosensitization in young pigs (Kral and
Schwartzman, 1964), sometimes considered to be of the hepatogenous type
(Forsyth, 1954). However, Lohrey et al. (1974) caused photosensitization
in albino rats by injecting protein concentrate made from *M. sativa* but could
not detect phylloerythrin in the blood of the photosensitized rats.

Certain molds and fungi are extremely potent hepatotoxic agents, and as
they proliferate on plants when conditions are ideal, the plant may be un-
justly labeled as an hepatotoxic plant that can cause secondary photosensitization. Certainly, when difficulty arises in detecting the hepatotoxic compound in a plant, first consideration should be given to any possible fungi it
may be harboring.

B. Primary Photosensitization

There are relatively few plant toxins that cause photosensitization of the primary type. In this type, the phototoxic substance is in the plant in its toxic
state. As the plant is ingested by an animal, the photosensitizing toxin is
absorbed from the digestive tract in its original state and is transported
through the peripheral circulation to the skin areas, where it is reached by
sunlight to cause the photosensitization.

Blum (1938) formulated three postulates to be used to determine if a substance was truly a photodynamic substance.

1. Photosensitization symptoms must be brought on by exposure to sunlight.
2. A photodynamic substance must be isolated in pure form, and this
   substance when injected into or fed to experimental animals must
   cause symptoms only when the animals are exposed to light.

3. It must be determined that the wavelengths that produce sensitivity in (1) are the same as those that produce sensitivity in (2).

Postulates 1 and 2 are relatively easy to accomplish, but postulate 3 is sometimes difficult.

1. *Hypericin*

One of the most interesting and well-known primary photosensitizing agents found in plants is hypericin. It is produced by several species of *Hypericum*. It is a red pigment that can be easily extracted from glandular spots on the petals, leaves, and stems of the plant with methanol or ethanol. Hypericin is a naphthodianthrone, hexahydroxydimethylnaphthodianthrone [3].

Many species of *Hypericum* occur from Africa and the Mediterranean area to Australia and the United States, but fewer than 10 species are known to cause photosensitization. *H. perforatum* and *H. crispum* are two of the better known species and are those on which much of the experimental work has been done.

Hypericum is an aggressive plant that establishes itself easily in dry ground, and more than 25 species may be found in the United States and Canada. *H. perforatum* (St. Johnswort, Klamath weed), the common toxic species in the United States, is presently less of a problem than it was because of biological control methods that have been instituted.

That *Hypericum* spp. cause photosensitization has been known for many years (Rogers, 1914; Marsh and Clawson, 1930; Henry, 1922; Dodd, 1920). These and other workers through observation and experimentation determined that symptoms in animals appeared only when the animals were exposed to sunlight, thus fulfilling the first postulate of Blum. They further noted and described the general symptoms of photosensitization as similar to those already described in this paper.

Ray (1914) first suggested that hypericism may be due to a photodynamic agent. Other workers, but especially Horsley (1934) and Brockmann and his co-workers (Brockmann et al., 1942), 1957; Brockmann and Kluge, 1951), extracted, studied the effects of, and characterized the compound hypericin. Horsley (1934) also suggested that another yellow substance, probably quercetin, found in *Hypericum* can enhance or potentiate the effects of hypericin. These workers made possible the fulfillment of Blum's second postulate.

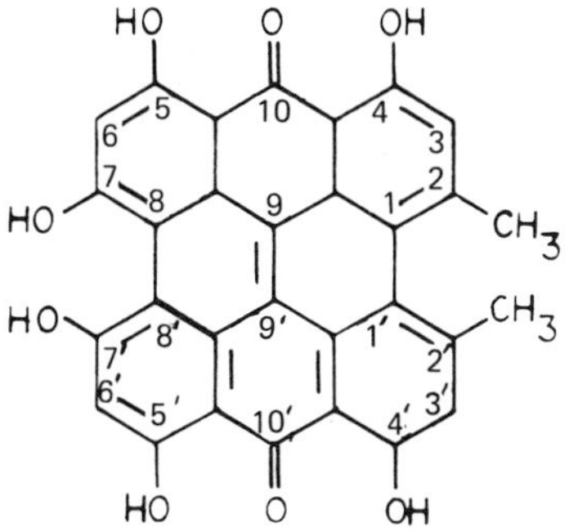

[3]

Blum's third postulate has not been totally fulfilled, but recent work of Araya and Ford (1981) definitely indicates that there is no disruption of liver function in hypericism; therefore, if the photosensitization is not secondary, it must be primary.

2. *Fagopyrin*

Fagopyrin [4] and its analogs are very closely related to hypericum both in their chemical structure and in their phototoxic abilities. Their absorption spectra is in the range 540-610 nm, as is hypericin. Fagopyrin occurs in the buckwheat group of plants, the *Fagopyrum* genus, and most commonly in *F. esculentum*.

Buckwheats have been grown and used in the past as grains and the whole plant used as livestock feed. The kernels contain very little fagopyrin, but the whole plant, either dried or green, can cause serious photosensitization problems in livestock that use it for feed. For this reason, it is grown and used much less than in past years.

The early work and historical aspects of fagopyrism have been recorded by Blum (1941) and much of the chemistry of fagopyrin has been worked out by Brockmann (1957) and Brockmann and Sanne (1953).

Both hypericism and fagopyrism require the presence of oxygen to induce their phototoxic reactions in the skin tissue.

3. *Psoralens*

Another group of primary photodynamic agents is the psoralens. Psoralens are furocoumarins, compounds composed of conjoined furan and coumarin rings.

Two of the isomeric forms of furocoumarins occur in nature, psoralens [5] and isopsoralens (angelicins) [6]. These and their substituted forms are those that occur in plants and are responsible for phototoxic reactions in livestock and humans. Psoralens absorb in the range 220-380 nm, but their action spectrum lies in the range 340-380 nm (Giese, 1971).

Psoralens are better known for their ability to cause contact photosensitization. Plants in several families cause this type of phototoxic reaction, but only four plant families have had psoralens extracted and identified from them: Umbelliferae (parsley family), Leguminosae (legume family), Rutaceae (rue family, including citrus), and Moraceae (fig family) (Pathak et al., 1962).

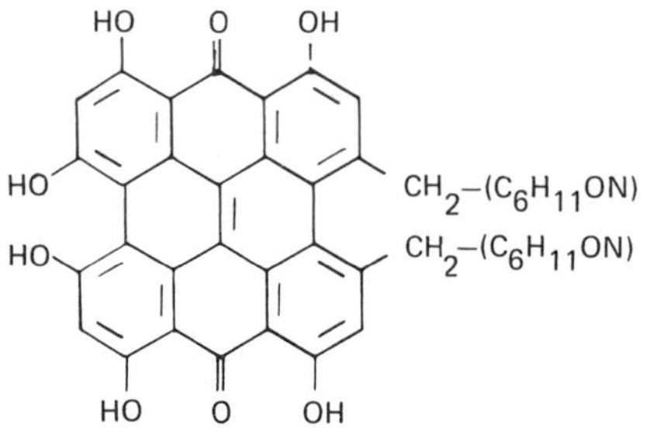

[4]

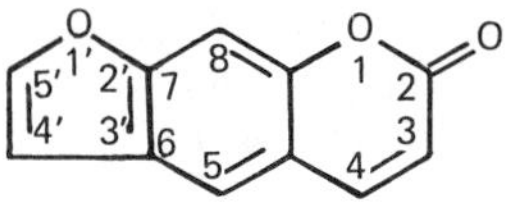

[5]

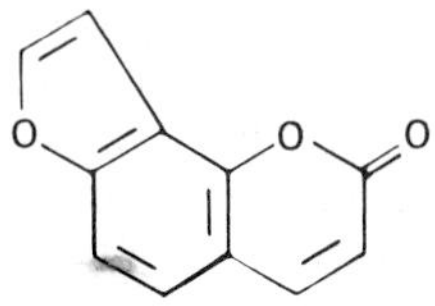

[6]

Extracts of the seeds of *Psoralea corylifera*, a legume, were used as early as 1400 B.C. in India and later in Egypt and China to repigment white patches of skin in humans (Giese, 1971). Bergamot oil, a citrus extract, contains psoralens (bergapten). Fig dermatitis is caused by contact with fig leaves containing psoralens, and plants in the Umbelliferae family are phototoxic to livestock (see below).

Symptoms of psoralen photosensitization are somewhat different from those noted in hepatogenous photosensitization and even from those caused by hypericin and fagopyrin in that they more closely resemble severe sunburn. Erythema is the first and most prominent symptom, followed by blistering and scaling if exposure is long enough. Pigmentation occurs also. Additionally, Giese (1971) reports that after ingestion of psoralens there may be gastric irritation, nausea, nervousness, and depression. This difference in symptoms exhibited is undoubtedly due to a difference in chemical reactions elicited, and to a difference in substrates involved. Psoralens do not require oxygen for reaction with the substrate, and they are known to react with flavin mononucleotides (Musajo, 1963) and with DNA and RNA (Pathak and Kraemer, 1969). Oxidative photosensitizers probably react more with certain amino acids.

Psoralen photosensitization in livestock can cause major localized problems. *Cymopterus watsonii* (spring parsley) was reported to cause sporadic outbreaks of photosensitization in sheep in some western areas (Binns et al., 1964). Williams (1970) later isolated 8-methoxypsoralen (xanthotoxin) [7] and 5-methoxypsoralen (bergapten) [8] from this plant. Another member of the parsley family (Umbelliferae), *Ammi visnaga* was fed to fowls by Trenchi (1960) to cause photosensitization. Harrison et al. (1971) isolated xanthotoxin from this plant. Seed of bishop's-weed, *Ammi majus*, also contains psoralens and has been determined by Egyed et al. (1975) and others to cause photosensitization in fowls. Witzel et al. (1978) fed finely ground seed of bishop's-weed to sheep and produced severe photosensitization.

Whereas there has been comparatively little interest in psoralen photosensitization in livestock, the actions of psoralens in humans have been under intensive investigation in recent years. Medical literature is replete with

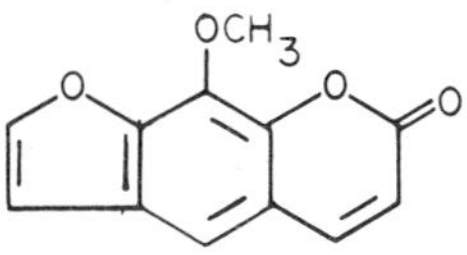

[7]

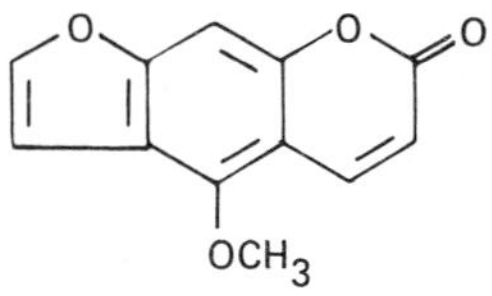

[8]

information on their use in tanning lotions, in skin depigmentation diseases (vitiligo), in treatment of psoriasis, in cancer research, and in the study of their actions on tissues and tissue components. The reader is referred to reviews by Pathak (1969), Giese (1971), Musajo and Rodighiero (1972), Pathak et al. (1974), Scheel (1973), and Ivie (1978).

Much of the interest in psoralens in human medicine has been generated by their ability to photosensitize when they are applied or contacted topically. There has been some concern, however, since Fitzpatrick et al. (1955) determined that psoralens taken internally could cause photosensitization, that some of the foods known to contain psoralens might cause phototoxic reactions. Pathak et al. (1962) reviewed the occurrence of psoralens in plants. Figs, parsely, caraway, and limes were among plants listed as sources of psoralens. It has not been shown that natural psoralen-containing foods cause photosensitization in humans, and Scheel (1973) points out that because coumarins steam distill, boiling foods might reduce their toxic psoralen content to less than toxic levels. However, Ivie et al. (1981) have determined that psoralen levels in parsnip roots were not reduced by boiling or microwave cooking.

There is much yet to be learned about phototoxic substances and their modes of action in both humans and animals. The medical use of the psoralens and concern regarding their occurrence in foods should provide sufficient impetus to sustain research activity on these and other photodynamic substances in humans, while photodynamic compounds in plants that cause severe losses of livestock should continue to generate research interest in animal phototoxicities.

REFERENCES

Araya, O. S., and Ford, E. J. H. (1981). An investigation of the type of photosensitization caused by the ingestion of St. Johnswort, *Hypericum perforatum*, by calves. *J. Comp. Pathol. 91*:135-142.

Bailey, E. M., Jr., Bridges, C. H., Livingston, C. W., Menzies, C. S., Tabor, R. A., Pettit, R. E., and Muchiri, D. (1977). Kleingrass-induced photosensitization in sheep. *Tex. Agric. Exp. Stn. Prog. Rep.*, pp. 3445-3470.

Bath, G. F., Van Tonder, E. M., and Basson, P. A. (1978). Geeldikkop: preservation of toxic material. *J. S. Afr. Vet. Assoc. 49*:23-25.

Binns, W., James, L. F., and Brooksby, W. (1964). *Cymopterus watsoni*: a photosensitizing plant for sheep. *Vet. Med./Small Anim. Clin. 59*:375-379.

Blum, H. F. (1938). Domestic animal diseases produced by light. *J. Am. Vet. Med. Assoc. 93*:185-191.

Blum, H. F. (1941). *Photodynamic Action and Diseases Caused by Light*. Reinhold, New York.

Brockmann, H. (1957). Photodynamically active plant pigments. *Proc. Chem. Soc. Lond.*, November, pp. 304-312.

Brockmann, H., and Kluge, F. (1951). Zur Synthese des Hypericins. *Naturwissenschaften 38*:14.

Brockmann, H., and Sanne, W. (1953). Zur Biosynthese des Hypericins. *Naturwissenschaften 40*:509-510.

Brockmann, H., Pohl, F., Maier, K., and Haschad, M. N. (1942). Über das Hypericin, den photodynamischen Forbstoff des Johanniskrautes (*Hypericum perforatum*). *Ann. Chem.* 553:1-52.

Brockmann, H., Kluge, F., and Muxfeldt, H. (1957). Totalsynthese des Hypericins. *Chem. Ber.* 90:2302-2318.

Case, A. A. (1954). Malnutrition complex. *Sheepbreeder Sheepman* 74(11): 9-10.

Case, A. A. (1957). Photosensitization syndrome in cattle, sheep, and swine. *N. Am. Vet.* 38:161-165.

Clare, N. T. (1952). *Photosensitization in Diseases of Domestic Animals.* Commonwealth Agricultural Bureaux, Farnham Royal, Bucks, England.

Dickie, C. W., and Berryman, J. R. (1979). Polioencephalomalacia and photosensitization associated with *Kochia scoparia* consumption in range cattle. *J. Am. Vet. Med. Assoc.* 175:463-465.

Dishington, I. W., and Laksesvela, B. (1976). The etiology of "Alveld" elucidated by the BSP-test. *Nord. Vet. Med.* 28:547-549.

Dodd, S. (1920). St. Johnswort and its effects on livestock. *Agric. Gaz. N.S. Wales* 31:265-272.

Egyed, M. N., Shlosberg, A., Eilat, A., and Malkinson, M. (1975). Acute and chronic manifestations of *Ammi majus*-induced photosensitization in ducks. *Vet. Rec.* 97:198-199.

Ender, F. (1955). Aetiology of photosensitization in lambs. *Nord. Vet. Med.* 7:329-377.

Fincher, M. G., and Fuller, H. K. (1942). Photosensitization–trifoliosis– light sensitization. *Cornell Vet.* 32:95-98.

Fitzpatrick, T. B., Hopkins, C. E., Blickenstaff, D. D., and Swift, S. (1955). Augmented pigmentation and other responses of normal human skin to solar pigmentation following oral administration of 8-methoxy- psoralen. *J. Invest. Dermatol.* 25:187-190.

Forsyth, A. A. (1954). British poisonous plants. *Ministry Agric. Fish. Food (Lond.) Bull. 161.*

Giese, A. C. (1971). Photosensitization by natural pigments. In *Photophysiology*, Vol. VI, A. C. Giese (Ed.). Academic Press, New York, pp. 77-129.

Gopinath, C., and Ford, E. J. H. (1969). The effect of *Lantana camara* on the liver of sheep. *J. Pathol.* 99:75-85.

Griffiths, M., Sistrom, W. R., Cohen-Bazire, G., and Stanier, R. Y. (1955). Function of carotenoids in photosynthesis. *Nature (Lond.)* 176:1211- 1214.

Hansen, A. A. (1930). Indiana plants injurious to livestock. *Purdue (Indiana) Agric. Exp. Stn. Circ. 175.*

Harrison, P. G., Bailey, B. K., and Steck, W. (1971). Biosynthesis of furanochromones. *Can. J. Biochem.* 49:964-967.

Heikel, T., Knight, B. C., Rimington, C., Ritchie, H. D., and Williams, E. J. (1960). Studies on biliary excretion in the rabbit. I. The effect of icterogenin and rehmannic acid on bile flow and the excretion of bilirubin, phylloerythrin, coproporphyrin, alkaline phosphatase, and bromsulphalein. *Proc. R. Soc. Lond. B* 153:47-79.

Henning, M. W. (1932). *Animal Diseases in South Africa*, Vol. II. Central News Agency, Ltd., Johannesburg, South Africa, p. 854.

Henry, M. (1922). Feeding and contact experiments with St. Johnswort. *Agric. Gaz. N.S. Wales* 33:205-207.

Horsley, C. H. (1934). Investigation into the action of St. Johnswort. *J. Pharmacol. Exp. Ther.* 50:310-322.

Ivie, W. (1978). Toxicological significance of plant furocoumarins. In *Effects of Poisonous Plants on Livestock*, R. F. Keeler, K. R. Van Kampen, and L. F. James (Eds.). Academic Press, New York, pp. 475-485.

Ivie, G. W., Holt, D. L., and Ivey, M. C. (1981). Natural toxicants in human foods: psoralens in raw and cooked parsnip root. *Science* 213:909-910.

Jennings, P. W., Reeder, S. K., Hurley, J. C., Caughlan, C. N., and Smith, G. D. (1974). Isolation and structure determination of one of the toxic constituents from *Tetradymia glabrata*. *J. Org. Chem.* 39:3392-3398.

Jennings, P. W., Reeder, S. K., Hurley, J. C., Robbins, J. E., Holian, S. K., Holian, A., Lee, P., Pribanic, J. A. S., and Hull, M. W. (1978). Toxic constituents and hepatotoxicity of the plant, *Tetradymia glabrata* (Asteroceae). In *Effects of Poisonous Plants on Livestock*, R. F. Keeler, K. R. Van Kampen, and L. F. James (Eds.). Academic Press, New York, pp. 217-228.

Johnson, A. E. (1974a). Predisposing influence of range plants on *Tetradymia*-related photosensitization in sheep: work of Drs. A. B. Clawson and W. T. Hoffman. *Am. J. Vet. Res.* 35:1583-1585.

Johnson, A. E. (1974b). Experimental photosensitization and toxicity in sheep produced by *Tetradymai glabrata*. *Can. J. Comp. Med.* 38:406-410.

Jungherr, E. (1931). Lechuguilla fever of sheep and goats; a form of swell-head in west Texas. *Cornell Vet.* 21:227-242.

Kingsbury, J. M. (1964). *Poisonous Plants of the United States and Canada*. Prentice-Hall, Englewood Cliffs, N.J., pp. 52-57.

Kral, F., and Schwartzman, R. M. (1964). *Veterinary and Comparative Dermatology*. Lippincott, Philadelphia, p. 126.

Lohrey, E., Tapper, B., and Hove, E. L. (1974). Photosensitization of albino rats fed on lucerne-protein concentrate. *Br. J. Nutr.* 31:159-166.

Marsh, C. D., and Clawson, A. B. (1930). Toxic effect of St. Johnswort (*Hypericum perforatum*) on cattle and sheep. *U.S. Dept. Agric. Tech. Bull. 202*.

Mathews, F. P. (1937). Lechuguilla (*Agave lecheguilla*) poisoning in sheep, goats, and laboratory animals. *Texas Agric. Exp. Stn. Bull. 554*.

Mathews, F. P. (1940). Poisoning in sheep and goats by sacahuiste (*Nolina texana*) buds and blossoms. *Texas Agric. Exp. Stn. Bull. 585*.

Musajo, L. (1963). Photoreactions between flavin coenzymes and skin-photosensitizing agents. *Pure Appl. Chem.* 6:369-384.

Musajo, L., and Rodighiero, G. (1972). Mode of photosensitizing action of furocoumarins. *Photophysiology* 7:115-147.

Pass, M. A., Seawright, A. A., Heath, T. J., and Gemmell, R. T. (1978). Lantana poisoning: a cholestatic disease of cattle and sheep. In *Effects of Poisonous Plants on Livestock*, R. F. Keeler, K. R. Van Kampen, and L. F. James (Eds.). Academic Press, New York, pp. 229-237.

Pathak, M. A. (1969). Basic aspects of cutaneous photosensitization. In *The Biologic Effects of Ultraviolet Radiation (with Emphasis on the Skin)*, F. Urbach (Ed.). Pergamon Press, Oxford, pp. 489-511.

Pathak, M. A., and Kraemer, D. M. (1969). Photosensitization of the skin in vivo by furocoumarins (psoralens). *Biochim. Biophys. Acta* 195:197-206.

Pathak, M. A., Daniels, F., Jr., and Fitzpatrick, T. B. (1962). The presently known distribution of furocoumarins (psoralens) in plants. *J. Invest. Dermatol. 39*:225-239.

Pathak, M. A., Kraemer, D. M., Fitzpatrick, T. B. (1974). Photobiology and photochemistry of furocoumarins psoralens. In *Sunlight and Man: Normal and Abnormal Photobiologic Responses*, M. A. Pathak (Ed.). Proc. Int. Conf. Photosensitization Photoprot. University of Tokyo Press, Tokyo, pp. 335-368.

Quin, J. I. (1928). Recent investigations into geeldikkop affecting sheep and goats in the Cape Province. *J. S. Afr. Vet. Med. Assoc. 1*:43-45.

Quin, J. I. (1933). Studies on the photosensitisation of animals in South Africa. I. The action of various fluorescent dye-stuffs. *Onderstepoort J. Vet. Sci. 1*:459-468.

Quin, J. I. (1936). Studies on the photosensitization of animals in South Africa. IX. The bile flow of the Merino sheep under various conditions. *Onderstepoort J. Vet. Sci. 7*:351-366.

Quin, J. I., and Rimington, C. (1935). Geeldikkop. A critical review of the problem as it affects sheep farming in the Karoo together with recommendations in the light of newer knowledge. *J. S. Afr. Vet. Med. Assoc. 6*:16-24.

Ray, G. (1914). Note sur les effets toxiques du millepertuis à feuilles crispées (*Hypericum crispum*). *Bull. Soc. Cent. Méd. Vét. 68*:39-42.

Rimington, C., and Quin, J. I. (1934). Studies on the photosensitisation of animals in South Africa. VII. The nature of the photosensitising agent in geeldikkop. *Onderstepoort J. Vet. Sci. Anim. Ind. 3*:137-157.

Rimington, C., and Quin, J. I. (1935). The isolation of an icterogenic principle from *Lippia rehmanni* pears. *S. Afr. J. Sci. 32*:142-151.

Rimington, C., and Quin, J. I. (1937). Dik-oor or Geel-dikkop on grassveld pastures. *J. S. Afr. Vet. Med. Assoc. 8*:141-146.

Rimington, C., Quin, J. I., and Roets, G. C. S. (1937). Studies upon the photosensitisation of animals in South Africa. X. The icterogenic factor in geel-dikkop. Isolation of active principles from *Lippia rehmanni*. *Onderstepoort J. Vet. Sci. 9*:225-255.

Rogers, T. B. (1914). On the action of St. Johnswort as a sensitizing agent for non-pigmented skin. *Am. Vet. Rev. 46*:145-162.

Scheel, L. D. (1973). Photosensitizing agents. In *Toxicants Occurring Naturally in Foods*, Vol. 2. National Academy of Sciences, Washington, D.C., pp. 558-572.

Sippel, W. L., and Burnside, J. E. (1954). Oat dermatitis. *Ga. Vet. 6*:3-4.

Smith, H. A., Jones, T. C., and Hunt, R. D. (1972). *Veterinary Pathology*, 4th ed. Lea & Febiger, Philadelphia, pp. 77-82.

Spikes, J. D. (1968). Photodynamic action. In *Photophysiology*, Vol. III, A. C. Giese (Ed.). Academic Press, New York, pp. 33-64.

Theiler, A. (1918). Geeldikkop in sheep (*Tribulus ovium*). *7th and 8th Rep. Dir. Vet. Res.*, Union of South Africa, pp. 1-55.

Trenchi, H. (1960). Ingestion of *Ammi visnaga* Seeds and photosensitization— the cause of vesicular dermatitis in fowls. *Avian Dis. 4*:275-280.

Wall, M. E., Warnock, B. H., and Willaman, J. J. (1962). Steroidal sapogens. 48. Their occurrence in *Agave lecheguilla*. *Econ. Bot. 16*:266-269.

Williams, M. C. (1970). Xanthotoxin and bergapten in spring parsley. *Weed Sci. 18*:479-480.

Witzel, D. A., Dollahite, J. W., and Jones, L. P. (1978). Photosensitization in sheep fed *Ammi majus* (bishops weed) seed. *Am. J. Vet. Res. 39*:319-320.

11

FUNGAL-TOXIN-INDUCED PHOTOSENSITIZATION

PETER H. MORTIMER and JOHN W. RONALDSON

Ruakura Agricultural Research Center, Hamilton, New Zealand

I. FUNGAL-TOXIN-INDUCED PHOTOSENSITIZATION

A. Introduction

In this context photosensitization generally refers to an injurious skin reaction or photodermatitis. It is produced when a photodynamic agent, aber-
rantly present in the skin or present in abnormally high quantities, is
activated by light energy (sunlight), causing cellular injury and tissue
inflammation (Clare, 1955). Photodynamic agents of exogenous origin causing
photodermatitis in mammals include drugs, chemicals, and plant components.
Under certain conditions endogenous catabolic porphyrin pigments may act as
photodynamic agents.

A contact phytophotodermatitis due to furocoumarins, metabolites of certain plants, especially of the orders Umbelliferae and Rutaceae, is well recognized (Ivie, 1978). Originally believed to be produced as a fungal metabolite,
xanthoxin accumulates in parsnips infected with *Cerratocystis fimbriata* and
both xanthoxin and trisoralen are produced in celery in response to *Sclerotinia sclerotiorum* infection ("pink rot"). Recent evidence indicates that both
these furocoumarins are produced by the plants as phytoalexins (Ivie, 1978)
and therefore that this photodermatitis is not of mycotoxic origin.

1. Photosensitization in Animals

Some photosensitizations in animals are of the primary type. In these the
photodynamic agents, often plant pigments not usually eaten, are absorbed
from the digestive tract and not being completely excreted or detoxified,
reach the peripheral circulation and the skin. Primary photosensitizing
agents are not known to be produced by fungi.

The majority of instances of photosensitization in animals are of the secondary or hepatogenous type (Clare, 1955). In these cases phylloerythrin, a
photodynamic porphyrin pigment derived from microbial breakdown of chlorophyll in the digestive tract (normally effectively excreted by the liver into
bile), accumulates in the peripheral circulation when hepatic excretory function is impaired. Such impairment can be caused by infective or toxic or
other liver or bile duct injury which impairs detoxification or efficient biliary
excretion. Under these circumstances photodermatitis develops rapidly when
the skin is exposed to sunlight for periods as short as several minutes. The
more prolonged the light exposure, the more severe is the skin injury.

2. Hepatogenous Photosensitization Diseases of Known Mycotoxic Etiology

Current knowledge recognizes only three groups of mycotoxins which produce photosensitization in animals, and in all three mycotoxicoses the resulting photodermatitis is a hepatogenous (secondary) photosensitization.

Two of these photosensitization mycotoxicoses occur spontaneously in
nature. The most commonly occurring and most extensively researched of
these is pithomycotoxicosis or "Facial eczema" of grazing ruminants. In this
disease the liver and bile duct injury is produced by the sporidesmins, toxins
known to be elaborated only by *Pithomyces chartarum* (Berk. and Curt.) M.B.
Ellis.

The other naturally occurring photosensitization disease is mycotoxic lupinosis of sheep, cattle, horses, and swine. Two hepatotoxic mycotoxins produced by *Phomopsis leptostromiformis* (Kühn) Bubák ex Lind. have been designated phomopsins A and B.

The third group of mycotoxins produce liver and bile duct lesions and photosensitization when dosed to laboratory animals, but they have not been known to produce spontaneous disease in laboratory or farm animals. These mycotoxins are xanthomegnin quinone metabolites, fungal pigments so far obtained from laboratory cultures of certain *Penicillium, Aspergillus, Trichophyton,* and *Microsporum* species (Carlton, 1980).

3. Hepatogenous Photosensitization Diseases with Suspected Mycotoxic Etiology

As well as the foregoing three established mycotoxicoses which may produce photosensitization as a symptom of hepatic derangement, there are several other naturally occurring photosensitization diseases of animals of unknown etiology. In some of these there are reasons to suspect a mycotoxic etiology for the hepatic injury, but evidence adequate for confirmation is not available. Detailed discussion on these diseases is beyond the scope of this review and the reader is referred to the informative review of Marasas and Kellerman (1978). Some contenders for mycotoxic diseases status are listed below.

Panicum photosensitization: Several species of *Panicum* grasses have been incriminated in outbreaks of photosensitization in sheep in various countries: *P. effusum* and *P. decompositum* in Australia (Hurst, 1942); *P. miliaceum* in New Zealand (Simpson, 1946); *P. maximum, P. laevifolium, P. coloratum* in South Africa (Rimington and Quin, 1937) where the disease is known as dikoor (thick ear); and *P. coloratum* (Kleingrass) in North America (Muchiri et al., 1980).

Icterus and some degree of liver injury has been described for most of these *Panicum*-associated photosensitizations, and phylloerythrin was established as the photodynamic agent at least in *P. miliaceum* (Clare, 1955). From descriptions available on *Panicum* photosensitization, histologic changes in livers and bile ducts have some general similarities. However, identification or isolation of the toxic icterogenic factors has not so far been accomplished with any of these *Panicum* spp. Furthermore, the hepatic derangement they produce in bile pigment and bile acid metabolism are also currently poorly understood.

Bermuda grass photosensitization: Sporadic outbreaks of hepatogenous photosensitization have been reported in Florida in cattle grazing frosted moldy Bermuda grass [*Cynodon dactylon* (L.) Pers.]. Prominent among the saprophytic fungi that developed on the frost-killed grass was *Periconia minutissima* Corda, but the feeding of mixtures of fungal spores collected from the toxic moldy grass did not reproduce the disease (Kidder et al., 1961). Cattle penned and forced to consume the moldy grass in situ developed photosensitization. In the outbreaks, icterus and abnormal granular livers were found. Histopathologic lesions in the liver included enlargement of the portal triads, proliferation of bile ducts, and cirrhosis. The mycotoxic etiology of Bermuda grass photosensitization still requires substantiation.

Geeldikkop (Tribulosis ovis): Serious outbreaks of a disease characterized by hepatogenous photosensitization occur sporadically in sheep grazing the plant *Tribulus terrestris* L. (Brown, 1959; van Tonder et al., 1972). Toxicity of this plant appears to be restricted to periods when specific climatic conditions occur (van Tonder et al., 1972). Although the disease has been reproduced on three occasions by feeding *T. terrestris*, other attempts have failed (Brown, 1959), as have all attempts to isolate from the plant hepatotoxic substances capable of causing hepatogenous photosensitization (Brown, 1959). Cultures of a variety of fungal isolated from *T. terrestris* also failed to produce photosensitization (Brown, 1964), although two of these isolates produced other toxic effects in sheep (Gouws, 1965).

Histopathologic changes in livers of affected sheep include necrosis of individual hepatocytes, bile duct proliferation, and fibrosis, together with depositions of crystalline material in hepatocytes, in Kupffer cells, in bile duct epithelial cells, and in the bile ducts (van Tonder et al., 1972; Kellerman et al., 1980).

Phylloerythrin has been determined as the photodynamic agent in geeldikkop (Rimington and Quin, 1934), but the hepatotoxic substances responsible for the disease remain undetected.

Alfalfa hay photosensitization: The feeding of hay made from alfalfa (*Medicago sativa* L.) which had been damaged by flooding prior to harvest resulted in a severe outbreak of hepatogenous photosensitization in cattle in eastern Oklahoma and neighboring states in the fall and winter of 1957-1958 (Monlux et al., 1963). The disease was fully reproduced by feeding the toxic hay to cattle. All became ill and most developed icterus and photosensitization (Glenn et al., 1964).

In the outbreak, anorexia and sudden drop in milk production occurred within 7 days of commencement of feeding the alfalfa hay, and icterus and photosensitization occurred 1-3 days later. Phylloerythrin was detected in serum. Histopathologic lesions revealed an acutely developing cholangitis, pericholangitis, and occlusion of the small bile ducts, lesions which were considered (Glenn et al., 1965) to be strikingly similar to those which occur in the disease known as facial eczema (pithomycotoxicosis).

Unfortunately, mycological investigations of the toxic hay gave no evidence to indicate a specific mycotoxic etiology, and the responsible hepatotoxins remain unknown.

II. PITHOMYCOTOXICOSIS

A. History

Facial eczema, a hepatogenous photosensitization disease which affects sheep, cattle, and farmed deer grazing on improved pastures, was first reported at the beginning of this century. The first recognition of the disease was in lambs showing exudation and swelling of the skin on the head and ears, from which the popular name was derived. Many years later the obvious facial lesions were recognized to be a photosensitization reaction to sunlight and an outward symptom of severe liver injury (Hopkirk, 1936). The correlation of severely jaundiced livers and conspicuous bile duct lesions with high serum levels of bilirubin and phylloerythrin led Cunningham et al. (1942) to suspect that the photodynamic agent responsible for photosensitization was phylloerythrin, and this was later confirmed by Clare (1944). The disease was found to be nontransmissible and, because of its strict seasonal incidence, there was a strongly held belief that an abnormal metabolic product in rapidly growing ryegrass was the cause. Toxic pasture material, collected, dried, and fed to lambs, produced the disease but used much of the material collected. By scaling down to a guinea pig feeding bioassay in 1951, a systematic laboratory isolation of the toxin became possible.

A colorless material, the basis of the "beaker test" (Sandos et al., 1959), was commonly found in toxic pasture samples. It was not the toxin being sought but it contained cyclic peptides (pithomycolide and sporidesmolides; see the review by Russell, 1966), which suggested a fungal association. Further evidence for a fungal etiology came soon after when a black dusty deposit was noticed on the blades of a mower cutting grass in a toxic pasture. The deposit contained an abundance of spores of *Pithomyces chartarum* (Berk. and Curt.) M. B. Ellis (then known as *Sporidesmium bakeri*). Grown on various media, fungal isolates of *P. chartarum* produced the beaker test materials (Percival and Thornton, 1958), and more important, when cultures were fed to guinea pigs and lambs, they produced characteristic liver and bile duct lesions (Thornton and Percival, 1959).

B. Mycological Aspects

1. *Source of Mycotoxin*

Sporidesmins, the toxins responsible for the animal disease pithomycotoxicosis (commonly called facial eczema), are produced only by the saprophytic mold *Pithomyces chartarum*. Cosmopolitan in distribution, *P. chartarum* is more commonly reported from tropical and warm temperate zones than from cooler zones (Ellis, 1960).

Occurrence of pithomycotoxicosis: Major outbreaks of the disease occur only under specific weather conditions, as when hot dry periods are terminated by warm rains and high humidity with grass-minimum temperatures of 12°C or greater on two or more consecutive nights (Mitchell et al., 1959). Under these conditions the saprophytic mold *Pithomyces chartarum* grows rapidly and sporulates freely. The substrate for growth of *P. chartarum* in the field is dead vegetative material. Compared with other forages, pasture litter provides a warm and humid ground surface mat with a particularly favorable microclimate for growth and sporulation of *P. chartarum*.

The incidence and severity of this disease in ruminants is related to the number of spores ingested, their sporidesmin content, and the susceptibility of the grazing animal. Close grazing of leafy, well-controlled pastures increases the intake of available *P. chartarum* spores in relation to amount of pasture material ingested. Consequently, the use of pasture management systems which have made possible higher rates of stocking and more efficient pasture utilization have also increased the likelihood of emergence of this disease. It appears to be this interplay of local factors which place New Zealand North Island flocks and herds at much greater risk to the disease than their counterparts in other warm-temperate regions of the world. The New Zealand summer/fall of 1981 again saw widespread severe outbreaks of pithomycotoxicosis and, although at the time of writing it is too early to quantify the resultant on-farm losses, authorities calculate that livestock and production losses may be in the region of $100 million.

Other countries where authentic outbreaks of pithomycotoxicosis have been recorded are in Victoria (Hore, 1960) and Western Australia states (Gardiner and Nairn, 1962; Edwards et al., 1981) and in South Africa (Marasas et al., 1972). Outbreaks of hepatogenous photosensitivity in cattle fed alfalfa hay occurred in Oklahoma (Monlux et al., 1963). The clinical, pathological, and blood chemical changes described were most typical of those found in pithomycotoxicosis in cattle in New Zealand. Furthermore, spores identical to those of *P. chartarum* were found in a hay sample, but on culture, they were not viable, so the mycological evidence for the involvement of *P. chartarum* in the Oklahoma outbreaks was incomplete.

Distribution of toxigenic strains of P. chartarum: Most New Zealand strains of *P. chartarum* have proved to be freely sporing and production of toxin is generally related to production of spores. It appears that not all sporing strains produce sporidesmins, for freely sporing Texas strains (Ueno et al., 1974) and South African strains (Kellerman et al., 1980) have failed to produce detectable amounts of toxin. On the other hand, a number of non-New Zealand strains have produced toxins for cultures of these strains fed to lambs in Australia (Hore, 1960) and in South Africa (Marasas et al., 1972) produced lesions of pithomycotoxicosis, and a strain from the United Kingdom reportedly produced a toxin resembling sporidesmin (Gregory and Lacey, 1964). The occurrence of toxigenic and nontoxigenic strains of *P. chartarum* in different geographical regions needs further examination.

Morphology of P. chartarum: The morphology and distinguishing characteristics of the imperfect mold were described by di Menna et al. (1977) and Sutton and Gibson (1977). Rough, dark, barrel-shaped septate aleurospores, 8-20 × 10-30 μm, are borne singly on short conidiophores (Fig. 1). A small piece of basal denticle remains attached to free spores. Transverse septa number none to five (usually three) and longitudinal none to three (usually two), and they distinguish *P. chartarum* spores from the rather similar spores of *P. karoo* Marasas and Schumann, which commonly have two transverse septa and one longitudinal septum.

Culture techniques and production of sporidesmins: In the laboratory, *P. chartarum* will grow on many substrates, but cultural conditions as well as strains influence both time and rate of sporulation. Since there is a general correlation between numbers of spores formed and amount of sporidesmin produced (di Menna et al., 1970), techniques that promote sporulation are preferable for toxin production. Sporulation has proved more abundant on non-defined than on synthetic media (Ross, 1960) and with most isolates but not all, sporulation is stimulated by ultraviolet irradiation.

Initially, methods for production of sporidesmin at the Ruakura Agricultural Research Center used potato-carrot broth. This method produced up to 3.9 mg of sporidesmin per liter in 7 days at 20°C (Ross and Thornton, 1962), and while it proved useful for animal dosing, it was inconvenient for bulk chemical extraction of sporidesmin because of its high water content. Moistened sterilized ryecorn (*Secale cereale*) proved to be a suitable substrate for *P. chartarum* (Lloyd and Clarke, 1959), and based on this a routine

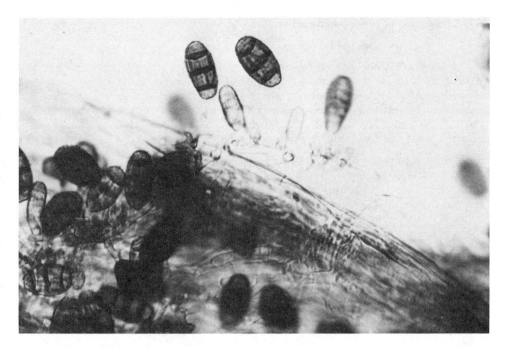

FIGURE 1 *P. chartarum* growing on dead leaf material in pasture (X750).

method for bulk sporidesmin production was developed (Done et al., 1961) and is currently employed at this center.

The routine procedure is as follows. Ryecorn or barley (75 g of whole grain plus 45 ml of distilled water per wide-necked 600-ml bottle) is soaked overnight and next morning the plugged bottles are autoclaved at 120°C for 30 min. Inoculated with a spore suspension of *P. chartarum*, the bottles are placed on their sides and incubated at 20°C in the dark for 4 weeks. A freely sporing laboratory isolate (strain C) yields about 100 mg of sporidesmin per kilogram of culture without ultraviolet irradiation. Aeration of the cultures is essential and foil caps reduce sporulation, but protective paper caps retained during incubation are satisfactory. To ensure even growth and to redistribute condensed water, cultures under incubation are shaken each week. Within 10 days of inoculation there is a blackening of cultures as spores appear, and cultures are completely black after 4 weeks of incubation and are suitable for harvesting. Most of the toxin is produced in the fourth week, and there is only slightly more after 5 weeks of incubation.

2. Mycotoxins Known to Be Produced

Sporidesmin (sometimes referred to as sporidesmin A) represents more than 90% of the total sporidesmins produced by cultures of *P. chartarum*, the remainder being small amounts of sporidesmins B, C, D, E, F, G, H, and J.

C. Chemistry of the Sporidesmins

1. *Structure and Derivatives*

Physical and chemical properties: In the original isolation of sproidesmin from *P. chartarum* cultures the presence of toxin in fractions was detected by biological assays. Initially, a guinea pig feeding test was utilized (see Sec. II.C.5). Subsequently, this work was expedited by a more rapid in vitro cytotoxic assay using tissue culture cells (see Sec. II.C.5), and finally, by the rabbit corneal opacity test (see Sec. II.C.5).

Structure and properties of sporidesmin and derivatives: There are only minimal differences in the positions and intensities of the ultraviolet (UV) absorptions for the various sporidesmins and their acetates. Relaxation of the strain on the dioxopiperazine ring by opening the -S-S- bridge results in halving the intensity of the absorption at about 300 nm. Using ether as solvent (less polar than alcohol or water), the peak at 254 nm is sharper. Mass spectral and nuclear magnetic resonance (NMR) data can be found in detailed papers referred to below.

Sporidesmin (Ronaldson et al., 1963): $C_{18}H_{20}ClN_3O_6S_2$, Str. [1] (R_1 = Cl, R_2,R_3 = OH, R_4,R_5 = S-S, R_6 = Me). The unsolvated crystals (from aqueous alcohols), mp 179°C, give $[\alpha]_D^{23}$ -45° (ca. 0.98 in MeOH), ν_{max} (CHCl₃) 1700 and 1670 (C=O) cm⁻¹, and λ_{max}(ether) 218.5, 254, and 302 nm (log ε = 4.60, 4.12, and 3.45). Sporidesmin crystallizes as solvates from benzene (see spectrum, Fig. 2), carbon tetrachloride, dibromomethane, and methyl iodide. The diacetate (from aqueous ethanol), mp 170-171°C, gives $[\alpha]_D^{22}$ -25° (ca. 1.2 in MeOH), and ν_{max}(paraffin) 1750, 1230 (acetate C=O), and 1690 cm⁻¹ (C=O). The monoacetate (Jamieson et al., 1969) from the diacetate in formic acid and methane-sulfonic acid gives mp 200-204°C and ν_{max}(KBr) 3350, 1760, 1715, and 1675 cm⁻¹. X-ray crystallography eluci- dated the structure of sporidesmin (Fridrichsons and Mathieson, 1962, 1965; Beecham et al., 1966), but detailed chemical studies are set out in: part 1, Ronaldson et al. (1963); part 2, Hodges et al. (1963); part 3, Hodges et al (1964); part 4, Hodges and Taylor (1964); part 5, Hermann et al. (1964); and parts 15 and 18, Ronaldson (1976, 1981). Total synthesis is given by Kishi et al. (1973).

Sporidesmin, as the benzene solvate or unsolvated, has been stored un- changed for years under refrigeration. Macro quantities of sporidesmin and derivatives may be concentrated in solution (e.g., rotary evaporation at 50°C)

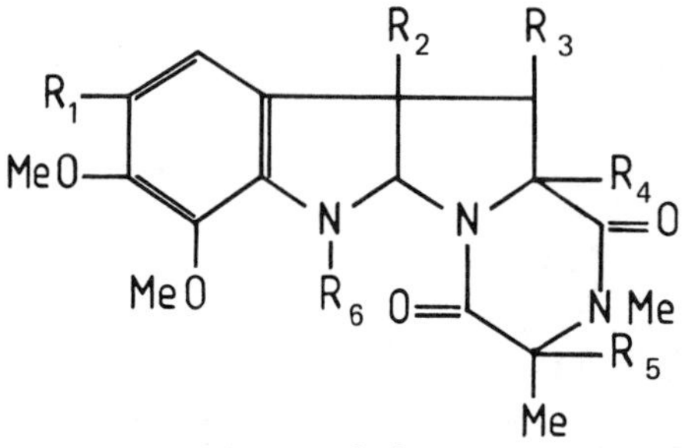

[1]

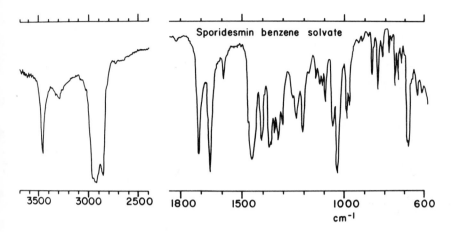

FIGURE 2 Infrared spectrum of sporidesmin benzene solvate (paraffin mull).

without serious loss provided that bright light is avoided. Recrystallization of sporidesmin from aqueous methanol solutions yields unsolvated sporidesmin, but the mother liquor decomposes, turning green. Solutions in other solvents could be evaporated without decomposition. Standard solutions should be used immediately the sporidesmin has dissolved. In dilute solution [e.g., in water (solubility 30 ppm; Clare and Gumbley, 1962)], loss of sporidesmin may be rapid (Clare and Mortimer, 1964; Brook and Matthews, 1960; Marbrook and Matthews, 1962). For experimental production of pithomycotoxicosis in farm animals, aqueous solutions of concentrates can be safely used because the toxin is protected by the naturally occurring antioxidants. Rain may remove sporidesmin from spores and aqueous solutions of sporidesmin are inactivated by sunlight (Clare and Gumbley, 1962; Marbrook and Matthews, 1962; Clare and Mortimer, 1964).

Sporidesmin B (Ronaldson et al., 1963; Hodges et al., 1966): $C_{18}H_{20}ClN_3O_5S_2$, Str. [1] (R_1 = Cl, R_2 = OH, R_3 = H, R_4,R_5 = S-S, R_6 = Me). The yield from *P. chartarum* cultures is less than 5% of total sporidesmins and it is readily separated from sporidesmin by partition chromatography. Crystals (from aqueous acetone), mp 183°C, give $[\alpha]_D^{21}$ -27° (ca. 1.0 in MeOH), +12° (ca. 0.75 in CHCl$_3$), and ν_{max}(CHCL$_3$) 1705 cm^{-1} (Figure 3 shows spectrum in paraffin mull). Synthesis of the d,1 form is given by Nakatsuka et al. (1974). The acetate is amorphous, mp 93-114°C.

Sporidesmin C diacetate (Hodges and Shannon, 1965): $C_{22}H_{24}ClN_3O_8S_3$, Str. [2]. Sporidesmin C per se has not been isolated, but its diacetate occurs as a minor component (0.2%) of sporidesmin diacetate, mp 230-240°C (decomp.), giving $[\alpha]_D$ -215° (ca. 0.46 in CHCl$_3$).

Sporidesmin D (Jamieson et al., 1969): $C_{20}H_{26}ClN_3O_6S_2$, Str. [1] (R_1 = Cl, R_2,R_3 = OH, R_4,R_5 = SMe, R_6 = Me). It is obtained from the ether extract of the routine sporidesmin concentrate (3.25 g) by chromatography on a silicic acid column (Benzene-ether, 44:6), arriving after sporidesmin (0.26 g).

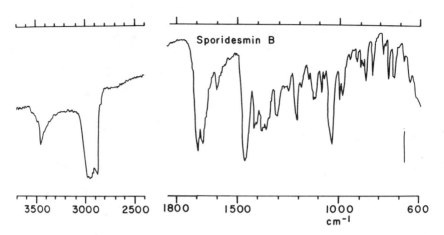

FIGURE 3 Infrared spectrum of sporidesmin B (paraffin mull).

It occurs as the etherate (from ether), mp 110-120°C, giving ν_{max}(KBr) 3450, 3330, 1680, 1665, and 1605 cm^{-1} and as the ethanolate (from ethanol), mp 105-107°C, giving $[\alpha]_D^{23}$ +58° (ca. 0.11 in MeOH).

Its diacetate (from ethanol), mp 202-204°C, gave $[\alpha]_D^{20}$ +6° (ca. 0.1 in CHCl$_3$), and ν_{max}(KBr) 1765, 1750, 1695, 1660, and 1603 cm^{-1}, and its monoacetate (from aqueous ethanol), mp 185-187°C, has ν_{max}(KBr) 3460, 3020, 3000, 1765, 1685, 1670, and 1605 cm^{-1}. Sporidesmin D is prepared from sporidesmin with pyridine, methanol, sodium borohydride and methyl iodide.

Sporidesmin E (Rahman et al., 1969): $C_{18}H_{20}ClN_3O_6S_3$, Str. [1] (R$_1$ = Cl, R$_2$,R$_3$ = OH, R$_4$,R$_5$ = S-S-S, R$_6$ = Me). This was separated from crude crystalline sporidesmin benzene solvate by chromatography on a silicic acid column and elution with t-butyl alcohol-light petroleum (3:17). This unstable compound (from ethyl acetate-acetic acid, Ronaldson, 1978c,d) mp 181-182.5°C, occurs also as the ethanolate (from ethanol), mp 180-187°C, giving $[\alpha]_D^{20}$ -131° (ca. 0.065 in CHCl$_3$), or the etherate (from ether), mp 144.5-

148°C (Ronaldson, 1978b,d), having $[\alpha]_D^{20}$ -131° (ca. 0.065 in $CHCl_3$), and ν_{max}(KBr) 3325, 1690, and 1655 cm^{-1}. The etherate (Ronaldson, 1978c,d) effloresced at 125-135°C (45 min), falling to a liquid and decomposing.

The diacetate is amorphous (from ethanol-water), mp 112-115°C, giving $[\alpha]_D^{20}$ -118° (ca. 0.11 in $CHCl_3$), and ν_{max}(KBr) 1760, 1690, and 1605 cm^{-1}. It may be converted to sporidesmin D by methyl iodide and sodium borohydride, and to sporidesmin by triphenylphosphine, or sodium borohydride and 5,5'-bis(2-nitrobenzoic acid). Sporidesmin gives sporidesmin E on treatment with sulfur and pyridine (Ronaldson, 1978b,d).

Sporidesmin F (Jamieson et al., 1969): $C_{19}H_{22}ClN_3O_6S$, Str. [3]. Chromatography of the sporidesmin D mother liquors with *t*-butyl alcohol-light petroleum (3:17) yielded, from 3.25 g of concentrate, 16 mg of an amorphous solid, mp 65-67°C, giving ν_{max}(KBr) 3430, 1690, 1615, and 1605 cm^{-1}.

Sporidesmin G (Francis et al., 1972): $C_{18}H_{20}ClN_3O_6S_4$, Str. [1] (R_1 = Cl, R_2,R_3 = OH, R_4,R_5 = S-S-S-S, R_6 = Me); for the stereochemistry of the sulfur bridge, see Przybylska et al. (1973). Preparative thin-layer chrmatography on sporidesmin E concentrates, using multiple solvent systems and development, segregated the slightly more polar sporidesmin G from sporidesmin E. The etherate (from ether at 4°C), mp 148-153°C, has $[\alpha]_D^{20}$ -217° (ca. 0.023 in $CHCl_3$), and ν_{max}($CHCl_3$) 3570, 3520, 1690, and 1660 cm^{-1}. It is prepared from sporidesmin or sporidesmin E by sulfur and pyridine treatment, converted to sporidesmin D by sodium borohydride with methyl iodide and pyridine, and to sporidesmin and sporidesmin E under a fluorescent lamp.

For the thin-layer chromatography (TLC) R_f values for the foregoing sporidesmins (except C and F) and dethiosporidesmin (see later) in various solvent systems (Rahman et al., 1970), see White et al. (1977).

Sporidesmin H (Rahman et al., 1978): $C_{18}H_{20}ClN_3O_4S_2$, Str. [1] (R_1 = H, R_2 = Cl, R_3 = H, R_4,R_5 = S-S, R_6 = Me), is obtained as an amorphous solid (ether-light petroleum) from silica gel chromatography of mixed sporidesmins by elution after sporidesmins D and G with increased concentration of ethyl acetate in benzene followed by TLC (high R_f), mp 150-152°C. It gives ν_{max}(KBr) 1700 cm^{-1}; and δ 6.98 (J = 8 Hz), 6.43 (J = 8 Hz), 3.27 (J = 16 Hz), and 2.70 (J = 16 Hz).

Sporidesmin J (Rahman et al., 1978): $C_{17}H_{18}ClN_3O_6S_2$, Str. [1] (R_1 = Cl, R_2,R_3 = OH, R_4,R_5 = S-S, R_6 = H), is obtained with sporidesmin H, but at low R_f on TLC, mp 168-169°C. It has $[\alpha]_D^{18}$ +43° (ca. 0.25 in $CHCl_3$), and δ 7.14, 5.17, exchangeable (J = 1.5 Hz).

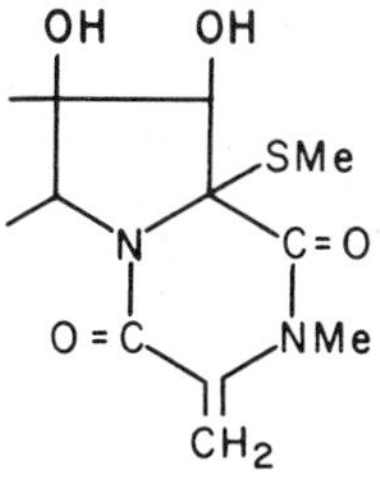

[3]

The diacetate, mp 185-187°C, gives ν_{max}(KBr) 1740 cm^{-1}.

Modified sporidesmins:

1. *Dimercaptosecosporidesmin* (Ronaldson, 1978a,d). $C_{18}H_{22}ClN_3O_6S_2$, Str. [1] (R_1 = Cl, R_2,R_3 = OH, R_4,R_5 = SH, R_6 = Me) is obtained by treating sporidesmin in methanol with sodium borohydride. The crystals mp 170-172.5°C (from aqueous methanol), give ν_{max}(KBr) 2540 (SH), 1650 br (C=O) cm^{-1}.

2. *Dethiosporidesmin* (Safe and Taylor, 1971). $C_{18}H_{20}ClN_3O_6S$, Str. [1] (R_1 = Cl, R_2,R_3 = OH, R_4,R_5 = S, R_6 = Me) has mp 72-74°C, but is not found in cultures.

3. *Sporidesmin antigens* (Ronaldson, 1975). Sporidesmin D was prepared (Jamieson et al., 1969) by opening the S-S bridge of sporidesmin with sodium borohydride and alkylating the sulfur atoms with methyl iodide. Ronaldson (1975), instead, alkylated with chloroacetic ester, sodium iodide, and pyridine and obtained three methyl ester derivatives and the corresponding ethyl ones:

Dimethyl secosporidesmin-S,S'-diacetate, $C_{24}H_{30}ClN_3O_{10}S_2$, Str. [1] ($R_1$ = Cl, R_2,R_3 = OH, R_4,R_5 = SCH$_2$COOMe, R_6 = Me), is a gum, giving R_{sdm} 0.59 (TLC, EtOEt-C_6H_6 2:3), and ν_{max} 1730 (ester C=O), 1670, and 1650 (C=O) cm^{-1}.

Methyl 11α-mercaptosecosporidesmin-3-S-acetate, $C_{21}H_{26}ClN_3O_8S_2$, Str. [1] (R_1 = Cl, R_2,R_3 = OH, R_4 = SH, R_5 - SCH$_2$COOMe, R_6 = Me), mp 150-156°C (from methanol), gives R_{sdm} 0.22 and ν_{max} 1720 (ester C=O), 1680, and 1650 (C=O) cm^{-1}.

Methyl 3-mercaptosecosporidesmin-11α-S-acetate, $C_{21}H_{26}ClN_3O_8S_2$, Str. [1] (R_1 = Cl, R_2,R_3 = OH, R_4 = SCH$_2$COOMe, R_5 - SH, R_6 = Me), mp 151-166°C (decomp.) (from acetone-ether) gives R_{sdm} 0.70 and ν_{max} 2520 (SH), 1725 (ester C=O), 1685, 1660, and 1640 cm^{-1} (C=O).

Each of these derivatives was linked to a protein through transacylation in alcohol-water (pH maintained at 9-10) to the ε-amino groups of lysine residues, so producing antigen complexes in which (by infrared measurements) the ester peaks were missing. Sporidesmin complexed to poly-L-lysine produced an immunological response in rabbits (Jonas and Ronaldson, 1974).

Taylor (1971) reviewed the chemistry and biological effects of sporidesmins and Sammes (1975) reviewed their chemistry.

2. Methods of Extraction and Isolation

Large-scale extraction of grain cultures: This method produces extracts of 5-15% sporidesmin. Highly sporulated grain cultures of *P. chartarum* are harvested from batches of 250 bottles (600-ml size) and are extracted with MeOH-H_2O (4:1). The filtered extract is concentrated in a cyclone evaporator to an aqueous suspension (ca. 8 liters). This is extracted with ether (4 × 1 liter), which is evaporated at 20°C per 10 mm to give a 12- to 35-g extract (Ronaldson et al., 1963). This concentrate may be used in animal dosing experiments.

Low-pressure liquid chromatography of concentrates: This method was described in detail (White et al., 1977) and is summarized here. The concentrate in ether is loaded onto three columns of 500 g of chromatographic alumina (washed with acetic acid-ether, 2%, 2400 ml, and then with ether, 14 liters) and developed with ether (800 ml each). The residue from this ether eluate is partitioned between methanol-water (5:1) and light petroleum. Partition chromatography (CS_2-MeOH-H_2O 50:12:3) of the residue from the methanol-water phase yields sporidesmolides, pithomycolide, sporidesmin B, then sporidesmin. The sporidesmin fractions are dissolved in benzene and the sporidesmin benzene solvate precipitated with light petroleum. If required, purer sporidesmin is obtained by silica gel chromatography (EtOEt-C_6H_6 1:4) (Ronaldson et al., 1963; Ronaldson and Fyvie, 1973).

Sporidesmin B is obtained from triturating the sporidesmin B fractions with ether. Pithomycolide precipitates and the residue from the solution is chromatographed on silica gel (EtOEt-C_6H_6 1:3). The active fractions are again partition chromatographed (CS_2-MeOH-H_2O 50:12:3) and sporidesmin B is crystallized from squeous acetone.

Preparative high-press liquid chromatography (HPLC) of concentrates: The concentrate is dissolved in a minimum of ether and the suspension centrifuged to remove pithomycolide and sporidesmolides. Sporidesmin is then extracted from the ether (2 liters) with concentrated hydrochloric acid (10 × 40 ml) until less than 40 mg (determined by Anal HPLC) of sporidesmin remains in the raffinate. Each lot of acid extract is separated into the same lot of water (1 liter), which is itself extracted with ether (350 ml). This ether is then extracted with 5% sodium carbonate-4% sodium hydroxide solution (5 × 25 ml), neutralized, and evaporated to dryness.

The chloroform solution (10-20 ml) of this residue (ca. 10 g) is injected into a Prep 500 HPLC silica cartridge and chromatographed with chloroform containing ca. 0.8% ethanol. A number (7-9) of arbitrary fractions (from front to k' 0.25, to k' 0.5, to k' 1.0, to k' 2.0 et seq.) are collected. The small fractions at the beginning serve to ascertain when the residual sporidesmolides have moved off the column so that the sporidesmin fractions may be collected and crystallized without contamination with sporidesmolides. These active fractions (3-4 g; 1.5-2.0 g sporidesmin by assay) are accumulated with ethyl acetate and the sporidesmin crystals (1.2-1.6 g), which are precipitated with hexane, are ca. 90% pure and contain ca. 5% of sporidesmin E (Ronaldson, 1982).

Other methods of isolation: The review by White et al. (1977) described methods for the isolation of sporidesmin from liquid cultures or from spores. Halder et al. (1980) described a benzene extraction method in which the benzene residue was extracted with methanol-water (7:3) and the solution extracted with hexanes. The methanol phase was diluted to 35:65 and the filtrate extracted with benzene, which solution was chromatographed on silica.

3. Methods of Estimation and Analysis by Analytical HPLC

a. Sporidesmin may be assayed by Anal HPLC in spores, concentrates, residues, fractions, or crystals.

(1) For spores, a 100-ml suspension of 10^6 spores per milliliter is extracted with chloroform (3 × 5 ml). The residue dissolved in methanol (0.6 ml) is diluted with water (0.4 ml) and passed through a SepPak (Waters) into a volumetric flask (5 ml).

Subsequent lots of 65% methanol are passed through the SepPak to make up to volume ready for injection into the HPLC apparatus.

(2) For concentrates, and so on, a small sample (2-5 mg) is taken up in methanol (0.6 ml) and diluted for SepPak treatment as under (1).

(3) Injection of 20 μl of ca. 800 μg of standard sporidesmin in 5 ml of 65% methanol-water into a C_{18} Zorbax column (25 cm) will give ca. 100 units at 254 nm in 0.8 ml/min, and 0.16 absorbance units full scale (AUFS) in ca. 8-10 min.

At these parameters sporidesmin B eluted at R_{sdm} 1.64, sporidesmin D at 0.93, and sporidesmin E at 1.35 (Ronaldson, 1982).

b. Halder et al. (1979) described an Anal HPLC method for estimation of sporidesmin in extracts of cultures grown on dried kleingrass or rye-corn. The extract (acetonitrile) was dissolved in methanol-water, then partitioned with hexane (discarded) followed by benzene. Aliquots of the benzene residue in acteonitrile were then dried and taken up in a standard naphthalene-methanol solution for Anal HPLC. They obtained a 91% recovery of sporidesmin added to 80-200 mg of fungal material in the range 1-2.5 μg. HPLC details: C_{18} column (Waters), 47% methanol-water, 254 nm, 2 ml/min, retention times 300 sec for sporidesmin and 600 sec for naphthalene.

4. *Detection and Estimation by Other Chemical Methods*

White et al. (1977) described the detection of sporidesmin in extracts. This was first done by paper chromatography and spraying with azide-iodine solution. With starch spray, the sporidesmin spots appeared white on a blue background. Later, TLC (silica gel GF_{254}) was used with benzene-ether (5:1) separating sporidesmin and sporidesmins B and E from D and G, or hexane-t-butyl alcohol (9:1) separating sporidesmin from sporidesmins B, D, E, and G.

The same authors described several chemical methods: azide-iodine reaction, a "penicillin"-type titration, UV absorption, and radiosulfur when the fungus was grown on a synthetic medium.

5. *Biological Methods: Detection and Estimation*

Generally, chemical methods have now supplanted the three biological methods which assisted in the detection and initial isolation of sporidesmin. These methods were reviewed by White et al. (1977) and are briefly as follows:

a. The guinea pig test, which requires the feeding of the test extracts dispersed onto ground grass meal (or similar) to recently weaned guinea pigs for 3 weeks (Perrin, 1957). The finding of characteristic changes, including marked bile-ductule proliferation, in the livers after 4 weeks of test indicates the presence of sporidesmin in the extract.

b. Assay on tissue culture cells, in which the test solutions are added in ethanol to cell culture media at increasing dilutions and are then placed in 24-hr test tube cultures. After further incubation, the characteristic cytopathogenic effects of sporidesmin appear after 24-48 hr. The test is sensitive to concentrations of sporidesmin as low as 1 ng/ml of culture medium (Done et al., 1961; Mortimer and Collins, 1968). With experience, a semiquantitative test procedure can be established.

c. In the rabbit corneal opacity test, the test fluids are instilled into the conjunctival sac of the rabbit eye. Within 5 days corneal opacity and edema are produced with a total sporidesmin application of about 5 µg, or within 2-3 days with 10-20 µg (Done et al., 1961). The method has been used to detect sporidesmin in biological fluids by Leaver (1968).

6. Biosynthesis

By culturing *P. chartarum* on synthetic substrates containing inorganic [^{35}S]sulfate, [^{35}S]sporidesmin has been produced (Brook and Matthews, 1960). Similarly, [^{35}S]sporidesmin has been obtained from L-[^{35}S]methionine or better from L-[^{35}S]cysteine (Towers and Wright, 1969). From L-[U-^{14}C]-alanine or L-[methyl-^{14}C]methionine ^{14}C has been incorporated, but D,L-[methylene-^{14}C]tryptophan or L-[U-^{14}C]serine gave a higher specific activity. By contrast, no ^{14}C was derived from D,L-[3-^{14}C]cysteine or L-[U-^{14}C]phenylalanine (Towers and Wright, 1969). When Kirby and Varley (1974) used (3S)-[3-^{3}H]tryptophan or (3R)-[3-^{3}H]tryptophan, the ^{3}H retained from the latter was <12% (against a ^{14}C reference label) of that from the former. Thus the S configuration of 3-C is retained in the formation of the sporidesmin 11-OH.

7. Pharmacological and Biochemical Studies

Available information on the absorption and excretion of sporidesmin in animals was reviewed by White et al. (1977).

Giving doses of [^{35}S]sporidesmin orally to guinea pigs and rats, Towers (1970a,b) found little difference between species in absorption and excretion rates or in distribution of counts in tissues. Excretion of counts in bile was effected within 7.5 hr in both species (Towers, 1972), and from two to five times more counts appeared in bile than in urine. Highest levels of activity were found in the liver, kidney, and spleen. Of the radioactivity excreted in feces, up to 40% was extractable in ether and was shown to be associated with intact [^{35}S]sporidesmin. Ether extracts of urine recovered 16-20% of counts, but extracts were not clearly identified as intact [^{35}S]sporidesmin. Precipitated glucuronide conjugates and other unidentified forms constituted the fractions not extracted by ether (Towers, 1970b).

In experiments using sheep with established bile duct reentrant cannulae which maintained homeostasis of bile secretion, Mortimer and Stanbridge (1968) recovered unchanged sporidesmin from the bile of two of five sporidesmin-dosed sheep. Using identically cannulated sheep and a cytotoxic assay method with cells in culture, sporidesmin activity was detected in bile taken at intervals from 10 min to 24 hr after dosing. Maximum cytotoxic activity (equal to 10-20 µg of sporidesmin per milliliter) was found in bile at 4 hr, but after 24 hr little sporidesmin activity was detected. The significance of this probable enterohepatic circulation of sporidesmin in bile in relation to severity of liver and bile duct lesions was first raised by Worker (1960) from studies using rabbits. On evidence presented by Mortimer and Stanbridge (1968), and by Leaver (1968), who independently examined the secretion of sporidesmin in the bile of sheep, both groups concluded that enterohepatic circulation of sporidesmin would add little to the severity of hepatobiliary lesions.

Sporidesmin activity was also detected in samples of urine from dosed sheep (Mortimer and Stanbridge, 1968). Maximum sporidesmin activity (equal to 2-4 μg of sporidesmin per milliliter) was detected 4 hr after oral dosing; it was still detectable at 12 hr, but not at 24 hr. In serum, activity equal to 0.2-0.4 μg of sporidesmin per milliliter was detectable up to 12 hr, but not after. No sporidesmin activity could be detected in milk from two lactating ewes orally dosed with 1.0 mg of sporidesmin per kilogram of body weight. This represents an intake of toxin several times higher than could be obtained by sheep or cattle grazing toxic pastures.

The effect of sporidesmin on the rate of flow of bile has been examined by several investigators. In sheep, a single oral dose of sporidesmin (1 mg/kg) produced a gradual progressive reduction in bile flow until cessation of flow occurred in most sheep after 10-14 days (Mortimer and Stanbridge, 1969). A similar sporidesmin-induced reduction in bile flow, followed by cessation of flow at 6-8 days, was reported for sheep with a different bile duct cannulation system (Leaver, 1968). In rats, intraperitoneal injections of sporidesmin (6 mg/kg) produced dramatic but transient decreases in bile flow rates, with cessation at 18-30 hr, and return to normal flow at 3 days (Slater and Griffiths, 1963).

Studies were made using the isolated perfused rat liver (Slater et al., 1968; Eakins et al., 1973; Bullock et al., 1974) and on rat liver plasma membranes in attempts to measure sporidesmin-induced cholestasis and to define its mechanisms (Eakins, 1978). From previous findings (Bullock et al., 1974), and on his further evidence, Eakins (1978) suggested that sporidesmin acts directly on canulicular membranes to disrupt bile secretion mechanisms, possibly through the interaction of the disulfide bridge of sporidesmin with membrane thiol groups, rather than by inhibition of canalicular ATPases. This same locus and mechanism for in vivo toxic action of sporidesmin was previously suggested by Slater (1972), whose electron micrographs of sporidesmin-infused rat livers show bile canalicular distortion and loss of microvilli, changes that became apparent 45 min after sporidesmin infusion. Middleton (1974b) also supported Slater's (1972) hypothesis of disulfide group/membrane interaction in discussing the action of sporidesmin on swelling and respiration of liver mitochondria and the inhibition of this toxic effect by dithiothreitol but not by a number of other mono- or dithiols. In contemporaneously reported findings, Middleton (1974a) confirmed that sporidesmin produced rapid and reversible swelling of guinea pig liver mitochondria and that sheep, rabbit, and rat liver mitochondria reacted similarly. His results helped to dispel apparently conflicting previous results of others in this field (Slater et al., 1964; Gallagher, 1964; Wright and Forrester, 1965).

Studies on the metabolism of sporidesmin in sheep are in progress and only preliminary results are currently available (Fairclough et al., 1978). Sheep liver microsomes, in which drug metabolizing enzymes had been induced by prior injection of sheep with hexachlorobenzene, were incubated with ^{3}H-, ^{14}C-, and ^{35}S-labeled sporidesmin. At least eight different metabolites were isolated, but a structure has been proposed for only two of these, monohydroxy and dihydroxy derivatives of sporidesmin, and neither of these ethyl acetate-extractable metabolites contains sulfur atoms. For this reason it was considered that the toxicity of these metabolites would be low.

D. Toxicity of Sporidesmins

Extracts of laboratory cultures of *P. chartarum* yield sporidesmin as the major toxic metabolite (ca. 90% of total sporidesmins). Sporidesmin B is present in small amounts and has less toxic activity than sporidesmin. Sporidesmin C has been obtained only as the diacetate, which has low toxic activity. Sporidesmin E has similar high toxic activity to that of sporidesmin. Also present in small amounts are sporidesmins D, F, G, H, and J. Of these, D and F have low toxicity, G and H appreciable toxicity, and no figures are available for the recently characterized sporidesmin J.

1. In Vitro Studies on Toxicity

Valuable information is available regarding the relative toxicities of the sporidesmins from several independent in vitro microassay studies (references in Table 1). Available figures on the least toxic doses which inhibit protein synthesis in HeLa cell cultures, or which produced clear-cut cytopathic effects when added to roller-culture cells, and data on the minimum inhibitory concentrations needed to increase the lag phase growth of *Bacillus subtilis* by 300%, are listed in Table 1. Notwithstanding differences in techniques between laboratories, there is good agreement, especially on relative toxic activities of the sporidesmins assayed.

The results suggest that biological activity is associated with the sulfur-containing ring; sporidesmins with an opened -S-S- bridge (D and F) have very low toxicity to cell cultures.

Phytotoxic effects of sporidesmin on germination of seeds and on growth of seedlings at concentrations in the range 5-45 µg/ml were reported (Wright, 1969), but no toxic effects of sporidesmin on ruminal protozoa in vitro were found (Shaw and Wright, 1972).

2. Toxicity to Animals

In a wide range of experiments, sporidesmin has been administered to a number of species over a period of years. Tabulated data, drawn from results of comparable experimental oral dosage conducted by one of us (P.H.M.) at this Research Center, indicate the range of toxicity and responses found in 10 animal species (Table 2). It is clear that, where information is available, ruminants are generally the most susceptible species, and of these, fallow deer (Mortimer and Smith, 1981) and sheep (Mortimer and Taylor, 1962) rank highest.

As well as sporidesmin, only the sporidesmins B and E have been available in quantities sufficient to dose laboratory animals and sheep. In sheep, sporidesmin and sporidesmin E, given at the same dose rate, produced a similar degree of liver injury, but to produce the same toxic effect with sporidesmin B required a higher (fivefold) dose rate (B. L. Smith, J. W. Ronaldson, and P. H. Mortimer, personal communication, 1975).

Toxic effects of sporidesmin in laboratory animals: The use of laboratory animals considerably assisted investigations into this mycotoxicosis. Initially, they played an essential role in the detection and bioassay of sporidesmin (see Sec. II.C). Subsequent to its isolation, several laboratory species were

TABLE 1 In Vitro Toxicities of the Sporidesmins

Compound	Growth inhibition of *Bacillus subtilis*, minimum concentration (μg/ml) Method of Brewer et al.[a]	Least toxic dose for tissue culture cells (ng/ml)	
		Method of Done et al.[b]	Method of Mortimer and Collins[c]
Sporidesmin	80[a]	3.0[b] 1.0[a]	0.4[c]
Sporidesmin diacetate	120[a]	6.0[d]	5.0[c]
Sporidesmin B	400[a]	3.0[a]	2.0[c]
Sporidesmin C diacetate	>1000[e]	–	500[c]
Sporidesmin D	–	>1000[e]	–
Sporidesmin E	10[e]	0.1[f] 0.04[e]	–
Sporidesmin F	–	>1000[e]	–
Sporidesmin G	–	2.0[g]	–
Sporidesmin H	–	10.0[g]	–
Sporidesmin J	–	–	–

[a] Brewer et al., 1966.
[b] Done et al., 1971.
[c] Mortimer and Collins, 1968.
[d] J. Done (quoted by Taylor, 1967).
[e] Review by Taylor, 1971.
[f] Brewer et al., 1968.
[g] Francis et al., 1972.

TABLE 2 Summary of Typical Clinical and Pathologic Responses of Animal Species to Oral Doses of Sporidesmin

Species	Time period of dosage (days)	Total dose (mg/kg)	Typical responses
Cattle			
Calves	3	3.0	100% mortality within 3-5 days
	3	0.8	50% photosensitized; icterus; low mortality
Gravid cows	3	1.0	Marked weight loss; 50% photosensitized; icterus with some mortalities; typical serious field outbreaks
	3	0.5	Transient weight loss; liver lesions, mainly asymptomatic
Sheep	3	3.0	100% mortality within 4 days
	3	1.0	Severe illness; 80% photosensitized; 70% mortality
	3	0.5	50% photosensitized; 10-15% mortality; typical of serious field outbreaks
	24	0.8	50% photosensitized; 35% mortality
	48	0.8	50% photosensitized; 7% mortality
Farmed red deer (*Cervus elaphus*)	3	1.8	No clinical responses; 25% sustained liver injury
	3	1.2	No clinical responses; 25% sustained liver injury
	3	0.6	No clinical responses; no liver injury
Farmed fallow deer (*C. damadama*)	3	1.2	Icterus; 60% photosensitized; 100% mortality within 10 days; severe pulmonary edema and emphysema
	3	0.6	Icterus; 60% photosensitized; 100% mortality within 14 days
	3	0.3	Icterus; 60% photosensitized; 100% mortality within 24 days

TABLE 2 (Continued)

Species	Time period of dosage (days)	Total dose (mg/kg)	Typical responses
Swine	3	3.0	Asymptomatic; mild bile duct lesions
	3	1.0	Asymptomatic; trivial lesions
Chickens	Continuous (5.0 mg/kg)	30 (max.)	100% mortality within 6 days; lesions of liver
	Continuous (0.5 mg/kg)	7 (max.)	100% mortality within 14 days; lesions of liver and gizzard
Rabbits	1	1.2	50% mortality within 14 days; severe lesions of bile ducts and liver; icterus
Guinea pigs	14	3.5 (max.)	100% mortality within 14-21 days; severe lesions of bile ducts and liver; icterus
	21	1-2	Hepatomegaly, marked liver fibrosis at 28 days; some mortalities
Rats	Continuous (0.02 mg/day)	4-8	Severe bile duct and liver lesions; icterus; pleural effusions and ascites; mortality at 2-4 months
Mice	Continuous (0.2 mg/day)	250-300	Average survival time 35 days; bile duct and liver lesions; icterus; anemia; some pleural effusions and ascites
	Continuous (0.1 mg/day)	200-250	Average survival time 57 days; lesions as above

utilized in studies on the comparative pathology of sporidesmin poisoning. The more important findings are reviewed here. Species are listed in decreasing order of susceptibility.

Effects on rabbits: New Zealand white rabbits were used for estimating the toxicity of dried pasture samples, extracts from grass, and spores of *P. chartarum* before sporidesmin was available (Clare, 1959). Early studies on the pathogenesis of the disease in rabbits using fungal extracts also predated the isolation of sporidesmin (Clare, 1959; Dodd, 1960). These and later experiments using sporidesmin (Clare and Mortimer, 1964) revealed the marked severity of hepatobiliary lesions in this species. A most florid picture of severe edema, inflammation, and necrosis, involving the major part of the bile duct system and gallbladder, is seen in this species. Obliterative cholangitis and icterus develop, and there are marked elevations in serum bilirubin, cholesterol, and glutamic oxalacetic transaminase levels.

Rabbits proved to be only marginally less susceptible to sporidesmin than sheep, and rabbits provide the closest similarity to sheep in relation to hepatobiliary lesions.

Effects on guinea pigs: This was the first laboratory animal species found to be susceptible to the toxin when it was initially fed as spontaneously toxic pasture material (Perrin, 1957). Subsequently, it was used extensively to detect toxic fractions (White, 1958).

Lesions produced by feeding cultures of *P. chartarum* over 3-4 weeks were those of marked liver enlargement and extensive hyperplasia of bile ductules. These changes were associated with edematous and degenerative lesions in the small bile ducts. The proliferated bile ductule cells often represented more than half the area of each liver lobule seen in sections (MacKinnon and Te Punga, 1961). Acute focal lesions were also found in which there were wedge-shaped areas of coagulative necrosis involving a few or many lobules. Necrosis of intrahepatic bile ducts with extension into surrounding tissues, sometimes including thrombosis of the portal vein, were histologic features noted in other livers.

Effects on chickens: Week-old chickens were found to be very susceptible to sporidesmin (Mortimer, 1967). Daily oral doses of 0.5 mg of sporidesmin per kilogram resulted in deaths, with an average survival time of 14 days. Deep bile-stained ulcerations were produced in the gizzard, and livers were fibrous and icteric.

Effects on rats: The rat proved more resistant to oral doses of sporidesmin than either the rabbit or guinea pig. Extended oral dosing of rats with 20 µg of sporidesmin daily (total dose in the range 4-8 mg/kg) produced severe liver and bile duct lesions after 2-4 months (Mortimer, 1965). Serum values for bilirubin, cholesterol, and transaminases were elevated. Pleural effusions and ascites were often found as terminal lesions. Rats that survived dosage beyond 4 months contracted glaucoma and blindness.

In contrast, Slater et al. (1964) produced no liver or bile duct lesions with single oral doses of 9 mg/kg, yet given by the intraperitoneal route, the same dose produced ascites, pleural effusions, lung edema, and death. The production by sporidesmin of vascular exudative reactions had previously been reported to occur in sheep (Mortimer, 1963). These manifestations of severe capillary injury as part of an inflammatory response illustrate a most important pharmacopathologic effect of sporidesmin, and they warrant further consideration.

Studying the mechanism of sporidesmin-induced edema and inflammation, Slater et al. (1964) used rats "blued" by intravenous injections of trypan blue dye. The rats were then given intradermal injections of weak solutions of sporidesmin which, within minutes, produced a local edematous blue weal where protein-bound blue dye had leaked into the dermis from capillary vessels. Two distinct inflammatory mechanisms were involved and it was suggested that the first, a rapid reaction (10-15 min) was mediated by histamine release because its appearance could be delayed by prior medication with antihistamine drugs. The second reaction was mediated by a slow-release substance, probably a kinin, because the second (delayed) reaction could be blocked by prior medication with sodium salicylate. These studies are very illuminating, for they examined sporidesmin-induced cellular injury in vascularized tissue at sites away from the liver and bile ducts, where other irritant substances such as bile acids confound the basic inflammatory and slower necrotizing effects of sporidesmin. The reaction in the rabbit corneal-opacity test (Done et al., 1961) is a further illustration of severe injury in nonhepatic sites initiated by microgram quantities of sporidesmin.

Hove and Wright (1969), in these laboratories, used daily oral doses of sporidesmin to challenge groups of rats fed different levels of protein in diets. They found that a protective effect with high protein was, for protein sources tested, specific for casein. Fractionation of the casein indicated that the phosphopeptone moiety was protective and that synthetic D,L-phosphoserine also protected against sporidesmin poisoning. The protective mechanism is not known, but it was suggested that maintenance of cell and membrane integrity by the diet factor may have counteracted the effusive pathology of sporidesmin poisoning.

Effects on mice: Mice were found to be considerably less susceptible to sporidesmin than rats. Daily oral doses of 0.1 and 0.2 mg per female mouse gave group average survival times of 57 and 35 days, respectively (Mortimer, 1970). Female mice showed significantly greater susceptibility than did males. In the same experiment the mice became severely icteric and progressively anemic, the latter change not being a feature in other laboratory animal species. Pulmonary edema and bile-stained pleural exudates (average protein content 3.8 g per 100 ml) were found at necropsy.

Histologic features included bizarre hepatocyte degeneration and discrete areas of coagulative necrosis in the liver. Typical exuberant concentric fibrosis and constriction of the larger bile ducts was present; some ducts were obliterated.

Pithomycotoxicosis (facial eczema) of farm animals. The spontaneous disease is restricted to animals grazing on improved pastures. Efficient controlled-grazing systems, which maximize both pasture utilization and production, favor the intake of toxic *P. chartarum* spores under warm, humid weather conditions. Sheep were regarded to be the most susceptible farm animal species, followed by cattle. Recent evidence shows that farmed fallow deer (*Cervus damadama*) are also very highly susceptible to sporidesmin and also readily contract spontaneous severe disease in the field (Mortimer and Smith, 1981).

Disease in sheep: The first comprehensive description was made in New Zealand by Cunningham et al. (1942), and further observations were added by McFarlane et al. (1950) and by Done et al. (1960). Outbreaks in Australia were first described by Hore (1960), and in South Africa by Marasas et al. (1972). Using oral doses of sporidesmin, the disease has been extensively reproduced and investigated (Mortimer and Taylor, 1962; Done et al., 1962; Mortimer, 1962; Mortimer, 1963). These and other relevant papers were recently reviewed (Mortimer et al., 1978b).

This account is restricted to the more important symptoms and pathological changes in sheep. Where species differences occur for cattle and deer, these are given later.

1. *Clinical signs*. Generally, in an outbreak the first signs are sudden and are related to skin photosensitization. Affected sheep rapidly develop edematous swelling of the skin on parts exposed to sunlight, especially the ears, eyelids, face, and lips (Fig. 4). The affected skin is erythematous,

FIGURE 4 Sheep showing facial lesions of photosensitization; with drooping edematous ears, exudative lesions, and scabs around eyes, on muzzle, nostrils, and lips.

hot, and inflamed. Irritation is evident from the head and ear shaking, scratching, rubbing, and general restlessness. Sheep seek shade and venture out only in overcast weather or in darkness. Serum exudation through the injured skin soon appears and the exudate dries to form scabs. Rubbing and scratching cause exoriation and bleeding on the ears, eyelids, and lips. Blowflies are attracted. Experiments show that onset of photosensitization postdates intake of sporidesmin by 10-20 days. Clinical examination of photosensitized sheep will reveal jaundice in the conjunctiva and membranes. The urine is deeply colored with bile pigments.

In severely affected flocks deaths occur from about 10 days after onset of photosensitization and may continue for several weeks. The severe debilitating stress of photosensitization and extensive superficial lesions contribute to the cause of death. In less severe outbreaks most sheep recover, jaundice clears, and skin lesions resolve remarkably well. Most recovered sheep are capable of normal production, a few succumb to the metabolic stresses of pregnancy, parturition, and/or lactation, and all are at greater risk to subsequent toxic challenges.

2. *Pathology and pathogenesis.* Sporidesmin solutions are highly cytotoxic per se and there is no evidence to indicate that metabolism enhances its toxicity. It is known that there is considerable sporidesmin toxic activity in bile and urine of dosed sheep (see Sec. II.C.7). Early hepatobiliary lesions, caused directly by sporidesmin, were evident on histological examination of orally dosed sheep (Mortimer, 1963). Over the first 48 hr, only apparently minor changes were produced in hepatocytes; mainly vacuolation and minor accumulations of triglycerides occurred (Mortimer et al., 1962). Electron microscope studies of these early liver changes were also made (Bertaud and Mortimer, 1966). At 48 hr, degenerative changes in the bile duct epithelium and marked inflammatory changes (edema and cellular infiltration) were present in the supportive tissues surrounding the bile ducts. Whereas the bile duct lesions became progressively more severe over succeeding days, changes previously seen in hepatocytes largely resolved. Sheep showed only mild transient symptoms of inappetance and diarrhea during this phase, and changes in serum enzyme values were not pronounced. From day 10 onward a severe necrotizing cholangitis and pericholangitis, involving intra- and extrahepatic bile ducts, was obvious in livers. Sections showed that the severe inflammatory reaction and necrosis had spread outward from bile ducts and had produced severe disorganization of surrounding tissues. Fibroblastic activity and granulation tissue deposition had spread inward to localize and repair the necrotic peribiliary tissues. This adventitious tissue, together with edematous exudates and necrotic detritus, was producing occlusion of bile ducts and preventing bile flow. The establishment of duct occlusion from day 10 onward in increasing numbers of sheep coincided with the onset of jaundice and photosensitization. Biliary obstruction was followed by a second and more severe phase of hepatocyte injury. This was reflected in marked increases in certain serum enzyme activities, in serum bilirubin, cholesterol, phospholipids, and bile acid values (Done et al., 1962; Peters and Mortimer, 1970; Ford, 1965, 1974).

Urinary cystitis commonly occurs in both the experimental and the spontaneous field disease in sheep. The mucosa and muscular wall of the bladder show edematous inflammatory changes characteristic of sporidesmin cytotoxic effect. Hemorrhagic ulcerations are produced which become deeply bile stained in jaundiced sheep. Animals so affected often show incontinence and urine-stained fleeces. These urinary lesions again illustrate that tissue specificity for toxic action of sporidesmin is low. Adrenal cortical hyperplasia is found (total adrenal weight more than doubled) in sheep with protracted photosensitization, and is likely related to the severe stress that photosensitization produces.

In sheep that recovery, bile ducts are reorganized, patency is reestablished, and liver excretory function is resumed. Excretion of accumulated bile pigments and phylloerythrin via the bile is reestablished, the signs of jaundice and photosensitivity disappear, and skin healing takes place. An increased rate of mitotic activity is evident in hepatocytes early in the disease. Newly regnerated nodes of hepatic tissue are formed which, superimposed on the original but now fibrosed hepatic lobes, produce a distortion in liver shape which persists throughout life (Fig. 5). For every sheep showing overt disease there are many more that contract milder spectrum of liver injury not detected by superficial examination. In these, a good measure of recent liver

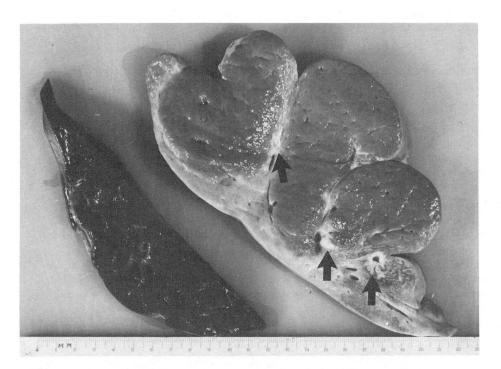

FIGURE 5 Pithomycotoxicosis (facial eczema) liver injury. These cross sections of livers show (left) the deeply colored liver of a nonaffected sheep and (right) the misshapen liver from a sheep after recovery. Superimposed on the pale, fibrous, atrophied tissues are large, deeper-colored masses of regenerated liver tissue. The small areas of white, glistening, fibrous tissue (arrowed) contain occluded or quasi-occluded bile ducts.

injury can be obtained from determination of serum γ-glutamyl transferase activity (Ford, 1974; Towers and Stratton, 1978).

Disease in cattle: Authenticated outbreaks in cattle have so far been reported only from New Zealand (Cunningham et al., 1942) and Australia (Hore, 1960; Flynn, 1962). In New Zealand, dairy cattle are more commonly affected than are beef cattle. The incidence is highest in lactating dairy cows, is lower in weaner calves, and lowest in yearling replacement heifers.

1. *Clinical signs.* The disease in cattle was described by Mortimer (1969) and Steffert (1970). In lactating herds the harbinger of outbreaks is often a sudden fall in milk production. Photosensitization soon follows and dosing experiments indicate that it occurs some 14-24 days postdosing (Mortimer, 1971). Skin lesions can vary in site according to pigment distribution of the skin, and nonpigmented areas are worst affected. The ears, face, lips, escutcheon, udder, teats, and coronets, with sparse hair cover, are particularly vulnerable. Udder and teat lesions are especially troublesome (Fig. 6), for the discomfort caused by milking makes the operation difficult or impossible, and is often abandoned. Photosensitized cattle show jaundice and bilirubinuria. A small proportion of affected cattle show toxic hemoglobinuria associated with massive intravascular hemolysis of red cells whose fragility has become abnormally high (Mortimer, 1967). The resulting severe anemia can cause extreme weakness, and cattle may collapse if driven. The marked reticulocytosis, seen in blood smears, indicates a prompt release of immature replacement cells and, given rest and care, most cattle recover from this complication.

The duration of jaundice and photosensitization is generally shorter in cattle than sheep, but recovery from general debility is equally protracted and may take 3-4 months. Mortality rate varies greatly between outbreaks but is generally less than in sheep. Necessary cullings of surviving cattle in dairy herds usually outnumber spontaneous deaths by 3:1. Subsequent milk production from recovered cows is generally equal to that in previous uninterrupted lactations.

2. *Pathology and pathogenesis.* Lesions in cattle are similar to those described previously for sheep. Livers of affected cattle show a marked diffuse fibrosis, but there is less pronounced peribiliary fibrosis than in sheep livers. Cystlike saccular dilatations are formed in the larger intrahepatic bile ducts and they impede bile flow. As in sheep, urinary bladder lesions are commonly found and, as a complication, suppurative cystitis and pyelonephritis may become established.

Disease in farmed deer: The utilization of well-established ryegrass/clover pastures for the farming of mainly red deer (*Cervus elaphus*), but also fallow deer (*Cervus damadama*), is a recent diversification in New Zealand. High stocking rates are used and there is efficient pasture utilization. Under continued warm, moist conditions the disease hazard that high fungal sporulation and this grazing regime could present to farmed deer was, until 1981, not known. Experiments were made to determine the susceptibility of both deer species to sporidesmin. Later in the same year, spontaneous outbreaks of pithomycotoxicosis (facial eczema) occurred in deer and the joint findings were reported (Mortimer and Smith, 1981).

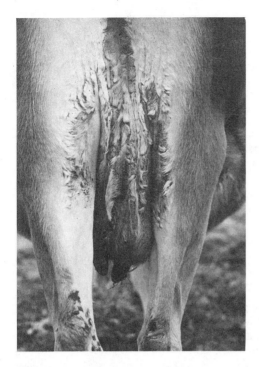

FIGURE 6 Dairy cow showing dry, necrotic, broken skin lifting from healing lesions on flanks, escutcheon, udder, and teats.

Red deer proved to be relatively resistant, and more so than sheep or cattle. Sporidesmin dosage produced no clinical signs, although minor liver injury was detected in some animals (see Table 2). In spontaneous field outbreaks, the incidence of disease in red deer was very low.

Fallow deer given comparable doses of sporidesmin reacted severely. At the highest dose (1.2 mg/kg) all died within 7-10 days after dosing, at the median dose (0.6 mg/kg) within 10-14 days, and at the lowest dose (0.3 mg/kg) within 14-24 days (see Table 2). In spontaneous field outbreaks, mortalities of more than 50% were reported in some herds. Fallow deer proved to be very highly susceptible to sporidesmin, and this contrasts strongly with the low susceptibility for red deer.

1. *Clinical signs in fallow deer.* The onset of photosensitization was 7-14 days after dosing and deer showed very severe distress, with violent shaking of the head and ears. There was frantic rubbing of the head and rapid tongue-licking movements. Deer were restless and dashed from place to place seeking shade. Intense jaundice was to be seen in the eyes, membranes, and in the urine, which was frequently voided.

2. *Pathological changes in fallow deer.* The early deaths occurring before day 12 were largely the result of severe pulmonary edema and interstitial emphysema. Livers were deeply bile-stained and typical severe bile duct lesions were present. In all deer necropsied the urinary bladder was severely

ulcerated and hemorrhagic. Exposing the tongue to sunlight while licking had produced severe deep-seated inflammation and necrotic ulceration deep into the musculature of the tip of the tongue. Stomach and intestinal ulcerations were found in several deer. Some ulcers had perforated, causing usually a local, but sometimes a general peritonitis.

Note on horses: Horses have shown no natural occurrence of pithomycotoxicosis. As a close-grazing herbivor with a high intake of feed in relation to body weight, it must ingest appreciable numbers of toxin-containing spores under New Zealand conditions. Experimental dosage with sporidesmin has not been attempted.

Disease control measures: There are at least seven major avenues in which prevention may theoretically be exploited. Since alone, none of these is likely to ensure absolute protection of livestock against high toxic challenges, a combination of measures needs to be considered. For New Zealand conditions, certain options for prevention have been established, and others are being investigated. However, because significant outbreaks may occur in a region only once in 3 or 5 or even 10 years, many farmers gamble on its nonoccurrence and are caught unprepared. The avenues are:

1. *Recognition of danger periods by on-farm counts of P. chartarum spores.* When weather conditions are conducive to fungal activity, spore numbers are monitored daily. Regional warning committees broadcast district warnings. An increasing number of farmers own or share a microscope to assess spore numbers and fluctuations on their own individual paddocks prior to grazing.

2. *Prevention of buildup of toxic spore numbers by timely spraying of pastures with substituted benzimidazole fungicides.* These fungicides bind to conidia of *P. chartarum* and during germination are released to inhibit germ tube elongation (Stutzenberger and Parle, 1972, 1973a,b). Data were presented (Stutzenberger and Parle, 1973b; Stutzenberger, 1974) to suggest that certain substituted benzimidazoles may act as a precursor in the formation of an inactive vitamin B_{12} coenzyme analog. Efficient applications of fungicide reduce spore numbers and control major buildup on pasture for up to 6 weeks (Sinclair and Parle, 1967). Effective rates and methods of application have been determined (Parle and di Menna, 1972a,b; Parle, 1973).

3. *Restrict the ingestion of spores by pasture management and alternative crop use.* Grazing of pastures to short length increases the intake of litter and spores, so overgrazing increases the toxic hazard. The strategic in situ grazing of fodder crops, such as turnips, swedes, and kale, is used at this center to feed lambs during toxic periods, but the practice is not widely accepted by grassland farmers.

4. *Increase inborn flock resistance to the toxin by selective breeding.* It was recognized that some sheep have a high resistance to sporidesmin liver injury and that resistance had a seemingly high heritability (Campbell et al., 1975). Since that finding, 160 Romney sires have been progeny tested for pithomycotoxicosis resistance status by sporidesmin challenge. Based on a total of 1455 progeny challenged, the heritability estimate (h^2) was 0.42 ± 0.09 (Campbell et al., 1981). The most resistant and susceptible sires have

been used to generate resistant and susceptible flocks. These flocks are being used in studies on sporidesmin metabolism in the hope of identifying a biochemical marker to aid field selection of sires for wider exploitation of resistance.

In relation to spontaneous field outbreaks, exposure of lambs to a severe field challenge did not swamp the resistance mechanism(s), and late after the challenge, figures for survival and live weight were considerably higher in resistant than in susceptible progeny groups. With this information, some commercial stud breeders are already attempting sire selection from resistance shown in spontaneous outbreaks.

5. *Increase resistance by immunoprophylactic measures.* Sporidesmin, being a small molecule, is nonantigenic. Attempts have been made to couple sporidesmin or its analogs to certain proteins and to inject the complexes formed to stimulate the production of antibodies in animals. A sporidesmin-poly-L-lysine complex antigen produced sporidesmin-specific low-titer antisera in rabbits (Jonas and Ronaldson, 1974). Antibodies that cross-react with sporidesmin have been produced in guinea pigs by injecting a derivative of 2-amino-5-chloro-3,4-dimethoxy benzyl alcohol (ACDMBA) coupled to certain heat-killed bacteria. ACDMBA is a synthetic molecule and is similar in structure to the aromatic moiety of sporidesmin. Mice immunized with this antigen and later challenged with sporidesmin were apparently more resistant to the bilirubinemic effects of sporidesmin than were control mice (Jonas and Erasmuson, 1977, 1979). There appears to be little prospect for the early field control of the disease by these means.

6. *To augment in-built detoxification mechanisms in grazing animals.* Hepatic drug-metabolizing enzyme activities in sheep are markedly induced by hexachlorobenzene (Turner and Green, 1974). Sheep pretreated with hexachlorobenzene and subsequently challenged, either with doses of sporidesmin or by grazing toxic pastures, had at the end of the trial significantly lower mean serum γ-glutamyl-transferase activities, and significantly lower liver injury scores than did control sheep (Mortimer et al., 1978a). The protection lasted throughout the toxic season. Hexachlorobenzene produced no toxic effects in the sheep, but tissue residues in food animals would present an unacceptable hazard to humans. Several accepted food additives, known to induce drug-metabolizing enzyme activities in laboratory animals, did not produce significant protection against sporidesmin in sheep. The experiments suggested that the hepatic mixed-function oxygenases may be partly responsible for metabolism and detoxification of sporidesmin.

7. *Protection afforded by zinc.* High oral intakes of zinc gave good protection from sporidesmin poisoning in rats (Towers, 1977), in sheep (Smith et al., 1977), and in dairy cattle (Towers and Smith, 1978). Zinc protection has also been reported in field trials in sheep (Towers et al., 1975) and in a field outbreak in dairy cows (Smith et al., 1978). For effective protection, zinc has to be given over the period when pastures are toxic and to coincide with sporidesmin intakes. Levels of zinc required are high (calculated as 20-25 mg of elemental zinc per kilogram of live weight per day), 20-30 times greater than nutritional requirements. Protracted dosage at these levels can produce toxic effects from excesses of zinc, mainly expressed as pancreatic necrosis and fibrosis in sheep and cattle (Smith, 1977, 1980).

The protection from sporidesmin toxic effects was not confined to hepato-biliary sites, for there was a lower incidence of lesions in the urinary bladder of sheep and cattle and a less marked drop in milk yield in cows when zinc was given. The mechanism(s) of zinc protection is not known. The stabiliza-tion of biomembranes by zinc, possibly by interference with lipid peroxida-tion, has been suggested from studies on zinc protection against carbon tetrachloride-induced hepatic injury in rats (Chvapil et al., 1973). Zinc also activates superoxide dismutase, an enzyme that McCord and Fridovitch (1969) suggest protects biological membranes from oxidative injury.

Practical regimes for the use of zinc in the prevention of pithomycotoxi-cosis in sheep and cattle have been investigated, and advisory information has been made available to farmers and their advisors.

III. MYCOTOXIC LUPINOSIS

A. History

Lupinosis is a mycotoxic disease causing severe liver injury, icterus, and sometimes photosensitization in livestock. It most commonly affects sheep, occasionally cattle, and rarely horses grazing or fed moldy lupin (*Lupinus* spp.) plant material (Gardiner, 1967a). It is caused by mycotoxins (so far designated phomopsins A and B) produced by the fungus *Phomopsis lepto-stromiformis* (Kühn) Bubák ex Lind (van Warmelo et al., 1970; Gardiner and Petterson, 1972).

The disease has been known for more than 100 years, and the earliest major investigations were made in the last century (ca. 1860-1880), in Germany where lupinosis became a threat to sheep farming. These first investigators clearly distinguished lupinosis from lupin alkaloid poisoning, which is an acute disease of sheep and horses fed large quantities of bitter lupin seed. Such feeding may quickly result in symptoms of confusion, head pushing, frenzy, falling convulsions, dyspnea (through respiratory paralysis), coma, and death, but not liver damage and icterus (Gardiner, 1967a).

Widespread and serious outbreaks of lupinosis next occurred in Western Australia (ca. 1950) and a further phase of lupinosis research was initiated (Bennetts, 1957, 1960; Gardiner, 1967a). Etiologies considered at that time included nutritional factors associated with the lupin diet; in areas where copper had been applied to correct deficiencies, excess copper was suspected. This suspicion was reinforced by the finding of excessive accumulations of copper in the livers of sheep with lupinosis (Gardiner, 1966b).

The belief that fungi were fundamentally involved in the development of toxicity in lupins (and suggested from the very outset by the early German workers) grew stronger in Western Australia in the decade commencing 1960 (Gardiner, 1967a) and research was so oriented. Investigations in South Africa followed similar directions after the severe 1969 outbreak of lupinosis in sheep (Marasas, 1974).

The mycotoxic etiology of the disease was independently established in South Africa by dosing fungal cultures of *P. leptostromiformis*, grown on white lupin seed, to sheep (van Warmelo et al., 1970), and in Western Australia by injection of mice with extracts from the same fungal species (syn. *P. rossiana*) grown on sterile lupin leaves and stalks (Gardiner and Petterson, 1972).

Currently, the growing of sweet lupins, once restricted to coastal areas of Western Australia, is now increasing in other areas, including the eastern states, where lupinosis has occurred. Sweet lupin seed is a valuable high-protein crop, and lupin stubbles provide a better alternative than dry annual pastures and cereal stubbles for summer grazing of sheep in Western Australia (Crocker et al., 1979). Lupinosis sporadically affects sheep grazing lupin stubbles, and this is an important factor limiting the exploitation of lupins in Western Australia (Allen et al., 1978). In mediterranean-climate areas of Australia, lupinosis is a growing problem, but the sparsity of the reported outbreaks from other countries (see Marasas, 1974) suggest that it is not a serious problem elsewhere.

B. Mycological Aspects

1. Morphology of P. leptostromiformis

This fungus is found in nature as a parasite and saprophyte on certain *Lupinus* species. On the stems, pods, and seeds of infected lupin plants it can be distinguished by its stromatic pycnidia. A-conidia are produced on slender tapering conidiophores, and are hyaline, one-celled, biguttulate, cylindrical, and straight, 5-12 × 1.5-2.5 µm. Also produced on lupin stems in separate pycnidia are B-conidia which are uninucleate, hyaline, filiform, strongly curved or hooked. Only A-conidia have been noted on artificial media (van Warmelo and Marasas, 1972; Marasas, 1977).

2. Isolation and Culture Techniques

The fungus is readily isolated in pure culture from infected lupin material on 1.5% malt extract agar at 18°C and irradiated with near-ultraviolet light. For inoculations of routine cultures in mycotoxin production, spore suspensions are made from 21-day-old sporulating cultures and maintained by periodic subculture.

3. Production of Phomopsins

A routine method, using solid media, has been described by van Rensberg et al. (1975) and Marasas (1977). Yellow corn kernels (e.g., 400 g in 180 ml of water) in Erlenmeyer flasks (2 liters) are autoclaved at 121°C on two consecutive days, then inoculated in the dark at 25-28°C for 4 weeks. Flask contents are then minced and air dried on trays at room temperature, milled to a fine powder, and stored at 4°C until fed or extracted. Others have used sterile lupin seed and stalk and leaf material as a substrate for toxin production (van Warmelo et al., 1970; Gardiner and Petterson, 1972; Culvenor et al., 1977).

The production of phomopsin A in liquid medium was investigated by Lanigan et al. (1979) in the hope of reducing the considerable amount of extraneous extractives associated with lupin seed or similar grain cultures. Using their method, stationary cultures of *P. leptostromiformis* (WA 1515) produced 75-150 mg of phomopsin A per liter in a Czapek-Dox medium supplemented with 5-10 g of yeast extract per liter. Alternative supplements were tried and an enzymatic digest of casein (oxoid tryptone) proved satisfactory. Approximately optima for pH and temperature were 6.0 to 25°C, respectively. No phomopsins were detected when shaken cultures were used. No phomopsin B was produced and the use of this liquid medium considerably facilitated the isolation of phomopsin A for chemical and toxicological studies. Phomopsin A seems to be the more important of the two phomopsins since its specific activity in the nursling rat test (Peterson, 1978) is four times that of phomopsin B.

4. Extraction of Phomopsins (Culvenor et al., 1977)

Milled corn cultures of *P. leptostromiformis* (see Sec. I.A.2) may also be extracted using this procedure. Cultures on lupin seed (5 kg) of proven toxicity to sheep were kibbled, then slurried with methanol into a 10-cm glass column. This column was washed with 80% methanol (25 liters). After raising the pH of the extract to 8.5, it was cycloned to the aqueous phase, which was ether-extracted. The aqueous phase, acidified to pH 3.5, was extracted with *n*-butyl alcohol, which was then diluted with an equal volume of light petroleum (30-40°C), yielding two phases. The upper phase was extracted with alkali (0.1 M NaOH) until the extracts were alkaline. These aqueous extracts were combined with the lower phase, the pH corrected to 8.5, and the mixture was concentrated to a brown gum (40-160 g) which contained 10-13 bovine kidney cell culture units per milligram (ccu/mg).

This crude concentrate was partitioned between water and tetrahydrofuran. The organic layer was then extracted with aqueous potassium chloride (15%) which was back extracted. The residue from evaporation of the tetrahydrofuran was an amber gum containing 57 ccu/mg.

In chromatographing 120,000 ccu (in 2 g) on a column of specially prepared macroreticular polystyrene beads (XAD-2), water eluted 4000 ccu (1.5 g), aqueous methanol (1:1), 7500 ccu (250 mg), and methanol 112,000 ccu (400 mg). The methanol residue dissolved in tetrahydrofuran-ethanol-methanol-water (31:3:5:1) deposited crystals (250 ccu/mg) at 0°C. The crystals obtained by crystallizing from 25% aqueous tetrahydrofuran contained phomopsins A and B and decomposed at 205°C without melting (evacuated tube).

Liquid cultures (see Sec. I.A.2), extracted by Lanigan et al. (1979). yielded crystalline phomopsin A but no phomopsin B. Their culture filtrate (10 liters, pH 6-8) was chromatographed (as above) on a column of polystyrene beads (Amberlite XAD-2, 1500 g) and, after water, the subsequent methanol (6 liters) eluate was evaporated. An aqueous solution of the residue (pH 7.5-8) was applied to another XAD column (800 g). The methanol eluate (after water) now left a residue containing 35-50% phomopsin A. Fine, colorless prisms were obtained from tetrahydrofuran-water (10:1).

5. Chemistry of Phomopsin A

Culvenor et al. (1978) have presented the UV, infrared, ^{1}H- and ^{13}C-NMR, and attempted mass spectra of phomopsin A and the results of hydrolysis. The microanalysis of their crystals gave: C, 51.7; H, 5.8; Cl, 4.6; N, 9.4; O, 27.4%. $C_{33}H_{44}ClN_5O_{13}$ requires C, 52.5; H, 5.9; Cl, 4.7; N, 9.3; O, 27.6%. This substance has a UV spectrum similar to that of tenuazonic acid ($C_{10}H_{15}NO_3$), which could also account for the ν_{OH}, ν_{NH}, and ν_{CO} absorptions. Tenuazonic acid also shows the same color reactions of yellow to white with $KMnO_4$.

For characterization of phomopsin A, in the ^{1}H-NMR spectrum there are these resonances: δ (D_2O, pH9-10) 1.07, t and s, 2CH$_3$s; 1.54, s, CH$_3$;

1.63, s, CH_3; 1.77, s, CH_3; 2.36, s, CH_3; 3.53, m; 4.95, m, 6Hs; 5.96, m; 6.24, s; 7.02, s.

6. Detection and Assay of Phomopsins

For confident stepwise advancement in the chemical isolation and identification of toxin from complex extract mixtures, a reliable bioassay using microgram quantities is essential. As refinement of the extract proceeds, chemical assays may be feasible and assume greater usefulness.

Bioassays: Initial toxicity was first established by force-feeding minced culture material (40 g per day for 4 days) to sheep of 30-35 kg live weight. This produced anorexia, jaundice, and death within 6 days. Methods used for bioassay in the isolation of phomopsins were listed by Culvenor et al. (1977). They were generally based on arrested mitotic effects in livers or in cultured cells.

Mouse liver assay: This was the only method used in the early stages of the investigation. Test extracts were injected into mice and they were killed at 6 days. On stained liver sections, parenchymal cell mitoses were counted in 50 representative microscope fields (×600). A mouse unit of toxin was the amount resulting in 2 mitotic figures per 100 parenchymal nuclei. Threefold dilutions of extracts were used. This test is a modification of an earlier method (Gardiner and Petterson, 1972).

Nursling rat liver assay: This method, using remarkably low doses of toxic extract, was described by Peterson (1978). Intraperitoneal injection of phomopsin material increased liver parenchymal cell mitosis. The mitotic cells contained a high proportion of abnormal figures of the scattered chromosomal type, indicative of spindle disruption. In the test, groups of five 15-day-old rats (25 ± 5 g) were injected using serial dilutions of test material and killed at 18 hr. Sections of liver were stained and microscopically examined (×630) for the readily recognizable abnormal mitotic figures. Control groups received either saline or a reference toxin standard. The minimum amount of toxin that would produce 15 abnormal figures in 20 fields was the nursling rat unit (nru) of toxin. In use, the test gave reliable results. Hepato-pathological changes produced in this test illustrated well the early sequence of nuclear and cell changes also characteristic of lupinosis in sheep.

In vitro cell assays: Three further assay methods used were based on the effects of addition of toxin test materials to cells growing in culture. The lamb kidney cell assay (Petterson and Coackley, 1973) method counted abnormal cells after 24 hr. The bovine kidney cell assay measured the reduction in protein increase in a standard cell sheet at 72 hr, compared with controls. The mouse mastocytoma cell assay was based on the reduction in number of live cells produced in 24 hr. Live cells were those impermeable to water-soluble nigrosin dye, and were counted in a hemocytometer. With crude materials for assay, these three cell tests probably gave less specific results than did in vitro tests, but they were more rapid.

By chemical methods: Lanigan et al. (1979) described an electrophoretic method to assay phomopsins in 25 ml of culture filtrate. The filtrate was first chromatographed on a small XAD-2 column. The residue from the methanol

eluate was serially diluted and applied to Whatman No. 4 paper saturated with
borate buffer (0.05 M, pH 9.2) for electrophoresis at 70 V/cm for 15 min,
followed by visualization at 254 nm. The mobilities of phomopsins A and B
were 5.7 cm/hr per kilovolt and 4.5 cm/hr per kilovolt at pH 9.2 (respectively)
(Culvenor et al., 1978).

The spots could also be detected by spraying with potassium permanganate-
chromium trioxide-sulfuric acid reagent when the phomopsins immediately
appeared as yellow spots on a pink background. When the background
changed to brown, the spots were white (Culvenor et al., 1977).

7. *Toxicity of Phomopsins*

Crystalline products containing approximately 80% phomopsin A and 20%
phomopsin B were obtained from extracts of *P. leptostromiformis* culture grown
on lupin seed (Culvenor et al., 1978). Small quantities of other apparently
related compounds were detected by electrophoresis and tentatively named
phomopsins C, D, and E. Introperitoneal injection of 1.2 mg of the crystalline mixture into a sheep of 19 kg live weight produced clinical, biochemical,
and histological changes characteristic of lupinosis.

Separated by electrophoresis and tested against nursling rats (Peterson,
1978), phomopsins A and B gave specific toxicities of 5000 and 1100 nru/mg,
respectively (Culvenor et al., 1977). *P. leptostromiformis* cultures grown on
a liquid medium gave only phomopsin A, and this prompted the suggestion
that phomopsin B is a breakdown product of phomopsin A produced during
extraction (Lanigan et al., 1979). No information is available on the toxicity
of the tentative phomopsins C, D, and E.

Toxicity of phomopsins to cells in vitro. Some information on the toxic
effects of phomopsins on cells in culture is available from the in vitro assay
procedures listed by Culvenor et al. (1977). Initial in vitro studies were made
by Petterson and Coackley (1973), who added toxin to 3-day-old monolayers of
passaged lamb kidney cells. Arrest of nuclear division at metaphase and the
production of bizarre-shaped formations (designated "Lup" cells) reached a
maximum by 24 hr. Many cells contained several nuclei, some much smaller
than normal. An assay method was developed using this information. A later
study, using standardized partly purified phomopsins, examined and compared the mitotic metaphase arrest produced in transformed sheep lymphocytes
by phomopsin and by colchicine (Petterson et al., 1979). Both agents produced results similar in several respects, although colchicine required shorter
exposure time on cells to produce the same severity. The duration of colchicine effects was also shorter. Phomopsins also decreased the mobility of
migrating leucocytes, and this effect was compared with a colchicine action
known to bind microtubular protein, prevent polymerization, and disrupt
microtubular formation.

Electron microscopy of transchromosomal sections of metaphase profiles in
cells treated with either agent showed no differences in the length or concentration of cytoplasmic microtubules. These and other findings suggested to
Petterson et al. (1979) that the microtubular apparatus is the primary target
for phomopsins and that, like colchicine, phomopsins also prevent the polymerization of microtubular proteins.

Toxicity of phomopsins to laboratory animals. Phomopsins in the form of toxic lupin material or extracts, or as laboratory cultures of *P. leptostromiformis* or extracts have, in recent years, produced experimental disease, akin to lupinosis, in rabbits (Dobberstein and Walkiewicz, 1933; Gardiner, 1961, 1967a), in mice (Gardiner, 1967a; Gardiner and Petterson, 1972; Papadimitriou et al., 1973, 1974; Peterson and Lanigan, 1976), in rats (Papadimitriou and Petterson, 1976; Peterson, 1978), in chicks (Kung et al., 1977), and in guinea pigs (Lanigan et al., 1979), for which no information was given.

Toxicity to rabbits: Spontaneously toxic lupin roughages were fed to rabbits for up to 31 days (Gardiner, 1961, 1967a). Periportal and midzone hepatocytes showed abnormal mitotic figures and centrilobular areas showed granules containing hemosiderin and lipofuscin-like material. Rabbits have not been used in numbers either for assays or for pathological studies.

Toxicity to chicks: Extracts of cultures of *P. leptostromiformis* grown on soybeans were dosed on alternate days to chicks for up to 42 days (Kung et al., 1977). A reported constant finding in livers from 14 to 42 days was biliary hyperplasia. Kupffer's cell and endothelial hyperplasia were present at 28 days. Regenerative hepatocyte hyperplasia with numerous mitotic figures, and cholangitis, were seen at 35 to 42 days. It was claimed that the lesions they described represent the sequential stages of liver injury in lupinosis.

Toxicity to mice: The occurrence of hepatocytes with nuclear aberrations, previously described in lupinosis in other species (rabbit: Dobberstein and Walkiewicz, 1933; sheep: Gardiner, 1965), was first reported in mice by Gardiner (1967a). Subsequent papers have reported complementary studies on the pathogenesis of phomopsin toxins in mice (Gardener and Petterson, 1972; Papadimitriou et al., 1973, 1974; Peterson and Lanigan, 1976).

In these studies, intraperitoneal injections of phomopsin extracts produced a consistent pattern of changes, some of which were dose dependent. Morphological changes in the liver included swelling of hepatocyte nuclei within a few hours. Increased mitotic activity occurred within 26 hr, and it reached a peak between 3 and 7 days. This early phase, which showed both cytotoxic manifestations and mitotic stimulation, was of short duration. Of longer duration was the widespread swelling of hepatocytes and nuclei, the formation of nuclear inclusions, and polykariocytes which contained clusters of small nuclei. Abnormal arrested metaphasic profiles were formed with scattered or condensed chromosomes. Considerable loss of parenchymal cells occurred through shrinkage necrosis.

Reported liver ultrastructural changes included the adherence of electron dense deposits to sinusoidal cells and hepatocytes within 30 min of dosage, the formation of autophagic vacuoles, dilated bile duct canaliculi with diminished numbers of microvilli, and more transient changes of lipid accumulation and mitochondrial swelling.

Histochemical examination confirmed that centrilobular zones of the hepatic lobules were most affected, and numbers of hepatocytes in these regions with nuclear changes were directly proportional to the dose of toxin given. At the highest dose, 26% of hepatocytes showed metaphase.

In severely affected mice, renal changes, qualitatively similar to those in hepatocytes, occurred in the proximal convoluted tubules. The wave of mitotic arrests commenced several hours later than in the liver.

Toxicity to rats: Toxic effects of single injections of crude extracts containing phomopsins have been studied in adult rats (Papadimitriou and Petterson, 1976) and in 15-day-old nursling rats (Peterson, 1978).

Pathological changes involved mainly hepatocytes, especially at centrilobular and midzonal sites. Bile ductal and sinusoidal cells were little affected. Histological findings were basically similar to those produced by phomopsins in mice.

The injury in adult rats (Papadimitriou and Petterson, 1976) was biphasic and the initial nonspecific toxic phase lasted for a few days. There was an accumulation of lipids in hepatocytes, some mitochondrial swelling, and autophagosome formation. Hepatic levels of certain enzyme activities fell, and serum levels increased, especially ornithine carbamyl transferase activity. Marked hypertrophy of smooth endoplasmic reticulum was a feature of rats, yet was not pronounced in mice similarly dosed.

The second phase was dominated by abnormalities in hepatocyte division and was more protracted. Both normal and abnormal metaphase profiles were present, and multinuclear hepatocytes with micronuclei were formed. Degeneration and necrosis of these aberrant hepatocytes became more pronounced and their numbers rapidly decreased, apparently after a very short life span.

The principal effect of low doses of phomopsins on nursling rats (Petersen, 1978) was metaphasic arrest in hepatocyte nuclei. With an already existing high rate of mitosis in nursling rats, metaphase arrest occurred as a single wave which commenced within 1 hr of injection of toxin. Peak arrest occurred at 2-4 days and then declined rapidly. Higher dose rates of phomopsins produced fatty changes and fibrosis in the liver and most of the rats became jaundiced. These are changes characteristic of lupinosis in sheep (Gardiner, 1965). With higher dose rates, mitotic arrests were produced in the kidney and acinar cells of the pancreas, but not in other tissues.

Toxicity of phomopsins in farm animals (lupinosis)

Lupinosis in sheep: The higher prevalence of lupinosis in sheep than in other farm species may be more related to opportunity than to species susceptibility. Reliable information on the latter is not available.

Various aspects of the symptoms and pathological changes have been described (Bennetts, 1957, 1960; Gardiner, 1961, 1965, 1966a,b; Gardiner and Parr, 1967; van Warmelo et al., 1970) in spontaneous outbreaks of lupinosis, in sheep fed fungal-infected lupin material, or latterly, in sheep dosed *P. leptostromiformis* fungal culture material. Detailed reviews of these investigations have been published (Gardiner, 1967a; Marasas, 1974, 1978a).

The first clinical signs of lupinosis are listlessness and inappetance. Complete anorexia and jaundice may soon follow and, in acute outbreaks, deaths can occur from 2 to 14 days after ingestion of toxic lupins. Secondary signs of liver injury sometimes seen in severely affected sheep are photosensitization (Bennetts, 1957) and ketosis (acidosis). Mortalities of more than 50% have been reported. In less acute outbreaks, partial inappetance, loss of weight, and dullness are general and some animals may show jaundice. There is slow recovery, and mortalities, which are spread over a period, may not be high.

Single doses of *P. leptostromiformis* cultures given to sheep produce marked serum elevations of bilirubin and γ-glutamyl and glutamic-oxalacetic transferases within 3-4 days and deaths within 5-10 days (Malherbe et al., 1977). Other chemical pathological changes reported are early elevations in serum lactate and glutamate dehydrogenases and in alkaline phosphatase activities.

In liver, elevations of copper and selenium values and decreases in zinc values have also been reported in sheep grazing toxic lupin stubbles for more than 2 weeks (Allen et al., 1979).

Marked anatomical changes show that the liver is the main target for phomopsins. With high intakes of toxin, which produce acute overt disease, livers are enlarged, orange-yellow in color, fatty on cutting, and easily disrupted. In less acute and in chronic cases, livers are often bronze or tan colored, firm, contracted in size and fibrosed, often distorted and likened in shape to a boxing glove (Bennetts, 1957), and nodes of regenerated tissue may be prominent. Kidneys are generally darker in color and the spleen is enlarged. Large volumes of transudates may occupy the thoracic and abdominal cavities and the pericardial sac.

Histopathological changes in the liver reflect the time course of the disease. Lipid phanerosis, mainly centrilobular, is most pronounced in acute cases, becoming less pronounced with time. Pigment granules, which stain as lipofuscins, and hemosiderin and hyaline droplets, become more profuse with time. In the active stages, pronounced nuclear changes and enlarged vesicular nuclei are present along with abnormal mitotic metaphase profiles, some showing bizarre chromosomal figures. A few polykariocytes and megalocytes are readily identified. More chronic cases show a higher proportion of normal mitotic figures. Neutrophiles are diffusely infiltrated within the lobules. Periportal fibrosis and bile ductule proliferation and centrilobular stellate fibrosis become more extensive with time. Kidney proximal tubular cells show vacuolation, pigmented inclusions, and degenerative changes.

Lupinosis in cattle: Lupinosis occurs infrequently in cattle. Over a 6-year period, six minor outbreaks were reported and investigated in Western Australia (Gardiner, 1967b). Elsewhere its occurrence is rare.

The early signs of bovine lupinosis are inappetance, listlessness, and loss of milk production (Marasas, 1974). Intense jaundice is usual and photosensitization occurs when there is access to green feed to provide the photodynamic pigments (Bennetts, 1960). Lachrymation and salivation are commonly noted. Cattle may become ketotic, followed by prostration and death.

In the acute disease, necropsy reveals pronounced jaundice in the carcase, body organs, and fat deposits. Acute cases show greatly enlarged fatty, friable, rich orange-yellow colored livers. Kidneys are deeply bile stained.

Histopathological changes found in 35 cows examined (Gardiner, 1967b) revealed predominantly fatty changes of the liver and kidney. In the liver, fatty infiltration produces rupture of adjacent hepatocytes to form fat cysts of mainly midlobular distribution. Granule-containing degenerating hepatocytes and focal cell necrosis are present in some livers. Abnormal mitotic hepatocytes are also commonly found. Proliferation of bile ductular cells and periportal fibrosis are present. The most chronic cases are characterized by extensive connective tissue depositions and bile duct proliferation, together with nodular regeneration. Renal and splenic siderosis is often present. No reports of experimentally produced lupinosis in cattle have been located.

Lupinosis in horses: Only rare reports of the disease are found in the literature (Marasas, 1974, 1978b). Isolated cases occurring in Australia were reported by Bennetts (1960), and findings in five cases were described by Gardiner and Seddon (1966).

Inappetance, depression, jaundice, conjunctival hemorrhages, and inability to stand are clinical signs recorded. Neurological ataxia and paralysis may also be present (Marasas, 1978b).

Pronounced jaundice is invariably found at necropsy. The liver is grossly enlarged, firm and greasy, and deep yellow in color. Urine may be reddish brown in color.

Histopathological changes in the liver are similar to those found in cattle with lupinosis. Fatty degeneration, pigmented granular masses, bile ductule proliferation, and generalized fibrosis are features to be seen. Hemoglobin casts may be numerous in renal tubules.

Apparently, horses are highly susceptible to lupinosis toxins and they show acute toxic hemorrhagic manifestations (Gardiner and Seddon, 1966) not shown by species mentioned previously.

Lupinosis in swine: The only report of spontaneous lupinosis in swine tells of a possible incident in which bitter lupin seed was fed to swine in Poland (Marczewski, 1955).

To replace soybeans in swine-feed formulations, seed from improved sweet lupin cultivars is being increasingly used in several countries. Since lupin seed could be a source of phomopsin toxins, the susceptibility of swine to these toxins was investigated (van Rensberg et al., 1975) by feeding toxic *P. leptostromiformis* cultures gown on autoclaved corn (*Zea mays* L).

Swine were found to be susceptible, and the cultures induced anorexia, marked loss of weight, and posterior paralysis. The early onset of anorexia prevented the voluntary intake of the toxic material. When this problem was overcome, marked pathological changes were induced. Generalized jaundice, atrophy of the liver, nephrosis, and hemorrhages into the intestine and into the peritoneal cavity were lesions found at necropsy.

Histopathological changes include severe necrosis of hepatocytes, swollen hepatocytes with nuclear changes of pyknosis, and karyorrhexis, the latter often resembling mitotic figures. Other hepatocytes were binucleate, and golden-brown pigmented globules were present in the cytoplasm. Kidney tubular epithelium also showed necrotic cells, and degenerative changes were recognized in the myocardium.

The possible dangers of feeding contaminated lupin seed were discussed.

Control of lupinosis. The use of sweet lupins either as a forage crop, or for grazing residues and stubble after grain harvesting, has considerable advantages over dry annual pastures for summer grazing of young sheep, or ewes prior to mating, in a mediterranean climate (Arnold and Charlick, 1976; Arnold et al., 1976). The seasonal danger of toxicity emerging on lupin plant material cannot, apparently, be gauged on weather factors alone. Toxicity can occur before summer rainfall and, after toxicity has developed, lupin material may remain significantly toxic for up to 17 weeks (Allen et al., 1980). Marasas (1978a) indicated that close surveillance of lupin paddocks, and the removal of sheep to alternative safe grazing as soon as fungal development becomes evident on lupin stubble, was the best recommendation. However, regular daily surveillance of sheep for the first 7 days of grazing lupin stubble, thereafter 3 times weekly, and if rain falls, to reinstate daily surveillance for a further 7 days, has also been recommended (Allen et al., 1980). More liver injury occurred in sheep grazed at high stocking rates than those grazed at lower rates, so outbreaks may be precipitated by grazing sheep at high stocking rates (Crocker et al., 1979). In the same trials, the monitoring of certain enzyme activities in serum of grazing sheep was examined for early signs of liver injury, but the enzyme tests were not considered reliable indicators.

The preparation of lupin hay can provide farmers who do not harvest seed with a safe means of utilizing lupin crops for summer feed (Allen et al., 1978). The crop is mown after pod formation and is left in an open swathe until it is bailed or rolled.

Development and use of lupin cultivars resistant to *P. leptostromiformis* offer great hope for control and have received some attention. *Lupinus mutabilis* Sweet, a cultivated crop plant in South America, has high resistance to both infection and to saprophytic colonization by *P. leptostromiformis* and is a strong contender for inclusion in lupin breeding programs (van Jaarsveld and Knox-Davies, 1974; Marasas, 1978a). In Australia, *Lupinus albus*-cv. Ultra has shown marked resistance when compared in trials with some other cultivars (Wood and Allen, 1980).

A protective effect from experimental phomopsin liver injury, with oral doses of zinc (0.5 g of zinc or more per day), has been demonstrated in sheep (Allen and Masters, 1980). Although the degree of protection varied between animals and between experiments, further trials seem warranted, using less toxic salts or oxide of zinc.

There is also a need for a rapid chemical method for detecting and assaying toxic levels of phomopsins in lupin plant material samples using, for example, HPLC methods.

IV. XANTHOMEGNIN AND RELATED QUINONE MYCOTOXINS

A. History

Xanthomegnin was first isolated by Blank et al. (1963) from the dermatophyte *Trichophyton megnini* Blanchard 1896. It has since been isolated from other *Trichophyton* and *Microsporum* spp. of dermatophytes and from a range of *Penicillium* and *Aspergillus* spp. which are commonly isolated from foodstuffs (Mislivec and Tuite, 1970; Pöhland and Mislivec, 1976). However, the detection of xanthomegnin quinone mycotoxins has not been reported from animal or human foodstuffs, yet the likelihood exists.

The dosing of mice with certain *Penicillium* and *Aspergillus* spp. cultures highlighted the presence of a mycotoxin with potent hepatotoxic activity (Carlton et al., 1970; Zimmermann et al., 1976).

Isolation of the xanthomegnin quinones from toxigenic cultures and feeding them to mice (Carlton et al., 1976) produced the evidence that the quinones were the mycotoxins responsible for the hepatotoxic changes repeatedly produced by toxigenic cultures. Only xanthomegnin and viomellein have so far been reported to have the hepatotoxic effect.

The hepatic pathology produced by these two mycotoxins features severe bile duct injury and occlusion, with consequent bile stasis and jaundice. Mice with identical liver lesions produced by feeding toxigenic *P. viridicatum* cultures showed photosensitization and photodermatitis when exposed to sunlight (Budiarso et al., 1970). On this evidence xanthomegnin and viomellein are considered to be mycotoxins which, fed in cultures to mice, produced severe hepatobiliary injury and photosensitization. There are no reports of these mycotoxins producing spontaneous hepatic lesions and photodermatitis in farm or laboratory animals.

B. Source of Mycotoxin

Xanthomegnin, viomellein, and related metabolites have so far been obtained from cultures of numerous mold species belonging to four fungal genera: *Microsporum, Trichophyton, Penicillium,* and *Aspergillus* (see Table 3 for metabolites produced and references). Data on morphologic features and identification of this broad range of toxin-producing molds is beyond the scope of this review (refer to Raper and Thom, 1949; Raper and Fennell, 1965).

Toxicity studies with xanthomegnin and the quinone metabolites have mainly used laboratory cultures of *P. viridicatum, P. ochraceum,* and some isolates of the *A. ochraceus* group (see the review by Carlton, 1980).

Isolates of *P. viridicatum* in particular produce numerous known toxic metabolites, and different strains of this mold apparently produce broadly different ranges of these metabolites (Stack et al., 1977). On data obtained from mouse toxicity experiments, xanthomegnin is produced as the major toxic metabolite by many Indiana isolates of *P. viridicatum* (Carlton, 1980).

C. Production of Xanthomegnin and Related Metabolites

Earlier studies on the toxigenicity of cultured products of *P. viridicatum* utilized several liquid substrates. Consistently highly toxic cultures were produced using corn-steep dextrose medium, potato dextrose broth, and a modified Czapek medium incubated at 23°C for 1 or 2 weeks under stationary conditions (Budiarso et al., 1971). A modified Czapek medium with 1% corn steep and 3% glucose apparently gave superior toxin production.

Cultures of *P. viridicatum, P. cyclopium,* and *A. ochraceus* grown on sterile polished rice were used for production of xanthomegnin and viomellein (Stack and Mislivec, 1978). Isolates were grown for 21 days at 23-26°C on 50 g of rice and 50 ml of water and were then extracted with chloroform. Yields of 0.4-1.6 mg of xanthomegnin and 0.2-0.4 mg of viomellein per gram of rice were obtained from three isolates of *P. viridicatum,* comparable yields were obtained from six isolates of *A. ochraceus,* but much lower yields were obtained from the one cultured isolate of *P. cyclopium.* Carlton (1980) lists oats, wheat, and corn as other suitable solid substrates, and Stack et al. (1978), using *P. viridicatum* cultures on corn, found production of increasing amounts of xanthomegnin up to day 13 (629 mg per kilogram of culture), at which time the experiment was terminated.

TABLE 3 Molds Producing Xanthomegnin and Related Metabolites

Mold species	Quinone metabolite			
	Xanthomegnin	Viomellein	Viopurpurin	Rubrosulfin
Penicillium viridicatum	+[a]	+[a]	+[a]	+[a]
P. cyclopium	+[b]	+[b]		
P. citreo-viride	+[c]	+[c]		
P. ochraceum	+[d]			
Aspergillus sulphureus	+[e]	+[e]	+[e]	+[e]
A. melleus	+[e]	+[e,f]	+[e]	
A. ostianus	+[f]	+[f]		
A. auricomus		+[f]		
A. ochraceus	+[b,f]	+[b,f]		
Trichophyton violaceum	+[g]		+[g,h]	
T. megnini	+[i]			
T. rubrum	+[j]			
Microsporum cookei	+[k]			

[a] Stack et al., 1977.
[b] Stack and Mislivec, 1978.
[c] Zeeck et al., 1979.
[d] Cited by Carlton, 1980.
[e] Durley et al., 1975.
[f] Robbers et al., 1978.
[g] Ng et al., 1969.
[h] Blank et al., 1966.
[i] Blank et al., 1963; Just and Day, 1963.
[j] Wirth et al., 1965.
[k] Ito et al., 1973.

D. Extraction and Isolation of Xanthomegnin and Related Metabolites

From larger-scale cultures of *P. viridicatum* (Purdue 66-68-2) grown at room temperature for 24 days on 19 × 500 g of rice per 500 ml of water (Stack et al., 1977), each flask was extracted with chloroform (1 liter). The residue (45 g) from the chloroform was put onto silica gel (500 g) and eluted with chloroform (8 liters). This chloroform residue (30 g) was rechromatographed on silica gel (500 g) but first eluted with hexane (5 liters), then with chloroform (5 liters). The red precipitate (15 g) from adding hexane (400 ml) to this chloroform, concentrated to 100 ml, was applied in chloroform to another column (250 g of SiO$_2$). Benzene-chloroform-acetic acid (25:25:1, 3 liters) eluted viomellein (1.5 g), then chloroform-acetic acid (50:1, 3 liters) eluted xanthomegnin (9 g) and rubrosulfin (0.5 g) together. On concentrating the chloroform-acetic acid solution and adding benzene, rubrosulfin crystallized out, leaving the xanthomegnin in solution.

Durley et al. (1975) describe also the extraction of the toxins from *Aspergillus sulphureus* cultures with chloroform and their separation by preparative layer chromatography.

E. Estimation of Xanthomegnin in Corn by HPLC (Stack et al., 1978)

Analytical methods for the detection of xanthomegnin in foodstuffs currently lack sensitivity. The following method is based on a technique developed and used for the detection of aflatoxin.

Finely ground corn was shaken with phosphoric acid and chloroform. An aliquot of the filtrate was chromatographed on silica gel. The residue from the chloroform-acetic acid fraction was analyzed by HPLC on a μPorasil column with chloroform-methanol-acetic acid. A recovery of 40% of 750 μg of xanthomegnin added to 1 kg of corn was achieved.

F. Structures of Xanthomegnin and Viomellein

Xanthomegnin [4] is a symmetrical dimer; each of its monomers is a substituted juglone residue fused to a δ-lactone. Viomellein [5] consists of one xanthomegnin monomer bonded to a dihydroxynaphthalene, which is also fused to a δ-lactone.

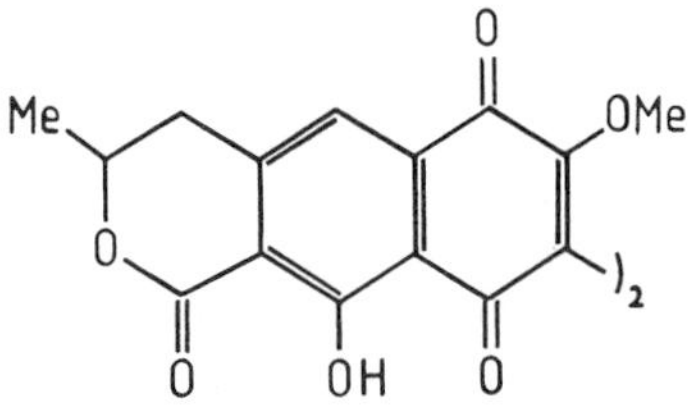

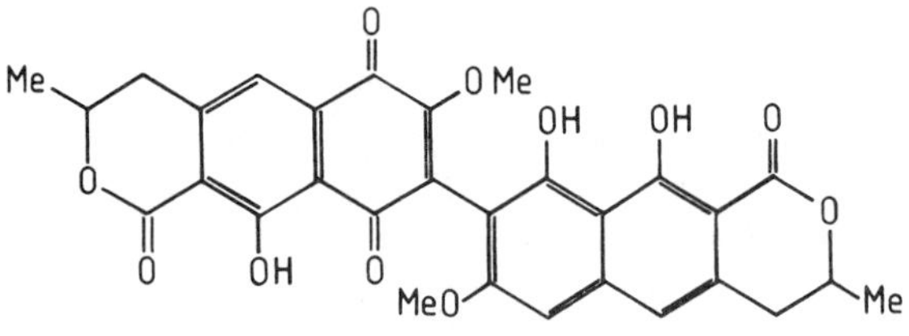

1. *Xanthomegnin* (Durley et al., 1975) $C_{30}H_{22}O_{12}$, Str. [4]. The orange plates from chloroform-benzene, sintering above 260°C, gave λ_{max} (methanol) 222, 264, and 380 nm (log ε 4.42, 4.29, 3.90); ν_{max}(Nujol) 3300br, 1712, 1672, 1618, and 1600 cm^{-1}; m/e 578 (5%, $M^+ + 4$), 577 (2, $M^+ + 3$), 576 (5, $M^+ + 2$), 575 (1.5, $M^+ + 1$), 574.116 (3, M^+ calc. 574.111), and 485.089 (100, calc. for $C_{27}H_{17}O_9$ 485.087); and δ (CDCl$_3$) 13.11, 7.49, 4.61, 4.15, 3.04, and 1.56.

2. *Viomellein* (Durley et al., 1975) $C_{30}H_{24}O_{11}$, Str. [5]. The brown beads (from chloroform-light petroleum) sintering above 260°C gave λ_{max}(methanol) 225, 264, and 395 nm (log ε 4.23, 4.31, 3.91); ν_{max} (Nujol) 3340, 2750br, 1725, 1675, 1656, 1640, 1630, and 1608 cm^{-1}; m/e 562 (0.4%, $M^+ + 2$), 561 (0.13, $M^+ + 1$), 560.137 (0.27, M^+ calc. 560.141), and 83 (100); and δ (CDCl$_3$) 13.88, 13.44, 9.80, 7.50, 6.96, 6.66, 4.63, 3.90, 3.84, 3.02, and 1.56.

The linear structure proposed for xanthomegnin (and viomellein) by Höfle and Röser (1978) compared with the earlier proposed angular one (Durley et al., 1975) is based (from ^{13}C-NMR studies) on a carbonyl and a bridgehead carbon in the acetate, showing long-range coupling, ^{3J}CH=4 and 6 Hz, respectively, to an aromatic proton.

G. Toxicity of Xanthomegnin and Viomellein

Spontaneous outbreaks of mycotoxic disease in domestic or laboratory animals caused by xanthomegnin, viomellein, or related quinone metabolites have not been reported. Both xanthomegnin and viomellein, isolated from laboratory cultures of a toxigenic strain of *Penicillium viridicatum*, were fed separately in diets to weanling male Swiss mice for 10 days (Carlton et al., 1976). The feed, containing either 448 mg of xanthomegnin per kilogram or 456 mg of viomellein per kilogram, was provided ad libitum.

Deaths occurred in each mycotoxin-fed group and body weight gains were reduced in the survivors. Gross lesions were similar for both groups and included jaundice and greenish foci in the liver and greenish discoloration of the kidneys.

Histopathologic lesions in both groups were similar and included periductal edema, pericholangitis, and necrotizing cholangitis of intrahepatic bile ducts, and corresponding lesions in the extrahepatic ducts and gallbladders. Bile ductule hypertrophy and focal hepatic necrosis were also noted. Kidney lesions included pigmented tubular casts, focal tubular dilation, and mineralization changes.

This was the first experiment in which these characteristic liver and kidney lesions had been produced by feeding xanthomegnin and viomellein, isolated from cultures of *P. viridicatum*, to laboratory animals. Previously, the feeding of whole fungal cultures of *P. viridicatum* (Carlton et al., 1968, 1970), of *P. ochraceum* (Carlton et al., 1972), and products of some cultured isolates of the *Aspergillus ochraceus* group (Zimmerman et al., 1976, 1977) had produced the same gross and histologic changes in mice. Furthermore, most of these mold isolates are known to produce xanthomegnin and other quinone metabolites (Stack and Mislivec, 1978; Robbers et al., 1978). It was on this good evidence that these mycotoxins were in retrospect held mainly responsible for the production of severe hepatobiliary and other lesions in mice, rats, guinea pigs, and swine fed fungal culture materials in a range of experiments carried out before xanthomegnin and related compounds had been incriminated. These studies have been previously reviewed (Carlton and Tuite, 1977; Carlton, 1980), and selected aspects are mentioned here.

1. Mouse

When fungal cultures containing xanthomegnin were fed at high dietary concentrations, mice became lethargic and deaths occurred during a feeding period of 1-2 weeks (Carlton et al., 1968, 1970; Budiarso et al., 1971). The most significant lesions were found at hepatobiliary sites, for livers showed bile-stained discoloration, and jaundice was generally widespread, the kidneys in particular showing green discoloration. The watery consistency of fluid in the gallbladder probably indicated eventual stasis of bile flow.

Histologically, intra- and extrahepatic bile ducts showed periductal edema, inflammation, and necrosis, sometimes involving the hepatic artery in the necrotizing cholangitis. Parenchymal cell necrosis was also evident, both as widespread small foci and as larger areas close to the portal structures. Less acute lesions were typified by fibroplasia and bile-ductule hypertrophy.

More protracted feeding resulted in more pronounced fibroplasia and proliferation of bile ducts, with dilatation and saccule formation in both intra- and extrahepatic ducts. There was marked individual variation in renal lesions, which included tubular dilatation, tubular necrosis, and mineralization.

When mice, fed cultures of *P. viridicatum*, were exposed to sunlight a photodermatitis developed (Budiarso et al., 1970). On the ears this lesion was characterized by prompt development of erythema and edema of the skin and the lesions sometimes progressed to necrosis of the skin and auricular cartilage. The photodermatitis also involved the muzzle, paws, and tail and, in some mice, edema and keratitis of the cornea developed.

2. Rat

Hepatic and renal lesions, closely similar to those reported in mice, were produced in rats by feeding cultures of *P. viridicatum* (Carlton and Tuite, 1970a; Rafiquzzaman, 1974), and *A. ochraceus* (Zimmermann et al., 1978).

In addition to these major primary lesions, McCracken et al. (1974a) described the development of occular lesions, lesions of the scrotum (McCracken et al., 1974b), and necrosis and ulceration of the gastric mucosa (McCracken et al., 1974c) in rats fed cultures of *P. viridicatum*. The occular lesions were first apparent as a mild corneal opacity, developing in a few days to marked keratitis with, in severely affected eyes, the development of corneal vascularization (pannus formation). The causal mechanisms of this lesion were not accounted for. The pathogenesis of scrotal lesions in rats was studied (McCracken et al., 1974b), and it was concluded that the necrosis and ulceration of the scrotal integument was probably due to ischemic changes, a consequence of vasculitis and thrombosis in the subcutis of the scrotum.

3. Guinea Pig

Guinea pigs fed corn cultures of *P. viridicatum* proved less susceptible than the rat (Carlton and Tuite, 1970a). The guinea pigs became lethargic and anorexic, with reduced weight gain. Histologic changes found in livers were hepatic cell focal necrosis, bile ductule proliferation, and periductal fibrosis.

4. Swine

Swine fed corn cultures of *P. viridicatum* (Carlton and Tuite, 1970b) and swine fed rice cultures of *A. ochraceus* (Zimmermann et al., 1979a,b) showed identical responses. The pathologic changes were, however, wholly different from those produced in mice, rats, and guinea pigs fed cultures of the same molds. In swine the principal organ affected was not the liver but the kidney. A composite clinicopathologic picture, drawn from the three references cited above, is given here.

Dependent on amounts of toxic culture in fed rations, swine either became lethargic, lost weight, or in some cases died. Most developed subcutaneous edema; effusions into the abdominal and thoracic cavities; pulmonary atelectasis; edema of the abdominal viscera, including the perirenal folds; and hemorrhages in the kidneys. Histologic examination showed proximal tubular changes in the kidney ranging from mild cell degeneration to frank necrosis and the presence of hyaline tubular casts and interstitial fibrotic changes. Feeding of lesser amounts of toxic fungal material for 8 weeks produced extensive renal interstitital fibrosis, but perirenal edema was not a feature. A range of clinical tests on serum and urine gave good indication of the nephrotoxic changes present (Zimmermann et al., 1979b).

H. Biosynthesis

There are no known reports regarding the biosynthetic pathways for xanthomegnin and related quinone metabolites.

I. Biochemical Effects

Early studies on the effect of xanthomegnin on rat liver mitochondrial respiration (Ito et al., 1970, 1973) showed that respiration was markedly accelerated when xanthomegnin was added to stage 4 respiration, but not when added to stage 3. It is a strong uncoupler of oxidative phosphorylation, and this mechanism was also investigated by Akita (1978), who suggested that the phenolic hydroxyl groups and hydrophobicity of xanthomegnin may contribute to this uncoupling action. Xanthomegnin also strongly induces mitochondrial swelling. Latent ATPase activity of freshly prepared rat liver mitochondria was strongly stimulated by xanthomegnin (Akita, 1978). The possibility of a xanthomegnin bypass to the mitochondrial electron transport system has been suggested (Kawai and Nozawa, 1979). Binding of xanthomegnin with protein and amino acids, with the production of spectral changes, has recently received attention (Kawai et al., 1977; Watanabe, 1978).

J. Antibiotic and Other Activity

Compounds of the xanthomegnin series were shown to limit the growth of gram-negative and gram-positive bacteria in dilution-series inhibition tests and in plate diffusion tests. Results in the plate diffusion test were strongly influenced by the diffusibility of the test substance. Xanthomegnin and viomellein showed an inhibitory effect on development and reproduction of the Mexican bean beetle (*Epilachna varivestis* Muls.) (Zeeck et al., 1979).

K. Control

Xanthomegnin and related quinone metabolites have been obtained only from laboratory cultures of producing molds, not from foodstuffs. For optimal growth and toxin production, high moisture content and temperatures near 25°C are required. Consequently, prevention of mold growth and contamination of foodstuffs would most likely be achieved by storage under controlled dry conditions.

To assist the surveillance of disease, further information is needed on the susceptibility of other farm animal species, sheep and cattle and especially poultry, to these mycotoxins. Known laboratory production and isolation procedures of these toxins will enable animal experimentation to proceed without the difficulties of interpretation resulting from the presence of multiple toxins in nonprocessed fungal cultures.

REFERENCES

Akita, T. (1978). Pigments from the pathogenic fungus, *Microsporum cookei*. The mechanism for the uncoupling action of xanthomegnin on the oxidative phosphorylation in liver mitochondria. *Gifu Daigaku Igakubu Kiyo 26*:593-620. *C.A. 91*:205062, 1979.

Allen, J. G., and Masters, H. G. (1980). Prevention of ovine lupinosis by the oral administration of zinc sulphate and the effect of such therapy on liver and pancreas zinc and liver copper. *Aust. Vet. J. 56*:168-171.

Allen, J. G., Wood, P. McR., and O'Donnell, F. M. (1978). Control of ovine lupinosis: experiments on the making of lupin hay. *Aust. Vet. J. 54*:19-22.

Allen, J. G., Masters, H. G., and Wallace, S. R. (1979). The effect of lupinosis on liver, copper, selenium and zinc concentrations in Merino sheep. *Vet. Rec. 105*:434-436.

Allen, J. G., Wallace, S. R., and Wood, P. McR. (1980). A study of the periods for which lupin stubbles infected with *Phomopsis leptostromiformis* remain toxic. *Aust. J. Exp. Agric. Anim. Husb. 20*:166-169.

Arnold, G. W. and Charlick, A. J. (1976). The use of sweet lupins in pastoral systems with breeding ewes. *Proc. Aust. Soc. Anim. Prod. 11*: 233-236.

Arnold, G. W., Maller, R. A., Charlick, A. J., and Hill, J. L. (1976). Lupin crops in pastoral systems for weaner sheep in a Mediterranean environment. *Agro. Ecosyst. 2*:99-116.

Beecham, A. F., Fridrichsons, J., and Mathieson, A. McL. (1966). The structure and absolute configuration of gliotoxin and the absolute configuration of sporidesmin. *Tetrahedron Lett. (27)*:3131-3138.

Bennetts, H. W. (1957). Lupin poisoning of sheep in Western Australia. *Aust. Vet. J. 33*:277-283.

Bennetts, H. W. (1960). Lupinosis. *J. Agric. West. Aust. 1*:47-52.

Bertaud, W. S., and Mortimer, P. H. (1966). Changes induced in liver cells by experimental doses of sporidesmin. *Proc. 6th Int. Congr. Elec. Micros.* Kyoto, Japan, Vol. 2, pp. 619-620.

Blank, F., Day, W. C., and Just, G. (1963). Metabolites of pathogenic fungi. II. The isolation of xanthomegnin from *Trichophyton megnini* Blanchard 1896. *J. Invest. Dermatol. 40*:133-137.

Blank, F., Ng, A. S., and Just, G. (1966). Isolation and tentative structures of vioxanthin and viopurpurin, two coloured metabolites of *Trichopyton violaceum. Can. J. Chem. 44*:2873-2879.

Brewer, D., Hannah, D. E., and Taylor, A. (1966). The biological properties of 3,6-epidithiadiketopiperazines. Inhibition of growth of *Bacillus subtilis* by gliotoxins, sporidesmins and chemotin. *Can. J. Microbiol. 12*:1187-1195.

Brewer, D., Rahman, R., Safe, S., and Taylor, A. (1968). A new toxic metabolite of *Pithomyces chartarum* related to the sporidesmins. *J. Chem. Soc. Chem. Commun.*, p. 1571.

Brook, P. J., and Matthews, R. E. F. (1960). The production of ^{35}S labelled sporidesmin. *N.Z. J. Sci. 3*:591-599.

Brown, J. M. M. (1959). Advances in "geeldikkop" (*Tribulosis ovis*) research. 1. The history of geeldikkop research. *J. S. Afr. Vet. Med. Assoc. 30*:97-111.

Brown, J. M. M. (1964). Advances in geeldikkop (*Tribulosis ovis*) research. 6. Studies on selected aspects of the biochemistry of geeldikkop and enzootic icterus. *J. S. Afr. Vet. Med. Assoc. 35*:507-552.

Budiarso, I. T., Carlton, W. W., and Tuite, J. F. (1970). Phototoxic syndrome induced in mice by rice cultures of *Penicillium viridicatum* and exposure to sunlight. *Pathol. Vet. 7*:531-546.

Budiarso, I. T., Carlton, W. W., and Tuite, J. (1971). The influence of some cultural conditions on the toxigenicity of *Penicillium viridicatum. Toxicol. Appl. Pharmacol. 20*:194-205.

Bullock, G., Eakins, M. N., Sawyer, B. C., and Slater, T. F. (1974). Studies on bile secretion with the aid of the isolated perfused rat liver. 1. Inhibitory action of sporidesmin and icterogenin. *Proc. R. Soc. Lond. B 186*:333-356.

Campbell, A. G., Mortimer, P. H., Smith, B. L., Clarke, J. N., and Ronaldson, J. W. (1975). Breeding for facial eczema resistance? *Proc. Ruakura Farmers' Conf.*, p. 62.

Campbell, A. G., Meyer, H. H., Henderson, H. V., and Wesselink, C. (1981). Breeding for facial eczema resistance—a progress report. *Proc. N.Z. Soc. Anim. Prod. 41*:273-278.

Carlton, W. W. (1980). Penicillic acid, citrinin and xanthomegnin quinone metabolites. A review. *Food Drug Admin., Bur. Vet. Med.* (Tech. Rep.) *80*:233-345.

Carlton, W. W., and Tuite, J. (1970a). Mycotoxicosis induced in guinea pigs and rats by corn cultures of *Penicillium viridicatum. Toxicol. Appl. Pharmacol. 16*:345-361.

Carlton, W. W., and Tuite, J. (1970b). Nephropathy and edema syndrome induced in miniature swine by corn cultures of *Penicillium viridicatum. Pathol. Vet. 7*:68-80.

Carlton, W. W., and Tuite, J. (1977). Metabolites of *P. viridicatum*: toxicology. In *Mycotoxins in Human and Animal Health*, J. V. Rodricks, C. W.

Hesseltine, and M. A. Mehlman (Eds.). Pathotox, Park Forest, Ill., pp. 525-541.

Carlton, W. W., Tuite, J., and Mislivec, P. (1968). Investigations of the toxic effects in mice of certain species of *Penicillium*. *Toxicol. Appl. Pharmacol.* *13*:372-387.

Carlton, W. W., Tuite, J., and Mislivec, P. (1970). Pathology of the toxicosis produced in mice by corn cultures of *Penicillium viridicatum*. *Proc. First U.S.-Japan Conference on Toxic Microorganisms*, M. Hertzberg (Ed.). U.S. Government Printing Office, Washington, D.C., pp. 94-196.

Carlton, W. W., Tuite, J., and Caldwell, R. W. (1972). Mycotoxicosis induced in mice by *Penicillium ochraceum*. *Toxicol. Appl. Pharmacol.* *21*:130-142.

Carlton, W. W., Stack, M. E., and Eppley, R. M. (1976). Hepatic alterations produced in mice by xanthomegnin and viomellein, metabolites of *Penicillium viridicatum*. *Toxicol. Appl. Pharmacol.* *38*:455-459.

Chvapil, M., Ryan, J. N., Elias, S. L., and Peng, Y. M. (1973). Protective effect of zinc on carbon tetrachloride-induced liver injury in rats. *Exp. Mol. Pathol.* *19*:186-196.

Clare, N. T. (1944). Photosensitivity disease in New Zealand. III. The photosensitizing agent in facial eczema. *N.Z. J. Sci. Technol.* *25A*:202-220.

Clare, N. T. (1955). Photosensitisation in animals. *Adv. Vet. Sci.* *2*:182-211.

Clare, N. T. (1959). Photosensitivity diseases in New Zealand. XIX. The susceptibility of New Zealand white rabbits to facial eczema liver damage. *N.Z. J. Agric. Res.* *2*:1249-1256.

Clare, N. T., and Gumbley, J. M. (1962). Some factors which may affect the toxicity of spores of *Pithomyces chartarum* (Berk. & Curt.) M.B. Ellis collected from pasture. *N.Z. J. Agric. Res.* *5*:36-42.

Clare, N. T., and Mortimer, P. H. (1964). The effect of mercury arc radiation and sunlight on the toxicity of water solutions of sporidesmin to rabbits. *N.Z. J. Agric. Res.* *7*:258-263.

Crocker, K. P., Allen, J. G., Petterson, D. S., Masters, H. G., and Frayne, R. F. (1979). Utilization of lupin stubbles by Merino sheep: studies of animal performance, rates and time of stocking, lupinosis, liver copper and zinc, and circulating plasma enzymes. *Aust. J. Agric. Res.* *30*:551-564.

Culvenor, C. C. J., Beck, A. B., Clarke, M., Cockrum, P. A., Edgar, J. A., Frahn, J. L., Jago, M. V., Lanigan, G. W., Payne, A. L., Peterson, J. E., Petterson, D. S., Smith, L. W., and White, R. R. (1977). Isolation of toxic metabolites of *Phomopsis leptostromiformis* responsible for lupinosis. *Aust. J. Biol. Sci.* *30*:269-277.

Culvenor, C. C. J., Smith, L. W., Frahn, J. L., and Cockrum, P. A. (1978). Lupinosis: Chemical properties of phomopsin A, the main toxic metabolite of *Phomopsin leptostromiformis*. In *Effects of Poisonous Plants on Livestock*, R. F. Keeler, K. R. Van Kampen, and L. F. James (Eds.). Academic Press, New York, pp. 565-573.

Cunningham, I. J., Hopkirk, C. S. M., and Filmer, J. F. (1942). Photosensitivity diseases in New Zealand. 1. Facial eczema: its clinical, pathological, and biochemical characterization. *N.Z. J. Sci. Technol.* *24A*: 185-198.

di Menna, M. E., Campbell, J., and Mortimer, P. H. (1970). Sporidesmin production and sporulation in *Pithomyces chartarum*. *J. Gen. Microbiol.* *61*: 87-96.

di Menna, M. E., Mortimer, P. H., and White, E. P. (1977). The genus *Pithomyces*. In *Mycotoxic Fungi, Mycotoxins, Mycotoxicoses*, Vol. 1, T. D. Wyllie and L. G. Morehouse (Eds.). Marcel Dekker, New York, pp. 99-103.

Dobberstein, J., and Walkiewicz, W. (1933). Die Leberveranderungen bei der chronischen Lupinose des Kaninchens. *Virchows Arch. Pathol. Anat.* 291:695-703.

Dodd, D. C. (1960). Photosensitivity diseases in New Zealand. XXI. The susceptibility of the rabbit to facial eczema. *N.Z. J. Agric. Res.* 3:491-497.

Done, J., Mortimer, P. H., and Taylor, A. (1960). Some observations on field cases of facial eczema: liver pathology and determination of serum bilirubin, cholesterol, transaminase and alkaline phosphatase. *Res. Vet. Sci.* 1:76-83.

Done, J., Mortimer, P. H., Taylor, A., and Russell, D. W. (1961). The production of sporidesmin and sporidesmolides by *Pithomyces chartarum*. *J. Gen. Microbiol.* 26:207-222.

Done, J., Mortimer, P. H., and Taylor, A. (1962). The experimental intoxication of sheep with sporidesmin, a metabolic product of *Pithomyces chartarum*. II. Changes in some serum constituents after oral administration of sporidesmin. *Res. Vet. Sci.* 3:161-171.

Durley, R. C., Macmillan, J., Simpson, T. J., Glen, A. T., and Turner, W. B. (1975). Fungal products. XIII. Xanthomegnin, viomellein, rubrosulphin and viopurpurin, pigments from *Aspergillus sulphureus* and *Aspergillus melleus*. *J. Chem. Soc. Perkins Trans.* 1, pp. 163-169.

Eakins, M. N. (1978). The effect of three triterpene acids and sporidesmin on the enzyme activities of rat liver plasma membranes. *Chem. Biol. Interact.* 21:117-124.

Eakins, M. N., Slater, T. F., Sawyer, B., and Bullock, G. (1973). Effects of icterogenin and sporidesmin on the isolated perfused rat liver and on the adenosine triphosphate of rat liver plasma membranes. *Biochem. Soc. Trans.* 1:170-172.

Edwards, J. R., Richards, R. B., Gwynn, R. V. R., and Love, R. A. (1981). A facial eczema outbreak in sheep. *Aust. Vet. J.* 57:392-394.

Ellis, M. B. (1960). Dematiaceous hyphomycetes. *Commonw. Mycol. Inst. Mycol. Paper 76.*

Fairclough, R. J., Sissons, C. H., Holland, P. T., and Ronaldson, J. W. (1978). Studies on sporidesmin metabolism in sheep. *Proc. N.Z. Soc. Anim. Prod.* 38:65-70.

Flynn, D. M. (1962). Facial eczema. Part 1. History of the disease in Victoria. *J. Dept. Agric., Victoria, Aust.* 60:49-50.

Ford, E. J. H. (1965). Changes in the activity of ornithine carbamyl transferase (O.C.T.) in the serum of cattle and sheep with hepatic lesions. *J. Comp. Pathol.* 75:299-308.

Ford, E. J. H. (1974). Activity of gamma-glutamyl transpeptidase and other enzymes in the serum of sheep with liver or kidney damage. *J. Comp. Pathol.* 84:231-243.

Francis, E., Rahman, R., Safe, S., and Taylor, A. (1972). Sporidesmins. Part XII. Isolation and structure of sporidesmin G, a naturally occurring 3,6-epitetrathiopiperazine-2,5-dione. *J. Chem. Soc. Perkin Trans.* 1: 470-472.

Fridrichsons, J., and Mathieson, A. McL. (1962). The structure of sprides-min: causative agent of facial eczema in sheep. *Tetrahedron Lett.*, No. 26, pp. 1265-1268.

Fridrichsons, J., and Mathieson, A. McL. (1965). The structure of the methylenedibromide adduct of sporidesmin at -150°. *Acta Crystallogr.* *18*: 1043-1052.

Gallagher, C. H. (1964). The effect of sporidesmin on liver enzyme systems. *Biochem. Pharmacol.* *13*:1017-1026.

Gardiner, M. R. (1961). Lupinosis—an iron storage disease of sheep. *Aust. Vet. J.* *37*:135-140.

Gardiner, M. R. (1965). The pathology of lupinosis in sheep. *Pathol. Vet.* *2*:417-445.

Gardiner, M. R. (1966a). Mineral metabolism in sheep lupinosis. II. Copper, *J. Comp. Pathol. Ther.* *76*:107-120.

Gardiner, M. R. (1966b). Fungus induced toxicity in lupinosis. *Br. Vet. J.* *122*:508-516.

Gardiner, M. R. (1967a). Lupinosis. *Adv. Vet. Sci.* *11*:85-138.

Gardiner, M. R. (1967b). Cattle lupinosis—A clinical and pathological study. *J. Comp. Pathol. Ther.* *77*:63-69.

Gardiner, M. R., and Nairn, M. (1962). Facial eczema in West Australian sheep. *J. Dept. Agric. W. Aust.* *4(Ser. 3)*:85-88.

Gardiner, M. R., and Parr, W. H. (1967). Pathogenesis of acute lupinosis of sheep. *J. Comp. Pathol. Ther.* *77*:51-62.

Gardiner, M. R., and Petterson, D. S. (1972). Pathogenesis of mouse lupino-sis induced by a fungus (*Cytospora* spp.) growing on dead lupins. *J. Comp. Pathol.* *82*:5-13.

Gardiner, M. R., and Seddon, H. D. (1966). Equine lupinosis. *Aust. Vet. J.* *42*:242-244.

Glenn, B. L., Monlux, A. W., and Panciera, R. J. (1964). A hepatogenous photosensitivity disease of cattle. I. Experimental production and clinical aspects of the disease. *Pathol. Vet.* *1*:469-484.

Glenn, B. L., Panciera, R. J., and Monlux, A. W. (1965). A hepatogenous photosensitivity disease of cattle. II. Histopathology and pathogenesis of the hepatic lesions. *Pathol. Vet.* *2*:49-67.

Gouws, L. (1965). Toxigenic dematiaceae. In *Symposium on Mycotoxins in Foodstuffs*, University of Pretoria, Sess. I: Human nutritional aspects. CSIR, Pretoria, pp. 15-23.

Gregory, P. H., and Lacey, M. E. (1964). The discovery of *Pithomyces chartarum* in Britain. *Br. Mycol. Soc. Trans.* *47*:25-30.

Halder, C. A., Taber, R. A., and Camp, B. J. (1979). High performance liquid chromatography of the mycotoxin, sporidesmin, from *Pithomyces chartarum* (Berk. & Curt.) M.B. Ellis. *J. Chromatogr.* *175*:356-361.

Halder, C. A., Hejtmancik, E., Camp, B. J., and Bridges, C. H. (1980). An alternative extraction procedure for the isolation of sporidesmin from *Pithomyces chartarum*. *N.Z. J. Agric. Res.* *23*:399-402.

Herrmann, H., Hodges, R., and Taylor, A. (1964). Sporidesmin. Part V. The stereochemistry of the bridged dioxopiperazine ring in sporidesmins and gliotoxin. *J. Chem. Soc.*, pp. 4315-4319.

Hodges, R., and Taylor, A. (1964). Sporidesmins. Part IV. The synthesis of 2-amino-5-chloro-3,4-dimethoxybenzoic acid and related compounds. *J. Chem. Soc.*, pp. 4310-4314.

Hodges, R., and Shannon, J. S. (1965). The isolation and structure of sporidesmin C. *Aust. J. Chem. 19*:1059-1066.

Hodges, R., Ronaldson, J. W., Taylor, A., and White, E. P. (1963). Sporidesmins. Part II. The structure of degradation products related to 5-chloro-6,7-dimethoxyisatin. *J. Chem. Soc.*, pp. 5332-5336.

Hodges, R., Ronaldson, J. W., Shannon, J. S., Taylor, A., and White, E. P. (1964). Sporidesmins. Part III. The degradation of sporidesmins under anhydrous acid conditions. *J. Chem. Soc.*, pp. 26-31.

Hodges, R., Shannon, J. S., and Taylor, A. (1966). Sporidesmins. Part VII. Oxindoles derived from sporidesmin B. *J. Chem. Soc. C*, pp. 1803-1805.

Höfle, G., and Röser, K. (1978). Structure of xanthomegnin and related pigments: reinvestigation by ^{13}C nuclear magnetic resonance spectroscopy. *J. Chem. Soc. Chem. Commun.*, pp. 611-612.

Hopkirk, C. S. M. (1936). Facial dermatitis in New Zealand. *N.Z. J. Agric. 52*:98-103.

Hore, D. (1960). Facial eczema. *Aust. Vet. J. 36*:172-176.

Hove, E. L., and Wright, D. E. (1969). Casein, phosphopeptones and phosphoserine protect rats against the mycotoxin, sporidesmin. *Life Sci. 8*: 545-550.

Hurst, E. (1942). *The Poison Plants of New South Wales*. N.S.W. Poison Plants Committee, Sydney.

Ito, Y., Nozawa, Y., and Kawai, K. (1970). Effect of naphthoquinone pigment, xanthomegnin from *Microsporum cookei* on the respiration of rat liver mitochondria. *Experientia 26*:826-827.

Ito, Y., Kawai, K., and Nozawa, Y. (1973). Biochemical studies of pigments of the pathogenic fungus, *Microsporum cookei*. Effect of the 1,4-naphthoquinone pigment, xanthomegnin, on oxidative phosphorylation in rat liver mitochondria. *J. Biochem. 74*:805-810.

Ivie, G. W. (1978). Toxicological significance of plant furocoumarins. In *Effects of Poisonous Plants on Livestock*, R. F. Keeler, K. R. Van Kampen, and L. F. James (Eds.). Academic Press, New York, pp. 475-485.

Jamieson, W. D., Rahman, R., and Taylor, A. (1969). Sporidesmins. Part VIII. Isolation and structure of sporidesmin D and sporidesmin F. *J. Chem. Soc. C*, 1564-1567.

Jonas, W. E., and Erasmuson, A. F. (1977). Immunological studies of sporidesmin: production of anti-bodies to azo-linked derivatives of 2-amino-5-chloro-3,4 dimethoxybenzyl alcohol. *N.Z. Vet. J. 25*:161-164.

Jonas, W. E., and Erasmuson, A. F. (1979). The effect of immunizing mice with a derivative of 2-amino-5-chloro-3,4-dimethoxybenzyl alcohol coupled to some bacteria on sporidesmin induced bilirubinaemia. *N.Z. Vet. J. 27*: 61-63.

Jonas, W. E., and Ronaldson, J. W. (1974). The production of rabbit antibodies to sporidesmin. *N.Z. Vet. J. 22*:111-116.

Just, G., and Day, W. C. (1963). Metabolites of pathogenic fungi. III. The structure of xanthomegnin. *Can. J. Chem. 41*:74-79.

Kawai, K., and Nozawa, Y. (1979). Biochemical studies of pigments from a pathogenic fungus; *Microsporum cookei*. VI. Formation of a xanthomegnin-bypass to the mitochondrial electron transport system. *Experientia 35*: 721-722.

Kawai, K., Atika, T., and Shimonaka, H. (1977). Interaction of xanthomegnin, a pigment from a dermatophyte *Microsporum cookei*, with hydrophobic
and basic amino acids. *Igaku To Seibutsugaku* 95:5-8. *Chem. Abstr.* 90:
67986, 1979.

Kellerman, T. S., van der Westhuizen, G. C. A., Coetzer, J. A. W., Roux,
C., Marasas, W. F. O., Minné, J. A., Bath, G. F., and Basson, P. A.
(1980). Photosensitivity in South Africa. II. The experimental production
of the ovine hepatogenous photosensitivity disease geeldikkop (*Tribulosis
ovis*) by the simultaneous ingestion of *Tribulus terrestris* plants and cultures of *Pithomyces chartarum* containing the mycotoxin sporidesmin.
Onderstepoort J. Vet. Res. 47:231-261.

Kidder, R. W., Beardsley, D. W., and Erwin, T. C. (1961). Photosensitization in cattle grazing frosted common Bermuda grass. *Univ. Fla. Agric.
Exp. Stn. Bull.* 630:1-121.

Kirby, G. W., and Varley, M. J. (1974). Synthesis of tryptophan stereo-
selectively labelled with tritium and deuterium in the β-methylene group;
the steric course of hydroxylation in sporidesmin biosynthesis. *J. Chem.
Soc. Chem. Commun.*, pp. 833-834.

Kishi, Y., Nakatsuka, S., Fukuyama, T., and Havel, M. (1973). A total
synthesis of sporidesmin A. *J. Am. Chem. Soc.* 95:6493-6495.

Kung, H.-C., Chipley, J. R., Latshaw, J. D., Kerr, K. M., and Wilson,
R. F. (1977). Chronic mycotoxicosis in chicks caused by toxins from
Phomopsis grown on soybeans. *J. Comp. Pathol.* 87:325-333.

Lanigan, G. W., Payne, A. L., Smith, L. W., Wood, McR. P., and Petterson,
D. S. (1979). Phomopsin A production by *Phomopsis leptostromiformis* in
liquid media. *Appl. Environ. Microbiol.* 37:289-292.

Leaver, D. D. (1968). Sporidesmin poisoning in the sheep. 1. Changes in
bile secretion and the excretion of sporidesmin in bile. *Res. Vet. Sci.* 9:
255-264.

Lloyd, A. B., and Clarke, R. T. J. (1959). Spore production by *Sporidesmin bakeri* on ryecorn (*Secale cereale*). *N.Z. J. Agric. Res.* 2:1084.

MacKinnon, M. M., and Te Punga, W. A. (1961). Pathological effects of feeding *Pithomyces chartarum* (Berk. & Curt.) M.B. Ellis to guinea pigs.
N.Z. J. Agric. Res. 4:141-146.

Malherbe, W. D., Kellerman, T. S., Kriek, N. P. J., and Haupt, W. H.
(1977). Gamma-glutamyl transpeptidase activity in sheep serum: normal
values and an evaluation of its potential for detecting liver involvement in
experimental lupinosis. *Onderstepoort J. Vet. Res.* 44:29-38.

Marasas, W. F. O. (1974). *Phomopsis leptostromiformis*. In *Mycotoxins*,
I. F. H. Purchase (Ed.). Elsevier, Amsterdam, pp. 111-127.

Marasas, W. F. O. (1977). The genus *Phomopsis*. In *Mycotoxic Fungi, Mycotoxins, Mycotoxicoses*, Vol. 1, T. D. Wyllie and L. G. Morehouse (Eds.).
Marcel Dekker, New York, pp. 111-118.

Marasas, W. F. O. (1978a). Lupinosis in sheep. In *Mycotoxic Fungi, Mycotoxins, Mycotoxicoses*, Vol. 2, T. D. Wyllie and L. G. Morehouse (Eds.).
Marcel Dekker, New York, pp. 213-217.

Marasas, W. F. O. (1978b). Lupinosis in horses. In *Mycotoxic Fungi, Mycotoxins, Mycotoxicoses*, Vol. 2, T. D. Wyllie and L. G. Morehouse (Eds.).
Marcel Dekker, New York, pp. 186-187.

Marasas, W. F. O., and Kellerman, T. S. (1978). Photosensitivity in cattle. In *Mycotoxic Fungi, Mycotoxins, Mycotoxicoses*, Vol. 2, T. D. Wyllie and L. G. Morehouse (Eds.). Marcel Dekker, New York, pp. 73-84.

Marasas, W. F. O., Adelaar, T. F., Kellerman, T. S., Minné, J. A., van Rensburg, I. B. S., and Burroughs, G. W. (1972). First report of facial eczema in sheep in South Africa. *Onderstepoort J. Vet. Res. 39*:107-112.

Marbrook, J., and Matthews, R. E. F. (1962). Loss of sporidesmin from spores of *Pithomyces chartarum* (Berk. & Curt.) M.B. Ellis. *N.Z. J. Agric. Res. 5*:223-236.

Marczewski, H. (1955). Przypadek masowego zatrucia swin lubenem gorzkin. *Med. Wet. (Warsaw) 11*:738. (Cited by Gardiner, 1967a).

McCord, J. M., and Fridovitch, I. (1969). Superoxide dismutase—an enzymic function for erythrocuprein (Hemocuprein). *J. Biol. Chem. 244*:6049-6055.

McCracken, M. D., Carlton, W. W., and Tuite, J. (1974a). *Penicillium viridicatum* mycotoxicosis in the rat. I. Ocular lesions. *Food Cosmet. Toxicol. 12*:79-88.

McCracken, M. D., Carlton, W. W., and Tuite, J. (1974b). *Penicillium viridicatum* mycotoxicosis in the rat. II. Scrotal lesions. *Food Cosmet. Toxicol. 12*:89-98.

McCracken, M. D., Carlton, W. W., and Tuite, J. (1974c). *Penicillium viridicatum* mycotoxicosis in the rat. Hepatic and gastric lesions. *Food Cosmet. Toxicol. 12*:99-105.

McFarlane, D., Evans, J. V., and Reid, C. S. W. (1959). Photosensitivity diseases in New Zealand. XIV. The pathogenesis of facial eczema. *N.Z. J. Agric. Res. 2*:194-200.

Middleton, M. C. (1974a). Effects of the mycotoxin sporidesmin on swelling and respiration of liver mitochondria. *Biochem. Pharmacol. 23*:801-810.

Middleton, M. C. (1974b). The involvement of the disulphide group of sporidesmin in the action of the toxin on swelling and respiration of liver mitochondria. *Biochem. Pharmacol. 23*:811-820.

Mislivec, P. B., and Tuite, J. (1970). Species of *Penicillium* occurring in freshly-harvested and in stored dent corn kernels. *Mycologia 62*:67-74.

Mitchell, K. J., Walshe, T. O., and Robertson, N. G. (1959). Weather conditions associated with outbreaks of facial eczema. *N.Z. J. Agric. Res. 2*:584-604.

Monlux, A. W., Glenn, B. L., Panciera, R. J., and Corcoran, J. E. (1963). Bovine hepatogenous photosensitivity associated with the feeding of alfalfa hay. *J. Am. Vet. Med. Assoc. 142*:989-994.

Mortimer, P. H. (1962). The experimental intoxication of sheep with sporidesmin; a metabolic product of *Pithomyces chartarum*. III. Some changes in cellular components and coagulation properties of the blood, in serum and in liver function. *Res. Vet. Sci. 3*:269-286.

Mortimer, P. H. (1963). The experimental intoxication of sheep with sporidesmin; a metabolic product of *Pithomyces chartarum*. IV. Histological and histochemical examinations of orally-dosed sheep. *Res. Vet. Sci. 4*:166-185.

Mortimer, P. H. (1965). Effects of sporidesmin on rats. *Ann. Rep., Animal Research, N.Z. Dept. Agric.*, p. 36.

Mortimer, P. H. (1967). Susceptibility of chickens to sporidesmin. *Ann. Rep. Res. Div. N.Z. Dept. Agric.*, p. 55.

Mortimer, P. H. (1969). Facial eczema: its effects within the animal. *N.Z. J. Agric. 119*: 22-25.

Mortimer, P. H. (1970). The toxic effects of sporidesmin on mice. *N.Z. J. Agric. Res. 13*: 437-447.

Mortimer, P. H. (1971). Sporidesmin poisoning in Jersey cows. *Ann. Rep. Res. Div. N.Z. Dept. Agric.*, p. 22.

Mortimer, P. H., and Collins, B. S. (1968). The in vitro toxicity of the sporidesmins and related compounds to tissue-culture cells. *Res. Vet. Sci. 9*: 136-142.

Mortimer, P. H., and Smith, B. L. (1981). Facial eczema in deer. *Proc. Ruakura Farmers' Conf.*, pp. 109-112.

Mortimer, P. H., and Stanbridge, T. W. (1968). Excretion of sporidesmin given by mouth to sheep. *J. Comp. Pathol. 78*: 505-512.

Mortimer, P. H., and Stanbridge, T. A. (1969). Changes in biliary secretion following sporidesmin poisoning in sheep. *J. Comp. Pathol. 79*: 267-275.

Mortimer, P. H., and Taylor, A. (1962). The experimental intoxication of sheep with sporidesmin; a metabolic product of *Pithomyces chartarum*. I. Clinical observations and findings at post-mortem examinations. *Res. Vet. Sci. 3*: 147-160.

Mortimer, P. H., Taylor, A., and Shorland, F. B. (1962). Early hepatic dysfunction preceding biliary obstruction in sheep intoxicated with sporidesmin. *Nature (Lond.) 194*: 550-551.

Mortimer, P. H., Manns, E., and Coe, B. D. (1978a). Manipulation of liver metabolism in relation to ruminant toxicology. *Proc. N.Z. Soc. Anim. Prod. 38*: 59-64.

Mortimer, P. H., White, E. P., and di Menna, M. E. (1978b). Pithomycotoxicosis "facial eczema" in sheep. In *Mycotoxic Fungi, Mycotoxins, Mycotoxicoses*, Vol. 2, T. D. Wyllie, L. G. Morehouse (Eds.). Marcel Dekker, New York, pp. 195-203.

Muchiri, D. J., Bridges, C. H., Ueckert, D. N., and Bailey, E. M. (1980). Photosensitization of sheep on kleingrass pasture. *J. Am. Vet. Med. Assoc. 177*: 353-354.

Nakatsuka, S., Fukuyama, T., and Kishi, Y. (1974). A total synthesis of d,l-sporidesmin B. *Tetrahedron Lett. (16)*: 1549-1552.

Ng, A. S., Just, G., and Blank, F. (1969). VII. On the structures and stereochemistry of xanthomegnin, vioxanthin and viopurpurin, pigments from *Trichophyton violaceum*. *Can. J. Chem. 47*: 1223-1227.

Papadimitriou, J. M., and Petterson, D. S. (1976). The pathogenesis of lupinosis in the rat. *J. Pathol. 118*: 35-47.

Papadimitriou, J. M., Walters, M. N.-I., Petterson, D. S., and Gardiner, M. R. (1973). Hepatic ultrastructural changes in murine lupinosis. *J. Pathol. 111*: 221-228.

Papadimitriou, J. M., Bradshaw, R. D., Petterson, D. S., and Gardiner, M. R. (1974). A histological, histochemical and biochemical study of the effect of the toxin of lupinosis on murine liver. *J. Pathol. 112*: 43-53.

Parle, J. N. (1973). Facial eczema. 1. Problems in control with fungicides. *Proc. Ruakura Farmers' Conf.*, pp. 46-49.

Parle, J. N., and di Menna, M. E. (1972a). Fungicides and the control of *Pithomyces chartarum*. I. Laboratory trials. *N.Z. J. Agric. Res. 15*: 48-53.

Parle, J. N., and di Menna, M. E. (1972b). Fungicides and the control of *Pithomyces chartarum*. II. Field trials. *N.Z. J. Agric. Res.* 15:54-63.

Percival, J. C., and Thornton, R. H. (1958). Relationships between the presence of fungal spores and a test for hepatotoxic grass. *Nature (Lond.) 182*:1095-1096.

Perrin, D. D. (1957). Photosensitivity diseases in New Zealand. X. The guinea pig as an experimental animal in the investigation of facial eczema. *N.Z. J. Sci. Technol. 38A*:669-681.

Peters, J. A., and Mortimer, P. H. (1970). Sporidesmin poisoning in sheep. The effects on lipid and bile acid metabolism as reflected in serum. *Res. Vet. Sci. 11*:183-187.

Peterson, J. E. (1978). *Phomopsis leptostromiformis* toxicity (Lupinosis) in nursling rats. *J. Comp. Pathol. 88*:191-203.

Peterson, J. E., and Lanigan, G. W. (1976). Effects of *Phomopsis rossiana* toxin on the cell cycle and on the pathogenesis of lupinosis in mice. *J. Comp. Pathol. 86*:293-306.

Petterson, D. S., and Coackley, W. (1973). Changes in cell cultures produced by toxic lupin extracts. *Aust. J. Exp. Biol. Med. Sci. 51*:513-520.

Petterson, D. S., Howlett, R. M., Robertson, T. A., and Papadimitriou, J. M. (1979). Alteration in cell division, morphology and motility induced by the toxic principle of lupinosis. *Aust. J. Exp. Biol. Med. Sci. 57*: 211-223.

Pöhland, A. E., and Mislivec, P. (1976). Metabilites of various *Penicillium* species encountered on foods. *Adv. Chem. Ser. 149*:110-143.

Przybylska, M., Gopalakrishna, E. M., Taylor, A., and Safe, S. (1973). X-ray crystallographic determination of the stereochemistry of the tetrathio-bridge in sporidesmin G. *J. Chem. Soc. Chem. Commun.*, pp. 554-555.

Rafiquzzaman, M. (1974). Experimental *Penicillium viridicatum* toxicosis in rats. *Acta Vet. Scand. Suppl. 47, 15*:1-36.

Rahman, R., Safe, S., and Taylor, A. (1969). Sporidesmins. Part IX. Isolation and structure of sporidesmin E. *J. Chem. Soc. C*, pp. 1665-1668.

Rahman, R., Safe, S., and Taylor, A. (1970). Separation of polythiadioxo-piperazine antibiotics by thin-layer chromatography. *J. Chromatogr. 53*: 592-594.

Rahman, R., Safe, S., and Taylor, A. (1978). Sporidesmins. Part 17. Isolation of sporidesmin H and sporidesmin J. *J. Chem. Soc. Perkin Trans. 1*, pp. 1476-1479.

Raper, K. B., and Fennell, D. I. (1965). *The Genus Aspergillus.* Williams & Wilkins, Baltimore.

Raper, K. B., and Thom, C. (1949). *A Manual of the Penicillia.* Williams & Wilkins, Baltimore.

Rimington, C., and Quin, J. I. (1934). Studies in the photosensitization of animals in South Africa. VII. The nature of the photosensitizing agent in geeldikkop. *Onderstepoort J. Vet. Sci. Anim. Ind. 3*:137-157.

Rimington, C., and Quin, J. I. (1937). Dikoor or geeldikkop on grassveld pastures. *J. S. Afr. Vet. Med. Assoc. 8*:141-146.

Robbers, J. E., Hong, S., Tuite, J., and Carlton, W. W. (1978). The production of xanthomegnin and viomellein by species of *Aspergillus* correlated with mycotoxicosis produced in mice. *Appl. Environ. Microbiol. 36*:819-823.

Ronaldson, J. W. (1975). Sporidesmins. XIV. Modifications to the opened
-S-S- bridge of sporidesmin for coupling to proteins by transacylation.
Aust. J. Chem. 28:2043-2050.

Ronaldson, J. W. (1976). Sporidesmins. XV. The ^{13}C nuclear magnetic
resonance spectra of sporidesmin and sporidesmin-D. The evidence in
the spectra for strain imposed by an epidithio bridge. *Aust. J. Chem. 29*:
2307-2314.

Ronaldson, J. W. (1978). Fungal metabolite chemistry: sporidesmin, crepido-
tine. D. Phil. thesis, Waikato University, New Zealand, (a) 13-16; (b) 101-
102; (c) 106-107; (d) 152-157.

Ronaldson, J. W. (1981). Sporidesmins. XVIII. The infrared solution spectra (4000-1600 cm^{-1}) of sporidesmin, sproidesmin-B, sporidesmin-D and
sporidesmin-E. *Aust. J. Chem. 34*:1215-1222.

Ronaldson, J. W. (1982). The purification of sporidesmin by preparative
high pressure liquid chromatography. *J. Chem. Tech. Biotechnol. 32*:
556-558.

Ronaldson, J. W., and Fyvie, A. A. (1973). Recovery of sporidesmin from
side fractions. *Lab. Pract. 22*:734-735.

Ronaldson, J. W., Taylor, A., White, E. P., and Abraham, R. J. (1963).
Sporidesmins. Part I. Isolation and characterisation of sporidesmin and
sporidesmin B. *J. Chem. Soc.*, pp. 3172-3180.

Ross, D. J. (1960). A study of the physiology of *Pithomyces chartarum*
(Berk. and Curt.) M.B. Ellis. 1. Nutrition. *N.Z. J. Sci. 3*:15-25.

Ross, D. J., and Thornton, R. H. (1962). Study of the physiology of
Pithomyces chartarum (Berk. and Curt.) M.B. Ellis. 3. Production of
toxin on some laboratory media. *N.Z. J. Sci. 5*:165-183.

Russell, D. W. (1966). Cyclodepsipeptides. *Q. Rev. (Lond.) 20*:559-576.

Safe, S., and Taylor, A. (1971). Sporidesmins. Part XI. The reaction of
triphenylphosphine with epipolythiodioxopiperazines. *J. Chem. Soc. C*,
pp. 1189-1192.

Sammes, P. G. (1975). Naturally occurring 2,5-dioxopiperazines and related
compounds. *Fortschr. Chem. Org. Naturst. 32*:51-118.

Sandos, J., Clare, N. T., and White, E. P. (1959). Improved procedure for
the beaker test for facial eczema toxicity. *N.Z. J. Agric. Res. 2*:623-626.

Shaw. B. A., and Wright, D. E. (1972). The action of sporidesmin on rumen
protozoa. *N.Z. J. Agric. Res. 15*:512-515.

Simpson, J. E. V. (1946). Danger in grazing broom corn millet. *N.Z. J.
Agric. 73*:243.

Sinclair, D. P., and Parle, J. N. (1967). Thiabendazole and facial eczema.
Proc. Ruakura Farmers' Conf., pp. 160-170.

Slater, T. F. (1972). *Free Radical Mechanisms in Tissue Injury*. Pion,
London, p. 254.

Slater, T. F., and Griffiths, D. B. (1963). Effects of sporidesmin on bile
flow rate and composition in the rat. *Biochem. J. 88*:60-61.

Slater, T. F., Sträuli, U. D., and Sawyer, B. (1964). Sporidesmin poisoning
in the rat. I. Chemical changes. *Res. Vet. Sci. 5*:450-472.

Slater, T. F., Sawyer, B. C., Delaney, V. B., and Bullock, G. (1968). The
effects of sporidesmin and icterogenin on bile flow in the isolated prefused
rat liver. *Biochem. J. 110*:15-16.

Smith, B. L. (1977). Toxicity of zinc in ruminants in relation to facial
eczema. *N.Z. Vet. J. 25*:310-312.

Smith, B. L. (1980). Effect of high concentrations of zinc sulphate in the drinking water of grazing yearling dairy cattle. *N.Z. J. Agric. Res. 23:* 175-178.

Smith, B. L., Embling, P. P., Towers, N. R., Wright, D. E., and Payne, E. (1977). The protective effect of zinc sulphate in experimental sporidesmin poisoning of sheep. *N.Z. Vet. J. 25:*124-127.

Smith, B. L., Coe, B. D., and Embling, P. P. (1978). Protective effect of zinc sulphate in a natural facial eczema outbreak in dairy cows. *N.Z. Vet. J. 26:*314-315.

Stack, M. E., and Mislivec, P. B. (1978). Production of xanthomegnin and viomellein by isolates of *Aspergillus ochraceus, Penicillium cyclopium* and *Penicillium viridicatum. Appl. Environ. Microbiol. 36:*552-554.

Stack, M. E., Eppley, R. M., Dreifuss, P. A., and Pöhland, A. E. (1977). Isolation and identification of xanthomegnin, viomellein, rubrosulphin and viopurpurin as metabolites of *Penicillium viridicatum. Appl. Environ. Microbiol. 33:*351-355.

Stack, M. E., Brown, N. L., and Eppley, R. M. (1978). High pressure liquid chromatographic determination of xanthomegnin in corn. *J. Assoc. Off. Anal. Chem. 61:*590-592.

Steffert, I. J. (1970). Facial eczema: the extent of damage in dairy cows. *Dairyfarming Ann.*, pp. 95-102.

Stutzenberger, F. J. (1974). Ribonucleotide reductase of *Pithomyces chartarum*: requirement for B_{12} coenzyme. *J. Gen. Microbiol. 81:*501-503.

Stutzenberger, F. J., and Parle, J. N. (1972). Binding of benzimidazole compounds to conidia of *Pithomyces chartarum. J. Gen. Microbiol. 73:*85-94.

Stutzenberger, F. J., and Parle, J. N. (1973a). Effect of 2-substituted benzimidazoles on the fungus *Pithomyces chartarum. J. Gen. Microbiol. 76:*197-209.

Stutzenberger, F. J., and Parle, J. N. (1973b). Binding of 2-(2-oxazolyl) benzimidazole by *Pithomyces chartarum* conidia and release during germination. *J. Gen. Microbiol. 78:*199-201.

Sutton, B. C., and Gibson, I. A. S. (1977). *Pithomyces chartarum. Commonw. Mycol. Inst. Description Pathog. Fungi Bact. 540.*

Taylor, A. (1967). The chemistry and biochemistry of sporidesmins and other 2,5-epidithia-3,6-dioxopiperazines. In *Biochemistry of Some Foodborne Microbiol Toxins*, R. I. Mateles and G. N. Wogan (Eds.). MIT Press, Cambridge, Mass., pp. 69-107.

Taylor, A. (1971). The toxicology of sporidesmins and other epipolythia-dioxopiperazines. In *Microbial Toxins*, Vol. 7, S. Kadis, A. Ciegler, and S. J. Ajl (Eds.). Academic Press, New York, pp. 337-376.

Thornton, R. H., and Percival, J. C. (1959). A hepatotoxin from *Sporidesmin bakeri* capable of producing facial eczema disease in sheep. *Nature (Lond.) 183:*63.

Towers, N. R. (1970a). Tissue distribution and excretion of sulphur 35 after administration of sporidesmin S[35] to guinea pigs. *N.Z. J. Agric. Res. 13:* 182-191.

Towers, N. R. (1970b). Tissue distribution and excretion of radioactivity in rats dosed with sporidesmin S[35]. *N.Z. J. Agric. Res. 13:*428-436.

Towners, N. R. (1972). Absorption and excretion of [35]sulphur by biliary fistulated rats and guinea pigs following administration of ^{35}S-sporidesmin. *Life Sci. 11:*691-698, part II.

Towers, N. R. (1977). Effect of zinc on the toxicity of the mycotoxin sporidesmin to the rat. *Life Sci. 20*:413-417.

Towers, N. R., and Smith, B.L. (1978). The protective effect of zinc sulphate in experimental sporidesmin intoxication of lactating dairy cows.
N.Z. Vet. J. 26:199-202.

Towers, N. R., and Stratton, G. C. (1978). Serum gamma-glutamyltrans-
ferase as a measure of sporidesmin induced liver damage in sheep. *N.Z.
Vet. J. 26*:109-112.

Towers, N. R., and Wright, D. E. (1969). Biosynthesis of sporidesmin from
amino acids. *N.Z. J. Agric. Res. 12*:275-280.

Towers, N. R., Smith, B. L., Wright, D. E., and Sinclair, D. P. (1975). Preventing facial eczema by using zinc. *Proc. Ruakura Farmers' Conf. 27*:57-61.

Turner, J. C., and Green, R. S. (1974). Effect of hexachlorobenzene on
microsomal enzyme systems. *Biochem. Pharmacol. 23*:2387-2390.

Ueno, K., Giam, C. S., and Taber, W. A. (1974). Absence of sporidesmin
in a Texas isolate of *Pithomyces chartarum*. *Mycologia 66*:360-362.

van Jaarsveld, A. B., and Knox-Davies, P. S. (1974). Resistance of lupins
to *Phomopsis leptostromiformis*. *Phytophylactia 6*:55-60.

van Rensberg, I. B. J., Marasas, W. F. O., and Kellerman, T. S. (1975).
Experimental *Phomopsis leptostromiformis* mycotoxicosis of pigs. *J. S. Afr.
Vet. Med. Assoc. 46*:197-204.

van Tonder, E. M., Basson, P. A., and van Rensburg, I. B. J. (1972).
Geeldikkop: experimental induction by feeding the plant *Tribulus terrestris* L. (Zygophyllaceae). *J. S. Afr. Vet. Med. Assoc. 43*:363-375.

van Warmelo, K. T., and Marasas, W. F. O. (1972). *Phomopsis leptostromiformis*: the causal fungus of lupinosis, a mycotoxicosis in sheep. *Mycologia
64*:316-324.

van Warmelo, K. T., and Marasas, W. F. O., Adelaar, T. F., Kellerman, T. S.,
van Rensberg, I. B. J., and Minné, J. A. (1970). Experimental evidence
that lupinosis of sheep is a mycotoxicosis caused by the fungus *Phomopsis
leptostromiformis* (Kühn) Bubák). *J. S. Afr. Vet. Med. Assoc. 41*:235-247.

Watanabe, R. (1978). Pigments from the pathogenic fungus, *Microsporum
cookei*. Protein binding and spectral changes of xanthomegnin. *Gifu
Daigaku Igakubu Kiyo 26*:654-670. *Chem. Abstr. 91*:204976, 1979.

White, E. P. (1958). Photosensitivity diseases in New Zealand. XII. Concentration of the facial eczema poison. *N.Z. J. Agric. Res. 1*:433-446.

White, E. P., Mortimer, P. H., and di Menna, M. E. (1977). Chemistry of
the sporidesmins. In *Mycotoxic Fungi, Mycotoxins, Mycotoxicoses*, Vol.
1, T. D. Wyllie and L. G. Morehouse (Eds.). Marcel Dekker, New York,
pp. 427-447.

Wirth, J. C., Beesley, T. E., and Anand, S. R. (1965). The isolation of
xanthomegnin from several strains of the dermatophyte, *Trichophyton
rubrum*. *Phytochemistry 4*:505-509.

Wood, P. McR., and Allen, J. G. (1980). Control of ovine lupinosis: use
of a resistant cultivar of *Lupinus albus*-cv. Ultra. *Aust. J. Exp. Agric.
Anim. Husb. 20*:316-318.

Worker, N. A. (1960). A hepatotoxin causing liver damage in facial eczema
of sheep. *Nature (Lond.) 185*:909-910.

Wright, D. E. (1969). Phytotoxicity of sporidesmin. *N.Z. J. Agric. Res. 12*:
275-280.

Wright, D. E., and Forrester, I. T. (1965). Some biochemical effects of sporidesmin. *Can. J. Biochem. 43*:881-888.

Zeeck, A., Russ, P., Laatsch, H., Loeffler, W., Wehrle, H., Zaehner, H., and Holst, H. (1970). Isolierung des Antibioticums *semi*-Vioxanthin aus *Penicillium citreo-viride* und Synthese des Xanthomegnins. *Chem. Ber. 112*:957-978.

Zimmermann, J. L., Carlton, W. W., and Tuite, J. (1976). Mycotoxicosis produced in mice by cultural products of an isolate of *Aspergillus ochraceus. Food Cosmet. Toxicol. 14*:571-575.

Zimmermann, J. L., Carlton, W. W., Tuite, J., and Fennell, D. I. (1977). Mycotoxic diseases produced in mice by species of the *Aspergillus ochraceus* group. *Food Cosmet. Toxicol. 15*:411-418.

Zimmermann, J. L., Carlton, W. W., and Tuite, J. (1978). Mycotoxicosis produced in rats by cultural products of an isolate of *Aspergillus ochraceus. Food Cosmet. Toxicol. 16*:449-461.

Zimmermann, J. L., Carlton, W. W., and Tuite, J. (1979a). Mycotoxicosis produced in swine by cultural products of an isolate of *Aspergillus ochraceus.* I. Clinical observations and pathology. *Vet. Pathol. 16*: 583-592.

Zimmermann, J. L., Carlton, W. W., and Tuite, J. (1979b). Mycotoxicosis produced in swine by cultural products of an isolate of *Aspergillus ochraceus.* II. Clinicopathologic changes. *Vet. Pathol. 16*:702-709.

12

ALLERGIC CONTACT DERMATITIS FROM PLANTS

HAROLD BAER

Food and Drug Administration, Bethesda, Maryland

I. INTRODUCTION

There are many types of allergic reactions, but they can be roughly categorized into immediate and delayed. When a sensitized individual is challenged by an antigen, immediate reactions may occur within minutes, and rarely later than an hour, while delayed reactions take at least 6-8 hr to begin to develop and reach a maximum between 24 and 72 hr. In humans whose last reaction occurred many years ago, as in poison ivy, the maximum reaction may take up to 5 days (Johnson et al., 1972). Individuals with contact sensitivity to plants may exhibit both types of reactions, although the delayed reactions are the most common type.

Immediate reactions result from the presence of antibodies in the serum portion of blood; therefore, the reaction may be transferred from sensitive to normal individuals using serum from the sensitive individual. Delayed reactions result from the presence of specifically sensitive cells and therefore can be transferred from sensitive to normal individuals using cells, usually lymph node cells. The mechanism of delayed contact sensitivity, including the variour types of cells involved, has been reviewed (Godfrey and Gell, 1978; Parker and Turk, 1974; Turk, 1980).

Contact allergic reactions from plants result from sensitization with substances of defined chemical structure and relatively low molecular weight, usually under 500 daltons. The immunologic mechanisms for these reactions have had very little study using plant-derived substances. The very extensive literature using chemically defined synthetic substances is undoubtedly applicable and can be used to give insights into the immunochemistry and immunologic mechanisms.

Any substance capable of combining with proteins is potentially immunologically reactive. Thus such diverse substances as dinitrofluorobenzene (Godfrey and Gell, 1978), 1-dodecene-1,3-sultone (Connor et al., 1975), nitroolefines (Josephson, 1966), and merthiolate (Godfrey and Gell, 1978) can induce contact sensitivity; the common structural feature of these substances is the presence of a reactive group capable of covalent binding to proteins, which usually occurs at the basic ε-amino groups of lysine or with free sulfhydryl groups. The capability of binding to proteins is a necessary but not sufficient criterion for a substance to be a sensitizer. To induce sensitivity by contact, the substance must be an irritant, or must be mixed with such a substance (Friedlaender et al., 1973). If it reacts too rapidly with proteins, it may induce circulating antibody and no delayed contact sensitivity, or there may be antibody and delayed contact sensitivity (Godfrey and Baer, 1971). Although immediate contact sensitivity is not usually considered to be common, some substances, such as oxazolone (Baer et al., 1976) and occasionally poison ivy as well as certain other naturally occurring contact sensitizers, are known to give both types of reactions. On the basis of model compound studies, this would imply antibody formation as well as cellular immunity. This has been demonstrated for oxazolone (Baer et al., 1976) but has not yet been proven by direct experimentation for the poison ivy active substance.

Some substances must first be converted to a second substance which is the sensitizer. Thus pentadecylcatechol, one of the sensitizers from poison ivy, cannot couple to proteins, but must first be oxidized to a quinone which is spontaneously reactive (Byers et al., 1978; Byk and Dawson, 1968; Liberato et al., 1980; Mason and Lada, 1954). The mechanism of this oxidation is not clear. Thus pentadecylcatechol can be oxidized with silver oxide to the quinone, which spontaneously couples to proteins. The oxidation also takes place slowly in the presence of oxygen. The slow darkening of solutions of these compounds is the result of air oxidation, followed by polymerization. The process can be catalyzed by enzymes, some of which occur in the plants themselves (McNair, 1917; Suminokura, 1936). Whether or not phenol oxidases that occur in animals can use the alkylcatechols of the Anacardiaceae as substrates is not certain, but would help to explain their reactivity. Other substances are altered by ultraviolet radiation and then become sensitizers. These substances are called photoallergic sensitizers, and if they are of plant origin, phytophotosensitizers (Willis, 1971). Once light has altered the substance, the mechanism of inducing the allergic response is the same as that for other sensitizers in that the transformed substance must be capable of binding to proteins. The various theories of the role of ultraviolet (UV) radiation in photosensitization by simple chemicals have been reviewed (Mitchell, 1975; Willis, 1971).

A number of other physicochemical properties are known to affect the sensitization process. Substances that induce intense delayed sensitivity are lipid rather than water soluble. Water-soluble substances that induce antibody can be made lipid soluble by introducing long hydrocarbon chains and then are found to induce cellular sensitivity of the delayed type (Dailey and Hunter, 1974). This has also been demonstrated for a series of n-alkylcatechols used as model compounds to study poison ivy sensitivity. It was found that those with n-alkyl side chains longer than 11 carbon atoms were very water insoluble and potent sensitizers (Baer et al., 1966, 1967).

The stereochemical configuration of the molecules may also play a role. The best example of this resulted from the study of sesquiterpene lactone sensitizers, where it was found that individuals reacting to parthenin did not react to its diasteromer hymenin (Subba et al., 1978). This was also investigated with synthetic alkylcatechols used as model compounds. Guinea pigs sensitized with catechols having linear and cyclic side chains with equal numbers of carbon atoms did not show reciprocal cross-reactivity. Animals sensitive to cyclohexylmethylcatechol reacted equally well to octylcatechol since the latter has a long alkyl side chain capable of bending to fit a variety of shapes. Since the catechol with the cyclic side chain has a more rigid structure and cannot alter its conformation, it has a much reduced activity in animals sensitized to the compound with the linear side chain (Baer et al., 1968). It has also been found that the d and l forms of usnic acid show little cross-reactivity (Mitchell, 1966).

The number of double bonds in the alkyl side chain of the catechol sensitizers also seems to play a role in antigenic specificity (Johnson et al., 1972) when these substances are tested in humans, but the effect is minimal or nonexistent in sensitive guinea pigs (Baer and Hooton, 1975).

In addition to chemical factors, biological factors can affect the immunologic response. It has been found that monkeys either do not become sensitive to poison ivy or occasionally can be made slightly sensitive (Bowser et al., 1964). On the other hand, guinea pigs are readily sensitized and become as sensitive as humans. That there are genetic factors within a species that can affect the sensitization process can be presumed from studies on model compounds in animals. Age of the individuals clearly plays a role. Young guinea pigs were readily sensitized to pentadecylcatechol, but aged guinea pigs were not (Baer and Bowser, 1963). Young humans are also readily sensitized to poison ivy, but older individuals are difficult to sensitize (Epstein, 1961; Zisserman, 1940). Even the season of the year in which contact with the sensitizer occurs may be important (Baer and Hooton, 1976).

The ability of some substances to be absorbed through the skin either directly or in the presence of other substances such as dimethyl sulfoxide (DMSO) undoubledly has some effect on the sensitization process (Stoughton,

1965). Studies with pentadecylcatechol in guinea pigs showed that this substance was rapidly absorbed through the unbroken skin and appeared in
serum, urine, and so on (Godfrey et al., 1971).

The route by which the individual first comes into contact with a substance
has a dramatic effect on the immunological outcome. Sensitizers placed directly
on the skin usually result in the development of contact sensitivity. This is
true whether they are in the pure state or dissolved in a solvent such as vegetable or mineral oils or acetone. In fact, the routine testing of humans to try
to determine the substance causing a dermatitis may result in sensitization to
the testing substance (Agrup et al., 1969). If they are introduced for the
first time by the oral route, the result is usually immune tolerance (Bowser
and Baer, 1963). The same may occur if the substance is injected subcutaneously in an oil solution (Bowser and Baer, 1963). This means that the individual not only fails to become sensitive, but that future attempts to sensitize
the individual by any route will fail. Chemical alteration of a sensitizer may
convert it to a substance inducing immune tolerance (Baer et al., 1977). The
cellular mechanism of this process is uncertain, but it has been shown that
sensitization probably involves Langerhans' cells of the skin and almost certainly the lumph nodes draining the skin area on which the sensitizer was
placed (Friedlaender et al., 1973).

One problem that always arises in studying contact sensitivity is whether
the substance is an irritant or a sensitizer. The appearance of the skin
lesion may be very similar in both cases. In animals it is easier to make this
determination than in humans. An irritant produces the reaction the first
time it is applied to the skin of normal animals. If the dose is large enough,
all individuals react, and as the dose decreases, fewer react. If a sufficient
number of animals are tested, a 50% irritant dose can be calculated. The sensitizers produce a reaction only after the individual has been sensitized, so
that the first contact with the substance produces no reaction. Some plants
have only irritants that produce a dermatitis that is sometimes confused with
allergic contact dermatitis. The Euphorbiaceae, Urticaceae, and the Thymelaceae, which include the stinging nettles, possess only irritants. They may
be in the form of silica-containing hairs that produce mechanical irritation, or
they may contain irritant chemical substances such as phorbol esters, histamine, hydroxytryptamine, and acetylcholine (Evans, 1978; Evans et al., 1975;
Hecker, 1977; Saxena et al., 1965; Thurston, 1974, 1976; Thurston and
Lersten, 1969). On the other hand, true allergic contact dermatitis may sometimes be misdiagnosed as an irritant reaction. Contact dermatitis from airborne
oleoresins has been confused with photodermatitis (Hjorth et al., 1976).

The presence of a substance in a plant that might be considered to be a
sensitizer does not automatically assure that sensitization will occur. It has
been claimed that wheat bran contains an alkylresorcinol (Wenkert et al., 1964)
but wheat bran is not known to cause dermatitis. The data in this one report
is limited, but it is possible that the alkylresorcinol is covalently linked to
some other substance rendering it inactive. However, many sensitizers are
also irritant and, at least for guinea pigs, a nonirritant may not sensitize by
the dermal route (Friedlaender et al., 1973). The distinguishing factor is
usually the dose. The dose required to elicit an immunologically mediated contact sensitization reaction in a previously sensitized animal is so low that
normal animals do not react. This is also true for humans, but the data are

frequently very difficult to obtain. Furthermore, in animals, the sensitization phenomenon is transferable from sensitive to normal animals with cells (Godfrey and Gell, 1978; Turk, 1980); this is not true for the irritant reaction.

II. *ANACARDIACEAE*

A. Biology

The Anacardiaceae family has the most familiar and common plants that cause contact dermatitis. The poison ivy, oak, and sumac (*Toxicodendrum radicans, diversilobum,* and *vernix*) are the most widespread and are found almost exclusively in North America, Japan, some of the East Indian Islands, China, and Taiwan. In the United States, a number of species and subspecies have been described that have a localized distribution. For example, *T. radicans* is largely an east coast plant, while *T. diversilobum* is limited to the west coastal region. There is a complete description of these plants with maps of their location (Gillis, 1971, 1975).

In addition to *Toxicodendrum*, there are several other genera that yield phenolic substances that can cause contact dermatitis. These include the cashew nut tree (*Anacardium occidentale*), the so-called marking-nut tree (*Semecarpus anacardium*), the mango (*Mangifera indica*), and the poison wood tree (*Metopium toxiferum*). Complete listings have been published (Evans and Schmidt, 1980). It is interesting that no members of this family are indigenous to Europe.

In addition to the poisonous members, the family also includes some very common nonpoisonous plants. Several species of sumac (*Rhus*) are widespread in the United States, as well as several species of *Cotinus* and *Schinus*.

There are two other families that contain substances chemically related to the Anacardiaceae that may occasionally induce contact dermatitis—the Ginkgoaceae, of which the common Ginkgo tree is most familiar, and the Proteaceae, of which several genera are involved (Evans and Schmidt, 1980).

The poisonous substances of poison ivy, oak, and sumac occur in poison canals (Fig. 1) that are found in the roots, stems, and leaves. Consequently, touching the unbroken plant parts causes no problems. Those plants old enough to have thick wood are safe on the outer woody portions. Any breaks resulting from mechanical trauma or insect attack causes leakage of the poison onto the outer surfaces. The cashew nut itself is perfectly safe; the poisonous oil occurs in the nut shell. The fruit of the mango is also safe, the poison occurring in the exocarp or skin of the fruit. Lacquer from the lacquer tree of China and Japan (*T. vernicifluum*) comes from the sap. The tree itself is a close relative of the poison sumac, *T. vernix*.

B. Chemistry

The poisonous substances of *Toxicodendron* and *Semecarpus* are alkyl catechols (Table 1). The alkyl side chains have 13-17 carbon atoms and exist in four forms (Fig. 2): a saturated side chain and side chains with one, two, or three unconjugated double bonds, except for *T. vernicifluum* (Japanese lac), in which two or three double bonds are conjugated (Baer et al., 1980; Billets

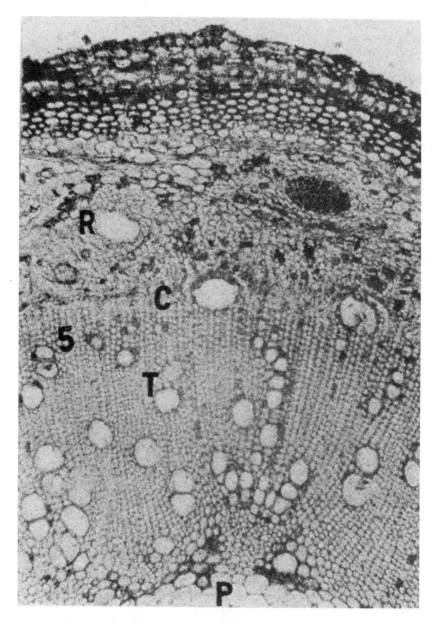

FIGURE 1 Transverse section through stem showing cork cambium, tracheal
tube (T), pericycle with sclerenchyma cells or bast fibers and thin-walled
pericycle parenchyma, phloem with resin duct (R), cambium (C), and pith
(P). (From J. B. McNair, ed., Rhus Dermatitis, University of Chicago Press,
1923, p. 28.)

TABLE 1 Composition of Urushiols from *Toxicodendra* sp. and *Metopium toxiferum*

Urushiol sample	Bis-TMSi-C$_{15}$-Catechols					Bis-TMSi-C$_{17}$-Catechols			
				Molecular weight					
	464	462	460	458	492	490	488	486	484
				Side chain					
	Saturated	Monoene	Diene	Triene	Saturated	Monoene	Diene	Triene	Tetraene
Poison ivy									
CI-1	0	25·8	67·8	3·3	0·2	0·7	1·8	0·4	0
CI-2	0	41·6	50·3	3·4	0	0·8	3·1*	0·8	0
CI-3	6·0	13·5	63·0	16·5	0	0·3	0·7	0	0
MI-1	0	15·4	17·9	62·5	0	0	0	4·2	0
FI-1	7·6	25·2	38·8	4·3	0	0·8	22·0*	1·3	0
FI-2	0	3·1	83·1	9·4	0	0·4	2·6*	1·4	0
FI-3	6·0	11·6	60·0	12·5	0	2·1	6·7*	1·1	0
Poison oak									
CO-1	22·3	8·4	2·7	0	2·1	10·7	18·7	35·1	0
MO-2	3·4	1·4	0·2	0·6	0	5·7	24·9	62·5	1·3
Poison sumac									
CS-1	15·3	41·4	32·7	10·6*	0	0	0	0	0
Poison wood									
CW-1	3·9	48·6	41·4	0	0	1·8	3·3†	1·0	0

* Two isomeric compounds of different R_t were observed for this M$^+$.
† Three isomeric compounds of different R_t were observed for this M$^+$.

Source: Gross et al., 1975.

Chemical Structure —

Molecular Weight —

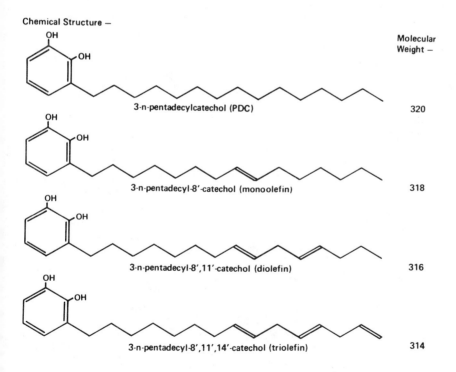

3-n-pentadecylcatechol (PDC)   320

3-n-pentadecyl-8'-catechol (monoolefin)   318

3-n-pentadecyl-8',11'-catechol (diolefin)   316

3-n-pentadecyl-8',11',14'-catechol (triolefin)   314

FIGURE 2 Chemical structure of the four pentadecylcatechols in poison ivy.

et al., 1976; Corbett and Billets, 1975; Dawson, 1956; Gross et al., 1975; Sunthankar and Dawson, 1954). The compound with a completely saturated side chain is generally in lowest concentration, and those with two or three double bonds in highest concentration. Recently, it was shown that there were two monoene compounds (Baer et al., 1980). An examination of individual leaves of poison ivy for urushiol content showed that the older lowest leaves had 10-15% as much as the newer upper leaves (Baer et al., 1980).

All of the long-chain alkylcatechols are extremely insoluble in water (Baer et al., 1967), but seem to increase in solubility in the presence of serum albumin. For purposes of skin testing they are usually applied in a solvent, such as acetone, alcohol, or oil. These substances become reactive when oxidized. They are then capable of binding to basic amino groups of proteins or to sulfhydryl groups (Byk and Dawson, 1968; Liberato et al., 1980). This oxidation is greatly enhanced by the presence of oxidases in the plants McNair, 1917; Suminokura, 1936). The oxidation is of great importance. When complete, it almost completely eliminates the toxic activity of the urushiol and it has great commercial importance in the conversion of the sap of the Jamanese lacquer tree to the black lacquer used on many decorative Japanese products. The same property of oxidation to a black insoluble substance accounts for the use of the poison of the fruit of the Indian marking nut tree (*S. anacardium*) as an indelible ink (Gillis, 1971, 1975).

Cashew nut shell liquid (*Anacardium occidentale*) is different from the previous two genera, in that the sensitizers are alkyl resorcinols instead of catechols. The side chains contain 15 carbons and are composed of a least four species: saturated and mono-, di-, and triolefins with unconjugated double bonds (Dawson, 1956; Symes and Dawson, 1953; Tyman, 1973; Tyman and Kiong, 1978). These substances also polymerize to yield polymers of commercial importance. The mechanism by which the alkyl resorcinols oxidize is not established, since they cannot form quinone-like substances. It is assumed that they are oxidized to a trihydroxy form, which then can yield quinone structures.

In addition to the resorcinol sensitizers, *Anacardium* species contain alkyl phenols, some with a carboxyl group in the ring (Dawson, 1956; Tyman and Kiong, 1978). The variation in composition of cashew nut shell liquid obtained from different sources has been studied (Tyman and Kiong, 1978).

The analysis for these substances in plant material is usually carried out by extracting fresh leaves with a solvent such as alcohol or acetone. Leaves that are dried or maintained cold for long periods have markedly reduced catechol content. The extracts can then be analyzed by gas chromatography, high-pressure liquid chromatography, or a combination of gas chromatography and mass spectrometric analysis (Baer et al., 1963, 1980; Billets et al., 1976; Gross et al., 1975; Lloyd et al., 1980; Tyman and Kiong, 1978).

III. *GINKGOACEAE*

The fruit and seeds of the ginkgo tree, which are produced on the female trees, contain several phenolic substances, including 6(pentadec-8-enyl)-2,4 dihydroxybenzoic acid, which is probably the sensitizer (Becker and Skipworth, 1975; Gellerman et al., 1976; Loev and Dawson, 1958; Morimoto et al., 1968), as well as an alkyl phenol and hydroxybenzoic acid (ginkolic acid) (Morimoto et al., 1968).

IV. PROTEACEAE

The only substances thus far isolated from several genera of this family are alkyl resorcinols, whose properties would be similar to those described above.

Table 2 is a list of plants from the foregoing three families that have been examined for the presence of catechols and resorcinols.

A. Miscellaneous Substances

In addition to the above, there are plant sensitizers of the hydroquinone, quinone, and naphthoquinone types. These include the sensitizer of *Primula obconica*, called primin, an alkyl methoxyquinone (Hjorth et al., 1969; Schultz et al., 1977), and the sensitizer of *Phacelia crenulata*, geranylhydroquinone, which occurs in the trichomes (Reynolds and Rodriguez, 1979; Reynolds et al., 1980). A hydroquinone, alliodorin, has been isolated from the leaves and flowers of *Wigandia kunthii* Gomez et al., 1980).

B. Clinical Immunology

These substances are the most important cause of contact dermatitis in humans, and can be important as occupational problems (Epstein, 1974; Klingman, 1955; Zisserman, 1940). Between 50 and 75% of the population of the United States are sensitive to poison ivy or oak (dermatitis to poison sumac is relatively uncommon because of its limited distribution) and this is entirely due to the extraordinarily wide distribution of these plants (Gillis, 1971); tests on 65 Eskimos showed none with sensitivity and these plants do not grow where Eskimos live (Heywood et al., 1977). That the catechols and resorcinols are the plant sensitizers has been demonstrated by sensitization of animals and humans with the substances isolated from plants and with synthetic substances. Cross-reactivity of sensitized individuals to various plants is at least partly dependent on the length of the alkyl side chain (Baer et al., 1967); the longer chains are the more potent sensitizers. It is interesting to note that most of the naturally occurring catechols and resorcinols have 15-17 carbon chains which were shown to be optimal for sensitization in guinea pigs, and which undoubtedly accounts for the extensive cross reactivity of sensitive individuals, not only to catechols from different sources, but also of catechol-sensitive individuals to resorcinal sensitizers (Howell, 1959; Hurtado, 1968; Keil et al., 1944, 1945; Olivera, 1953). This extensive cross-reactivity is another reason for the widespread dermatitis due to these compounds, because once an individual is sensitized to any one of them, reactivity to others is certain; thus sensitization to poison ivy of the United States makes an individual sensitive to Japanese lac and the Indian marking-nut tree.

The importance of the double bonds in the alkyl side chain was demonstrated by a study of the reactivity of naturally poison ivy-sensitive humans to pentadecylcatechols isolated from leaves having a saturated chain and compounds

TABLE 2 Plants Producing Phenolic Sensitizing Substances

| Plant | Sensitizer[a] | | References |
	Catechol	Resorcinol	
Anacardiaceae			
Anacardium occidentale L.		x	Symes and Dawson, 1953 a, b; Tyman, 1973; Tyman Kiong, 1978
Camposperma auriculata[b]			Lamberton, 1959b
Gluta renghas	x		Backer and Haack, 1941a
Holigarna arnottiana	x		Majima, 1922
Melanorrhoea usitata	x		Majima, 1922
Metopium toxiferum	x		Gross et al., 1975
Pentaspadon motleyi	c		Backer and Haack, 1941b
Pentaspadon officinalis	c		Lamberton, 1959a
Semecarpus anacardium	x		Pillay and Siddiqui, 1931
Semecarpus heterophylla	x		Backer and Haack, 1938
Semecarpus travancorica	x		Majima, 1922
Toxicodendron diversilobum	x		Corbett and billets, 1975; Gross et al., 1975
Toxicodendron radicans	x		Billets et al., 1976; Gross et al., 1975; Symes and Dawson, 1954
Toxicodendron striatum	x		Nakano et al., 1970
Toxicodendron succedaneum	x		Majima, 1922
Toxicodendron vernicifluum	x		Majima, 1922
Toxicodendron vernix	x		Gross et al., 1975
Ginkgoaceae			
Ginkgo biloba		x	Gellerman et al., 1976; Loev and Dawson, 1958
Proteaceae			
Cardwellia sublimis		x	Cirigottis et al., 1974
Grevillea hilliana		x	Cirigottis et al., 1974
Grevillea banksii		x	Cirigottis et al., 1974
Grevillea pteridifolia		x	Cirigottis et al., 1974
Grevillea pyramidalis		x	Occolowitz and Wright, 1962
Grevillea robusta		x	Ridley et al., 1968; Ritchie et al., 1965
Hakea persiehana		x	Cirigottis et al., 1974
Opisthiolepis heterophylla		x	Cirigottis et al., 1974
Persoonia elliptica		x	Brongersman-Oosterhoff, 1967
Persoonia linearis		x	Cirigottis et al., 1974
Petrophila shirleyae		x	Cirigottis et al., 1974

[a]All plants seem to produce multiple substances which usually differ in the number of carbon atoms and double bonds in the side chain. In a few cases, monophenolic substances may be present. For detailed structures, see the references.

[b]This plant had alkyl phenols and was the only plant of these three genera that contained an alkyl hydroquinone, which is probably the sensitizer.

[c]Thus far, only monophenols have been isolated from these plants. Since these have not been shown to be sensitizers by direct test, it is likely that further studies would yield additional substances.

with one, two, or three double bonds. The compound with two double bonds reacted in all individuals, with three double bonds in all but one, with one double bond in about 70%, and with none in about 40% of those tested. This may possibly be due to the fact that the compound with two double bonds is most frequently in highest concentration in leaves (Baer et al., 1980; Gross et al., 1975).

Because this type of dermatitis is so prevalent, methods of treatment have been sought. For those who acquire the disease on rare occasions, various lotions or topical steroid drugs provide relief until the symptoms subside, which usually takes 1-2 weeks. For those who suffer regular bouts because of their occupation (i.e., foresters) or because of other activities (i.e., nature walks, camping, etc.), methods of prevention are important. Attempts have been made to apply barrier creams to prevent the plant sap from coming in contact with the skin or to apply chemical oxidizing substances or even an oxidase enzyme (McNair, 1917); these methods have met with no success.

Attempts at immunological intervention appear to provide more hope. The oral administration of pure urushiol in capsules reduced sensitivity in many individuals (Epstein et al., 1974), but it required large doses and the reduction in sensitivity was not permanent, and therefore required continuing doses. Products with large quantities of urushiol are not yet regularly available commercially. Injection of urushiol-containing extracts is generally unacceptable in sensitive individuals, since it induces large reactions at the injection site.

A second approach to prevention is the induction of immune tolerance. This must be done in individuals who are not yet sensitive but who anticipate becoming sensitive because of their activities. This has been accomplished in children (Epstein et al., 1981), exactly as it was accomplished in guinea pigs (Bowser and Baer, 1963).

Another approach to the induction of a tolerant state is to employ urushiols that have been modified chemically. It was shown in guinea pig studies that the introduction of various groups into position 6 of the benzene ring of pentadecylcatechol resulted in substances that were very poor sensitizers, but instead induced a tolerant state so that the animals could not be sensitized with the related potent sensitizer pentadecylcatechol (Baer et al., 1977). This approach of chemical modification of sensitizers may open the way for effective therapeutic products for humans.

V. SESQUITERPENE LACTONES

In addition to dermatitis resulting from contact with members of the Anacardiaceae, there are a relatively small number of individuals who get the disease from contact with other plants (Grater, 1975; Roed-Peterson and Hjorth, 1976; Shelmire, 1940; Towers, 1979). Chemical studies of plants (other than Anacardiaceae) that cause contact dermatitis revealed that most of these contain a group of compounds that are classified as sesquiterpene lactones (Evans and Schmidt, 1980; Heywood et al., 1977; Mitchell and Rook, 1979; Rodriquez et al., 1977). Most of the plants that have these sensitizers are from the Compositae, but some are from Lauraceae, Magnoliaceae, and Jubilaceae (Mitchell et al., 1972; Mitchell, 1975; Mitchell and Rook, 1979). These sensitizers are present in such diverse plants as chrysanthemums and liverworts. This has caused some confusion in determining the source of the offending plant. For example, sensitivity to forest workers, at first thought to be due to the trees, was actually caused by liverworts growing on the trees (Mitchell et al., 1972).

In at least some, if not all Compositae, the sensitizer is present in the trichomes and capitale sessile glands found on leaves and stems (Fig. 3) (Reynolds and Rodriguez, 1979; Rodriguez et al., 1976). The substances responsible for ragweed dermatitis may also be found in pollen, since it has been claimed that pollen oil can cause the dermatitis (Brown et al., 1931; Fromer and Burrage, 1953).

The purpose of these substances in the plants appears to be the obvious. The trichomes and their poisons protect the plant from predation by herbivores and pathogens (Levin, 1973; Rodriguez, 1977). Where the pressure is

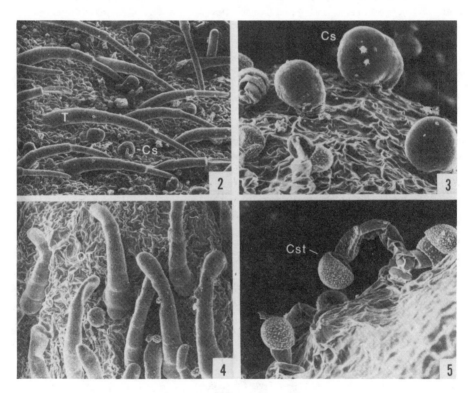

FIGURE 3 Glandular and nonglandular trichomes on leaf, phyllary, and palea of *Parthenium hysterophorus*. (2) Portion of mature leaf with numerous capitate-sessile glands and multicellular, glandular trichomes (X200). (3) Bulbous, capitate-sessile gland on disk-floret pale (X500). (4) Portion of phyllary with oblong, nonglandular trichomes (X297). (5) Capitate-stalked, nonglandular trichome, present on ray-floret pale and achene (X500). All specimens were air dried and coated with gold-paladium (40:60). Cs, capitate-sessile; T. multicellular trichome; OT, oblong, nonglandular trichome; Cst, capitate-stalked trichome. (From Rodriguez et al., 1976b.)

greatest, as in tropical plants, the concentration is highest. In the tropical members of the genus *Parthenium*, the quantity of lactones per species may be 1-2% of dry weight, whereas in temperate members it is as low as 0.001% (Pillay and Siddique, 1931).

A. Chemistry

Many hundreds of sesquiterpene lactones have been isolated and structures established (Fischer et al., 1979; Heywood et al., 1977). However, it has been found that certain structural features are regularly associated with those lactones capable of causing contact dermatitis; examples of the various classes are shown in Fig. 4. As with other simple chemical substances, the prime requisite for a sensitizer is that it possess one or more reactive groups capable of binding to proteins. Those sesquiterpene lactones that are sensitizers have an excocyclic α-methylene group conjugated to the γ-lactone ring. These substances have been shown to react with the sulfhydryl group of cysteine and, almost certainly, the ε-amino group of lysine (Dupuis et al., 1974, 1980). The importance of the lactone ring with an exocyclic double bond is shown by the fact that α-methylene butyrolactone, the simplest lactone of this type, is the sensitizer that occurs in tulip bulbs (Bergman et al., 1967; Brongersman-Oosterhoff, 1967; Jhorth and Wilkinson, 1968). It is not a potent sensitizer since the dermatitis occurs only in workers who handle large numbers of tulip bulbs. Furthermore, the concentration of this sensitizer seems to vary in the different varieties, some varieties being more potent than others (Rook, 1961). Stereochemical configuration is also important since it was shown that individuals in India sensitive to *Parthenium* react to parthenin—the major sesquiterpene lactone present in this plant—but are unreactive to hymenin, its diasteriomer. The latter is present in some South American species of *Parthenium*. Whether hymenin is capable of inducing sensitivity is not known and has not been determined by direct test (Subba et al., 1978).

Analysis for these substances is usually carried out by extracting plant material with an organic solvent such as acetone, petroleum ether, or benzene, followed by column or thin-layer chromatography. Individual compounds have

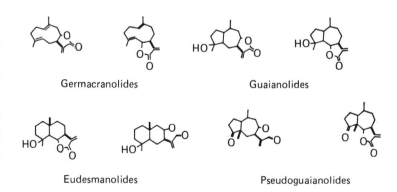

Germacranolides   Guaianolides

Eudesmanolides   Pseudoguaianolides

FIGURE 4 Basic skeletal classes of sesquiterpene lactones.

distinctive infrared and especially nuclear magnetic resonance (Fischer et al.,
1979; Sohi et al., 1979). Recently, some new reagents have been developed
for the visualization of sesquiterpene lactones on thin-layer chromatograms
(Picman et al., 1980).

B. Immunology and Clinical Application

Although the majority of cases of contact dermatitis due to plants results
from the Anacardiaceae (almost exclusively poison ivy and oak in the United
States), other plants have long been known to be involved, especially in
selected population groups. Plants containing sesquiterpene lactones account
for most of these. Thus some forest workers may get dermatitis from liverworts (Knoche et al., 1969; Mitchell et al., 1972), farmers from ragweeds
(Mitchell et al., 1971), and florists from chrysanthemums (Bleumink et al.,
1973; Mitchell and Rook, 1979).

Unlike poison ivy, which sensitizes a large part of the population, even
those who appear to have had only casual contact, plants containing sesquiterpene lactones require long and intimate contact before sensitization occurs;
the only instance where there appears to be a severe problem is sensitization
to *Parthenium hysterophorus* in India. Even in the latter case, most of the
individuals affected appear to be males who work in the fields, and therefore
receive prolonged exposure.

The reason for the low sensitization rate is not known. Some of the plants,
such as the ragweeds, are so prevalent that numerous individuals must come
into contact with them. Consequently, it would appear that sesquiterpene
lactones are poor sensitizers. Unfortunately, almost all of the information concerning the sensitizing properties of these substances comes from tests on
naturally sensitized humans. Extensive studies in animals would provide the
controlled, detailed information required to understand thoroughly the immunology of these substances. One such study has been carried out (Dupuis
et al., 1974), but many more are required.

At this time, it seems clear that these substances sensitize some individuals,
and that sensitivity can be diagnosed by patch testing (Fisher, 1973; Mitchell,
1975). Because these compounds are so widely distributed, patch testing with
crude oleoresin or pure chemicals may not necessarily prove the source of the
problem. This may require a detailed examination of all plants in the environment of the affected individual, as was shown by forest workers who thought
they were sensitive to trees but were really sensitive to the liverworts growing
on them.

There is no recommended specific immunologic treatment for this sensitivity
since it has not been shown that either injecting or feeding these substances
is effective in reducing or eliminating the sensitive state. In fact, administering sesquiterpene lactones in large quantities may be hazardous, due to
their rather profound biological properties. They are known to have poisoned
livestock feeding on the plants, have antimicrobial activity, and are inhibitors
of cellular metabolism (Lee et al., 1977; Rodriguez et al., 1976). It has been
claimed that they are antitumor agents (Lee et al., 1977). Avoidance, therefore, is the only solution to the problem at this time.

VI. MISCELLANEOUS

Occasional instances of contact dermatitis due to a variety of plants are known (Fisher, 1973; Grater, 1975; Shelmire, 1940). Sensitivity to the radish, *Raphanus sativus*, has been reported and is presumably due to the presence of alkyl isothiocyanate, a substance or closely related substances that occur in many members of the mustard family, Cruciferae (Mitchell and Jordan, 1974).

Perfumes have long been known to sensitize some individuals. Heptin-carboxylic acid methylester has been identified as one such substance and its sensitizing properties as well as that of related substances has been thoroughly studied in guinea pigs (Griepentrog, 1959). Atranorin, present in *Evernia prunastri*, is also found in perfumes and is a sensitizer (Dalquist and Fregart, 1980). An oxidation product of carene sensitizes a few individuals. The carene itself is present in turpentine, but the sensitizer appears only following oxidation to the peroxide form (Hellerstrom et al., 1955).

REFERENCES

Agrup, G., Fregert, S., and Rorsman, H. (1969). Sensitization by routine patch testing with ether extracts of *Primula obconica*. *Br. J. Dermatol.* *81*:897.

Backer, H. J., and Haack, N. H. (1938). Le principe toxique des fruits de renghas (*Semecarpus heterophylla* B1). *Rec. Trav. Chim.* 57:225-232.

Backer, H. J., and Haack, N. H. (1941a). Le principe toxique due Gluta renghas. *Rec. Trav. Chim.* *60*:656-660.

Backer, H. J., and Haack, N. H. (1941b). La structure de l'acide peland-janique. *Rec. Trav. Chim.* *60*:678-688.

Baer, H., and Bowser, R. T. (1963). Antibody production and the development of contact skin sensitivity in guinea pigs of various ages. *Science* *140*:1211-1212.

Baer, H., and Hooton, M. (1975). The relationship between human and guinea pig sensitivity to poison ivy and poison oak. In *Animal Models in Dermatology*, H. Maibach (Ed.). Churchill Livingstone, Edinburgh, pp. 53-55.

Baer, H., and Hooton, M. (1976). Effect of season of immunization of the induction of delayed contact sensitivity in guinea pigs. *Int. Arch. Allergy* *51*:140-143.

Baer, H., Srinivasan, S., Bowser, R. T., and Karman, A. (1963). The quantitative assay of the active principles of poison ivy by biological and gas chromatographic methods. *J. Allergy 34*(3):221-224.

Baer, H., Watkins, R. C., and Bowser, R. T. (1966). Delayed contact sensitivity to catechols and resorcinols. The relationship of structure and immunization procedure to sensitizing capacity. *Immunochemistry 3*:479-485.

Baer, H., Watkins, R. C., Kurtz, A. P., Byck, J. S., and Dawson, C. R. (1967). Delayed contact sensitivity to catechols. III. The relationship of side-chain length to sensitizing potency of catechols chemically related to the active principles of poison ivy. *J. Immunol.* *99*:370-375.

Baer, H., Watkins, R. C., and Bowser, R. T. (1966). Delayed contact sensitivity to catechols and resorcinols. The relationship of structure and immunization procedure to sensitizing capacity. *Immunochemistry 3*: 479-485.

Baer, H., Stone, S. H., and Malik, F. (1976). Early and late contact sensitivity reactions in guinea pigs sensitized to oxazolone. *J. Immunol. 117*: 1159-1163.

Baer, H., Hooton, M. L., Dawson, C. R., and Lerner, D. I. (1977). The induction of immune tolerance in delayed contact sensitivity by the use of chemically related substances of low immunogenicity. *J. Invest. Dermatol. 69*: 215-218.

Baer, H., Hooton, M., Fales, H., Wu, A., and Schaub, F. (1980). Catecholic and other constituents of the leaves of *Toxicodendron radicans* and variation of urushiol concentrations within one plant. *Phytochemistry 19*: 799-802.

Becker, L. E., and Skipworth, G. B. (1975). Ginkgo-tree dermatitis, stomatitis and proctitis. *JAMA 231*: 1160-1163.

Bergman, H. H., Beijersbergen, J. C. M., Overmeen, J. C., and Sijpesteijn, A. K. (1967). Isolation and identification of α-methylene butyrolactone, a fungitoxic substance from tulip bulbs. *Rec. Trav. Chim. 86*: 709-714.

Billets, S., Craig, J. C., Corbett, M. D., and Mickery, J. F. (1976). Component analysis of the urushiol content of poison ivy and oak. *Phytochemistry 15*: 533-535.

Bleumink, E., Mitchell, J. C., and Nater, J. P. (1973). Contact dermatitis to chrysanthemums. *Arch. Dermatol. 108*: 220-222.

Bowser, R. T., and Baer, H. (1963). Contact sensitivity and immunologic unresponsiveness in adult guinea pigs to a component of poison ivy extract, 3-*n*-pentadecylcatechol. *J. Immunol. 91*: 791-794.

Bowser, R. T., Kirschstein, R. L., and Baer, H. (1964). Contact sensitivity in *Rhesus* monkeys to poison ivy extracts and fluorodinitrobenzene. *Proc. Soc. Exp. Biol. Med. 117*: 763-766.

Brongersman-Oosterhoff, U. W. (1967). Structure determination of the allergenic agent isolated from tulip bulbs. *Rec. Trav. Chim. 86*: 705-708.

Brown, A., Milford, E. L., and Coca, A. F. (1931). Studies on contact dermatitis. I. The nature and etiology of pollen dermatitis. *J. Allergy 2*: 301-309.

Byers, V. S., Epstein, W. L., Castagnoli, N., and Baer, H. (1979). In vitro studies of poison oak immunity. I. In vitro reaction of human lymphocytes to urushiol. *J. Clin. Invest. 64*: 1437-1448.

Byk, J. S., and Dawson, C. R. (1968). Assay of protein-quinone coupling involving compounds structurally related to the active principle of poison ivy. *Anal. Biochem. 25*: 123-135.

Cannon, J. R., and Metcalf, B. W. (1971). Phenolic constituents of *Persoonia elliptica* (Proteaceae). *Aust. J. Chem. 24*: 1925-31.

Champlin, R., and Hunter, R. L. (1975). Studies of the composition of adjuvants which selectively enhance delayed-type hypersensitivity to lipid conjugated protein antigens. *J. Immunol. 114*: 76-80.

Cirigottis, K. A., Cleaver, L., Corrie, J. E. T., Grasby, R. G., Green, G. H., Mock, J., Ningirawath, S., Read, R. W., Ritchie, E., Taylor, W. C., Vadasz, A., and Webb, U. R. G. (1974). Chemical studies of the

Proteaceae. VII. An examination of the woods of 17 species for resorcinol derivatives. *Aust. J. Chem.* 27:345-355.

Connor, D. S., Ritz, H. L., Ampulski, R. S., Kowollik, H. G., Lim, P., Thomas, D. W., and Parkhurst, R. (1975). Identification of certain sultones as the sensitizers in an alkyl ethoxy sulfate. *Fette Seifen Anstrichm.* 77:25-29.

Corbett, M. D., and Billets, D. (1975). Characterization of poison oak urushiol. *J. Pharm. Sci.* 64:1715-1718.

Dahlquist, I., and Fregart, S. (1980). Contact allergy to atranorin in lichens and perfumes. *Contact Dermatitis* 6:111-119.

Dailey, M. O., and Hunter, R. L. (1974). The role of lipid in the induction of hapten-specific delayed hypersensitivity and contact sensitivity. *J. Immunol.* 112:1526-1534.

Dawson, C. R. (1956). The chemistry of poison ivy. *Trans. N.Y. Acad. Sci.* 18:427-433.

Dupuis, G., Mitchell, J. C., and Towers, G. H. M. (1974). Reaction of alantolactone, an allergenic sesquiterpene lactone, with some amino acids. Resultant loss of immunologic reactivity. *Can. J. Biochem.* 52:575-581.

Dupuis, G., Benezra, C., Schleiver, G., and Stampf, J. (1980). Allergic contact dermatitis to α-methylene-γ-butyrolactones. Preparation of alantolactone-protein conjugates and induction of contact sensitivity in the guinea pig by an alantolactone-skin protein conjugate. *Mol. Immunol.* 17:1045-1051.

Epstein, W. L. (1961). Contact-type delayed hypersensitivity in children and infants: induction of *Rhus* sensitivity. *Pediatrics* 27:51-53.

Epstein, W. L. (1974). Poison oak and poison ivy dermatitis as an occupational problem. *Cutis* 13:544-548.

Epstein, W. L., Baer, H., Dawson, C. R., and Khurana, R. G. (1974). Poison oak hyposensitization. *Arch. Dermatol.* 109:356-360.

Epstein, W. L., Byers, V. S., and Baer, H. (1981). Induction of persistent tolerance to urushiol in humans. *J. Allergy Clin. Immunol.* 68:20-25.

Evans, F. J. (1978). The irritant toxins of blue *Euphorbia* (*E. coerulescens* Haw.). *Toxicon* 16:51-57.

Evans, F. J., and Schmidt, R. J. (1980). Plants and plant products that induce contact dermatitis. *Planta Med.* 38:289-316.

Evans, F. J., Kinghorn, A. D., and Schmidt, R. J. (1975). Some naturally occurring skin irritants. *Acta Pharmacol. Toxicol.* 37:3250-256.

Fischer, N. H., Olivier, E. J., and Fischer, H. K. (1979). The biogenesis and chemistry of sesquiterpene lactones. *Prog. Chem. Org. Nat. Prod.* 38:47-390.

Fisher, A. E. (1973). *Contact Dermatitis*. Lea & Febiger, Philadelphia.

Friedlaender, M. H., Chisari, F. V., and Baer, H. (1973). The role of the inflammatory response of skin and lymph nodes in the induction of sensitization to simple chemicals. *J. Immunol.* 111:164-170.

Fromer, J. L., and Burrage, W. S. (1953). Ragweed oil dermatitis. *J. Allergy* 24:425.

Gellerman, J. L., Anderson, W. H., and Schenk, H. (1976). 6-(Pentadec-8-enyl)-2,4 dihydroxybenzoic acid from seeds of *Ginkgo biloba*. *Phytochemistry* 15:1959-1961.

Gillis, W. T. (1971). The systematics and ecology of poison-ivy and the poison oaks (*Toxicodendron,* Anacardiaceae). *Rhodora* 73:72-159; 161-237; 370-443; 465-540.

Gillis, W. T. (1975). Poison ivy and its kin. *Arnoldia* 35:73-123.

Godfrey, H. P., and Baer, H. (1971). The effect of physical and chemical properties of the sensitizing substance on the induction and elicitation of delayed contact sensitivity. *J. Immunol.* 106:431-441.

Godfrey, H. P., and Gell, P. G. H. (1978). Cellular and molecular events in the delayed-onset sensitivities. *Rev. Physiol. Biochem. Exp. Pharmacol.* 84:1-92.

Godfrey, H. P., Baer, H., and Watkins, R. C. (1971). Delayed hypersensitivity to catechols. V. Absorption and distribution of substances related to poison ivy extracts and their relation to the induction of sensitization and tolerance. *J. Immunol.* 106:71-102.

Gomez, F., Quijano, L., Calderon, J. S., and Rios, T. (1980). Terpenoids isolated from *Wigandia kunthii. Phytochemistry* 19:2202-2203.

Grater, W. C. (1975). Hypersensitivity dermatitis from American weeds other than poison ivy. *Ann. Allergy* 35:159-164.

Griepentrog, F. (1959). Allergiestudien mit einfachen chemischen Substanzer. 3. Mitteilungen: Heptin- und Octin Carbonsauremethylester. *Allergie Asthma* 5:224-226.

Gross, M., Baer, H., and Fales, H. M. (1975). Urushiols of poisonous Anacardiaceae. *Phytochemistry* 14:2261-2266.

Hecker, E. (1977). New toxic, irritant and cocarcinogenic diterpene esters from Euphorbiaceae and from Thymelaceae. *Pure Appl. Chem.* 49:1423-1431.

Heinbecker, P. (1928). Studies in hypersensitiveness. XXXIV. The susceptibility of Eskimos to an extract from *Toxicodendron radicans* (L). *J. Immunol.* 15:365-367.

Hellerstrom, S., Thyresson, N., Blohm, S., and Widmark, G. (1955). On the nature of the eczematogenic component of oxidized Δ^3-carene. *J. Invest. Dermatol.* 24:217-224.

Heywood, V. H., Harborne, J. B., and Turner, B. L. (1977). *The Biology and Chemistry of the Compositae.* Academic Press, New York.

Hjorth, N., and Wilkinson, D. S. (1968). Contact dermatitis. IV. Tulip fingers, hyacinth itch and lily rash. *Br. J. Dermatol.* 80:696-698.

Hjorth, N., Fregert, S., and Schildknecht, H. (1969). Cross-sensitization between synthetic primin and related quinones. *Acta Dermato-Venereol.* 49:552-555.

Hjorth, N., Roed-Petersen, J., and Thomsen, K. (1976). Airborne contact dermatitis from Compositae oleoresins simulating photodermatitis. *Br. J. Dermatol.* 95:613-620.

Howell, J. B. (1959). Cross-sensitization in diverse poisonous members of the sumac family (Anacardiaceae). *J. Invest. Dermatol.* 32:21-25.

Hurtado, I. (1968). Studies on the biological activity of *Rhus striata* (Manzanillo). II. Skin response to patch test in humans. *Int. Arch. Allergy* 33:209-216.

Johnson, R. A., Baer, H., Kirkpatrick, C. H., Dawson, C. R., and Khurana, R. G. (1972). Comparison of the contact allergenicity of the four pentadecyl-catechols derived from poison ivy urushiol in human subjects. *J. Allergy Clin. Immunol. 49*(1):27-35.

Josephson, A. S. (1966). Immunologic response to the nitro-olefins. *J. Immunol. 96*:699-706.

Keil, H., Wasserman, D., and Dawson, C. R. (1944). The relation of chemical structure in catechol compounds and derivatives to poison ivy hypersensitiveness in man as shown by the patch test. *J. Exp. Med. 80*:275-287.

Keil, H., Wasserman, D., and Dawson, C. R. (1945). The relation of hypersensitiveness to poison ivy and to the pure ingredients in cashew nut shell liquid and related substances. *Inc. Med. 14*:825-830.

Kinghorn, A. D. (1979). *Toxic Plants*, Columbia University Press, New York.

Kligman, A. M. (1955). Poison ivy dermatitis. *Arch. Dermatol. 77*:149-180.

Knoche, H., Ourisson, G., Perold, W., Maleville, J., and Fousseraau, J. (1969). Allergenic component of liverwort: a sesquiterpene lactone. *Science 166*:239-240.

Lamberton, J. A. (1959a). Studies on the optically active compounds of Anacardiaceae exudates. VI. The exudate from *Pentaspadon officinalis* Holmes. *Aust. J. Chem. 12*:234-239.

Lamberton, J. A. (1959b). Studies of the optically active compounds of Anacardiaceae exudates. V. Further investigation of the exudate from *Camposperma auriculata* Hook f. *Aust. J. Chem. 12*:224-233.

Langman, R. E. (1978). Cell mediated immunity and the major histocompatibility complex. *Rev. Physiol. Biochem. Pharmacol. 81*:2-37.

Lee, K., Hall, I., Mar, E., Starnes, C. O., El-Gebaly, S. A., Waddell, T. G., Hadgraft, R. I., Ruffner, C. G., and Weidner, I. (1977). Sesquiterpene antitumor agents: inhibitors of cellular metabolism. *Science 196*:533-535.

Levin, D. A. (1973). The role of trichomes in plant defense. *Q. Rev. Biol. 48*:3-15.

Liberato, D. J., Byers, V. S., Dennick, R. G., and Castagnoli, N., Jr. (1980). Regiospecific attack of nitrogen and sulfur nucleophiles on quinones derived from poison oak/ivy catechols (urushiols) and analogues as models for urushiol-protein conjugate formation. *J. Med. Chem. 24*:28-33.

Lloyd, H. A., Denny, C., and Krishna, G. (1980). A simple liquid chromatographic method for the analysis and isolation of the unsaturated components of anacardic acid. *J. Liq. Chromatogr. 3*:1497-1504.

Loev, B., and Dawson, C. R. (1958). An investigation of the geometrical configuration of the olefinic components of cardanol and some observations concerning ginkgol. *J. Am. Chem. Soc. 80*:643-645.

Loukar, A., Mitchell, J. C., and Caluan, C. D. (1974). Contact dermatitis from *Parthenium hysterophorus*. *Trans. St. Johns Hosp. Dermatol. Soc. 60*:43-53.

Majima, R. (1922). Über den Hauptbestandteil des Japan-Lacks. IX. Mit.: Chemische Untersuchung den verschiedenen naturlichen Lack arten, die den Japan-Lack nahe verwandt sind. *Ber. Dtsch. Chem. Ges. 55*:191-124.

Mason, H. S. (1945). The structure of bhilawanol. *J. Am. Chem. Soc. 67*:418-420.

Mason, H. S., and Lada, A. (1954). Allergenic properties of poison ivy. VIII. Immunological properties of a hydrourushiol-albumin conjugate. *J. Invest. Dermatol. 22*:457-461.

McNair, J. B. (1917). The oxidase of *Rhus diversiloba. J. Infect. Dis. 20*: 485-498.

Mitchell, J. C. (1966). Stereoisomeric specificity of usnic acid in delayed hypersensitivity. *J. Invest. Dermatol. 47*:167-168.

Mitchell, J. C. (1975). Contact allergy from plants. *Recent Adv. Phytochem. 9*:119-150.

Mitchell, J. C., and Jordan, W. P. (1974). Allergic contact dermatitis from the radish, *Raphanus sativus. Br. J. Dermatol. 91*:183-189.

Mitchell, J. C., and Rook, A. (1979). *Botanical Dermatology.* Greengrass, Vancouver.

Mitchell, J. C., Roy, A. K., Dupuis, G., and Towers, G. H. N. (1971). Allergic contact dermatitis from ragweeds (*Ambrosia* species). *Arch. Dermatol. 104*:73-76.

Mitchell, J. C., Dupuis, G., and Geissman, T. A. (1972). Allergic contact dermatitis from sesquiterpenoids of plants. Additional allergenic sesquiterpene lactones and immunological specificity of compositae, liverworts, and lichens. *Br. J. Dermatol. 87*:235-240.

Morimoto, H., Kawamatsu, Y., and Sugihara, H. (1968). Sterische Struktur der Giftstoffe aus dem fruchtfleisch von *Ginkgo biloba. Chem. Pharm. Bull. 16*:2282-2286.

Nakano, T., Medina, J. D., and Hurtado, I. (1970). The chemistry of *Rhus striata* (Manzanillo). *Planta Med. 18*:260-265.

Occolowitz, J. L., and Wright, A. S. (1962). 5-(10-pentadecynl)resorcinol from *Grevillea pyramidalis. Aust. J. Chem. 15*:858-861.

Olivera, Lima A. (1953). Über das antigene Verhalten der Olharze einiger Gattungen der Familie Anacardiaceae. *Int. Arch. Allergy 4*:169-174.

Ollodart, R., and Rose, N. R. (1962). Antibodies to 1,2-naphthoquinone. *Cancer Res. 22*:689-695.

Parker, D., and Turk, J. L. (1974). *Contact Hypersensitivity in Experimental Animals.* S. Karger, New York.

Picman, A. K., Ranieri, R. L., Towers, G. H. N., and Lam, J. (1980). Visualization reagents for sesquiterpene lactones and polyacetylenes on thin-layer chromatograms. *J. Chromatogr. 189*:187-198.

Pillay, P. P., and Siddiqui, S. (1931). Chemical examination of the marking nut (*Semecarpus anacardium*, Linn). *J. Int. Chem. Soc. 8*:517-525.

Rasmussen, M., Ridley, D. D., Ritchie, E., and Taylor, W. C. (1968). Chemical studies of the Proteaceae. III. The structure determination and synthesis of striatol, a novel phenol from *Grevillea striata. Aust. J. Chem. 21*:2989-2999.

Reynolds, G., and Rodriguez, E. (1979). Geranylhydroquinone: a contact allergen from trichomes of *Phacelia crenulata. Phytochemistry 18*:1567-1568.

Reynolds, G., Epstein, W., Terry, D., and Rodriguez, E. (1980). A potent contact allergen of *Phacelia* (Hydrophyllaceae). *Contact Dermatol. 6*:272-274.

Ridley, D. D., Ritchie, E., and Taylor, W. C. (1968). Chemical studies of the Proteaceae. II. Some further constituents of *Grevillea robusta*

A. cunn.; experiments on the synthesis of 5-*n*-tridecyl resorcinol (grevillal) and related substances. *Aust. J. Chem. 21*:2979-2988.

Ritchie, E., Taylor, W. C., and Vautin, S. T. K. (1965). Chemical studies of the Proteaceae. I. *Gravillea robusta* A. cunn. and *Orites excelsa* R. Br. *Aust. J. Chem. 18*:2015-2020.

Rodriguez, E. (1977). Ecographic distribution of secondary constituents in *Parthenium* (Compositae). *Biochem. Syst. Ecol. 5*:207-218.

Rodriguez, E., Dillon, M. O., Mabry, T. J., Mitchell, J. C., and Towers, G. H. N. (1976a). Dermatologically active sesquiterpene lactones in trichomes of *Parthenium hysterophorus*, L. (Compositae). *Experientia 32*: 236-237.

Rodriguez, E., Towers, G. H. N., and Mitchell, J. C. (1976b). Biological activities of sesquiterpene lactones. *Phytochemistry 15*:1573-1580.

Rodriguez, E., Epstein, W. L., and Mitchell, J. C. (1977). The role of sesquiterpene lactones in contact hypersensitivity to some North and South American species of fever few (*Parthenium*-Compositae). *'Contact Dermatol. 3*:155-162.

Roed-Petersen, J., and Jhorth, N. (1976). Compositae sensitivity among patients with contact dermatitis. Value of Compositae oleoresins in a standard test series. *Contact Dermatol. 2*:271-281.

Rook, A. (1961). Plant dermatitis. The significance of variety specific sensitization. *Br. J. Dermatol. 73*:283-287.

Saxena, P. R., Paut, M. C., Kishor, K., and Bhargave, K. P. (1965). Identification of pharmacologically active substances in the Indian stinging nettle, *Purtica parviflora* (Roxb.). *Can. J. Physiol. Pharmacol. 43*:869-876.

Schultz, K. H., Garbe, I., Hausen, B. M., and Simatupang, M. H. (1977). The sensitizing capacity of naturally occurring quinones. Experimental studies in guinea pigs. *Arch. Dermatol. Res. 258*:41-52.

Shelmire, B. (1940). Contact dermatitis from vegetation. *South. Med. J. 33*:337-346.

Sohi, A. S., Tiware, V. D., Loukar, A., Rangachar, S. K., and Nagasampagi, B. A. (1979). Allergenic nature of *Parthenium hysterophorus*. *Contact Dermatol. 5*:133-136.

Stoughton, R. B. (1965). Percutaneous absorption. *Toxicol. Appl. Pharmacol. 7*:1-6.

Subba Rao, P. V., Mangala, A., Towers, G. H. N., and Rodriguez, E. (1978). Immunological activity of parthenin and its diasteriomer in persons sensitized by *Parthenium hysterophorus*, L. *Contact Dermatol. 4*:199-203.

Suminokura, K. (1936). Über die Laccase des Japanischen Lacks. II. Über Polyphenole und Diamine. *Biochem. Z. 11*:299-309.

Sunthankar, S. V., and Dawson, C. R. (1954). The structural identification of the olefinic components of Japanese lac urushiol. *J. Am. Chem. Soc. 76*:5070.

Symes, W. F., and Dawson, C. R. (1953a). Cashew nut shell liquid. IX. The chromatographic separation and structural investigation of the olefinich components of methycardanol. *J. Am. Chem. Soc. 75*:4952.

Symes, W. F., and Dawson, C. R. (1953b). Separation and structural determination of the olefinic components of poison ivy urushiol, cardanol and cardol. *Nature (Lond.) 171*:841.

Symes, W. F., and Dawson, C. R. (1954). Poison ivy "urushiol." *J. Am. Chem. Soc.* 76:2959.

Thurston, E. L. (1974). Morphology, fine structure and ontogeny of the stinging emergence of *Urtica divica*. *Am. J. Bot.* 61:809-817.

Thurston, E. L. (1976). Morphology and ontogency of the stinging emergence of *Tragia ramosa* and *T. saxicola* (Euphorbiaceae). *Am. J. Bot.* 63:710-718.

Thurston, E. L., and Lersten, N. R. (1969). The morphology and toxicology of plant stinging hairs. *Bot. Rev.* 35:393-412.

Towers, G. H. N. (1979). Contact sensitivity and photodermatitis evoked by Compositae. In *Toxic Plants*, A. D. Kinghorn (Ed.). Columbia University Press, New York.

Turk, J. L. (1980). *Delayed Hypersensitivity*. Elsevier/North-Holland Biomedical Press, New York.

Tyman, J. H. P. (1973). Long chain phenols. Part III. Identification of the components of a novel phenolic fraction in Anacardium occidentale (cashew nut-shell liquid) and synthesis of the saturated member. *J. Chem. Soc. Perkin Trans. 1* 15:1639-1647.

Tyman, J. H., and Kiong, L. S. (1978). Long chain phenols. Part XI. Composition of natural cashew nut shell liquid (*Anacardium occidentale*) from various sources. *Lipids* 13:525-532.

Wenkert, E., Loeser, E. M., Mahapatra, S. N., Schenker, F., and Wilson, T. M. (1964). Wheat bran phenols. *J. Org. Chem.* 29:435-439.

Willis, I. (1971). Photoallergic contact dermatitis. In *Immunology and the Skin*, W. Montagna and R. E. Billingham (Eds.). Appleton-Century-Crofts, New York.

Zisserman, L. (1940). Susceptibility to poison ivy dermatitis. *J. Allergy 11*: 600-603.

Part V

TOXINS INDUCING
PSYCHIC OR NEUROTOXIC EFFECTS

13

NEUROTOXINS AND OTHER TOXINS FROM *ASTRAGALUS* AND RELATED GENERA

LYNN F. JAMES

U.S. Department of Agriculture, Logan, Utah

I. INTRODUCTION

The 368 species and 184 varieties of *Astragalus* listed in North America (Barneby, 1964) make the genus one of the largest of the legume family in North America (Kingsbury, 1964). Species of *Astragalus* are abundant throughout Europe and Asia. The recorded history of the *Astragalus* goes back to at least the year A.D. 1. The name itself may be much older. *Astragalus* is a Greek word meaning ankle bone, which was used in ancient times as a form of dice. It is suspected that the name was given to these plants because of the rattle of the seeds in the pods of some species of *Astragalus* growing in the Mediterranean area (Barneby, 1964).

Closely related to the genus *Astragalus* is the genus *Oxytropis*. Some taxonomists have felt that the genus *Oxytropis* should be included with the *Astragalus*. Early literature lists *Aragallus* as a synonym for the genus *Oxytropis*. However, the name *Oxytropis* now seems secure. The taxonomy and proper identification of these two groups of plants is difficult, even for a plant taxonomist. For a complete discussion of the taxonomic features of the genera *Astragalus* and *Oxytropis*, the reader is referred to Barneby's *Atlas of North American Astragalus*, Parts I and II (Barneby, 1964), and *A Revision of North American Species of Oxytropis DC* (Barneby, 1952). Whatever name is applied to these two genera of plants, they are extremely fascinating, not only because of their taxonomic complexity, but also because they exert so many and varied toxicological effects on animals that consume them. In addition to their close taxonomic relationship, they have almost identical toxicological effects on animals and therefore can be considered as a unit by those concerned with their poisonous effects on animals (Marsh, 1909).

Astragalus species are mostly perennial (stemmed or stemless) herbs. Leaves are alternate and pinnately compound. Flowers are leguminous. The fruit is a legume pod that varies in size and shape. Barneby (1964) points out that one of the most remarkable characteristics of the genus *Astragalus* is that there are hardly two species, even those that are closely related, that do not differ from one another in the form or structure of their fruit. The seeds are kidney-shaped. The seeds of some species of *Astragalus* may maintain their viability for 40 years or longer (Barneby, 1964). *Oxytropis* can be distinguished from *Astragalus* in that the keel (lowermost petal) of the species of *Oxytropis* is prolonged into a long, distinct point, while the keel petal of each *Astragalus* is blunt. Barneby (1952) discusses the differentiating features of *Oxytropis* species in detail.

The genus *Swainsona*, closely related to the *Astragalus*, is native to Australia except for one species, which has no toxicological impact, found in New Zealand (Everist, 1972). The more than 50 species are herbs or subherbs with compound leaves. The mostly large and showy pea-shaped flowers are in shades of blue, mauve, purple, or red, and occasionally yellow or white. The pods are thin-walled and mostly inflated. For a detailed taxonomic description, the reader is referred to *The Poisonous Plants of Australia* by Everist (1972). It was suggested as early as 1909 that *Swainsona* had the same toxicological effects on livestock as certain species of *Astragalus* and *Oxytropis* (Marsh, 1909). More recently this suspected relationship has been demonstrated (James et al., 1970a; Hartley, 1978).

Poisonous species of the genera *Oxytropis* and *Astragalus* and probably the most destructive to livestock of all poisonous plants (Marsh, 1929; Butcher et al., 1953). Not all species of these genera are necessarily toxic to livestock. Some nontoxic species can provide nutritious forage. The nontoxic species will be discussed only briefly.

The toxic species of *Astragalus* can be divided into three general categories according to their effects on livestock. The toxic syndromes include (1) the acute and chronic toxicosis of the nitro-containing *Astragalus*, (2) the acute and chronic toxicosis of the selenium-accumulating *Astragalus*, and (3) the toxic effects of locoweeds. Included with the discussion on the *Astragalus* that have as a common name "locoweed" will be the locoweed of the genus *Oxytropis* and the toxic species of *Swainsona*.

II. NONTOXIC *ASTRAGALUS* PLANTS

Many species of *Astragalus* are nontoxic and useful forage for livestock. *A. nuttallianus*, for example, is a low-growing succulent plant of the southwest United States that is relished by livestock.

Plant breeders and plant introduction centers have been interested in such *Astragalus* plants not only as potential forage producers, but also because of their soil-building characteristics and drought resistance. *A. cicer*, a plant introduced into the United States, has gained some acceptance as a forage plant. Other species, such as *A. tenellus*, are nontoxic and nonpalatable to livestock and big game (James, 1971).

III. NITRO-CONTAINING *ASTRAGALUS*

Species of this group of *Astragalus* grow throughout much of the world. Williams and Barneby (1977a,b) and Williams (1981) have surveyed much of the world's *Astragalus* for the nitro compounds. Plants of this group have been especially troublesome in western Canada, western United States, and northern Mexico. Examples include *A. emoryanus* in Texas, Arizona, and New Mexico (Fig. 1); *A. tetrapterus* in Utah, Arizona, and Nevada; *A. pterocarpus* in central Nevada; *A. miser* var. *serotinus* in western Canada; *A. miser* var. *hylophilus* in Utah, Wyoming, Colorado, and Montana; and *A. canadensis* in most of the United States and western Canada. *Astragalus*

FIGURE 1 *Astragalus emoryanus*, a nitro-containing poisonous plant of the southwestern United States.

falcatus is an introduced plant now growing in certain parts of the western United States. Scientists have expressed concern that other toxic species of *Astragalus* may be inadvertently introduced into the United States. Other countries interested in plant introductions share this concern.

Poisoning of livestock by the nitro-containing *Astragalus* was first reported by Marsh and Clawson (1920) when they demonstrated that *A. tetrapterus* was toxic to cattle and sheep. Bruce (1927), in western Canada, later showed that *A. campestris* (now identified as *A. miser* var. *serotinus*) was poisonous to cattle, sheep, and horses. Beath et al. (1932) reported a similar type of poisoning in cattle grazing *A. miser* var. *hylophilus* in Wyoming. Newsom et al. (1936) observed comparable poisoning in animals using the same plant in Colorado. In 1940, Mathews described poisoning in sheep, goats, and cattle by *A. emoryanus* in Texas. MacDonald (1952) in Canada reported poisoning in cattle grazing *A. miser* var. *serotinus*. The descriptions given by these authors of livestock intoxicated by these plants are similar.

Beath et al. (1932) reported that the toxicity of *A. hylophilus* correlated with specific geological formations and suggested tin as the toxic agent. Stermitz et al. (1969) later showed that plants in this group of *Astragalus* synthesize aliphatic nitro compounds. The compound miserotoxin (3-nitro-1-propyl-β-D-glucopyranoside) (Fig. 2) was first isolated from *A. miser* var. *oblongifolius*. This compound is metabolized to the highly toxic 3-nitro-1-propanol in the digestive tract of ruminants (Williams et al., 1970). Other plants of this group synthesize nitropropionic acid glycoside, which is to the less toxic 3-nitropropionic acid in the digestive tract of ruminants. The 3-nitropropionic acid is degraded much more rapidly in the rumen than is 3-nitro-1-propanol (Majak and Clark, 1981); however, it is much less toxic than the alcohol form. Many of the plants that synthesize the acid form accumulate higher levels than those that synthesize the alcohol form. Thus in many cases these two groups of plants have nearly the same toxicity per unit of plant material (Williams and James, 1978).

The levels of toxin in these plants decline as they mature and dry. The same happens following the treatment of these plants with the phenoxy herbicides (Williams, 1970; Williams and James, 1978).

The consumption of either of these two groups of plants by cattle and sheep can result in either acute or chronic intoxication. The type of intoxication is apparently related to the level of toxin in the plant and the rate at which the plant is consumed. The lethal dose of 3-nitro-1-propanol is approximately 57 mg/kg body weight (Williams and James, 1978).

In the rumen, miserotoxin is thought to break down into two different fractions—the inorganic nitrate and a three-carbon side chain. This theory

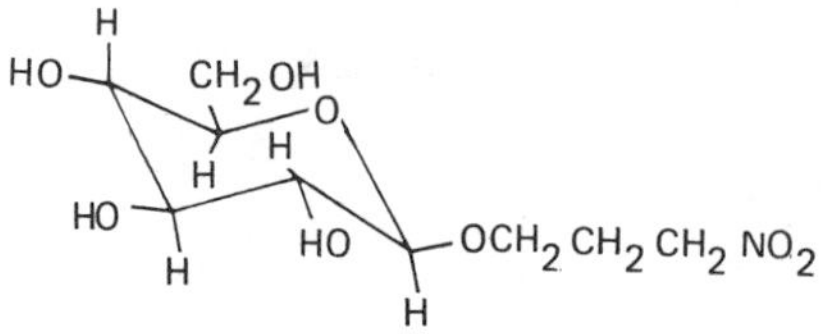

FIGURE 2 Miserotoxin, 3-nitro-1-propyl-β-D-glucopyranoside, is the aliphatic nitro compound that is responsible for the toxicity of many of the *Astragalus* plants.

is based on the fact that animals poisoned on the nitro compound develop methemoglobinemia. Absorbed nitrite could complex with blood hemoglobin to form the methemoglobin. Production of methemoglobin by cattle fed a lethal dose of plants containing 3-nitro-1-propanol can be prevented by administering methylene blue, but the animals will still die. Thus the nitrate may contribute to the characteristics of the syndrome, but it apparently is not the primary cause of death (Cronin et al., 1974).

Cattle are more susceptible than sheep to intoxication by nitro-containing *Astragalus*. The signs of acute poisoning in cattle include general body weakness, knuckling at the fetlocks, interference between the hind limbs as the animal moves, respiratory distress, cyanotic mucous membranes, and sudden collapse of the animal accompanied by difficulty in rising. If a lethal dose of plant has been consumed, the animal will die within 4-25 hr. The formation of the methemoglobin may contribute to the respiratory distress. Sudden exertion may result in instant death. Sheep show fewer signs of poisoning than do cattle and often die suddenly.

Gross lesions in acute poisoning include pulmonary congestion, edema, petechial hemorrhage on the epithelium of the heart, and microhemorrhages in the central nervous system (Cronin et al., 1974; James et al., 1980).

Chronic poisoning follows the grazing of nitro-containing *Astragalus* at a much lower level than in acute poisoning and over an extended period of time.

The signs of chronic poisoning are related to the effects of the toxin on the respiratory system and the central nervous system (Mosher et al., 1971; James et al., 1980). The first signs of poisoning include general depression and rough, dry hair coat, followed by cocked fetlocks and perhaps a knuckling over of the fetlocks (Fig. 3). Some animals become blind or partially blind. Goose-stepping and varying degrees of incoordination may develop, the animal's back may sag in the loin area, and the hind quarters sway when the animal walks. The hind legs may interfere when walking, causing a clicking sound. As intoxication advances, the animal may become partially paralyzed in the hind quarters and, in severe cases, paralysis may cause the animal to drag its hind quarters. The respiratory distress is manifested by a distinct roaring sound that can be heard for some distance. If poisoning continues, the animal becomes emaciated and dies. If affected animals are removed from access to the plants and given good feed, they will make an apparent recovery. Many may die later, however, if placed under stress. Animals that are disturbed or excited may fall and have difficulty in rising. Exertion may cause death. The clicking of the dew claws has caused the condition to be called "cracker heels" in some areas, and the respiratory difficulty has caused it to be called "roaring disease" in others.

Affected sheep show fewer central nervous system disturbances and more respiratory problems. Sheep tend to die quite suddenly after eating the plant for a few days.

Horses are also adversely affected by this group of plants. Signs of poisoning are similar but more acute.

The signs of intoxication may persist for up to a year or longer after the animal no longer has access to the plant. Death can occur over an equal length of time.

There are no spectacular gross lesions in sheep or cattle chronically poisoned on nitro-containing *Astragalus* (James et al., 1980). Congestion and

FIGURE 3   Cow poisoned on *Astragalus emoryanus*.   Note the cocked fetlocks.

swelling of the liver are found in most poisoned animals. In animals that have
ingested the plant for more than a few days, varying degrees of emphysema
and pneumonia may be present. Perhaps the most interesting observation is
that most animals have an excessive amount of cerebrospinal fluid. There are
ulcerations of the mucosa in the cardiac region of the abomasum of some
animals.

The principal microscopic lesions in sheep and cattle include pulmonary
alveolar emphysema, bronchiolar constriction, interlobular edema, and fibrosis, wallerian-type degeneration in the spinal cord and peripheral nerves,
and focal hemorrhages in the brain (James et al., 1980).

Bloat may occur in cattle and sheep grazing lush stands of *A. emoryanus*
(Mathews, 1940).

Poisoning of livestock by the nitro-containing *Astragalus* can be minimized
by controlling the plant with herbicides, using grazing programs that favor
more desirable plants, and removing animals from access to the plants.
Plants of this group of *Astragalus* do not grow every year, so the problem is
sporadic. Also, some plants such as *A. emoryanus* are somewhat palatable,
whereas others of this group are grazed only when desirable forage is
deficient.

Intramuscular injections of thiamine hydrochloride at the rate of 400 mg per
adult cow and 100 mg per sheep or lamb have had some therapeutic value to
cattle and sheep poisoned on *A. miser* var. *serotinus*. The mechanism for
this effect is now known (Nicholson, 1963).

IV. SELENIUM–ACCUMULATING PLANTS

Selenium, an element that occurs in some soils, may be taken up by some plants in quantities that render them toxic. Selenates may occur in alkaline soils of semiarid regions in high concentrations and are quite stable. They are water soluble, easily leached, and available to plants (NRC, 1976).

The distribution of selenium in the soil profile is important to the selenium content of the plants. If the selenium has been leached into the deeper soil profiles, the deeper-rooted plants may absorb more selenium than will shallow-rooted ones. Even shallow-rooted plants such as the grasses, however, may absorb some selenium under these circumstances. Reports in the literature indicate that selenium is absorbed at least in trace quantities by all members of the plant kingdom. Perhaps the most notorious of the selenium-accumulating plants are certain species of the *Astragalus* genus. Among the selenium-accumulating plants, *Astragalus* shows the greatest diversity of form and the most extensive geographic distribution. About 24 North American species of *Astragalus* are known to accumulate high levels of selenium. Some of the more notable of the selenium-accumulating *Astragalus* are *A. bisulcatus*, two-grooved milkvetch (Fig. 4); *A. pattersonii*, Patterson's milkvetch; *A. prae-longus*, stinking milkvetch; *A. pectinatus*, five-leaved milkvetch; and *A. racemosus*, alkali milkvetch. This group of plants can accumulate selenium at concentrations of up to several thousand parts per million. These plants are often referred to as indicator plants because they characteristically grow on soils high in selenium, and therefore help identify such soils (Rosenfeld and

FIGURE 4 *Astragalus bisulcatus*, two-grooved milkvetch, a selenium-accumulating plant.

Beath, 1964). These plants were called converter plants since it appeared that they could absorb selenium from relatively insoluble compounds and convert them to forms more available to plants such as grass, corn, and wheat. The term "converter plant" has fallen into disuse because the function was never demonstrated (NRC, 1976). Deep-rooted plants may bring selenium from the subsoil to the surface, making it available to shallow-rooted plants. Many plants that accumulate high levels of selenium have an unpleasant odor and are unpalatable.

Three types of selenium poisoning have been described: (1) acute, (2) chronic of the alkali disease type, and (3) chronic of the blind staggers type (Rosenfeld and Beath, 1964).

Acute selenium poisoning results when livestock graze plants principally of species of *Astragalus* that are high in selenium (usually over 40 ppm), such as *A. bisulcatus* or *A. praelongus*. Although these plants are usually unpalatable, acute poisoning can and does occur on occasion. Signs of poisoning may include abnormal movement and posture, elevated temperature, diarrhea, increased heart rate, labored breathing, prostration, and death. In sheep the signs are primarily depression and sudden death (NRC, 1976).

V. CHRONIC SELENIUM POISONING (ALKALI DISEASE)

Chronic selenium poisoning, alkali disease, results from the prolonged consumption by livestock of forage containing 5-40 ppm selenium (NRC, 1976). This condition may be produced in livestock that graze forage such as the grasses and small grains. The time an animal must graze seleniferous forage before signs of poisoning occur is highly variable. The responses of classes of livestock also vary. Signs of poisoning in affected cattle, horses, and hogs may include rough hair coat, lack of vitality, lameness, depressed appetite, and emaciation, and the long hair of the mane and tail may fall out. Signs of poisoning in chickens include depressed growth rate and roughened features. Sheep show no typical signs of alkali disease. Perhaps the most significant effect of selenium poisoning in livestock is depressed reproductive functions (Olsen, 1978).

Chronic selenium poisoning of livestock is difficult to determine visually. Analysis of hair samples of cattle grazing seleniferous forage has been helpful in determining the status of each animal (Moxon and Olsen, 1974).

The following guide has been used by the South Dakota State Experiment Station in evaluating hair samples from cattle suspected of selenium poisoning (Olsen, 1978):

ppm Se in hair	
Less than 5.0	Chronic selenosis not expected
5.0-10.0	Borderline selenosis suspected
More than 10.0	Selenium intake excessive, and selenosis can be expected, but not necessarily with signs of poisoning

VI. CHRONIC SELENIUM POISONING (BLIND STAGGERS)

Blind staggers as described in Wyoming livestock is said to be caused by animals grazing a limited amount of selenium indicator plants over a period of time (Rosenfeld and Beath, 1964). The selenium in these plants occurs in organic compounds with or without small amounts of selenates that are readily extractable with water. The plants incriminated here include those of the selenium-accumulating *Astragalus* and *Stanleya* spp. The presence of other toxic substances in the plants may alter the course of the disease; however, these same investigators stated that *A. bisulcatus* contains no toxic material other than selenium (Rosenfeld and Beath, 1964). If so, other selenium-accumulating species of *Astragalus* would be expected to produce similar intoxification.

The signs of poisoning in blind staggers include wandering, stumbling, going in circles, apparent blindness, and anorexia. Paralysis and respiratory distress may precede death (Rosenfeld and Beath, 1964).

Researchers in Colorado described a condition they called forage poisoning that had similar lesions to those of blind staggers (Jensen et al., 1956). The lesions were described as those of polioencephalomalacia. They could find no relationship to selenium consumption. Blind staggers has not been observed in livestock in South Dakota, where much of the early research on selenium poisoning was done (Moxon, 1937). Also, there are other plants that are suspected of causing polioencephalomalacia, such as *Kochia scoparia* (Dickie and Bergman, 1979) and *Descurainia pinnata* (Kingsbury, 1964).

Recent work suggests that *A. bisulcatus* contains the same toxin as the locoweeds (discussed in Sect. VII) (Van Kampen and James, 1978). *A. bisulcatus* was listed as a locoweed in the early literature. The listing of *A. bisulcatus* as a locoweed and the later finding suggesting that it accumulates selenium have caused some to think that selenium is the toxic principle of the locoweeds.

There is some question as to whether blind staggers, as described in Wyoming, is associated with selenium consumption or, in fact, is part of the disease syndrome polioencephalomalacia (Jensen et al., 1956; NRC, 1976).

VII. *ASTRAGALUS* AND *OXYTROPIS* PLANTS AS A CAUSE OF LOCOWEED POISONING

"Loco" is a Spanish word meaning crazy and has been used not only as a name for this group of plants, but also to describe animals poisoned on them (Butcher et al., 1953). The term "loco" is properly applied only to the intoxication that occurs when livestock graze plants of this group. The following is a list of some of the more important species of *Astragalus* and *Oxytropis* that are called locoweeds: *A. lentiginosus*, 36 varieties; *A. mollissimus*, 11 varieties; *A. wootonii*, 2 varieties; *A. thurberi*; *A. nothoxys*; *O. sericea*, 2 varieties; *O. lambertii*, 3 varieties; and *O. besseyi*, 7 varieties. The dried skeletons of the *Astragalus* spp. of locoweed are toxic (James, 1981).

The most destructive of all the poisonous plants are those going under the general name "locoweed" (Butcher et al., 1953). Of the locoweeds, the *Oxytropis* species are probably the most destructive (Fig. 5). This is not because

FIGURE 5 *Oxytropis sericea*, one of the principal locoweeds in the western United States.

they are more toxic than the other locoweed, but because of their wider geographic distribution (Marsh et al., 1936).

Locoweeds were one of the first poisonous plants to be encountered by livestock owners as they began to drive their herds of cattle, sheep, and horses into the West. The first recorded account of locoweed poisoning was published by the U.S. Commissioner of Agriculture in 1873 (Marsh, 1909). In 1883, 25,000 cattle died in one area in Kansas (Crawford, 1908). However, it was not until 1909 that Marsh presented conclusive evidence that it was indeed locoweed that was responsible for this and many other similar losses. Until that time many doubted. Since that time many investigators have described the disease and the plants, and discussed ways to prevent it. Locoweed continues to be a serious problem to livestock producers of the West.

Almost simultaneously with Marsh's research on the cause of locoweed disease, Crawford undertook to identify the toxic principle of these plants. In 1908, Crawford published that barium was the toxic principle of locoweed, but in 1912 it was shown not to be the case (Marsh et al., 1912). Since that time much research has been done in an effort to isolate the locoweed toxin. Swainsonine, $8\alpha,\beta$-indolizidine-$1\alpha,2\alpha,8\beta$-triol (Fig. 6), recently isolated from *Swainsona canescens* in Australia, has been shown to be the toxic constituent of that plant (Colgate et al., 1979). This same compound has been found in *A. lentiginosus* and *Oxytropis serecia*, two of the locoweeds (Molyneux and James, 1982). Because of the multiple effects of locoweed on an animal, it is quite possible that there are other toxic compounds in the plant.

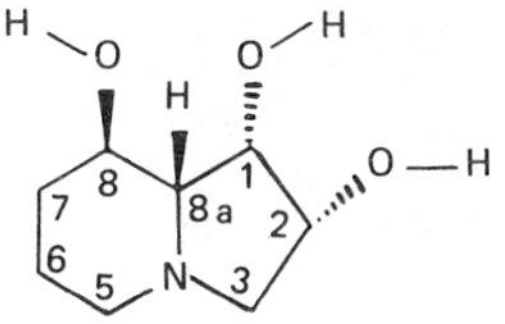

FIGURE 6 Swainsonine, 8α,β-indolizidine-1α,2α,8β-triol, the toxic constituent of *Swainsona canescens*.

Locoweed causes a multitude of problems in livestock. These include neurologic changes, reproductive alterations, emaciation, habituation (James et al., 1969), congestive right-heart failure (James et al., 1983), and probably other physiopathologic changes.

The word "loco" was applied to this group of plants because of their neurological effects on livestock. Most of the signs observed are related to the effect of the toxin on the nervous system. The signs of poisoning include depression, a slow, awkward gait, rough hair coat, dull-appearing eyes with a staring look, muscular incoordination, and nervousness, especially when stressed (Mathews, 1932; Marsh, 1909; James et al., 1967). The animal may become stiff in the legs. In addition, affected animals may become solitary and hard to handle, and they may have difficulty in eating and drinking. Conjunctivitis, keratitis, and excessive lacrimation may be associated with locoism (Mathews, 1932). Prolonged consumption of locoweed leads to recumbency and eventual death. Sexual activity may become suppressed. Spermatogenesis (James and Van Kampen, 1971b) and oogenesis (Van Kampen and James, 1971) both cease.

Horses that have been poisoned with locoweed are useless for draft or riding purposes because of permanent injury to the animal. Signs of poisoning will be observed if an animal that has been removed from loco is subjected to stress. A full appreciation of the effects of locoweed poisoning on an animal can only be had by observing poisoned animals.

Locoweed poisoning is a chronic type of intoxication. Acute intoxication has not been observed in association with locoweed poisoning. Livestock must graze this plant for long periods before signs of poisoning become obvious. Lambs when first fed locoweed have a slight weight gain advantage over those not fed locoweed (James et al., 1970b); however, as the feeding continues, the animals fed the locoweed lose that advantage, lose weight, become emaciated, and eventually die. Some livestock owners have reported that their animals do well for a short period of time after they start grazing locoweed. Continued grazing always brings emaciation and death. The length of time an animal must graze locoweed before signs of poisoning occur depends on the rate of consumption of the plant. Signs may appear within 10 days to 2 weeks, or as long as 60 days after the animal starts grazing locoweed. If feeding or grazing of locoweed is discontinued before intoxication becomes too advanced and good nutritious feed given, the animal will recover. However, such animals will exhibit signs of poisoning when placed under stress.

The principal effect of locoweed on pregnant animals is abortion (Fig. 7) (James et al., 1967). The abortion may occur at any time during the gestation

FIGURE 7 One of the principal effects of locoweed poisoning in pregnant
animals is abortion. Shown here is a ewe that has aborted due to locoweed
poisoning.

period. Abortions may approach 100% of the pregnant animals grazing on locoweed. In addition, the offspring of animals poisoned on locoweed may be
small and weak at birth and fail to survive; poisoned pregnant cows and ewes
may give birth to offspring with skeletal malformations (James et al., 1967).
Placentation is delayed in ewes fed locoweed early in gestation (James, 1972).
Fetal lambs from ewes fed locoweed have been edematous and had enlarged
right ventricles of the heart (James, 1972). Sexual activity ceases in both
sexes of livestock poisoned on locoweed but returns when the ingestion of
locoweed ceases. The locoweed toxin is excreted in the milk (James, 1977).

Recent evidence indicates that when the locoweed *O. sericea* is grazed or
fed to cattle at high altitudes, congestive right-heart failure develops (James
et al., 1983). This leads to accumulation of fluid under the jaw and brisket,
cirrhosis of the liver, accumulation of fluid in the thorax, the microscopic
lesions characteristic of locoweed poisoning (see below), and congestive right-
heart failure (Fig. 8).

Hydrops amnii has been observed in pregnant cows grazing locoweed
(Mathews, 1932). This condition is characterized by a distended abdomen
due to the accumulation of fluid in the uterus.

In both of these conditions, congestive right-heart failure and hydrops
amnii, affected cattle do not show typical signs of poisoning. They seem,
however, to have increased susceptibility to disease conditions such as
pneumonia and foot rot.

FIGURE 8 There is evidence that when calves graze locoweed at high altitudes, they develop congestive right-heart failure.

Locoweed is generally considered nonpalatable to livestock, although under some conditions they graze it readily, and after they start to graze it, they seem to develop a liking or craving for it (Marsh, 1909). Not all locoweeds, however, are equally unpalatable. Most locoweeds stay green late in the fall or become green during the winter or early spring. Green locoweed may be more palatable than the associated dried and dormant grasses and shrubs. Locoweed is deep rooted; therefore, it withstands drought and other adverse conditions better than do the more shallow-rooted grasses. Locoweed intoxication thus may take place under conditions of drought.

Although locoweed may be less palatable than grasses, forbs, and shrubs, when these are grazed off, livestock will graze locoweed. Livestock that once start grazing locoweed will graze it much more readily than before. It is as if they acquire the habit.

Gross lesions in locoweed poisoning are not spectacular. There may be a slight accumulation of fluid in the thoracic cavity and ulceration in the cardiac region of the abomasum may be found, and enlarged thyroid glands may occur (Van Kampen and James, 1969; James and Van Kampen, 1971a).

Serum glutamic oxalacetic transaminase, alkaline phosphatase, and lactic dehydrogenase may all increase. There is a marked decrease in the white blood cell count, which may account for the apparently increased susceptibility to infection (James et al., 1970b).

The microscopic lesions of locoweed poisoning in livestock can be described as neurovisceral foamy cytoplasmic vacuolation (Fig. 9) (Van Kampen and James, 1969; Hartley and James, 1973). The vacuolization can be observed as early as 4 days after the feeding of locoweed is initiated, and Purkinje cells

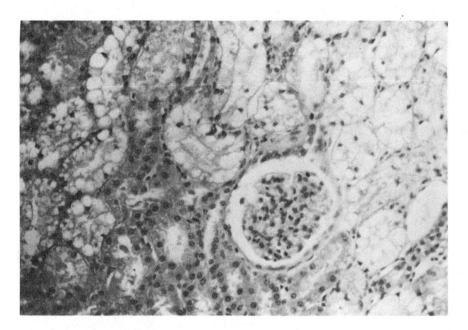

FIGURE 9 The microscopic lesions of locoweed poisoning are neurovisceral foamy cytoplasmic vacuolation. This is a photomicrograph of kidney from a sheep poisoned on locoweed.

and other neurons of the central nervous system become vacuolated after about 8 days (Van Kampen and James, 1970). Vacuolation disappears shortly after locoweed consumption ceases. If feeding or grazing is extended, however, Purkinje cells may be permanently lost from the cerebellum. Placental and fetal tissues develop microscopic lesions similar to those of adult animals that are locoweed poisoned (Hartley and James, 1973).

Electron microscopic studies have revealed that the cellular vacuolar lesions produced by locoweed resulted primarily from continued enlargement of single membranelike structures presumed to be lysosomes (Rhees et al., 1978). At later stages of intoxication, mitrochondrial swelling and degeneration may be observed. In the central nervous system, vacuolar degeneration was observed in neurons and glial cells. Fluid accumulations separated the myelin sheaths of axons. Axons degenerated as poisoning progressed.

Vacuoles were observed in other organs, such as the thyroid gland, adrenal, pancreas, and lymph nodes, as well as in the lymphocytes. Axonal spheroids were observed in many areas of the brain (Hartley and James, 1973; Van Kampen and James, 1970).

Swainsona, darling pea, produces a disease condition known as pea struck in cattle, sheep, and horses (Hartley, 1971a,b; Gardiner et al., 1969). The condition is similar, if not identical, to locoweed poisoning in livestock (Hartley, 1978). The signs of poisoning are very similar to those of locoweed poisoning, as are the gross and microscopic lesions. However, congenital malformations have not been observed in livestock poisoned on it (Hartley, 1978).

Ingestion of *Swainsona* by livestock may induce lysosomal storage disease, which is biochemically and morphologically similar to the genetically induced mannosidosis (Colegate et al., 1979; Dorling et al., 1978). The compound swainsonine, isolated from *Swainsona*, is thought to be the plant's toxin (Colegate et al., 1979).

REFERENCES

Barneby, R. C. (1952). A revision of the North American species of *Oxytropis* DC. *Calif. Acad. Sci. Mem. 27*:177-312.

Barneby, R. C. (1964). *Atlas of North American Astragalus*, Parts I and II. New York Botanical Garden, The Bronx, New York.

Beath, O. A., Draize, J. A., and Eppson, H. F. (1932). Three poisonous vetches. *Wyo. Agric. Exp. Stn. Bull. 189.*

Bruce, E. A. (1927). *Astragalus campestris* and other stock poisoning plants of British Columbia. *Dom. Can. Dept. Agric. Bull. 88.*

Butcher, J. E., Bronson, F. A., Booth, W. E., and Warden, A. L. (1953). Loco—the taxonomy of the plants, a review of literature, and a report of current Montana research. *Mont. Agric. Exp. Stn. Circ. 72.*

Colegate, S. M., Dorling, P. R., and Huxtable, C. R. (1979). A spectroscopic investigation of swainsonine: an induced monosidase inhibitor isolated from *Swainsona canescens. Aust. J. Chem. 32*:2237-2264.

Crawford, A. C. (1908). Barium, a cause of the locoweed disease. *U.S. Dept. Agric. BAI Bull. 129.*

Cronin, E. H., Williams, M. C., Van Kampen, K. R., and Holmgren, A. H. (1974). The poisonous timber milkvetches. *U.S. Dept. Agric. Handbook 459.*

Dorling, P. R., Huxtable, C. R., and Vogel, P. (1978). Lysosomal storage disease in *Swainsona* spp. toxicosis: an induced monosidosis. *Neuropathol. Appl. Neurobiol. 4*:285-295.

Dickie, C. W., and Bergman, J. R. (1979). Polioencephalomalacia and photosensitization associated with *Kochia scoparia* consumption in range cattle. *J. Am. Vet. Med. Assoc. 175*:463-465.

Everist, S. L. (1972). *Poisonous Plants of Australia.* Angus & Robertson, Sydney.

Gardiner, M. R., Linto, A. C., and Aplin, T. E. H. (1969). Toxicity of *Swainsona canescens* for sheep in Western Australia. *Aust. J. Agric. Res. 20*:87-97.

Hartley, W. J. (1971a). Observations of *Swainsona galegifolia* poisoning in cattle in northern New South Wales. *Aust. Vet. J. 47*:300-305.

Hartley, W. J. (1971b). Some observations on the pathology of *Swainsona* spp. poisoning in farm livestock in eastern Australia. *Acta Neuropathol. 18*:342-355.

Hartley, W. J. (1978). A comparative study of Darling pea (*Swainsona* spp.) poisoning in Australia with locoweed (*Astragalus* and *Oxytropis* spp.) poisoning in North America. In *Effects of Poisonous Plants on Livestock,* R. F. Keeler, K. R. Van Kampen, and L. F. James (Eds.). Academic Press, New York, pp. 363-369.

Hartley, W. J., and James, L. F. (1973). Microscopic lesions in fetuses of ewes ingesting locoweed (*Astragalus lentiginosus*). *Am. J. Vet. Res. 34*:209-211.

James, L. F. (1971). The effects of *Astragalus tenellus* in sheep. *J. Range Manage.* 24:161.

James, L. F. (1972). Effects of locoweed on fetal development: preliminary study in sheep. *Am. J. Vet. Res.* 33:835-841.

James, L. F. (1977). Effects of milk from animals fed locoweed on kittens, calves, and lambs. *Am. J. Vet. Res.* 38:1263-1265.

James, L. F. (1981). Syndromes of *Astragalus* poisoning in livestock. *J. Am. Vet. Med. Assoc.* 178:146-150.

James, L. F., and Van Kampen, K. R. (1971a). Acute and residual lesions of locoweed poisoning in cattle and horses. *J. Am. Vet. Med. Assoc.* 158: 614-618.

James, L. F., and Van Kampen, K. R. (1971b). Effects of locoweed intoxication on the genital tract of the ram. *Am. J. Vet. Res.* 32:1293-1295.

James, L. F., Shupe, J. L., Binns, W., and Keeler, R. F. (1967). Abortive and teratogenic effects of locoweed on sheep and cattle. *Am. J. Vet. Res.* 27:1379-1388.

James, L. F., Van Kampen, K. R., and Staker, G. (1969). Locoweed (*Astragalus lentiginosus*) poisoning in cattle and horses. *J. Am. Vet. Med. Assoc.* 155:525-530.

James, L. F., Van Kampen, K. R., and Hartley, W. J. (1970a). Comparative pathology of *Astragalus* (locoweed) and *Swainsona* poisoning in sheep. *Pathol. Vet.* 7:116-125.

James, L. F., Van Kampen, K. R., and Johnson, A. E. (1970b). Physiopathologic changes in locoweed poisoning of livestock. *Am. J. Vet. Res.* 31:663-672.

James, L. F., Hartley, W. J., Williams, M. C., and Van Kampen, K. R. (1980). Field and experimental studies in cattle and sheep poisoned by nitro-bearing *Astragalus* or their toxins. *Am. J. Vet. Res.* 41:377-382.

James, L. F., Hartley, W. F., Van Kampen, K. R., and Nielsen, D. (1983). Rlationship between ingestion of the locoweed *Oxytropis sericea* and congestive right-sided heart failure in cattle. *Am. J. Vet. Res.* 44(2):254-259.

Jensen, R., Grier, L. A., and Adams, O. R. (1956). Polioencephalomalacia of cattle and sheep. *J. Am. Vet. Med. Assoc.* 129:311-321.

Kingsbury, J. M. (1964). *Poisonous Plants of the United States and Canada.* Prentice-Hall, Englewood Cliffs, N.J., p. 305.

MacDonald, M. A. (1952). Timber milkvetch poisoning on British Columbia ranges. *J. Range Manage.* 5:16-20.

Majak, W., and Clark, L. J. (1981). Metabolism of aliphatic nitro compounds in bovine rumen fluid. *Can. J. Anim. Sci.* 60:319-325.

Marsh, C. D. (1909). The locoweed disease of the plains. *U.S. Dept. Agric. Bur. Anim. Ind. Bull. 112.*

Marsh, C. D. (1929). Stock-poisoning plants of the range. *U.S. Dept. Agric. Bull. 1249.*

Marsh, C. D., and Clawson, A. B. (1920). *Astragalus tetrapterus.* A new poisonous plant of Utah and Nevada. *U.S. Dept. Agric. Circ. 81.*

Marsh, C. D., Alsberg, C. L., and Black, O. F. (1912). The relation of barium to the locoweed disease. *U.S. Dept. Agric. Bur. Anim. Ind. Bull. 246.*

Marsh, C. D., Clawson, A. B., and Eggleston, W. W. (1936). The locoweed disease. *U.S. Dept. Agric. Farmers Bull. 1054.*

Mathews, F. P. (1932). Locoism in domestic animals. *Tex. Agric. Exp. Stn. Bull. 456.*

Mathews, F. P. (1940). The toxicity of red-stemmed peavine (*Astragalus emoryanus*) for cattle, sheep, and goats. *J. Am. Vet. Med. Assoc.* 97: 125-134.

Molyneux, R., and James, L. F. (1982). Loco Intoxication: Indolizidine Alkaloids fo Spotted Locoweed (*Astragalus lentiginosus*). *Science 216*:190-191.

Mosher, G. A., Kirshnamurate, C. R., and Ketla, W. D. (1971). Physiological effects of timber milkvetch, *Astragalus miser* var. *serotinus* on sheep. *Can. J. Anim. Sci.* 51:465-474.

Moxon, A. L. (1937). Alkali disease or selenium poisoning. *South Dakota Agric. Exp. Stn. Bull. 311.*

Moxon, A. L., and Olsen, O. E. (1974). Selenium in Agriculture. In *Selenium.* R. A. Zingaro and W. C. Cooper (Eds.). Van Nostrand Reinhold, New York, p. 679.

National Research Council (1976). *Selenium.* National Academy of Sciences, Washington, D.C.

Newsom, I. E., Cross, F., McCrory, B. R., Groth, A. H., Tobiska, J. W., Balis, E., Durrell, L. W., and Smith, E. C. (1936). Timber milkvetch as a poisonous plant. *Colo. Agric. Exp. Stn. Bull. 425.*

Nicholson, H. H. (1963). The treatment of timber milkvetch poisoning among cattle and sheep. *Can. J. Anim. Sci.* 43:237-240.

Olsen, O. E. (1978). Selenium in plants as a cause of livestock poisoning. In *Effects of Poisonous Plants on Livestock*, R. F. Keeler, K. R. Van Kampen, and L. F. James (Eds.). Academic Press, New York.

Rhees, R. W., James, L. F., and Van Kampen, K. R. (1978). Ultra-structural observations following locoweed poisoning in sheep. *Fed. Proc.* 37:501.

Rosenfeld, I., and Beath, O. A. (1964). Selenium. *Geobotany, Biochemistry, Toxicity, and Nutrition.* Academic Press, New York.

Stermitz, F. R., Norris, F. A., and Williams, M. C. (1969). Miserotoxin, a new naturally occurring nitro compound. *J. Am. Chem. Soc.* 91:4599.

Van Kampen, K. R., and James, L. F. (1969). Pathology of loco poisoning in sheep. *Pathol. Vet.* 6:413-423.

Van Kampen, K. R., and James, L. F. (1970). Pathology of locoweed poisoning in sheep: sequential development of cytoplasmic vacuolation in tissue. *Pathol. Vet.* 7:503-508.

Van Kampen, K. R., and James, L. F. (1971). Ovarian and placental lesions in sheep caused by ingestion of locoweed (*Astragalus lentiginosus*). *Pathol. Vet.* 8:193-199.

Van Kampen, K. R., and James, L. F. (1978). Manifestations of intoxication by selenium-accumulating plants. In *Effects of Poisonous Plants on Livestock*, R. F. Keeler, K. R. Van Kampen, and L. F. James (Eds.). Academic Press, New York.

Williams, M. C. (1970). Detoxication of timber milkvetch by 2,4,5-T and silvex. *J. Range Manage.* 23:400-402.

Williams, M. C. (1981). Nitro-compounds in foreign species of *Astragalus*. *Weed Sci. 29*:261-269.

Williams, M. C., and Barneby, R. C. (1977a). The occurrence of nitro-toxins in North American *Astragalus* (Fabaceae). *Brittonia 29*:310-326.

Williams, M. C., and Barneby, R. C. (1977b). The occurrence of nitro-toxins in the Old World and South Carolina *Astragalus* (Fabaceae). *Brittonia 29*:327-331.

Williams, M. C., and James, L. F. (1978). Livestock poisoning from nitro-bearing *Astragalus*. In *Effects of Poisonous Plants on Livestock*, R. F. Keeler, K. R. Van Kampen, and L. F. James (Eds.). Academic Press, New York.

Williams, M. C., Norris, F. A., and Van Kampen, K. R. (1970). Metabolism of miserotoxin to 3-nitro-1-propanol in bovine and ovine ruminal fluids. *Am. J. Vet. Res. 31*:259-262.

14

CYCAD POISONING

PETER T. HOOPER

James Cook University of North Queensland, Queensland, Australia

I. INTRODUCTION

This chapter describes the poisonings caused by the cycad palms, with particular emphasis on the neurological diseases in domestic animals. The cycads can cause either neurological disease characterized by ataxia and partial hind limb paralysis, or a syndrome of hepatic and to a lesser extent, gastrointestinal disease (Hooper, 1978).

II. THE CYCADS AND THEIR TOXICITY

Cycads are easily recognized by their palmlike appearance (Fig. 1) (Whiting, 1963; Birdsley, 1972; Everist, 1974). They occur in tropical and subtropical environments throughout the world with fairly well defined geographical locations for the genera, for example, *Encephalartos* spp. in Africa, *Diion* spp. and *Zamia* spp. in central America, *Cycas* spp. in Oceania, and *Cycas* spp.,

FIGURE 1   Cycad palms near Darwin, Australia, *Cycas armstrongii*. The cycads are the only plants that regrew after a fire. (From J. R. Maconochie.)

Bowenia spp., *Lepidozamia* spp., and *Macrozamia* spp. in Australia. There have been reports of poisonings with most genera and it would be reasonable to assume that all species of these genera are potentially dangerous. However, toxicity problems are not present in all cycad-infested areas. In some areas, there is apparent freedom of disease although there is an abundance of plant which seems to be readily eaten.

In many tropical areas where cycads grow, there are very pronounced wet and dry seasons. However, poisonings can occur at any time, irrespective of the season. Individual animals may develop particular preferences for the plants even though there is an abundance of other green feed in the rainy season. The dry season is probably more dangerous in that many cycads rapidly regenerate after bushfires have burned and denuded the ground. All that remains in some pastures are the regenerated palms with their attractive green fronds. Further, the plants are often more dangerous because they develop their more toxic nuts at that time.

Cattle in both good and poor nutritional condition seem equally susceptible. The disease occurs commonly in cattle newly introduced to an area but also in stock born in the area.

Wesley-Smith (1973) summarized techniques used for the control of *C. armstrongii* (previously believed to be *C. media* in that area). Chemical control is practical, although he warns that care must be taken for at least 2 years, as regeneration is possible within that time. He believed that the frequent fires which are a common feature of the area may reduce the plant populations over the longer term. On the other hand, in the short term, they cause immediate problems because of prompt regeneration of fronds on the stumps.

III. THE DISEASES

Early information on cycad poisoning was comprehensively reviewed by Whiting (1963). There have been reviews on later experimental work (Whiting et al., 1966; Laqueur and Spatz, 1968; Spatz, 1969) and a series of short papers published from a cycad research conference (Sixth International Cycad Conference, 1972; Yang et al., 1972).

From early times, cycads have been recognized as natural hazards in tropical areas. In domestic animals, there are two distinct field syndromes. One is characterized by hepatic and gastrointestinal disease and the other is the neurotoxic syndrome characterized by ataxia. In addition, in laboratory animals, the cycad toxins that cause the hepatic and gastrointestinal syndrome, glucoside derivatives of methylazoxymethanol (MAM), produce radiomimetic and/or carcinogenic effects. The effects include neural disease in fetuses and newborn animals, which are not necessarily related to the ataxia syndrome seen in domestic animals.

The results of experimental feeding with plant material can vary depending on which species is being tested, what part of the plant is used, and how it is processed. Both nuts and fronds of the cycads can be either hepatotoxic and enterotoxic, or neurotoxic.

It seems that the hepatotoxic components of the cycads, the MAM derivatives, are the least stable and often do not survive storage, particularly in the leaves. Multiple washings of cycad flour render it safer because of the instability of the MAM derivatives in the plants and their solubility in water. These techniques have been employed by many people in different tropical regions (Whiting, 1963). The instability of the toxins could also explain the wide disparities in results of different experimental feeding trials.

The animal species most frequently affected are cattle and sheep, but disease in horses and pigs has also been observed (Whiting, 1963). Cattle are more frequently affected by neurotoxic disease, while sheep are more frequently affected by the hepatic and gastrointestinal disease. Sheep, however, can develop spinal disease (J. M. Armstrong, personal communication, 1975) and cattle also can develop the hepatic disease after being fed comparatively low quantities of fresh *M. pauli-guilielmi* quickly but ataxia when fed larger quantities of dried plant (Hall, 1957).

There are numerous occurrences of disease in humans with symptoms corresponding to the hepatic/gastrointestinal syndrome and there is evidence that a paralytic or ataxic syndrome, similar to amyotrophic lateral sclerosis, occurs in people who consume the plants in a number of tropical areas Mickelsen, 1972).

A. The Neurotoxic Syndromes

1. *The Neural Disease Seen in Domestic Animals*

Hall (1957) described the clinical signs of *Macrozamia* sp. and *Bowenia* sp. poisoning in cattle as "those of uncontrolled extension and flexion of the hind quarters, which may swing to either side. In the early stages there is over extension of the lower hind limbs, and a 'goose-stepping' gait. Later there is incomplete extension, and the beast 'knuckles over' on its phalanges. Affected animals become worse when exercised and very severe cases will fall with the hind legs extended behind them dragging themselves along with their forelimbs which are never affected. There is never complete recovery." The cattle, although permanently paretic, could still remain viable. Other organs were not affected.

Microscopic changes were recognized in areas of the central nervous system (CNS) which were consistent with these clinical signs (Hall and McGavin, 1968; Mason and Whiting, 1968). These were degenerations of nerve fibers seen bilaterally in the upper spinal cords stained by the Marchi method, in the fasciculus gracilis, the dorsal spinocerebellar tract, and possibly the corticospinal tract (Fig. 2). In acute experimental cases, and very recent field cases, these changes could be recognized under hematoxylin and eosin staining as spheroids, pink staining masses, which have been interpreted as axonal swellings (Fig. 3) (Hooper et al., 1974). In transverse sections of the cervical and anterior thoracic spinal cords between C3 and T4, these masses, 10-50 μm in diameter, were most frequently visible in the superficial dorsolateral areas of the white matter, under the dorsal nerve roots. In the lumbar areas they were present most frequently in the ventromedial white matter. There was also some equivocal evidence of degeneration in the dorsal root ganglia of affected steers.

There have been few descriptions of the neurotoxic syndrome in other species, as they seem more likely to develop the hepatic/gastrointestinal syndrome. J. M. Armstrong (Personal communication, 1977) fed the freshly collected tips of old fronds of *M. riedlei* to sheep. These sheep became paralyzed and showed demyelination in their spinal cords on histological examination.

The changes described in the spinal cords of the cattle are probably axonal degenerations in the long spinocerebellar and corticospinal tracts, at the ends of the tracts most distant to the neurones. Hence the degenerations

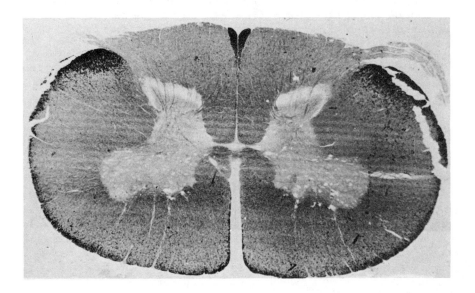

FIGURE 2 Degeneration in the cervical spinal cord in a steer poisoned by *Cycas armstrongii*, demonstrated by the Marchi method. (From Hall and McGavin, 1968.)

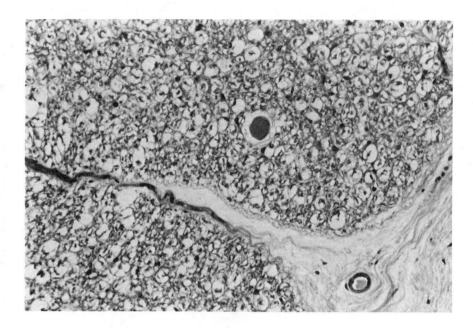

FIGURE 3 Spheroid or axonal swelling visible in the spinal cord of the same steer as Fig. 2, stained by hematoxylin and eosin.

could be interpreted as a dying-back process similar to that discussed by Cavanagh (1964).

The process has also been recognized in other intoxications, including those induced by some organophosphorus compounds in humans (Johnson, 1975), cattle (Williams et al., 1976; Beck et al., 1977), and pigs (Kruckenberg, et al., 1973), and by some poisonous plants—*Solanum esuriale* (O'Sullivan, 1976; McMeniman, 1976) and in part by *Karwinskia humboldtiana* (Charlton et al., 1970).

2. Experimental Teratogenic and Newborn Neural Diseases in Laboratory Animals

An experimental neural condition in laboratory animals is that of ataxia and hindlimb paralysis. This can be produced in newborn rats and mice or as a teratogenic affect by experimental poisonings by cycasin, a compound extracted from a number of the cycads, or by MAM, the active aglycone derivative of cycasin or cycasinlike compounds (Hirono and Shibuya, 1967; Hirono et al., 1968; Fushimi and Hirono, 1973; Spatz and Laqueur, 1968). The effect is specific against those cells dividing in the central nervous system at the time (e.g., those in the cerebellum in neonatal laboratory animals).

This destruction of dividing cells in the central nervous system (CNS) is a common feature of a number of animal species given prenatal or perinatal treatments of MAM (Haddad et al., 1972) and is believed to be a form of "granuloprival cerebellar hypoplasia" (Jones et al., 1972; Kurland, 1972), similar to that caused by radiation and radiomimetic viruses and poisons. This CNS effect, which is one primarily directed at the nuclei of dividing cells, seems more closely related to the carcinogenic and hepatotoxic effects of MAM and its derivatives than to the neurotoxic effects of cycads seen in field diseases.

3. *Hepatic Encephalopathy*

Another form of neural disease associated with cycad intoxication is that in animals affected with severe hepatic intoxication. Some of these may show hepatic encephalopathy. Sheep so poisoned have shown clinical signs of depression, head pressing, and even coma. They have also shown CNS spongy degeneration, which was probably associated with hyperammonemia during hepatic encephalopathy (Hooper, 1975).

B. The Hepatic and Gastrointestinal Syndromes

1. *The Hepatic and Gastrointestinal Syndromes Seen in Domestic Animals*

Sheep, pigs, horses, cattle, and rats poisoned by *Macrozamia* spp. (Gardiner, 1970; Gabbedy et al., 1975) and *Cycas* spp. (Hooper, 1978) develop liver cirrhosis with occlusion of central veins and proliferative gastro-enteritis. If large quantities of the cycads are consumed quickly, the more acute, haemorrhagic necrosis develops along with gastric or abomasal ulceration. Over a longer term, there is moderate hepatocyte megalocytosis and venoocclusive disease (Hooper, 1978).

2. *Experimental Hepatic and Gastrointestinal Diseases and Neoplasia in Laboratory Animals*

Methylazoxymethanol and its derivatives cause hepatic diseases in laboratory animals similar to those caused by whole plants in domestic animals. Over a longer term, they are extremely carcinogenic (Spatz, 1969; Hirono, 1972; Watanabe et al., 1975) and can produce a variety of radiomimetic effects other than those previously mentioned neural diseases in the newborn. Radiomimetic effects involve mutagenesis (Smith, 1966; Teas and Dyson, 1967), chromosomal aberrations (Teas et al., 1965), and other aberrations in dividing cells in the intestinal crypts, spleen, thymus, and testis (Zedeck et al., 1970).

IV. THE TOXINS

Poisons that have been isolated include glucoside derivatives of methylazoxymethanol (MAM), for example, cycasin and an amino acid, α-amino-β-methylaminopropionic acid. The MSM derivatives are toxic only if administered by the oral route. They require enzymic metabolism to the active aglycone MAM in the intestines for toxicity (Spatz et al., 1967).

There is evidence that extracted plant material is still neurotoxic when fed over a long period of time (exceeding 7 weeks). This may be due to the very low concentrations of the MAM derivatives remaining in the plant, or be due to the amino acid or to some other as yet unidentified toxin.

Although the MAM derivatives have been shown to be neurotoxic to prenatal and perinatal animals, this radiomimetic effect is only coincidentally clinically similar to the neural syndrome of cycad poisoning in the field. The pathogeneses of the two syndromes do not resemble one another. The activity of the MAM derivatives seems to be directed at the nuclei, particularly in rapidly dividing cells (Matsumoto and Higa, 1966; Shank and Magee, 1967), whereas the neurotoxic action of the plants in the field seems to start at the distal ends of axons.

The nonprotein amino acid has been measured only at very low levels, possibly too low to be of significance (Polsky et al., 1972). Chemically, it resembles the neurolathyrogens.

There is some evidence of a neurotoxic compound with a molecular weight exceeding 1000. Louw and Oelofsen (1975) dosed guinea pigs with retentate remaining after ultrafiltration of an extract of *Encephalartos altensteini* at a molecular retentivity of 1000. Their guinea pigs developed paralyses after a latent period of 15-30 days. These guinea pigs also developed "swelling of the eyes" within 1 day of dosage. Similar results were obtained with *Cycas armstrongii* (P. T. Hooper, I. J. Hamdorf and D. R. McEwan, unpublished observations, 1976) but the work needs further refining, as there has been difficulty in completely separating the hepatotoxic fraction.

REFERENCES

Beck, B. E., Wood, C. D., and Whenham, G. R. (1977). Triaryl phosphate poisoning in cattle. *Vet. Pathol.* 14:128-137.

Birdsley, M. B. (1972). A brief description of the cycads. *Fed. Proc. 31*: 1467-1469.

Cavanagh, J. B. (1964). The significance of the "dying back" process in experimental and human neurological disease. *Int. Rev. Exp. Pathol. 3*: 219-267.

Charlton, K. M., Pierce, K. R., Stortz, R. W., and Bridges, C. H. (1970). A neuropathy in goats caused by experimental coyotillo (*Karwinskia humboldtiana*) poisoning. IV. Lesions in the central nervous system. *Pathol. Vet. 7*:435-447.

Everist, S. L. (1974). In *Poisonous Plants of Australia*. Angus & Robertson, Sydney, p. 165.

Fushimi, K., and Hirono, I. (1973). Induction of a retinal disorder with cycasin in newborn mice and rats. *Acta. Path. Jpn.* 23:307-314.

Gabbedy, B. J., Meyer, E. P., and Dickson, J. (1975). Zamia palm (*Macrozamia riedlei*) poisoning of sheep. *Aust. Vet. J.* 51:303-305.

Gardiner, M. R. (1970). Chronic ovine hepatosis following feeding of *Macrozamia riedlei* nuts. *Aust. J. Agric. Res.* 21:519-526.

Haddad, R. K., Rabe, A., and Dumas, R. (1972). Comparison of effects of methylazoxymethanol on brain development in different species. *Fed. Proc. 31*:1520-1522.

Hall, W. T. K. (1957). Toxicity of the leaves of *Macrozamia* spp. for cattle. *Qld. J. Agric. Sci.* 14:41-52.

Hall, W. T. K., and McGavin, M. D. (1968). Clinical and neuropathological changes in cattle eating the leaves of *Macrozamia lucida* in *Bowenia serrulata* (family Zamiaceae). *Pathol. Vet. 5*:26-34.

Hirono, I. (1972). Carcinogenicity and neurotoxicity of cycasin with special reference to species differences. *Fed. Proc.* *31*:1493-1499.

Hirono, I., and Shibuya, C. (1967). Induction of a neurological disorder by cycasin in mice. *Nature (Lond.)* *216*:1311-1312.

Hirono, I., Shibuya, C., and Hayashi, K. (1969). Induction of a cerebellar disorder with cycasin in newborn mice and hamsters. *Proc. Soc. Exp. Biol. Med.* *131*:593-599.

Hooper, P. T. (1975). Spongy degeneration in the central nervous system of domestic animals. Part 1. Morphology. *Acta Neuropathol (Berl.).* *31*: 325-334.

Hooper, P. T. (1978). Cycad poisoning in Australia—etiology and pathology. In *Effects of Poisonous Plants on Livestock*, R. F. Keeler, K. R. Van Kampen, and L. F. James (Eds.). Academic Press, New York, pp. 337-347.

Hooper, P. T., Best, S. M., and Campbell, A. (1974). Axonal dystrophy in the spinal cords of cattle consuming the cyacad palm, *Cycas media*. *Aust. Vet. J.* *50*:146-149.

Johnson, M. K. (1975). Organophosphorus esters causing delayed neurotoxic effects. *Arch. Toxicol.* *34*:259-288.

Jones, M., Yang, M., and Nickelsen, O. (1972). Effects of methylazoxymethanol acetate on the cerebellum of the postnatal Swiss albine mouse. *Fed. Proc.* *31*:1508-1511.

Kobayashi, A. (1972). Cycasin in cycad materials used in Japan. *Fed. Proc.* *31*:1476-1477.

Kruckenberg, S. M., Strafuss, A. C., Marsland, W. P., Blanch, B. S., Marsh, R. T., Vestweber, J. G. E., and Winter, W. G. (1973). Posterior paresis in swine due to dermal absorption of the neurotoxic organophosphate tri-*o*-tolyl phosphate. *Am. J. Vet. Res.* *34*:403-404.

Kurland, L. T. (1972). An appraisal of the neurotoxicity of cycad and the etiology of amyotrophic lateral sclerosis on Guam. *Fed. Proc.* *31*:1540-1542.

Laqueur, G. L., and Spatz, M. (1968). Toxicology of cycasin. *Cancer Res.* *28*:2262-2267.

Louw, W. W. A., and Oelofsen, W. (1975). Carcinogenic and neurotoxic components in the cycad *Encephalartos altensteinii* Lehm. (Family Zamiaceae). *Toxicon* *13*:447-452.

Mason, M. M., and Whiting, M. G. (1968). Caudal motor weakness and ataxia in cattle in the Caribbean area following ingestion of cycads. *Cornell Vet.* *58*:541-554.

Matsumoto, H., and Higa, H. H. (1966). Studies on methylazoxymethanol, the aglycone of cycasin: methylation of nucleic acids in vivo. *Biochem. J.* *98*:20C-22C.

McMeniman, N. P. (1976). *Solanum esuriale*, a possible cause of humpy back in sheep. *Aust. Vet. J.* *52*:432-433.

Mickelsen, O. (1972). Introductory remarks. *Fed. Proc.* *31*:1454-1466.

O'Sullivan, B. M. (1976). Humpy back of sheep, clinical and pathological observations. *Aust. Vet. J.* *52*:414-418.

Polsky, F. I., Nunn, P. B., and Bell, E. A. (1972). Distribution and toxicity of α-amino-β-methylaminopropionic acid. *Fed. Proc.* *31*:1473-1475.

Shank, R. C., and Magee, P. N. (1967). Similarities between the biochemical actions of cycasin and dimethylnitrosamine. *Biochem. J.* *105*:521-527.

Sixth International Cycad Conference (1972). *Fed. Proc. 31*:1454-1546.

Smith, P. W. E. (1966). Mutagenicity of cycasin aglycone (methylazoxymethanol), a naturally occurring carcinogen. *Science 152*:1273-1274.

Spatz, M. (1969). Toxic and carcinogenic alkylating agents from cycads. *Ann. N.Y. Acad. Sci. 163*:848-859.

Spatz, M., and Laqueur, G. L. (1968). Transplacental chemical induction of microencephaly in two strains of rats. I. *Proc. Soc. Exp. Biol. Med. 129*:705-708.

Spatz, M., Smith, D. W. E., McDaniel, E. G., and Laqueur, G. L. (1967). Role of intestinal microorganisms in determining cycasin toxicity. *Proc. Soc. Exp. Biol. Med. 124*:691-697.

Teas, H. J., and Dyson, J. G. (1967). Mutation in *Drosophila* by methylazoxymethanol the aglycone of cycasin. *Proc. Soc. Exp. Biol. Med. 125*: 988-990.

Teas, H. J., Sax, H. J., and Sax, K. (1965). Cycasin: radiomimetic effect. *Science 149*:541-542.

Watanabe, K., Iwashita, H., Muta, K., Hamada, Y., and Hamada, K. (1975). Hepatic tumours of rabbits induced by cycad extract. *Gann 66*:335-340.

Wesley-Smith, R. N. (1973). Cycads and cattle in the Northern Territory. *J. Aust. Inst. Agric. Sci. 39*:233-236.

Whiting, M. G. (1963). Toxicity of cycads. *Econ. Bot. 17*:271-301.

Whiting, M. G., Spatz, M., and Matsumoto, H. (1966). Research progress on cycads. *Econ. Bot. 20*:98-108.

Williams, J. F., Dade, A. W., and Benne, R. (1976). Posterior paralysis associated with anthelmintic treatment of sheep. *J. Am. Vet. Med. Assoc. 169*:1307-1309.

Yang, M. G., Kobayashi, A., and Mickelsen, O. (1972). Bibliography of cycad research. *Fed. Proc. 31*:1543-1546.

Zedeck, M. S., Sternberg, S. S., Poynter, R. W., and McGowan, J. (1970). Biochemical and pathological effects of methylazoxymethanol acetate, a potent carcinogen. *Cancer Res. 30*:801-812.

15

CHEMISTRY, TOXICOLOGY AND PSYCHIC EFFECTS OF CANNABIS *

COY W. WALLER

University of Mississippi, University, Mississippi

I. INTRODUCTION

The cannabis plant and its toxic constituents have contributed to the stimulus in the early development of organic chemistry. The thesis of 1821 by Tscheepe (1821) on chemical research on the leaves of Indian hemp is probably the first recorded chemical effort on the plant. This work was done shortly after

---

* The plant, *Cannabis sativa* L., and marihuana, the leaves and flowering parts of it, should not be used as synonymous terms.

Serturner (1805) in 1817 had isolated morphine from opium. Since morphine is a nitrogenous base (alkaloid) and can be manipulated by dissolving it in organic solvent as its free base and into water as its salts, its isolation in a pure form was facilitated. Δ^9-Tetrahydrocannabinol (THC), the main psychic substance in cannabis, is not an alkaloid and isolation in a pure form was delayed about 130 years. The correct structure for Δ^9-THC had to await the development for methodology and instrumentation until Gaoni and Mechoulam (1964a) published their work. The contributions from early works on plant toxins (especially alkaloids) and the development of chemistry was acknowledged by Gates and Tschudi (1956) when they reported on the first complete synthesis of morphine. Several complete syntheses of Δ^9-THC were reported subsequent to Mechoulam's work (Gaoni and Mechoulan, 1964a). The most practical synthesis was published by Petrzilka and Sikemeier (1967). It was Petrzilka's synthetic method that made it possible to provide large quantities of Δ^9-THC for a formal research program that is still ongoing.

Since alkaloids have been frequently the toxin in plants and since Δ^9-THC is not an alkaloid, it needs to be said early that the cannabis plant does contain alkaloids. However, the cannabis alkaloids do not appear to occur in sufficiently high concentration to contribute to the toxicity of the plant. These cannabis alkaloids will be discussed later.

The cannabis plant is an unstable *genus* and many species have been described and named. A United Nations (Working) Committee of world experts considered the question of separate species and decided that only one species was justified and specified it to be the monotypic, *Cannabis sativa* L. (U.N., 1976). There are a large number of "varients" of *C. sativa* and a wide variation of constituents in these, particularly as to quantities of each, which will be discussed later.

The hemp plant is a synonymous term for the cannabis plant and has been used over the millennia. Its use as a economic crop (Li, 1974) for fiber, food, and medicine has spread all over the world with human migration. It has been an important source of fiber, but producers of fiber were not aware of the psychic effects of the leaves and their constituents. Yet its therapeutic effects were recorded in the ancient literature of China.

Therapeutic claims have been made for cannabis in treatment of nearly every known malady. Thus it has been an item of folk medicine and in the earliest pharmacopeia and related compendia. With development of more potent drugs with better specificity and standardization, its use in medicine declined. In 1940, it was dropped entirely and did not appear in the *Pharmacopeia of the United States XII*, which is the official compendium, and this issue was recognized by the Congress on November 1, 1942. The standardization of the strength of cannabis and its preparation was so difficult that the dog assay last appeared in the *U.S. Pharmacopeia X* of 1926. Reasons for this difficulty are discussed later.

The question of potency of cannabis and its preparation has been the underlying cause of debates for or against safety. Results reported in the Cannabis literature are no better than the potency of the material used in the study. Fortunately, since 1970, scientists could have, if desired, obtained material of known potency for their research.

This report endeavors to review the toxicity and psychic effects with relevance to chemical standards that support the results. In a recent report by the Council of the American Medical Association on "Marijuana, Its Health

Hazards and Therapeutic Potentials," the following conclusion was reached: "Marijuana is a dangerous drug. . . . The fact that marijuana may prove to have therapeutic value in medical practice does not indicate that it is a safe drug for recreational use" (*JAMA*, 1981).

II. LITERATURE AND RESEARCH PROGRAMS

Since antiquity hemp has been an important item of commerce but less so after cotton and synthetic fibers have become known. Also, since the medicinal properties of the plant have been included in all recorded compendia of remedies, and research has been done since early 1800s, the literature on cannabis is massive. In 1965, Nathan B. Eddy assisted the U.S. Commission on Narcotic Drugs with a cannabis bibliography (Eddy, 1965) which listed 1860 citations in the technical literature. The literature during the 10 years following the U.N. publication was covered by Waller et al. (1976) with annotated abstracts of 3045 references. Also, Waller et al. (1982) has in press an annotated bibliography covering additional 2670 citations. Thus the more than 7500 references on cannabis are mostly technical articles and do not include those technical aspects of the cultivation and processing of the plant or its parts as food, as a textile, or for other industrial purposes. Also excluded are the lay press articles. There are other bibliography sources such as that of Abel (1979), who lists 8177 citations through 1977, and the STASH publication (Gamage and Zerkin, 1969). Books and chapters in books will be referenced when used the same as journal articles; however, the book by Walton (1938) deserves special consideration because of its extensive coverage of the pharmacological and toxicological literature and its timeliness. With the title of his book, Walton caught the trend of the time in the Americas by using marihuana instead of cannabis. He states that marihuana is derived from the generic word "maraguango," meaning any substance producing an intoxication. Walton devotes an entire chapter of his book to nomenclature with extensive references. He refers to Herodotus (486-406 B.C.) for the early use of cannabis (kannabis) as an ancient Greek and Latin term. The United Nations in its *Multilingual List of Narcotic Drugs Under International Control* (U.N., 1968) has an extensive list of terms of cannabis and cannabis resin (without references). It is almost inconceivable that there is no listing or reference to any constituent of cannabis or of cannabis resin in the U.N. list, whereas even chemical structures are given for morphine, codeine, cocaine, and many of their synthetic surrogates.

Cannabis as defined by the U.S.P. XI (*U.S. Pharmacopeia XI*, 1935) "consists of the dried flowering tops of the pistillate plants of *Cannabis sativa* Linné (Fam. Moraceae). [It] contains not more than 10 percent of its fruits, large foliage leaves and stems over 3 mm in diameter, and not more than 2 percent of other foreign organic matter [nor] yields more than 5 percent of acid-insoluble ash." "Marihuana," the item of illicit trade, does not necessarily follow this definition for cannabis. When the marijuana preparation is composed of smaller leaves, it can come from the male as well as the female plant. The respective potencies of the male and female plants will be discussed later. A very recent form of marihuana with increased potency is "sinsemilla." It comprises the flowering tops of the female plant, which are grown with great care to eliminate any male plants before pollination can occur.

Thus is contains no seeds. It can be up to seven times as potent as most "street" marihuana (C. E. Turner and M. Russell, unpublished observations, 1982).

A staggering increase in the number of meaningful publications followed the availability of standardized materials and monetary support for research from the National Institute of Mental Health (NIHM) and later the National Institute on Drug Abuse (NIDA). The formalized program on marihuana by NIHM was an outgrowth of a recommendation made by this author to a program review committee chaired by D. X. Freedman on March 8, 1968. The necessity for NIMH to supply materials and money to accomplish its mission was described by Morten G. Miller (1970) on February 16, 1969. Since marihuana and its constituents are still schedule I drugs, to obtain materials for research required approval by NIMH, which functioned mainly through the FDA-PHS Psychotomimetic Agents Advisory Committee. This committee was guided by its executive secretary, John A. Scigliano, from its inception until 1981, when Dr. Scigliano retired. Each marihuana researcher was required to obtain registration with the Bureau of Narcotics and later with DEA. This critical symbiotic tie between NIMH (NIDA) and the research community is still functional. By the end of the first year of this expanded research program, NIMH was supporting 58 grants and 7 contracts with a yearly budget of over 42 million (Miller, 1970). In addition to the NIDA support of a marihuana research program, today NCI and the Eye Institute are supporting research on the theraputic use of cannabis and/or its constituents in their cancer and eye programs. These U.S. programs are supportive of the worldwide cannabis program of the Division of Narcotic Drugs, United Nations.

III. CONSTITUENTS OF *CANNABIS SATIVA* L.

From a chemotaxonomical consideration, the cannabis plant is unique. Taxonomists earlier classified *Cannabis* (hemp) and *Humulus* (hops) in the same family, Moraceae, but more recently *Cannabis* was placed in the Cannabaceae family. However, the cannabinoid constituents of cannabis have been found only in the cannabis plant. The cannabinoids were not found even in hops (Fenselau et al., 1976).

The botany of cannabis (Quimby et al., 1973) is not discussed here; however, in describing the origin of the plant, Schultes (1970) made reference ot Linnaeus' "Species plantarum" of 1753 for the name *Cannabis sativa* L.

A. Cannabinoids

Cannabinoid is a term used by Mechoulam (Mechoulam and Gaoni, 1967; Mechoulam, 1973) for the C_{21} compounds, their carboxylic acids, analogs, and transformation products, which are typical of and present in the cannabis plant. More than 60 cannabinoid compounds have been isolated from the cannabis plant and have been structurally characterized. A recent review listed 421 chemicals of various classes that occur in the plant (Turner et al., 1980).

The first cannabinoid to be isolated in a pure form from cannabis or, more correctly, from hashish (cannabis resin) was cannabinol [1]. The correct

[1]
Cannabinol

empirical formula, $C_{21}H_{26}O_2$, was assigned to it. Its crystalline acetate was prepared and melted at 75°C. This work on cannabinol by Wood et al. (1899) was terminated probably because of a series of tragic accidents: Wood was seriously injured when he took some cannabinol and barely escaped from a zinc ethyl fire; Esterfield was killed in an explosion while performing hydrogenation of cannabinol; and Spivey was killed during the nitration of cannabinol (Walton, 1938). It was not until 1933, when Cahn (1933) showed cannabinol to be a dibenzopyran, and not until 1940, when B. R. Baker in his doctoral work under Roger Adams (Adams et al., 1940b) synthesized cannabinol by an unequivocal method, that the total structure was established.

More than a century had passed from the first chemical publication (Tscheepe, 1821) until the structure of the first isolated cannabinoid was established. Yet ironically, we know today that the cannabis plant does not produce cannabinol. It is an artifact produced by an oxidative process from Δ^9-THC (Levine, 1944; Razdan et al., 1972). When fresh plant material of cannabis was analyzed the cannabinol content approached zero percentage (Fetterman et al., 1971a). Furthermore, cannabinol is only about one-tenth as active biologically as Δ^9-THC. Still, cannabinol and some of its analogs occur in marihuana and products from the leaves of cannabis. The cannabinols so far isolated from cannabis are shown in Table 1.

The synthesis of cannabinol by Adams' group (Adams et al., 1940b) and the testing of it confirmed that it was not responsible for the intense biological active of cannabis. This resulted in a renewed effort to isolate the active substance. At about the same time, two groups of researchers isolated the new crystalline substance, cannabidiol, $C_{21}H_{30}O_4$. Adams and coworkers (1940a) used American hemp, which contained a large amount of cannabidiol. The other group, Todd and co-workers (Jacob and Todd, 1940) in England, used the Indian variety, which contained very little, and Egyptian material, which contained a intermediate amount of cannabidiol (Todd, 1946). This new substance was also found to be biologically inactive, but cyclization converted it to tetrahydrocannabinols which were highly biologically active. The cyclization of cannabidiol to THC with acidic catalysts gave products with rotation varying with the strength of acidic substance used (Adams et al., 1940c). The structure of cannabidiol [2] was not fully established until 1963 by Mechoulam and Shvo (1963). In subsequent work, Mechoulam and collaborators established the structures of the tetrahydrocannabinols Δ^9-THC [3] and Δ^8-THC [4] (Mechoulam, 1973, and references therein).

TABLE 1

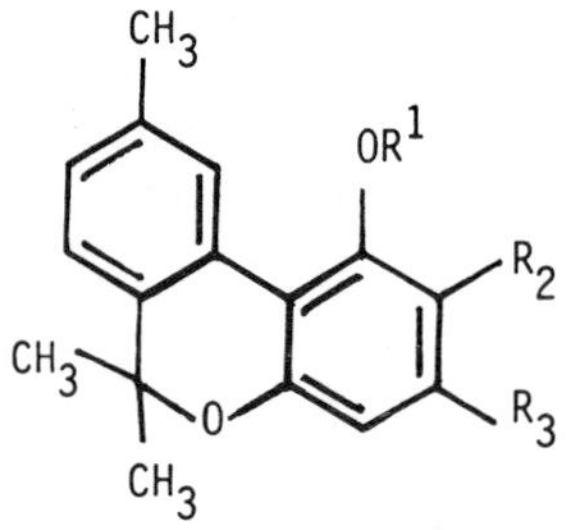

6,6,9-Trimethyl-6H-Dibenzo[b,d]pyran-1-ol type

Cannabinol type

Name and abbreviation	R_1	R_2	R_3	CAR number[a]	Reference
Cannabinol, CBN	H	C_5H_{11}	H	521-35-7	Adams et al., 1940b
Cannabinolic acid, CBNA	H	C_5H_{11}	COOH	2808-39-1	Mechoulam, 1973
Cannabinol methylether, CBNOM	CH_3	C_5H_{11}	H		Turner et al., 1980
Cannabinol-C_4 CBN-C_4	H	C_4H_9	H		Turner et al., 1980
Cannabivarin, (cannabivarol), CBN-C_3	H	C_3H_7	H	33745-21-0	Turner et al., 1980
Cannabiorcol, CBN-C_3	H	CH_3	H	19825-73-1	Turner et al., 1980

[a]Chemical Abstracts Substance Registry Number.

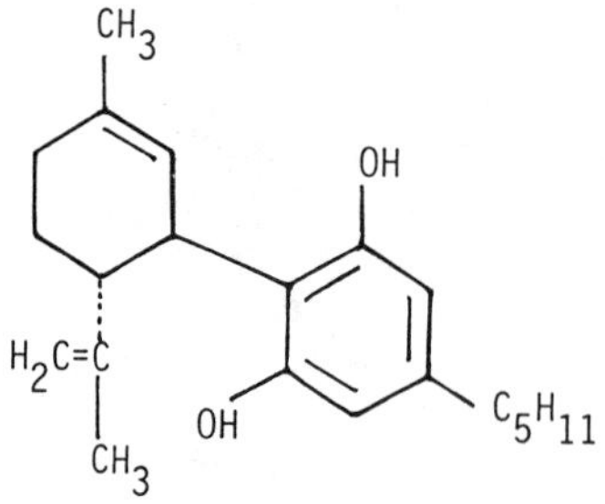

[2]
Cannabidiol (CBD)

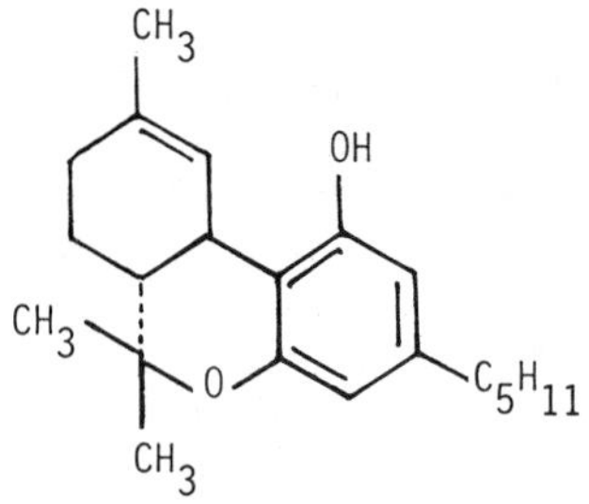

[3]
Δ^9-trans-Tetrahydrocannabinol
Δ^9-THC=Δ^1-THC

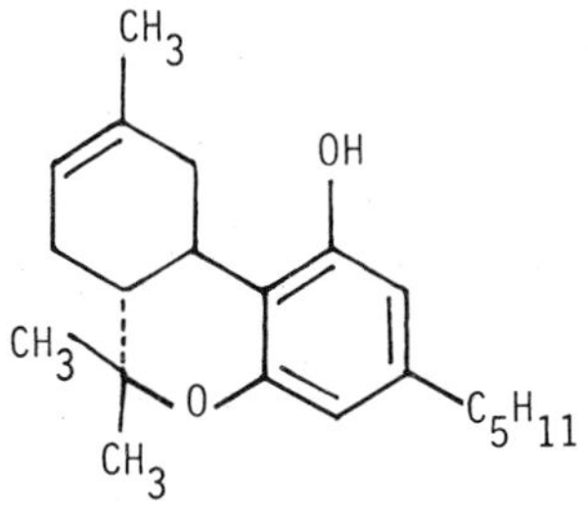

[4]

Δ⁸-trans-Tetrahydrocannabinol
Δ^8-THC=Δ^6-THC

Table 2 lists the cannabidiol type of compounds which have been isolated from cannabis. These resorcinol and β-resorcinolic acid derivatives have the terpenoid moeity in a trans configuration with the phenolic ring and are optically active.

The cannabinodiol [5] that was isolated from Lebanese hashish (Lousberg et al., 1977) is probably not a constituent of the cannabis plant but is formed exogenously by an oxidative dehydrogenation of the terpenoid ring.

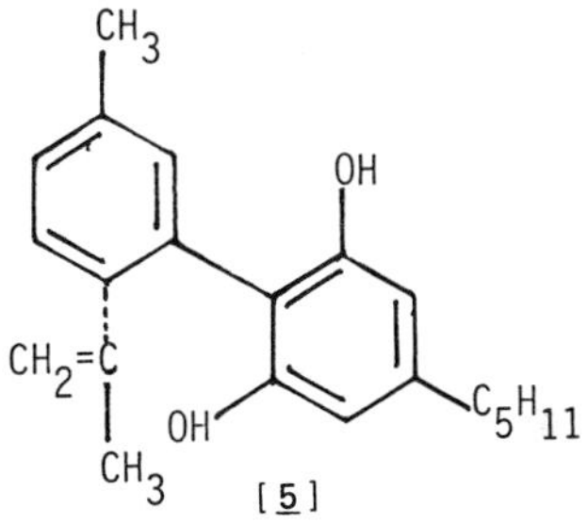

Cannabinodiol
5'-methyl-2'-(1-methylethenyl)-4-pentyl-1,1'-biphenyl-2,6-diol
CAR No.   39624-81-2

Δ^9-*trans*-tetrahydrocannabinol [3] is the most active of the constituents in cannabis on the central nervous system (CNS); therefore, it and its isomer Δ^8-THC [4], which is also active on the CNS (75%), have been the subject of many research reports. Since these compounds are also called Δ^1 and Δ^6, respectively, it is necessary to recognize the different number systems.

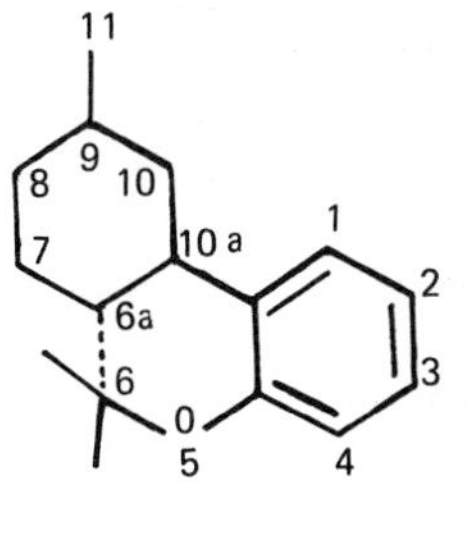

Dibenzopyran-CA System

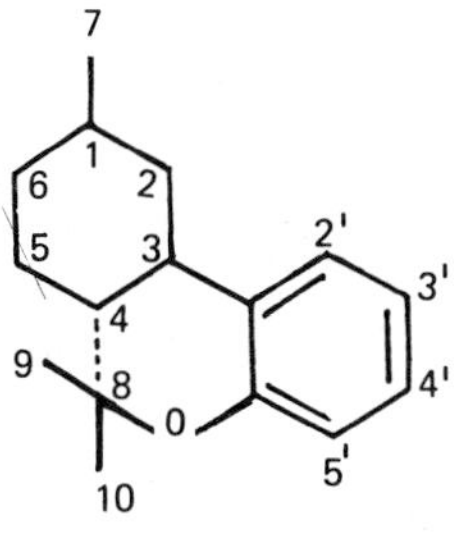

Terpenoid System

TABLE 2

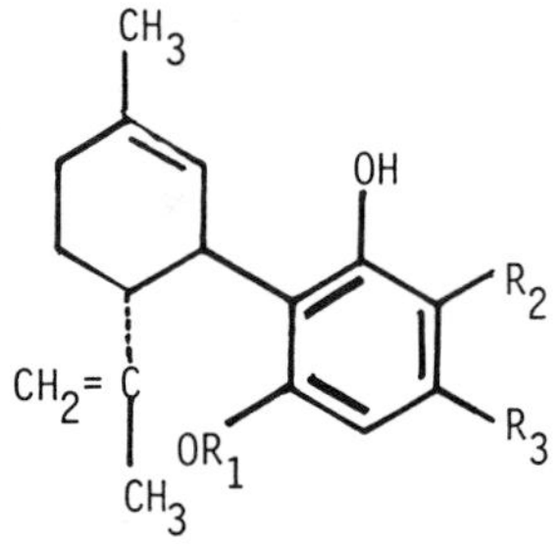

Cannabidiol type

2-(3-methyl-6-(1 methylethenyl)-2-cycolhexen-1-yl)-
5-alkyl-, (IR-<u>trans</u>)-1-benzeneol   type

Name and abbreviation	R_1	R_2	R_3	CAR number	Reference
Cannabidiol, CBD	H	H	C_5H_{11}	13956-29-1	Mechoulam and Shvo, 1963
Cannabidiolic acid (cannabidiolcarboxylic acid), CBDA	H	COOH	C_5H_{11}	1244-58-2	Mechoulam and Gaoni, 1965
Cannabidiol methyl ether, CBDM	CH_3	H	C_5H_{11}		Turner et al., 1980
Cannabidiol-C_4, CBD-C_4	H	H	C_4H_9		Turner et al., 1980
Cannabidivaric acid, CBDVA	H	COOH	C_3H_7		Shoyama et al., 1977
Cannabidivarin, CBDV	H	H	C_3H_7	24272-48-4	Turner et al., 1980
Cannabidorcol, CBDO	H	H	CH_3	35482-50-9	Turner et al., 1980
Cannabidorcolic acid	H	COOH	CH_3	64147-52-0	Turner et al., 1980

In Table 3 are listed the Δ^9-*trans*-tetrahydrocannabinol compounds found in cannabis with the exception of the Δ^9-*cis*-THC, which was reported by Smith and Kempfert (1977). The presence of this cis compound needs to be confirmed. In any event the cis isomers of the THCs are inactive as CNS agents (Mechoulam, 1973). The neutral compounds appear to be absent in fresh plant material. When the analysis of fresh plant material for both Δ^9-THC and Δ^9-THCA was performed (Fetterman et al., 1971b; Turner et al.,

TABLE 3

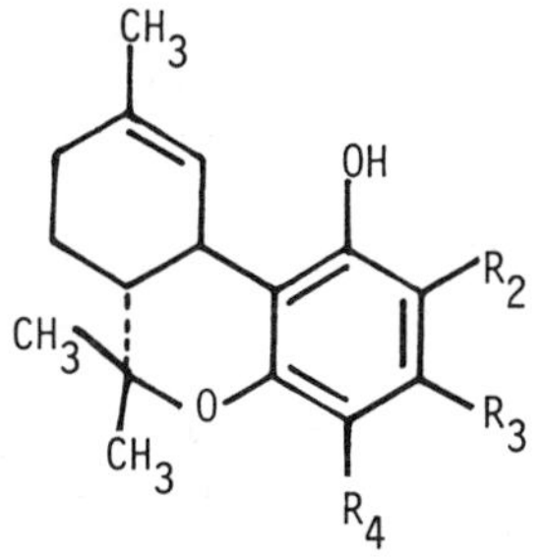

Δ^9-<u>trans</u>-tetrahydrocannabinol type

6a,7,8,10a-Tetrahydro-6,6,9-trimethyl-3-alkyl-(6aR-<u>trans</u>)-
6H-dibenzo(b,d)pyran-1-ol type

Name and abbreviation	R_2	R_3	R_4	CAR number	Reference
Δ^9-Tetrahydrocannabinol, Δ^9-THC, Δ^1-THC	H	C_5H_{11}	H	1972-08-3	Mechoulam and Shvo, 1963
Δ^9-Tetrahydrocannabinolic acid A, Δ^9-THCA(A)	COOH	C_5H_{11}	H	23978-85-0 4326-55-0	Mechoulam and Shvo, 1963
Δ^9-Tetrahydrocannabinolic acid B, Δ^9-THCA(B)	H	C_5H_{11}	COOH	23978-84-9	Mechoulam and Shvo, 1963
Δ^9-Tetrahydrocannabinol-C_4, Δ^9-THC-C_4	H	C_4H_9	H		Turner et al., 1980
Δ^9-Tetrahydrocannabinolic acid-C_4, Δ^9-THCA-C_4	COOH	C_4H_9	H		Turner et al., 1980
Δ^9-Tetrahydrocannabivarol (Δ^9-Tetrahydrocannabivarin), Δ^9-THCV	H	C_3H_7	H	31262-37-0	Turner et al., 1980
Δ^9-Tetrahydrocannabivarolic acid, Δ^9-THCVA	COOH	C_3H_7	H	39986-26-0	Turner et al., 1980
Δ^9-Tetrahydrocannabiorcol, Δ^9-THCO	H	CH_3	H	22972-65-2	Mechoulam, 1973
Δ^9-Tetrahydrocannabiorcolic acid, Δ^9-THCOA	COOH	CH_3	H		Mechoulam, 1973

1973a) Δ^9-THC approached a very low level. Δ^9-THCV also exists in fresh plant material as the Δ^9-THCVA (Turner et al., 1973b).

Thus Δ^9-THC is not produced in cannabis in the cannabinoid biosynthetic pathway, but through the decarboxylation of Δ^9-THCA. This decarboxylation takes place readily on heating, smoking, and more slowly by aging.

The analysis of fresh plant material shows at best a very low content of Δ^8-THC [4], Chemical Abstracts Substance Registry (CAR) number 5957-75-5, and Δ^8-THCA, CAR 23988-89-4 (Fetterman et al., 1971b). This is particularly true when samples are handled using good analytical practices (Turner et al., 1973a). Acid catalysts even at very low levels convert Δ^9-THC to Δ^8-THC. The synthesis of Δ^9-THC frequently goes through Δ^8-THC, where it is purified and then converted back to Δ^9-THC by procedures that avoid acids. Razdan recently reviewed the total synthesis of cannabinoids (Razdan, 1981), so this area is not covered here.

However, since almost all of the Δ^9-THC [3] that has been used in research was totally synthetic and not of natural origin, it should be stated that the synthesis used was published by Petrzilka et al. (1967, 1969). Adaptation of the Petrzilka process to large-scale work was done at Arthur D. Little, Inc., under NIMH contract PH 43-68-1339 (see *Technical Report 3*, January 1972). Some of the individuals who made major contribution to the success of this effort and achieved synthesis of several batches each in the 3- to 5-kg sizes were: R. Razdan, G. R. Handrick, H. G. Pars, and the project engineer, L. K. R. Woodland, who carried out the pilot plant operation. The starting materials are (+)-*cis*- or (+)-*trans*-p-mentha-2,8-dien-1-ol and olivetol using p-tolnenesulfonic acid as a catalyst.

The $\Delta^{9,11}$-THC and Δ^8-THC were removed in the final product by chromatography on silver nitrate-silica gel to give essentially pure (-)-Δ^9-*trans*-THC, identical with the natural Δ^9-THC [3].

(+) <u>cis</u>   (+) <u>trans</u>   Olivetol
       p-mentha-2,8-dien-ol

(-) Δ^8-<u>trans</u>-THC ⟶ (-) Δ^9-<u>trans</u>-THC 90% + (-) $\Delta^{9,11}$-<u>trans</u>-THC 3% +
                                (-) Δ^8-<u>trans</u>-THC 5%

Razdan (1981) and co-workers studied the details of the Petrzilka reaction and identified several abnormal products therein. One of these products was in the iso-THC series and identified as compound [6]. Recently, a propyl analog [7] in the iso-THC series was isolated from an Indian variant of cannabis (Turner er al., 1981).

The synthesis of (-)-Δ^8-*trans*-THC [4] and (-)-Δ^9-*trans*-THC [3] from 1970 to date provided the natural THCs (although from synthetic sources) for research. However, the synthetic THC [8] that was available from the methods of Roger Adams' group from the 1930s to 1960 was a different THC. Unfortunately, the literature during the period 1930-1960 is frequently unclear as to which tetrahydrocannabinol is used in research. Adams et al. (1949) prepared many derivatives of [8] in which R_3, R_6, and R_9 were varied.

Cannabigerol [9] and its acids, ethers, and analogs which have been isolated from cannabis are listed in Table 4. Cannabigerolic acid is the first

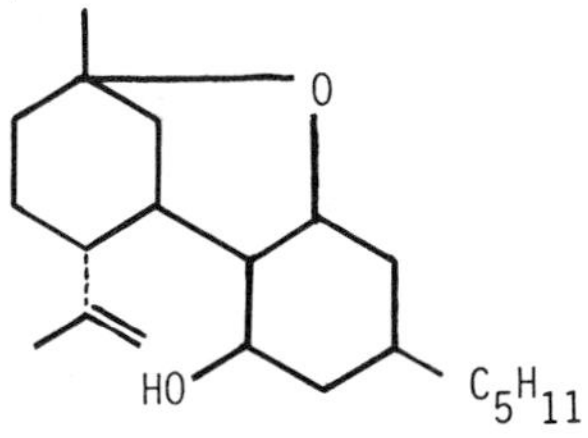

[6]

Iso-THC Series
CAR No. 23050-47-7

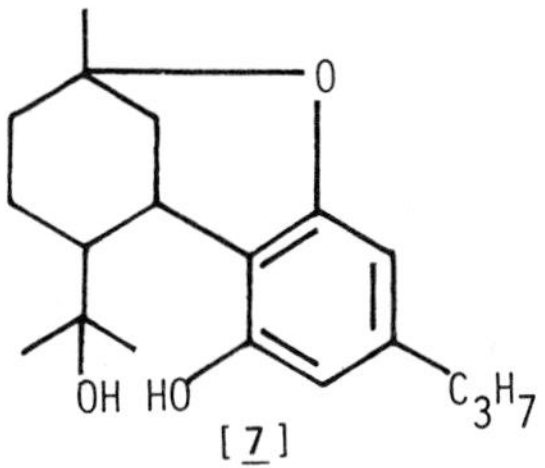

[7]

Cannabiglendol

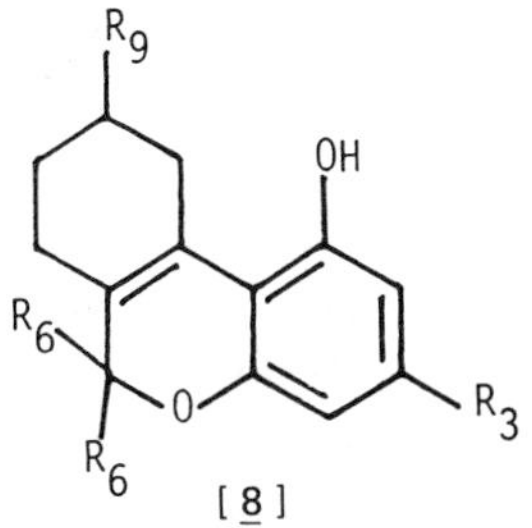

[8]

$\Delta^{6,10a}$-THC

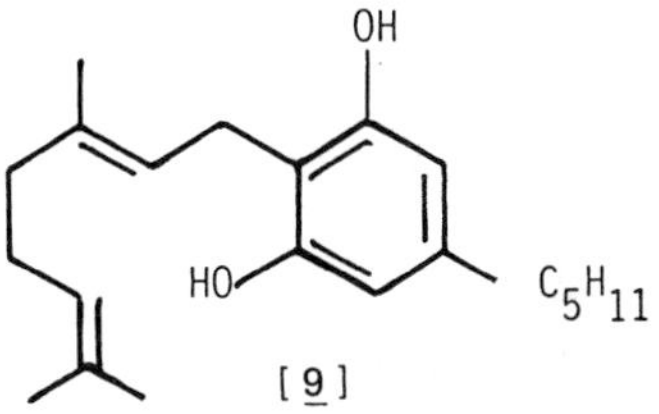

[9]

Cannabigerol

cannabinoid in the biosynthetic pathway. This biosynthetic pathway was determined by Shoyama et al. (1975a) by using labeled malonic acid, mevalonic acid, geraniol, and/or nerol in feeding experiments.

Cannabigerol and its acid were first isolated by Mechoulam and Gaoni (Shoyama et al., 1970; Gaoni and Mechoulam, 1964b) from hashish. Nishioka and co-workers (Shoyama et al., 1970, 1975b, 1977) used fresh plant materials to isolate cannabigerol and its acids, analogs, and monomethyl ethers.

Cannabichromenic acid (CBCA) was isolated from cannabis (Shoyama et al., 1968) and decarboxylated with heat to cannabichromene [10]. Earlier, [10] had been isolated from hashish (Clausen et al., 1966; Gaoni and Mechoulam, 1966). Nishioka, working with fresh plant material, also isolated CBCV (Shoyama et al., 1975b) and its acid (Shoyama et al., 1977). Table 5 shows the CBC derivatives that have been isolated from cannabis.

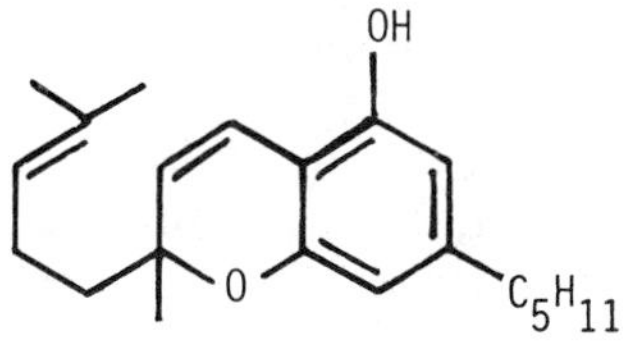

[10]

Cannabichromene

TABLE 4

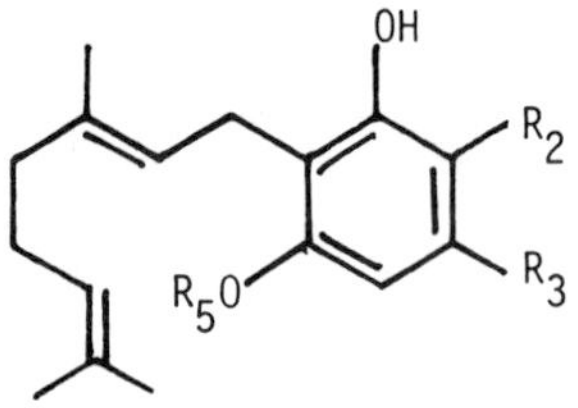

Cannabigerol   type

2-(3,7-Dimethyl-2,6-octdienyl)-5-alkyl,(E)-1,3-benzendiol - type

Name and abbreviation	R_2	R_3	R_5	CAR number	Reference
Cannabigerol, CBG	H	C_5H_{11}	H	25654-31-3	Shoyama et al, 1975b
Cannabigerolic acid, CBGA	COOH	C_5H_{11}	H	25555-57-1	Mechoulam and Gaoni, 1965
Cannabigerol methyl ether, CBGM	H	C_5H_{11}	CH_3	29106-17-0	Yamauchi et al., 1968
Cannabigerolic acid methyl ether, CBGAM CBGAM	COOH	C_5H_{11}	CH_3	29624-08-6	Shoyama et al., 1970
Cannabigerovarin, CBGV	H	C_3H_7	H	55824-11-8	Shoyama et al., 1975b
Cannabigerovarinic acid	COOH	C_3H_7	H	64924-07-8	Shoyama et al., 1977

    Cannabichromene is not a primary biosynthetic cannabinoid (its acid, CBCA, is the precusor) from cannabis, but it is present in dried plant material such as marihuana and hashish. Since CBC and CBD both have the same retention time on most columns used in gas chromatographic (GC) analysis, frequently the analysis of marihuana reported a content of CBD that was really a content of CBD + CBC. Turner et al. (Turner and Hadley, 1973; Turner et al., 1975) developed procedures of analysis by GC to analyze each of these and found some variants (Turner and Hadley, 1973) of cannabis that had no CBD content, while the content of CBC varied widely in the different variants (Holley et al., 1975). An improved synthesis of CVC over earlier methods was recently published (ElSohly et al., 1978a). Decomposition products of CBC, CBCA, and CBCV often found in old and abused plant materials and extracts are the cannabinoids cannabicyclol [11], its acid (CBLA), and cannabicyclovarin (CBLV).

TABLE 5

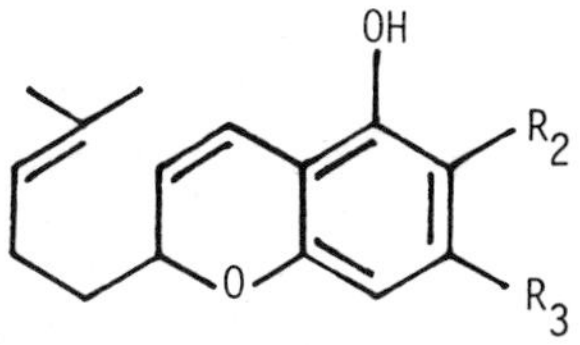

Cannabichromene   type

2-Methyl-2-(4-methyl-3-pentenyl-1)-7-alkyl-2H-1-benzopyran-5-ol
type

Name and abbreviation	R_2	R_3	CAR number	Reference
Cannabichromene CBC	H	C_5H_{11}	20675-51-8	Claussen et al., 1966
Cannabichromenic acid, CBCA	COOH	C_5H_{11}	20408-52-0	Shoyama et al., 1968
Cannabichromevarin, CBCV	H	C_3H_7	57130-04-8	Shoyama et al., 1977
Cannavichromevarinic, CBCVA	COOH	C_3H_7	64898-02-8	Shoyama et al., 1975b

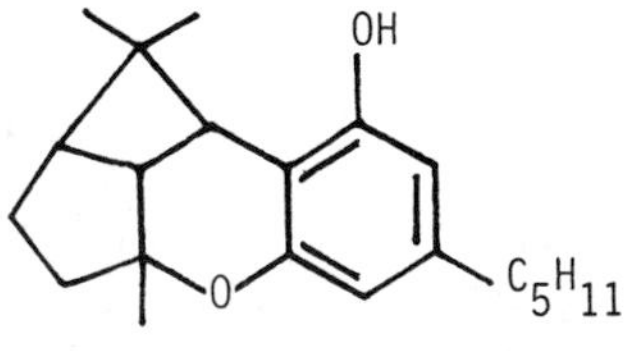

[11]

Cannabicyclol, CBL, CAR No. 21366-63-2

Korte et al. (1965), in a review of compounds from hashish, called CVC (mp 146°C) THC II and CBL (mp 128°C) THC III. Nishioka and co-workers isolated CBLA and determined it to be an artifact (Shoyama et al., 1972).

Cannabielsoin is formed by pyrrolysis of CBD (Kuppers et al., 1973), and its acids A and B have been isolated from hashish (Shani and Mechoulam, 1974). Even the propyl analog of cannabielsoin and its acid B were detected in cannabis extract (Grote and Spiteller, 1978). Still, these are considered to be artifacts and to be formed by oxidation of the double bond of the cyclohexenyl ring of CBD [2] and of CBDA followed by cyclization to form the

TABLE 6

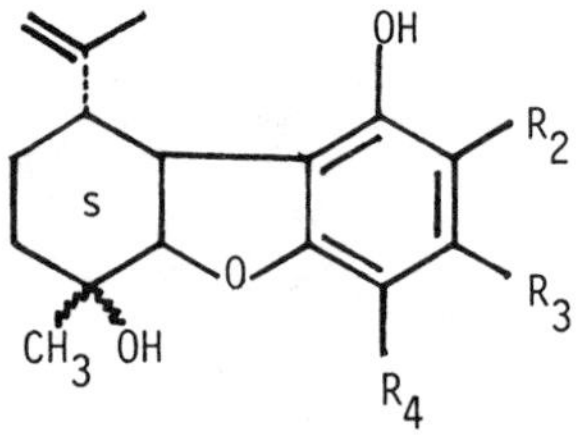

Cannabielsoin type

5α,6,7,8,9,9α-Hexahydro-6-methyl-9-(1-methylethenyl-1)-3-alkyl
[5aS(5aα,6α,9α,9aα)]-1,6-dibenzofurandiol type

Name and abbreviation	R_2	R_3	R_4	CAR number	Reference
Cannabielsion, CBE	H	C_5H_{11}	H	52025-76-0	Kuppers et al., 1973
Cannabielsoic acid A, CBEA-A	COOH	C_5H_{11}	H	26674-02-2	Shani and Mechoulam, 1974
Cannabielsoic acid B, CBEA-B	H	C_5H_{11}	COOH	55652-62-5	Shani and Mechoulam, 1974
C_3-Cannabielsoic acid A	COOH	C_3H_7	H	67664-29-6	Grote and Spiteller, 1978
C_3-Cannabielsoic acid B	H	C_3H_7	COOH	67665-30-9	Grote and Spiteller, 1978

hexahydrodibenzofuran moiety. In Table 6 are listed the cannabielsoin of compounds isolated from cannabis.

Two related furans, cannabifuran [12] and dehydrocannabifuran [13], have been isolated from cannabis by Friedrich-Fiechtl and Spiteller (1975). These authors (Friedrich-Fiechtl and Spiteller, 1975) also reported the isolation of cannabichromanon [14] and 10-oxo-$\Delta^{6,10a}$-THC [15].

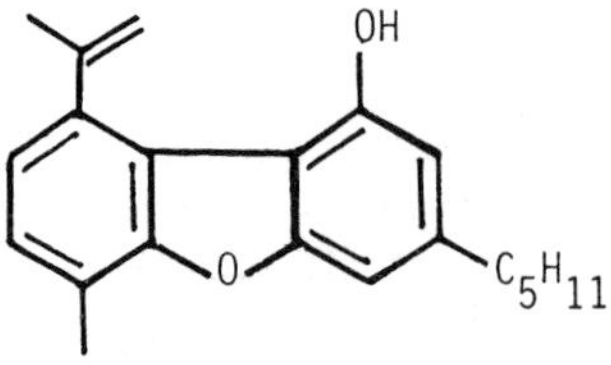

[13]

Dehydrocannabifuran

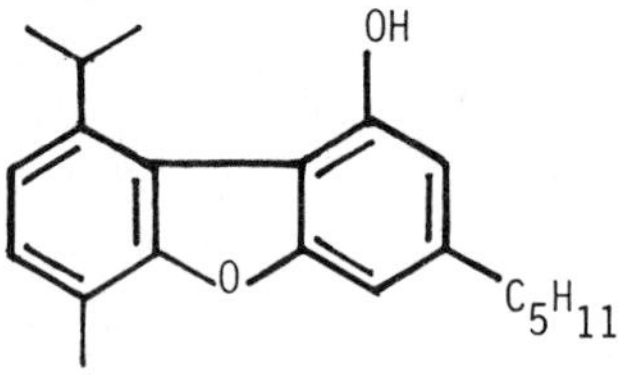

[12]

Cannabifuran

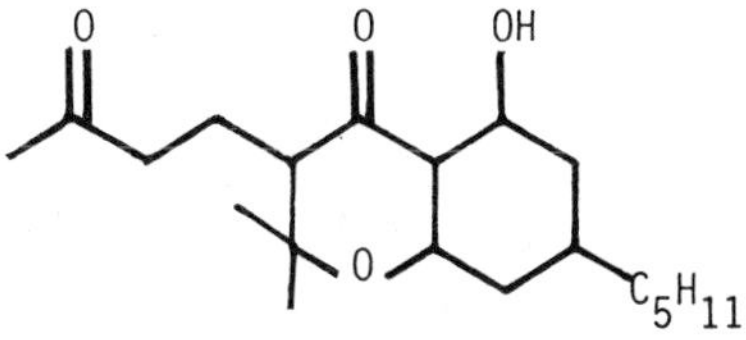

[14]
Cannabichromanon

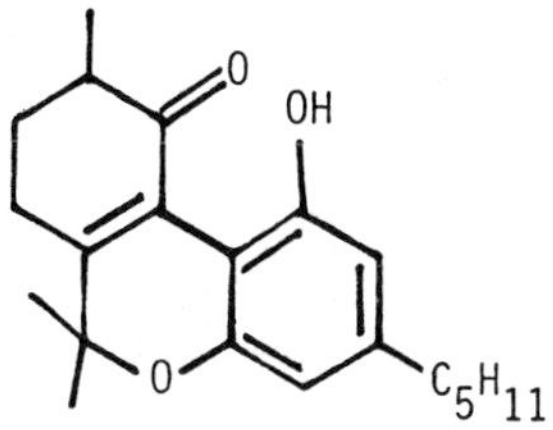

[15]
10-oxo-$\Delta^{6a,10a}$-Tetrahydrocannabinol

Razdan et al. (1972), in studying the stability of Δ^9-THC [3], found that a diene was formed which went to CBN [1]. The instability of this diene (a), which ws produced by oxidation dehydrogenation, would itself form poly-hydroxylated compounds. Some of these that have been found in hashish and marijuana extracts are shown below:

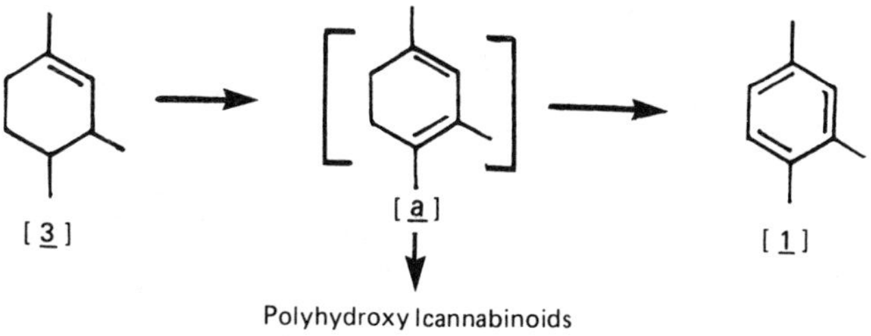

[3]

[a]

[1]

Polyhydroxy lcannabinoids

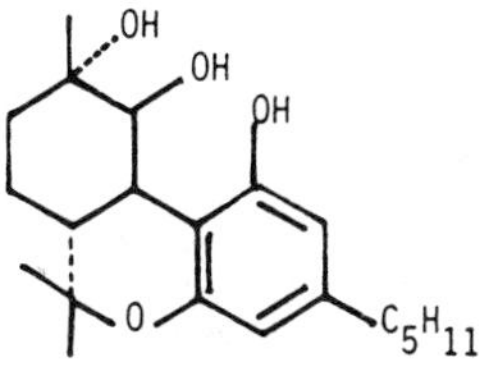

Cannabiripsol
CAR No. 72236-32-9
(Boeren et al., 1979)

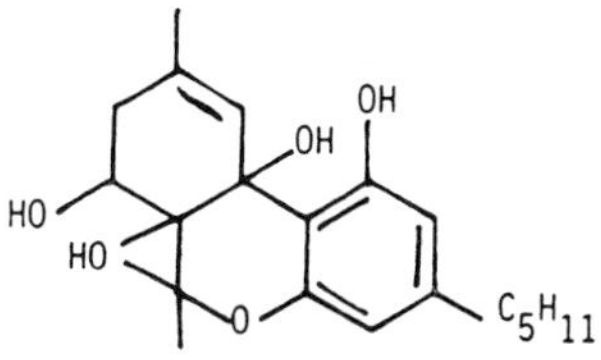

6a,7,10a-Trihydroxy-Δ^9-THC
(Boeren et al., 1979)

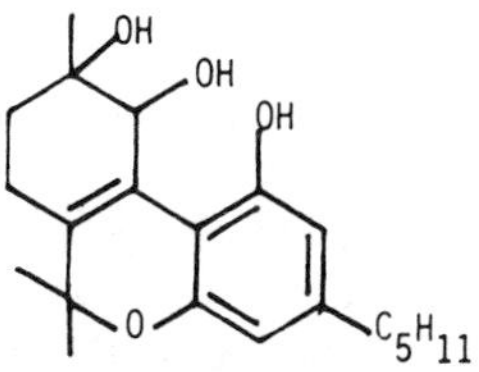

Cannabitriol
CAR No. 11003-36-4
(ElSohly et al., 1977)

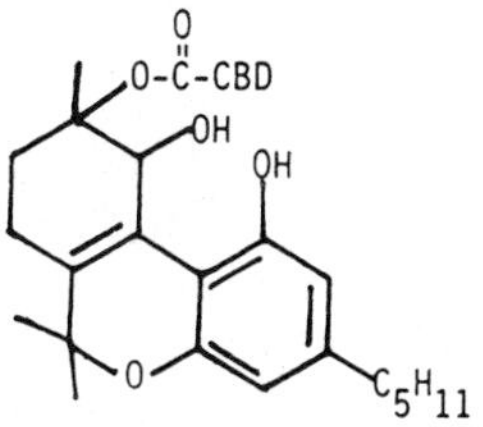

CBD acid ester of cannabitriol
(von Spulak et al., 1968)

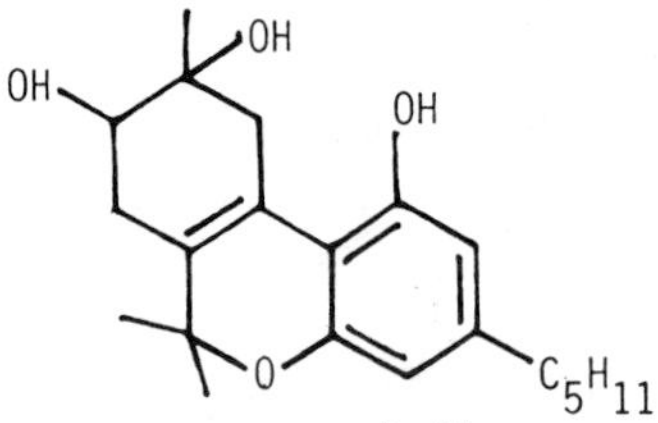

8,9 - Dihydroxy - $\Delta^{6a,10a}$- THC
(ElSohly et al., 1978)

B. Cannabispirans ([16] and [17])

These nonalkaloidol phenols are not classified as cannabinoids, yet in the GC analysis of marihuana these compounds can confound the THC results. El-Feraly et al. (1977) found that the relative retention time of cannabispiran was identical to Δ^8-THC [4] and stated that the peak for Δ^8-THC in fresh cannabis is mostly, if not all, cannabispiran. For a review of the isolation, structural determination, and synthesis, see Turner et al. (1980).

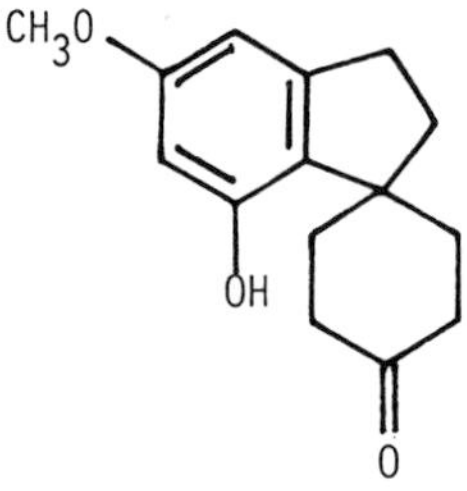

[16]

Cannabispiran
Cannabispirone
CAR No. 61262-81-5

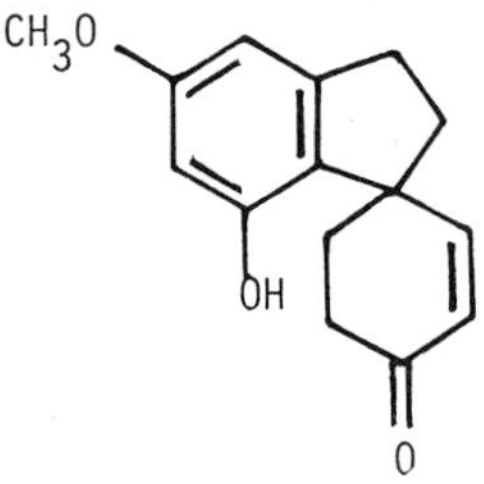

[17]

Dehydrocannabispiran
Cannabispirenone
CAR No. 63213-00-3

C. Alkaloids—Cannabisativine Type

Considering the pharmacological action of cannabis, it is not surprising that a great deal of effort has gone into the investigations of cannabis for alkaloids. In spite of this effort very little could be concluded in 1968 when the U.S. cannabis program expanded. At most it was known that cannabis contrained chloine and trigonelline. Turner et al. (1980) recently reviewed the literature on alkaloids in cannabis. In 1975, El-Feraly and Turner isolated hordenine, a β-arylethylamine, and cannabisativine [18] (Lotter et al., 1975). Anhydrocannabisativine [19] was also isolated from Mexican cannabis (ElSohy et al., 1978c).

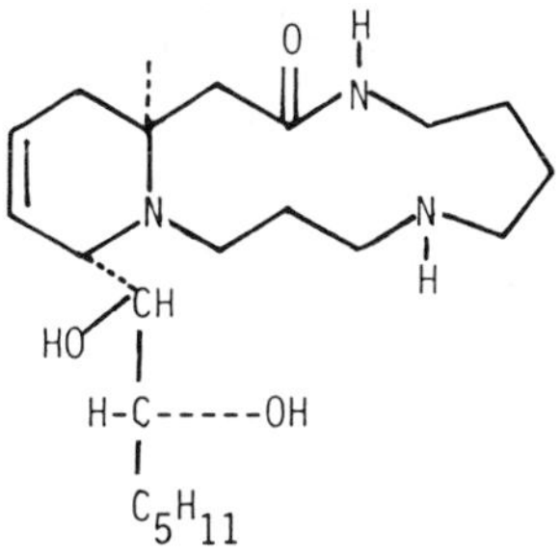

[**18**]

Cannabisativine
CAR No. 57682-64-1

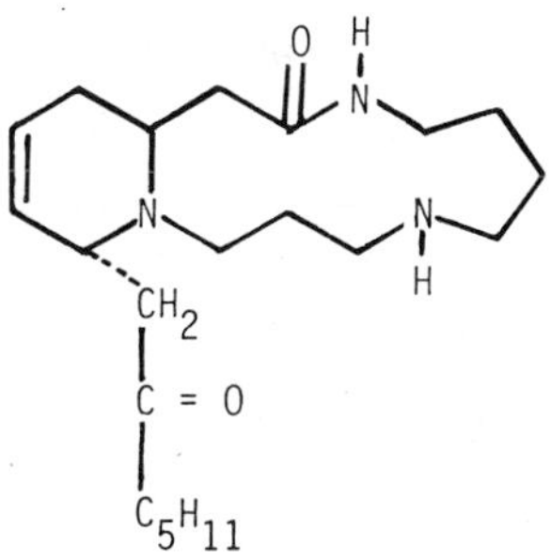

[**19**]

Anhydrocannabisativine
CAR No. 65664-79-1

IV. ANALYSIS OF CANNABIS CONSTITUENTS

In the national program on cannabis and in determination of its effects on humans and animals, the analysis of its cannabinoid constituents in the plant materials and in body tissues and fluids became of paramount importance. Color reactions were useful to a limited extent in plant extracts, but were useless in body tissues and fluids. The concentrations of cannabinoids and their metabolites were below the levels in which the color tests would detect them. Therefore, the highly sensitive methods of gas chromatography, mass spectrometry, high-pressure chromatography, and thin-layer chromatography were employed. The analysis of natural, pyrolytic, and metabolic products of marihuana was recently reviewed (Gudzinowicz and Gudzinowicz, 1980) and will not be covered here. There has been a great need for a method that could be used as a forensic tool to detect usage of marihuana. Recent developments on radioimmunoassay (RIA) methodology appear to satisfy this need. Rogers et al. (1978) used a homogeneous enzyme immunoassay (EMIT) for cannabinoids in urine. O'Connor and Rejent (1981) compared the EMIT method with the RIA and the GC/mass spectrometry methods and confirmed the EMIT's usefulness in urine. In blood and plasma both the EMIT method (Peel and Perrigo, 1981) and the RIA methods (Yeager et al., 1981; Bergman et al., 1981; Owens et al., 1981) have also been found to be useful.

Metabolism of Δ^8- and Δ^9-THC have been studied extensively and some of the major metabolites are listed below (Waller et al., 1976, 1982).

Δ^9-*trans*-THC	Δ^8-*trans*-THC
11-OH-Δ^9-THC	11-OH-Δ^8-THC
8α- or 8β-OH-Δ^9-THC	7,11-Di-OH-Δ^8-THC
11-Oxo-Δ^9-THC	1'-, 2'-, 3'-, 4'-, or 5'-OH-Δ^8-THC
11-*Nor*-9-COOH-Δ^9-THC	11-Oxo-Δ^9-THC
1'-, 2'-, 3'-, and 4'-OH-Δ^9-THC	
9α,10α-Epoxy-HHC	

These metabolites have also been produced by microbial transformation (Binder and Popp, 1980).

V. BIOLOGICAL ACTIVITY—THERAPEUTIC AND ADVERSE EFFECTS

A. Marihuana

The "therapeutic use of marihuana," except in smoking as cigarettes, is a misnomer. With the use of the cigarettes come inflammatory and/or neoplastic changes, particularly when the lung has to deal with the double burden of tobacco and cannabis smoke. The pyrolitic changes on plant material and cannabinoid constituents is so great that the cigarettes could not be considered as a dosage form. The U.S. Food and Drug Administration (FDA) and its advisory committees has never seriously considered approval of marihuana itself as a drug substance. The approval of research studies at the IND stage using the marihuana cigarettes was not intended to convey approval. Such studies were pilot scale and a prelude to studies on the constituents of cannabis. The folk medicine of cannabis provides a lead as to areas for research, not as arguments for its use per se. Chemists and biologists with their outstanding research on the constituents and their synthetic surrogates, have negated the use of marihuana in medicine. A quotation of Mikuriya (1969) is still most appropriate:

> Cannabis—has a long history characterized by usefulness, euphoria or evil—depending on one's point of view. To the agriculturists, cannabis is a fiber crop; to the physician of a century ago, it was a valuable medicine; to the physician of today, it is an enigma; to the user, a euphoriant; to the police, a menace; to the trafficker, a source of profitable danger; to the convict or parolee and his family, a source of sorrow.

The growing of C. sativa and the many variants using seed stock from many parts of the world (Turner et al., 1980) resulted in the selection of the Mexican seed stock as the most appropriate for extensive study in the Marihuana Program. Therefore, the plant material used in cigarettes and in the preparation of extracts were from this variant. The project at the University of Mississippi has continued to grow and has analyzed many variants of cannabis and also "street" marihuana. Turner (1980) recently reviewed these findings. Prior to 1975 (Turner, 1980), confiscated materials rarely exceeded 1% THC, whereas the "street" sinsemilla is above 6%. Also, the available marihuana extract has as much as 28% THC. The marihuanas contain not only THC but also many of the cannabinoids listed herein and other noncannabinoid constituents (Turner et al., 1980). The "street" marihuanas are often contaminated with other drugs of abuse and even microorganisms (Kagen, 1981). Recently, outbreaks of salmonellosis in Ohio and in Michigan were reported in which marihuana was found to be the source of the infection (H.H.S. Publications, 1981).

The effect of marihuana can be related to the Δ^9-trans-THC content to some extent; however, other cannabinoids may potentiate or suppress these effects. THC is a central nervous system depressant in which the intoxication is biphasic: euphoria followed by depression. In the chronic use this continuous state of intoxication brings on the clinical condition called the amotivational syndrome. This syndrome, called "burnout," is characterized by people who are mentally dull, emotionally blunt, and without drive or goals. Since the process of reversing the amotivational syndrome in the chronic user requires many weeks or years, brain damage is postulated. Heath et al. (1979) has found brain damage in the monkey in chronic smoking experiments. "Marijuana is potentially damaging to health in a variety of ways; but it can be especially harmful when used by children and adolescents, by persons who are psychologically vulnerable or by those already physically or mentally ill" (*JAMA*, 1981).

The loss of short-term memory is a common effect of THC and marihuana. This, in conjunction with intoxication, would seem to rule out their use in any society.

Before leaving marihuana to discuss the therapeutic uses and adverse effects of the individual constituents of cannabis, the following statement should be made and clarified. Since the fresh and growing plant of cannabis and its variants do not contain THC in at least appreciable concentrations, "cannabis is not a poisonous plant."

The cannabinoids are present in fresh plant material almost exclusively as their acids. In 1967, Yamauchi et al. (1967) isolated $(-)-\Delta^9$-*trans*-THCA and found it to be inactive in the catalepsy test in mice. In 1970, Mechoulam et al. (1970) independently found that THC acids were inactive in monkeys. It has also been found that Δ^9-THCA B is inactive in humans (E. A. Carlini, M. Perez-Reyes, C. E. Turner, unpublished observation; see Turner et al., 1980).

Cannabis is not a poisonous plant when consumed fresh but becomes poisonous when damaged by drying, heating, smoking, and/or aging of it or its extracts and plant parts. For example, a booklet (Roffman, 1979) was prepared on the use of marihuana in nausea during treatment with chemotherapy using illicit, "street" marihuana. The recommended dosage forms included cookies (where the heat during baking would have decomposed the THC acids) and suppositories or capsules in which the marihuana was heated in butter and water to extract the freed THC into the butter, which was then made into the suppositories or capsules.

The dog bioassay in the early 1900s became less useful as the cannabis plant material became fresher and supposedly better.

B. Cannabinoids

Shortly after $(-)-\Delta^9$-*trans*-THC became available, Hollister (1970) found that orally in humans it produced increased pulse rate, lowered blood pressure, increased conjuntival injection, caused muscle weakness, showed euphoria followed by sleepiness, and caused behavoral impairment of visual distortions, depersonalization, difficulty in thinking, and dreamlike states. The dosages were 50 to 70 mg. Later (Miller and Drew, 1972) it was shown in smoking experiments that the impairment included loss of short-term memory.

Hollister's high dosage was questioned and it was supposed that because of early inexperience with analysis dosage, formulation, and stability, the dosage might have been about half the amount stated (Braude, 1972a). In any event, 40 mg per person produces a psychotic reaction. (The FDA considered 30 mg orally per person to be the high range.)

Hollister and co-workers (Hollister, 1974) have reported relative activity of cannabinoids in humans to ascertain some structure-activity relationships by parameters given above. These activities are as follows:

	Relative activity
(-)-Δ^9-*trans*-THC [3]	100
Δ^8-THC [4]	75
$\Delta^{6a,10a}$-THC [8]	30
CBD	0
CBN	0
THCV	25
11-OH-Δ^9-THC	120
11-OH-Δ^8-THC	90
8β-OH-Δ^9-THC	20
8α-OH-Δ^9-THC	25

These relative activities seem to be valid regardless of the route of administration (Hollister and Gillespie, 1973). Recently, Ohlsson et al. (1980) and Lindgren et al. (1981) related plasma concentration in humans of Δ^9-THC with various biological effects when given oral, intravenously, or by smoking.

From the brief discussion of the bioactivity of the major cannabinoids in marihuana and hashish, it is obviously an impossible situation to relate structure to activity without careful analysis of research materials and to some extent just before usage. Even this is not sufficient, without a full history of the research material, contamination can occur and negate results.

Before continuing the discussion of (-)-Δ^9-*trans*-THC, its activities, its medical uses, and its toxicity, a reflection on its metabolism is needed. Their formation in many species of animals of the metabolic products of THC and the isolation and characterization of their structure have been carried nearer completion than those of most any other drug except alcohol. In humans the metabolite that might be selected as the only one of great importance is (-)-11-OH-Δ^9-THC. The other metabolites are either inactive or appear to be in low concentrations and can be discounted. Present data appear to support this conclusion. On the other hand, subsequent data may prove the conclusion to be a mere "cop-out."

From the work on 11-OH-Δ^8-THC (Watanabe et al., 1981), it can be extrapolated that 11-OH-Δ^9-THC gets into the brain faster than Δ^9-THC itself. This may or may not explain why 11-OH-Δ^9-THC is more active than Δ^9-THC itself. However, with existing data this quantitative difference is not overly significant in view of the lack of qualitative difference in this biological activity. Since 11-OH-Δ^9-THC is difficult to produce, not much has been available for extensive research—and so on to Δ^9-THC itself.

The pharmacology has been reviewed often and recently (Turner, 1980; Hollister and Gillespie, 1973; Braude and Szara, 1976; Nahas et al., 1976;

Nahas and Paton, 1979; Kettenes-Van Den Bosch et al., 1980; Harris et al., 1977), so only brief statements on a few areas will be given.

The *CNS activities* of Δ^9-THC in animals carry over reasonably well to humans; these have already been stated (Hollister, 1974). Psychotic episodes are seen in humans with confusion and panic reactions at high doses which are difficult to simulate in animals. Tolerance to this effect in both animals and humans occurs when high dosages are given and when blood levels are maintained. Hypothermia in animals has been shown to be due to a decrease in heat production, not to heat loss (Haavik et al., 1980). This hypothermia has been demonstrated in humans to some extent.

Physical dependence on THC was thought not to occur and the literature, especially the lay literature, is replete with the statement that marihuana causes only psychological dependence. This is not true; THC does cause physical dependence, with irritability, sleep disturbances, decreased appetite, nausea, vomiting, sweating, salivation, and tremors (Jones and Benowitz, 1976).

The mechanism of action of THC on the central nervous system has been the subject of many reports. The research approach is mainly through the neurotransmitters. The biogenic amines which are most studied to ascertain their uptake, release, transport, and storage are norepinephrine, epinephrine, serotonin, acetylcholine, their precursors, and frequently inhibitors of these. Biphasic effects are often found with one dose showing opposite effects to another dosage. This confusion in data appears even in recent reports. Thus the mechanism of action of THC in the brain is still uncertain. In spite of the frequent contradictory findings, some of the various biogenic amine systems are implicated in the mechanism of THC action on the CNS.

The cardiovascular effect of THC in the early phase of intoxication is a clear-cut clinical fact, whereas predicting chronic pathology from these acute effects is more problematic. The acute effects are summarized in nearly every review on cannabis research. The heart can increase up to 160 beats per minute. This tachycardia is often the opposite in animals, where the heart rate is reduced and shows bradycardia. The bradycardia can be noted in humans when very high doses are given. The reviews by Harris et al. (1977) and Jones and Benowitz (1976) should be noted. Blood pressure changes are slight in a supine position but when standing, blood pressure decreases (orthostatic hypotension). Subjects who have hypertension and coronary diseases are predisposed and can have serious problems with THC and marihuana. A serious loss in blood pressure has been reported clinically (Merritt et al., 1981). With a decrease in myocardial oxygen delivery THC could precipitate angina or a myocardial infraction in predisposed individuals. The long-term effect of THC administration on the cardiovascular system needs further research.

The disposition of Δ^8-THC, 11-OH-Δ^8-THC, and 11-oxo-Δ^8-THC in mouse blood, liver, and brain indicates that the two metabolites are distributed more readily from blood to brain in mice than is Δ^8-THC itself. This was stated to explain the greater bioactivity of these metabolites over Δ^8-THC (Watanabe et al., 1981).

The structure-activity relationships were reviewed by Mechoulam (1973) for a number of compounds, mostly in monkeys. He reported that (+)-Δ^9-THC, (+)-Δ^8-THC, (-)-Δ^7-THC, (-)-$\Delta^{9,11}$-THC, and (+)-*cis*-Δ^9-THC were inactive. A reexamination of the bioactivity of some of the stereoisomers of

Δ^9-THC and Δ^8-THC was recently reported by Martin et al. (1981). Again the (-)-*trans*-Δ^9-THC was found to be 100 times more potent than (+)-*trans*-Δ^9-THC. The low activities of (+)-*trans*-Δ^9-THC, (+)-*cis*-Δ^9-THC, and (+)-*trans*-Δ^8-THC were measured by increased dosage. The dog ataxia, mouse hypothermia, and mouse spontaneous activity tests were used in addition to the monkey test. The conclusion was that the magnitude of the relative activity of the unnatural stereoisomers depends on the animal model. These findings of Martin et al. (1981) will need validation before the earlier conclusion—that a (-) *trans* stereoisomer of Δ^8 and Δ^9-THC is required for cannabis-type bioactivity—can be changed.

In addition to metabolites that are formed due to allylic hydroxylation at positions 11 and 8 and epoxy formation on the double bond of Δ^9-THC and at positions 11 and 7 and epoxy formation of Δ^8-THC, there is a hydroxyl group formed on each of the carbons in the pentyl side chain. Thus there are 1'-hydroxy-Δ^9-THC, 1'-hydroxy-Δ^8-THC, and 2'-hydroxy-Δ^9-THC to 5'-hydroxy-Δ^8-THC. These were detected in dog tissue first by Maynard et al. (1971), but most of the compounds were detected in lung tissue by Agurell et al. (1976). The 3'-OH-Δ^8-THC and 3'-OH-Δ^9-THC were more active in monkeys than was the parent compound.

Animals used in testing for cannabis-type bioactivity include the cat, dog, rabbit, rat, mouse, monkey, humans, gerbil, and pigeon (Waller et al., 1976, 1982). The rabbit corneal areflexia and the dog ataxia assay (Loewe, 1945) were the two tests used in the first half of this century. The dog ataxia and monkey activity tests are still used to detect cannabis activity, but the mouse catalepsy test is used most (Christensen et al., 1971). Review articles and books cover this area (Mechoulam, 1973; Turner, 1980; Braude and Szara, 1976; Nahas et al., 1976; Nahas and Paton, 1979). The most recent review of the biological activity of the THCs is by Kettenes-Van Den Bosch et al. (1980).

A most intriguing bioassay system was recently reported by Consroe and Fish (1980). A selected line of genetically unique rabbits exhibited nonfatal convulsions when given 0.05 mg/kg of Δ^9-THC intravenously and with increasing severity at doses of 0.1, 0.5, and 0.9 mg/kg. These workers reported later that only psychoactive cannabinoids caused convulsions; also, tolerance could be produced to Δ^9-THC that was cross-tolerant to CBN (Consroe and Fish, 1981).

The pharmacology of cannabinoids, individual and collective, and the pharmacology of marihuana is badly confused in the literature because the terms are used so interchangeably. Another factor in this confusion is the separation of cannabinoids into psychoactive cannabinoids and nonpsychoactive cannabinoids, while assuming that the nonpsychoactive cannabinoids would not have any pharmacological or biological action. Any chemical compound has some biological effects at the appropriate concentration level.

Marihuana is a preparation produced from the cannabis plant. The type or variant or subspecies used determines its cannabinoid content. Thus a drug type of cannabis will produce a marihuana high in THC content with the second preponderant cannabinoid being CBC. A fiber type of cannabis will produce a marihuana with high CBD content, with little or no THC, and some CBC. Each of these cannabis types (and marihuana) contain varying amounts of other cannabinoids. Of course, these cannabinoids in fresh plant material will be present as their acids. Rarely is a cannabis plant a pure drug type or a fiber type but may be anywhere in between.

Hashish, the cannabis resin, has the same problem as marihuana as to cannabinoid content. But hashish has at least two additional problems. The first is that it is exposed to oxidative changes; the THC content goes down as CBN goes up. In old and exposed hashish there is often very little THC and the other cannabinoids go up, particularly in their ratio to the total cannabinoid content. Of course, the pharmacology and toxicity then change. Hashish is often not the main product but is a by-product from fiber-type cannabis, where the main product from the plant is hemp fiber and/or seeds. Most "street" marihuana today comes from the drug type and is high in THC content, whereas most hashish has a high content of CBD and most, or at least a larger portion of it, probably comes from fiber-type cannabis.

The importance of analysis of hashish and marijuana was recognized by NIDA (see Turner, 1980).

From the data in Table 7 it can be seen that CBD, CBC, Δ^9-THC, and CBN occur in concentrations high enough to consider the activity of each and interactions on each other.

The activity of CBN was found to be about one-tenth as active as Δ^9-THC (Perez-Reyes et al., 1973). It does not appreciably affect the action of Δ^9-THC (Siemens et al., 1976).

CBD was found to be inactive in cannabis activity (Hollister, 1974), but it does antagonize THC behavoral effects (Siemens et al., 1976; Karniol et al., 1974). CBD does have anticonvulsant properties and has been studied as an antiepileptic drug by Karler and Turkanis (1979).

The biological effect of CBC is the least studied of the major cannabinoids occurring naturally in drug-type cannabis, yet it is more abundant than CBD in this strain. The Mississippi group has shown more interest in CBC than have other researchers (Turner et al., 1975, 1980; Hatoum et al., 1981). The first bioactivity for CBC, reported by Gaoni and Mechoulam (1966), showed sedation and ataxia in a dog, but in the monkey CBC has no cannabis-like activity (Mechoulam et al., 1970). It could be said that it has its own pharmacologic profile, with a depressant action on spontaneous motor activity. CBC

TABLE 7

Cannabis preparation	CBDV	Δ^9-THCV	CBL	CBD	CBC	Δ^8-THC	Δ^9-THC	CBN
Sinsemilla (fiber)	0.07	0.02	0.03	4.68	0.47	0.09	0.21	0.06
Sinsemilla (intermediate)	t	0.08	0.01	3.69	0.61	0.07	3.58	0.21
Sinsemilla (drug)	-	0.15	0.02	0.02	0.20	-	6.28	0.22
Hashish (UN standard)	0.03	0.03	0.03	2.89	0.38	0.22	2.2	2.50
NIDA Cig. 1	0.01	t	t	t	0.12	0.09	0.84	0.30
Cig. 2	-	0.02	0.02	0.02	0.28	0.15	1.86	0.13

prolongs the hexobarbital sleep time and synergizes the effects of Δ^9-THC on this parameter. CBC also induced a hypothermic response and synergizes Δ^9-THC-induced analgesia (Hatoum et al., 1981). Thus in general, CBC enhances the activity of Δ^9-THC.

The *reproductive systems and functions* are affected by THC ingestion. THC effects on neuroendocrine functions were recently reviewed by Smith (1980). The effects on reproduction and development were reviewed by Harclerode (1980). Marihuana and THC affect the peripheral endocrine organs that are regulated by the hypothalamic-pituitary axis, and alter reproductive hormones, affect spermatogensis, and affect ovulation. The decrease in gonadotropin levels by THC is not dose related as to the magnitude, but the duration of the depression in levels is dose related. THC depresses the release of prolactin even at low dosage (0.5 mg/kg, i.v.) levels in the ovariectomized rat (Hughes et al., 1981). Thus fertility and milk production are adversely effected by THC and marihuana.

Testosterone production is decreased by THC and marihuana in both rodents and humans. THC decreases organ weights and functions in the rat. There is a decreased number of sperm and an increase in the percentage of abnormal sperm.

The cannabinoids, THCs, CBD, and CBN inhibit prostaglandin synthesis. This hormonal substance is necessary in both males and females for proper tone of the circulatory system associated with the reproductive organs.

C. Toxicity of (-)-Δ^9-*trans*-Tetrahydrocannabinol

Reported deaths from overdosage with THC in humans has not been reported, nor has Δ^9-THC been found as a single substance on the "street." Older literature (Walton, 1938) reports a few deaths due to the use of Indian hemp.

The acute toxicity of pure Δ^9-THC was reported by Forney and Kiplinger (1971) as the following LD_{50} values (mg/kg):

Route	Rat	Mouse
i.v.	28.6	42.47
i.p.	372.9	454.5
Orally	666.1	481.9

Rosenkrantz et al. (1974) and Braude (1972b) considered intravenous injection and inhalation to be equivalent routes of administration in rats, where the LD_{50} value was approximately 40 mg/kg. The minimum lethal dose by intravenous injection was 63 mg/kg in monkeys and 100 mg/kg in dogs. The minimum lethal oral dose in monkeys and dogs was greater than 3 g/kg. These lethal oral doses are at least 1000 times the behaviorally effective oral doses in these species and about 4000 times the oral psychoactive dose in humans. The therapeutic index or safety ratio for Δ^9-THC is very great.

In view of these high lethal doses in animals, it is interesting to note that 1 mg of 11-OH-Δ^9-THC given intravenously in humans produced tachycardia in 2-3 min, and the "high" is more intense than that reported by any of the volunteers from smoking marihuana. The psychological effects during this "high" were so intense as to be unpleasant (Lemberger et al., 1972).

Rosenkrantz and Braude (1976) have performed chronic toxicity studies of Δ^9-THC in rats by oral and inhalation administration. In the inhalation, standardized marihuana was used, but the THC in the smoke was determined by analysis. The dose of THC in both forms was equivalent to that which humans would get from one, three, or six marihuana cigarettes.

CNS depression followed by CNS stimulation was found as tolerance to the depression developed. Fighting aggression was seen after the depression phase. The unusual neurotoxicity (popcorn response, tremors, convulsions) and some lethal toxicity were reported. Weight losses of organs were limited to the tests. Morphological changes were seen in the lungs. It was later reported from this same group (Rosenkrantz and Fleischman, 1979) that long-term (up to 360 days) exposure to marihuana (THC) smoke produced dose-related intense morphological changes in the respiratory system. There were large numbers of macrophage aggregates and cholesterol clefts. These pathological lung changes did not reverse during a 30-day recovery period.

The effect of THC and marihuana smoke (containing measured THC content) in rodents on fetal development was reported by Rosenkrantz (1979). Unequivocal embryocidal effects and resorptions were found. The embryotoxicity was correlated with vaginal bleeding and excessive fetal resorption.

The aggressiveness (with fighting often to death) after tolerance to the sedation which Rosenkrantz et al. described in rats with THC has been confirmed in monkeys by Chapman et al. (1979). With the monkeys, dosages (in food) of 0.6, 1.2, and 2.4 mg/kg per day were used and calculated to be equivalent to smoking one to three marihuana cigarettes per day in humans. In the monkey, social behavioral patterns were changed and hierarchial or the "pecking" order was frequently reversed, showing individual variability. One female monkey was fatally injured during fighting.

In a recent review on the effects of marihuana in humans, Jones (1978) discusses criminal and aggressive behavior. Marihuana intoxication and hostile behavior has been a topic of great interest and discussion. In India, violent behavior is commonly associated with cannabis psychosis (Thacore and Shukla, 1976). Yet in the laboratory situation, chronic cannabis intoxication, even at high dosages (210 mg in 2 hr in a divided dosage), produced tolerance to the pleasurable effects, followed by unpleasant, lethargic, sedated states. Cessation of the drug resulted in symptoms of withdrawal or dependence, with restlessness, irritability, and insomnia (Jones and Benowitz, 1976). In monkeys, the aggression was more intense in a stressful situation. Until stress is ruled out, criminality associated with marihuana use remains an open question. In most laboratory studies, the subjects are screened, and those with predisposed violent tendencies may be eliminated.

In two reviews by Jones (1978, 1980), some of the topics covered were:

Acute effects
Metabolism of cannabinoids and biochemistry
Cardiovascular effects
Sexual functioning
Effects on cell-mediated immunity
Psychopathology
Acute panic anxiety
Toxic delirium

Prolonged reactions—flashbacks, amotivational syndrome
Dependence
Chronic studies—field and laboratory
Neurological effects
Driving performance

The ultimate parameter of toxicity is death in validated and repeatable cases. Such finds would come with humans only in clinical cases. The adverse effects with crude, unsterilized preparations taken by injection is not a meaningful toxic finding (Payne and Brand, 1975). The cerebral atrophy found by Campbell et al. (1971) by injecting air into the brain of a supposed cannabis user needs validation, but such techniques are considered unethical for laboratory experiments. The damage to the brain in the monkey with Δ^9-THC but not with CBD or CBC as shown when viewed with the electron microscope (Heath et al., 1979) has not been confirmed in humans. Attempts to see structural changes in the brain of cannabis users with computerized axial tomography indicated no cerebral atrophy. It is easy to see that improved techniques are needed.

In the review of health hazards and the therapeutic potential of marihuana by the Council on Scientific Affairs of the American Medical Association, it was concluded that "marijuana is a dangerous drug" (*JAMA*, 1981).

Who or what body is the ultimate authority on medical affairs? Consideration of the adverse effects of marihuana on humans would not be complete without reference to social, political, and legal climates. Two major organizations have advocated the legalization of marihuana for recreational usage. Books on each of these organizations have appeared recently.

The book *Facts about Drug Abuse* from the Drug Abuse Council (1980) reports the activities and sources of financial support of the Drug Abuse Council. A review of essentially the same areas of effects of marihuana on humans as Jones (1978) covered reached the conclusion that marihuana was not damaging enough to rule out its use. The reviewers were Robert T. Carr (affiliation not given) and Erik J. Meyer (a lawyer). The language in this review approaches that of a prosecuting attorney. If one has been an expert witness in a court of law, then he or she realizes that the opposing attorney can pervert technical facts to fit the intended purpose.

The other book (Anderson, 1981) is *High in America: The True Story Behind NORML and the Politics of Marijuana*, by Patrick Anderson. NORML, the National Organization for the Reform of Marijuana Laws, has clearly publicized its objectives on the legalization of marihuana.

When one examines the laws of the United States or the individual states as related to marihuana, it is obvious that none of these laws legalize or decriminalize* cannabis, marihuana, or any of their preparations. Permission

---

*
The Drug Abuse Council (1980) used the word "decriminalization" to indicate changes in laws in more than 20 states without a definition of the word. NORML also used this word in lobbying for changes in state laws. Thus the prodrug groups coined the word to describe these changes in some state laws as decriminalization and the lay press was quick to pick it up. This word is not in the dictionary. To the average person "de" means "without," so to them, decriminalization is equivalent to legalization. Cuskey et al., in a

for a research organization to use a substance in research does not carry rights for individuals to use it.

Unfortunately, the report of the Council on Scientific Affairs of the American Medical Association was entitled, "Marijuana, Its Health Hazards and Therapeutic Potentials." As stated earlier, marihuana has no therapeutic use as a dosage form and this title only continues the confusion. It would be a public service to divorce the medical usage of cannabis plant parts from the potential of its constituents. The cannabinoids, mainly Δ^9-THC and CBD, are chemical leads to new structures in search of new medicinal agents. Δ^9-THC may or may not be approved for medical use as a drug for the treatment of extreme nausea for patients who undergo cancer chemotherapy. Even if approved, it does not mean that marihuana itself is approved or that marihuana is safe for use as a recreational drug. A number of publications even confuse compound [8] with marihuana and Δ^9-THC. It is presently inconceivable that even a purified mixture of the cannabinoids as they occur in cannabis would ever be approved for medicinal use.

D. Therapeutic Areas

In this review on cannabis as a plant toxin, it seems appropriate to list the potential chemical leads for therapeutic agents that have evolved from studies on its constituents. A book, *The Therapeutic Potential of Marihuana* (Cohen and Stillman, 1976), publishing the papers presented at a conference in 1976, and a listing of the potential areas by Cohen (1978), are two summary reports.

1. *Antinauseant in cancer patients receiving chemotherapy.* Clinical studies are under the sponsorship of the National Cancer Institute, and preparations of Δ^9-THC for this research are available from NCI. This will probably be the first area of use of Δ^9-THC to be approved for medicinal usage.

2. *Glaucoma.* The lowering of intraocular pressure by Δ^9-THC has been validated in humans and animals (Hepler and Frank, 1971; Green et al., 1978; ElSohly et al., 1981). The activity of CBD in rabbits by Green et al. (1978) was not confirmed by ElSohly et al. (1981). There may be a problem with CNS effects and orthostatic hypotension when Δ^9-THC is used in glaucoma.

3. *Epilepsy.* The anticonvulsant activity in animals indicates that CBD should be considered as a potential antiepileptic agent. CBD does not have psychoactivity. Δ^9-THC should not be considered as an antiepileptic drug because it has CNS activity and provokes seizures in epileptic beagles. Epileptics should be discouraged from using marihuana since there is some risk of provoking seizures (Feeney, 1979).

---

special report, *Contemporary Drug Problem*, Winter 1978 (Federal Legal Publications, Inc., 1979), found it necessary to define decriminalization before they could review the impact of these changes in some state laws on patterns of marihuana usage. Cuskey et al., defined "decriminalization" to mean that the possession of small amounts of marihuana (between 1 and 3-1/2 ounces) constitutes a civil offense, punishable by a fine, rather than a criminal act, punishable by incarceration.

4. *Asthma.* The bronchodilation of Δ⁹-THC in normal and asthmatic subjects has been reported (Vachon et al., 1976). Better dosage forms of Δ⁹-THC are needed than oral capsules or marihuana cigarettes. It is still a lead for the further study of new compounds that would not have psychoactivity.

5. *Analgesia.* The relief of pain by marihuana and Δ⁹-THC does not appear to be of morphine type, since narcotic antagonists do not block THC analgesia. The potency of THC varies with test systems. Synthetic compounds under study for analgesia include 9-nor-9β-hydroxy-hexahydrocannabinol (Wilson and May, 1975), Nabilone-Lilly, Levon-antradol-Pfizer (Milne et al., 1981), and Nibitan (sp-106)−Burroughs-Wellcome (Razdan et al., 1972). These compounds are also being examined for uses as antinauseants, antidepressants, anticonvulsants, and so on.

6. *Other areas.* Other areas of usefulness for the natural cannabinoids that have been alluded to but have very little or no validation are: antidepressant, hypnotic, antibacterial, antifertility, and antitumor.

ACKNOWLEDGMENTS

The author acknowledges the assistance of B. S. Urbanek with literature searches and M. G. Taylor for secretarial services in the preparation of the manuscript.

REFERENCES

Abel, E. L. (1979). *A Comprehensive Guide to the Cannabis Literature.* Greenwood Press, Westport, Conn.

Adams, R., Hunt, M., and Clark, J. H. (1940a). Structure of cannabidiol, a product isolated from marihuana extract of Minnesota wild hemp. I. *J. Am. Chem. Soc. 62*:196-200.

Adams, R., Baker, B. K., and Wearn, R. B. (1940b). Cannabinol. III. Synthesis of cannabinol, 1-hydroxy-3-n-amyl-6,6,9-trimethyl-6-di-benzopyran. *J. Am. Chem. Soc. 62*:2204-2207.

Adams, R., Pease, D. C., Cain, C. K., and Clark, J. H. (1940c). Structure of cannabidiol. IV. Isomerization of cannabidiol to tetrahydro-cannabinol, a physiologically active product. Conversion of cannabidiol to cannabinol. *J. Am. Chem. Soc. 62*:2402-2405.

Adams, R., Harfenist, M., and Loewe, S. (1949). New analogs of tetra-hydrocannabinol. XIX. *J. Am. Chem. Soc. 71*:1624-16288.

Agurell, S., Binder, M., Fonseka, K., Lindgren, J. E., Lander, K., Martin, B., Nilsson, I. M., Nordqvist, M., Ohlsson, A., and Widman, M. (1976). Cannabinoids: metabolites hydroxylated in the pentyl side chain. In *Marihuana: Chemistry, Biochemistry, and Cellular Effects*, G. G. Nahas, W. D. M. Paton, and J. E. Idänpään-Heikkilä (Eds.). Springer-Verlag, New York, pp. 141-148.

Bergman, R. A., Lubaszewski, T., and Wang, S. Y. S. (1981). The detection of tetrahydrocannabinol in blood: a comparative study. *J. Anal. Toxicol. 5*:85-89.

Binder, M., and Popp, A. (1980). Microbial transformation of cannabinoids. *Helv. Chim. Acta 63*:2515-2518.

Boeren, E. G., ElSohly, M. A., and Turner, C. E. (1979). Cannabiripsol: a novel cannabis constituent. *Experientia 35*:1278.

Braude, M. C. (1972a). In *Cannabis and Its Derivatives*, W. D. M. Paton and J. Crown (Eds.). Oxford University Press, London, p. 183.

Braude, M. C. (1972b). Toxicology of cannabinoids. In *Cannabis and Its Derivatives*, W. D. M. Paton and J. Crown (Eds.). Oxford University Press, London, pp. 88-98.

Braude, M. C., and Szara, S. (1976). *Pharmacology of Marihuana*, Vols. 1 and 2. Raven Press, New York.

Cahn, R. S. (1933). *Cannabis indica* resin. Part IV. The synthesis of some 2:2-dimethyldibenzopyrans and confirmation of the structure of cannabinol. *J. Chem. Soc.*, pp. 1400-1405.

Campbell, A. M. G., Evans, M., Thomson, J. L. G., and Williams, M. J. (1971). Cerebral atrophy in young cannabis smokers. *Lancet 2*:1219-1225.

Chapman, L. F., Sassenrath, E. N., and Goo, G. P. (1979). Social behavior of rhesus monkeys chronically exposed to moderate amounts of delat-9-tetrahydrocannabinols. In *Marihuana: Biological Effects*, G. G. Nahas and W. D. M. Paton (Eds.). Pergamon Press, Elmsford, N.Y., pp. 693-712.

Christensen, H. D., Freudenthal, R. I., Gidley, J. T., Rosenfeld, R., Boegli, G., Testino, L., Brine, D. R., Pitt, C. G., and Wall, M. E. (1971). Activity of Δ^8 and Δ^9-tetrahydrocannabinol and related compounds in the mouse. *Science 172*:165-167.

Claussen, U., von Spaluk, F., and Korte, F. (1966). Zur chemischen Klassifizierung von Pflanzen. XXXI. Haschisch-X. Cannabichromen, ein neuer Haschischinhaltsstoff. *Tetrahedron 22*:1477-1479.

Cohen, S. (1978). Marijuana: Does it have a possible therapeutic use? JAMA *240*:1761-1763.

Cohen, S., and Stillman, R. C. (1976). *The Therapeutic Potential of Marihuana*. Plenum Press, New York.

Consroe, P., and Fish, B. S. (1980). Behavioral pharmacology of tetrahydrocannabinol convulsions in rabbits. *Commun. Psychopharmacol. 4*: 287-291.

Consroe, P., and Fish, B. S. (1981). Rabbit behavioral model of marijuana psychoactivity in humans. *Med. Hypotheses 7*:1079-1090.

The Drug Abuse Council. (1980). *The Facts About Drug Abuse*. Free Press, Ne York.

Eddy, N. E. (1965). *The Question of Cannabis, Cannabis Bibliography*. United Nations Economic and Social Council E/CN.7/479.

El-Feraly, F. S., and Turner, C. E. (1975). Alkaloids of *Cannabis sativa* leaves. *Phytochemistry 14*:2304.

El-Feraly, F. S., ElSohly, M. A., Boeren, E. G., Turner, C. E., Ottersen, T., and Aasen, A. (1977). Crystal and molecular structure of cannabispiran and its correlation to D hydrocannabispiran. Two novel cannabis constituents. *Tetrahedron 33*:2373-2387.

ElSohly, M. A., El-Feraly, F. S., and Turner, C. E. (1977). Isolation characterization of (+)cannabitriol and (-)-10-ethoxy-9-hydroxy-$\Delta^{6a(10a)}$-tetrahydrocannabinol: two new cannabinoids from *Cannabis sativa* L. extract. *Lloydia (Cinci.) 40*:275-280.

ElSohly, M. A., Boeren, E. G., and Turner, C. E. (1978a). An improved method for the synthesis of dl-cannabichromene. J. Heterocycl. Chem. 15: 699-700.

ElSohly, M. A., Boeren, E. G., and Turner, C. E. (1978b). (+)9,10-Dihydroxy-$\Delta^{6a(10b)}$-tetrahydrocannabinol and (+)8,9-dihydroxy-$\Delta^{6a(10a)}$-tetrahydrocannabinol: 2 new cannabinoids from Cannabis sativa L. Experientia 34: 1127-1128.

ElSohly, M. A., Turner, C. E., Phoeba, C. H., Knapp, J. E., Schiff, P. L., and Slatkin, D. J. (1978c). Anhydrocannabisativine, a new alkaloid from Cannabis sativa L. J. Pharm. Sci. 67: 124.

ElSohly, M. A., Harland, E., Murphy, J. C., Wirth, P., and Waller, C. W. (1981). Cannabinoids in glaucoma: a primary screening procedure. J. Clin. Pharmacol. 21: 4725-2785.

Feeney, D. M. (1979). Marihuana and epilepsy: paradoxical anticonvulsant and convulsant effects. In Marihuana: Biological Effects, G. G. Nahas and W. D. M. Paton (Eds.). Pergamon Press, Elmsford, N.Y., pp. 643-657.

Fenselau, C., Kelly, S., Salmon, M., and Billets, S. (1976). The absence of tetrahydrocannabinol from hops. Food Cosmet. Toxicol. 14: 35-39.

Fetterman, P. S., Keith, E. S., Waller, C. W., Guerrero, O., Doorenbos, N. J., and Quimby, M. W. (1971a). Mississippi-grown Cannabis sativa L. preliminary observation on chemical definition of phenotype and variations in tetrahydrocannabinol content versus age, sex and plant part. J. Pharm. Sci. 60: 1246-1249.

Fetterman, P. S., Doorenbos, N. J., Keith, E. S., and Quimby, M. W. (1971b). A simple gas liquid chromatography procedure for determination of cannabinoidic acids in Cannabis sativa L. Experientia 27: 988-990.

Forney, R. B., and Kiplinger, G. F. (1971). Toxicology and pharmacology of marihuana. Ann. N.Y. Acad. Sci. 191: 74-82.

Friedrich-Fiechtl, J., and Spiteller, G. (1975). Neue Cannabinoide. I. Tetrahedron 31: 479-487.

Gamage, J. R., and Zerkin, E. L. (1969). A Comprehensive Guide to the English Language Literature on Cannabis. Stash Press, Beloit, Wis.

Gaoni, Y., and Mechoulam, R. (1964a). Isolation, structure and partial synthesis of an active constituent of hashish. J. Am. Chem. Soc. 86: 1646-1647.

Gaoni, Y., and Mechoulam, R. (1964b). The structure and synthesis of cannabigerol, a new hashish constituent. Proc. Chem. Soc., p. 82.

Gaoni, Y., and Mechoulam, R. (1966). Cannabichromene, a new active principle in hashish. J. Chem. Soc. Chem. Commun. 1: 20-21.

Gates, M., and Tschudi, G. (1956). The synthesis of morphine. J. Am. Chem. Soc. 78: 1380-1393.

Green, K., Wynn, H., and Bowman, K. A. (1978). A comparison of topical cannabinoids on intraocular pressure. Exp. Eye Res. 27: 239-246.

Grote, H., and Spiteller, G. (1978). Neue Cannabinoide. II. J. Chromatogr. 154: 13-23.

Gudzinowicz, B. J., and Gudzinowicz, M. J. (1980). Analysis of Drugs and Metabolites by Gas Chromatography-Mass Spectrometry, Vol. 7. Marcel Dekker, New York, pp. 271-489.

Haavik, C. D., Hardman, H. F., Collins, F. G., and Skibba, J. (1980). The effects of Δ^9-tetrahydrocannabinol and heat loss. Res. Commun. Subst. Abuse L.: 381-405.

Harclerode, J. (1980). The effect of marihuana on reproduction and development. In *NIDA Research Monograph 31*. U.S. Government Printing Office, Washington, D.C., pp. 137-166.

Harris, L. S., Dewey, W. L., and Razdan, R. K. (1977). Cannabis: its chemistry, pharmacology and toxicology. In *Handbook of Experimental Pharmacology*, Vol. 45: *Drug Addiction 2*, W. R. Martin (Ed.). Springer-Verlag, New York, pp. 371-429.

Hatoum, N. S., Davis, W. M., ElSohly, M. A., and Turner, C. E. (1981). Cannabichromene and Δ^9-tetrahydrocannabinol: interactions relative to lethality, hypothermia and hexabarbital hypnosis. *Gen. Pharmacol. 12*: 357-362.

Heath, R. G., Fitzjarrell, A. R., Garey, R. E., and Mayers, W. A. (1979). Chronic marihuana smoking: its effects on function and structure of the primate brain. In *Marihuana: Biological Effects*, G. G. Nahas and W. D. M. Paton (Eds.). Pergamon Press, Elmsford, N.Y., pp. 713-730.

Hepler, R. S., and Frank, I. R. (1971). Marihuana smoking and intraocular pressure. *JAMA 217*:1391.

H. H. S. Publications (1981). No. (CDC) 81-8017. *Morbid. Mortal. Wkly. Rep. 30*:77-79.

Holley, J. H., Hadley, K. W., and Turner, C. E. (1975). Constituents of *Cannabis sativa* L. XI. Cannabidiol and cannabichromene in samples of known geographical origin. *J. Pharm. Sci. 64*:892-895.

Hollister, L. E. (1970). Recent research on the effects of marihuana in man. *Drug Depend. (NIMH) 4*:38.

Hollister, L. E. (1974). Structure-activity relationships in man of cannabis constituents, and homologs and metabolites of Δ^9-tetrahydrocannabinol. *Pharmacology 11*:3-11.

Hollister, L. E., and Gillespie, H. K. (1973). Delta-8- and delta-9-tetrahydrocannabinol. Comparison in man by oral and intravenous administration. *Clin. Pharmacol. Ther. 14*:353-357.

Hughes, C. L., Everett, J. W., and Tyrey, L. (1981). Δ^9-Tetrahydrocannabinol suppression of prolactin secretion in the rat: lack of direct pituitary effect. *Endocrinology 190*:876-880.

Jacob, A., and Todd, A. R. (1940). *Cannabis radica*. Part II. Isolation of cannabidiol from Egyptian hashish. Observations on the structure of cannabinol. *J. Chem. Soc.*, pp. 649-653.

Jones, R. T. (1978). Marihuana: human effects. In *Handbook of Pharmacology*, Vol. 12: *Drug Abuse*, L. L. Iverson, S. D. Iverson, and S. H. Snyder (Eds.). Plenum Press, New York, pp. 373-412.

Jones, R. T. (1980). Human effects: an overview. In *NIDA Research Monograph 31*. U.S. Government Printing Office, Washington, D.C., pp. 54-80.

Jones, R. T., and Benowitz, N. (1976). The 30-day trip: clinical studies of cannabis tolerance and dependence. In *Pharmacology of Marihuana*, M. C. Braude and S. Szara (Eds.). Raven Press, New York, pp. 627-642.

Journal of the American Medical Association (1981). Council Report. *246*: 1823-1827.

Kagen, S. L. (1981). *Aspergillus*: an inhalable contaminant of marihuana. *N. Engl. J. Med. 304*:483-484.

Karler, R., and Turkanis, S. A. (1979). Cannabis and epilepsy. In
Marihuana: Biological Effects, G. G. Nahas and W. D. M. Paton (Eds.).
Pergamon Press, Elmsford, N.Y., pp. 619-641.

Karniol, I. G., Shirakawa, I., Kasinski, N., Pfeferman, A., and Carlini,
E. A. (1974). Cannabidiol interferes with the effects of Δ^9-tetrahydro-
cannabinol in man. *Eur. J. Pharmacol. 28*:172-177.

Kettenes-Van Den Bosch, J. J., Salemink, C. A., Van Noordwijk, J., and
Khan, I. (1980). Biological activity of the tetrahydrocannabinoids. *J.
Ethanopharmacol. 4*:197-204.

Korte, F., Sieper, H., and Tira, S. (1965). New results on hashish-specific
constituents. *Bull. Narc. 17*:35-43.

Kuppers, F. J. E. M., Lousberg, R. J. J. Ch., Bercht, C. A. L., Salemink,
C. A., Terlouw, J. K., Heerman, W., and Laven, A. (1973). Cannabis.
Viii. Pyrolysis of cannabidiol. Structure elucidation of the main pyrolytic
product. *Tetrahedron 29*:27-97-2802.

Lemberger, L., Crabtree, R. E., and Rowe, H. M. (1972). 11-Hydroxy-Δ^9-
tetrahydrocannabinol: pharmacology, disposition, and metabolism of a
major metabolite of marihuana in man. *Science 177*:62-64.

Levine, J. (1944). Origin of cannabinol. *J. Am. Chem. Soc. 66*:1868-1869.

Li, H. L. (1974). An archaeological and historical account of cannabis in
China. *Econ. Bot. 28*:437-48.

Lindgren, J. E., Ohlsen, A., Agurell, S., Hollister, L. E., and Gillespie,
H. K. (1981). Clinical effects and plasma levels of Δ^9-tetrahydrocannabi-
nol in heavy and light users of cannabis. *Psychopharmacology (Berl.) 74*:
208-212.

Loewe, S. (1945). The chemical basis of marihuana activity. *J. Pharmacol.
Exp. Ther. 84*:78-81.

Lotter, H. L., Abraham, D. J., Turner, C. E., Knapp, J. E., Schiff, P. L.,
and Slatkin, D. J. (1975). Cannabisativine, a new alkaloid from *Cannabis
sativa* L. root. *Tetrahedron Lett. 33*: 2815-2818.

Lousberg, R. J. J. Ch, Bercht. C. A. L., van Ooyen, R., and Spronck,
H. J. W. (1977). Cannabinoidiol: conclusive identification and synthesis
of a new cannabinoid from *Cannabis sativa. Phytochemistry 16*:595-597.

Martin, B. R., Balster, R. L., Razdan, R. K., Harris, L. S., and Dewey,
W. L. (1981). Behavioral comparison of the stereoisomers of tetrahydrocannabinols. *Life Sci. 29*:565-574.

Maynard, D. E., Gurny, O., Pitcher, R. G., and Kierstead, R. W. (1971).
(-)Δ^8-Tetrahydrocannabinol: two novel in vitro metabolites. *Experientia
27*:1154-55.

Mechoulam, R. (1973). *Marijuana: Chemistry, Pharmacology, Metabolism,
and Clinical Effects.* Academic Press, New York.

Mechoulam, R., and Gaoni, Y. (1965). Hashish. IV. The isolation and structure of cannabinolic, cannabidiolic and cannabigerolic acids. *Tetrahedron
21*:1223-1229.

Mechoulam, R., and Gaoni, Y. (1967). Recent advances in the chemistry of
hashish. *Fortschr. Chem. Org. Naturst. 25*:175-213.

Mechoulam, R., and Shvo, Y. (1963). Hashish. I. The structure of
cannabidiol. *Tetrahedron 19*:2073-2078.

Mechoulam, R., Shani, A., Edery, H., and Grunfield, Y. (1970). Chemical
basis of hashish activity. *Science 169*:611-612.

Merritt, J. C., Olsen, J. L., Armstrong, J. R., and McKinnon, S. M. (1981). Topical Δ 9-tetrahydrocannabinol in hypertensive glaucomas. *J. Pharm. Pharmacol. 33*: 40-41.

Mikuriya, T. H. (1969). Marijuana in medicine: past, present and future. *Calif. Med. 110*: 34-40.

Miller, L. L., and Drew, W. C. (1972). Marihuana induced impairment of recent memory and mental set shifting. *Fed. Proc., Fed. Am. Soc. Exp. Biol. 31*: 515 (abstr.).

Miller, M. G. (1970). The National Institute of Mental Health Marihuana Research Program. In *Committee on Problems of Drug Dependence*. National Academy of Science, National Research Council, Division of Medical Sciences, Washington, D.C., pp. 6386-6393.

Milne, G. M., Johnson, M. R., Wiseman, E. H., and Hutcheon, D. E. (1981). Therapeutic progress in cannabinoid research. *J. Clin. Pharmacol. 21*: 1S-494S.

United Nations (1968). *Multilingual List of Narcotic Drug Under International Control*. U.N. number E/CN.7/513.

Nahas, G. G., and Paton, W. D. M. (1979). *Marihuana: Biological Effects*. Pergamon Press, Elmsford, N.Y.

Nahas, G. G., Paton, W. D. M., and Idänpään-Heikkilä, J. E. (1976). *Marihuana: Chemistry, Biochemistry and Cellular Effects*. Springer-Verlag, New York.

O'Conner, J. E., and Rejent, T. A. (1981). EMIT cannabinoid assay. *J. Anal. Toxicol. 5*: 168-173.

Ohlsson, A., Lindgren, J. E., Wahlen, A., Agurell, S., Hollister, L. E., and Gillespie, H. K. (1980). Plasma levels of Δ 9-tetrahydrocannabinol after intravenous, oral and smoke administration. In *NIDA Research Monograph 34*. U.S. Government Printing Office, Washington, D.C., pp. 250-256.

Owens, S. M., McBay, A. J., Reisner, H. M., and Perez-Reyes, M. (1981). Iodine-125 radioimmunoassay of delta-9-tetrahydrocannabinol in blood and plasma with a solid-phase second antibody separation method. *Clin. Chem. 27*: 619-624.

Payne, J. R., and Brand, S. N. (1975). The toxicity of intravenously used marijuana. *JAMA 233*: 351-354.

Peel, H. W., and Perrigo, B. J. (1981). Detection of cannabinoids in blood using EMIT. *J. Anal. Toxicol. 5*: 165-7.

Perez-Reyes, M., Timmons, M. C., Davis, K. H., and Wall, M. E. (1973). A comparison of the pharmacological activity in man of intravenous administered Δ 9-tetrahydrocannabinol, cannabinol, and cannabidiol. *Experientia 29*: 1368-1369.

Petrzilka, T., and Sikemeier, C. (1967). Über Inhaltsstoffe des Hashisch. III. Umwandlung von (-)Δ 1,2-3,4-*trans*-Tetrahydrocannabinol. *Helv. Chim. Acta 50*: 2111-2113.

Petrzilka, T., Haefliger, W., Sikemeier, C., Ohloff, G., and Eschenmoser, A. (1967). Synthese und Chiralitat des (-)cannabidiol. *Helv. Chim. Acta 50*: 719-723.

Petrzilka, T., Haefliger, W., and Sikemeier, C. (1969). Synthese von Haschischinhaltsstoffen. *Helv. Chim. Acta 52*: 1102-1134.

The Pharmacopeia of the United States of America, Vol. XI (1935). Lippincott, Philadelphia, p. 104.

Quimby, M. W., Doorenbos, N. J., Turner, C. E., and Masoud, A. (1973). Mississippi grown marihuana—*Cannabis sativa*: cultivation and observed morphological variations. *Econ. Bot. 27*:117-127.

Razdan, R. K. (1981). The total synthesis of cannabinoids. In *Total Synthesis of Natural Products*, Vol. 4, J. Apsimon (ed.). Wiley, New York, pp. 186-262.

Razdan, R. K., Putlick, A. J., Zitko, B. A., and Handrick, G. R. (1972). Hashish. IV. Conversion of (-)$\Delta^{1(6)}$-tetrahydrocannabinol to (-)$\Delta^{1(7)}$-tetrahydrocannabinol. Stability of (-)Δ^{1}- and (-)$\Delta^{1(6)}$-tetrahydrocannabinols. *Experientia 28*:121-122.

Roffman, R. A. (1979). *Using Marijuana in the Reduction of Nausea Associated with Chemotherapy*. Murry, Seattle.

Rogers, R., Crowl, C. P., Eimstad, W. M., Hu, M. W., Kam, J. K., Ronald, R. C., Rawley, G. L., and Ullman, E. F. (1978). Homogeneous enzyme immunoassay for cannabinoids in urine. *Clin. Chem. 24*:95-100.

Rosenkrantz, H. (1979). Effects of cannabid on fetal development of rodents. In *Marihuana: Biological Effects*, G. G. Nahas and W. D. M. Paton (Eds.). Pergamon Press, Elmsford, N.Y., pp. 479-499.

Rosenkrantz, H., and Braude, M. C. (1976). Comparative chronic toxicities of Δ^{9}-tetrahydrocannabinol administered orally or by inhalation in rats. In *The Pharmacology of Marihuana*, M. C. Braude and S. Szara (Eds.). Raven Press, New York, pp. 571-584.

Rosenkrantz, H., and Fleischman, R. W. (1979). Effects of cannabis on lungs. In *Marihuana: Biological Effects*, G. G. Nahas and W. D. M. Paton (Eds.). Pergamon Press, Elmsford, N.Y., pp. 279-299.

Rosenkrantz, H., Heyman, I. A., and Braude, M. C. (1974). Inhalation, parenteral and oral LD50 values of Δ^{9}-tetrahydrocannabinol in Fischer rats. *Toxicol. Appl. Pharmacol. 28*:18-27.

Schultes, R. E. (1970). Random thoughts and queries on the botany of cannabis. In *The Botany and Chemistry of Cannabis*, C. R. B. Joyce and S. H. Curry (Eds.). J. & A. Churchill, London, pp. 11-38.

Serturner, F. W. (1805). *Trommsdorf's J. Pharm. 13*:1234. (See Gates and Tschudi, 1956.)

Shani, A., and Mechoulan, R. (1974). Cannabielsoic acids. Isolation and synthesis by a novel oxidative cyclization. *Tetrahedron 30*:2437-2446.

Shoyama, Y., Fujita, T., Yamauchi, T., and Nishioka, I. (1968). Cannabis. II. Cannabichrominic acid, a genuine substance of cannabichromene. *Chem. Pharm. Bull. 16*:1157-1158.

Shoyama, Y., Yamauchi, T., and Nishioka, I. (1970). Cannabis. V. Cannabigerolic acid monomethyl ether and cannabinolic acid. *Chem. Pharm. Bull. 18*:1327-1332.

Shoyama, Y., Oku, R., Yamauchi, T., and Nishioka, I. (1972). Cannabis. VI. Cannabicyclolic acid. *Chem. Pharm. Bull. 20*:1927-1930.

Shoyama, Y., Yagi, M., Nishioka, I., and Yamauchi, T. (1975a). Biosynthesis of cannabinoid acids. *Phytochemistry 14*:2189-2192.

Shoyama, Y., Hirano, H., Oda, M., Somehara, T., and Nishioka, I. (1975b). Cannabichromevarin and cannabigerovarin, two new propyl homologues of cannabichromen and cannabigerol. *Chem. Pharm. Bull. 23*:1894-1895.

Shoyama, Y., Hirano, H., Makino, H., Umekita, N., and Nishioka, I. (1977). Cannabis. X. The isolation and structure of four new propyl cannabinoid acids, cannabidivarinic acid, tetrahydrocannabivarinic acid, cannabi-chromevarinic acid and cannabigerovarinic acid, from Thai cannabis, 'Meao variant.' *Chem. Pharm. Bull.* 25:1206-2311.

Siemens, A. J., Kalant, H., and deNie, J. C. (1976). Metabolic interactions between Δ^9-tetrahydrocannabinol and other cannabinoids in rats. In *The Pharmacology of Marihuana*, M. C. Braude and S. Szara (Eds.). Raven Press, New York, pp. 77-92.

Smith, C. S. (1980). Effects of marijuana on neuroendocrine function. In *NIDA Research Monograph 31*. U.S. Government Printing Office, Washington, D.C., pp. 120-136.

Smith, R. M., and Kempfert, K. D. (1977). Δ^1-*cis*-Tetrahydrocannabinol in *Cannabis sativa*. *Phytochemistry* 16:1088-1089.

Thacore, F. S., and Shukla, S. R. P. (1976). Cannabis psychosis and para-noid schizophrenia. *Arch. Gen. Psychiatry* 33:383-386.

Todd, A. R. (1946). Hashish. *Experientia* 11:55-60.

Tscheepe (1821). *Chemische Untersuchung der Hanfblatter*. Tubingen. (See Eddy, 1965.)

Turner, C. E. (1980). Chemistry and Metabolism. In *Marijuana Research Findings (NIDA Research Monograph)*, R. C. Peterson (Ed.). U.S. Government Printing Office, Washington, D.C., pp. 81-97.

Turner, C. E., and Hadley, K. W. (1973). Constituents of *Cannabis sativa* L. II. Absence of cannabidiol in an African variant. *J. Pharm. Sci.* 62: 251-255.

Turner, C. E., Hadley, K. W., and Davis, K. H. (1973a). Constituents of *Cannabis sativa* L. V. Stability of an analytical sample extracted with chloroform. *Acta Pharm. Jugoslav.* 23:89-94.

Turner, C. E., Hadley, K. W., and Fetterman, P. S. (1973b). Constituents of *Cannabis sativa* L. VI. Propyl homologs in samples of known geo-graphical origin. *J. Pharm. Sci.* 62:1739-1741.

Turner, C. E., Hadley, K. W., Holley, J. H., Billets, S., and Mole, M. L. (1975). Constitutents of *Cannabis sativa* L. VIII. Possible biological application of a new method to separate cannabidiol and cannabichromene. *J. Pharm. Sci.* 64:810-815.

Turner, C. E., ElSohly, M. A., and Boeren, E. G. (1980). Constituents of *Cannabis sativa* L. XVII. A review of the natural constituents. *J. Nat. Prod.* 43:169-234.

Turner, C. E., Mole, M. L., Hanus, L., and ElSohly, H. N. (1981). Con-stituents of *Cannabis sativa* L. XIX. Isolation and structure elucidation of cannabiglendol, a novel cannabinoid from an Indian variant. *J. Nat. Prod.* 44:27-33.

United Nations, Division of Narcotics (1976). *Botany and Chemotaxonomy of Cannabis*. NMAR/15. 77-1388.

Vachon, L., Mikus, P., Morrissey, W., Fitzgerald, M., and Gaensler, E. (1976). Bronchial effect of marihuana smoke in asthma. In *Pharmacology of Marihuana*, M. C. Braude and S. Szara (Eds.). Raven Press, New York, pp. 777-784.

von Spulak, F., Claussen, U., Fehdhaber, H. W., and Korte, F. (1968). Haschisch. XIX. Cannabidiolcarbonsaure-tetrahydrocannabitriol-ester, ein neuer Haschisch-inhaltsstoff. *Tetrahedron* 24:5378-5383.

Waller, C. W., Johnson, J. J., Buelke, J., and Turner, C. E. (1976). *Marihuana, an Annotated Bibliography*. Macmillan, New York.

Waller, C. W., Nair, R. S., McAllister, A. F., Urbanek, B. S., and Turner, C. E. (1982). *Marihuana, an Annotated Bibliography*, Vol. II. Macmillan, New York,

Walton, R. P. (1938). *Marihuana, America's New Drug Problem*. Lippincott, Philadelphia.

Watanable, K., Yamamoto, I., Oguri, K., and Yoshimura, H. (1981). Metabolic disposition of Δ^8-tetrahydrocannabinol and its active metabolites, 11-hydroxy-Δ^8-tetrahydrocannabinol, in mice. *Drug Metab. Dispos.* 9:261-264.

Wilson, R. S., and May, E. L. (1975). Analgesic properties of the tetrahydrocannabinols, their metabolites and analogs. *J. Med. Chem.* 18:700-703.

Wood, T. B., Spivey, W. T. N., and Easterfield, M. A. (1899). Cannabinol. Part I. *J. Chem. Soc.* 75:20-36.

Yamauchi, T., Shoyama, Y., Aramaki, H., Azuma, T., and Nishioka, I. (1967). Tetrahydrocannabinolic acid, a genuine substance of tetrahydrocannabinol. *Chem. Pharm. Bull.* 15:1075-1076.

Yamauchi, T., Shoyama, Y., Matsuo, Y., and Nishioka, I. (1968). Cannabigerol monomethyl ether, a new component of hemp. *Chem. Pharm. Bull.* 16:1164-1165.

Yeager, E. P., Goebelsmann, U., Soares, J. R., Grant, J. D., and Gross, S. J. (1981). Δ^9-Tetrahydrocannabinol by GLC-MS validated radioimmunoassays of hemolyzed blood or serum. *J. Anal. Toxicol.* 5:81-84.

Part VI

TOXINS INDUCING
GASTROINTESTINAL OR HEPATIC EFFECTS

16

COMPARATIVE FEATURES OF LANTANA, MYOPORUM AND PIMELEA TOXICITIES IN LIVESTOCK

ALAN A. SEAWRIGHT, SELWYN L. EVERIST[†], and JANA HRDLICKA

University of Queensland, Brisbane, Australia

[†]Deceased.

I. INTRODUCTION

Lantana spp. have their origin in the tropical and subtropical regions of the
western hemisphere, but in recent times the plants have become established
as pasture pests and poisonous plants throughout most of the warmer areas
of the world. Poisoning of livestock due to *L. camara* was first described in
northern Queensland, Australia (Tucker, 1910; Pound, 1913), and subsequently studied by Seddon et al. (1927) and Seddon and Carne (1929) in
Australia and by Steyn and Van der Walt (1941) and Reimerschmidt and Quin
(1941) in South Africa. It was not until later that intoxication of cattle was
first reported in areas where *L. camara* was native, namely Florida (Sanders,
1946), Mexico (de Aluja, 1970), and Brazil (Da Silva and Couto, 1971).
Lantana poisoning of all classes of domesticated grazing ruminants is a particular problem on the Indian subcontinent (Lal and Kalra, 1960).

The genus *Myoporum* is confined to the Pacific Basin, mainly Australia and
New Zealand. Although *Myoporum* intoxication of grazing livestock occurs
only in Australasia, the North American *Tetradymia* spp. contain similar toxic
principles (Jennings et al., 1978) and cause identical syndromes in livestock
in parts of the United States where the plants are prevalent (Johnson, 1978).

Pimelea spp. are restricted mainly to Australasia, but a few species are
found in Malaysia, New Guinea, and the Philippines. Although all species
tested for toxicity in livestock are acutely irritant to the gastrointestinal
tract, the cardiopulmonary syndrome known as St. George disease, which is
due to chronic intake of particular *Pimelea* spp., occurs only in certain parts
of Australia.

In the following sections descriptions are presented of the respective main
toxic plants involved, the relevant chemical toxicants they contain, and the
clinical, pathological, and pathogenetic features of the poisonings resulting
from consumption of the plants by grazing livestock. Finally, significant similarities and differences in various aspects of the intoxications are indicated.

II. TOXIN SOURCES

A. Lantana

Lantana camara sens. lat. is a complex of many closely related plant taxa. The
taxonomy of the group is extremely complicated and the nomenclature is even
more confusing. Different authors have proposed several specific and varietal
epithets for some of the taxa. In addition, some 300 cultivar names have been
used for these plants. The precise application of these names is difficult and
sometimes impossible, and they are discussed here as one species complex.

All of them are shrubs with coarse woody stems that are square in cross
section, have a pithy center, and have small prickles along the angles. The
leaves are opposite, dull green, rough to the touch, 3.5-7.5 cm long, 2.5-
4.5 cm broad, with toothed margins, acute tips, and short basal stalks. The
flowers occur in dense clusters (shortened racemes) at the ends of long, stiff
stalks in the axils of the upper leaves: each individual flower has a small
green calyx at the base and a conspicuous, narrow tubular corolla that
spreads out abruptly at the tip into a four-lobed limb. The flowers open
successively from the outside of the cluster inward to the center and in most
forms change color as they mature. The fruits are small and succulent, with

a single hard seed. They are clustered on the swollen, knoblike tip of the stalk. Intensive field and herbarium studies showed that more than 20 recognizable taxa of *Lantana camara* are naturalized in eastern Australia, and others are grown in gardens. The taxonomic and nomenclatural status was not established and for convenience they were treated as cultivars. Each recognizable taxon was assigned a distinguishing English name. These taxa were characterized on the basis of size, shape, and color of the flowers; size, shape, and hairiness of the leaves; nature of the stem prickles; and habit of growth. Most of them were found to be tetraploids; there was one known diploid, one pentaploid, and several triploids (Seawright, 1965a; Everist, 1974).

Toxicological studies showed that, although recognizable taxa differ in toxicity, there was no consistent correlation between toxicity and ploidy or flower color per se. Toxicity is apparently determined by genetic constitution and not by environmental factors (Seawright, 1965a). Taxa in Australia that are toxic to livestock include most red-flower forms, especially the "pink-edged red" of southeastern Queensland and northeastern New South Wales and the "northern toxic pink" or "Townsville red-centered pink" from certain wet areas in northern coastal districts (Fig. 1a,b).

Nontoxic taxa include the "common pink" (Fig. 1c), which occurs throughout the range of the species in eastern Australia, and a "small-flowered pink" from a local area in southeastern Queensland. The "Townsville prickly orange," a pentaploid form from northern Queensland, is also virtually nontoxic (Seawright, 1965a; Everist, 1974).

The general distribution of naturalized taxa of *Lantana camara* in eastern Australia is shown in Fig. 2a.

B. Myoporum

The genus *Myoporum* comprises about 35 species. Four species have been recorded as toxic to livestock: *M. deserti*, *M. acuminatum*, and *M. tetrandrum* in Australia and *M. laetum* in New Zealand. A few species are regarded as useful fodder trees or shrubs in some semiarid areas in inland Australia. Although some of them may grow into small trees, most species are shrubs with thick, succulent, elongated leaves that contain oil glands embedded within the leaf tissues. The flowers are small, on individual stalks, either solitary or in small groups in the axils of the leaves. They have a five-lobed calyx and a bell-shaped, five-lobed corolla, usually white in color and enclosing four stamens. The fruits are small and more or less succulent on the outside, each containing a single hard seed. In *M. deserti*, the most important toxic species in Australia (Fig. 3), there is considerable morphological variation in size and shape of leaves and in the habit of the plant. There are also differences in the nature of the essential oils from different plants. However, no consistent correlation has been established between morphological characters on the one hand and chemical composition on the other. Figure 2b shows the distribution of known toxic species in Australia.

C. Pimelea

The genus *Pimelea* includes more than 80 species. The Australian species are the most important from a toxicity viewpoint and occur mainly in the southern half of the continent, although a few are found in northern and eastern tropical regions.

FIGURE 1 *Lantana camara*: (a) "pink-edged red," (b) "northern toxic pink," and (c) "common pink." (Queensland Herbarium photo.)

All have slender stems with thin, very tough bark, narrowly tubular, four-lobed flowers, and small fruits, some dry, some succulent. Most of them are shrubs but a few are annual herbs. Like other members of the family Thymelaeaceae, virtually all species are unpalatable to livestock in the green state. The known toxic plants include seven shrubby species from forest country in eastern and southern Australia and two red-flowered annual species from semiarid parts of tropical Queensland. The intoxication is similar to that produced by the related genera *Daphne* and *Wikstroemia* (Everist, 1974, 1981). At least four closely related annual species are more palatable when dry. The standing dry stalks or broken-down fragments on the soil, if ingested by cattle, can produce the chronic syndrome known as

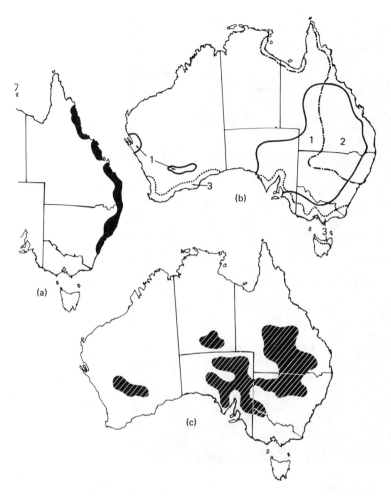

FIGURE 2 Approximate limits of distribution in Australia of: (a) naturalized taxa of *Lantana camara*; (b) known toxic species of *Myoporum*: 1, *M. deserti*, 2, *M. acuminatum*, 3, *M. tetrandrum-M. adscendens-M. insulare* complex; (c) *Pimelea trichostachya-P. simplex-P. continua-P. elongata* group.

St. George disease, as well as acute toxicity (Everist, 1974, 1981; Kelly and Seawright, 1978). These annual species are erect herbs with thin, branched stems bearing small alternately arranged leaves and with small, hairy, tubular flowers in the axils of the upper leaves. In *P. simplex* and *P. continua*, the flowering tips of the stems remain short, so that the flower spikes appear to be compact heads; in *P. trichostachya* and *P. elongata*, the stems elongate as the flowers mature and appear to be slender terminal spikes. These species are fairly widely distributed in arid and semiarid regions in inland subtropical Australia. Figure 2c shows the generalized distribution of the group. Figure 4 illustrates *P. trichostachya*, the most widespread species.

FIGURE 3 *Myoporum deserti*. (Queensland Herbarium photo.)

III. TOXIC PRINCIPLES

A. Lantana

The only known biologically active compounds found in *L. camara* are pentacyclic triterpene acids. The first two of these compounds isolated were
lantadene A (rehmannic acid) (Fig. 5a) and lantadene B (Fig. 5b) (Louw,
1943, 1948), and these were later characterized as 22β-angeloyloxyoleanonic
and 22β-dimethylacroyloxyoleanonic acids by Barton and de Mayo (1954),
Barton et al. (1956) and Barton et al. (1954), respectively. These compounds are present in toxic lantana at from 0.5 to 2.2% of the dry matter of
the leaves, with A and B in the ratio 3:1. Additional significant triterpene
acids later isolated from the plants included the C-3 hydroxy forms of lantadenes A and B, 22β-angeloyloxyoleanolic (Fig. 5c) and 22β-dimethylacroyl-
oxyoleanolic acids as well as traces of icterogenin (22β-angeloyloxy-24-

FIGURE 4 *Pimelea trichostachya*. (Queensland Herbarium photo.)

hydroxyoleanonic acid) (Fig. 5d) (Hart et al., 1976a). These compounds are present only in trace amounts and contribute little, if anything, to the toxicity of the plants. Lantadenes A and B have an oral toxicity for sheep of 60-80 and 200-300 mg/kg, respectively, while for C-3 hydroxy (or reduced) lantadene A, the toxic dose is about 40 mg/kg (Hart et al., 1976b). Nontoxic "common pink" lantana contains small amounts of the new lantadene A and B analogs, 22β-angeloyloxylantanolic and 22β-dimethylacroyloxylantanolic acids. The latter have an oxide bridge from C-25 to C-3, with an acetal system at C-3.

More recently, Beeby (1977) has described a partial synthesis of lantadene A and subsequently employed this procedure to synthesize a series of 22β-ester analogs of the compound (Beeby, 1978). In toxicity studies in sheep the 22β-cinnamoyl ester was toxic but the corresponding acetate was not. This suggests that a 22β-ester moiety of sufficiently large size associated with

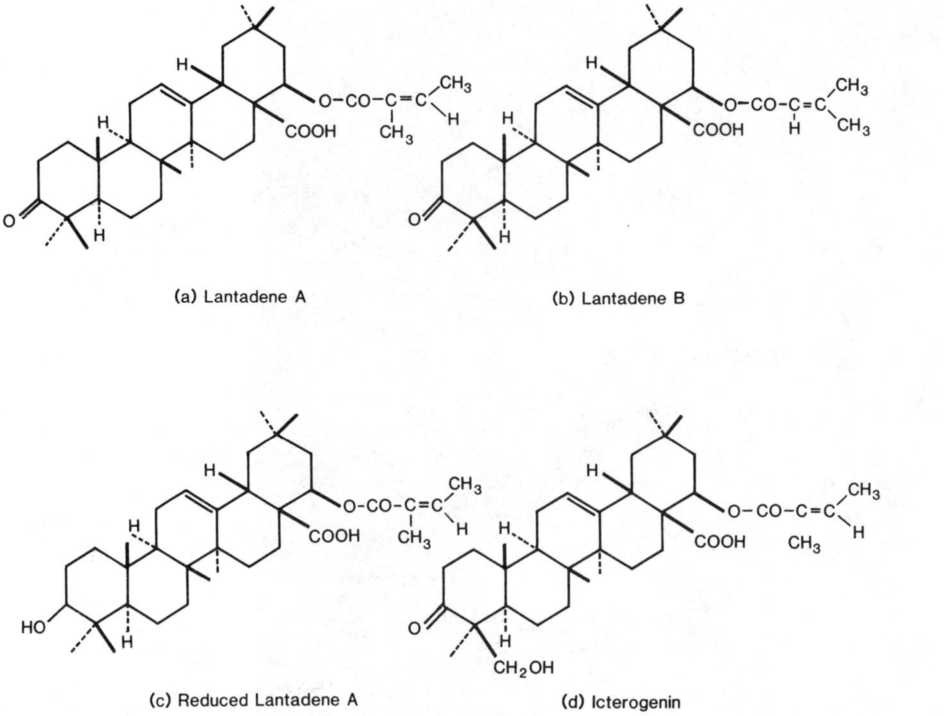

(a) Lantadene A   (b) Lantadene B

(c) Reduced Lantadene A   (d) Icterogenin

FIGURE 5 Toxic pentacyclic triterpene acids found in *Lantana camara*.

oleanonic or oleanolic acid was necessary for toxicity, at least in this species. Oleanonic and oleanolic acids themselves are common constituents of plants and are not toxic.

B. Myoporum

The toxic constituents of the various species of *Myoporum* are furanosesqui-terpenoid essential oils, the best known of which is ngaione (McDowall, 1925). The latter oil is the diastereoisomer of ipomeamarone (Birch et al., 1954), a toxic essential oil previously isolated from mold-affected sweet potatoes (*Ipomoea batatas*) (Hiura, 1943).

Of the representatives of this genus, *M. deserti* is the most common cause of poisoning of livestock in Australia and the essential oils of this plant have been the most extensively investigated. So far, 12 such furanoid oils have been characterized (Fig. 6), and apart from ngaione (Hegarty et al., 1970) include myodesmone and isomyodesmone (Blackburne et al., 1971), dehydro- and dehydroisomyodesmone (Blackburne and Sutherland, 1972), myoporone and dehydromyoporone (Blackburne et al., 1972), epingaione, dehydro-ngaione, dehydroepingaione and deisopropylngaione (Hamilton et al., 1973), and "ketol A" (M. D. Sutherland, personal communication, 1981). All these

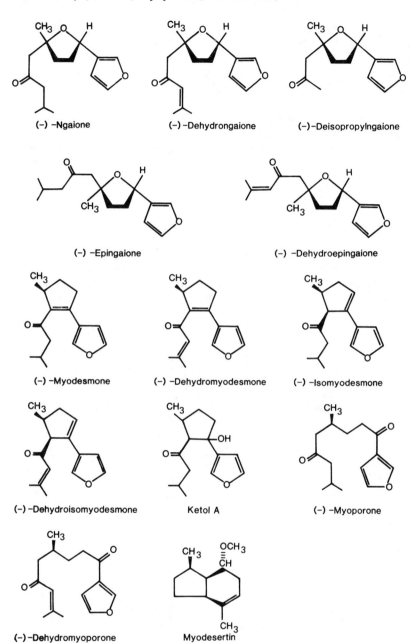

FIGURE 6 Terpenoid essential oils isolated from *Myoporum deserti*.

compounds are acutely hepatotoxic for laboratory animals, and deisopropyl-ngaione is also nephrotoxic and pulmonary toxic (Seawright et al., 1981).

Several years ago samples of foliage from *M. deserti* from 29 separate localities in eastern inland Australia where livestock poisonings occurred were investigated chemically, and from these samples nine different mixtures of essential oils were distinguished. Six of these oils contained one or more of the various furanosesquiterpenes as the major constituents, while three contained iridoid monoterpenes, myodesertin, and myodesertal as sole or major constituents, usually free of furanosesquiterpenes. An account of the chemistry of myodesertin (IR-I-methoxymyodesert-3-ene) is given by Grant et al. (1980). These monoterpenoid oils are nontoxic for both domestic and laboratory animals.

C. Pimelea

The highly irritant toxic principles present in *Pimelea* spp. are diterpene esters based on tigliane and daphnane (Hecker, 1978; Evans and Soper, 1978). The tigliane derivatives include 12-deoxy,13-esters of phorbol (Fig. 7a). The weak irritant 13-O-acetyl-12-deoxyphorbol (prostratin) (Fig. 7b) was the first

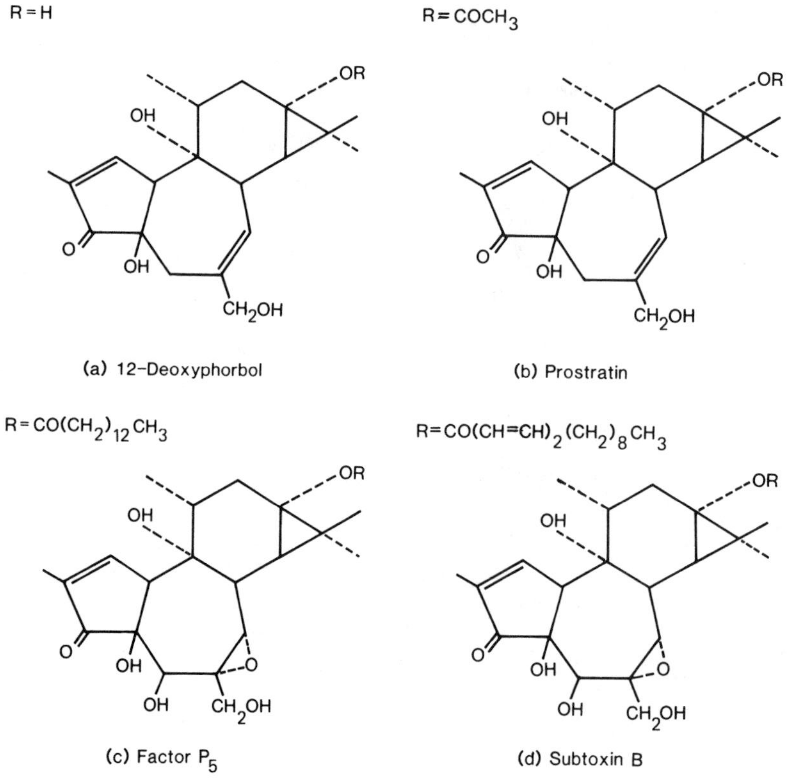

FIGURE 7 Irritant tigliane diterpenoid esters from *Pimelea* spp.

compound of this type isolated and described from the toxic New Zealand plant *P. prostrata* (Cashmore et al., 1976). Recently, a more powerful tigliane irritant, the 13-tetradecanoate of 12-deoxy-5β-hydroxyphorbol-6α,7α-oxide (Factor P5) (Fig. 7c), has been isolated from this plant (Zayed et al., 1977b), and a similar compound, subtoxin B (Fig. 7d), has been found in Australian *P. simplex* (Freeman et al., 1979).

The main class of irritant found in toxic *Pimelea* spp. belongs to the daphnane esters. The prototype parent alcohol of this group is resiniferonol (Fig. 8a), the functionality of which is almost identical with 12-deoxyphorbol. In addition, the compound contains a secondary 14α-hydroxyl group and a stable 9α,13α,14α-orthoester function apart from the 13β-isopropyl group. The prototype orthoester irritant of this series in *Pimelea* spp. is simplexin (Fig. 8b) (Roberts et al., 1975), the base alcohol of which is 5β-hydroxy-resiniferonol-6α,7α-oxide. A further compound of this type recently found in *P. simplex* and *P. elongata* is 12β-acetoxyhuratoxin (Fig. 7c) (Freeman et al., 1979). Simplexin, as well as its active homolog, factor P4 (Fig. 8d), have also been found in *P. prostrata* (Zayed et al., 1977b).

A further group of highly active irritants found in *Pimelea* spp. is the 1α-alkyldaphnane series of compounds (Zayed et al., 1977a). The compound found in *P. simplex* is based on 1,2,6,7-tetrahydro-5-hydroxy-21-methyl-6α,7α-epoxy-1α-nonylresiniferonol-30-oic acid, with the carboxyl group of the parent acid located at the tail end (ωposition) of the 1α-alkyl chain, forming an orthoester group with hydroxyls 9, 13, and 14 of the six-membered ring (factor P6) (Fig. 8e). This gives an intramolecular ring on the α side of the diterpene molecule. A similar compound, pimelea factor P2, isolated from *P. prostrata*, varies from the latter in that there is a benzoyl group in lieu of the ketone function at C-3 as well as a modification to the 1α-alkyl side chain (Fig. 8f).

The daphnane orthoesters, apart from being highly irritant, are also potent tumor promotors in mouse skin and resemble croton oil in this respect (Hecker, 1978). They therefore represent an important group of cocarcinogens. *P. simplex* extracts also have antileukemia activity (Howard and Howden, 1975). Both simplexin and crude plant alcoholic extracts, injected intravenously into cattle, have been shown to be able to produce typical St. George disease (Roberts et al., 1975; Clark, 1973). Apart from *P. simplex* (McClure and Farrow, 1971; Roberts and Healy, 1971), *P. trichostachya* (Clark, 1973), *P. elongata* (Freeman et al., 1979; Threlfall, 1980), *P. continua* (Kelly, 1975a), and *P. altior* (Rogers and Roberts, 1976) have also been shown to cause St. George disease in Australia.

IV. CLINICAL FEATURES AND PATHOLOGY

A. Lantana

All types of domesticated ruminants are equally susceptible to lantana poisoning. Animals become sick within a day of ingestion of the plant, and death or recovery from intoxication normally takes place within 1-3 weeks (Seawright and Sferco, 1966). The syndrome is characterized by intense jaundice with photosensitization (Fig. 9) and, depending on the severity of intoxication, also by a profuse black scour, the irregular passage of dark feces coated with blood or mucus (Sanders, 1946), or constipation with straining (Seawright,

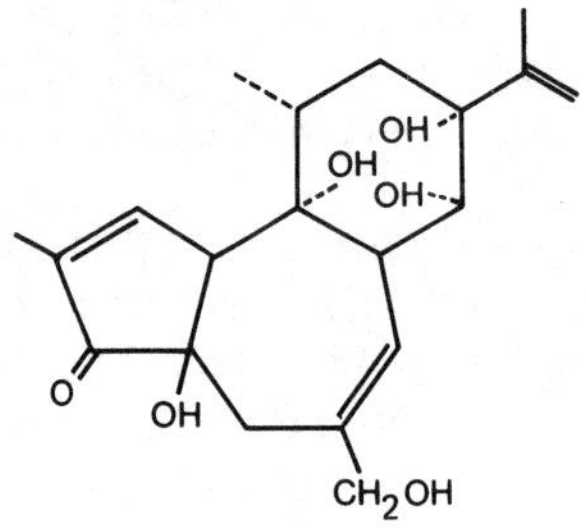

(a) Resiniferonol

$R = (CH_2)_8CH_3$

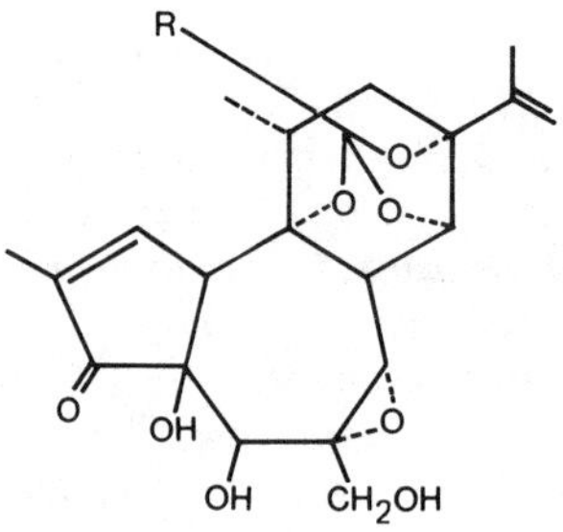

(b) Simplexin

$R = (CH{=}CH)_2(CH_2)_4CH_3$

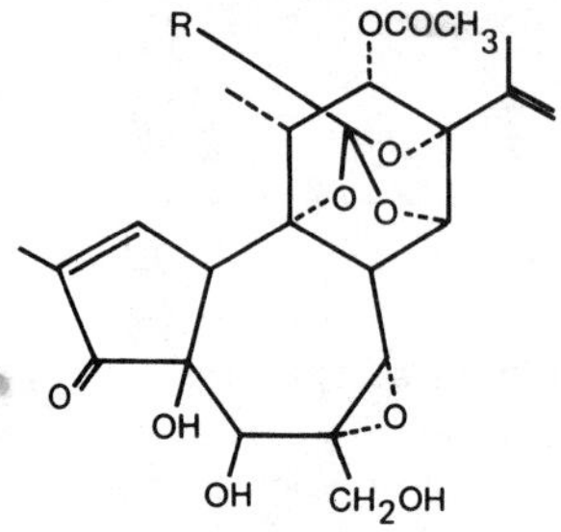

(c) 12 β-acetoxyhuratoxin

$R = (CH_2)_{12}CH_3$

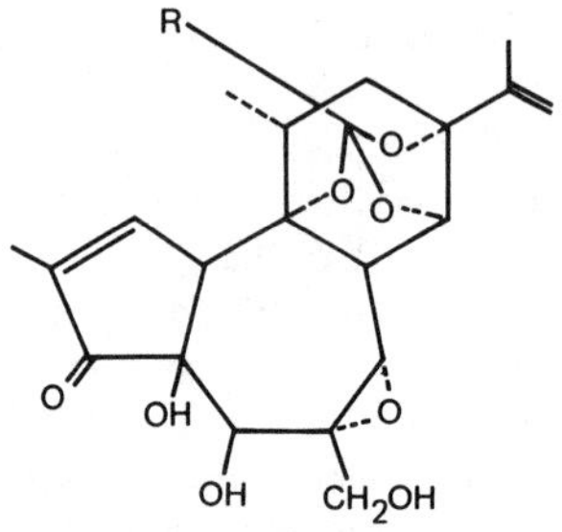

(d) Pimelea Factor P_4

$R = CH(CH_3)\text{-}(CH_2)_7$
$R' = COC_6H_5$

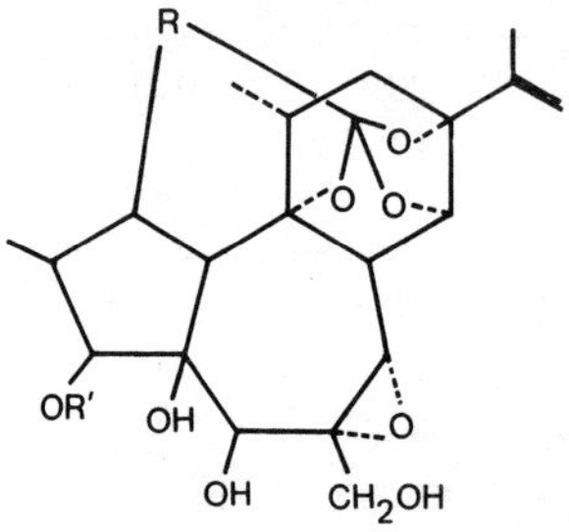

(f) Pimelea Factor P_2

$R = CH(CH_3)\text{-}(CH_2)_6\text{-}CH(OCOC_6H_5)$

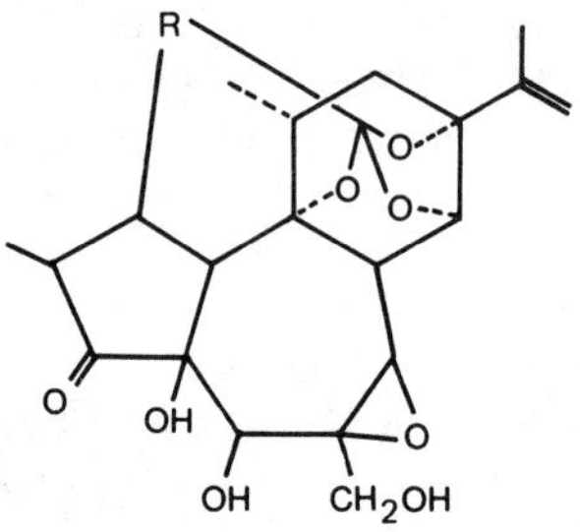

(e) Pimelea Factor P_6

FIGURE 8 Irritant daphnane diterpenoid esters from *Pimelea* spp.

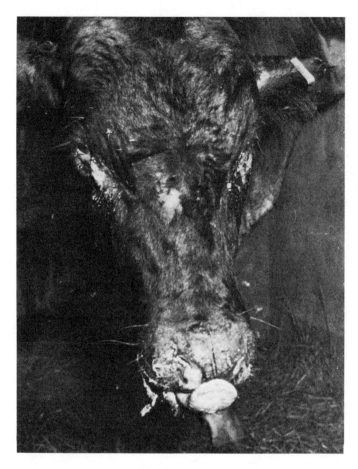

FIGURE 9 Photosensitization of the muzzle characteristic of lantana poisoning.

1963, 1964). In the early stages of poisoning animals pass urine frequently, develop transient, mild proteinuria, and become quite dehydrated. There is loss or reduction of ruminal movements, persistent anorexia and infrequency of drinking, and elevated levels of serum phylloerythrin, bilirubin, and liver-derived enzymes (Gopinath and Ford, 1969). Pulmonary edema sometimes develops during the agonal phase of the condition (Seawright, 1964).

Gross changes observed at necropsy usually reveal an intensely jaundiced carcase, sometimes with profuse subcutaneous hemorrhages, with a yellowish, ocher, sometimes mottled, swollen liver, and with a grossly distended gall-bladder (Fig. 10) filled with watery yellow to dark green bile. There may also be copious amounts of yellow serous fluid in the peritoneal and pleural cavities. Kidneys are usually swollen, pale, and yellowish, turning green on prolonged exposure to air (Seawright, 1964; Seawright and Allen, 1972). When death occurs from 2 to 3 weeks after poisoning, the rumen contents are abnormally dry and those of the large intestine firm, dark, dry, and

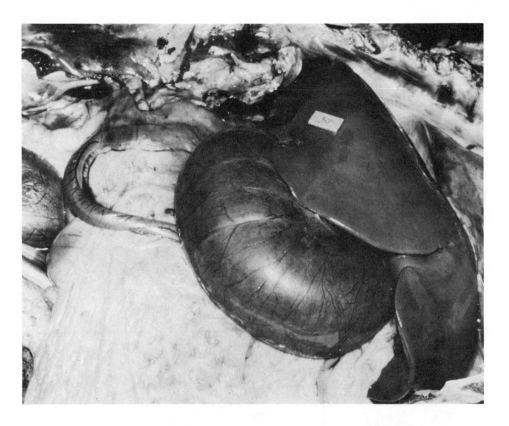

FIGURE 10 Marked distension of the gallbladder of a sheep with lantana
poisoning.

associated with copious amounts of glairy mucus due to prolonged stasis.
Sheep may show white, patchy areas of the myocardium and moist, heavy
lungs due to pulmonary edema (Seawright, 1964).

Significant microscopic changes in cattle are confined to the liver and
kidneys (Seawright and Allen, 1972). With developing cholestasis there is a
gradual swelling of the hepatocytes with vesiculation of the nuclei. Small
portal tracts become more prominent than normal due to biliary ductular
hyperplasia, portal fibrosis, and mild lymphocytic infiltration. Mitotic
figures may be seen in hepatocytes and multinucleated pleomorphic hepatocytes, and bile canalicular plugs are frequent. Due to prolonged biliary
retention, a feathery vacuolation mainly of the periportal hepatocytes may be
present (Fig. 11). In other livers there may be midzonal foci of coagulative
necrosis with infiltration by inflammatory cells, and marked disorganization
and fragmentation of the hepatic cell plates (Seawright and Allen, 1972).

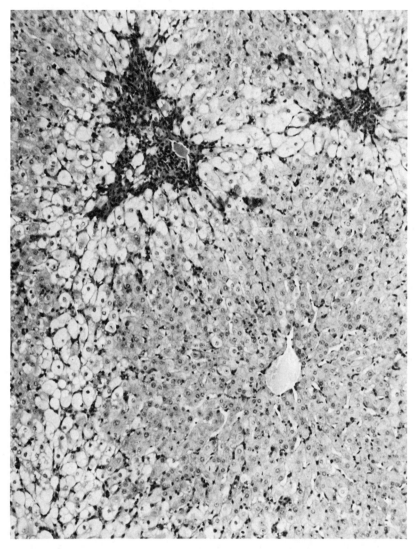

FIGURE 11 Periportal vacuolar and feathery degeneration of hepatocytes in a sheep with lantana poisoning (hematoxylin and eosin, X150).

Depending on the duration and severity of intoxication, the changes in the kidneys range from widespread hydropic vacuolation of the epithelium of the convoluted tubules with presence of hyaline and leucocyte casts and focal interstitial leucocytic infiltration, to extensive coagulative necrosis and fatty change affecting most of the tubular epithelium (Seawright, 1963, 1964; Seawright and Allen, 1972). In sheep with myocardial lesions there is interstitial edema, granular degeneration, and coagulative necrosis of myocardial fibers and focal inflammatory infiltration and fibrosis (Seawright, 1964).

B. Myoporum

Toxicity due to ingestion of *Myoporum* leaves has been recorded in cattle,
sheep, goats, pigs, and horses (Cunningham and Hopkirk, 1945). After a
toxic dose of *Myoporum* leaves or essential oils, animals become depressed
and anorexic between 6 and 10 hr. By 12 hr there is usually a tympanic
swelling in the left flank in ruminants together with signs of abdominal discomfort. At this stage there is a copious seromucoid nasal discharge which
persists for 24-48 hr. Animals tend to become constipated and any feces
passed are usually hard, dry, and blood stained. Up to this stage affected
animals tend to be abnormally aggressive. Signs of acute liver injury usually
become apparent within about 2 days and at this stage jaundice and photosensitization may be present. In the terminal stages of illness animals become
prostrate and comatose and usually die within 2-5 days of intoxication. In
other animals liver injury and photosensitization tend to subside within 2-3
days of appearance of skin lesions, and recovery is normally complete by 6-10
days (Allen et al., 1978). In some sheep however, an acute pulmonary edema
develops within 3-24 hr of ingestion of the plant, and in such animals there
may be little opportunity for signs of acute hepatic injury to appear before
death occurs (Allen and Seawright, 1973).

Gross pathological findings include jaundice with subcutaneous hemorrhages, and excessive yellow serous fluid containing strands of fibrin in the
peritoneal cavity. The liver is swollen, pale or reddened, and may be mottled
in a red and yellow zonal pattern, occasionally with a discrete patchiness
apparent on the surface of the organ (Fig. 12). Sometimes large blood lakes
may be present in the parenchyma and the gallbladder wall is markedly
thickened due to edema. There is usually hyperemia of the mucosa of the
lower alimentary tract, sometimes with petechiae, and free blood staining of
the contents of the duodenum, ileum, and cecum. There is usually intestinal
stasis with the presence of firm, dry feces in the large intestine. The kidneys may be swollen and pale but are usually unremarkable. The myocardium
is flabby and pale and there is excess serous fluid in the pericardial and
pleural sacs. When pulmonary edema is present, the lungs are swollen and
reddish and the bronchi and trachea contain copious white, stable froth.

Microscopically, lesions are seen consistently in the liver and include fatty
infiltration with massive or zonal necrosis. The latter lesions may be midzonal, centrolobular (Fig. 13), or periportal (Fig. 14) in location (Allen and
Seawright, 1973)--most frequently centrolobular in cattle and periportal in
sheep and goats (Allen et al., 1978). Within 2 days of intoxication biliary
ductular hyperplasia, portal fibrosis, and lymphocytic infiltration are usually
marked. In some livers there is fibrinoid necrosis of some arteriolar walls,
causing the discrete patchiness shown in Fig. 12. In the kidneys there is
often extensive fatty infiltration of the tubular epithelium and cortical and
medullary congestion, but coagulative necrosis is rare. In the myocardium
there is a diffuse fine-droplet fatty degeneration and interstitial and sub-
endocardial hemorrhages. Affected lungs show alveolar capillary hyperemia,
distension of lymphatics, and protein coagulum in alveolar and lymphatic
spaces.

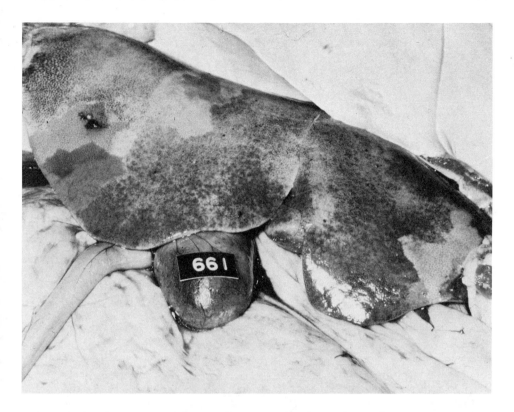

FIGURE 12 Discrete patchiness with necrosis of the liver of a sheep dead from myoporum poisoning.

C. Pimelea

Following consumption of *Pimelea* spp. animals rapidly develop profuse diarrhea (Everist, 1974; Hill, 1970), while horses consuming *P. prostrata* have vesicular and necrotic lesions of the oral and pharyngeal mucosa as well (Connor, 1977). Except where diarrhea is associated with St. George disease, episodes of typical poisoning following consumption of these plants is uncommon. St. George disease occurs following chronic daily intake of small quantities (40-60 mg/kg) of the dried whole plant over several weeks (Kelly, 1975a) and is characterized by severe cardiopulmonary dysfunction similar to that seen in high mountain or Brisket disease (Clark, 1973). Diarrhea may not necessarily be present in affected cattle (Dodson, 1965).

Intoxicated cattle are characterized by marked anasarca of the head, neck, and pectoral region (Fig. 15), with a marked jugular pulse indicative of raised

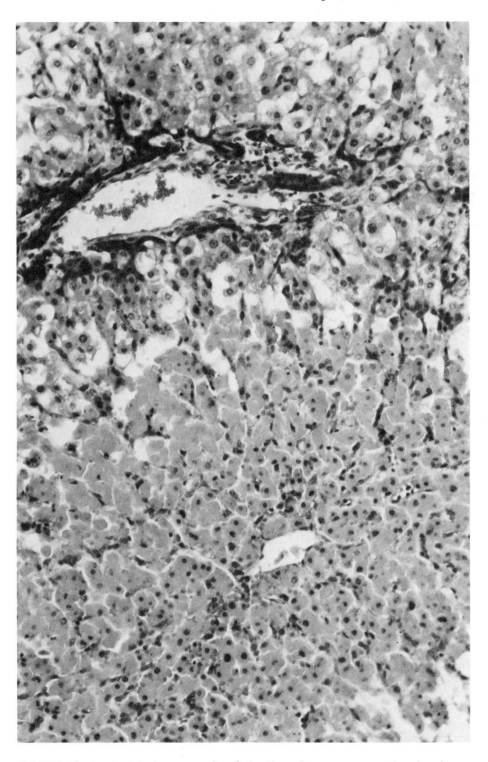

FIGURE 13 Centrolobular necrosis of the liver in myoporum poisoning in
sheep (hematoxylin and eosin, X200).

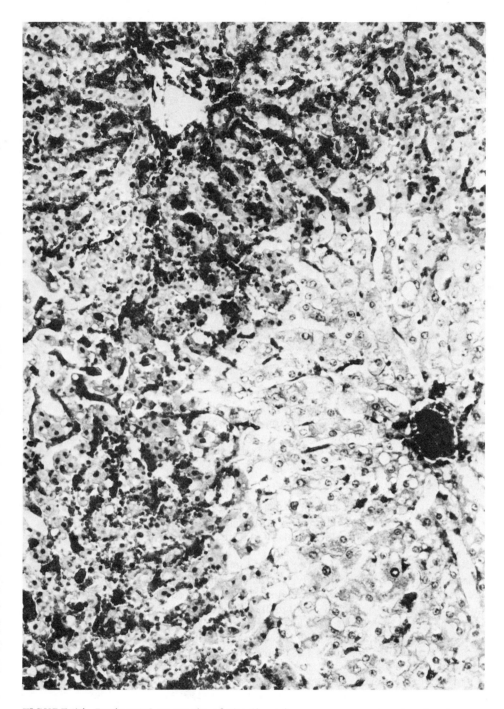

FIGURE 14 Periportal necrosis of the liver in myoporum poisoning in sheep (hematoxylin and eosin, X200).

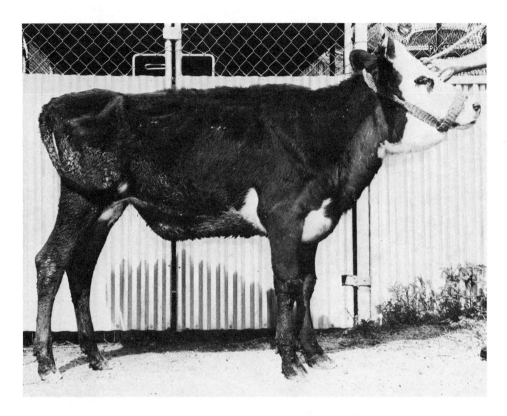

FIGURE 15 Calf with anasarca of the head and neck due to cardiopulmonary
dysfunction in chronic pimelea poisoning (St. George disease). (Courtesy of
W. R. Kelly, University of Queensland, Brisbane, Australia.)

central venous pressure (Clark, 1973; Kelly, 1975a). The condition is associated with a normochromic, normocytic anemia and panleucopenia due mainly to
lymphocytopenia. There is a marked weight loss and reduced exercise
tolerance.

Necropsy findings in affected cattle include marked subcutaneous ventral
edema and hydrothorax, usually with a notable absence of ascites (Seawright
and Francis, 1971). In acute, severe intoxication the lungs are swollen,
edematous, and red, but in chronically affected cattle they are not remarkable
except for ventral atelectasis caused by the hydrothorax. The right ventricle
and the trunks of the pulmonary arteries are dilated (Clark, 1973) and lymph
nodes remote from edematous areas are moist and swollen. The liver is enlarged and blue-black, with the contents of abnormally friable consistency.
In freshly dead animals copious amounts of blood ooze from the cut surface.
Other organs, as well as urine, are unremarkable.

Histologically, in the lungs there may be a variable degree of congestion,
with erythrophagocytosis and leucocytic sequestration in alveolar capillaries,
siderosis, interstitial fibrosis, and collapse, but except in acute poisoning the

tissues are unremarkable. In the heart there are foci of myocardial fiber degeneration and necrosis, with foci of fibrosis and macrophage accumulation (Kelly, 1975a). In the livers there is distension of the smallest portal venules, leading to dilation of periportal sinusoids. This gives rise in some areas to large cavities lined with endothelium and often containing blood clots, a phlebectatic peliosis hepatis (Fig. 16). In other areas there is breakdown of the parenchyma (Fig. 17), particularly at the limiting plate under Glisson's capsule, allowing sludging of large pools of blood immediately beneath the capsule (Fig. 18). Hepatocyte cords and plates become atrophic and there may be fatty change of the centrolobular parenchyma. Centrolobular necrosis is unusual but may occur if the cardiopulmonary disturbance is rapid in onset. In recovered animals there is substantial organization of subcapsular thrombi, with the development of a form of hepatoportal sclerosis. Lymph nodes show marked lymphoid depletion and distended vascular spaces may be present in the spleen, bone marrow, adrenal cortex, and kidneys (Kelly, 1975a). Changes in the alimentary mucosa are usually unremarkable.

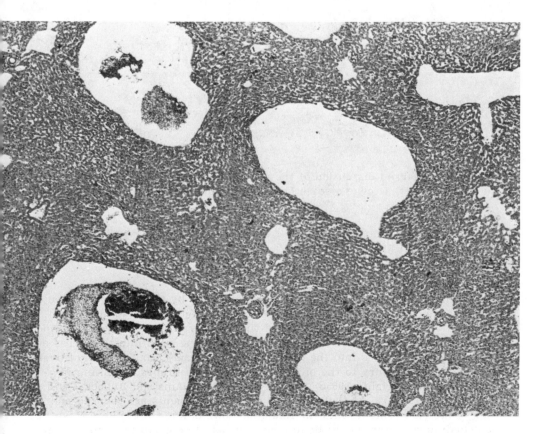

FIGURE 16 Circumscribed cavities in the liver of cattle characteristic of peliosis hepatis in chronic pimelea poisoning (St. George disease) (hematoxylin and eosin, X40).

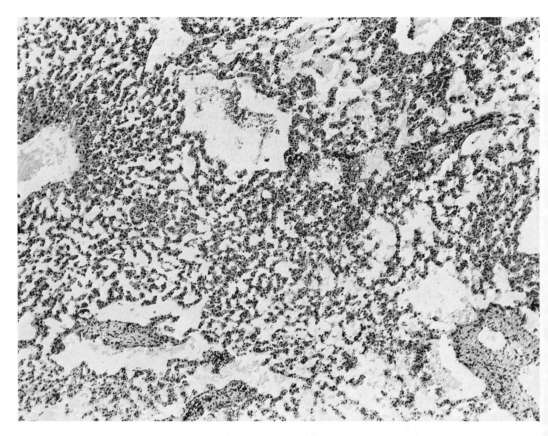

FIGURE 17 Marked distension of the portal venules and sinusoids in the liver of cattle with St. George disease (hematoxylin and eosin, X75).

V. PATHOGENETIC MECHANISMS

A. Lantana

Lantadene A is about 20 times more toxic for sheep by the intravenous route than by the oral route (Pass et al., 1979b), suggesting that alimentary absorption of the toxin after ingestion of the plant is comparatively slow. Once ingested the toxin is absorbed most rapidly from the small intestine (Pass et al., 1981a) and passes into the portal blood. An early manifestation of liver damage is rapid injury to the biliary canalicular membranes, resulting in cholestasis (Seawright, 1965b; Pass et al., 1978; Eakins and Bullock, 1978; Pass et al., 1981b). At the same time ruminal movements are reduced and flow of stomach contents into the intestine slows down. Nevertheless, sufficient toxin does enter the duodenum continuously to sustain the intoxication since complete evacuation of the rumen up to 6 days after intoxication and its replacement by normal rumen contents results in rapid recovery from the condition (C. S. McSweeney and M. A. Pass, unpublished data, 1982).

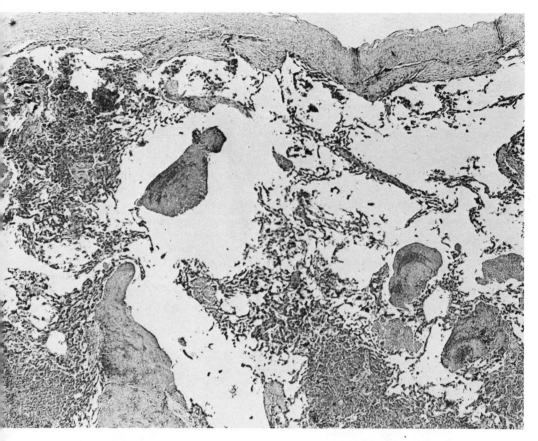

FIGURE 18 Subcapsular rupture of the parenchyma with cavity formation and thrombosis in the liver in St. George disease (hematoxylin and eosin, X40).

Studies of the effect of lantana poisoning on intestinal motility reveal that this function is not affected (Pass and Heath, 1978), while the intoxication does in fact result in paralysis of the gallbladder wall (Pass and Heath, 1977), as previously suggested by Seawright (1964). Reduced lantadene A but not lantadene A itself is toxic by the oral route for the female but not the male rat, causing typical cholestasis and alimentary stasis (Pass et al., 1979a).

The hepatotoxicity is enhanced by pretreatment with spironolactone but not with prior dosing with phenobarbitone, SKF525A, or carbon disulfide, suggesting that a metabolite of the triterpene produced by the microsomal cytochrome P_{450} oxygenases is the actual toxic agent. The toxicity is not sex dependent in domestic animals, and studies of the effects on the kidney and myocardium have yet to be made.

B. Myoporum

Mechanism studies with *Myoporum* oils have been concerned mainly with the explanation of hepatic midzonal necrosis (Fig. 19) which occurs consistently in mice (Seawright and O'Donahoo, 1972). Seawright and Hrdlicka (1972) showed that both inhibition and induction of the hepatic microsomal mixed-function oxygenases (HMFO) with SKF525A and phenobarbitone, respectively, protected mice against ngaione toxicity. With the former treatment the liver lesion present was centrolobular necrosis, while in the latter periportal necrosis occurred. The response of the animals to the oil returned to normal (midzonal necrosis) when the altered state of the HMFO returned to pretreatment levels. A similar transposition of the liver lesion was demonstrated in rats (Seawright et al., 1978), sheep (Allen and Seawright, 1973), and calves (Allen et al., 1978). Seawright and Hrdlicka (1972) proposed that alternative routes of HMFO metabolism were available to the oils, toxic and nontoxic, and whether injury occurred in particular hepatocytes depended on the level and type of the HMFO metabolism at the time of dosing with the oil or ingestion of the plants. This was partly demonstrated by the study in mice by Lee et al. (1979), in which the lesion caused by ngaione given in the inhibitory phase of phenobarbitone induction of HMFO was centrolobular necrosis, whereas in animals given the oil several hours later when the enzyme levels

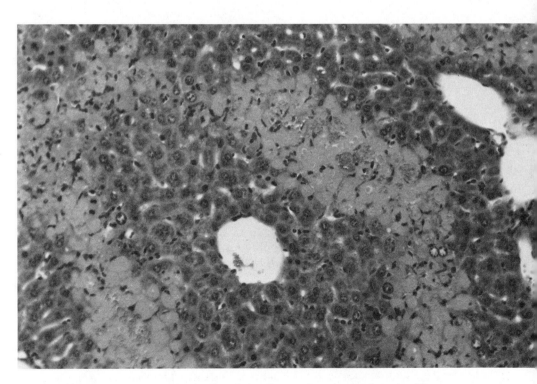

FIGURE 19 Midzonal necrosis of the liver of a mouse dosed with a myoporaceous furanosesquiterpenoid essential oil (hematoxylin and eosin, X230).

were increased twofold the lesion was periportal. The zonal location of liver lesions in animals poisoned by *Myoporum* spp. in the field depends on the HMFO activity at the time of ingestion of the plants.

Tolerance to chronic daily dosing with the oils was demonstrated in rats and this was shown to be due to 75% reduction of the activating cytochrome P_{450} which occurred in the liver microsomes (Lee et al., 1981).

Mechanism studies of the mode of action of 3-substituted furans have been most extensively studied by M. R. Boyd using 4-ipomeanol from moldy sweet potatoes. Extrahepatic lesions caused by such compounds occur through their similar toxic metabolism in organs such as the kidneys and lungs (Boyd et al., 1979), and species and strain differences in the injurious effects due to the oils can be explained on this basis (Dutcher and Boyd, 1979).

C. Pimelea

Although the biological mechanism of action of the diterpene ester irritants has been investigated in the last decade (Evans and Soper, 1978), the relevance of these studies to the mechanism of St. George disease has so far not been assessed. It has been claimed that all the clinical and pathological manifestations of the latter condition can be produced with simplexin (Roberts et al., 1975).

Studies are handicapped by the fact that the disease occurs only in cattle. The latter animals have thick spiral muscles in the walls of the pulmonary veins (Alexander and Jensen, 1963); and Clark (1973) has proposed that the marked constricting action of the toxin on this structure is responsible for the cardiopulmonary dysfunction of St. George disease. In vitro studies with *Pimelea* extracts by Kelly and Bick (1976) using isolated bovine pulmonary vein strips supported Clark's in vivo observations. Extracts of *Euphorbia tirucalli* which contain tigliane esters (4-deoxyphorbol) (Furstenberger and Hecker, 1977) do not appear to have such effects in vivo or in vitro (W. R. Kelly, unpublished observations, 1981).

Hematological studies in calves by Kelly (1975b) showed that the anemia of St. George disease was due to hemodilution. Immediately after beginning of chronic daily dosing with the plants, there was a progressive increase in the plasma volume of up to 178% without increase in the red cell mass, to give an overall increase in the blood volume by about 100% within 40-80 days. The additional blood was accommodated mainly in the liver, which responded by passive dilation of portal venules and sinusoids, resulting in typical phlebectatic peliosis hepatis. The mechanism by which this massive enhancement of the plasma volume occurred is not understood, but a possible direct effect of the toxin on the microcirculation of the liver, causing loss of tone of the vessel walls with consequent unimpeded distention of sinusoids, would seem to warrant further investigation.

VI. COMPARATIVE FEATURES

An important difference between lantana and myoporum poisoning of ruminants is that uptake of the toxin from the gut in the former is slow and continuous (Pass et al., 1981a), whereas absorption of essential oils in the latter condition is comparatively rapid (A. A. Seawright and A. R. Mattocks, unpublished observations, 1972). In both diseases zonal liver necrosis occurs if the dose of toxin reaching the liver is high enough. In lantana poisoning the liver lesion in sheep after intravenous injection of the toxin can be centrolobular necrosis (Pass et al., 1979b), but in earlier experimental studies of oral dosing with lantana extracts in the guinea pig, midzonal necrosis also occurred (Seawright, 1965c). In both conditions, activation of the liver in the HMFO system appears to be necessary for toxic action, at least in some species, but the respective cytochrome enzyme species involved are clearly different. The peripheral feathery degeneration of the liver parenchyma in lantana poisoning is a secondary characteristic of severe cholestasis, whereas the periportal necrosis of myoporum toxicity is the result of a primary intrahepatocytic injury.

In both these diseases constipation and gut stasis are common features of the intoxication. Recent studies by C. S. McSweeney and M. A. Pass (unpublished observations, 1982) have shown that disruption of hepatic function of almost any kind has a direct depressant effect on the rumen, causing its frequency and amplitude of contraction to be greatly reduced, leading to stasis in the lower alimentary tract. It is not unusual, therefore, that constipation and large intestinal stasis commonly occur in hepatotoxic diseases of ruminants except where there is direct toxic injury to the gastrointestinal mucosa, possibly stimulating enhanced motility of these organs, and diarrhea.

In lantana poisoning, a primarily cholestatic disease under natural conditions, photosensitization is always comparatively severe and prolonged even though histological evidence of hepatocellular injury is slight. In zonal necrosis of the liver in which the biliary excretory reserve may not be exceeded, evidence of biliary excretory failure is often slight and transient, and photosensitization may not even occur. In the latter condition photosensitization would indicate extensive parenchymal necrosis and a relatively grave prognosis, whereas in lantana poisoning and other primarily cholestatic conditions this is not necessarily so.

Chronic pimelea intoxication has little in common with either lantana or myoporum toxicities, notwithstanding the fact that there is a liver lesion in the former and sometimes scouring and blood in the gut contents in one or the other of the latter. Diterpene esters are not usually metabolized in the liver for toxicity in the animal, and the subcutaneous edema due to photosensitization is not difficult to differentiate from that of heart failure. St. George disease superficially resembles high mountain disease of cattle, and apart from differences in the pulmonary dysfunction and pathology in these two conditions, the liver in the former is characterized mostly by phlebectatic peliosis hepatis, whereas in the latter there is always the expected chronic venous congestion (Alexander and Will, 1963).

REFERENCES

Alexander, A. F., and Jensen, R. (1963). Normal structure of bovine pulmonary vasculature. *Am. J. Vet. Res. 24*:1083-1093.

Alexander, A. F., and Will, D. H. (1963). High mountain (Brisket) disease. In *Diseases of Cattle*, 2nd rev. ed., W. J. Gibbons (Ed.). American Veterinary Publications, Wheaton, Ill., pp. 110-116.

Allen, J. G., and Seawright, A. A. (1973). The effects of prior treatment with phenobarbitone, dicophane (DDT) and β-diethyl-aminoethyl phenyl-propyl acetate (SKF525A) on experimental intoxication of sheep with the plant *Myoporum deserti* Cunn. *Res. Vet. Sci. 15*:167-179.

Allen, J. G., Seawright, A. A., and Hrdlicka, J. (1978). The toxicity of *Myoporum tetrandrum* (boobialla) and myoporaceous furanoid essential oils for ruminants. *Aust. Vet. J. 54*:287-292.

Barton, D. H. R., and de Mayo, P. (1954). Terpenoids. Part XVI. The constitution of rehmannic acid. *J. Chem. Soc.*, pp. 900-903.

Barton, D. H. R., de Mayo, P., Warnhoff, E. W., Jeger, O., and Perold, G. W. (1954). Terpenoids. Part XIX. The constitution of lantadene B. *J. Chem. Soc.*, pp. 3689-3692.

Barton, D. H. R., de Mayo, P., and Orr, J. C. (1956). Terpenoids. Part XXIII. Nature of lantadene A. *J. Chem. Soc.*, pp. 4160-4162.

Beeby, P. J. (1977). Angeloyl chloride: synthesis and utilization in the partial synthesis of lantadene A (rehmannic acid). *Tetrahedron Lett.*, pp. 3379-3382.

Beeby, P. J. (1978). Chemical modification of triterpenes from *Lantana camara*. 22β-ester analogues of lantadene A. *Aust. J. Chem. 31*:1313-1322.

Birch, A. J., Massy-Westropp, R. A., Wright, S. E., Kubota, T., Matsuura, T., and Sutherland, M. D. (1954). Ipomeamarone and ngaione. *Chem. Ind.*, p. 902.

Blackburne, I. D., and Sutherland, M. D. (1972). Terpenoid chemistry. XIX. Dehydro- and dehydroiso-myodesmone, toxic furanoid sesquiterpene ketones from *Myoporum deserti*. *Aust. J. Chem. 25*:1779-1786.

Blackburne, I. D., Park, R. J., and Sutherland, M. D. (1971). Terpenoid chemistry. XVIII. Myodesmone and isomyodesmone, toxic furanoid ketones from *Myoporum deserti* and *M. acuminatum*. *Aust. J. Chem. 24*: 995-1007.

Blackburne, I. D., Park, R. J., and Sutherland, M. D. (1972). Terpenoid chemistry. XX. Myoporone and dehydromyoporone, toxic furanoid ketones from *Myoporum* and *Eremophila* species. *Aust. J. Chem. 25*:1787-1796.

Boyd, M. R., Dutcher, J. S., Buckpitt, A. R., Jones, R. B., and Statham, C. N. (1979). Role of metabolic activation in extrahepatic target organ alkylation and cytotoxicity by 4-ipomeanol, a furan derivative from mouldy sweet potatoes: possible implications for carcinogenesis. In *Naturally Occurring Carcinogens—Mutagens and Modulators of Carcinogenesis*, E. C. Miller, J. A. Miller, I. Hirono, T. Sugimura, and S. Takayama (Eds.). University Park Press, Baltimore, pp. 35-56.

Cashmore, A. R., Seelye, R. N., Cain, B. F., Moch, H., Schmidt, R., and Hecker, E. (1976). The structure of prostratin. A toxic tetracyclic diterpene ester from *Pimelea prostrata*. *Tetrahedron Lett.*, p. 1737-1738.

Clark, I. A. (1973). The pathogenesis of St. George disease of cattle. *Res. Vet. Sci. 14*:341-349.

Connor, H. E. (1977). Thymelaeaceae. In *The Poisonous Plants in New Zealand*. E. C. Keating, Government Printer, Wellington, New Zealand, pp. 172-175.

Cunningham, I. J., and Hopkirk, C. S. M. (1945). Experimental poisoning of sheep by ngaio (*Myoporum laetum*). *N.Z. J. Sci. Technol. A26*:333-339.

Da Silva, F. M., and Couto, E. S. (1971). Experimental *Lantana camara* poisoning in cattle in the State of Pernambuco. *Arq. Esc. Vet. (Univ. Fed. Minas Gerais. 23*:77-89.

de Aluja, A. S. (1970). *Lantana camara* poisoning in cattle in Mexico. *Vet. Rec. 87*:628.

Dodson, M. E. (1965). A disease of cattle in South Australia resembling St. George disease. Part I. Field Investigation. *Aust. Vet. J. 41*:65-67.

Dutcher, J. S., and Boyd, M. R. (1979). Species and strain differences in target organ alkylation and toxicity by 4-ipomeanol. Predictive value of covalent binding in studies of target organ toxicities by reactive metabolites. *Biochem. Pharmacol. 28*:3367-3372.

Eakins, M. N., and Bullock, G. R. (1978). Studies on bile secretion with the aid of the isolated perfused rat liver. II. The effect of two further pentacyclic triterpenes, asiatic acid and 22β-angeloyloxyoleanolic acid. *Chem.-Biol. Interact. 21*:79-87.

Evans, F. J., and Soper, C. J. (1978). The tigliane, daphnane and ingenane diterpenes, their chemistry, distribution and biological activities. A review. *Lloydia (Cinci.) 41*:193-233.

Everist, S. L. (1974). *Poisonous Plants of Australia.* Angus & Robertson, Sydney, pp. 379-382, 488-499, 520-527.

Everist, S. L. (1981). *Poisonous Plants of Australia,* rev. ed. Angus & Robertson, Sydney, pp. 537-542, 692-707, 738-747.

Freeman, P. W., Ritchie, E., and Taylor, W. C. (1979). The constituents of Australian *Pimelea simplex.* I. The isolation and structure of the toxin of *Pimelea simplex* and *P. trichostachya* Form B responsible for St. George disease of cattle. *Aust. J. Chem. 32*:2495-2506.

Furstenberger, G., and Hecker, E. (1977). The new diterpene 4-deoxyphorbol and its highly unsaturated irritant diesters. *Tetrahedron Lett.,* pp. 925-928.

Gopinath, C., and Ford, E. J. H. (1969). The effect of *Lantana camara* on the liver of sheep. *J. Pathol. 99*:75-85.

Grant, H. G., O'Regan, P. J., Park, R. J., and Sutherland, M. D. (1980). Terpenoid chemistry. XXIV. (IR)-1-Methoxymyodesert-3-ene, an iridoid constituent of *Myoporum deserti* (Myoporaceae). *Aust. J. Chem. 33*: 853-878.

Hamilton, W. D., Park, R. J., Perry, G. J., and Sutherland, M. D. (1973). Terpenoid chemistry. XXI. (-)Epingaione, (-)dehydrogaione, (-)dehydroepingaione and (-)deisopropylngaione, toxic furanoid sesquiterpenoid ketones from *Myoporum deserti. Aust. J. Chem. 26*:375-387.

Hart, N. K., Lamberton, J. A., Sioumis, A. A., and Suares, H. (1976a). New triterpenes of *Lantana camara.* A comparative study of the constituents of several taxa. *Aust. J. Chem. 29*:655-671.

Hart, N. K., Lamberton, J. A., Sioumis, A. A., Suares, H., and Seawright, A. A. (1976b). Triterpenes of toxic and nontoxic taxa of *Lantana camara. Experientia 32*:412-413.

Hecker, E. (1978). Structure-activity relationships in diterpene esters irritant and cocarcinogenic to mouse skin. In *Carcinogenesis,* Vol. 2: *Mechanisms of Tumour Promotion and Cocarcinogenesis,* T. J. Slaga, A. Swak, and R. K. Boutwell (Eds.). Raven Press, New York, pp. 11-48.

Hegarty, B. F., Kelly, J. R., Park, R. J., and Sutherland, M. D. (1970). Terpenoid chemistry. XVII. (-)Ngaione, a toxic constituent of *Myoporum deserti*. The absolute configuration of (-)ngaione. *Aust. J. Chem. 23*: 107-117.

Hill, M. W. M. (1970). Toxicity of *Pimelea decora* in horses. *Aust. Vet. J. 46*: 287-289.

Hiura, M. (1943). Studies on storage and rot of sweet potato (2). *Rept. Gifu Agric. Coll. 50*: 5-9.

Howard, H. T. C., and Howden, M. E. H. (1975). Antitumour activity in *Pimelea simplex*. *Cancer Chemother. Rep. 59*: 585-586.

Jennings, P. W., Reeder, S. K., Hurley, J. C., Robbins, J. E., Holian, S. K., Holian, A., Lee, P., Pribanic, J. A. S., and Hull, M. (1978). Toxic constituents and hepatotoxicity of the plant *Tetradymia glabrata* (Asteraceae). In *Effects of Poisonous Plants on Livestock*, R. F. Keeler, K. R. Van Kampen, and L. F. James (Eds.). Academic Press, New York, pp. 217-228.

Johnson, A. E. (1978). Tetradymia toxicity—a new look at an old problem. In *Effects of Poisonous Plants on Livestock*, R. F. Keeler, K. R. Van Kampen, and L. F. James (Eds.). Academic Press, New York, pp. 209-216.

Kelly, W. R. (1975a). The pathology and haematological changes in experimental *Pimelea* spp. poisoning in cattle (St. George disease). *Aust. Vet. J. 51*: 233-243.

Kelly, W. R. (1975b). ^{59}Fe utilisation and excretion in anaemia of cattle caused by *Pimelea trichostachya* intoxication. *Aust. Vet. J. 51*: 504-510.

Kelly, W. R., and Bick, I. R. C. (1976). Some in vivo and in vitro properties of various fractions of *Pimelea trichostachya*. *Res. Vet. Sci. 20*: 311-315.

Kelly, W. R., and Seawright, A. A. (1978). *Pimelea* spp. poisoning of cattle. In *Effects of Poisonous Plants on Livestock*, R. F. Keeler, K. R. Van Kampen, and L. F. James (Eds.). Academic Press, New York, pp. 293-300.

Lal, M., and Kalra, D. B. (1960). Lantana poisoning in domesticated animals. *Indian Vet. J. 37*: 263-269.

Lee, J. S., Hrdlicka, J., and Seawright, A. A. (1979). The effect of size and timing on a pretreatment dose of phenobarbitone on the liver lesion caused by ngaione in the mouse. *J. Pathol. 127*: 121-127.

Lee, J. S., Seawright, A. A., and Hrdlicka, J. (1981). The effect of daily oral dosing with ngaione for seven weeks on the liver of the rat. *J. Appl. Toxicol. 1*: 165-173.

Louw, P. G. J. (1943). Lantanin, the active principle of *Lantana camara* L. Part I. Isolation and preliminary results on the determination of its constitution. *Onderstepoort J. Vet. Sci. Anim. Ind. 18*: 197-202.

Louw, P. G. J. (1948). Lantadene A, the active principle of *Lantana camara* L. Part II. Isolation of lantadene B and the oxygen functions of lantadene A and lantadene B. *Onderstepoort J. Vet. Sci. Anim. Ind. 23*: 233-238.

McClure, T. J., and Farrow, B. R. H. (1971). Chronic poisoning of cattle by desert rice flower (*Pimelea simplex*) and its resemblance to St. George disease as seen in northwestern New South Wales. *Aust. Vet. J. 47*: 100-102.

McDowall, F. H. (1925). Constituents of *Myoporum laetum*, Forst. (the "ngaio"). Part I. *J. Chem. Soc. 127*: 2200-2207.

Pass, M. A., and Heath, T. J. (1977). Gall bladder paralysis in sheep during lantana poisoning. *J. Comp. Pathol.* 87:301-306.

Pass, M. A., and Heath, T. J. (1978). The effect of *Lantana camara* on intestinal motility in sheep. *J. Comp. Pathol.* 88:149-156.

Pass, M. A., Gemmell, R. T., and Heath, T. J. (1978). Effect of lantana on the ultrastructure of the liver of sheep. *Toxicol. Appl. Pharmacol.* 43: 589-596.

Pass, M. A., Findlay, L., Pugh, M. W., and Seawright, A. A. (1979a). Toxicity of reduced lantadene A (22β-angeloyloxyoleanolic acid) in the rat. *Toxicol. Appl. Pharmacol.* 51:515-521.

Pass, M. A., Seawright, A. A., Lamberton, J. A., and Heath T. J. (1979b). Lantadene A toxicity in sheep. A model for cholestasis. *Pathology* 11:89-94.

Pass, M. A., McSweeney, C. S., and Reynoldson, J. A. (1981a). Absorption of the toxins of *Lantana camara* L. from the digestive system of sheep. *J. Appl. Toxicol.* 1:38-41.

Pass, M. A., Pugh, M. W., and Findlay, L. (1981b). Studies on the mechanism of reduced lantadene A toxicity in rats. *Biochem. Pharmacol.* 30:1433-1437.

Pound, C. J. (1913). In *Annual Report Department of Agriculture and Stock, Queensland.*, p. 109.

Reimerschmidt, G., and Quin, J. I. (1941). Studies on the photosensitization of animals in South Africa. XI. The reaction of the sensitised merino skin to radiation in different regions of the spectrum. *Onderstepoort. J. Vet. Sci.* 17:89-104.

Roberts, H. B., and Healy, P. J. (1971). *Pimelea simplex* and St. George disease of cattle. *Aust. Vet. J.* 47:123-124.

Roberts, H. B., McClure, T. J., Ritchie, E., Taylor, W. C., and Freeman, P. W. (1975). The isolation and structure of the toxin of *Pimelea simplex* responsible for St. George disease of cattle. *Aust. Vet. J.* 51:325-326.

Rogers, R. J., and Roberts, K. H. (1976). *Pimelea altior* poisoning of cattle. *Aust. Vet. J.* 52:193-194.

Sanders, D. A. (1946). Lantana poisoning in cattle. *J. Am. Vet. Med. Assoc.* 109:139-141.

Seawright, A. A. (1963). Studies on experimental intoxication of sheep with *Lantana camara*. *Aust. Vet. J.* 39:340-344.

Seawright, A. A. (1964). Studies on the pathology of experimental lantana (*Lantana camara* L.) poisoning of sheep. *Pathol. Vet.* 1:504-529.

Seawright, A. A. (1965a). Toxicity of *Lantana* species in Queensland. *Aust. Vet. J.* 41:235-238.

Seawright, A. A. (1965b). Electronmicroscopic observations of the hepatocytes of sheep in lantana poisoning. *Pathol. Vet.* 2:175-196.

Seawright, A. A. (1965c). Toxicity for the guinea pig of an extract of *Lantana camara*. *J. Comp. Pathol.* 75:215-221.

Seawright, A. A., and Allen, J. G. (1972). Pathology of the liver and kidney in lantana poisoning of cattle. *Aust. Vet. J.* 48:323-331.

Seawright, A. A., and Francis, J. (1971). Peliosis hepatis—a specific liver lesion in St. George disease of cattle. *Aust. Vet. J.* 47:91-99.

Seawright, A. A., and Hrdlicka, J. (1972). The effect of prior dosing of phenobarbitone and β-diethylaminoethyl phenylpropyl acetate (SKF525A) in the toxicity and liver lesion caused by ngaione in the mouse. *Br. J. Exp. Pathol.* _53:242-252.

Seawright, A. A., and O'Donahoo, R. M. (1972). Light and electron-microscopic observations of the liver of the mouse after a single oral dose of ngaione. *J. Pathol. 106:251-259.*

Seawright, A. A., and Sferco, F. G. (1966). Progressive microscopic changes in the liver after lantana poisoning in calves. *Proc. IVth Int. Meet. World Assoc. ̈Buiatrics*, Zurich, pp. 196-200.

Seawright, A. A., Lee, J. S., Allen, J. G., and Hrdlicka, J. (1978). Toxicity of *Myoporum* spp. and their furanosesquiterpenoid essential oils. In *Effects of Poisonous Plants on Livestock*, R. F. Keeler, K. R. Van Kampen, and L. F. James (Eds.). Academic Press, New York, pp. 241-250.

Seawright, A. A., Hrdlicka, J., Lee, J. S., and Ogunsan, E. A. (1982). Toxic substances in the food of animals. Some recent findings of Australian poisonous plant investigations. *J. Appl. Toxicol. 2:75-82.*

Seddon, H. R., and Carne, H. R. (1929). Feeding tests with *Lantana camara* (pink-flowered lantana). *Vet. Res. Rep. 5*, Dept. Agric., New South Wales, pp. 91-94.

Seddon, H. R., Carne, H. R., and McGrath, T. T. (1927). Feeding trials with *Lantana crocea* (red- and yellow-flowered lantana). *Vet. Res. Rep. 4*, Dept. Agric., New South Wales, pp. 91-96.

Steyn, D. G., and Van der Walt, S. J. (1941). Recent investigations into the toxicity of known and unknown poisonous plants in the Union of South Africa. XI. *Onderstepoort J. Vet. Sci. 16:121-147.*

Threlfall, S. (1980). *Pimelea elongata* (Thymelaeaceae) a new species from inland Australia. *Telopea 2:55-56.*

Tucker, G. (1910). In *Annual Report, Department of Agriculture and Stock, Queensland*, p. 25.

Zayed, S., Adolf, W., Hafez, A., and Hecker, E. (1977a). New highly irritant 1-alkyldaphnane derivatives from several species of Thymelaeaceae. *Tetrahedron Lett.*, pp. 3481-3482.

Zayed, S., Hafez, A., Adolf, W., and Hecker, E. (1977b). New tigliane and daphnane derivatives from *Pimelea prostrata* and *Pimelea simplex. Experientia 33:1544-1555.*

17

SESQUITERPENE LACTONES: STRUCTURE, BIOLOGICAL ACTION, AND TOXICOLOGICAL SIGNIFICANCE

G. WAYNE IVIE and DONALD A. WITZEL

U.S. Department of Agriculture, College Station, Texas

I. INTRODUCTION

Plants elaborate a seemingly endless number of organic compounds that vary tremendously in chemical complexity and functional significance. Among these are a large assemblage of structurally complex, biogenetically related compounds that are commonly referred to as sesquiterpene lactones. Studies during the past several decades have shown that many higher and lower plant species contain sesquiterpene lactones, and these chemicals have been the focus of intense interdisciplinary research efforts for several reasons. These include the rather formidable challenges associated with their chemical characterization, their chemotaxonomic significance, their potential roles in plant-animal coevolution, and their spectrum of biological activities that may lead to potentially useful agricultural and medicinal applications or to undesired toxicological effects on humans and animals. Our purpose here is to review the available information regarding structural and biological activity aspects of sesquiterpene lactones, with particular emphasis on their potential toxicological interactions with humans and domestic animals.

II. STRUCTURE, BIOGENETIC RELATIONSHIPS, AND DISTRIBUTION

Sesquiterpene lactones are C_{15} terpenoids (or their derivatives) that arise biogenetically through the mevalonate pathway. In their pure state, these compounds are crystalline, nonvolatile, colorless, relatively stable, and usually of intermediate polarity. Sesquiterpene lactones have historically been referred to as "bitter principles" because most are intensely bitter. For many years prior to the definition of their chemical nature, these compounds were known to be highly irritating to the mucous membranes of the nose, eyes, and throat (Lamson, 1913; Reeb, 1910), a characteristic that accounts for many sesquiterpene lactone-containing plants being referred to as "sneezeweeds."

Because the germacranolides are generated by rather straightforward cyclization and subsequent oxidative modifications of *trans,trans*-farnesyl pyrophosphate, they are considered to represent the most primitive type of sesquiterpene lactones, and all other classes of these lactones can be said to be derived from the germacranolide type (Fig. 1). The five-membered lactone ring of sesquiterpene lactones, which is generally but not always α,β-unsaturated, may be fused at the C_6-C_7 or C_8-C_7 positions, and the fusion may be either cis or trans. In Fig. 1, the most common mode of lactone closure is indicated for the various lactone types (Herz, 1977).

Naturally occurring sesquiterpene lactones possess varying stereochemistry, degrees of unsaturation, and functional groups, which may include hydroxyl (one or more, free or esterified), ketone, aldehyde, carboxylic acid, ether, epoxide, hemiketal, hemiacetal, an additional five- or six-membered lactone ring, and even sulfhydryl and halogen. Conjugation of some of these

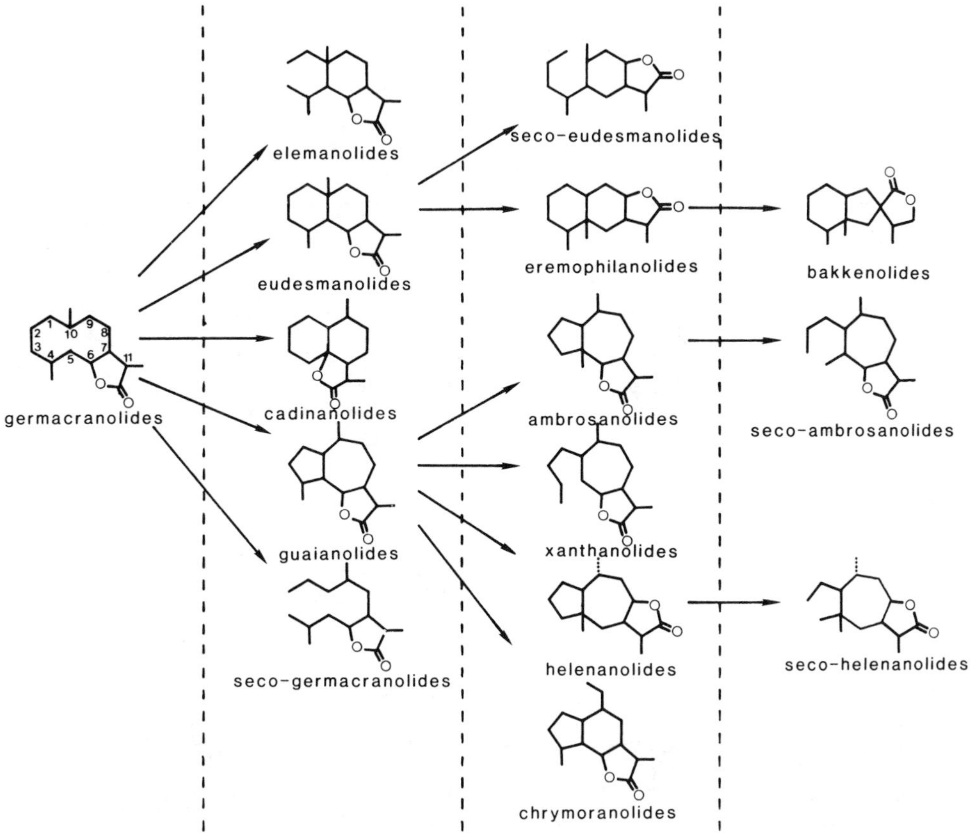

FIGURE 1 Basic carbon skeletons for major classes of sesquiterpene lactones. The most common mode of lactone closure is indicated for the various skeletal types. The arrangement indicates presumed relationships in order of increasing biogenetic complexity from left to right. (From Herout, 1973; Herz, 1977, 1978; Rodriguez et al., 1976; and Burnett et al., 1978b.)

compounds with sugars is known to occur. The variety of functional groups in naturally occurring sesquiterpene lactones is illustrated by the compounds shown in Fig. 2.

Of the 700 or so sesquiterpene lactones known from nature, the vast majority (perhaps 90%) are from species of the family Compositae (Herz, 1977; Burnett et al., 1978b; Mabry and Bohlmann, 1977). Sesquiterpene lactones are, in fact, considered to be characteristic constituents of this very large family of flowering plants, and they are known to occur in most tribes and genera of Compositae (Burnett et al., 1978b; Mabry and Bohlmann, 1977). Sesquiterpene lactones occur sporadically in genera of several other, mostly angiosperm families, including Umbelliferae, Lauraceae, Burseraceae, Magnoliaceae, Hepaticae, Amaranthaceae, Aristolochiaceae, and Cannellaceae (Herz, 1977). With only a few exceptions; however, the more biogenetically advanced types of sesquiterpene lactones occur exclusively in the Compositae, which is in accord with the generally held view that Compositae is an advanced angiosperm family. Paradoxically, Compositae taxa with advanced morphological features tend either to lose the more biogenetically advanced types of sesquiterpene lactones or to lose these compounds altogether (Burnett et al., 1978b; Mabry and Bohlmann, 1977).

Sesquiterpene lactones are usually concentrated in glands on the leaves, inflorescences, involucral bracts, and achenes (where they are often associated with glandular trichomes), but these lactones have also been found to occur in the roots, root bark, wood, and flowers of some species (Burnett et al., 1978b; Rodriguez et al., 1976). Given species may contain as much as

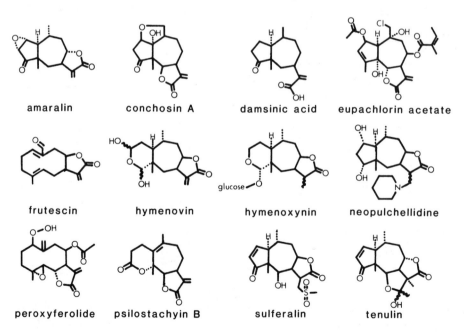

FIGURE 2 Structures of some naturally occurring sesquiterpene lactones, indicating the variety of functional groups that may be present in these compounds.

5% of their dry weight as sesquiterpene lactones (Rodriguez, et al., 1976). It appears that in most cases, individual plant species elaborate only one skeletal type of sesquiterpene lactone, but where wide geographical ranges are involved, considerable infraspecific variation in skeletal types may occur (Rodriguez et al., 1976). Several other factors, such as season, growth stage, climate, altitude, and stress agents, may affect the qualitative and/or quantitative elaboration of sesquiterpene lactones by a single species (Amo and Anaya, 1978; Boughton and Hardy, 1937; Calhoun et al., 1978; Hill et al., 1979; Kelsey et al., 1973).

III. CYTOTOXIC AND ANTITUMOR PROPERTIES

In recent years a great amount of effort has been directed toward the evaluation of natural products as potential antitumor agents, and sesquiterpene lactones have been quite extensively researched with regard to their cytotoxic and antitumor activity. A number of sesquiterpene lactones are, in fact, quite cytotoxic and some exhibit rather potent antitumor activity in vivo. Early work (Kupchan et al., 1971) noted that all of the cytotoxic sesquiterpenes known at that time were lactones, almost all were α,β-unsaturated. and the α,β-ethylenic linkage in the lactone was exocyclic in every case. Compounds found to possess both significant cytotoxicity and antitumor activity in vivo contained, in addition to the α-methylene-γ-lactone moiety, other reactive functional groups, such as epoxide, chlorohydrin, unsaturated ester, unsaturated lactone, or unsaturated ketone groups (Kupchan et al., 1970, 1971). It is now known, however, that the α-methylene-γ-lactone moiety is not an absolute requirement for either cytotoxicity or in vivo antitumor activity. The lactones tenulin and plenolin (Fig. 3), which contain neither an α-methylene-γ-lactone moiety nor an exocyclic methylene of any type, possess both cytotoxicity and significant in vivo antitumor activity (Lee et al., 1977a; Waddell et al., 1979a). Structures of several cytotoxic sesquiterpene lactones that also exhibit in vivo antitumor activity are shown in Fig. 3.

In early attempts to explain the cytotoxicity of sesquiterpene lactones, it was postulated that these compounds might be active as a result of alkylation of nucleophilic centers in biological systems (Kupchan et al., 1970). The fact that cytotoxicity is associated with sesquiterpene lactones containing an α-methylene-γ-lactone moiety supported this hypothesis because unsaturated lactones are well known to be reactive toward thiols (Cavallito and Haskell, 1945; Jones and Young, 1968) and amines (Mitchell, 1975; Unde et al., 1968). It has been confirmed that thiols do undergo "Michael-type" additions to not only the exocyclic methylene of sesquiterpene lactones containing α-methylene-γ or δ-lactone groups, but also to the α,β-unsubstituted cyclopentenone moiety as well. Thus elephantopin and eupatundin form adducts with L-cysteine in vitro and the dilactone vernolepin yields a biscysteine adduct (Fig. 4) (Kupchan et al., 1970). Tenulin, which lacks an exocyclic methylene but does contain an α,β-unsaturated cyclopentenone group, likewise forms a cysteine adduct by reaction with the endocyclic methylene (Fig. 4) (Hall et al., 1977; Lee et al., 1977a). Helenalin, which is potent both in cytotoxicity (Lee et al., 1971) and antitumor activity (Lee et al., 1977a), contains both α-methylene-γ-lactone and cyclopentenone moieties. This compound forms a

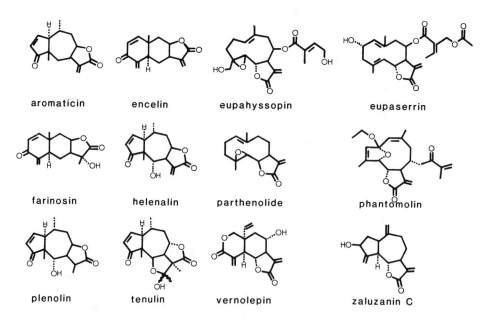

aromaticin
encelin
eupahyssopin
eupaserrin
farinosin
helenalin
parthenolide
phantomolin
plenolin
tenulin
vernolepin
zaluzanin C

FIGURE 3 Some sesquiterpene lactones known to possess both cytotoxic and antitumor properties.

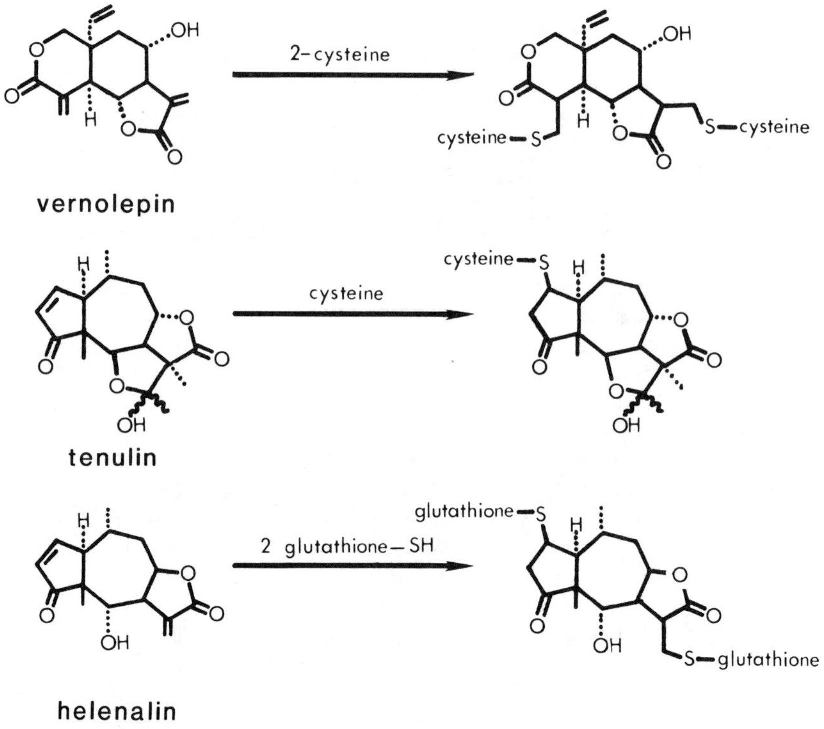

2-cysteine
cysteine—S
S—cysteine
vernolepin
cysteine
cysteine—S
tenulin
2 glutathione—SH
glutathione—S
S—glutathione
helenalin

FIGURE 4 Reaction of sesquiterpene lactones with biological nucleophiles through Michael-type addition.

bis adduct with reduced glutathione (Fig. 4) (Lee et al., 1977a). Vernolepin and euparotin acetate inhibit the sulfhydryl enzyme phosphotofructokinase, apparently by reaction with the sulfhydryl groups of the enzyme (Hanson et al., 1970). Glycogen synthetase is similarly inhibited by such lactones (Smith et al., 1972).

Studies with helenalin and tenulin injected intraperitoneally into male CF_1 mice bearing Ehrlich ascites tumors have shown that these compounds inhibit the activity of several enzymes in the ascites fluid, including deoxyribonuclease, ribonuclease, cathepsin, and HMG-coenzyme A (CoA) reductase (Hall et al., 1977; Lee et al., 1977a). Helenalin inhibited the activity of phosphofructokinase and hexokinase, and decreased the levels of serum cholesterol (the effects of tenulin on these three parameters was not studied). Both compounds resulted in increased adenosine 3',5'-cyclic monophosphate (cyclic AMP) concentrations, and both decreased the rate of incorporation of thymidine into DNA. The incorporation of uridine into RNA was unaltered by helenalin administration. Leucine incorporation into protein was significantly reduced in ascites fluid from helenalin-treated mice, whereas tenulin had no significant effect. Both helenalin and tenulin inhibited the in vitro aerobic respiration [basal and adenosine triphosphate (ATP)-stimulated] of Ehrlich ascites cells. In separate studies, a number of other sesquiterpene lactones similarly inhibited respiration of Ehrlich ascites cells (Hall et al., 1978). Although helenalin and tenulin inhibit the incorporation of thymidine into DNA, they apparently do not act through alkylation of purine bases of nucleic acids (Lee et al., 1977a). Recent studies (Hall et al., 1980b) have shown that the germacranolide eupaformosanin inhibits a number of enzymes in Ehrlich ascites cells. Among these are DNA polymerase, thymidylate synthetase, messenger and ribosomal polymerase, and a number of glycolytic and Krebs cycle enzymes. All of the inhibited enzymes are known to bear thiol groups (Hall et al., 1980b). Studies to date thus indicate that the cytotoxic and antitumor properties of sesquiterpene lactones are probably the result of the inhibition of a number of vital metabolic functions within the cell, and that alkylation of sulfhydryl groups through the occurrence of a Michael-type reaction is probably a major inhibitory mechanism. Consistent with the proposed mechanisms of cytotoxic action is the view that most if not all sesquiterpene lactones that exhibit antitumor properties have general cytotoxic effects rather than specific antitumor activity (Burnett et al., 1978b).

IV. MUTAGENICITY

At least one sesquiterpene lactone, hymenovin (Fig. 5), is mutagenic in certain histidine-deficient *Salmonella typhimurium* strains when tested in the Ames plate incorporation assay (MacGregor, 1977). Hymenovin is quite mutagenic in strains TA 100 and TA 98, where 100 μg of hymenovin per plate resulted in >900 and >100 revertant colonies/plate for the two strains, respectively. Hymenovin was not mutagenic in strains TA 1535 or TA 1537. Inclusion of an in vitro rat liver enzyme-metabolizing system did not significantly alter the mutagenicity of hymenovin in these tests. Hymenovin's response patterns with respect to tester strain specificity are similar to those of the known mutagens and carcinogens 4-nitroquinoline 1-oxide and methylmethanesulfonate. Tests with helenalin and tenulin showed that neither of these

hymenovin   diethylhymenovin   dihydrohymenovin

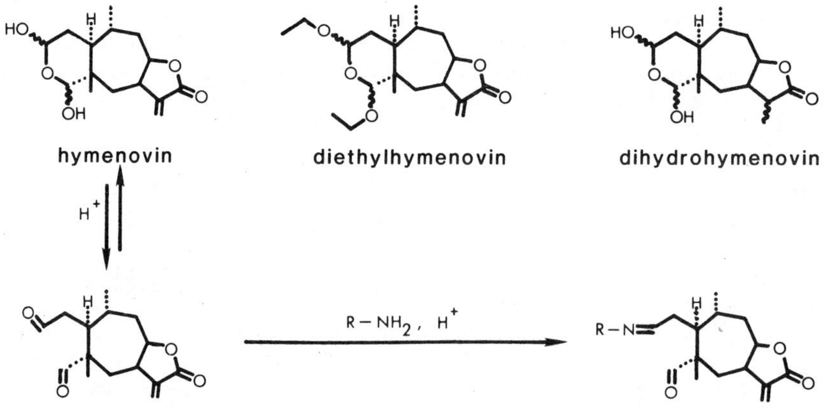

FIGURE 5 Structure of the bacterial mutagen, hymenovin, and certain of
its derivatives, with a proposed reaction sequence for alkylation of DNA
that could account for its mutagenic properties. (Adapted from Manners
et al., 1978.)

compounds was appreciably mutagenic in the strains used, even though both
contain an α,β-unsaturated ketone and helenalin in addition contains an α-
methylene-γ-lactone moiety (Fig. 3). Of the three lactones studied, however,
helenalin was the most cytotoxic to the *Salmonella* tester strains, and levels
of 300 µg or more of helenalin per plate resulted in growth inhibition of the
background lawn in every case (MacGregor, 1977).

Subsequent studies in *S. typhimurium* strain TA 100 demonstrated that
the principal reactive site responsible for the mutagenicity of hymenovin to
S. typhimurium is the unusual bishemiacetal moiety and not the α-methylene-
γ-lactone. Hymenovin retained 41% of its mutagenic potency after hydrogenation of the exocyclic methylene group, but less than 3% after diethyla-
tion of the bishemiacetal (Fig. 5) (Manners et al., 1978). On the basis of
these observations, it was postulated that in biological systems the bishemiacetal group probably exists in equilibrium with the noncyclic dialdehyde,
which can react with amines to form imine or diimine derivatives (Fig. 5).
Thus such a reaction between the dialdehyde and DNA could explain the observed bacterial mutagenicity of hymenovin (Manners et al., 1978).

Studies on the toxic effects of tenulin and hymenovin on *Bacillus thuringiensis* (Norman et al., 1976) have shown that hymenovin apparently has a
mutagenic effect on *B. thuringiensis* cultures that results in the production
of several variant strains. Both tenulin and hymenovin inhibited the growth
of *B. thuringiensis*, with tenulin being considerably the more toxic of the
two, but tenulin treatment did not result in any apparent mutagenic effects.
In hymenovin-treated bacterial cultures, some of the variant strains observed
had lost their ability to form spores and crystals, and hymenovin also induced *B. thuringiensis* to produce significant levels of bacteriophage. All
variant strains observed were as susceptible to phage as was the parent
strain (Norman et al., 1976).

It is known that helenalin and tenulin do not bind or intercalate with
deoxyguanosine 5'-phosphate, deoxyguanosine 5'-triphosphate, or DNA, and
neither compound alkylates the N_7 position of guanosine (Hall et al., 1977;

Lee et al., 1977a). These observations, when considered with the structure-activity data related to the bacterial mutagenicity of hymenovin, helenalin, and tenulin (MacGregor, 1977), suggest that the presence of an α-methylene-γ-lactone, α,β-unsaturated ketone, or hemiketal function in sesquiterpene lactones does not itself confer appreciable mutagenic potential. However, it is clear that the bishemiacetal moiety of hymenovin is associated with mutagenic activity, at least against *Salmonella* (Manners et al., 1978) and probably *Bacillus* (Norman et al., 1976). The potential mutagenicity of sesquiterpene lactones that contain other functional groups (epoxide, chlorohydrin, etc.) has apparently not been investigated. Clearly, compounds such as hymenovin that are not only mutagenic (MacGregor, 1977) but have potential for alkylation of sulfhydryl groups as well could possibly exhibit mechanisms of action and effects on biological systems that differ from those of more "typical" sesquiterpene lactones. Thus the potential medical or toxicological significance of such compounds may merit particular attention.

V. CONTACT DERMATITIS/ALLERGENIC RESPONSES

Sesquiterpene lactones are among the most important causative agents of allergic contact dermatitis in humans, and their allergenic effects have been discussed in several recent reviews (Burnett et al., 1978b; Mitchell, 1975; Rodriguez et al., 1976; Towers et al., 1977). Current theory of the pathogenesis of allergic contact dermatitis is that low-molecular-weight chemicals, designated haptens, become bound to skin proteins to form higher-molecular-weight complexes, designated allergens or antigens, which induce sensitization of lymphocytes (Mitchell, 1975). Among a large number of sesquiterpene lactones tested for allergenic potential in human skin patch tests, the α-methylene-γ-lactone moiety was found to be the principal group responsible for allergenic effects (Mitchell, 1975, and references therein). Reduction of the α-methylene group attached to the γ-lactone renders the compounds inactive as haptens. Studies with alantolactone (Fig. 6) and other sesquiter-

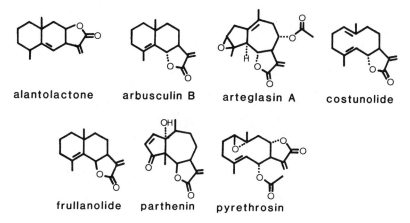

alantolactone   arbusculin B   arteglasin A   costunolide

frullanolide   parthenin   pyrethrosin

FIGURE 6 Some sesquiterpene lactones that induce allergic contact dermatitis in humans. (Adapted from Burnett et al., 1978b, and Rodriguez et al., 1976.)

pene lactones have shown that these compounds form adducts with the sulfhydryl groups of cysteine, with the imidazole group of histidine, and with
the ε-amino group of lysine but not with the guanido group of arginine, the
hydroxyl group of serine, or the thioether moiety of methionine (Dupuis et
al., 1974; Mitchell, 1975).

  Cross-sensitivity patterns are also observed in sesquiterpene lactone-
induced allergenic responses. Individuals sensitized to alantolactone have
shown positive patch test reactions to numerous other sesquiterpene lactones,
and cross-sensitivity between lactones of different skeletal types is known to
occur (Mitchell, 1975). More than 50 plant species that contain sesquiterpene lactones have been reported to cause allergic contact dermatitis in
humans, including representatives from several families but, particularly,
Compositae (Mitchell, 1975; Rodriguez et al., 1976). Structures of several
sesquiterpene lactones implicated as causative agents in allergic contact
dermatitis in humans are shown in Fig. 6.

VI. ANTIMICROBIAL EFFECTS

A number of sesquiterpene lactones possess antimicrobial activity (Fig. 7).
The pseudoguanilide carpesiolin inhibits the growth of *Xanthomonas oryzae*
(Maruyama and Omura, 1977), and certain growth parameters of *Sclerospora*

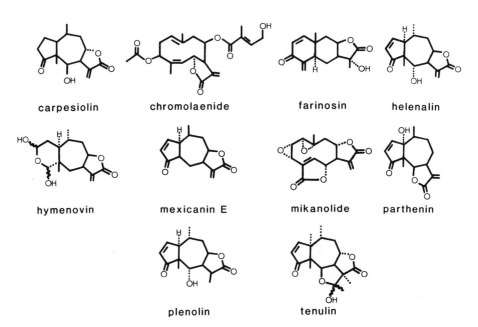

FIGURE 7 Structures of some sesquiterpene lactones that are active as
antimicrobial agents.

graminicola are inhibited by the lactone parthenin (Char and Shankarabhat, 1975). Sesquiterpene lactones have been reported to inhibit the growth of the pathogenic protozoans *Trichomonas vaginalis* and *Entamoeba histolytica* Rubinchik et al., 1976; Vichkanova et al., 1971). The germacranolide mikanolide (Mathur et al., 1975) and several other sesquiterpene lactones (Lee et al., 1977b) inhibit the growth of yeast *Candida albicans*, but a rather large number of additional sesquiterpene lactones had no noticeable effects on this organism (Calzada et al., 1980; Lee et al., 1977b). Although sesquiterpene lactones have been considered to be toxic primarily to gram-positive bacteria (Calzada et al., 1980; Lee et al., 1977b), studies with *Salmonella typhimurium* (MacGregor, 1977) indicate that these lactones are toxic to at least some gram-negative bacteria as well.

It might be predicted that antimicrobial activity would be associated with the reactive α-methylene-γ-lactone moiety present in many sesquiterpene lactones, and studies by Calzada et al. (1980) have shown with a series of heliangolides that such a correlation does indeed exist when *Staphylococcus aureus* was used as a test bacteria. However, in a study of structure-antibacterial relationships among 36 sesquiterpene lactones and related compounds, Lee et al. (1977b) found that the α,β-unsaturated cyclopentenone moiety was the major structural requirement for antibacterial activity against *S. aureus* and *Bacillus subtilis*. Substitution at the β-position of the cyclopentenone moiety greatly reduced antibacterial activity, probably because such modifications would be expected to hinder alkylation reactions. It is of considerable interest that significant antibacterial activity in these studies appeared to be independent of the presence or absence of an α-methylene-γ-lactone moiety (Lee et al., 1977b). It thus appears that studies to date are less than definitive as regards the precise structural requirements for anti-microbial activity among the sesquiterpene lactones. The great diversity of structural features seen among this group of compounds, the diversity of the organisms affected, and the possibility of multiple mechanisms of action may well preclude the assignment of specific structural requirements for anti-microbial activity.

VII. ANTHELMINTHIC PROPERTIES

Much of the early work on the potential value of sesquiterpene lactones as therapeutic agents was focused on the santanolide santonin (Fig. 8) and certain of its derivatives which are well known for their anthelminthic and ascaricidal activities (Haynes, 1948; Lee et al., 1971). More recently, the lactones eremanthine, costunolide, α-cyclocostunolide, and goyazensolide (Fig. 8), isolated from the wood oils of some Brazilian trees, have been shown to inhibit skin penetration by cercarie of the trematode, *Schistosoma mansoni* (Baker et al., 1972; Garcia et al., 1976; Vichnewski and Gilbert, 1972; Vichnewski et al., 1976). It was proposed that the mechanism of anti-schistosomal action of these compounds probably involved the α-methylene-γ-lactone moiety, on the basis of observations that dihydro-β-cyclo-costunolide, which lacks the exocyclic methylene on the γ-lactone moiety, was inactive (Baker et al., 1972).

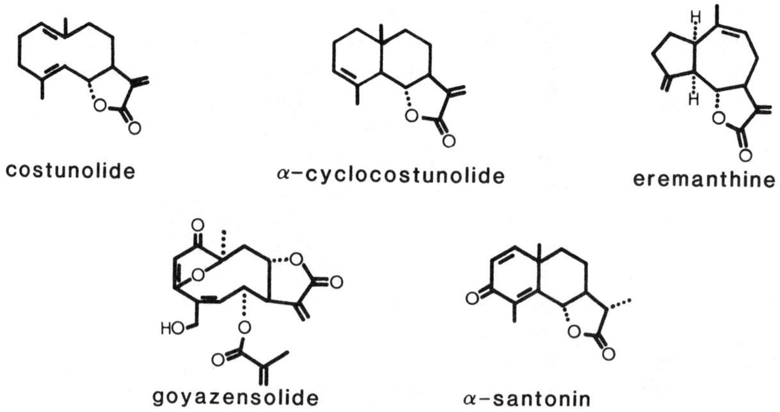

costunolide   α-cyclocostunolide   eremanthine

goyazensolide   α-santonin

FIGURE 8 Sesquiterpene lactones with anthelminthic properties.

VIII. ANALGESIC AND ANTI-INFLAMMATORY ACTIONS

Some plants known to contain sesquiterpene lactones have historically been used as anti-inflammatory and antipyretic herbal remedies (Hall et al., 1979; Towers et al., 1977). In laboratory studies, crude extracts of *Helenium amarum* when injected subcutaneously into mice exhibited analgesic activity in the mouse tail flick test, and also inhibited the writhing syndrome in mice induced by the intraperitoneal injection of acetic acid. However, the extract demonstrated no anticonvulsant properties and did not cause alterations in spinal reflexology in cats (Lucas et al., 1964). The sesquiterpene lactones amaralin and tenulin were shown to be present in the extract, and both exhibited analgesic actions. Nine of eleven sesquiterpene lactones tested exhibited some analgesic activity, the order of potency being amaralin > helenalin > tenulin, isotenulin, dihydroamarlin > helenalin oxide, isotenulin oxide, dihydroisotenulin > aromaticin oxide. Aromaticin and tetrahydro-helenalin oxide were inactive (Lucas et al., 1964).

Recently, Hall et al. (1979) noted that several known anti-inflammatory drugs, such as ethacrynic acid and *N*-ethylmaleimide, are known to bind with sulfhydryl groups and that a number of clinically useful anti-inflammatory agents have effects on cell metabolism that are quite similar to those caused by sesquiterpene lactones. They therefore subjected a number of these lactones to a battery of in vivo rodent inflammatory tests (Hall et al., 1979). In the enema-induced carragenan inflammatory screen and writhing reflex analgesic screen, the α-methylene-γ-lactone moiety was required for inhibitory activity, and helenalin was the most active compound studied in both tests. Several lactones tested against induced pleurisy, including helenalin and tenulin, were marginally active. Delayed hypersensitivity was suppressed, and immunoglobulin synthesis was slightly stimulated by some of these lactones. None of the compounds tested afforded any protection against induced hyperpyrexia in rats, and they were essentially inactive as antiallergy agents. The lactones studied did not exhibit analgesic activity in the mouse tail flick test, findings that do not necessarily conflict with

with those of Lucas et al. (1964) because higher dosages were used in the earlier work.

In the chronic adjuvant arthritic screen in rats, several sesquiterpene lactones afforded significant inhibition at relatively low treatment rates. Structural requirements for chronic arthritic inhibition were not as rigid as those for general anti-inflammatory activity, and arthritic inhibition was associated with compounds containing the α-methylene-γ-lactone, α,β-unsubstituted cyclopentenone, or α-epoxycyclopentenone moiety. It was speculated that sesquiterpene lactones may offer an alternative to existing anti-inflammatory therapy for arthritic patients, on the basis of the observed dose-activity relationships of these compounds with respect to those of existing drugs (Hall et al., 1979).

The mode of action of sesquiterpene lactones as anti-inflammatory drugs appears to be at multiple sites. Detailed studies by Hall et al. (1980c) have shown that a number of these lactones containing an α-methylene-γ-lactone moiety effectively uncouple oxidative phosphorylation of human polymorphonuclear neutrophils and elevate cyclic AMP levels in rat and mouse liver cells. Lysosomal enzymatic activity was increased by these lactones in both rat and mouse liver and rat and human neutrophils. The structure-activity relationships for the stabilization of lysosomal membrane for rat liver cathepsin activity followed the same structural requirements necessary for anti-inflammatory activity; that is, the α-methylene-γ-lactone moiety contributed the most activity, whereas the β-unsubstituted cyclopentenone and α-epoxycyclopentenone contributed only minor activity. Human polymorphonuclear neutrophil chemotaxis and prostaglandin synthetase activity were inhibited by these lactones (Hall et al., 1980c).

Structures of some sesquiterpene lactones that exhibit analgesic or anti-inflammatory activity are indicated in Fig. 9.

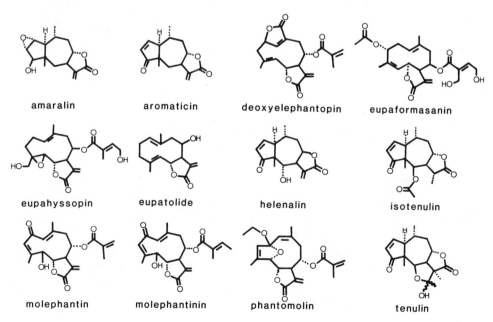

amaralin   aromaticin   deoxyelephantopin   eupaformasanin

eupahyssopin   eupatolide   helenalin   isotenulin

molephantin   molephantinin   phantomolin   tenulin

FIGURE 9 Some sesquiterpene leactones that exhibit analgesic or anti-inflammatory activities.

IX. ANTIHYPERLIPIDEMIC ACTIVITY

Initial indications that some sesquiterpene lactones reduced serum cholesterol levels in rodents (Lee et al., 1977a) led to a more definitive evaluation of these compounds as antihyperlipidemic agents (Hall et al., 1980a). Intraperitoneal dosing of CF_1 male mice with several lactones at 20 mg/kg per day or less significantly lowered serum cholesterol levels. The most active compound was helenalin, but tenulin, 2,3-epoxytenulin, isotenulin, 2,3-epoxyisotenulin, deoxyelephantopin, and eupahyssopin (Fig. 10) also reduced serum cholesterol levels by at least 30% after 16 days (Hall et al., 1980a). Structure-activity considerations indicated that the α-methylene-γ-lactone, the β-unsubstituted cyclopentenone, and the α-epoxycyclopentenone moieties contributed to antihypercholesterolemic activity. Enzyme inhibition studies in vitro showed that these lactones are potent inhibitors of several key regulatory enzymes involved in fatty acid and cholesterol synthesis, including acetyl-CoA synthetase, citrate-lyase, acetyl-CoA carboxylase, fatty acid synthetase, and β-hydroxy-β-methylglutaryl-CoA reductase. It was concluded that sesquiterpene lactones possess sufficient ability to block lipogenesis to warrant further investigations of their activities as antihyperlipidemic agents (Hall et al., 1980a).

X. EFFECTS ON PLANT GROWTH

A number of sesquiterpene lactones affect plant growth, although the nature and extent of the effects produced depend on a number of factors, including the lactone tested, its concentration, and the species on which it acts. The lactones arbusculin-A, achillin, desacetoxymatricarin, and viscidulin-B and -C (Fig. 11) inhibit radicle, hypocotyl, and lateral root growth of *Cucumis sativis* (McCahon et al., 1973). Heliangine (Shibaoka et al., 1967) and vernolepin (Sequeira et al., 1968) inhibit the elongation of wheat coleoptiles, but heliangine exhibits a stimulatory effect on adventitious root formation of bean cuttings (Shibaoka et al., 1967). Alantolactone inhibits the growth of *Chlorella* cells and has been shown to restrict seed germination and seedling growth in several species (Burnett et al., 1978b; Dalvi et al., 1971; Kwon et al., 1973).

Sesquiterpene lactones from *Ambrosia cumanensis* were tested against seeds of several tropical plants, including *Mimosa pudica, Achyranthes asperia, Bidens pilosa,* and *Crusea calocephala* (Amo and Anaya, 1978). The effects on germination and root and stem growth varied both with respect to compound and species, and in some cases different levels of exposure resulted in opposite effects. Exposure of *Bidens pilosa* seeds to seven sesquiterpene lactones, including psilostachyin B, psilostachyin C, cumambrin A, cumambrin B, cumanin, confertin, and peruvin (Fig. 11) resulted in significant stimulation of stem growth, root growth, or both when the seeds were germinated in medium containing the individual lactones at 100 ppm. However, both stem and root growth were greatly inhibited by each of these lactones at 250 ppm (Amo and Anaya, 1978). Several other naturally occurring sesquiterpene lactones, including chrysartemin-A, chrysartemin-B, and pyrethrosin, have also been shown to possess plant growth regulating activity (Gross, 1975).

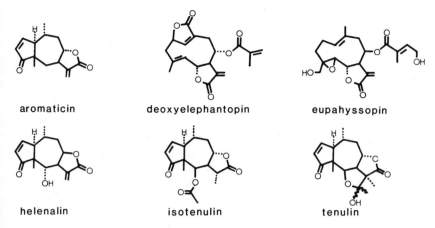

aromaticin　　deoxyelephantopin　　eupahyssopin

helenalin　　isotenulin　　tenulin

FIGURE 10　Sesquiterpene lactones with antihyperlipidemic properties.
(Adapted from Hall et al., 1980a.)

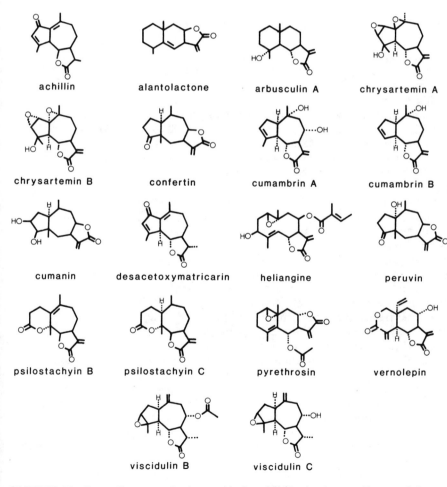

achillin　　alantolactone　　arbusculin A　　chrysartemin A

chrysartemin B　　confertin　　cumambrin A　　cumambrin B

cumanin　　desacetoxymatricarin　　heliangine　　peruvin

psilostachyin B　　psilostachyin C　　pyrethrosin　　vernolepin

viscidulin B　　viscidulin C

FIGURE 11　Sesquiterpene lactones that exhibit plant growth regulator
activities.

It is likely that plant growth regulator activity of at least some sesquiter-pene lactones is related to the presence of the α-methylene-γ-lactone moiety, because Shibaoka et al. (1967) have shown that cysteine reduced the promo-tion effects of heliangine on adventitious root formation in bean cuttings, and that hydrogenation of the exocyclic methylene group of heliangine gives similar reduction in root-promoting activity. It has been speculated, quite reasonably but without direct experimental evidence, that some of the plant growth regulator actions of sesquiterpene lactones may be related to the inhi-bition of crucial cellular enzymes (Dalvi et al., 1971).

On the basis of studies to date, it is clear that a large number of sesqui-terpene lactones may be expected to exhibit effects on plant growth, although these effects would be predicted to be quite variable with respect to com-pound, species, and dose. It therefore would be very difficult at this time to arrive at generalizations regarding the plant growth regulator activity of these compounds.

XI. INSECTICIDAL/INSECT ANTIFEEDANT PROPERTIES

Sesquiterpene lactones apparently exhibit little contact insecticidal activity. More than 60 years ago, de Waal (1920) tested the powdered flowers of *Helenium autumnale* against flies, ants, fleas, and roaches, and found the preparation to exhibit no appreciable insecticidal activity. Some 20 years later, McGovran and Mayer (1942) tested helenalin, tenulin, and isotenulin as contact insecticides against nymphs and adults of the green peach aphid, larvae and adults of the Mexican bean beetle, and adult houseflies and American cockroaches. These three lactones, topically applied, were not significantly toxic to any of the insects studies, except that second instar larvae of the Mexican bean beetle treated with about 125 μg of lactone per insect experienced at least 50% mortality within 3 days. Helenalin (about 250 μg per insect) caused some mortality in fourth instar larvae and adult beetles, but tenulin and isotenulin were not appreciably toxic under these circum-stances (McGovran and Mayer, 1942).

Although sesquiterpene lactones are not significantly active as contact insecticides, some are potent insect antifeedants, inhibitors of larval insect growth, and oviposition deterrents. These properties of sesquiterpene lac-tones have received considerable study, and are the subject of a recent re-view that specifically considers the potential role of sesquiterpene lactones in plant-animal coevolution (Burnett et al., 1978b).

Feeding studies with two species of *Vernonia* (*V. gigantea* and *V. glauca*) that produce the lactone glaucolide-A (Fig. 12) and one (*V. flaccidifolia*) that does not produce sesquiterpene lactones have provided experimental evidence that sesquiterpene lactones confer resistance to feeding by the larvae of some lepidopterous insects. Larvae of the fall armyworm, southern armyworm, and saddleback caterpillar (insects not observed to feed on glaucolide-A containing plants in the field) exhibited strong tendencies to avoid feeding on the fresh foliage of *Vernonia* species that contain glaucolide-A, although they readily consume *V. flaccidifolia* that does not contain sesqui-terpene lactones (Burnett et al., 1978b). On the other hand, the insect species that had been observed to feed in nature on *Vernonia* that contained

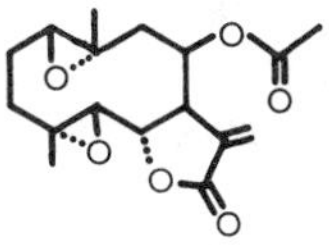

epitulipinolide diepoxide

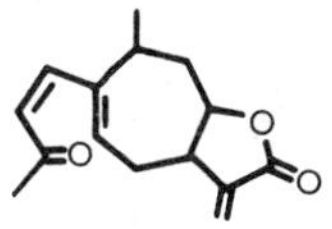

8-epi-xanthatin

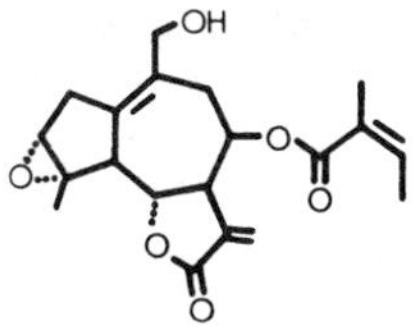

euponin

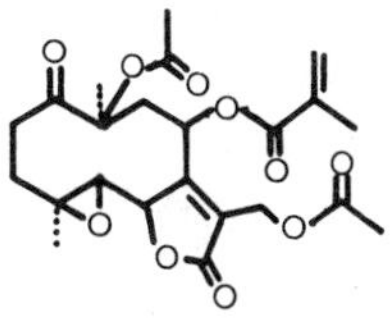

glaucolide A

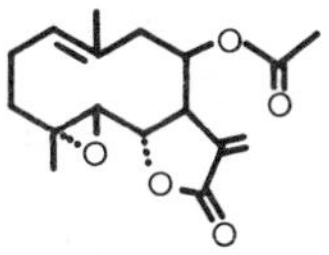

lipiferolide

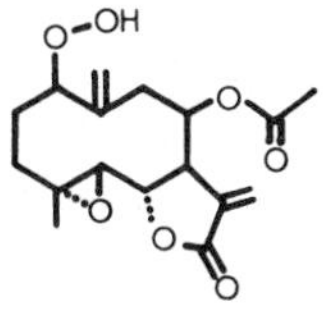

peroxyferolide

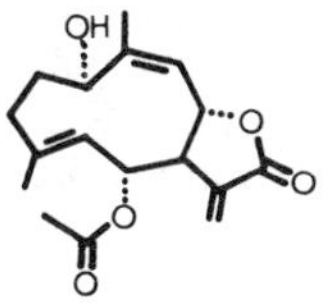

tulirinol

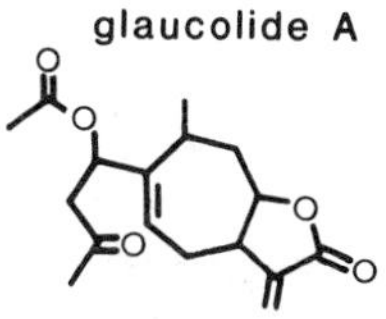

xanthumin

FIGURE 12 Sesquiterpene lactones active as insect antifeedants, insect development inhibitors, or insect oviposition deterrents.

glaucolide-A tended to exhibit the same feeding preferences in the laboratory. Feeding studies with powdered *Vernonia* foliage added to disks of agar diet showed generally the same patterns observed with fresh foliage but, significantly, when larvae were offered agar diets of *V. flaccidifolia*, with or without added glaucolide-A, all the insect species studied preferred the medium without glaucolide-A.

Incorporation of glaucolide-A into larval diets resulted in growth inhibition of the fall, southern, and yellow-striped armyworms, and increased the number of days to pupation in most of the insect species studied (Burnett et al., 1977a, 1978b; Jones et al., 1979). These data suggest that even in insects that prefer to forage in nature on plants that contain sesquiterpene lactones, extension of the insect's life cycle could result in an increased exposure to pathogenic microorganisms, predators, parasites, and adverse climatic conditions that might lower survival. It was also speculated that because pupal weight has been correlated with the number of eggs laid by adults of some lepidoptera, reproductive capacity might be lowered as a result of dietary consumption of plants that contain sesquiterpene lactones (Burnett et al., 1978b).

Studies with the tulip poplar (*Lirodendron tulipifera*) have shown that the leaves contain at least four sesquiterpene lactones (lipiferolide, epitulipinolide diepoxide, peroxyferolide, and tulirinol, Fig. 12) that are antifeedants for larvae of the gypsy moth (Doskotch et al., 1975, 1977a,b, 1980). Recently, Japanese workers (Kawazu et al., 1979, Nakajima and Kawazu, 1980) studied the sesquiterpene lactones, xanthumin, 8-epixanthatin, and euponin as growth inhibitory agents against the fruit fly, *Drosophila melanogaster*. When eggs were exposed to these lactones, the resulting larvae experienced reduced growth, pupation was delayed, and both larvae and pupae were smaller than normal. Significantly, studies with one of the lactones, euponin, showed that the compound is not a feeding deterrent per se for fruit fly larvae, and data indicated that euponin is capable of inhibiting the larval growth of *D. melanogaster* only when administered at the egg stage. The authors concluded that euponin is an insect development inhibitor rather than an antifeedant.

Euponin exhibited essentially no ovicidal activity against *D. melanogaster* (Nakajima and Kawazu, 1980).

Studies by Burnett and colleagues have also been directed toward evaluating the potential of sesquiterpene lactones as insect oviposition deterrents. Sesquiterpene lactones are not volatile; thus it would be predicted that they would not repel (or attract) insects in search of oviposition sites. However, once insects come in contact with plants that contain sesquiterpene lactones, these chemicals might exert an influence on egg laying because they might readily be detected by the insects' chemoreceptor organs. Data from studies with three *Vernonia* species, adults of several lepidoptera species, and glaucolide-A indicated that the presence of sesquiterpene lactones does not deter oviposition by lepidoptera that normally prefer sesquiterpene lactone-containing *Vernonia* as a food source. However, at least one generalist feeder, the fall armyworm, exhibits a clear oviposition preference for plants that do not contain these lactones (Burnett et al., 1978a,b).

On the basis of data from studies to date, it is clear that at least some sesquiterpene lactones can function as insect oviposition deterrents, antifeedants, and/or growth inhibitory agents. However, the role of these chemicals in protecting plants from the damaging effects of insect feeding are far from clear. The presence of sesquiterpene lactones may deter oviposition and feeding by at least some generalist insect feeders (e.g., some armyworm species). Other insects, however, perhaps through coevolution or simply tolerance to the toxic effects of sesquiterpene lactones, are adapted to survive very well on sesquiterpene lactone-containing plants, and in fact appear to prefer these plants as food sources (Burnett et al., 1978b).

XII. MAMMALIAN ANTIFEEDANT PROPERTIES

It is reasonable to assume that some plant chemicals have evolved as defense mechanisms against browsing mammals, and Burnett et al. (1978b) have suggested that the mammalian antifeedant properties of sesquiterpene lactones may represent their major role in plant-animal coevolution. *Vernonia* and other Compositae that contain high concentrations of sesquiterpene lactones apparently remain essentially untouched by cattle in heavily grazed pastures in eastern North America, an observation that led to studies of the feeding deterrent properties of glaucolide-A against mammalian herbivores (Burnett et al., 1977b).

Wild eastern cottontail rabbits were observed to feed heavily on *Vernonia flaccidifolia*, which lacks sesquiterpene lactones, but they sampled and rejected *V. gigantea* and *V. glauca*, both of which contain glaucolide-A. Caged rabbits exhibited the same feeding preferences, and tests also established that they preferred *V. flaccidifolia* over glaucolide-A-coated *V. flaccidifolia* (Burnett et al., 1977b, 1978b). Feeding studies with white-tail deer gave data essentially identical to those with rabbits. Deer given choices between *V. gigantea* and *V. flaccidifolia* or *V. flaccidifolia* and glaucolide-A-coated *V. flaccidifolia* would smell, lick, and sample the plants, but invariably chose to feed on plants lacking glaucolide-A (Burnett et al., 1977b, 1978b). These authors have noted that the two species of *Vernonia* in North America that lack sesquiterpene lactones, *V. flaccidifolia* and *V. pulchella*, are also species with restricted ranges. On the basis of studies with mammalian herbivores, it was suggested that heavy mammalian feeding pressure may account for the restricted ranges and small population sizes of these two *Vernonia* species (Mabry et al., 1977).

Although Burnett et al. (1977b, 1978b) observed no evidence of mammalian feeding on sesquiterpene lactone-containing *Vernonia* species in nature, it is clear that not all such plants are immune from mammalian grazing pressure. A considerable number of plants that contain these lactones are potent livestock poisons, and the lactones themselves are known in several cases to be the toxic agents (vide infra). It is generally believed that most instances of livestock poisoning by toxic plants results from the lack of other palatable forages, and such appears to be true in most cases of livestock poisoning by sesquiterpene lactone plant toxins (Dollahite et al., 1972; Marsh et al., 1921; Rimington and Roets, 1936; Roew et al., 1973). However, the suggestion by others (Burnett et al., 1978b; Mabry et al., 1977) that mammals avoid these plants because of their "bitter" taste appears to us to be tenuous and anthropormorphic. We are unaware of any evidence that mammalian herbivores perceive or react to bitterness per se in the same way as do humans. We suggest that rather than avoidance of food sources on account of bitterness itself, mammals have adapted through coevolution with these plants to *recognize*, by taste or resulting physiological effects, the unpalatable and toxic nature of sesquiterpene lactone plants. This hypothesis is more in accord with the fact that certain foods (e.g., acorns) considered to be quite bitter and unpalatable from the human perspective are nevertheless highly desirable and nutritious foodstuffs for many mammalian herbivores. For the same reasons, the proposal that the insect antifeedant properties of sesquiterpene lactones is due to their bitterness (Burnett et al., 1978b) seems particularly speculative.

XIII. ACUTE MAMMALIAN TOXICITY

Other than passing mention of the acute toxicity of sesquiterpene lactones to fish (Clark, 1936) and frogs (Rimington et al., 1936), mammals are apparently the only vertebrates in which the toxicity of sesquiterpene lactones has been evaluated to any appreciable extent.

The toxicologic aspects of livestock plant poisons that are known or suspected to contain sesquiterpene lactones as their toxic principles will be discussed in detail in the section to follow. However, an appreciable number of studies have considered the acute toxicity effects of individual sesquiterpene lactones to mammals, and these will be considered here. Acute toxicity data for nine lactones against a number of mammalian species are shown in Table 1. Several conclusions can be drawn from these data. First, some sesquiterpene lactones are potent mammalian poisons. Second, these lactones appear to be considerably more toxic, by an order of magnitude or more, when administered intravenously or intraperitoneally as compared to oral treatment. This is probably the result of higher concentrations of lactone at the site(s) of toxic action caused by more rapid entrance into the bloodstream, but it is also possible that the orally administered lactones undergo some degradation in the digestive tract to nontoxic products before absorption. Third, structure-activity considerations indicate that high acute toxicity is associated with compounds that possess two reactive functional groups, such as helenalin, mexicanin-E, and hymenovin (see Sec. XIV.B). Fourth, although

TABLE 1 Acute Toxicity of Sesquiterpene Lactones to Mammals

Lactone	Species	Route	LD_{50} (mg/kg)	Reference
Geigerinin	Guinea pig	s.c.	250 (LD_{100})	de Kock et al., 1968
				Vermeulen et al., 1978
Dihydrogriesenin	Guinea pig	s.c.	250 (LD_{100})	de Kock et al., 1968
				Vermeulen et al., 1978
Helenalin	Dog	i.v.	<< 60	Szabuniewicz and Kim, 1972
	Hamster	Oral	85	Witzel et al., 1976
	Mouse	Oral	150	Witzel et al., 1976
		i.p.	10	Kim, 1980
	Rabbit	Oral	90	Witzel et al., 1976
	Rat	Oral	125	Witzel et al., 1976
	Sheep	Oral	100–125	Witzel et al., 1976
Hymenolane	Mouse	i.p.	>200	Kim, 1980
Hymenovin (hymenoxon)	Dog	i.v.	30–50	Kim et al., 1974a

	Animal	Route	Dose	Reference
	Goat	Oral	<300	Kim et al., 1974a
	Hamster	Oral	250	Ivie et al., 1975a
	Mouse	Oral	150	Ivie et al., 1975a
		i.p.	16	Kim, 1980
	Rabbit	Oral	<1000	Kim et al., 1974a
		i.v.	25-50	Kim et al., 1974a
	Sheep	Oral	50-100	Ivie et al., 1975a
			50-75	Terry et al., 1981
		i.p.	7	Kim et al., 1975
			5-9	Rowe et al., 1980
Hymenovin dimethyl ether	Mouse	i.p.	141	Kim, 1980
Mexicanin-E	Mouse	i.p.	3	Kim, 1980
Psilotropin (= floribundin)	Mouse	i.p.	112	Kim, 1980
Tenulin	Hamster	Oral	1200	Ivie et al., 1975b
	Mouse	i.p.	185	Kim, 1980
	Sheep	Oral	500-1000	Ivie et al., 1975b

each of the most toxic lactones contain an α-methylene-γ-lactone moiety, the presence of this group per se does not appear to be the major requirement for high mammalian toxicity. Thus hymenovin dimethyl ether, which retains the unsaturated lactone, is about an order of magnitude less toxic than hymenovin itself, an observation supporting the hypothesis that the high biological activity of hymenovin may be related more to the bishemiacetal moiety than to the α-methylene-γ-lactone (Manners et al., 1978). Psilotropin similarly contains as a reactive center only the α-methylene-γ-lactone (the saturated δ-lactone of this compound would be predicted to confer little reactivity), and psilotropin is relatively low in acute mammalin toxicity (Table 1). Hymenolane, which contains no reactive functional groups, is apparently the least toxic of the compounds tested, and tenulin, which contains the α,β-unsaturated ketone but not α-methylene-γ-lactone, is relatively low in acute mammalian toxicity (Table 1). It is interesting that the most toxic of the compounds in Table 1 is mexicanin-E, a norpseudoguainolide that differs from helenalin only in the absence of the C-5 methyl, the C-6 hydroxyl, and opposite stereochemistry of the C-10 methyl (Fig. 13).

Studies with helenalin- and hymenovin-poisoned dogs, rabbits, and sheep have quite clearly defined the toxicological syndromes associated with acute mammalian poisoning by these lactones (Kim et al., 1974a,b; Szabuniewicz and Kim, 1972; Szabuniewicz et al., 1974; Terry et al., 1981). These effects are discussed in detail in a section to follow, but it is emphasized here that the toxic effects elicited by these individual lactones are consistent with effects seen in animals poisoned by the whole plants that contain helenalin or hymenovin as their toxic agents.

XIV. SESQUITERPENE LACTONES AS LIVESTOCK POISONS

The poisoning of livestock as a result of grazing toxic forage plants has historically been a major obstacle to livestock production in many areas of the world (Kingsbury, 1964). Studies in recent years have shown that a number of livestock plant poisons contain sesquiterpene lactones, and in several instances these lactones are now known to be the major if not the only toxic chemicals involved. Burnett et al. (1978b) and Rodriguez et al. (1976) have briefly reviewed the role of sesquiterpene lactones as livestock poisons, and Herz (1978) has considered this subject somewhat more thoroughly. We will review here the known information on sesquiterpene lactone containing plant poisons, particularly as regards the chemistry of the toxic agents, the nature of the clinical, physiopathological, and pathological responses elicited, and structure activity and mode of action considerations.

Livestock scientists began to realize no later than the early 1900s that some Compositae species were capable of causing poisonings in grazing livestock. The poisonous nature of these plants had no doubt existed for millenia but, in usual cases, overgrazed ranges were invaded by these plants and in the absence of more desirable forage, consumption by livestock and poisoning resulted. Although plants that are known, or that might reasonably be predicted to contain sesquiterpene lactones have been studied as livestock poisons for many years, it is largely on the basis of studies conducted during the past decade that these chemicals have been shown to be the major toxic agents in a number of plant species.

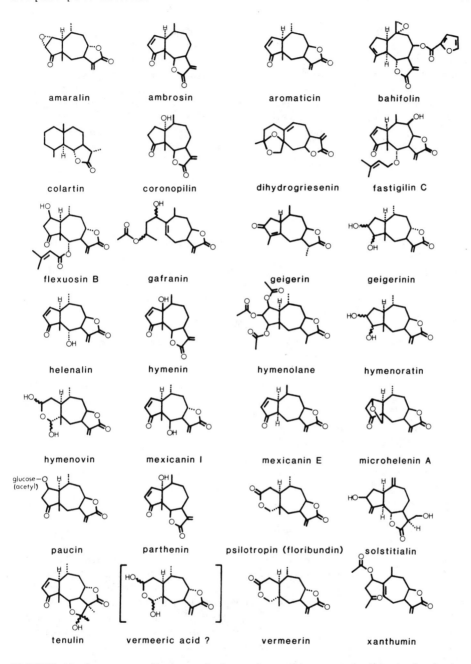

FIGURE 13 Some sesquiterpene lactones known to occur in livestock plant poisons. Acute toxicity data (see Table 1) confirm the roles of some of these lactones as agents responsible in part or whole for the toxicity of some range plants (see Table 2).

Table 2 is a list of 25 species of Compositae that have been reported to be poisonous to livestock. In several cases, sesquiterpene lactones are known to be the major if not the only toxic agents of significance in these plants, while others are known to contain sesquiterpene lactones, but their possible contribution to the toxicity of these plants has yet to be defined. Still other plants (e.g., *Artemisia spinescens*, *Centaurea repens*, *Hymenoxys lemmoni*, *Psilostrophe* spp., *Viguiera annua*, and some *Xanthium* spp.) are livestock poisons that have apparently not been reported to contain sesquiterpene lactones but, on the basis of phytochemical investigations with related species in the same genera, sesquiterpene lactones might reasonably be predicted to be present in these plants.

A. *Helenium* Species

Helenium autumnale (common sneezeweed) was apparently the first livestock plant poison from which sesquiterpene lactones were isolated and shown to be mammalian poisons. Reeb (1910) obtained helenalin from the plant, and subsequent studies by Lamson (1913) showed helenalin to be highly irritating to mucous membranes and toxic to dogs. A number of additional sesquiterpene lactones have now been reported from *H. autumnale* (Table 2; see also Herz, 1973). Because of the high acute toxicity of helenalin to mammals (Table 1), it is reasonable to speculate that this lactone is primarily responsible for the toxicity of the plant. However, *H. autumnale* is quite variable as to sesquiterpene lactone content, and some collections have been studied that apparently do not contain helenalin (Herz and Subramaniam, 1972).

Helenium microcephalum (smallhead sneezeweed) may well be the most toxic of the *Helenium* species; as little as 2.5 g of green plant per kilogram of body weight administered orally causes acute poisoning and death in cattle, sheep, and goats (Dollahite et al., 1964). Although *H. microcephalum* contains sesquiterpene lactones other than helenalin (Kim, 1980; Lee et al., 1976), studies with a Texas collection of the plant (which contained 3.6% helenalin on a dry weight basis) indicated that helenalin was the primary constituent that accounted for the toxicity of *H. microcephalum* (Witzel et al., 1976).

Helenium amarum (= *H. tenuifolium*) (bitter sneezeweed) is widely distributed in the eastern and southern United States and is of economic importance to the dairy industry because, when eaten by cattle, it imparts a bitter taste to the milk and renders it unpalatable. *H. amarum* is also a livestock poison that seems to affect horses and mules to a greater extent than other species (Dollahite et al., 1972; Kingsbury, 1964; West and Emmel, 1952). Laboratory studies with sheep have shown that fresh *H. amarum* causes death when force fed at 2% of body weight per day for 2 days (Dollahite et al., 1972). Studies with *H. amarum* have shown that tenulin is the major sesquiterpene lactone constituent that accounts for the toxicity of the plant (Ivie et al., 1975b), although the mammalian toxicity of tenulin is quite low (Table 1). Studies with tenulin administered orally to a lactating cow indicated that unmetabolized tenulin was secreted into the milk, but of the total administered dose, not more than 0.1% appeared in the milk (Ivie et al., 1975b). Because humans detect the bitter taste of tenulin in milk at levels as low as 1 ppm (Herzer, 1942; Ivie et al., 1975b), the bitter properties of tenulin guard

TABLE 2 Livestock Plant Poisons in which Sesquiterpene Lactones May Account in Part or Whole for Toxic Effects Observed

Plant	Sesquiterpene lactones present[a]	References
Artemisia filifolia Torr.	Colartin	Herz, 1978 Kingsbury, 1964 Schmutz et al., 1968
Artemisia spinescens DC Eat.	—	Herz, 1978 Kingsbury, 1964 Schmutz et al., 1968
Bahia oppositifolia (Nutt.) DC	Bahifolin	Herz, 1978 Kingsbury, 1964 Schmutz et al., 1968
Baileya multiradiata Harv. Gray	Hymenovin (hymenoxon) Baileyin Baileyolin Fastigilin C	Herz, 1978 Hill et al., 1977 Schmutz et al., 1968 Sperry et al., 1964
Centaurea solstitials L.	Solstitialin Solstitialin acetate	Cordy, 1978 Herz, 1978
Centaurea repens	—	Schmutz et al., 1968
Dugaldia hoopesii Rhydb. (= *Helenium hoopesii* Gray)	Hymenovin (hymenoxon)	Herz, 1978 Hill et al., 1977 Ivie et al., 1976 Schmutz et al., 1968
Geigeria aspera Harv.	"Vermeeric acid"	Anderson et al., 1967

TABLE 2 (Continued)

Plant	Sesquiterpene lactones present[a]	References
Geigeria filifolia Mattf. (= *G. africana* Gries)	Vermeerin Geigerin Geigerinin Gafranin Griesenin Dihydrogriesenin	de Kock et al., 1968 Herz, 1978 Rimington et al., 1936
Helenium amarum (Raf.) Rock (= *H. tenui-folium* Nutt.)	Tenulin Amaralin Aromaticin Heleniamarin Mexicanin I	ElSohly et al., 1979 Herz, 1978 Ivie et al., 1975b Sperry et al., 1964 Waddell et al., 1979b
Helenium autumnale L.	Helenalin Tenulin Mexicanin I Flexuosin A	Herz, 1973, 1978
Helenium flexuosum Raf. (= *H. nudiflorum* Nutt.)	Flexuosin A Flexuosin B	Herz, 1973, 1978 Kingsbury, 1964
Helenium microcephalum DC	Helenalin Mexicanin-E Microhelenin-A	Kim, 1980 Lee et al., 1976 Sperry et al., 1964 Szabuniewicz and Kim, 1972 Witzel et al., 1976
Hymenoxys lemmoni (Greene) Ckll	—	Herz, 1978 Kingsbury, 1964

Species	Compounds	References
Hymenoxys odorata DC	Hymenovin (hymenoxon) Hymenoratin Paucin Hymenoxynin Hymenolide Hymenolane	Herz, 1978 Ivie et al., 1975a Kim et al., 1975 Petersen and Kim, 1976 Schmutz et al., 1968 Sperry et al., 1964
Hymenoxys richardsonii (Hook.) Ckll. var. *floribunda*	Hymenovin (hymenoxon) Psilotropin (floribundin) Vermeerin	Herz, 1978 Ivie et al., 1976 Schmutz et al., 1968
Parthenium hysterophorus L.	Parthenin Hymenin Ambrosin	Towers et al., 1977
Iva acerosa (Nutt.) Jackson (= *Oxytenia acerosa* (Nutt.))	Coronopilin	Herz, 1978 Kingsbury, 1964
Psilostrophe gnaphalodes DC	—	Herz, 1978
Psilostrophe sparsiflora (Gray) A. Nels.	—	Kingsbury, 1964
Psilostrophe tagetina (Nutt.) Greene	—	Schmutz et al., 1968 Sperry et al., 1964
Viguiera annua (Jones) Blake[b]	—	Herz, 1978 Kingsbury, 1964 Schmutz et al., 1968
Xanthium strumarium L.[b]	Xanthumin	Schmutz et al., 1968
Xanthium saccharatum L.[b]	—	Sperry et al., 1964
Xanthium spinosum L.[b]	—	Rodriguez et al., 1976

[a] For some species, only a partial listing of sesquiterpene lactones known to occur in the plant is given.

[b] Although these species are known or suspected to contain sesquiterpene lactones, available data suggest that other agents may be primarily or solely responsible for their toxicity to livestock (see the text).

against its consumption by humans in appreciable amounts, and it was concluded that the consumption of milk from dairy animals grazing on *H. amarum* will probably not constitute a significant human health hazard (Ivie et al., 1975b).

Helenium flexuosum (= *H. nudiflorum*) (sneezeweed) is reported to be a livestock poison (Kingsbury, 1964), but no data are available regarding the nature of its toxic agents. However, the lactones flexuosin-A and -B have been isolated from *H. flexuosum* (Herz, 1973), and Herz (1978) speculated that these compounds might be responsible for the toxicity of this plant.

B. *Hymenoxys* Species

Hymenoxys odorata (western bitterweed) has for decades been a significant obstacle to livestock production in the southwestern United States. Economic losses (particularly sheep, but to a lesser extent goats and cattle) amount to millions of dollars annually (Jaggi, 1962; Rowe et al., 1973), and certain ranges are rendered unusable because of the presence of this plant. The poisonous nature and economic implications of *H. odorata* have been fully reviewed (Rowe et al., 1973). Several sesquiterpene lactones were isolated from *H. odorata* by Herz et al. (1970), who speculated that these lactones might account for the toxicity of *H. odorata*. Subsequent studies (Ivie et al., 1975a; Kim et al., 1975) demonstrated that, in fact, the sesquiterpene lactone hymenovin is the major toxic agent of *H. odorata*. Hymenovin is a mixture of two hemiacetals epimeric at C-2 or C-4 (Ivie et al., 1975a; Fig. 13), and one of the hymenovin epimers has been identified by x-ray crystallography and is referred to as hymenoxon (Pettersen and Kim, 1976). Although the term "hymenoxon" has been used in a number of reports on the chemical and toxicological behavior of the toxic principle of *H. odorata* (Hill and Camp, 1979; Hill et al., 1977, 1979, 1980; Kim, 1980; Rowe et al., 1980; Terry et al., 1981), it is not clear whether the toxicant used was in fact the hymenoxon epimer or the epimeric mixture hymenovin—the latter seems most probable. We use the term hymenovin throughout this report, as have others (Herz et al., 1980), in what would appear to be a more proper designation of the toxin as normally isolated.

Hymenoxys richardsonii var. *floribunda* (pingue) is quite closely related to *H. odorata* and is similarly of considerable importance as a livestock poison in certain of the Rocky Mountain states, particularly Colorado, New Mexico, and Arizona (Aanes, 1961; Kingsbury, 1964). Subsequent to the identification of hymenovin as the major poisonous principle of *H. odorata*, studies with a Utah collection of *H. richardsonii* conclusively showed that hymenovin accounts for the toxicity of this species as well (Ivie et al., 1976).

Hymenoxys lemmoni is poisonous to sheep (Kingsbury, 1964), and although nothing is known of the nature of its toxic principle, it seems quite probable that hymenovin or other sesquiterpene lactones are responsible for its toxicity.

C. *Dugaldia* (= *Helenium*) *hoopesii* (Orange Sneezeweed)

This plant has been a major obstacle to sheep production on high-altitude summer ranges in most of the Rocky Mountain states (Doran and Cassady, 1944; Kingsbury, 1964; Marsh et al., 1921). Consumption of *D. hoopesii* by sheep results in the syndrome known as "spewing sickness" (vomiting), and extensive death losses may result. A definitive study of the poisonous nature of this species was undertaken early in this century, and it was reported that the active constituent of *D. hoopesii* was a poorly defined glycoside to which the name "dugaldin" was given (Marsh et al., 1921). However, work by Ivie et al. (1976), confirmed by others (Hill et al., 1977), has shown that the active principle of *D. hoopesii* is, in fact, hymenovin. Whether or not "dugaldin" is involved at all in the toxicity of *D. hoopesii* is unclear, but the reported isolation procedures and solubility characteristics of dugaldin (Marsh et al., 1921) are consistent with the compound being similar to or identical with hymenovin (Ivie et al., 1976).

D. *Baileya multiradiata* (Desert Baileya)

Mortality as high as 25% has occurred in bands of sheep grazing *B. multiradiata* in certain areas of Texas (Dollahite, 1960; Kingsbury, 1964). Although the green plant is relatively distasteful, the flower heads are quite palatable and may be consumed even when other, more acceptable forage is available (Dollahite, 1960). A number of sesquiterpene lactones have been isolated from this plant (Table 2), and a recent report that hymenovin is present in *B. multiradiata* (Hill et al., 1977) indicates that hymenovin accounts, at least in part, for the poisonous nature of this species. However, comparative toxicity tests with various *B. multiradiata* fractions have not been done; thus the possibility that toxins other than hymenovin may also contribute to the toxicity of this plant cannot be ruled out.

E. *Geigeria* Species

The term "vermeerbos" (vomiting bush) is applied to several *Geigeria* species that have caused heavy losses of sheep in South Africa during at least the past 100 years (Rimington and Roets, 1936). The toxic syndrome is referred to as "vermeersiekte" (vomiting disease) and is quite similar to that produced by *Hymenoxys* spp. or *Dugaldia hoopesii*. Almost 50 years ago, studies by Rimington et al. (1936) resulted in the isolation from *G. aspera* of a substance, "vermeeric acid," determined by toxicity tests to be the poisonous agent in the plant. Vermeeric acid was a poorly defined compound which gradually transformed to another compound, vermeerin, on standing. Much more recently, Anderson et al. (1967) isolated vermeerin from *G. filifolia* (= *G. africana*), although all efforts at isolation of vermeeric acid were unsuccessful. Vermeeric acid has, in fact, not been isolated since Rimington's work, but vermeerin has been studied extensively and was characterized as a sesquiterpene dilactone (Fig. 13) (Anderson et al., 1967; Herz et al., 1970). Although Rimington implied that vermeeric acid, not vermeerin, is primarily responsible for the toxicity of *Geigeria* spp., the nature of vermerric acid remains obscure. The reported chemical properties are not compatible with the earlier deduction that it is simply the dicarboxylic acid of vermeerin (Rimington et al., 1936), now that the structure of vermeerin is known. It is plausible that vermeeric acid is a hymenovin isomer that is related to vermeerin as hymenovin is related to psilotropin (= floribundin) (Fig. 13) (Ivie et al., 1975a). Hymenovin readily dehydrates to psilotropin upon treatment with base (Hill and Camp, 1979), and if the structure shown for vermeeric acid in Fig. 13 is correct, this compound is obviously even more prone than hymenovin to undergo a similar dehydration reaction. This would account for its instability in Rimington's early work and possibly for failures to isolate vermeeric acid in later studies. We emphasize, however, that the structure indicated for vermeeric acid in Fig. 13 is speculative and is not supported by direct experimental data.

At least six sesquiterpene lactones are known from *Geigeria* (Table 2), and recent studies with *G. aspera* have shown that the lactones dihydrogriesenin and geigerinin (Fig. 13) account in part for the mammalian toxicity of this species (Vermeulen et al., 1978). Studies with these two compounds injected subcutaneously into guinea pigs showed that both are mammalian poisons, although their toxic potential appears to be less than that of some other, more thoroughly studied sesquiterpene lactones (Table 1). Comparative toxicity tests with various fractions from *G. aspera* showed that dihydrogriesenin and geigerinin contribute about 30% and 4%, respectively, of the toxicity of an ethanol extract of the plant (Vermeulen et al., 1978).

F. *Artemesia* Species

A. filifolia (sand sagebrush) is reported to cause a syndrome in horses known as "sage sickness" (Kingsbury, 1964). The eudesmanolide colartin (Fig. 13) has been isolated from this plant (Herz, 1978), but its toxic properties have not been evaluated. Because colartin contains no reactive functional groups, it seems likely that this compound would not be appreciably toxic. *A. spinescens* has also been suspected as a livestock poison (Kingsbury, 1964), but no phytochemical data are available on this species that would implicate sesquiterpene lactones as causative agents. However, Herz (1978) has noted that almost all *Artemesia* species elaborate sesquiterpene lactones and has speculated that these compounds might be involved in *Artemisia* toxicity.

G. *Bahia oppositifolia* (Plains Bahia)

Poisoning under range conditions by *B. oppositifolia* rarely if ever occurs (Kingsbury, 1964; Schmutz et al., 1968), but feeding and analysis studies have shown that the plant is toxic to livestock and that the major toxic agent is probably a cyanogenic glycoside (Kingsbury, 1964). However, the guaianolide bahifolin (Fig. 13) has been isolated from *B. oppositifolia*. Most if not all other *Bahia* species elaborate sesquiterpene lactones (Herz, 1978), and thus it is possible that sesquiterpene lactones contribute to the toxicity of this plant.

H. *Centaurea* Species

The prolonged consumption of *C. solstitialis* (yellow starthistle) or *C. repens* (Russian knapweed) produces a disease in horses known as equine nigro-pallidal encephalomalacia (Cordy, 1978). The production of germacranolides and guaianolides in characteristic of *Centaurea* species (Herz, 1978), and two lactones, solstitialin and solstitialin acetate, have been isolated from California

collections of *C. solstitialis* (Herz, 1978). The toxic syndrome in horses results from necrosis and softening in the globus pallidus and substantia nigra
of the brain (Cordy, 1978), and such lesions are not suggestive of sesquiterpene lactone poisoning, on the basis of current knowledge regarding toxic syndromes associated with these compounds in mammals (vide infra). However,
the horse appears to be uniquely susceptible to this disease and it is possible
that sesquiterpene lactones are involved in equine nigropallidal encephalomalacia through biochemical mechanisms of toxicity peculiar to this species.

I. *Iva* (= *Oxytenia*) *acerosa* (Copperweed)

This species is toxic to cattle and sheep, and the symptoms of poisoning described by Kingsbury (1964) are consistent with the hypothesis that sesquiterpene lactones may be responsible for the plant's toxicity. The ambrosano-
lide coronopilin has been isolated from *I. acerosa* (Herz, 1973, 1978), but experimental data are lacking that might directly implicate this or related lactones as the toxic agents of *Iva* poisoning (Herz, 1978).

J. *Parthenium hysterophorus*

This aggressive weed is endemic to the Americas but within the last 100 years
has become established in Africa, Australia, and Asia (Towers et al., 1977).
P. hysterophorus is apparently consumed readily by at least some livestock
species, and water buffalo developed toxic symptoms within a few weeks after
being fed a diet that contained 50% *P. hysterophorus* foliage. Clinical and
pathological observations on poisoned animals as summarized by Towers et al.
(1977) are fully consistent with those elicited by plants in which sesquiterpene lactones are known to be the toxic agents (vide infra). *P. hysterophorus* contains the lactones parthenin, hymenin, and ambrosin (Fig. 13),
each of which contains two reactive functional groups. It therefore seems
highly likely that these lactones account for the toxicity of the plant.

    P. hysterophorus is responsible for serious outbreaks of human allergic
contact dermatitis, particularly in India (Towers et al., 1977), and it is of
considerable interest that buffalo fed foliage of the plant developed severe
dermatitis. Within 7 days after feeding began, the animals developed itching,
followed by erythematous eruptions involving the tip and base of the ears,
the neck, the thoracic and brisket regions, and the front legs. A few weeks
later, the affected areas were mostly devoid of hair. In addition, surviving
animals developed edema around the eyelids and the muzzle (Towers et al.,
1977). To our knowledge, these studies with *P. hysterophorus* represent the
only instance in which livestock plant poisons that contain sesquiterpene lactones have been shown to cause dermatitis in any livestock species.

K. *Psilostrophe* Species

Three species of paperflowers, *P. gnaphalodes, P. sparsiflora,* and *P. tage-
tina,* have caused economically important losses of sheep on ranges in the
southwestern United States (Kingsbury, 1964). The toxic syndrome associated with these plants is quite similar to that caused by *Hymenoxys* or
Dugaldia, and although these three *Psilostrophe* species have apparently not

been investigated chemically, the related *P. cooperi* (Gray) Greene is known to contain psilotropin. On the basis of these observations, and of the close chemical relationship between psilotropin and hymenovin, Herz (1978) has speculated that hymenovin or closely related compounds may be responsible for the toxicity of *Psilostrophe* species.

L. *Viguiera annua* (Annual Goldeneye)

Cattle have been killed after grazing large amounts of this plant, but the toxic symptoms associated with poisoning by *V. annua* are suggestive of either cyanide or nitrate poisoning (Kingsbury, 1964). It has been noted by Herz (1978) that other species of *Viguiera* are known to contain sesquiterpene lactones, but their possible occurrence in *V. annua* has not been established. It is therefore not known at this time whether sesquiterpene lactones contribute to the toxicity of this plant.

M. *Xanthium* Species

Cockleburs are widely distributed throughout the world and can be poisonous to livestock (Cole et al., 1980; Kingsbury, 1964). The poisonous properties of cockleburs are apparently associated only with the seed and with young plants at the cotyledon stage, and young cocklebur plants are particularly hazardous to pastured swine (Cole et al., 1980). The sesquiterpene lactone xanthumin has been reported from *X. strumarium* (Yoshioka et al., 1973) and other *Xanthium* species are known to contain these lactones (Kawazu et al., 1979). Recent studies (Cole et al., 1980) have resulted in the identification of carboxyatractyloside as a highly toxic agent occurring in *Xanthium* seed or seedlings. The possible roles that sesquiterpene lactones might play in *Xanthium* poisoning are unknown at this time.

XV. CLINICAL, PHYSIOPATHOLOGIC, AND PATHOLOGIC EFFECTS OF SESQUITERPENE LACTONES

A number of studies have shown that the toxic syndromes are quite similar among livestock plant poisons in which sesquiterpene lactones are known or strongly suspected to be the toxic agents. In addition, studies with helenalin and hymenovin indicate that the toxic effects of these two lactones on mammals are consistent with the effects observed in animals poisoned by plants that contain these compounds as major toxic agents.

Animals that consume *Helenium* spp., *Hymenoxys* spp., *Dugaldia hoopesii*, *Geigeria* spp., and other sesquiterpene lactone-containing plants may exhibit considerable time variation with respect to the onset of toxic symptoms. Poisoning may occur within a day or less, or symptoms may not be observed for a month or more. The amount of plant consumed, the levels of toxicant present, the availability of other forage, the nutritional state of the exposed animals, and probably other factors are no doubt related to such variability. Livestock poisoning under range conditions does, however, tend to be associated more with exposures of at least a few weeks' duration.

Rowe et al. (1973) have described in detail the clinical signs of *Hymenoxys odorata* poisoning in sheep, and these, summarized below, are very similar to

clinical symptoms reported for livestock poisoning by several other sesquiterpene lactone-containing plants (Aanes, 1961; Dollahite et al., 1964, 1972;
duToit, 1928; Kingsbury, 1964; Marsh et al., 1921). Early clinical signs of
acute poisoning may include bloating, central nervous system (CNS) depression, loss of appetitie, and an apparent cessation of rumination. Animals may
exhibit arched backs and grinding of the teeth, suggestive of abdominal pain.
With time, CNS depression becomes more severe, and swallowing movements,
a mucous nasal discharge, muscular tremor, and labored breathing may occur.
Vomiting is sometimes associated with *H. odorata* poisoning, but is particularly
characteristic in many animals poisoned with *Dugaldia hoopesii* (Marsh et al.,
1921) or *Geigeria* spp. (duToit, 1928). Terminally, there may be intermittent
walking movements of the limbs, tonic convulsions, and severe dyspnea. Regurgitation of rumen ingesta at the time of death sometimes occurs. Chronically poisoned animals suffer loss of appetite and become very weak and re-
cumbent. Starvation and dehydration probably contribute to the death of
some of these animals due to insufficient strength to obtain food and water.
Many chronically poisoned animals can recover if removed from access to the
toxic plants and provided with food and water soon after the onset of signs
of intoxication.

Marked physiopathologic effects occur in animals poisoned with sesquiterpene lactone-containing plants or the lactones isolated therefrom. Acute
poisoning of sheep with *Hymenoxys odorata* resulted in hypoglycemia, metabolic acidosis, lactacidemia, and pyruvemia (Witzel et al., 1974). Additional
studies in sheep treated orally with either *H. odorata* or its toxic principle,
hymenovin, showed that both treatments produced dose-related elevations in
blood urea nitrogen (BUN), creatinine, inorganic phosphorus, alkaline phosphatase, serum glutamic-pyruvic transaminase (SGPT), serum glutamic-
oxalic transaminase (SGOT), lactic dehydrogenase (LDH), and creatinine
phosphokinase (CPK) (Terry et al., 1981). Sheep subacutely poisoned with
Dugaldia hoopesii similarly exhibited elevated SGOT levels, but unlike *H.
odorata*, SGPT activities in *D. hoopesii* poisoned sheep were not appreciably
affected (Buck et al., 1961). Sheep poisoned with *H. odorata* (Steel et al.,
1976) or *H. richardsonii* (Aanes, 1961) also exhibit abnormalities in certain
other hematological parameters.

Necropsy observations from animals poisoned by sesquiterpene lactones or
by plants that contain them often show pronounced lesions in the digestive,
respiratory, cardiac, renal, and hepatic systems (Aanes, 1961; Dollahite et
al., 1964, 1972; Marsh et al., 1921; Rimington et al., 1936; Rowe et al., 1973;
Terry et al., 1981; Towers et al., 1977; Witzel et al., 1977). Acute congestion of the lungs is usually observed, and this may be accompanied by edema
and hemorrhage. Epicardial hemorrhages are common. The intestinal tract,
particularly the forestomach and abomasal mucosa, may show congestion and
hemorrhage. Accumulation of edematous fluids may be seen along the rumino-
reticular folds and ruminal sulci. Renal effects tend to be pronounced and
include severe glomerulonephrosis characterized by proteinaceous casts,
swollen and degenerated glomerular tufts, and degeneration and necrosis in
the inner renal cortex and outer medulla. The liver may be congested, and
toxic hepatosis, characterized by vacuolar degeneration of hepatocytes, may
be seen (Witzel et al., 1977). The observed elevations of BUN, creatinine,
and inorganic phosphorus seen in poisoned animals are compatible with the
renal lesions observed morphologically. Similarly, elevations in alkaline

phosphatase, SGPT, SGOT, and LDH are consistent with the toxic hepatic effects (Terry et al., 1981).

XVI. STRUCTURE ACTIVITY AND MODE OF ACTION CONSIDERATIONS

On the basis of structure-activity relationships with respect to the toxic action of sesquiterpene lactones on a number of biological systems, including mammals (Table 1), it seems clear that appreciable biological activity of these lactones is related to the presence of reactive functional groups that can form covalent bonds with critical biological nucleophiles. Although biological activity among the sesquiterpene lactones is usually, but not always, associated with an α-methylene-γ-lactone moiety, the presence of an additional reactive center, such as unsaturated ketone, chlorohydrin, epoxide, hemiacetal, or hemiketal, is generally required for a high degree of biological activity. That these lactones can efficiently alkylate thiols and amines in biological systems is firmly established; thus it follows that their toxic action can be ascribed to general cytotoxic effects on any number of cellular, tissue, and organ systems. Consistent with this view are the clinical, physiopathologic, and pathologic observations in sesquiterpene lactone-poisoned animals, which indicate that a number of body systems are adversely affected.

Due to the reactivity of sesquiterpene lactones toward sulfhydryl groups, it might be predicted that mercaptans would show some antagonistic properties against sesquiterpene lactone poisoning in mammals. Studies with *Hymenoxys odorata* and its toxic constituent, hymenovin, have shown that such is indeed the case (Ivie et al., 1975a; Kim et al., 1974b; Rowe et al., 1980). Cysteine effectively prevents poisoning if it is administered orally or intravenously soon after mammals are exposed to *H. odorata* or hymenovin. However, such treatment probably has little or no practical value, because even though mercaptans can inactivate sesquiterpene lactone toxins within the digestive tract or bloodstream, they cannot be expected to reverse the biochemical lesions once reaction of the lactones with critical body constituents has occurred (Ivie et al., 1975a).

An additional consideration in livestock poisoning by plants that contain sesquiterpene lactones as toxic agents is the possibility that these lactones may alter the microbial composition of the rumen and thus affect its vital metabolic functions (Ivie et al., 1975a). Cessation of rumination appears to occur in some sesquiterpene lactone-poisoned animals, and because many such lactones are known to be potent antimicrobial agents, it may well be that sesquiterpene lactone-induced rumen dysfunction contributes to the toxicity of these compounds in ruminants.

XVII. DISPOSITION AND FATE OF SESQUITERPENE LACTONES IN MAMMALS

There is apparently very little information available relative to the fate of sesquiterpene lactones in animal systems. Hill et al. (1980) orally administered tritiated hymenovin (hymenoxon) to rabbits and observed that the compound was absorbed, rapidly metabolized, and was excreted in the bile and urine. None of several metabolites observed were chemically characterized, but the major metabolites appeared to be glucuronides of at least two different hymenovin transformation products. Although some hymenovin per se was excreted in the urine, its relative concentration was low and most of the absorbed dose was metabolized prior to elimination. Data were obtained which indicated that hymenovin rapidly binds, both in vitro and in vivo, to cellular components of the blood. However, studies with sulfhydryl inhibitors and hymenovin indicated that the binding of hymenovin with blood apparently did not involve sulfhydryl groups as reactive sites (Hill et al., 1980).

XVIII. CONCLUSIONS

On the basis of studies to date, it is apparent that many sesquiterpene lactones are highly biologically active compounds that may affect a wide variety of microbial, plant, and animal systems. It may be that additional research will lead to practical applications of sesquiterpene lactones or their analogs as antimicrobial, antiparasitic, plant or insect growth regulator, or insect or mammalian antifeedant agents. Certain pharmacological activities of these compounds, particularly their actions as anthelminthic, anti-inflammatory, and antihyperlipidemic drugs, merit more detailed study. The potential antitumor properties of some sesquiterpene lactones are encouraging, although it can be argued that their general cytotoxic effects rather than specific antitumor actions, which results in relatively poor "selectivity" and high mammalian toxicity, may well preclude the effective use of these lactones in cancer chemotherapy.

It is fortuitious that sesquiterpene lactones are intensely bitter compounds, because such a circumstance no doubt greatly minimizes the potential for human poisoning. Although sesquiterpene lactones are of considerable human health significance as contact allergens, human toxicity resulting from the consumption of these compounds is apparently exceedingly rare. We are aware of only a single such incident, possibly anecdotal, in which consumption of bread made with flour that contained large quantities of seeds of *Helenium amarum* was said to have caused poisoning in humans (Kingsbury, 1964). The consumption of bitter milk (Ivie et al., 1975b; Towers et al., 1977) and possibly meat of livestock that have grazed on sesquiterpene lactone-containing plants has been of historical significance and no doubt continues to occur. However, it is not likely that sufficient amounts of these toxicants would be consumed under such circumstances to constitute a significant threat to human health, at least from an acute toxicity standpoint.

The potential effects related to chronic exposures of both humans and domestic animals to sesquiterpene lactones have, for all practical purposes, not been considered. The rather potent mutagenic activity of at least one of these compounds (hymenovin) has led to speculation that the ingestion of sesquiterpene lactones by livestock might result in adverse genetic or reproductive effects that could lead to long-term economic losses (MacGregor, 1977). Of even more importance are the potential mutagenic and carcinogenic risks to humans that may be associated with exposure to some of these lactones, whether exposure be by contact or orally through the consumption of contaminated foodstuffs of plant or animal origin. It seems prudent to us to research the long-range health implications of sesquiterpene lactones to both humans and animals.

XIX. ACKNOWLEDGMENT

We thank Douglas L. Holt of this laboratory for excellent librarial assistance in support of this project.

REFERENCES

Aanes, W. A. (1961). Pingue (*Hymenoxys richardsonii*) poisoning in sheep. *Am. J. Vet. Res.* 22:47-52.

Amo, S. D., and Anaya, A. L. (1978). Effect of some sesquiterpenic lactones on the growth of certain secondary tropical species. *J. Chem. Ecol.* 4:305-313.

Anderson, L. A. P., de Kock, W. T., and Pachler, K. G. R. (1967). The structure of vermeerin, a sesquiterpenoid dilactone from *Geigeria africana* Gries. *Tetrahedron 23*:4153-4160.

Baker, P. M., Fortes, C. C., Fortes, E. G., Gazzinelli, G., Gilbert, B., Lopes, J. N. C., Pelligrino, J., Tomassini, T. C. B., and Vichnewski, W. (1972). Chemoprophylactic agents in schistosomiasis: eremanthine, costunolide, α-cyclocostunolide and bisabolol. *J. Pharm. Pharmacol. 24*: 853-857.

Boughton, I. B., and Hardy, W. T. (1937). Toxicity of bitterweed (*Actinea odorata*) for sheep. *Tex. Agric. Exp. Stn. Bull. 552.*

Buck, W. B., James, L. F., and Binns, W. (1961). Changes in serum transaminase activities associated with plant and mineral toxicity in sheep and cattle. *Cornell Vet. 51*:568-585.

Burnett, W. C., Jones, S. B., and Mabry, T. J. (1977a). Evolutionary implications of herbivory on *Vernonia* (Compositae). *Plant Syst. Evol. 128*:227-286.

Burnett, W. C., Jones, S. B., and Mabry, T. J. (1977b). Evolutionary implications of sesquiterpene lactones in *Vernonia* (Compositae) and mammalian herbivores. *Taxon 26*:203-207.

Burnett, W. C., Jones, S. B., and Mabry, T. J. (1978a). Influence of sesquiterpene lactones of *Vernonia* (Compositae) on oviposition preferences of Lepidoptera. *Am. Midl. Nat. 100*:242-246.

Burnett, W. C., Jr., Jones, S. B., and Mabry, T. J. (1978b). The role of sesquiterpene lactones in plant-animal coevolution. In *Biochemical Aspects of Plant and Animal Coevolution*, J. B. Harborne (Ed.). Academic Press, New York, pp. 233-257.

Calhoun, M. C., Ueckert, D. N., Livingston, C. W., and Camp. B. J. (1978). Effect of spraying with 2,4-D on hymenoxon concentration and toxicity of harvested bitterweed fed to sheep. *Tex. Agric. Exp. Stn. Prog. Rep. PR-3511*, pp. 87-93.

Calzada, J., Ciccio, J. F., and Echandi, C. (1980). Antimicrobial activity of the heliangolide chromolaenide and related sesquiterpene lactones. *Phytochemistry 19*:967-968.

Cavallito, C. J., and Haskell, T. H. (1945). Mechanism of action of antibiotics. Reaction of unsaturated lactones with cysteine and related compounds. *J. Am. Chem. Soc. 67*:1991-1994.

Char, M. B. S., and Shankarabhat, S. (1975). Parthenin, a growth inhibitor behavior in different organisms. *Experimentia 31*:1164-1165.

Clark, E. P. (1936). Helenalin. I. Helenalin, the bitter sternutative substance occurring in *Helenium autumnale*. *J. Am. Chem. Soc. 58*:1982-1983.

Cole, R. J., Stuart, B. P., Lansden, J. A., and Cox, R. H. (1980). Isolation and redefinition of the toxic agent from cocklebur (*Xanthium strumarium*). *J. Agric. Food Chem. 28*:1330-1332.

Cordy, D. R. (1978). *Centaurea* species and equine nigropallidal encephalomalacia. In *Effects of Poisonous Plants on Livestock*, R. F. Keeler, K. R. Van Kampen, and L. F. James (Eds.). Academic Press, New York, pp. 327-336.

Dalvi, R. R., Singh, B., and Salunkhe, D. K. (1971). Phytotoxicity of alantolactone. *Chem.-Biol. Interact. 3*:13-18.

deKock, W. T., Pachler, K. G. R., and Wessels, P. L. (1968). Griesenin and dihydrogriesenin, two new sesquiterpenoid lactones from *Geigeria africana* Gries. II. Nuclear magnetic resonance studies and conformation. *Tetrahedron 24*:6045-6052.

de Waal, M. (1920). Study of the insecticidal power of Compositae, especially of *Helenium autumnale*. *Pharm. Weekbl. 57*:1100-1107.

Dollahite, J. W. (1960). Desert baileya poisoning in sheep, goats, and rabbits. *Tex. Agric. Exp. Stn. Prog. Rep.* PR-2149.

Dollahite, J. W. Hardy, W. T., and Henson, J. B. (1964). Toxicity of *Helenium microcephalum* (smallhead sneezeweed). *J. Am. Vet. Med. Assoc. 145*:694-696.

Dollahite, J. W., Rowe, L. D., Kim, H. L., and Camp, B. J. (1972). Toxicity of *Helenium amarum* (bitter sneezeweed) to sheep. *Southwest. Vet. 26*:135-137.

Doran, C. W., and Cassady, J. T. (1944). Management of sheep on range infested with orange sneezeweed. *U.S. Dept. Agric. Circ. 691*.

Doskotch, R. W., Keely, S. L., Hufford, C. D., and El-Feraly, F. S. (1975). Antitumor agents 8. New sesquiterpene lactones from *Liriodendron tulipifera*. *Phytochemistry 14*:769-773.

Doskotch, R. W., El-Feraly, F. S., Fairchild, E. H., and Haung, C. T. (1977a). Isolation and characterization of peroxyferolide, a hydroperoxy sesquiterpene lactone from *Liriodendron tulipifera*. *J. Org. Chem. 42*: 3614-3618.

Doskotch, R. W., Odell, T. M., and Godwin, P. A. (1977b). Feeding responses of gypsy moth larvae, *Lymantria dispar*, to extracts of plant leaves. *Environ. Entomol. 6*:563-566.

Doskotch, R. W., Fairchild, E. H., Huang, C. T., and Wilton, J. H. (1980). Tulirinol, an antifeedant sesquiterpene lactone for the gypsy moth larvae from *Liriodendron tulipifera*. *J. Org. Chem. 45*:1441-1446.

Dupuis, G., Mitchell, J. C., and Towers, G. H. N. (1974). Reaction of alantolactone an allergenic sesquiterpene lactone with some amino acids. Resultant loss of immunologic reactivity. *Can. J. Biochem. 52*:575-581.

duToit, P. J. (1928). Investigations into the cause of vomeersiekte in sheep. *13th and 14th Rep. Dir. Vet. Res. Educ. Union S. Afr.*, pp. 109-153.

ElSohly, M. A., Craig, J. C., Turner, C. E., and Sharma, A. S. (1979). Constituents of *Helenium amarum*. II. Isolation and characterization of heleniamarin and other constituents. *J. Nat. Prod. 42*:450-454.

Garcia, M., DaSilva, A. J. R., Baker, P. M., Gilbert, B., and Rabi, J. A. (1976). Chemical transformations of abundant natural products. Part 1.

Absolute stereochemistry of eremanthine, a schistosomicidal sesquiterpene lactone from *Eremanthus elaeagnus*. *Phytochemistry* *15*:331-332.

Gross, D. (1975). Growth regulating substances of plant origin. *Phytochemistry* *14*:2105-2112.

Hall, I. H., Lee, K. H., Mar, E. C., and Starnes, C. O. (1977). Antitumor agents. 21. A proposed mechanism for inhibition of cancer growth by tenulin and helenalin and related cyclopentenones. *J. Med. Chem.* *20*:333-337.

Hall, I. H., Lee, K. H., Starnes, C. O., El Gebaly, S. A., Ibuka, T., Wu, Y. S., Kimura, T., and Haruna, M. (1978). Antitumor agents. XXX. Evaluation of α-methylene-γ-lactone containing agents for inhibition of tumor growth, respiration, and nucleic acid synthesis. *J. Pharm. Sci.* *67*:1235-1239.

Hall, I. H., Lee, K. H., Starnes, C. O., Sumida, Y., Wu, R. Y., Waddell, T. G., Cochran, J. W., and Gerhart, K. G. (1979). Anti-inflammatory activity of sesquiterpene lactones and related compounds. *J. Pharm. Sci.* *68*:537-542.

Hall, I. H., Lee, K. H., Starnes, C. O., Muraoka, O., Sumida, Y., and Waddell, T. G. (1980a). Antihyperlipidemic activity of sesquiterpene lactones and related compounds. *J. Pharm. Sci.* *69*:694-697.

Hall, I. H., Lee, K. H., Williams, W. L., Kimura, T., and Hirayama, I. (1980b). Antitumor agents. XLI. Effects of eupaformosanin on nucleic acid, protein, anaerobic and aerobic glycolytic metabolism of Ehrlich ascites cells. *J. Pharm. Sci.* *69*:294-297.

Hall, I. H., Starnes, C. O., Lee, K. H., and Waddell, T. G. (1980c). Mode of action of sesquiterpene lactones as anti-inflammatory agents. *J. Pharm. Sci.* *69*:537-543.

Hanson, R. L., Lardy, H. A., and Kupchan, S. M. (1970). Inhibition of phosphofructokinase by quinone methide and α-methylene lactone tumor inhibitors. *Science* *168*:378-380.

Haynes, L. J. (1948). Physiologically active unsaturated lactones. *Q. Rev. Chem. Soc. (Lond.)* *2*:46-72.

Herout, V. (1973). A chemical compound as a taxonomic character. In *Chemistry in Botanical Classification, Nobel Symposium 25*, G. Bendz and J. Santesson (Eds.). Academic Press, New York, pp. 55-62.

Herz, W. (1973). Pseudoguaianolides in Compositae. In *Chemistry in Botanical Classification, Nobel Symposium 25*, G. Bendz and J. Santesson (Eds.). Academic Press, New York, pp. 153-172.

Herz, W. (1977). Biogenetic aspects of sesquiterpene lactone chemistry. *Is. J. Chem.* *16*:32-44.

Herz, W. (1978). Sesquiterpene lactones from livestock poisons. In *Effects of Poisonous Plants on Livestock*, R. F. Keeler, K. R. Van Kampen, and L. F. James (Eds.). Academic Press, New York, pp. 487-497.

Herz, W., and Subramaniam, P. S. (1972). Pseudoguaianolides in *Helenium autumnale* from Pennsylvania. *Phytochemistry* *11*:1101-1103.

Herz, W., Aota, K., Holub, M., and Samek, Z. (1970). Sesquiterpene lactones and lactone glycosides from *Hymenoxys* species. *J. Org. Chem.* *35*:2611-2624.

Herz, W., Govindan, S. V., Bierner, M. W., and Blount, J. F. (1980). Sesquiterpene lactones of *Hymenoxys insignis*. X-ray analyses of hymenograndin and hymenosignin. *J. Org. Chem.* *45*:493-497.

Herzer, F. H. (1942). Bitterweed studies. *Assoc. South. Agric. Workers, Proc. Annu. Conv. 43*:112-113.

Hill, D. W., and Camp, B. J. (1979). Reactions of hymenoxon: Base conversion to psilotropin and greenein and formation of "Michael adduct" with cysteine. *J. Agric. Food Chem. 27*:882-885.

Hill, D. W., Kim, H. L., Martin, C. L., and Camp. B. J. (1977). Identification of hymenoxon in *Baileya multiradiata* and *Helenium hoopsii*. *J. Agric. Food Chem. 25*:1304-1307.

Hill, D. W., Kim, H. L., and Camp, B. J. (1979). Quantitative analysis of hymenoxon in plant tissue. *J. Agric. Food Chem. 27*:885-887.

Hill, D. W., Bailey, E. M., and Camp, B. J. (1980). Tissue distribution and disposition of hymenoxon. *J. Agric. Food Chem. 28*:1269-1273.

Ivie, G. W., Witzel, D. A., Herz, W., Kannan, R., Norman, J. O., Rushing, D. D., Johnson, J. H., Rowe, L. D., and Veech, J. A. (1975a). Hymenovin: major toxic constituent of western bitterweed (*Hymenoxys odorata* DC). *J. Agric. Food Chem. 23*:841-845.

Ivie, G. W., Witzel, D. A., and Rushing, D. D. (1975b). Toxicity and milk bittering properties of tenulin, the major sesquiterpene lactone constituent of *Helenium amarum* (bitter sneezeweed). *J. Agric. Food Chem. 23*: 845-8459.

Ivie, G. W., Witzel, D. A., Herz, W., Sharma, R. P., and Johnson, A. E. (1976). Isolation of hymenovin from *Hymenoxys richardsonii* (pingue) and *Dugaldia hoopesii* (orange sneezeweed). *J. Agric. Food Chem. 24*:281-282.

Jaggi, F. P. (1962). Committee Report on Animal Disease Losses of Economic Importance to the Livestock in Texas. Texas A&M University, School of Veterinary Medicine, College Station, Tex., pp. 56-62.

Jones, J. B., and Young, J. M. (1968). Carcinogenicity of lactones. III. The reaction of unsaturated γ-lactones with L-cysteine. *J. Med. Chem. 11*:1176-1182.

Jones, S. B., Burnett, W. C., Coile, N. C., Mabry, T. J., and Betkouski, M. F. (1979). Sesquiterpene lactones of *Vernonia*—Influence of glaucolide-A on the growth rate and survival of lepidopterous larvae. *Oecologia 39*: 71-77.

Kawazu, K., Nakajima, S., and Ariwa, M. (1979). Xanthumin and 8-epi-xanthatin as insect development inhibitors from *Xanthium canadense* Mill. *Experientia 35*:1294-1295.

Kelsey, R. G., Morris, M. S., Bhadane, N. R., and Shafizadeh, F. (1973). Sesquiterpene lactones of *Artemisia*: TLC analysis and taxonomic significance. *Phytochemistry 12*:1345-1350.

Kim, H. L. (1980). Toxicity of sesquiterpene lactones. *Res. Commun. Chem. Pathol. Pharmacol. 28*:189-192.

Kim, H. L., Rowe, L. D., Szabuniewicz, M., Dollahite, J. W., and Camp, B. J. (1974a). The isolation of a poisonous lactone from bitterweed, *Hymenoxys odorata* DC (Compositae). *Southwest. Vet. 27*:84-87.

Kim, H. L., Szabuniewicz, M., Rowe, L. D., Camp, B. J., Dollahite, J. W., and Bridges, C. H. (1974b). L-Cysteine, an antagonist to the toxic effects of an α-methylene-γ-lactone isolated from *Hymenoxys odorata* DC (bitterweed). *Res. Commun. Chem. Pathol. Pharmacol. 8*:381-384.

Kim, H. L., Rowe, L. D., and Camp, B. J. (1975). Hymenoxon, a poisonous sesquiterpene lactone from *Hymenoxys odorata* DC (bitterweed). *Res. Commun. Chem. Pathol. Pharmacol. 11*:647-650.

Kingsbury, J. M. (1964). *Poisonous Plants of the United States and Canada.* Prentice-Hall, Englewood Cliffs, N. J.

Kupchan, S. M., Fessler, D. C., Eakin, M. A., Giacobbe, T. J. (1970). Reactions of alpha methylene lactone tumor inhibitors with model biological nucleophiles. *Science 168*:376-378.

Kupchan, S. M., Eakin, M. A., and Thomas, A. M. (1971). Tumor inhibitors. 69. Structure-cytotoxicity relationships among the sesquiterpene lactones. *J. Med. Chem. 14*:1147-1152.

Kwon, Y. M., Woo, W. S., Woo, L. K., and Lee, M. J. (1973). Effect of *Inula* sesquiterpene lactones on the respiration of plants. *Han'guk Saenghwahakhoe Chi 6*:85-94.

Lamson, P. D. (1913). On the pharmacological action of helenin, the active principle of *Helenium autumnale*. *J. Pharm. Exp. Ther. 4*:471-489.

Lee, K. H., Huang, E. S., Piantadosi, C., Pagano, J. S., and Geissman, T. A. (1971). Cytotoxicity of sesquiterpene lactones. *Cancer Res. 31*: 1649-1654.

Lee, K. H., Imakura, Y., and Sims, D. (1976). Antitumor agents. XVII. Structure and stereochemistry of microhelenin-A, a new antitumor sesquiterpene lactone from *Helenium microcephalum*. *J. Pharm. Sci. 65*:1410-1412.

Lee, K. H., Hall, I. H., Mar, E. C., Starnes, C. O., ElGebaly, S. A., Waddell, T. G., Hadgraft, R. I., Ruffner, C. G., and Weidner, I. (1977a). Sesquiterpene antitumor agents: inhibitors of cellular metabolism. *Science 196*:533-535.

Lee, K. H., Ibuka, T., Wu, R. Y., and Geissman, T. A. (1977b). Structure-antimicrobial activity relationships among the sesquiterpene lactones and related compounds. *Phytochemistry 16*:1177-1181.

Lucas, R. A., Rovinski, S., Kiesel, R. J., Dorfman, L., and MacPhillamy, H. B. (1964). A new sesquiterpene lactone with analgesic activity from *Helenium amarum* (Raf.) H. Rock. *J. Org. Chem. 29*:1549-1554.

Mabry, T. J., and Bohlmann, F. (1977). Summary of the chemistry of the Compositae. In *Chemistry and Biology of the Compositae*, V. Heywood, J. B. Harborne, and B. L. Turner (Eds.). Academic Press, New York, pp. 1097-1104.

Mabry, T. J., Gill, J. E., Burnett, W. C., and Jones, S. B. (1977). Antifeedant sesquiterpene lactones in the Compositae. In *Host–Plant Resistance to Insects*, P. A. Hedin (Ed.). *ACS Symp. Ser. 62*, pp. 179-184.

MacGregor, J. T. (1977). Mutagenic activity of hymenovin, a sesquiterpene lactone from western bitterweed. *Food Cosmet. Toxicol. 15*:225-227.

Manners, G. D., Ivie, G. W., and MacGregor, J. T. (1978). Mutagenic activity of hymenovin in *Salmonella typhimurium*: association with the bishemiacetal functional group. *Toxicol. App. Pharmacol. 45*:629-633.

Marsh, C. D., Clawson, A. B., Couch, J. F., and Marsh, H. (1921). Western sneezeweed (*Helenium hoopesii*) as a poisonous plant. *USDA Bull. 947.*

Maruyama, M., and Omura, S. (1977). Carpesiolin from *Carpesium abrotanoides*. *Phytochemistry 16*:782-783.

Mathur, S. B., Tello, P. G., Fermin, C. M., and Mora-Arellano, V. (1975). Terpenoids of *Milcania monagasensis* and their biological activities. *Rev. Latinoam. Quim. 6*:201-205.

McCahon, C. B., Kelsey, R. G., Sheridan, R. P., and Shafizadeh, F. (1973). Physiological effects of compounds extracted from sagebrush. *Bull. Torrey Bot. Club 100*: 23-28.

McGovran, E. R., and Mayer, E. L. (1942). The toxicity of the natural bitter substances, quassin, tenulin, helenalin, and picrotoxin, and some of their derivatives to certain insects. *U.S. Dept. Agric., Bur. Entomol. Plant Quarantine. Entomol. Tech. E-572.*

Mitchell, J. C. (1975). Contact allergy from plants. *Recent Adv. Phytochem. 9*: 119-138.

Nakajima, S., and Kawazu, K. (1980). Coumarin and euponin, two inhibitors for insect development from leaves of *Eupatorium japonicum*. *Agric. Biol. Chem. 44*: 2893-2899.

Norman, J. O., Johnson, J. H., Mollenhauer, H. H., and Meola, S. M. (1976). Effects of sesquiterpene lactones on the growth of *Bacillus thuringiensis*. *Antimicrob. Agents Chemother. 9*: 535-539.

Petersen, R. C., and Kim, H. L. (1976). X-ray structures of hymenoxon and hymenolane: pseudoguaianolides isolated from *Hymenoxys odorata* DC (bitterweed). *J. Chem. Soc. Perkins Trans. 2*, pp. 1399-1403.

Reeb, E. (1910). *Helenium autumnale* et son principe actif. Mulhouse Imprimerie, J. Brinkmann.

Rimington, C., and Roets, G. C. S. (1936). Chemical studies upon the vermeerbos, *Geigeria aspera* Harv. I. Isolation of a bitter principle, "Geigerin." *Onderstepoort J. Vet. Sci. Anim. Ind. 7*: 485-506.

Rimington, C., Roets, G. C. S., and Steyn, D. G. (1936). Chemical studies upon the vermeerbos, *Giegeria aspera*, Harv. II. Isolation of the active principle, "veermeeric acid." *Onderstepoort J. Vet. Sci. Anim. Ind. 7*: 507-520.

Rodriguez, E., Towers, G. H. N., and Mitchell, J. C. (1976). Biological activities of sesquiterpene lactones. *Phytochemistry 15*: 1573-1580.

Rowe, L. D., Dollahite, J. W., Kim, H. L., and Camp, B. J. (1973). *Hymenoxys odorata* (bitterweed) poisoning in sheep. *Southwest. Vet. 26*: 287-293.

Rowe, L. D., Kim, H. L., and Camp, B. J. (1980). The antagonistic effect of L-cysteine in experimental hymenoxon intoxication in sheep. *Am. J. Vet. Res. 41*: 484-486.

Rubinchik, M. A., Rybalko, K. S., Evstratova, R. I., and Konovalova, O. A. (1976). Sesquiterpene lactones of higher plants as a possible source of new antiprotozoal drugs. *Rastit. Resur. 12*: 170-181.

Schmutz, E. M., Freeman, B. N., and Reed, R. E. (1968). *Livestock-Poisoning Plants of Arizona*. University of Arizona Press, Tucson.

Sequeira, L., Kupchan, S. M., and Hemingway, R. J. (1968). Vernolepin: a new reversible plant growth inhibitor. *Science 161*: 789-790.

Shibaoka, H., Shimokoriyama, M., Iruichijima, S., and Tamura, S. (1967). Promoting activity of terpenic lactones in *Phaseolus* rooting and their reactivity toward cysteine. *Plant Cell Physiol. 8*: 297-305.

Smith, C. H., Larner, J., Thomas, A. M., and Kupchan, S. M. (1972). Inactivation of glycogen synthase by the tumor inhibitor vernolepin. *Biochim. Biophys. Acta 276*: 94-104.

Sperry, O. E., Dollahite, J. W., Hoffman, G. O., and Camp, B. J. (1964). Texas plants poisonous to livestock. *Tex. Agric. Exp. Stn. Rep. B-1028.*

Steel, E. G., Witzel, D. A., and Blanks, A. (1976). Acquired coagulation factor X activity deficiency connected with *Hymenoxys odorata* DC (Compositae), bitterweed poisoning in sheep. *Am. J. Vet. Res.* 37:1383-1386.

Szabuniewicz, M., and Kim, H. L. (1972). The pharmacodynamic and toxic action of *Helenium microcephalum* extract and helenalin. *Southwest. Vet.* 25:305-311.

Szabuniewicz, M., Kim, H. L., Rowe, L. D., Bridges, C. H., Dollahite, J. W., Camp, B. J., and Storts, R. W. (1974). The lesions and treatment of poisoning in dogs produced by a lactone isolated from bitterweed (*Hymenoxys odorata* DC). *Southwest. Vet.* 27:238-243.

Terry, M. K., Kim, H. L., Corrier, D. E., and Bailey, E. M. (1981). The acute oral toxicity of hymenoxon in sheep. *Res. Commun. Chem. Pathol. Pharmacol.* 31:181-184.

Towers, G. H. N., Mitchell, J. C., Rodriguez, E., Bennett, F. D., and Subba Rao, P. V. (1977). Biology and chemistry of *Parthenium hysterophorus* L., a problem weed in India. *Biochem. Rev.* 48:65-77.

Unde, N. R., Hiremath, S. V., Kulkarni, G. H., and Kelkar, G. R. (1968). Terpenoids. CXXVII. Preparation of some inaccessible conjugated α-methylene γ-lactones. *Tetrahedron Lett.* 1968:4861-4862.

Vermeulen, N. M. J., Vogelzang, M. E., and Potgieter, D. J. J. (1978). Dihydrogriesenin in *Geigeria aspera* Harv. *Agrochemophysica* 10:1-3.

Vichkanova, S. A., Rubinchik, M. A., and Adgina, V. V. (1971). Antimicrobial activity of sesquiterpene lactones from *Compositae. Tr. Vses. Nauchno.-Issled. Inst. Lek. Rast.* 14:230-308.

Vichnewski, W., and Gilbert, B. (1972). Schistosomicidal sesquiterpene lactone from *Eremanthus elaeagnus. Phytochemistry* 11:2563-2566.

Vichnewski, W., Sarti, S. J., Gilbert, B., and Herz, W. (1976). Goyazensolide, a schistosomicidal heliangolide from *Eremanthus goyazensis. Phytochemistry* 15:191-193.

Waddell, T. G., Austin, A. M., Cochran, J. W., Gerhart, K. G., Hall, I. H., and Lee, K. H. (1979a). Antitumor agents: structure-activity relationships in the tenulin series. *J. Pharm. Sci.* 68:715-718.

Waddell, T. G., Ridley, M. B., Evans, K. D., and Green, M. E. (1979b). Antitumor agents. Constituents of an east Tennessee population of *Helenium amarum* (Raf.) H. Rock. *J. Tenn. Acad. Sci.* 54:103-105.

West, E., and Emmel, M. W. (1952). Poisonous plants in Florida. *Fla. Agric. Exp. Stn. Bull. 510.*

Witzel, D. A., Rowe, L. D., and Clark, D. E. (1974). Physiopathologic studies on acute *Hymenoxys odorata* (bitterweed) poisoning in sheep. *Am. J. Vet. Res.* 35:931-934.

Witzel, D. A., Ivie, G. W., and Dollahite, J. W. (1976). Mammalian toxicity of helenalin, the toxic principle of *Helenium microcephalum* DC (smallhead sneezeweed). *Am. J. Vet. Res.* 37:859-861.

Witzel, D. A., Jones, L. P., and Ivie, G. W. (1977). Pathology of subactue bitterweed (*Hymenoxys odorata*) poisoning in sheep. *Vet. Pathol.* 14:73-78.

Yoshioka, H., Mabry, T. J., and Timmermann, B. N. (1973). *Sesquiterpene Lactones: Chemistry, NMR, and Plant Distribution.* University of Tokyo Press, Tokyo.

18

PEPTIDE TOXINS FROM *AMANITA*

THEODOR WIELAND and HEINZ FAULSTICH

Max-Planck-Institute for Medical Research, Heidelberg, Federal Republic of Germany

I. INTRODUCTION

Fungi of the genus *Amanita* account for nearly all fatal mushroom intoxications in humans. This has prompted chemists, biochemists, toxicologists, and pharmacologists all over the world, but mainly in Europe, to isolate and analyze the biologically active components and to try to understand the mechanisms by which they display their activity. The starting hour was 43 years ago when Lynen and Ulrich Wieland (1938) crystallized the first homogeneous toxin from *A. phalloides* called phalloidin. The next milestone, crystallization of amanitin, by Heinrich Wieland et al. (1941) marked at the same time the approaching end of efforts in Munich until after the war, when research was taken up in the senior author's laboratories. Several review articles describe the progress of chemical, biochemical, toxicological, and molecular biological work: Wieland and Wieland (1959), Wieland (1967, 1968), Wieland and Wieland (1971), Litten (1975), Wieland and Faulstich (1978), and Faulstich et al. (1980a,c). In this chapter, in addition to our present state of knowledge on the chemistry and toxicology of phallotoxins and amatoxins, the virotoxins will be considered, which have been discovered, isolated, and structurally elucidated in recent years (Faulstich et al., 1980b).

Phallolysins, a mixture of three hemolytically active proteins in *A. phalloides* (Seeger et al., 1973) and also under investigation in the author's laboratory (Faulstich and Weckauf-Bloching, 1974), will not be considered. They are labile against acids and heat, and do not contribute to human death cap poisoning.

II. OCCURRENCE, CHARACTERIZATION, AND ANALYSIS OF THE TOXINS

A. Mushroom Species Containing Toxic Peptides

Toxic peptides were detected in several species of the genera *Amanita*, *Galerina*, and *Lepiota* (Table 1). The most convenient source for the toxic peptides described in this chapter is the green death cap, *A. phalloides*. This is true for two reasons: the mushroom is abundant in central Europe, and more recently, also in North America. Second, no other mushroom contains the toxic peptides in such a variety and in such high concentrations as this toadstool. Accordingly, most of the toxic *Amanita* peptides elucidated in structure were isolated from that green species.

A similar spectrum of toxins is apparently also present in the white species *A. verna*, which from chemotaxonomical and morphological similarities may be regarded as a subspecies, *A. phalloides* var. *verna*. However, this mushroom is rare and in the few examples analyzed so far for their toxin composition, the concentration of toxins was in the average lower than in the green species (see Table 2).

White *A. virosa* is an individual species different from *A. verna* both in morphology and chemotaxonomy. For example, all specimens analyzed so far lacked β- and γ-amanitin, most of them lacked the acidic phallotoxins, and several were even devoid of α-amanitin, but had a new toxin, amaninamide, instead (Buku et al., 1980b). *A. virosa* is the only species containing a new group of toxic peptides, the virotoxins (Faulstich et al., 1980b). *A. virosa* is abundant in some parts of Europe (e.g., in central France and Scandinavia) and also in North America.

Some of the toxic *Amanita* spp. described are found exclusively on the North American continent (e.g., *A. bisporigera*, *A. tenuifolia*, and *A. ocreata*). Possibly, some of them are not individual species and may be identical or closely related to others. Chemotaxonomic studies should be used to find out close relationships of species, but so far only *A. bisporigera* has been analyzed (Table 2).

TABLE 1 Fungi Accumulating Amatoxins and Phallotoxins

Amanita phalloides	(Vaill. ex Fr.) Secr.	Lynen and Wieland, 1938; Wieland et al., 1941; Wieland et al., 1948; Wieland and Dudensing, 1956; Wieland and Mannes, 1957; Wieland and Schnabel, 1962; Wieland and Buku, 1968; Faulstich et al., 1973; Faulstich et al., 1975a
A. phalloides var. verna	Bull.	Tyler et al., 1966; Wieland et al., 1966; Seeger and Stijve, 1979
A. virosa	Lamm ex Secr. = *Amanita verna* Bull. ex Fr. Pers. ex Vitt. (Fr. 1821)	Tyler et al., 1966; Faulstich et al., 1974; Yocum and Simons, 1977; S. A. Mahoney and R. W. Holton, personal communication, 1978
A. bisporigera	AtK (possibly = *A. virosa*)	Tyler et al., 1966; Yocum and Simons, 1977; S. A. Mahoney and R. W. Holton, personal communication, 1978
A. tenuifolia	Murr	Tyler, 1963
A. ocreata	Peck	Horgen et al., 1976; Ammirati et al., 1977
A. suballiacea		J. F. Preston et al., personal communication, 1982
Galerina marginata	(Fr.) Kühn	Rinaldi and Tyndailo, 1974
G. autumnalis		Smith, 1974
G. venenata	A. H. Smith	Tyler and Smith, 1963; Tyler et al., 1963
Lepiota brunneoincarnata	Chod. et Mart.	Gerault and Girre, 1975
L. helveola	Bres.	Gerault and Girre, 1975

TABLE 2 Amounts of Amatoxins and Phallotoxins in Poisonous *Amanita*, *Galerina*, and *Lepiota* Species[a]

Species	Σ ama tox.	α	β	γ	Others	Σ phallo tox.	Acidic	Neutral	Method	References
A. *phalloides*[b]	2500	950	1150	100	300	4650	3675	975	Chrom. + spectroph.	Faulstich et al., 1973
	3440	1880	1260	300	-[c]	-	-	-	Chrom. + spectroph.	Andary et al., 1977
	3290	1100	2000	190	-	5260	3550	1710	Chrom. + spectroph.	Yocum and Simons, 1977
	4420	1900	2170	350	-	2400	-	2400	High-perform. thin-layer chrom.	Stijve and Seeger, 1979
	3650	2000	1400	250	-	-	-	-	Chrom. + colorim.	Andary et al., 1979
	4000	-	-	-	-	11000	-	-	Radioimmunoassay	Bodenmüller et al., 1980; N. Stewart and H. Faulstich, unpublished observations, 1979
	2010	650	1200	160	-	-	-	-	High-perform. thin-layer chrom.	Seeger and Stijve, 1980
A. *verna* or *phall.* var. *verna*	450	250	200	-	-	2400	1900	500	Chrom. + spectroph.	Faulstich et al., 1974a
	0-1050	-	-	-	-	-	-	-	RNA-polym. inhib.	Preston et al., 1975
	0	-	-	-	-	0	-	-	Chrom. + spectroph.	Yocum and Simons, 1977
	1550	600	950	-	-	-	-	-	Chrom. + colorim.	Andary et al., 1979
	3210	1550	1290	370	-	-	-	-	High-perform. thin-layer chrom.	Seeger and Stijve, 1980; Seeger and Stijve, 1979
A. *virosa*	1150	1150	-	-	-	3600	1550	2050	Chrom. + spectroph.	Faulstich et al., 1974
	2250	2250	-	-	-	0-470	-	0-470	Chrom. + spectroph.	Yocum and Simons, 1977

Species							Method	Reference
	560	–	–	–	875	875	Chrom. + spectroph.	Andary et al., 1979
	2600	2600	–	–	–	–	Chrom. + colorim.	–
	1860	–	–	–	–	–	Radioimmunoassay	N. Stewart and H. Faulstich, unpublished observations, 1979
A. bispoigera	2400	1800	600	560[d]	1230	1230	Chrom. + spectroph.	Yocum and Simons, 1977
G. marginata	400	400	–	–	–	–	Chrom. + spectroph.	Faulstich et al., 1974
	1100	600	500	–	–	–	Chrom. + colorim.	Andary et al., 1979
	240	–	–	–	–	–	Radioimmunoassay	N. Stewart and H. Faulstich, unpublished observations, 1979
	–	150–440	–	–	–	–	High-perform. thin-layer chrom.	Stijve, 1981
G. autumnalis	1200	–	–	–	–	–	RNA-polym. inhib.	Johnson and Preston, 1976
L. brunneo-incarnata	1300	900	300	100	–	–	Chrom. + colorim.	Andary et al., 1979

[a] All toxins in μg/g dry weight.
[b] Amanin
[c] –, means: not available
[d] Amaninamide

Amatoxins in genera other than *Amanita* were first detected in 1963 in *Galerina* species (Tyler and Smith, 1963). Today, amatoxins have been quantitated in several *Galerina* species (e.g., *G. marginata* and *G. autumnalis*) (Table 2). The concentration of amatoxins in these mushrooms is, on the average, ca. 25% of that of *A. phalloides*. Similar concentrations of amatoxins were found in species of small *Lepiotas*, such as *L. brunneoincarnata* and *L. helveola* (Gerault and Girre, 1975).

The yellowish *A. citrina*, formerly considered as poisonous, does not contain any toxic peptides, but contains bufotenine (5'-hydroxytryptamine), according to Wieland et al. (1953).

1. Distribution of Toxins in Mushroom Tissue

The distribution of the peptides in the fruiting body of *A. phalloides* is not homogeneous, neither with respect to the concentration of the toxins in the various tissues, nor with respect to the ratio amatoxins: phallotoxins. When pileus and stipe are analyzed separately, the amatoxin concentration in the cap was found to be three times higher than in the stipe (Andary et al., 1977). When gills, cap, stipe, and volva were analyzed separately (Bodenmüller et al., 1980), the highest amount of amatoxins was found in the gills, which contain ca. 46% of the total amount of amatoxins. Pileus and stipe represent about 23% each, while the volva contains only 9% of the total amount of amatoxins. This result was recently confirmed by radioimmunoassay (Faulstich et al., 1981c), by which the following concentrations of amatoxins were determined for the different tissues (in milligrams per gram of dry weight): gills 2.6; pileus 1.9; stipe 1.7; and volva 0.6.

The low content of amatoxins in the volva has been corroborated recently by a semiquantitative analysis (S. A. Mahoney and R. W. Holton, personal communication, 1978). Remarkably, however, the phallotoxins are the predominant components in the volva, while in the gills and in the pileus the amatoxins prevail (H. Bodenmüller, H. Faulstich, and S. Zobeley, unpublished observations, 1981).

B. Methods for Identification of Toxins

1. Physical Properties

Description of Compounds: Most of the toxic peptides of *Amanita* mushrooms can be obtained in the crystalline state. Amatoxins crystallize with 3-4 mol, phallotoxins with 5 mol of water:

α-Amanitin: $C_{39}H_{54}N_{10}O_{14}S \cdot 4H_2O$, MW 990
β-Amanitin: $C_{39}H_{53}N_9O_{15}S \cdot 3H_2O$, MW 973
γ-Amanitin: $C_{39}H_{54}N_{10}O_{13}S \cdot 4H_2O$, MW 974
Phalloidin: $C_{35}H_{48}N_8O_{11}S \cdot 5H_2O$, MW 878
Phalloin: $C_{35}H_{48}N_8O_{10}S \cdot 5H_2O$, MW 862
Phallisin: $C_{35}H_{48}N_8O_{12}S \cdot 5H_2O$, MW 894

The other toxins (Tables 5-7) have so far been obtained in the lyophilized form only, including the acidic phallotoxins and the related virotoxins; for example:

Phallacidin: $C_{37}H_{50}N_8O_{12}S$, MW 831
Viroidin: $C_{38}H_{56}N_8O_{15}S$, MW 897

TABLE 3 UV-Absorption Maxima and Molar Extinction Coefficients of the Toxic Peptides of *Amanita* Fungi

Toxin family	Chromophore	Maximum absorbance (nm)	Molar extinction coefficients used for characterization (mol^{-1} liter^{-1})
Phallotoxins	2'-(SR)trp	292	$\varepsilon_{300} = 10.100$
Amanin some virotoxins	2'-(SOR)trp	287	$\varepsilon_{287} = 12.500$
Amatoxins	2'-(SOR),6'-(OH)trp		
pH 7		305	$\varepsilon_{310} = 13.500$
pH 11	(anion)	330	$\varepsilon_{330} = 14.70$
Virotoxins	2'-(SO$_2$R)trp		
pH 7		276	$\varepsilon_{276} = 13.400$
pH 11	(anion)	298	$\varepsilon_{298} = 10.300$

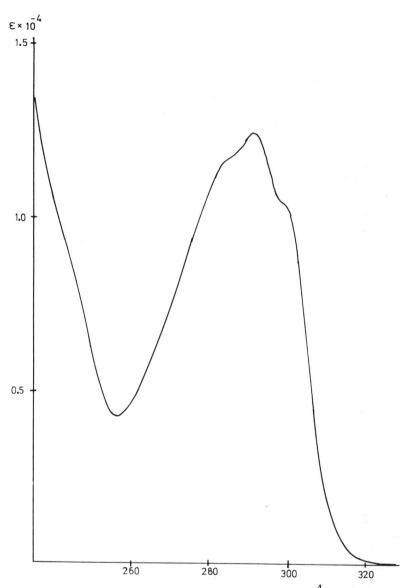

FIGURE 1 Ultraviolet absorption spectrum ($\varepsilon \times 10^{-4}$) of phalloidin (or other phallotoxins) in water.

Spectroscopy

Ultraviolet (UV) absorption: All toxins contain tryptophan, substituted by sulfur functions in position 2'. By the substitution the absorption maximum of tryptophan (around 280 nm) is shifted to longer wavelengths and modulated in a way typical for each toxin group. In the phallotoxins (Fig. 1) it is located around 290 nm, in the amatoxins the 6'-hydroxy group causes a

further shift to longer wavelengths of about 10 nm (Fig. 2), while oxidation
at the sulfur atom (e.g., in the sulfone moiety of virotoxins) (Fig. 3) shifts
the absorption maximum to shorter wavelengths. Two groups of toxins, the
amatoxins and the virotoxins, form anions in alkaline solutions (Figs. 2 and
3). UV absorbance can be used to characterize the different types of toxins
by the typical form of their UV spectra in neutral or alkaline solution. Some
of the spectra are presented in Figs. 1 to 3. For quantitation, the molar
extinction coefficients can be used, which for some characteristic maxima are
listed in Table 3. The maxima at longest wavelengths, for phallotoxins ($\sim$300
nm) and amatoxins ($\sim$310 nm), are located at wavelengths beyond the usual
absorption maximum of proteins. Accordingly, these toxins can be determined also in the presence of, or after attachment to, proteins.

Circular Dichroism (CD) spectra, Nuclear Magnetic Resonance (NMR), and
x-ray structure analysis: Valuable characteristics are the optical rotatory dispersion (ORD) and CD spectra of the toxins. The inherently dissymmetric
chromophore is, at least in the bicyclic peptides, fixed in the peptide ring
and can be used as a probe for conformational changes in the backbone.
Accordingly, in the amatoxin and phallotoxin series, all biologically active
components exhibit unique ORD or CD spectra (Figs. 4 and 5), while most
of the inactive monocyclic or linear derivatives do not. Most probably they
have lost their biological activity by adopting a different conformation, which

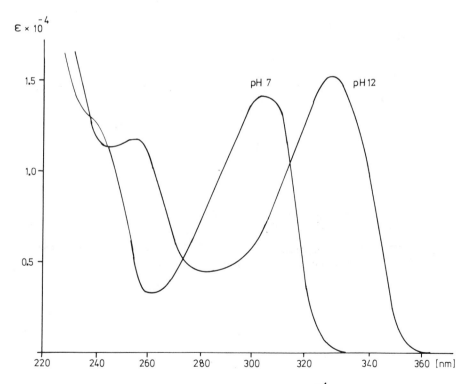

FIGURE 2 Ultraviolet absorption spectrum ($\varepsilon \times 10^{-4}$) of α-, β-, γ-amanitin in
water at pH 7 and pH 12 (abstraction of the 6'-OH proton).

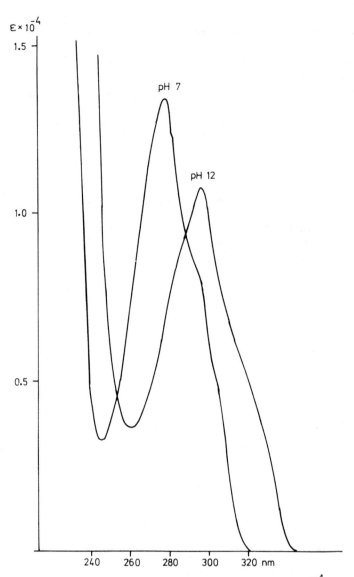

FIGURE 3 Ultraviolet absorption spectrum ($\varepsilon \times 10^{-4}$) of viroidin in water of pH 7 and pH 12 (abstraction of the indole NH proton).

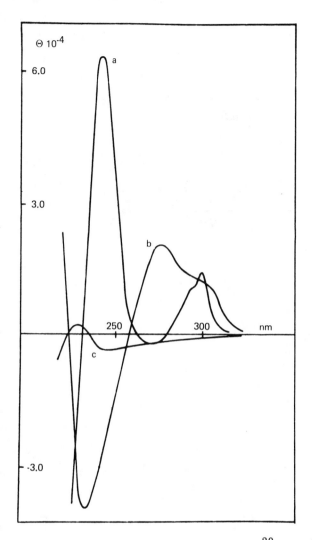

FIGURE 4 Circular dichroism spectra ($\theta^{20} \times 10^{-4}$) in water of (a) phalloidin, (b) *seco*-phalloidin, and (c) dethiophalloidin.

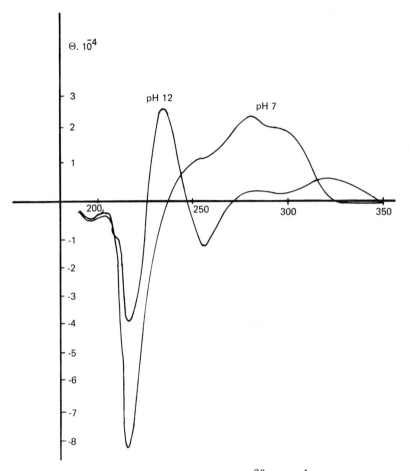

FIGURE 5 Circular dichroism spectra ($\theta^{20} \times 10^{-4}$) of α-amanitin in water at pH 7 and pH 72.

no longer fits into the binding site of the target proteins. A few inactive peptides still have the conformation of active compounds. In these cases functional groups (e.g., hydroxy groups) could be pinpointed, which must be essential for binding the toxins to their target proteins (see Sec. IV,A,2; IV,B,4. By ^{1}H-NMR analysis the active conformations of phalloidin (Fig. 6) (Patel et al., 1973) and α-amanitin (Tonelli et al., 1978) were elucidated in dimethylsulfoide solution. In the case of β-amanitin, crystals could be obtained which were suitable for x-ray analysis (Kostansek et al., 1977, 1978). The conformation of the amatoxin in the crystal is closely similar to that of α-amanitin in solution. It is shown in Fig. 7.

2. Chemical Reactions

Color reactions: Several color reactions have been described to visualize the toxins on paper or silica thin layers, even in crude mushroom extracts. Most commonly used is their reaction with cinnamaldehyde, developed in the

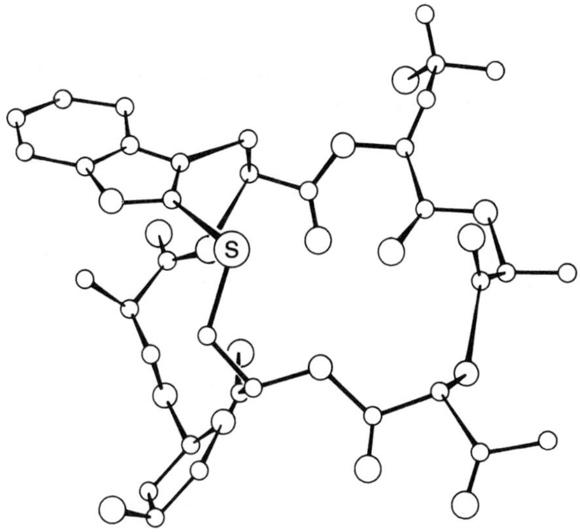

FIGURE 6 Spatial structure of phalloin designed from NMR data. Thioether moiety is plus-helical.

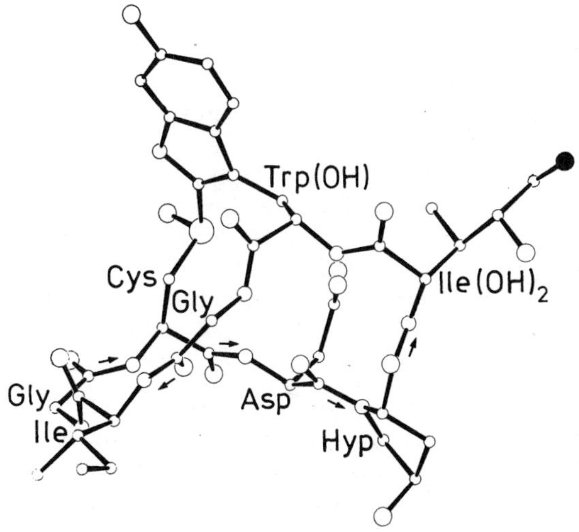

FIGURE 7 Spatial structure of β-amanitin revealed by x-ray analysis. (From Kostansek et al., 1977.)

vapor of hydrochloric acid (Wieland et al., 1949). Amatoxins give a bright purple color, whereas phallotoxins stain brownish, turning to a light blue after a few minutes. The limit of detection is 0.05 and 0.5 µg, respectively. The reaction with amanin or with the sulfoxides among the virotoxins is much retarded, yielding a faint blue color, while the virotoxins as sulfones do not react. They can be detected, like all the other toxins, by their strong absorbance of UV light on thin layers doped with fluorescent dyes.

A very simple and convenient staining method makes use of unsaturated aldehydes obviously present in or formed from lignin in crude papers. After drying a drop of amatoxin-containing solution on a newspaper, a blue color develops on contact with 10 N hydrochloric acid in about 10 min. The limit of detection is ca. 20 µg of amatoxins per milliliter. Other reagents for the toxic peptides are croconic acid (Faulstich et al., 1975a), which gives a wine-red color with amatoxins and a yellow color with phallotoxins, as well as Ehrlich's reagent, which stains amatoxins red.

3. Separation Techniques

In the early phase of *Amanita* research, partition of toxic peptides between two phases, mainly *n*-butanol-water, played an important role. Today this method has been abandoned and substituted by chromatographic procedures.

Chromatography

Thin-layer chromatography: Many efforts have been made to detect amatoxins in crude mushroom extracts by a single-step chromatography on thin-layer plates. Although brown water-soluble components often badly disturb the chromatography, efficient separation can be achieved with some solvents listed in Table 4. The most sensitive identification of the toxins has been established recently by high-performance thin-layer chromatography (Stijve and Seeger, 1979).

Liquid chromatography: A very effective separation of the various toxins, also at preparative scale, is achieved by adsorption chromatography on Sephadex LH-20 with water as solvent (Faulstich et al., 1973). If long columns are used (2.50 m), a single run separates acidic phallotoxins from acidic amatoxins, neutral phallotoxins, α-amanitin, and γ-amanitin, which can be obtained with about 80% purity (Fig. 8). A further resolution of toxin families, whose members differ only by single hydroxy groups, can be achieved by high-pressure liquid chromatography. Using the reverse-phase technique, a single run, for example, separated the six virotoxins into five peaks, with complete resolution of the main components (Faulstich et al., 1980b). The same technique has also been used to determine the different toxins in crude extracts of mushrooms by calibrating the areas of the toxin peaks (Yocum and Simons, 1977).

Paper chromatography: At the time when the toxic peptides were elucidated in structure (Wieland and Wieland, 1972) chromatography on special filter paper was of highest value. Today the technique is nearly abandoned, but it is worthwhile to note that the method is remarkably efficient, especially for the fine analysis of prepurified mixtures of toxic peptides.

TABLE 4 Solvent Mixutres for the Chromatography of *Amanita* Peptides on Silica Thin Layers or Paper (all ratios are v/v)

Solvent	Reference	Remarks
Ethylene glycol monobutyl ether/ammonia (25%) (7:3)	Palyza, 1972	Especially for acidic toxins
sec-Butanol/ethyl acetate/water (14:12:5)	Faulstich et al., 1974	
Chloroform/methanol/water (65:25:4)	A. Buku, personal communication, 1978	
sec-Butanol/ammonia (3%) (100:44)	Faulstich et al., 1974	Especially for acidic toxins
Chloroform/methanol/acetic acid (100%)/water (75:33:5:7.5)	Andary, et al., 1977	
Butanone/acetone/water (30:3:5)	Wieland and Boehringer, 1960	Especially for paper chromatography

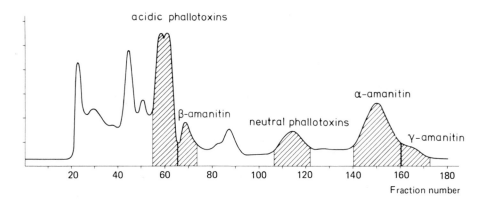

FIGURE 8   Elution diagram of a chromatographic separation of an extract of *Amanita phalloides* on Sephadex LH-20 with water.

4. *Biological Assays*

For the detection of very small amounts of the toxic peptides, macro-molecules with high affinity for amatoxins or phallotoxins ($K_D = 10^{-8}$ M) were used. Assays were established using either the target proteins or immuno-globulin antibodies against the two toxin families. In general, the assays determine the total amount of a toxin family because there is large-scale cross-reaction between the members of one toxin family. However, no cross-reactivity has ever been reported for amatoxin-binding proteins with phallo-toxins, or vice versa.

Reactions with target proteins: RNA polymerases II have been used to determine the concentration of amatoxins in carpophores (Preston et al., 1975), and recently also for medical use (Brown and Garrity, 1980). Using rabbit muscle actin, the phallotoxins could be measured down to concentra-tions of 160 μg/ml (Schäfer and Faulstich, 1977). More convenient is the use of immunoglobulins, antibodies against amatoxins (Faulstich et al., 1975b; Fiume et al., 1975) and phallotoxins (Faulstich et al., 1980a). The highest sensitivity in these assays was ca. 0.5 ng of amatoxin per milliliter and ca. 2 ng of phallotoxin per milliliter. To meet the request of clinics for a simple and rapid assay for amatoxins by which cases of *Amanita* poisoning can be diagnosed immediately after hospitalization, a radioimmunoassay for ama-toxins on nylon support has been developed which allows detection of about 3 ng of amatoxins per milliliter (e.g., in the urine of a patient) within 2 hr (Faulstich et al., 1982).

III. CHEMISTRY OF THE PEPTIDES

A. Phallotoxins

1. *Structure and Conformation*

The first phallotoxin isolated from *A. phalloides* in 1937 (Lynen and Wie-land, 1938) was named phalloidin. It is the main representative of the phallo-toxins upon which the elucidation of the general structure of the phallotoxins has been performed (Wieland and Schön, 1955). Thereafter, seven additional phallotoxins have been isolated from *Amanita* mushrooms and characterized. They are compiled in Table 5. The spatial structure of phalloidin as determined by ^{1}H-NMR analysis (Patel et al., 1973) is shown in Table 5.

As shown in Fig. 6, all the phallotoxins are derived from the same cyclic peptide backbone, consisting of seven amino acids and cross-linked by a 2'-indolyl thioether bridge. There are two groups of phallotoxins: neutral and acidic. These (phallacin, phallacidin, phallisacin) contain D-erythro-α-amino-β-hydroxy-succinic acid (D-β-hydroxyaspartic acid) in position 2 instead of D-threonine in the neutral toxins (phalloin, phalloidin—the main component, phallisin, and prophalloin). A second fundamental difference between the two groups is the occurrence of L-valine in position 1 of the acidic peptides instead

TABLE 5 Structural Formulas of the Naturally Occurring Phallotoxins and of Demethylpalloin, and LD$_{50}$ Values (mg/kg, white mouse)

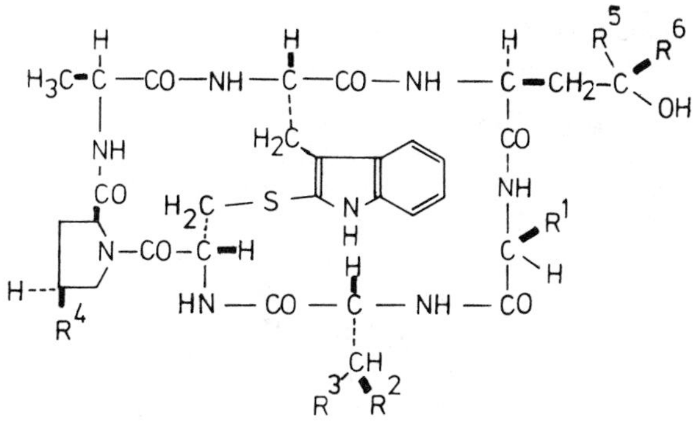

Name	R^1	R^2	R^3	R^4	R^5	R^6	LD$_{50}$
Phalloin (PHN)	CH$_3$	CH$_3$	OH	OH	CH$_3$	CH$_3$	1.5
Phalloidin (PHD)	CH$_3$	CH$_3$	OH	OH	CH$_2$OH	CH$_3$	2
Phallisin (PHS)	CH$_3$	CH$_3$	OH	OH	CH$_2$OH	CH$_2$OH	2
Prophalloin (PPN)	CH$_3$	CH$_3$	OH	H	CH$_3$	CH$_3$	>20
Phallacin (PCN)	CH(CH$_3$)$_2$	OH	CO$_2$H	OH	CH$_3$	CH$_3$	1.5
Phallacidin (PCD)	CH(CH$_3$)$_2$	OH	CO$_2$H	OH	CH$_2$OH	CH$_3$	1.5
Phallisacin (PSC)	CH(CH$_3$)$_2$	OH	CO$_2$H	OH	CH$_2$OH	CH$_2$OH	4.5
Demethylphalloin (DMPHN)	CH$_3$	CH$_3$	OH	OH	CH$_3$	H or ^{3}H	2

of L-alanine in the neutral ones. Common features are L-alanine in position 5, and the thioether moiety linking the cysteine residue with the 2'-position of the indole nucleus of L-tryptophan denoted as tryptathionine. The inherently dissymetric P-helical chromophore is responsible for the maximum of UV absorption around 290 nm (Fig. 1) and for the positive Cotton effects in this wavelength region, in the CD spectrum (Wieland et al., 1981b) (Figs. 4 and 6). Common also is an L-proline that bears a cis-standing hydroxyl group (allohydroxyproline) in all but one case, and a L-γ-hydroxyleucine in position 7 which additionally can be hydroxylated once in the δ-position or twice (in the δ,δ'-position).

The exact name of the bis-hydroxylated leucine which has a second chiral center is (2S,4R)-2-amino-4,5-dihydroxyisocaproic acid (Wieland and Schöpf, 1959). The toxicity values (LD50) per kilogram of the white mouse after intraperitoneal or intravenous injection are also given in Table 5. Prophaloin, which lacks the hydroxyl group at the proline ring, is not toxic in doses up to 50 mg/kg. None of the toxins has proven lethal after peroral administration in mice or rats. Included in Table 5 is a phallotoxin not occurring in nature, demethylphalloin, which has been obtained by oxidative degradation of side chain 7 of phalloidin (Wieland and Schöpf, 1959) and subsequent hydrogenation of the keto compound so formed (Wieland and Rehbinder, 1963).

Radioactive phallotoxins: Radioactively labeled phallotoxins have been obtained by introducing ^{3}H or ^{35}S into appropriate precursors.

^{3}H-Demethylphalloin: This compound, included in Table 5, was obtained from ketophalloidin by hydrogenation using ^{3}H-containing NaBH$_4$ (Puchinger and Wieland, 1969a). As it behaves pharmacokinetically almost like phalloidin, it could be utilized in numerous biological experiments (see Sec. IV,A,2 and 3).

$$-CH_2-\underset{\underset{CH_2-OH}{|}}{\overset{\overset{OH}{|}}{C}}-CH_2 \xrightarrow{IO_4^-} -CH_2-\overset{\overset{O}{||}}{C}-CH_3 \xrightarrow{[^3H]NaBH_4} -CH_2-\underset{\underset{H}{\overset{|}{3}}}{\overset{\overset{OH}{|}}{C}}-CH_3$$

Side chain 7   Keto-PHD   Demethyl-PHN

^{35}S-Phallodithiolan: By reaction of keto-PHD with ^{35}S-containing ethanedithiol, Wieland and Rehbinder (1963) prepared a radioactive dithiolane of similar toxicity as phalloidin.

$$\text{Ketophalloidin} + \text{HS-}\overset{*}{C}H_2\text{-}\overset{*}{C}H_2\text{-SH} \longrightarrow -CH_2-\underset{\underset{*S\ S*}{/\backslash}}{C}-CH_2$$

Side chain 7

2. Correlation of Toxicity and Structural Features of the Phallotoxins

The structural features important for the toxicity (and a strong binding to actin, see Sec. IV,B,2) of the phallotoxins have been recognized by chemical modifications and by synthesis of numerous analogs. It is not possible to enumerate all of them here, but we will mention only the most illustrative ones.

Dependence on the molecular shape: The bicyclic system of the phallo-toxins can be converted to a monocyclic one by two reactions:

Seco compounds: Since the peptide bond between the γ-hydroxylated side chain 7 and amino acid 1 is split preferentially on treatment with mild acids (water-free trifluoroacetic acid, 2 hr at room temperature), monocyclic seco compounds can be obtained. They proved nontoxic. The easy splitting of one peptide bond is due to the tendency to form a γ-lactone.

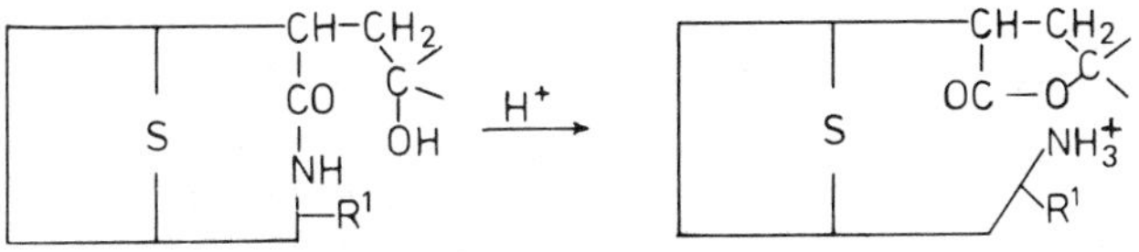

The alteration of the shape of the bicyclic peptide as a consequence of the opening of one ring can be observed readily by circular dichroism measurements. All of the bicyclic, toxic parent compounds show a characteristic CD curve with positive Cotton effects around 240 and 300 nm, whereas the seco compounds exhibit only one positive Cotton effect around 270 nm and a negative one in the short-wavelength region (Fig. 4).

Dethio compounds: The sulfur atom forming the bridge in the bicyclic peptides is removed hydrogenolytically on treatment with Raney nickel. The dethio compounds so formed are also completely nontoxic.

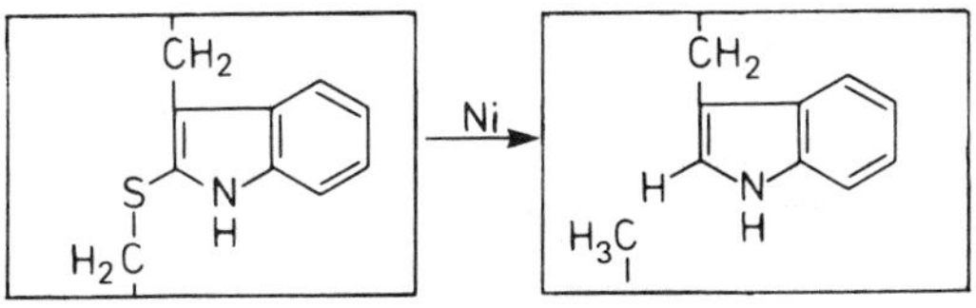

The removal of the sulfur bridge also causes a strong change of the molecular shape as evident from the CD spectrum (Fig. 4).

Influence of the side chains: As results of chemical degradation and modification experiments at the side chain 7 of phalloidin and of syntheses of many analogs (for a review, see Munekata, 1981), the contributions of the diverse side chains to the toxicity (and binding strength to actin) can be summarized as follows:

Side chain 1: The methyl group (alanine) of phalloidin, for example, can be replaced by an isopropyl group (valine) as in the acidic phallotoxins, without any diminution of toxicity. However, the absence of a methyl group as in Gly[1]-phalloidin (Munekata et al., 1977a) leads to a significant loss of toxicity.

Side chain 2: The D-threonyl side chain can be replaced by that of β-hydroxyaspartic acid (as in the acidic toxins) or even by that of D-α-aminobutyric acid without loss of toxicity. Derivatization of the carboxyl group as in phallacidin, for example, leads to a diminution of the toxicity: LD_{50} of monomethylamide 6.3, dimethylamide 8.3 mg/kg in the white mouse (Faulstich et al., 1975a). A fluorescent ethylenediamide (with a 7-nitrobenz-2-oxa-1,3-diazole group) still has some affinity to F-actin (Barak et al., 1980).

D-Alanine instead of D-threonine in position 2 leads to the disappearance of the affinity to actin of the toxin (Wieland et al., 1982).

Side chain 4: As nature shows with the nontoxic prophalloin, the cis standing hydroxyl group in 4'-position of the L-proline ring is essential. A synthetic product containing 4'-*trans*-hydroxy-L-proline was nontoxic.

Side chain 5: A methyl group (alanine) in this position is essential. A synthetic Gly^5-phallotoxin was not toxic and showed minimal affinity to actin (Munekata et al., 1977a).

Side chain 7: The hydroxyl groups of this side chain are permissible and the number of the carbon atoms can be reduced to at least three (norphalloin, Wieland and Jeck, 1968) without loss of toxicity (see also demethylphalloin). Side chain 7 must possess the L-configuration. A synthetic phalloin which had γ-hydroxy-D-leucine in position 7 was not toxic up to 30 mg/kg in the white mouse (Munekata et al., 1977b).

Fluorescent phallotoxins. Side chain 7 can also be utilized for introduction of an amino group via dithiolanamine to provide fluorescing derivatives by reaction with fluoresceine isothiocyanate (Wieland et al., 1980) or rhodamine isothiocyanate (Faulstich et al., 1983). Both are suitable for visualizing F-actin by fluorescence microscopy (Wulf et al., 1979). The reaction of the iso-thiocyanates with δ-aminophalloin, obtained from Wieland and Rehbinder's (1963) δ-tosylphalloidin, with ammonia, also yields useful fluorescent phalloidins (Wieland et al., 1983).

$$-CH_2-\overset{\overset{\displaystyle CH_3}{|}}{C}=O + H_2\overset{\overset{}{\underset{\underset{\displaystyle SH}{|}}{C}}}{} - \overset{}{\underset{\underset{\displaystyle SH}{|}}{CH}}-CH_2NH_2 \longrightarrow -CH_2-\overset{\overset{\displaystyle CH_3}{|}}{C}\overset{S-CH_2}{\underset{S-CH-CH_2NH_2}{|}}$$

$$-CH_2-\overset{\overset{\displaystyle CH_3}{|}}{\underset{\underset{\displaystyle OH}{|}}{C}}-CH_2OTos + NH_3 \longrightarrow -CH_2\overset{\overset{\displaystyle CH_3}{|}}{\underset{\underset{\displaystyle OH}{|}}{C}}-CH_2-NH_2 \qquad \text{or}$$

$$+ \quad \begin{array}{c}\text{fluorescent} \\ \text{isothiocyanates}\end{array} \longrightarrow \begin{array}{c}\text{fluorescent} \\ \text{phallotoxins}\end{array}$$

For the total synthesis of phalloin, see Munekata et al., 1977b.

B. Virotoxins

1. *Structure and Nomenclature*

The virotoxins discovered as components of the mushroom *A. virosa* produce the same toxicological symptoms as the phallotoxins. The elucidation of the structure of viroidin surprisingly led to the fact that the virotoxins are not bicyclic peptides (as the phallotoxins are) but monocyclic hepta-peptides (Faulstich et al., 1980b). The general formula, the variations of side chains leading to the six different virotoxins, and their proportions are given in Table 6. As shown in the formula in Table 6, there exist in the amino acid compositions some common features with the phallotoxins: alanine

TABLE 6 Structural Formulas of the Virotoxins

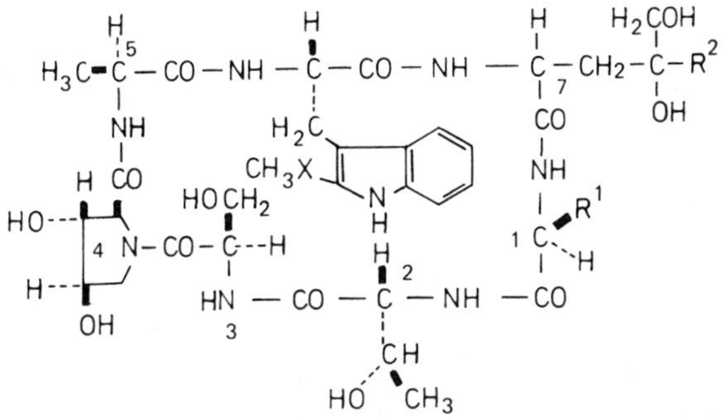

Name	X	R^1	R^2	Percent of total
Viroidin	SO_2	$CH(CH_3)_2$	CH_3	18
Desoxoviroidin	SO	$CH(CH_3)_2$	CH_3	4
Ala^1-viroidin	SO_2	CH_3	CH_3	10
Ala^1-desoxoviroidin	SO	CH_3	CH_3	
Viroisin	SO_2	$CH(CH_3)_2$	CH_2OH	49
Desoxoviroisin	SO	$CH(CH_3)_2$	CH_2OH	19

(respectively, valine) in position 1, D-threonine (position 2), L-alanine (position 5), and L-leucine containing a varying number of hydroxy groups in position 7. Amino acids different from those of phallotoxins are: D-serine instead of the L-cysteine as part of tryptathionine in position 3: 2,3-*trans*-3,4-*trans*-3,4-dihydroxy-L-proline (formula A) instead of 2,4-*cis*-4-hydroxy-L-proline in position 4, and 2'-(methylsulfinyl)-L-tryptophan or 2'-(methylsulfonyl)-L-tryptophan instead of the tryptophan part of tryptathionine in position 6. The 3,4-dihydroxyproline (A) was unknown before its detection as a component of the virotoxins (Buku et al., 1980a), while the diastereomeric 2,3-*cis*-3,4-*trans*-3,4-dihydroxy-L-proline (B) was recognized as a component of a protein in diatoms several years ago (Nakashima and Volcani, 1969).

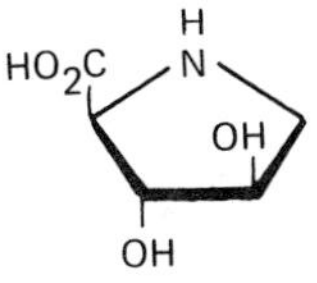

A 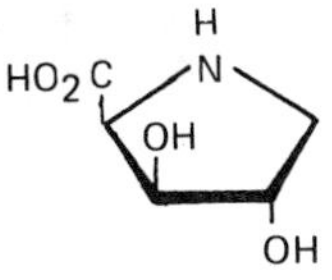   B

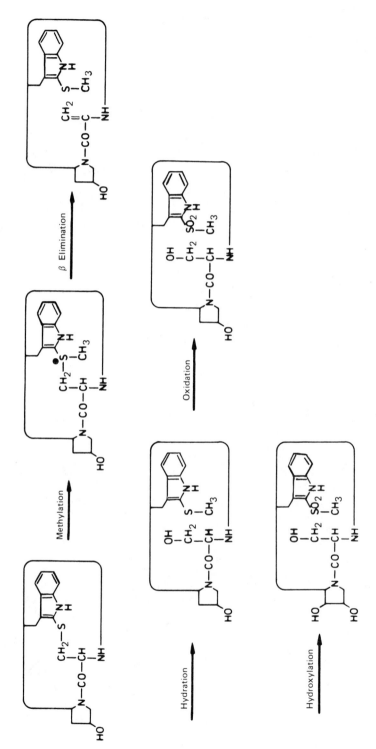

FIGURE 9   Plausible biochemical transformation of phalloidin to viroidin.

2. Presumptive Derivation from Phallotoxins

The nomenclature of the virotoxins corresponds to that of the phallotoxins. The ending "-din" denotes a twofold hydroxylated side chain of residue 7 (e.g., phalloidin, viroidin), and "-sin" a threefold hydroxylated one (e.g., phallisin, viroisin). It is highly probable that the virotoxins are derived from the phallotoxins or from a common precursor molecule. A plausible biochemical transformation of phalloidin to viroidin is formulated in Fig. 9. The toxicity of the virotoxins corresponds to that of the phallotoxins (LD$_{50}$ ∿2 mg/kg, white mouse). They also bind strongly to F-actin. This is surprising since a similar monocyclic structure in the group of the phallotoxins as in their dethio compounds does not show any affinity to the protein. Features different from dethiophalloidin which could be responsible for such a strong interaction with F-actin are D-serine instead of L-alanine, the additional *trans*-hydroxyl group in the 3'-position of allo-hydroxy-L-proline, and the methylsulfinyl group at the tryptophan moiety. A methyl sulfinyl group has been introduced into dethiohalloidin by A. Buku and J. U. Kahl (personal communication, 1980); however, the product exhibited no affinity for F-actin.

C. Amatoxins

The amatoxins are by far more potent poisons than the phallotoxins. All symptoms described for poisoning by *Amanita* phalloides can be produced by pure amatoxins. It is concluded, therefore, that amatoxins are the sole cause of fatal human poisoning. For a recent review, see Faulstich (1979, 1980a).

1. Structure and Conformation

The amatoxins are bicyclic octapeptides. The nine individuals isolated so far, of which α-amanitin is the most frequently studied, are derived from one parent molecule and differ only by the number of hydroxyl groups and by an amide *versus* carboxyl function (Table 7). The amatoxins consist exclusively of L-amino acids and glycine. Uncommon building blocks are the bridging 6'-hydroxytryptathionine-(R)-sulfoxide (numbers 4-8), which is responsible for the UV absorption at about 300 nm (Fig. 2) and for the positive Cotton effects in the same region of the circular dichroism spectrum (Fig. 5). On addition of alkali the UV maximum is shifted by 30 nm to longwave lengths as a consequence of phenolate ion formation at the 6'-hydroxyl group. This is also expressed by the CD spectrum (Fig. 5). The OH group is also responsible for the very intensive color reaction with cinnamaldehyde plus hydrochloride vapors (deep violet). In position 3 is an L-isoleucine which is monohydroxylated in the γ position (γ-amanitin, ε-amanitin) or bishydroxylated in the γ and δ positions (α-amanitin, β-amanitin). The exact stereochemistry of this side chain is (2S,3R,4R)-2-amino-3-methyl-4,5-dihydroxyvaleric acid (Gieren et al., 1974). Amanullin has a normal L-isoleucine side chain instead. Amanin and amaninamide differ from β-(respectively, α-) amanitin only by the lack of the phenolic 6'-hydroxyl group at the tryptophan moiety. All of the amatoxins are (R)-sulfoxides. For the conformation of the amatoxins in solution and crystal, see Sec. II, B, 1.

TABLE 7 Structural Formulas of the Amatoxins

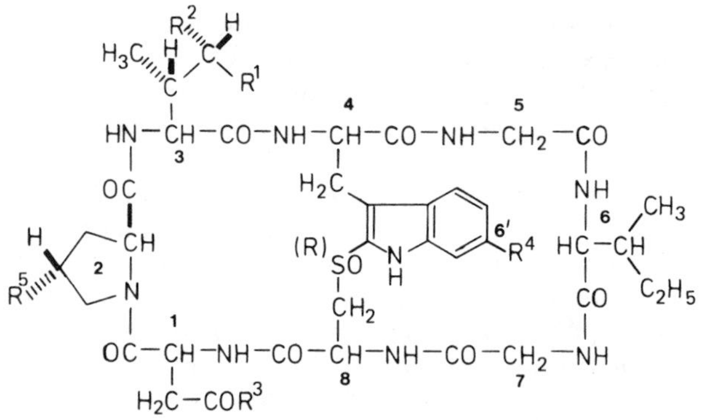

Name	R^1	R^2	R^3	R^4	R^5	LD$_{50}$ (mg/kg, white mouse)
α-Amanitin	CH_2OH	OH	NH_2	OH	OH	0.3
β-Amanitin	CH_2OH	OH	OH	OH	OH	0.5
γ-Amanitin	CH_3	OH	NH_2	OH	OH	0.2
ε-Amanitin	CH_3	OH	OH	OH	OH	0.3
Amanin	CH_2OH	OH	OH	H	OH	0.5
Amanin amide[a]	CH_2OH	OH	NH_2	H	OH	0.3
Amanullin	CH_3	H	NH_2	OH	OH	>20
Amanullinic acid	CH_3	H	OH	OH	OH	>20
Proamanullin	CH_3	H	NH_2	OH	H	>20

[a]In *A. virosa* only.

2. *Structure and Toxicity*

Molecular shape: A preferential cleavage of one peptide bond (between the γ-hydroxylated isoleucine and the tryptophan moiety) is achieved under mild acidolytic conditions yielding monocyclic, nontoxic seco compounds from both the phallotoxin and amatoxin series. Hydrogenolytic removal of the sulfoxide bridge occurs by treatment of amatoxins with Raney nickel, the monocyclic dethio products being nontoxic. Since both reactions are accompanied by a change of conformation, one must conclude that the shape of the molecule plays an important role in toxicity. The S-deoxo compounds bicyclic thioethers obtained as intermediates by deoxygenation with Raney nickel or with K_3MoCl_6 are as toxic as the parent (R)-sulfoxides. This is also true for the sulfone obtained from 6'-methoxy-α-amanitin by oxidation with an excess of peroxyacetic acid (Buku et al., 1974).

Contributions of the side chains: Insight into the role of side chains has been gained by chemical modifications of native amatoxins and by synthesis of several analogs. Since the toxic effect of the amatoxins is intrinsically connected with their inhibitory capacity of DNA-dependent RNA polymerase II (or B) of eukaryotic cells (see p. 622), structure-activity studies can clearly be performed in vitro by comparing the inhibitory constants K_i of the amatoxins to be investigated with purified enzyme preparations (e.g., from calf thymus) (Cochet-Meilhac and Chambon, 1974 and Wieland et al., 1981a).

Naturally occurring variants: Two members of the amatoxin family give information on the contribution to toxicity of hydroxyl groups: the OH group of hydroxyproline (side chain 2) is essential, as proamanullin which contains proline instead of hydroxyproline is nontoxic and has practically no affinity to RNA polymerase II. In contrast, the OH group in position 6' of the tryptophan moiety is not important, as amanin which lacks this group is as toxic as all amatoxins. From the spatial structure (Fig. 7) it appears that the indole part of the molecule protrudes out of that region which bears the groups specific for binding to the enzyme. The nonparticipation of the indole part permits the introduction of various residues without seriously affecting the inhibitory capacity of the amatoxin moiety (see below).

Chemical modifications: Accessible for chemical modifications are the side chains of aspartic acid (number 1) in β-amanitin, of γ,δ-dihydroxyisoleucine (number 3) in α-amanitin, and the 6'-OH group of the indolyl side chain of residue 6.

1. *Side chain 1.* The aspartic acid side chain of β-amanitin was derivatized as methyl ester, thiophenyl ester, and as amides of various structures (Wieland and Boehringer, 1960), all derivatives remaining toxic. The carboxylic group provided a site of attachment of amatoxin to serum albumin, yielding toxic conjugates (Cessi and Fiume, 1969).

2. *Side chain 3.* The γ,δ-dihydroxyisoleucine residue of the 6'-O-methyl ether of α-amanitin was oxidized with periodate, IO_4^-, into an aldehyde (O-methylaldoamanitin) which is nontoxic. By hydrogenation with $NaBH_4$ an alcoholic function is generated which restores a part of the toxicity (about 10%, O-methyldehydroxymethyl-α-amanitin, γ-hydroxyvaline-7 analog).

α-amanitin	aldo compound	dehydroxymethyl-
100% toxic	nontoxic	10% toxic
100% affinity	1% affinity	25% affinity

Removal of the hydroxy group of the γ-hydroxyvaline analog causes a drop of the affinity to less than 1%. So one can conclude that to produce strong inhibition of the polymerase, side chain 3 must contain at least four carbon atoms, with branched β position, and at least one hydroxyl group in the γ position.

3. Biologically Important Derivatives

Radioactively labeled amatoxins: As Table 8 shows, radioactive atoms have been introduced mostly into the indole nucleus of amatoxins, or into side chain 3. Using the hydrogenation of 6'-O-methylaldoamanitin (Wieland and Fahrmeir, 1970), one tritium atom was firmly bound. [^{14}C]Methyl or [^{3}H]methyl was introduced as an ether into the 6'-hydroxy group and, concomitantly, the indole nitrogen was substituted in some cases. An N-phenyltetrazolyl ether of β-amanitin was hydrogenolyzed by introducing ^{3}H into the 6'-position of the indole ring. Finally, ^{125}I iodination of α-amanitin yielded a 7'-iodo derivative. The radioactive derivatives are summarized in Table 8.

6'-Ethers of α-amanitin

Ether of monocaproyl-ethylenediamine [6'-O(CH$_2$)$_5$CONHCH$_2$-CH$_2$NH$_2$]: For the preparation, see Faulstich et al. (1981b). The amanitin derivative was used to attach the toxin to an activated gel matrix or, after succinylation and activation, to several proteins.

Fluorescent amatoxin: FAMA was obtained by reaction of the amino derivative (see above) with fluoresceine isothiocyanate (Wulf et al., 1980).

Biotin-amatoxin: Similarly, after reaction with activated β-biotin, a conjugate was obtained with high affinity to both avidin and RNA-polymerase II, which could be used for affinity chromatography (Faulstich, unpublished results).

Macromolecular conjugates: Prior to coupling via 6'-ethers, Cessi and Fiume (1969) used carbodiimide for coupling β-amanitin by its carboxyl group to proteins. In our laboratory amatoxins have been conjugated to macromolecular compounds not only via 6'-ethers but also by an azo linkage (Faulstich and Trischmann, 1973). For enhanced toxicity of the conjugates, see Cessi and Fiume (1969).

D. Antamanide

1. Structure and Properties

Besides toxic peptides, *A. phalloides* contains a number of nontoxic monocyclic peptides, cycloamanides (Gauhe and Wieland, 1977), among them one cyclic decapeptide which protects the white mouse from death by phalloidin (Wieland et al., 1969; review: Wieland, 1972; Burgermeister and Winkler-Oswatitsch, 1977) (see Fig. 10). In water-free solvents antamanide forms rather strong complexes with Li$^+$, Na$^+$, and Ca^{2+} but is not an ionophore like gramicidine S. A great number of analogs has been obtained by synthesis. By comparing their antitoxic capability, the following molecular features have been recognized as essential:

Two pairs of proline in the 2,3- and 7,8-positions.
Amino acid 2 (valine) must be lipophilic.
Amino acid 4 (alanine) must contain at least one methyl, can be longer.
Amino acids 9 (phenylalanine) and 10 may not be replaced, not even by
    tyrosine or hexahydrophenylalanine, whereas the phenylalanines No.
    5 and 6 may be replaced by various amino acids.

TABLE 8 Radioactively Labeled Amatoxins

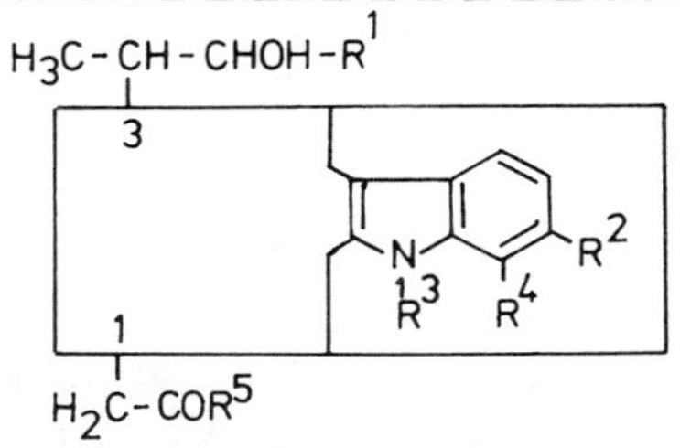

Name	R^1	R^2	R^3	R^4	R^5	Specific activity of preparation (Ci/mol)	Notes	References
[^{3}H]O-Methyldehydroxy-methyl-α-amanitin	[^{3}H]	OCH_3	H	H	NH_2	3.8	K_i 3 times higher than K_i of α-amanitin	Wieland and Fahrmeir, 1970
[^{14}C]O-Methyl-γ-amanitin	CH_3	$O[^{14}C]H_3$	H	H	NH_2	0.07	Labile, partly removable by microsomal oxidation	Govindan, 1969
[^{3}H]O-Methyl-α-amanitin	CH_2OH	$OC[^3H]_3$	H	H	NH_2	4.5		Faulstich et al., 1981b
[^{3}H]Dimethyl-α-amanitin	CH_2OH	$OC[^3H]_3$	$C[^3H]_3$	H	NH_2	8.3	K_i $\sim$2 times higher than K_i of α-amanitin	Faulstich et al., 1981
[^{3}H]-6'-amanin	CH_2OH	[^{3}H]	H	H	OH	1.4		Wieland and Brodner, 1976
[^{125}I] 7'-α-amantin	CH_2OH	OH	H	[^{125}I]	NH_2	0.3	K_i $\sim$5 times lower than that of α-amanitin	Morris et al., 1978

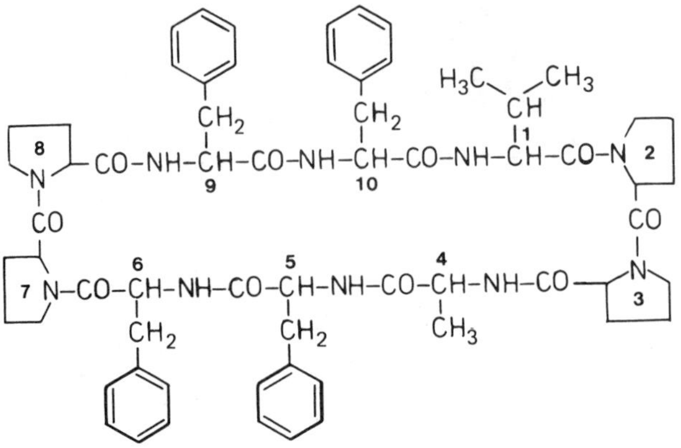

FIGURE 10 Formula of antamanide.

For further biologically active and inactive analogs, see Wieland and Faulstich (1978).

IV. PATHOLOGY AND MOLECULAR TOXICOLOGY OF THE TOXINS

A. Phallotoxins (Virotoxins)

1. Symptoms of Intoxication

The toxicity of phalloidin has been under investigation since 1938 (Vogt, 1938), when the crystalline toxin became available. It has fascinated both pharmacologists and biochemists since that time. Most of the studies have been done in the white mouse and in the rat. Here, phalloidin causes a severe swelling of the liver; the weight of the liver may reach two to three times that of a normal rat liver. Such swelling is due to an excessive accumulation of blood in this organ, the blood content being five times higher than that in controls. The animals bleed to death into their liver. The depletion of fluid in peripheral circulation probably leads to hemodynamic shock. Such toxic lesions were observed exclusively in the liver. This is a consequence of the specific accumulation of the toxin in this organ (Faulstich et al., 1977b). In less than 10 min, the rat liver takes up more than 70% of the toxin present in the circulatory system. For comparison, the rate of uptake for amatoxins is about 20 times **slower.** In addition to this, phallotoxins cannot be excreted into the bile like the amatoxins, because the bile flow stops a few minutes after administration of phalloidin (Wieland, 1965). The intake of phallotoxins into liver cells is mediated by a few proteins which also are responsible for the transport of antamanide (competitive protection, see above), and of bile salts (M. Nassal, G. Kurz, personal communication, 1982).

In search of the cellular events preceding the accumulation of blood in the liver, a study by electron microscopy established that the hepatocytes become interspersed with vacuoles. These vacuoles develop from invaginations of the membranes near the sinusoidal space, as indicated by erythrocytes and other blood components present in the interior of the vacuoles. An event very similar to this occurs when isolated hepatocytes are incubated with phalloidin. In this case exvaginations develop instead of invaginations, giving rise to numerous bulbous protrusions or blebs on the cell surface (Weiss et al., 1973), (Fig. 11). In both cases the plasma membrane of the hepatocytes presumably has lost its inherent elasticity. As a consequence, the membrane can be deformed by even low-pressure gradients such as the moderate pressure of the portal vein in vivo or a slight intracellular pressure apparently present in the isolated hepatocytes.

In the perfused rat liver, the vacuolization induced by phalloidin is succeeded by a series of other lesions. One of the earliest events is a loss of K^+ ions from the hepatocytes, which is followed by a loss of cytoplasmic enzymes. About one hour later the liver is obviously depleted from adenosine triphosphate (ATP) and glycogen, effects that finally destroy the organ.

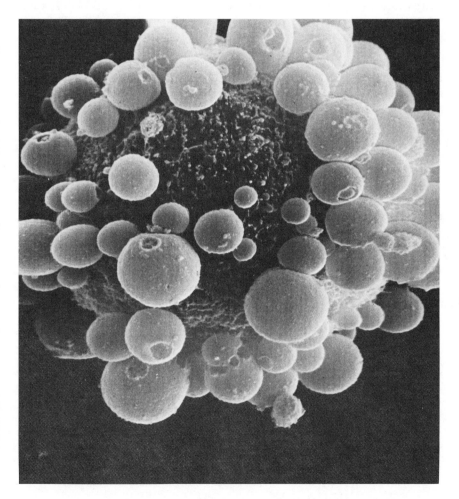

FIGURE 11 Isolated rat hepatocyte after 10 min incubation in 10^{-3} M phalloidin containing Tyrode medium at 37°C. (By courtesy of M. Frimmer.) This effect is entirely counteracted by ca. 10^{-4} M Na-O-carboxymethyl-Tyr[6]-antamanide, according to Faulstich et al. (1974b).

All the disturbances cited here seem to run parallel with the observed vacuolization rather than being the cause of vacuolization.

2. Interaction of Phallotoxins with Actin

To determine the reasons for the apparent alteration of the cytoplasmic membrane of the liver caused by phalloidin, electron microscopic studies have been made on membrane preparations from livers of poisoned rats (Govindan et al., 1972). Here, bundles of microfilaments were observed which were associated with the membrane fragments. Such filaments were visible only occasionally in membrane preparations of control animals. Most interestingly, similar filamentous structures could also be produced by phalloidin in vitro in preparations of cytoplasmic membranes from livers of unpoisoned rats, and their in vitro formation could be prevented by a water-soluble derivative of antamanide (on antamanide see Sec. III.D). These filaments consisted of actin, as documented by their reaction with heavy meromyosin forming the typical arrowhead-like structures (Lengsfeld et al., 1974). Since rat liver cell actin corresponds to rabbit muscle actin in most aspects (e.g., in its reaction with phallotoxins) (Govindan and Wieland, 1975), all following experiments were performed with the muscle protein as a model. The results obtained are summarized in the following paragraphs (see also the review by Wieland, 1977).

Stabilization of F-actin: In the F-actin-phalloidin complex (Ph-actin) phallotoxins strongly stabilize the structure of F-action. The equilibrium of monomeric G-actin (protomers) and polymeric F-actin, which is independent from the F-actin concentration, contains about 40 μg G-actin per milliliter. By binding the toxins, this equilibrium is shifted toward the polymeric form (Fig. 12).

Since phallotoxins probably also bind to and stabilize trimers, which are regarded to be the nuclei of polymerization, the rate of polymerization of G-actin is accelerated by phalloidin. The stoichiometry of the stabilizing influence of some phallotoxins against depolymerization of F-actin by 0.6 M potassium iodide has been measured initially by viscosimetry (Löw and Wieland, 1974; Dancker et al., 1975). Figure 13 demonstrates that the viscosity of the solution will drop to $\eta_{spec} = 0$ on addition of KI in the absence of phalloidin,

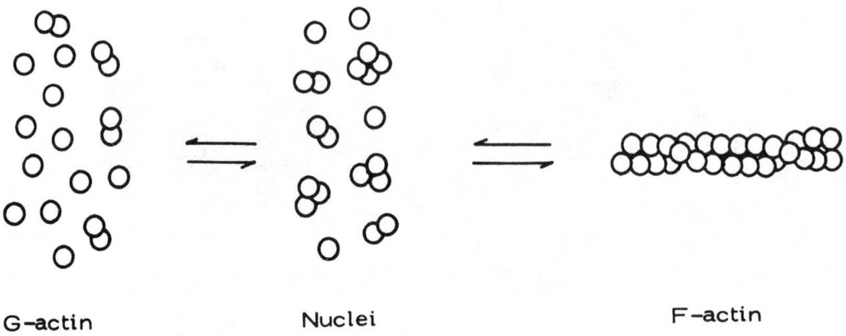

G-actin   Nuclei   F-actin

FIGURE 12 Equilibrium of G-actin with F-actin, oligomers (nuclei) as intermediates. This equilibrium is shifted entirely to the right in the presence of an excess of phalloidin.

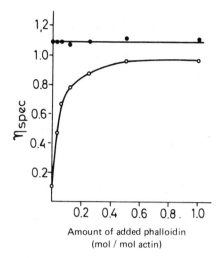

Amount of added phalloidin
(mol / mol actin)

FIGURE 13 Stabilizing effect of phalloidin on F-actin against depolymerization by 0.6 M KI. Viscosity in this medium is retained if more than 1 mol of
phalloidin per 2 mol of G-actin is present.

but will be conserved increasingly on the addition of increasing amounts of
the toxin. Complete stabilization is achieved if about 0.5 mol of phalloidin
per one G-actin unit is present. Although on saturation a 1:1 ratio has
been stated, complexation of one out of two or three protomers is sufficient
to stabilize the double chain of F-actin. The drug not only stabilizes but
even effects a slow polymerization of G-actin in 0.6 M KI, also in the presence of various cytochalasins and chaetoglobosins (Wieland and Löw, 1980).
Another depolymerizing agent is pancreatic DNAase I. This enzyme forms
1:1 adducts with actin monomers and, as a consequence, shifts the actin
equilibrium toward the monomers. There is no depolymerization by this
enzyme in the presence of phalloidin, because the interaction of phalloidin
with the filamentous form is apparently stronger than that of DNAase I with
the monomers.

   Phalloidin not only inhibits the depolymerization reaction but also restrains
local ruptures in the filaments. Such ruptures can be induced by ultrasonic
vibration, by cytochalasin B, or by low pH values. The subsequent healing
of such perturbations is accompanied by splitting of ATP. Hence the measurement of this weak ATPase activity allows one to follow the extent of local
ruptures. Phalloidin suppresses completely such ATPase activities induced
by ultrasonication as well as by cytochalasin B. Also, the ATPase induced
by protons is totally inhibited at pH 6 (see Wieland, 1977). Phalloidin also
protects actin from denaturation by heat. While F-actin will be denatured by
100% after 3 min heating to 70°C, Ph-actin is denatured by only 15%. Full
protection by phalloidin was achieved when Ph-actin was heated 3 min to 60°C.
Similarly, phalloidin protects F-actin filaments from proteolytic digestion, as
observed with subtilisin, trypsin, or α-chymotrypsin. Full protection in this
case needs the presence of 0.1 M K^+ ions (de Vries and Wieland, 1977).

   As already pointed out, phalloidin shifts the actin equilibrium toward the
filamentous form. As a consequence, the concentration of actin monomers in

equilibrium with actin polymers decreases. The concentration of monomers in equilibrium with filaments in 0.1 M KCl is about 10^{-6} M. In the presence of 1 equivalent of phalloidin, this value decreases to 3.6×10^{-8} M, which is about 30 times lower (Faulstich et al., 1977a). In the presence of 2 equivalents of phalloidin, the concentration of the monomers decreases further, to a value about 100 times lower than without phalloidin. The low concentration of monomers can probably account for most of the stabilization effects observed with muscle actin. For example, proteolytic action may predominantly affect monomers or terminal protomers in the filaments, which can be expected to be better substrates for proteases. Also, heat denaturation probably proceeds via actin monomers.

The protective capacity of all phallotoxins investigated so far runs grossly parallel to their affinity toward actin. The highest affinity to actin is achieved by phalloidin and phallacidin, indicating that the two hydroxyl groups at the leucine side chain play a certain role in binding to the protein. In all cases, the affinity to actin of phallotoxins could be correlated with their toxicity in the white mouse. Derivatives, with dissociation constants up to 10 times higher than that of phalloidin (3×10^{-8} M), were still toxic, whereas those with dissociation constants between 10 and 100 times that of phalloidin turned out to be nontoxic. It is, therefore, reasonable to assume that the binding of the phallotoxins to cell actin probably represents a possible mechanism for phalloidin toxicity by the concomitant decrease of the concentration of monomeric actin.

Modified actin and liver injury: The question of how the phallotoxin-induced stabilization of liver F-actin leads to the fatal swelling of the organ has not fully been elucidated. The endocytotic formation of vacuoles under normal blood pressure after phallotoxin administration in vivo and in the perfused rat liver points to a loss of the elasticity of the cytoplasmic membrane. The early finding of bundles of actin filaments in membrane fractions of rat liver after administration of the toxin (p. 614) has been supplemented by visualization of the distribution of F-actin in phalloidin-poisoned liver cells (Jahn et al., 1980a). After removal of the toxin by washing with dimethyl sulfoxide-water the cryosections were stained in parallel with control sections with a fluorescent phallotoxin (FL-PHD, p. 605). As the fluorescence microscopy in Fig. 14 shows, in the normal liver the polygonal hepatocytes are separated by bright fluorescent seams of the filamentous F-actin underlying the membranes, whereas in the poisoned liver this layer appears interrupted, with the F-actin being clustered in discrete bundles. The actin-free parts may well be the entering sites for the endocytotic vacuolization. An explanation for the spotwise accumulation of F-actin filaments could be that there exists a permanent equilibration between G- and F-actin, also underneath the membrane, and that at places where phalloidin accumulates modification of actin starts converting the surrounding actin into Ph-actin filaments. Fluorescent phallotoxins are also very appropriate for visualizing actin cables in cultured cells (Fig. 15). Vacuoles artificially generated in perfused rat livers by rising the pressure of the medium will after releasing the pressure, be eliminated within 30 min. Electron microscopic examination of such "normal" vacuoles revealed that these vacuoles are surrounded by a filamentous layer of F-action (+ myosin?) which, apparently as a contractile system, presses the vacuoles out of the cytoplasm. In contrast, the vacuoles in phalloidinized liver cells are lacking such a contractile seam

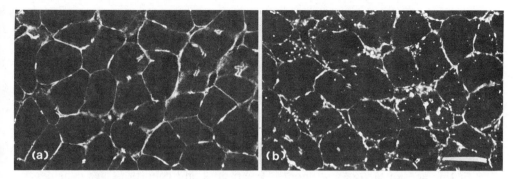

FIGURE 14 Cryosections of rat liver stained with fluorescent phallotoxin.
(a) Normal rat liver; F-actin along the cell borders and around bile canaliculi.
(b) Liver perfused for 2 hr with a medium containing 0.5 mg of phalloidin.
Layer along cell borders disrupted, part of F-actin concentrated in clusters.
(Length of bar = 30 μm.)

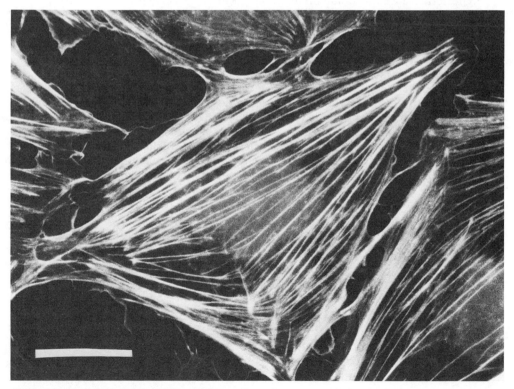

FIGURE 15 Rat kangaroo fibroblast cell stained with fluorescent phallotoxin
after fixation with 4% formaldehyde. (Length of bar = 10 μm.) (From Wulf
et al., 1979.)

(Jahn, 1977). Based on the reasoning discussed above, there is insufficient G-actin to build up a meshwork of filaments around the vacuoles in the interior.

Virotoxins may act similarly: Much less is known about the toxicology of the virotoxins. However, since they cause in animals symptoms analogous to those produced by phallotoxins and bind to actin in a very similar way, one may assume that generally their mechanism of intoxication is very similar to that of the phallotoxins.

3. Protection and Tolerance Against Phallotoxins

In this section drugs and agents, treatments and conditions are considered under which doses of phallotoxins, normally lethal, do not lead to death of the animals. Early in phalloidin research, Vogt (1938) reported on the survival after lethal phalloidin doses of mice in which sublethal doses of phalloidin had been injected several times before. We will come back to this phenomenon shortly, but describe now some drugs that reduce phallotoxin potency when given shortly before application of the toxin.

Drugs and agents

Antamanide: The classical drug preventing the white mouse from death by hyperlethal doses of phallotoxins is antamanide, a component of *A. phalloides* itself (see p. 610, Wieland, 1972; Burgermeister and Winkler-Oswatitsch, 1977). Antamanide acts only if it is injected not earlier than 1-2 hrs before the toxin. A single dose of 0.5-1 mg/kg body weight by this protocol will protect a white mouse from certain death by 5 mg of phalloidin per kilogram. When applied 10 min after the toxin, protection can be achieved by much higher doses, but it disappears 20 min after the toxin. The mechanism of the rather specific antiphallotoxin effect of antamanide has been shown to be a competition for the membrane proteins involved in the uptake of the phallotoxins (M. Nassal, G. Kurz, personal communication, 1982). It does not compete for the target protein actin, but strongly reduces the uptake of phalloidin by the liver in vivo and in vitro, thus preventing the accumulation of a noxious concentration of the toxin inside the cell. It has been suggested that the cyclic peptide, by virtue of its Ca^{2+} complex (Wieland et al., 1972), is bound to phosphoric acid residues in the cell membrane, thereby preventing entry of toxin (Ivanov, 1975). However, since several analogs of antamanide with equally strong Ca^{2+}-complexing property are biologically inactive, a general membrane-tightening effect seems too simple an explanation.

Silymarin: Silymarin, a mixture of components of the milk thistle *Silybum marianum*, Gaertn., reduces the LD_{50} value of phalloidin in mice to a LD_0 when administered in doses of 15 mg/kg. An even more effective inhibition was exerted by disilybin (Vogel and Trost, 1975), which protected the mice with a dose of 5 mg/kg against lethal doses of phalloidin. As with antamanide, the protection by silymarin derivatives lasts for only a finite time, due to the excretion of the compounds through the bile.

Rifampicin: According to Floersheim (1971), the antibiotic rifampicin also has a protective effect (10 mg/kg versus 5 mg/kg phalloidin, white mouse), which is due to a reduction of the rate of uptake of the toxin in the liver, as with the other drugs. This effect was much higher in a perfused liver system (Frimmer et al., 1975).

Bile salts: Considerable retardation of phalloidin uptake in isolated rat hepatocytes by bile salts was stated by Petzinger and Frimmer (1980) using [^{3}H]demethylphalloin as a tracer. In a suspension containing 6 µg of phallotoxin per milliliter, the toxin uptake was reduced to 30% if about 5 µg of cholate or taurocholate per milliliter was present, and totally inhibited by 40 µg/ml. Cholecystographic compounds which competitively inhibit the uptake of bile acids by isolated hepatocytes also inhibit the uptake of phalloidin. This fact is considered as further support of the theory of Frimmer that bile salts and phallotoxins are taken up by the same mechanism (Frimmer et al., 1980). This theory has been substantiated by the observation that affinity-labeled phallotoxins, antamanide and bile acids bind to identical membrane proteins (M. Nassal, G. Kurz, personal communication, 1982). Bile salts, however, have no protective effect against phalloidin in vivo. We found that not even an intravenous dose of 100 mg of cholate per kilogram of weight could protect rats from death by 2 mg/kg phalloidin (Th. Wieland and A. Schmitz, unpublished observations, 1980).

Liver-damaging substances: Next, several substances will be considered which in the course of days alter the liver of animals in such a way as to render it insensitive against the noxious effect of phallotoxins. Many substances have been described which increase the resistance of mice, rats, or rabbits to phalloidin. Today, most of them are regarded as drugs which damage or simply cover those surface structures of hepatocytes that are responsible for the extraordinarily rapid uptake of the toxin. For a review of these substances, see Frimmer et al. (1975).

Some of the earliest observations in this field were made by Floersheim (1966), who found that pretreatment of mice with carbon tetrachloride or Na$^+$-cinchophen protects the animals against phalloidin. The doses needed to protect against LD$_{95}$ of phalloidin were 0.2 ml/kg 24 hr before and 400 mg/kg 30 min before, respectively. Protection in these cases was achieved by severe damage of hepatocyte structures. Kroker and Frimmer (1974) reported a decreased binding of phalloidin to isolated hepatocytes after treatment with carbon tetrachloride.

Since drug-metabolizing enzymes could have been destroyed during the damage of liver, an earlier suggestion (van der Decken et al., 1960; Fiume, 1965), was seized on by Floersheim (1966), that phallotoxins might not be toxic by themselves but only after toxification by microsomal enzymes. However, this theory had to be abandoned after Puchinger and Wieland (1969b) extracted more than 95% of a radioactively labeled phallotoxin unmetabolized from the homogenates of poisoned rats. Further proof of the fact that the toxic activity of phalloidin is independent of the metabolizing enzyme system was given by T. M. Guenther and A. W. Nebert (unpublished observations, 1976) who found that phalloidin neither induces P$_{450}$-dependent enzymes nor requires protein synthesis for its lethal activity in mice. Finally, no metabolites of phalloidin could be detected in rat livers, neither in the perfusion medium nor in the bile after application of low doses of toxin (Rehbinder et al., 1963).

Damaging the liver of rats by galactosamine also produces resistance in animals against phallotoxins (Agostini et al., 1977). The same is true for rats with liver tumors induced by diethylnitrosamine (Agostini et al., 1976). Since isolated liver tumor cells (ascites hepatoma) in suspension do not take up [^{3}H]phallotoxin (Petzinger and Frimmer, 1980), and decreased phalloidin uptake has also been stated for hepatocytes prepared from rats chronically (up to 90 weeks) treated with diethylnitrosamine (Ziegler et al., 1980), reduced uptake of the toxin in vivo seems an obvious explanation. Apparently, such cells have lost binding sites on the outside of the plasma membrane. Mice also acquire tolerance against LD_{100} of phalloidin 5-6 days after implantation of Ehrlich ascites tumors (Agostini et al., 1978a). As foci of tumor cells were observed in the livers, a similar situation as in the liver of tumor-bearing rats may be supposed.

Tolerance

Tolerance is induced

1. *By phalloidin.* As already mentioned, Vogt (1938) found that tolerance to phalloidin is achieved by prolonged administration of sublethal doses of the toxin. The effect was confirmed later (Szabados, 1971; Floersheim, 1976) and followed more intensively by Gabbiani et al. (1975). These authors observed that under permanent treatment with small doses of phalloidin, hepatocytes of rats developed a hyperplasia of filamentous material and an extensive formation of tight junctions between the liver cells (Montesano et al., 1976). Liver cells isolated from such animals despite their high actin content, incorporated much less phalloidin than do those from control animals (Faulstich et al., 1977b). Accordingly, tolerance here can be explained by a slow resorption rate of the toxin and a "neutralization" by the high content of F-actin. Tight junctionlike filamentous seams at contact sites between normally single growing cells could be induced by phalloidin even in lymphocyte cultures (Saxon et al., 1978).
2. *By bile duct ligation (cholestasis).* Tolerance of rats against phallotoxins can also be induced by bile duct ligation, which leads to a severe cholestasis (Agostini et al., 1978b). Using [^{3}H]demethylphalloin it was shown by Walli et al. (1981) that as early as 6 hr after the operation the in vivo uptake of the liver was reduced to about 15% of the normal, and that hepatocytes isolated from such animals take up the toxin slower than control cells, but not slow enough to account for the tolerance in vivo. However, since serum of cholestatic rats protects normal hepatocytes from phallotoxin incorporation, a factor (mixture) has been traced which, like an intrinsic antamanide, protects the animal (E. Wieland, 1981).

Tolerance of newborn animals: A natural tolerance to phalloidin was observed by Fiume (1965) in newborn rats, which survive a manifold lethal dose up to the age of 18 days. This supported the theory of a metabolic toxification of the natural product (see p. 619), which, however, has been disproved. The inborn tolerance is correlated with a reduced rate of phallotoxin uptake by livers in vivo (Walli et al., 1979) as well as by isolated hepatocytes of the young animals (Ziegler et al., 1979). The livers of 12-day-old rats, after application of a dose 80 times that of an LD_{50} dose for adults, without being extremely swollen, contained 30 μg of toxin per gram compared with 20 μg/g in the adult after a dose of 5 mg/kg, which in the latter causes severe lesions and death. According to Siess et al. (1970), high doses of phalloidin in young rats, mice, or rabbits cause vacuoles in the liver as in adults, but to a distinctly lesser extent. This suggests that the tolerance of young animals against phallotoxins cannot be explained solely by the reduced rate of uptake of the toxin, but also by an unique property of the plasma membrane, which is more resistant against invagination than in adult animals.

B. Amatoxins

1. Symptoms of Intoxication

The symptoms of human poisoning with green and white *Amanita* spp. have frequently been described. They correspond well to those in most laboratory animals caused by pure amatoxins. We therefore are certain that human *Amanita* poisoning is brought about solely by the amatoxins. The symptoms develop in four phases. The first is a latency period of 6-12 hr after ingestion. This latency period is specific for amatoxins and therefore of great diagnostic value. It is followed by the gastrointestinal phase, which is usually the first indication of *Amanita* poisoning, characterized by cholera-like diarrhea with concurrent dehydration, vomiting, and abdominal pains. At this stage there may occur already a rapid fall of coagulation factors, which confirms the diagnosis. The third phase is another latency period. The patient feels better; however, this recovery is delusive, because the hepatic lesions continue to develop. During the final phase the liver enzymes in the serum [serum glutamic-oxalic transaminase (SGOT), serum glutamic-pyruvic transaminase (SGPT), and lactic dehyrogenase (LDH)] increase rapidly, and the hepatic failure often causes encephalopathy. High values of creatinine and urea indicate additional damage to the kidney cells.

In most cases the patients die between the fourth and eighth days in hepatic coma combined with renal failure.

2. Doses and Pharmacodynamics

In humans the lethal dose of amatoxins can be estimated from accidents to be about 0.1 mg/kg body weight, or even lower. The toxins must be readily absorbed by the intestine. A similar situation is given in the guinea pig, where the lethal dose is also low (0.1 mg/kg), and identical for oral and intravenous or intraperitoneal administration, indicating a strong intestinal absorption. This is surprising because other rodents, such as the mouse (0.35 mg/kg, i.v.) or the rat (3.5 mg/kg, i.v.), cannot be poisoned orally, even by high doses of amatoxins. Among other animals, only the dog is known to absorb amatoxins from the gut (Fiume et al., 1973; Faulstich, 1980b). Here, however, the intestinal absorption is slow and limits the intoxication; the oral LD_{50} dose for α-amanitin is about five times higher (0.5 mg/kg) than the intravenous dose (0.1 mg/kg). In accordance with the capability of intestinal absorption, the gut cells of humans and dog seem to be the first cells to be affected. In both species the intestinal phase begins about 9 hr after administration of the toxins. In the dog, the intestinal symptoms occur after oral as well as parenteral administration (Faulstich, 1980b), possibly another indication for the enterohepatic circulation (see below), which brings the intestinal cells into direct contact with the toxins.

Uptake into liver cells is strictly proportional to the toxin concentration in the serum. In the perfused rat liver, after an initial phase of rapid uptake (0.5 hr, in which 8% of the applied dose was incorporated), the uptake was determined to be 4% of the dose per hour. Immediately after incorporation has begun the liver cells excrete amatoxins by bile, resulting in a steady-state concentration of the toxins corresponding to about 5% of the dose administered (Jahn et al., 1980b). The high excretion rate of the toxin (about two-thirds of the incorporated amount is excreted) emphasizes the importance of the enterohepatic circulation in animal species capable of intestinal absorption (e.g., in dog and humans) (Fauser and Faulstich, 1973).

Amatoxins are highly dialysable. In vitro, the dialysis rate is only three to four times slower than for salts. Correspondingly, in dogs, after a single intravenous dose of radioactively labeled amatoxin, about 90% of the bolus is excreted after 5 hr (Faulstich and Fauser, 1973), despite renal reabsorption. The existence of renal reabsorption of amatoxins has been concluded from the damage of proximal tubuli observed in the mouse poisoned with low doses of amatoxins (Fiume et al., 1969).

3. Treatments

The most crucial parameter in amatoxin poisoning seems to be the exposure time of liver cells or nuclei to the toxin. The concentration of RNA-polymerase II in the rat liver is about 5×10^{-8} M (Cochet-Meilhac et al., 1974). According to the high affinity of α-amanitin to these enzymes ($K_D \sim 10^{-9}$ M), an intracellular toxin concentration as low as the above can inhibit more than 90% of the mRNA transcription. Since the intracellular toxin concentration is directly dependent on the concentration of the toxin in the serum, the first precautions to be taken should be to lower the toxin concentration in serum. During the first and second days any extracorporal purification method (hemoperfusion over ion exchange resin, hemodialysis, hemofiltration, or plasmapheresis—or if not feasible, forced diuresis) can be considered.

The next aim should be the interruption of the enterohepatic circulation of the toxins, which permanently sustains the intracellular concentration of the toxins in the liver cells. This can be achieved mechanically by insertion of a special duodenal tube (Faulstich, 1979) as well as by some drugs such as silymarin (Vogel, 1980), or—less effective—penicillin, which have been shown to inhibit the incorporation of the toxins into hepatocytes (Faulstich et al., 1980e). Beside these drugs other substances with unknown mechanism (e.g., thioctic acid) have also been used, apparently with good results (Bartter et al., 1980; Bastien, 1980). As yet unknown is the possible beneficial effect of immunoglobulins specific for amatoxins. In antisera thus far produced the concentration of the amanitin-binding antibodies was too low to be tested for their antitoxic activity.

4. Amatoxins Inhibit RNA-Polymerases II (B)

The interaction of amatoxins with RNA-polymerases was detected in 1966 by Fiume and Stirpe, who found structural lesions in nuclei of mouse liver cells and concomitantly significant decrease of RNA synthesis as soon as 1 hr after administration of α-amanitin (Fiume and Stirpe, 1966). More recently, three laboratories independently confirmed that α-amanitin inhibits RNA synthesis (Seifart and Sekeris, 1969; Kedinger et al., 1970; Lindell et al., 1970).

The biological aspects of this amatoxin activity have recently been reviewed by Wieland and Faulstich (1978) and Faulstich (1980a). With the differentiation of RNA-polymerases into three classes, I, II, and III (or A, B, and C), it became evident that the three classes of enzymes have different sensitivities to amatoxins (Fig. 16).

Among the mammalian enzymes, RNA-polymerase II is highly sensitive ($K_i \sim 10^{-8}$ M) while RNA-polymerase III transcribing tRNA and low-molecular-

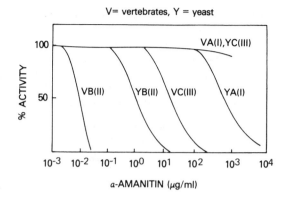

FIGURE 16 Inhibition curves of the different RNA-polymerases A (or I),
B (or II), and C(or III) of vertebrates (V) and yeast (Y) by α-amanitin.

weight RNA is about 10^3 times less sensitive ($K_i \sim 10^{-5}$ M); RNA-polymerase
I, transcribing the rRNA, is completely insensitive ($K_i > 10^{-2}$ M). Also insensitive are the RNA-polymerases of *Escherichia coli* (Seifart and Sekeris,
1969) mitochondria and chloroplasts; only the RNA-polymerases of mitochondria in mammalian cells can be inhibited by high amatoxin concentrations. In order to understand the course of human *Amanita* poisoning, it
may be of importance that in contrast to the in vitro experiments, the in
vivo activity of RNA-polymerases I can be completely blocked by amanitin
(e.g., in the liver of the amanitin-poisoned rat and mouse) (Jacob et al.,
1970; Niessing et al., 1970; Hadjiolov et al., 1974).

Differences between species: The sensitivity of RNA-polymerases II of
eukaryotic cells varies widely with the species. Apparently, the sensitivity
to amatoxins increases with the evolutionary hierarchy. For example, the
RNA-polymerases II of the frog are about five times and those of a slime
mold about 10 times less sensitive than the enzymes isolated from mammalian
tissues (Table 9). However, several mammalian cell lines grown under $\sim 10^{-6}$
M amatoxins can develop resistance against the toxins. Such cells survive
by producing RNA-polymerases up to 1000-fold less sensitive than the enzymes of the wild type. RNA-polymerases II with even lower sensitivity to
amatoxins have been found in the amatoxin-accumulating *Amanita* mushrooms.

Molecular mechanism: Different from most of the inhibitors of transcription, the amatoxins bind to the enzyme, not to the template. Most probably,
binding itself represents the inhibitory activity, because the K_i values are
very close to the equilibrium dissociation constants (K_D). These values are
about 10^{-8}-10^{-10} M, depending on the assay conditions (Cochet-Meilhac and
Chambon, 1974). The toxins form 1:1 complexes with the enzyme. Since
RNA-polymerase II consists of several subunits, it was of great interest to
know which of the subunits binds the amatoxins. By affinity labeling experiments it was shown that the 140,000-dalton polypeptide is the target
subunit (Brodner and Wieland, 1976a).

For the mechanism of inhibition of RNA synthesis it has been suggested
that amatoxins generally block the formation of phosphodiester bonds, since

TABLE 9 Concentrations of α-Amanitin ($\times$ 10^{-8} M) Causing 50% Inhibition of RNA-Polymerases II of Various Eukaryotic Cells

		References
Various mammalian tissues	1	Zylber and Penman, 1971; Seifart and Benicke, 1975
HeLa cells	1	Hossenlopp et al., 1975; Weil and Blatti, 1976
Bombyx mori	3	Sklar and Roeder, 1975
Drosophila melanogaster	3	Greenleaf et al., 1976
Dictyostelium discoideum	3	Pong and Loomis, 1973; Yagura et al., 1976
Xenopus leavis	5	Roeder, 1974; Wilhelm et al., 1974
Zea mays	5	Strain et al., 1971; Jendrisak and Guilfoyle, 1978
Wheat germ	5	Jendrisak and Guilfoyle, 1978
Brassica oleracea	5	Jendrisak and Guilfoyle, 1978
Saccharomyces cerevisiae	80	Hager et al., 1976; Schultz and Hall, 1976
Physarum polycephalum	100	Grant, 1972; Hildebrandt and Sauer, 1973
Tetrahymena pyriformis	300	Higashinakagawa and Mita, 1973
Agaricus bisporus	650	Vaisius and Horgen, 1979
Resistant mammalian cells	100-1000	Reviewed in: Faulstich, 1980a
Amanita brunnescens (nontoxic)	900	Johnson and Preston, 1979
Amanita hygroscopica (toxic)	180.000	Johnson and Preston, 1979
Amanita suballiacea (toxic)	300.000	Johnson and Preston, 1979
Aspergillus nidulans	40.000	Stunnenberg et al., 1981

no pyrophosphate exchange could be observed (Cochet-Meilhac and Chambon, 1974). This mechanism would affect the formation of the first phosphodiester bond (initiation) and of the following phosphodiester bonds (elongation) as well. However, in a more detailed study, Vaisius and Wieland (1982) found that with RNA polymerase B from calf thymus a first internucleotide bond is catalyzed also in the presence of high concentrations of α-amanitin (10^{-6} M), but that the next step, translocation, is inhibited due to the stabilization by the drug of the ternary complex of template, enzyme, and nascent ribonucleotide chain. This complex could be visualized by gel electrophoresis.

5. Interaction of Amatoxins with Proteins Other Than RNA Polymerases

Amatoxins do not bind to serum albumin (Fiume, 1977) or to any other serum protein (H. Faulstich, 1979). Such speculations have occasionally been put forth in the past to try to understand the mode of action of anti-dotes. A protein with amatoxin-binding property has been isolated from calf thymus during the preparation of RNA-polymerases (Brodner and Wieland, 1976b). In column chromatography radioactively labeled amatoxins comigrate with this amatoxin-binding protein (ABP), indicating an affinity comparable to that of RNA-polymerase. As yet its function is unknown. From experiments with fluorescent-labeled amatoxins the possibility arises that ABP may be present in centrioles and spindle apparatus during the mitosis of cells (Wulf et al., 1980). Recently, an ABP has been isolated from wheat germ by Brodner and Sabbagh in the author's laboratory (Brodner et al., 1982).

Another class of proteins binding amatoxins are the immunoglobulines produced as antibodies against amatoxin-protein conjugates in rabbits. They consist of 95% IgG and about 5% IgM. Their affinity as determined by equilibrium dialysis is comparable to that of RNA-polymerases II (K_D ca. 10^{-8} M) (Faulstich et al., 1980d).

6. Biochemical and Biological Applications

The applications of amatoxins in biochemistry and biology have been summarized in two recent reviews Wieland and Faulstich, 1978; Faulstich, 1980a). Use of the amatoxins in biological science relies on the possibility of differentiating the activities or transcription products of the three classes of RNA-polymerases. This section presents a brief summary of that work.

In experiments with cell cultures the amatoxins helped elucidate the regulation of DNA, RNA, and protein synthesis. Several cell lines resistant to amatoxins have been produced, and studies with the isolated RNA-polymerases of these cells gave insight into the physical background of the resistance phenomenon. When amatoxins were attached to carrier proteins, the conjugates exhibited specific toxicity for protein-consuming cells.

Amatoxins were also applied in virus research. Since the virion-coded RNA-polymerases are not inhibited by amatoxins, their application allowed identification of those viruses which depend on the transcription apparatus of the host cell. In general, all viruses that replicate in the nucleus were found to be inhibited by amatoxins. In contrast, among the species multiplying in the cytoplasm, the few investigated so far were found to be insensitive to amatoxins (see also the brief review of Campadelli-Fiume, 1978). Hormones and drugs can induce the synthesis of enzymes. This process can be inhibited by amatoxins, demonstrating that the induction involves a transcription step. In combination with other inhibitors the kinetics of the different steps could be elucidated. In some cases the amatoxins helped to establish the existence of long-lasting mediators of the hormonal signal. Similar conditions as after hormone induction exist in developing systems, such as eggs and embryos. The amatoxins were also useful to detect the participation of transcription and to study the time course of gene expression.

Finally, several behavioral processes apparently depend on the synthesis of mRNA. For example, intracerebral administration of α-amanitin in rats and mice can inhibit the consolidation of memory, especially of long-term memory.

Apparently, this process depends on protein synthesis or, at least, on the synthesis of RNA.

REFERENCES

Agostini, B., Wieland, Th., Ivankovic, S., and Hofmann, W. (1976). Phalloidin tolerance in rats with liver carcinoma induced by diethyl-nitrosamine. *Naturwissenschaften 63*:438.

Agostini, B., Wieland, Th., and Lesch, R. (1977). Decreased phalloidin toxicity in rats pretreated with D-galactosamine. *Naturwissenschaften 64*:649.

Agostini, B., Ivankovic, S., and Granzow, C. (1978a). Development of phalloidin tolerance in mice bearing Ehrlich ascites tumors. *Naturwissenschaften 65*:602.

Agostini, B., Walli, A. K., and Wieland, H. (1978b). Protection of rats against phalloidin by ligation of bile duct. *Naturwissenschaften 65*:664.

Ammirati, J. F., Thiers, H. D., and Horgen, P. (1977). Amatoxin-containing mushrooms: *Amanita ocreata* and *amanita phalloides* in California. *Mycologia 69*:1095.

Andary, C., Enjalbert, F., Privat, G., and Mandrou, B. (1977). Dosage des amatoxines par spectrophotométrie directe sur chromatogramme chez *Amanita phalloides* Fries (Basidiomycetes). *J. Chromatogr. 132*:525.

Andary, C., Privat, G., Enjalbert, F., and Mandrou, B. (1979). Teneur comparative en amanitines de différentes agaricales toxiques d'Europe. *Doc. Mycol. 10*:61.

Barak, L. S., Yocum, R. R., Nothnagel, E. A., and Webber, W. W. (1980). Fluorescence staining of the actin cytoskeleton in living cells with 7-nitrobenz-2-oxa-1,3-diazole-phallacidin. *Proc. Natl. Acad. Sci. USA 77*:980.

Barak, L., and Yocum, R. (1981). 7-Nitrobenz-2-oxa-1,3-diazole (NBD) phallacidin: Synthesis of a fluorescent actin probe. *Anal. Biochem. 110*:31.

Bartter, F., Berkson, B., Gallelli, J., and Hiranaka, P. (1980). Thioctic acid in the treatment of poisoning with alpha-amanitin. In *Amanita Toxins and Poisoning*, H. Faulstich, B. Kommerell, and Th. Wieland (Eds.). Witzstrock, Baden-Baden, p. 197.

Bastien, P. (1980). A general practitioner's experience of *Amanita phalloides* poisoning. In *Amanita Toxins and Poisoning*, H. Faulstich, B. Kommerell, and Th. Wieland (Eds.). Witzstrock, Baden-Baden, p. 211.

Bautz, F. K., Wulf, E., Deboben, A., Faulstich, H., and Wieland, Th. (1979). Fluorescent phallotoxin, a tool for the visualization of cellular actin. *Proc. Natl. Acad. Sci. USA*.

Bodenmüller, H., Faulstich, H., and Wieland, Th. (1980). Distribution of amatoxins in *Amanita phalloides* mushrooms. In *Amanita Toxins and Poisoning*, H. Faulstich, B. Kommerell, and Th. Wieland (Eds.). Witzstrock, Baden-Baden, p. 18.

Brodner, O. G., and Wieland, Th. (1976a). Identification of the amatoxin-binding subunit of RNA polymerase B by affinity labeling experiments. Subunit B3—the true amatoxin receptor protein of multiple RNA polymerase B. *Biochemistry 15*:3480.

Brodner, O. G., and Wieland, Th. (1976b). Die Isolierung eines Amatoxin-bindenden Proteins, das von der RNA-Polymerase B und C verschieden ist. *Hoppe-Seyler's Z. Physiol. Chem. 357*:89.

Brodner, O. T., Sabbagh, M., and Wieland, Th. (1982). Isolierung und Charakterisierung eines Amatoxin-bindenden Proteins aus Weizenkeimen. *Hoppe-Seyler's Z. Physiol. Chem. 363*:273.

Brown, A., and Garrity, G. M. (1980). Detection and quantitation of amanitin-using an RNA-polymerase competition binding assay. *Toxicon 18*:702.

Buku, A., Altmann, R., and Wieland, Th. (1974). Über die Inhaltsstoffe des grünen Knollenblätterpilzes. XLVI. Das zum giftigen O-Methyl-amanitin diastereomere ungiftige Sulfoxid. *Liebigs Ann. Chem.*, p. 1580.

Buku, A., Faulstich, H., Wieland, Th., and Dabrowski, J. (1980a). 2,3-*trans*-3,4-*trans*-3,4-Dihydroxy-proline: an amino acid in toxic peptides of *Amanita virosa* mushrooms. *Proc. Natl. Acad. Sci. USA 77*:2370.

Buku, A., Wieland, Th., Bodenmüller, H., and Faulstich, H. (1980b). Amaninamide, a new toxin of *Amanita virosa* mushrooms. *Experientia (Basel) 36*:33.

Burgermeister, W., and Winkler-Oswatitsch, R. (1977). Complex formation of monovalent cations with biofunctional ligands. In *Topics in Current Chemistry*, Vol. 69. Springer-Verlag, Berlin, p. 93.

Campadelli-Fiume, G. (1978). Amanitins in virus research. *Arch. Virol. 58*:1.

Cessi, C., and Fiume, L. (1969). Increased toxicity of β-amanitin when bound to a protein. *Toxicon 6*:309.

Cochet-Meilhac, M., and Chambon, B. (1974). Animal DNA-dependent RNA polymerases. 11. Mechanism of the inhibition of RNA polymerases B by amatoxins. *Biochim. Biophys. Acta 353*:160.

Cochet-Meilhac, M., Nuret, P., Courvalin, J. C., and Chambon, B. (1974). Animal DNA-dependent RNA polymerases. 12. Determination of the cellular number of RNA polymerase B molecules. *Biochim. Biophys. Acta 353*:185.

Dancker, P., Löw, I., Hasselbach, W., and Wieland, Th. (1975). Inter-action of actin with phalloidin: polymerization and stabilization of F-actin. *Biochim. Bioplys. Acta 400*:407.

Decken, A. v. d., Low, H., and Hultin, T. (1960). The primary effect of phalloidin on liver cells. *Biochem. Z. 332*:503.

de Vries, J., and Wieland, Th. (1977). Influence of phallotoxins and metal ions on the rate of proteolysis of actin. *Biochemistry 17*:1965.

Faulstich, H. (1979). New aspects of *Amanita* poisoning. *Klin. Worchenschr. 57*:1143.

Faulstich, H. (1980a). The amatoxins. In *Progress in Molecular and Sub-cellular Biology*, Vol. 7, F. E. Hahn (Ed.). Springer-Verlag, Berlin, pp. 88-134.

Faulstich, H. (1980b). Pharmacokinetics of amatoxins in the dog. In *Amanita Toxins and Poisoning*, H. Faulstich, B. Kommerell, and Th. Wieland (Eds.). Witzstrock, Baden-Baden, p. 88.

Faulstich, H., and Fauser, U. (1973). Untersuchungen zur Frage der Hämodialyse bei der Knollenblätterpilzvergiftung. *Dtsch. Med. Wochen-schr. 98*:2258.

Faulstich, H., and Trischmann, H. (1973). Toxicity of and inhibition of RNA polymerase by α-amanitin bound to macromolecules by an azo linkage. *Hoppe-Seyler's Z. Physiol. Chem. 384*:1395.

Faulstich, H., and Weckauf-Bloching, M. (1974). Isolation and toxicity of two cyclolytic glycoproteins from *Amanita phalloides* mushrooms. *Hoppe-Seyler's Z. Physiol. Chem. 355*:1489.

Faulstich, H., Georgopoulos, D., and Bloching, M. (1973). Quantitative chromatographic analysis of toxins in single mushrooms of *Amanita phal-loides*. *J. Chromatogr. 79*:257.

Faulstich, H., Georgopoulos, D., Bloching, M., and Wieland, Th. (1974a).
Analysis of the toxins of amanitin containing mushrooms. *Z. Naturforsch.*
29c:86.

Faulstich, H., Wieland, Th., Walli, A. K., and Birkmann, K. (1974b). Antamanide protects hepatocytes from phalloidin destruction. *Hoppe-Seyler's*
Z. Physiol. Chem. 355:1162.

Faulstich, H., Brodner, O., Walch, St., and Wieland, Th. (1975a). Über
die Inhaltsstoffe des grünen Knollenblätterpilzes. XLIX. Über Phallisacin
und Phallacin, zwei neue saure Phallotoxine und einige Amide des Phallacidins. *Liebigs Ann. Chem.*, p. 2324.

Faulstich, H., Trischmann, H., and Zobeley, S. (1975b). A radioimmunoassay for amanitin. *FEBS Lett. 56*:312.

Faulstich, H., Schäfer, A. J., and Weckauf, M. (1977a). The dissociation of
the phalloidin-actin complex. *Hoppe-Seyler's Z. Physiol. Chem. 358*:181.

Faulstich, H., Wieland, Th., Schimassek, H., Walli, A. K., and Ehler, N.
(1977b). Mechanism of phalloidin intoxication. II. Binding studies. In
Membrane Alterations as Basis of Liver Injury, Falk Symp. No. 22 MTP
Press, Lancaster, England, p. 301.

Faulstich, H., Deboben, A., and Zobeley, S. (1980a). A radioimmunoassay for
phallotoxins. To be published.

Faulstich, H., Buku, A., Bodenmüller, H., and Wieland, Th. (1980b). Virotoxins: actin-binding cyclic peptides of *Amanita virosa* mushrooms.
Biochemistry 19:3334-3343.

Faulstich, H., Kommerell, B., and Wieland, Th. (Eds.) (1980c). *Amanita*
Toxins and Poisoning. Witzstrock, Baden-Baden.

Faulstich, H., Trischmann, H., and Zobeley, S. (1980d). Raising and use of
antibodies against amatoxins. In *Amanita Toxins and Poisoning*, H. Faulstich,
B. Kommerell, and Th. Wieland (Eds.). Witzstrock, Baden-Baden, p. 37.

Faulstich, H., Jahn, W., and Wieland, Th. (1980e). Silybin-inhibition of amatoxin uptake in the perfused rat liver. *Arzneimittelforschung (Drug Res.)*
30:3.

Faulstich, H., Trischmann, H., and Vaisius, A. C. (1981a). To be published.

Faulstich, H., Trischman, H., Wieland, Th., and Wulf, E. (1981b). Ether
derivatives of α-amanitin. Introduction of spacer moieties, lipophilic residues and radioactive labels. *Biochemistry 20*:6498.

Faulstich, H., Zobeley, S., and Trischmann, H. (1982). A rapid radioimmunoassay, using a nylon support for amatoxins from amanita mushrooms.
Toxicon 20:913.

Faulstich, H., Trischmann, H., and Mayer, D. (1983). Preparation of tetra-
methylrhodaminylphalloidin and uptake of the toxin into short-term
cultured hepatocytes by endocytosis. *Experm. Cell Res. 114*:73.

Fauser, U., and Faulstich, H. (1973). Beobachtungen zur Therapie der
Knollenblätterpilzvergiftung. *Dtsch. Med. Wochenschr. 98*:2259.

Fiume, L. (1977). Amanitins do not bind to serum albuminum. *Lancet*, p. 1111.

Fiume, L., and Stirpe, F. (1966). Decreased RNA content in mouse liver
nuclei after intoxication with α-amanitin. *Biochim. Biophys. Acta 123*:643.

Fiume, L., Marinozzi, V., and Nardi, F. (1969). The effect of amanitin
poisoning on the mouse kidney. *Br. J. Exp. Pathol. 50*:270.

Fiume, L., Derenzini, M., Marinozzi, V., Petazzi, F., and Testoni, A.
(1973). Pathogenesis of gastro-intestinal symptomatology during
poisoning by *Amanita phalloides*. *Experientia 29*:1520.

Fiume, L., Busi, C., Campadelli-Fiume, G., and Franceschi, C. (1975). Production of antibodies to amanitins as the basis for their radioimmunoassay. *Experientia* 31:1233.

Floersheim, G. L. (1966). Schutzwirkung hepatotoxischer Stoffe gegen letale Dosen eines Toxins aus *Amanita phalloides* (Phalloidin). *Biochem. Pharmacol.* 15:1589.

Floersheim, G. L. (1971). Antagonistic effects to phalloidin, α-amanitin and extracts of *Amanita phalloides*. *Agents Actions* 2:141.

Floersheim, G. L. (1976). Protection by phalloidin against lethal doses of phalloidin. *Agents Actions* 6:490.

Frimmer, M. (1979). Phalloidin, ein leberspezifisches Pilzgift. *Biol. Zeit* 9:147.

Frimmer, M., Petzinger, H., and Homann, J. (1975). Phalloidin-Antagonisten. 4. Mitteilung: Thioctsäure, SH-Verbindungen, Rifampicin, Choleretika, Dexamethason, Östradiol, unspezifische Hemmstoffe und unwirksame Verbindungen. *Arzneim.-Forsch.* 25:1881.

Frimmer, M., Petzinger, E., and Ziegler, K. (1980). Protective effect of anionic cholecystographic agents against phalloidin on isolated hepatocytes by competitive inhibition of the phallotoxin uptake. *Naunyn-Schmiedeberg's Arch. Pharmacol.* 313:85.

Gabbiani, G., Montesano, R., Tuchweber, B., Salas, M., and Orci, L. (1975). Phalloidin induced hyperplasia of actin filaments in rat hepatocytes. *Lab. Invest.* 33:562.

Gauhe, A., and Wieland, Th. (1977). Über die Inhaltsstoffe des grünen Knollenblätterpilzes. LI. Die Cycloamanide, monocyclische Peptide; Isolierung und Strukturaufklärung eines cyclischen Hepatopeptids (CyA B) und zweier cyclischer Oktapeptide (CyA C und CyA D). *Liebigs Ann. Chem.*, p. 859.

Gerault, A., and Girre, L. (1975). Recherches toxicologiques sur le genre *Lepiota* FR. *C. R. Acad. Sci. Ser. D* 280:2841.

Gieren, A., Narayanan, P., Hoppe, W., Hasan, M., Michl, K., Wieland, Th., Smith, H. O., Jung, G., and Breitmaier, E. (1974). Über die Inhaltsstoffe des grünen Knollenblätterpilzes. XLIV. Die Konfiguration der hydroxylierten Isoleucine der Amatoxine. *Justus Liebigs Ann. Chem.*, p. 1561.

Govindan, V. M. (1969). Vorbereitung der radioaktiven Markierung der Amanitine. Diploma work, University of Heidelberg.

Govindan, V. M., and Wieland, Th. (1975). Isolation and identification of an actin from rat liver. *FEBS Lett.* 59:117.

Govindan, V. M., Faulstich, H., Wieland, Th., Agostini, B., and Hasselbach, W. (1972). In-vitro effect of phalloidin on a plasma membrane preparation from rat liver. *Naturwissenschaften* 59:521.

Grant, W. D. (1972). Effect of alpha-amanitin and ammonium sulfate on RNA synthesis in nuclei and nucleoli isolated from *Physarum polycephalum* at different times during the cell cycle. *Eur. J. Biochem.* 29:94.

Greenleaf, A. L., Krämer, A., and Bautz, E. K. F. (1976). DNA-dependent RNA polymerases from *Drosophila melanogaster* larvae. In *RNA Polymerase*, R. Losick and M. Chamberlin (Eds.)., Cold Spring Harbor Laboratory, Cold Spring Harbor, New York, p. 793.

Hadjiolov, A. A., Dabeva, M. D., and Mackedonski, V. V. (1974). The action of α-amanitin in vivo on the synthesis and maturation of mouse liver ribonucleic acids. *Biochem. J.* 138:321.

Hager, G., Holland, P., Valenzuela, F., Weinberg, F., and Rutter, W. (1976). RNA polymerases and transcriptive specifity in *Saccharomyces cerevisiae.* In *RNA Polymerase*, R. Losick and M. Chamberlin (Eds.). Cold Spring Harbor Laboratory, Cold Spring Harbor, New York, p. 745.

Higashinakagawa, T., and Mita, T. (1973). DNA-dependent RNA polymerase of eukaryotic cells: a study with protozoon *Tetrahymena pyriformis. Gunma Symp. Endocrinol. 10*:41.

Hildebrandt, A., and Sauer, H. W. (1973). DNA dependent-RNA polymerases from *Physarum polycephalum. FEBS Lett. 35*:41.

Horgen, P. A., Ammirati, J. F., and Thiers, H. D. (1976). Occurrence of amatoxins in *Amanita ocreata. Lloydia 39*:368.

Hossenlopp, P., Wells, D., and Chambon, P. (1975). Animal DNA-dependent RNA polymerase. Partial purification and properties of three classes of RNA polymerases from uninfected and adenovirus-infected HeLa cells. *Eur. J. Biochem. 58*:237.

Ivanov, V. T. (1975). "Sandwich" complexation in cyclopeptides and its implications in membrane processes. *Ann. N.Y. Acad. Sci. 264*:221.

Jacob, S. T., Sajdel, E. M., Muecke, W., and Munro, H. N. (1970). Soluble RNA polymerases of rat liver nuclei. Properties, template specificity, and amanitin responses in vitro and in vivo. *Cold Spring Harbor Symp. Quant. Biol. 35*:681.

Jahn, W. (1977). Phalloidin hemmt die Bildung eines filamentösen Netzwerkes an der Membran endocytotischer Vakuolen in Leberparenchymzellen. *Cytobiologie 15*:452.

Jahn, W., Faulstich, H., Deboben, A., and Wieland, Th. (1980a). Formation of actin clusters in rat liver parenchymal cells on phalloidin poisoning as visualized by a fluorescent phallotoxin. *Z. Naturforsch. 35c*:467.

Jahn, W., Faulstich, H., and Wieland, Th. (1980b). Pharmacokinetics of [^{3}H-]-methyl-dehydroxymethyl-α-amanitin in the isolated perfused rat liver, and the influence of several drugs. In *Amanita Toxins and Poisoning*, H. Faulstich, B. Kommerell, and Th. Wieland (Eds.). Witzstrock-Verlag, Baden-Baden, p. 79.

Jendrisak, J., and Guilfoyle, T. I. (1978). Eukaryotic RNA polymerases: Comparative subunit structures, immunological properties, and α-amanitin sensitivities of the class II enzymes from higher plants. *Biochemistry 17*:1322.

Johnson, B. E. C., and Preston, J. F. (1976). Quantitation of amanitins in *Galerina autumnalis. Mycologia 68*:1248.

Johnson, B. E. C., and Preston, J. F. (1979). Unique amanitin resistance of RNA-synthesis in isolated nuclei from *Amanita* species accumulating amanitins. *Arch. Microbiol. 122*:161.

Jupille, T. H. (1977). High-performance thin-layer chromatography: a review of principles, practice and potential. *CRC Crit. Rev. Anal. Chem. 6*(4), 325.

Kedinger, C., Gniazdowski, M., Jr., Gissinger, F., and Chambon, P. (1970). a-Amanitin: a specific inhibitor of one of two DNA-dependent RNA polymerase activities from calf thymus. *Biochem. Biophys. Res. Commun. 38*:165.

Kostansek, E. C., Lipscomb, W. N., Yocum, R. R., and Tiessen, W. E. (1977). The crystal structure of the mushroom toxin β-amanitin. *J. Am. Chem. Soc. 99*:1273.

Kostansek, E. C., Lipscomb, W. N., Yocum, R. R., and Tiessen, W. E. (1978). Conformation of the mushroom toxin β-amanitin in the crystalline state. *Biochemistry 17*:3790.

Kroker, R., and Frimmer, M. (1974). Decrease of binding sites for phalloidin on the surface of liver cells during carbon tetrachloride intoxication. *Naunyn-Schmiedeberg's Arch. Pharmakol. 282*:109.

Lengsfeld, A. M., Löw, I., Wieland, Th., Dancker, P., and Hasselbach, W. Interaction of phalloidin with actin. *Proc. Natl. Acad. Sci. USA 71*:2803.

Lindell, Th. J., Weinberg, F., Morris, P. W., Roeder, R. G., and Rutter, W. I. (1970). Specific inhibition of nuclear RNA polymerase II by α-amanitin. *Science 170*:447.

Litten, W. (1975). The most poisonous mushrooms. *Sci. Am. 232*:90.

Löw, I., and Wieland, Th. (1974). The interaction of phalloidin, some of its derivatives and of other cyclic peptides with muscle actin as studied by viscosimetry. *FEBS Lett. 44*:340.

Lynen, F., and Wieland, U. (1938). Über die Giftstoffe des Knollenblätter-pilzes. IV. Kristallisation von Phalloidin. *Liebigs Ann. Chem. 533*:93.

Montesano, R., Gabbiani, G., Perrelet, A., and Orci, L. (1976). In vivo induction of tight junction proliferation in rat liver. *J. Cell Biol. 68*:973.

Morris, P. W., Venton, D. L., and Kelley, K. M. (1978). Biochemistry of the amatoxins: preparation and characterisation of stably iodinated α-amanitin. *Biochemistry 17*:690.

Munekata, E. (1981). New advances in phallotoxin chemistry. In *Perspectives in Peptide Chemistry*, A. Eberle, R. Geiger, and Th. Wieland (Eds.). S. Karger, Basel, p. 129.

Munekata, E., Faulstich, H., and Wieland, Th. (1977a). Rapid access to analogues of phalloidin by replacing alanine-1 in the natural toxin by other amino acids. *J. Am. Chem. Soc. 99*:6151.

Munekata, E., Faulstich, H., and Wieland, Th. (1977b). Totalsynthese von Phalloin und [Leu[7]]-phalloin. *Liebigs Ann. Chem.*, p. 1758.

Munekata, E., Faulstich, H., and Wieland, Th. (1978). Die Isolierung, Charakterisierung und Totalsynthese von Prophalloin (Pro[4]-phalloin), einem ungiftigen vermutlichen Vorläufer der Phallotoxine. *Liebigs Ann. Chem.*, p. 776.

Nakashima, T., and Volcani, B. (1969). 3,4-Dihydroxyproline: a new amino-acid in diatom cell walls. *Science 164*:1400.

Niessing, J., Schnieder, B., Kunz, W., Seifart, K. H., and Sekeris, C. E. (1970). Inhibition of RNA synthesis by α-amanitin in vivo. *Z. Naturforsch. 25b*:1119.

Palyza, V. (1972). Chromatographie der Amanita-Toxine. II. Neue Methode zur Identifizierung von Amanitatoxinen durch Dünnschichtchromatographie. *J. Chromatogr. 64*:317.

Palyza, V., and Kulhanek, V. (1970). Über die chromatographische Analyse von Toxinen aus *Amanita phalloides*. *J. Chromatogr. 53*:545.

Patel, D. J., Tonelli, A. E., Pfaender, P., Faulstich, H., and Wieland, Th. (1973). Experimental and calculated conformational characteristics of the bicyclic heptapeptide phalloidin. *J. Mol. Biol. 79*:185.

Petzinger, E., and Frimmer, M. (1980). Comparative studies on the uptake of ^{14}C-bile acids and ^{3}H-demethylphalloin in isolated rat liver cells. *Arch. Toxicol. 44*:127.

Pong, S. S., and Loomis, W. F., Jr. (1973). Multiple nuclear ribonucleic acid polymerases during development of *Dictyostelium discoideum*. *J. Biol. Chem. 218*:3833.

Preston, J. F., Stark, H. J., and Kimbrough, J. W. (1975). Quantitation of amanitins in *Amanita vera* with calf thymus RNA polymerase B. *Lloydia 38*: 153.

Puchinger, H., and Wieland, Th. (1969a). ^{3}H-Demethylphalloin. *Liebigs Ann. Chem. 725*:238.

Puchinger, H., and Wieland, Th. (1969b). Suche nach einem Metaboliten bie Vergiftung mit Desmethylphalloin (DMP). *Europ. J. Biochem. 11*:1.

Rehbinder, D., Löffler, G., Wieland, O., and Wieland, Th. (1963). Studien über den Mechanismus der Giftwirkung des Phalloidins mit radioaktiv markierten Giftstoffen. *Hoppe-Seyler's Z. Physiol. Chem. 331*:132.

Rinaldi, A., and Tyndalo, V. (1974). *Pilzatlas*. Hörnemann, Bonn.

Roeder, R. G. (1974). Multiple forms of deoxyribonucleic acid-dependent ribonucleic acid polymerase in *Xenopus laevis*: isolation and partial characterization. *J. Biol. Chem. 249*:249.

Saxon, M. E., Popov, V. I., Kirkin, A. H., Allakhverdov, B. L., Kovalenko, V. A., and Miroshnikov, A. I. (1978). De-novo formation of tight-like junctions induced with phalloidin between mouse lymphocytes. *Naturwissenschaften 65*:62.

Schäfer, A. J., and Faulstich, H. (1977). A protein-binding assay for phallotoxins using muscle actin. *Anal. Biochem. 83*:720.

Schultz, L. D., and Hall, B. D. (1976). Transcription in yeast: α-Amanitin sensitivity and other properties which distinguish between RNA polymerase I and III. *Proc. Natl. Acad. Sci. USA 73*:1029.

Seeger, R., and Stijve, T. (1979). Amanitin content and toxicity of *Amanita verna* Bull. *Z. Naturforsch. 34c*:330.

Seeger, R., Kraus, H., and Wiedmann, R. (1973). Zum Vorkommen von Hämolysinen in Pilzen der Gattung *Amanita*. *Arch. Toxicol. 30*:215.

Seifart, K. H., and Benecke, B. J. (1975). DNA-dependent RNA polymerase C. Occurrence and localization in various animal cells. *Eur. J. Biochem. 53*:293.

Seifart, K. H., and Sekeris, C. E. (1969). α-Amanitin, a specific inhibitor of transcription by mammalian RNA-polymerase. *Z. Naturforsch. 24b*:1538.

Siess, E., Wieland, O., and Müller, F. (1970). Elektronenmikroskopische Untersuchungen zur Phalloidintoleranz neugeborener Ratten, Mäuse und Kaninchen. *Virchows Arch., Abt. B. Zellpathol. 6*:151.

Sklar, V. E. F., and Roeder, R. G. (1975). Purification, characterization and structure of class III RNA polymerases. *Fed. Proc. Fed. Am. Soc. Exp. Biol. 34*:650.

Stijve, T. (1981). High performance thin-layer chromatographic determination of the toxic principles of some poisonous mushrooms. *Mitt. Geb. Lebensmittelunters. Hyg. 72*:44.

Stijve, T., and Seeger, R. (1979). Determination of α-, β-, and γ-amanitin by high performance thin-layer chromatography in *Amanita phalloides* (Vaill. ex Fr.) secr. from various origins. *Z. Naturforsch. 34c*:1133.

Strain, G. C., Mullinix, K. P., and Bogorad, L. (1971). RNA polymerase of maize: nuclear RNA polymerases. *Proc. Natl. Acad. Sci. USA 68*:2647.

Stunnenberg, H. G., Wennekes, L. M. J., Spierings, T., and Broek, H. A. v. d. (1981). An α-amanitin-resistant DNA-dependent RNA polymerase II from the fungus *Aspergillus nidulans*. *Eur. J. Biochem. 117*:121.

Szabados, A. (1971). Zum Phänomen der Phalloidintoleranz neugeborener Ratten — ein Beitrag zur Pathologie der Knollenblätterpilzvergiftung. M.D. thesis, University of Munich.

Tonelli, A. E., Patel, D. J., Wieland, Th., and Faulstich, H. (1978). The structure of α-amanitin in dimethylsulfoxide solution. *Biopolymers* 17:1973.

Tyler, V. E., Jr. (1953). Poisonous mushrooms. *Prog. Chem.Toxicol.* 1:339.

Tyler, V. E., and Smith, A. H. (1963). Chromatographic detection of amanita toxins in *Galerina venenata*. *Mycologia* 55:358.

Tyler, V. E., Brady, L. R., Benedict, R. G., Khanna, J. M., and Malone, M. H. (1963). Chromatographic and pharmacologic evaluation of some toxic Galerina species. *Lloydia* 26:154.

Tyler, V. E., Benedict, R. G., Brady, L. R., and Robbers, J. E. (1966). Occurrence of amanita toxins in American collections of deadly amanitas. *J. Pharm. Sci.* 55:590.

Vaisius, A. C., and Horgen, P. A. (1979). Purification and characterization of RNA polymerase II resistant to α-amanitin from the mushroom *Agaricus bisporus*. *Biochemistry* 18:795.

Vaisius, A. C., and Wieland, Th. (1982). Formation of a single phosphodiester bond by RNA polymerase B from calf thymus is not inhibited by α-amanitin. *Biochemistry* 21:3097.

Vogel, G. (1980). The anti-*Amanita* effect of silymarin. In *Amanita Toxins and Poisoning*, H. Faulstich, B. Kommerell, and Th. Wieland (Eds.). Witzstrock, Baden-Baden, p. 180.

Vogel, G. and Trost, W. (1975). Zur Anti-Phalloidin-Aktivität der Silymarine Silybin und Disilybin. *Arzneim. Forsch. (Drug Res.)* 25:392.

Vogt, M. (1938). Pharmakologische Untersuchung des kristallisierten Giftes "Phalloidin" des Knollenblätterschwammes. *Arch. Exp. Pathol. Pharmakol.* 190:406.

Walli, A. K., Wieland, E., Faulstich, H., and Wieland, Th. (1979). Reduced phallotoxin uptake by livers of young compared with adult rats. *Naunyn-Schmiedeberg's Arch. Pharmacol.* 307:283.

Walli, A. K., Wieland, E., and Wieland, Th. (1980). Phalloidin uptake by the liver of cholestatic rats in vivo in isolated perfused liver and isolated hepatocytes. *Naunyn-Schmiedeberg's Arch. Pharmacol.* 316:257.

Weil, P. A., and Blatti, St. P. (1976). HeLa cell deoxyribonucleic acid-dependent RNA polymerases: function and properties of the class III enzymes. *Biochemistry* 15:1500.

Weiss, E., Sterz, I., Frimmer, M., and Kroker, R. (1973). Electron microscopy of isolated rat hepatocytes before and after treatment with phalloidin. *Beitr. Pathol.* 150:345.

Wieland, E. (1981). Toleranz gegen Phalloidin, ein hepatotropes Pilzgift. M.D. thesis, University of Heidelberg.

Wieland, H., Hallermayer, R., and Zilg, W. (1941). Über die Giftstoffe des Knollenblätterpilzes. VI. Amanitin, des Hauptgift des Knollenblätterpilzes. *Liebigs Ann. Chem.* 548:1.

Wieland, O. (1975). Changes in liver metabolism induced by the poisons of *Amanita phalloides*. *Clinical Chemistry* 2:323.

Wieland, O., and Szabados, A. (1968). On the nature of phalloidin tolerance in newborn rats. *VIth Int. Congr. Clin. Chem.*, Munich. S. Karger, Basel, p. 59.

Wieland, Th. (1967). The toxic peptides of *Amanita phalloides*. In *Progr. Chem. of Organic Natural Products*, Vol. 25, L. Zechmeister (Ed.). Springer-Verlag, Vienna, p. 214.

Wieland, Th. (1968). Poisonous principles of mushrooms of the genus *Amanita*. *Science* 159:946.

Wieland, Th. (1972). Properties of antamanide and some of its analogues. In *Chemistry and Biology of Peptides, Proc. 3rd Am. Peptide Symp.*, J. Meienhofer (Ed.). Ann arbor science Publishers, Ann Arbor, Mich., p. 377.

Wieland, Th. (1977). Modification of actin by phallotoxins. *Naturwissenschaften 64*:303.

Wieland, Th., and Schön, W. (1955). Giftstoffe des grünen Knollenblätterpilzes. X. Die Konstitution des Phalloidins. *Liebigs Ann. Chem. 593*:157.

Wieland, Th., and Dudensing, Ch. (1956). Über die Giftstoffe des grünen Knollenblätterpilzes. XI. γ-Amanitin, eine weitere Giftkomponente. *Liebigs Ann. Chem. 600*:156.

Wieland, Th., and Mannes, K. (1957). Über die Giftstoffe des grünen Knollenblätterpilzes. 13. Mitt.: Phalloin, ein weiteres Toxin. *Angew. Chem. 11"389.*

Wieland, Th., and Schöpf, A. (1959). Über die Giftstoffe des grünen Knollenblätterpilzes. XVIII. Ergänzungen zur Phalloidinformel, Ketophalloidin. *Liebigs Ann. Chem. 626*:174.

Wieland, Th., and Wieland, O. (1959). Chemistry and toxicology of the toxins of *Amanita phalloides*. *Pharmacol. Rev. 11*:87.

Wieland, Th., and Boehringer, W. (1960). Über die Giftstoffe des grünen Knollenblätterpilzes. XIX. Umwandlung von β-Amanitin in α-Amanitin. *Liebigs Ann. Chem. 635*:178.

Wieland, Th., and Schnabel, H. W. (1962). Über die Giftstoffe des grünen Knollenblätterpilzes. XXI. Die Konstitution des Phallacidins. *Liebigs Ann. Chem. 657*:218.

Wieland, Th., and Rehbinder, D. (1963). Über die Giftstoffe des grünen Knollenblätterpilzes. XXIII. 35 S-Markierung und chem. Umwandlungen an einer Seitenkette des Phalloidins. *Liebigs Ann. Chem. 670*:149.

Wieland, Th., and Buku, A. (1968). Über die Inhaltsstoffe des grünen Knollenblätterpilzes. XXXVIII. Die Konstitution von ε-Amanitin and Amanullin. *Liebigs Ann. Chem. 717*:215.

Wieland, Th., and Jeck, R. (1968). Über die Inhaltsstoffe des grünen Knollenblätterpilzes. XXXV. Umwandlung des Phalloidins in das ebenfalls giftige Deoxydesmethylphalloin (Norphalloin). *Liebigs Ann. Chem. 713*:196.

Wieland, Th., and Fahrmeir, A. (1970). Über die Inhaltsstoffe des grünen Knollenblätterpilzes. XL. Oxidation und Reduktion an der γ-Dihydroxyisoleucin-Seitenkette des O-Methyl-α-amanitins. Methyl aldoamanitin, ein ungiftiges Abbauprodukt. *Leibigs Ann. Chem. 736*:95.

Wieland, Th., and Wieland, O. (1971). The toxic peptides of *Amanita* species. In *Microbial Toxins*, Vol. 8. Academic Press, New York, p. 249.

Wieland, Th., and Brodner, O. (1976). Über die Inhaltsstoffe des grünen Knollenblätterpilzes. L. Herstellung von [6^{ind}-^{3}H]Amanin, einem radioaktiven Amatoxin mit Carboxyfunktion. *Liebigs Ann. Chem.*, p. 1412.

Wieland, Th., and Faulstich, H. (1978). Amatoxins, phallotoxins, phallolysin and antamanide: the biologically active components of poisonous *Amanita* mushrooms. *Crit. Rev. Biochem. 5*:185.

Wieland, Th., and Löw, I. (1980). Concentration-dependent influence of various cytochalasins and chaetoglobusines on the phalloidin-induced polymerization of G-actin in 0.6 M potassium iodide. *Biochemistry 19*:3363.

Wieland, Th., Wirth, L., and Fischer, E. (1949). Über die Giftstoffe des Knollenblätterpilzes. 7. Mitt.: β-Amanitin, eine 3. Komponente. *Liebigs Ann. Chem. 564*:152.

Wieland, Th., Motzel, W., and Merz. H. (1953). Giftstoffe des Knollenblätterpilzes. 9. Mitt.: Über das Vorkommen von Bufotenin im gelben Knollen-
blätterpilz. *Liebigs Ann. Chem. 581*:10.

Wieland, Th., Schiefer, H., and Gebert, U. (1966). Giftstoffe von *Amanita
verna. Naturwissenschaften 53*:39.

Wieland, Th., Lüben, G., Ottenheym, H., and Schiefer, H. (1969). Über die
Inhaltsstoffe des grünen Knollenblätterpilzes. XXXIX. Isolierung und Charakterisierung eines antitoxischen Cyclopeptids, Antamanid, aus der lipo-
philen Extrakfraktion von *Amanita phalloides. Liebigs Ann. Chem. 722*:173.

Wieland, Th., Faulstich, H., and Burgermeister, W. (1972). Antamanide and
analogs. Studies on selectivity and stability of complexes. *Biochem.
Biophys. Res. Commun. 47*:984.

Wieland, Th., de Vries, J., Schäfer, A. J., and Faulstich, H. (1975). Spectroscopic evidence for the interaction of phalloidin with actin. *FEBS Lett. 54*:73.

Wieland, Th., Deboben, A., and Faulstich, H. (1980). Einige vom Ketophalloidin abgeleitete, in der biochemischen Forschung verwendbare Dithiolane.
Liebigs Ann. Chem., p. 416.

Wieland, Th., Götzendörfer, Ch., Zanotti, G., and Vaisius, A. C. (1981a).
The effect of the chemical nature of the side chains of amatoxins in the
inhibition of eukaryotic RNA polymerase. *Eur. J. Biochem. 117*:161-164.

Wieland, Th., Beijer, B., Seeliger, A., Dabrowski, J., Zanotti, G., Tonelli,
A. E., Gieren, A., Dederer, B., and Lamm, V. (1981b). Über die
Inhaltsstoffe des grünen Knollenblätterpilzes. LIX. Die Raumstruktur
der Phallotoxine. *Liebigs Ann. Chem.*, p. 2318.

Wieland, Th., Miura, T., and Seeliger, A. (1983). Analogs of phalloidin:
D-Abu[2]-lys[7]-phalloin, an F-actin binding analog, its rhodamine conjugate
(RLP) a novel fluorescent F-actin-probe, and D-Ala[2]-Leu[7]-phalloin, an
inert peptide. Submitted to *Int. J. Pep. Prot. Res. 21*:7.

Wieland, Th., Hollosi, M., and Nassal, M. (1983). Das (2', 5'-Diamino-4'-
hydroxy-4'-methylvaleriansäure)-7-analoge des Phalloidins (δ-Amino-
phalloin) und biochemisch nützliche Derivate. *Liebigs Ann. Chem.*, in press.

Wilhelm, I., Dina, D., and Crippa, M. (1974). A special form of deoxyribonucleic acid dependent ribonucleic acid polymerase from oocytes of
Xenopus laevis. Isolation and characterization. *Biochemistry 13*:1200.

Wulf, E., Deboben, A., Bautz, F. A., Faulstich, H., and Wieland, Th.
(1979). Fluorescent phallotoxin, a tool for the visualization of cellular
actin. *Proc. Natl. Acad. Sci. USA 76*:4498.

Wulf, E., Bautz, F. A., Faulstich, H., and Wieland, Th. (1980). Distribution of fluorescent α-amanitin (FAMA) during mitosis in cultivated rat
kangaroo (PtK1) cells. *Exp. Cell Res. 130*:415.

Yaguara, T., Yanagisawa, M., and Iwabuchi, M. (1976). Evidence for two
α-amanitin-resistant RNA polymerases in vegetative amobae of Dictyostelium discoideum. *Biochem. Biophys. Res. Commun. 68*:483.

Yocum, R. R., and Simons, D. M. (1977). Amatoxins and phallotoxins in
Amanita species of the northeastern United States. *Lloydia 40*:178.

Ziegler, K., Grundmann, E., and Frimmer, M. (1979). Decreased sensitivity
of isolated hepatocytes from baby rats from regnerating and from poisoned
livers to phalloidin. *Naunyn-Schmiedeberg's Arch. Pharmacol. 306*:295.

Ziegler, K., Petzinger, E., and Frimmer, M. (1980). Decreased phalloidin
response, phallotoxin uptake and bile acid transport in hepatocytes prepared from Wistar rats treated chronically with diethylnitrosamine.
Naunyn-Schmiedeberg's Arch. Pharmacol. 310:245.

Zylber, E. A., and Penman, Sh. (1971). Products of RNA polymerases in
HeLa cell nuclei. *Proc. Natl. Acad. Sci. USA 68*:2861.

19

HEPATOTOXIC PYRROLIZIDINE ALKALOIDS

J. E. PETERSON and C. C. J. CULVENOR

Commonwealth Scientific and Industrial Research Organization, Melbourne, Victoria, Australia

I. INTRODUCTION AND PLANT SOURCES

Pyrrolizidine alkaloids, characteristically, are constituents of herbs present
in grasslands and uncultivated areas as weeds or small shrubs. Suspicion of
poisoning due to pyrrolizidine alkaloid plants was first recorded for *Senecio
jacobaea* in England in the eighteenth century and for *Crotalaria* species in
the United States in the nineteenth century. The first proof that a chronic
liver disease was due to *S. jacobaea* came in 1903 (Gilruth, 1903) and a similar
demonstration on related species soon followed in a number of countries.
Poisoning of humans through the consumption of contaminated bread was
recognized in South Africa in 1920 (Willmott and Robertson, 1920) and in the
USSR from 1930 (Bourkser, 1947; Milenkov and Kizhaikin, 1952), the sources
of alkaloid being *Senecio* spp. and *Heliotropium lasiocarpum*, respectively.
The Soviet studies provided the first incrimination of species in the family
Boraginaceae.

Investigations over the last 30 years have confirmed these three plant
groups, *Crotalaria* species, *Senecio* and related species, and the family
Boraginaceae, as the main sources of the hepatotoxic pyrrolizidine alkaloids
(Smith and Culvenor, 1981). Another genus of composites, *Eupatorium*, is
of growing importance and there have been other significant additions to the
source list (Table 1). These principal genera contain a large number of
species, some of which are found in almost all regions of the world. The
Senecio genus is cosmopolitan, *Crotalaria* is largely a tropical to subtropical
genus, *Eupatorium* is centered in the Americas, and the Boraginaceae are
mainly tropical and temperate and especially suited to mediterranean climates.

Toxic species typically contain 0.2-3% alkaloid, although levels above 10%
have been recorded, for example, in seed of *Crotalaria retusa* (Kumari et al.,
1966) and leaf of *Senecio riddellii* (Molyneux et al., 1979; R. J. Molyneux,
personal communication, 1982). A feature of pyrrolizidine alkaloids is that
they may be present in large proportion as *N*-oxides which are neither basic
nor bitter and are not rejected in feed by animals.

II. CHEMICAL STRUCTURE AND PROPERTIES

About one-half of the approximately 250 known pyrrolizidine alkaloids are
hepatotoxic. They are esters of the unsaturated amino alcohols, heliotridine, retronecine, otonecine, crotanecine, and supinidine (Fig. 1), usually
with a unique type of branched-chain hydroxylated acid but also angelic,
tiglic, and acetic acids. Pyrrolizidine alkaloids which are derived from

TABLE 1 Plant Families and Genera Containing Hepatotoxic Pyrrolizidine Alkaloids[a]

Family	Subfamily/tribe	Genera
Apocynaceae	Echitoideae	*Parsonsia*
Boraginaceae	Heliotropioideae	*Heliotropium, Tournefortia*
	Boraginoideae	*Amsinckia, Anchusa, Cynoglossum, Lappula, Lindelofia, Myosotis, Paracynoglossum, Rindera, Solenanthus, Symphytum, Trachelanthus, Trichodesma*
Celastraceae		*Bhesa*
Compositae	Eupatorieae	*Conoclinium, Eupatorium*
	Senecioneae	*Adenostyles, Brachyglottis, Cacalia, Doronicum, Emilia, Erechtites, Farfugium, Ligularia, Petasites, Senecio, Syneilesis, Tussilago*
Leguminosae	Genisteae	*Crotalaria*
Ranunculaceae		*Caltha*
Scrophulariaceae		*Castilleja*

[a]See Smith and Culvenor, 1981.

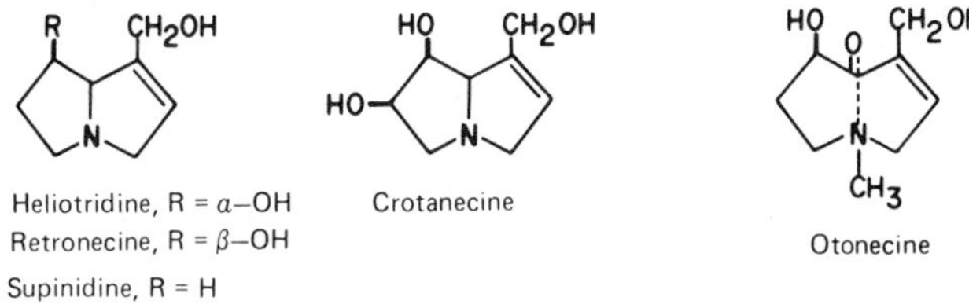

Heliotridine, R = α—OH
Retronecine, R = β—OH
Supinidine, R = H

Crotanecine

Otonecine

FIGURE 1 Amino alcohols from hepatotoxic pyrrolizidine alkaloids.

saturated amino alcohols or are not esters are not hepatotoxic. Typical hepatotoxic alkaloids are shown in Fig. 2. They may be mono- or diesters with monocarboxylic acids (e.g., heliotrine, lasiocarpine) or macrocyclic diesters formed from dicarboxylic acids (e.g., monocrotaline, retrorsine). Supinidine esters are only weakly hepatotoxic. A more detailed account of structure-toxicity relationships is given by Culvenor et al. (1976a); see also Mattocks (1972a).

As esters, the alkaloids are susceptible to hydrolysis either chemically or enzymatically. The hydrolysis products are nonhepatotoxic. Another significant reaction is oxidation, which includes a ready conversion of the

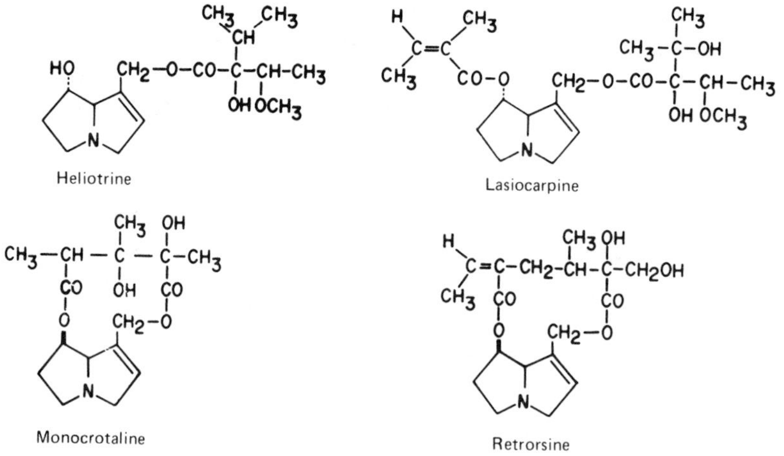

FIGURE 2 Typical hepatotoxic pyrrolizidine alkaloids.

pyrroline ring to a pyrrolic derivative. This may be effected by chloranil
and related reagents, by manganese dioxide, and at a slow rate even by
molecular oxygen (Culvenor et al. , 1970a,b). Isolation of the pyrrolic
products can be difficult because of the high alkylating reactivity engendered
by the acyloxymethylpyrrole groupings (shown in heavy outline in Fig. 3;
note that diester alkaloids produce a difunctional alkylating agent).

An effective method of preparing the pyrrolic analogs of the alkaloids,
the dehydro alkaloids, is by reaction of the alkaloid *N*-oxides with acetic
anhydride (Culvenor et al. , 1970a; Mattocks, 1969) or with ferrous sulfate
in methanol (Mattocks, 1968a).

Metabolic activation of the alkaloids to the pyrrolic dehydroalkaloids is the
main basis of alkaloid toxicity in animals (Sec. VIII.B). The dehydro alkaloids
are apparently also formed in pyrrolizidine alkaloid-containing plants. A reaction product of dehydroheliotrine and heliotrine was detected in *Heliotropium europaeum* (Culvenor and Smith, 1969) and 5-keto derivatives of
dehydro alkaloids (Fig. 4) have been found in nonbasic extracts of some
Senecio species (e.g., Bohlmann et al. , 1977). The possible contribution of
the 5-keto compounds to toxicity of *Senecio* species has not yet been investigated.

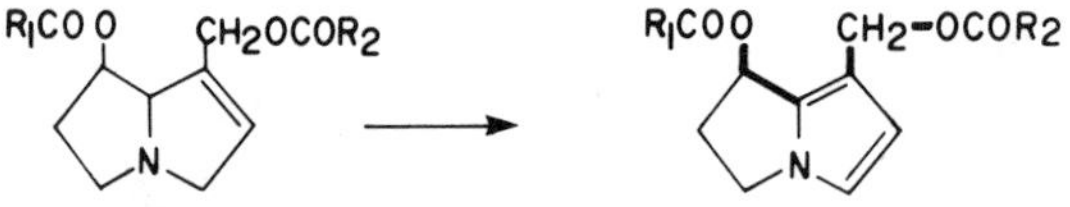

FIGURE 3 Oxidation of pyrrolizidine alkaloids to pyrrolic dehydro alkaloids.

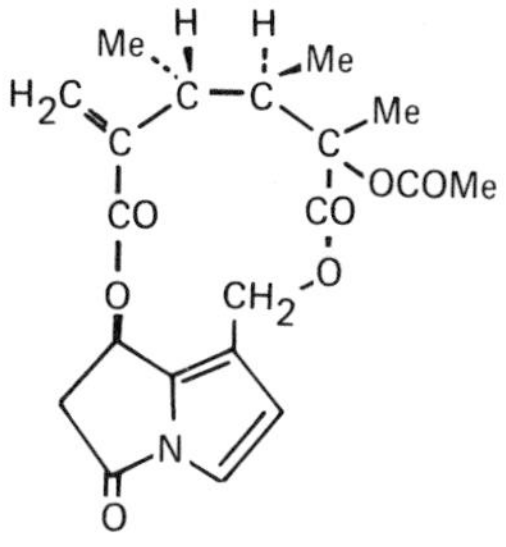

FIGURE 4 Pterophorine, a 5-oxo derivative of a pyrrolizidine dehydro alkaloid, from *Senecio pterophorus*.

III. DETECTION AND ASSAY

Pyrrolizidine alkaloids may be detected in, or extracted from plants or animal tissues by the usual alkaloid procedures with incorporation of a reduction step to convert N-oxides into the tertiary bases (Culvenor and Smith, 1955). The usual reduction procedure, zinc dust in sulfuric acid, is tedious and subjects the alkaloids to mildly hydrolyzing conditions. The use of a column of oxidation-reduction resin for this step (Huizing and Malingré, 1979) is a welcome recent development which needs wider trial.

Detection of the alkaloids on thin layers and pherograms may be achieved with reagents such as iodine vapor and Draggendorff's reagent. In a specific method for hepatotoxic pyrrolizidine alkaloids (Mattocks, 1967a), the alkaloids are converted to N-oxides with hydrogen peroxide and then reacted with acetic anhydride to form the pyrrolic analogs, which give a strong purple color with Ehrlich's reagent. Iodine partially oxidizes the alkaloid directly to the pyrrolic derivative, and the use of Ehrlich's reagent after iodine vapor achieves a similar, although probably less sensitive, result. o-Chloranil has recently been introduced to oxidize the alkaloids directly to pyrroles for detection on thin layers (Molyneux and Roitman, 1980; Huizing et al., 1980).

Quantitative estimation of each alkaloid in a mixture of alkaloids can be performed by gas chromatography-mass spectrometry (e.g., Culvenor et al., 1980a, 1981). This technique is sensitive enough to measure levels below 0.1 ppm in foodstuffs or to 0.1 ppm in blood. Procedures that measure total hepatotoxic alkaloids have also been developed:

1. An alkaloid mixture is converted to the pyrrolic analogs by Mattocks' oxidation technique, treated with Ehrlich's reagent, and estimated photometrically (Mattocks, 1967b, 1968b). The method has been adapted to measure pyrrolic metabolites bound to animal tissue (Mattocks and White, 1970).

2. The alkaloids are hydrolyzed to retronecine (or heliotridine), which is converted into the bis-trifluoroacetate or bis-heptafluorobutyrate and estimated by electron-capture gas chromatography (Deinzer et al., 1978). This method has high sensitivity and can measure nanogram amounts of alkaloid. It is suited to measuring low-level alkaloid contamination of foodstuffs.

3. A nuclear magnetic resonance spectrum is taken on the total crude
    alkaloid and the hepatotoxic alkaloid estimated from the integral of the
    olefinic hydrogen atom in the pyrrolizidine ring (Molyneux et al.,
    1979). This method was developed for assaying plant samples and
    gives additional information about the identity of the alkaloids.

High-pressure liquid chromatography has been used for the analysis of
pyrrolizidine alkaloids, in particular the mixtures present in *Senecio jacobaea,
S. longilobus*, and *S. vulgaris* (Segall and Molyneux, 1978; Segall, 1979a;
Segall and Krick, 1979). Good resolution is obtained with a C_{18} reversed-
phase column and ultraviolet (UV) detection is satisfactory even with
Amsinckia alkaloids (Dimenna et al., 1980), which do not contain an α, β-
unsaturated ester system. Preparative applications are also successful
(Segall, 1979b).

IV. PATHOLOGICAL EFFECTS

Toxicity of pyrrolizidine alkaloids may be manifested in peracute, acute, or
chronic form, distinguished in rats by the time elapsing between a single
lethal injection of alkaloid and death. Peracute deaths occur within 24 hr,
usually within 2 hr; acute deaths occur mostly in 2-5 days, seldom in less
than 2 or more than 8 days; chronic deaths are infrequent in less than 2
weeks. If repeated sublethal doses are administered, the changes characteristic of both acute and chronic effects may be present together.

A. Peracute Toxicity

A number of the alkaloids possess pharmacological activities that are not related to their hepatotoxicity (Pomeroy and Raper, 1971, 1972). One such
activity is the neuromuscular blockade exhibited by heliotrine and lasiocarpine
(Gallagher and Koch, 1959), which may result in respiratory failure. At
present this seems the likely cause of the deaths that occur within a few
hours of the administration of alkaloid without recognized morphological
changes in any tissues. With hepatotoxic alkaloids, the dose rate required
for peracute death is seldom less than twice that required to produce acute
deaths (Culvenor et al., 1976a). Nonhepatotoxic alkaloids show similar peracute toxicity.

B. Acute Toxicity

The hallmark of acute pyrrolizidine toxicity is hepatic zonal necrosis that
tends to be centrilobular, although lesions restricted to midlobular or periportal zones have been described (Harris et al., 1942, 1957; Wakim et al.,
1946). This distribution is influenced by factors such as lobular patterns
of blood flow and microsomal enzyme distribution. When a single lethal dose
is administered the death that follows within a few days is due to acute liver
failure as a result of extensive necrosis, which may involve all but a narrow
rim of periportal hepatocytes. In domestic animals acute deaths are more
liable to be associated with the feeding of heavily contaminated hay, silage,
or grain than with the free grazing of plants, where lack of palatability is

likely to restrict the amount consumed. Acute toxicity in humans is most
likely to be associated with the drinking of extracts of toxic plants for medicinal purposes.

The development of hepatic necrosis has been described by Christie
(1958a) and Kerr (1969a) and variations among animal species have been reviewed by McLean (1970), who also discusses the nature and time sequence
of the biochemical events involved. There is an early, rapid decline in the
synthesis of protein (Svoboda and Soga, 1966; Demale and Moulé, 1971) and
RNA (Frayssinet and Moulé, 1969). Shortly before necrosis becomes recognizable the tricarboxylic acid cycle becomes disorganized due to loss of NAD-
dependent respiratory enzyme activity (Christie et al., 1962; Gallagher,
1960).

Morphological changes that have been described include mitochondrial
damage (Kerr, 1969a), ribosomal alterations (Monneron, 1969; Harris et al.,
1969), nucleolar segregation (Svoboda and Soga, 1966), and increased lysosomal activity with, later, formation of giant cytosegresomes (Kerr, 1967,
1969b). Impairment of liver cell membrane function has been indicated in
histochemical studies (Persaud et al., 1970), and Tsuji (1980) has suggested
that lipid peroxidation might be involved.

1. Acute Toxicity Levels of Pure Alkaloids

The available figures are mostly for rats by intraperitoneal injection. Alkaloids are grouped in the following tables to indicate approximate relative
toxicities; individual values can be found in the cited references.

1. Acute LD_{50} values for adult male rats are given as dose per kilogram
    of body weight; intraperitoneal injections, based on deaths within 3
    days (Bull et al., 1968):

    60-100 mg: heliosupine, lasiocarpine, seneciphylline, senecionine
    125-200 mg: latifoline, heleurine, monocrotaline, echimidine
    220-300 mg: spectabiline, cynaustine, heliotrine
    350-550 mg: echinatine, rinderine
    >1000 mg: europine, heliotridine, heliotrine *N*-oxide

    Acute toxicities of retrorsine and monocrotaline in male and female
    rats, normal and pretreated with an inducer or an inhibitor of drug-
    metabolizing enzymes, have been reported by Mattocks (1972b).
2. Minimum acute lethal doses in 2-week-old rats; intraperitoneal injec-
    tions, expressed as millimoles per kilogram of body weight with
    approximate mg/kg ranges in parentheses (Culvenor et al., 1976a):

    0.05 mmol (17-21 mg): lasiocarpine, retusamine, seneciphylline
    0.1 mmol (30-40 mg): heliosupine, crispatine, retrorsine, senecionine,
        usaramine
    0.2 mmol (60-70 mg): fulvine, anacrotine
    0.4 mmol (120-140 mg): heliotrine, monocrotaline, madurensine
    3.2 mmol (960 mg): rinderine

    Except with a small number of alkaloids (e.g., monocrotaline and
    rinderine), young rats are more susceptible than adults. Minimum
    dose levels for peracute and chronic toxic effects in 2-week-old rats
    are also given by Culvenor et al. (1976a).

3. Acute LD$_{50}$ values for mice, for some alkaloids, have also been published (*Research Today*, 1949). The values are not greatly different from those for adult rats.

Outside the hepatocytes the same process of acute cellular necrosis may also affect endothelial cells and may be the basis of later extensive chronic vascular damage. For reasons explained later (Sec. VIII.B), this primary vascular damage is usually restricted to hepatic sinusoids or central and sublobular veins, or to pulmonary arterioles and capillaries.

In acute toxicity there may be little to see at necropsy other than the mottling of the liver caused by hemorrhagic zonal necrosis. If the dose has not been lethal, repair processes come into operation and the progressive chronic disease becomes established. However, the liver remains fully susceptible to further acute reactions if alkaloid is again consumed, although the lobular distribution may not be readily apparent in advanced disease when hepatic architecture has been severely disrupted.

C. Chronic Toxicity

Two classes of reaction can be distinguished in the development of chronic pyrrolizidine alkaloid poisoning. There is first a primary reaction between specific tissue components and pyrrolic metabolites of the alkaloids (Sec. VIII.B). Secondary reactions, in which the alkaloid metabolites are not directly involved, then develop in response to the initial damage, and these, in turn, may lead to more widely distributed pathology. It is characteristic of these alkaloids that a single dose that is not acutely toxic may induce a chronic disease that progresses over a long period without further contact with the initiating toxin (Schoental and Magee, 1957; Nolan et al., 1966). This would suggest either the continued presence of firmly bound metabolite in affected cells or permanent alteration of cell control mechanisms.

The primary biochemical and morphological changes seen in acute necrosis no doubt occur at a lower level of intensity when a nonnecrogenic dose of alkaloid is administered, and in this respect there may be a dose-effect relationship between acute and chronic toxicity. Some of the effects described, such as polyribosomal disaggregation and nucleolar segregation, may be of short duration in cells that survive.

There is also a primary reaction between pyrrolic metabolite and cell nucleus that is fundamental to the development of the chronic disease and is expressed as a strong antimitotic effect that persists for several months after a single injection (Peterson, 1965; Downing and Peterson, 1968). However, although mitosis is blocked at the end of the S phase or early in G2 (Samuel and Jago, 1975), the cell is able to bypass the mitotic phase and to continue repeatedly through the remainder of the cycle, including the DNA-synthetic phase. The stimulus for hepatocytes to enter this activity may be the demands for either normal growth or regeneration (Jago, 1969). The final result is the development of large cells with polyploid nuclei, referred to as megalocytes (Bull, 1955). As a large proportion of hepatocytes can be affected by a single injection of alkaloid, it is clear that the cells are susceptible throughout most, if not the whole, of the cell cycle. In the absence of a demand for an increase in hepatic tissue the damage may persist for long periods in latent form. There has been considerable speculation on the significance of megalocytes, but Svoboda et al. (1971), on the basis of bio-chemical and ultrastructural studies, concluded that they were functionally normal except for hypertrophy and the inability to divide. Megalocytosis is a relatively slowly developing phenomenon that is best observed in the liver since this is the site of production of the responsible pyrroles, but may occur in other tissues, usually those with a normal slow rate of cell turnover and lacking a defined compartment of generative cells.

The megalocytic response is not the only possible consequence of mitotic inhibition. In some tissues, such as intestinal epithelium and skin, characterized by a clearly defined generative compartment of cells and a high rate of cell turnover, neither mitosis nor megalocytes develop, even though cells continue to be lost at the normal rate. DNA replication may also be blocked in these tissues. This type of inhibition, in general, does not persist as long as that associated with megalocytosis. Reactions of this type are well seen in young rats treated with the pyrrolic metabolites, dehydroheliotridine (Peterson et al., 1972), and dehydroretronecine (Allen and Hsu, 1974), and have also been observed, but less consistently, in other animals treated with the parent alkaloids (Hooper, 1975; Castles et al., 1974).

Secondary responses that follow the initial damage in the liver are the normal repair mechanisms characteristic of that organ. Acute necrosis is followed by the development of fibrosis, accentuated by stromal collapse as the necrotic tissue is removed. At the same time, acute endothelial damage may lead to thrombosis and occlusion of central veins. The combination of these reactions eventually leads to extensive nonportal cirrhosis. Some portal fibrosis may also develop, particularly when there is a prolonged intake of alkaloid at a low dose level. In advanced chronic disease, lobular architecture may be further disrupted by bands of dissecting fibrosis.

Proliferation of bile ductule cells and the formation of new bile ducts can be extensive. The stimulus for this reaction is possibly related to the disruption of normal canalicular or ductular bile flow. It does not occur early enough to suggest any direct association with the alkaloids or their metabolites.

Loss of functional liver tissue gives rise to nodular hyperplasia. These nodules of regeneration tissue may eventually account for a large proportion of the surviving hepatocytes. Nodule formation tends to be favored by intermittent alkaloid intake and suppressed when the administration is continuous (Bull and Dick, 1959).

As this chronic damage develops, the functional capacity of the liver progressively deteriorates and in most animals death is due in large measure to hepatic failure. At this stage the liver is commonly atrophic and fibrotic. Failing liver function is indicated by the usual biochemical aberrations and the consequences can be observed clinically. For example, reduced albumen synthesis, causing a lowering of plasma protein levels, may affect capillary permeability and contribute to the development of edema and the accumulation of fluid in serous cavities; reduction of clotting factors promotes hemorrhage; reduced capacity to conjugate and excrete bile and other pigments contributes to the development of jaundice and photosensitization; reduced urea synthesis may lead to hyperammonemia, cerebral damage, and disturbances of behavior (Hooper et al., 1974).

In the lung two distinct types of damage may occur. Vascular damage, which originates as an acute primary reaction between pyrrolic metabolite and arteriolar or capillary endothelium, progresses to more extensive and chronic damage with considerable vascular occlusion (Allen et al., 1967, 1969; Allen and Carstens, 1970a,b; Brooks et al., 1970; Lalich et al., 1977; Heath and Kay, 1967). This results in the development of pulmonary edema and in severe cases to right-ventricular hypertrophy and to cor pulmonale syndrome, although this extreme progression is seldom seen under field conditions. The second type of lung damage that may occur is a cellular reaction in which a variable increase in interstitial tissue may develop, within which large cells, analogous to the hepatic megalocytes and probably derived from type 2 pneumocytes, may be prominent (Barnes et al., 1964; Butler et al., 1970). In addition, lung alveolar epithelium may undergo the cuboidal or low columnar metaplasia that is often referred to as "epithelialization." The pathogenesis of the lesion is not known. It has been seen in several species of animals exposed to various alkaloids under both field and experimental conditions. It is not a specific reaction to pyrrolizidine alkaloids. The lesions in advanced cases, as in jaagsiekte of horses eating *C. dura* (Steyn, 1934), can resemble pulmonary adenomatosis and interference with gas exchange can result in severe dyspnea.

The development of lung disease depends largely, if not fully, on pyrrolic metabolites produced in the liver. Consequently, it is always accompanied by liver disease. The balance between the severity of the damage to the two organs and the resulting clinical disease depends on many factors, of which animal species and type of alkaloid seem of particular importance, and while respiratory disease can be the most obvious manifestation, it occurs in only a small proportion of outbreaks of pyrrolizidine alkaloid poisoning.

Alimentary tract damage is common but not constant in the chronic disease. Early ulceration, hemorrhage, and diarrhea may follow from epithelial loss and villous atrophy that result from the direct action of pyrrolic metabolites, but chronic diarrhea and tenesmus in the later, terminal stages are probably related to the extensive edema that develops in the mesentery and in the wall of the stomach and intestine.

Numerous lesions in other organs have been described inconsistently, including nephritis, splenomegaly, lymphadenopathy, thymic cortical necrosis, anemia, and widespread hemorrhage. Most seem to be secondary in nature and not caused by direct action of the pyrrolic metabolites.

Alkaloids administered to pregnant animals may also have embryotoxic or teratogenic effects (Peterson and Jago, 1980). Such responses are readily produced in rats, but evidence is lacking on their importance in domestic animals or humans.

V. FACTORS INFLUENCING THE TOXICITY OF PYRROLIZIDINE ALKALOIDS

Animal species: Variations between species in their susceptibility to pyrrolizidine alkaloids are known and utilized. For example, because of their relatively low susceptibility, sheep have been used extensively to control *S. jacobaea*. There are marked differences in the levels of activity of hepatic pyrrole-producing enzymes in different species, and such variations are often a major factor in the susceptibility differences between species (White et al., 1973); Chesney and Allen, 1973; Shull et al., 1976). Poor absorption of alkaloids or their destruction in the intestine appears to be responsible for low susceptibility of rabbits (Pierson et al., 1977) and Japanese quail (Buckmaster et al., 1977). Differences between breeds or strains within a species have been reported in rats (Goldenthal et al., 1964; Lalich, 1964) and sheep (Bull et al., 1956), in the latter case thought to be based on selectivity of grazing.

Age: Susceptibility to both acute and chronic disease, especially megalocytosis, is high in young, rapidly growing animals but declines as the growth rate is reduced (Jago, 1969, 1970, 1971). The difference is due partly to the rate of cell division in the liver of young animals and partly to changes in the level of drug-metabolizing enzymes. The rat fetus is generally regarded as deficient in these enzymes (Kato et al., 1964; Henderson, 1971), although cultures of human embryonic liver cell are capable of converting alkaloids to the pyrrolic dehydro alkaloids (Armstrong and Zuckerman, 1971). After birth, rat liver enzyme levels increase rapidly, reaching adult levels at about 5 days (Mattocks and White, 1973).

Sex: Interactions affecting toxicity of alkaloids have been described between sex and type of alkaloid, animal species, age, and diet (Smit, 1952; Goldenthal et al., 1964; Jago, 1969, 1971; Mattocks, 1972a). With some alkaloids (e.g., retrorsine, monocrotaline, and heliotrine) males are more susceptible than females, but with other alkaloids (e.g., lasiocarpine and senecionine) males are more resistant. Toxicity can be influenced by steroidal sex hormones (Campbell, 1957a,b).

Diet: Deficiencies of protein (Mattocks, 1972a) and lipotropic factors (Newberne et al., 1971) reduce the activity of the hepatic pyrrole-producing microsomal enzymes, but this may either reduce or enhance toxicity, depending on the alkaloid (Newberne, 1968; Newberne et al., 1971; Cheeke and Gorman, 1974; Schoental and Magee, 1957; Selzer et al., 1951; Mattocks, 1972a). More general dietary restriction has been reported to inhibit the development of cardiopulmonary lesions (Hayashi et al., 1979). Dietary constituents rich in sulfydryl compounds may have some protective value by competing with tissue nucleophilic sites for alkylation by pyrroles (Hayashi and Lalich, 1968; Cheeke and Gorman, 1974; Buckmaster et al., 1976).

VI. PLANTS KNOWN OR SUSPECTED OF CAUSING NATURAL OUTBREAKS

*Horse**

Amsinckia intermedia (McCulloch, 1940).
Crotalaria crispata[†] (Gardiner et al., 1965); *C. dura*[†] (Steyn and van der Walt, 1945); *C. globifera*[†] (Steyn and van der Walt, 1945); *C. retusa* (Rose et al., 1957); *C. spectabilis* (Gibbons et al., 1953); *C. sagittalis* (Gibbons et al., 1950).

---

*
Eupatorium adenophorum (O'Sullivan, 1979) may cause acute respiratory disease but probably not due to its pyrrolizidine alkaloid content.
[†]Indicates species associated with clinical respiratory disease.

Senecio aquaticus (Dybing and Hjelle, 1958); *S. erraticus* (Araya and
Gonzalez, 1979); *S. erraticus* subsp. *barbaraeifolius* (Vanek, 1956);
S. integerrimus (Clawson, 1933); *S. jacobaea* (Gilruth, 1903); *S. lati-
folius* (Verney, 1911); *S. longilobus* (Clawson, 1933); *S. riddellii*
(Bukey and Cunningham, 1933); *S. vernalis* (Köhler, 1950); *S. vul-
garis* (Gulick et al., 1980).
Trichodesma incanum (Smirnov et al., 1959a).
    See also Smirnov and Stoljarova (1959), Kalkus et al. (1925), and
    Hill and Martin (1958) for further clinical and pathological reports.

Cattle

Amsinckia intermedia (McCulloch, 1940).
Crotalaria spectabilis (Piercy and Rusoff, 1946).
Heliotropium europaeum (Bull et al., 1961).
Senecio alpinus (von Pohlenz et al., 1980); *S. aquaticus* (Jalving, 1930);
    S. bipinnatisecticus (Mortimer and White, 1975); *S. brasiliensis* var.
    tripartitus (Podesta et al., 1977); *S. burchellii* (Steyn, 1934); *S.
    erraticus* (Araya and Gonzalez, 1979); *S. isatideus* (Steyn, 1934); *S.
    jacobaea* (Gilruth, 1903; Pethick, 1906); *S. latifolius* (Robertson,
    1906); *S. longilobus* (Dollahite, 1972); *S. moorei* (Mugera, 1970); *S.
    quadridentatus* (Dickson and Hill, 1977); *S. retrorsus* (Steyn, 1934);
    S. riddellii (Mathews, 1933); *S. sceleratus* (Wild, 1947); *S. spartioides*
    (Dollahite, 1972); *S. tweedii* (Carrillo et al., 1976); *S. vulgaris*
    (Fowler, 1968).
Trichodesma incanum (Smirnov et al., 1959b).
    See also Donald and Shanks (1956), Thorpe and Ford (1968), and
    Pearson (1977) for further clinical and pathological reports.

Sheep

Echium plantagineum (St. George-Grambauer and Rac, 1962).
Heliotropium dasycarpum (Absalyamov and Muratov, 1971); *H. europaeum*
    (Bull et al., 1956); *H. lasiocarpum* (Bourkser, 1947).
Senecio jacobaea (Gilruth, 1904); *S. longilobus* (Clawson, 1933).
Crotalaria mucronata (Laws, 1968) may cause acute respiratory disease,
    probably not due to its pyrrolizidine alkaloid content.

Pig

Amsinckia intermedia (McCulloch, 1940).
Crotalaria retusa (Hooper and Scanlan, 1977); *C. spectabilis* (Emmel et al.,
    1935a,b).
Echium plantagineum (Hunt, 1975).
Heliotropium lasiocarpum (Bourkser, 1947).
    See also Harding et al. (1964) for additional pathological reports.

Poultry

Crotalaria goreensis (Norton and O'Rourke, 1979); *C. mucronata* var.
    giant striata (Bierer et al., 1960); *C. retusa* (Hooper and Scanlan,
    1977); *C. spectabilis* (Thomas, 1934).
Heliotropium europaeum (Pass et al., 1979); *H. lasiocarpum* (Bourkser,
    1947).

Trichodesma incanum (Shevchenko and Ibadullaev, 1974).

See also Gopinath and Ford (1977) for additional pathological report.

Humans

Grain toxicity

Crotalaria nana (Tandon et al., 1978).
Heliotropium lasiocarpum (Khanin, 1956); *H. popovii* (Mohabbat et al., 1976).
Senecio burchellii, S. ilicifolius (Willmot and Robertson, 1920).

Medicinal or other use

Crotalaria fulva (Bras et al., 1961); *C. juncea* (Lyford et al., 1976).
Heliotropium eichwaldii (Datta et al., 1978).
Senecio longilobus (Huxtable, 1980); *S. barbellatus, S. burchellii, S. coronatus, S. deltoideus, S. ilicifolius, S. retrorsus* (Rose, 1972).
Numerous authors imply that many other toxic species may be used in herbal teas or refer to specific plants that are known to be used by particular communities without evidence of causing sickness.

VII. SYNDROMES ASSOCIATED WITH PYRROLIZIDINE ALKALOIDS

Natural outbreaks of pyrrolizidine alkaloid poisoning tend to result from a unique combination of circumstances in which a particular species of plant or its seeds are consumed by one or a very few species of animals. All plants containing hepatotoxic pyrrolizidine alkaloids must be regarded as potentially toxic for all species of animals. The limited number of plant species that have been associated with animal toxicity under natural conditions is an indication of the likelihood of their consumption. When animals have unrestricted grazing, palatability is an important factor determining the amount of toxic plant eaten, but when plant or seed are included in chaff, silage, food pellets, or meals fed as a sole item of diet, there is no opportunity for voluntary restriction of the amount of alkaloid consumed.

Because the pattern of intake of alkaloid has an important bearing on the pathological changes and clinical disease and because outbreaks of disease in any single species have been observed under only a limited range of circumstances, generalizations about the differences among species in their reaction to the alkaloids must be accepted with some caution.

A. Domestic Animals

1. Clinical Disease

Acute death from a large intake of toxic plant over a short period may occur under conditions of natural grazing but a longer time course is more usual, resulting from a number of sublethal episodes that may extend over a period of weeks or several years. Most commonly, all the clinical signs relate to hepatic insufficiency. Sluggishness, weakness, loss of appetite, wasting, ascites, jaundice, photosensitization, and behavioral abnormalities have been observed, although their frequency tends to vary in different species.

Horses seem particularly prone to develop a terminal hepatic encephalo-pathy, with resulting disturbances of behavior ranging from abnormal stance to aimless wandering or periods of frenzy. Similar signs, and amaurosis, have been described in cattle and, less severely, in pigs. In sheep the condition has been associated with deaths without premonitory signs.

Alimentary tract disturbances, including severe diarrhea with tenesmus that may lead to rectal eversion, are commonly seen in cattle, especially after grazing *S. jacobaea*. In other species diarrhea occurs less consistently and less severely.

Respiratory syndromes have been described most frequently in horses, under field conditions, but have been seen in occasional outbreaks in cattle and pigs. Although only a few species of plants have been associated with lung disease, they contain neither a common alkaloid nor alkaloids unique to plants causing the condition. Moreover, lung disease is not a constant result of the consumption of a particular plant species. Clearly, other undefined factors are also involved.

Compared with other domestic animals, pigs respond to the alkaloids with a wider range of clinical signs, and in addition to hepatic and respiratory disease they may also develop signs of renal failure.

Because of their ability to store large amounts of copper, sheep are subject also to the toxaemic jaundice syndrome (Bull et al., 1956). The storage of copper is favored by the alkaloid-induced damage. Its subsequent rapid release into the blood precipitates an acute hemolytic crisis with rapid death from renal failure.

2. Pathology

Considerable variation may be seen in all species in the character and severity of liver damage under field conditions. In acute toxicity extensive hemorrhagic necrosis of the liver may be the only abnormality seen. In the chronic disease, especially if it has progressed intermittently over 2-3 years, lesions of various ages may be discernible in the one animal, with terminal acute damage superimposed on a background of extensive fibrosis, bile ductule proliferation, nodular regeneration, and parenchymal megalocytosis. At the other extreme, very susceptible sheep under experimental conditions have died of hepatic failure showing only minimal liver damage on histological examination (Lanigan and Peterson, 1979). Fibrosis tends to develop more extensively in cattle than in other animals.

Pigs seem less liable than other animals to exhibit extensive bile ductule proliferation or nodular hypertrophy, possibly because outbreaks are more likely to be associated with the feeding of contaminated milled grain or pellets over a single continuous period.

Ascites and edema are likely to develop in all species as a consequence of liver insufficiency, but the rapid accumulation of large volumes of ascitic fluid has received particular mention as a characteristic of the late stages of the clinical disease in cattle. Obliteration of central and sublobular veins caught up in the extensive fibrosis may be a significant contributing factor. Although the results are the same, the development of this is somewhat different from the venous occlusion that occurs at an earlier stage from direct endothelial damage by pyrrolic metabolites and which is particularly prominent in the disease in humans. The extensive edema of the intestinal wall that accompanies the ascites in cattle is probably the main cause of the severe tenesmus that is seen in these animals.

Megalocytosis of renal tubule epithelium occurs occasionally in all species, but is most readily produced in pigs and seen in most outbreaks of pyrrolizidine alkaloid poisoning. It may be indicative of the excretion of a larger proportion of ingested alkaloid as pyrrolic metabolite in the urine or of more efficient concentration of this material in the renal tubule.

In other organs, including lungs, the pathology seen is within the range already described, with no feature displaying any particular species-related emphasis.

B. Poultry

It would be unusual for poultry to ingest pyrrolizidine alkaloids other than through seed contamination of grain fed directly or as milled grain products. Most reports deal with experimental poisoning, but it can be assumed that the few toxic plant species tested are likely sources of grain contamination. Little detail has been published on the clinical disease. Depression of growth rate and egg production seem general and depression, diarrhea, and anemia have been recorded.

Liver pathology is essentially similar to that in mammals, although megalocytosis has not been referred to in many reports. Ascites and a gelatinous or fibrinous coating of the liver are common. Widely distributed hemorrhages, necrotic enteritis, splenic atrophy, anemia, pulmonary and subcutaneous edema, and minor renal changes, including megalocytosis, have been inconsistently described.

C. Humans

Contamination of grain used for meal or bread is the usual cause of outbreaks of pyrrolizidine poisoning of humans. In some affected communities there has been an ignorance of the danger associated with the contaminating seeds; in others inspection for known toxic seeds has been neglected. Serious outbreaks have occurred in South Africa, the USSR, Afghanistan, and India. Contamination has also occurred in flax seed used as a source of oil for cooking (Braginskii and Bobokhodzhaev, 1965).

Grain toxicity usually has a well-defined epidemiology, with outbreaks occurring in largely closed communities using locally grown grain when seasonal and other environmental conditions have favored heavy growth of toxic weeds in the crops.

The other source of pyrrolizidine alkaloid poisoning comes from the use of infusions of toxic plants, including the so-called "bush teas," in the traditional folk medicine practiced in some countries. Cases are likely to be sporadic and the pattern of intake of alkaloid, and consequently the clinical disease, is likely to vary widely. The consumption of whole plants is unusual, but an exception is comfrey (*Symphytum* spp.), which has some popularity as a salad item (Culvenor et al., 1980b).

A further potential but undefined risk lies in the consumption of foods derived from animals that have had access to alkaloid-containing plants and excrete unchanged alkaloid. The products most likely to be affected are milk and honey (Deinzer et al., 1979; Dickinson et al., 1976; Dickinson and King, 1978; Culvenor et al., 1981), but as yet no evidence of toxicity in humans has been reported. The possible risk of alkaloid in the meat of animals that have grazed toxic plants shortly before slaughter has not been investigated.

1. Clinical Disease

Acute toxicity can occur, particularly in infants given a single large dose of toxic herbal medicine. Some authors have said that all human cases are acute, but the reference seems to be to acute clinical disease that more often is a late manifestation of hepatic failure or circulatory obstruction resulting from chronic pathology that has developed over a period of weeks or months of low-level intake of alkaloid.

The classical symptoms and signs in the grain toxicity syndrome are abdominal pain and rapidly developing ascites. Lassitude, anorexia, nausea, vomiting, diarrhea, edema, emaciation, hepatomegaly, splenomegaly, and mild jaundice have also been observed.

2. Pathology

The expected hemorrhagic central necrosis of the liver occurs in acute poisoning, while the chronic disease is characterized by failing liver function and hepatic vascular damage. Occlusion of the central and sublobular veins is histologically the most prominent lesion and considerable emphasis has been placed on it as the essential lesion underlying the clinical disease, which has accordingly been called venoocclusive disease. Attention has been drawn to the similarity to the Budd-Chiari syndrome, in which the main hepatic veins are obstructed (Selzer and Parker, 1951). Occluded vessels may be recannulated and perivascular fibrosis may progress to nonportal cirrhosis, but bile duct proliferation and nodular hyperplasia are not prominent responses and hepatic parenchymal megalocytosis has not been recorded. The liver failure and hepatic veno-occlusion are the cause of the rapidly developing ascites.

There thus appear to be some differences in the responses of humans to pyrrolizidine alkaloid poisoning compared with other species. Whether these differences are absolute or a matter of degrees of susceptibility is open to speculation, as human poisoning has necessarily been examined only within a limited context. The scarcity of references to some of the characteristic chronic lesions seen in other animals suggests that the circumstances seldom operate where human survival is possible in the presence of comparable damage. On the other hand, there are reports, some supported by serial biopsy, that indicate clinical recovery and tissue repair after moderate liver damage. This has not been observed in other species.

VIII. METABOLISM

A. Liver Metabolism

Metabolic studies have been carried out on a small number of pyrrolizidine alkaloids, mainly heliotrine, lasiocarpine, retrorsine, monocrotaline, and indicine N-oxide. The principal metabolic changes observed—hydrolysis, N-oxidation, and oxygenation adjacent to N, leading to pyrrolic derivatives—are applicable to the whole group of hepatotoxic alkaloids. Liver and blood concentrations of the alkaloids appear to peak about 5 min after intraperitoneal injection, subsequently declining to zero over several hours (Bull et al., 1968a).

1. Hydrolysis

Hydrolysis by esterases in liver and possibly elsewhere leads to the amino alcohol and the esterifying acid, both of which are nontoxic (Fig. 5). The esterases are effective enough to afford substantial protection against simple esters of retronecine but are much less effective against the natural alkaloids, on which attack is inhibited or prevented by the highly branched nature of the esterifying acids. Thus Mattocks (1970) has shown that diacetylretronecine is hepatotoxic if administered with an esterase inhibitor but not if given alone. Even with the natural alkaloids, a significant degree of protection may be afforded by esterases, since Dann found 18% of heliotrine and 25% of rinderine excreted as heliotridine and its N-oxide. For the lipophilic lasiocarpine, the excretion of hydrolysis products fell to about 8% (Bull et al., 1968b).

2. N-Oxidation

The alkaloids are partially converted into water-soluble N-oxides (Fig. 5) in the liver, the change being effected by the mixed-function oxidases of the microsomal fraction (Mattocks and White, 1971). The N-oxides so formed are largely excreted. They are nontoxic themselves and require reduction back to the parent alkaloid before they can be activated. Reduction is effected mainly by the bacterial flora of the gut (Mattocks, 1971; Powis et al., 1979), so that N-oxides are nearly as toxic as the parent alkaloids when administered orally, but much less so after parenteral injection. Powis et al. (1979) have shown that liver microsomes are capable of reducing N-oxides to the parent alkaloid under anaerobic conditions, but there is no evidence for this occurring in vivo.

N-Oxides make up 20% of the urinary pyrrolizidine derivatives produced from heliotrine in the rat. Heliotrine N-oxide is excreted to the extent of 62% (Bull et al., 1968b) and retrorsine N-oxide to more than 40% (Mattocks, 1971) after intraperitoneal injection in rats.

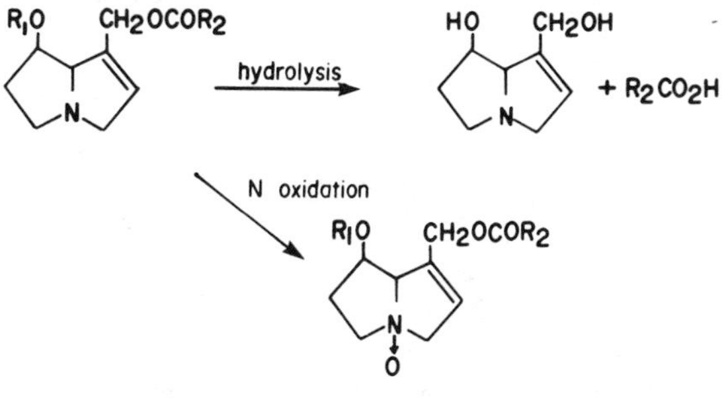

FIGURE 5 Hydrolysis and N-oxidation of pyrrolizidine alkaloids.

3. Oxygenation Adjacent to N, Leading to Pyrrolic Metabolites

Another major metabolic reaction for the hepatotoxic alkaloids is conversion into the reactive pyrrolic analog of the alkaloids (Sec. II). This change is also brought about by mixed-function oxidases in the liver. The initiating step is probably hydroxylation on carbon adjacent to the N atom (e.g., C-3 or C-8), normally the first stage in metabolic demethylation but followed in this instance by the ready elimination of water to give an aromatic pyrrole ring (Fig. 6). The overall process is equivalent to dehydrogenation and the product is called a "dehydro alkaloid" or "ester pyrrole."

The change is difficult to study in vivo because of secondary reactions of the dehydroalkaloids, but pyrrolic products are detectable in liver tissue, other organs, and urine after administration of pyrrolizidine alkaloids (Mattocks, 1968c, 1972b). The alkaloids are readily converted into pyrrolic products in liver microsomal preparations (Jago et al., 1970; Mattocks and White, 1971). The dehydro alkaloids have not been isolated directly but their formation is inferred from their secondary reaction products such as those formed with added cysteine (M. V. Jago, personal communication, 1971), with unchanged alkaloid, and with the water medium (Jago et al., 1970). Thus dehydroheliotridine was isolated as the main soluble pyrrolic product in an incubation mixture of liver microsomes and heliotrine, but is not formed similarly from heliotridine. The proportion of alkaloid converted into pyrrolic metabolites in vivo has not been estimated, but the relevant experimental data indicate that it could be in the range 15-50%, with water-soluble alkaloids giving the lower percentages and lipid-soluble alkaloids the higher percentages (e.g., Bull et al., 1968b; Hayashi, 1966; Mattocks, 1968c; White, 1977). Both the dehydro alkaloids and the dehydroamino alcohols may be prepared synthetically from the parent alkaloids (Culvenor et al., 1970a; Mattocks, 1969).

The alkaloids which are esters of otonecine, with a C-8-oxo-quaternary-N-methyl structure, are also capable of forming the same types of pyrrolic metabolites (Culvenor et al., 1971). In this case it is probable that N-demethylation occurs, initiated by hydroxylation on the *N*-methyl group, and is followed by tautomeric change to give a C-8-OH intermediate which eliminates water as before (Fig. 7). The product is expected to be the dehydro alkaloid of the corresponding retronecine ester, although this has not been demonstrated.

Being powerful alkylating agents, the dehydro alkaloids attack nucleophilic centers (SH, NH, and OH groups) in proteins and other cell constituents.

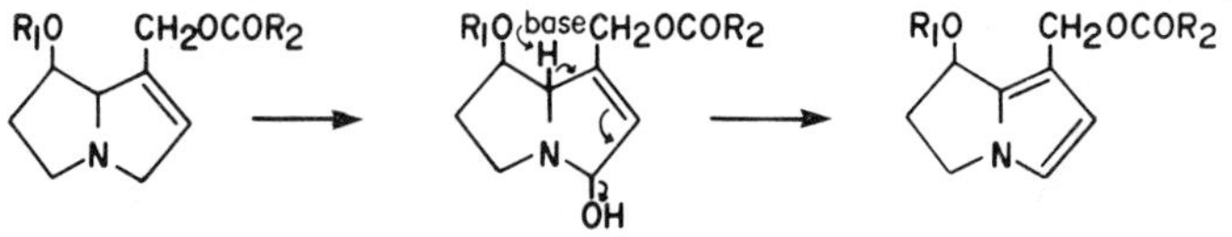

FIGURE 6 Metabolic conversion of heliotridine and retronecine ester alkaloids to pyrrolic dehydro alkaloids.

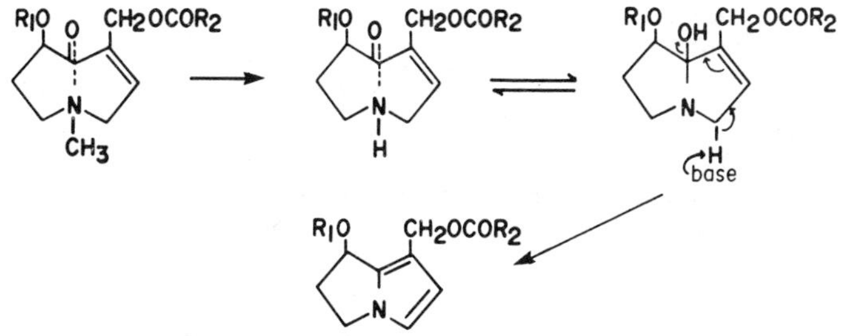

FIGURE 7 Metabolic conversion of otonecine-ester alkaloids to pyrrolic derivatives.

The amount of pyrrolic residue which becomes bound to the insoluble components of liver cells, the "bound pyrroles," has been measured and shown to be proportional to the acute toxicity of the alkaloid administered (Mattocks, 1972b). The conclusion drawn from studies such as this is that a high proportion of the dehydroalkaloid reacts and is converted into other products within the liver cell in which it was formed. Some may escape and reach the next organ, the lung, which is consistent with the development of major lung lesions when certain alkaloids are administered to rats. The available evidence is that lung tissue is unable to convert the alkaloids into pyrrolic metabolites in situ to a significant extent (Mattocks and White, 1971).

B. Pyrrolic Metabolites: Toxic Properties and Role in Pathogenesis

The two known toxic forms of pyrrolic metabolites are the dehydro alkaloids and the dehydroamino alcohols. The inability to study the effects of the former in the absence of the latter has meant difficulty in clarifying their respective contributions to pyrrolizidine alkaloid effects.

The reactive dehydro alkaloids are highly cytotoxic and react with animal tissues near the site of administration. They are regarded as the main cause of necrosis in acute pyrrolizidine poisoning. When formed in the hepatocyte, a high proportion of dehydro alkaloid is bound within the cell, but sufficient apparently escapes from the hepatocytes to initiate the vascular injury associated with some alkaloids. There is little evidence that they progress beyond the lung or penetrate beyond the vascular endothelium.

Injection of dehydromonocrotaline or dehydroretrorsine into the portal vein causes acute endothelial necrosis of the small radicles of the portal vein. However, injection into the tail vein, a route such that the liver is bypassed and the lung is the first vascular bed encountered, gives a distribution of lesions essentially the same as that induced by the intraperitoneal injection of alkaloid (Butler et al., 1970).

The dehydroamino alcohols are moderately reactive, circulate in the blood, and are excreted in urine. Dehydroheliotridine has an acute LD_{50} value of 76.5 mg/kg in nursling rats. It does not cause necrosis but it possesses marked antimitotic activity, which results in acute lesions in tissues with a high rate of cell renewal. In nursling rats a single dose causes arrest of growth, alopecia, defects in incisor teeth, and atrophic changes in hair folli- cles, gastrointestinal tract mucosa, thymus, spleen, bone marrow, and testis. Megalocytosis develops slowly in the hepatic parenchyma and in renal tubular epithelium (Peterson et al., 1972). Similar effects, with emphasis on gastric ulcers and hemorrhage in the stomach, were reported for dehydroretronecine (Hsu et al., 1973; Allen and Hsu, 1974; Shumaker et al., 1976).

Dehydroretronecine, given subcutaneously to rats, has been shown to produce turmors at the site of injection (Allen et al., 1975). Dehydrohelio- tridine, given intraperitoneally to rats, produced tumors in a variety of organs (Peterson et al., 1983). When given to pregnant females, dehydro- heliotridine is teratogenic at a medium dose level and embryotoxic at higher dose levels (Peterson and Jago, 1980).

Thus most of the chronic effects of the parent alkaloids can be produced by the dehydroamino alcohols. The dehydroalkaloids might also exert an anti mitotic effect, but the simpler hypothesis is that this activity is a function of the dehydroamino alcohols. Thus the development of a megalocytic liver fol- lowing alkaloid administration is seen as resulting from the antimitotic action of the dehydroamino alcohol combined with the stimulus of growth or regenera- tion after necrosis by the dehydroalkaloid. Beyond the liver and lung, the dehydroamino alcohol would appear to be a sufficient cause of lesions and effects. The embryotoxicity and teratogenicity of the alkaloids are probably very largely due to the dehydroamino alcohol originating in the maternal liver. Dehydroheliotridine has been demonstrated in the fetus after the injection of alkaloid into the dam, but without the vascular or necrotic lesions indicative of dehydro alkaloid activity in the placenta or fetus (Peterson and Jago, 1980).

The relative reactivities and partition behavior of the dehydro alkaloids might be expected to underlie some differences in the toxic effects of alka- loids, but these influences are not clear at the present time.

C. Rumen Metabolism

In the rumen of sheep, the hepatotoxic alkaloids undergo a reductive elimina- tion of the esterifying acid to form a 1-methylenepyrrolizidine derivative (Dick et al., 1963; Bull et al., 1968b). Heliotrine, for example, gives 7α- hydroxy-1-methylenepyrrolizidine (Fig. 8), and lasiocarpine gives the

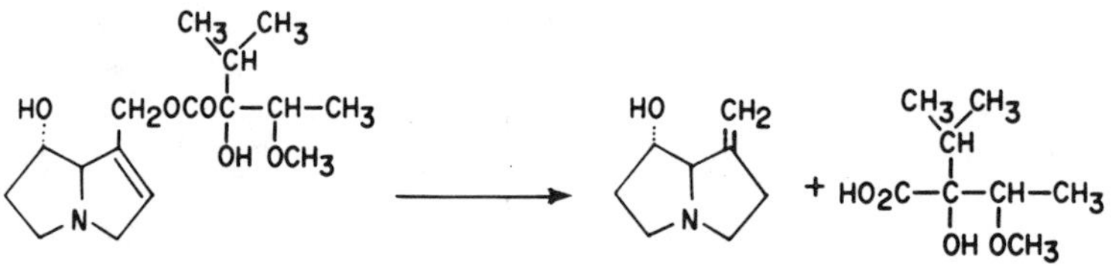

FIGURE 8 Rumen metabolism of pyrrolizidine alkaloids to 1-methylenepyr- rolizidine derivatives.

corresponding 7α-angeloxy compound (Lanigan and Smith, 1970). The reaction appears to be general for the hepatotoxic alkaloids (Bull et al., 1968b) and since the products are nonhepatotoxic (as shown for the 7α-angeloxy compound, Culvenor et al., 1976a), it is an important detoxification process. Heliotrine N-oxide is rapidly reduced to heliotrine by rumen liquor, so the reaction is equally applicable to ingested N-oxides (Lanigan, 1970).

In rumen liquor in vitro, 1-methylenepyrrolizidine formation may proceed almost to completion, with partial further reduction to a 1-methylpyrrolizidine derivative (Lanigan and Smith, 1970). However, the degree of conversion in vivo is uncertain and recent studies indicate that there may be other reasons for the comparative insensitivity of sheep to pyrrolizidine alkaloids (Shull et al., 1976; Swick et al., 1979).

An organism responsible for the change, *Peptococcus heliotrinreducans* (Lanigan, 1976), is usually present at a low level in the rumen and multiplies rapidly if heliotrope is added to the diet. It uses molecular hydrogen as a hydrogen source but competes poorly for this substrate with methanogenic bacteria (Lanigan, 1971). Inhibition of the latter with iodoform, chloral hydrate, or related substances results in an increased rate of metabolism of heliotrine (Lanigan, 1972). However, regular administration of iodoform to sheep gives only a small measure of protection against heliotrope (Lanigan et al., 1978). Under certain conditions in vitro, vitamin B_{12} also appeared to facilitate the reduction of heliotrine (Dick et al., 1963), but a continuous supplement of cobalt from pellets in the rumen failed to protect sheep eating heliotrope (Lanigan and Whittem, 1970).

IX. CYTOTOXIC EFFECTS, MUTAGENICITY, AND CARCINOGENICITY

A. Cell Culture Effects

Pyrrolizidine alkaloids produce toxic effects in cell cultures, probably via production of pyrrolic metabolites. Human embryonic liver cells show inhibition of nucleic acid and protein synthesis, possible chromosomal damage and mutation, and a long-term inhibition of mitosis leading to megalocytes (Sullman and Zuckerman, 1969; Armstrong and Zuckerman, 1971; Armstrong et al., 1972). These cells give color reactions for bound pyrrolic products (Armstrong and Zuckerman, 1970).

The alkaloids cause chromosome breakage in plant (Avanzi, 1961) and mammalian cells (Ameda and Saito, 1971; Martin et al., 1972; Bick et al., 1975) and in *Drosophila* (Brink, 1969). In leucocytes of the potoroo (*Potorous tridactylus*) the breaks were mostly simple chromatid and isochromatid deletions, with heliotrine producing some chromatid exchanges (Bick et al., 1975). Cells affected by monocrotaline showed more extensive damage than those affected by other alkaloids, but the proportion of cells affected was less.

Bick et al. (1975) showed the presence of pyrrolic products, primarily dehydroheliotridine, in incubation mixtures of heliotrine and homogenized leucocytes and in leucocyte cultures containing heliotrine. Dehydroheliotridine is more active than heliotrine in damaging leucocyte chromosomes, and dehydroheliotrine may also play a part as the initial product formed within the leucocyte.

B. Mutagenicity

Several alkaloids have been shown to be mutagenic in *Drosophila melanogaster* (Clark, 1960) and in *Aspergillus nidulans* (Alderson and Clark, 1966). In their ability to induce sex-linked lethals in *D. melanogaster*, monocrotaline, lasiocarpine, and heliotrine were stronger than echinatine, echimidine, senecionine, and supinine, while jacobine was weaker still. *N*-Oxides of the alkaloids were more weakly mutagenic than the parent bases.

The hepatotoxic alkaloids also give positive results in short-term mutagenicity tests such as with *Salmonella typhimurium* in the presence of rat liver microsomes (Koletsky et al., 1978; Yamanaka et al., 1979; Wehner et al., 1979), a mammalian cell transformation test (Styles et al., 1980) and a transplacental micronucleus test (Stoyel and Clark, 1980).

C. Carcinogenicity

Several pyrrolizidine alkaloids have been shown to be carcinogenic under appropriate dosing conditions, namely heliotrine, lasiocarpine, a mixture of intermedine and lycopsamine, monocrotaline, retrorsine, retrorsine *N*-oxide, petasitenine (fukinotoxin), senkirkine, and symphytine (IARC, 1976; Culvenor and Jago, 1979; Hirono, 1981). It must be expected that many more will prove to be carcinogenic when adequately tested. The main target organ is the liver, with tumors also recorded in a wide range of other organs. The carcinogenic action is rather weak, and tumor development seems to be dependent, at higher dose rates, on the test animals being without alkaloid for an interval of several months. Continuous feeding at low dose rates (e.g., lasiocarpine at 7, 15, or 30 ppm in the diet) (NCI, 1978), may result in a substantial tumor yield.

Plant materials containing the alkaloids have also been shown to induce tumors when fed to rats. Given the wide distribution of pyrrolizidine alkaloid plants throughout the world as herbs and weeds in grain crops and the demonstrated presence of the alkaloids at levels of about 0.5-3 ppm in certain milk and honey samples in the United States (Dickinson et al., 1976; Dickinson and King, 1978; Deinzer et al., 1977) and Australia (Culvenor et al., 1981), it is clear that these alkaloids rank with the mycotoxins as environmental hazards. Whether they add appreciably to the human carcinogen load has yet to be established, but evidence that this might be so has increased substantially in recent years.

D. Reaction with DNA and Protein

The metabolic conversion of the alkaloids into strongly alkylating pyrrolic metabolites, their effects on chromosomes, and their mutagenicity lead to an expectation of chemical modification of cell nucleic acids and protein. The details of such changes are not yet established. Dosing a rat with randomly ^{14}C-labeled lasiocarpine led to binding of 0.6% activity in the liver protein and 0.005% activity in the liver nucleic acids (Culvenor et al., 1969). The labeling of the nucleic acids was of the order of 1 pyrrolizidine nucleus per 10^5 nucleotides. Dehydroretrorsine and dehydromonocrotaline were found to cross-link *Escherichia coli* and rat liver DNA in vitro (White and Mattocks,

1972). It is probable that cross linking occurs in vivo, but an attempt to demonstrate it gave an indication only, without proof (White and Mattocks, 1972).

Dehydroheliotridine has been shown to combine with calf thymus DNA in vitro, but the rate is appreciable only below pH 7.5 (Black and Jago, 1970). Tritiated dehydroretronecine, administered subcutaneously to infant rhesus monkeys, was found to bind preferentially to the protein fraction of the gastric mucosa (Hsu et al., 1975).

X. DIRECT ALKALOID TOXICITY AND THE POSSIBILITY OF OTHER TOXIC METABOLITES

Peracute toxicity of the alkaloids is believed to be due to the pharmacological effects of the intact alkaloids (Sec. IV.A), but acute and chronic effects are well explained on the basis of causation by the pyrrolic metabolites (Sec. VIII.B). Cytotoxic effects evinced in cell culture experiments are also probably due to pyrrolic metabolites (Sec. IX.A) rather than the unchanged alkaloids.

1,2-Epoxides of the alkaloids have been suggested as possible toxic metabolites (Bull et al., 1968c; Schoental, 1970). No evidence has been reported which indicates their metabolic formation, and the two diastereoisomeric epoxides prepared from monocrotaline did not possess cytotoxic or related properties (Culvenor et al., 1971).

Studies of the antitumor activity of indicine N-oxide, particularly the finding that indicine has lower activity than indicine N-oxide, have led to the suggestion that indicine N-oxide is acting by an unknown mechanism which does not involve reduction to indicine and conversion into dehydroindicine (Powis et al., 1979). It is possible that unexpected differences in distribution of the drug and availability at the tumor site are involved, or that the tumor cells have previously unknown pathways for producing the pyrrolic metabolites from alkaloid N-oxides.

REFERENCES

Absalyamov, I. F., and Muratov (1971). Toxicity of *Heliotropium dasycarpum* alkaloids for farm animals. *Tr. Uzb. Nauchno.-Issled. Inst. Vet. 19*:102-104. *Chem. Abstr. 78*:848 (1973).

Alderson, A., and Clark, A. M. (1966). Interlocus specificity for chemical mutagens in *Aspergillus nidulans. Nature (Lond.) 210*:593-595.

Allen, J. R., and Carstens, L. A. (1970a). Pulmonary vascular occlusions initiated by endothelial lysis in monocrotaline-intoxicated rats. *Exp. Mol. Pathol. 13*:159-171.

Allen, J. R., and Carstens, L. A. (1970b). Clinical signs and pathologic changes in *Crotalaria spectabilis*-intoxicated rats. *Am. J. Vet. Res. 31*: 1059-1070.

Allen, J. R., and Hsu, I. C. (1974). Antimitotic effects of dehydroretronecine pyrrole. *Proc. Soc. Exp. Biol. Med. 147*:546-550.

Allen, J. R., Carstens, L. A., and Olson, B. E. (1967). Veno-occlusive disease in *Macaca speciosa* monkeys. *Am. J. Pathol. 50*:653-667.

Allen, J. R., Carstens, L. A., and Katagiri, G. J. (1969). Hepatic veins of monkeys with veno-occlusive disease. Sequential ultrastructural changes. *Arch. Pathol. 87*:279-289.

Allen, J. R., Hsu, I. C., and Carstens, L. A. (1975). Dehydroretronecine-induced rhabdomyosarcoma in rats. *Cancer Res. 35*:997-1002.

Araya, O., and Gonzalez, S. (1979). *Senecio erraticus* poisoning in four horses. *Gac. Vet. (B. Aires) 41*:743-745. *Vet. Bull. 50*:7691 (1980).

Armstrong, S. J., and Zuckerman, A. J. (1970). Production of pyrroles from pyrrolizidine alkaloids by human embryo tissue. *Nature (Lond.) 228*:569-570.

Armstrong, S. J., and Zuckerman, A. J. (1971). The effects of lasiocarpine, retrorsine and retronecine pyrrole on human embryo lung and liver cells in culture. *Br. J. Exp. Pathol. 53*:138-144.

Armstrong, S. J., Zuckerman, A. J., and Bird, R. G. (1972). Induction of morphological changes in human embryo liver cells by the pyrrolizidine alkaloid, lasiocarpine. *Br. J. Exp. Pathol. 53*:145-149.

Avanzi, S. (1961). Chromosome breakage by pyrrolizidine alkaloids and modification of the effect by cysteine. *Caryologia 14*:251-261.

Barnes, J. M., Magee, P. N., and Schoental, R. (1964). Lesions in the lungs and livers of rats poisoned with the pyrrolizidine alkaloid fulvine and its *N*-oxide. *J. Pathol. Bacteriol. 88*:521-531.

Bick, Y. A. E., Culvenor, C. C. J., and Jago, M. V. (1975). Comparative effects of pyrrolizidine alkaloids and related compounds on leucocyte cultures from *Potorous tridactylus*. *Cytobios 14*:151-160.

Bierer, B. W., Vickers, C. L., Rhodes, W. H., and Thomas, J. B. (1960). Comparison of the toxic effects of *Crotalaria spectabilis* and *Crotalaria giant striata* as complete feed contaminants. *J. Am. Vet. Med. Assoc. 142*:318-322.

Black, D. B., and Jago, M. V. (1970). Interaction of dehydroheliotridine, a metabolite of heliotridine-based pyrrolizidine alkaloids, with native and heat-denatured deoxyribonucleic acid in vitro. *Biochem. J. 118*:347-353.

Bohlmann, F., Knoll, K.-H., Zdero, C., Mahanta, P. K., Grentz, M., Suivita, A., Ehlers, D., Le Van, N., Abraham, W.-R., and Natu, A. A. (1977). Terpen-derivate aus *Senecio*-arten. *Phytochemistry 16*:965-985.

Bourkser, G. V. (1947). On the question of the aetiology and pathogenesis of toxic hepatitis with ascites (heliotrope toxicosis). *Hyg. Sanit. 1947*(6): 24-26.

Braginskii, B. M., and Bobokhodzhaev, I. Y. (1965). The clinical picture of hepatomegalia of heliotropic medicine. *Klin. Med. (Mosc.) 28*:42-45.

Bras, G., Brooks, S. E. H., and Watler, D. C. (1961). Cirrhosis of the liver in Jamaica. *J. Pathol. Bacteriol. 82*:503-512.

Brink, N. G. (1969). The mutagenic activity of the pyrrolizidine alkaloid heliotrine in *Drosophila melanogaster*. III. Chromosome rearrangements. *Mutat. Res. 8*:139-146.

Brooks, S. E. H., Miller, C. G., McKenzie, K., Audretsch, J. J., and Bras, G. (1970). Acute veno-occlusive disease of the liver. Fine structure in Jamaican children. *Arch. Pathol. 89*:507-520.

Buckmaster, G. W., Cheeke, P. R., and Shull, L. R. (1976). Pyrrolizidine alkaloid poisoning in rats: protective effect of dietary cysteine. *J. Anim. Sci. 43*:464-473.

Buckmaster, G. W., Cheeke, P. R., Arscott, G. H., Dickinson, E. O., Pierson, M. L., and Shull, L. R. (1977). Response of Japanese quail to dietary and injected pyrrolizidine (*Senecio*) alkaloid. *J. Anim. Sci. 45*:1322-1325.

Bukey, F. S., and Cunningham, R. W. (1933). A study of *Senecio riddellii*. *J. Am. Pharm. Assoc.* 22:399-401.

Bull, L. B. (1955). The histological evidence of liver damage from pyrrolizidine alkaloids: megalocytosis of the liver cells and inclusion globules. *Aust. Vet. J.* 31:33-40.

Bull, L. B., and Dick, A. T. (1950). The chronic pathological effects on the liver of the rat of the pyrrolizidine alkaloids, heliotrine, lasiocarpine, and their *N*-oxides. *J. Pathol. Bacteriol.* 78:483-502.

Bull, L. B., Dick, A. T., Keast, J. C., and Edgar, G. (1956). An experimental investigation of the hepatotoxic and other effects on sheep of the consumption of *Heliotropium europaeum* L: heliotrope poisoning of sheep. *Aust. J. Agric. Res.* 7:281-332.

Bull, L. B., Scott Rogers, E., Keast, J. C., and Dick, A. T. (1961). Heliotropium poisoning in cattle. *Aust. Vet. J.* 37:37-43.

Bull, L. B., Culvenor, C. C. J., and Dick, A. T. (1968). *The Pyrrolizidine Alkaloids*. North-Holland, Amsterdam, (a) p. 215; (b) p. 217; (c) p. 214.

Butler, W. H., Mattocks, A. R., and Barnes, J. M. (1970). Lesions in the liver and lungs of rats given pyrrole derivatives of pyrrolizidine alkaloids. *J. Pathol.* 100:169-175.

Campbell, J. G. (1957a). Studies on the influence of sex hormones on the avian liver. 2. Acute liver damage in the male fowl, and the protective effect of oestrogen, as determined by a liver function test. *J. Endocrinol.* 15:346-350.

Campbell, J. G. (1957b). Studies on the influence of sex hormones on the avian liver. 3. Oestrogen-induced regeneration of the chronically damaged liver. *J. Endocrinol.* 15:351-354.

Carrillo, B. J., Casaro, A., Ruksan, B., and Okada, K. A. (1976). Intoxication of cattle with *Senecio tweediei* H. et A. *Rev. Med. Vet. (B. Aires)* 57:205-214.

Castles, T. R., Snyer, J. L., Lee, C.-C., Folk, R. M., and Cooney, D. A. (1974). Toxicity of indicine *N*-oxide (NSC-132,319) in mice, dogs and monkeys. *Gov. Rep. Announce. Index* (U.S.) 75:57 (1975); U.S. NTIS, PB Rep. *Chem. Abstr.* 84:54223c (1976).

Cheeke, P. R., and Gorman, G. R. (1974). Influence of dietary protein and sulfur amino acid levels on the toxicity of *Senecio jacobaea* (tansy ragwort) to rats. *Nutr. Rep. Int.* 9:197-207.

Chesney, C. F., and Allen, J. R. (1973). Resistance of the guinea pig to pyrrolizidine alkaloid intoxication. *Toxicol. Appl. Pharmacol.* 26:385-392.

Christie, G. S. (1958a). Liver damage in acute heliotrine poisoning. 1. Structural changes. *Aust. J. Exp. Biol. Med. Sci.* 36:405-412.

Christie, G. S. (1958b). Liver damage in acute heliotrine poisoning. 2. Biochemical changes. *Aust. J. Exp. Biol. Med. Sci.* 36:413-424.

Christie, G. S., Bailie, M. J., and Le Page, R. N. (1962). Acute toxic liver injury. *Biochem. J.* 84:364-358.

Clark, A. M. (1960). The mutagenic activity of some pyrrolizidine alkaloids in *Drosophila*. *Z. Vererbungsl.* 91:74-80.

Clawson, A. B. (1933). The American groundsels species of *Senecio* as stock poisoning plants. *Vet. Med. (Kansas City, Mo.)* 28:105-110. *Vet. Bull.* 3:440 (1933).

Culvenor, C. C. J., and Jago, M. V. (1979). Carcinogenic plant products and DNA. In *Chemical Carcinogens and DNA*, P. L. Grover (Ed.). CRC Press, Boca Raton, Fla., pp. 161-186.

Culvenor, C. C. J., and Smith, L. W. (1955). The alkaloids of *Erectites quadridentata* D.C. *Aust. J. Chem. 8*:556-561.

Culvenor, C. C. J., and Smith, L. W. (1969). A quaternary *N*-dihydro-pyrrolizinomethyl derivative of heliotrine from *Heliotropium europaeum*. *Tetrahedron Lett.* No. 41, p. 3603.

Culvenor, C. C. J., Downing, D. T., Edgar, J. A., and Jago, M. V. (1969). Pyrrolizidine alkaloids as alkylating and antimitotic agents. *Ann. N.Y. Acad. Sci. 163*:837-847.

Culvenor, C. C. J., Edgar, J. A., Smith, L. W., and Tweeddale, H. J. (1970a). Dihydropyrrolizines. III. Preparation and reactions of derivatives related to pyrrolizidine alkaloids. *Aust. J. Chem. 23*:1853-1867.

Culvenor, C. C. J., Edgar, J. A., Smith, L. W., and Tweeddale, H. J. (1970b). Dihydropyrrolizines. IV. Manganese dioxide oxidation of 1,2-dehydropyrrolizidines. *Aust. J. Chem. 23*:1869-1879.

Culvenor, C. C. J., Edgar, J. A., Smith, L. W., Jago, M. V., and Peterson, J. E. (1971). Active metabolites in the chronic hepatotoxicity of pyrrolizidine alkaloids including otonecine esters. *Nature (Lond.) 229*:255-256.

Culvenor, C. C. J., Edgar, J. A., Jago, M. V., Outteridge, A., Peterson, J. E., and Smith, L. W. (1976a). Hepato- and pneumo-toxicity of pyrrolizidine alkaloids and derivatives in relation to molecular structure. *Chem.-Biol. Interact. 12*:299-324.

Culvenor, C. C. J., Edgar, J. A., Smith, L. W., and Hirono, I. (1976b). The occurrence of senkirkine in *Tussilago farfara*. *Aust. J. Chem. 29*: 229-230.

Culvenor, C. C. J., Edgar, J. A., Frahn, J. C., and Smith, L. W. (1980a). The alkaloids of *Symphytum* x *uplandicum* (Russian comfrey). *Aust. J. Chem. 33*:1105-1113.

Culvenor, C. C. J., Clarke, M., Edgar, J. A., Frahn, J. L., Jago, M. V., Peterson, J. E., and Smith, L. W. (1980b). Structure and toxicity of the alkaloids of Russian comfrey (*Symphytum* x *uplandicum* Nyman), a medicinal herb and item of human diet. *Experientia 36*:377-379.

Culvenor, C. C. J., Edgar, J. A., and Smith, L. W. (1981). Pyrrolizidine alkaloids in honey from *Echium plantagineum* L. *J. Agric. Food Chem. 29*:958-960.

Datta, D. V., Khuroo, M. S., Mattocks, A. R., Aikit, B. K., and Chhuttani, P. N. (1978). Herbal medicines and veno-occlusive disease in India. *Postgrad. Med. J. 54*:511-515.

Deinzer, M. L., Thomson, P. A., Burgett, D. M., and Isaacson, D. L. (1977). Pyrrolizidine alkaloids: their occurrence in honey from tansy ragwort (*Senecio jacobaea* L.). *Science 195*:497-499.

Deinzer, M. L., Thomson, P., Griffin, D., and Dickinson, E. (1978). A sensitive analytical method for pyrrolizidine alkaloids. The mass spectra of retronecine derivatives. *Biomed. Mass Spectrom. 5*:175-179.

Deinzer, M. L., Thomson, P. A., Griffin, D. A., and Burgett, D. M. (1979). The analysis of pyrrolizidine alkaloids in agricultural food products. In *Symposium on Pyrrolizidine (Senecio) Alkaloids*, P. R. Cheeke (Ed.). Nutrition Research Institute, Oregon State University, Corvallis, pp. 95-106.

Demale, A., and Moulé, Y. (1971). Effect of lasiocarpine on protein synthesis in rat liver. *Int. J. Cancer 8*:86-96.

Dick, A. T., Dann, A. T., Bull, L. B., and Culvenor, C. C. J. (1963). Vitamin B$_{12}$ and the detoxification of hepatotoxic pyrrolizidine alkaloids in rumen liquor. *Nature (Lond.)* 197:207-208.

Dickinson, J. O., and King, R. R. (1978). The transfer of pyrrolizidine alkaloids from *Senecio jacobaea* into the milk of lactating cows and goats. In *Effects of Poisonous Plants on Livestock*, R. F. Keeler, K. R. Van Kampen, and L. F. James (Eds.). Academic Press, New York, p. 201-208.

Dickinson, J. O., Cooke, M. P., King, R. R., and Mohamed, P. A. (1976). Milk transfer of pyrrolizidine alkaloids in cattle. *J. Am. Vet. Med. Assoc.* 169:1192-1196.

Dickson, J., and Hill, R. (1977). Cotton fireweed (*Senecio quadridentatus*) poisoning of cattle. *Proc. 54th Annu. Conf. Aust. Vet. Assoc.*, p. 92.

Dimenna, G. P., Krick, T. P., and Segall, H. J. (1980). Rapid high performance liquid chromatography isolation of mono-esters, diesters and macrocyclic diester pyrrolizidine alkaloids from *Senecio jacobaea* and *Amsinckia intermedia*. *J. Chromatogr.* 192:474-478.

Dollahite, J. W. (1972). The use of sheep and goats to control *Senecio* poisoning in cattle. *Southwest. Vet.* 25:223-226.

Donald, L. G., and Shanks, P. L. (1956). Ragwort poisoning from silage. *Br. Vet. J.* 112:307-311.

Downing, D. T., and Peterson, J. E. (1968). Quantitative assessment of the persistent antimitotic effect of certain hepatotoxic pyrrolizidine alkaloids on rat liver. *Aust. J. Exp. Biol. Med. Sci.* 46:493-502.

Dybing, O., and Hjelle, A. (1958). Toxic action of *Senecio aquaticus*. *Nord. Vet. Med.* 10:719-729. *Vet. Bull.* 29:1559 (1959).

Emmel, M. W., Sanders, D. A., and Henley, W. W. (1935a). *Crotalaria spectabilis* Roth seed poisoning in swine. *J. Am. Vet. Med. Assoc.* 39:43-54.

Emmel, M. W., Sanders, D. A., and Henley, W. W. (1935b). Additional observations on the toxicity of *Crotalaria spectabilis* (Roth) for swine. *J. Am. Vet. Med. Assoc.* 40:175-176.

Fowler, M. E. (1968). Pyrrolizidine alkaloid poisoning in calves. *J. Am. Vet. Med. Assoc.* 152:1131-1137.

Frayssinet, C., and Moulé, Y. (1969). Effect of lasiocarpine on transcription in liver cells. *Nature (Lond.)* 223:1269-1270.

Gallagher, C. H. (1960). The effect of pyrrolizidine alkaloids on liver enzyme systems. *Biochem. Pharmacol.* 3:220-230.

Gallagher, C. H., and Koch, J. H. (1959). Action of pyrrolizidine alkaloids on the neuro-muscular junction. *Nature (Lond.)* 183:1124-1125.

Gardiner, M. R., Royce, R., and Bokor, A. (1965). Studies on *Crotalaria crispata*, a newly recognized cause of Kimberley horse disease. *J. Pathol. Bacteriol.* 89:43-55.

Gibbons, W. J., Hokanson, J. F., Wiggins, A. M., and Schmitz, M. B. (1950). (1950). Cirrhosis of the liver in horses. *N. Am. Vet.* 31:229-233.

Gibbons, W. J., Durr, E. H., and Cox, S. A. (1953). An outbreak of cirrhosis of the liver in horses. *N. Am. Vet.* 34:556-558.

Gilruth, J. A. (1903). Hepatic cirrhosis affecting horses and cattle (so-called "Winton disease"): a report on its nature, cause, treatment, distribution, etc. *11th Annu. Rep., N.Z. Dept. Agric.*, pp. 228-278.

Gilruth, J. A. (1904). Hepatic cirrhosis due to ragwort (*Senecio jacobaea*). *Div. Vet. Sci., N.Z. Dept. Agric., Bull. 9.*

Goldenthal, E. J., D'Aguanno, W., and Lynch, J. F. (1964). Hormonal
    modification of sex differences following monocrotaline administration.
    Toxicol. Appl. Pharmacol. 6:434-441.

Gopinath, C., and Ford, E. J. H. (1977). The effect of ragwort on the
    liver of the domestic fowl (*Gallus domesticus*): a histopathological and
    enzyme histochemical study. *Br. Poult. Sci. 18*:137-141. *Vet. Bull.* 47:
    4611 (1977).

Gulick, B. A., Liu, I. K., Qualls, C. W., Gribble, D. H., and Rogers, Q. R.
    (1980). Effect of pyrrolizidine alkaloid-induced hepatic disease on plasma
    amino acid patterns in the horse. *Am. J. Vet. Res. 41*:1894-1898.

Harding, J. D. J., Lewis, G., Done, J. T., and Allcroft, R. (1964). Ex-
    perimental poisoning by *Senecio jacobaea* in pigs. *Pathol. Vet. 1*:204-
    220.

Harris, C., Reddy, J., Chiga, M., and Svoboda, D. (1969). Polyribosome
    disaggregation in rat liver by lasiocarpine. *Biochim. Biophys. Acta 182*:
    587-589.

Harris, P. N., Anderson, R. C., and Chen, K. K. (1942). The action of
    senecionine, integerrimine, jacobine, longilobine, and spartioidine,
    especially on the liver. *J. Pharmacol. Exp. Ther. 75*:69-77.

Harris, P. N., Rose, C. L., and Chen, K. K. (1957). Hepatotoxic and
    pharmacologic properties of heliotrine. *AMA Arch. Pathol. 64*:152-157.

Hayashi, Y. (1966). Excretion and alteration of monocrotaline in rats after
    a subcutaneous injection. *Fed. Proc. 25*:668.

Hayashi, Y., and Lalich, J. J. (1968). Protective effect of mercaptoethyl-
    amine and cysteine against monocrotaline intoxication in rats. *Toxicol.
    Appl. Pharmacol. 12*:36-43.

Hayashi, Y., Kato, M., and Otsuka, H. (1979). Inhibitory effects of diet-
    reduction on monocrotaline intoxication in rats. *Toxicol. Lett. 3*:151-155.

Heath, D., and Kay, J. M. (1967). Medial thickness of pulmonary trunk in
    rats with cor pulmonale induced by ingestion of *Crotalaria spectabilis*
    seeds. *Cardiovasc. Res. 1*:74-79.

Henderson, P. Th. (1971). Metabolism of drugs in rat liver during the peri-
    natal period. *Biochem. Pharmacol. 20*:1225-1232.

Hill, K. R., and Martin, H. M. (1958). Hepatic veno-occlusive disease and
    megalocytosis in *Senecio* poisoning of horses. *Br. Vet. J. 114*:345-350.

Hirono, I. (1981). Natural carcinogenic products of plant origin. *CRC
    Crit. Rev. Toxicol. 8*:235-277.

Hooper, P. T. (1975). Experimental acute gastrointestinal disease caused by
    the pyrrolizidine alkaloid, lasiocarpine. *J. Comp. Pathol. 85*:341-349.

Hooper, P. T., and Scanlan, W. A. (1977). *Crotalaria retusa* poisoning of
    pigs and poultry. *Aust. Vet. J. 53*:109-114.

Hooper, P. T., Best, S. M., and Murray, D. R. (1974). Hyperammonaemia
    and spongy degeneration of the brain in sheep affected with hepatic
    necrosis. *Res. Vet. Sci. 16*:216-222.

Hsu, I. C., Allen, J. R., and Chesney, C. F. (1973). Identification and
    toxicological effects of dehydroretronecine, a metabolite of monocrotaline.
    Proc. Soc. Exp. Biol. Med. 144:834-838.

Hsu, I. C., Robertson, K. A., Shumaker, R. C., and Allen, J. R. (1975).
    Binding of tritiated dehydroretronecine to macromolecules. *Res. Commun.
    Chem. Pathol. Pharmacol. 11*:99-106.

Huizing, H. J., and Malingré, Th. M. (1979). Reduction of pyrrolizidine N-oxides by the use of a redox polymer. *J. Chromatogr.* 173:187-189.

Huizing, H. J., De Boer, F., and Malingré, Th. M. (1980). Chloranil, a sensitive detection reagent for pyrrolizidine alkaloids on thin layer chromatograms. *J. Chromatogr.* 195:407-411.

Hunt, E. R. (1975). Plant alkaloids poison livestock. *Agric. Gaz. N.S. Wales* 86:20-21.

Huxtable, R. J. (1980). Herbal teas and toxins: novel aspects of pyrrolizidine poisoning in the United States. *Perspect. Biol. Med.* 24:1-14.

IARC (1976). *Monographs on the Evaluation of Carcinogenic Risk of Chemicals to Man*, Vol. 10: *Some Naturally Occurring Substances*. International Agency for Research on Cancer, Lyon.

Jago, M. V. (1969). The development of the hepatic megalocytosis of chronic pyrrolizidine alkaloid poisoning. *Am. J. Pathol.* 56:405-421.

Jago, M. V. (1970). A method for the assessment of the chronic hepatotoxicity of pyrrolizidine alkaloids. *Aust. J. Exp. Biol. Med. Sci.* 48:93-103.

Jago, M. V. (1971). Factors affecting the chronic hepatotoxicity of pyrrolizidine alkaloids. *J. Pathol.* 105:1-11.

Jago, M. V., Edgar, J. A., Smith, L. W., and Culvenor, C. C. J. (1970). Metabolic conversion of heliotridine-based pyrrolizidine alkaloids to dehydroheliotridine. *Mol. Pharmacol.* 6:402-406.

Jalving, H. (1930). Een Platen-intoxicatie veroorzaakt door Ingestie van *Senecio aquaticus* Huds. en *Senecio jacobaea* Lin. *Tijdschr. Diergeneeskd.* 57:328-342.

Kalkus, J. W., Trippeer, H. A., and Fuller, J. R. (1925). Enzootic hepatic cirrhosis of horses (walking disease) in the Pacific Northwest. *J. Am. Vet. Med. Assoc.* 68:285-298.

Kato, R., Vassanelli, P., Frontino, G., and Chiesara, E. (1964). Variation in the activity of liver microsomal drug-metabolizing enzymes in rats in relation to their age. *Biochem. Pharmacol.* 13:1037-1051.

Kerr, J. F. R. (1967). Lysosome changes in acute liver injury due to heliotrine. *J. Pathol. Bacteriol.* 93:167-174.

Kerr, J. F. R. (1969a). An electron-microscope study of liver cell necrosis due to heliotrine. *J. Pathol.* 97:557-562.

Kerr, J. F. R. (1969b). An electron-microscopic study of giant cytosegresomes in acute liver injury due to heliotrine. *Pathology* 1:83-94.

Khanin, M. N. (1956). The etiology and pathogenesis of toxic hepatitis with ascites. *Arkh. Patol.* 18:35-45.

Köhler, H. (1950). Uber ungewöhnlich grosse Leberzellen beim Leberkoller des Pferdes (Schweinsberger Krankheit). *Zentralbl. Allg. Pathol. Pathol. Anat.* 86:282-285. *Vet. Index* 20(3):84 (1952).

Koletsky, A., Oyasu, R., and Reddy, J. K. (1978). Mutagenicity of the pyrrolizidine (*Senecio*) alkaloid, lasiocarpine, in the *Salmonella*/microsome test. *Lab. Invest.* 38:352.

Kumari, S., Kapur, K. K., and Atal, C. K. (1966). Occurrence of a high monocrotaline-yielding strain of *Crotalaria retusa*. *Curr. Sci. (India)*, No. 21, 546-547.

Lalich, J. J. (1964). Influence of rat strain on pulmonary vascular responses to monocrotaline feeding. *Pathol. Microbiol.* 27:965-973.

Lalich, J. J., Johnson, W. D., Raczniak, T. J., and Schumaker, R. C. (1977). Fibrin thrombosis in monocrotaline pyrrole-induced cor pulmonale in rats. *Arch. Pathol. Lab. Med. 101*:69-73.

Lanigan, G. W. (1970). Metabolism of pyrrolizidine alkaloids in the ovine rumen. II. Some factors affecting rate of alkaloid breakdown by rumen fluid in vitro. *Aust. J. Agric. Res. 21*:633-639.

Lanigan, G. W. (1971). Metabolism of pyrrolizidine alkaloids in the ovine rumen. 3. The competitive relationship between heliotrine metabolism and methanogenesis in rumen fluid in vitro. *Aust. J. Agric. Res. 22*:123-130.

Lanigan, G. W. (1972). Metabolism of pyrrolizidine alkaloids in the ovine rumen. 4. Effects of chloral hydrate and halogenated methanes on rumen methanogenesis and alkaloid metabolism in fistulated sheep. *Aust. J. Agric. Res. 23*:1085-1091.

Lanigan, G. W. (1976). *Peptococcus heliotrinreducans*, sp. nov., a cytochrome-producing anaerobe which metabolises pyrrolizidine alkaloids. *J. Gen. Microbiol. 94*:1-10.

Lanigan, G. W., and Peterson, J. E. (1979). Bromosulphothalein clearance rates in sheep with pyrrolizidine liver damage. *Aust. Vet. J. 55*:220-224.

Lanigan, G. W., and Smith, L. W. (1970). Metabolism of pyrrolizidine alkaloids in the ovine rumen. I. Formation of 7α-hydroxy-1α-methyl-8α-pyrrolizidine from heliotrine and lasiocarpine. *Aust. J. Agric. Res. 21*: 493-500.

Lanigan, G. W., and Whittem, J. H. (1970). Cobalt pellets and *Heliotropium europaeum* poisoning in penned sheep. *Aust. Vet. J. 46*:17-21.

Lanigan, G. W., Payne, A. L., and Peterson, J. E. (1978). Antimethanogenic drugs and *Heliotropium europaeum* poisoning in penned sheep. *Aust. J. Agric. Res. 29*:1281-1292.

Laws, L. (1968). Toxicity of *Crotalaria mucronata* to sheep. *Aust. Vet. J. 44*:453-455.

Lyford, C. L., Vergara, G. G., and Moeller, D. D. (1976). Hepatic veno-occlusive disease originating in Ecuador. *Gastroenterology 70*:105-108.

Martin, P. A., Thorburn, M. J., Hutchinson, S., Bras, G., and Miller, C. G. (1972). Preliminary findings of chromosomal studies on rats and humans with veno-occlusive disease. *Br. J. Exp. Pathol. 53*:374-380.

Mathews, F. P. (1933). Poisoning of cattle by species of groundsel. *Texas Agric. Exp. Stn. Bull. 481*, October.

Mattocks, A. R. (1967a). Detection of pyrrolizidine alkaloids on thin-layer chromatograms. *J. Chromatogr. 27*:505-508.

Mattocks, A. R. (1967b). Spectrophotometric determination of unsaturated pyrrolizidine alkaloids. *Anal. Chem. 39*:443-447.

Mattocks, A. R. (1968a). Iron (II)-catalysed conversion of unsaturated pyrrolizidine alkaloid N-oxides to pyrrole derivatives. *Nature (Lond.) 219*:480.

Mattocks, A. R. (1968b). Spectrophotometric determination of pyrrolizidine alkaloids--some improvements. *Anal. Chem. 40*:1749-1750.

Mattocks, A. R. (1968c). Toxicity of pyrrolizidine alkaloids. *Nature (Lond.) 217*:723-728.

Mattocks, A. R. (1969). Dihydropyrrolizine derivatives from unsaturated pyrrolizidine alkaloids. *J. Chem. Soc. C 1969*:1155-1162.

Mattocks, A. R. (1970). Role of the acid varieties in the toxic actions of pyrrolizidine alkaloids on liver and lung. *Nature (Lond.) 228*:174-175.

Mattocks, A. R. (1971). Hepatotoxic effects due to pyrrolizidine alkaloid N-oxides. *Xenobiotica 1*:563-565.

Mattocks, A. R. (1972a). Toxicity and metabolism of *Senecio* alkaloids. In *Phytochemical Ecology*, J. B. Harborne (Ed.). Academic Press, London, pp. 179-200.

Mattocks, A. R. (1972b). Acute hepatotoxicity and pyrrolic metabolites in rats dosed with pyrrolizidine alkaloids. *Chem.-Biol. Interact. 5*:227-242.

Mattocks, A. R., and White, I. N. H. (1970). Estimation of metabolites of pyrrolizidine alkaloids in animal tissues. *Anal. Biochem. 38*:529-535.

Mattocks, A. R., and White, I. N. H. (1971). The conversion of pyrrolizidine alkaloids to N-oxides and to dihydropyrrolizine derivatives by rat-liver microsomes in vitro. *Chem.-Biol. Interact. 3*:383-396.

Mattocks, A. R., and White, I. N. H. (1973). Toxic effects and pyrrolic metabolites in the liver of young rats given the pyrrolizidine alkaloid, retrorsine. *Chem.-Biol. Interact. 6*:297-306.

McCulloch, E. C. (1940). Hepatic cirrhosis of horses, swine and cattle due to the ingestion of seeds of the tarweed, *Amsinckia intermedia. J. Am. Vet. Med. Assoc. 96*:5-18.

McLean, E. K. (1970). The toxic actions of pyrrolizidine (*Senecio*) alkaloids. *Pharmacol. Rev. 22*:429-483.

Milenkov, S. M., and Kizhaikin, Y. (Eds.) (1952). Toxic hepatitis with ascites. *Proc. Symp. V. M. Molotov Med. Inst.*, Tashkent.

Mohabbat, O., Younos, M. S., Merzad, A. A., Srivastava, R. N., Sediq, G. G., and Aram, G. N. (1976). An outbreak of hepatic veno-occlusive disease in north-western Afghanistan. *Lancet 2*:269-271.

Molyneux, R. J., and Roitman, J. N. (1980). Specific detection of pyrrolizidine alkaloids on thin-layer chromatograms. *J. Chromatogr. 195*:412-415.

Molyneux, R. J., Johnson, A. E., Roitman, J. N., and Benson, M. E. (1979). Chemistry of toxic range plants. Determination of pyrrolizidine alkaloid content and composition in *Senecio* species by nuclear magnetic resonance spectroscopy. *J. Agric. Food. Chem. 27*:494-499.

Monneron, A. (1969). Experimental induction of helical polysomes in adult rat liver. *Lab. Invest. 20*:178-183.

Mortimer, P. H., and White, E. P. (1975). Toxicity of some composite (*Senecio*) weeds. *Proc. 28th N.Z. Conf. Weed Pest Control*, pp. 88-91.

Mugera, G. M. (1970). Toxic and medicinal plants of East Africa. Part 2. *Bull. Epizoot. Dis. Afr. 18*:389-403.

National Cancer Institute (1978). Bioassay of lasiocarpine for possible carcinogenicity. *Publ. DHEW (NIH) 78-839. Dept. Health, Education and Welfare*, Bethesda, Md.

Newberne, P. M. (1968). The influence of a low lipotrope diet on response of maternal and fetal rats to lasiocarpine. *Cancer Res. 28*:2327-2337.

Newberne, P. M., Wilson, R. B., and Rogers, A. E. (1971). Effects of a low-lipotrope diet on the response of young male rats to the pyrrolizidine alkaloid, monocrotaline. *Toxicol. Appl. Pharmacol. 18*:387-397.

Nolan, J. P., Scheig, R. L., and Klatskin, G. (1966). Delayed hepatitis and cirrhosis in weanling rats following a single small dose of the senecio alkaloid, lasiocarpine. *Am. J. Pathol. 49*:129-151.

Norton, J. H., and O'Rourke, P. J. (1979). Toxicity of *Crotalaria goreensis* for chickens. *Aust. Vet. J. 55*:173-174.

O'Sullivan, B. M. (1979). Crofton weed (*Eupatorium adenophorum*) toxicity in horse. *Aust. Vet. J. 55*:19-21.

Pass, D. A., Hogg, G. G., Russell, R. G., Edgar, J. A., Tence, I. M., and Rikard-Bell, L. (1979). Poisoning of chickens and ducks by pyrrolizidine alkaloids of *Heliotropium europaeum*. *Aust. Vet. J. 55*:284-288.

Pearson, E. G. (1977). Clinical manifestations of tansy ragwort poisoning. *Mod. Vet. Pract. 58*:421-424.

Persaud, T. V. N., Putzke, H.-P., Tessmann, D., and Bienengräber, A. (1970). Microscopic and enzyme histochemical studies on the liver and kidney of rats treated with fulvine, the toxic alkaloidal constituent of *Crotalaria fulva*. *Acta Histochem. 37*:369-378.

Peterson, J. E. (1965). Effects of the pyrrolizidine alkaloid, lasiocarpine N-oxide, on nuclear and cell division in the liver of rats. *J. Pathol. Bacteriol. 89*:153-171.

Peterson, J. E., and Jago, M. V. (1980). Comparison of the toxic effects of dehydroheliotridine and heliotrine in pregnant rats and their embryos. *J. Pathol. 131*:339-355.

Peterson, J. E., Jago, M. V., Reddy, J. K., and Jarrett, R. G. (1983). Neoplasia and chronic disease associated with the prolonged administration of dehydroheliotridine to rats. *J. Natl. Cancer Inst. 70*:381-386.

Peterson, J. E., Samuel, A., and Jago, M. V. (1972). Pathological effects of dehydroheliotridine, a metabolite of heliotridine-based pyrrolizidine alkaloids, in the young rat. *J. Pathol. 107*:175-189.

Pethick, N. H. (1906). *Special Report on Pictou Cattle Disease*. Can. Dept. Agric., Ottawa.

Piercy, P. L., and Rusoff, L. L. (1946). Crotalaria spectabilis poisoning in Louisiana livestock. *J. Am. Vet. Med. Assoc. 108*:69-73.

Pierson, M. L., Cheeke, P. R., and Dickinson, E. O. (1977). Resistance of the rabbit to dietary pyrrolizidine (*Senecio*) alkaloids. *Res. Commun. Chem. Pathol. Pharmacol. 16*:561.

Podesta, M., Tortora, J. L., Moyna, P., Izaguirre, P. R., Arrillaga, B., and Altamirano, J. (1977). *Senecio* intoxication of cattle in Uruguay. *Veterinaria (Uruguay) 13*:98-112. *Vet. Bull. 48*:3815 (1978).

Pomeroy, A. R., and Raper, C. (1971). Pyrrolizidine alkaloids: actions on muscarinic receptors in the guinea pig ileum. *Br. J. Pharmacol. 41*:683-690.

Pomeroy, A. R., and Raper, C. (1972). Pyrrolizidine alkaloids: analysis of spasmolytic activity in the guinea-pig ileum. *Arch. Int. Pharmacodyn. Ther. 199*:5-15.

Powis, G., Ames, M. M., and Kovach, J. S. (1979). Metabolic conversion of indicine N-oxide to indicine in rabbits and humans. *Cancer Res. 39*:3564-3570.

Research Today (Eli Lilly and Co., Indianapolis, Ind.) (1949). Senecio and related alkaloids. *5*:55-73.

Robertson, W. (1906). Cirrhosis of the liver in stock in Cape Colony, produced by two species of *Senecio* (*Senecio burchelli* and *Senecio latifolius*). *J. Comp. Pathol. Ther. 19*:97-110.

Rose, A. L., Gardner, C. A., McConnell, J. D., and Bull, L. B. (1957). Field and experimental investigation of "walk-about" disease of horses (Kimberley horse disease) in northern Australia: *Crotalaria* poisoning in horses. *Aust. Vet. J. 33*:25-33.

Rose, E. F. (1972). *Senecio* Species: toxic plants used as food and medicine in the Transkei. *South Afr. Med. J. 46*:1039-1043.

Samuel, A., and Jago, M. V. (1975). Localization in the cell cycle of the antimitotic action of the pyrrolizidine alkaloid, lasiocarpine, and of its metabolite, dehydroheliotridine. *Chem.-Biol. Interact. 10*:185-197.

Schoental, R. (1959). Liver lesions in young rats suckled by mothers treated with the pyrrolizidine (*Senecio*) alkaloids, lasiocarpine and retrorsine. *J. Pathol. Bacteriol. 77*:485-495.

Schoental, R. (1970). Hepatotoxic activity of retrorsine, senkirkine and hydroxysenkirkine in newborn rats, and the role of epoxides in carcinogenesis by pyrrolizidine alkaloids and aflatoxins. *Nature (Lond.) 227*: 401-402.

Schoental, R., and Magee, P. N. (1957). Chronic liver changes in rats after a single dose of lasiocarpine, a pyrrolizidine (*Senecio*) alkaloid. *J. Pathol. Bacteriol. 74*:305-318.

Segall, H. J. (1979a). Reverse phase isolation of pyrrolizidine alkaloids. *J. Liq. Chromatogr. 2*:429-436.

Segall, H. J. (1979b). Preparative isolation of pyrrolizidine alkaloids derived from *Senecio vulgaris*. *J. Liq. Chromatogr. 2*:1319-1323.

Segall, H. J., and Krick, T. P. (1979). Pyrrolizidine alkaloids: organohalogen derivative ioslated from *Senecio jacobaea* (tansy ragwort). *Toxicol. Lett. 4*:193-198.

Segall, H. J., and Molyneux, R. J. (1978). Identification of pyrrolizidine alkaloids (*Senecio longilobus*). *Res. Commun. Chem. Pathol. Pharmacol. 19*:545-548.

Selzer, G., and Parker, R. G. F. (1951). *Senecio* poisoning exhibiting as Chiari's syndrome. *Am. J. Pathol. 27*:885-901.

Selzer, G., Parker, R. G. F., and Sapeika, N. (1951). An experimental study of *Senecio* poisoning in rats. *Br. J. Exp. Pathol. 32*:14-20.

Shevchenko, N. Kh., and Ibadullaev, F. I. (1974). Prophylactic measures against *Trichodesma* toxicosis. *Veterinariya (Mosc.) 1974*(part 10):113-115.

Shull, L. R., Buckmaster, G. W., and Cheeke, P. R. (1976). Factors influencing pyrrolizidine (*Senecio*) alkaloid metabolism: species, liver sulphydryls and rumen fermentation. *J. Anim. Sci. 43*:1247-1253.

Shumaker, R. C., Hsu, I. C., and Allen, J. R. (1976). Localisation and tissue effects of tritiated dehydroretronecine in young rats. *J. Pathol. 119*:21-28.

Smirnov, F. E., and Stoljarova, A. G. (1959). Pathogensis of *Trichodesma* poisoning (suiljuk) in horses. *Uzbek. Inst. Vet. Sborn. Nauch. Tr. 13*: 186-194.

Smirnov, F. E., Grosheva, G. A., and Stoljarova, A. G. (1959a). Role of the plant *Trichodesma incanum* in the aetiology of "suiljuk" disease of horses. *Uzbek. Inst. Vet. Sborn. Nauch. Tr. 13*:173-185.

Smirnov, F. E., Grosheva, G. A., and Stoljarova, A. G. (1959b). Action of the leaves and stems of *Trichodesma incanum* on cattle. *Uzbek. Inst. Vet. Sborn. Nauch. Tr. 13*:195-198.

Smit, J. D. (1952). Preliminary report on the toxicity of retrorsine, isatidine and pteropherine on rats, with special reference to the influence of sex on susceptibility. *J. S. Afr. Vet. Med. Assoc. 23*:94-102.

Smith, L. W., and Culvenor, C. C. J. (1981). Plant sources of hepatotoxic pyrrolizidine alkaloids. *J. Nat. Prod. 44*:129-152.

Steyn, D. G. (1934). *The Toxicology of Plants in South Africa.* South Africa Central News Agency Ltd., Johannesburg, South Africa.

Steyn, D. G., and van der Walt, S. J. (1945). Jaagsiekte in horses, and sunn-hemp poisoning in stock. *Farming S. Afr. 20*:445-447.

St. George-Grambauer, T. D., and Rac, R. (1962). Hepatogenous chronic copper poisoning in sheep in South Australia due to the consumption of *Echium plantagineum* L. (Salvation Jane). *Aust. Vet. J. 38*:290-293.

Stoyel, C. J., and Clark, A. M. (1980). The transplacental micronucleus test. *Mutat. Res. 74*:393-398.

Styles, J., Ashby, J., and Mattocks, A. R. (1980). Evaluation in vitro of several pyrrolizidine alkaloid carcinogens: observations on the essential pyrrolic nucleus. *Carcinogenesis 1*:161-164.

Sullman, S. F., and Zukerman, A. J. (1969). The effect of heliotrine, a pyrrolizidine alkaloid, on human liver cells in culture. *Br. J. Exp. Pathol. 50*: 361-368.

Svoboda, D., and Soga, J. (1966). Early effects of pyrrolizidine alkaloids on the fine structure of rat liver cells. *Am. J. Pathol. 48*:347-373.

Svoboda, D., Reddy, J., and Bunyaratvey, S. (1971). Hepatic megalocytosis in chronic lasiocarpine poisoning. *Am. J. Pathol. 65*:399-407.

Swick, R. A., Cheeke, P. R., and Buhler, D. R. (1979). Factors affecting the toxicity of dietary tansy ragwort to rats. In *Symposium on Pyrrolizidine Alkaloids: Toxicity, Metabolism and Poisonous Plant Control Measures*, P. R. Cheeke (Ed.). Nutrition Research Institute, Oregon State University, Corvallis, 1979, pp. 115-123.

Tandon, B. N., Puri, B. K., Tandon, A., Koshi, A., and Joshi, Y. K. (1978). Ultra-structure of liver in veno-occlusive disease due to *Crotalaria nana* Burm. *Ind. J. Med. Res. 68*:790-797.

Thomas, E. F. (1934). The toxicity of certain species of *Crotalaria* seed for the chicken, quail, turkey and dove. *J. Amer. Vet. Med. Assoc. 38*:617-622.

Thorpe, E., and Ford, E. J. H. (1968). Development of hepatic lesions in calves fed with ragwort (*Senecio jacobaea*). *J. Comp. Pathol. 78*:195-205.

Tsuji, T. (1980). Lipid peroxidation in experimental interstitial pneumonitis induced by monocrotaline. *Nihon Univ. J. Med. 21*:221-234.

Umeda, M. and Saito, M. (1971). The effect of monocrotaline, a pyrrolizidine alkaloid, on HeLa cells and primary cultured cells from rat liver and lung. *Acta Pathol. Jap. 21*:507-514.

Vanek, J. (1956). Comparative study of experimentally caused *Senecio erraticus* poisoning with Zd'ár disease of horses. *Vet. Med. (Prague) 29*:705-724.

Verney, F. A. (1911). Dunsickness. *J. Comp. Pathol. Ther. 24*:226-229.

von Pohlenz, J., Luthy, J., Minder, M. P., and Bivetti, A. (1980). Enzootische Leberzirrhose beim Rind, verursacht durch Pyrrolizidinalkaloide nach Aufnahme von *Senecio alpinus* (Alpenkreuzkraut). *Schweiz. Arch. Tierheilkd. 122*:183-193.

Wakim, K. G., Harris, P. N., and Chen, K. K. (1946). The effects of senecionine on the monkey. *J. Pharmacol. Exp. Ther. 87*:38-41.

Wehner, F. C., Thiel, P. G., and Van Rensburg, S. J. (1979). Mutagenicity of alkaloids in the *Salmonella*/microsome system. *Mutat. Res. 66*:187-190.

White, I. N. H. (1977). Excretion of pyrrolic metabolites in the bile of rats given the pyrrolizidine alkaloid retrorsine or the bis-*N*-ethylcarbamate of synthanecine A. *Chem.-Biol. Interact. 16*:169-180.

White, I. N. H., and Mattocks, A. R. (1972). Reaction of dihydropyrroli- zines with deoxyribonucleic acids in vitro. *Biochem. J. 128*: 291-297.

White, I. N. H., Mattocks, A. R., and Butler, W. H. (1973). The conversion of the pyrrolizidine alkaloid retrorsine to pyrrolic derivatives in vivo and in vitro and its acute toxicity to various animal species. *Chem.-Biol. Interact. 6*: 207-218.

Wild, H. (1947). The Eastern Districts *Senecio* problem. *Rhodesia Agric. J. 44*: 164-171.

Willmot, F. C., and Robertson, G. W. (1920). *Senecio* disease or cirrhosis of the liver due to *Senecio* poisoning. *Lancet 2*: 848-849.

Yamanaka, H., Nagao, M., and Sugimura, T. (1979). Mutagenicity of pyrrolizidine alkaloids in the *Salmonella*/mammalian-microsome test. *Mutat. Res. 68*: 211-216.

Part VII

TOXINS IN
EVOLUTION AND SPECIES INTERACTION

20

THE EVOLUTIONARY AND ECOLOGICAL SIGNIFICANCE OF PLANT TOXINS

JOHN M. KINGSBURY

Cornell University, Ithaca, New York

I. PURPOSE

The purpose of this chapter is to review the evolutionary origin and current ecological roles of plant toxins and to consider some practical consequences.

II. HISTORY AND REVIEWS

Plant toxins are a subset of that heterogeneous array of molecules in plants which are commonly but loosely termed secondary compounds or secondary substances. The concept embodied in these two terms is a very early one in the study of plant chemistry (see the reviews by Fraenkel, 1959; Hegnauer, 1975; Rhoades, 1979). A precise definition distinguishing primary and secondary compounds was provided by Czapek in 1913 and 1921, but the general distinction between compounds involved in the primary metabolism of plants and the remainder had been made by 1888 (Stahl; see Fraenkel, 1959) or earlier, and as early as 1813, DeCandolle had recognized that plants produced noncycling (therefore "waste") substances (Muller, 1970).

As for many early generalizations, the idea of separating the entire spectrum of plant chemistry into just two categories has proved too simple to stand unmodified under the accumulation of increasing detail. Primary and secondary compounds are now better defined by examples. Primary compounds are those, especially sugars, acids (e.g., amino acids, citric acid cycle acids, fatty acids), purines, and pyrimidines, and their active polymers, involved in the creation of structure and the production and use of chemical energy by plants. Secondary compounds include those in categories such as phenolics, alkaloids, saponins, free amino acids, steroids, essential oils, glycosides, terpenes, and resins (Hegnauer, 1975; Levin, 1976b). Compounds such as vitamins, coenzymes, some sterols, tetrapyrroles, and carotenoids do not fit the increasingly functional implicit definitions of secondary compounds even though they may be generated from the same basic reaction pathways (Hegnauer, 1975) and what have long been considered typical plant secondary compounds are now being found also in animals, especially sedentary marine species (Jones, 1972). Aside from the question of whether secondary compounds are waste products in incomplete cycles, or useful compounds in their own right, the evidence now is that many of them (especially those containing nitrogen) are scavenged or otherwise returned to the basic metabolic pathways of the plant rather than being ultimately excreted into the environment (Jones, 1972; Robinson, 1974). This fact by itself blurs the original distinction between primary and secondary compounds.

At first, secondary compounds were generally viewed as plant wastes, to be gotten rid of before they damaged the producer (Muller, 1969). As knowledge of their number, variety, and amount accumulated, it became necessary either to consider plants wildly profligate with their resources ("leaky") (Muller, 1966) or that the secondary substances performed one or another useful role in the life of the plant producing them. Most plants influence the feeding behavior of one or more animals in some manner associated with secondary compounds. The preponderance of evidence today supports the conclusion of a useful role for most if not all secondary compounds.

The rapid advance of technique and instrumentation in the field of organic chemistry over the past two decades has led to an absolute deluge of specific information on the identity of plant secondary compounds (Mabry, 1972). By 1976 some 12,000 molecules of molecular weight between 20 and 2,000 had been identified (Levin, 1976a; Reese, 1979). Information is now accumulating, although less rapidly, on the reactions by which secondary compounds appear and disappear in plants and what happens to them if ingested by an animal. On another front, information (both experimental and observational) is also slowly accruing on the role of particular secondary compounds in the ecology of the organisms possessing them. Some could argue that not enough of the latter two kinds of information has yet been gathered to allow establishing solid unifying generalizations concerning the roles of secondary compounds, but recent authors have not hesitated to put forth ideas, some predictive, hence ultimately testable, others purely speculative, and study of the co-evolution of plants and animals, interrelated as it is by secondary compounds, has become extraordinarily productive of thought and experimentation. References in the following review are not intended to represent the totality of this massive literature, but to provide access to pertinent sections of it. Useful reviews include: Arnold and Hill, 1972; Beck and Reese, 1976; Brower and Brower, 1964; Chambers, 1970; Culvenor, 1970; Deverall, 1977; Feeny, 1976; Fraenkel, 1959, 1969; Freeland and Janzen, 1974; Gilbert and Raven, 1975; Harbourne, 1972b, 1977; Laycock, 1978; Levin, 1971, 1976a,b; Muller, 1970; Rhoades and Cates, 1976; Robinson, 1974; Rosenthal and Janzen, 1979; Schoonhoven, 1972; Sondheimer and Simeone, 1970; Southwood, 1973; Swain, 1977; Whittaker and Feeny, 1971; Zahorik and Houpt, 1977.

III. EVOLUTION OF SECONDARY COMPOUNDS

A discussion of secondary compounds in general and plant toxins in particular directs attention at the molecular level to the differences between plants and animals, and tends to obscure the greater generalization of the similarities. Let us remind ourselves at the outset that production and consumption of primary compounds is remarkably similar among all plants and animals. Basic chemistry and ultrastructure is closely comparable among all cellular organisms. Important consequences of this generalization include the fact that a green leaf, as far as its primary constituents go, is acceptable and adequate food for herbivorous animals from protozoa through mammals. Herbivores differ little in general nutritional needs from one end of the animal kingdom to the other, and plants, in general, contain all the nutrients necessary to sustain animal life. It is this fundamental nutritional compatibility that makes possible (indeed, encourages) the variety of interrelationships among organisms recognized in the terms "herbivore," "carnivore," "pathogen," and "parasite."

It is the secondary compounds that prevent, modify, or mediate these relationships (Beck and Reese, 1976).

Life originated in the oceans. Most products of incomplete cycling among primary compounds within an organism are water soluble. Primitive organisms were acellular or unicellular and microscopic in size. Most secondary compounds are dangerous to the producer if allowed to accumulate (Muller, 1970). Many affect a fundamental process in some way (McKey, 1979). It is probable, therefore, that the first secondary compounds appearing on the evolutionary scene were wastes to the producing organism and were excreted by direct solution into the surrounding aqueous medium. Even today, primitive plants such as blue-green algae (Cyanophyta) seem unusually "leaky" compared with more advanced forms. In general, plants and animals that are phylogenetically "native" to water (as opposed to those, like marine mammals, that have come on land and returned to water secondarily), have less complex means and systems of excretion. Fishes excrete ammonia from gill surfaces, for example, and need no recourse to the formation of urea or other complex nitrogenous compounds for this purpose (Brodie and Maickel, 1961). Similarly, the number of detoxifying enzymes present in a tadpole rises significantly in the transition to toad, and mixed-function oxidases, absent in the neonatal mammal, appear and increase dramatically shortly after birth

(Brattsten, 1979). Evidence like this supports the idea that the evolutionary removal of plants and animals to the land environment was accompanied by a sudden requirement for new methods of excretion of wastes. The evolution of increasingly complex means of excretion of nitrogenous wastes by terrestrial animals from birds to mammals has been long and widely known and has tended to color the early thinking about plant secondary compounds (Robinson, 1974), reinforcing the early idea that they, too, were wastes and end products of otherwise-directed chemical pathways.

Whatever their function, secondary compounds usually are active compounds that are difficult for the protoplasm of living plants to accommodate without damage except at very low concentrations (Muller, 1970). The means by which plants protect themselves against injurious effects of the secondary compounds they produce are numerous (Fowden and Lea, 1979). They include: removal into dead tissue (heartwood, peripheral bark, or deciduous leaves, or isolation in microscopic islands of tissue composed of dead cells killed by the secondary compound), removal by volatization or secretion onto surfaces subject to washing by rain (Tukey, 1966), secretion to the surface via special glandular organs, secretion into schizogenous or lysigenous canals within the plant, or sequestering into the vacuole or other nonsusceptible organelles within the cell. At the molecular level, secondary compounds may be concentrated into crystals, or made physiologically inactive by complexing with other molecules into a "bound" form. Or an endangered reaction within the protoplasm may be protected by a specific mechanism such as the cyanide-detoxifying enzymes associated with the mitochondria of cyanide-producing plants (Hegnauer, 1975; McKey, 1979).

Every plant species from primitive to most advanced forms some secondary compounds, often many. Nearly every species produces one or more than are unique to it (Cates and Orians, 1975; Mabry, 1972). Sometimes they are produced in great abundance. Douglas fir (*Pseudotsuga menziesii*), for example, elaborates at least 40 different compounds in its needles (Freeland and Janzen, 1974). Some secondary compounds seem to be end products of separate special chemical pathways. Others, perhaps more numerous, appear as side reactions in other general pathways (Hegenauer, 1975; McKey, 1979). Patterns in the distribution of secondary compounds of plants are beginning to appear and eventually promise to give considerable insight into their evolution and function.

Primitive plants, existing before the advent of insects or other terrestrial animals, already encountered the activities of parasitic and pathogenic microorganisms (Chapman and Blaney, 1979). All extant plants contain secondary compounds that function as effective defenses against most microorganisms at the present time (Ingham, 1972). It is possible that secondary compounds originating as the result of incomplete cycling in early evolution (wastes) became useful initially in fending off attack by microorganisms. It is also likely, as terrestrial plants and animals evolved and became more complex, creating and occupying increasingly narrow niches and suffering increasing competitive pressures, that the "luxury" of discarding energy-rich wastes, or spending additional resources on detoxifying or sequestering them, became too costly. Plants that found beneficial use for these potentially embarrassing substances placed themselves doubly at a definite competitive advantage. Similarly, without effective means (by innovative use of secondary compounds) of accomplishing attraction, reproduction, dispersion, defense, and other necessary functions in an increasingly competitive and complex biotic world, particular species were doomed to early extinction.

The production or capability for production of secondary compounds is ultimately under genetic control. Presence, amount, and differential distribution among parts of a given plant, may, however, relate at any given moment most obviously to ontogenetic change, environmental variables, or in the case of inducible substances (such as phytoalexins), to the appropriate initiating stimulus (McKey, 1979). There is as yet little specific information on the exact genetics of secondary compounds. Most comes from the exploitation of crop plants. Human selection in the case of most crop plants speedily affects both the presence and amount of specific compounds (McKey, 1979). Irradiation of potato (*Solanum tuberosum*), for example, resulted in a 60% increase in its solanine content (Spiher, 1974). Cultivars or clones of particular crops may vary widely from one another in their secondary compound complement (Williams et al., 1971). Hybridization may result in the appearance of entirely new alkaloids, unknown in either parental species. The genetics of alkaloid formation seems to be particularly complex (Robinson, 1979). Cyanogens vary with plant families, genera, and species, but the same cyanogen is sometimes found in two species from widely separated families, and also in some invertebrate species (Jones, 1972). These examples reinforce an idea of general plasticity among plants in the creation of secondary compounds.

On the other hand, there is evidence for genetic stability of many secondary compounds. The recent science of chemotaxonomy relies on a fundamental evolutionary stability of this sort, and its success is an indirect measure of that stability (Mabry, 1972). Certain groups tend to specialize in certain types of secondary compounds: thus terpenes in conifers, alkaloids in Ranunculaceae (buttercups), tannins in Fagaceae (oaks and beeches), essential oils in Labiatae (mints) (Cronin et al., 1978). Alkaloids are mostly (but not entirely) lacking in gymnosperms. Their distribution among the orders of angiosperms is intriguing. Sixty-nine percent of the species in the primitive orders Magnoliales and Ranales contain alkaloids, whereas only 40% of the species in the remaining orders are so endowed (Levin, 1976a). Compared with gymnosperms, angiosperms are provided with a richer assortment of secondary compounds (Culvenor, 1970).

Many, probably most, of the herbivore-plant interrelationships which are controlled by the effects of secondary compounds at the present time occur between insect herbivores and plants. Insects and angiosperms appeared in the evolutionary record at about the same time (lower Carboniferous, some 200,000 million years ago) (Bowers et al., 1976; Southwood, 1973). Ehrlich and Raven (1964) have gone so far as to suggest that angiosperms were able to evolve explosively at first under the shield of insect-inhibiting secondary compounds already elaborated in the more prevalent gymnosperms. Similar speculation involves several other facets of the coevolution of insects and plants, including hypotheses for carpel closure, the transition from wind to insect pollination, and the role of nectar in a further transition to pollination and dispersal by birds and mammals, all of which are mediated through particular secondary compounds (e.g., Mulcahy, 1979; Sussman and Raven, 1978). But even primitive plants such as ferns (those species still extant, at least) possess secondary compounds such as juvenile hormones that are effective specifically only against insects (Culvenor, 1970).

Whatever the plausibility or subtlety of evolutionary generalizations such as those suggested above, the end point of evolution to date is the present situation. Enough relationships have been established for particular secondary compounds in recent years that a spectrum of functions can now be generalized (e.g., Chew and Rodman, 1979). Some secondary compounds, such as alkaloids and cyanogenic glycosides, have turnover rates of sufficient magnitude that, whatever else they may be doing, they are also contributing to the metabolic processes of the plant (Jones, 1972; Robinson, 1979). Some are clearly involved in internal regulation (growth hormones, etc.) in the producer (McKey, 1979). Some, for which no other positive functions can be discovered, are probably simple wastes, of no positive benefit to the plant producing them. Together, these functions are probably in the minority. Most secondary compounds are involved with external regulation, mediating interrelationships between plants and animals, or between plants and other plants (Hsiao, 1969). These relationships may be either positive (attraction, stimulation, etc.) or negative (defensive). Among positive relationships between plants and animals are those involved in the attraction or encouragement of pollinators, the dispersal or germination (passage through gut) of seeds, or the attraction to the plant of animals for plant defense (ants, etc.) (Bentley, 1977; Janzen, 1966), either individually or as populations, through the presentation of secondary compounds (tastes, odors, colors, etc.) perceived as attractive by the target animal.

Negative secondary compounds include allelopathic compounds (Muller, 1966; Muller and Del Moral, 1971; Whittaker and Feeny, 1971) inhibitors, estrogens or hormones (upsetting the normal reproductive patterns of consumers) (Berger et al., 1977), phytoalexins (induced defensive compounds against disease organisms) (Janzen, 1979), refractory compounds impairing normal digestion, deterrents, compounds that "mask" the action or effect of other compounds, or toxicants.

Rarely does a secondary compound act alone. Rather, it is one of many to which an animal is exposed in the process of normal feeding, and the signals from which the animal must integrate in the process of reacting (Schoonhoven, 1972).

Plant toxins are a subset of this complex, heterogeneous array of secondary compounds. The following discussion is limited to plant toxins alone, and thereby to the single functional definition implied in that designation. Even so, the definition is inclusive, and plant toxins are a large (perhaps the largest) and diverse group of secondary compounds, the specific functions of which depend as much on the consumer as the producer. To place this in somewhat broader context, I include brief mention of additional mechanical ways in which plants deny herbivores their nutritional benefits through the agency of injury.

IV. SUBSET OF PLANT TOXINS

Chemical heterogeneity is a dominant characteristic of plant toxins. Switching from a fundamentally chemical orientation implicit in the foregoing discussion of secondary compounds to a fundamentally functional one for plant toxins emphasizes that heterogeneity. For purposes of the following, a plant toxin may broadly be defined as any compound present in plants which, under natural (but possibly abnormal) circumstances of production or consumption (or exposure) causes or is capable of causing an injurious consequence in one or more individual animals. This definition excludes animal injury as the result of drugs from plants or malnutrition resulting solely from an inadequate diet (although it includes malnutrition as the result of a nutrient antagonist in the diet). It perhaps begs a further definition of "injury" which I do not attempt.

Nearly every secondary compound (as is, or as transformed in the exposed animal) if sufficiently concentrated, and if not detoxified or excreted effectively by the exposed animal, is probably capable of causing injury under some set of conditions. Some discussions limit themselves to compounds that have actually caused injury under known conditions of exposure. Much of the literature of plant toxins, however, is based on postulated toxicity on the basis of known biochemical characteristics of the compounds under discussion and the known physiology of animal systems.

Directing attention by examples to certain characteristics of secondary compounds of particular interest in reviewing their toxicity to animals seems more useful than presenting a catalog of such compounds. Size and solubility of the molecule determines such things as whether it can be transported within the plant or must be manufactured at the location of use, and also on an ultrastructural level, largely determines the degree to which organelle membranes will serve as barriers. Thus alkaloids are commonly formed in the roots and transported to leaves or seeds (Robinson, 1974). Terpenoids, on the other hand, are only slightly soluble and not easily transported in the plant. They must generally either be manufactured in the leaves they serve to protect, or in some cases, plants provide special transport machanisms (resin canals) for their movement. Many terpenes are readily volatized directly from the plant leaf surface (Muller, 1966). Tannins are formed in leaves and are not remobilized or scavenged on senescence of the leaf (McKey, 1979). Solubility characteristics also influence the distribution, course of excretion, or destruction of the compound when it enters an animal. Lipophilic compounds are excreted with difficulty (Brattsten et al., 1977) and may tend to accumulate in particularly sensitive or easily damaged locations within the animal. Chemically refractory compounds in plants often present problems of excretion or high cost of handling to animals. Disulfide linkages in protease inhibitors and lectins give them this kind of protection within the animal and, in a sense, even cellulose and lignins, primarily structural molecules from the plant perspective, are also simultaneously inimical (or at least not beneficial) to most animals (Janzen, 1979; Rhoades, 1979).

Cost of toxin production is necessarily of importance in the competitive world of plants. Unnecessary virulence is produced only at the expense of something else the plant might have done, and is selected against (Atsatt and O'Dowd, 1976). Most defensive compounds are carbohydrates in composition and cost of construction can be measured primarily in terms of the energy required in their formation (Rhoades, 1979). Phenolics are second only to the primary carbohydrates in abundance in plants and are very stable compounds. They are readily polymerized into larger, less active substances (Miles, 1969), and they are generally not scavenged or returned to metabolic use before the death of the plant. The additional cost of such mobilization would not be returned in any commensurate benefit to the plant (Levin, 1971). Alkaloids and cyanogens, in comparison, demand for manufacture not only the energetic costs, but also the use of a scarce material (nitrogen). These are reactive compounds which may readily be complexed into less dangerous "bound" forms within the plant, and equally easily taken apart again (McKey, 1974). The benefit of returning the nitrogen in such compounds to other uses apparently exceeds the cost of taking them apart. Availability of sulfur and possibly other elements may influence the cost of compounds containing them. The level of isothiocyanates, for example, is higher in crucifers when they are gown on high-sulfur soils than otherwise (Chew and Rodman, 1979).

Benefit, hence presence or potential abundance, of a particular substance may relate to a function other than defense, even though the compound is also toxic. The high oxalate content of halogeton (*Halogeton glomeratus*) and other oxalate-rich plants of dry habitats, for example, relates primarily to its role in sodium balance and water relations within the plant (Cronin et al., 1978). The presence of toxic selenium compounds in plants on seleniferous soils may relate more simply to the fact that selenium is taken up in place of sulfur in some plant compounds rather than to the toxic nature of the resulting compounds. In situations where animal attack on a plant species is gradual, there would be obvious cost-benefit to a defense strategy based on the rapid creation of a toxic compound on the stimulus of injury. The inducibility of phytoalexins has already been mentioned. Similarly, protease inhibitors can accumulate rapidly (within hours) in herbage after injury (Janzen, 1979).

Perception of toxic compounds in plants must obviously relate to their reactivity, but not necessarily directly to their toxicity. Alkaloids and cyanogens are almost universally bitter, but the toxic reactions they cause vary widely (Bate-Smith, 1972). The flavonoids include many highly pigmented or odiferous compounds, and some terpenes are especially malodorous, but these characteristics are probably not related to their toxicity (Harbourne, 1972a; Rothschild, 1975). On the other hand, tannins precipitate proteins, including those of the tongue, and the sensation of astringency thus engendered is directly a consequence of the substance and is also responsible for its toxicity (Rhoades, 1979).

Newly formed plant tissues are usually soft, and if defended, secondary compounds are usually involved. Hairiness can also be effective (Johnson, 1975) against insects. Older tissues tend to toughen and in this process become generally less acceptable to animal herbivores (Feeny, 1970). Many plants form specific hard structures which are clearly defensive. An advantage of such structures is that their cost is entirely in the construction; they are essentially maintenance free (Janzen, 1979). Mechanical defenses extend all the way from concentration of calcium and silica compounds in surficial tissues (McKey, 1979) to the spatial defense of plants that place their foliage higher than most mammals can reach (Nichol, 1938). Spatial defense can be entirely effective when coevolved in concordance with the habits of particular large herbivores but may be useless or even detrimental against introduced herbivores. The camels early introduced into central Australia for transportation purposes, and now feral, for example, do tremendous damage to the elevated natural vegetation by pulling down and breaking off branches and small trees in order to reach the leaves (personal observation). Fine hairiness on the surface of leaves is known to prevent insects from reaching the leaf surface to feed (Levin, 1973; Piltemer and Tingey, 1976). Hooked hairs can trap foraging insects and immobilize them (Gilbert, 1971). On the other hand, sawflies harvest hairs on some plants like hay (Johnson, 1975), and some butterfly larvae can form a scaffolding of web material over the hairs from which to operate effectively on the plant itself (Rathke and Poole, 1975). Some hairs are not simply a mechanical defense. Many are glandular, secreting toxins of various kinds to the outer surfaces of the plant (Levin, 1973; Thurston and Lersten, 1969). Stinging hairs, found in four plant families, are effective against larger animals and can sometimes provoke a serious systemic effect (Kingsbury, 1964). Other morphological defenses of plants include mimicry. Plants (e.g., *Passiflora*) which produce superficial growths that mimic clusters of insect eggs are less likely than others to have insect eggs laid on them (Williams and Gilbert, 1981).

Spines, thorns, prickers, and stickers are effective against larger animals (including man) (Arnold, 1964). (An old saying about the plants of the inimical Texas plateau country is that they all "stick, stink, or sting.") The seeds of spiny-trunked plants tend to be dispersed not by climbing mammals, but by birds. Longer spines are produced on shorter internodes, therefore in greater density, by African acacias after they have been browsed (Janzen, 1979). Thorns and spines may inflict wounds that subsequently become infected, or they may become embedded in animal tissue in such a way as to result in immobilization or diminution in function of the affected part (usually the mouth). Grass awns, with their backward-pointed sawteeth, will work their way relentlessly through the fur, skin, and even bone of large animals, propelled by body movements (Kingsbury, 1964). Some acacias have hollow thorns which house ants (ant domatia) in a relationship that now seems nearly obligate on both partners. The ants create a vigorous defense against mammals if disturbed (Bentley, 1977).

Thorns are not always an effective defense. Anyone who has watched a goat avidly consume black locust (*Robinia pseudoacacia*), branches, thorns, and all, will be forced to that conclusion. Despite their general effectiveness in defense, Ehrlich (1970) has concluded that as a category, thorns and similar mechanical defenses are less useful than secondary compounds in response to the pressures of herbivores.

V. ANIMAL PRESSURE AND PLANT DEFENSE STRATEGIES

The pressure that herbivores place on plants is pervasive, constant, and in an evolutionary sense, opportunistic. Pressure is exerted on individuals, species, populations, or communities of plants, and may be related also to the environmental particularities or niches within which the plants grow and compete. The emergence of a powerful rationalizing generalization has been most useful for ordering the disparate data already available, allowing certain ecological predictions, and suggesting fruitful lines of investigation. This is the concept of apparency (Feeny, 1976; Rhoades and Cates, 1976). Briefly put, it states that short-lived, cryptic plants or plant parts (unapparent to herbivores) will be minimally defended or defended against attack by only one or a small number of specialist (targeted) secondary compounds. Long-lived plants or plant parts (which inevitably will be found at some time during their lifetime by a variety of herbivores and therefore are apparent) can be expected to mount a general defense involving a larger number or a higher concentration of less specifically directed secondary compounds. Unapparent plants may include species in which the individuals are spatially scattered and difficult to find, as well as temporally evanescent plants or parts of plants. Temporally unapparent plants typically occur in dense populations, grow rapidly, reproduce copiously (and usually simultaneously), possess an effective means of dissemination, and die immediately after reproducing. In its essentials, the strategy of unapparent plants is to accomplish a successful life cycle before they are found by herbivores, or if found, to reproduce so speedily and abundantly that the herbivore population cannot itself increase rapidly enough to extinguish the resource before successful reproduction has taken place. Unapparent plants tend to invest heavily in structures and functions necessary to completing a rapid and fecund life cycle, and only lightly in defenses. This strategy thus does not deter herbivores but rather satiates the instant herbivore pressure with enough resource left over to accomplish the plant's purposes.

A. Unapparent Plants with Specialist Defense

Unapparent plants may not be able to escape all herbivores. In the competition of herbivore species for an adequate food supply, it is likely that some species of herbivore will have adapted its life cycle to "find" an unapparent plant species. In such case, for successful continuation of the newly "discovered" species, the latter may draw on its chemosynthetic resources to elaborate a secondary compound targeted (in an evolutionary sense) against the adapted discoverer. Generalizing this, unapparent species of plants can be expected to spend a significant portion of their resources on specific, targeted, effective defensive compounds. Such compounds will usually be metabolically expensive, requiring diversion of some of the plant's resources to produce. They should be effective in small quantity (highly virulent), and if possible, easily scavenged back into the plant's basic metabolism to recover some of the cost. Specialist secondary compounds occur in quantities typically less than 2% of the plant by weight (Laycock, 1978). Plants with moderate concentrations of any one specialist compound rarely have much of any other (Bate-Smith, 1972). Specialist secondary compounds that are formed only at the stimulus of a particular insult to the plant require no investment of resources until actually needed. Examples of such induced defensive compounds include phytoalexins which are formed in response to attack by pathogenic microorganisms in a manner analogous to antibodies in animals (Ingham, 1972) and protease inhibitors which accumulate rapidly in herbage in response to damage (Janzen, 1979).

Herbivores, especially insect herbivores, have reacted with great genetic ingenuity to the strategy of unapparency and the presence of specialist defense compounds in unapparent plants. If a plant elaborates a specific defensive compound, and an herbivore species then evolves a means of detoxifying that compound, the animal at one stroke provides itself with a copious food resource and removes itself from competition with other herbivores for that resource. The compound in the plant which was initially defensive now becomes an attractant for the herbivore that can handle it (Fraenkel, 1959). Even more, by sequestering in some way the toxic compound, the animal may gain a measure of protection against its own predators. The elegant relationships of butterflies with milkweeds and birds as worked out by Brower and Brower (1964) and co-workers are familiar. Certain butterflies sequester certain glycosides from some species of milkweeds (*Asclepias*) in a way harmless to themselves. These insects, in addition to having an improved food supply, now become toxic to certain birds. Birds learn to avoid them. Other butterflies have evolved a pattern of coloration similar to the glycoside-sequestering ones. They too, although harmless, are avoided by birds. However, in the complexity of the natural world, a beautiful series of interrelationships such as this inevitably has its own complications. Some butterflies sequester the toxic glycosides but still remain palatable to birds (Duffey, 1970). Insects that are toxic for whatever reason stand to profit from being easily recognized by potential predators. Typically, they are sharply patterned and brightly colored, especially close up. Nontoxic species of insects tend to be cryptic and not easily found either close up or at a distance (Rothschild, 1971).

Two more examples show additional ways in which sequestered secondary compounds can protect the sequestering animal. Sawfly larvae sequester a terpenoid from pines (*Pinus* spp.) and release it as a defense when disturbed (Eisner et al., 1974). Insects can attain protection against parasites this way, too. A wasp that is a serious pest of tomatoes (*Lycopersicon lycopersicum*) is itself attacked by a parasite. Agronomic efforts to increase the tomatine content of tomato plants (to improve their natural defense) results in an increased level of tomatine sequestered by the wasp. This in turn results in a decrease in the effectiveness of the parasite as a natural control of the wasp (Campbell and Duffey, 1979).

Some specialist defense compounds interfere with the successful completion of the herbivore life cycle. Metamorphoses among the morphological stages in a typical insect's life cycle and initiation of sexual reproduction are mediated by a series of hormones. Many plants produce secondary compounds that are chemically or functionally identical with one or more of these hormones. These include juvenile hormones, antijuvenile hormones, and molting hormones or antagonists (Bowers et al., 1976; Bowers and Nishida, 1980). By presenting these compounds to insects at "wrong" times in the life cycle, plants force abnormal consequences. Most insects make their own pheromones (sex attractants), but some have evolved to depend entirely on plant sources for them (Chapman and Blaney, 1979). This puts such insects at the mercy of a continued supply. The actual situation can be more complex and ingenious than these extremes. The normal pheromone response in the western pine beetle, for example, is adequate for successful reproduction on any adequate food resource. It is enhanced, however, by a synergistic relationship with a terpene the insect obtains when feeding on Pondorosa pine (*Pinus ponderosa*), a preferred diet. This "sets up" the pine for a mass infestation by the resulting population explosion of the beetle after it finds the pine (Bedard et al., 1969).

Plants can affect the reproduction of animals in additional ways. The Colorado beetle, for example, continually assays the concentration of secondary compounds in its host (potato). By this means it times its reproduction to occur just before the senescence and disappearance of its food resource (DeWilde et al., 1969). Some plants manufacture compounds that act as estrogens in mammals. These occasionally cause reproductive difficulties in livestock (Kingsbury, 1964), but are probably more important in reducing the damage of rodent populations to plants containing them (Berger et al., 1977) than they are to larger vertebrate herbivores in the wild.

Two recent papers by Berenbaum (1978) and Berenbaum and Feeny (1981) describe a series of relationships between decondary compounds in certain plant families and insects which give further insight into the delicate balances that may exist between herb and herbivore as one creates new defensive toxins and the other new ways of dealing with them. Ubelliferone, found in many plant families, is the precursor of certain linear furanocoumarins restricted to some eight plant families. Umbelliferone is moderately toxic to insects but the linear derivatives are even more so and thus are assumed to represent an advanced evolutionary stage in the defenses of those plants which began with umbelliferone. Furanocoumarins are most toxic in the presence of ultra-violet light, which enhances their ability to disrupt the herbivore's DNA (per-haps suggesting why leaf rollers, which seem well adapted to foraging on plants with linear furanocoumarins, have taken up that habit). Taking the story one step further, Berenbaum and Feeny have found that certain mem-bers of just two plant families (Leguminosae and Umbelliferae) are able to make angular furanocoumarins. These are therefore scarcer, sometimes occur in the same plants with linear furanocoumarins (never the converse), and are toxic to some insects that can live successfully on plants containing linear furanocoumarins. (Even so, a few species of insects are able to handle even the angular furanocoumarins.) Berenbaum and Feeny view these several realtionships as brief plateaus in a series of genetic moves and countermoves in the escalation of the evolutionary arms race between plants and insect herbivores.

B. Apparent Plants with Generalist Defense

Long-lived woody plants, whether deciduous or evergreen, expose them-selves inevitably to a variety of herbivore pressures through their longevity. In comparison with the variety of specialist compounds evolved by unapparent plants to meet specific needs, the demand of apparent plants is for a general defense. This has led to a functional convergence of defensive compounds in such plants (Rhoades and Cates, 1976). Defensive compounds of apparent plants commonly affect palatability or digestibility. A compound that reduces the digestibility of associated nutrients has a good chance of working effec-tively against most kinds of herbivores, from microorganisms to mammals, since the ultimate processes of nutrient breakdown are fundamentally the same across that entire spectrum. Compounds affecting palatability may or may not be toxic in themselves, and their effectiveness depends heavily on the sensory reaction of the herbivore (Laycock, 1978). Tannins, sterols, and simple hardening of plant tissues are the prime digestibility reducers (McKey, 1974). Tannins precipitate proteins, inactivating digestive enzymes and making the precipitated proteins refractory to breakdown (Harbourne, 1972a). Essential oils tend to disrupt normal rumen function in ruminants (Longhurst et al., 1964).

Generalist defensive compounds are both more widespread in woody plants and more abundant in amount than is typical of specialist compounds. Tannins are found in 80% of woody dicotyledonous plants compared with only about 15% of annual herbs (Rhoades, 1979). Under some circumstances, the amount of a generalist defensive compound in a plant may reach as much as 60% by weight (Laycock, 1978). Generalist compounds are usually manufactured in situ and are not scavenged with the death of the plant or with leaf fall. They accumulate instead in the litter on the soil surface (where they continue to decay only very slowly).

As described earlier, unapparent plants may use a strategy of satiation to escape serious losses. Apparent plants may also use that technique in the following way (McKey, 1974). Elaboration of a high concentration of digestibility reducers in foliage, or hardening of tissues, are processes that take appreciable time. In the interim, the new foliage is vulnerable. It may be protected temporarily by a specialist secondary compound which is scavenged when no longer needed (Ikada et al., 1977), or by the synchrony of production of foliage. If new foliage is produced continuously in time and undefended, an insect population will enlarge in time to take advantage of it at a sustained level equal to or greater than that production. If on the other hand the new foliage appears all at the same time, it may well satiate the low-level instant population of insect herbivores that might utilize it, and a generalized defense can then be elaborated before that population increases enough to cause serious damage (McKey, 1974).

In summary, generalist defensive compounds evolve slowly and convergently; specialist compounds appear rapidly (in evolutionary context) and are diverse.

The foregoing discussion has related mostly to plants as a whole. Defensive strategy may be directed more specifically to the protection of the most vulnerable or valuable parts of a plant. Thioglycosides, for example, vary greatly in concentration from one plant part to another in the genus *Brassica* (Joseffson, 1967). Roots generally need little protection against insect or mammalian herbivores, but must be protected against microbial pathogens. Foliage, growing points, seeds, and fruits are most generally susceptible to herbivore attack. Damage to foliage reduces the plant's ability to nourish itself and subjects it to stress. Damage to the growing point delays and may drastically alter the shape of future growth. Damage to seeds obviously reduces the plant's reproductive capacity, while damage to fruit may not only do that but may also reduce the effectiveness of dispersion and change the pattern of distribution. General or specialist defensive compounds may be deployed selectively to protect parts according to their apparency and value in completing the life cycle of the plant on the one hand and cost to the plant on the other (Rhoades and Cates, 1976). A similar spatial strategy is to concentrate defenses, both chemical and mechanical, at the surface or in the peripheral tissues of a plant (McKey, 1979).

In some cases, specialist compounds may be mobilized from one plant part and sent to another as the life cycle progresses. Thus defensive compounds in seeds (often highly concentrated) may be transferred into the tender seedling (McKey, 1979). Like new tissues, ephemeral tissues and seeds (unapparent) may be protected by the strategy of satiation or by specialist defensive compounds. Thus it can be predicted that undefended seeds depending on satiation as a strategy will be small, numerous, produced in synchrony, and readily dispersed, whereas those possessing defensive compounds will be larger, fewer, capable of delayed germination (dormancy), and "artfully placed" (McKey, 1975) by selective, costly, but effective methods of dispersion.

As well as separate parts, ontogenetic stages in the life cycle of a species and successional stages in the progression to a climax vegetation may profit

from varying defensive strategies. Annual plants, in general, have twice the
concentration of specialist defensive compounds as perennials (Levin, 1976a).
Defensive compounds may be scavenged (or abandoned) during senescence
at the end of an ontogenetic cycle. Solanine content of potato foliage decreases markedly, for example, as it ages (Schoonhoven, 1972). In the case
of L-dopa and plant lectins generally, commonly present in legume seeds, the
nitrogen is scavenged at the time of germination (Rehr et al., 1973). Climax
vegetation is more heavily defended than are earlier stages (Cates and Orians,
1975). Early successional stages tend to be characterized instead by high
fecundity, quick and effective dispersion, and rapid completion of the life
cycle (McKey, 1979).

Defensive strategies may vary with environmental circumstances. Plants
under mineral stress tend to produce greater concentrations of defensive
compounds than do plants growing on richer soils (Chew and Rodman, 1979).
Trees on nonfertile forest soils contained twice the concentration of phenols
as comparable trees on more fertile soils (McKey et al., 1978). The cost of
defending what has been elaborated seems less than the cost of obtaining
more of the scarce resource in these instances. On the other hand, it can be
argued that plants under other kinds of stress, such as injury, may divert
a greater portion of their resources to dealing with the consequences of that
stress, with a reciprocal reduction in the proportion of resources devoted to
the production of defensive compounds (Rhoades, 1979). Manuring and
clipping are agricultural practices that enhance the succulence of vegetation.
However, sheep and other animals reject lush forage growing from the spots
where they have dropped feces (Marten and Donker, 1966; Simons and Marten,
1971), and clipping may result in major changes in chemistry as, for example,
the immediate change in odor of a field of hay when mown (Arnold, 1970).
The concentration of alkaloids increases at points of wounding in some plants
(McKey, 1974).

Some relationships mediated by defensive secondary compounds can be
very subtle. Insects may accumulate at the growing tips and on the south
side of trees in temperate northern latitudes in winter because these parts
of trees are able to photosynthesize and manufacture nutrients more successfully than the parts that receive less light (Longhurst et al., 1968). Given
its short life span compared with that of a tree, a given species of insect can
evolve populations that are adapted to the characteristics of an individual
tree. A single long-lived tree may support 200 or more successive generations of aphids, for example, and this is sufficient for considerable dietary
specialization to take place in the herbivore before any genetic change takes
place in the host (Edmunds and Alstad, 1978). It can be observed that certain trees in a population of Douglas fir, for that reason or another, are
consistently subject to more intesive insect attack than are others (Longhurst
et al., 1968).

In summary, the relation of unapparent plants and plant parts with insect
herbivores has coevolved over some 200,000 million years and is finely tuned
and constantly adjusted by both parties. In the final analysis, since the complete success of an herbivore in an immediate sense can only be at the total
expense of its resource, it is an undesirable consequence for both parties.
The "best" coevolutionary consequence for the herbivore is one in which the
herbivore and resource are more or less evenly matched at population levels

that approximate the maximum sustainable yield of the resource. In populations capable of rapid change, skirmishes of advantage and counteradvantage are waged constantly between the protagonists, yet they must balance around that point over evolutionary time. Even in more slowly responding systems, when serious inbalance occurs it must be corrected. Otherwise, not only the resource species will be lost but also the herbivores that depend on it.

Dutch elm disease (with subsequent phloem necrosis) and the American chestnut blight are conspicous examples (using pathogens) of one-sided change and its consequences. To a degree, these may be self-correcting perturbations. As elms become less common, the insect vectors which carry the disease from one to another find the remaining elms with increasing difficulty and eventually the population level of the disease organism, having no host, may become infinitely small and in practical terms disappear altogether. If some few isolated elms survive at that point, the elm population can then begin to rebuild (other things equal—which they never are). A common way in which communities of organisms become unbalanced is through the introduction of foreign speces that have not suffered the checks and balances of coevolutionary history within the community. Introduced mammals press first the nutritious, palatable, nontoxic resources. When those have been depleted, the herbivores will turn to less palatable, less nutritious, and eventually toxic species, or will starve. That, in turn, results in an abrupt diminution in the herbivores and (other things equal) an opportunity for the original plant community to make a comeback. Until coevolution returns a balance between herbivore and resource or one partner succumbs, there will be an alternating boom or bust relationship between them.

Intense short-term genetic skirmishes are characteristic of insects and short-lived plants. Perturbations from steady state are less obvious when apparent plants with generalized defenses, or long-lived vertebrates, are involved.

Subtleties of defense may be abstracted to another level of complexity. In a varied population of plants, the defensive strategy may become so well tuned to the pressures against it that in some measure the population reacts as a whole against unselective pressure. Individual species may develop defensive characteristics that are better suited to defending other species than to defending themselves. Such use of individual resource is not altruistic; the species that defends another receives some specific benefit in return directly or in some other way related to the interactions within the community. Such defensive associations among the plant species of specialized communities have been called "plant defense guilds" (McNaughton, 1978).

VI. CONSUMER STRATEGY

Whatever the designed benefit to the plant, secondary compounds are present in every species, often in substantial molecular variety, and often in considerable amount. All herbivores must contend with them successfully. Basic mechanisms for dealing with potentially poisonous secondary compounds are limited in number, but the strategies of avoiding a toxic consequence are numerous and often complex. Just as the idea of apparency was useful in organizing the spectrum of plant defense strategies, so a parallel idea relating to animal exposure is useful in organizing consumer strategies (Freeland and Janzen, 1974).

Consumers vary from minute insects (or plant parasites and pathogens, to extend the range downward in size) to large herbivorous or omnivorous vertebrates. Within this range, some will be specialist organisms, feeding on a single species of plant, a single plant part (e.g., leaf, seed) of a single species, or even a single part in a single state of maturity. Specialist consumers of this sort need capacity to deal only with whatever dangerous or difficult secondary compounds may exist in a limited diet. Typically, they will possess just one or two specific detoxification mechanisms (Cates and Orians, 1975). Most insect species pursue a specialist consumer strategy. Most large vertebrate herbivores, on the other hand, consume plants from a wide botanical spectrum, and of necessity must adopt the strategy of a generalist, with ability to deal with all the diversity of potentially troublesome secondary compounds such a diet brings.

There are some important exceptions to these generalizations. Some insects are generalists in the context of the usual insect herbivore diet. They can be recognized by observation of their dietary nonspecificity and also by the fact that they possess an unusually well developed enzymatic array for dealing with a variety of secondary compounds (Krieger et al., 1971). Similarly, a few large vertebrate herbivores are specialists. An example is the koala, which feeds exclusively on the leaves of certain species of eucalyptus (Brattsten, 1979). Despite these exceptions, insects generally are specialists and vertebrate herbivores normally are generalists in consumer strategy.

Information on consumer strategy has been largely derived from observation of species (both plant and animal) of economic importance to man and from experimentation with an immediate economic motivation. The built-in biases of this kind of experimentation, at present unavoidable, must be kept in mind through the following discussion.

With minor exception what does not enter the body cannot poison it. Much consumer protective strategy is erected on this maxim. Marked preferences and aversions in diet are displayed both by insects (Feeny, 1975) and by vertebrate herbivores (Ivins, 1952). The preferences and aversions of insects seem mostly to be genetically programmed, whereas those of vertebrates are apt to be learned (phenotypic) (McKey, 1979). The distinction is not clear cut, however, because in experiments, responses with certain characteristics that satisfy parts of some definitions of learning have been shown in a variety of invertebrates and in some insects (Gilbert, 1975), while some vertebrates display a strong innate dislike for some food materials proffered for the first time.

Except as an evolutionary adaptation to toxicity, learned food aversions would not be necessary (Zahorik and Houpt, 1977). Experimentation with sensory reactions and with learning in vertebrates has been extensive and the resulting literature contains much information on avoidance. The cytology of chemosensory cells of vertebrates and invertebrates is similar. Such cells will respond individually to more than one stimulus and detection or recognition of a particular secondary compound thus seems to require the integration of individual cellular reactions (Chapman and Blaney, 1979). In vertebrates, stimuli that integrate into the sense of taste (or its equivalent if that term is reserved for a human response) provide the strongest mechanism selecting for or against a particular food (Krueger et al., 1974). Other senses that are involved in selection are smell, sight, and touch. Odor may act by association with flavor. In a learned response, it may act in lieu of flavor as a premonitory warning that will produce avoidance before the food material is actually tasted (Eisner and Grant, 1981). Sight aversions are usually weaker than either odor or taste aversions, but are still effective as premonitory warnings in learned responses. Touch is more subtle than simple perception of thorns or hairs. Anyone who closely watches a horse graze in mixed vegetation will recognize that the lips are intimately involved in selection of what goes into the mouth. One of the major aims of the product scientist in a toothpaste manufactory is to get the right "mouth feel" (as separate from taste or odor) in the product. Such persons are well aware of the very sensitive human ability to discriminate against a product that does not "feel" right, and there is no reason to assume other vertebrate herbivores are any less sensitive to this kind of stimulus.

Learning, in vertebrates, is associated with the same chemosensory stimuli that produce a direct oral response, probably in the same order of strength. Learned taste aversions are easily created, requiring only an initial exposure in small amount—often only a single such exposure—and are persistent over long times without reinforcement. In experiments, vertebrates commonly have the ability to discriminate a single novel food which causes sickness from the background of a spectrum of familiar foods, whether the novel food is given before or after the other components of the experimental feeding, and to reject it selectively in the future (Revusky and Bedarf, 1967). The unpleasant consequences of a novel food may be delayed by hours or even days after ingestion without affecting the subject's ability to associate the consequences with the source. Aversions of this sort are easily created by negative sequellae that are only accidentally associated with a novel food (or if strong enough, even with a familiar food). Thus in man, a bacterial or viral enteritis following upon a meal with a novel food may produce a long-lasting aversion to that food. The nausea associated in the human being with some types of cancer treatments produces similar aversions, and clinicians must take diet into consideration in treatment regimens in order not to evoke unnecessary food aversions in the patient's future (Wallace, 1976).

Domestic livestock have been subjected to vigorous genetic selection over many scores of generations and what natural selectivity for diet may remain (Westoby, 1974) is further modified by the highly specialized feeding and agronomic practices of modern animal science. The habits of wild mammalian vertebrates, although less studied, are more useful for demonstrating the fundamental effectiveness of palatability and food aversions, direct and learned, in preventing poisonings. When placed on new pasturage, deer are cautious, tasting new plants in small amounts (Nichol, 1938). In familiar situations, deer use olfaction for initial selection or rejection (Longhurst et al., 1968). An innate characteristic (unlearned) of deer foraging is a requirement for variety. Deer will lose weight if confined to a single palatable species supplied in ample quantity and of a quality capable of sustaining weight (Nichol, 1938). In natural circumstances, they forage by a "bite and skip" technique that diversifies the diet botanically and dilutes the amount of any single component in relation to the whole. It also tends to delay the addition of incremental amounts of any particular dangerous species to the gut. All of these consequences are beneficial in preventing poisoning or reducing its severity. This protective type of foraging cannot, however, be generalized to all wild vertebrates. Many have much more specialized feeding habits. Similarly, artificial pressures on deer produce unusual responses. Under conditions of semimanagement, deer will seek out "islands" of vegetation in single-plant fields (wheat and clover)—perhaps those growing on the more fertile spots (Swift, 1948). Penned wild deer cannot be made to ingest mountain laurel (*Kalmia* spp.) or rhododendron (*Rhododendron maximum*) most of which unpenned animals in the same location will eat in small amount. The penned deer will starve first (Forbes and Bechdel, 1931).

Despite the difficulties of drawing fundamental conclusions from the reactions of managed animals (wild or livestock), the following observations are pertinent. Sheep are able to select low-cyanide varieties of sudangrass (*Sorghum sudanense*) and sudan-sorghum (*Sorghum vulgare*) hybrids (Rabas et al., 1970), and also low-alkaloid populations of reed canarygrass (*Phalaris arundinacea*) (Simons and Marten, 1971; Williams et al., 1971). Sweet lupines (*Lupinus* spp.) (low in alkaloidal content) are preferred over others by sheep (Arnold and Hill, 1972). Cattle will not eat high cyanide sudangrass if other good forage is available (Stuckey, 1955). The palatability of sericea lespedeza (*Lespedeza cuneata*) to sheep varies with its tannin content (Donnelly and Anthony, 1973; Wilkins et al., 1953). The food preference of sheep can be modified twofold by learned responses (Arnold, 1964). Burrowweed (*Aplopappus heterophyllus*) aversion is a learned response (Cronin et al., 1978).

Palatability and potential toxicity are often but not always synonymous, nor constant from one species to another. Bitterbrush (*Purshia tridentata*), which is bitter to human taste, is not only palatable to range livestock, but often a very important food resource under natural conditions (Laycock, 1978). Young scrub oak (*Quercus* spp.) vegetation is usually palatable to cattle and occasionally dangerous. Larkspur (*Delphinium* spp.) is always palatable and always dangerous. But bitterweed (*Helenium* sp.) is always unpalatable and always dangerous (Merrill and Schuster, 1978).

The basic relationships of food aversions can be modified in many ways. Hunger lowers sensory barriers (Arnold, 1964). Simple fatigue of mouthparts or distension of the gut may result in cessation of feeding, as may subtle changes in body chemistry or various kinds of stress such as thirst (Arnold, 1970; Arnold and Hill, 1972). Also influential are the choices available to an animal. Some items may be eaten in larger amount than normal in the absence of sufficient alternative. On the other hand, deer may seek out conifers, almost as a man uses a condiment, against the background of an otherwise bland diet (Nichol, 1938).

Examples such as these show that aversions, learned and direct, may be highly effective in protecting large vertebrate herbivores from poisoning by preventing the entry of potentially poisonous material into the animal. Although insects show dietary preferences (often strongly selective) and some aspects of learning (Gilbert, 1975), the true learned food aversion is a characteristic of vertebrates.

A second means of protection is the ability to rid the gut rapidly of offensive or dangerous material. Two major routes are possible, vomition (or anterior regurgitation in insects) and speeded or intensified defecation. Insects commonly regurgitate or defecate when handled (Eisner et al., 1974), but the degree to which this response can be stimulated by ingestion of potentially dangerous food materials is not clear. Regurgitation or defecation in insects by handling is often defensive against a carnivorous predator, and secondary compounds of plants may have been sequestered by the insect against such use.

The vomition reflex, on the other hand, is common among vertebrates (although absent in rodents—Oehme, 1970). Although present in horses, the vomition reflex if stimulated usually results in aspiration of ingesta, because of the structure of the oral cavity, and is typically fatal. Bluejays vomit readily (Brower, 1970), and the intensity of vomition and ease with which it is stimulated in most mammalian vertebrates is impressive. Even ruminants can vomit (not the same as normal eructation) and thus rapidly clear much of the contents of the rumen from the alimentary tract. Vomition may be stimulated directly by the foodstuff itself during passage through the esophagus, or compounds absorbed from the gut may stimulate a vomition center in the brain stem (Chapman and Blaney, 1979), as for example by parenteral administration of the digitalis glycosides. Explosive diarrhea also is characteristic of vertebrates and is effective in ridding the lower gut of offensive or dangerous materials before absorption of secondary compounds has been completed.

It has even been postulated (Drummond, 1979) that the nausea and vomition of seasickness (which has no discernible evolutionary advantage for those who go to sea) is a consequence of the evolution of vomition as a protection against neurotoxins. The coordination of eye, head, and body motion, and perception of position, is easily upset by neurotoxins, and may be the most sensitive mechanism by which vertebrates can detect them. If vomition is an evolved mammalian means of protection against the action of ingested neurotoxins, it can also be expected to result from the "unnatural" motions of the sea affecting the inner ear in a manner similar to dysfunctions by neurotoxins.

A third method of protection is by detoxication of dangerous secondary compounds within the gut before absorption. This may be accomplished in insects by mixed-function oxidases present or induced in the gut (Brattsten, 1979), especially the midgut, or possibly to a lesser extent by other kinds of enzymes (Fox and Morrow, 1981).

In comparing alimentary detoxication among vertebrate herbivores, it is helpful to distinguish certain fundamental differences in gut structure and function found across this spectrum of organisms. Specialized organs such as the crop in birds, the rumen in ruminants, and the cecum in the horse modify the general process of digestion considerably and help account for major differences in susceptibility to toxins among these different kinds of herbivores. Much of the effectiveness of the rumen and cecum depends on the kinds and function of the flora of microorganisms housed in these fermentative vessels. Many of the secondary compounds that enter the vertebrate diet have been elaborated in plants as a defense against microbial attack. They can be expected also to have an effect on the microflora of the gut, which in turn can be expected to have evolved defensive capacities against such effects. A ruminal flora well adjusted to the existing diet will have specific capacities for inactivating or nulling much of the danger to the host animal of the secondary compounds in it. A novel food, or a sudden change in diet, however, may present the ruminal flora an insult to which it cannot respond effectively on the instant. The same diet, introduced slowly, allowing the microflora to change to meet the new conditions, may be ingested and digested entirely with beneficial result. This ability to adapt is the basis of long-recognized practices of practical animal management, and is apt to approximate what happens among wild ruminants as the nature of the diet changes with the seasons. Ruminal adaptation is similar in its time course to the induction of microsomal enzymes (Arnold, 1970; Freeland and Janzen, 1974).

Simple changes in pH from one part of the gut to another, particularly noticeable in ruminants, may have major consequence for the potential toxicity of specific compounds as they pass from one milieu to another.

Possession of a rumen is not always an advantage to a herbivore, however. Already mentioned has been the lesser ability of a ruminant to clear the gut by vomition compared with a monogastric animal. Essential oils in plants are particularly troublesome to ruminants by depressing the normal ruminal function (Longhurst et al., 1968). Certain small molecules, such as cyanide, are readily released from plant materials by enzymatic action enhanced by the microflora of the rumen and may be absorbed directly through the ruminal wall (Kingsbury, 1964). Toxic secondary compounds in small, heavy seeds may be concentrated in the abomasum by the mechanical sorting by density that takes place in the rumen; they would remain dispersed in the ingesta in the monogastric animal. Finally, the ruminant is more susceptible to bloat than is the monogastric herbivore.

Detoxication by enzymes in insects is greatest in the midregion of the gut, but has not been further localized. The vertebrate herbivore possesses additional ways of detoxifying compounds after absorption through the gut wall. These include the filtering action of the kidney, and especially an assortment of enzymatic reactions in various locations of the body. Principal among the latter is the role of mixed-function oxidases. These enzymes, associated with smooth endoplasmic reticulum ("microsomes") of cells, always present in some concentration, may be increased rapidly ("induced") following increased concentration of a stimulating substrate. Mixed-function oxidases are known in all phyla of animals. Experimentally, the period of induction may be as brief as a few hours. Under natural conditions, the titer of a specific mixed-function oxidase may increase to effective protective levels over a time span in the order of a day to a week (Brattsten, 1979).

Mixed-function oxidases are effective against a wide variety of potentially dangerous substrate molecules. In general, the oxidative reactions they catalyze make secondary compounds more polar, more soluble, and hence more easily excreted. Although they have been shown to have certain other functions in animals, their principal function seems invariably to be the mobilization or detoxication of secondary compounds from plants (Brattsten, 1979). In the mammal they are most common in the liver, then in the kidney, lung, placenta, and small intestine. While mixed-function oxidases are at the moment the most conspicuous and best understood of the detoxifying enzymes, vertebrate herbivores also possess conjugating and other types of enzymes that occasionally or commonly catalyze detoxifying reactions.

The reactions of mixed-function oxidases are not always beneficial. Quercetin (a phenol), a common mutagen, becomes more mutagenic after reacting with liver microsomes (Bjeldanes and Chang, 1977). Pyrrolizidine alkaloids are converted into hepatotoxic pyrroles by mixed-function oxidases (Brattsten et al., 1977).

Man, as a vertebrate herbivore, has developed certain unique protections against poisoning. These include the processes of fermentation (of which some animals take fortuitous advantage, but do not promote), and cooking. Cooking, especially, is an effective detoxicant both by heat which destroys proteinacious and certain other toxins (Janzen et al., 1976), and when water is used, by leaching dangerous compounds from the food materials. It has been suggested (Leopold and Ardrey, 1972) that the invention of cooking allowed the spread of man into latitudes and areas where the toxicity of the native vegetation would have posed a serious hazard otherwise. However, while wild plants were important in the human diet before the invention of cooking, uncooked wild plants remained important afterward (Dornstreich, 1973).

VII. COMPLEX HERB-HERBIVORE RELATIONSHIPS

Relationships between plants and herbivores mediated by toxic secondary compounds can become more complex than those described above, as modifying circumstances shape the final outcome. No class of secondary compound is toxic solely to insects or to vertebrates alone (Janzen, 1969). Any kind of secondary compound may have toxic consequences in almost any species of animal if specific mechanisms for detoxifying specialist compounds, or a generalist defense against toxic compounds fails, or under conditions of abnormal exposure. Vertebrates in poor general condition will have less ability to withstand environmental toxicity, other things being equal, than healthy animals (Cronin et al., 1978). Young animals, naive in learning and scantily equipped as yet with detoxifying enzymes or competent ruminal microflora, may react to toxic compounds quite differently from mature animals (Oehme, 1970).

Among large herbivores there is probably no such thing any longer as a truly wild animal whose diet is not influenced in some measure by the actions of humans. Deer in eastern North America, for example, are no longer subject to natural selective factors controlling population size, are restricted in movement by habitations, fences, and domestic dogs, and must forage in vegetation that is being managed or has been changed by human activities. In some places the foraging of deer is so strenuous as to produce a line, like a high-water mark, at the highest level to which deer can reach (Forbes and Bechdel, 1931). Deer populations commonly increase after logging opens the forest to renewed lower growths (Cates and Orians, 1975). Despite food pressures suggested by observations such as these, and recognizing that animals like deer are less well monitored than domestic vertebrates, it appears nonetheless that they are infrequently poisoned by plants. The detoxifying resources of wild herbivores may be greater than those of domesticated livestock and they may possess specialist defenses well matched in evolution to their normal diet's potential toxicity. Certainly, many wild vertebrates can handle specific plant material that would kill domestic vertebrates in equivalent dosage (Freeland and Janzen, 1974). Even so, they can be poisoned under extreme circumstances. For example, wild antelope have been poisoned by chokecherry (*Prunus* sp.) (Ogilvie, 1955), and drought has forced pronghorn antelope eventually to consume enough tarbrush (*Flourensia cernua*) to result in a mortality of about 60% (Hailey et al., 1966).

The reactions of domesticated vertebrate herbivores to potentially toxic secondary compounds are largely influenced, determined, or controlled by human management practices. If wild native animals display a high degree of nutritional wisdom (seeking out a nutritionally adequate diet, by selection of suitable plant species or vegetation with particular nutrient content to satisfy deficiences, from among those available), then domesticated vertebrates have lost some of that ability (Arnold and Hill, 1972; Coppock et al., 1974). Penned animals to which all forage is delivered, and animals pastured closely on seeded pastures, have little ability to excerise what natural selectivity may remain to them, and also have little chance to learn which plants are troublesome by experiencing a variety of materials in small amounts. Such animals running free (as by a fence break) may get themselves rapidly into serious dietary trouble.

Even under range conditions in which domesticated animals appear relatively unmanaged, the influence of man is strongly present. Ranges are always fenced and commonly overgrazed. Animals may be prevented from access to available nutritious forage in ways that are not always immediately obvious, such as by terrain that is too steep, too rocky, too slippery (after a rain), too marshy, or contains dense thorny vegetation which impedes free movement. Trailing, branding, shearing, restricted water, cold temperatures, and giving birth are factors that further restrict the free movement of grazing animals (and hence free choice) among available vegetation (Heady, 1964; Krueger and Sharp, 1978). Practices such as close herding, trucking from one range to another, or unprotected severe winter exposure, obviously also affect an animal's innate selectivity or capacity to detoxify. This effect may vary even with the individual animal. In herded sheep, for example, laggard animals eat more toxic species than those in front (Doran, 1943).

Other agricultural practices also influence the toxicity of plants to vertebrate herbivores. The application of herbicides, particularly, may have a large effect on the level of certain secondary compounds in forage and may or may not affect palatability. Alkaloids increase in larkspur (*Delphinium* spp.) after its treatment with herbicides, but the palatability to cattle (which for larkspurs is already high) remains good (Williams and Cronin, 1963). In some cases, palatability of dangerous forage increased after herbicide application (Laycock, 1978). Mechanical control of larkspurs by clipping, on the other hand, reduces the concentration of alkaloids in the regrowth (perhaps by requiring the plant to spend its resources elsewhere).

Seed distribution by animals arose early in the coevolution of vascular plants and herbivores (Snow, 1971). Developing fruits and seeds require protection not only from vertebrate herbivores but also from insects until mature and usually are richly provided with toxic secondary compounds. If the mature fruits are to attract the kinds of animals that can distribute them (larger than insects), they must be detoxified in the ripening process. Tannins in immature fruits may be polymerized into nontoxic compounds (McKey, 1979). Similarly, alkaloids disappear either by being bound into an inactive form or more commonly by being remobilized out of the fruit (McKey, 1974). Another way in which fruits are adequately protected, yet well distributed, is by an evolved relationship in which the particular distributing animal has developed a specialist capacity to handle the given toxic compounds protecting that fruit. Commonly, birds are able to eat fruits (such as belladonna, *Atropa belladonna*) with impunity that would be lethal to most vertebrates (Snow, 1971). In fact, some seeds require passage through the digestive system of a particular bird or other vertebrate in order for successful germination to follow. When cooperative pairing of this sort has been refined to an obligate relationship, as may happen over a particular coevolutionary history, one member of the pair is in serious trouble if something happens to the other. *Calvaria major* seeds normally require passage through the gut of the dodo in order to germinate. This plant is now endangered by the extinction of the dodo (Temple, 1977). It may be possible to predict the dispersal/ protective strategy of fruits by their appearance. Fruits designed for distribution by birds should be easily found and not too large to carry. Ripening may be accompanied not only by a diminution in toxicity but also by a change in color or pattern from cryptic to showy. Insect pressure would, on the contrary, be expected to select for cryptic fruits and size would be of little consequence (Snow, 1971).

Toxic plants are not usually confined to one or a few niches in the ecological pattern of any given area (Cronin et al., 1978; Whittaker, 1970). Niches seem to be filled by the jostling of species (Force, 1974) and specialist relationships among plants and consumer species serve to create greater relief and more sharply defined boundaries among niches. It may be possible for larger lacunae to develop in evolutionary time when competing populations of plants and animals are characterized largely by generalist strategies of protection and consumption.

Thus monophagy among insects with specialist capacities has probably evolved repeatedly (Smiley, 1978). It may start as a response to availability of a particular plant, and evolve to an increasingly specific one-to-one relationship as specialist detoxifying mechanisms are evolved by the insect to meet specialist defensive compounds developed by the plant and by the evolution of greater digestive efficiency against any generalist defensive compounds that may be present. On the other hand, large herbivores mostly disappeared from North America during the Pleistocene glaciation. West of the Rockies, before the advent of domestic animals, the vegetation was largely unpressurized by large herbivores (although perhaps by increasing dryness). Sagebrush (*Artemisia* spp.) is among the dominants, and it is interesting that the one remaining large native herbivore (the pronghorn antelope) relishes sagebrush despite its heavy concentration of essential oils (Cronin et al., 1978). None of the introduced domestic livestock can handle sagebrush as effectively. Nor is the pressure of the few remaining native vertebrates sufficient to account, in an evolutionary sense, for the unusual spininess of range plants of the southwestern desert at the present time (Janzen, 1979). Its evolution dates to a time before the Pleistocene. The flora of New Zealand developed without large vertebrates present on either North or South Island, and was quickly decimated when they arrived (Longhurst et al., 1968). Islands, in general, have fewer poisonous plants than do continents (Carlquist, 1974), presumedly because islands cannot support the variety or density of large vertebrate herbivores (or any kind of large vertebrates) that larger land-masses can.

Just as the concept of apparency has brought order to much of the diversity of secondary compounds in plants, so can the concept of relative size and longevity organize some of the observations on the toxic relationships between plants and animals. A single leaf is large to a typical insect, but small to a vertebrate herbivore. A consequence is that a typical insect is apt to be adapted to the specificities of a botanically monotonous diet while a vertebrate herbivore normally consumes a botanically varied diet. On the other hand, boring insects may attack seeds without regard to size, while a typical vertebrate herbivore cannot eat a coconut (*Cocos nucifera*) or may strangle on an osage orange (*Maclura pomifera*) if it tries to eat one. Based on relative volumes and resources, it is less costly for large animals to produce a unit volume of a detoxifying enzyme than it is for small animals (Brattsten, 1979; Whittaker, 1970).

In any herb-herbivore relationships relative size may be predictive of the consumer strategy and the plant defense. In a similar way, temporal relationships also provide insight into coevolved strategies between pairs of dependent organisms. The ability of a population of short-lived insects to adapt genetically to the peculiarities of a single long-lived host tree has already been mentioned. In general, the short-lived feeding stages of insects are temporally limited to a diet that is not only botanically monotonous but may be restricted further to a single stage of seasonal growth. The diet of long-lived vertebrate herbivores, in contrast, must change drastically from season to season in temporal or boreal latitudes, and from wet to dry seasons in many tropical locations.

VIII. CONCLUSIONS

Secondary compounds are produced by plants in bewildering variety and sometimes in large amount. Perhaps originating as wastes in the evolution of primitive organisms, they soon came to serve useful functions both within the organism (metabolism, regulation, etc.) and by mediating positive and negative relationships with other individuals of the same species and among adjacent species. No species of plant is without a complement of useful secondary compounds, some usually unique to that species, and few relationships take place among plants or between plant and animal species except as controlled or modified by secondary metabolites of plants. Perhaps the greatest single use of secondary compounds is in defense or denial of resource to potential herbivores by means of toxicity.

Actual toxic relationships are myriad; many are complex. The genetic skirmishes that have been deduced from present circumstances give an impression of great coevolutionary ingenuity. Specific toxic relationships usually involve the specialist defenses of unapparent plants and the "targeted" attack of limited consumer herbivores. For reasons of relative size and longevity, these specialist relationships occur primarily between plants and herbivore insects. Yet vertebrate herbivores are also exposed inevitably to the same specialist defensive compounds by the nature of their varied diets. They are normally exposed in addition to the generalized defensive compounds in apparent plants, some of which may have coevolved primarily to confer protection against large herbivores. Thus the consumer strategy of large herbivores must typically be based on a spectrum of abilities to avoid or detoxify this array of potentially toxic secondary compounds.

Under appropriate (often abnormal) circumstances specific secondary substances of plants can damage nearly any organ or system in a vertebrate animal from skin to mouth, gut, liver, circulatory system, kidney, heart, bone, lung, thyroid, eye, nervous systems, embryo, milk, and hair (Bailey, 1978; Kingsbury, 1980). Similarly, toxicity can vary with plant family, genus, species, variety, or even individual or sex of plant as well as with ecological or agronomic variables and stage of growth (Kingsbury, 1960).

Few authors have speculated on the degree to which toxins in plants have been directed specifically against large vertebrate herbivores (e.g., Culvenor, 1970; Freeland and Janzen, 1974; Kingsbury, 1978; Rothschild, 1972; Swain, 1977), and the conclusions vary. Nevertheless, whatever the evolutionary source of toxic secondary compounds in vegetation, wild vertebrate herbivores (especially those living under conditions little disturbed or pressured by man) are rarely poisoned. Their defensive strategies against secondary compounds are, by and large, remarkably effective. If they are poisoned, it is generally through the failure for one reason or another (usually associated with human activities) of their defenses. In other words, poisoning of large vertebrate herbivores appears rare and truly accidental under natural conditions. The toxic secondary compounds were there for some other evolved function than to kill or injure large vertebrates, against which they are normally ineffective.

The situation with man, their domestic livestock, and their pets, is quite different. Incidents involving toxic plants rank first or second among categories of potentially toxic ingestions reported in children under age 5 by poison control centers in the United States (Oehme, 1978). Even more striking are responsible figures (James, 1978) for poisonings of range livestock in the western states. Mortality is estimated at 3-5% annually. Estimates of illness associated with toxic plants (assessed by determining the cause of illness among all sick animals in field surveys of all animals in a specific area), not including mortality, are around 9%. No reliable figures are available either for eastern livestock or pets, but experience shows that significant trouble occurs in these animals as well.

Human activity in the agronomic practices of growing crops, in the genetic manipulation of domesticated animals, and in practices of animal management has severely reduced the ability of domesticated animals to defend themselves effectively against toxic secondary compounds in plants. In general, little anterior attention has been given consciously to potential toxicity as practices have changed. We either muddle through empirically, or deal with the consequences of toxicity after the fact.

Of particular concern at present is the extensive research effort to improve the defenses of commercial plant crops against insects and pathogens. The principal way in which this can be accomplished is by manipulating the complement of secondary substances in these species. Only lately have regulatory agencies recognized that such manipulation may have toxic consequences to domestic livestock, pets, and man by presenting additional poisons to which natural defenses are inadequate or no longer extant. Given the past intense public concern about "chemicals" added to our foods, it might now be wise to consider how the general public will react to the idea that scientists are now at work breeding "chemicals" into our foodstuffs.

REFERENCES

Arnold, G. W. (1964). Some principles in the investigation of selective grazing. *Proc. Aust. Soc. Anim. Prod.* 5:258-271.

Arnold, G. W. (1970). Regulation of food intake in grazing ruminants. In *Physiology of Digestion and Metabolism in the Ruminant*, A. T. Phillipson (Ed.). Oriel Press, Newcastle on Tyne, England, pp. 264-276.

Arnold, G. W., and Hill, J. L. (1972). Chemical factors affecting selection of food plants by ruminants. In *Phytochemical Ecology*, J. B. Harbourne (Ed.). Academic Press, New York, pp. 71-101.

Atsatt, P. R., and O'Dowd, D. J. (1976). Plant defense guilds. *Science 193*:24-29.

Bailey, E. M., Jr. (1978). Physiologic responses of livestock to toxic plants. *J. Range Manage. 31*:343-347.

Bate-Smith, E. C. (1972). Attractants and repellents in higher animals. In *Phytochemical Ecology*, J. B. Harbourne (Ed.). Academic Press, New York, pp. 45-56.

Beck, S. D., and Reese, J. C. (1976). Insect-plant interactions: nutrition and metabolism. *Rec. Adv. Phytochem. 10*:41-92.

Bedard, W. D., Tilden, P. E., Wood, D. L., Silverstein, R. M., Brownlee, R. G., and Rodin, J. O. (1969). Western pine beetle: field response to its sex pheromone and a synergistic host terpene, myrcene. *Science 164*: 1284-1285.

Bentley, B. L. (1977). Extrafloral nectaries and protection by pugnacious bodyguards. *Annu. Rev. Ecol. Syst. 8*:407-427.

Berenbaum, M. (1978). Toxicity of a furanocoumarin to army worms: a case of biosynthetic escape from insect herbivores. *Science 201*:532-534.

Berenbaum, M., and Feeny, P. P. (1981). Toxicity of angular furano-coumarins to swallowtail butterflies: escalation in an evolutionary arms race? *Science 212*:927-929.

Berger, P. J., Sanders, E. H., Gardner, P. D., and Negus, N. C. (1977). Phenolic plant compounds functioning as reproductive inhibitors in *Microtus montanus*. *Science 195*: 575-577.

Bjeldanes, L. F., and Chang, G. W. (1977). Mutagenic activity of quercetin and related compounds. *Science 197*:577-578.

Bowers, W. S., and Nishida, R. (1980). Juvocimenes: potent juvenile hormone mimics from sweet basil. *Science 209*:1030-1032.

Bowers, W. S., Ohta, T., Cleere, J. S., and Marsella, P. A. (1976). Discovery of insect anti-juvenile hormones in plants. *Science 193*:542-547.

Brattsten, L. B. (1979). Biochemical defense mechanisms in herbivores against plant allelochemics. In *Herbivores: Their Interaction with Secondary Plant Metabolites*, G. A. Rosenthal and D. H. Janzen (Eds.). Academic Press, New York, pp. 199-270.

Brattsten, L. B., Wilkinson, C. F., and Eisner, T. (1977). Herbivore-plant interactions: mixed function oxidases and secondary plant substances. *Science 196*:1349-1352.

Brodie, B. B., and Maickel, R. P. (1961). Comparative biochemistry of drug metabolism. *Proc. First Int. Pharmacol. Meet. 6*:299-324.

Brower, L. P. (1970). Plant poisons in a terrestrial food chain and implications for a mimicry theory. In *Biochemical Coevolution*, K. L. Chambers (Ed.). Oregon State University Press, Corvallis, pp. 69-82.

Brower, L. P., and Brower, J. VanZ. (1964). Birds, butterflies, and plant poisons: a study in ecological chemistry. *Zoologica 49*:137-159.

Campbell, B. C., and Duffey, S. S. (1979). Tomatine and parasitic wasps: incompatibility of plant antibiosis with biological control. *Science 205*:700-702.

Carlquist, S. (1974). *Island Biology*. Columbia University Press, New York.

Cates, R. G., and Orians, G. H. (1975). Successional status and the palatability of plants to generalized herbivores. *Ecology 56*:410-418.

Chambers, K. L. (Ed.) (1970). *Biochemical Coevolution.* Oregon State University Press, Corvallis.

Chapman, R. F., and Blaney, W. M. (1979). How animals perceive secondary compounds. In *Herbivores: Their Interaction with Secondary Plant Metabolites,* G. A. Rosenthal and D. H. Janzen (Eds.). Academic Press, New York, pp. 161-198.

Chew, F. S., and Rodman, J. E. (1979). Plant resources for chemical defense. In *Herbivores: Their Interaction with Secondary Plant Metabolites,* G. A. Rosenthal and D. H. Janzen (Eds.). Academic Press, New York, pp. 271-307.

Coppock, C. E., Everett, R. W., Smith, N. E., Slack, S. T., and Harner, J. P. (1974). Variations in forage preference in dairy cattle. *J. Anim. Sci. 39*:1170-1179.

Cronin, E. H., Ogden, P. A., Young, J. A., and Laycock, W. (1978). The ecological niche of poisonous plants in range communities. *J. Range Manage. 31*:328-334.

Culvenor, C. C. J. (1970). Toxic plants—a reevaluation. *Search 1*:103-110.

Deverall, B. J. (1977). *Defence Mechanisms of Plants.* Cambridge University Press, Cambridge.

DeWilde, J., Bongers, W., and Schooneveld, H. (1969). Effects of host plant age on phytophagous insects. *Entomol. Exp. Appl. 12*:714-720.

Donnelly, E. D., and Anthony, W. B. (1973). Relationship of sericea lespedeza leaf and stem tannin to forage quality. *Agron. J. 65*:993-994.

Doran, C. W. (1943). Activities and grazing habits of sheep on summer ranges. *J. Forestry 41*:253-258.

Dornstreich, M. D. (1973). Food habits of early man: balance between hunting and gathering. *Science 179*:306.

Drummond, A. J., Jr. (1979). Motion sickness. *Sea Front.*, January-February, pp. 39-43.

Duffey, S. S. (1970). Cardiac glycosides and distastefulness: some observations on the palatability spectrum of butterflies. *Science 169*:78-79.

Edmunds, G. F., Jr., and Alstad, D. N. (1978). Coevolution in insect herbivores and conifers. *Science 199*:941-945.

Ehrlich, P. R. (1970). Coevolution and the biology of communities. In *Biochemical Coevolution,* K. L. Chambers (Ed.). Oregon State University Press, Corvallis, pp. 1-11.

Ehrlich, P. R., and Raven, P. H. (1964). Butterflies and plants: a study in coevolution. *Evolution 18*:586-608.

Eisner, T., and Grant, R. P. (1981). Toxicity, odor aversion, and "olfactory aposematism," *Science 213*:476.

Eisner, T., Johnessee, J. S., Carrel, J., Hendry, L. B., and Meinwald, J. (1974). Defensive use by an insect of a plant resin. *Science 184*:996-999.

Feeny, P. P. (1970). Seasonal changes in oak leaf tannins and nutrients as a cause of spring feeding by winter moth caterpillars. *Ecology 51*:565-581.

Feeny, P. P. (1975). Biochemical coevolution between plants and their insect herbivores. In *Coevolution of Animals and Plants,* L. E. Gilbert and P. H. Raven (Eds.). University of Texas Press, Austin, pp. 3-19.

Feeny, P. P. (1976). Plant apparency and chemical defense. *Phytochemistry 10*:1-40.

Forbes, E. B., and Bechdel, .S I. (1931). Mountain laurel and rhododendron as foods for the white-tailed deer. *Ecology 12*:323-333.

Force, D. C. (1974). Ecology of insect host-parasitoid communities. *Science 184*:624-632.

Fowden, L., and Lea, P. J. (1979). Mechanism of plant avoidance of auto-toxicity by secondary metabolites, especially by nonprotein amino acids. In *Herbivores: Their Interactions with Secondary Plant Metabolites*, G. A. Rosenthal and D. H. Janzen (Eds.). Academic Press, New York, pp. 135-160.

Fox, L. R., and Morrow, P. A. (1981). Speciation: species property or local phenomenon? *Science 211*:887-893.

Fraenkel, G. S. (1959). The raison d'être of secondary plant substances. *Science 129*:1466-1470.

Fraenkel, G. (1969). Evaluation of our thoughts on secondary plant substances. *Entomol. Exp. Appl. 12*:473-486.

Freeland, W. J., and Janzen, D. H. (1974). Strategies in herbivory by mammals: the role of plant secondary compounds. *Am. Nat. 108*:269-289.

Gilbert, L. E. (1971). Butterfly-plant coevolution: Has *Passiflora adenopoda* won the selectional race with heliconiine butterflies? *Science 172*:585-586.

Gilbert, L. E. (1975). Ecological consequences of a coevolved mutualism between butterflies and plants. In *Coevolution of Plants and Animals*, L. E. Gilbert and P. H. Raven (Eds.). University of Texas Press, Austin, pp. 210-240.

Gilbert, L. E., and Raven, P. H. (Eds.) (1975). *Coevolution of Animals and Plants*. University of Texas Press, Austin.

Hailey, T. L., Thomas, J. W., and Robinson, R. M. (1966). Pronghorn die-off in Trans-Pecos, Texas. *J. Wildl. Manage. 30*:488-496.

Harbourne, J. B. (1972a). Evolution and function of flavonoids in plants. *Recent Adv. Phytochem. 4*:107-147.

Harbourne, J. B. (Ed.) (1972b). *Phytochemical Ecology*. Academic Press, New York, 286 pp.

Harbourne, J. B. (Ed.) (1977). *Biochemical Aspects of Plant and Animal Coevolution*. Ann. Proc. Phytochem. Soc. Eur. 15. Academic Press, New York.

Heady, H. F. (1964). Palatability of herbage and animal preference. *J. Range Manage. 17*:76-82.

Hegnauer, R. (1975). Secondary metabolites and crop plants. In *Crop Genetic Resources for Today and Tomorrow*, O. H. Frankel and J. G. Hawkes (Ed.). Cambridge University Press, Cambridge, pp. 249-265.

Hsiao, T. H. (1969). Chemical basis of host selection and plant resistance in oligophagous insects. *Entomol. Exp. Appl. 12*:777-788.

Ikada, T., Matsumura, F., and Benjamin, D. M. (1977). Chemical basis for feeding adaptation of pine sawflies *Neodiprion rugifrons* and *N. swainei*. *Science 197*:497-498.

Ingham, J. L. (1972). Phytoalexins and other natural products as factors in plant disease resistance. *Bot. Rev. 38*:343-424.

Ivins, J. D. (1952). The relative palatability of herbage plants. *J. Br. Grassl. Soc. 7*:43-54.

James, L. F. (1978). Overview of poisonous plants problems in the United States. In *Effects of Poisonous Plants on Livestock*, R. F. Keeler, K. R. Van Kampen, and L. F. James (Eds.). Academic Press, New York, pp. 3-5.

Janzen, D. H. (1966). Coevolution of ants and acacias in Central America. *Evolution 20*:249-275.

Janzen, D. H. (1969). Seed-eaters versus seed size, number, toxicity, and dispersal. *Evolution* 23:1-27.

Janzen, D. H. (1979). New horizons in the biology of plant defenses. In *Herbivores: Their Interaction with Secondary Plant Metabolites*, G. A. Rosenthal and D. H. Janzen (Eds.). Academic Press, New York, pp. 331-350.

Janzen, D. H., Juster, H. B., and Liener, I. E. (1976). Insecticidal action of the phytohaemagglutinin in black beans on a bruchid beetle. *Science* 192:795-796.

Johnson, H. B. (1975). Plant pubescence: an ecological perspective. *Bot. Rev.* 41:233-258.

Jones, D. A. (1972). Cyanogenic glycosides and their function. In *Phytochemical Ecology*, J. B. Harbourne (Ed.). Academic Press, New York, pp. 103-124.

Joseffson, E. (1967). Distribution of thioglucosides in different parts of *Brassica* plants. *Phytochemistry* 6:1617-1627.

Kingsbury, J. M. (1960). Poisonous plants of particular interest to animal nutritionists. *Proc. Cornell Nutr. Conf. Feed Manuf.* 1960:14-23.

Kingsbury, J. M. (1964). *Poisonous Plants of the United States and Canada*. Prentice-Hall, Englewood Cliffs, N.J.

Kingsbury, J. M. (1978). Ecology of poisoning. In *Effects of Poisonous Plants on Livestock*, R. F. Keeler, K. R. Van Kampen, and L. F. James (Eds.). Academic Press, New York, pp. 81-91.

Kingsbury, J. M. (1980). Phytotoxicology. In *Casarett and Doull's Toxicology*, 2nd ed., J. Doull, C. D. Klaassen, and M. O. Amdur (Eds.). Macmillan, New York, pp. 578-590.

Krieger, R. I., Feeny, P. P., and Wilkinson, C. F. (1971). Detoxification enzymes in the guts of caterpillars: an evolutionary answer to plant defense? *Science* 172:579-581.

Krueger, W. C., and Sharp, L. A. (1978). Management approaches to reduce livestock losses from poisonous plants on rangeland. *J. Range Manage.* 31:347-350.

Krueger, W. C., Laycock, W. A., and Price, D. A. (1974). Relationships of taste, smell, sight, and touch to forage selection. *J. Range Manage.* 27:258-262.

Laycock, W. A. (1978). Coevolution of poisonous plants and large herbivores on rangelands. *J. Range Manage.* 31:335-342.

Leopold, A. C., and Ardrey, R. (1972). Toxic substances in plants and the food habits of early man. *Science* 176:512-513.

Levin, D. A. (1971). Plant phenolics: an ecological perspective. *Am. Nat.* 105:157-181.

Levin, D. A. (1973). The role of trichomes in plant defense. *Q. Rev. Biol.* 48:3-16.

Levin, D. A. (1976a). Alkaloid bearing plants: an eco-geographic perspective. *Am. Nat.* 110:261-284.

Levin, D. A. (1976b). The chemical defenses of plants to pathogens and herbivores. *Ann. Rev. Ecol. Syst.* 7:121-159.

Longhurst, W. M., Oh, H. K., Jones, M. B., and Kepner, R. E. (1968). A basis for the palatability of deer forage plants. *Trans. N. Am. Wildl. Nat. Res. Conf.* 33:181-192.

Mabry, T. J. (1972). Major frontiers in phytochemistry. *Rec. Adv. Phytochem.* 4:273-306.

Marten, G. C., and Donker, J. D. (1966). Animal excrement as a factor influencing acceptibility of grazed forage. *Proc. Int. Grassl. Congr.* 10: 524-527.

McKey, D. (1974). Adaptive patterns in alkaloid physiology. *Am. Nat.* 108: 305-320.

McKey, D. (1975). The ecology of coevolved seed dispersal systems. In *Coevolution of Animals and Plants*, L. E. Gilbert and P. H. Raven (Eds.). University of Texas Press, Austin, pp. 159-191.

McKey, D. (1979). The distribution of secondary compounds within plants. In *Herbivores: Their Interaction with Secondary Plant Metabolites*, G. A. Rosenthal and D. H. Janzen (Eds.). Academic Press, New York, pp. 55-133.

McKey, D., Waterman, P. G., Mbi, C. N., Gartlan, J. S., and Struhsaker, T. T. (1978). Phenolic content of vegetation in two African rain forests: ecological implications. *Science* 202:61-64.

McNaughton, S. J. (1978). Serengeti ungulates: feeding selectivity influences the effectiveness of plant defense guilds. *Science* 199:806-807.

Merrill, L. B., and Schuster, J. L. (1978). Grazing management practices affect livestock losses from poisonous plants. *J. Range Manage.* 31:351-354.

Miles, P. W. (1969). Interactions of plant phenols and salivary phenolases in the relationship between plants and hemiptera. *Entomol. Exp. Appl.* 12: 736-744.

Mulcahy, D. L. (1979). The rise of the angiosperms: a genecological factor. *Science* 206:20-23.

Muller, C. H. (1966). The role of chemical inhibition (allelopathy) in vegetational composition. *Bull. Torrey Bot. Club* 93:332-351.

Muller, C. H. (1969). The "co-" in coevolution. *Science* 164:197.

Muller, C. H. (1970). Phytotoxins as plant habitat variables. In *Recent Adances in Phytochemistry*. Vol. 3, C. Steelink and V. C. Runeckles (Eds.). Appleton-Century-Crofts, New York, pp. 105-121.

Muller, C. H., and Del Moral, R. (1971). Role of animals in suppression of herbs by shrubs. *Science* 173:462-463.

Nichol, A. A. (1938). Experimental feeding of deer. *Univ. Ariz. Agric. Exp. Stn. Tech. Bull.* 75:1-39.

Oehme, F. W. (1970). Species differences: the basis for and importance of comparative toxicology. *Clin. Toxicol.* 3:5-10.

Oehme, F. W. (1978). The hazard of plant toxicities to the human population. In *Effects of Poisonous Plants on Livestock*, R. F. Keeler, K. R. Van Kampen, and L. F. James (Eds.). Academic Press, New York, pp. 67-80.

Ogilvie, S. (1955). Chokecherry toxic to an antelope. *J. Mammal.* 36:146.

Piltemer, E. A., and Tingey, W. M. (1976). Hooked trichomes: a physical plant barrier to a major agricultural pest. *Science* 193:482-484.

Rabas, D. L., Schmid, A. R., and Marten, G. C. (1970). Relationship of chemical composition and morphological characteristics to palatability in sudan grass and sorghum x sudan grass hybrids. *Agron. J.* 62:762-763.

Rathke, B. J., and Poole, R. W. (1975). Coevolutionary race continues: butterfly larval adaptation to plant trichomes. *Science* 187:175-176.

Reese, J. C. (1979). Interactions of allelochemics with nutrients in herbivore food. In *Herbivores: Their Interaction with Secondary Plant Metabolites*, G. A. Rosenthal and D. H. Janzen (Eds.). Academic Press, New York, pp. 309-350.

Rehr, S. S., Janzen, D. H., and Feeny, P. P. (1973). L-Dopa in legume seeds: a chemical barrier to insect attack. *Science 181*:81-82.

Revusky, S. H., and Bedarf, E. W. (1967). Association of illness with prior ingestion of novel foods. *Science 155*:219-220.

Rhoades, D. F. (1979). Evolution of plant chemical defenses against herbivores. In *Herbivores: Their Interaction with Secondary Plant Metabolites*, G. A. Rosenthal and D. H. Janzen (Eds.). Academic Press, New York, pp. 3-54.

Rhoades, D. F., and Cates, R. G. (1976). Toward a general theory of plant anti-herbivore chemistry. *Rec. Adv. Phytochem. 10*:168-213.

Robinson, T. (1974). Metabolism and function of alkaloids in plants. *Science 184*:430-435.

Robinson, T. (1979). The evolutionary ecology of alkaloids. In *Herbivores: Their Interaction with Secondary Plant Metabolites*, G. A. Rosenthal and D. H. Janzen (Eds.). Academic Press, New York, pp. 413-448.

Rosenthal, G. A., and Janzen, D. H. (Eds.) (1979). *Herbivores: Their Interaction with Secondary Plant Metabolites*. Academic Press, New York.

Rothschild, M. (1971). Speculation about mimicry with Henry Ford. In *Ecological Genetics and Evolution*, R. Creed (Ed.). Blackwell, Oxford, pp. 202-223.

Rothschild, M. (1972). Some observations on the relationship between plants, toxic insects, and birds. In *Phytochemical Ecology*, J. B. Harbourne (Ed.). Academic Press, New York, pp. 2-12.

Rothschild, M. (1975). Remarks on carotenoids in the evolution of signals. In *Coevolution of Animals and Plants*, L. E. Gilbert and P. H. Raven (Eds.). University of Texas Press, Austin, pp. 20-50.

Schoonhoven, L. M. (1972). Secondary plant substances and insects. *Rec. Adv. Phytochem. 5*:197-224.

Simons, A. B., and Marten, C. G. (1971). Relationship of indole alkaloids to palatability of *Phalaris arundinacea* L. *Agron. J. 63*:915-919.

Smiley, J. (1978). Plant chemistry and the evolution of host specificity: new evidence from *Heliconius* and *Passiflora*. *Science 201*:745-747.

Snow, D. W. (1971). Evolutionary aspects of fruit eating by birds. *Ibis 113*:194-202.

Sondheimer, E., and Simeone, J. B. (Eds.) (1970). *Chemical Ecology*. Academic Press, New York.

Southwood, T. R. E. (1973). The insect/plant relationship—an evolutionary perspective. *Symp. R. Entomol. Soc. Lond. 6*:3-30.

Spiher, A. T., Jr. (1974). FDA regulations: a new development in agriculture. In *The Effect of FDA Regulations (GRAS) on Plant Breeding and Processing*, C. H. Hasen et al. (Eds.). Crop Science Society of America, Madison, Wis., pp. 1-6.

Stuckey, I. H. (1955). Animals don't usually eat poison plants. *Rhode Island Agric.*, Summer 1955:5-6.

Sussman, R. W., and Raven, P. H. (1978). Pollination by lemurs and marsupials: an archaic coevolutionary system. *Science 200*:731-736.

Swain, T. (1977). Secondary compounds as protective agents. *Annu. Rev. Plant Physiol. 28*:479-500.

Swift, R. W. (1948). Deer select most nutritious forages. *J. Wildl. Manage. 12*:109-110.

Temple, S. A. (1977). Plant-animal mutualism: coevolution with dodo leads to near extinction of plant. *Science 197*:885-886.

Thurston, L. E., and Lersten, N. R. (1969). The morphology and toxicology of plant stinging hairs. *Bot. Rev. 35*:393-412.

Tukey, H. B., Jr. (1966). Leaching of metabolites from above-ground plant parts and its implications. *Bull. Torrey Bot. Club 93*:385-401.

Wallace, P. (1976). Animal behavior: the puzzle of flavor aversion. *Science 193*:989-991.

Westoby, M. (1974). An analysis of diet selection by large generalist herbivores. *Am. Nat. 108*:290-304.

Whittaker, R. H. (1970). The biochemical ecology of higher plants. In *Chemical Ecology*, E. Sondheimer and J. B. Simeone (Eds.). Academic Press, New York, pp. 43-70.

Whittaker, R. H., and Feeny, P. P. (1971). Allelochemics: chemical interactions between species. *Science 171*:757-770.

Wilkins, H. L., Bates, R. P., Henson, P. R., Lindahl, I. L., and Davis, R. E. (1953). Tannin and palatability in sericea lespedeza, *L. cuneata*. *Agron. J. 45*:335-336.

Williams, K. S., and Gilbert, L. E. (1981). Insects as selective agents on plant vegetative morphology: egg mimicry reduces egg laying by butterflies. *Science 212*:467-469.

Williams, M., Barnes, R., and Cassady, J. (1971). Characterization of alkaloids in palatable and unpalatable clones of *Phalaris arundinacea* L. *Crop Sci. 11*:213-217.

Williams, M. C., and Cronin, E. H. (1963). Effect of Silvex and 2,4,5-T on alkaloid content of tall larkspur. *Weeds 11*:317-319.

Zahorik, D. M., and Houpt, K. A. (1977). The concept of nutritional wisdom: applicability of laboratory learning models to large herbivores. In *Learning Mechanisms in Food Selection*, L. M. Barker, M. R. Best, and M. Domjan (Eds.). Baylor University Press, Waco, Tex., pp. 45-67.

21

THE ROLE OF TOXINS IN PLANT-PLANT INTERACTIONS

LAWRENCE G. STOWE* and BONG-SEOP KIL†

University of Massachusetts, Amherst, Massachusetts

Present affiliations
*Choate Rosemary Hall, Wallingford, Connecticut
†Dept. of Biology, Won Kwang University, Iri, Korea

I. INTRODUCTION

The hypothesis that plants influence each other by means of releasing toxic substances into the soil has been discussed since the time of De Candolle (1832), and the past 20 years have witnessed an enormous swell of interest in this phenomenon. "Allelopathy," as it was named by Molische (1937), has been postulated to explain patterns of distribution, succession and growth in a great variety of ecosystems, from deserts to tropical rain forests, from rice paddies to heath scrub (Muller, 1974; Rice, 1974, 1979; Fisher, 1977; Horsley, 1977b; Putnam and Duke, 1978). Not only have plant shoots and roots (living and decomposing) been found to release phytotoxins, but so have fern gametophytes (Peterson and Fairbrothers, 1980), pollen (Sukhada and Jayachandra, 1980), and seeds (Gressel and Holm, 1964). The scope of these allelopathy investigations is too broad to be encompassed in a brief review; here we focus on the direct chemical interactions among higher plants, thus passing over interactions among lower organisms (Owens, 1969; Turner, 1971; Keating, 1978; Wolfe and Rice, 1979) and plant-plant interactions occurring indirectly, through mycorrhizae (Handley, 1963; Olsen et al., 1971; Robinson, 1972) and nitrifying and nitrogen-fixing bacteria (Blum and Rice, 1969; Rice and Pancholy, 1972, 1973; Lodhi, 1979c; Lodhi and Killingbeck, 1980).

The field of allelopathy remains somewhat in turmoil, largely because of the great difficulty of distinguishing it from competition in the field (Harper, 1964; Muller, 1969a). Although hundreds of investigations have turned up evidence supporting the role of allelopathy by one species or another (Rice, 1974, 1979), various scientists have expressed misgivings about some of the methods used in these studies (Harper, 1961, 1964, 1977; Cannon et al., 1962; del Moral and Cates, 1971; Kaminsky and Muller, 1977; Newman, 1978; Putnam and Duke, 1978; Montes and Christensen, 1979; Stowe, 1979; Klein and Muller, 1980). In this chapter, after a brief discussion of the evolution of allelopathy, we will elucidate the allelopathic potential of plants by reviewing the phytotoxins which they produce and the effects of those toxins on other plants. We will then approach the question of whether these allelopathic agents are effective in natural communities by considering the factors which mitigate their toxicity in the field, and the sorts of evidence which have been used to document allelopathy.

II. EVOLUTIONARY CONSIDERATIONS

The ability of plants to produce phytotoxic compounds may have evolved in some cases as a means for plants to inhibit or kill their neighbors and thus eliminate the competition for limited resources (Rice, 1967). Such a mechanism would be analogous to territoriality in animals and might be considered an example of "interference competition" (Pianka, 1978, p. 174; not to be confused with Harper's (1964) term, "interference").

However, Newman (1978) argues that allelopathy does not appear to have evolved because of any advantages it might confer on the plants releasing phytotoxins. This he concludes from three generalizations taken from the allelopathy literature to date: (1) In bioassays, plant-produced phytotoxins appear to be just as detrimental to the producer species as to other species. (2) Increased competition, with resultant deficiencies, does not tend to stimulate increased toxin production. (3) Native species that have grown together for centuries have evolved no greater resistance to each other's toxins than have introduced species. All three of these generalizations tend to refute the hypothesis that plants have evolved an ability to release phytotoxins because of the advantages of poisoning their competition. Of course, these three points are to some extent based on the results of laboratory and greenhouse bioassays, the validity of which has been questioned (Harper, 1977; Stowe, 1979).

Another hypothesis for the origin of allelopathic agents is that they are waste compounds, by-products of other chemical pathways (Muller, 1966, 1969b, 1970; Whittaker, 1970). Swain (1977) has argued cogently against this idea, however, for plant secondary compounds in general. He believes that waste products (1) would be relatively simple; (2) would be stable, with very low amounts of turnover, in the plant tissue; and (3) would be fairly uniform and unchanging through the course of evolution. None of these characteristics can be applied to the secondary substances found in plants. Of course, it is dangerous to generalize about such a diverse array of chemicals. Yet if the biosynthetic pathways of phenolic compounds (Brown, 1961; Neish, 1964; Harborne, 1980) and terpenes (Charlwood and Banthorpe, 1978; Banthorpe and Charlwood, 1980) are examined, the substances seem to be, not by-products, but end products of various energy-requiring, biosynthetic pathways. Far from trying to excrete phenolics and terpenes, plants are going to considerable effort to produce them.

Alternatively, allelopathic agents may have evolved in plants for some function other than allelopathy. Swain (1977) and Bell (1980) point out the growing evidence that secondary compounds protect plants against herbivores and pathogens; clearly, chemicals which arose for this function could also poison neighboring plants upon release (Whittaker, 1970). Phytotoxins might also be useful to their producers as storage compounds, regulatory compounds, or intermediates in the synthesis of more complex molecules.

III. TOXINS

A. Phenolic Compounds

Phenolic compounds have been identified as allelopathic agents more commonly than any other substances. As has been pointed out (Horsley, 1977b; Putnam and Duke, 1978), their importance may be overemphasized because

of the fact that they are often the first or the only secondary compounds
sought by the ecological chemist investigating allelopathy. They are virtually
always present, whether the plant or the soil is analyzed.

Phenolics are ubiquitous in the plant kingdom. Gentisic and p-hydroxybenzoic acids are present in 97% of the large number of species tested, and
p-coumaric, protocatechuic, ferulic, and caffeic acids also occur in most
species tested (Tomaszewski, 1960; Harborne and Simmonds, 1964). Vanillic
and syringic acids are universally present in lignin-containing plants (Harborne and Simmonds, 1964). In plant tissues these phenolics usually occur in
conjugated form, bound with sugars or other substances, probably because
their toxicity is thereby reduced (Towers, 1964; Harborne, 1980).

The first suggestion that phenolic compounds might be involved in plant-
plant interactions came from agricultural studies by Schreiner and co-workers
(Schreiner and Reed, 1908; Schreiner and Skinner, 1912), who felt that some
types of "soil sickness" might be caused by organic toxins. Although this
work was later criticized (see the review by Loehwing, 1937), agronomic
interest in plant-produced phenolic toxins is still strong and has shifted its
focus to crop residues. It appears that buried plant residues can indeed
produce phenolic substances during their decomposition, particularly under
conditions of poor aeration (Welbank, 1960, 1963; Guenzi and McCalla, 1966;
Patrick, 1971; Chou and Patrick, 1976; Shindo and Kuwatsuka, 1977). The
importance of this phenomenon in agronomy depends on just how local and how
temporary is the occurrence of these toxins (Patrick et al., 1964).

Ecological investigations implicating phenolic substances in plant-plant
interactions really began with the studies of Rice and co-workers (Abdul-
Wahab and Rice, 1967; Wilson and Rice, 1968; Parenti and Rice, 1969) and
Muller and co-workers (del Moral and Muller, 1969; Chou and Muller, 1972).
Since those experiments phenolics have been identified in extracts or leach-
ates of many plants suspected of allelopathy, most recently in shrubs of the
Spanish heathland (Ballester et al., 1977; Carballeira, 1980), in two species
that colonize mine spoils (Lodhi, 1979a,b), in a colonizer of abandoned agri-
cultural fields (Rasmussen and Einhellig, 1979), in several seagrasses
(Zapata and McMillan, 1979), and in coffee (Chou and Waller, 1980).

Although the simple phenolics (Fig. 1) are most commonly extracted from
plants and soil, more complex phenols and polyphenols have also been implicated as allelopathic agents. Chlorogenic acid, isochlorogenic acid, and
neochlorogenic acid, phenolic compounds bound with quinic acid or closely
related compounds, are found in or released by a variety of plants (Williams,
1960; Rice, 1965; Abdul-Wahab and Rice, 1967; del Moral and Muller, 1969,
1970; Al-Naib and Rice, 1971). The coumarins esculetin, scopoletin, and
aesculetin are also fairly common (Winter, 1961; Al-Naib and Rice, 1971; Lodhi
and Rice, 1971; McCahon et al., 1973; Rice and Pancholy, 1974). Phlorizin
is a flavonoid that occurs at high concentration in apple root bark and may be
responsible for apple replant failures (Börner, 1959; but see Savory, 1969).
Other flavonoids—quercetin, chalcone, and flavonol—are exuded from fronds
of *Pityrogramma* ferns (Star, 1980). Finally, tannins are polyphenols which
are certainly abundant throughout the plant kingdom, and which might
possibly be phytotoxic (Mitin, 1970, as discussed in Rice, 1974; Corcoran
et al., 1972).

Phenolics appear to have several different functions in plants, in addition
to their possible allelopathic activity. They may play a role in hormone

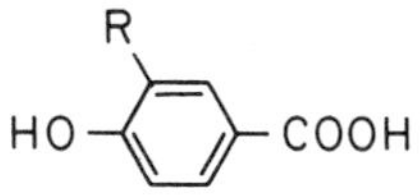

R
HO—⟨benzene⟩—COOH

R = H   *p*-hydroxybenzoic acid
R = OCH₃ vanillic acid

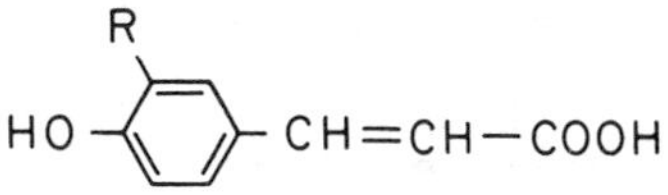

R
HO—⟨benzene⟩—CH=CH—COOH

R = H   *p*-coumaric acid
R = OH   caffeic acid
R = OCH₃ ferulic acid

FIGURE 1 Some simple compounds occurring commonly in plants and soils.

regulation through their influence on indoleacetic acid (IAA) oxidation (Zenk and Muller, 1963; Tomaszewski and Thimann, 1966), although this possibility has never been adequately examined in vivo (Harborne, 1980). Some phenolics serve as precursors in the synthesis of lignins, tannins, and other compounds (Brown, 1969; Steelink, 1972). Finally, phenolics have in many instances been found to deter herbivores (McKey et al., 1978; Swain, 1979), and especially, pathogens (Uritani, 1961; Friend, 1979). Even phenolics released into the soil may play a protective role. Tomato plants exposed to phenolics through soil amendments acquired resistance against the root-pathogenic nematode *Meloidogyne javanica*, probably because the roots themselves contained larger quantities of phenolics (Sitaramaiah and Pathak, 1979; Sitaramaiah and Singh, 1979). Phenolics released from roots as exudates may also play a role in deterring pathogenic bacteria and fungi (Lingappa and Lockwood, 1962; Schroth and Hildebrand, 1964).

The concentration of phenolic compounds in plants is probably influenced by environmental conditions, judging from a variety of experiments performed with sunflower and tobacco. Stress in these species seems generally to enhance the production of chlorogenic acid and scopolin, whether the stress is caused by ultraviolet light, x-rays, low temperatures, or deficiencies of various nutrients (Wender, 1970; Koeppe et al., 1971; del Moral, 1972; Lehman and Rice, 1972). There are exceptions to this rule, however (see Newman, 1978, for a review), and little work has been done with other species. The subject may be of some ecological importance, because stress can cause greater concentrations of these phenolics not only in the plant tissues but also in the leachates of those tissues (Koeppe et al., 1976).

B. Terpenoids

Terpenes and their relatives are not as widespread in the plant kingdom as phenolic compounds, but they have a high potential as allelopathic agents because they volatilize readily from intact leaves and because they can be phytotoxic at concentrations as low as $1-3 \times 10^{-6}$ M (Asplund, 1968). They appear to be especially abundant in plants of arid regions, and have been implicated in allelopathy for *Salvia leucophylla*, *S. apiana*, *S. mellifera*, *Artemisia californica*, *A. tridentata*, *Eucalyptus camaldulensis*, *E. globulus*, and *Sassafras albidum* (W. Muller and Muller, 1964; Baker, 1966; Asplund, 1969; del Moral and Muller, 1970; McCahon et al., 1973; Tyson et al., 1974; Gant and Clebsch, 1975; Halligan, 1975; Weaver and Klarich, 1977).

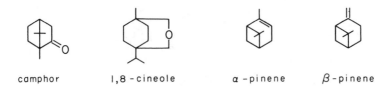

camphor   1,8-cineole   α-pinene   β-pinene

FIGURE 2 Several of the most common terpenes.

Camphor, 1,8-cineole, and α- and β-pinene (Fig. 2) are the most common terpenes identified in plants suspected of allelopathic suppression. Camphor is a particularly phytotoxic compound, perhaps owing to its functional ketone group, and cineole is also highly toxic (C. Muller and Muller, 1964; Asplund, 1968; Halligan, 1975).

Like phenolic compounds, terpenes may have various functions in plants, in addition to their hypothesized allelopathic role. Some are known to turn over rapidly (Seigler and Price, 1976), and they may in fact act as substrates for metabolism when other sources are depleted (Charlwood and Banthorpe, 1978). Terpenes also appear to be toxic to herbivores. Many, including 1,8-cineole, α-pinene, and thujone, have been shown to be effective feeding deterrents against both insects and mammals, although monophagous herbivores may alternatively be attracted by the terpenes of their host (Charlwood and Banthorpe, 1978). Terpenes can also serve as digestibility-reducing substances by interfering with bacterial metabolism in the digestive tracts of ruminant mammals (Oh et al., 1967; Schwartz et al., 1980).

C. Juglone

The characteristic wilting of plants grown near, or with root bark of, walnut trees (Cook, 1921; Massey, 1925) may be due to juglone (5-hydroxy-naphthaquinone; Fig. 3). This quinone is present in the bark of *Juglans nigra, J. cinerea, J. regia,* and some species of *Carya.* Juglone can inhibit seedling growth at concentrations as low as 10^{-7} M, and can be lethal at 10^{-4} M (Case and Funk, 1978). It is also highly toxic to animals. It is a strong deterrent to feeding of the bark beetle, *Scolytus multistriatus,* which feeds only on those members of the Juglandaceae which do not possess juglone (Gilbert et al., 1967). It also has been observed to poison earthworms when applied to soil (MacDaniels and Pinnow, 1976).

Plants seem to vary in their response to exposure to black walnut roots. While some species gorwing near walnut trees show wilting and browning of the vascular tissue for which no pathogen can be found responsible (Massey, 1925; Gries, 1943), other species growing near it are unharmed (MacDaniels and Pinnow, 1976). Even those species which are susceptible, like tomato, sometimes suffer no deleterious effects from walnut. MacDaniels and Muenscher (1941) obtained normal growth from tomatoes grown with walnut roots or watered with the brownish leachate of those roots. A U.S. Department of Agriculture experiment also gave a negative result (USDA, 1948). It is possible that the toxicity of walnut root leachates depends on soil conditions. Fisher (1978) found that young *Pinus strobus* and *P. resinosa* trees growing near walnut trees were inhibited or killed in two poorly drained sites,

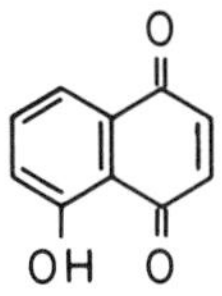

juglone

FIGURE 3   The structure of juglone.

but not in a well-drained site. Correspondingly the inhibition due to juglone, as assayed with *P. resinosa* seedlings, disappeared fairly rapidly in dry soil in an open container, but in a closed container of wet soil, persisted with little diminution throughout 90 days.

D. Other Phytotoxic Compounds

Plants contain and release many other phytotoxic compounds which may act as allelopathic agents. Cyanogenic compounds have been found in several species: amygdalin, in the root bark of peach trees, may be responsible for replant problems (Patrick et al., 1964); and dhurrin, a constituent of Johnson grass (*Sorghum halepense*), is thought to impair the growth of other early successional species in Oklahoma old fields (Abdul-Wahab and Rice, 1967). These compounds hydrolyze upon tissue disruption to form toxic HCN and toxic benzoic acids (Patrick, 1955). Gant and Clebsch (1975) identified a phenylpropene, eugenol, in the throughfall of *Sassafras albidum*, and it proved to be phytotoxic. Patrick (1971) attributed the deleterious effects of decomposing rye residues partially to simple organic acids—butyric and acetic. Amino acids are thought to explain the depression of crop seed germination by seeds of *Abutilon theophrasti* (Gressel and Holm, 1964; Elmore, 1980). An indole compound may be responsible for the inhibition of gametophytes by *Thelypteris normalis* (Davidonis and Ruddat, 1973).

C_{10}-polyacetylenes have recently been shown to have a high potential for allelopathic activity (Kobayashi et al., 1980). These substances are released from *Solidago altissima* and from several *Erigeron* species, and are deleterious at very low concentrations; *cis*-dehydromatricaria ester and *cis*-lacnophyllum ester, for example, inhibited germination at only 1 ppm. *trans*-Dehydromatricaria ester, a somewhat less toxic polyacetylene, was present in soil at a concentration of approximately 6 ppm. These compounds have a limited distribution in the plant kingdom, occurring primarily in the Compositae (Bohlmann et al., 1973). They may conceivably explain other apparent cases of allelopathy by plants of the genera *Erigeron* and *Solidago* (Keever, 1950; Horsley, 1977a; Fisher et al., 1978; but see Raynal and Bazazz, 1975).

Even sodium chloride has potential as an agent of "allelopathy." *Mesenbryanthemum crystallinum*, a coastal annual that accumulates salt throughout its lifetime, releases so much salt upon its death that it can triple the osmolality of the soil beneath it. This change in salt concentration has the effect of inhibiting several grassland species and thereby allowing *Mesenbryanthemum* itself to invade (Vivrette and Muller, 1977).

IV. PHYSIOLOGICAL MECHANISMS OF PHYTOTOXICITY

A. Effects of Phenolic Compounds

Phenolic compounds appear to influence a number of plant processes, and it is difficult to distinguish between their primary effects and their secondary effects. One effect that appears to be primary and ecologically meaningful is the impairment of nutrient uptake by phenolics. Schreiner and Skinner (1912) observed that in a water culture experiment the inhibition of wheat plants by coumarin was the most drastic at low phosphorus levels. Glass (1973), working with excised barley roots, found that phosphorus uptake was drastically curtailed in the presence of any of 10 phenolics. Further experimentation showed that potassium uptake was likewise reduced by phenolic compounds, and that the degree of inhibition caused by each phenolic was correlated with its lipid solubility (Glass, 1973, 1974). Phenolics such as 2,4-dinitrophenol (DNP) apparently depolarize the membrane potentials of root cells and thereby interfere with the active transport of nutrients. The effectiveness of different phenolics in depolarizing the root cell membranes was again correlated with their lipid solubility and therefore with inhibition of nutrient uptake (Glass and Dunlop, 1974), indicating that the depolarization is probably a primary, not secondary, effect of phenolic compounds. Membrane changes in response to salicylic acid have since been observed in yeast as well as root cells (Scharff and Perry, 1976). Soybean roots, whether excised or intact, also show reduced phosphate uptake in the presence of ferulic acid (McClure et al., 1978).

Taking a more ecological viewpoint, Glass (1976) and Stowe and Osborn (1980) have shown that the growth reduction caused by phenolics is much more drastic under low-nutrient stress than in the presence of ample nutrients. Thus in those cases in which plants appear to inhibit their neighbors' growth and nutrient uptake, despite ample nutrients in the soil (Chambers and Holm, 1965; Moore and Keraitis, 1966; Bevege, 1968; Elmstrom and Howard, 1970; Buchholtz, 1971; Miner, 1974; Newman and Miller, 1977), one might suspect phenolic compounds as possible allelopathic agents.

The depolarization of root cell membranes might be related to the observable effects of phenolics on roots. Coumarin and its derivatives, and scopoletin, inhibit root growth in cereals partly by decreasing the elasticity of cell walls (Goodwin and Taves, 1950; Avers and Goodwin, 1956; Thimann, 1972). It may be incorrect to try to generalize about these effects, however, because coumarin and scopoletin appear to affect different root tip cells in different ways (Pollock et al., 1954; Avers and Goodwin, 1956). Phenolics may also be responsible (Chou and Patrick, 1976) for root lesions and browning of the root-apical meristem in lettuce and spinach plants grown in contact with decomposing plant debris (Patrick et al., 1963).

In addition to impairing root growth and nutrient uptake, phenolics influence many plant processes, and it is unclear which effects are direct and which are indirect. Net photosynthesis in several species is reduced by scopoletin (Einhellig et al., 1970). Respiration of coleoptiles or seeds can be stimulated or impaired by phenolics (Morinos and Heinberg, 1960; Van Sumere et al., 1972); yet Demos et al. (1975) showed that several phenolics were able to inhibit hypocotyl growth without influencing the rate of respiration or the rate of Ca^{2+} transport by mitochondria. It has also been suggested that phenolics, like 2,4-dinitrophenol, may block adenosine triphosphate (ATP) synthesis by uncoupling oxidation from phosphorylation in mitochondria (Kefeli and Kadyrov, 1971; Mayer and Poljakoff-Mayber, 1975). However, ferulic acid and coumarin influence the respiration rates and germination probability in seeds of lettuce and barley without any apparent uncoupling ability (Van Sumere et al., 1972). Pilet (1966) observed that the auxin content of lentil stem sections was inversely proportional to the concentration of p-hydroxybenzoic acid in the medium, and seemed to be related to the rate of elongation. He thus argued that phenolics may influence growth through their inhibition or stimulation of IAA oxidation. However, Machackova and Zmrhal (1976) have been unable to find any correlation between the effects of various phenolics on IAA-oxidase activity and their effects on coleoptile section growth. Protein synthesis was reduced by various phenolics in plant cell suspensions (Danks et al., 1975), and in lettuce seedlings (Cameron and Julian, 1980). Einhellig and Rasmussen (1979) investigated the chlorophyll content of soybean and sorghum treated with phenolic acids. Growth in dry weight was inhibited in both species, but only soybean showed a corresponding reduction in leaf chlorophyll. Finally, phenolics seem to cause low water potentials and stomatal closure, but these conditions may be only temporary (Einhellig and Kuan, 1971; Colton and Einhellig, 1980).

B. Effects of Terpenes

The anatomical and physiological effects of terpenes on plants are known primarily through several studies by W. Muller and co-workers. The volatiles from *Salvia leucophylla*, primarily terpenes, were first noticed to cause cucumber radicles and hypocotyls to be short and thick, due to reduced cell elongation and cell division. Also the cells of inhibited roots frequently contain abnormal nuclei and globules of fats or oils which presumably accumulated because new cells were not being produced (Muller, 1965). Excessive amounts of cutinlike matter also forms on the outer surface of root-epidermal cells exposed to these terpenes, and the inhibited roots tend to produce few lateral roots (Muller and Hauge, 1967). The number of intact organelles, including mitochondira, is reduced by exposure to terpenes, and the membranes surrounding nuclei, mitochondria, and dictyosomes are frequently disrupted (Lorber and Muller, 1976).

As with phenolic compounds, there is some evidence that terpenes may inhibit plant processes by blocking respiration. Cineole at 1.2×10^{-3} M and dipentene at comparable concentrations cause a decrease in oxygen uptake by mitochondrial suspensions, although other terpenes (α-pinene, β-pinene, camphene) do not appear to possess this inhibitory capacity. The inhibition appears to be localized in the conversion of succinate to fumarate or fumarate to malate, since no inhibition is observed when malate is used as a substrate (Muller et al., 1969).

C. Effects of Juglone

There are several suggestions regarding the physiological effects upon plants of juglone, an important secondary compound in the bark of walnut trees (*Juglans nigra*, *J. cinerea*, *J. regia*). Jensen and Welbourne (1962) found that mitosis in excised roots of *Vicia faba* and *Pisum sativum* was inhibited by juglone. Perry (1967) sought the cause of this inhibition in respiration rates, using leaf disks of tomato and bean. Both species indeed showed reduced oxygen uptake with juglone concentrations as low as 7.3×10^{-5} M. Since walnut trees in the field inhibit tomato, and not bean, Perry speculated that bean plants may have a mechanism that protects its respiratory function. Koeppe (1972) found that excised corn roots also showed a dramatic reduction in oxygen consumption upon treatment with juglone. Perhaps by acting as an electron acceptor, juglone inhibits the state 3 oxidation rates of succinate and of malate and pyruvate (Koeppe, 1972).

D. Effects of Unidentified Toxins

In some cases plants have shown specific responses to inhibitors which have not yet been characterized. Ballantyne (1962) observed that decomposing chrysanthemum roots seemed to inhibit chrysanthemum growth by reducing the number of internodes and the internode length. He proposed the existence of a compound released by the decomposing roots which was antagonistic to the action of giberellic acid, and showed that the addition of exogenous giberellic acid could indeed offset these growth reductions, in some respects. Lodhi and Nickell (1973) found that the leaf extract of *Celtis laevigata* not only impaired shoot and root growth in several grasses, but also reduced their photosynthetic CO_2 uptake. Effects on dark respiration and shoot water content were temporary and not consistently correlated with inhibition.

Retig et al. (1972) have observed some striking changes in the root anatomy of crop plants grown in intimate contact with weeds. They germinated crop seeds and weed seeds together in test tubes of agar, and found that the roots of the crop seedlings showed such symptoms as inhibited cell elongation, disrupted epidermal tissue, and enlarged parenchyma cells. Extracts of *Kalmia angustifolia* cause similar root-epidermal necrosis in black spruce seedlings (Peterson, 1965).

V. FACTORS MITIGATING TOXICITY IN NATURE

A. Dilution

Both rainwater and residual soil water dilute any water-soluble toxins released by plants, and this must decrease their effectiveness in the field. McPherson and Muller (1969), Halligan (1975), and Gliessman and Muller (1978) have noted the rapid loss of toxicity from aboveground plant parts on the first few rains of the wet season in southern California. Phenolics appear to leach readily from the soil surface into the soil, where they are generally adsorbed by soil particles and/or degraded by bacteria and fungi (Shindo and Kuwatsuka, 1975). The minimal leaching and dilution of toxins in arid regions might favor the operation of allelopathy (Muller, 1966). Few field studies have examined this aspect of the chemical interactions among plants.

B. Degradation by Soil Microorganisms

Soil microorganisms can metabolize a great variety of organic compounds, and probably the most effective allelopathic agents will be compounds that cannot be rapidly degraded. Such may be the case for juglone, which, in very wet soil, seems to retain its toxicity for over 90 days (Fisher, 1978).

Phenolics, on the other hand, seem to disappear fairly rapidly from un-sterilized soils, as first discovered by Bonner and co-workers. Bonner and Galston (1944) investigated the apparent autotoxicity of guayule (*Parthenium argentatum*) and found that these plants produced a highly toxic leachate which contained a large quantity of cinnamic acid. However, they deferred any judgments about whether this phytotoxin was active under field condi-tions until a study of cinnamic acid in soils could be undertaken. Bonner (1946) later found that the soil's toxicity and its cinnamic acid largely dis-appeared from incubated soil within 2 weeks if concentrated, and within 2 days if dilute. Sterilized soil, on the other hand, retained its toxicity and its cinnamic acid, implying that soil microorganisms were responsible for its disappearance. Even soils that had supported guayule plants for as long as 8 years were, on the average, no more toxic than other soils. Bonner con-cluded that in soils with a suitable microflora, any accumulation of cinnamic acid would be slight.

Turner and Rice (1975) and Shindo and Kuwatsuka (1975) also noted the rapid loss of phenolic compounds to soil bacteria and fungi. Turner and Rice (1975) allowed leaves of *Celtis laevigata*, which contain large quantities of ferulic acid (Lodhi and Rice, 1971), to decompose in nylon mesh bags on the soil surface. The leaves lost 65% of their extractable ferulic acid in the first 100 days of decomposition, yet the soil beneath them showed little variability in ferulic acid concentration over the course of a year. There was a signifi-cant increase in extractable soil ferulic acid in March, but Turner and Rice attributed this increase to living or decomposing roots rather than to leaf leachates. Addition of synthetic ferulic and vanillic acids to soil resulted in rapid CO_2 evolution, reaching a peak approximately 45-60 hr after addition. The concentration of microorganisms in the soil increased dramatically during this period. Bacteria from the genera *Cephalosporium* and *Rhodotorula* showed a particularly high ability to utilize ferulic acid as their sole carbon source.

An isotopic labeling experiment carried out by Haider and Martin (1975) provided more detailed information concerning the rate of degradation of various phenolic compounds, and of various portions of the phenolic mole-cules. They synthesized a number of simple phenolics in such a way that specific carbon atoms were labeled with ^{14}C. These compounds were then incubated at 100 and 1000 ppm with soil, and the $^{14}CO_2$ evolution was followed. $^{14}COOH$ carbons of the phenolic molecules tended to be metabolized quickly—within a week, over 50% had been lost from caffeic acid, over 81% from vanillic acid and over 94% from *p*-hydroxybenzoic acid. Ring carbons were slower to be metabolized—30-80% were lost within a week, and 65-85% within 12 weeks. The ^{14}C which was not released by microbial degradation of these phenolics was largely bound in humic and fluvic acids. These phenolic polymers are metabolized only gradually, as evidenced by the slow decomposition of ^{14}C-labeled, model polymers added by Haider and Martin.

Both bacteria and fungi probably participate in the degradation of free phenolic compounds. Many taxonomically diverse soil bacteria are capable of growing in media in which phenolic compounds are the sole carbon source (Evans, 1947; Rogoff, 1961; Sundman, 1964). Fungi also appear very capable of degrading phenolics (Henderson and Farmer, 1955; Henderson, 1956; Black and Dix, 1976), as well as their polymers, lignin and tannin (Lewis and Starkey, 1969; Trigiano and Fergus, 1979).

C. Adsorption and Polymerization in Soil

Very little is known about the physical and chemical changes that befall potential allelopathic substances in the soil, with the exception of the phenolic compounds. Dry clay soils do have an ability to adsorb terpenes from the air,
and they retain their toxicity, if kept dry, for months (Muller and del Moral,
1966). Terpenes could thus remain in the soil throughout a dry period, and
inhibit germinating seeds or seedlings at the beginning of a subsequent wet
season; this hypothesis could be tested by examining the changes in soil
toxicity throughout a year.

Phenolic compounds, when added to soils, show some changes which at
first seem very puzzling. Wang et al. (1971) performed an experiment in
which they added simple phenolics to two tropical soils, waited for 3 hr, and
then extracted the phenolics with alcoholic sodium hydroxide. They found
that ferulic and syringic acids could not be extracted unless they were added
at rates of over 8 μmol per 10 g soil. p-Hydroxybenzoic, p-coumaric, and
vanillic acids added to these soils were also fixed, but other phenolics were
released, such that the total quantity of phenolics recovered was about the
same as the amount added. These changes seemed to be fairly instantaneous,
and did not require microorganisms, since they occurred even when the
experiment was conducted in alcohol. Isotopic labeling confirmed the hypothesis that the phenolics that became bound by the soil were fixed by humic
complexes (Wang et al., 1971).

Organic molecules in soil tend to polymerize, particularly in the presence
of clay, to form large complexes, generally called "humic acids" (Hurst and
Burges, 1967). Although their structure and composition vary widely, humic
acids are principally composed of aromatic monomers, particularly phenolic
compounds (Haider et al., 1975). In fact, the types and quantities of phenolics contained in humic acid reflect the vegetation (particularly the type of
lignin decomposition) which the soil has supported (Burges et al., 1964; Tate
and Anderson, 1978).

Although they appear to exchange peripheral monomers readily with the
soil (Wang et al., 1971), humic acids persist for remarkably long periods of
time in natural soils. Carbon-14 dating techniques have shown that the mean
residence time for humic acid molecules can vary from several hundred to
over 2000 years (Hurst and Burges, 1967). Clearly, polymerization affords
the phenolic compounds a remarkable degree of protection from microbial
degradation, since they would last only a few weeks as monomers.

The role of clay in the polymerization of phenolics has recently been clarified. It had previously been hypothesized that, by adsorbing various organic
and nonorganic substances, clay particles concentrated the resources necessary for fungal activity, and that fungi were primarily responsible for humic
acid formation (Sarkanen and Ludwig, 1971). However, Wang et al. (1978)
recently tried adding a solution of phenolic compounds to pure, sterilized
montmorillonite, illite, or kaolinite; they observed the formation of humic
acids closely resembling those of natural soils. Apparently, the synthesis of
humic acids from phenolic compounds is purely a chemical reaction catalyzed
by clay particles and does not require microorganisms. The reaction proceeds
more quickly with montmorillonite or illite than with kaolinite or quartz sand,
and more quickly at alkaline or neutral than at acid pH (Wang et al., 1980).
Using montmorillonite or illite, most phenolics were quickly withdrawn from

solution by polymerization, and could barely be detected after 1-6 days (Wang et al., 1978).

It has been speculated that phenolics that become bound in organic complexes might nonetheless be active in allelopathy, and might thus be concentrated by soil particles (Muller and Chou, 1972; Turner and Rice, 1975). However, humic acids do not seem to have deleterious effects on plants, and in fact are generally beneficial (Whitehead, 1963; Kononova, 1966). They may stimulate plant growth by enhancing nutrient uptake or by maintaining soluble iron in the soil (de Kock, 1955; Whitehead, 1963; Kononova, 1966). Alternatively, they may release phenolic compounds into the soil in quantities sufficient to inhibit fungal plant pathogens but insufficient to affect plant growth (Lingappa and Lockwood, 1962).

D. Resultant Concentrations in Soil

There are clearly various soil processes which remove or inactivate phytotoxins, yet if these substances are continually added to soils, they might temporarily accumulate or reach an equilibrium at a concentration that would be inhibitory to plants (Patrick et al., 1964; Turner and Rice, 1975). The straightforward approach to testing this idea is simply to analyze soil beneath plants suspected of allelopathy for its content of toxins, and compare the amount found in the soil to the amount necessary to cause inhibition.

This approach has several problems, however. One is that plants grown hydroponically, with known concentrations of toxins, may respond differently to those toxins than plants grown in soil. For example, phytotoxins in nature may have a very patchy distribution (Patrick et al., 1964; McCalla and Norstadt, 1974), whereas those in water culture or sand culture would be uniform. A second and more critical problem is the choice of soil extraction procedures, as pointed out by Kaminsky and Muller (1977, 1978). The techniques used to extract soil phenolics, for example (Morrison, 1963; Whitehead, 1964; Morita, 1965; Guenzi and McCalla, 1966; Wang et al., 1967; Al-Naib and Rice, 1971; Patrick, 1971; Lodhi, 1975, 1978; Shindo et al., 1979), vary tremendously, and probably have a substantial effect on the results. The large quantities and unusual composition of the phenolics extracted by Lodhi (1975, 1978), in comparison with those extracted by others, may be related to his technique of autoclaving the soil during extraction.

Phenolic compounds are generally extracted from soil using alkaline solutions, yet it is known that alkaline conditions result in partial degradation of humic complexes in the soil (Tinsley and Salam, 1961; DeSerra and Schnitzer, 1972). Kaminsky and Muller (1978) showed that alkaline conditions can alter phenolic compounds even in the absence of soil, and that mild extraction procedures release fewer organic compounds from soil than does alkaline extraction.

Clearly, we must examine the procedures for extracting phenolics, and other phytotoxins as well, to ensure that they yield all of, but no more than, the substances available to roots or seeds. Kaminsky (1980) has undertaken this task for the phenolic compounds, in relation to seed germination. He first ascertained the dose-response curve for lettuce seeds in the presence of phenolics, both in soil and in the absence of soil. The inhibitory effect of the phenolics was moderated in soil, presumably because of adsorption by soil particles and incorporation into humic acids. Microbial degradation was not considered to be important, since the seeds germinated within 48 hr; however, in light of the rapid stimulation of microbial activity observed by Turner and Rice (1975), it would be comforting to verify this assumption by using sterilized soil. Kaminsky found that soil extraction with (ethylenedinitrilo)tetraacetic acid (EDTA) yielded a quantity of phenolics which corresponded well with the measured inhibition of the lettuce seeds; alkali extraction, on the other hand, apparently released compounds from the soil which were not available to plant roots, as it yielded a quantity of phenolics which would have caused more severe inhibition of germination than that observed. EDTA extraction thus appeared to be more realistic than alkali extraction, for two different soils and for both pure p-hydroxybenzoic acid and a mixture of three phenolic acids. This promising approach may eventually give us a better knowledge of the phytotoxin concentrations preceived by plants in various natural soils.

VI. EXPERIMENTAL METHODS USED TO DOCUMENT ALLELOPATHY

There is considerable debate over what constitutes sufficient evidence to claim allelopathy (Welbank, 1961, 1963; Bartholomew, 1970; Harper, 1977; Newman, 1978; Stowe, 1979). Unfortunately, the most convincing data are also the most difficult to obtain. In this section various types of observations and experiments are examined and evaluated with respect to their potential for demonstrating allelopathic interactions between higher plants.

A. Elimination of Alternatives

Often the first step in an investigation of the role of allelopathy in the apparent suppression of some plants by others is an attempt to rule out the other possibilities. Usually examined are the possibility that the suspected plant is competing intensely with its neighbors for light, water, or nutrients, and the possibility that soil properties near the suspected plant are intrinsically different from those in surrounding areas and are inhibitory to many species (Curtis and Cottam, 1950; Wilson and Rice, 1968; del Moral and Muller, 1970; Al-Naib and Rice, 1971; Neill and Rice, 1971; Chou and Muller, 1972; Davidonis and Ruddat, 1973; Lodhi, 1976; Hull and Muller, 1977). Although it is comforting to find no obvious differences in light and soil characteristics when one is entertaining the idea that allelopathy may explain certain distribution patterns, this approach can never convincingly document allelopathy because of the innumerable ways in which plants may interact (Harper, 1977, p. 380).

It is impossible to examine all the environmental factors which might result in the observed inhibition of one plant by another. Wilson and Rice (1968), for example, examined soils near *Helianthus annuus* and found no significant differences in nutrient levels to explain the observed suppression of neighbors; but they were later criticized (Harper, 1977) for choosing to analyze *total* nitrogen and phosphorus rather than *available* nitrate, ammonium, phosphorus, and potassium. Webb et al. (1967) found dramatic self-inhibition in the tropical tree *Grevillea robusta*, which could not be overcome even when ample light, water, and nutrients were supplied; but the possibility of a root pathogen was not adequately tested and could explain the observations (Harper, 1977).

A more complex example of the difficulty of demonstrating allelopathy by process of elimination is to be found in the investigations of certain shrub species (*Artemisia californica, Salvia leucophylla, S. mellifera*) which invade annual grassland in southern California. Bare areas surrounding the enlarging stands of these shrubs were first attributed to volatile terpene inhibitors released by the shrubs (Muller et al., 1964). Wells (1964) argued that the patterns of these bare areas instead suggested the trampling effects of cattle which were avoiding the dense shrubs, so C. Muller and Muller (1964) were obliged to rule out this alternative by a study of the distribution of cow feces. Muller (1966) also did a number of soil tests in an attempt to rule out competition for soil factors. Next, Bartholomew (1970) tested a different hypothesis —that the bare areas result from predation by birds and small mammals, which are especially numerous in and near the shrub stand because of the protection it affords. By constructing wire mesh exclosures, Bartholomew was able to eliminate this predation in certain portions of the bare zone; the resulting herb growth nearly equaled that of the uninhibited grassland. Muller and del Moral (1971) replied that, although animal activities do exert some influence on herbs in the bare zone, allelopathy is still necessary to explain the pattern of inhibition. Finally, Halligan (1973) has performed further experiments which implicate not only predation and allelopathy but also competition for moisture. Although Muller (1966) found no substantial differences in soil moisture beneath shrub stand and grassland, he may not have analyzed the soils during the beginning of the growing season, at which time the shurb zone soil is drier than the grassland soil (Halligan, 1973). This long series of experiments, beginning in 1964, illustrates nothing as clearly as it illustrates the difficulty of eliminating all the alternatives to allelopathy. This is not to say that observations and experiments designed to rule out other factors are worthless; but the primary effort should probably be directed toward gathering solid evidence that allelopathy is occurring, rather than evidence that other factors are unimportant.

B. Laboratory and Greenhouse Bioassays

Biological assays for phytotoxicity have long been the principal tool for demonstrating allelopathic potential. In these bioassays plant materials (e.g., leaves, roots, stems) or the leachings from those materials are in some way tested on seed germination or plant growth of one or more test species. A great many variants of this general procedure have been employed, and some are more likely than others to give positive results (Stowe, 1979). In Table 1 are presented the results of a survey of 84 papers in which 96 qualitatively different bioassays were examined for positive and negative results. The bioassays were classified according to the plant material tested for toxicity, the method used for preparing it, and the situation in which test species were exposed to it. The table reveals that some plant materials tend to be more toxic than others and that the method of assaying toxicity greatly influences the result. Extraction and testing procedures that mimic natural processes— such as the leaching of intact leaves by rain or the leaching of decomposing roots by percolating water—are clearly desirable if the results are to have

TABLE 1 Relationships Between Methods and Results of Bioassays Reported in the Literature

	Number of investi-gations	Number of positive results	Number of negative results	Percent of positive results
Material assayed				
Shoots	108	709	414	63
Roots	72	617	400	61
Litter	50	265	98	73
Soil	18	77	76	50
Treatment of plant material				
Left intact	71	394	329	54
Cut up	16	54	24	69
Dried and ground	88	930	440	68
Ground fresh in blender	25	113	76	60
Test procedure				
Seed germination in petri dishes	39	459	353	56
Seed germination in soil or sand	25	101	61	62
Radicle elongation in petri dishes	53	619	226	73
Growth in soil or sand	59	227	216	51
Growth in water culture	18	85	13	87

ecological meaning. Yet to duplicate the timing and amounts of rainfall, the soil properties, the distribution of plants, and the rate of release of phyto-toxins which occur in natural ecosystems is clearly impossible, and most bio-assays are crude simulations of the processes that take place in the field.

We must therefore address the question of whether the results of these simulations have any relationship to the presence or absence of allelopathy in nature. Some undoubtedly do have such a relationship, but a great many do not, as can be deduced from several observations:

1. In those cases in which bioassays have been repeated under more natural conditions, or with more appropriate controls, they have often been found to be invalid for some reason (Howard, 1925; Bonner, 1946; Keever, 1950; Welbank, 1960, 1963; Cannon et al., 1962; Ohman and Kommedahl, 1964; Brown, 1967; Chou and Muller, 1972; Gregg, 1972; June, 1976; Buchanan et al., 1978).

2. A large proportion of species tested for phytotoxicity show positive bioassay results, even when there is no ecological evidence for sus-pecting allelopathic interactions (Pickering, 1917; Jameson, 1961; Lawrence and Kilcher, 1962; Welbank, 1963; Grant and Sallans, 1964; Quarterman, 1973; Stowe, 1979). Apparently, as Harper (1961) stated, "it is relatively easy, perhaps too easy, to extract from a plant a substance which may prove toxic to its own seedlings or to the seedlings of other species."

3. Autotoxicity is commonly noted in the results of bioassays (Benedict, 1941; Curtis and Cottam, 1950; Keever, 1950; Guyot, 1957; Muller, 1966; Webb et al., 1967; Abdul-Wahab and Rice, 1967; Wilson and Rice, 1968; del Moral and Muller, 1969; Quinn, 1974; Bendall, 1975; Newman and Rovira, 1975) and is, on the average, just as strong as "allotoxicity" (Newman, 1978; Stowe, 1979). Since a plant that releases phytotoxins will generally be closer to the source than its neighbors, this observation suggests that plants would be inhibiting themselves more than any other plants. Selection would presumably favor either the cessation of toxin release or the development of resistance if autotoxicity were operating in the field.

4. The distribution and succession of plants generally does not seem to correspond with the results of bioassays, as discussed in the next section.

These points suggest that bioassays are very unreliable as indicators of allelopathy, and should probably be used only for comparisons of phytotoxicity or susceptibility.

C. Bioassays Related to Succession or Distribution

In some cases it might be found that the results of a biological assay show a statistically significant correlation with the sequence of succession or the patterns of distribution. The early successional species, for example, might be susceptible to their own and each other's phytotoxins, while later species were more resistant. Or species occurring in the vicinity of a certain suspected species might be unharmed by its extracts, while those occurring farther away were inhibited by them. Such correlations, although not really proving any causal relationship, would strongly support the claims that the bioassay was indeed an adequate simulation of natural processes and that allelopathy at some point influenced the vegetation dynamics of the community. Unfortunately, most studies in which bioassay results have been compared with successional sequence or spatial distribution pattern show little evidence of any correlation [Keever, 1950; W. Muller and Muller, 1956; Jameson, 1970; del Moral and Cates, 1971; Neill and Rice, 1971; Hoffman and Hazlett, 1977; Weaver and Klarich, 1977; Stowe, 1979; see Stowe (1979) for a more detailed review].

The sequence of succession has sometimes shown a tantalizing relationship to bioassay results, but the number of species examined has been insufficient for any definite conclusions. Rice and co-workers (Abdul-Wahab and Rice, 1967; Wilson and Rice, 1968; Parenti and Rice, 1969; Neill and Rice, 1971) found that a number of early successional species of abandoned fields in Oklahoma were susceptible to their own and each other's extracts, whereas *Aristida oligantha*, a later successional species, was generally resistant. Phenolic compounds were suspected as toxins, and Olmsted and Rice (1970) confirmed that *A. oligantha* was markedly more resistant to phenolics than two early successional species. Jackson and Willemsen (1976) also discovered that two early successional species were more sensitive to their own and each other's extracts than two later successional species. However, without testing a greater number of species one cannot with statistical justification infer a correlation, and certainly not causation, between inhibition by extracts and successional sequence.

Spatial distribution has also shown some tantalizing relationships with bioassay results. Halligan (1973) found that two of the herbaceous species able
to grow in the vicinity of *Artemisia californica* were resistant to the volatiles,
rain drip, and soils of that shrub, whereas two species unable to grow near
A. californica were significantly inhibited by all three. Friedman et al. (1977)
similarly found that the germination of three species which were infrequently
found near *Artemisia herba-alba* was suppressed by its extracts, whereas
that of another species, including both a variety which did and a variety
which did not grow near *A. herba-alba*, was unaffected. In these studies the
results are suggestive enough to warrant more investigation, but the sample
size of species or varieties is too small for firm conclusions.

Agronomists interested in the practical importance of allelopathy have also
tried to relate bioassay results with ability to suppress weed growth. Cucumber varieties exhibit a striking range of abilities to inhibit the growth of
Panicum miliaceum in sand culture, and for three varieties whose leachates
were bioassayed, leachate toxicity was related to *Panicum* suppression (Putnam
and Duke, 1974). Field evaluation of this preliminary finding gave inconsist-
ent results, however; of two varieties expected to inhibit weed growth, only
one variety did so, and only on the drier of two years (Lockerman and Putnam,
1979). In similar experiments with different oat cultivars, toxicity in sand
culture was possibly related to scopoletin production, but did not seem to be
effective in soil (Fay and Duke, 1977). Small sample sizes are again a problem
with these interesting studies.

In another experiment the germination of jack pine (*Pinus banksiana*)
seeds in the field was strongly correlated with the results of a laboratory biological assay. Brown (1967), wishing to understand the patchy regeneration
of this species, decided to investigate allelopathy by first bioassaying 56
species for water-soluble phytotoxins. He macerated the fresh leaves or other
organs of these plants in a blender and then tested these extracts on jack
pine seeds in petri plates. Although most species had no significant effect,
some did reduce germination and a few stimulated it. For a field comparison,
he cleared litter from small plots beneath 11 species and planted jack pine
seeds in them. Despite the fact that his method for preparing extracts with
a blender was a poor imitation of natural leaching, Brown obtained a remarkably good correlation between germination in the laboratory bioassay and
emergence in the field. (If the numbers 1, 2, and 3 are assigned to Brown's
results of inhibitory, neutral, and stimulatory, the correlation coefficient obtained is $r = .892$; $p < 0.01$). This correlation, although not absolute proof
that allelopathy was occurring, is far more convincing than the bioassays
alone.

Christensen and Muller (1975) also found an excellent relationship between
bioassay results and field distribution in their studies of *Adenostoma fasciculatum*, a chaparral shrub. McPherson and Muller (1969) had verified the
allelopathic potential of the leaves of this species, but Christensen and Muller
tested leaf leachate on 10 chaparral species, two of which grew commonly near
A. fasciculatum and eight of which grew only in artificial clearings or burned-
over areas of the chaparral. The two tolerant species had a similar resistance
to the phytotoxins in the leaf leachate, while all eight of the other species
were inhibited in germination and radicle elongation; the difference between
the two groups was significant ($p < 0.05$), as determined by t-tests.

The experiments of Brown (1967) and Christensen and Muller (1975) illustrate a powerful tool in the study of allelopathy. Of course, a correlation between bioassay results and distribution does not itself prove causation. One alternative explanation, for example, might be that those species growing near the suspected species have certain differences from those growing more distant from it, and one of these differences happens to be related to susceptibility to phytotoxins. Another might be that resistance to phytotoxins is not the determinant of distribution, but only a minor adaptation, and that the agent presently enforcing the distribution pattern is some soil property or competitive interaction. The latter explanation would at least imply that allelopathy had operated in the past to alter fitness and encourage adaptation. Overall, the method can provide good evidence for allelopathy, although it requires diligence, in the testing of many species, and honesty, in reporting results for each species examined.

D. Field Bioassays

In some cases, experiments have been designed in which naturally produced extracts are tested in the field on one or more test species. Brown's (1967) study of *Pinus banksiana*, described above, is essentially such a field bioassay, strengthened by the comparison of field results with laboratory results. Field bioassays can produce striking evidence for allelopathy if they correlate with laboratory studies or if they are designed in such a way that the only difference between control and treatment is the extract of a suspected toxin producer.

Turner and Quarterman (1975) performed a field bioassay in their investigations of the effects of *Petalostemon gattingeri*, a legume of cedar glades, on the winter annual *Arenaria patula*. They assayed the toxicity of *Petalostemon* leaf leachates by enclosing *Arenaria* seeds in cloth bags and placing these bags either under *Petalostemon* plants or on bare rock. Unfortunately, conditions during the 2 weeks of deployment were not appropriate for germination, but when the seeds were subsequently incubated in the laboratory, those which had lain on the bare rock germinated significantly better than those beneath *Petalostemon*. This test is weakened not only by the necessity of germinating the seeds under laboratory conditions but also by the possible effects of cloth bags on the seed environment and the possible differences in environmental conditions between bare rock and herb understory. The results are suggestive, nonetheless.

Volatile phytotoxins can also be assayed using this field approach, either by placing potted plants of the producer species in the field near the test plants, as Harper (1977, p. 377) has suggested, or by placing potted plants of the test species among plants of the producer species, as Weaver and Klarich (1977) have done. *Artemisia tridentata*, big sagebrush, releases potentially phytotoxic terpenes as volatiles, and these terpenes had the effect of accelerating leaf respiration in laboratory experiments. Weaver and Klarich (1977) deployed wheat plants to three locations: small openings in a sagebrush stand, a grassy site 30 m from the same stand, and a lawn 2 km distant. They observed mean respiration rates in these plants of 63, 65, and 40 μl of O_2/min per gram of dry weight, respectively, and found these means to be significantly different ($p < 0.01$). Apparently, the concentrations of volatiles occurring in the field near sagebrush stands are indeed capable of affecting metabolism in wheat plants; it would now be instructive to determine their effects on the growth of native species with a similar experimental approach, and to compare the sensitivity of different species to the known distribution patterns (Hazlett and Hoffman, 1975).

When standing dead plants are suspected of releasing phytotoxins, the field bioassay can again be useful, because the dead plants may be transported. Vivrette and Muller (1977) made use of this technique in their investigations of *Mesembryanthemum chrystallinum*, which releases large quantities of sodium chloride upon its death. But perhaps the best illustration of the use of field bioassays is a study of bracken fern, *Pteridium aquilinum*, carried out by Gliessman and Muller (1978). Previous investigations (Gliessman and Muller, 1972; Stewart, 1975; Gliessman, 1976) had documented bracken's strong interference with other plants and its allelopathic potential. Standing dead fronds seemed to be particularly toxic, so Gliessman and Muller (1978) designed an experiment in which the dead fronds were suspended in a twine grid, at a density similar to that occurring naturally, over an adjacent grassland area. They used cheesecloth to provide similar shading of a control plot. After two seasons of this treatment, species composition beneath the suspended ferns had changed substantially in the direction one might predict based on the susceptibility of the different grassland species. This experiment, combined with the contributing evidence in Gleissman and Muller's investigation, constitutes one of the most convincing demonstrations of the occurrence of allelopathy in the field.

E. Recognition of Symptoms

Harper (1977) suggested an approach to allelopathy reminiscent of Koch's approach to human disease: If specific symptoms of phytotoxicity can be identified in the laboratory and then found or produced in the field, the conditions under which allelopathy occurs might be carefully documented.

This approach was adopted for research on agricultural systems by Patrick and co-workers to examine the phytotoxicity of plant residues. Patrick and Koch (1958) had shown that, particularly with high soil water content, plant material incorporated into the soil could produce substances which inhibited plant growth and caused browning of apical meristems and other radicle injury. Patrick et al. (1963) then examined lettuce and spinach in fields in the Salinas Valley, California, where growers had reported that crop establishment was poor when they planted too soon after turning under barley or rye cover crops. The lettuce and spinach roots indeed exhibited apical browning and other injuries similar to those observed in the laboratory, and in addition possessed discolored or sunken lesions at those points which had come into contact with fragments of decomposing plant material. These observations clearly confirmed the phytotoxicity of decomposing plant residues. Patrick et al. (1963) concluded that toxicity was localized in the immediate vicinity of the decomposing plant material, and therefore was rather patchy in the soil.

The success of this method depends on the possibility of finding specific symptoms of poisoning in the test species. *Grevillea robusta* causes blackening of leaf tips in its own seedlings (Webb et al., 1967) and *Juglans nigra* induces nearby herbs to wilt dramatically; if these symptoms are indeed

brought about by plant-produced phytotoxins, they are ideal for field studies
of allelopathy. Unfortunately, symptoms that appear in the root cells (Pollock
et al., 1954; Avers and Goodwin, 1956) are difficult to observe in the field;
changes in tissue nutrient concentrations (Walters and Gilmore, 1976) might
be attributable to a number of conditions unless exhaustively tested; and some
types of tissue browning and corking are caused not only by plant extracts
but also by a number of other irritants (Bogdan, 1971; Grodzinsky and Bogdan,
1972).

F. Analysis of Toxin Dynamics

Undoubtedly, a thorough understanding of allelopathy will come only through
detailed chemical examination of the release, the movement, the persistence,
and the physiological effects of plant-produced toxins. Potential phytotoxins
could be identified in naturally produced leachates, and their concentrations
in both leachates and soil could be followed throughout the year. Soil extractions could be calibrated using the bioassay method of Kaminsky (1980). Test
species of differing susceptibility to these phytotoxins could be grown in field
soil containing known amounts of the substances.

These chemical studies could be followed and validated by a field experiment which, surprisingly, has not been tried. The phytotoxins could be
added at natural rates to field plots containing various test species, and the
productivity of those plots could be compared with control plots receiving an
equal amount of water. Similar field tests could be devised for the study of
volatile compounds (e.g., terpenes) suspected of acting as allelopathic agents.
These sorts of experiments would indicate what concentrations or rates of
release are necessary to give rise to inhibition in the field.

This chemical approach to allelopathy, although ultimately providing a very
complete understanding of the processes involved, is rather complex, as
illustrated by the work to date on phenolic compounds. Studies of toxin dynamics should therefore not precede, but follow convincing demonstrations
that allelopathy is occurring in the field.

VII. CONCLUDING REMARKS

It is readily apparent that the research on allelopathy to date has been very
extensive and not sufficiently intensive. Hundreds of species have been
demonstrated to have a potential for allelopathy; a great many phytotoxic
compounds are known to be produced and released by plants; and the physiological effects of these substances on other plants are gradually becoming
understood; yet studies which convincingly demonstrate that allelopathy
occurs, in either natural or agricultural systems, are deplorably few.

Field experiments are needed to determine whether the allelopathic potential of plants is realized in nature. Four experimental approaches discussed
above—relating bioassay results to distribution, bioassaying naturally produced leachates in the field, recognizing specific symptoms of phytotoxicity,
and finally, examining the dynamics of the chemical substances involved—
have provided fairly convincing evidence for allelopathy in a few instances.
Hopefully, these methods will be used to reevaluate critically the information

on allelopathy and to reveal both laboratory artifacts and genuine cases of chemical interactions among plants.

REFERENCES

Abdul-Wahab, A. S., and Rice, E. L. (1967). Plant inhibition by Johnson grass and its possible significance in old-field succession. *Bull. Torrey Bot. Club 94*:486-497.

Al-Naib, F. A., and Rice, E. L. (1971). Allelopathic effects of *Platanus occidentalis*. *Bull. Torrey Bot. Club 98*:75-82.

Asplund, R. O. (1968). Monoterpenes: relationships between structure and inhibition of germination. *Phytochemistry 7*:1995-1997.

Asplund, R. O. (1969). The phytotoxicity of essential oils of different species of sagebrush. *Who. Range Manage.*, No. 270, pp. 40-44.

Avers, C. J., and Goodwin, R. H. (1956). Studies on roots. IV. Effects of coumarin and scopoletin on the standard root-growth pattern of *Phleum pratense*. *Am. J. Bot. 43*:612-620.

Baker, H. G. (1966). Volatile growth inhibitors produced by *Eucalyptus globulus*. *Madroño 18*:207-210.

Ballantyne, D. J. (1962). A growth inhibitor from the roots of *Chrysanthemum morifolium* Ramat. and its influence upon the gibberellic acid response. *Can. J. Bot. 40*:1229-1235.

Ballester, A., Albo, J. M., and Vieitz, E. (1977). The allelopathic potential of *Erica scoparia* L. *Oecologia 30*:55-61.

Banthorpe, D. V., and Charlwood, B. V. (1980). The terpenoids. In *Secondary Plant Products. Encyclopedia of Plant Physiology*, Vol. 8, E. A. Bell and B. V. Charlwood (Eds.). Springer-Verlag, New York, pp. 183-220.

Bartholomew, B. (1970). Bare zone between California shrub and grassland communities: the role of animals. *Science 170*:1210-1212.

Bell, E. A. (1980). The possible significance of secondary compounds in plants. In *Secondary Plant Products. Encyclopedia of Plant Plysiology*, Vol. 8, E. A. Bell and B. V. Charlwood (Eds.). Springer-Verlag, New York, pp. 11-21.

Bendall, G. M. (1975). The allelopathic activity of California thistle (*Cirsium arvense* (L.) Scop.) in Tasmania. *Weed Res. 15*:77-81.

Benedict, H. M. (1941). The inhibitory effect of dead roots on the growth of bromegrass. *Agron. J. 33*:1108-1109.

Bevege, D. I. (1968). Inhibition of seedling hoop pine (*Araucaria cunninghamii* Ait.) on forest soils by phytotoxic substances from the root zones of *Pinus, Araucaria* and *Flindersia*. *Plant Soil 29*:263-273.

Black, R. L. B., and Dix, N. J. (1976). Utilization of ferulic acid by microfungi from litter and soil. *Trans. Br. Mycol. Soc. 66*:313-317.

Blum, U., and Rice, E. L. (1969). Inhibition of symbiotic nitrogen-fixation by gallic and tannic acid and possible roles in old-field succession. *Bull. Torrey Bot. Club 96*:531-544.

Bogdan, G. P. (1971). Anatomical study of affections in plant conducting system by the allelopathic substances. *Ukr. Bot. Z. 28*:703-707.

Bohlmann, F., Burkhardt, T., and Zdero, C. (1973). *Naturally Occurring Acetylenes*. Academic Press, New York.

Bonner, J. (1946). Further investigations of toxic substances which arise from guayule plants: relation of toxic substances to the growth of guayule in soil. *Bot. Gaz. 107*:343-351.

Bonner, J., and Galston, S. W. (1944). Toxic substances from the culture media of guayule which may inhibit growth. *Bot. Gax. 106*:185-198.

Börner, H. (1959). The apple replant problem. I. The excretion of phlorizin from apple root residues and its role in the soil sickness problem. *Contrib. Boyce Thompson Inst. 20*:39-56.

Brown, R. T. (1967). Influence of naturally occurring compounds on germination and growth of jack pine. *Ecology 48*:542-546.

Brown, S. A. (1961). Biosynthesis of plant phenols. In *Symposium on Biochemistry of Plant Phenolic Substances*, G. Johnson and T. A. Geissman (co-directors). Colorado State University, Fort Collins, pp. 9-46.

Brown, W. A. (1969). Biochemistry of lignin formation. *BioScience 19*:115-121.

Buchanan, B. A., Harper, K. T., and Frischknecht, N. C. (1978). Allelopathic effects of bur buttercup tissue on germination and growth of various grasses and forbs in vitro and in soil. *Great Basin Nat. 38*:90-96.

Buchholtz, K. P. (1971). The influence of allelopathy on mineral nutrition. In *Biochemical Interactions Among Plants*. National Academy of Sciences, Washington, D.C., pp. 86-89.

Burges, N. A., Hurst, H. M., and Walkden, B. (1964). The phenolic constituents of humic acid and their relation to the lignin of the plant cover. *Geochim. Cosmochim. Acta 28*:1547-1554.

Cameron, H. J., and Julian, G. R. (1980). Inhibition of protein synthesis in lettuce (*Lactuca sativa* L.) by allelopathic compounds. *J. Chem. Ecol. 6*:989-995.

Cannon, J. R., Corbett, N. H., Haydock, K. P., Tracey, J. G., and Webb, L. J. (1962). An investigation of the effect of the dehydroangustione present in the leaf litter of *Backhousia angustifolia* on the germination of *Araucaria cunninghamii*—an experimental approach to a problem in rainforest ecology. *Aust. J. Bot. 10*:119-128.

Carballeira, A. (1980). Phenolic inhibitors in *Erica australis* L. and in associated soil. *J. Chem. Ecol. 6*:593-596.

Case, P. J., and Funk, D. T. (1978). The complex effects of juglone on the growth of pine, spruce and larch seedlings. *Proc. 5th N. Am. For. Biol. Workshop*, C. A. Hollis and A. E. Squillace (Eds.). Abstract, p. 422.

Chambers, E. E., and Holm, L. G. (1965). Phosphorus uptake as influenced by associated plants. *Weeds 13*:312-314.

Charlwood, B. V., and Banthorpe, D. V. (1978). The biosynthesis of monoterpenes. In *Progress in Phytochemistry*, Vol. 5, L. Reinhold, J. B. Harborne, and T. Swain (Eds.). Pergamon Press, Elmsford, N.Y., pp. 66-118.

Chou, C.-H., and Muller, C. H. (1972). Allelopathic mechanisms of *Arctostaphylos glandulosa* var. *zacaensis*. *Am. Midl. Nat. 88*:324-347.

Chou, C.-H., and Patrick, AZ. A. (1976). Identification and phytotoxic activity of compounds produced during decomposition of corn and rye residues in soil. *J. Chem. Ecol. 2*:369-387.

Chou, C.-H., and Waller, G. R. (1980). Isolation and identification by mass spectrometry of phytotoxins in *Coffea arabica*. *Bot. Bull. Acad. Sin. 21*:25-34.

Christensen, N. L., and Muller, C. H. (1975). Effects of fire on factors controlling plant growth in *Adenostoma* chaparral. *Ecol. Monogr. 45:* 29-55.

Colton, C. E., and Einhellig, F. A. (1980). Allelopathic mechanisms of velvetleaf (*Abutiolon theophrasti* Medic., Malvaceae) on soybean. *Am. J. Bot. 67:*1407-1413.

Cook, M. T. (1921). Wilting caused by walnut trees. *Phytopathology 11:* 346.

Corcoran, M. R., Geissman, T. A., and Phinney, B. O. (1972). Tannins as gibberellin antagonists. *Plant Physiol. 49:*323-330.

Curtis, J. T., and Cottam, G. (1950). Antibiotic and autotoxic effects in prairie sunflower. *Bull. Torrey Bot. Club 77:*187-191.

Danks, M. L., Fletcher, J. S., and Rice, E. L. (1975). Effects of phenolic inhibitors on growth and metabolism of glucose-UL-^{14}C in Paul's scarlet rose cell-suspension cultures. *Am. J. Bot. 62:*311-317.

Davidonis, G. H., and Ruddat, M. (1973). Allelopathic compounds, thelypterin A and B in the fern *Thelypteris normalis*. *Planta 111:*27-32.

De Candolle, A. P. (1832). *Physiologie végétale*, Vol. III. Béchet Jeune, Lib. Fac. Med., Paris.

de Kock, P. C. (1955). Influence of humic acids on plant growth. *Science 121:*473.

del Moral, R. (1972). On the variability of chlorogenic acid concentration. *Oecologia 9:*289-300.

del Moral, R., and Cates, R. G. (1971). Allelopathic potential of the dominant vegetation of western Washington. *Ecology 52:*1030-1037.

del Moral, R., and Muller, C. H. (1969). Fog drip: a mechanism of toxin transport from *Eucalyptus globulus*. *Bull. Torrey Bot. Club 96:*467-475.

del Moral, R., and Muller, C. H. (1970). The allelopathic effects of *Eucalyptus camaldulensis*. *Am. Midl. Nat. 83:*254-283.

Demos, E. K., Woolwine, M., Wilson, R. H., and McMillan, C. (1975). The effects of ten phenolic compounds on hypocotyl growth and mitochondrial metabolism of mung bean. *Am. J. Bot. 62:*97-102.

DeSerra, M. O., and Schnitzer, M. (1972). Extraction of humic acid by alkali and chelating resin. *Can. J. Soil Sci. 52:*365-374.

Einhellig, F. A., and Kuan, L.-Y. (1971). Effects of scopoletin and chlorogenic acid on stomatal aperture in tobacco and sunflower. *Bull. Torrey Bot. Club 98:*155-162.

Einhellig, F. A., and Rasmussen, J. A. (1979). Effects of three phenolic acids on chlorophyll content and growth of soybean and grain sorghum seedlings. *J. Chem. Ecol. 5:*815-824.

Einhellig, F. A., Rice, E. L., Risser, P. G., and Wender, S. H. (1970). Effects of scopoletin on growth, CO_2 exchange rates, and concentration of scopoletin, scopolin, and chlorogenic acids in tobacco, sunflower and pigweed. *Bull. Torrey Bot. Club 97:*22-23.

Elmore, C. D. (1980). Free amino acids of *Abutilon theophrasti* seed. *Weed Res. 20:*63-64.

Elmstrom, G. W., and Howard, F. O. (1970). Promotion and inhibition of iron accumulation in soybean plants. *Plant Physiol. 45:*327-329.

Evans, W. C. (1947). Oxidation of phenol and benzoic acid by some soil bacteria. *Biochem. J. 41:*373-382.

Fay, P. K. and Duke, W. B. (1977). An assessment of allelopathic potential in *Avena* germ plasm. *Weed Science 25:*224-228.

Fisher, R. F. (1977). Allelopathic interference among plants. I. Ecological significance. *Proc. 4th N. Am. For. Biol. Workshop*, H. E. Wilcox and A. F. Hammer (Eds.). Coll. of Environ. Sci. For., Syracuse, N.Y., pp. 73-92.

Fisher, R. F. (1978). Juglone inhibits pine growth under certain moisture regimes. *Soil Sci. Soc. Am. J. 43*:801-803.

Fisher, R. F., Woods, R. A., and Glavicic, M. R. (1978). Allelopathic effects of goldenrod and aster on young sugar maple. *Can. J. For. Res. 8*:1-9.

Friedman, J., Orshan, G., and Ziger-Cfir, Y. (1977). Suppression of annuals by *Artemisia herba-alba* in the Negev Desert of Israel. *J. Ecol. 65*:413-426.

Friend, J. (1979). Phenolic substances and plant disease. In *Biochemistry of Plant Phenolics. Recent Advances in Phytochemistry*, Vol. 12, T. Swain, J. B. Harborne and C. F. Van Sumere (Eds.). Plenum Press, New York, 557-588.

Gant, R. E., and Clebsch, E. E. C. (1975). The allelopathic influences of *Sassafras albidum* in old-field succession in Tennessee. *Ecology 56*:604-615.

Gilbert, B. L., Baker, J. E., and Harris, D. M. (1967). Juglone (5-hydroxy-1,4-naphthoquinone) from *Carya ovata*, a deterrent to feeding by *Scolytus multistriatus*. *J. Insect Physiol. 13*:1453-1459.

Glass, A. D. M. (1973). Influence of phenolic acids on ion uptake. I. Inhibition of phosphate uptake. *Plant Physiol. 51*:1037-1041.

Glass, A. D. M. (1974). Influence of phenolic acids upon ion uptake. III. Inhibition of potassium absorption. *J. Exp. Bot. 25*:1104-1113.

Glass, A. D. M. (1976). The allelopathic potential of phenolic acids associated with the rhizosphere of *Pteridium aquilinum*. *Can. J. Bot. 54*:2440-2444.

Glass, A. D. M., and Dunlop, J. (1974). Influence of phenolic acids on ion uptake. IV. Depolarization of membrane potentials. *Plant Physiol. 54*:855-858.

Gliessman, S. R. (1976). Allelopathy in a broad spectrum of environments as illustrated by bracken. *Bot. J. Linn. Soc. 73*:95-104.

Gliessman, S. R., and Muller, C. H. (1972). The phytotoxic potential of bracken, *Pteridium aquilinum* (L.) Kuhn. *Madroño 21*:299-304.

Gliessman, S. R., and Muller, C. H. (1978). Allelopathic mechanisms of dominance in bracken. *J. Chem. Ecol. 4*:337-362.

Goodwin, R. H., and Taves, C. (1950). The effect of coumarin derivatives on the growth of *Avena* roots. *Am. J. Bot. 37*:224-231.

Grant, E. A., and Sallans, W. G. (1964). Influence of plant extracts on germination and growth of eight forage species. *J. Br. Grassl. Soc. 19*:191-197.

Gregg, W. P., Jr. (1972). Ecology of the annual grass *Setaria lutescens* on old-fields of the Pennsylvania piedmont. *Proc. Acad. Nat. Sci. Phil. 124*:135-196.

Gressel, J. B., and Holm, L. G. (1964). Chemical inhibition of crop germination by weed seeds and the nature of inhibition by *Abutilon theophrasti*. *Weed Res. 4*:44-53.

Gries, G. A. (1943). Juglone, the active agent in walnut toxicity. *Rep. Proc. Northern Nut Growers' Assoc. 32*:52-55.

Grodzinsky, A. M., and Bogdan, G. P. (1972). Histochemical study of pectins, lignin, suberin and melanines in plants subjected to the action of allelopathically active substances. *Ukr. Bot. Zh.* 29:137-143.

Guenzi, W. D., and McCalla, T. M. (1966). Phenolic acids in oats, wheat, sorghum and corn residues and their phytotoxicity. *Agron. J.* 58:303-304.

Guyot, A. L. (1957). Les microasscoiations végétales au sein du *Brometum erecti*. *Vegetatio* 7:321-354.

Haider, K., and Martin, J. P. (1975). Decomposition of specifically carbon-14 labelled benzoic and cinnamic acid derivatives in soil. *Soil Sci. Soc. Am. Proc.* 39:657-662.

Haider, K., Martin, J. P., and Filip, Z. (1975). Humus biochemistry. In *Soil Biochemistry*, Vol. 4, E. A. Paul and A. D. McLaren (Eds.). Marcel Dekker, New York, pp. 195-244.

Halligan, J. P. (1973). Bare areas associated with shrub stands in grassland: the case of *Artemisia californica*. *BioScience* 23:429-432.

Halligan, J. P. (1975). Toxic terpenes from *Artemisia californica*. *Ecology* 56:99-1003.

Handley, W. R. C. (1963). Mycorrhizal associations and *Calluna* heathland afforestation. *Gr. Br. For. Comm. Bull. 36.*

Harborne, J. B. (1980). Plant phenolics. In *Secondary Plant Products*, E. A. Bell and B. V. Charlwood (Eds.). Springer-Verlag, New York, pp. 329-402.

Harborne, J. B., and Simmonds, N. W. (1964). The natural distribution of the phenolic aglycones. In *Biochemistry of Phenolic Compounds*, J. B. Harborne (Ed.). Academic Press, New York, pp. 77-127.

Harper, J. L. (1961). Approaches to the study of plant competition. *Symp. Soc. Exp. Biol.* 15:1-39.

Harper, J. L. (1964). The nature and consequence of interference among plants. Proc. XI Int. Congr. Genet. *Genetics Today*, S. J. Gerts (Ed.). Pergamon Press, Oxford, pp. 465-482.

Harper, J. L. (1977). *Population Biology of Plants*. Academic Press, New York.

Hazlett, D. L. and Hoffman, G. R. (1975). Plant species distributional patterns in *Artemisia tridentata-* and *Artemisia cana-*dominated vegetation in western North Dakota. *Bot. Gaz.* 136:72-77.

Henderson, M. E. K. (1956). A study of the metabolism of phenolic compounds by soil fungi using spore suspensions. *J. Gen. Microbiol.* 14:684-691.

Henderson, M. E. K., and Farmer, V. C. (1955). Utilization by soil fungi of p-hydroxybenzaldehyde, ferulic acid, syringaldehyde and vanillin. *J. Gen. Microbiol.* 12:37-46.

Hoffman, G. R., and Hazlett, D. L. (1977). Effects of aqueous *Artemisia* extracts and volatile substances on germination of selected species. *J. Range Manage.* 30:134-137.

Horsely, S. B. (1977a). Allelopathic inhibition of black cherry by fern, grass, goldenrod and aster. *Can. J. For. Res.* 7:205-216.

Horsley, S. B. (1977b). Allelopathic interference among plants. II. Physiological modes of action. *Proc. 4th N. Am. For. Biol. Workshop*, H. E. Wilcox and A. F. Hamer (Eds.). Coll. Environ. Sci. For., Syracuse, N.Y., pp. 93-136.

Howard, A. (1925). The effect of grass on trees. *Proc. R. Soc. Lond.*, *Ser. B* 97:284-321.

Hull, J. C., and Muller, C. H. (1977). The potential for dominance by *Stipa pulchra* in a California grassland. *Amer. Midl. Nat. 97*:147-175.

Hurst, H. M., and Burges, N. A. (1967). Lignin and humic acids. In *Soil Biochemistry*, A. D. McLaren and G. H. Peterson (Eds.). Marcel Dekker, New York, pp. 260-286.

Jackson, J. R., and Willemsen, R. W. (1976). Allelopathy in the first stages of secondary succession on the piedmont of New Jersey. *Am. J. Bot. 63*: 1015-1023.

Jameson, D. A. (1961). Growth inhibitors in native plants of northern Arizona. *Res. Note 61.* Rocky Mt. For. Range Exp. Stn., Fort Collins, Colo.

Jameson, D. A. (1970). Degradation and accumulation of inhibitory substances from *Juniperus osteosperma* (Torr.) Little. *Plant Soil 33*:213-224.

Jensen, T. E., and Welbourne, F. W. (1962). The cytological effect of growth inhibitors on excised roots of *Vicia faba* and *Pisum sativum*. *Proc. S. Dak. Acad. Sci. 41*:131-136.

June, S. R. (1976). Investigations on allelopathy in a red beech forest. *Mauri Ora 4*:87-91.

Kaminsky, R. (1980). The determination and extraction of available soil organic compounds. *Soil Sci. 130*:118-123.

Kaminsky, R., and Muller, W. H. (1977). The extraction of soil phytotoxins using a neutral EDTA solution. *Soil Sic. 124*:205-210.

Kaminsky, R., and Muller, W. H. (1978). A recommendation against the use of alkaline soil extractions in the study of allelopathy. *Plant Soil 49*:641-645.

Keating, K. I. (1978). Blue-green algal inhibition of diatom growth: transition from mesotrophic to eutrophic community structure. *Science 199*:971-973.

Keever, C. (1950). Causes of succession on old fields of the Piedmont, North Carolina. *Ecol. Mongr. 20*:229-250.

Kefeli, V. I., and Kadyrov, C. S. (1971). Natural growth inhibitors, their chemical and physiological properties. *Annu. Rev. Plant Physiol. 22*: 185-196.

Klein, R. R., and Miller, D. A. (1980). Allelopathy and its role in agriculture. *Commun. Soil Sci. Plant Anal. 11*:43-56.

Kobayashi, A., Morimoto, S., Shibata, Y., Yamashita, K., and Numata, M. (1980). C_{10}-polyacetylenes as allelopathic substances in dominants in early stages of secondary succession. *J. Chem. Ecol. 6*:119-132.

Koeppe, D. E. (1972). Some reactions of isolated corn mitochondria influenced by juglone. *Physiol. Plant. 27*:89-94.

Koeppe, D. E., Rohrbaugh, L. M., and Wender, S. H. (1971). The effect of environmental stress conditions on the concentration of caffeoylquinic acids and scopolin in tobacco and sunflower. In *Biochemical Interactions Among Plants*. National Academy of Sciences, Washington, D.C., pp. 102-108.

Koeppe, D. E., Southwick, L. M., and Bittell, J. E. (1976). The relationship of tissue chlorogenic acid concentrations and leaching of phenolics from sunflowers grown under varying phosphate nutrient conditions. *Can. J. Bot. 54*:593-599.

Kononova, M. M. (1966). *Soil Organic Matter*, 2nd Engl. ed. Trans. from Russian by T. Z. Nowakowski and A. C. D. Newman. Pergamon Press, Elmsford, N.Y., 544 pp.

Lawrence, T., and Kilcher, M. R. (1962). The effect of fourteen root extracts upon germination and seedling length of fifteen plant species. *Can. J. Plant Sci. 42*:308-313.

Lehman, R. H., and Rice, E. L. (1972). Effect of deficiencies of nitrogen, potassium and sulfur on chlorogenic acids and scopolin in sunflower. *Am. Midl. Nat. 87*:71-30.

Lewis, J. A., and Starkey, R. L. (1969). Decomposition of plant tannins by some soil microorganisms. *Soil Sci. 107*:235-241.

Lingappa, B. T., and Lockwood, T. L. (1962). Fungitoxicity of lignin monomers model substances, and decomposition products. *Phytopathology 52*: 295-299.

Lockerman, R. H., and Putnam, A. R. (1979). Evaluation of allelopathic cucumbers (*Cucumis sativus*) as an aid to weed control. *Weed Sci. 27*: 54-57.

Lodhi, M. A. K. (1975). Soil-plant phytotoxicity and its possible significance in patterning of herbaceous vegetation in a bottomland forest. *Am. J. Bot. 62*:618-622.

Lodhi, M. A. K. (1976). Role of allelopathy as expressed by dominating trees in a lowland forest in controlling the productivity and pattern of herbaceous growth. *Am. J. Bot. 63*:1-8.

Lodhi, M. A. K. (1978). Allelopathic effects of decaying litter of dominant trees and their associated soil in a lowland forest community. *Am. J. Bot. 65*:340-344.

Lodhi, M. A. K. (1979a). Allelopathic potential of *Salsola kali* and its possible role in rapid disappearance of weedy stage during revegetation. *J. Chem. Ecol. 5*:429-438.

Lodhi, M. A. K. (1979b). Germination and decreased growth of *Kochia scoparia* in relation to its autoallelopathy. *Can. J. Bot. 57*:1083-1088.

Lodhi, M. A. K. (1979c). Inhibition of nitrifying bacteria, nitrification and mineralization in spoil soils as related to their successional stages. *Bull. Torrey Bot. Club 106*:284-289.

Lodhi, M. A. K. , and Killingbeck, K. T. (1980). Allelopathic inhibition of nitrification and nitrifying bacteria in a ponderosa pine (*Pinus ponderosa* Dougl.) community. *Am. J. Bot. 67*:1423-1429.

Lodhi, M. A. K., and Nickell, G. L. (1973). Effects of leaf extracts of *Celtis laevigata* on growth, water content and carbon dioxide exchange rates of three grass species. *Bull. Torrey Bot. Club 100*:159-165.

Lodhi, M. A. K., and Rice, E. L. (1971). Allelopathic effects of *Celtis laevigata*. *Bull. Torrey Bot. Club 98*:83-89.

Loehwing, W. F. (1937). Root interactions of plants. *Bot. Rev. 3*:195-239.

Lorber, P., and Muller, W. H. (1976). Volatile growth inhibitors produced by *Salvia leucophylla*: effects on seedling root tip ultrastructure. *Am. J. Bot. 63*:196-200.

MacDaniels, L. H., and Muenscher, W. C. (1941). Black walnut toxicity. *Rep. Proc. Northern Nut Growers' Assoc. 31*:172-179.

MacDaniels, L. H., and Pinnow, D. L. (1976). Walnut toxicity, an unsolved problem. *67th Annu. Rep. Northern Nut Growers' Assoc.*, pp. 114-122.

Machackova, I., and Zmrhal, Z. (1976). Comparison of the effect of some phenolic compounds on wheat coleoptile section growth with their effect on IAA-oxidase activity. *Biol. Plant. 18*:147-151.

Massey, A. B. (1925). Antagonism of the walnuts (*Juglans nigra* and *J. cinerea*) in certain plant associations. *Phytopathology 15*:773-784.

Mayer, A. M., and Poljakoff-Mayber, A. (1975). *The Germination of Seeds*, 2nd ed. Pergamon Press, Elmsford, N.Y.

McCahon, C. B., Kelsey, R. G., Sheridan, R. P., and Shafizadeh, F. (1973). Physiological effects of compounds extracted from sagebrush. *Bull. Torrey Bot. Club 100*:23-28.

McCalla, T. M., and Norstadt, F. A. (1974). Toxicity problems in mulch tillage. *Agric. Environ. 1*:153-174.

McClure, P. R., Gross, H. D., and Jackson, W. A. (1978). Phosphate absorption by soybean varieties: the influence of ferulic acid. *Can. J. Bot. 56*:764-767.

McKey, D., Waterman, P. G., Mbi, C. N., Gartlan, J. S., and Struhsaker, T. T. (1978). Phenolic content of vegetation in two African rain forests: ecological implications. *Science 202*:61-64.

McPherson, J. K., and Muller, C. H. (1969). Allelopathic effects of *Adenostoma fasciculatum*, "Chamise," in the California chaparral. *Ecol. Monogr. 39*:177-198.

Miner, J. (1974). The effect of couch grass on the growth and mineral up-take of wheat. *Folia Fac. Sci. Nat. Univ. Purkynianae Brun. Biol. 15*:1-84.

Mitin, V. V. (1970). On study of chemical nature of growth inhibitors in the dead leaves of hornbean and beech. In *Physiological-Biochemical Basis of Plant Interactions in Phytocenoses*, Vol. 1, A. M. Grodzinsky (Ed.). Nanhova Dumka, Kiev, pp. 177-181. (In Russian.)

Molische, H. (1937). *Der Einfluss einer Pflanze auf die Andere—Allelopathie*. Gustav Fischer, Jena.

Montes, R. A., and Christensen, N. L. (1979). Nitrification and succession in the piedmont of North Carolina. *For. Sci. 25*:287-297.

Moore, C. W. E., and Keraitis, K. (1966). Nutrition of *Grevillea robusta*. *Aust. J. Bot. 14*:151-163.

Morinos, N. G., and Heinberg, T. (1960). Observations on a possible mechanism of action of the inhibitor-β complex. *Physiol. Plant. 13*:571-581.

Morita, H. (1965). The phenolic acids in organic soils. *Can. J. Biochem. 43*:1277-1280.

Morrison, R. I. (1963). Products of the alkaline nitrobenzene oxidation of soil organic matter. *J. Soil. Sci. 14*:201-216.

Muller, C. H. (1966). The role of chemical inhibition (allelopathy) in vegetational composition. *Bull. Torrey Bot. Club 93*:332-351.

Muller, C. H. (1969a). Allelopathy as a factor in ecological processes. *Vegetatio 18*:348-357.

Muller, C. H. (1969b). The "co-" in coevolution. *Science 164*:197-198.

Muller, C. H. (1970). The role of allelopathy in the evolution of vegetation. In *Biochemical Coevolution*, K. L. Chambers (Ed.). Oregon State University Press, Corvallis, pp. 13-31.

Muller, C. H. (1974). Vegetation and environment. In *Handbook of Vegetation Science*, B. R. Strain and W. D. Billings (Eds.). Dr. W. Junk, The Hague, pp. 73-87.

Muller, C. H., and Chou, C.-H. (1972). Phytotoxins: an ecological phase of phytochemistry. In Phytochemical Ecology, J. B. Harborne (Ed.). Academic Press, New York, pp. 201-216.

Muller, C. H., and del Moral, R. (1966). Soil Toxicity induced by terpenes from Salvia leucophylla. Bull. Torrey Bot. Club 93:130-137.

Muller, C. H., and del Moral, R. (1971). Role of animals in suppression of herbs by shrubs. Science 173:462-463.

Muller, C. H., and Muller, W. H. (1964). Antibiosis as a factor in vegetation patterns. Science 144:889-890.

Muller, D. H., Muller, W. H., and Haines, B. L. (1964). Volatile growth inhibitors produced by shurbs. Science 143:471-473.

Muller, W. H. (1965). Volatile materials produced by Salvia leucophylla: effect on seedling growth and soil bacteria. Bot. Gaz. 126:195-200.

Muller, W. H., and Hauge, R. (1967). Volatile growth inhibitors produced by Salvia leucophylla: effect on seedling anatomy. Bull. Torrey Bot. Club 94:182-191.

Muller, W. H., and Muller, C. H. (1956). Association patterns involving desert plants that contain toxic products. Am. J. Bot. 43:354-361.

Muller, W. H., and Muller, C. H. (1964). Volatile growth inhibitors produced by Salvia species. Bull. Torrey Bot. Club 91:327-330.

Muller, W. H., Lorber, P., Haley, B., and Johnson, K. (1969). Volatile growth inhibitors produced by Salvia leucophylla: effect on oxygen uptake by mitochondrial suspensions. Bull. Torrey Bot. Club 96:89-95.

Neill, R. L., and Rice, E. L. (1971). Possible role of Ambrosia psilostachya on pattern and succession in old-fields. Am. Midl. Nat. 86:344-357.

Neish, A. C. (1964). Major pathways of biosynthesis of phenols. In Biochemistry of Phenolic Compounds, J. B. Harborne (Ed.). Academic Press, New York, pp. 295-359.

Newman, E. I. (1978). Allelopathy: adaptation or accident? In Biochemical Aspects of Plant and Animal Coevolution, J. B. Harborne (Ed.). Academic Press, New York, pp. 327-342.

Newman, E. I., and Miller, M. H. (1977). Allelopathy among some British grassland species. II. Influence of root exudates on phosphorus uptake. J. Ecol. 65:399-411.

Newman, E. I., and Rovira, A. D. (1975). Allelopathy among some British grassland species. J. Ecol. 63:727-737.

Oh, H. K., Sakai, T., Jones, M. B., and Longhurst, W. M. (1967). Effects of various essential oils isolated from Douglas fir needles upon sheep and deer rumen microbial activity. Appl. Microbiol. 15:77-784.

Ohman, J. H., and Kommedahl, T. (1960). Relative toxicity of extracts from vegetative organs of quackgrass to alfalfa. Weeds 8:666-670.

Ohman, J. H. and Kommedahl, T. (1964). Plant extracts, residues and soil minerals in relation to competition of quackgrass with oats and alfalfa. Weeds 12:222-231.

Olmsted, C. E., III, and Rice, E. L. (1970). Relative effects of known plant inhibitors on species from first two stages of old-field succession. Southwest. Nat. 15:165-173.

Olsen, R. A., Odham, G., and Lindeberg, G. (1971). Aromatic substances in leaves of Populus tremula as inhibitors of mycorrhizal fungi. Physiol. Plant. 25:122-129.

Owens, L. D. (1969). Toxins in plant disease: structure and mode of action. *Science 165*:18-25.

Parenti, R. L., and Rice, E. L. (1969). Inhibitional effects of *Digitaria sanguinalis* and possible role in old-field succession. *Bull. Torrey Bot. Club 96*:70-78.

Patrick, Z. A. (1955). The peach replant problem in Ontario. II. Toxic substances from microbial decomposition products of peach root residues. *Can. J. Bot. 33*:461-486.

Patrick, Z. A. (1971). Phytotoxic substances associated with the decomposition in soil of plant residues. *Soil Sci. 111*:13-18.

Patrick, Z. A., and Koch, L. W. (1958). Inhibition of respiration, germination and growth by substances arising during the decomposition of certain plant residues in the soil. *Can. J. Bot. 36*:621-647.

Patrick, Z. A., Toussoun, T. A., and Snyder, W. S. (1963). Phytotoxic substances in arable soils associated with decomposition of plant residues. *Phytopathology 53*:152-161,

Patrick, Z. A., Toussoun, T. A., and Koch, L. W. (1964). Effect of crop residue decomposition products on plant roots. *Annu. Rev. Phytopathol. 2*:267-291.

Perry, S. F. (1967). Inhibition of respiration by juglone in *Phaseolus* and *Lycopersicon*. *Bull. Torrey Bot. Club 94*:26-30.

Peterson, E. B. (1965). Inhibition of black spruce primary roots by a water-soluble substance in *Kalmia angustifolia*. *For. Sci. 11*:473-479.

Peterson, R. L., and Fairbrothers, D. E. (1980). Reciprocal allelopathy between the gametophytes of *Osmunda cinnamomea* and *Dryopteris intermedia*. *Am. Fern J. 70*:73-78.

Pianka, E. R. (1978). *Evolutionary Ecology*, 2nd ed. Harper & Row, New York.

Pickering, S. (1917). The effect of one plant on another. *Ann. Bot. 31*:181-187.

Pilet, P. E. (1966). Effect of *p*-hydroxybenzoic acid on growth, auxin content and auxin catabolism. *Phytochemistry 5*:77-82.

Pollock, B. M., Goodwin, R. H., and Green, S. (1954). Studies on roots. II. Effects of coumarin, scopoletin and other substances in growth. *Am. J. Bot. 41*:521-529.

Putnam, A. R., and Duke, W. B. (1974). Biological suppression of weeds: Evidence for allelopathy in accessions of cucumber. *Science 185*:370-372.

Putnam, A. R., and Duke, W. B. (1978). Allelopathy in agroecosystems. *Annu. Rev. Phytopathol. 16*:431-451.

Quarterman, E. (1973). Allelopathy in cedar glade plant communities. *J. Tenn. Acad. Sci. 48*:147-150.

Quinn, J. A. (1974). *Convolvulus sepium* in old field succession on the New Jersey piedmont. *Bull. Torrey Bot. Club 101*:89-95.

Rasmussen, J. A., and Einhellig, F. A. (1979). Allelochemic effects of leaf extracts of *Ambrosia trifida* (Compositae). *Southwestern Naturalist 24*:637-644.

Raynal, D. J., and Bazzaz, F. A. (1975). Interference of winter annuals with *Ambrosia artemisiifolia* in early successional fields. *Ecol. 56*:35-49.

Retig, B., Holm, L. G., and Struckmeyer, B. E. (1972). Effects of weeds on the anatomy of roots of cabbage and tomato. *Weed Sci. 20*:33-49.

Rice, E. L. (1965). Inhibition of nitrogen-fixing and nitrifying bacteria by seed plants. IV. The inhibitors produced by *Ambrosia elatior* L. and *Ambrosia psilostachya* D. C. *Southwest. Nat. 10*:248-255.

Rice, E. L. (1967). Chemical warfare between plants. *Bios (Mr. Vernon, Iowa) 38*:67-74.

Rice, E. L. (1974). *Allelopathy*. Academic Press, New York.

Rice, E. L. (1979). Allelopathy—an update. *Bot. Rev. 45*:15-109.

Rice, E. L., and Pancholy, S. K. (1972). Inhibition of nitrification by climax ecosystems. *Am. J. Bot. 59*:1033-1040.

Rice, E. L., and Pancholy, S. K. (1973). Inhibition of nitrification by climax ecosystems. II. Additional evidence and possible role of tannins. *Am. J. Bot. 60*:691-702.

Rice, E. L., and Pancholy, S. K. (1974). Inhibition of nitrification by climax ecosystems. III. Inhibitors other than tannins. *Am. J. Bot. 61*: 1095-1103.

Robinson, R. K. (1972). The production by roots of *Calluna vulgaris* of a factor inhibitory to growth of some mycorrhizal fungi. *J. Ecol. 60*:219-224.

Rogoff, M. H. (1961). Oxidation of aromatic compounds by bacteria. In *Advanced Applied Microbiology*, Vol. 3, W. W. Umbreit (Ed.). Academic Press, New York, pp. 192-221.

Sarkanen, K. V., and Ludwig, C. H. (Eds.) (1971). *Lignins: Occurrence, Formation, Structure and Reactions*. Wiley-Interscience, New York.

Savory, B. M. (1969). Evidence that toxins are not the causal factors of the specific apple replant disease. *Ann. Appl. Biol. 63*:225-231.

Scharff, T. G., and Perry, A. C. (1976). The effects of salicylic acid on metabolism and potassium ion content in yeast. *Proc. Soc. Exp. Biol. Med. 151*:72-77.

Schreiner, O., and Reed, H. S. (1908). The toxic action of certain organic plant constituents. *Bot. Gaz. 45*:73-102.

Schreiner, O., and Skinner, J. J. (1912). The toxic action of organic compounds as modified by fertilizer salts. *Bot. Gaz. 54*:31-48.

Schroth, M. N., and Hildebrand, D. C. (1964). Influence of plant exudates on root-infecting fungi. *Annu. Rev. Phytopathol. 2*:101-132.

Schwartz, C. C., Nagy, J. G., and Regelin, W. L. (1980). Juniper oil yield, terpenoid concentration, and antimicrobial effects on deer. *J. Wildl. Manage. 44*:107-113.

Seigler, D., and Price, P. W. (1976). Secondary compounds in plants: primary functions. *Am. Nat. 110*:101-105.

Shindo, H., and Kuwatsuka, S. (1975). Behavior of phenolic substances in the decaying process of plants. III. Degradation pathway of phenolic acids. *Soil Sci. Plant Nutr. 21*:227-238.

Shindo, H., and Kuwatsuka, S. (1977). Behavior of phenolic substances in the decaying process of plants. VI. Changes in quality and quantity of phenolic substances in the decaying process of rice straw in a soil. *Soil Sci. Plant Nutr. 23*:319-332.

Shindo, H., Marumoto, T., and Higashi, T. (1979). Behavior of phenolic substances in the decaying process of plants. X. Distribution of phenolic acids in soils of greenhouses and fields. *Soil Sci. Plant Nutr. 25*:591-600.

Sitaramaiah, K., and Pathak, K. N. (1979). Effect of phenolics and an aromatic acid on *Meloidogyne javanica* infecting tomato (*Lycopersicon esculentum*). *Nematologica* 25:281-287.

Sitaramaiah, K., and Singh, R. S. (1979). Effect of organic amendment on phenolic content of soil and plant and response of *Meloidogyne javanica* and its host to related compounds. *Plant Soil* 50:671-680.

Star, A. E. (1980). Found exudate flavonoids as allelopathic agents in *Pityrogramma*. *Bull. Torrey Bot. Club* 107:146-153.

Steelink, C. (1972). Biological oxidation of lingnin phenols. In *Recent Advances in Phytochemistry*, Vol. 4, V. C. Runeckles and J. E. Watkins (Eds.). Appleton-Century-Crofts, New York, pp. 237-271.

Stewart, R. E. (1975). Allelopathic potential of western bracken. *J. Chem. Ecol.* 1:161-169.

Stowe, L. G. (1979). Allelopathy and its influence on the distribution of plants in an Illinois old-field. *J. Ecol.* 67:1065-1085.

Stowe, L. G., and Osborn, A. (1980). The influence of nitrogen and phosphorus levels on the phytotoxicity of phenolic compounds. *Can. J. Bot.* 58:1149-1153.

Sukhada, K., and Jayachandra. (1980). Pollen allelopathy—a new phenomenon. *New Phytol.* 84:739-746.

Sundman, V. (1964). A description of some lignanolytic soil bacteria and their ability to oxidize simple phenolic compounds. *J. Gen. Microbiol.* 36:171-183.

Swain, T. (1977). Secondary compounds as protective agents. *Annu. Rev. Plant Physiol.* 28:479-501.

Swain, T. (1979). Tannins and lignins. In *Herbivores, Their Interaction with Secondary Plant Metabolites*. G. A. Rosenthal and D. H. Janzen (Eds.). Academic Press, New York, pp. 657-682.

Tate, K. R., and Anderson, H. A. (1978). Phenolic hydrolysis products from gel chromatographic fractions of soil humic acids. *J. Soil Sci.* 29:76-83.

Thimann, K. V. (1972). The natural plant hormones. In *Plant Physiology*, Vol. VIB, F. C. Steward (Ed.). Academic Press, New York, pp. 3-332.

Tinsley, J., and Salam, M. (1961). Extraction of soil organic matter with aqueous solvents. *Soils Fert.* 24:81-84.

Tomaszewski, M. (1960). The occurrence of p-hydroxybenzoic acid and some other simple phenols in vascular plants. *Bull. Acad. Pol. Sci.* 8:61-63.

Tomaszewski, M., and Thimann, K. V. (1966). Interactions of phenolic acids, metallic ions and chelating agents on auxin-induced growth. *Plant Physiol.* 41:1443-1454.

Towers, G. H. N. (1964). Metabolism of phenolics in higher plants and microorganisms. In *Biochemistry of Phenolic Compounds*, J. B. Harborne (Ed.). Academic Press, New York, pp. 247-294.

Trigiano, R. N., and Fergus, C. L. (1979). Extracellular enzymes of some fungi associated with mushroom culture. *Mycologia* 71:908-917.

Turner, B. H., and Quarterman, E. (1975). Allelochemic effects of *Petalostemon gattingeri* on the distribution of *Arenaria patula* in cedar glades. *Ecology* 56:924-932.

Turner, J. A., and Rice, E. L. (1975). Microbial decomposition of ferulic acid in soil. *J. Chem. Ecol.* 1:41-58.

Turner, W. B. (1971). *Fungal Metabolites*. Academic Press, New York.

Tyson, B. J., Dement, W. A., and Mooney, H. A. (1974). Volatilization of terpenes from *Salvia mellifera*. *Nature (Lond.)* 252:119-120.

Uritani, I. (1961). The role of plant phenolics in disease resistance and immunity. In *Symposium on Biochemistry of Plant Phenolic Substances*, G. Johnson and T. A. Geissman (co-directors). Colorado State University, Fort Collins, pp. 98-124.

U.S. Department of Agriculture (1948). Clip Sheet 2375-48. Test clears walnut's reputation.

Van Sumere, C. F., Cottenie, J., DeGreef, J., and Kint, J. (1972). Biochemical studies in relation to the possible germination regulatory role of naturally occurring coumarin and phenolics. *Recent Adv. Phytochem.* 4: 165-221.

Vivrette, N. J., and Muller, C. H. (1977). Mechanism of invasion and dominance of coastal grassland by *Mesembryanthemum crystallinum*. *Ecol. Monogr.* 47:301-318.

Walters, D. T., and Gilmore, A. R. (1976). Allelopathic effects of fescue on the growth of sweetgum. *J. Chem. Ecol.* 2:469-479.

Wang, T. S. C., Yang, T.-K., and Chuang, T. T. (1967). Soil phenolics acids as plant growth inhibitors. *Soil Sci.* 103:239-246.

Wang, T. S. C., Yeh, K.-L., Cheng, S.-Y., and Yang, T.-K. (1971). Behavior of soil phenolic acids. In *Biochemical Interactions Among Plants*. National Academy of Sciences, Washington, D.C., pp. 113-120.

Wang, T. S. C., Li, S. W., and Ferng, Y. L. (1978). Catalytic polymerization of phenolic compounds by clay minerals. *Soil Sci.* 126:15-21.

Wang, T. S. C., Kao, M.-M., and Huang, P. M. (1980). The effect of pH on the catalytic synthesis of humic substances by illite. *Soil Sci.* 129: 333-338.

Weaver, T. W., and Klarich, D. (1977). Allelopathic effects of volatile substances from *Artemisia tridentata* Nutt. *Am. Midl. Nat.* 97:508-512.

Webb, L. J., Tracey, J. G., and Haydock, K. P. (1967). A factor toxic to seedlings of the same species associated with living roots of the non-gregarious subtropical rain forest tree *Grevillea robusta*. *J. Appl. Ecol.* 4:13-25.

Welbank, P. J. (1960). Toxin production from *Agropyron repens*. In *Biology of Weeds*. Br. Ecol. Sco. Symp., Vol. 1, J. L. Harper (Ed.). Blackwell, Oxford, pp. 158-164.

Welbank, P. J. (1961). A study of the nitrogen and water factors in competition with *Agropyron repens*)L.) Beauv. *Ann. Bot.* 25:116-137.

Welbank, P. J. (1963). Toxin production during decay of *Agropyron repens* (couch grass) and other species. *Weed Res.* 3:205-214.

Wells, P. V. (1964). Antibiosis as a factor in vegetation patterns. *Science* 144:889.

Wender, S. H. (1970). Effects of some environmental stress factors on certain phenolic compounds in tobacco. In *Recent Advances in Phytochemistry*, Vol. 3. C. Steelink and V. C. Runeckles (Eds.). Appleton-Century-Crofts, New York, pp. 1-30.

Whitehead, D. C. (1963). Some aspects of the influence of organic matter on soil fertility. *Soils Fert.* 26:217-223.

Whitehead, D. C. (1964). Identification of p-hydroxybenzoic, vanillic, p-coumaric and ferulic acids in soils. *Nature (Lond.)* 202:417-418.

Whittaker, R. H. (1970). The biochemical ecology of higher plants. In *Chemical Ecology*, E. Sondheimer and J. B. Simeone (Eds.). Academic Press, New York, pp. 43-70.

Williams, A. H. (1960). The distribution of phenolic compounds in apple and pear trees. In *Phenolics in Plants in Health and Disease*, J. B. Pridham (Ed.). Pergamon Press, Oxford, pp. 3-7.

Wilson, R. E., and Rice, E. L. (1968). Allelopathy as expressed by *Heli-] anthus annuus* and its role in old-field succession. *Bull. Torrey Bot. Club 95*:432-448.

Winter. A. G. (1961). New physiological and biological aspects in the inter-relationships between higher plants. *Symp. Soc. Exp. Biol. 15*:229-244.

Wolfe, J. M., and Rice, E. L. (1979). Allelopathic interactions among algae. *J. Chem. Ecol. 5*:533-542.

Zapata, O., and McMillan, C. (1979). Phenolic acids in seagrasses. *Aquat. Bot. 7*:307-318.

Zenk, M. H., and Müller, G. (1963). In vivo destruction of exogenously applied indolyl-3-acetic acid as influenced by naturally occurring phenolic acids. *Nature (Lond.) 200*:761-763.

22

TOXINS OF PLANT-FUNGAL INTERACTIONS

JEFFREY B. HARBORNE

University of Reading, Reading, England

I. INTRODUCTION

Although the interactions between higher plants and fungi may take many forms, most fungi are inimical to higher plant survival in that they are frequently parasitic on them and depend on them for nutrients. In establishing an association with a host plant, a pathogenic fungus is capable of causing considerable damage in host tissue and indeed death of the plant may eventually ensue. Saprophytic fungi may then enter the plant tissue and complete the breakdown and recycling of the organic matter of the plant.

In this coevolutionary struggle for survival between plant and fungus, chemical agents or toxins may be involved at various stages of the interaction. Two major types of toxin may be discerned: those produced by the host plant to ward off fungal attack, and those produced by the fungus to enable it to establish itself within the host tissues.

In the present account of these toxins, the host plant defense toxins will first be considered. Such natural fungitoxic substances are of many types and may be present in different sites within the plant. Conveniently, pre-formed toxins may be divided into those which are generally lipid soluble and are present at the plant surface and those which are usually water soluble and are located within the plant tissue. Toxins are also formed de novo in plants in response to microbial infection; these are called phytoalexins and are by definition toxic to a fungal organism. Toxins produced by fungi as part of their pathogenic association with plants will then be considered and the different phytotoxins described according to the degree of complexity in their chemical structures. Major references which have been consulted in the preparation of this chapter include Friend and Threlfall (1976), Heitefuss and Williams (1976), Horsfall and Cowling (1979, 1980), and Wood et al. (1972).

II. FUNGITOXINS PRODUCED BY PLANTS

A. Toxins at the Plant Surface

1. Introduction

The first hurdle that a fungal organism has to overcome in establishing itself on a host plant is at the plant surface. The spores have to germinate and grow, so that eventually the fungal haustoria can penetrate into the epidermal cells. There is significant circumstantial evidence in the plant pathological literature indicating that toxins are often present on the plant surface, particularly on the upper surface of the leaf, which are inimical to the development of fungi. Such substances may either prevent spore germination or interfere with fungal growth. For example, Purnell (1971) was able to show that leaching the leaves of the swede plant with water removed unidentified toxins; such prewashed leaves were much more readily infected with *Erysiphe cruciferarum* than were control leaves. Also, Topps and Wain (1957) collected aqueous leaf washings from woodland tree species and demonstrated fungitoxicity in some of the ether-soluble fractions of these washings. Again, Schneider and Sinclair (1975) reported unidentified toxins in leaf surface diffusates of young leaves of *Vigna unguiculata* which inhibited conidial germination and germ tube growth of the fungal pathogen *Cercospora canescens*. Older leaves, by contrast, did not contain any active material.

Other indirect evidence that toxins may be present on leaf surfaces is provided by phytochemical studies of leaf waxes (Martin and Juniper, 1970). In addition to the normal hydrocarbons universally present in these waxes, many plants also contain lipid-soluble secondary constituents which may vary in structure and class from species to species or genus to genus. These secondary substances in the wax are generally phenolics or terpenoids and include a number of structures which are potentially inimical to fungal growth. The actual wax itself, if present in high concentration as in certain *Eucalyptus* species, may create such a predominantly hydrophobic environment that fungal spores which require moisture for germination are unable to develop.

In addition to toxic constituents in the wax, there is the possibility of toxins in leaf surface hairs or trichomes. Such toxins, if volatile, may be released continuously in small amounts so that whenever fungal spores land on the leaf, they will be unable to get established. Toxins may also be present in pigment incrustations on leaves or as farinaceous dusts present on some leaves (as in *Primula*).

In spite of the wealth of circumstantial evidence that plants gain a certain immunity from fungal attack by secreting toxins at the leaf or other surface, evidence that this defense is actually effective in vivo is often lacking. Experiments must be carried out, in any given interaction, in order to establish the exact location of the toxin at the leaf surface, its effectiveness as a fungitoxin and its presence in sufficient amount to represent a real barrier to infection. The number of examples where these requirements have been met are quite few. These are discussed in turn to indicate the range and type of toxin involved.

2. Terpenoids of Leaf Surfaces

The sesquiterpene lactones are one of the most biologically active classes of terpenoid known in higher plants. Nearly 1000 different structures have been described, largely from a single plant family, the Compositae (Heywood et al., 1977). Individual sesquiterpene lactones have variously been shown to be allelopathic, schistosomicidal, antiherbivore, cytotoxic, allergenic, or poisonous to mammals (Rodriguez et al., 1976). A number have antimicrobial activity and one particular structure, parthenolide [1], has recently been shown to be an important leaf surface toxin of *Chrysanthemum parthenium*.

[1]

In this plant, parthenolide occurs in special leaf surface glands and is the major antimicrobial constituent of the crude extract obtained by dipping leaves into chloroform. It also occurs on the surface of the seeds, where it has a similar antimicrobial role (Blakeman and Atkinson, 1979). Parthenolide is differentially toxic to microbes, a number of gram-positive bacteria and yeasts being killed when grown in a nutrient broth containing it. On the other hand, gram-negative bacteria and filamentous fungi are relatively insensitive to its toxic properties.

Field experiments have shown that parthenolide is potentially a useful protective agent, since when added to spore-containing droplets of two facultative pathogens on chrysanthemum petals and bean leaves, it reduced the level of infection. Several other toxins are present in the trichomes of *C. parthenium* leaves, but these have yet to be identified. Parthenolide is a germacranolide, and other sesquiterpene lactones of this type as well as the guaianolide parthenin have been shown to be antifungal (Rodriguez et al., 1976). A number of plants of the Compositae besides *C. parthenium* contain sesquiterpene lactones in leaf trichomes, so it is likely that yet other lactones will be found to have fungitoxic properties.

Diterpenoids have been identified as leaf surface toxins in members of another plant family, the Solanaceae, in fact in two tobacco species. Thus it has been found that *Nicotiana glutinosa* secretes two highly fungitoxic bicyclic diterpenes in the leaf wax. These occur in high concentration in liquid droplets which are visible under the microscope on the upper surface of the leaves. They have been identified as sclareol [2] and its 13-epimer, episclareol (Bailey et al., 1974). At a concentration of between 5 and 100 μg/ml these diterpenes inhibited the growth of 16 of 18 fungal species tested. In the case of the

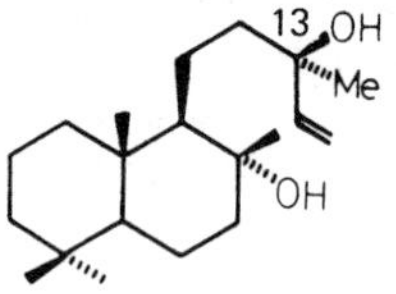

[2]

fungus *Alternaria longipes*, germ tube growth was restricted largely through an increase in hyphal branching. Thus sclareol could interfere with fungal growth as a hormonal analog and this is made more likely when one considers that sclareol has a structure which is at least distantly related to the gibberellin class of natural hormone, which are also diterpene-based. Finally, it may be noted that both sclareol and episclareol are toxic to fungi, so that the stereochemistry at carbon-13 cannot be involved in fungitoxicity.

The cultivated tobacco plant, *Nicotiana tabacum*, has also yielded a diterpene toxin from the leaf surface. It is structurally distinct from the diterpenes of *N. glutinosa*, since it is macrocyclic. The substance has been identified as 4,8,13-duvatrien-1,3-diol [3], which is present as a mixture of α- and

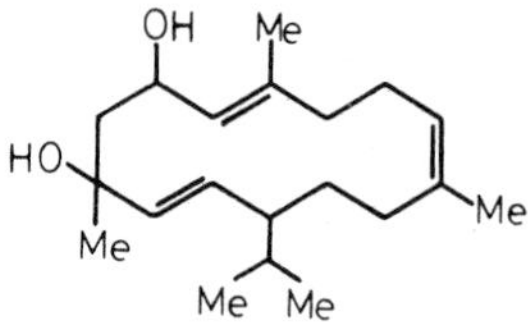

[3]

β-isomers (Cruickshank et al., 1977). This diterpene mixture accounts for more than 0.1% of the weight of fresh tobacco leaf and the ED_{50} value toward *Peronospora hyoscyami* f.sp. *tabacina* is about 20 μg/ml. There is evidence that the diterpene diol [3] plays a role in the epidemiology of blue mold on tobacco, since *N. debneyi*, which is much more susceptible to the disease than *N. tabacum*, lacks this diterpene in the wax. However, it is not the only source of blue mold resistance in the tobacco, since other factors are also clearly involved (Cruickshank et al., 1977).

3. Flavones and Isoflavones of Leaf Surfaces

Two fungistatic flavones, tangeretin [4] and nobiletin [5], have been isolated from *Citrus* leaves (Rutaceae) and shown to be responsible in part for resistance in citrus to the pathogenic fungus (*Deuterophoma tracheiphila*)

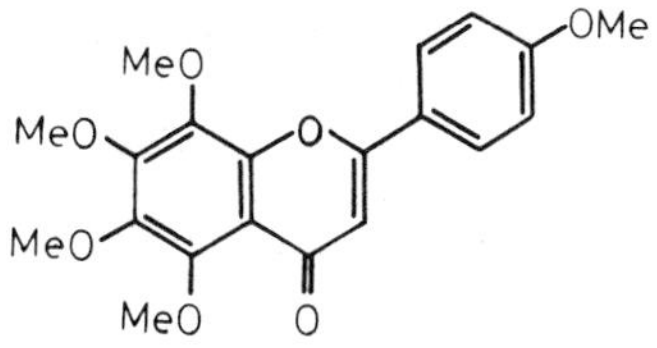

[**4**]

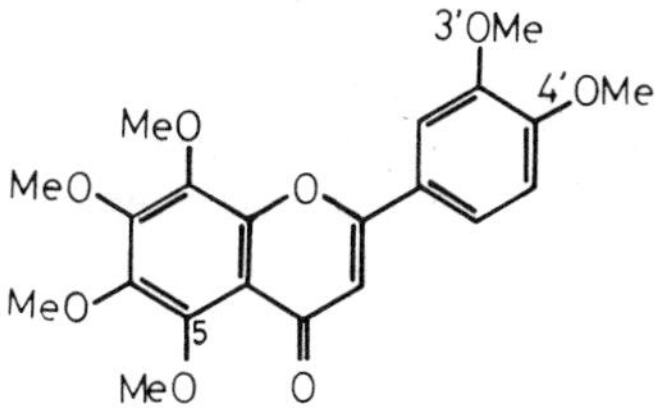

[**5**]

(Piattelli and Impellizzeri, 1971). This organism causes the disease known as mal-secco, which leads to the loss of foliage followed eventually by death of the trees. These two flavones are present in high concentrations in species such as *C. reticulata*, *C. sinensis*, and *C. volkamericana*, which are resistant to the disease, while being absent from the susceptible citrus species, *C. medica* and *C. aurantifolia*. However, they are not the only factors controlling the spread of this fungus in citrus, since another species, *C. aurantium*, contains both flavones but is yet susceptible to fungal attack.

The fungistatic properties of the two flavones [4] and [5] are presumably related to their high degree of methylation, since the closely similar 5-hydroxy-6,7,8,3',4'-pentamethoxyflavone or 5-demethylnobiletin is biologically inactive. Complete O-methylation per se is not, however, essential for activity, since 5,4'-demethylnobiletin is as fungistatic as nobiletin. Various methylated flavones, including [4] and [5], also occur in the peel of citrus fruits, where they may also have a protective function against fungal invasion.

The occurrence of tangeretin and nobiletin in the leaf wax, rather than within the leaf in the vacuoles, of citrus plants is still only presumptive. It is highly likely in view of the solubility properties of these flavones, but it has not been definitively proved by experiment. However, in the case of the two isoflavones, luteone [6] and wighteone [7] of *Lupinus albus*, they have

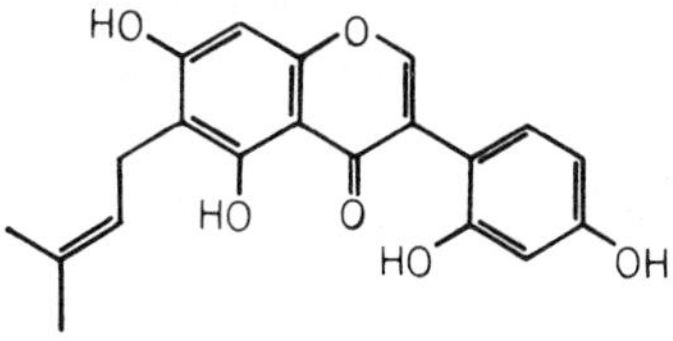

[**6**]

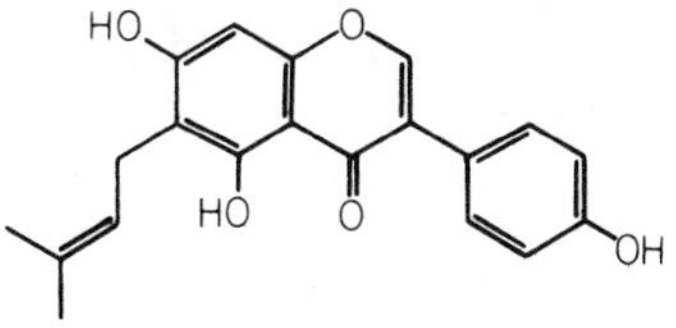

[**7**]

been positively identified as being leaf surface constituents (Harborne et al., 1976; Ingham et al., 1977). Fungitoxicity was established for these isoflavones when it was found that luteone has an ED_{50} value of 35-40 µg/ml on mycelial growth of *Helminthosporium carbonum*. Toxic effects on germ tube development and curled, distorted, and highly branched germ tubes were observed as this fungus developed on lupin leaves. Structurally, isoflavones take up a molecular shape which is similar to certain steroids, particularly the estrogens, and it is possible that the fungitoxicity associated with luteone is due to a disturbance of the permeability of the fungal cell membrane, which is known to be sensitive to the external application of steroid derivatives.

The isopentenyl side chain in luteone seems to be essential for fungitoxicity, since simple isoflavones such as formononetin (7-hydroxy-4'-methoxyisoflavone) are much less inhibitory to fungal growth.

Unlike the diterpenoids of *Nicotiana* (see Sec. II.A.2), these two isoflavones of lupin were also detected within the leaf, at similar concentrations to their occurrence in the wax. They are presumably synthesized in the epidermal cells and then extruded out onto the surface during leaf growth. Luteone also occurs in unripe seeds of *Lupinus albus*, where it may also have a protective role. Luteone and related isopentenyl isoflavones have been detected in leaf washings of about 11 other *Lupinus* species (Harborne et al., 1976), so that it is likely that there are similar fungitoxins throughout this genus. This type of microbial defense is, however, rather specialized within the Leguminosae as a whole, since the great majority of legume genera tested respond strongly to the induction of phytoalexins following fungal infection (see Sec. II.C).

4. *Miscellaneous Surface Toxins*

Other toxins have been isolated from plant tissues where there is some presumptive evidence of their presence in or near the plant surface. For example, Kawazu et al. (1973) isolated two antifungal compounds from the leaves of *Dendropanax trifidus* (Araliaceae), which were extracted into ethyl acetate from fresh leaves and thus they might have been in the wax. The compounds were identified as the acetylenic derivatives: *cis*-9,17-octadecadien-12,14-diyn-1,16-diol and R-(-)-16-hydroxy-*cis*-9,17-octadecadien-12,14-diynoic acid. They cause total inhibition of germination in *Cochliobolus miyabeanus* at concentrations of 12.5 and 50.0 μg/ml, respectively. Again, Tripathi et al. (1978) reported lawsone (2-hydroxynaphthoquinone) as being the fungitoxic principle of leaves of *Lawsonia inermis*; since it is essentially lipid soluble, lawsone might be expected to occur at the surface. Its role in preventing disease in this plant is suspect in view of the high concentration (1000 ppm) required before toxic symptoms appear in fungi in in vitro tests.

Plant roots often show significant resistance to fungal disease and it is possible that substances in or near the surface may provide such resistance. Van Wyck and Koeppen (1974) have reported, for example, that p-hydroxybenzoylcalleryanin, a phenolic, occurs to the extent of 5-8% dry weight in root bark of *Protea cynaroides* and is active against the root bark fungus *Phytophthora cinnamomi*.

Another root constituent, the lactone borbonol, has been reported in species of *Persea*, where it also provides toxicity to the same fungus that attacks *Protea* (Zaki et al., 1980). Borbonol inhibits vegetative growth of *Phytophthora cinnamonii* at concentrations as low as 1 μg/ml. Alternatively, it is possible that fungitoxic substances may be exuded from root hairs as a protection against infection. Burden et al. (1974) have detected several fungitoxic isoflavans and pterocarpans in pea and bean root exudates, in plants grown in a nonsterile aqueous medium.

Perennial plants with subterranean storage organs are liable to microbial attack and there is evidence of protective substances being present in the outer layers of such organs. It has long been known that catechol and protocatechuic acids, two highly active antimicrobial phenols, occur in the outer scales of onions. When present in sufficient concentration, they provide resistance in some onion varieties to the smudge disease *Colletotrichum cir-cinans* (Walker and Stahmann, 1955). More recently, the peel of yam tubers, *Dioscorea rotundata*, have yielded two antifungal phenanthrenes, hircinol [8] and 7-hydroxy-2,4,6-trimethoxyphenanthrene [9]. These substances do not

[8]

[9]

generally prevent spore germination but instead they inhibit germ tube growth, with ED_{50} values between 16 and 100 $\mu g/ml$. Their specific localization in the peel, but not the flesh, of the yam tuber provides resistance to soft rot fungi, such as *Cladosporium cladosporioides*, which are only able to infect the plants after tissue damage (Coxon et al., 1982).

It has recently been shown that phenanthrenes related to [9], called batatosins, are widespread in tubers of members of the Dioscoreaceae (Ireland et al., 1981). It is also to be noted that the dihydrophenanthrene hircinol and several related structures have been detected in orchid bulbs, but only after infection (i.e., as phytoalexins) (see Sec. II.C). Clearly, the hydroxy-phenanthrene structural basis is an important means of providing fungitoxicity, under varying circumstances, in plants of both the Dioscoreaceae and Orchida-ceae.

One further situation where substances present in surface layers may be active against fungal attack is that of gymnosperm trees, where the bark normally yields an oleoresin rich in terpene materials. How far the volatile mono- and sesquiterpenes present in these oleoresins provide fungitoxicity is not yet clear. However, Rockwood (1974) found some evidence that high levels of β-phellandrene, a sesquiterpene present in the oleoresin of slash pine *Pinus elliottii*, were correlated with resistance to fusiform rust, *Cro-nartium fusiforme*, in this tree. By contrast, in vitro tests with other fungi, such as *Fomes annosus* and *Ceratocystis* spp., suggest that the monoterpenes myrcene and limonene are generally more inhibitory to fungal growth than β-phellandrene (Cobb et al., 1968).

B. Bound Toxins

1. Chemical Nature

A familiar concept in ecological biochemistry is the storage within plant cells of toxins needed for their protection, in safe, inactive, bound form. It applies to toxins produced to deter herbivores such as cyanogenic glycosides, and to toxins serving as allelopathic agents. There is increasing evidence that disease resistance in plants is afforded by a similar mechanism. The toxins involved are sometimes referred to as postinhibitins (Ingham, 1973), although the bound form is normally present prior to infection. Microbial invasion triggers an enzymic modification of the bound form with the release, around the site of invasion, of the active toxin. A number of examples of postinhibitins have now been reported (Table 1), involving several different types of toxin (see Schönbeck and Schlösser, 1976). Enzymes concerned in the release of the toxin are mainly specific glucosidases, but at least one oxidase and one amino acid lyase have also been implicated.

A most convincing, but complex example of bound toxins is that of the 1-tuliposides A and B [10, 11], which occur in young tulip bulbs and provide

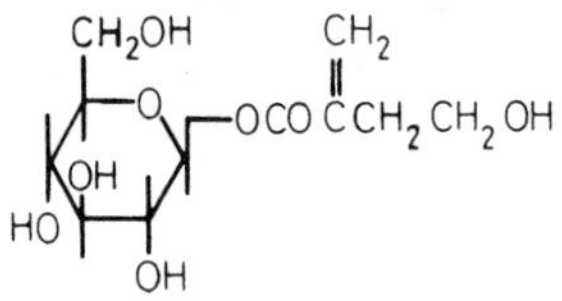

[10] 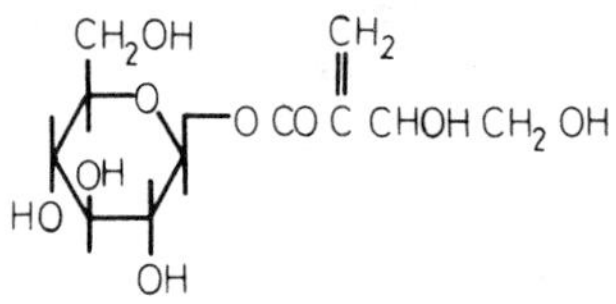   [11]

them with resistance to the pathogen *Fusarium oxysporum* until harvest time. The two glucosides are themselves only mildly antibiotic but on fungal invasion, rearrange to the isomeric but inactive 6-tuliposides. These, in turn, undergo enzymic hydrolysis, followed by cyclization, to yield the highly active tulipalins A and B [12, 13]. The fungitoxicity of these two lactones

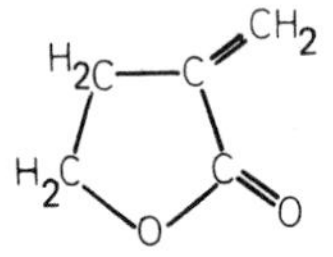

[12] 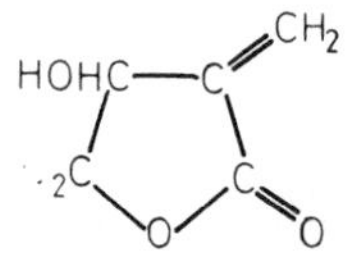   [13]

is thought to be related to their ability to complex with SH groups and thus to inhibit enzymic activities within the attacking fungus (Beijersbergen and Lemmers, 1972). The tuliposides are relatively widespread in Liliaceae (Slob et al., 1975) and hence may contribute to disease resistance in other plants besides tulip, although this has yet to be demonstrated.

Another example of the release of a bound toxin is that of the leek oils found in *Allium* species. Intact tissues contain the nonvolatile sulfur amino acid alliin [14]. On maceration, this amino acid is rapidly attacked by the enzyme allicin-lyase with the production within a few minutes of pyruvic acid, ammonia, and the antimicrobial toxin allicin [15]. Allicin then undergoes further modification with loss of oxygen to give diallyl disulfide [16], the major

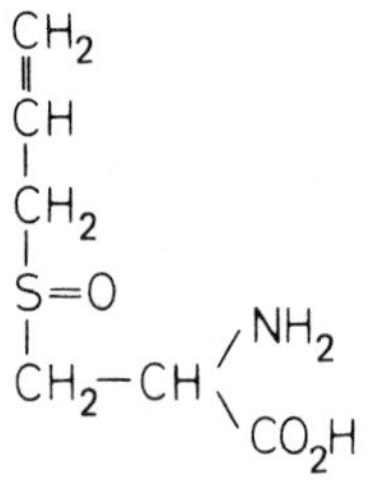

[14] 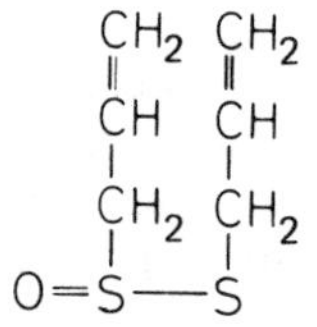   [15] 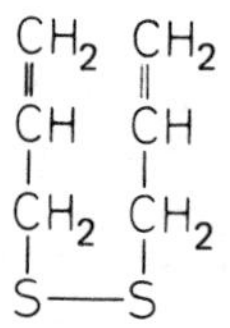   [16]

TABLE 1 Toxins Present in Plants in Bound Form

Bound Form	Toxin	Enzyme(s) controlling release	Plant and potential pathogen(s)
1. Glucosinolates Sinigrin	Allyl isothiocyanate	Myrosinase	*Brassica oleracea* (*Peronospora parasitica*)
2. Cyanogenic glycosides Linamarin	HCN	Linamarase	*Lotus corniculatus* (various)
3. Lactonic glucosides 1-Tuliposides A, B	Tulipalin A, B	Hydrolase	*Tulipa* cvs. (*Fusarium oxysporum*)
Ranunculin	Protoanemonin	Glucosidase	*Ranunculus* spp.
4. Saponins Hederasaponins B, C	α- and β-Hederin	Glycosidase	*Hedera helix* (various)
Avenacosides A, B	26-Desglucoavenacosides	Glycosidase	*Avena sativa* (*Fusarium avenaceum*)

5. Benzoxazolinones
   2-Glucosyloxy-4-hydroxy-
   1,4-benzoxazin-3-one
	2-Benzoxazolinone	Glucosidase	*Secale cereale*
			(*Fusarium nivale*)

6. Sulfur compounds Alliin	Allicin, diallyl disulfide	Alliin lyase	*Allium* spp. (various)
7. Phenolic quinones Phloridzin	Hydroxyphloretin quinone	Phenolase	*Malus pumila* (*Venturia inaequalis*)

volatile of garlic oil. Both allicin and the disulfide are antifungal (Durbin and Uchytil, 1971). Diallyl disulfide inhibits *Phytophthora cactorum* in liquid culture at a concentration of 0.01%. Most *Penicillium* species are also inhibited by either the disulfide or allicin. The role of these sulfur toxins in disease resistance in *Allium* is not yet clear, and there is good evidence that certain soil fungi are actually stimulated to grow in their presence. Thus the sclerotia of *Sclerotium cepivorum* are stimulated to germinate by the release of diallyl disulfide from the roots of onion and garlic (Coley-Smith, 1976).

2. *Mode of Enzymic Release of Toxin*

The active toxin is normally released enzymically, with the enzyme being either of plant origin or derived from the invading microorganism. In most cases that have been studied, the former seems to be true. Certainly, with both glucosinolates and cyanogenic glycosides, specific enzymes for the release of the respective toxins—isothiocyanate and hydrogen cyanide—always accompany the glycosides wherever they occur in plants. Thus simple disruption of the plant cellular material, as must occur when fungal haustoria penetrate within the leaf, is probably enough to bring the bound toxin in contact with the requisite enzyme and hydrolytic cleavage of the sugar residue then rapidly occurs.

In the case of the various saponin postinhibitins (Table 1), the situation is more complex, since the toxic principle is usually an intermediate glycoside rather than a completely sugar-free molecule. Most plant saponins have several sugars attached to the triterpenoid skeleton, with substitution of sugars at one, two, or more reactive positions. The hederasaponins B and C of ivy leaves (*Hedera helix*) are no exception. Schlösser (1973) has shown that these two substances [17] and [18] provide disease resistance, because

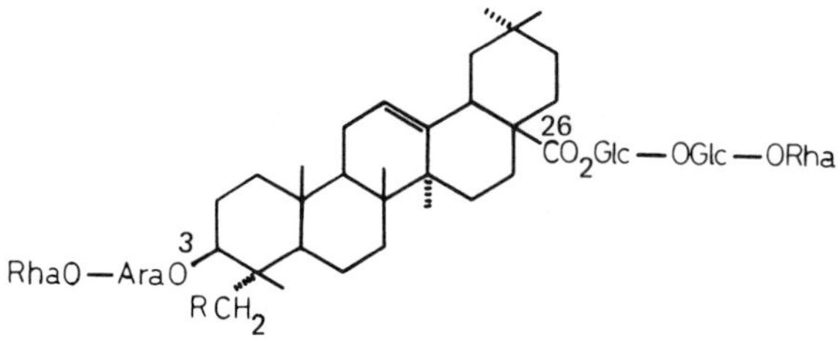

$$[17] \quad R = H$$

$$[18] \quad R = OH$$

when ivy leaves are damaged by penetration of a parasite, the two saponins undergo partial hydrolysis, with specific loss of the sugars attached to the 26-carboxylic acid group. The α- and β-hederins thus formed are highly toxic to a range of fungi. The in vivo concentration in plant juice can reach 150 mg/ml, which is a powerful barrier considering that most fungi are completely inhibited by a α-hederin at 50-250 μg/ml. Significantly, if further hydrolysis of the ivy leaf saponin occurs and the sugars in the 3-position are lost, fungitoxicity completely disappears. While α-hederin provides protection

from most fungi, *Pythium* and *Phytophthora* are exceptional in being resistant
to these saponins. Since pythiaceous fungi differ from most others in having
membranes free of sterols, it appears that the toxicity of α- and β-hederin to
the majority of fungi is due to their interference with fungal cell membranes
at particular sites where sterols are attached.

The *Hedera helix* example explains why chemical barriers to fungal para-
sitism are rarely complete and why infection may occur in spite of the presence
of resistance factors. Saponin protection is only effective against fungi with
a certain common type of cell wall membrane and mutation to a modified membrane may be sufficient to impart immunity from the saponin's toxic effects.
From the structure of the hederins, it is apparent that the requirements for
fungitoxicity lie in a degree of water solubility linked to the presence of active
sites (e.g., a free 26-carboxylic acid group) elsewhere in the molecule.

A wide range of saponins are known in plants (Tschesche and Wulff, 1973),
with many different sugar residues and some with nitrogen atoms (also called
steroidal alkaloids) are also important in disease resistance. Undoubtedly,
some of these occur naturally in a form that is highly fungitoxic without enzymic modification. They may well be barriers to fungal infection, as seems
to be true of cyclamin, a saponin present in *Cyclamen* tissues, which is glyco-
sylated only at the 3-hydroxyl position (Schlösser, 1971). It is probably true
of the alkaloids tomatine of tomato leaves and solanine of potato leaves. In the
case of these two alkaloids, fungitoxicity, at least in vitro, is highly dependent
on the pH of the medium and modification of pH in vivo may determine whether
or not they are able to prevent fungal invasion (Schlösser, 1975). Both
tomatine and solanine are toxic at fairly low concentrations to a range of fungal organisms.

Phenolic compounds occur widely in plants in glycosidic form and are potentially an important class of bound toxin, since the free phenols in in vitro
tests show considerable fungitoxicity. Convincing evidence that natural
phenols are important in protecting plants from fungal invasion is generally
lacking. One substance with which fungitoxicity is definitely associated is
phloridzin, the dihydrochalcone glucoside [19] which is characteristically

[19]

present in the apple genus, *Malus* (Rosaceae). Toxicity is associated not
with the aglycone phloretin released by β-glucosidase hydrolysis but with the
oxidation product, hydroxyphloretin o-quinone, produced by phenolase activity on phloretin. This quinone is undoubtedly fungitoxic to the apple scab
organism, *Venturia inaequalis*. Attempts to prove it has an in vivo toxic role
in protecting apples from scab have not been successful, but this is largely
because of its great instability (see Hunter, 1975).

As with certain saponins such as cyclamin (see above) certain natural
phenolics may be fungitoxic in bound form as glucosides. This is apparently

true of hordatine A and its methoxy analog hordatine B, two phenolics synthesized by young barley shoots (Stoessl, 1967) and which are present in gluco-
sidic form [20]. These compounds inhibit spore germination of *Monilinia*

[20]

fructicola at 10 ppm. Curiously, these substances lose their effectiveness in
warding off the barley pathogen *Helminthosporium sativum* as the seedlings
develop. This is due to the accumulation of Ca^{2+} and Mg^{2+} ions in the seedlings and inactivation of the antifungal agents as metal complexes.

3. Detoxification by Fungi

Fungi are known to have the ability to detoxify some of the toxins mentioned in Table 1. Thus HCN released from cyanogenic leaf glycosides of
Lotus corniculatus does not prevent invasion and colonization by the leaf-
spot pathogen *Stemphylium loti* because this organism produces the enzyme
formamide hydrolyase, which detoxifies HCN by converting it to formamide,
$HCONH_2$ (Fry and Millar, 1971, 1972). Attempts to relate resistance in flax,
Linum usitatissimum, to *Fusarium lini* to the concentration of linamarin present have also failed (Trione, 1960). Nevertheless, considering the toxicity
of HCN to many microorganisms, it is difficult to believe that these cyanogens do not provide the plant with some protection from fungal infection.

Presumably, isothiocyanates produced from glucosinolates must also be
susceptible to detoxification by microbes, although this has yet to be established. Certainly, mustard oils are effective antifungal agents in *Brassica*
cultivars. Thus allyl isothiocyanate released from sinigrin in *Brassica
oleracea* protects the plant from downy mildew *Peronospora parasitica* as long
as the concentration at the site of infection is above 450 μg per gram of dry
weight (Greenhalgh and Mitchell, 1976).

Detoxification of saponins may occur through fungal β-glycosidases, which
are either produced within the fungal tissue or else diffuse out in advance of
the developing hyphae. A dramatic example of fungal detoxification of saponin
is the case of *Avena* leaves, which contain a bound toxin with sugars attached
both at C-3 and C-26 [21]. Release of the single glucose at C-26 by a specific
plant enzyme gives rise to a saponin molecule still carrying a sugar at C-3
which is highly active against fungal development. This intermediate may provide protection to oat leaves from many potential pathogens. It is not,

[21]

however, a complete panacea since no active saponin can be detected when oat leaves have been infected with the pathogenic fungus *Helminthosporium avenae*. In this case, it has been clearly shown that the fungus produces, both in vivo and in vitro, a highly specific β-glucosidase which removes the sugar at C-3, leaving behind a water-insoluble and hence inactive toxin. Pathogenicity of of *H. avenae* must be at least partially dependent on the presence of this enzyme of fungal manufacture (Lüning and Schlösser, 1976). Several other examples of enzymic hydrolysis of saponins by pathogenic fungi are also known (see Schönbeck and Schlösser, 1976).

In conclusion, one may observe that bound toxins are probably more effective antifungal agents than substances at the leaf surface. They differ in being water rather than lipid soluble. To overcome such barriers, fungi have to develop special detoxifying enzymes. A considerable degree of enzyme specificity is implied in the case of the saponins, since the sugar residues that have to be removed may be oligosaccharides in which the nature and number of sugars, as well as their linkages, can vary considerably. A simple change in structure within the sugar moiety can immediately produce a saponin, which the fungus can no longer detoxify. The amount of resistance to fungal modification that can be programmed by the plant into complex molecules such as saponins is thus very considerable.

C. Phytoalexins

1. Definitions

Phytoalexins are toxins which are synthesized de novo by higher plants in response to fungal attack and which accumulate in plant tissues around the site of infection. They are, by definition (Müller and Börger, 1941), fungitoxic at relatively low concentrations to a wide range of organisms. Under favorable circumstances, they are capable of preventing the colonization of the plant by the fungus. Phytoalexins are occasionally formed in plants in response to bacterial infection and are also sometimes produced in tissues subjected to stress (e.g., illumination by ultraviolet radiation, treatment with metal salts, low-temperature shock, etc.). Nevertheless, they are characteristically produced in quantity in response to fungal invasion and their primary role in plant defense would seem to be against fungal parasites.

Phytoalexin formation has been shown to take place in a wide range of plants from over 20 different families and it is a reasonable assumption that most higher plants are probably capable of producing these phytoalexins as antifungal agents. The chemistry of the phytoalexin response, however, has only been fully explored in a relatively small number of species. Phytoalexin synthesis is a response of living plant cells, and although mainly observed in leaf tissues, it can be induced in hypocotyl, stem, root, and seed. Synthesis is quite rapid, being detectable within a few hours of infection and reaching a maximum within 1-2 days. To avoid further metabolism of the phytoalexin produced (see Sec. II.C.5), it is preferable to use a nonpathogenic fungus for induction and *Helminthosporium carbonum* and *Botrytis cinerea* are particularly useful test organisms for studying phytoalexin accumulation. A single fungitoxic substance may be produced, but more usually several different substances of varying fungitoxicities are formed in any given plant.

Much has been written elsewhere on phytoalexins (see, e.g., Ingham, 1972; Harborne and Ingham, 1978; VanEtten and Pueppke, 1976) and a monograph dealing with these has recently appeared (Bailey and Mansfield, 1981). Only certain aspects of these toxic compounds are discussed here, in particular their chemical nature, natural distribution, fungitoxicity, and further metabolism.

2. Chemical Variation

About 100 phytoalexins have been characterized in plants to date. The first to be identified as such was the pterocarpan pisatin [22] from the pea

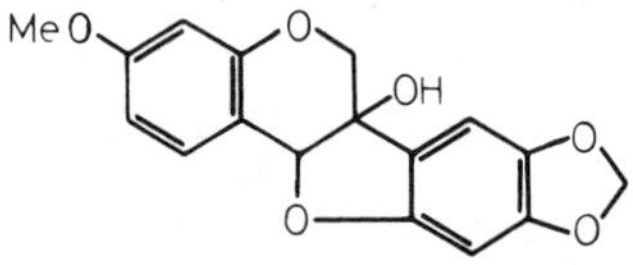

[22]

Pisum sativum (Cruickshank and Perrin, 1960) and even today pterocarpans and related isoflavonoids comprise the largest single group of known phytoalexins. Many related shikimate-derived structures have also been described (Table 2). The great majority of isoflavonoids and related compounds are phenolic, but in general, they have a low polarity and lipid solubility, since most of the potential phenolic groups in the molecule are masked by O-methylation or methylenedioxy group formation. Lipid solubility may be enhanced in the isoflavonoid and benzofuran series by isopentenyl substitution, as for example in the phytoalexin phaseollidin [23], from *Phaseolus vulgaris* (Leguminosae). Such an isopentenyl substitution may be present alternatively in

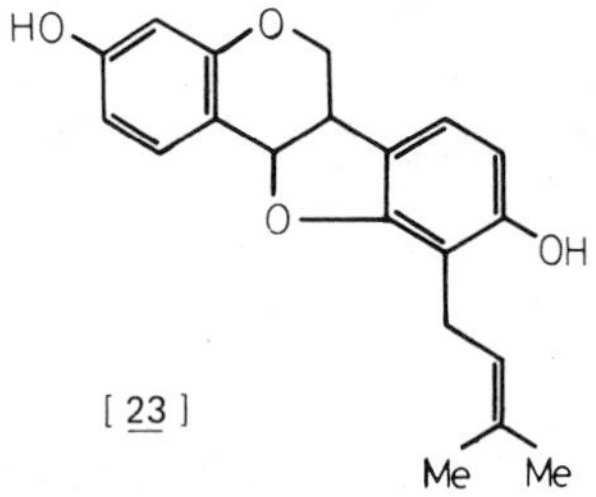

[23]

TABLE 2 Structural Classes Among Phytoalexins

Shikimate derived	Isoprene derived
Isoflavonoid	Sesquiterpenoid
Isoflavone	Sesquiterpene
Isoflavone	Sesquiterpene furanolactone
Pterocarpan	Sesquiterpene dimer
Isoflavan	
	Diterpenoid
Flavan	Diterpene
Benzofuran	
Stilbene	
Phenanthrene	Fatty acid derived
Phenylpropanoid	Polyacetylene
1,3-Diphenylpropanoid	Furanoacetylene
Lignan	
Chromone	
Isocoumarin	Amino acid derived
Furanocoumarin	Benzoxazin-4-one
Phenolic acid	

ring-closed form to produce a furano- or pyranoisoflavonoid. Glyceollin I [24], a phytoalexin of the soya bean *Glycine max*, is one such example. However, within the Leguminosae, the most common phytoalexin is a pterocarpan with simple hydroxy-methoxy substitution, namely medicarpin [25]. This pterocarpan was first described from lucerne *Medicago sativa* but it has subsequently been detected in many other legumes (Ingham, 1981).

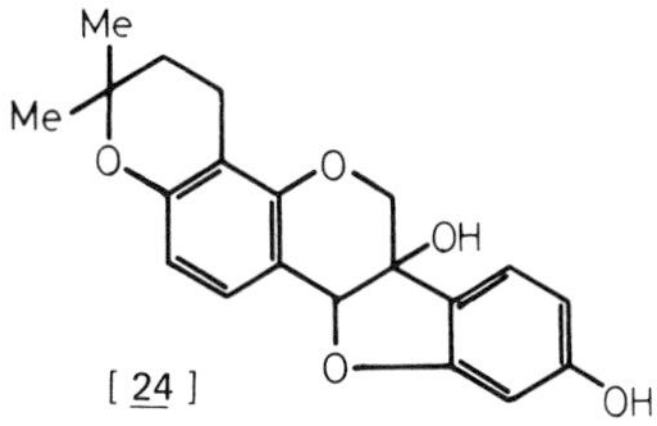

[24] 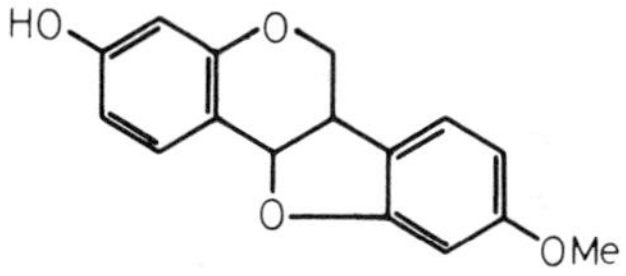   [25]

With so many different structural types known among the phytoalexins, no direct relationship between chemical structure and fungitoxicity is apparent. However, within the shikimate-derived molecules, there are some clear structural analogs. One may note, for example, that hydroxystilbenes such as resveratrol [26], recently recognized as a phytoalexin in *Arachis hypogaea*

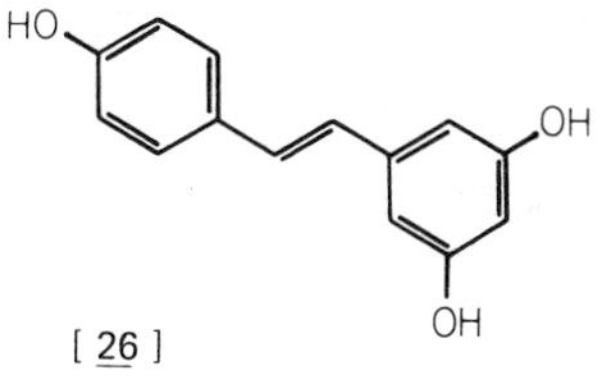

[26]

(Leguminosae) and in *Vitis vinifera* (Vitaceae), show a close structural relationship in their naturally occurring trans forms to the isoflavonoid skeleton
(compare medicarpin [25]). Also, benzofurans such as vignafuran [27] from
Vigna unguiculata are structurally related to the isoflavonoids. Again, the
dihydrophenanthrenes such as orchinol [28], a phytoalexin of *Orchis militaris*

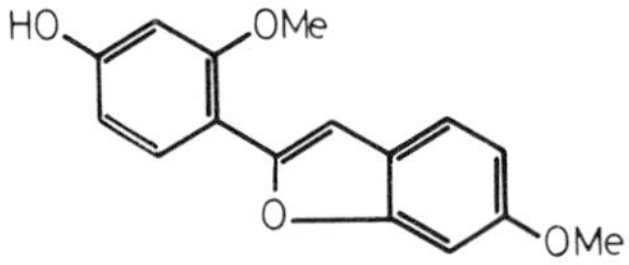

[27]

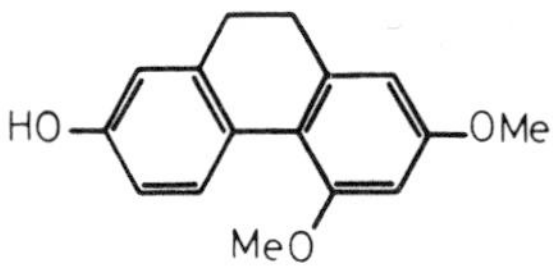

[28]

(Orchidaceae), are closely related biosynthetically to the hydroxystilbenes
(Gorham, 1980).

Quite simple structures are formed on occasion as phytoalexins. For example, coniferyl alcohol [29] and the related coniferaldehyde are reported to
be induced fungitoxins in the flax plant *Linum usitatissimum*. The simple
5,7-dihydroxy-3-ethylchromone [30] is formed, together with pisatin [22] as

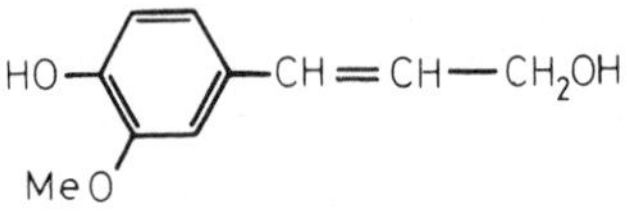

[29]

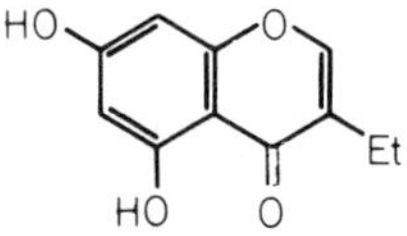

[30]

a phytoalexin in *Lathyrus odoratus*. The simplest of all phytoalexins is benzoic acid [31], the synthesis of which is induced in apple fruit following
infection with the storage rot organism, *Nectria galligena*.

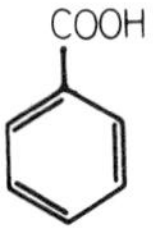

[31]

A second major group of phytoalexins are those that are terpenoid-based
(Table 2). They are characteristically formed and are widespread in members of the Solanaceae (Stoessl et al., 1976). A variety of oxygenated sesquiterpenes have been detected in the potato *Solanum tuberosum*, tobacco
Nicotiana tabacum, and tomato *Lycopersicon esculentum*. They range from
the bicyclic sesquiterpene alcohol rishitin [32] from potato tubers and tomato
plants to the structurally more unusual spiro derivative, solavetivone [33],
also from the potato.

Among the diterpenoid phytoalexins, casbene [34] is distinctive in its
macrocyclic structure. It is unique among known phytoalexins in that fungal
infection of castor bean seedlings induces not the phytoalexin itself but the
production of an enzyme system for its synthesis. It is necessary to supply
the seedlings with mevalonate as substrate before casbene can be isolated in
any quantity (Sitton and West, 1975).

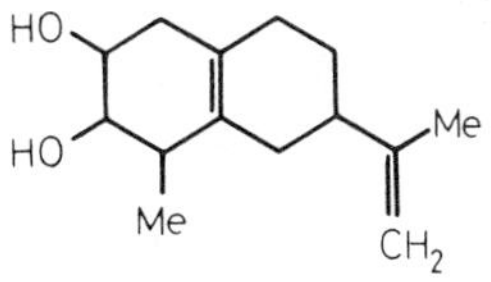

[32]

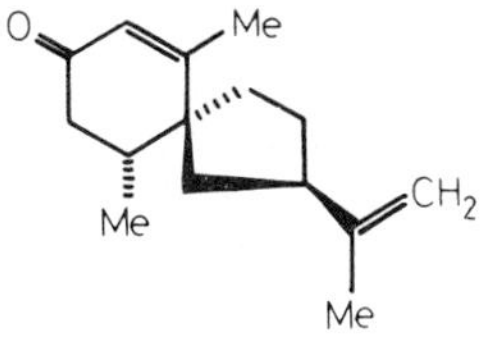

[33]

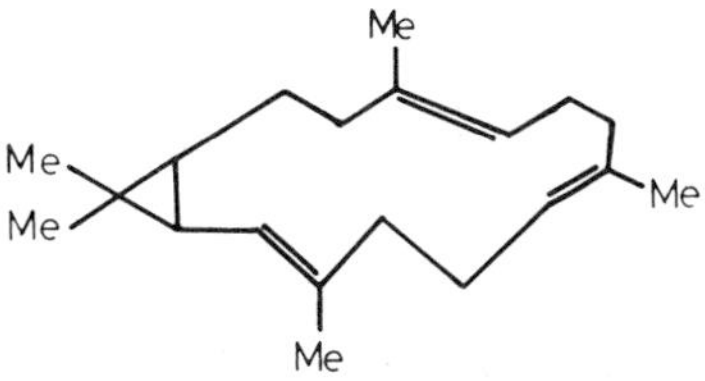

[34]

A third, small group of phytoalexins are the fatty acid derived acetylenic compounds. Simple acetylenes, such as safynol [35], from safflower,

$$Me-CH-CH-(C \equiv C)_3-CH=CH-CHOH-CH_2OH$$

[35]

Carthamus tinctorius, may be characteristic phytoalexins of the Compositae, although this has yet to be established by surveys. Otherwise, acetylenes seem to be mainly formed anomalously in families where the major phytoalexin response produces other structures. Thus the production of wyerone acid [36] and the related methyl ester wyerone and epoxide (at the terminal double

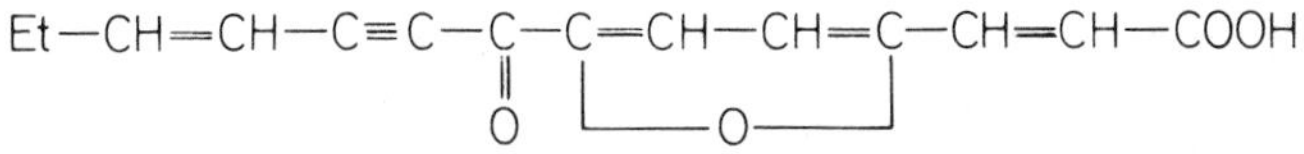

$$Et-CH=CH-C \equiv C-\underset{\substack{\|\\O}}{C}-\underset{\substack{|\\ \underline{\hspace{1em}}O\underline{\hspace{1em}}}}{C}=CH-CH=C-CH=CH-COOH$$

[36]

bond) in *Vicia faba* is unusual in a family (the Leguminosae) where isoflavonoids are the main phytoalexins. Wyerone derivatives are, in fact, produced widely in the genus *Vicia* and the related *Lens*, but not by *Lathyrus* or *Pisum* species, although they are members of the same tribe of the Leguminosae as *Vicia* and *Lens* (Robeson and Harborne, 1980). Recently, three acetylenic phytoalexins have been reported in tomato inoculated with *Cladosporium fulvum* (DeWit and Kodde, 1981), namely falcarinol [37] and falcarindiol, two

$$CH_2=CH-CHOH-(C \equiv C)_2-CH_2-CH=CH-(CH_2)_6-Me$$

[37]

known compounds, and the new compound tetradeca-6-en-1,3-diyn-5,8-diol. This is the first report of acetylenics in a family, the Solanaceae, which otherwise produces only sesquiterpene-based phytoalexins.

Finally, mention must be made of the first nitrogen-containing phytoalexins, avenalumins I, II, and III, from fungally infected barley leaves
(Mayama et al., 1981). Avenalumin I [38] is a benzoxazin-4-one, derived bio-

[38]

biogenetically from hydroxyanthranilic acid condensed with a p-coumaroyl
residue. The discovery of these nitrogenous phytoalexins in barley *Avena
sativa* contrasts with the earlier report of diterpene phytoalexins, momilactones A and B, from rice, *Oryza sativa* (Cartwright et al., 1981). Clearly,
there may be several different types of phytoalexin response in the family
which barley and rice belong (i.e., the Gramineae). Other grasses, however,
have yet to be studied in this respect.

3. Natural Distribution

Phytoalexin induction has now been studied in at least 15 plant families
and, from chemical studies of the substances formed on fungal infection, it
is apparent that every family differs in the nature of the characteristic fungitoxins formed (Table 3). The toxins produced in this way are typical secondary compounds in every known case. Furthermore, they are likely to be
similar, at least in biosynthetic type, to substances which occur constitu-
tively in that family or in a related family. In nearly every case, however,
the actual phytoalexins formed differ substantially from the characteristic
naturally occurring substances known to accumulate in that family.

This is true, for example, in the Leguminosae. Although isoflavonoids
are widespread as natural products in this family, the major type to accumulate are isoflavones, while pterocarpans and isoflavans are comparatively rare.
On the other hand, in respect to the phytoalexins produced, isoflavones are
quite rare and indeed are generally ineffective as fungitoxins. It is the
pterocarpans and isoflavans which provide the most usual phytoalexin response, these two classes of isoflavonoid being extremely effective against
many fungal pathogens.

While many of the substances that have been characterized as phytoalexins
have been described as naturally occurring (but always in a different plant
from that in which they are induced by fungal infection), a significant number are quite novel and are not known as natural products. The furanoacetylene, wyerone, is not known from any other source, for example, and it
is particularly distinctive since acetylenics as a class are not reported to
occur naturally in members of the Leguminosae. The chromone, lathodoratin,
is also a unique structure since it has never been recorded as a natural plant
product.

The data of Table 3 are based on only limited plant surveys and it is quite
likely that other responses may be recorded in any of the families listed.

TABLE 3 Characteristic Phytoalexin Responses According to Plant Family

Family	Phytoalexin type	Example[a]
Amaryllidaceae	Flavan	7-Hydroxyflavan from daffodil, *N. pseudonarcissus* (*Narcissus*)
Chenopodiaceae	Isoflavonoid	Betavulgarin from sugar beet, *B. vulgaris* (*Beta*)
Compositae	Polyacetylene	Safynol from safflower, *C. tinctoria* (*Carthamus, Dahlia*)
Convolvulaceae	Sesquiterpenoid furanolactone	Ipomeamarone from sweet potato, *I. batatas* (*Ipomoea*)
Euphorbiaceae	Diterpene	Casbene from castor bean, *R. communis* (*Ricinus*)
Gramineae[b]	Diterpene	Momilactone A from rice, *O. sativa* (*Oryza*)
Leguminosae[b]	Isoflavonoids	Pisatin from peas, *Pisum sativum* (*Many*)
Linaceae	Phenylpropanoid	Coniferyl alcohol from flax, *L. usitatissimum* (*Linum*)
Malvaceae	Terpenoid polyphenol	Gossypol from cotton, *G. barbadense* (*Gossypium*)
Moraceae[b]	Benzofuran	Moracin C from mulberry, *M. alba* (*Broussonetia, Morus*)
Orchidaceae	Hydroxyphenanthrene	Orchinol from military orchid, *O. militaris* (*Loroglossum, Orchis*)
Rosaceae	Benzoic acids	Benzoic acid from apple, *M. pumila* (*Malus, Prunus*)
Solanaceae[b]	Oxygenated sesquiterpenes	Rishitin from potato, *S. tuberosum* (*Capsicum, Datura, Lycopersicon, Nicotiana, Solanum*)
Umbelliferae[b]	Isocoumarins	6-Methoxymellein from carrot, *D. carota* (*Daucus*)
Vitaceae	Hydroxystilbene	Resveratrol from grape vine, *V. vinifera* (*Vitis*)

[a]Literature references can be found in Ingham (1972), in Harborne and Ingham (1978), or in the text.
[b]Other types of response have been recorded in these families.

Indeed, only one family, the Leguminosae, has been thoroughly examined for
its phytoalexins (Ingham, 1981). Over 500 species have been surveyed,
representing the great majority of tribes in this large plant group. It has
been found that most species produce pterocarpans and/or isoflavans, and
these isoflavonoids are entirely typical for the legumes. Nevertheless, a number of minor structural types have been isolated as phytoalexins, namely isoflavones, isoflavanones, stilbenes, benzofurans, and furanoacetylene. In
some cases, the phytoalexins produced are only of the minor type; in *Cajanus
cajan* fungal infection causes the production of only isoflavones and isoflavanones. More frequently, minor types accompany the production of a pterocarpan or isoflavan. For example, in *Lathyrus odoratus*, the chromones
lathodoratin and its 7-methyl ether are formed, together with large amounts
of pisatin, during phytoalexin induction.

There is increasing evidence that other plant families besides the Leguminosae form more than one type of phytoalexin. Thus in the Moraceae, only
two taxa have been examined, the mulberry tree *Morus alba* and the paper
mulberry, *Broussonetia papyrifera* (Takasugi et al., 1978, 1979, 1980). In
spite of this, 10 phytoalexins representing four different structural types
have been recorded: stilbenes, benzofurans, flavans, and the related ring-
opened 1,3-diphenylpropane derivatives. The complexity of phytoalexin synthesis in this case is further complicated by the presence of isopentenyl substituents in some but not all the compounds produced.

Two other families in which more than one type of phytoalexin response
have been recorded are the Solanaceae and the Gramineae; their phytoalexins
have already been outlined in Sec. IIC.2. One other family appears also to
be variable in its response, the Umbelliferae. The evidence for this, however,
is still not certain, as there have been considerable difficulties in determining,
in the case of umbellifer species, whether a particular substance is a genuine
phytoalexin or a natural product of either the fungus or the higher plant.
At one time, considerable controversy centered around whether the isocoumarin 6-methoxymellein formed in carrot root infected by *Ceratocystis
fimbriata* was a genuine phytoalexin, since closely related isocoumarins were
obtained from pure cultures of this and related fungi. However, the production of the isocoumarin abiotically in carrot tissues supports its classification
as a phytoalexin, and this view has now been generally accepted (see Harborne and Ingham, 1978). Other antifungal compounds produced in carrots
which may also be phytoalexins in this plant include the phenolic eugenin (5-
hydroxy-7-methoxy-2-methylchromone) and the acetylenic, falcarinol. Yet
another structural class—furanocoumarin—has been implicated as a phytoalexin type in celery *Apium graveolens* and parsnip, *Pastinaca sativa*.
Wider surveys of the family are, however, needed to establish precisely
what range of phytoalexins are formed in umbelliferous plants.

4. Fungitoxicity and Molecular Structure

Fungitoxicity of phytoalexins is usually measured against spore germination, germ tube growth, or radial growth of the mycelium on a solid medium.
Toxicity is commonly expressed as the concentration at which 50% inhibition
of germination or growth occurs (ED_{50} value). Such measurements have
indicated that the known phytoalexins vary considerably in fungitoxicity and
ED_{50} values between 1 and 100 $\mu g/ml$ have been recorded. The toxicity may

vary slightly according to the test microorganism, and it is preferable to test any given phytoalexin against several fungi in order to establish its fungitoxic capacity.

Attempts to relate fungitoxicity with molecular structure have had only limited success. In the isoflavonoid series, there is clear evidence of increasing toxicity in the biosynthetically related series of structural types: isoflavone-isoflavanone-pterocarpan-isoflavan. This can be seen clearly in the data shown in Table 4. Within a given structural type, the number, nature, and position of substituents can affect activity. For example, the isoflavan sativan, the 2'-methyl ether of vestitol, is more fungitoxic than vestitol itself. The introduction of isopentenyl substituents may also enhance fungitoxicity in this series.

The molecular shape of the isoflavonoid molecule might be expected to be crucial for fungitoxicity, but attempts to establish correlations between shape and activity have not been successful so far. Perrin and Cruickshank (1969) suggested that aplanarity was essential, since the aplanar pterocarpans were antifungal, whereas the planar coumestans and pterocarpenes were not. However, in a more extensive survey of isoflavonoid structures, VanEtten (1976) was not able to confirm this hypothesis. Thus three planar pterocarpenes were highly inhibitory to mycelial growth and one of them was more active than the corresponding aplanar pterocarpan. It is possible that molecular shape is indeed significant in determining fungitoxicity but that some physicochemical property of the molecule, such as its degree of lipid solubility and ability to penetrate cell membranes, may have overriding importance.

Fungitoxicity in the sesquiterpenes of the Solanaceae has also been examined (Ward et al., 1974). Capsidiol [39], the principal phytoalexin of

[39]

Capsicum frutescens, is one of the most active compounds here. Various modified derivatives (Table 4) are significantly less toxic. The exocyclic double bond seems to increase fungitoxicity, since the reduced dihydrocapsidiol is three times less active. The stereochemistry at C-6 is also important, since the epimer of capsidiol at this position is 12 times less active. Such observations, however, have little predictive value, since among other fungitoxic terpenoids (see Sec. II.A) changes in sterochemistry do not apparently alter the activity.

The fungitoxicities of the *Vitis vinifera* phytoalexins have also been compared, with conflicting results. The stilbene, resveratrol [26], is accompanied in this phytoalexin response by three oligomeric derivatives, the α-, β-, and γ-viniferins. α-Viniferin, for example, is a cyclic trimer of resveratrol and is produced in major amount in infected vine leaves (Langcake and Price, 1977). Fungitoxic tests show that the parent stilbene is less active than the oligomeric forms. However, vine leaves also yield after infection

TABLE 4 Relationships Between Phytoalexin Structure and Fungitoxicity

Phytoalexins	Fungitoxicity
Isoflavonoids vs. *Helminthosporium carbonum*	ED_{50} (μg/ml)
Pterocarpan: medicarpin	25
Isoflavan: vestitol (related 2'-OH flavan)	17
Isoflavan: sativan (vestitol 2'-Me ether)	10
Isoflavonoids vs. *Aphanomyces cuteiches*	Percent inhibition of radial growth at 0.1 mM
Isoflavone: formononetin	17
Pterocarpan: phaseollin	86
Isoflavan: 2'-methoxyphaseollin isoflavan	89
Isoflavan: phaseollin isoflavan	97
Sesquiterpenes vs. *Monilinia fructicola*	ED_{50} (μg/ml)
Epicapsidiol (epimer at C-6)	106
Capsenone (ketone)	46
Dihydrocapsidiol (reduction product)	33
Capsidiol	9

another stilbene, pterostilbene, which is the 3,5-dimethyl ether of [26] (Langcake et al., 1979). This turns out to be more fungitoxic than the three viniferins. Thus oligomerization of the stilbene nucleus does not necessarily guarantee increased fungitoxicity, and the higher fungitoxicity of α-viniferin compared to resveratrol may simply reflect the fact that fewer phenolic hydroxyls are present in the molecule.

5. Further Metabolism

While plants are developing, by natural selection, more effective phytoalexins to ward off fungal attack, the fungi are themselves producing detoxification mechanisms which inactivate these antimicrobial agents. Certainly, there is increasing evidence that these compounds are further metabolized, principally by pathogenic fungi, to substances with decreased fungitoxicity. In the case of isoflavonoids, such detoxification appears to be an essential prerequisite for fungal pathogenicity in some interactions. It should be pointed out that nonpathogenic fungi may also have the enzymic machinery for detoxification.

A variety of mechanisms have been detected for detoxification. For isoflavonoids, two principal steps are hydroxylation of the nucleus and demethylation of methoxyl groups; both processes at once reduce lipid solubility and increase water solubility. The products of such reactions are immediately more susceptible to oxidative cleavage of the aromatic rings and undoubtedly the ultimate fate of these phytoalexins is to be broken down by well-described aromatic-cleavage pathways to eventually yield CO_2. The first step in the metabolism of medicarpin, the sweet clover phytoalexin, by *Botrytis cinerea*, is the production of 6a-hydroxymedicarpin [40] (Ingham, 1976). Loss in fungitoxicity is quite dramatic: thus medicarpin is highly active (ED_{50} = 25 μg/ml against mycelial growth of *Helminthosporium carbonum*),

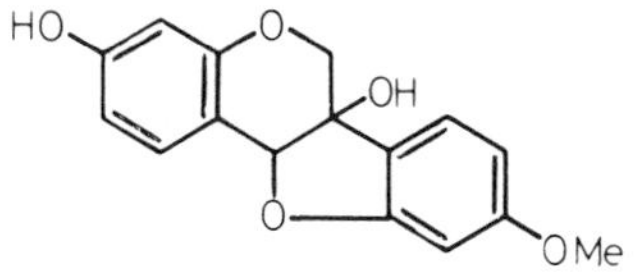

[40]

whereas its fungus-derived hydroxylation product (6α-hydroxymedicarpin) has only weak antifungal properties (ED_{50} < 100 μg/ml). A second oxidation, which is carried out by *Colletotrichum coffeanum* but not by *Botrytis*, produces 6α,7-dihydroxymedicarpin [41], which is virtually inactive as a fungitoxin. Yet another step in pterocarpan metabolism, shown in the case of

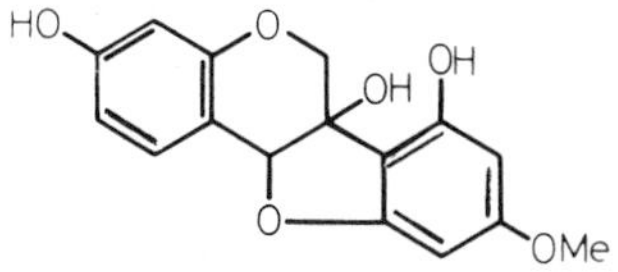

[41]

pisatin and *Fusarium oxysporum*, is ring cleavage of the five-membered furan ring to give the corresponding isoflavene (Harborne and Ingham, 1978).

The detoxification of the acetylenic phytoalexin wyerone acid [36] of *Vicia faba* by *Botrytis fabae* leads first to a hexahydroxy derivative in which the acetylenic bond is completely reduced and where the keto group adjacent to it becomes an alcohol (Hargreaves et al., 1976). By contrast, wyerone and wyerone epoxide are first metabolized to the corresponding alcohols, wyerol and wyerol epoxide, without the acetylenic triple bond being affected. Wyerol epoxide is then further reduced at the epoxide carbons with production of the related diol. All these reduction products are much less fungitoxic than the original phytoalexins. It is also interesting that while the organism *B. fabae*, which is pathogenic on the broad bean, is capable of these detoxifications, the related nonpathogenic *B. cinerea* lacks some of the enzymes to carry out all these conversions.

Some work has also been done on the detoxification of sesquiterpene phytoalexins of the Solanaceae and capsidiol [39] from *Capsicum frutescens* is known to be oxidized by pathogens (e.g., *Phytophthora capsici*) to the corresponding hydroxyketone, capsenone, which is considerably less fungitoxic than capsidiol (Ward and Stoessl, 1972). Less is known about the detoxification of other phytoalexins, although some of the steps of the pathway are predictable from known pathways of metabolism.

For the fungus, the timing at which detoxifying enzymes are produced is clearly crucial to survival. Although it is easy to demonstrate eventual turnover of many phytoalexins by fungi in culture, the important matter is whether turnover occurs in vivo at the time when the fungus first enters host cells. One can envisage a dynamic interaction occurring in biochemical terms between host and parasite before the ultimate outcome is decided—resistance or susceptibility. Our knowledge of these complexities is at present very limited and the understanding of the biochemistry of these interactions is an important goal for future research efforts.

III. PHYTOTOXINS PRODUCED BY FUNGI

A. Definitions

Many different organic substances are synthesized by fungi (Turner, 1971) and a significant number of them are capable of causing toxic symptoms when applied to other organisms. They have been variously defined according to their mode of action or the organisms they affect. Some are mammalian poisons, such as the cyclic peptides of *Amanita* species, which may be fatal to humans who eat the fruiting bodies. Again, there are mycotoxins, produced usually by fungal infection of foodstuffs, which are toxic to vertebrates rather than to the host plant. Certain mycotoxins have been shown to be damaging to plant tissues in vitro, but it is unlikely that they are concerned in plant pathogenesis (see Rudolph, 1976).

Here, we are concerned principally with fungal metabolites which are directly or indirectly responsible for disease symptoms in higher plants, products which are sometimes called phytotoxins. The term "phytotoxin" could include substances produced in pure culture, which might not necessarily be formed as a consequence of an interaction between that fungus and a higher plant. In order to refer to substances which are indeed formed as a result of the infection of living plant cells by fungi, Dimond and Waggoner (1953) have proposed the term "vivotoxin."

Even the term "vivotoxin" has its disadvantages, since it does not necessarily imply that the substance so formed is highly phytotoxic. Yet another term, "pathotoxin," has been proposed (Wheeler and Luke, 1963) to describe a host-specific toxin, which is capable of inducing typical disease symptoms at very low concentrations and which is highly selective in damaging a susceptible, but not a resistant variety of that host.

Although such distinctions between different toxins of fungal origin are important to the plant pathologist, it is often difficult in practice to determine whether a particular natural product identified from a given interaction fulfills all the requirements of one or other definition. In the present account, attention is focused principally on those compounds formed by fungi which are capable of producing significant damage to the higher plant and which are responsible for some, at least, of the symptoms of plant disease. Phytotoxin will be used as a general term, while pathotoxin will be reserved for those substances that are selective in their action on higher plants.

The chemical literature on phytotoxins is often confusing because of the many difficulties associated with the isolation and characterization of fungal products which are often highly toxic at very low concentrations. Some are very labile or become so during the process of purification. A named toxin from a specific host-parasite interaction isolated and described by one researcher may be reisolated by a second and found to have quite different characteristics. Such confusion may be due to the fact that toxins of several different types but with similar effects on the host may be present. Alternatively, a group of several closely related toxins with almost identical chemical properties may be encountered. Again, different strains of the same fungus can vary in their secondary metabolism.

Several previous accounts of phytotoxins have been consulted in the preparation of this section (Rudolph, 1976; Strobel, 1976; Wood et al., 1972).
More recent research reports have been included, but because of the large
volume of literature on the subject, the present review is selective rather than
comprehensive in nature. With few exceptions, toxins that have been only
partly characterized have been omitted from this treatment. The names of
certain well-known fungi have undergone changes in nomenclature recently;
here, the more familiar names are generally retained.

B. Low-Molecular-Weight Toxins

1. Phenolics and Quinonoids

Fungi are capable of synthesizing a variety of phenolic and quinonoid compounds, some of the latter of which may be highly colored. The phytotoxicities of such compounds are generally limited and there is some doubt in many
cases whether they are the causal agents of plant diseases with which their
production is associated.

One typical phenol of this class is zinniol [42], which was first detected in
Alternaria zinniae, the causal agent of leaf spot on zinnia, sunflower, and

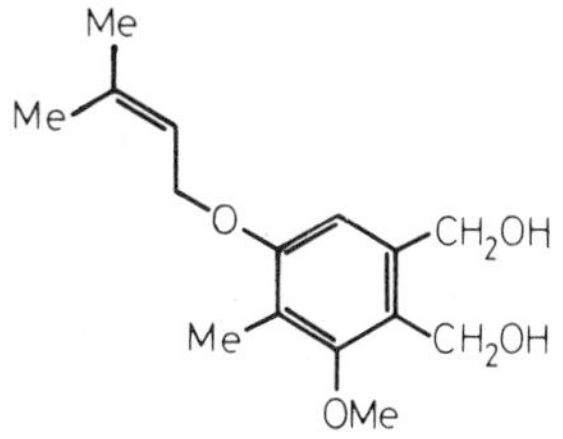

[42]

marigold (White and Starratt, 1967). More recently, zinniol has been obtained
from *A. dauci*, a fungus responsible for carrot blight. Although it produces
brown necrotic spots within 1 hr of its application on carrot leaves, the lowest phytotoxic dose is only 1 mM and it is doubtful whether it is alone responsible for carrot blight symptoms (Barash et al., 1981).

A series of three closely related phenols, including the ketone [43], have
been isolated from *Ceratocystis ulmi*, the fungus responsible for Dutch elm

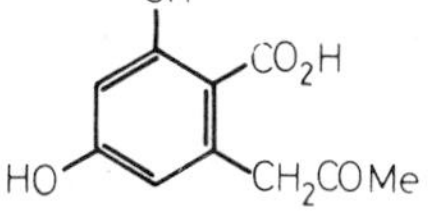

[43]

disease (Claydon et al., 1974). The ketone [43] has also been isolated from
two other unrelated fungal pathogens, *Alternaria kikuchiana* and *Pyricularia
oryzae*. It is not clear how far the three phenols of *Ceratocystis* are involved
in causing Dutch elm disease, although they are capable of producing necrotic

lesions in elm leaves. It may also be significant that they are formed in higher concentrations in the virulent "fluffy" strain of the fungus than in the aviru-lent "waxy" strain. However, the high-molecular-weight toxins also isolated from *Ceratocystis* appear to be more important in producing wilting symptoms (see Sec. III.C). The phytotoxicity associated with ketone [43] also appears in the related phenolic acid [44], which is one of several toxins present in culture of *Pyrenochaeta terrestris*, the causal agent of onion pink root disease (Sato et al., 1981). A more complex phenolic toxin is radicinin [45], a pyrano-

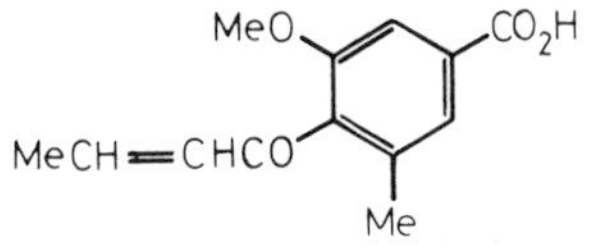

[**44**] 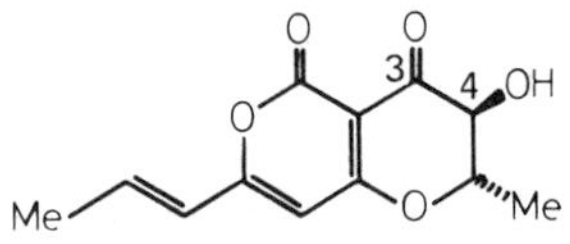   [**45**]

[4,3-*b*]pyran first characterized from *Alternaria radicina* (Grove, 1964). It also occurs in *Alternaria chrysanthemi*, a pathogen of *Leucanthemum maximum*, along with the related 3-alcohol, radicinol (Robeson et al., 1982). The 4-deoxy derivative of radicinin is a metabolite of *Alternaria helianthi*, which is pathogenic on sunflower (Robeson and Strobel, 1982). Both radicinin and its 4-deoxy derivative are phytotoxic at concentrations of 330 ppm and there is evidence that radicinin is the pathotoxin of *A. chrysanthemi*, since it produces the characteristic necrotic lesions of the disease in chrysanthemum leaves. Both compounds are, incidentally, growth inhibitors of other fungal species (e.g., of *Cladosporium*).

Typical among a number of quinonoids reported to be phytotoxic are marti-cin and isomarticin [46], two red pigments of *Fusarium solani* f.sp. *pisi*, a

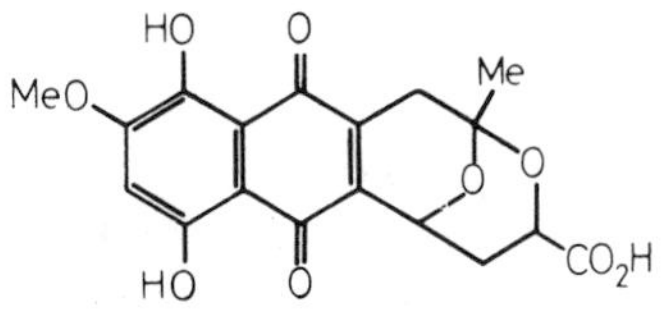

[46]

pathogen of *Pisum sativum* causing a root and stem rot. The two pigments have an unusual naphthazarin skeleton and differ in structure only in the conformation of the ketal ring system. The two pigments have been extracted from diseased pea plants in quantities sufficient to be responsible for the typical tissue damage caused by the fungus (i.e., 34 µg of pigment per gram of fresh weight of whole plant) (Kern, 1972).

One further example of a quinonoid-based toxin is cercosporin, a benzo-perylene derivative synthesized by *Cercospora beticola*, the leaf spot disease of sugar beet, and *C. kikuchii*, the purple speck disease of soya bean. Cer-cosporin is interesting as a photodynamic substance, capable of killing bac-terial and animal cells as well as whole plants. Its poisonous effect is en-hanced by light and oxygen and it is thus phototoxic in its biological activity (Towers, 1980).

2. Pyridine Derivatives

Two different low-molecular-weight toxins have been found in *Fusarium oxysporum* cultures and both have been implicated as wilting agents in the tomato. They are a small peptidelike substance lycomarasmin and a pyridine derivative, fusaric acid (5-*n*-butylpicolinic acid) [47]. The role of

HO$_2$C–N —(CH$_2$)$_3$Me

[47]

lycomarasmin in pathogenicity is still not entirely clear because although readily formed in culture, it has yet to be detected unequivocally in infected plants. The role of fusaric acid in causing wilting is, however, reasonably well established. Thus it has been detected in plants after infection and is present in much higher concentrations in plants infected by virulent strains than those treated with avirulent strains. A given virulent strain of *Fusarium* produces up to 80 mg/liter in in vitro culture and as much as 100 μg of fusaric acid per liter of fresh weight has been found in badly infected tomato plants. As part of the coevolution of *Fusarium* with the tomato, some varieties have developed the ability to resist attack. Such varieties apparently resist infection or the effects of infection because they are able to conjugate the fusaric acid with glycine, the resulting conjugate being inactive. This is not an all-or-nothing effect, since susceptible varieties are able to conjugate between 5 and 10% of the toxin in this way. However, the resistant varieties conjugate up to 25%, the greater efficiency in conjugating ability being apparently sufficient to avoid wilting.

Fusaric acid has been detected in other plants (cotton, flax, and banana) besides tomato after inoculation with wilt pathogens and it seems to be fairly widely produced by *Fusarium* species. A comparison of fusaric acid with a range of synthetic pyridine derivatives has shown that the carboxyl group in the α position to the nitrogen is essential for toxicity. The aliphatic side chain in the β position is also important, since its presence improves water permeability (Kern, 1972). It may be noted, however, that the side chain is not necessary for toxicity per se, since the parent compound, picolinic acid, which lacks any β substituent, causes necrosis in rice. Indeed, it is a major toxin of the agriculturally important disease "blast of rice," which is caused by the fungus *Pyricularia oryzae*.

Picolinic acid, like fusaric, is a metal chelator and, in the rice blast disease, it acts largely by scavenging vital iron and copper ions from within the plant tissue; its toxic effects can be reversed by supplying these metal ions back to the plant. Picolinic acid is detoxified by the host plant by conversion to the methyl ester and the *N*-methyl ether and resistant rice varieties have been shown to have a greater capacity for detoxification than susceptible forms. The pathogenicity of *Pyricularia* in rice is partly due to the synthesis by the fungus of a second toxin, piricularin, $C_{18}H_{14}N_2O_3$, a compound still only partly characterized. It is interesting that this second pathotoxin is actually toxic to the conidia of the parasite and is capable of preventing germination of the spores at a concentration of 0.25 ppm. This inhibitory effect is circumvented in vivo by the fact that piricularin exists as a complex with protein, the complex being nontoxic to the fungus but still highly lethal to the host plant.

3. Terpenoids

Two groups of phytotoxin which are particularly damaging to plant tissues are those which are terpenoid or peptide based. Both groups of toxin appear to act at the cellular level and affect the function of the plasmalemma. Among the terpenoid-based toxins, Marré (1980) has distinguished three types according to their major mode of action. One type, including fusicoccin and the cotylenins, induces an early hyperpolarization of the electric potential difference of the membrane and also enhance active H^+/K^+ exchange across the membrane. A second type, including the ophiobilins, increases diffusion or leakage of electrolytes and other solutes associated with a rapid depolarization of the potential difference. A third type, including helminthosporal, appears to act directly with the membranes of organelles within the cell, namely those of mitochondria or chloroplasts. An account of these different types of terpenoid toxin follows.

Fusicoccin is undoubtedly one of the most widely studied of all phytotoxins because of its easily measurable effects on water uptake, growth, metabolism, and transport in the higher plant cell. It is the major toxin of *Fusicoccum amygdali*, the causal agent of leaf wilting and necrosis in almond and peach trees. It is a tricyclic diterpene with a substituted glucose moiety [48]

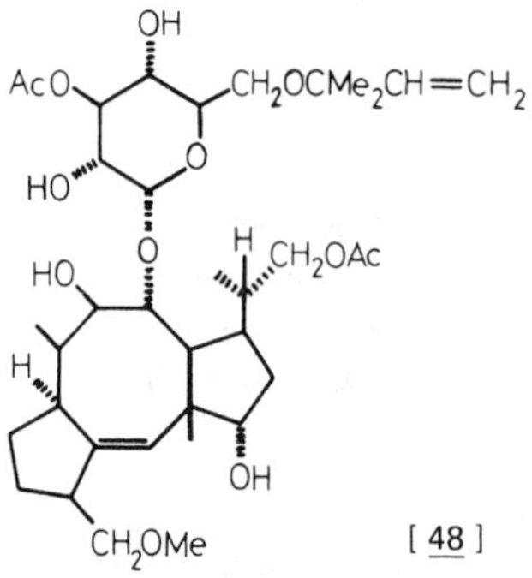

[48]

(Barrow et al., 1968). Fusicoccin, its aglycone, and several acetyl derivatives all seem to have similar pathological effects on plants. Far from being host specific, this diterpene glucoside induces the same general symptoms in all higher plants investigated. However, unlike the quinonoid toxin cercosporin (see Sec. III.B.1), fusicoccin is completely inactive in animal or bacterial tissues. The minimal concentration inducing detectable responses in plants is between 10^{-7} and 10^{-8} M, saturation being observed at 10^{-5} M. Its main toxic effects in leaves are due to the very rapid induction of stomatal opening, leading to uncontrollable increases in transpiration and hence wilting.

The cotylenins A-G comprise a group of seven closely related diterpene glucosides all obtainable together from the culture filtrate of a fungus believed to be a *Cladosporium* species (Sassa and Takahama, 1975). All seven cotylenins have a common aglycone cotylenol which is chemically close to that of fusicoccin. The similarities in chemical structure to fusicoccin are mirrored by the almost identical phytotoxicities.

Another closely related group of terpenoid toxins are the six ophiobilins, which are non-host-specific toxins of *Helminthosporium oryzae*, a fungus responsible for brown spot disease in rice. The ophiobilins are C_{25} terpenoids containing a carbotricyclic ring similar to that of fusicoccin and the cotylenins. Ophiobilin A has structure [49] and the other five compounds have related

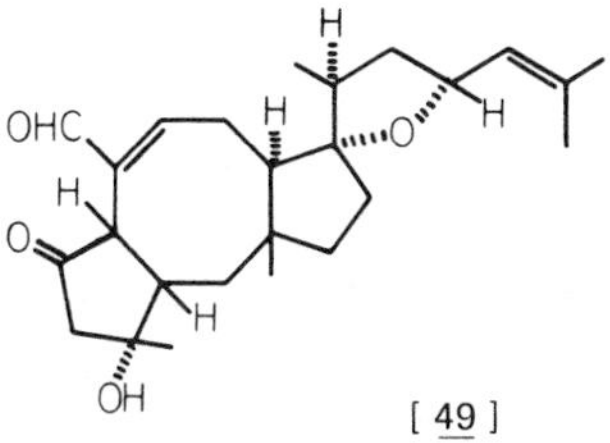

[49]

structures in this series (Canonica et al., 1966). At concentrations around 10^{-5} M, ophiobilin A inhibits elongation of rice coleoptile segments and strongly antagonizes the stimulating effect of fusicoccin on cell enlargement in maize coleoptiles. It thus appears that ophiobilin A competes with fusicoccin at the same membrane system, which is inhibited by the former and stimulated by the latter. Besides inhibiting the growth of cereal plants, the ophiobilins induce necrotic lesions.

Helminthosporal, which represents the third type of terpenoid toxin acting on the plasmalemma (Marré, 1980), is simpler in structure to those already described, since it is a sesquiterpene dialdehyde. Its structure [50] was

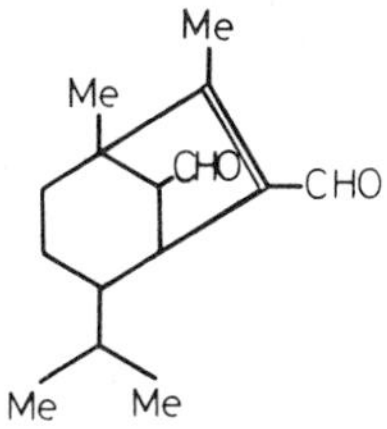

[50]

established by De Mayo et al. (1962). It is synthesized by *Helminthosporium sativum*, which causes seedling blight, root rot, and leaf spot in several cereal crops. Helminthosporal is an artifact of its isolation, formed by a rearrangement from the true phytotoxin, prehelminthosporal [51]. The related

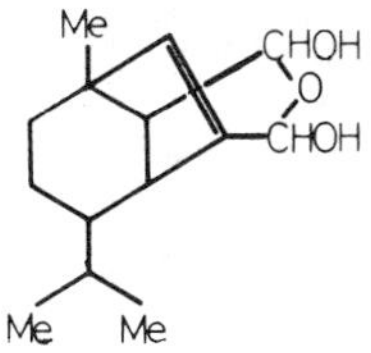

[51]

alcohol helminthosporol, which can also be isolated from *H. sativum*, is likewise a rearranged product derived from prehelminthosporol. The rearranged metabolites are highly phytotoxic and are presumably similar in action to the natural precursors.

Unlike the terpenoid toxins so far described, helminthosporol induces both necrosis and chlorosis in plant tissues. It is capable of inhibiting mitochondrial oxidative phosphorylation by acting at a site between flavoprotein and the cytochromes. Its toxic effect on the plasmalemma and tonoplast can be readily observed in beetroot disks since it induces the leakage of the betacyanin pigment of the vacuoles within 15 min of its application.

4. Peptides

A number of peptidelike toxins have been characterized from fungi, some of which are undoubtedly responsible for certain disease symptoms observed in the relevant host plants. The simplest of the group, lycomarasmin, has already been mentioned as being formed by the wilt fungus *Fusarium oxysporum*. It is not strictly speaking a peptide, since its two amino acid residues, aspartic acid and alanine, are linked by an $NH-CH_2$ group rather than by a peptide bond. In addition, the amino group of the alanyl residue carries an acetamido substituent. The complete structure of lycomarasmin, which was established by synthesis (see Barbier, 1972) is

$$HO_2CCH_2CH(CO_2H)NHCH_2CH(CO_2H)NHCH_2CONH_2$$

Most of the other toxins in this group have true peptide links, but differ in some feature or other from a peptide based simply on the common protein amino acids (Table 5). Thus they may contain one or more nonprotein amino

TABLE 5 Peptide Toxins of Fungal Origin

Name	Source	Amino acid and other components	Reference
Tentoxin	*Alternaria tenuis*	Leu, Gly, N-Me Ala, N-Me-dehydro Phe (cyclic)	Meyer et al., 1974
Alternario-lide	*Alternaria mali*	α-Hydroxyisovaleryl-α-amino-(p-methoxyphenylvaleryl-α-acrylalanyl)lactone (cyclic)	Okuno et al., 1974
Malformin	*Aspergillus niger*	Ile, Cys (2 mols.), Val, Leu (cyclic)	Anzai and Curtis, 1965
Victorin[a]	*Helminthosporium victoriae*	Asp, Glu, Gly, Val, Leu + sesquiterpenoid amine (victotoxinine)	Dorn and Arigoni, 1972
HC toxin[a]	*Helminthosporium carbonum*	Ala (2 mols.), Pro, unsat. Leu, hydroxamic acid	Pringle, 1973

[a]Structures still incompletely known.

acids (tentoxin, alternariolide) or contain a nonamino acid component (victorin, HC toxin). In many cases, the peptide is cyclic, but this is not an essential requirement for toxicity. The structures of most peptide toxins are still chemically insecure. In particular, the structure of victorin is uncertain and some studies of this toxin have suggested that it may be a mixture of extended peptides. Although it is reasonably stable in partly purified fungal extracts, victorin becomes quite labile on purification.

In fact, victorin is the most active peptide toxin known. In long-term experiments, it can be shown to inhibit the growth of oat roots of susceptible varieties at concentrations as low as 10^{-11} M. It is also the most highly selective toxin so far reported, since its effects on resistant oat varieties can only be demonstrated at concentrations 400,000 times greater than this. The effects of victorin on plants resemble those of Ca^{2+} deficiency and it is possible that its selective phytotoxicity on susceptible oat varieties are due to the induction of Ca^{2+} deficiency (Saftner et al., 1976). A toxin similar to victorin in its selectivity has been reported in culture filtrates of *Periconia circinata*, the causal agent of milo disease in *Sorghum*. This toxin is probably a mixture of three or more related compounds, but it is peptidal, since it yields alanine, aspartic acid, glutamic acid, and serine in a 3:3:1:1 ratio on acid hydrolysis (Pringle and Scheffer, 1967).

The only other peptide toxin to have been well studied is tentoxin, a cyclic tetrapeptide from *Alternaria tenuis* (Table 5). It contrasts with victorin in its nonhost specificity. It will produce chlorosis in seedlings of cotton, citrus, cucumber, and most other plants tested. Its mode of action is complex in that it causes stomatal closure as well as inhibition of K^+ uptake and of photophosphorylation.

5. Long Chain Alkanols

Novel structures based on long-chain hydrocarbon skeleta have recently been assigned to host-specific toxins isolated from *Helminthosporium maydis* and *Alternaria alternata*. Two such derivatives have been characterized from race T of *H. maydis* which produce a blighting effect when applied to *Zea mays* leaves (Kono et al., 1980). They are C_{41} hydrocarbons and differ in structure only as to whether there is a carbonyl or alcoholic function at C-24. The structure of the ketone has been established as

$$CH_3(CH_2)_4CHOHCH_2CHOHCH_2CO(CH_2)_3COCH_2CHOHCH_2CO(CH_2)_5$$

$$COCH_2CHOHCH_2CO(CH_2)_3COCH_2CHOHCH_2CO(CH_2)_3COCH_3$$

Two phytotoxins of *Alternaria alternata* f.sp. *lycopersici*, the fungus causing stem canker disease in tomato, are C_{19} hydrocarbons with a terminal amino group. Extracts containing the two toxins produce disease symptoms at concentrations of less than 10 ng/ml. The two compounds characterized have the same basic structure:

$$CH_3CH_2CH(CH_3)(CHOH)_2CH_2CHCH_3(CH_2)_5(CHOH)_2CH_2CHOHCH_2NH_2$$

This structure is alternatively substituted at the C-4 or C-5 alcohol group with the tricarboxylic acid, $HO_2C \cdot CH_2CH(CO_2H)CH_2CO_2H$ (Bottini and Gilchrist, 1981). Two further related structures are also present in the culture filtrate.

It will be interesting to see if further toxins of these unusual types are discovered in extracts of other fungal pathogens. They are unique as natural products. Their mode of action is as yet unclear.

6. Miscellaneous Toxins

A variety of other toxins, not classifiable under the earlier headings, have been characterized from fungi pathogenic on plants. One that at first appeared to have a very simple structure is helminthosporoside (not to be confused with helminthosporol or helminthosporal; see above). This is the toxin of the sugar cane pathogen *Helminthosporium sacchari*, which is responsible for the eyespot symptom. Helminthosporoside produces the characteristic reddish brown stripes, called runners, when injected into sugar cane plants (Strobel, 1976).

This toxin was originally thought to be 2-hydroxycyclopropyl α-galactoside, but this structure was never confirmed by rigorous chemical procedures. More recent reexamination (Macko et al., 1981) indicates that it is in fact a mixture of three isomeric glycosides $C_{39}H_{64}O_{22}$, each of which is active at concentrations of 2×10^{-11} M. The three isomers are galactosides but have four galactose moieties linked to a so far uncharacterized aglycone $C_{15}H_{24}O_2$.

Helminthosporoside is of special biochemical interest because of its selective mode of action on cell membranes. It binds to a single membrane protein in susceptible but not in resistant sugar cane varieties. Indeed, the difference between disease resistance and susceptibility in sugar cane seems to be based on the availability or otherwise of a particular binding site on the cell membrane. The protein concerned has also been isolated and defined (Strobel, 1976).

Many fungal metabolites are polyketide-derived, so that it is not surprising that a few such structures are among the phytotoxins. One such compound is zearalenone [52] (Mirocha et al., 1971), a non-host-specific macrocyclic

[52]

substance produced by several *Fusarium* species growing on maize and other cereals. At concentrations of 1-10 μg/ml, it will kill young plants and inhibit seed germination and embryo growth. It appears to act at the plasmalemma and like the terpenoid ophiobilins, it inhibits electrogenic proton transport.

Toxins of other biosynthetic origins have been described from fungi, including triglycerides, steroids, cyclic diketones, and indole derivatives (see Rudolph, 1976). In general, however, little is known about the way they cause damage in host plant tissues.

C. Macromolecular Toxins

A number of macromolecular toxins have been described from fungi, some of which are pure protein, others pure polysaccharide, and yet others glycoprotein or glycopeptide. In terms of the toxic effects produced, they all seem to act in a similar manner in that the major symptom produced in the host plant is wilting. The most extensively studied toxic activity is that of the Dutch elm disease pathogen, *Ceratocystis ulmi*, on the elm tree. At least two different macromolecules have been described from this source.

Strobel et al. (1978) have reported from *C. ulmi* culture filtrates a peptidorhamnomannan, which is 83% sugar and 7% amino acid. This toxin has three components, the major (80%) polydisperse fraction of molecular weight 270,000, a higher-molecular-weight fraction of less than 5% of the total, and the remainder with molecular weight 70,000. All the common amino acids, except histidine, cysteine, and methionine, are present in the peptide backbone. Linked to this through threonyl and seryl residues are a variety of very long rhamnomannan chains, together with some much shorter mannose-based oligosaccharide chains. The polysaccharide chains are composed of α-$(1 \rightarrow 6)$-linked mannose units, with a 3-linked terminal rhamnosyl residue attached to almost every mannosyl residue. In bioassay on elm cuttings, the toxin is active at 400 μg/ml, but its wilting effect is not specific to elm, since it similarly affects alfalfa, geranium, petunia, and other test plants.

Another fraction from the culture filtrate of *Ceratocystis ulmi* has been studied by Stevenson et al. (1979), who have described a small protein, which they call cerato-ulmin. This is active at the low concentration of 0.4 μg/ml, and it causes necrosis and chlorosis in the elm, as well as wilting. Cerato-ulmin has 128 amino acid residues and a molecular weight of 13,000. It is unusual in its amino acid composition in having high amounts of proline (12 residues) and seven cystine disulfide bridges; it also lacks cysteine, methionine, or tryptophan. It is unusually soluble in ethanol, highly hydrophobic, and appears to have a compact globular structure. Its relationship to the peptidorhamnomannan of Strobel et al. (1978) is not entirely clear, although there are some similarities between the peptide backbone of the glycopeptide and the pure protein of cerato-ulmin. On the face of it, it would seem that cerato-ulmin is a more likely candidate for the specific virulence that has been produced in *C. ulmi* in recent years. Aided by the elm bark beetle as vector, this fungus has caused the complete demise of the elm tree from the southern English countryside and has also killed millions of trees in North America.

Several other phytotoxic glycopeptides have also been described in the plant pathological literature, but they have rarely been fully characterized. One from lemon leaves infected with *Phoma tracheiophila* has been named malseccin, and shown to contain mannose, galactose, and glucose as well as most of the standard amino acids (Nachmias et al., 1979). The glycopeptide from the infected lemon tree was compared to that isolated from the pure fungal culture and they were found to be essentially similar in their properties and wilting activities. Another glycopeptide from *Verticillium albo-atrum* and *V. dahliae* was found to be distinctive in containing 15% lipid, as well as protein (15%) and polysaccharide (70%). The sugars present are glucose, galactose, mannose, and galacturonic acid (Keen and Long, 1972).

Fungal polysaccharides are also capable of producing wilting symptoms in higher plants. Most that have been isolated have only been partly characterized (Table 6). The simple idea that they act as toxins by physical obstruction of the water vessels is an attractive one and is supported by their water solubility. Their mode of action is more likely to be at the cellular level, particularly since very similar polysaccharides from other fungi are capable of causing other disease symptoms, such as necrosis and chlorosis.

TABLE 6 Polysaccharide Toxins of Fungal Origin

Source	Sugars	Linkages	Reference
Cephalosporium gramineum	Glc, Man, Rha	–	Spalding et al., 1961
Ceratocystis fagacearum	Man	α-(1 → 6), with branching	McWain and Gregory, 1972
Fusarium solani	Glc (?)	–	Thomas, 1949
Phytophthora cinnamoni P. *cryptogea* P. *nicotianae*	Glc	β-(1 → 3), β-(1 → 6), with branching	Woodward et al., 1980

There does not seem to be any obvious structural factor in terms of sugar content or linkage to explain their ability to cause wilting in host plants (Table 6). Their action, indeed, seems to be relatively non-host specific. The *Phytophthora* polysaccharides will produce their effects in *Eucalyptus* seedlings (Woodward et al., 1980). Their significance vis-à-vis other wilting factors is not clear. Further structural studies are needed before it is profitable to speculate about their effectiveness as toxins and their mode of attack.

IV. CONCLUSION

A great variety of different organic structures are elaborated as toxins at the interface between higher plants and fungi in order to limit predation of the latter on the former or to cause damage to the former by the latter. In the case of preformed fungitoxins produced by plants, practically all the compounds so far characterized (diterpenoids, lactones, methylated flavones, phenanthrenes, etc.) are typical secondary metabolites and hence are almost certain to fulfill more than one role in plants. Many fungitoxic compounds may be active deterrents of herbivore feeding (Levin, 1976) or be useful as allelopathic agents against other plants (Rice, 1974). Their ability to suppress fungal spore germination or mycelial growth may be accidental to the overall economy of plant growth and metabolism.

Even with the phytoalexins, substances specifically designed to resist fungal invasion, there is no obvious structural rationale in the type of compound formed. So many unrelated structures have been recorded as being produced in the phytoalexin response in different plants. Again, the compounds are characteristically secondary in terms of their biosynthesis, mainly isoflavonoids, terpenoids, or acetylenic compounds. Furthermore, there is evidence that some, if not all, phytoalexins can be formed in plants subjected to other stresses besides microbial invasion. Some phytoalexins at low concentrations are highly damaging to fungal cell growth and further exploration of their mode of action might indicate a greater degree of specificity in the structures synthesized. It is at least conceivable that they are able to produce damage at different target sites within the fungus.

In the case of the phytotoxins synthesized by the fungi, it is possible to distinguish some compounds which are damaging to a number of different host plants from those which are highly host specific, and may be capable of tissue damage in only susceptible varieties of that host. In chemical terms, there is some overlap between the two classes of toxin and it is difficult to pinpoint particular structural features which lead to host specificity. A number of phytotoxins exert their toxicity through interference in the transport system of the cell membrane. Further exploration of the effects of these particular pathotoxins on higher plant cells may lead to the enunciation of structure-activity relationships. While the role of low-molecular-weight pathotoxins in causing disease symptoms in plants is beginning to become clear, that of the macromolecular toxins is still relatively obscure. Further chemical, ultra-structural, and pathological studies of purified glycopeptide, protein, and polysaccharide fractions from pathogenic fungi would seem to be particularly desirable at the present time. Even with Dutch elm disease pathogen, which has been extensively studied in recent years, it is not yet sure which of several toxins characterized is the most important in causing all the symptoms produced in the elm host plant. With most crop plant pathogens, too, our knowledge of their pathotoxins is still fragmentary and the subject needs much further interdisciplinary research effort.

REFERENCES

Anzai, K., and Curtis, R. W. (1965). Structure of malformin A. *Phytochemistry 4*:263-271.

Bailey, J. A., and Mansfield, J. (Eds.) (1981). *The Phytoalexins*. Blackie, Glasgow, pp. 334.

Bailey, J. A., Vincent, G. G., and Burden, R. S. (1974). Diterpenes from *Nicotiana glutinosa* and their effect on fungal growth. *J. Gen. Microbiol. 85*:57-64.

Barash, I., Mor, H., Netzer, D., and Kashman, Y. (1981). Production of zinniol by *Alternaria dauci* and its phytotoxic effect on carrot. *Physiol. Plant Pathol. 19*:7-16.

Barbier, M. (1972). The chemistry of some amino acid derived phytotoxins. In *Phytotoxins in Plant Diseases*, R. K. S. Wood, A. Ballio, and A. Graniti (Eds.). Academic Press, London, pp. 91-903.

Barrow, K. D., Barton, D. H. R., Chain, E. B., Ohnsorge, U. F. W., and Thomas, R. (1968). The structure of fusicoccin. *J. Chem. Soc. Chem. Commun.*, p. 1197.

Beijersbergen, J. C. M., and Lemmers, C. B. G. (1972). Enzymic and non-enzymic liberation of tulipalin A in extracts of tulip. *Physiol. Plant Pathol. 2*:265-270.

Blakeman, J. P., and Atkinson, P. (1979). Antimicrobial properties of parthenolide, a sesquiterpene lactone from glands of *Chrysanthemum parthenium*. *Physiol. Plant Pathol. 15*:183-192.

Bottini, A. T., and Gilchrist, D. G. (1981). Phytotoxins from *Alternaria alternata* f.sp. *lycopersici*. *Tetrahedron Lett. 22*:2719-2726.

Burden, R. S., Rogers, P. M., and Wain, R. L. (1974). Fungicides. XVI. Natural resistance of plant roots to fungal pathogens. *Ann. Appl. Biol. 78*:59-63.

Canonica, L., Fiecchi, A., Kienle, M., and Scala, A. (1966). The constitu-
tion of cochliobilin. *Tetrahedron Lett. 11*:1211-1218.
Cartwright, D. W., Langcake, P., Pryce, R. J., Leworthy, D. P., and Ride,
J. P. (1981). Isolation and characterisation of two phytoalexins from rice
as momilactones A and B. *Phytochemistry 20*:535-537.
Claydon, N., Grove, J. F., and Hosken, M. (1974). Phenolic metabolites of
Ceratocystis ulmi. *Phytochemistry 13*:2567-2571.
Cobb, F. W., Krstic, M., Zavarin, E., and Barber, H. W. (1968). Inhibi-
tory effects of volatile oleoresin components on *Fomes annosus* and four
Ceratocystis species. *Phytopathology 58*:1327-1335.
Coley-Smith, J. R. (1976). Some interactions in soil between plants,
sclerotium-forming fungi and other micro-organisms. In *Biochemical
Aspects of Plant-Parasite Relationships*, J. Friend and D. R. Threlfall
(Eds.). Academic Press, London, pp. 11-24.
Coxon, D. T., Ogundana, S. K., and Dennis, C. (1982). Antifungal com-
pounds in yam tubers. *Phytochemistry 21*:1389-1392.
Cruickshank, I. A. M., and Perrin, D. R. (1960). Isolation of a phytoalexin
from *Pisum sativum*. *Nature (Lond.) 187*:799-800.
Cruickshank, I. A. M., Perrin, D. R., and Mandryk, M. (1977). Fungitoxicity of duvatrienediols associated with the cuticular wax of tobacco
leaves. *Phytopathol. Z. 90*:243-249.
De Mayo, P., Spencer, E. Y., and White, R. W. (1962). The constitution of
helminthosporal. *J. Am. Chem. Soc. 84*:494-495.
DeWit, P. J. G., and Kodde, E. (1981). Induction of acetylenic phytoalexins
in tomato inoculated with *Cladosporium fulvum*. *Physiol. Plant Pathol. 18*:
143-148.
Dimond, A. E., and Waggoner, P. E. (1953). On the nature and role of vivo-
toxins in plant disease. *Phytopathology 43*:229-235.
Dorn, F., and Arigoni, D. (1972). The structure of victoxinine. *J. Chem.
Soc. Chem. Comm.* 1342-1343.
Durbin, R. D., and Uchytil, T. F. (1971). The role of allicin in the resist-
ance of garlic to *Penicillium* spp. *Phytopathol. Mediterr. 10*:227-230.
Friend, J., and Threlfall, D. R. (Eds.) (1976). *Biochemical Aspects of
Plant-Parasite Relationships*. Academic Press, London, pp. 354.
Fry, W. E., and Millar, R. L. (1971). Cyanide tolerance in *Stemphylium loti*.
Phytopathology 61:494-500.
Fry, W. E., and Millar, R. L. (1972). Cyanide degradation by an enzyme
from *Stemphylium loti*. *Arch. Biochem. Biophys. 151*:468-474.
Gorham, J. (1980). The stilbenoids. *Prog. Phytochem. 6*:203-252.
Greenhalgh, J. R., and Mitchell, N. D. (1976). The involvement of flavour
volatiles in the resistance to downy mildew of wild and cultivated forms of
Brassica oleracea. *New Phytol. 77*:391-398.
Grove, J. (1964). Metabolic products of *Stemphylium radicinum*. *J. Chem.
Soc.*, pp. 3234-3241.
Harborne, J. B., and Ingham, J. L. (1978). Biochemical aspects of the co-
evolution of higher plants with their fungal parasites. In *Biochemical
Aspects of Plant and Animal Coevolution*, J. B. Harborne (Ed.). Academic
Press, London, pp. 343-405.
Harborne, J. B., Ingham, J. L., King, L., and Payne, M. (1976). The iso-
pentenylisoflavone luteone as a pre-infectional antifungal agent in the
genus *Lupinus*. *Phytochemistry 15*:1485-1487.

Hargreaves, J. A., Mansfield, J. W., Coxon, D. T., and Price, K. R. (1976). Wyerone epoxide as a phytoalexin in *Vicia faba* and its metabolism by *Botrytis cinerea* and *B. fabae*. *Phytochemistry 15*:1119-1121.

Heitefuss, R., and Williams, P. H. (Eds.) (1976). *Physiological Plant Pathology*, Vol. 4, *Encyclopedia of Plant Physiology, New Seies*. Springer-Verlag, Berlin.

Heywood, V. H., Harborne, J. B., and Turner, B. L. (Eds.) (1977). *The Biology and Chemistry of the Compositate*. Academic Press, London, 2 vols.

Horsfall, J. G., and Cowling, F. B. (Eds.) (1979). *Plant Disease*, Vol. IV: *How Pathogens Induce Disease*. Academic Press, New York.

Horsfall, J. G., and Cowling, F. B. (Eds.) (1980). *Plant Disease*, Vol. V: *How Plants Defend Themselves*. Academic Press, New York.

Hunter, L. D. (1975). Phloridzin and apple scab. *Phytochemistry 14*:1519-1522.

Ingham, J. L. (1972). Phytoalexins and other natural products as factors in plant disease resistance. *Bot. Rev. 38*:343-424.

Ingham, J. L. (1973). Disease resistance in plants: the concept of pre-infectional and post-infectional resistance. *Phytopathol. Z. 78*:314-335.

Ingham, J. L. (1976). Fungal modification of pterocarpan phytoalexins from *Melilotus alba* and *Trifolium pratense*. *Phytochemistry 15*:1489-1495.

Ingham, J. L. (1981). Phytoalexin induction and its taxonomic significance in the Leguminosae, subfamily Papilionoideae. In *Advances in Legume Systematics*, R. M. Polhill and P. H. Raven (Eds.). HMSO, London, pp. 599-626.

Ingham, J. L., Keen, N. T., and Hymowitz, T. (1977). A new isoflavone phytoalexin from fungus-inoculated stems of *Glycine wightii*. *Phytochemistry 16*:1943-1946.

Ireland, C. R., Schwabe, W. W., and Coursey, D. G. (1981). The occurrence of batatasins in the Dioscoreaceae. *Phytochemistry 20*:1569-1572.

Kawazu, K., Noguchi, H., Fujishita, K., Iwasa, J., and Egawa, H. (1973). Two new antifungal compounds from leaves of *Dendropanax trifidus*. *Tetrahedron Lett.*, pp. 3131-3134.

Keen, N. T., and Long, M. (1972). Isolation of a protein-lipopolysaccharide complex from *Verticillium albo-atrum*. *Physiol. Plant Pathol. 2*:307-315.

Kern, H. (1972). Phytotoxins produced by *Fusaria*. In *Phytotoxins in Plant Diseases*, R. K. S. Wood, A. Ballio, and A. Graniti (Eds.). Academic Press, London, pp. 35-48.

Kono, Y., Takeuchi, S., Kawarada, A., Daly, J. M., and Knoche, H. W. (1980). Structure of the host-specific pathotoxins produced by *Helminthosporium maydis*, race T. *Tetrahedron Lett. 21*:1537-1540.

Langcake, P., and Price, R. J. (1977). A new class of phytoalexins from grapevines. *Experientia 33*:151.

Langcake, P., Cornford, C. A., and Pryce, J. R. (1979). Identification of pterostilbene as a phytoalexin from *Vitis vinifera* leaves. *Phytochemistry 18*:1025-1027.

Levin, D. A. (1976). The chemical defenses of plants to pathogens and herbivores. *Annu. Rev. Ecol. Syst. 7*:121-159.

Lüning, H. V., and Schlösser, E. (1976). Role of saponins in antifungal resistance. VI. Interactions *Avena sativa-Drechslera avenacea*. *Z. Pflanzenkr. Pflanzenschutz 83*:317-327.

Macko, V., Goodfriend, K., Wachs, T., Renwick, J. A. A., Acklin, W., and Arigoni, D. (1981). Characterisation of the host-specific toxins of *Helminthosporium sacchari*. *Experientia 37*:923-924.

McWain, P., and Gregory, G. F. (1972). A neutral mannan from *Cerato-cystis fagacearum* culture filtrate. *Phytochemistry 11*:2609-2612.

Marré, E. (1980). Mechanism of action of phytotoxins affecting plasma-lemma functions. *Prog. Phytochem. 6*:253-284.

Martin, J. T., and Juniper, B. E. (1970). *The cuticles of Plants*. Edward Arnold, London.

Mayama, S., Tani, T., Ueno, T., Hirabayashi, K., Nakashima, T., Fukami, H., Mizuno, Y., and Irie, H. (1981). Isolation and structure elucidation of a genuine oat phytoalexin avenalumin I. *Tetrahedron Lett. 22*:2103-2106.

Meyer, W. L., Kuyper, L. F., Lewis, R. B., Templeton, G. E., and Wood-head, S. H. (1974). The amino acid sequence and configuration of ten-toxin. *Biochem. Biophys. Res. Commun. 56*:234-240.

Mirocha, C. J., Christensen, C. M., and Nelson, G. H. (1971). In *Microbial Toxins*, Vol. VII, S. Kadis, A. Ciegler and S. J. Ajl (Eds.). Academic Press, New York, pp. 107-

Müller, K., and Börger, H. (1941). Experimentelle Unterschungen über die Phytophthora-Resistenz der Kartoffel. *Arb. Biol. Reichsanstalt. Lander. Forstw. Berl. 23*:189-231.

Nachmias, A., Barash, I., Buckner, V., Solel, Z., and Strobel, G. A. (1979). A phytotoxic glycopeptide from lemon leaves infected with *Phoma tracheiphila*. *Physiol. Plant Pathol. 14*:135-140.

Okuno, T., Ishita, Y., Sawai, K., and Matsumoto, T. (1974). Characterisation of alternariolide, a host-specific toxin produced by *Alternaria mali*. *Chem. Lett.*, pp. 635-639.

Perrin, D. R., and Cruickshank, I. A. M. (1969). The antifungal activity of pterocarpans towards *Monilinia fructicola*. *Phytochemistry 8*:971-978.

Piattelli, M., and Impellizzeri, G. (1971). Fungistatic flavones in the leaves of *Citrus* species resistant and susceptible to *Deuterophoma tracheiphila*. *Phytochemistry 10*:2657-2659.

Pringle, R. B. (1973). Abolishment of specific toxicity of host-specific toxin of *Helminthosporium carbonum* by electrolytic reduction. *Plant Physiol. 51*:403-404.

Pringle, R. B., and Scheffer, R. P. (1967). Isolation of the host-specific toxin and a related substance with non-specific toxicity from *Helmintho-sporium carbonum*. *Phytopathology 57*:1169-1172.

Purnell, T. J. (1971). Effects of preinoculation washings of leaves with water on subsequent infections by *Erysiphe cruciferarum*. In *Ecology of Leaf Surface Micro-organisms*, T. F. Preece and C. H. Dickinson (Eds.). Academic Press, London, pp. 269-275.

Rice, E. L. (1974). *Allelopathy*. Academic Press, New York.

Robeson, D. J., and Harborne, J. B. (1980). A chemical dichotomy in phyto-alexin induction within the tribe Vicieae of the Leguminosae. *Phytochemistry 19*:2359-2366.

Robeson, D. J., and Strobel, G. A. (1982). Deoxyradicinin, a novel phyto-toxin from *Alternaria helianthi*. *Phytochemistry 21*:1821-1823.

Robeson, D. J., Gray, G. R., and Strobel, G. A. (1982). Radicinin and radicinol production by *Alternaria chrysanthemi*. *Phytochemistry 21*:2359-2362.

Rockwood, D. J. (1974). Cortical monoterpenes and fusiform rust resistance relationships in slash pine. *Phytopathology 64*:976-979.

Rodriguez, E., Towers, G. H. N., and Mitchell, J. C. (1976). Biological activities of sesquiterpene lactones. *Phytochemistry 15*:1573-1580.

Rudolph, K. (1976). Non-specific toxins. In *Physiological Plant Pathology*, R. Heitefuss and P. H. Williams (Eds.), Vol. 4, *Encyclopedia of Plant Physiology, New Series*. Springer-Verlag, Berlin, pp. 270-315.

Saftner, R. A., Evans, M. L., and Hollander, P. B. (1976). Specific binding of victorin and calcium: evidence for calcium binding as a mediator of victorin activity. *Physiol. Plant Pathol. 8*:21-34.

Sassa, T., and Takahama, A. (1975). Isolation and identification of cotylenins F and G. *Agric. Biol. Chem. 39*:2213-2215.

Sato, H., Konoma, K., and Sokamura, S. (1981). Three new phytotoxins produced by the fungus *Pyrenochaeta terrestris*, pyrenochaetic acids A-C. *Agric. Biol. Chem. 45*:1675-1679.

Schlösser, E. (1973). Role of saponins in antifungal resistance. II. The hederasaponins in leaves of *Hedera helix*. *Z. Pflanzenkr. Pflanzenschutz 80*:704-710.

Schlösser, E. (1975). Role of saponins in antifungal resistance. III. Tomatine dependent development of fruit rot organisms on tomato fruits. *Z. Pflanzenkr. Pflanzenschutz 82*:476-484.

Schönbeck, F., and Schlösser, E. (1976). Preformed substances as potential protectants. In *Physiological Plant Pathology*, R. Heitefuss and P. H. Williams (Eds.), Vol. 4, *Encyclopedia of Plant Physiology, New Series*. Springer-Verlag, Berlin, pp. 656-678.

Schneider, R. W., and Sinclair, J. B. (1975). Inhibition of conidial germination and germ tube growth of *Cercospora canescens* by cowpea leaf diffusates. *Phytopathology 65*:63-65.

Sitton, D., and West, C. A. (1975). Casbene, an antifungal diterpene produced in cell-free extracts of *Ricinus communis* seedlings. *Phytochemistry 14*:1921-1925.

Slob, A., Jekel, B., de Jong, B., and Schlatmann, E. (1975). On the occurrence of tuliposides in the Liliiflorae. *Phytochemistry 14*:1997-2005.

Spalding, D. H., Bruehl, G. W., and Foster, R. J. (1961). Possible role of pectinolytic enzymes and polysaccharide in pathogenesis by *Cephalosporium gramineum* in wheat. *Phytopathology 51*:227-235.

Stevenson, K. J., Slater, J. A., and Takai, S. (1979). Cerato-ulmin, a wilting toxin of Dutch elm disease fungus. *Phytochemistry 18*:235-238.

Stoessl, A. (1967). Antifungal factors in barley: constitutions of hordatines A and B. *Can. J. Chem. 45*:1745-1760.

Stoessl, A., Stothers, J. B., and Ward, E. W. B. (1976). Sesquiterpenoid stress compounds of the Solanaceae. *Phytochemistry 15*:855-872.

Strobel, G. A. (1976). Toxins of plant pathogenic bacteria and fungi. In *Biochemical Aspects of Plant-Parasite Relationships*, J. Friend and D. R. Threlfall (Eds.). Academic Press, London, pp. 135-160.

Strobel, G. A., van Alfen, N., Hapner, K. D., McNeill, M., and Albersheim, P. (1978). Some phytotoxic glycopeptides from *Ceratocystis ulmi*, the Dutch elm disease pathogen. *Biochim. Biophys. Acta 538*:60-75.

Takasugi, M., Nagao, S., Ueno, S., and Masamune, T. (1978). Phytoalexins of Moraceae, moracin C and D from diseased mulberry. *Chem. Lett.*, pp. 1239-1240.

Takasugi, M., Nagao, S., and Masumune, T. (1979). Moracins E, F, G and H, new phytoalexins from diseased mulberry. *Tetrahedron Lett.*, pp. 4675-4678.

Takasugi, M., Yumagai, Y., and Nagao, S. (1980). Cooccurrence of flavan and 1,3-diphenylpropane derivatives in wounded paper mulberry. *Chem. Lett.*, pp. 1459-1460.

Thomas, C. A. (1949). A wilt-inducing polysaccharide from *Fusarium solani*. *Phytopathology 39*:572-579.

Topps, J. H., and Wain, R. L. (1957). Fungistatic properties of leaf exudates. *Nature (Lond.) 179*:652-653.

Towers, G. H. N. (1980). Photosensitizers from plants and their photodynamic action. *Prog. Phytochem. 6*:183-202.

Trione, E. J. (1960). Extracellular enzyme and toxin production by *Fusarium oxysporum* f. *lini*. *Phytopathology 50*:480-482.

Tripathi, R. D., Srivastava, H. S., and Dixit, S. N. (1978). Fungitoxic principle from the leaves of *Lawsonia inermis*. *Experientia 34*:51-52.

Tschesche, R., and Wulff, G. (1973). Chemie und Biologie der Saponine. *Fortschr. Chem. Org. Naturst. 30*:461-606.

Turner, W. B. (1971). *Fungal Metabolites*. Academic Press, London, 446 pp.

VanEtten, H. D. (1976). Antifungal activity of pterocarpans and other selected isoflavonoids. *Phytochemistry 15*:655-659.

VanEtten, H. D., and Pueppke, S. G. (1976). Isoflavonoid phytoalexins. In *Biochemical Aspects of Plant-Parasite Relationships*, J. Friend and D. R. Threlfall (Eds.). Academic Press, London, pp. 239-290.

Van Wyk, P. S., and Koeppen, B. H. (1974). *p*-Hydroxybenzoyl-calleryanin, antifungal compound in root bark of *Protea cynaroides*. *S. Afr. J. Sci. 70*:121-122.

Walker, J. C., and Stahmann, M. A. (1955). Chemical nature of disease resistance in plants. *Annu. Rev. Plant Physiol. 6*:351-366.

Ward, E. W. B., and Stoessl, A. (1972). Detoxification of capsidiol, an antifungal compound from peppers. *Phytopathology 62*:1186-1187.

Ward, E. W. B., Unwin, C. H., and Stoessl, A. (1974). Fungitoxicity of the phytoalexin capsidiol and related sesquiterpenes. *Can. J. Bot. 52*:2481-2488.

Wheeler, H. and Luke, H. H. (1963). Microbial toxins in plant disease. *Annu. Rev. Microbiol. 17*:223-242.

White, G. A., and Starratt, A. N. (1967). The production of a phytotoxic substance by *Alternaria zinniae*. *Can. J. Bot. 45*:2087-2090.

Wood, R. K .S, Ballio, A., and Graniti, A. (Eds.) (1972). *Phytotoxins in Plant Diseases*. Academic Press, London.

Woodward, J. R., Keane, P. S., and Stone, B. A. (1980). Structures and properties of wilt-inducing polysaccharides from *Phytophthora* spp. *Physiol. Plant Pathol. 16*:439-454.

Zaki, A. I., Zentmyer, G. A., Pettus, J., Sims, J. J., Keen, N. T., and Sing, V. O. (1980). Borbonol from *Persea* spp. chemical properties and antifungal activity against *Phytophthora cinnamomi*. *Physiol. Plant Pathol. 16*:205-212.

Part VIII

TOXINS USEFUL IN MEDICINE

23

CONTRIBUTIONS OF HERBOLOGY TO MODERN MEDICINE AND DENTISTRY

WALTER H. LEWIS

Washington University, St. Louis, Missouri

MEMORY P. F. ELVIN-LEWIS

Washington University, School of Dental Medicine, St. Louis, Missouri

I. INTRODUCTION

A few years ago, books on herbology in relation to human health began appearing in every-increasing numbers. Although some contained well-documented herbal recipes from traditional medicine, many were merely lists of plants accompanied by supposed efficacies. Little attempt was made to relate their contents to scientific data and only rarely were they correlated with current use in modern medicine. However, it was unusual for these

herbals to discuss plants known to be injurious; consequently, cross references to indicate remedial, as well as toxic properties, which so often depend
on either dosage or mode of preparation, were usually absent. It became
clear that although numerous systems, syndromes, or bodily functions were
glibly mentioned, the reasons why the plant worked, and for that matter what
the plant substances affected specifically, were wanting. Often it was difficult to understand the implication of presumed value and thus the average
reader found most data meaningless. This concern is real for an increasing
number of readers are using such information for self-medication without a
recent cultural background of herbology and therefore without a full appreciation of the potential toxicities that such preparations might possess.

On the other hand, many plant-derived products were incorporated into
modern pharmacopeias without full appreciation of their original sources.
Because of this, related plants were often ignored for their medicinal potential. Moreover, new investigations of plants were rarely conceived through
the empirical method of indigenous medicine. It became more the rule for
scientists in search of a particular medicinal property to manipulate a synthetic compound rather than to search the folk literature or to explore known
cures among indigenous peoples for clues that might lead ultimately to a more
efficacious product. Unfortunately, this approach has often served only to
delay the application of many potential remedies. For example, it was unfortunate that the first cosmopolitan tranquilizer, derived from *Rauvolfia*, did
not come into general use until 1952, despite an ancient history of its application in Ayarvedic (Hindu) medicine in India.

The search today for medically useful elements seldom combines the ability
to distinguish chemical compounds for medical value with the ability to recognize the relationship of plants used medically by different cultures. Clearly,
empirical selection has led to studies resulting in the isolation and use of
important active principles and these, together with semisynthetic and synthetic derivatives of natural products, are an important source of the modern
therapeutic armament. Nature is still humankind's greatest chemist, and many
compounds that remain undiscovered in plants are beyond the imagination of
even our best researchers.

However, the development from traditional medicine of therapeutic agents
for assimilation into modern medicine should not be the only goal of current
research. Existing remedies from traditional medicine should be rigorously
examined for potential relevance in the treatment of disease, especially in
societies and third-world countries where modern medicine is scarce or unavailable for various reasons. Promotion of traditional medicine is not to advo-
cate a return to domestic medicine, but only where shown efficacious and valid
should it be retained and where proved invalid discarded. Should herbalists
acquire diagnostic skills and scientific knowledge, their treatment of disease
would be even more effective. Indeed, traditional medicine ought to complement rather than substitute for modern medicine. A good illustration is the
practice of the New Chinese Medicine, where a harmonious interplay exists
between traditional medical techniques involving herbology and acupuncture
and modern Western-oriented medicine. This approach has contributed significantly to the high level of health care enjoyed in China today. Other
developing nations need to utilize the best available from both traditional and
modern medicine if in the foreseeable future adequate health is to be enjoyed
by their peoples.

II. PLANTS IN MODERN MEDICINE AND DENTISTRY

The common denominator of all traditional medicine has been the wide use of plants for healing. Have plants or their derivatives that trace their relevance to indigenous practitioners been applied to modern medicine and dentistry? Have long-ranged experiments utilizing empiricism contributed substantially to the current materia medica? Have biomedical researchers been able to augment traditional herbal medicine by their unique discoveries? Selected answers to these questions will be discussed using examples from the eighteenth to the current century.

A. Circulatory System

Digitalis or foxglove (Fig. 1) is a genus of European plants used for centuries to treat edema by farmers and housewives. The home remedies contained up to 20 different herbs, but it remained for William Withering to select *Digitalis* as the plant probably possessing the active ingredients. In 1785 he published his monumental account describing the diuretic effect of foxglove and associating it with a remarkable power over the heart. Since that time the glycosides from *Digitalis*, and also the African *Strophanthus*, remain the most important compounds in the treatment of congestive heart failure, atrial flutter, and fibrillation. They act by increasing cardiac output, relieving pulmonary congestion, and, as observed by Withering, relieving peripheral edema.

FIGURE 1 *Digitalis purpurea* (foxglove).

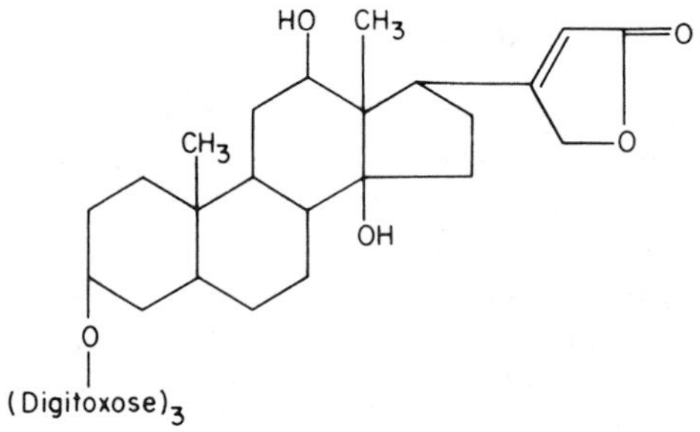

(Digitoxose)$_3$

FIGURE 2 Digoxin.

The prototype and most widely used glycoside is digoxin (Fig. 2) with an intermediate duration of action (36-hr biological half-life). It is extracted from the leaves of *Digitalis purpurea* and *D. lanata*. Digitoxin is also obtained from the former species, a less used glycoside with prolonged action time (5- to 7-day biological half-life) and more toxic side effects. In contrast, ouabain from the seeds of *Strophanthus gratus* has a rapid onset and short duration of action, with maximal effects being obtained 30-60 min after intravenous administration (21-hr biological half-life). It is used in emergencies and for research purposes.

Intoxication with *Digitalis* glycosides is common and hazardous, and in general hospitals they are the most frequent of all adverse drug reactions (Goth, 1981). Recently, an elderly couple died after drinking tea prepared from the leaves of *D. purpurea* (*Morbidity Mortality Weekly Report*, 1977), and Dickstein and Kunkel (1980) reported an additional case of poisoning from tea made from the leaf. Toxic symptoms included anorexia, ventricular ectopic beats, and bradycardia in mild to moderate intoxication that may progress to nausea, headache, malaise, and ventricular premature beats. In severe intoxication, symptoms are characterized by blurred vision, disorientation, diarrhea, ventricular tachycardia, and sinoatrial block that may lead to ventricular fibrilation. These symptoms often terminate in death.

These data show that the toxic dose of cardioactive glycosides is close to the therapeutic dose, and consequently that all therapy must be individualized. Nevertheless, the magnitude of the need for these drugs is suggested by the estimate that more than 3 million cardiac sufferers in the United States routinely use digoxin (Lewis and Elvin-Lewis, 1977), a tremendous mandate for the success of an herbal medicine in modern medicine.

Whenever Mahatma Gandhi felt the need to induce a state of philosophical detachment, he sipped tea brewed from the leaves of the plant that grows wild in India and in most of the world's tropical lands. For centuries, the plant was widely used for its calming effect. Holy men chewed it while meditating. Native medicine men employed it to treat highly agitated mental patients; it was even used to soothe fretful babies. For a long time the plant was ignored by Western scientists, for what plant could possibly live up to all the extravagant claims made for it? But in 1952, a root extract from the plant made its debut in American medicine as the first modern tranquilizer. It was the indole alkaloid reserpine (Fig. 3) from *Rauvolfia serpentina* (Monachino, 1954). The discovery of reserpine, together with the later introduction of chlorpromazine,

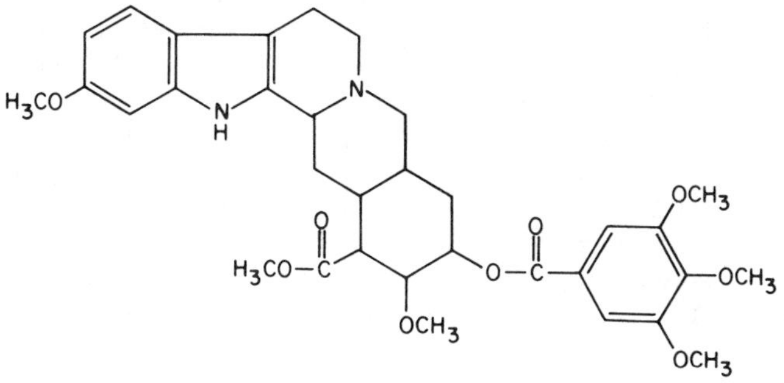

FIGURE 3 Reserpine.

opened up the entire field of psychopharmacology, for heretofore there was no effective pharmacotherapeutic agent available for the treatment of major mental disease. A vast improvement in the management of psychotic patients was now possible because of the central tranquilizing effects of reserpine that caused depletion of the brain monoamines—serotonin, norepinephrine, and dopamine. Subsequently, it was shown that reserpine also decreased sympathetic functions, that were hypotensive and bradycardiac, because of catecholamine depletion at adrenergic neurons. Today the drug is not much used in mental health therapy, but it remains one of the most important drugs employed to relieve high blood pressure.

Drug therapy now controls about 80% of all cases of hypertention. Drugs do not cure, but their control of this disease marks a tremendous change in the outlook for patients whose inflexible fate until the 1950s was a stroke, heart failure, or kidney failure. Today, just a few decades later, the ability to live a reasonably normal and healthy life represents one of the great advances of medicine in the twentieth century, an advancement attributable directly to the historic use in traditional medicine of *Rauvolfia* (Lewis and Elvin-Lewis, 1977).

B. Musculoskeletal System

1. *Neuromuscular System*

When Danile Bovet received the Nobel Prize for Physiology and Medicine in the 1950s for his production of a synthetic form of curare, that accomplishment recognized what has been described as one of the greatest advances of the past 20 years in clinical medicine—the introduction of muscle relaxants into anesthetic practice. It is no exaggeration to suggest that the relaxant drugs revolutionized the techniques of anesthesiology.

What preceeded Bovet's discovery of a synthetic curare? The stage was set in the late 1930s when a crude extract of an arrow poison called curare was brought to the United States and Europe from northern South America. Curare had been prepared from the young bark of *Chondodendron tomentosum* (Menispermaceae) and usually one of several species of *Strychnos* (Loganiaceae), mixed with other substances, and boiled in water and strained, or extracted by crude percolation with water. It was evaporated to a paste or syrup of bitter brownish or black resin. The Indians dipped arrows in the paste for hunting and killing game, for they had long realized how effective it was as a paralyzer.

The major ingredient of the arrow poison is *d*-tubocurarine, a quaternary compound having a bis(benzyliso)quinoline structure (Fig. 4). Tubocurarine chloride is now used as a skeletal muscle relaxant to secure relaxation in surgical procedures without deep anesthesia, for the compound combines with acetylcholine receptor sites at neuromuscular junctions and effectively blocks all messages to the muscles. The muscles then lose their tone and become paralyzed for various periods of time.

Tubocurarine is especially useful in abdominal surgery because it produces profound surgical relaxation (comparable to spinal anesthesia) with a minimal concentration of anesthetic agent. There is then prompt postanesthetic recovery and a radical reduction of the possibilities of postoperative complications. The drug is also employed as a relaxant in oral intubation, tetanic seizures, and lumbar and scaroiliac manipulations, and is useful in the diagnosis of myasthenia gravis. It is also used to relieve spastic paralysis, in the reduction of fractures and dislocations by eliminating muscle pull during setting or reduction, and in other areas where profound muscle relaxation is desired.

Many synthetic muscle relaxant agents are now available, but their synthesis was prompted by the discovery of the action of natural *d*-tubocurarine. Although all act at the neuromuscular junction, the mode of action of most synthetics, such as decamethonium and succinylcholine, differs from the blocking action (competitive antagonist) of *d*-tubocurarine by effecting depolarization of the muscle just as acetylcholine does (Carrier, 1970).

Menispermum canadense (moonseed), a common plant of eastern North America, possesses a similar isoquinoline alkaloid, dauricine, that has curarelike action. Recently, an Osage Indian told us (Lewis and Elvin-Lewis, 1983) that the probable reason why extracts from these plants had not been used as poisons by native Americans was that rattlesnake venom, dried at the end of an arrow tip, had been widely utilized and thus the search for plant poisons was not a viable alternative. The empirical method gave the South American Indians a number of effective muscle relaxant poisons that have contributed significantly to modern medicine, but a parallel selective process did not occur in North America and as a result the curarelike *Menispermum* alkaloid has not entered the modern pharmacopeia.

FIGURE 4 *d*-Tubocurarine.

2. Skeletal System

There has been essentially no incorporation of knowledge from traditional medicine into orthopedic medicine. Even so, plants exist that may be pertinent to the healing of bones. *Ehretia cymosa* (Boraginaceae), for example, is used by the herbalists of Ghana for treating fractures of the leg by wrapping the site in a leaf poultice that is covered with whole leaves and strengthened with splints (Lewis and Elvin-Lewis, 1979). Inflammation is reduced rapidly and healing of the fracture is quicker than by conventional methods, presumably because of the presence of both anti-inflammatory and healing compounds in the leaves. However, a much better indication of the resistance to herbal innovation by many orthopedic surgeons is illustrated by the controversy surrounding the use of the proteolytic enzyme chymopapain.

Chymopapain and the related sulfhydryl protease papain are obtained from the unripe fruit of papaya (*Carica papaya*, Fig. 5). Papain is extensively used in tenderizing commercial meats and in instant meat tenderizers for domestic use, and for stabilizing and chillproofing beer (by hydrolyzing proteins that normally are soluble at room temperature but likely to precipitate from chilled beer). The crude latex has traditional uses in treating a wide variety of illnesses or conditions, including infected wounds, sores, ulcers, chronic diarrhea, tumors, hayfever, esophogeal obstruction, catarrh, and psoriasis (Leung, 1980).

In 1964, Smith reported on his experiences with injection of chymopapain into the intervertebral disks of patients who had herniated disk syndrome. He coined the term "chemonucleolysis" for the injection of chymopapain into the intervertebral disk. The enzyme was subsequently used in the United

FIGURE 5 Fruit of *Carica papaya* (papaya).

States and Canada for about 17,000 patients and widely reported as effective
(Ford, 1977; Javid, 1980; McCulloch, 1980). Yet based on a single study of
104 patients that has been criticized for its small number of poorly selected
patients, lack of experience of the surgeons performing the injections, and
possible pharmacologic activity of the placebo, the U.S. Food and Drug
Administration (FDA) prohibited its use for treating herniated lumbar disks
in 1975. Currently, chemonucleolysis is an accepted treatment in Canada and
many other countries.

Chymopapain liquifies that part of the slipped disk that pinches a patient's
spinal nerve. The classical treatment in severe cases is surgical removal of
the herniated part, and in properly selected cases there is a 70-75% success
rate following surgical removal. But many individuals fear open-back surgery.
It is not without risks, and it requires hospitalization for 5-10 days with concomitant trauma and major expense. In contrast, the more conservative enzymatic treatment requires only a brief hospital stay (maximum 3 days) or none
at all (many individuals are treated as outpatients), results in low incidence
of complications (3%, mostly allergic sensitivity that can be alleviated by including steroids in the injection), and allows for greatly reduced trauma and
cost. Success rate is similar as with surgery, and, therefore, injections
would eliminate the need for 70-75% of about 200,000 annual disk operations in
the United States. Failure would still permit patients the option of surgery.
Clearly, chemonucleolysis is a viable procedure to bridge the wide gulf between standard conservative treatment and surgery (McCulloch, 1980). It is
unfortunate for so many patients that some surgeons and the FDA believe
otherwise.

The U.S. manufacturer of chymopapain is Trevenol Laboratories Inc., who
have recently (1981) resubmitted documentation to the FDA for approval to
market new products for chemonucleolysis.*

C. Skin

The skin is the largest human organ and, as much of it is exposed, humans
have plied it with many plant extracts for cosmetic purposes. Most important
medically has been the selection and use of plants to heal wounds and burns,
to stop bleeding, and to fight infections. Two examples will illustrate their
significance.

1. Burns

Modern medical reports using aloe leaves (Fig. 6) date from 1935, when a
patient suffering from facial x-ray burns was treated. Itching and burning
subsided in 24 hr, and within 5 weeks the 4 × 8 cm area had completed regeneration with no scar. After 3 months the area was pigmenting normally along
with other exposed skin surfaces (Collins and Collins, 1935). A similar experience of x-ray hand burns was reported by Loveman (1937). Brown (1964,
1967) and Brown et al. (1963) advocated the use of aloe gel in radiation burns
for relief from pain and itching, and to keep down keratosis and ulceration,
thus slowing any possible change toward malignancy. Zawahry et al. (1973)
successfully treated leg ulcers and dermatoses using fresh gel from aloe leaves.

---

*Chymodiactin approved for use in the United States November 12, 1982.

FIGURE 6 *Aloe barbadensis* (aloe) leaves. (Photo by V. E. Zenger.)

In domestic medicine the popularity of aloes as an aid to itching, minor cuts, and first- and second-degree thermal and other burns is spreading rapidly. No household should be without an aloe plant for treating these skin afflictions, essentially as did the ancient Egyptians, who advocated aloe leaf pulp for treating burns, ulcers, and parasitic infections of the skin (Zawahry et al., 1973).

2. Infections

Under development at the University of California (Berkeley) and Exover Corporation are antiherpetic substances derived from a number of red algae (e.g., *Cryptosyphonia woodii* and *Farlowia mollis*, Fig. 7) (Richards et al., 1978). One of these antiviral agents has been identified as a carbohydrate-protein complex that appears to affect the herpes virus by interference with its absorption to the host cell. Subsequent investigations have shown that the protein is required for antiviral action in vivo, whereas the polysaccharide fraction alone is sufficient for activity in vitro. Nonetheless, enhanced activity is affected if the protein is combined with the polysaccharide. Studies with mice have shown that systemic and skin infections of herpes simplex virus can be controlled when this substance is given parentally or topically, respectively. The fact that this agent is somewhat more effective against herpes simplex virus 2 rather than herpes simplex virus 1 infections also suggests that it may be particularly valuable in the treatment of genital herpes.

FIGURE 7 Red alga (*Farlowia mollis*). (Specimen courtesy of H. W. Nichols, Washington University, St. Louis, Mo., photo by V. E. Zenger.)

D. Gastrointestinal System

A bewildering array of plant extracts are available for the alleviation of most symptoms involving the gastrointestinal tract. Countless plants known to traditional medicine are used today for stimulation of digestion and indigestion, and for emetics, antiemetics, and purgatives. The same ones are employed in many parts of the world to control diarrhea, flatulence, and spasms, to kill worms and to destroy amebic infections, and to treat liver complains and hemorrhoids.

1. Purgatives

Human preoccupation with the bowels has a long history, although the laxative abuse syndrome is fortunately on the wane. Nevertheless, some 700 purgatives are available for purchase in the United States alone, the most

commonly used ones being the irritant cathartics containing anthraquinone compounds (e.g., emodin, Fig. 8).

Even though widely used in self-medication today and with historic use
dating from Alexander the Great and Dioscorides, the anthraquinone glycosides are excessively irritating and are probably responsible for many cases
of toxicity. The leaves of the garden rhubarb (*Rheum rhabarbarum*), for
example, contain glycosides of anthraquinones, which probably are responsible
for the poisonings associated with leaf ingestion. Most plant cathartics (senna,
buckthorn, aloe, dock) are also available as herbal teas and, although long
shelf life destroys activity by oxidation, plant material kept up to a year is
sufficiently active to cause diarrheal illness after drinking two cups of tea
(*Morbidity Mortality Weekly Report*, 1978). According to this report six
individuals developed severe abdominal cramps and profuse watery diarrhea,
and two women experienced palpitations.

2. Peptic Ulcers

The incidence of peptic ulcers in the United States has declined dramatically over the past 20 years, although an estimated 150,000 operations are
still conducted annually and about 8000 Americans die annually from bleeding,
perforation of the stomach, and other complications of ulcer attack. Ulcers
occur in the stomach or more commonly the duodenum, possibly because of
excess production of stomach acid.

Until the mid-1960s there was no drug available to assist in healing the
ulcer once developed; rest in bed, no smoking, and bland diets were the only
forms of therapy. From that time on there have been dramatic improvements
in healing ulcers: at least nine drugs, with apparently different pharmacological properties, have been reported to heal approximately 75% of patients
within a month or so (Wormsley, 1977). The most popular in the United States
are the histamine blockers cimetidine and ranitidine, but healing rates achieved
are no greater than those following treatment with carbenoxolone sodium
(synthesized from the aglycone glycerrhetinic acid obtained from glycyrrhizic
acid of licorice roots, *Glycyrrhiza glabra*), colloidal bismuth agents, sucral-
fate, and indeed antacids (Marks, 1980). In fact, use of placebo among
American sufferers of peptic ulcers gives a surprisingly high rate of healing,
up to 60%, compared, for example, to a maximum of 29% in the United Kingdom
(Wormsley, 1977).

Of the drugs discussed, only the licorice-derived carbenoxolone sodium
(Fig. 9) is prohibited from general use by the FDA in the United States. It
and deglycyrrhinized licorice can on the average reduce the size of an ulcer
by 70-90% after 1 month of treatment, but side effects, such as headache,

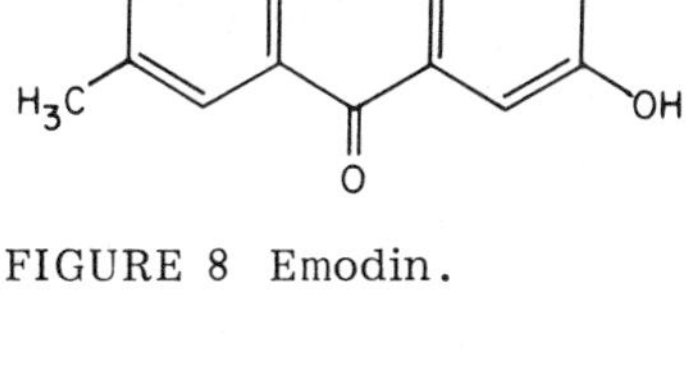

FIGURE 8 Emodin.

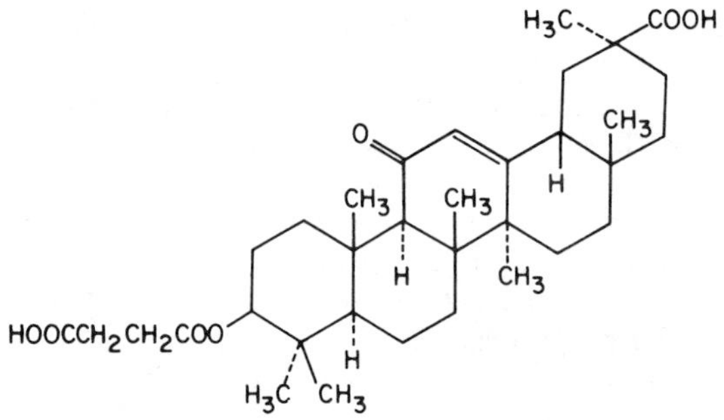

FIGURE 9 Carbenoxolone.

hypertension, edema, and potentially lethal hypokalemia in about 20% of patients restrict its usefulness. Yet, according to Sircus (1972), benefits
derived from the use of licorice extracts in the treatment of peptic ulcers far
outweigh these undesirable effects.

Ingestion of as much as a quarter pound of licorice candy daily may induce
abnormal cardiac behavior through hypokalemia and incipient kidney failure
(Robinson et al., 1971), and even lead to cardiac arrest (Bannister et al.,
1977). Chewing tobacco is soaked in licorice extract to improve its flavor and
swallowing too much of the tobacco juice can also lead to hypokalemia (Blackley
and Knochel, 1980). Nevertheless, the use of licorice root in moderate
amounts to treat adverse stomach and digestive symptoms has long been
valued in traditional medicine and in seventeenth-century Europe was believed
capable of clearing inflamed stomachs (Lewis and Elvin-Lewis, 1977).

Any amount of licorice root may be purchased from retail stores even
though it is now clear that a deleterious cardiovascular condition could be
initiated or exasperated by excessive use of licorice. Herein exists an obvious paradox. The FDA limits the use of licorice extracts under prescriptions
by physicians, yet allows the root to be sold freely as food or in tobacco products. Obviously, the health of the public is not being served by either
approach.

E. Oral Hygiene

The selection of plant species used for teeth cleaning throughout the world
is not made indiscriminately. Wherever they are now used, or have been used
in the past, these species are carefully chosen for their appealing flavor and
texture, and with the belief that their use maintains good dental health (Elvin-
Lewis, 1979, 1980). In folk dentistry, it is assumed that some also have added
therapeutic value for the purpose of healing and strengthening affected gums,
to ease pain and heal toothache, or to cure oral mucosal infections. This
empirical rationale is with foundation. For example, in many parts of West
Africa, where the technique of utilizing sticks (roots or twigs), fibrous
sponges, fruits, and gums for teeth cleaning is still a popular practice, caries
rates are often extremely low. For example, in examining a population of
chewing stick users in southern Ghana, it became apparent that their

preferred choices also conferred a better state of dental health than those less popular (Elvin-Lewis et al., 1980). Moreover, these species exhibited antibiotic activity, especially against odontopathic microorganisms, that could be correlated with their clinical efficacy (Elvin-Lewis, 1979). Also in these particular species, or those closely related to them, other active substances related to either astringent, healing, analgesic, or foaming properties have been identified (Elvin-Lewis, 1982). The examination of chewing stick species used throughout the world for teeth cleaning has revealed similar patterns, and thus observations made through folk dentistry can serve to identify plants, and their products, that have the potential to be utilized in the development of dental hygienic products (Elvin-Lewis, 1983).

The placing of known species into phylogenetic groups has revealed the presence of active compounds that are shared by a number of disparent members, thus making the selection of candidate species easier, especially when availability or associated toxic substances can be a factor. For example, in phylogenetic group 1, *Sassafras albidum* used as an Appalachian chewing stick, is also the source of an oil once used to disinfect root canals and to treat toothache. Among those active ingredients that have been identified is the antibiotic eugenol (Fig. 10) and a number of alkaloids with possible analgesic properties (Elvin-Lewis, 1983; Leung, 1980). However, since sassafras oil contains large amounts of the potentially hepatocarcinogenic compound, safrole (Fig. 11) (Segelman et al., 1976), eugenol that is still used in root canal therapy is usually obtained from *Syzgium aromaticum* (of phylogenetic group 8), which is the source of clove oil.

In phylogenetic group 2, another American chewing stick, sweet gum (*Liquidambar styraciflua*) is the source of storax, used commercially as a stimulating expectorant (Friar's Balsam) and in proprietary drugs for the treatment of aphthous stomatitis (Canker sores), acute necrotizing gingivitis (trench mouth), and herpes simplex (Leung, 1980). Storax contains a number of compounds that could contribute to its healing capacity; for example, the glycoside phenolic arbutin (Fig. 12), may act like other phenolics that possess antimicrobial and antiherpetic properties, and salicin (Fig. 13) is known to have anti-inflammatory and analgesic qualities.

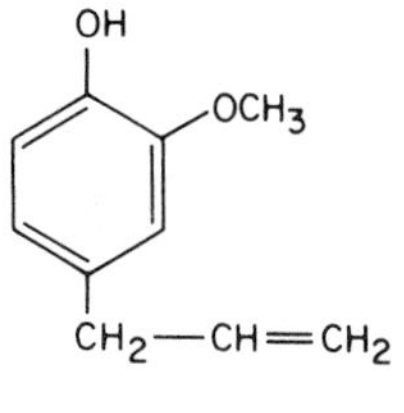

FIGURE 10 Eugenol.

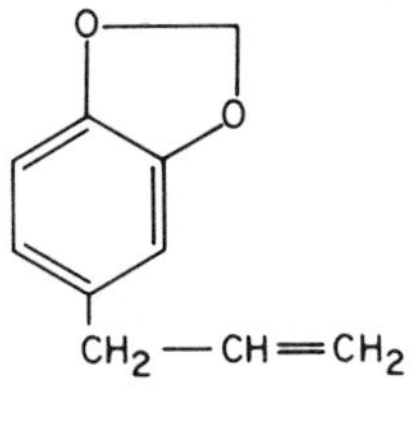

FIGURE 11 Safrole.

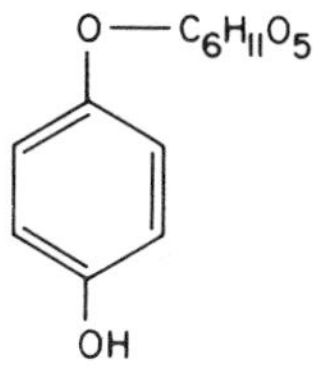

FIGURE 12 Arbutin.

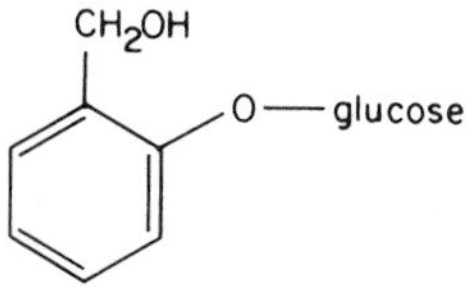

Figure 13 Salicin.

Achyranthes aspera, as a representative of phylogenetic group 3, is not only used for teeth cleaning in North Africa and China, but is used medically for its astringent, hemostatic, analgesic, and healing properties. A number of compounds with possible medicinal value have been isolated from the plant and include achyranthin, oleanolic acid, a sapogenin, and alkaloids.

In phylogenetic group 4, a number of West African species of *Garcinia* are not only very popular for teeth cleaning, but their use has been closely correlated with good dental health they impart (Elvin-Lewis et al., 1980). They also possess a broad spectrum of antibiotic activity (Elvin-Lewis, 1983) which is probably related to their possession of antibiotics such as morellin (Fig. 4), and the α- and β-gutifferins (Fig. 15) that have been isolated from an Indian species, *G. morella*. Also, a number of *Diospyros* species are popular in this region as well, and one, *D. tricolor*, exhibits an antibiotic spectrum that suggests that it may contain, like other species of phylogenetic group 4 (*D. canaliculata, D. mespiliformis*, and *Plumbago europaea*), an antibiotic-like compound plumbagin (Fig. 16).

The American Indian once used a number of *Populus* species that belong to phylogenetic group 5 for teeth cleaning. Today, *Populus* is used as commercial toothpicks and as the source of the balm of Gilead buds used in cough medicine (Leung, 1980). *Populus* contains a number of potentially active compounds, including the anti-inflammatory salicin and populin, and antibiotics such as cinnamic acid, α-d-bisabolol, and trichocarpin (Fig. 17).

Within phylogenetic group 6, a number of active principles have been isolated from *Hibiscus* and *Gossypium* species and include salicyclic acid, with known analgesic and anti-inflammatory properties, and polyphenols, with potential antibiotic activity. *Hibiscus rosa-sinensis* extracts are an ingredient of a Chinese toothpaste, although its efficacy as a dentrifrice is unknown.

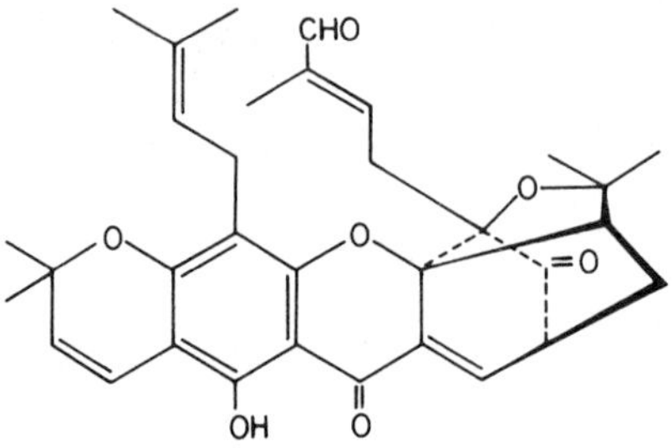

FIGURE 14 Morellin.

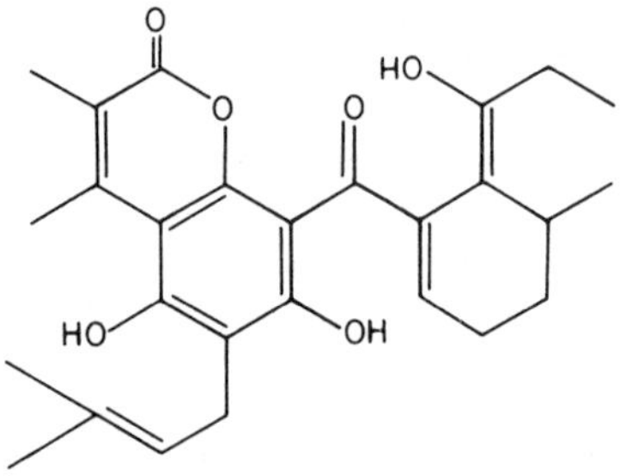

FIGURE 15 α-Gutifferen.

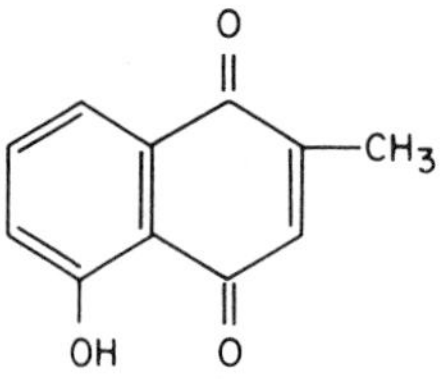

FIGURE 16 Plumbagin.

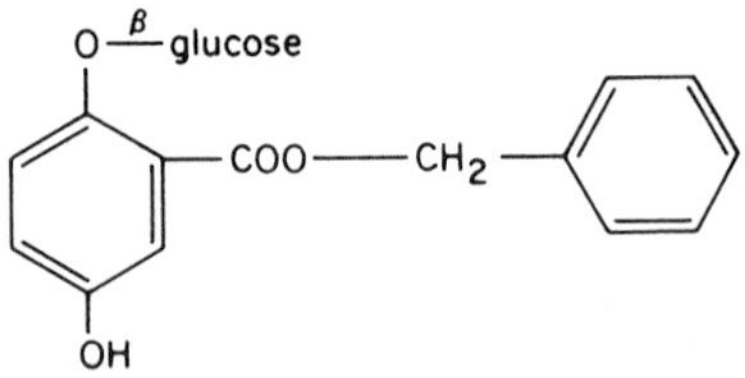

FIGURE 17 Trichocarpin.

Also, a Caribbean member of this group, *Gouania lupiloides*, long used for teeth cleaning, has been incorporated into a dentrifrice now under development in Jamaica. This species has broad-spectrum antibiotic activity that may be associated with its foamy, detergentlike saponin. Antibiotic activity is also widespread in the Euphorbiaceae, including those used for chewing sticks, such as the African *Alchornia cordifolia*. Those antibiotics that have already been identified in this family include selobicin from *Croton sellowii* and urensin from *Cnidoscolus urens*, although neither species are used for teeth cleaning.

Antibiotic activity is common in plants found in phylogenetic group 7 and a number of antibiotics have been identified, although not specifically from species used for teeth cleaning (Lewis and Elvin-Lewis, 1977; Elvin-Lewis, 1983). Ethyl gallate has been isolated from several *Acacia* species, and others used for teeth cleaning have been found to exhibit antibiotic activity. It has been suggested that the selection of *Acacia* among individuals with existing periodontal disease may be related to the presence of cachotannic or catechol tannis and organic fluorides, that could exert both an astringent and anticarciogenic effect, respectively (Elvin-Lewis, 1983).

Perhaps the best example of plants used in dentistry in phylogenetic group 8 is *Syzygium aromaticum*, the source of clove oil used for its eugenol in root canal therapy, and to treat toothache. The antibiotic spectrum of *Terminalia* species, used as chewing sponges in West Africa, suggests the possession of a similar active compound (Elvin-Lewis, 1983).

West African *Fagara* species, of phylogenetic group 9, contain a number of compounds that can be associated with their use not only for teeth cleaning but also for their analgesic and counterirritant properties in the treatment of toothache, aphthae, rheumatism, snake bits, and cough. These include a muscle irritant, atarine, an anticancer agent 8 methyldihydronitidine, and antibiotics such as the antimicrobial phenol, 7-dimethyl suberosin, chelerythrine, berberine 1,6-canthine, and others not yet identified. Berberine (Fig. 18) has also been isolated from *Zanthoxylum clava-herculis*, known as the toothache tree.

In this same group, the widespread popularity of the Asian *Azadirachta indica* in Africa and India for teeth cleaning and other medicinal uses has led to a few limited studies on its medicinal and dental value. Its oil contains antianthelmintic, insecticidal, healing, and anti-inflammatory properties. In folk medicine, it is used for its healing and antimalarial properties, and is known

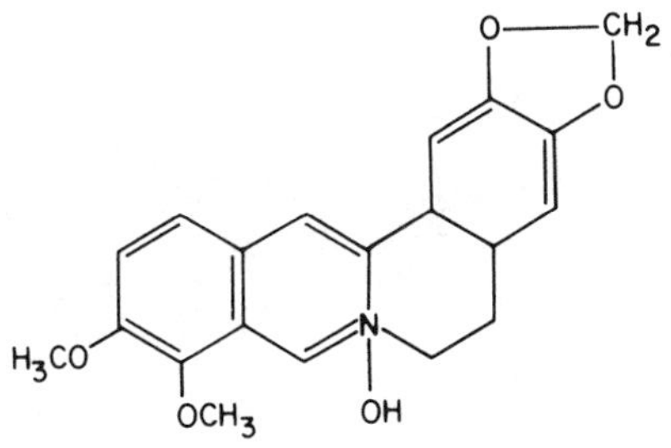

FIGURE 18 Berberine.

to contain a hypoglycemic capacity similar to that of the sulphylureas (Elvin-Lewis, 1983). For a number of years, extracts of its oil have been used in an Indian toothpaste that has been found in limited clinical trials to reduce gingival inflammation (Rathje, 1971; M. Elvin-Lewis, unpublished observations, 1979). It has been proposed that components of neem oil, known to raise the oxidation potential of erythrocytes and thus prevent the propagation of malarial parasites within them (Etkin, 1979), may act in a similar manner in the gingival sulcus, so that anaerobic microorganisms responsible in part for the development of periodontal disease cannot proliferate in numbers large enough to cause gingival inflammation (Elvin-Lewis, 1979).

Salvadora persica of phylogenetic group 10 is another popular Afro-Asian chewing stick whose extracts have been incorporated into a toothpaste. Isothiocyanates are an important component of these species, and may, through enzymatic action of saliva, release unstable products that exert antimicrobial activity that has yet to be detected by standard procedures (Elvin-Lewis, 1983).

Although antibiotic activity among genera in phylogenetic group 11 is widespread, the only specific antibiotics that have been identified in those used for teeth cleaning are the phenols, eugenol and thymol (Fig. 19) that have been isolated from species of Lamiaceae in Africa (Watt and Breyer-Brandwijk, 1962).

Within phylogenetic group 13, it is probable that the use of the pedicel or peduncle of the coconut palm (*Cocos nucifera*), date palm (*Phoenix dactylifera*), and *Pandanus* is related to their availability and the bristle that is formed. Nothing is known of antibiotic potential.

However, the pounded peduncle and inflorescence of plantain (*Musa sapientum*) of phylogenetic group 15, with charcoal added, used by Ashanti (Ghanian) women and children is particularly efficacious (Elvin-Lewis et al., 1980). Nonetheless, although the plant contains broad-spectrum antibiotic activity, it is not clear if those parts used for teeth cleaning share this potential. Should this be the case, the widespread occurrence of the species in the tropics would make it an excellent candidate for use among those that cannot afford the toothbrush and toothpaste, particularly in the neotropics.

Therefore, these studies have shown that plants used for teeth cleaning have a great potential for yielding compounds that can be used in modern dentistry. Once ethnobotanical investigations are complete it is likely that the best candidates will be those that have been identified by their popularity and efficacy.

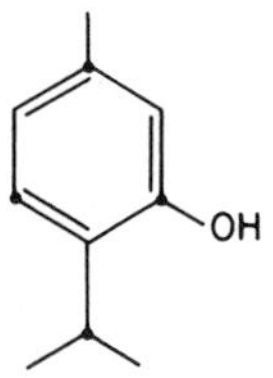

FIGURE 19 Thymol.

F. Reproductive System

The development in the early 1950s of semisynthetic steroids opened the way
for the mass use of these compounds as oral contraceptives. The standard
female contraceptive pill combines an estrogen with a progestin. These hormones have two major functions: to suppress the pituitary hormones through
negative feedback, and to stimulate the endometrium of the uterus, initiating
a menstrual cycle.

Most steroidal drugs, whether destined for contraceptive pills or topical
corticosteroids, are commonly based on several different sapogenins, the most
frequent starting material being diosgenin from various *Dioscorea* (yam)
species. There is an enormous medical demand for human sex hormones and
for cortisone and its analogs, the steroid nucleus almost exclusively originating from plants and particularly yams (Lewis and Elvin-Lewis, 1977).

Although the widespread use of the female contraceptive pill has to some
extent revolutionized social standards and behavior in the latter part of the
twentieth century, the search for contraceptives need not be restricted to
female use. Predictably, men should share the contraceptive burden with
women, so that both would get the health benefits of periodically discontinuing
the drug should an effective one be found for the male.

Because a physician in eastern China during the late 1950s observed the
high incidence of childless marriages among couples consuming large quantities
of crudely processed cottonseed oil in cooking, and also because doctors in
central China found that women raised on a cottonseed oil diet who married
outside the region had families at the normal rate whereas those who wed men
from the same area often remained childless, Chinese researchers decided to
test the oil for contraceptive purposes (Wen, 1980). Further studies conducted in the 1960s showed that males were more susceptible to antifertility
effects than females. Animal studies completed in 1971 confirmed that gossypol
(Fig. 20) was the effective nontoxic antifertility constituent in crude cottonseed oil (*Chinese Medical Journal*, 1978).

Clinical trials of gossypol began in 1972 in China and, to date, more than
10,000 men have been studied (Maugh, 1981). Each received a daily oral dose
of 20 mg until sperm count was sufficiently reduced—about 2 months—with subsequent maintenance doses of 75-100 mg taken twice monthly. Among the first
4000 men who received the drug for periods ranging from 6 months to 4 years
it was found 99.89% effective. Complete recovery after discontinuation of
gossypol took about 3 months. Side effects at effective doses were minimal.
A few individuals failed to produce normal frequencies of sperm for as long as
2 years after they stopped taking the pill (Wen, 1980), and a small number of

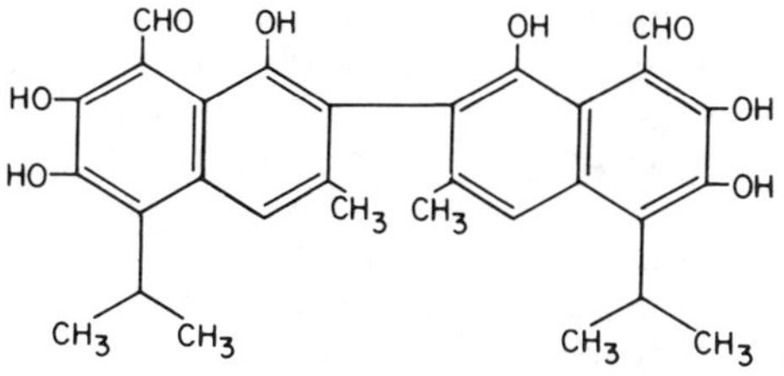

FIGURE 20 Gossypol.

men developed hypokalemia (*Chinese Medical Journal*, 1978) (percentages were not given), disturbing consequences requiring further study. More transient during the first few weeks of the trials were general weakness (12.8%), increased (3.0%) and decreased (2.4%) appetites, gastric discomforts (2.0%), nausea (1.0%), and about 6% complained of decreased libido. Recovery from these symptoms usually followed reassurances of symptomatic treatments.

The mechanism of gossypol action is poorly understood. In an appropriate chemical form (e.g., gossypol-polyvinylpyrrolidone), it is spermicidal in humans (Waller et al., 1980) and mice (Coulson et al., 1980). Serum testosterone levels are significantly reduced in rats after 1 week of treatment, suggesting that reduction or absence of sperm induced by gossypol may be secondary to decreased testosterone biosynthesis (Lin et al., 1981).

Following this Chinese discovery, research is now under way elsewhere to better understand the action of gossypol, for more data are needed before the compound could be used as a male contraceptive in the West. In time, however, it may represent another example of a plant product introduced into modern medicine from current observations and experimentation that will greatly alter human lifestyles.

G. Multiple-System Affectors

1. Lectins-Mitogens

In the folk medicine of India, Egypt, and China, castor oil from *Ricinus communis* (Fig. 21) was used for centuries as a cathartic, and externally to treat sores and abscesses. The seeds, from which the oil is derived, are known to be poisonous to cattle, and ingestion of only two to four seeds by humans can cause nausea, muscle spasms, and purgations, whereas as few as eight seeds can result in convulsions and death. Before the turn of the century (1888), a phytotoxic protein, ricin, was extracted from the seeds and found to have the curious ability to agglutinate erythrocytes. This same agglutinating property was also associated with the protein, abrin, isolated from seeds of *Abrus precatorius* (the rosary beads used for jewelry) (Fig. 22). Also like *R. communis*, *A. precatorius* was used in folk medicine; in Panama and Nepal, for example, natives ingest the cooked (or otherwise treated) seeds as an aphrodisiac, although eating only four raw seeds can cause severe gastrointestinal symptoms, circulatory collapse, coma, and death (Lewis and Elvin-Lewis, 1981).

Their toxic and associated agglutinating properties caused Paul Ehrlich in 1891 to use ricin and abrin in fundamental immunologic studies to demonstrate the concept of immunospecificity of toxins and thus began the use of plant toxins in modern medicine (Sharon, 1977). Since then these compounds have been further investigated and the toxic and nontoxic, but lectinic, components identified. The mechanisms of these two phytotoxins are identical in that both bind to the cell surface by their lectinic moiety and then their toxic portion is pinocytosed where it inhibits chain elongation in protein synthesis. Also, their lectinic affinities are somewhat similar in that they both have an affinity for β-D galactopyranosyl groups, although only *Abrus* lectins are blood group specific. Thus, in spite of other properties, which may be toxic in nature, a lectin is defined as a protein with carbohydrate-binding properties, but is not enzymatic, an antibody, nor a substance that participates in carbohydrate

FIGURE 21 *Ricinus communis* (castor bean). (Photo by V. E. Zenger.)

transport. The term "lectin" was derived from the Latin *legere*, which means to select or pick out. Although screening techniques have detected thousands of potentially useful compounds, only a few of these phytohemagglutinins have been characterized and purified. Also, since a lectin's activity can be inhibited by simple sugars, these proteins have also been found in bacteria, fungi, lichens, fish roe, snails, eels, and even mammals (Goldstein and Hayes, 1978). A common nomenclature of lectins has yet to be derived, although it has been proposed that each should be designated by the genus and species from which it was obtained followed by the sugar-binding specificity in parenthesis. This would exclude the popular use of generic (e.g., ricin, concanavalin A, abrin, and favin), common (pea, soybean, lima bean) or nonsense derivations (PHA-phytohemagglutinin from *Phaseolus vulgaris*), or the blood type it may agglutinate (e.g., A_{DB} = *Dolichos biflorus* agglutinates blood type A). Table 1 is a list of plant lectins whose blood group specificity and blood group binding

TABLE 1 Plants with Lectinic (Including Mitogenic) Properties[a]

Organism	Common name	Blood group specificity [b]	Carbohydrate specificity	Ligand
Fungi				
Agaricus bisporus	Commercial mushroom	Nonspecific	Sialoglycoprotein	Nonspecific
A. compestris	Meadow Mushroom	Nonspecific	Complex	
Angiosperms: dicotyledons				
Phylogenetic Group 3				
Phytolaccaceae				
Phytolacca americana	Poke, pokeweed	Nonspecific; leucoagglutination; mitogenic	Unknown	
(5 mitogens, Pa-1 to Pa-5)				
P. esculenta (2 mitogens)	Shoriku		Unknown	
Chenopodiaceae				
Bassia decurrens		Mitogenic		
Rhagodia crassifolia		Mitogenic		
Amaranthaceae				
Amaranthus caudatus	Inca wheat		β-D-Galp	
Phylogenetic Group 4				
Brassicaceae				
Iberis spp.				
Ericaceae				
Gaultheria procumbens	Wintergreen	M		
Phylogenetic Group 5				
Cucurbitaceae				
Momordica charantia	Balsam pear		Polysaccharide	

		Specificity	Sugar hapten	Best inhibitor
Phylogenetic Group 6				
Sterculiaceae				
Sterculia foetida		Mitogenetic (sterculic acid)		
Moraceae				
Maclura pomifera	Osage orange	Nonspecific	D-Galp, D-GalNAcp	D-Galactose
Euphorbiaceae				
Ricinus communis[c]	Castor bean[c]	Nonspecific	β-D-Galp > α-D-Galp	Lactose; *p*-nitro-phenyl-β-D-galactopyranoside
Phylogenetic Group 7				
Fabaceae:Caesalpinoideae				
Griffonia (Bandeiraea) simplicifolia (I)		B, AB >> A$_1$	α-D-Galp > α-D-GalNAcp	Methyl-D-galacto-pyranoside
(II)		T$_k$	β-D-GlcNAcp = α-D-GlcNAcp	
Bauhinia carronii	Queensland bean			
B. purpurea	Camel's foot tree	β-D-Galp		
Fabaceae:Faboideae				
Abrus precatorius	Rosary or jequirity bean[c]	B, O > A	β-D-Galp > α-D-Galp	Lactose
Arachis hypogaea		Neuramidase-digested A, B, O, or T antigen	β-D-Galp-(1→3)-D-GalNAcp	
Canavalia ensiformis (Con A)	Jack bean	Nonspecific; mitogenic	α-D-Manp > α-D-GLcp > α-D-GLc-NAcp	Methyl-D-galactose Pyranoside 2-Acetamido-2 Deoxy-D-galactose Lactose?
Caragana aborescens	Pea tree (I) Pea tree (II)	A, B, O (weak)		
Crotalaria juncea	Sunn or san hemp		D-Galactose	
Cytisus sessifolius	Broom	O > A$_2$ > A$_1$	β-D-GlcNAcp-(1→4)	N,N'-Diacetylchito-biose?

TABLE 1 (Continued)

Organism	Common name	Blood group specificity	Carbohydrate specificity[b]	Ligand
Dolichos bifloris	Horse gram	A1 >> A2	α-D-GalNAcp >> α-D-Galp	2 Acetamido-2-deoxy-D-galactose
Glycine max	Soybean	A > O > B	α-D-GalNAcp > GalNAcp >> α-D-Galp	
Lathyrus odoratus	Sweet pea		Polysaccharide	
Lens culinaris	Lentil	Nonspecific	α-D-Manp > α-D-Glcp, α-D-GlcNAcp	D-Glycopyranoside
Lotus tetragonolbus	Winged (asparagus) pea	O >> A2	α-L-Fucp, 2-O-Me-D-Fucp	L-Fucose
Phaseolus lunatus	Lima bean	A1 > A2 >> B	α-D-GalNAcp > α-D-Galp	III-Methyl-2-acetamido-2 II-Deoxy-D-galacto-pyranoside
P. vulgaris (PHA)	Kidney bean	Nonspecific, mitogenic	β-D-Galp-(1→4)-β-D-GlcNAcp-(1→2)-α-D-Manp β-D-Galp-(1→4)-β-D-GlcNAcp-(1→2)-α-D-Manp	
Pisum sativum	Garden pea	Nonspecific	α-D-Manp > α-D-Glcp > α-D-GlcNAc Sialoglycoprotein	D-Mannose; methyl-D-glucopyranoside
Robinia pseudoacacia	Black locust			
Sophora japonica	Japan pagoda tree	A > B >> O	β-D-GalNAcp > β-D-Gal > α-D-Galp	
Ulex europaeus	Gorse or furze seed	I-O >> A2 II-O >> A2	α-L-Fucp β-D-GlcNAcp-(1→4)-β-D-GlcNac	L-Fucose
Vicia cracca	Gerard vetch	A1 > A2 Mitogenic		
V. ervilia			α-D-GalpWAc	
V. faba	Fava or broad bean[c]	Nonspecific, mitogenic	α-D-Manp > α-D-Glcp >> α-D-GlcNAcp	

Species	Common name	Specificity	Structure
V. graminea	Vetch	N (N$_{vg}$)	2-Acetamido-deoxy-D-galactose
Wisteria floribunda	Japanese wisteria	Nonspecific	D-Galactosyl-
W. sinensis	Chinese wisteria		
Phylogenetic Group 9			
Juglandaceae			
Juglans nigra	Black walnut	Mitogenic	
Phylogenetic Group 10			
Celastraceae			
Euonymus europaeus	Spindle tree	B + A, A$_2$	α-D-Galp-[α-D-Fucp-(1→2)]-(1→3)-β-D-Galp-(1→3 or 4)-β-D-GlcNac
Loranthaceae			
Viscum album	European mistletoe[c]		D-Galactosyl-
Phylogenetic Group 11			
Solanaceae			
Datura stramonium	Jimsonweed		
Solanum tuberosum	Potato	Nonspecific	β-D-GlcNAcp-(1→4)-[β-D-GlcNAcp-(1→4)]$_2$-β-D-GlcNAc
Lamiaceae		A, N	D-Galactosyl-
Moluccella laevia		T$_n$	
Salvia sclarea			

TABLE 1 (Continued)

Organism	Common name	Blood group specificity	Carbohydrate specificity[b]	Ligand
Angiosperms:monocotyledons				
Phylogenetic Group 14				
Poaceae				
Triticum aestivum (*vulgaris*)	Common wheat (wheat germ)	Nonspecific	β-D-GlcNAcp-(1→4)-β-D-GlcNAcp-(1→4)-GlcNAc > β-D-GlcNAcp-(1→4)-β-D-GlcNac	2-Acetamido-2-deoxy-D-glucose; N,N',N'',N'''-tetraacetyl chito-tetaitol

[a] After Lewis and Elvin-Lewis, 1977 and Goldstein and Hayes, 1978.
[b] Abbreviations: Fucp = fucopyranoside, Galp = galactopyranoside, Glcp = glucopyranoside, Manp = mannopyranoside.
[c] Toxins: *Abrus precatorius* = abrin; *Ricinus communis* = ricin; *Viscum album* toxin; *Vicia faba* toxin.

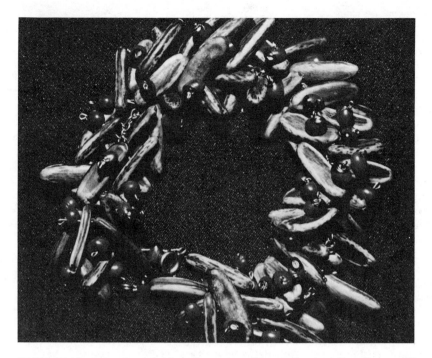

FIGURE 22 Bracelet of *Abrus precatorius* seeds (small and round) and *Delonix regia* seeds.

sites have been deliniated; among these are many that have been chemically and physically defined and are described in detail elsewhere (Goldstein and Hayes, 1978).

In nature it has been proposed that plant lectins have several purposes. For example, lectins in the root hairs of leguminous species may serve to attract specific bacterial species of the genus *Rhizobium* that are found in the nitrogen-fixing nodules of this group; in contrast, lectins may exert a specific inhibitory effect on bacteria, fungi, and beetle larvae that otherwise can destroy their plant seeds. Also, since the concentration of lectins often varies during the life cycle of the plant, and may only be found in the developing seed, involvement in germination and maturation have been suggested. In this respect, concanavalin A and the lectin from *Phaseolus vulgaris* can stimulate germination of pollen (Barondes, 1981).

Bacterial lectins have also been found to be important in the selective adherence to animal cells that they infect. By inference, this may be the way that certain bacteriophage utilize the somatic O antigen (polysaccharide) of *Escherichia coli* as their host receptor, and thus lectinlike proteins may also be identified in viruses. In slime molds it is assumed lectins are responsible for cell-to-cell adhesion during colony formation. In a similar fashion it has been proposed that lectins in developing vertebrates play a role in intercellular interactions during tissue differentiation. Also, in phagocytic cells they may serve as specific receptors to mediate pinocytosis and subsequent degradation of glycoproteins. Specifically, lectins have been implicated in myoblast fusion and nervous system development in the chick embryo and the secretion

of mucin from goblet cells into the intestinal lumen in the adult chicken.
Multiple lectins have been located on lymphocyte cell surfaces as well, where
they are associated with immune response such as lymphoblastic transformation, and on tetracarcinoma stem cells, where they may participate in cell
adhesion. Barondes (1981) concluded that the variation in location and function of one lectin is not dictated by the nature of its binding site alone, but
by opportunistic factors that determine which of its many receptors of adequate complementarity are available in much the same way as a neurotransmitter functions. Since more than one lectin may be present, their interaction could result in a highly specific binding property that is probably
structural in nature.

Originally, lectins were used to determine blood types and secretor status,
and still have these functions in addition to being used to separate leukocytes
from erythrocytes, and to agglutinate red blood cells from heparanized blood
in the preparation of plasma. Other carbohydrate macromolecules may be
detected, isolated, and characterized by their selective use, especially if
conjugated to column substrates such as Sephadex, chitin, agarose, or
Sepharase. Moreover, carbohydrate structures within membranes may be
located and determined by their use of lectins when the latter are conjugated
to ferritin, fluorescein, or radiolabeled isotopes. Since lectins also serve to
aggregate surface glycoproteins, particularly during cell division, they have
been useful in determining the distribution and mobility of these compounds
in normal and malignant cells and on the surfaces of certain viruses. In this
respect, certain lectins can distinguish normal from malignant cells; it has
been proposed that some be used in cancer chemotherapy. Moreover, they
have been useful as models in studying carbohydrate specific antibodies.
Also, the ability of certain mitogenic lectins to stimulate lymphocyte division
in vitro is helpful in studying the growth, differentiation, and division associated with such a phenomenon, especially as it relates to immunogenesis in
T and B cells (Sharon, 1976). Some lectins also have the ability to probe
certain membrane-bound hormone receptors, and thus have been as effective
as insulin in enhancing transport of D-glucose and in inhibiting epinephrine-
stimulated lipolysis in isolated adiopocytes (Goldstein and Hayes, 1978).

Lectins have been further studied from the point of view of how their
toxic properties affect the nutritional value of beans (Liener, 1976) and the
toxicity of mistletoe (Stirpe et al., 1980), which is sometimes consumed as an
herbal tea. Lectinic mitogens, such as pokeweed mitogen, have been found
responsible for inducing severe illness and causing death following either
purposeful or accidental ingestion in humans (Lewis and Elvin-Lewis, 1983).
However, the value of pokeweed extracts (*Phytolacca americana*). in dilute
form, in homeopathic remedies for treating rheumatism and arthritis has yet
to be evaluated.

2. *Allergy*

Bronchial asthma is characterized by paroxysms of expiratory dyspnea and
wheezing, overinflation of the lungs, cough, and rhonchi. Onset may be
sudden or insidious; duration may be brief or last for weeks. Symptoms are
due to bronchial airway obstruction, which occurs as a result of contraction
of bronchial smooth muscle, hypertrophy of the bronchial wall, edema of the
bronchial mucosa, and accumulation of secretions in the bronchial lumens.

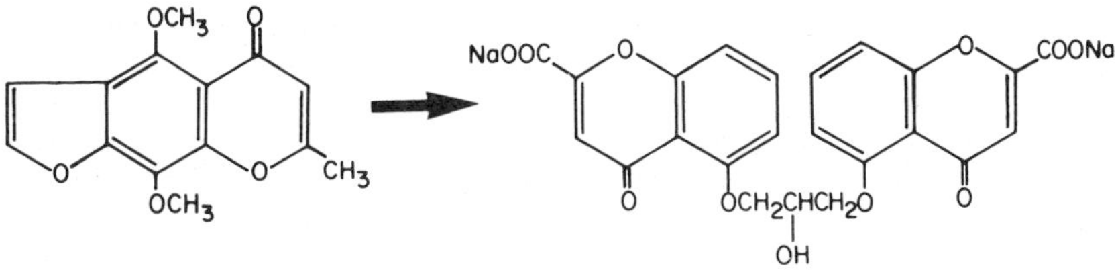

FIGURE 23 Khellin to cromolyn sodium.

No other new development in the treatment of asthma has singled out such glowing forecasts as the new compound, cromolyn sodium (disodium cromogly-cate). Altounyan (1967) and Kennedy (1967) first reported that the prior inhalation of cromolyn sodium prevented immediate asthmatic attacks against provocative inhalation of mixed pollen allergens.

Cromolyn sodium was derived after 9 years of molecular engineering by the Fisons Corporation (1973) from khellin (Fig. 23), a chromone extracted from the seeds of *Ammi visnaga* (Apiaceae) (Fig. 24). This plant had been used for centuries in folk medicine of the eastern Mediterranean region as an anti-spasmodic, for relieving the pain of renal colic, dental caries, and angina pectoris, and for relieving bronchial congestion (Quimby, 1953). Its primary action may be as a vasodilator. However, toxic reactions to khellin, which included nausea, constipation, dizziness, diarrhea, insomnia, and dermatitis,

FIGURE 24 Umbelliferous *Ammi visnaga*.

were high in trials among selected patients with bronchial asthma (Rosenman et al., 1950) and angina pectoris (Hultgren, 1952), but these were largely removed by manipulation of the molecule.

The new compound specifically inhibits the degranulation of IgE sensitized mast cells which occurs after exposure to specific allergenic antigens. Consequently, histamine and SRS-A (the slow-reacting substance of anaphylaxis) are inhibited. Bronchial asthma induced by the inhalation of specific antigens can therefore be prevented to varying degrees by *pretreatment* with cromolyn sodium, but because of this prophylactic mechanism of action, it has no role during an acute attack of asthma. Cromolyn sodium is widely used in modern medicine today, particularly for treating children systematically before the onset of acute symptoms as a major alternative to the use of **corticosteroids.**

The frequency of adverse reactions to cromolyn sodium was 2% among 375 asthmatic patients and included **dermatitis, myositis, and gastroenteritis** (Settipane et al., 1979). These reactions were not life threatening and were completely reversible, so that the safety of the drug for the treatment of asthma is not a problem.

III. CONCLUSIONS

In many instances the development of new agents has followed insights from traditional medicine and dentistry. A number of examples, such as the use of *Digitalis* glycosides to treat congestive heart failure, reserpine to alleviate hypertension, *d*-tubocurarine as a surgical relaxant, aloe for treating skin afflictions, licorice extracts for healing peptic ulcers, dental chewing sticks with anti-inflammatory, antibiotic, and anticariogenic potentials to treat periodontal disease, gingival inflammation, and caries, and highly modified molecules from *Ammi visnaga* as a prophylactic treatment for bronchial asthma all illustrate the importance of empiricism to medicine as practiced today. In addition, recent epidemiological data leading to the discovery of gossypol from cottonseed oil as a potential male birth control pill and current screening procedures resulting in novel antiviral agents specific against herpes simplex emphasize the importance of research involving natural plant products to the practice of medicine and dentistry, which may also have far-reaching applications to modern medicine.

ACKNOWLEDGMENTS

We thank Melvin T. Hatch, Exover Corporation, for information regarding the development of an antiherpetic agent, H. Wayne Nichols for a specimen of *Farlowia mollis*, and Vincent E. Zenger for the photographic prints.

REFERENCES

Altounyan, R. E. C. (1967). Inhibition of experimental asthma by a new compound—disodium cromoglycate "Intal." *Acta Allergol.* 22:487.
Bannister, B., Ginsburg, R., and Shneerson, J. (1977). Cardiac arrest due to liquorice-induced hypokalaemia. *Br. Med. J.* 2:738.

Barondes, S. H. (1981). Lectins: their multiple endogenous cellular functions. *Annu. Rev. Biochem. 50*:207-231.

Blackley, J. D., and Knochel, J. P. (1980). Tobacco chewer's hypokalemia:
licorice revisited. *N. Engl. J. Med. 302*:784-785.

Brown, J. B. (1964). Prevention and treatment of radiation-induced cancer,
including pure atomic and cathode-ray lesions. *CA—Cancer J. Clin. 14*:
14-15.

Brown, J. B. (1967). Management of cancer of the skin of the face and neck.
In *Cancer of the Head and Neck*, J. Conley (Ed.). Butterworth, Washington, D. C., pp. 91-98.

Brown, J. B., Freyer, M. P., and Killiopoulos, P. (1963). Long-term control
of cancer of head and neck with planned use of natural survival and healing tendencies. *Arch. Surg. 86*:945-954.

Carrier, O., Jr. (1970). Curare and curareform drugs. In *Medicinal Chemistry*, Part 2, 3rd ed., A. Burger (Ed.). Wiley-Interscience, New York,
pp. 1581-1599.

Chinese Medical Journal (1978). Gossypol—a new antifertility agent for males.
4(new series):417-428.

Collins, C. E., and Collins, C. (1935). Roentgen dermatitis treated with
fresh whole leaf of *Aloe vera*. *Am. J. Roentgen 33*:396-397.

Coulson, P. B., Snell, R. L., and Parise, C. (1980). Short term metabolic
effects of the anti-fertility agent, gossypol, on various reproductive
organs of male mice. *Int. J. Androl. 3*:507-518.

Dickstein, E. S., and Kunkel, F. W. (1980). Foxglove tea poisoning. *Am.
J. Med. 69*:167-169.

Elvin-Lewis, M. (1979). Empirical rationale for teeth cleaning plant selection.
Med. Anthropol. 3:431-454.

Elvin-Lewis, M. (1980). Plants used for teeth cleaning throughout the world.
J. Prev. Dent. 6:61-70.

Elvin-Lewis, M. (1983). The antibiotic and anticariogenic potential of chewing-
sticks. In *The Anthropology of Medicine*, L. Romanucci-Ross, D. E. Moer-
man, and L. R. Tancredi (Eds.). Praeger, New York, pp. 201-220.

Elvin-Lewis, M., Hall, M., Adu-Tute, M., Afful, Y., Asante-Appiah, K., and
Lieberman, D. (1980). The dental health of chewing stick users of
southern Ghana: preliminary findings. *J. Prev. Dent. 6*:151-159.

Etkin, N. L. (1979). An indigenous medical system among the Hausa of
northern Nigeria: laboratory evaluation for potential therapeutic efficacy
of antimalarial plant medicinals. *Med. Anthropol. 3*:401-430.

Fisons Corporation (1973). *Intal (Cromolyn Sodium): A Monograph.* Bedford,
Mass.

Ford, L. T. (1977). Chymopapain—past, present and future? *Clin. Orthop.
122*:367-373.

Goldstein, I. J., and Hayes, C. E. (1978). The lectins: carbohydrate-
binding proteins of plants and animals. In *Advances in Carbohydrate
Chemistry and Biochemistry*, Vol. 35, R. S. Tipson and D. Horton (Eds.).
Academic Press, New York, pp. 127-340.

Goth, A. (1981). *Medical Pharmacology*, 10th ed. Mosby, St. Louis.

Hultgren, H. M., Robertson, H. S., and Stevens, L. E. (1952). Clinical and
experimental study of use of khellin in treatment of angina pectoris. *JAMA
148*:465-469.

Javid, N. J. (1980). Treatment of herniated lumbar disk syndrome with chymopapain. *JAMA 243*:2043-2048.

Kennedy, M. C. S. (1967). Preliminary results of a double-blind cross-over trial on the value of FPL670 in the treatment of asthma. *Acta Allergol. 22*: 487-489.

Leung, A. Y. (1980). *Encyclopedia of Common Natural Ingredients.* Wiley-Interscience, New York.

Lewis, W. H., and Elvin-Lewis, M. P. F. (1977). *Medical Botany: Plants Affecting Man's Health.* Wiley-Interscience, New York.

Lewis, W. H., and Elvin-Lewis, M. P. F. (1979). Systematic botany and medicine. In *Systematic Botany, Plant Utilization and Biosphere Conservation*, I. Hedberg (Ed.). Almqvist & Wiksel, Stockholm, pp. 24-31.

Lewis, W. H., and Elvin-Lewis, M. P. F. (1983). Efficacious plants of the neotropics. In *Flora of Panama*, W. D'Arcy (Ed.) (in press).

Liener, I. E. (1974). Phytohemagglutinings: their nutritional significance. *J. Agric. Food Chem. 22*:17-22.

Lin, T., Murono, E. P., Osterman, J., Nankin, H. R., and Coulson, P. B. (1981). Gossypol inhibits testicular steroidogenesis. *Fertil. Steril. 35*: 563-566.

Loveman, A. B. (1937). Leaf of *Aloe vera* in treatment of roentgen ray ulcers. *Arch. Dermatol. Syphilol. 36*:838-843.

Marks, I. N. (1980). Current therapy in peptic ulcer. *Drugs 20*:283-299.

Maugh, T. H., II. (1981). Research news: male "pill" blocks sperm enzyme. *Science 212*:314.

McCulloch, J. A. (1980). Chemonucleolysis: experience with 2000 cases. *Clin. Orthop. 146*:128-135.

Monachino, J. (1954). *Rauwolfia serpentina*—its history, botany and medical use. *Econ. Bot. 8*:349-365.

Morbidity Mortality Weekly Report (1977). Poisoning associated with herbal teas—Arizona, Washington. *26*:257-259.

Morbidity Mortaility Weekly Report 1978). Diarrhea from herbal tea—New York, Pennsylvania. *27*:248-249.

Quimby, M. W. (1953). *Ammi visnaga* Lam.—a medicinal plant. *Econ. Bot. 7*: 89-92.

Rathje, R. (1971). Influence of neem tree extracts on inflammatory changes on the gingiva. *Quintessenz 22*:5.

Richards, J. T., Kern, E. R., Glasgow, L. A., Overall, J. C., Jr., Deign, E. F., and Hatch, M. T. (1978). Antiviral activity of extracts from marine algae. *Antimicrob. Agents Chemother. 14*:24-30.

Robinson, H. E., Harrison, F. S., and Nicholson, J. T. L. (1971). Cardiac abnormalities due to licorice intoxication. *Pa. Med. 74*:51-54.

Rosenman, R. H., Fishman, A. P., Kaplan, S. R., Levin, H. G., and Katz, L. N. (1950). Observations on the clinical use of fisammin (khellin). *JAMA 143*:160-165.

Segelman, A. B., Segelman, F. P., Karliner, J., and Sofia, R. D. (1976). Sassafras and herb tea: potential health hazards. *JAMA 236*:477.

Settipane, G. A., Klein, D. E., Boyd, G. K., Sturma, J. H., Freye, H. B., and Weltman, J. K. (1979). Adverse reactions to cromolyn. *JAMA 241*: 811-813.

Sharon, N. (1976). Lectins as mitogens. In *Mitogens in Immunobiology*, J. J. Oppenheim and D. L. Rosenstreich (Eds.). Academic Press, New York, pp. 31-41.

Sharon, N. (1977). Lectins. *Sci. Am. 236*(6):108-119.

Sircus, W. (1972). Progress report: carbenoxolone sodium. *Gut 13*:816-824.

Smith, L. (1964). Enzyme dissolution of the nucleus polposus in humans. *JAMA 187*:137-140.

Stirpe, F., Legg, R. F., Onyon, L. J., Ziska, P., and Franz, H. (1980). Inhibition of protein synthesis by a toxic lectin from *Viscum album* L. (mistletoe). *Biochem. J. 190*:843-845.

Waller, D. P., Zaneveld, L. J. D., and Fong, H. H. S. (1980). In vitro spermicidal activity of gossypol. *Contraception 22*:183-187.

Watt, J. M., and Breyer-Brandwijk, M. B. (1962). *Medicinal and Poisonous Plants of Southern and Eastern Africa*, 2nd ed. Churchill Livingstone, Edinburgh.

Wen, W. (1980). China invents male birth control pill. *Am. J. Chin. Med. 8*:195-197.

Wormsley, K. G. (1977). Testing anti-ulcer drugs. *Lancet 2*:719.

Zawahry, M. E., Hegazy, M. R., and Helal, M. (1973). Use of aloe in treating leg ulcers and dermatoses. *Int. J. Dermatol. 12*:68-73.

24

ANTIBIOTICS FROM FUNGI

STEPHEN M. HAMMOND

Leeds University, Leeds, England

PETER A. LAMBERT

University of Aston, Birmingham, England

I. INTRODUCTION

Antibiotics may be defined as compounds produced by microorganisms which at high dilution inhibit the multiplication and growth of other microorganisms. Many hundreds of antibiotic-producing strains are described in the scientific literature each year, but for an antibiotic to be useful as a chemotherapeutic agent in the treatment of disease it must selectively inhibit the growth or kill the infective agent without any significant deleterious effect(s) on the host. Fewer than 1 in 1000 of the described antimicrobial agents possess sufficient selective toxicity to be tolerated by an animal host and of these fewer than 1 in 100 will make the transition to become a bona fide chemotherapeutic agent.

Antibiotic production shows a restricted taxanomic distribution among microorganisms. Not only is the production limited to specific microbial groups, but the predominance of certain genera is readily apparent. It must, however, be stressed that this distribution may reflect the areas and methodology of searching for new antimicrobial drugs rather than the true situation in nature. In the first decade of the antibiotic era fungi and to a lesser extent bacteria produced the greatest number of antimicrobial compounds recorded (Fig. 1), but between 1955 and 1964 the Actinomycetes were responsible for almost 80% of all antibiotics described (Berdy, 1974). Since 1965 fewer antibiotics have been obtained from the Actinomycetes while the number of new compounds from other organisms, particularly fungi has increased. Eubacteria produce about 10% of the antibiotics described. Almost half of the bacterial strains producing antibiotics and all those producing clinically useful drugs belong to the genus *Bacillus*, while the bulk of the remaining strains are members of the Pseudomonadacea. About 65% of all antibiotics described are produced by Actinomycetes, and 95% of these by the genus *Streptomyces*. These myceliated bacteria provide the greatest part of the antibiotics used in current clinical practice.

   Although fungi are responsible for only one-fourth of the antibiotics described, they produce compounds of great historical, clinical, and commercial importance. Analysis of the phylogenic distribution of antibiotics among the fungi is made difficult since many of the compounds produced break the rule of specificity (i.e., a wide variety of fungi produce the same antibiotics) (Berdy, 1974; Lechevalier, 1975). However, several generalizations can be

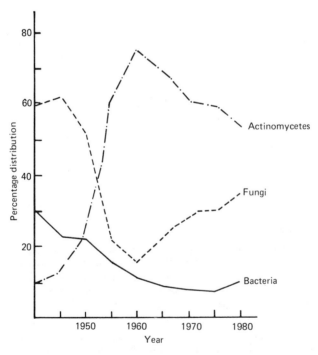

FIGURE 1 Percentage distribution of antibiotic-producing microorganisms [adapted from Berdy (1974) but now containing data up to 1980]. Only well-characterized antibiotics are included, but if all published data were used, the total antibiotic number, although greater, would not significally change the relative distribution.

made. Antibiotic production is limited to certain fungal groups (Hesseltine and Ellis, 1975). Members of the Aspergilliaceae family of the Ascomycetes, in particular members of the genera *Aspergillus* and *Penicillium*, have an important place in the antibiotic story, by producing the first clinically useful antibiotics, the penicillins. For nearly 25 years after their discovery, the penicillins and their semisynthetic derivatives held an unchallenged position as the most important antibiotics in terms of clinical usage and commercial value. The importance of the Aspergillales was reinforced by the discovery and clinical use of griseofulvin produced by *Penicillium*. The Aspergilliacea account for about 40% of the described antibiotics, divided almost equally between the genus *Penicillium* and *Aspergillus*. However, over the last 15 years interest has turned away from Aspergillales to the fungi imperfecti (Fig. 2), particularly the Moniliales (e.g., *Cephalosporium*, *Trichoderma*, and *Oospora*). Of particular importance was the discovery of cephalosporin C produced by *Cephalosporium acremonium* (Brotzu, 1948), derivatives of which have rapidly overtaken the penicillins as the most important antibiotics of fungal origin. The fungi imperfecti produce about 45% of the fungal antibiotics of which over three-fourths belong to the order Moniliales. Although the Basidiomycetes are responsible for about 15% of antibiotic-producing fungi, none has any commercial importance. Similarly, the large number of compounds isolated

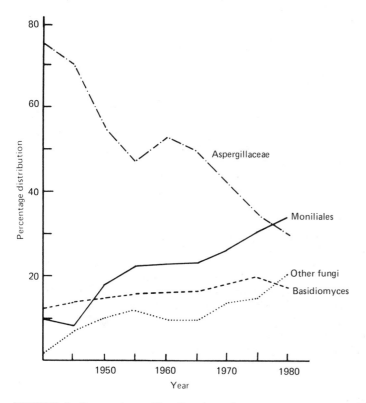

FIGURE 2 Percentage distribution of antibiotic-producing fungi [adapted from Berdy (1974) but now containing data up to 1980].

TABLE 1 Antibiotics Produced by Fungi Which Have Clinical Use

Antibiotic	Producing organism	Antimicrobial spectrum
Penicillins	*Penicillium chrysogenum* (order Aspergillales)	Gram-positive bacteria, derivatives active against gram negatives
Cephalosporins	*Cephalosporium acremonium* (order Moniliales)	Gram-positive and gram-negative bacteria
Griseofulvin	*Penicillium griseofulvum* (order Aspergillales)	Antifungal, mainly against dermatophyte infections
Fusidic acid	*Fusidium coccineum* (fungi imperfecti)	Gram-positive bacteria

from yeasts, Phycomycetes, Dermatophyta, and *Mycelia steralia*, have had little impact. Slightly over 1000 distinct antibiotic-producing fungal strains have been isolated, but only four of these (Table 1) have sufficient selective toxicity to allow their clinical use.

II. ANTIBIOTICS OF FUNGAL ORIGIN

A. Penicillins

The discovery of penicillin is generally attributed to Fleming in 1929, although it is clear that the antibacterial properties of low-molecular-weight extracts from *Penicillium glaucum* had been observed many years earlier. Fleming observed the lysis of colonies of staphylococci around a colony of a contaminating fungus, *Penicillium notatum* (Fleming, 1929). He showed that the active compound produced by the fungus, penicillin, was unstable and of low molecular weight; it inhibited a wide range of gram-positive bacteria and some gram-negatives (*Neisseria* and *Haemophilus*).

Between 1939 and 1943, Florey and Chain increased the purity of penicillin and demonstrated its therapeutic effectiveness and low toxicity in the systemic treatment of septicemia in humans (Chain et al., 1940). The period from 1944-1947 saw vast improvements in the production of penicillin, resulting from intensive work in the United States by Moyer and Coghill. A strain of *Penicillium chrysogenum* was selected from many different fungi as the optimum penicillin producer. Media and culturing techniques were improved, giving a 5000-fold increase in yield from 50,000-gallon cultures (Moyer and Coghill, 1946).

Purification of the fermentation products showed there to be a mixture of penicillins bearing different side groups, identified in the following way:

Benzyl penicillin, G
p-Hydroxybenzyl penicillin, X
2-Pentenyl penicillin, F
n-Amyl penicillin, dihydro F
n-Heptyl penicillin, K

At this stage it was realized that the proportions of the different penicillins depended on the nature of the medium. Penicillin produced in the United Kingdom contained mainly F and K, whereas that produced in the United States contained mainly G. Penicillin G was recognized to have the highest intrinsic activity, whereas K was markedly serum bound. The therapeutic value of penicillin G in treatment of bacterial infections was firmly established on military personnel during World War II. The yield of penicillin G was boosted by addition of phenylacetic acid to the culture of *Penicillium chrysogenum*. Labeling studies by Behrens showed that the phenylacetic acid was incorporated into penicillin G (Behrens et al., 1948). In a similar manner, addition of an alternative precursor, phenoxyacetic acid, led to the biosynthesis of phenoxymethyl penicillin (V). This compound had comparable activity to penicillin G but was acid stable and could be used orally, whereas penicillin G was acid labile and had to be administered by injection.

In 1949 the structure of the penicillin G was finally established (Clarke et al., 1949). It consists of a β-lactam ring fused with a thiazolidine ring. The β-lactam ring is a four-membered cyclic amide and the β-lactam bond is the linkage between the carbonyl and amino groups. The benzyl side chain derived from phenylacetic acid is joined via an amide linkage to an amino group on the 6-position of the penicillin nucleus (Fig. 3). Penicillin V contains a phenoxymethyl group in place of the benzyl group of penicillin G.

The emergence of resistance to penicillin G by the production of microbial enzymes which break down the β-lactam ring structure was first described in 1940 (Abraham and Chain, 1940). These penicillinases, or β-lactamases, cleave the β-lactam bond, converting penicillins to inactive penicilloic acids (Fig. 4). They posed a serious threat to the therapeutic value of penicillin G. In England the incidence of penicillin G-resistant staphylococci rose from 15% to 60% between 1946 and 1948 and was as high as 75% in hospitals in the United States. Methods were needed to prepare a wide range of penicillins with varying structures in the hope that some would be resistant to hydrolysis by the β-lactamases.

Although it was possible to achieve total synthesis of penicillins (Sheehan and Henery-Logan, 1957), it was not practical to prepare novel penicillins in this way. The breakthrough came from work carried out by Beechams Research Ltd. in England in 1959 (Batchelor et al., 1959). They found that starving cultures of *Penicillium chrysogenum* of side-chain precursors, such as phenylacetic acid, led to the production of the penicillin nucleus, 6-amino-

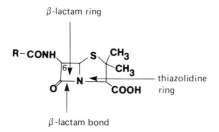

FIGURE 3 Structure of penicillins G and V. For the R equivalents for penicillin G (benzyl penicillin) and penicillin V (phenoxymethyl penicillin), see Table 2.

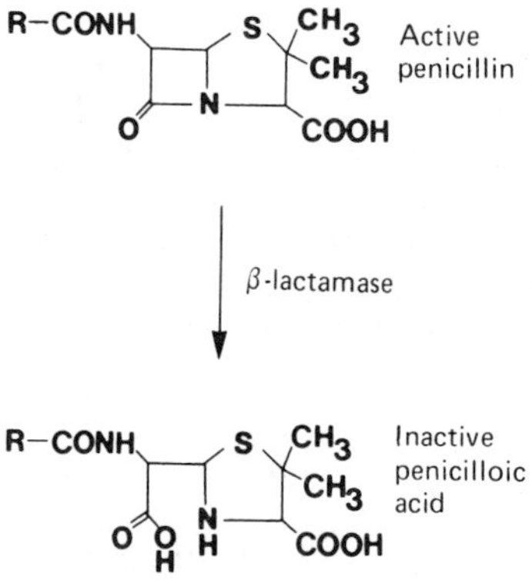

FIGURE 4 Inactivation of penicillins by the action of β-lactamase.

penicillanic acid (6-APA). Reaction of 6-APA with phenylacetyl chloride or
phenoxyacetyl chloride yielded penicillin G and penicillin V, respectively.
This route was used for the synthesis of many "semisynthetic" penicillins with
novel side chains. However, the yields of 6-APA from starved cultures were
low and its extraction from the fermentation medium was difficult. The isolation of microbial amidases which cleave the side chains from penicillin G and
penicillin V to give 6-APA provided the answer to the large-scale production
of 6-APA (Fig. 5).

Penicillin G and V are the only naturally produced penicillins in use today.
All of the others are made by the semisynthetic route, involving acylation of
6-APA. Fermentation of *Penicillium chrysogenum* is still the most efficient way
of producing penicillins G and V both for direct use and as a source of 6-APA
for production of the semisynthetic antibiotics.

Table 2 lists the principal penicillins in clinical use today. Methicillin,
introduced in 1960, was the first penicillin to show useful resistance to staphylococcal β-lactamases (Rolinson et al., 1960). The bulky methoxy groups on
the side chain are thought to protect the β-lactam ring from attack by β-
lactamases. Methicillin is acid labile and must be given parenterally for good
therapeutic results. The isoxazolyl penicillins, cloxacillin, oxacillin, dicloxacillin, and flucloxacillin (Marcy and Klein, 1970), and the related antibiotic
nafcillin (Klein and Finland, 1963) are resistant to many β-lactamases and to
acid. Ampicillin (Bear et al., 1970) and amoxycillin (Sutherland et al., 1972)
are broad-sprectrum penicillins. Although they are less active than penicillin G against gram-positive bacteria, they are active against many gram-
negative bacteria. For this reason they are widely used antibiotics, despite
their sensitivity to many β-lactamases.

Talampicillin (Clayton et al., 1974) has no antibacterial activity itself but
after absorption it is rapidly hydrolyzed to release ampicillin. The blood and
urine levels achieved are higher than with ampicillin alone.

Carbenicillin (Hewitt and Winters, 1973) was developed particularly for
its activity against *Pseudomonas*, *Proteus*, and *Serratia* species, although it
is less effective than ampicillin against other gram-negatives. It is sensitive
to β-lactamases and acid labile.

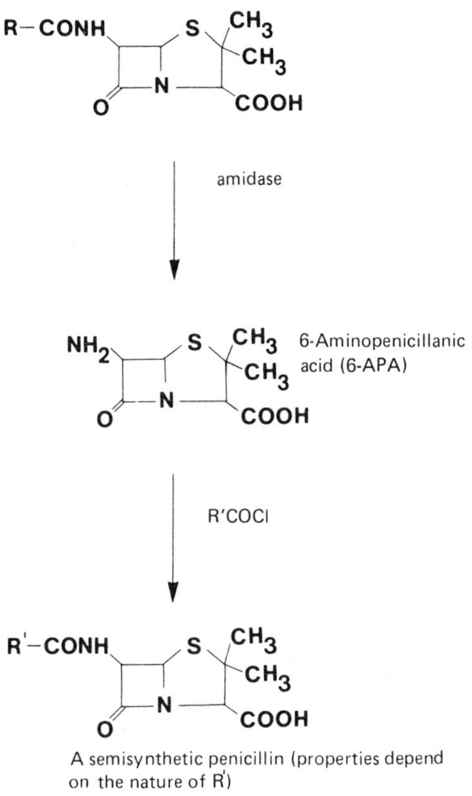

FIGURE 5 Production of semisynthetic penicillins.

Mecillinam (Williams et al., 1976) is a recently introduced β-lactam antibiotic produced semisynthetically via 6-APA from penicillin G. Unlike all the other penicillins, the side chain is linked by an amidino group to the penicillin nucleus. Mecillinam is unusual in being more active against gram-negative bacteria than gram-positives. It is sensitive to β-lactamases and is not orally absorbed. Its pivaloyloxymethyl ester, pivmecillinam, is absorbed from the gastrointestinal tract and subsequently hydrolyzed, giving high serum levels of mecillinam (Verrier Jones and Asscher, 1975).

B. Cephalosporins

In 1945 a strain of *Cephalosporium acremonium* was isolated from the sea off Sardinia by Brotzu (Van Heyningen, 1967). The mold produced a number of antibacterial substances which were investigated in the United Kingdom (Burton and Abraham, 1951). The major components were designated cephalo-sporins P (a group of closely related steroids) and N (an unstable peptide). A third compound, cephalosporin C, was isolated while studying the structure of cephalosporin N (Newton and Abraham, 1955). Its antibacterial activity was too low for it to have been detected as an antibiotic if it had occurred

TABLE 2 Some Penicillins Currently in Use

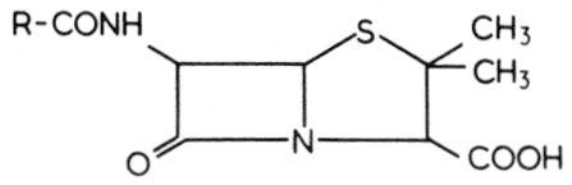

R	Name	Properties
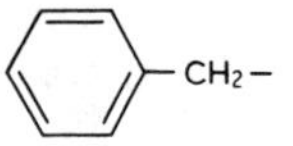	Benzyl penicillin (penicillin G)	Highly active, but lactamase-sensitive, acid-labile, narrow spectrum
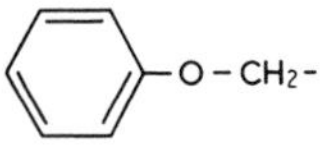	Phenoxymethyl penicillin (penicillin V)	Lactamase-sensitive, acid-stable, narrow spectrum
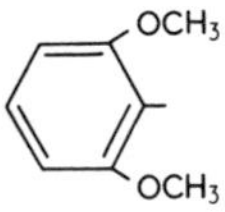	Methicillin	Lactamase-resistant, acid-labile, narrow spectrum
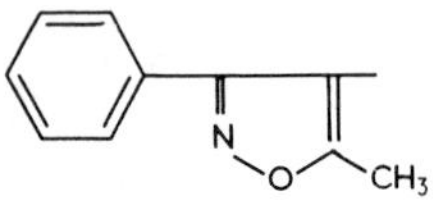	Oxacillin	Lactamase-resistant, acid-stabile, narrow spectrum
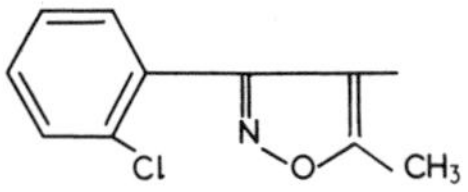	Cloxacillin	Lactamase-resistant, acid-stable, narrow spectrum
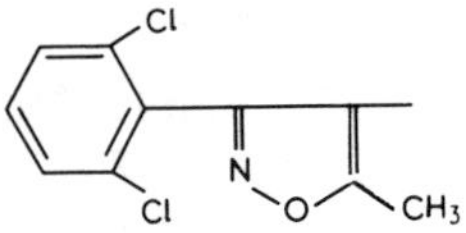	Dicloxacillin	Lactamase-resistant, acid-stable, narrow spectrum
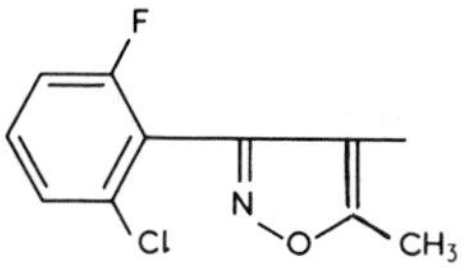	Flucloxacillin	Lactamase-resistant, acid-stable, narrow spectrum
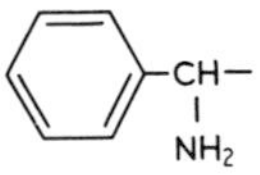	Ampicillin	Lactamase-sensitive, acid-stable, broad spectrum
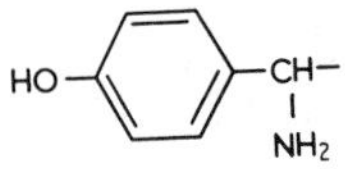	Amoxycillin	Lactamase-sensitive, acid-stable, broad spectrum

TABLE 2 (Continued)

R		Properties
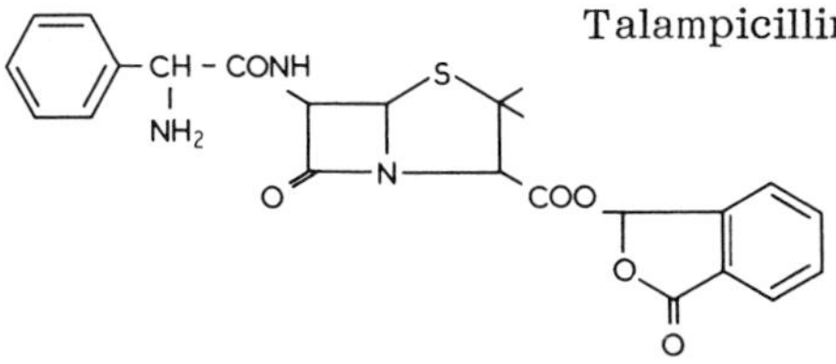	Talampicillin	Phthalidyl ester of ampicillin hydrolyzed after absorption, releasing ampicillin; achieves higher blood and urine levels than ampicillin alone
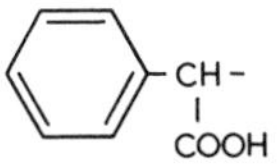	Carbenicillin	Lactamase-sensitive, acid-labile, active against some gram-negatives
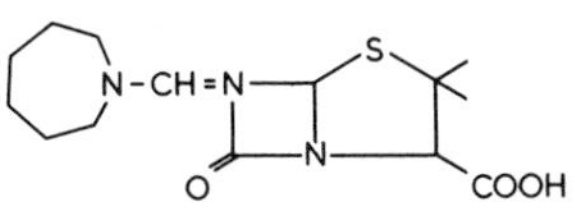	Mecillinam	Lactamase-sensitive, very active against some gram-negatives, less active against gram positives
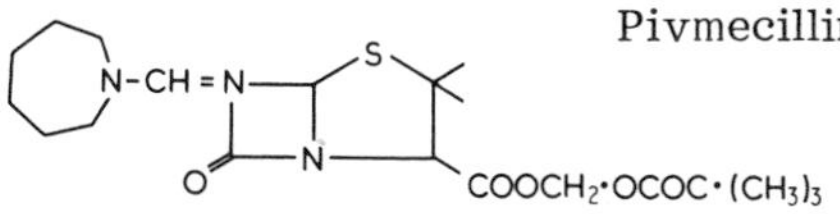	Pivmecillinam	Pivoloyloxymethyl ester of mecillinam; hydrolyzed after oral absorption, releasing mecillinam

alone in the fermentation products of *C. acremonium*. Cephalosporin C attracted interest because of its resistance to staphylococcal β-lactamases. It had extremely low toxicity and was of potential use for individuals who were allergic to penicillins. Problems of purification due to its low yield and hydrophilic properties were eventually solved and its structure was determined (Abraham and Newton, 1961). It contains a β-lactam ring fused with the six-membered dihydrothiazine ring with substituents on the 3,7-position (Fig. 6). Cephalosporin C would have been used to treat infections due to β-lactamase-producing staphylococci had not methicillin been introduced in 1960.

7-Aminocephalosporanic acid (7-ACA), the cephalosporin analog of 6-APA, can be made in very low yield by gentle acid hydrolysis of cephalosporin C. No enzymic method is available for removing the 7-amino side chain, but a chemical method has been devised (involving treatment with nitrosyl chloride in anhydrous formic acid) which gives a high yield of 7-ACA (Morin et al., 1962). Chemical and enzymic methods can be used to remove the 3-substituent from cephalosporin C (Jeffrey et al., 1961), so semisynthetic cephalosporins can be prepared with different substituents on both the 3- and 7-positions (Fig. 6).

The first two such cephalosporins, cephalothin and cephaloridine (both injectable), were introduced in 1964, followed by the orally absorbed cephalexin in 1967 (Boniece et al., 1962; Muggleton et al., 1964, 1969). Since

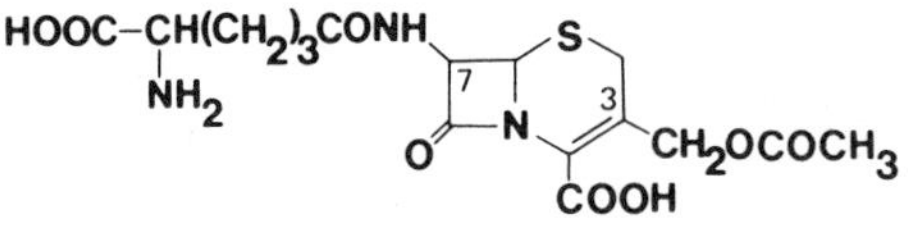

HOOC–CH(CH$_2$)$_3$CONH
|
NH$_2$

(structure of cephalosporin C with S, ring positions 7 and 3, O, N, CH$_2$OCOCH$_3$, COOH)

FIGURE 6   Structure of cephalosporin C.

then many semisynthetic cephalosporins have been synthesized and evaluated (Table 3). The guiding principle is that the 7-substituent controls the antibacterial activity, while the 3-substituent controls the pharmacokinetic properties. The properties conferred by the two substituents do interact to some extent.

The ever-expanding number of cephalosporins reaching clinical trials and the diversity of structures, properties, and names has resulted in attempts to classify them according to their antibacterial and pharmacokinetic properties (O'Callaghan, 1979). The cephalosporins available for human clinical use are first divided according to their route of administration (i.e., parenteral and oral). Further division is then made on the basis of their sensitivity or resistance to β-lactamases.

In group 1 are the metabolically unstable, β-lactamase-sensitive, parenterally administered compounds: cephalothin, cephapirin (Chisholm et al., 1970), and cephacetrile (Knusel et al., 1971). Group 2 contains the equivalent compounds which are metabolically stable: cephaloridine and cefazolin (Nishida et al., 1970). Group 3 contains the orally administered, β-lactamase-sensitive cephalosporins: cephalexin and cephradine (Bill and Washington, 1977). Group 4 is reserved for β-lactamase-resistant, orally administered compounds, of which none is available at present. Group 5 contains the β-lactamase-resistant parenteral compounds: cefuroxime (O'Callaghan et al., 1976), cefamandole (Neu, 1974) and cefoxitin (Birnbaum et al., 1978). (Note that the names of all new compounds derived from cephalosporin and developed since 1975 are spelled with "f" instead of "ph.") Cefoxitin, strictly speaking, should not be included in a list of semisynthetic cephalosporins derived from fungi because it is the first representative of the cephamycins, a name chosen for the new group of β-lactams from Streptomycetes (Gadebusch et al., 1978). Although the ring structure is the same as the cephalosporins, the cephamycins bear a 7α-methoxy group on the β-lactam ring.

The compounds in group 5 are probably the best antibiotics available at present when considered on the basis of their spectrum, activity, and low toxicity. Newer compounds to be included in this group, such as cefotaxime (Masuyoshi et al., 1980) and ceftazidime (O'Callaghan et al., 1980), show even greater promise, with broad-spectrum activity, β-lactamase resistance, antibacterial activity approaching that of the aminoglycosides and very low toxicity.

C. Mode of Action of β-Lactam Antibiotics

Penicillins and cephalosporins exert their antibacterial action by the same mechanism. The target enzymes which they inhibit are transpeptidases (Blumberg and Strominger, 1974). The function of the transpeptidases is to insert cross-

TABLE 3 Some Examples of Cephalosporins Currently in Use[a]

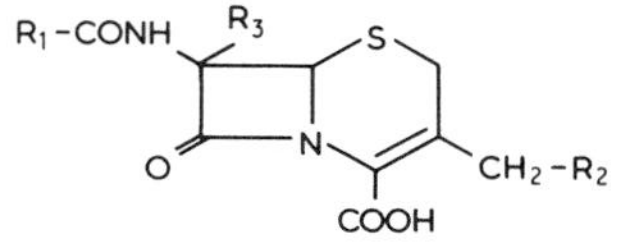
R1-CONH
R3
S
N
O
COOH
CH2-R2

R_1	R_2	R_3	Name	Properties
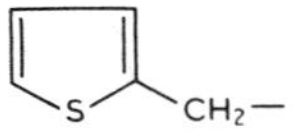	$-OCOCH_3$	-H	Cephalothin	Group 1: β-lactamase-sensitive, metabolically unstable, parenteral administration
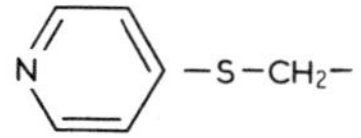	$-OCOCH_3$	-H	Cephapirin	
$NC-CH_2-$	$-OCOCH_3$	-H	Cephacetrile	
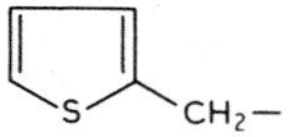	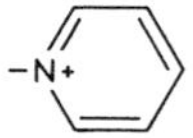	-H	Cephaloridine	Group 2: β-lactamase-sensitive, metabolically stable, parenteral administration
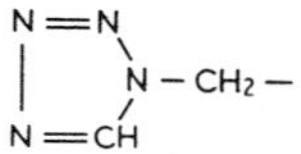	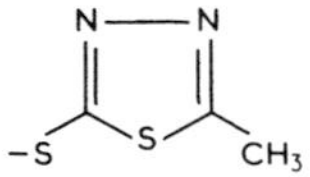	-H	Cefazolin	
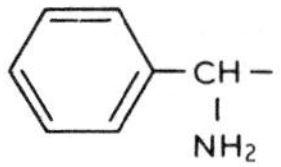	-H	-H	Cephalexin	Group 3: β-lactamase-sensitive, oral administration
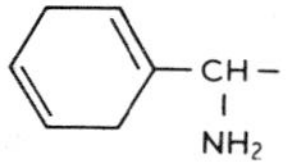	-H	-H	Cephradine	

Table 3 (Continued)

R$_1$	R$_2$	R$_3$	Name	Properties
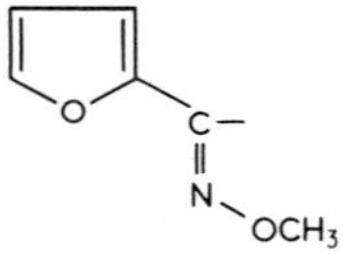 (furan)—C(=N—OCH$_3$)—	-OCOHN$_2$	-H	Cefuroxime	
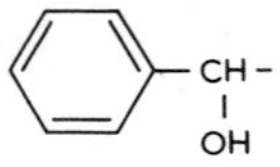 (phenyl)—CH(OH)—	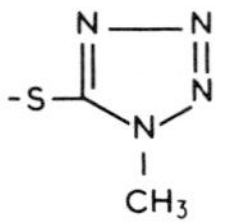 (1-methyltetrazol-5-yl-thio)	-H	Cefamandole	
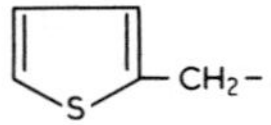 (thiophene)—CH$_2$—	-OCONH$_2$	-OCH$_3$	Cefoxitin	Group 5: β-lactamase-resistant, parenteral administration
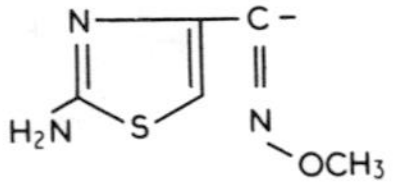 (2-amino-thiazolyl)—C(=N—OCH$_3$)—	-OCOCH$_3$	-H	Cefotaxime	
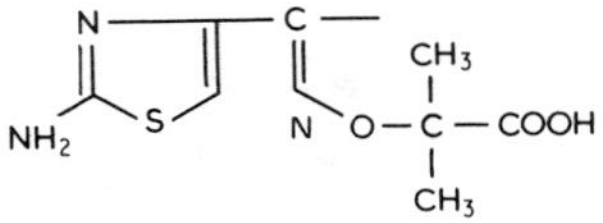 (2-amino-thiazolyl)—C(=N—O—C(CH$_3$)$_2$—COOH)—	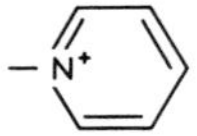 (pyridinium)	-H	Ceftazidime	

[a]See O'Callaghan (1979) for a more extensive list.

links between the linear strands of peptidoglycan (Ghuysen, 1976; Scheifer and Kandler, 1972), the macromolecule responsible for the shape, strength, and integrity of the bacterial cell wall and essential for the survival of virtually all bacteria (Fig. 7). The cross-linking of the glycan strands of peptidoglycan via short oligopeptide chains is the final stage in peptidoglycan synthesis and it occurs at growing points in the cell wall. The transpeptidases are located in the cytoplasmic membrane, presumably on the outer face, close to the sites of wall growth. The natural substrate of transpeptidase is the carboxy terminal D-alanyl-D-alanine of the pentapeptide, which is attached to each muramic acid residue of newly synthesized peptidoglycan. Having bound to D-alanyl-D-alanine, the enzyme cleaves the peptide bond, releasing the terminal D-alanine residue. The energy generated by the cleavage is then used in the formation of a new peptide bond between the carboxyl group of the remaining D-alanine and the amino group of the *meso*-diaminopimelic acid residue of the oligopeptide on an adjacent glycan strand. A cross-link between peptides on adjacent glycan strands is generated, giving the peptidoglycan polymer its mechanical strength (Fig. 8).

A number of points must be clarified at this stage. Although most bacterial peptidoglycans contain *meso*-diaminopimelic acid as the vital third amino acid in the oligopeptide chain linked to muramic acid, there are some exceptions (Scheifer and Kandler, 1972). For example, in *Staphylococcus aureus*

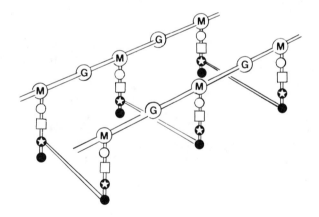

Ⓜ = N̲-acetylmuramic acid

Ⓖ = N̲-acetylglucosamine

◯ = L̲-alanine

▢ = D̲-glutamic acid

✪ = meso̲-diaminopimelic acid

● = D̲-alanine

FIGURE 7 Structure of the peptidoglycan component of bacterial cell walls.

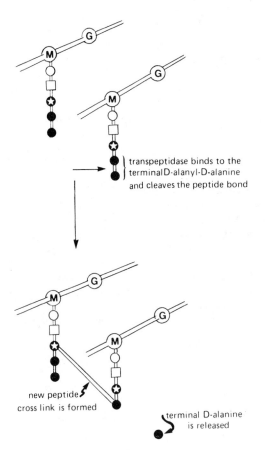

FIGURE 8 Cross linking of peptidoglycan catalyzed by transpeptidase. Peptidoglycan biosynthesis involves the release of the terminal D-alanine of the
pentapeptide synthesized initially. A new peptide bond is then formed between
the carboxy-terminal D-alanine and a free amino group of a diaminopimelic acid
residue on an adjacent glycan chain.

the third amino acid in L-lysine and the cross-link is not formed directly with
D-alanine on an adjacent chain but through a short bridging chain consisting
of five glycine residues. There are other exceptions, but the mechanism by
which the cross-link is formed remains essentially the same, with a transpeptidase binding first to the D-alanyl-D-alanine group.

In most bacteria not all the oligopeptides are cross-linked. The degree of
cross-linking can vary from almost 100% in *Staphylococcus aureus* to around
30% in bacilli (Scheifer and Kandler, 1972). A second enzyme or group of
enzymes, called carboxypeptidases, are thought to be involved in the regulation of the extent of cross-linking (Blumberg and Strominger, 1974; Ghuysen,
1976). Carboxypeptidases are similar to transpeptidase in their action. They
bind to the D-alanyl-D-alanine groups of newly synthesized peptidoglycan,
cleave the peptide bond, and release the terminal D-alanine. The carboxypeptidase is then released from the oligopeptide, leaving a terminal D-alanine

residue which is not cross-linked. In the completed peptidoglycan of all bacteria, each muramic acid bears a tetrapeptide rather than the pentapeptide which is synthesized initially. The fifth residue is removed either by transpeptidase action in forming a cross-link, or by carboxypeptidase action.

Penicillins and cephalosporins inhibit both transpeptidases and carboxypeptidases by binding covalently to the active sites of the enzymes. Tipper and Strominger proposed that the β-lactam antibiotics are structural analogs of D-alanyl-D-alanine (Tipper and Stominger, 1965). The antibiotics are mistakenly recognized by the enzymes as the natural substrates; the β-lactam bond is cleaved instead of the peptide bond between the two D-alanine residues. In the process the cleaved β-lactam becomes covalently bound to an amino acid residue at the active site of the enzyme and effectively inactivates the enzyme. In cleaving the terminal D-alanine from the natural substrate the enzymes are transiently bound to the substrate by a covalent linkage, forming reactive acyl-enzyme intermediates. The β-lactam antibiotics form analogous acyl-enzyme intermediates with much longer half-lives, effectively trapping the enzymes in inactive states (Fig. 9).

The close structural similarity between the β-lactams and D-alanyl-D-alanine may not be immediately obvious. However, measurements of bond angles and lengths supports the view. The β-lactam bond of the antibiotics is labile because of its strained, cyclic planar nature. The labile bond lies in a similar position in the antibiotic molecules to the peptide bond of D-alanyl-D-alanine; it is therefore in just the right position to acylate the active site of the enzyme.

Although both transpeptidase and carboxypeptidase activities are inhibited by β-lactams, it is thought that transpeptidase is the "killing site" of antibiotic action. The reasons are, first, that the concentrations of various penicillins and cephalosporins required to inhibit bacterial growth closely match

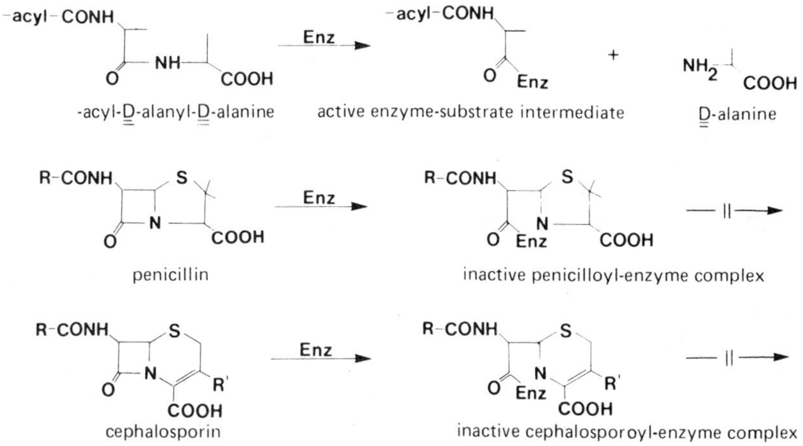

FIGURE 9 Inactivation of transpeptidase by penicillins and cephalosporins. The β-lactam antibiotics are structural analogs of the transpeptidase substrate acyl-D-alanyl-alanine. They act as alternative substrates, binding covalently to the active site to form inactive enzyme complexes.

those required to inhibit transpeptidase; and second, that transpeptidase fulfills a vital function, whereas carboxypeptidase activity can be almost entirely dispensed of without affecting the viability of the cells.

It has been known since 1972 that bacteria contain several different proteins which each covalently bind benzyl penicillin (Blumberg and Strominger, 1974). Many such penicillin-binding proteins (PBPs) have now been isolated; the enzyme activity of some has been established and their cellular functions have been probed using powerful genetic techniques (Spratt, 1977a). It appears that in rod-shaped organisms up to 80% of the bound penicillin is linked to a group of carboxypeptidases. These penicillin-binding proteins are not essential for cell viability; consequently, inhibition of their enzymic activity through binding penicillin does not result in cell death. The remaining 20% of bound penicillin is distributed between PBPs, which are probably transpeptidases (Strominger, 1980). In *Escherichia coli*, the most extensively studied organism, it is possible to ascribe specific functions to those PBPs. They are identified as PBP 1a, 1b, 2 and 3 (PBP 4, 5, and 6 are the carboxypeptidases). If PBP 1a and/or 1b are inhibited by binding penicillin, the cells lyse. When PBP 2 is selectively inhibited, the cells adopt an oval or spherical shape but are osmotically stable. Selective inhibition of PBP 3 results in cell elongation without division, filaments without septa being formed. This surprising range of different morphological effects has been established largely through study of mutants with temperature-sensitive PBPs (Spratt, 1977a).

The PBP patterns of bacteria are established using labeled benzyl penicillin. Competition experiments using labeled benzyl penicillin and unlabeled penicillins or cephalosporins have established that the relative affinities of different β-lactams for the PBPs vary. For example, at low concentrations, cephalexin binds selectively to PBP 3 of *E. coli* and causes filamentation of the cells (Spratt, 1975). Mecillinam binds almost exclusively to PBP 2 of *E. coli* and produces characteristic round forms (Spratt, 1977b). Low concentrations of ampicillin and cephaloridine bind to 1a and/or 1b with resulting lysis (Curtis et al., 1979). Most β-lactams bind to several or all of the PBPs at higher concentrations, so their action on cells in a therapeutic situation is probably a combination of effects.

A complete explanation of how β-lactams kill bacteria is still a long way off; many observations need to be explained. A simple explanation would be that the antimicrobial action results from unbalanced growth. Lack of peptidoglycan cross-linking through the inhibition of transpeptidases results in weak spots in the cell wall. Since other metabolic and biosynthetic activities are not inhibited, the cell mass continues to increase and, eventually, the cell wall bursts under pressure from the expanding cytoplasm. We must now add to this model the complication that there are several different transpeptidases, each with distinct functions. In *E. coli*, 1a and 1b are involved in maintaining the integrity of the cell, 2 is involved in controlling the shape, and 3 is concerned with cell division. It is clear that other factors also contribute to the mode of action, in particular the control of autolytic enzymes (Tomasz, 1979a,b). Peptidoglycan-hydrolyzing enzymes are present in bacterial cell walls, their function presumably being to effect cell separation during division and possibly to create zones for insertion of new peptidoglycan during wall growth. During normal growth and division the activity of these autolytic

enzymes must be regulated and there is evidence to suggest that during penicillin treatment the control is disturbed.

Although penicillins and cephalosporins are of fungal origin, antibiotics
containing the β-lactam ring are also produced by other microorganisms, in
particular streptomycetes and actinomycetes. Antibiotics such as clavulanic
acid, thienamycin, the olivanic acids and the nocardicins all show potential
usefulness as therapeutic antimicrobial agents (Brown, 1981).

D. Griseofulvin

In 1947, Grove and McGowan isolated a strain of *Penicillium janczewski* (= *P.
nigricans*) from soils in which conifers grew poorly. This stunted growth was
found to be the result of an agent produced by the *Penicillium* which inhibited
the symbiotic mycorrhizal fungi of the tree's roots, which allow the tree to
grow in nutrient-deficient soils. This compound was given the name "curling
factor" and shown to be identical with a substance purified in 1939 from *P.
griseofulvum* and given the name griseofulvin (Oxford et al., 1939). It is
now known that the antibiotic is produced by several species of *Penicillium*,
commercial production being from a high-yielding strain of *P. patulum*. It
was later shown to have the structures (2S-*trans*)-7-chloro-2',4,6-trimethoxy-
6'-methylspiro [benzofuran-2(3H)-1'-(2)-cyclohexene-3,4'-dione] (Grove et
al., 1952) (Fig. 10).

Although griseofulvin exhibits no effects against bacteria, including
actinomycetes, most fungi are inhibited to some degree. Sensitivity to the
antibiotic appears to be greatest among filamentous fungi, while yeasts and
oomycete fungi are unaffected even at high dilutions (Brian, 1949; Napier et
al., 1956; Roth et al., 1959; El Nakeeb et al., 1965). Dermatophyte fungi
appear to be particularly sensitive to low levels of griseofulvin. It has also
been reported that griseofulvin possesses activity against slime molds (Scholes,
1962). Griseofulvin is fungistatic rather than fungicidal and its action appears
limited to actively growing cells (Foley and Greco, 1960).

High concentrations of griseofulvin do not inhibit spore germination (Brian,
1949, 1960), but growing fungal filaments actively accumulate the drug by an
energy-dependant process (El Nakeeb et al., 1965). At high concentrations
(1/10 µg/ml) the antibiotic causes extensive branching, swelling, and other
gross abnormalities in the hyphal wall growing point, while the rate of hyphal
growth falls almost to zero. The morphological effects of the drug appear
localized; older cells remote from the growing point and arial hyphae appear
unaffected by the drug (Brian, 1960). Hyphal distortion may be so severe
that the cell wall bursts, allowing extrusion of the protoplast (Aytoun, 1956).
Treatment of growing hyphae with lower levels of griseofulvin (0.1-0.2 µg/ml)

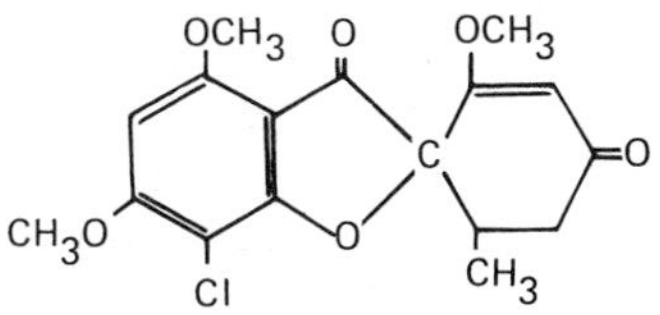

FIGURE 10 Structure of griseofulvin.

inhibits net protein, carbohydrate, lipid, and RNA content without affecting DNA synthesis. These levels also cause a characteristic regular curling of the hyphae and the formation of binucleate and multinucleate cells (Huber and Gottleib, 1968; Gull and Trinici, 1973). The drug also appears to be able to induce mutations in *Nannizzia incurvita* and *Microsporium gypseum* (Lenhart, 1969a). These observations have led to the suggestion that griseofulvin acts by directly inhibiting fungal mitosis.

Confirmation of the mechanism of action of griseofulvin has come from studies with mammalian cells. The antibiotic causes metaphase arrest of mammalian cells grown in vitro through disruption of the mitotic spindle (Malawista et al., 1968). The mitotic spindle fibers are long strands consisting of bundles of microtubular connecting the two centrioles of the dividing cell. The fibers of the spindles determine the segregation of the dividing chromatids to opposite poles during anaphase. As the nuclear membrane forms during telophase the microtubules disappear. Several drugs, for example colchicine and vinblastine, inhibit mitosis by binding to microtubule precursors, thereby inhibiting spindle assembly. Griseofulvin rapidly penetrates the cytoplasmic membrane of mammalian cells and binds to a soluble protein (Creasey et al., 1971). Griseofulvin binds on the spindle at a site distinct from other inhibitors of mitosis (e.g., colchicine and vinblastine), and unlike other mitotic inhibitors only appears to affect the function of polymerized microtubules (Wilson, 1970; Grisham et al., 1973). It is postulated that during the segregation of the daughter chromosomes to the opposite poles of the cell, adjacent microtubules slide over one another and that the interaction of griseofulvin inhibits this sliding, thereby preventing spindle contraction. Microtubules have also been implicated in other aspects of cellular physiology and the binding of the drug to this organelle may result in the observed changes in hyphal morphology.

Oral administration of griseofulvin is effective in the treatment of dermatophyte infection of the skin, hair, and nails caused by *Microsporium, Trichophyton* and *Epidermophyton*. The inhibition of other pathogenic and opportunistic fungi would require concentrations of griseofulvin in the bloodstream, which would prevent host mitosis. Dermatophyte fungi exclusively inhabit the surface keratinized tissue and it is of great importance to the effectiveness of griseofulvin that the drug appears to be concentrated in such tissue (Epstein et al., 1975). This selective localization is reversible since as the drug levels in the blood decline, tissue levels also fall. Griseofulvin has been shown to bind strongly to keratin in the skin, high concentrations developing in the outer layers of the epidermis, in particular in the stratum corneum (Epstein et al., 1972). Griseofulvin is also readily incorporated into hair and nails. Although some of the drug may be removed from hair and nails by washing, a significant amount is so strongly associated with the keratin that it can only be extracted with methanol (Gentiles et al., 1959).

Treatment of ringworm infection of experimental animals with griseofulvin has revealed that although the fungus is lost from the hair follicles, the tips remain infected, leading to the suggestion that it is only as the drug becomes associated with the keratin of the growing point that fungal growth is arrested (Gentiles, 1958). Many dermatophytes produce keratin-degrading enzymes which allow the fungus to invade host tissue. It has been demonstrated that the association of griseofulvin with keratin greatly diminishes the activity of these enzymes (Yu and Blank, 1973), thereby preventing the fungus from invading and parasitizing host tissues. It must be stressed that griseofulvin merely halts fungal growth, without killing, and treatment must be continued until all the mycelium has been removed by outgrowth or by desquamation of the tissue concerned.

Griseofulvin-resistant fungi have been isolated (Vidmar-Cvjetanovic et al., 1966; Lenhart, 1968, 1969b) but have never proved a problem in clinical practice. This may be explained in part by the observation that griseofulvin-resistant strains of *Trichophyton* possessed reduced pathogenic potential.

E. Fusidic Acid

A wide range of fungi produce steroidlike antibiotics based on a cyclopenta-phenanthrene skeleton. The most important of these antibiotics is fusidic acid (Godtfredsen et al., 1962), first isolated from *Fusidium coccidium*, a member of the fungi imperfecti. The production of fusidic acid occurs widely among other fungal groups notably by the order Moniliales (e.g., *Cephalosporium* species and the Phycomycete *Mucor ramannius*). The structure of fusidic acid (Godtfredsen and Vangedal, 1962; Godtfredsen et al., 1965) has been shown to be similar to that of cephalosporin P (Halsall et al., 1966) and helvolic acid (Iwaki et al., 1970) (Fig. 11). Cephalosporin P was isolated from a culture of a *Cephalosporium* species cultivated from the sea off Sardinia (Brotzu, 1948; Burton and Abraham, 1951). Cephalosporin P is distinct from and not to be confused with cephalosporin C. Crude cephalosporin P contains at least five components, P_1, P_2, P_3, P_4, and P_5, of which P_1 is the major active substance. Helvolic acid was first isolated from filtrates of *Aspergillus fumigatus* (Waksman et al., 1943). Although the three antibiotics share a common spectrum of antimicrobial activity, cephalosporin P_1 and helvolic acid possess approximately one-tenth of the potency of fusidic acid (on a w/w basis). Fusidic acid-resistant mutants arise frequently in both clinical isolates and in the laboratory; such mutations also confer resistance to cephalosporin P_1 and helvolic acid (Barber and Waterworth, 1962; Godtfredsen et al., 1962). Fusidic acid has been used extensively in medicine, particularly in the treatment of infections caused by penicillin-resistant *Staphylococci* (Ball et al., 1975).

Bacteriocidal levels of fusidic acid rapidly halt protein synthesis in gram-positive cells (Yamaki, 1965; Harvey et al., 1966). The failure of fusidic to inhibit gram-negative bacteria may be due to the inability of the drug to penetrate the gram-negative envelope since the compound rapidly halts protein synthesis in cell-free preparations derived from both gram-negative and gram-positive bacteria. Fusidic acid also arrests in vitro protein synthesis by ribosome preparations from yeast and mammalian cells (Malkin and Lipmann, 1969; Tanaka et al., 1969a, 1970). Techniques permitting in vitro protein synthesis have proved invaluable in determining the exact mechanism of action of fusidic acids. The ribosomes of prokaryotes and eukaryotes may be differentiated by their sedimentation coefficient; 80S ribosomes are apparently confined to eukaryotes, while in general, bacteria possess 70S ribosomes. Reduction of the magnesium concentration dissociates the 80S particle into 60S and 40S subunits and the 70S ribosomes into 50S and 30S components. The subunits consist entirely of protein and ribonucleic acid (RNA). Although differing in detail, protein synthesis in prokaryotes and eukaryotes follows a

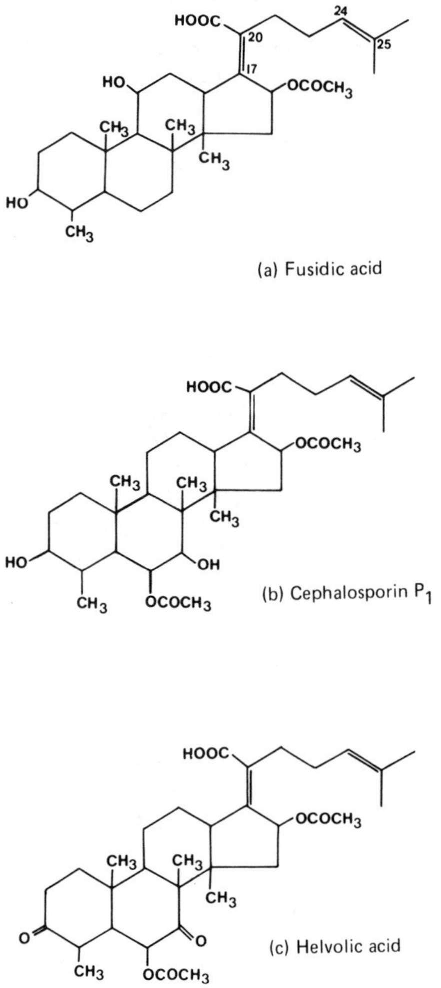

(a) Fusidic acid

(b) Cephalosporin P₁

(c) Helvolic acid

FIGURE 11 Structures of steroidal antibiotics from fungi.

similar path. Ribosomes engaged in protein synthesis are bound to a strand
of messenger RNA (mRNA) and carry a single developing polypeptide. Each
particle is capable of binding two transfer (tRNA) molecules, each charged
with an amino acid, at two distinct sites. The first site, the A (aminoacyl)
site, is concerned with codon-specific binding of a monomeric aminoacyl tRNA
and the second the P (peptidyl) site, which holds the peptidyl tRNA bearing
the developing polypeptide chain. Entry to the P site may be achieved only
through the A site. The enzyme peptidyl transferase catalyzes the formation
of the peptide bond. In addition to the peptidyl transferase, which constitutes an integral part of 50S subunit, several soluble proteins have been
demonstrated to play a vital role in chain initiation, elongation, and termination.

During translocation the peptidyl tRNA moves into the P site and in doing so opens up the A site by expelling the deacylated mRNA. The mRNA is translocated relative to the ribosome, in effect advancing three nucleotides, and the ribosome undergoes certain conformational changes. In *E. coli* translocation requires the obligatory participation of guanosine triphosphate (GTP) and a soluble tetrameric protein termed elongation factor G (EF G), shown to have a molecular weight of 80,000 (Parmeggiani, 1968; Leder et al., 1969). A similar protein, elongation factor 2 (EF 2), has been demonstrated in mammalian systems and shown to be somewhat smaller (molecular weight 70,000). During translocation in *E. coli* EF G and GTP interact with the ribosome to form a highly labile complex (Brot et al., 1969; Parmeggiani and Gottschalk, 1966). This complex is capable of GTP hydrolysis, presumably releasing the energy required for translocation (Nishizuka and Lipmann, 1969). This complex then dissociates, leaving the ribosome free for the next round of protein synthesis (Fig. 12).

Low concentrations of fusidic acid rapidly inhibit in vitro protein synthesis without affecting activation of the amino acids, formation of the aminoacyl tRNA, binding of tRNA to ribosome-mRNA complex or inhibiting the peptidyl transferase reaction (Gordon, 1969). Addition of fusidic acid to 70S ribosomes in vitro prevents the translocation of peptidyl tRNA from the A to the P site and indirectly inhibits the cleavage of GTP (Bodley et al., 1970a) through selective inhibition of EF G (Nishizuka and Lipmann, 1966). Radiolabeled fusidic acid has been shown to bind to EF G in a molar ratio of 1:1,

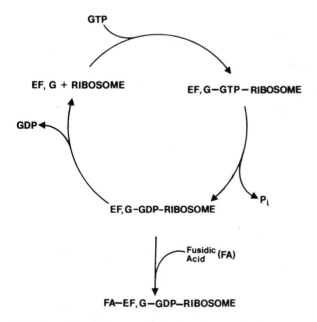

FIGURE 12 Diagram representing the role of elongation factor G (EF G) in bacterial protein synthesis and the mechanism of action of fusidic acid. The drug stabilizes the EF G:GDP:ribosome complex, preventing its dissociation and involvement with subsequent rounds of peptide biosynthesis.

and this binding is strongly stimulated by ribosomes and GTP. In in vitro systems of protein synthesis the inhibitory action of fusidic acid may be overcome by adding excess EF G. It appears that fusidic acid interacts with the previously labile EF G:GDP:ribosomes complex (Fig. 12) to form a complex stable enough to be isolated by sucrose density centrifugation. The stabilized complex has been shown to consist of fusidic acid:ribosome:EF G:GDP in the ratio 1:1:1:1 (Okura et al., 1970). While the fusidic acid:EF G:GTP complex remains associated with the ribosome access of nucleotides and additional EF G is prevented, the A site is not cleared and is therefore not available to permit the next round of polypeptide biosynthesis. The drug also prevents the conformational changes in the ribosome that normally accompany translocation (Chuang and Simpson, 1971). EF G-dependant GTP hydrolysis may be carried out by 50S subunits in vitro, although with reduced efficiency as compared to intact ribosomes, whereas 30S particles show no such activity (Modelell et al., 1971). Since fusidic acid is capable of inhibiting GTP hydrolysis by EF G:50S complexes, it appears that the larger subunit constitutes the active site of the drug (Okura et al., 1971).

Studies with wild-type and fusidic acid resistant (*fus*[r]) mutants of E. coli have shown that the drug binds only weakly to EF G from wild-type cells, but that this binding is greatly stimulated by GTP or ribosomes. Resistance to the drug was demonstrated to be associated with the EF G, not the ribosome. EF G from *fus*[r] bacteria was found to bind fusidic acid only weakly, but this binding was not enhanced by the addition of ribosomes or GTP (Kinoshita et al., 1968; Tocchini-Valenti and Mattoccia, 1968; Bernardi and Leder, 1970; Okura et al., 1970). At higher concentrations fusidic acid will also inhibit chain initiation (Sala and Ciferri, 1970) through an interaction with initiation factor 1.

In mammalian cells elongation factor (EF 2) combines with the 80S ribosome, forming a complex capable of GTP hydrolysis. Fusidic acid forms a stable complex with EF 2, GTP, and the 80S ribosome in cell-free systems (Bodley et al., 1970b; Richter, 1973), and it is believed that the drug inhibits mammalian in vitro protein synthesis in a manner analogous to that of fusidic acid in bacterial ribosomes. Surprisingly, fusidic acid has no in vivo activity against protein synthesis in mammalian cells, and it is believed that the drug fails to develop intracellular concentrations capable of interacting with EF 2.

The antimicrobial activity of a large number of fusidic and helvolic acids have been examined. It appears that the 17,20-double bond is essential for activity but the 24,25-double bond is dispensable (Fig. 11). The diacetyl lactone and the 21-methyl derivatives lack activity, suggesting that a free carboxyl group at C-20 is essential. Derivatives without the C-16 acetoxyl group retain their potency, but all activity is lost by replacement with a hydroxyl group (Godtfredsen et al., 1966). Similar structural criteria (i.e., a free carboxyl group at C-20, the 17,20-double bond, as well as 16,21-*cis* stereochemistry) are necessary for fusidic acid derivatives to bind to EF G (Tanaka et al., 1969b; Bodley and Godtfredsen, 1972).

Fusidic acid and some related terpenoid acids are uncouplers of oxidative phosphorylation (i.e., inhibit the synthesis of adenosine triphosphate by mitochondria), without affecting the rate of respiration. The compounds appear to act by binding to key lysyl groups participating in mitochondrial phosphorylation. Although the related drugs glycyrrhetic acid and polyporenic acid are relatively potent uncouplers, it requires levels of fusidic acid several hundred

times greater to uncouple the mitochondria than to inhibit cellular protein synthesis (Whitehouse et al., 1967). This suggests that uncoupling has little
importance in the mechanism of action of the drug.

A fusidic acid like antibiotic, named viridin, has been isolated from *Glio-
cladium virens* and *Trichoderma viridae* and shown to have antifungal properties (Brian and McGowan, 1945; Grove et al., 1965).

III. IN CONCLUSION

During the past 40 years a remarkable change has occurred in the treatment
of microbial disease. Before World War II the clinician used symptoms to
diagnose the infection and the microbiologist used the laboratory to confirm
the presence of the disease-causing organism, without effective treatments
being available for most infections. With the advent of chemotherapy, names
such as cholera, bubonic plague, syphilis, and typhoid have ceased to have
the ring of a death sentence, but have become the names of conditions that
respond to the appropriate therapy. This change, arguably the most important contribution of the biological sciences to humanity, began with a fungal
metabolite, penicillin. This drug began the antibiotic revolution, and when its
progress was threatened by the development of β-lactamase-producing strains,
particularly *Staphylococci*, the synthesis of lactamase-resistant penicillins
and the discovery of another fungal metabolite, fusidic acid, saved the day.
If a clinician were asked to describe an ideal antibacterial antibiotic, he or
she would specify that it be highly effective in killing pathogenic bacteria, be
able to be given orally, and have no deleterious effects on the host. There
is no ideal antibiotic, but many of the cephalosporins come very near to this
ideal. Indeed, some have stated that it is unlikely that some of the newer
cephalosporins will ever be bettered. If this were true, it would be appropriate that antibacterial chemotherapy began and ended with the fungal
metabolites. Griseofulvin, although useful in the treatment of dermatophyte
infections, cannot be used in systemic mycoses. Antifungal chemotherapy
has consistantly failed to mirror the progress made in the fight against bacteria, and as our attention becomes increasingly focused on this problem, it
would perhaps not be too chauvanistic to hope that fungi would again provide
the answer.

REFERENCES

Abraham, E. P., and Chain, E. (1940). An enzyme from bacteria able to
    destroy penicillin. *Nature (Lond.) 146*:837.
Abraham, E. P., and Newton, G. C. F. (1961). The structure of cephalo-
    sporin C. *Biochem. J. 79*:377-393.
Aytoun, R. S. C. (1956). The effect of griseofulvin on certain phytopatho-
    genic fungi. *Ann. Bot. Lond. 20*:297-305.
Ball, A. P., Gray, J. A., and Murdoch, J. M. (1975). Antibacterial drugs
    today. *Drugs 10*:1-55.
Barber, M., and Waterworth, P. W. (1962). Antibacterial activity in vitro of
    fucidin. *Lancet*, 931-932.
Batchelor, F. R., Doyle, F. P., Nayler, J. H. C., and Rolinson, G. N.
    (1959). Synthesis of penicillin: 6-aminopenicillanic acid in penicillin
    fermentations. *Nature (Lond.) 183*:257-258.

Bear, D. M., Turck, M., and Petersdorf, R. G. (1970). Ampicillin. *Med. Clin. N. Am. 54*:1145-1159.

Behrens, O. K., Corse, J., Jones, R. G., Mann, M. J., Soper, Q. F., Van Abeele, F. R., and Chiang, M.-C. (1948). Biosynthesis of penicillins. I. Biological precursors for benzyl penicillin (penicillin G). *J. Biol. Chem. 175*:751-764.

Berdy, J. (1974). Recent developments of antibiotic research and classification of antibiotics according to chemical structure. *Adv. Appl. Microbiol. 18*:309-406.

Bernardi, A., and Leder, P. (1970). Protein biosyntheses in *E. coli*: purification and characterization of a mutant G-factor. *J. Biol. Chem. 245*:4265-4268.

Bill, N. J., and Washington, J. A. (1977). Comparison of in vitro activity of cephalexin, cephradine and cefaclor. *Antimicrob. Agents Chemother. 11*: 470-474.

Birnbaum, J., Stapley, E. O., Miller, A. K., Wallick, H., Hedlin, D., and Woodruff, H. B. (1978). Cefoxitin, a semisynthetic cephamycin, a microbiological overview. *J. Antimicrob. Chemother. 4(Suppl. B)*:15-32.

Blumberg, P. M., and Strominger, J. L. (1974). Interaction of penicillin with the bacterial cell, penicillin-binding proteins and penicillin-sensitive enzymes. *Bacteriol. Rev. 38*:291-335.

Bodley, J. W., and Godtfredsen, W. O. (1972). Studies on translocation. XI. Structure function relationship of fusidane-type antibiotics. *Biochem. Biophys. Res. Commun. 46*:871-877.

Bodley, J. W., Zieve, F. J., and Lin, L. (1970a). Studies on translocation. IV. The hydrolysis of a single round of guanosine triphosphate in the presence of fusidic acid. *J. Biol. Chem. 245*:5662-5667.

Bodley, J. W., Lin, L., Salas, M. L., and Tao, M. (1970b). Studies on translocation. V. Fusidic acid stabilization of a eukaryotic ribosome-translocation factor—GDP complex. *FEBS Lett. 11*:153-156.

Boniece, W. S., Wick, W. E., Holmes, D. H., and Redman, C. E. (1962). In vitro and in vivo laboratory evaluation of cephalothin, a new broad spectrum antibiotic. *J. Bacteriol. 84*:1292-1296.

Brian, P. W. (1949). Studies on the biological activity of griseofulvin. *Ann. Bot. Lond. 13*:59.

Brian, P. W. (1960). Griseofulvin. *Trans. Br. Mycol. Soc. 43*:1-13.

Brian, P. W., and McGowan, J. C. (1945). Viridin: a highly fungicidal substance produced by *Trichoderma viride*. *Nature (Lond.) 156*:144-145.

Brot, N., Spears, C., and Weissbach, H. (1969). The formation of a complex containing ribosomes, transfer factor G and guanosine nucleotide. *Biochem. Biophys. Res. Commun. 34*:843-848.

Brotzu, G. (1948). Ricerche su di un nuovo antibiotico. Labori dell' Instituto d'Igiene di Calgari.

Brown, A. G. (1981). New naturally occurring β-lactam antibiotics and related compounds. *J. Antimicrob. Chemother. 7*:15-48.

Burton, H. S., and Abraham, E. P. (1951). Isolation of antibiotics from a species of *Cephalosporium*. Cephalosporins P_1, P_2, P_3, P_4 and P_5. *Biochem. J. 50*:168-174.

Chain, E., Florey, H. W., Gardner, A. D., Heatley, H. G., Jennings, M. A., Orr-Ewing, J., and Saunders, A. G. (1940). Penicillin as a chemotherapeutic agent. *Lancet 2*:226-228.

Chisholm, D. R., Leitner, F., Misiek, M., Wright, G. E., and Price, K. E. (1970). Laboratory studies with a new cephalosporanic acid derivative. *Antimicrob. Agents Chemother. 1969*:244-246.

Chuang, D. M., and Simpson, M. V. (1971). A translocation-associated ribosomal conformational change detected by hydrogen exchange and sedimentation velocity. *Proc. Natl. Acad. Sci. USA 68*:1474-1478.

Clarke, H. T., Johnson, J. R., and Robinson, R. (1949). *The Chemistry of Penicillin*. Princeton University Press, Princeton, N.J.

Clayton, J. P., Cole, M., Elson, S. W., and Ferres, H. (1974). BRL 8988 (Talampicillin), a well-absorbed oral form of ampicillin. *Antimicrob. Agents Chemother. 5*:670-671.

Creasey, W. A., Bensch, K. G., and Malawista, S. E. (1971). Colchicine, vinoblastine and griseofulvin—pharmacological studies with human leucocytes. *Biochem. Pharmacol. 20*:1579-1588.

Curtis, N. A. C., Orr, D., Ross, G. W., and Boulton, M. G. (1979). Affinities of penicillins and cephalosporins for the penicillin-binding proteins of *Escherichia coli* K-12 and their antibacterial activity. *Antimicrob. Agents Chemother. 16*:533-539.

El Nakeeb, M. A., McLellan, W. L., and Lampen, J. O. (1965). Antibiotic action of griseofulvin on dermatophytes. *J. Bacteriol. 89*:557-563.

Epstein, W. L., Shah, V. P., and Riegelman, S. (1972). Griseofulvin levels in stratum corneum. *Arch. Dermatol. 106*:344-352.

Epstein, W. L., Shah, V. P., Jones, H. E., and Riegelman, S. (1975). Topically applied griseofulvin in the prevention and treatment of *Trichophyton mentagrophytes*. *Arch. Dermatol. 111*:1293-1297.

Fleming, A. (1929). The antibacterial action of cultures of a *Penicillium*, with special reference to their use in the isolation of *B. influenzae*. *Br. J. Exp. Pathol. 10*:226-236.

Foley, E. J., and Greco, G. A. (1960). Studies in the mode of action of griseofulvin. *Antibiot. Ann. (1959-1960)*:670-673.

Gadebusch, H. H., Schwind, R., Lukaszow, P., Whitney, R., and McRipley, R. J. (1978). Cephamycin derivatives: comparison of the in vitro and in vivo antibacterial activities of SQ 14359, CS-1170 and cefoxitin. *J. Antibiot. 31*:1046-1058.

Gentiles, J. C. (1958). Experimental ringworm in guinea pigs: oral treatment with griseofulvin. *Nature (Lond.) 182*:476-477.

Gentiles, J. C., Barnes, M. J., and Fantes, K. H. (1959). Presence of griseofulvin in hair of guinea pigs after oral administration. *Nature (Lond.) 183*:256-257.

Ghuysen, J.-M. (1976). *The Bacterial DD-Carboxypeptidase-Transpeptidase Enzyme System*. University of Tokyo Press, Tokyo.

Godtfredsen, W. O., and Vangedal, S. (1962). The structure of fusidic acid. *Tetrahedron 18*:1029-1048.

Godtfredsen, W. O., Jahnson, S., Lorck, H., Roholt, H., and Tybring, L. (962). Fusidic acid, a new antibiotic. *Nature (Lond.) 193*:987.

Godtfredsen, W. O., Von Daehne, W., Vangedal, S., Maquet, M., Arigoni, D., and Melera, A. (1965). The stereochemistry of fusidic acid. *Tetrahedron 21*:3505-3530.

Godtfredsen, W. O., Von Daehne, W., Tybring, L., and Vangedal, S. (1966). Fusidic acid derivatives. I. Relationship between structure and antibacterial activity. *J. Med. Chem. 9*:15-22.

Gordon, J. (1969). Hydrolysis of guanosine 5'-triphosphate associated with binding of amino acyl transfer ribonucleic acid to ribosomes. *J. Biol. Chem.* *244*:5680-5686.

Grisham, L. M., Wilson, L., and Bensch, K. G. (1973). Antimitotic action of griseofulvin does not involve disruption of microtubules. *Natue (Lond.)* *244*:294-296.

Grove, J. F. (1951). The structure of griseofulvin. *Chem. Ind.*, pp. 219-220.

Grove, J. F., and McGowan, J. C. (1947). Identity of griseofulvin and curling factor. *Nature (Lond.)* *160*:574-575.

Grove, J. F., MacMillan, J., Mulholland, T. P. C., and Rodgers, M. A. (1952). Griseofulvin. IV. Structure. *J. Chem. Soc.*, pp. 3977-3987.

Grove, J. F., Moffat, J. S., and Vischer, E. B. (1965). Viridin. I. Isolation and characterisation. *J. Chem. Soc.*, pp. 3803-3811.

Gull, K., and Trinici, A. J. P. (1973). Griseofulvin inhibits fungal mitosis. *Nature (Lond.)* *244*:292-294.

Halsall, T. G., Jones, R. H., Lowe, G., and Newall, C. E. (1966). Cephalosporin P. *Chem. Commun.*, pp. 685-687.

Harvey, C. L., Sih, C. J., and Knight, S. G. (1966). The mode of action of fusidic acid. *Biochemistry* *5*:3320-3327.

Hesseltine, C. W., and Ellis, J. (1975). Antibiotic producing fungi: current status of nomenclature. *Adv. Appl. Microbiol.* *19*:47-57.

Hewitt, W. L., and Winters, R. E. (1973). The current status of parenteral carbenicillin. *J. Infect. Dis.* *127*:120-129.

Huber, F. M., and Gottleib, D. (1968). The mechanism of action of griseofulvin. *Can. J. Microbiol.* *14*:111-118.

Iwaki, S., Sair, M. I., Igarashi, H., and Okuda, S. (1970). Revised structure of helvolic acid. *J. Chem. Soc. Chem. Commun.*, pp. 1119-1120.

Jeffrey, J. D'A, Abraham, E. P., and Newton, G. G. F. (1961). Deacetylcephalosporin C. *Biochem. J.* *81*:591-596.

Kinoshita, T., Kuwana, G., and Tanaka, N. (1968). Association of fusidic acid sensitivity with G. factor in a protein-synthesizing system. *Biochem. Biophys. Res. Commun.* *33*:769-773.

Klein, J. O., and Finland, M. (1963). Nafcillin; antibacterial action in vitro and absorption and excretion in normal young men. *Am. J. Med. Sci.* *246*:10-26.

Knusel, F., Konopkam, E. A., Glezer, J., and Rosselet, A. (1971). Antimicrobial studies in vitro with CIBA 36278-Ba, a new cephalosporin derivative. *Antimicrob. Agents Chemother.* *1970*:140-149.

Lechevalier, H. A. (1975). Production of the same antibiotics by members of different genera of micro-organisms. *Adv. Appl. Microbiol.* *19*:25-44.

Leder, P., Skogerson, L. E., and Neu, M. M. (1969). Translocation of mRNA codons. *Proc. Natl. Acad. Sci. USA* *62*:454-460.

Lenhart, K. (1968). Spontaneous mutants of *Microsporium gypseum* resistant to griseofulvin. *Mycopathol. Mycol. 1ppl.* *36*:150-160.

Lenhart, K. (1969a). Mutagenic effect of griseofulvin. *Mycosen* *12*:687-693.

Lenhart, K. (1969b). Griseofulvin-resistant mutants in dermatophytes. *Mycosen* *12*:655-660.

Malawista, S. E., Sato, H., and Bensch, K. G. (1968). Vinblastine and griseofulvin reversibly disrupt the living mitotic spindle. *Science 160*:770-771.

Malkin, M., and Lipmann, F. (1969). Fusidic acid: inhibition of factor T2 in reticulocyte protein synthesis. *Science 164*:71-72.

Marcy, S. M., and Klein, J. O. (1970). The isoxazolyl penicillins: oxacillin, cloxacillin and dicloxacillin. *Med. Clin. N. Am.* 54:1127-1143.

Masuyoshi, S., Arai, S., Miyamoto, M., and Mitsuhashi, S. (1980). In vitro antimicrobial activity of cefotaxime, a new cephalosporin. *Antimicrob. Agents Chemother.* 18:1-8.

Modelell, J., Vazquez, D., and Monro, R. E. (1971). Ribosomes, G factor and Siomycin. *Nature New Biol.* 230:109-112.

Morin, R. B., Jackson, B. G., Flynn, E. H., and Roeske, R. W. (1962). Chemistry of cephalosporin antibiotics. I. 7-Aminocephalosporanic acid from cephalosporin C. *J. Am. Chem. Soc.* 84:3400-3401.

Moyer, A. J., and Coghill, R. D. (1946). Penicillin. VIII. Production of penicillin in surface cultures. *J. Bacteriol.* 51:57-78.

Muggleton, P. W., O'Callaghan, C. H., and Stevens, W. K. (1964). Laboratory evaluation of a new antibiotic—cephaloridine (Ceporin). *Br. Med. J.* 2:1234-1237.

Muggleton, P. W., O'Callaghan, C. H., Foord, R. D., Kirby, S. M., and Ryan, D. M. (1969). Laboratory appraisal of cephalexin. *Antimicrob. Agents Chemother.* 1968:353-360.

Napier, E. J., Turner, D. I., and Rhodes, A. (1956). The in vitro action of griseofulvin against pathogenic fungi and plants. *Ann. Bot. Lond.* 20: 461-466.

Neu, H. C. (1974). Cefamandole, a cephalosporin antibiotic with an unusual wide spectrum of activity. *Antimicrob. Agents Chemother.* 6:177-182.

Newton, G. G. F., and Abraham, E. P. (1955). Cephalosporin C, a new antibiotic containing sulphur and D-α-aminoadipic acid. *Nature (Lond.)* 175:548.

Nishida, M., Martubara, T., Murakawa, T., Mine, Y., Yokota, Y., Kuwahara, S., and Goto, S. (1970). In vitro and in vivo evaluation of cefazolin, a new cephalosporin C derivative. *Antimicrob. Agents Chemother.* 1969: 236-243.

Nishizuka, Y., and Lipmann, F. (1966). The inter-relationship between guanosine triphosphate and amino acid polymerization. *Arch. Biochem. Biophys.* 116:344-351.

O'Callaghan, C. H. (1979). Description and classification of the newer cephalosporins and their relationships with the established compounds. *J. Antimicrob. Chemother.* 5:635-671.

O'Callaghan, C. H., Sykes, R. B., Griffiths, A., and Thornton, J. C. (1976). Cefuroxime, a new cephalosporin antibiotic: activity in vitro. *Antimicrob. Agents Chemother.* 9:511-519.

O'Callaghan, C. H., Acred, P., Harper, P. B., Ryan, D. M., Kirby, S. M., and Harding, S. M. (1980). GR20263, a new broad-spectrum cephalosporin with anti-pseudomonal activity. *Antimicrob. Agents Chemother.* 17:876-883.

Okura, A., Kinoshita, T., and Tanaka, N. (1970). Complex formation of fusidic acid with G factor, ribosome and guanosine nucleotide. *Biochem. Biophys. Res. Commun.* 41:1545-1550.

Okura, A., Kinoshita, T., and Tanaka, N. (1971). Formation of fusidic acid G factor-GDP-ribosome complex and the relationship to the inhibition of GTP hydrolysis. *J. Antibiot. (Tokyo)* 24:655-661.

Oxford, A. E., Raistrick, H., and Simonart, P. (1939). Studies on the biochemistry of micro-organisms 60. Griseofulvin a metabolic product of *Penicillin griseofulvum*. *Biochem. J.* 33:240-248.

Parmeggiani, A. (1968). Crystalline transfer factors from *Escherichia coli*. *Biochem. Biophys. Res. Commun. 30*:614-619.

Parmeggiani, A., and Gottschalk, E. M. (1969). Properties of the crystalline amino acid polymerization factors from *Escherichia coli*: binding of G to ribosomes. *Biochem. Biophys. Res. Commun. 35*:861-867.

Richter, D. (1973). Competition between the elongation factors 1 and 2 and phenylalanyl transfer RNA for the ribosomal binding site in a polypeptide-synthesizing system from brain. *J. Biol. Chem. 248*:2853-2857.

Rolinson, G. N., Stevens, S., Batchelor, F. R., Wood, J. C. and Chain, E. B. (1960). Bacteriological studies with a new penicillin, BRL 1241. *Lancet 2*:1176-1178.

Roth, F. J., Sallman, B., and Blank, H. (1959). In vitro studies of the antifungal antibiotic griseofulvin. *J. Invest. Dermatol. 33*:403-425.

Sala, F., and Ciferri, O. (1970). Inhibition of peptide chain initiation in *E. coli* by fusidic acid. *Biochim. Biophys. Acta 224*:199-205.

Scheifer, K., and Kandler, O. (1972). Peptidoglycan types of bacterial cell walls and their taxonomic implications. *Bacteriol. Rev. 36*:407-472.

Scholes, P. M. (1962). Some observations on the cultivation, fruiting and germination of *Fuligo septica*. *J. Gen. Microbiol. 29*:137-148.

Sheehan, J. C., and Henery-Logan, K. R. (1957). The total synthesis of penicillin V. *J. Am. Chem. Soc. 79*:1262-1263.

Spratt, B. G. (1975). Distinct penicillin-binding proteins involved in the division, elongation and shape of *Escherichia coli* K12. *Proc. Natl. Acad. Sci. USA 72*:2999-3003.

Spratt, B. G. (1977a). Penicillin-binding proteins of *Escherichia coli*: general properties and characterization of mutants. In *Microbiology 1977*, D. Schlessinger (Ed.). American Society for Microbiology, Washington, pp. 182-194.

Spratt, B. G. (1977b). The mechanism of action of mecillinam. *J. Antimicrob. Chemother. 3 (Suppl. B)*:13-19.

Strominger, J. L. (1980). Interaction of penicillin with its receptors in bacterial membranes. *Trends Biochem. Sci.*, April, pp. 97-101.

Sutherland, R., Croydon, E. A. P., and Rolinson, G. N. (1972). Amoxycillin: a new semi-synthetic penicillin. *Br. Med. J. 3*:13-16.

Tanaka, N., Kinoshita, T., and Masukawa, H. (1969a). Mechanism of inhibition of protein synthesis by fusidic acid and related steroidal antibiotics. *J. Biochem. (Tokyo) 65*:459-464.

Tanaka, N., Nishimura, T., Kinoshita, T., and Umezawa, H. (1969b). The effect of fusidic acid on protein synthesis in mammalian systems. *J. Antibiot. (Tokyo) 22*:181-182.

Tanaka, N., Nishimura, T., and Kinoshita, T. (1970). Inhibition by fusidic acid of transferase 11 in reticulocyte protein synthesis. *J. Biochem. (Tokyo) 67*:459-463.

Tipper, D. J., and Strominger, J. L. (1965). Mechanism of action of penicillins: a proposal based on the structural similarity to acyl-D-alanyl-D-alanine. *Proc. Natl. Acad. Sci. USA 54*:1133-1141.

Tocchini-Valenti, G. P., and Mattoccia, E. (1968). A mutant of *E. coli* with an altered supernatant factor. *Proc. Natl. Acad. Sci. USA 61*:146-151.

Tomasz, A. (1979a). The mechanism of the irreversible effects of penicillins. How the β-lactam antibiotics kill and lyse bacteria. *Annu. Rev. Microbiol. 33*:113-137.

Tomasz, A. (1979b). From penicillin-binding proteins to the lysis and death of bacteria: a 1979 view. *Rev. Infect. Dis. 1*:434-467.

Van Heyningen, E. (1967). Cephalosporins. *Adv. Drug Res. 4*:1-70.

Verrier Jones, E. R., and Asscher, A. W. (1975). Treatment of recurrent bacteriuria with pivmecillinam (FL1039). *J. Antimicrob. Chemother. 1:* 193-196.

Vidmar-Cvjetanovic, B., Gaudin, P., Lozeron, H., and Jadassohn, W. (1966). Induced griseofulvin resistance of *Trichophyton quinckeanum* in vitro and in vivo. *Experientia 22*:737-738.

Waksman, S. A., Horning, E. A., and Spencier, E. L. (1943). Two antagonistic fungi *Aspergillus fumigatus* and *Aspergillus clavatus* and their antibiotic substances. *J. Bacteriol. 45*:233-248.

Whitehouse, M. W., Dean, P. D. G., and Halsall, T. G. (1967). Uncoupling of oxidative phosphorylation by glycyrrhetic acid, fusidic acid and some related triterpenoid acids. *J. Pharm. Pharmacol. 19*:533-544.

Williams, J. D., Andrews, J., Mitchard, M., and Kendall, M. J. (1976). Bacteriology and pharmacokinetics of the new amidino penicillin-mecillinam. *J. Antimicrob. Chemother. 2*:61-69.

Wilson, L. (1970). Properties of colchicine binding proteins from chick embryo brain. *Biochemistry 9*:4999-5007.

Yamaki, H. (1965). Inhibition of protein synthesis by fusidic and helvolinic acids. *J. Antibiot. (Tokyo) Ser. A 18*:228-232.

Yu, R. J., and Blank, F. (1973). On the mechanism of action of griseofulvin in dermatophytosis. *Sabouraudia 11*:274-278.

AUTHOR INDEX

Numbers are page numbers and indicate that an author's work is cited in the text. Underlined numbers give the page on which the complete reference is listed.

A

Aamodt, O. S., 130, <u>147</u>
Aanes, W. A., 570, 575, <u>578</u>
Aasen, A., 488, <u>501</u>
Abdel-Kader, M. M., 300, 303, <u>313</u>
Abdulla, R. F., <u>34</u>
Abdull-Wahab, A. S., 710, 713, 723, <u>728</u>
Abe, F., 49, 75, <u>83</u>
Abedo, A. H., 303, <u>312</u>
Abel, E. L., 475, <u>500</u>
Abisch, E., 44, <u>76</u>
Abraham, D. J., 488, <u>504</u>
Abraham, E. P., 821, 823, 825, 835, <u>837</u>, <u>838</u>, <u>840</u>, <u>841</u>
Abraham, R. J., 368, 369, 372, 372, 373, <u>416</u>
Abraham, W.-R., 640, <u>660</u>
Abrahamov, A., 133, <u>150</u>
Abramovich, A., 183, <u>189</u>
Absalyamov, I. F., 648, <u>659</u>
Achiwa, K., 31, <u>35</u>
Acklin, W., 774, <u>779</u>
Acred, P., 826, <u>841</u>
Adam, S. E. I., 57, <u>80</u>
Adams, J. H., 139, <u>146</u>
Adams, O. R., 453, <u>460</u>
Adams, R., 477, 478, 482, <u>500</u>
Adamson, R., 182, <u>190</u>
Adamson, R. H., 182, <u>198</u>, 246, 293, <u>294</u>, 306, <u>312</u>
Adekunle, A., 308, <u>313</u>

Adekunle, A. A., 203, 211, <u>227</u>, <u>229</u>, <u>233</u>
Adelaar, T. F., 366, 382, 390, 396, <u>413</u>, <u>418</u>
Adeuja, A. O. C., 143, <u>151</u>
Adgina, V. V., 553, <u>584</u>
Adolf, W., 242, 264, <u>268</u>, 269, 270, <u>276</u>, <u>298</u>, 521, <u>541</u>
Adu-Tute, M., <u>813</u>
Adye, J. C., 302, <u>318</u>
Afful, Y., <u>813</u>
Aftergood, L., 272, <u>297</u>
Agnihotri, V. P., 302, <u>315</u>
Agostini, B., 614, 619, <u>620</u>, <u>626</u>, <u>629</u>
Agrup, G., 424, <u>435</u>
Agurell, S., 492, <u>494</u>, <u>500</u>, <u>504</u>, <u>505</u>
Aikit, B. K., 649, <u>662</u>
Aitken, M. M., 88, <u>89</u>, <u>106</u>
Akai, S., 17, <u>38</u>
Akazawa, T., 17, <u>34</u>, <u>35</u>
Akazawa, Y., 17, <u>35</u>
Akinrele, I. A., 133, <u>146</u>
Akita, T., 405, <u>406</u>
Alary, J., 169, <u>190</u>
Albersheim, P., 775, <u>781</u>
Albo, J. M., 710, <u>728</u>
Alcañiz, J., 258, <u>281</u>
Alderson, A., 658, <u>659</u>
Alexander, A. F., 535, 536, <u>536</u>
Alexander, J. C., 135, <u>147</u>
Aleyas, N. M., 188, <u>189</u>
Alfin-Slater, R. B., <u>272</u>, <u>297</u>
Alikutty, K. M., 188, <u>189</u>
Allakhverdov, B. L., 620, <u>632</u>

847

M

SUBJECT INDEX

A

AB$_1$, 210, 211
ABPE, 86
Abdomen, 62
Abortifacient activity, 264
Abortion, 351, 455
Abrin, 802, 803
Abrus precatorius, 802
Absorption, 92
Absorption spectra, 354
Acacia, 799
   catechu (L. fil.) Willd., 248
   suma (Roxb.), 248
   villosa, 255
Acetylaminofluorene, 254
Acetic acids, 713
12 β-Acetoxyhuratoxin, 521
13-O-Acetyl-12-deoxyphorbol, 520
Acetylenic derivatives, 748
Achyranthes aspera, 798
Achyranthin, 798
Acorus calamus, 252
Actin, 614, 620
Actinomycin D, 224
Activated molecules, 347
Active tumor-promoting agent, 273
Acute aflatoxicosis in humans,
   suspected cases, 308
Acute bovine pulmonary edema and
   emphysema (ABPE), 5, 86
Acute necrotizing gingivitis, 797
Acute poisoning, 652
Acute pulmonary edema, 90, 96
Acute toxicity, 561, 642, 650

Acute toxicity levels (acute LD$_{50}$), 643,
   655
Adduct formation, 326, 330
Adenostoma fasciculatum, 724
Adenostyles, 639
Adrenal gland, 261
Adriamycin, 224-225
Adsorption, 718
Affinity labeling, 623
Affinity toward actin, 616
Aflatoxicol (AFL), 305
Aflatoxical AFR$_0$, 305, 311
Aflatoxin, 202-212, 299
   AFB$_1$ hemiacetal, 305
   AFB$_{2a}$, 305
   AFD$_1$, 305
   AFD$_2$, 305
   AFG$_{2a}$, 305
   AFGM$_1$, 305
   Aflatoxin B$_1$, 271, 303, 304
   Aflatoxin B$_2$, 303, 304
   Aflatoxin G$_1$, 303, 304
   Aflatoxin G$_2$, 303, 304
   Aflatoxin M, 271
   Aflatoxin Q$_1$, 211
   Aflatoxin W, 305
AFM$_1$, 303
AFM$_2$, 303
AFP$_1$, 305
AFQ$_1$, 305
Afrogenin, 73
Afroside, 47, 48, 49, 70, 73
Agaricus bisporus, 624
Agave lecheguilla, 351
Agglutinate erythrocytes, 802